Nonlinear Systems Tracking

Nonlinear Systems Tracking

LYUBOMIR T. GRUYITCH

CRC Press
Taylor & Francis Group
Boca Raton London New York

CRC Press is an imprint of the
Taylor & Francis Group, an **informa** business

CRC Press
Taylor & Francis Group
6000 Broken Sound Parkway NW, Suite 300
Boca Raton, FL 33487-2742

First issued in paperback 2017

ISBN-13: 978-1-4987-5325-8 (hbk)
ISBN-13: 978-1-138-74951-1 (pbk)

Visit the Taylor & Francis Web site at
http://www.taylorandfrancis.com

and the CRC Press Web site at
http://www.crcpress.com

Contents

List of Figures

Part I

Preface

Systems, control, and computers

0.1 Dynamical systems

Is there a real (dynamical) physical system that is not nonlinear? Probably not. The boundedness of energy and matter sources available for the systems work cause the limitation nonlinearity. There are often geometric and kinematic limitations. The path and its length of every displacement are limited. The maximal speed and acceleration of every technical object are bounded. Some physical processes are inherently nonlinear (e.g., friction). Sophisticated control algorithms are mainly nonlinear. Linear control algorithms cannot ensure a finite reachability time. The study of nonlinear dynamical systems has been attracting more and more research efforts. Their study is challenging for their complexity and for their theoretical and practical importance.

Note 1 ***The three classes of the dynamical systems treated herein***
The classes of the dynamical systems treated herein are:

- *the input-output (IO)* ***systems*** *described by the m-th order time-varying nonlinear vector differential equation (3.62) in the output vector* $\mathbf{Y}$,
- *the input-state-output (ISO)* ***systems*** *described by the first-order time-varying nonlinear vector differential equation (3.65) in the state vector* $\mathbf{X}$, *and by the time-varying nonlinear vector algebraic output equation (3.66) that defines the dependence of the output vector* $\mathbf{Y}$ *on the state vector* $\mathbf{X}$ *and on the input vector* $\mathbf{I}$,
- *the input-internal dynamics-output (IIDO)* ***systems*** *described by the α-th-order time-varying nonlinear vector differential equation [the first equation (3.12)] in the internal dynamics vector* $\mathbf{R}$, *and by the time-varying nonlinear vector algebraic output equation, [the second equation (3.12)], which determines the dependence of the output vector* $\mathbf{Y}$ *on the internal dynamics vector* $\mathbf{R}$ *and on the input vector* $\mathbf{I}$.

Note 2 ***The unifying class of the systems***
We prove how the IIDO system (3.12) reduces to:

- *The IO systems (3.62),*
- *The ISO systems (3.65), (3.66).*

The IO systems (3.62) and the ISO systems (3.65), (3.66) represent special subclasses of the IIDO systems (3.12). The study of the (stability, tracking, trackability) properties of the IIDO systems (3.12) incorporates the study of the same properties of the IO systems (3.62) and the ISO systems (3.65), (3.66). The results (concepts, definitions, criteria, conditions, control algorithms) obtained for the IIDO systems (3.12) simultaneously hold for, and are applicable to, the IO systems (3.62) and the ISO systems (3.65), (3.66).

The IIDO systems (3.12) are the unifying systems. We call them simply ***the systems (3.12).***

0.2 Dynamical systems and computers

The computer techniques and technology have been very successfully developing. Problems that need the computer's big memory, fast operations, and small computer space are now solvable by computer engineers, computer designers, and by software specialists. The computers enable the development of artificial intelligence. The digital computer now succeeds in carrying out simulations of dynamical processes practically equally well as the analog computer.

The nonlinear systems can abruptly change the character of their dynamical behavior due to a very small, sometimes infinitesimal, variation of initial conditions and/or of the external actions. Their qualitative properties (e.g., controllability, observability, optimality, stability and trackability [279]) can be valid only for initial conditions belonging to a connected bounded infinite set (i.e., can be only local). Such largest set is *the domain* $\mathcal{D}_{(.)}$ *of the corresponding qualitative dynamical property*. It can be open, or closed, but need not be either. The problem of how to determine the shape, the size, and the boundary of the property domain can be very difficult and complex. The computer simulations are very illustrative, but they cannot resolve the problem of the stability domain determination [287]. In order to cope effectively, the corresponding theory should be well developed. It serves as the basis for computer simulations, for engineering design, for its applications, and for further research.

0.3 Dynamical systems and control

A (technical, economical, biological, or social) *dynamical physical system* that cannot realize by itself either its demanded behavior or a real behavior sufficiently close to its requested behavior in the real environment, under arbitrary initial conditions and under (usually unknown, unpredictable) external influences, is a **plant** (also called an **object**). It can be, for example, a submarine, ship, car, train, tool machine, industrial robot, mobile robot, production line, turbine, motor, hydraulic process, thermal process, chemical process, power plant, aircraft, missile, space vehicle, electrical network, factory, human organ, market, or society.

0.4 Control goal

The very control goal, the primary control purpose, is for control to force a plant to behave exactly as demanded (which is the ideal case), *or at least sufficiently closely to the demanded behavior over some (bounded or unbounded) time interval.* The object desired (internal and/or output) behavior expresses its demanded (internal and/or output) dynamical behavior, respectively. Its desired output dynamical behavior is mathematically described by the desired *time* evolution $\mathbf{Y}_d(t)$ of the plant output vector $\mathbf{Y}$. This means that the plant real output response $\mathbf{Y}(t)$ should **follow/track** its desired output response $\mathbf{Y}_d(t)$ sufficiently closely; i.e., control is to force the object to realize/to exhibit an appropriate kind of ***tracking***. Such a control is ***tracking control***. It is clear that **tracking** and **tracking control synthesis** are the fundamental control issues.

Tracking studies essentially started as a *servomechanism* or *servosystem* theory in the broad sense. The pioneering contributions are due to L. A. MacColl, 1945 [402]; H. Lauer, R. Lesnick, and L. E. Matson, 1947 [375]; G. S. Brown and D. P. Campbell,1948 [61]; J. C. West, 1953 [553]; H. Chestnut and R. W. Mayer, 1955 [82]; I. Flügge-Lotz and C. F. Taylor, 1956 [132]; and J. C. Lozier, 1956 [396]. A. I. Talkin [535] constructed (1961) the term "servo tracking". The name *servomechanism* or *servosystem* means the controller that should force the plant output, or forces the controlled plant output, *to follow*, i.e., *to track*, its *time*-varying desired output rather than to track only a constant desired output. The latter is the purpose of the feedback controller called classically the *regulator*.

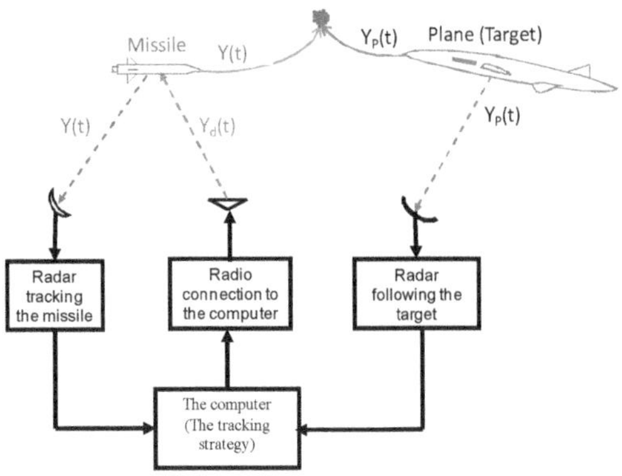

Figure 1: Target (plane) tracking.

0.5 Tracking control tasks

The work by Y. Bar-Shalom [26] was probably the pioneering one that opened the research on the target tracking, Figure 1. Further developments of target tracking are multitarget tracking and the related estimation and data acquisition theories due to Y. Bar-Shalom and T. E. Fortmann [28]; Y. Bar-Shalom and X. R. Li [29], [30]; Y. Bar-Shalom, X. R. Li, and T. Kirubarajan [31]; Y. Bar-Shalom, P. K. Willet, and X. Tian [32]; S. S. Blackman [52]; F. E. Daum [96]-[98]; R. P. S. Mahler [404]; D. B. Reid [496]; and L. D. Stone, C. A. Barlow, and T. L. Corwin [528].

Task 3 ***Estimation and data acquisition to determine the plant desired output behavior***

One tracking control task is to estimate and to gather data about the target motion [the enemy plane flight $\mathbf{Y}_P(t)$*] in order to define the desired output behavior* $\mathbf{Y}_d(t)$ *of the plant (the missile).*

The preceding task occupied (multi)target tracking research. What follows does not deal with this tracking control task 3. This means that we assume the following condition holds:

Condition 4 ***Knowledge of the plant desired output behavior***

The plant desired output behavior $\mathbf{Y}_d(t)$ *is well defined and known.*

The purpose, the goal, and the aim of the plant determine its desired output behavior $\mathbf{Y}_d(t)$. In the case of the missile its goal is to reach and destroy the enemy plane.

Task 5 ***Control synthesis for tracking quality***

Another control task is that control $\mathbf{U}(t)$ *acts on the plant so that its real output behavior* $\mathbf{Y}(.)$ *tracks (follows) sufficiently closely, with a requested tracking quality, its well-defined desired output behavior* $\mathbf{Y}_d(.)$*, and that* $\mathbf{Y}(t)$ *reaches* $\mathbf{Y}_d(t)$ *at a (possibly given, predetermined) finite instant* t_R *after the initial instant* t_0*,* $t_R > t_0$*, to stay equal because* t_R *until a final instant* t_F*,* $t_R < t_F \leq \infty$*,*

$$\mathbf{Y}(t) = \mathbf{Y}_d(t), \forall t \in [t_R, t_F),\ t_0 < t_R < t_F \leq \infty, \tag{1}$$

or at least that $\mathbf{Y}(t)$ *converges asymptotically to* $\mathbf{Y}_d(t)$ *as time t goes to infinity:*

$$t \longrightarrow \infty \Longrightarrow \mathbf{Y}(t) \longrightarrow \mathbf{Y}_d(t). \tag{2}$$

On the book

0.6 Goals of the book

Various purposes of different plants and the great variety of the requirements on the quality of the plant behavior lead to various tracking demands and characterizations. In order to establish their systematic study we meet the following.

Question 6 *Can we establish the tracking theory in the general framework of time-varying nonlinear dynamical systems (plants and their control systems)?*

By relying on:

- The pioneering works [245], [246], [487], [488], which deal with the absolute tracking of Lurie systems, on the comparative or joint stability and tracking studies in [163], [210], [221], [277], which explain the differences among stability and tracking phenomena and concepts and how they both can be achieved simultaneously
- On the works [150]-[154], [173], [177], [180], [181], [183]-[188], [194]-[197], [210], [212]-[215], [218], [220], [225], [232]-[235], [237], [238], [241], [242], [253], [263], [267], [277]-[283], which contain studies of Lyapunov type tracking concept and properties
- The contributions [183], [185], [191], [196], [212], [234], [236], [240], [239], [243], [269], [284], [286], [467], which established the finite-*time* tracking concept and tracking control synthesis, by referring to the concept of the high-gain tracking [445]-[449]
- On the papers [190], [192], [219], [220], [270], which deal with the guaranteed performance index tracking, by following [194], [195], which introduced and developed the tracking domains concept
- The concept of the practical tracking established in [183], [184], [196], [197], [212], [214], by recalling natural tracking control applications [449]-[453]
- The generalization of both the high-quality tracking theory of linear systems tracking control [279] and the theory of the stability domains of *time*-invariant and *time*-varying nonlinear systems [155]-[164], [170]-[172], [175], [174], [182], [199]-[205], [207]-[209], [217], [216], [287]

we state:

Reply 7 *Yes, we can establish the tracking theory in the general framework of time-varying nonlinear dynamical systems.*

This imposes the following goal.

Goal 8 ***The tracking theory***

The first goal of the book is to establish the tracking theory for time-varying nonlinear plants and their control systems.

The theory should comprise the characteristic tracking concepts, the definitions of the main tracking properties involved in the corresponding tracking concept, the corresponding conditions for the controlled plant to exhibit the tracking property, and various algorithms for tracking control synthesis.

The tracking theory will not only extend stability theory but it will introduce novel qualitative dynamical systems and control concepts that will incorporate various new system properties. It will effectively cope with more complex dynamical problems resulting from the real systems' environmental conditions and from the demands for increasing the quality of the systems and of their controls. It will discover new problems, novel systems properties, and will establish new systems and control methods. It will be applicable to the real (time-varying, hence also to time-invariant, nonlinear, and linear) dynamical and control systems. Its main tools are physics, mathematics, the theory of dynamical and control systems. The engineering is the world of its vivid applications.

Studying the tracking phenomena, concepts, and properties, and attacking the tracking control synthesis problems we meet the following fundamental control problem [279].

Problem 9 ***The fundamental control problem***

a) Do the properties of the plant enable the existence of tracking control for all initial conditions from a neighborhood $\mathfrak{N}(t_0)$ *of* $\mathbf{Y}_{d0}$*, [i.e., of* $\mathbf{Y}_d(t)$ *at the initial moment* $t = t_0$*], for all permitted disturbances* $\mathbf{D}(.)$ *belonging to a functional family* $\mathfrak{D}$ *and for every object desired output behavior* $\mathbf{Y}_d(.)$ *belonging to another functional family* $\mathfrak{Y}_d$*?*

b) If, and only if, they do then the plant is ***trackable*** *over* $\mathfrak{D}\times\mathfrak{Y}_d$*. What are the conditions on the plant to be trackable?*

c) What are the size, the shape, and the boundary of the trackability domain?

Goal 10 ***The trackability theory***

The second goal of the book is to establish the trackability theory for time-varying nonlinear plants and their control systems.

The theory should comprise the perfect and imperfect trackability properties.

Once the first and the second goals of the book are achieved then we meet the core control goal.

Goal 11 ***The tracking control synthesis***

The core control goal is the effective synthesis of the tracking control, under the action of which, the plant behavior satisfies all the imposed tracking requirements resulting from its purpose.

The book should present the crucial tracking control concepts that comprise effective tracking control algorithms.

We can now summarize that the book is on a part of the qualitative theory of the nonlinear control systems; i.e., it is on *the tracking theory*, on *the trackability theory,* and on *the tracking control synthesis for time-varying nonlinear plants* as parts of control theory.

The general aim of this book is to treat and effectively solve the problem of achieving the tracking control task whatever the physical (biological, chemical, economical, electrical, electromechanical, mechanical, social) nature of the plant by using appropriate mathematical data and description of the plant.

The aim of the book is to contribute to the creation and the establishment of the tracking and trackability theories of nonlinear plants and of their control systems, and to contribute to the development of the synthesis of tracking control of nonlinear plants.

This, it is hoped, should enrich essentially the corresponding university control systems courses, should open new directions for research in control theory and should enable fruitful new various control engineering applications.

What follows represents a further tracking control aimed development:

- of the nonlinear systems stability theory [6]-[12], [20]-[25], [37], [38], [45] -[48], [56], [66], [67], [71], [78], [83]-[86], [92]-[94], [99], [107], [113], [123], [125], [149], [155]-[164], [160]-[164], [170]-[172], [175], [176], [178], [182], [198]-[211], [217], [216], [222]-[224], [227], [231], [266], [280], [281], [287], [292]-[297], [300], [304]-[306], [311], [316], [322], [346], [356], [358], [359], [364], [363], [367]-[374], [383], [384], [400], [401], [406], [405], [410]-[417], [419]-[421], [427]-[433], [440], [454], [455], [463], [473], [479], [497], [504], [506], [511], [512], [531]-[534], [542]-[548], [562]-[565], [578],

- and of the nonlinear control systems theory including works on tracking [1]-[5], [10], [13], [14], [16], [19], [33]-[35], [39], [43], [51], [53]-[56], [58]-[65], [68]-[70], [72]-[77], [79]-[82], [87]-[91], [95], [100]-[106], [108]-[110], [112], [114]-[124], [126]-[128], [130], [132], [134]-[138], [141]-[147], [150]-[154], [165]-[169], [181]-[180], [183]-[197], [212]-[215], [218]-[221], [225], [226], [232]-[247], [260], [263]-[265], [267]-[270], [277]-[279], [281]-[286], [288]-[291], [298], [307]-[310], [312]-[315], [317]-[321], [323]-[344], [345]-[355], [360]-[362], [366], [376]-[382], [385]-[393], [394]-[399], [402], [403], [407]-[409], [418], [423], [424], [437]-[439], [441]-[453], [456]-[457], [465]-[474], [475]-[478], [483]-[492], [493], [495], [498], [499], [500]-[505], [507]-[513], [515]-[521], [524]-[530], [535]-[541], [549]-[561], [566]-[577]

as well as the development of the trackability theory of *time*-invariant linear or nonlinear plants [165]-[169], [232]. [233], [235]-[241], [279], [282]-[286] to *time*-varying nonlinear plants.

There is not in this book an available space for examples, simulations, or applications. They are left open to everybody for future works.

0.7 The book structure and composition

There are in the sequel eleven parts, eight of which compose the main body of the book. The parts mainly contain chapters that incorporate sections. Some sections contain subsections. The systems defined by the α-th-order nonlinear vector differential equation that describes the system state behavior and by the nonlinear algebraic vector equation that determines the system output dependence on the system internal dynamics, on the control action, and on the external disturbances are **the input-internal dynamics-output** ($IIDO$) **systems**. Their special classes are **the input-state-output** (ISO) **systems** and **the input-output** (IO) **systems**. The first-order nonlinear vector differential equation defines the internal dynamics, i.e., the state dynamics, and the nonlinear algebraic vector equation defines the output behavior of the ISO systems. The mathematical model of the IO systems is the m-th-order nonlinear vector differential equation that shows directly how the system output behavior depends on the external (control and disturbance) actions on the system. The book treats the $IIDO$ systems so that all the results are directly applicable to both IO systems and ISO systems. For the sake of simplicity the $IIDO$ systems are simply called *the systems*, if not stated otherwise.

The *time*-varying and nonlinear nature of the systems ensures the generality of the concepts, definitions, conditions, criteria, and control algorithms.

The main body begins with the essential topics for dynamical systems and their control. The next parts start with new concepts that incorporate definitions of new systems and control properties and end with the criteria or with the conditions on the system.

The fact that the book contributes new concepts, definitions, and system and control features, the author suggests to the reader to pass through the text continuously, without omitting any part, chapter, section, or subsection.

In order to describe the systems, to define their various dynamical and control properties, and to cope effectively with the dynamical and control problems caused by the systems complexity, the book uses the new simple vector and matrix notation analogous to the scalar notation, [252], [279].

0.8 In gratitude

The author is grateful to Mr. George Pearson with MacKichan Company for his very kind and effective assistance to resolve various problems related to the SWP application among which are the problem of the adaptation of the trim size of the book typed by using Scientific Work Place and the problem of effective application of SWP to generate simultaneously the Author Index and Subject Index.

The author is also indebted to
Mr. *Marcus Fontaine, LaTeX Project Editor, for his careful editing*
Ms. *Florence Kizza, Editor, Engineering & Environmental, for her fair leading the review process*
Ms. Nora Konopka, Publisher of Engineering & Environmental Sciences, for leading and organizing the publication process elegantly and effectively
Ms. Michele Smith, Editorial Assistant – Engineering, for very useful assistance
Ms. Jessica Vakili, Senior Project Coordinator, Editorial Project Development, for very useful assistance and for her endeavor to get the trim size of the book suitably adjusted
all of CRC Press/Taylor & Francis.

Belgrade, November 26, 2012 through May 6, 2015, July 8, and September 28 - October 18, November 7, 2015.

Lyubomir T. Gruyitch

Part II

SYSTEMS AND CONTROL BASIS

Chapter 1

Introduction

1.1 *Time* and systems

All processes, motions, and movements, all behaviors of dynamical systems and their responses, as well as all external actions on the systems, occur and propagate in *time*. All tracking processes take place in *time*. It is natural from the physical point of view to study the systems' behavior and its dynamical properties directly in the temporal domain. This requires us to be clear how we understand what *time* is and what are its properties, which we explain in brief as follows [252], [279] (for the more complete analysis of the *time* phenomenon see: [261], [272]-[274]).

Definition 12 ***Time****[252], [279]*

Time *(i.e.,* ***the temporal variable****) denoted by* t *or by* τ *is an independent scalar physical variable such that:*

- *Its value called* ***instant*** *or* ***moment*** *determines uniquely* ***when*** *somebody or something started/interrupted to exist*
- *Its values determine uniquely* ***since when and until when*** *somebody or something existed/exists or will exist*
- *Its values determine uniquely* ***how long*** *somebody or something existed/exists or will exist*
- *Its values determine uniquely whether an event* E_1 *occurs* ***then when*** *another event* E_2 *has not yet happened, or the event* E_1 *took/takes/will take place just* ***then when*** *the event* E_2 *was/is/will be happening, or the event* E_1 *occurs* ***then when*** *the event* E_2 *has already happened*
- *Its value* ***occupies (covers, encloses, imbues, impregnates, is over and in, penetrates) equally*** *everybody and everything (i.e., beings, objects, energy, matter, and space)* ***everywhere and always***
- *Its value has been, is, and will be* ***permanently changing smoothly, strictly monotonously continuously, equally*** *in all spatial directions and their senses, in and around everybody and everything,* ***independently*** *of everybody and everything (i.e., independently of beings, objects, energy, matter, and space),* ***independently*** *of all other variables,* ***independently*** *of all happenings, movements, and processes.*

Continuity, permanent monotonous motion of the instants, and independence are among the basic properties of *time*.

Time is a basic constituent of the existence of everybody and of everything. All human trials for millennia have failed to explain, to express, the nature, the phenomenon, of *time* in terms of other well-defined notions [273], [274]. The nature of *time*, the physical content of it, cannot be explained in terms of other basic constituents of existence (in terms of energy, matter, space) or in terms of other physical phenomena or variables. *Time* has its own, original, self-contained, nature that we can only call *the nature of time, i.e., the temporal nature* or *the time nature* [261], [272]-[274] [252], [279].

The value of *time* t (τ), i.e., **instant** or **moment**, is denoted also by t (or by τ), respectively. It is an **instantaneous** (**momentous**) and **elementary *time* value**. It is the temporal value. It does not contain either energy or matter or space. It can happen *exactly once* and then it is *the same everywhere for, and in, everybody and everything* (i.e.,, for, and in, beings, energy, matter, objects, and space), for all other variables, for all happenings, for all movements, for all processes, and for all biological, physical, and social systems. It is *not repeatable. Nobody and nothing can influence the flow of instants,* [261], [272]-[274] [252], [279].

The physical dimension of *time* is denoted by [T], where T stands for *time dimension.* It *cannot be expressed in terms of the physical dimension of another variable.* Its physical dimension is one of the basic physical dimensions. It is used to express the physical dimensions of most of the physical variables. A selected unit 1_t of *time* can be arbitrarily chosen and then fixed. If it is *second* s then $1_t = s$, which we denote by $t\ \langle 1_t \rangle = t\ \langle s \rangle$ [261], [272]-[274], [252], [279].

There can be assigned *exactly one* (which is denoted by $\exists!$) real number to every moment (instant), and vice versa. The numerical value **num** t of the moment t is a real number, num $t \in \mathfrak{R}$, where $\mathfrak{R}$ is the set of all real numbers, [261], [272]-[274] [252], [279].

Theorem 13 ***Universal time speed law*** *[261], [272]-[274], [252], [279]*
Time is the unique physical variable such that the speed v_t (denoted also with v_τ) of the evolution of its value (of the flow of its values), and of its numerical value, (of their numerical values):

a) Is invariant with respect to a choice of a relative zero moment t_{zero}, of an initial moment t_0, of a time scale and of a time unit 1_t, i.e., invariant relative to a choice of a time axis, invariant relative to a selection of spatial coordinates, and invariant relative to everybody and everything

and

b) Its value (its numerical value) is invariant relative to everybody and everything, and equals one arbitrary time unit per the same time unit (equals one), respectively,

$$v_t = \frac{dt}{dt} = 1[\mathrm{TT}^{-1}]\left\langle 1_t 1_t^{-1} \right\rangle = 1[TT^{-1}]\left\langle 1_\tau 1_\tau^{-1} \right\rangle = \frac{d\tau}{d\tau} = v_\tau,$$
$$num\ v_t = num\ v_\tau = 1, \tag{1.1}$$

relative to arbitrary time axes T and T_τ; i.e., its numerical value equals 1 (one) with respect to all time axes (with respect to any accepted relative zero instant t_{zero}, any chosen initial instant t_0, any time scale and any selected time unit 1_t), with respect to all spatial coordinate systems, and with respect to all beings and all objects.

Comment 14 *The derivative of every variable x over itself equals identically one:*

$$\frac{dx}{dx} \equiv 1,$$

but such derivative is the speed of the variation of the value of the variable x if, and only if, the variable $\times$ is time t (or τ). Otherwise, dx/dx is not the speed of the variable x value variation. Its speed is dx/dt (or $dx/d\tau$). This explains why time is the unique variable with such speed properties. This is so obvious. Physics ignores this time speed property so much that it has become able to accept since 1905 the meaningless claim on the existence of different times with different time speeds (for more on this see [261], [272]-[274], [252], [279]).

Time set $\mathfrak{T}$ is the set of all moments. It is an open, unbounded, and connected set. It is in biunivoque (one-to-one) correspondence with the set $\mathfrak{R}$ of all real numbers,

$$\mathfrak{T} = \{t : \text{num } t \in \mathfrak{R},\ dt > 0,\ \ v_t = \frac{dt}{dt} = t^{(1)} \equiv 1\ [TT^{-1}]\},$$
$$\forall t \in \mathfrak{T},\ \exists! x \in \mathfrak{R} \Longrightarrow x = \text{num } t,\ \textit{and}$$
$$\forall \times \in \mathfrak{R},\ \exists! t \in \mathfrak{T} \Longrightarrow \text{num } t = x,$$
$$\text{num inf } \mathfrak{T} \ = \text{num } t_{\text{inf}} = -\infty \ \textit{and} \text{ num sup } \mathfrak{T} = \text{num } t_{\text{sup}} = \infty. \tag{1.2}$$

The rule of the correspondence determines an **accepted relative zero numerical *time* value** t_{zero}, a ***time* scale**, and a ***time* unit** denoted by 1_t (or by 1_τ). The *time* unit can be ... , millisecond, second, minute, hour, day, ..., as well noted by Isaac Newton [462, p. 8: Scholium]. He clearly explained therein the sense and the meaning of *relative time* in the same sentence in which he explained the sense and meaning of *absolute time*. Newton's *absolute time* is *the temporal variable*: *time*. Newton's *relative time* is *the value of the temporal variable*: *the value of time* measured in a chosen time scale T_t with the corresponding *time* unit 1_t [261], [272]-[274], [252], [279].

Although the *time* set $\mathfrak{T}$ is in biunivoque (one-to-one) correspondence with *the set* $\mathfrak{R}$ *of the real numbers*, they are inherently different. The crucial difference between them is the dynamical property of the *time* set $\mathfrak{T}$ and *the set* $\mathfrak{R}$ is static. The permanent, monotonous, continuous flow of the instants (which are the elements of the *time* set $\mathfrak{T}$) expresses the dynamics of the *time* set $\mathfrak{T}$. The real numbers are immobile, [261], [272]-[274], [252], [279].

We accept in this book, as in [261], [272]-[274], [252], [279], the *relative zero moment* t_{zero} to have the zero numerical value, num $t_{zero} = 0$. However, we need not adopt it to be also *the initial moment* t_0, $t_0 \neq t_{zero}$, num $t_0 \neq 0$, in view of the *time*-varying nature of the systems to be studied. This determines the subset $\mathfrak{T}_0$ of $\mathfrak{T}$,

$$\mathfrak{T}_0 = \{t : t \in \mathfrak{T},\ \text{num } t \in [\text{num } t_0, \infty[\}. \tag{1.3}$$

If the system properties do not depend on the choice of the initial instant t_0, which means that the properties are uniform (relative to t_0), then we can accept $t_0 = t_{zero}$, num $t_0 = 0$. This holds for all *time*-invariant systems.

Various properties of *time*-varying systems can vary in *time*. For example, stability properties can depend essentially on the initial moment t_0 [156], [170], [256], [257]. The set of the initial moments t_0 of our interest is $\mathfrak{T}_i$. It is a subset of $\mathfrak{T}$,

$$\mathfrak{T}_i = \{t : t[T]\,\langle s\rangle\,,\ t \in \mathfrak{T}\} \subseteq \mathfrak{T},\ \inf\ \mathfrak{T}_i \geq -\infty,\ \ \sup\ \mathfrak{T}_i \leq \infty. \tag{1.4}$$

Note 15 *[252], [273], [274] It has been the common custom to use the letters t and τ to denote not only time itself, and an arbitrary instant, but also the numerical value of the arbitrary instant relative to the chosen zero instant, e.g., $t = 0$ is used in the sense num $t = 0$. From the physical point of view this is incorrect. The numerical value num t of the instant t is a real number without a physical dimension, whereas the instant t is a temporal value that has the physical dimension, the temporal dimension T of time. We overcome this by using the normalized, dimensionless, mathematical temporal variable, denoted by $\overline{t}$ and defined by*

$$\overline{t} = \frac{t}{1_t}\ [-],$$

so that the time set $\mathfrak{T}$ is to be replaced by

$$\overline{\mathfrak{T}} = \{\overline{t}\ [-] : \overline{t} = num\ \overline{t} = num\ t \in \mathfrak{R},\ d\overline{t} > 0,\ \overline{t}^{(1)} \equiv 1\}.$$

With this in mind we use in the sequel the letter t also for $\overline{t}$, and $\mathfrak{T}$ also for $\overline{\mathfrak{T}}$. Hence,

$$t\ [-] = num\ t\ [-].$$

Between any two different instants $t_1 \in \mathfrak{T}$ and $t_2 \in \mathfrak{T}$ there is a third instant $t_3 \in \mathfrak{T}$, either $t_1 < t_3 < t_2$ or $t_2 < t_3 < t_1$. The *time* set $\mathfrak{T}$ is **continuum**. It is also called **the continuous-*time* set**.

Let $\tau \in \mathfrak{T}$ or $\tau = \infty$, $\tau = t_0$ is permitted. The instant τ determines the *time* set $\mathfrak{T}_\tau$,

$$\mathfrak{T}_\tau = \{t : t \in \mathfrak{T}, t \geq \tau\} \subseteq \mathfrak{T}, \ t_0 \in \mathfrak{T}_\tau, \ \mathfrak{T}_0 \subseteq \mathfrak{T}_\tau. \tag{1.5}$$

If $\tau = t_0$ then $\mathfrak{T}_\tau = \mathfrak{T}_0$.

Being a physical variable, *time* possesses some general properties valid for all physical variables. We summarize them in the form of the principles, [261], [272]-[274], [252], [279]:

1.2 *Time*, physical variables, and systems

Common qualitative properties of the physical variables are important for the study of control systems. We summarize some of them in the following principles by using the books [261], [272]-[274], [252], [279]:

Principle 16 ***Physical Continuity and Uniqueness Principle (*****PCUP*****): scalar form***

A physical variable can change its value from one value to another one only by passing through every intermediate value, and it possesses a unique local instantaneous real value in any place (in any being or in any object) at any moment.

Principle 17 ***Physical Continuity and Uniqueness Principle (*****PCUP*****): matrix and vector form***

A vector physical variable or a matrix (vector) of physical variables can change, respectively, its vector or matrix (vector) value from one vector or matrix (vector) value to another one only by passing elementwise through every intermediate vector or matrix (vector) value, and it possesses a unique local instantaneous real vector or matrix (vector) value in any place (i.e., in any being or in any object) at any moment, respectively.

Principle 18 ***Physical Continuity and Uniqueness Principle (*****PCUP*****): system form***

The system physical variables (including those of their derivatives and integrals that appear in the system, as well as system solutions) can change, respectively, their (scalar or vector or matrix) values from one (scalar or vector or matrix) value to another one only by passing elementwise through every intermediate (scalar or vector or matrix) value, and they possess unique local instantaneous real (scalar or vector or matrix) values in any place at any moment.

The *PCUP* is inherent to accurate modeling physical systems.

Corollary 19 ***Mathematical model of a physical variable, mathematical model of a physical system and PCUP***

a) For a mathematical (scalar or vector) variable to be, respectively, an adequate description of a physical (scalar or vector) variable it is necessary that it obeys the physical continuity and uniqueness principle.

b) For a mathematical model of a physical system to be an adequate description of the physical system it is necessary that all its system variables obey the physical continuity and uniqueness principle; i.e., that the mathematical model obeys the physical continuity and uniqueness principle.

The synthesis of the properties of *time* and of the common properties of the physical variables expressed by *PCUP* (Principle 16 through Principle 18) results in:

Principle 20 ***Time Continuity and Uniqueness Principle (TCUP)***
Any (scalar or vector) physical variable and any vector / matrix of physical variables can change, respectively, its scalar / vector / matrix value from one scalar / vector / matrix value to another one only continuously in time by passing (elementwise) through every intermediate scalar / vector / matrix value, and it possesses a unique local instantaneous real scalar / vector / matrix value in any place (in any being or in any object) at any moment.

Definition 21 ***The system form of the*** *TCUP*
The system form of the TCUP means that all system variables (including their derivatives and integrals that appear in the system, as well as system solutions) satisfy the TCUP.

The effective application of the *TCUP* to the stability study of dynamical systems and to their control synthesis is in [262], [255], [155], [268], [275]-[277].

Corollary 22 *[273], [274], [252]* ***Mathematical representation of a physical variable, mathematical model of a physical system and TCUP***
a) For a mathematical (scalar or vector) variable to be, respectively, an adequate description of a physical (scalar or vector) variable it is necessary that it obeys the time continuity and uniqueness principle TCUP.
b) For a mathematical model of a physical system to be an adequate description of the physical system it is necessary that its system variables obey the time continuity and uniqueness principle; or equivalently, that the mathematical model obeys the time continuity and uniqueness principle.
c) For a mathematical model of a physical system to be an adequate description of the physical system it is necessary that its solutions are unique and continuous in time.

The books [259], [274], [273] contain the complete study of *time*, the establishment of the novel, consistent, physical, and mathematical relativity theory, its relationship to systems, and its importance for control.

Comment 23 ***Time relativity theory and controller design***
It might seem at first glance that there is not any importance of relativity theory for control science and engineering.
There exist different theories on time relativity, two of which are well known: Newton's and Einstein's, and the new, consistent, relativity theory [261], [272]-[274], [252], [279].

Question 24 *Which theory, Galilean-Newtonean or Einstein's, is correct and should be applied to design very high speed objects (vehicles)?*

Comment 25 *The answer to the question is very important for control designers of very fast rockets, projectiles, and space vehicles.*
For explanation and reply see [259], [261], [272]-[274], [252], [279].

1.3 Notational preliminaries

Lower-case ordinary letters denote scalars, bold (lower-case and capital, Greek and Roman) block letters signify vectors, capital italic letters stand for matrices, and we use capital $\mathfrak{Fraktur}$ letters for sets and spaces. Motions (solutions) or sets are denoted by capital $\mathcal{CALLIGRAPHIC}$ letters.

Let us explain the meaning of the basic notation and symbols.

In what follows:

$\mathfrak{C}^{ki} = \mathfrak{C}^{k}(\mathfrak{R}^{i})$ is *the family of all functions defined and k-times continuously differentiable on* $\mathfrak{R}^{i}$.

$\mathfrak{C}^k(\mathfrak{T}_0)$ is *the family of all functions defined, continuous, and k-times continuously differentiable on* $\mathfrak{T}_0$, $\mathfrak{C}^k = \mathfrak{C}^k(\mathfrak{T})$ and $\mathfrak{C}^0 = \mathfrak{C}$.

$\mathfrak{C}^{k-}(\mathfrak{R}^N)$ is *the family of all functions defined and continuous everywhere on* $\mathfrak{R}^N$, $(k-1)$ *times continuously differentiable on* $\mathfrak{R}^N$, *and k-times continuously differentiable on* $\mathfrak{R}^N \setminus \{\mathbf{0}_N\}$, which have defined the left and the right k-th-order derivative at the origin $\mathbf{0}_N$.

$\mathfrak{C}^{-}(\mathfrak{R}^N)$ is *the family of all functions defined everywhere on* $\mathfrak{R}^N$,

$i, j, k, l, K, m, N, n, \alpha, \rho$ are natural numbers.

$\mathbf{I}(.)$ is *the input vector function.*

$\mathcal{J}^k$ is *the family of all bounded k-times continuously differentiable on* $\mathfrak{T}_0$ *permitted input vector functions* $\mathbf{I}(.)$.

K can be any of the following combinations:

$$K \in \{N, \rho, (k+1)N, \alpha\rho\}, \quad k \in \{0, 1, \cdots, \alpha-1\}. \tag{1.6}$$

$\mathfrak{L} \subseteq \mathfrak{T}$ is *the largest temporal set of the initial moments* t_0 *for which the system has a solution,* $\mathfrak{L} = (l_m, l_M)$.

$\mathfrak{R}$ is *the set of all real numbers.* The division of two real numbers x and y different from zero, is well-defined in $\mathfrak{R}$:

$$x \in \mathfrak{R}, \quad y \in \mathfrak{R}, \quad xy \neq 0 \Longrightarrow \frac{x}{y} \in \mathfrak{R} \ \textit{and} \ \frac{y}{x} \in \mathfrak{R}.$$

$\mathfrak{R}^i$ is *the i-dimensional real vector space,* the zero vector in which is $\mathbf{0}_i$. Notice that $\mathfrak{R}^1 \neq \mathfrak{R}$. The division of two vectors (x) and (y) from $\mathfrak{R}^1$ is not defined in $\mathfrak{R}^1$ despite that the divisions xy^{-1} and yx^{-1} are well-defined in $\mathfrak{R}$ for $xy \neq 0$.

$\mathfrak{R}^K$ is the *K-dimensional real vector space,*

$$\mathfrak{R}^K \in \left\{\mathfrak{R}^N, \ \mathfrak{R}^\rho, \mathfrak{R}^{(k+1)N}, \ \mathfrak{R}^{\alpha\rho}\right\}. \tag{1.7}$$

$\mathfrak{R}^{M\times N}$ is $M\times N$ matrix space, the zero matrix in which is $O_{M\times N}$. If $M{=}N$ then, for short $O_{N\times N} = O_N$.

$\mathfrak{R}^N \setminus \{\mathbf{0}_N\}$ *is the set difference between* $\mathfrak{R}^N$ *and the singleton* $\{\mathbf{0}_N\}$, *i.e., the complement of* $\{\mathbf{0}_N\}$ *to* $\mathfrak{R}^N$.

$\mathbf{R}^{\alpha-1} \in \mathfrak{R}^{\alpha\rho}$ is *the total internal dynamics vector of the system (3.12).*

$\mathfrak{Y} \subseteq \mathfrak{R}^N$ is *the total output vector space of the system.*

$\mathbf{Y} \in \mathfrak{Y}$ is *the total output vector of the system.*

$\mathcal{Y}(.)$ is a mapping called *the output response* that describes the output dynamics behavior of the system,

$$\mathcal{Y}(.) : \mathfrak{T}_0 \times \mathfrak{L} \times \mathfrak{Z}_0 \times \mathcal{J}^i \rightarrow \mathfrak{Y}, \quad \mathcal{Y}(t_0; t_0; \mathbf{Z}_0; \mathbf{I}) \equiv \mathbf{Y}_0.$$

$\mathfrak{Z} \subseteq \mathfrak{R}^K$ is *the general total internal dynamics vector space.*

$\mathfrak{Z}_0$ is *the set of initial conditions* $\mathbf{Z}_0$ or $\mathbf{z}_0$ for which the system has a solution.

$\mathcal{Z}(.)$ is a mapping called *motion (solution)* that describes the internal dynamics behavior of a system,

$$\mathcal{Z}(.) : \mathfrak{T}_0 \times \mathfrak{L} \times \mathfrak{Z}_0 \times \mathcal{J}^i \rightarrow \mathfrak{Z}, \quad \mathcal{Z}(t_0; t_0; \mathbf{Z}_0; \mathbf{I}) \equiv \mathbf{Z}_0.$$

Note 26 ***On the new notation***

In the sequel

$$\alpha, \rho, N \in \{1, 2, \cdots\}, \quad k \in \{0, 1, \cdots, \alpha - 1\}, \quad j, l \in \{0, 1, \cdots\}. \tag{1.8}$$

Notice that for $k = 0$:

$$\mathbf{Y}^k \mid_{k=0} = \mathbf{Y}^0 = \mathbf{Y}.$$

In order to define, and to use effectively, the system properties we need new, simple, and elegant notation proposed in [159]. For example, instead of using the output vector extended by its derivatives:

$$\begin{bmatrix} \mathbf{Y} \\ \mathbf{Y}^{(1)} \\ \cdots \\ \mathbf{Y}^{(j)} \end{bmatrix}$$

we can use $\mathbf{Y}^j$:

$$\mathbf{Y}^j = \begin{bmatrix} \mathbf{Y} \\ \mathbf{Y}^{(1)} \\ \cdots \\ \mathbf{Y}^{(j)} \end{bmatrix} = \begin{bmatrix} \mathbf{Y}^{(0)} \\ \mathbf{Y}^{(1)} \\ \cdots \\ \mathbf{Y}^{(j)} \end{bmatrix} \in \mathfrak{R}^{(j+1)N}, \quad \mathbf{Y}^0 = \mathbf{Y}, \tag{1.9}$$

by introducing this general compact vector notation that is different from

$$\mathbf{Y}^{(j)} = \frac{d^j \mathbf{Y}}{dt^j} \in \mathfrak{R}^N, \quad j \in \{0, 1, \ \cdots\}.$$

The compact notation yields

$$\mathbf{D}^\eta(t) = \left[\mathbf{D}^T(t) \vdots \mathbf{D}^{(1)T}(t) \vdots \ldots \vdots \mathbf{D}^{(\eta)T}(t)\right]^T \in \mathfrak{R}^{(\eta+1)d}, \tag{1.10}$$

$$\mathbf{U}^\mu(t) = \left[\mathbf{U}^T(t) \vdots \mathbf{U}^{(1)T}(t) \vdots \ldots \vdots \mathbf{U}^{(\mu)T}(t)\right]^T \in \mathfrak{R}^{(\mu+1)r}, \tag{1.11}$$

$$\mathbf{Y}^\alpha(t) = \left[\mathbf{Y}^T(t) \vdots \mathbf{Y}^{(1)T}(t) \vdots \ldots \vdots \mathbf{Y}^{(\alpha)T}(t)\right]^T \in \mathfrak{R}^{(\alpha+1)N}. \tag{1.12}$$

The superscripts α, η, μ, and kp are not in the parentheses in $\mathbf{R}^{\alpha-1}(t)$, $\mathbf{D}^\eta(t)$, in $\mathbf{U}^\mu(t)$, and in $\mathbf{Y}^k(t)$, respectively, because $\mathbf{R}^{(\alpha-1)}(t)$, $\mathbf{D}^{(\eta)}(t)$, $\mathbf{U}^{(\mu)}(t)$, and $\mathbf{Y}^{(k)}(t)$ denote the $(\alpha-1)$-th derivative $d^{\alpha-1}\mathbf{R}(t)/dt^{\alpha-1}$ of $\mathbf{R}(t)$, the η-th derivative $d^\eta \mathbf{D}(t)/dt^\eta$ of $\mathbf{D}(t)$, the μ-th derivative $d^\mu \mathbf{U}(t)/dt^\mu$ of $\mathbf{U}(t)$ and the k-th derivative $d^k \mathbf{Y}(t)/dt^k$ of $\mathbf{Y}(t)$, respectively,

$$\begin{aligned}
\mathbf{R}^{\alpha-1}(t) &= \left[\mathbf{R}^T(t) \vdots \mathbf{R}^{(1)T}(t) \vdots \ldots \vdots \mathbf{R}^{(\alpha-1)T}(t)\right]^T \neq \mathbf{R}^{(\alpha-1)}(t) = \frac{d^{\alpha-1}\mathbf{R}(t)}{dt^{\alpha-1}}, \\
\mathbf{D}^\eta(t) &= \left[\mathbf{D}^T(t) \vdots \mathbf{D}^{(1)T}(t) \vdots \ldots \vdots \mathbf{D}^{(\eta)T}(t)\right]^T \neq \mathbf{D}^{(\eta)}(t) = \frac{d^\eta \mathbf{D}(t)}{dt^\mu}, \\
\mathbf{U}^\mu(t) &= \left[\mathbf{U}^T(t) \vdots \mathbf{U}^{(1)T}(t) \vdots \ldots \vdots \mathbf{U}^{(\mu)T}(t)\right]^T \neq \mathbf{U}^{(\mu)}(t) = \frac{d^\mu \mathbf{U}(t)}{dt^\mu}, \\
\mathbf{Y}^k(t) &= \left[\mathbf{Y}^T(t) \vdots \mathbf{Y}^{(1)T}(t) \vdots \ldots \vdots \mathbf{Y}^{(k)T}(t)\right]^T \neq \mathbf{Y}^{(k)}(t) = \frac{d^k \mathbf{Y}(t)}{dt^p}.
\end{aligned}$$

The desired vector value $\mathbf{R}_d^{\alpha-1}$ of $\mathbf{R}^{\alpha-1}$ and the induced *deviation vector* $\mathbf{r}$ are defined by

$$\mathbf{r} = \mathbf{R} - \mathbf{R}_d \Longrightarrow \mathbf{r}^{\alpha-1} = \mathbf{R}^{\alpha-1} - \mathbf{R}_d^{\alpha-1}. \tag{1.13}$$

The desired vector value $\mathbf{Y}_d$ of $\mathbf{Y}$ and the induced *deviation vector* $\mathbf{y}$ are defined by

$$\mathbf{y} = \mathbf{Y} - \mathbf{Y}_d, \tag{1.14}$$

which is opposite to *the output error vector* ε,

$$\varepsilon = \mathbf{Y_d} - \mathbf{Y} = -\mathbf{y}, \tag{1.15}$$

$\mathbf{Y}^{k-1} \in \mathfrak{R}^{(k+1)N}$ is *the total output vector of the systems, its desired vector value* is $\mathbf{Y}_d^{k-1}$, and the induced *deviation vector* $\mathbf{y}^{p-1}$ is defined by

$$\mathbf{y}^{k-1} = \mathbf{Y}^{k-1} - \mathbf{Y}_d^{k-1}. \tag{1.16}$$

$\mathbf{Z} \in \mathfrak{R}^K$ is *the general (unifying) total vector, and* $\mathbf{Z}_d$ *is its desired vector value,*

$$\mathbf{Z} \in \left\{\mathbf{Y}, \mathbf{Y}^k, \mathbf{R}^{\alpha-1}\right\},\ \mathbf{Z} = [Z_1\ Z_2\ ...Z_K]^T \in \mathfrak{R}^K,$$
$$\mathbf{z} = \mathbf{Z} - \mathbf{Z}_d,\ \mathbf{z} \in \left\{\mathbf{y}, \mathbf{y}^k, \mathbf{r}^{\alpha-1}\right\},\ \mathbf{z} \in \mathfrak{R}^K. \tag{1.17}$$

$\mathbf{z}$ is the *deviation vector.*

Note 27 *The higher the system order and/or the higher the system dimension, the more advantageous the new notation.*

We use symbolic vector notation and operations in the elementwise sense as follows:

- *The zero and unit vectors,*

$$\mathbf{0}_N = [0\ \ 0\ \ \cdots\ \ 0]^T \in \mathfrak{R}^N,\ \ \mathbf{1}_N = [1\ \ 1\ \ \cdots\ \ 1]^T \in \mathfrak{R}^N.$$

- *The diagonal matrix* V *associated elementwise with a vector* $\mathbf{v}$,

$$\mathbf{v} = [v_1\ \ v_2\ \ \cdots\ \ v_N]^T \Longrightarrow V = diag\{v_1\ \ v_2 \cdots\ \ v_N\},$$

- *The vector and matrix absolute values,*

$$|\mathbf{v}| = [|v_1|\ \ |v_2| \cdots\ \ |v_N|]^T,\ \ |V| = diag\{|v_1|\ \ |v_2| \cdots\ \ |v_N|\},$$

- *the elementwise vector inequality,*

$$\mathbf{w} \neq \mathbf{v} \Longleftrightarrow w_i \neq v_i,\ \ \forall i = 1, 2, \cdots, N.$$

We define the following sign function:

- $sign(.) : \mathfrak{R} \to \{-1, 0, 1\}$ *the scalar signum function,*

$$sign\ (v) = |v|^{-1} v\ if\ v \neq 0,\ and\ sign\ (0) = 0.$$

$\nexists \mathbf{Y}^{-1}$ signifies that $\mathbf{Y}^{-1}$ does not exist in general. However, if all entries Y_i of $\mathbf{Y}$ are nonzero real numbers, $Y_i \neq 0$, then we can define $\mathbf{Y}^{-1}$ as follows.

Definition 28 ***The vector inversion***

a) The vector $\mathbf{Y}$,

$$\mathbf{Y} = [Y_1\ \ Y_2 \cdots\ \ Y_N]^T \in \mathfrak{R}^{N\times 1},\ Y_i \in \mathfrak{R},\ \forall i = 1, 2, \cdots, N, \tag{1.18}$$

has its inverse vector $\mathbf{Y}^{-1}$ *if, and only if, all its entries* Y_i *are nonzero numbers,*

$$\exists \mathbf{Y}^{-1} = \left[Y_1^{-1}\ \ Y_2^{-1} \cdots\ \ Y_N^{-1}\right] \in \mathfrak{R}^{1xN} \Longleftrightarrow Y_i \neq 0,\ \ \forall i = 1, 2, \cdots, N.$$

b) If, and only if, the vector $\mathbf{Y}$, *(1.18), has its inverse vector* $\mathbf{Y}^{-1}$,

$$Y_i \in \mathfrak{R},\ Y_i \neq 0,\ \ \forall i = 1, 2, \cdots, N, \tag{1.19}$$

then

$$\mathbf{Y}^{-1} = \left[Y_1^{-1} \ \ Y_2^{-1} \cdots \ Y_N^{-1}\right] \in \mathfrak{R}^{1\text{x}\mathsf{N}} \tag{1.20}$$

satisfies:

$$\frac{1}{N}\mathbf{Y}^{-1}\mathbf{Y} = 1,$$

and for

$$\begin{gathered} diag\ \mathbf{Y} = diag\ \{Y_1 \ \ Y_2 \cdots \ Y_N\} \in \mathfrak{R}^{N\times N}, \\ diag\ \mathbf{Y}^{-1} = diag\ \left\{Y_1^{-1} \ \ Y_2^{-1} \cdots \ Y_N^{-1}\right\} \in \mathfrak{R}^{N\times N} \end{gathered} \tag{1.21}$$

it obeys

$$\begin{gathered} diag\ \left(\mathbf{Y}\mathbf{Y}^{-1}\right) = diag\ \mathbf{Y}\ \ diag\ \mathbf{Y}^{-1} = \begin{bmatrix} 1 & 0 & \ldots & 0 \\ 0 & 1 & \ldots & 0 \\ \ldots & \ldots & \ldots & \ldots \\ 0 & 0 & \ldots & 1 \end{bmatrix} = I_N \in \mathfrak{R}^{N\times N}, \\ \left(diag\ \mathbf{Y}\right)\left(\mathbf{Y}^{-1}\right)^T = \left(diag\ \mathbf{Y}^{-1}\right)\mathbf{Y} = \mathbf{1}_N = [1 \ \ 1 \cdots \ 1]^T \in \mathfrak{R}^{N\text{x}1}, \\ \mathbf{Y}^T\left(diag\ \mathbf{Y}^{-1}\right) = \mathbf{Y}^{-1}\ diag\ \mathbf{Y} = [1 \ \ 1 \cdots \ 1] \in \mathfrak{R}^{1\text{x}\mathsf{N}}. \end{gathered} \tag{1.22}$$

Other new notation is defined at its first use and in Appendix A.

1.4 Classes of the systems

This book concerns the three general classes of nonlinear systems.

Note 29 ***System classes***
The new notation enables us to unify the qualitative (e.g., stability, tracking) theory of all three classes of the dynamical systems treated herein.

- *The IO* ***systems*** *described by the m-th-order time-varying nonlinear vector differential equation (3.62) in the output vector* $\mathbf{Y}$*, which is the first subvector of the internal dynamics vector* $\mathbf{Y}^{m-1}$
- *The ISO* ***systems*** *described by the first-order time-varying nonlinear vector differential equation (3.65) in the state vector* $\mathbf{X}$*, and by the time-varying nonlinear vector algebraic output equation (3.66) that defines the dependence of the output vector* $\mathbf{Y}$ *on the state vector* $\mathbf{X}$ *and on the input vector* $\mathbf{I}$
- *The IIDO* ***systems*** *described by the* α*-th-order time-varying nonlinear vector differential equation [the first equation (3.12)] in the subvector* $\mathbf{R}$ *of the internal dynamics vector* $\mathbf{R}^{\alpha-1}$*, and by the time-varying nonlinear vector algebraic output equation [the second equation (3.12)] that determines the dependence of the output vector* $\mathbf{Y}$ *on the internal dynamics vector* $\mathbf{R}^{\alpha-1}$ *and on the input vector* $\mathbf{I}$*. The internal dynamics vector* $\mathbf{R}^{\alpha-1}$ *is the* ***state vector*** *of the IIDO system.*

Comment 30 The results on the IIDO systems hold directly for both the IO systems and the ISO systems
We show in the sequel (Claim 82 and Conclusion 83) that both the IO systems (3.62) and the ISO systems (3.65), (3.66) represent special cases of the IIDO systems (3.12). The

tracking and trackability studies of the IIDO systems incorporate the corresponding studies of both the IO systems and the ISO systems. The results on the former are directly valid for and applicable to the latter. After proving this (Claim 82) we deal with the IIDO system(s) that are for short called simply ***the system(s).***

Chapter 2

Sets

2.1 Set basis

The meaning and the notion of the term **set** is in the following sense:

Definition 31 *Set*

***Set**, which is denoted by $\mathfrak{S}$, is **separateness** of some beings, objects, symbols, numbers, letters, ..., which are its **members** (i.e., **elements**). The nature of its members can be any (biological, chemical, economic, financial, mathematical, physical, social,...).*

Set is mathematically denoted by the brackets $\{\}$ between which are the symbols in biunivoque correspondence with the set members.

If, and only if, x is member of $\mathfrak{S}$ then it is denoted by $x \in \mathfrak{S}$ meaning "x belongs to $\mathfrak{S}$", i.e., "x is an element of $\mathfrak{S}$"; i.e., "x is a member of $\mathfrak{S}$".

If, and only if, y is not a member of $\mathfrak{S}$ then it is denoted by $y \notin \mathfrak{S}$ meaning "y does not belong to $\mathfrak{S}$"; "y is not element of $\mathfrak{S}$"; "y is not an member of $\mathfrak{S}$".

For the basic set properties, operations and relations see, for example, [390], and [426].

A set can be *time*-invariant, (*time*-independent), i.e., constant. This is the classical treatment of sets. For example, the set of all people who wrote this book is constant, but the set of all people who are reading this book at a moment τ depends on τ. In the former case the set is *time*-invariant, i.e., constant, whereas in the latter case the set is *time*-varying.

For stability (and for tracking) properties, and for their studies, it is crucial whether their validity is related to *time*-invariant sets or to *time*-varying sets. Assuming that the reader is familiar with the basis of the theory of constant sets (e.g., [390], [426]), let us briefly explain several interesting phenomena of *time*-varying sets [175], [227, pp. 33-42].

The singleton (the set that has exactly one element) containing only the origin $\mathbf{0}_K$ of the vector space $\mathfrak{R}^K$ is denoted by $\mathfrak{O}$ for any natural number K,

$$\mathfrak{O} = \{\mathbf{0}_K\}, \quad K \in \{N, \rho, (k+1)N, \alpha\rho\}, \tag{2.1}$$

where (1.8), $\alpha, \rho, N \in \{1, 2, \cdots\}, k \in \{0, 1, \cdots, \alpha - 1\}$.

Let ϕ be *the empty set* meaning the set without any element.

The set of all subsets of a set $\mathfrak{S}$ including the empty set, $\mathfrak{S} \subseteq \mathfrak{R}^K$, is denoted by $2^{\mathfrak{S}}$ or by $\mathfrak{P}(\mathfrak{S})$,

$$\mathfrak{P}(\mathfrak{S}) = 2^{\mathfrak{S}}. \tag{2.2}$$

It is *the power set of the set* $\mathfrak{S}$. The power set of $\mathfrak{R}^K$ is

$$\mathfrak{P}\left(\mathfrak{R}^K\right) = 2^{\mathfrak{R}^K}. \tag{2.3}$$

For example, the power set of the empty set ϕ is the nonempty set $\mathfrak{P}(\phi) = 2^{\phi}$, $\mathfrak{P}(\phi) = 2^{\phi} = \{\phi\} \neq \phi$. Because the empty set ϕ is a subset of every set, then the power set $\mathfrak{P}(\mathfrak{S})$ of any singleton $\mathfrak{S} = \{z\}$ is the two member set, $\mathfrak{S} = \{z\} \Longrightarrow \mathfrak{P}(\mathfrak{S}) = \{\phi, \mathfrak{S}\}$.

A set mapping $\mathfrak{S}(.) : \mathfrak{T} \longrightarrow \mathfrak{P}(\mathfrak{R}^K)$ is a *time*-varying *set-valued function.* Its set value $\mathfrak{S}(t)$ at a moment $t \in \mathfrak{T}$ is a *time-varying set* (at the moment $t \in \mathfrak{T}$). If, and only if, $\mathfrak{S}(t) = \mathfrak{S}$ for all $t \in \mathfrak{T}$ then it is a *constant set*; i.e., a *time-invariant set (time-independent set).*

A *time*-varying set $\mathfrak{S}(t; t_0)$ is *bounded on* $\mathfrak{T}_0$ if, and only if, there is a constant bounded nonempty set $\mathfrak{A}$ such that

$$\mathfrak{S}(t; t_0) \subseteq \mathfrak{A}, \quad \forall t \in \mathfrak{T}_0.$$

If, and only if, this holds also for all $t_0 \in \mathfrak{T}_i$ and $\mathfrak{T}_i \subseteq \mathfrak{T}$ then the set $\mathfrak{S}(t; t_0)$ is *bounded on* $\mathfrak{T}_i$. If, and only if, this is valid for $\mathfrak{T}_i = \mathfrak{T}$; i.e., if, and only if, t_0 can be arbitrary in $\mathfrak{T}$, *then* the set $\mathfrak{S}(t; t_0) \equiv \mathfrak{S}(t)$ is *bounded.*

If, and only if, t_0 can be arbitrary in $\mathfrak{T}$ then $\mathfrak{S}(t; t_0) \equiv \mathfrak{S}(t)$.

Let $\|.\| : \mathfrak{R}^K \to \mathfrak{R}_+$ be *an accepted norm on* $\mathfrak{R}^K$, which can be the *Euclidean norm* $\|.\|_2$ *on* $\mathfrak{R}^K$ in a special case, $||\mathbf{z}||_2 = \left(\mathbf{z}^T\mathbf{z}\right)^{1/2} = \left(z_1^2 + z_2^2 + \cdots + z_K^2\right)^{1/2}$.

In$\mathfrak{S}(t; t_0)$ is *the interior of the set* $\mathfrak{S}(t; t_0)$ *at a moment* $t \in \mathfrak{T}_0$, which is the set of all points $\mathbf{z}$ such that there is an open hyperball $\mathfrak{B}_\mu(t, \mathbf{z})$,

$$\mathfrak{B}_\mu(t, \mathbf{z}) = \{\mathbf{x} : \mathbf{x} \in \mathfrak{S}(t; t_0), \ \ \|\mathbf{x} - \mathbf{z}\| < \mu\},$$

centered at $\mathbf{z}$ at the moment t so that it is a subset of the set $\mathfrak{S}(t; t_0)$, $\mathfrak{B}_\mu(t, \mathbf{z}) \subseteq \mathfrak{S}(t; t_0)$. The point $\mathbf{z}$ is *the interior point* of the set $\mathfrak{S}(t; t_0)$ *at the moment* t.

$\partial\mathfrak{S}(t; t_0)$ is *the boundary of the set* $\mathfrak{S}(t; t_0)$ *at a moment* $t \in \mathfrak{T}_0$, which is the set of all points $\mathbf{z}$ such that in every open hyperball $\mathfrak{B}_\eta(t, \mathbf{z})$ centered at $\mathbf{z}$ at the moment t there is a point $\mathbf{y}$ belonging to the set $\mathfrak{S}(t; t_0)$, $\mathbf{y} \in \mathfrak{S}(t; t_0)$, and a point $\mathbf{w}$ that is not in the set $\mathfrak{S}(t; t_0)$, $\mathbf{w} \notin \mathfrak{S}(t; t_0)$. The point $\mathbf{z}$ is *the boundary point of the set* $\mathfrak{S}(t; t_0)$ *at the moment* t.

A set $\mathfrak{S}(t; t_0)$ is an *open set at a moment* $t \in \mathfrak{T}_0$ if, and only if, it is its own interior In $\mathfrak{S}$, $\mathfrak{S} =$In $\mathfrak{S}$, *at the moment* t.

Cl $\mathfrak{S}(t; t_0)$ is *the closure of the set* $\mathfrak{S}(t; t_0)$ *at a moment* $t \in \mathfrak{T}_0$, which is the set of all interior and boundary points *of the set* $\mathfrak{S}(t; t_0)$ *at the moment* t,

$$\text{Cl}\mathfrak{S}(t; t_0) = \partial\mathfrak{S}(t; t_0) \ \cup \ \text{In}\mathfrak{S}(t; t_0).$$

A set $\mathfrak{S}(t; t_0)$ is a *closed set at a moment* $t \in \mathfrak{T}_0$ if, and only if, it is its own closure Cl $\mathfrak{S}(t; t_0)$, $\mathfrak{S}(t; t_0) =$ Cl $\mathfrak{S}(t; t_0)$, *at the moment* t.

The set $\mathfrak{S}_l(\mathfrak{T}_\tau)$ is *the lower set limit* on $\mathfrak{T}_\tau$ of the set $\mathfrak{S}(t; t_0)$,

$$\mathfrak{S}_l(\mathfrak{T}_\tau) = \cap\left[\mathfrak{S}(t; t_0) : (t; t_0) \in \mathfrak{T}_0 \times \mathfrak{T}_\tau\right].$$

The set $\mathfrak{S}^u(\mathfrak{T}_\tau)$ is *the upper set limit* on $\mathfrak{T}_\tau$ of the set $\mathfrak{S}(t; t_0)$,

$$\mathfrak{S}^u(\mathfrak{T}_\tau) = \cup\left[\mathfrak{S}(t; t_0) : (t; t_0) \in \mathfrak{T}_0 \times \mathfrak{T}_\tau\right].$$

Let $\mathbf{z} = [z_1 \ z_2 \ \ldots \ z_K]^T \in \mathfrak{R}^K$. The *set difference* between the set $\mathfrak{A}(t; t_0)$ and the set $\mathfrak{B}(t; t_0)$ *at a moment* $t \in \mathfrak{T}_0$ is the set $\mathfrak{A}(t; t_0) \backslash \mathfrak{B}(t; t_0)$. It is the set of all vectors $\mathbf{z}$ in $\mathfrak{A}(t; t_0)$ that do not belong to $\mathfrak{B}(t; t_0)$,

$$\mathfrak{A}(t; t_0) \backslash \mathfrak{B}(t; t_0) = \{\mathbf{z} : \ \mathbf{z} \in \mathfrak{A}(t; t_0), \ \mathbf{z} \notin \mathfrak{B}(t; t_0)\}. \tag{2.4}$$

$\mathfrak{A}(t; t_0)\Delta\mathfrak{B}(t; t_0)$ is *the symmetric set difference* between the set $\mathfrak{A}$ and the set $\mathfrak{B}$ *at a moment* $t \in \mathfrak{T}_0$. It is the set of all vectors $\mathbf{z}$ in $\mathfrak{A}(t; t_0)$, which do not belong to $\mathfrak{B}(t; t_0)$, and of all vectors $\mathbf{z}$ that are in $\mathfrak{B}(t; t_0)$ but not in $\mathfrak{A}(t; t_0)$,

$$\mathfrak{A}(t; t_0)\Delta\mathfrak{B}(t; t_0) = \left[\mathfrak{A}(t; t_0) \backslash \mathfrak{B}(t; t_0)\right] \cup \left[\mathfrak{B}(t; t_0) \backslash \mathfrak{A}(t; t_0)\right]. \tag{2.5}$$

Let $\mathfrak{B}_\xi(\mathbf{a};t;t_0)$ be an open hyperball with the radius $\xi > 0$ centered at the point $\mathbf{a}$ in the space $\mathfrak{R}^K$ at $t \in \mathfrak{T}_0$,

$$\mathfrak{B}_\xi(\mathbf{a};t;t_0) = \{\mathbf{z} : \mathbf{z} \in \mathfrak{R}^K, \ \ \|\mathbf{z} - \mathbf{a}(t;t_0)\| < \xi\} \subseteq .\mathfrak{R}^K.$$

If, and only if, $\mathbf{a} = \mathbf{0}_K$ then we omit $(\mathbf{0}_K)$ from the notation $\mathfrak{B}_\xi(\mathbf{0}_K)$,

$$\mathfrak{B}_\xi = \mathfrak{B}_\xi(\mathbf{0}_K) = \left\{\mathbf{z} : \mathbf{z} \in \mathfrak{R}^K, \ \ \|\mathbf{z}\| < \xi\right\}. \tag{2.6}$$

Definition 32 ***Set diameter***

a) A nonnegative real number $\mathfrak{d} \in \mathfrak{R}_+$, $\mathfrak{d} = \mathfrak{d}(\mathfrak{S})$, or $\mathfrak{d} = \mathfrak{d}(\mathfrak{S}) = \infty$, is:

a-1. The ***inner diameter of a time-invariant set*** *$\mathfrak{S}$, which is denoted by $\mathfrak{d}_m = \mathfrak{d}_m(\mathfrak{S})$, if, and only if, it is the diameter of the closure $Cl\mathfrak{B}_{\mathfrak{d}_m/2}$ of the largest hyperball $\mathfrak{B}_{\mathfrak{d}_m/2}$ such that $Cl\mathfrak{B}_{\mathfrak{d}_m/2} \subseteq Cl\mathfrak{S}$*

a-2. The ***outer diameter of a time-invariant*** *$\mathfrak{S}$, which is denoted by $\mathfrak{d}_M = \mathfrak{d}_M(\mathfrak{S})$, if, and only if, it is the diameter of the closure $Cl\mathfrak{B}_{\mathfrak{d}_M/2}$ of the smallest hyperball $\mathfrak{B}_{\mathfrak{d}_M/2}$such that $Cl\mathfrak{B}_{\mathfrak{d}_M/2} \supseteq Cl\mathfrak{S}$*

b) A nonnegative real number $\mathfrak{d}$, $\mathfrak{d} = \mathfrak{d}(t;\mathfrak{S}) \in \mathfrak{R}_+$, or $\mathfrak{d}(t;\mathfrak{S}) = \infty$, is:

b-1. ***The inner diameter of a time-varying set*** *$\mathfrak{S}(t)$* ***at a moment*** *$t \in \mathfrak{T}$, which is denoted by $\mathfrak{d}_m[\mathfrak{S}(t)] = \mathfrak{d}_m(t;\mathfrak{S})$, if, and only if, it is the diameter of the closure $Cl\mathfrak{B}_{\mathfrak{d}_m/2}(t)$ of the largest hyperball $\mathfrak{B}_{\mathfrak{d}_m/2}(t)$ at the moment t such that $Cl\mathfrak{B}_{\mathfrak{d}_m/2}(t) \subseteq Cl\mathfrak{S}(t)$*

b-2. ***The outer diameter of a time-varying set*** *$\mathfrak{S}(t)$* ***at a moment*** *$t \in \mathfrak{T}$, which is denoted by $\mathfrak{d}_M[\mathfrak{S}(t)] = \mathfrak{d}_M(t;\mathfrak{S})$, if, and only if, it is the diameter of the closure $Cl\mathfrak{B}_{\mathfrak{d}_M/2}(t)$ of the smallest hyperball $\mathfrak{B}_{\mathfrak{d}_M/2}(t)$ at the moment t such that $Cl\mathfrak{B}_{\mathfrak{d}_M/2}(t) \supseteq Cl\mathfrak{S}(t)$*

c) A nonnegative real number $\mathfrak{d}$, $\mathfrak{d} = \mathfrak{d}(\mathfrak{T}_\tau;\mathfrak{S}) \in \mathfrak{R}_+$, or $\mathfrak{d}(\mathfrak{T}_\tau;\mathfrak{S}) = \infty$, is:

c-1. ***The inner diameter of a time-varying set*** *$\mathfrak{S}(t)$* ***on*** *$\mathfrak{T}_\tau$, which is denoted by $\mathfrak{d}_m = \mathfrak{d}_m(\mathfrak{T}_\tau;\mathfrak{S})$, if, and only if,*

$$\mathfrak{d}_m(\mathfrak{T}_\tau;\mathfrak{S}) = \inf\left[\mathfrak{d}_m(t;\mathfrak{S}) : t \in \mathfrak{T}_\tau\right] = \mathfrak{d}_m(\mathfrak{S}_{m\tau}), \quad \mathfrak{S}_{m\tau} = \cap\left[\mathfrak{S}(t) : t \in \mathfrak{T}_\tau\right]$$

c-2. ***The outer diameter of a time-varying set*** *$\mathfrak{S}(t)$* ***on*** *$\mathfrak{T}_\tau$, which is denoted by $\mathfrak{d}_M = \mathfrak{d}_M(\mathfrak{T}_\tau;\mathfrak{S})$, if, and only if,*

$$\mathfrak{d}_M(\mathfrak{T}_\tau;\mathfrak{S}) = \sup\left[\mathfrak{d}_M(t;\mathfrak{S}) : t \in \mathfrak{T}_\tau\right] = \mathfrak{d}_M(\mathfrak{S}_{M\tau}), \quad \mathfrak{S}_{M\tau} = \cup\left[\mathfrak{S}(t) : t \in \mathfrak{T}_\tau\right].$$

The expression "**on** *$\mathfrak{T}_\tau$*" *is to be omitted if, and only if, $\mathfrak{T}_\tau = \mathfrak{T}$. Then, and only then, $\mathfrak{d}_{(.)}(\mathfrak{T}_\tau;\mathfrak{S}) = \mathfrak{d}_{(.)}(\mathfrak{S})$.*

If, and only if, $\mathfrak{S}$ is a time-varying set $\mathfrak{S}(t)$, $\mathfrak{S} = \mathfrak{S}(t)$, then the previous definitions hold at a moment $t \in \mathfrak{T}_\tau$.

A set $\mathfrak{S}(t;t_0)$ is *a bounded set at a moment* $t \in \mathfrak{T}_0$ if, and only if, its outer diameter is a positive real number at the moment t, $\mathfrak{d}_M(t;\mathfrak{S}) \in \mathfrak{R}^+$. It is *bounded on* $\mathfrak{T}_\tau$ if, and only if, its outer diameter on $\mathfrak{T}_\tau$ is a positive real number, $\mathfrak{d}_M(\mathfrak{T}_\tau;\mathfrak{S}) \in \mathfrak{R}^+$, or, equivalently, a *time*-varying set $\mathfrak{S}(t)$ is bounded on $\mathfrak{T}_\tau$ if, and only if, there is a bounded *time*-invariant set $\mathfrak{A}$ such that it is the upperset of $\mathfrak{S}(t)$ for every $t \in \mathfrak{T}_\tau$,

$$\mathfrak{S}(t) \subset \mathfrak{A}, \quad \forall t \in \mathfrak{T}_\tau. \tag{2.7}$$

2.2 Neighborhood

Definition 33 ***Neighborhood of the origin***

A set $\mathfrak{N}$ is a ***neighborhood*** *of the origin $\mathbf{a} = \mathbf{0}_K$ if, and only if, there is an open hyperball $\mathfrak{B}_\zeta$ centered at the origin $\mathbf{a} = \mathbf{0}_K$ with the radius $\zeta \in \mathfrak{R}^+$ such that*

$$\mathfrak{B}_\zeta \subseteq \mathfrak{N}. \tag{2.8}$$

Note 34 *We use neighborhoods of the origin* $\mathbf{a} = \mathbf{0}_K$ *denoted, for example, by* $\mathfrak{N}_\delta$, $\mathfrak{N}\varepsilon$, *or* $\mathfrak{N}_\Delta$. *In general, they are different from the hyperballs* $\mathfrak{B}_\delta$, $\mathfrak{B}\varepsilon$, *and* $\mathfrak{B}_\Delta$, *respectively,*

$$\mathfrak{N}_\delta \neq \mathfrak{B}_\delta, \mathfrak{N}_\varepsilon \neq \mathfrak{B}_\varepsilon, \text{ and } \mathfrak{N}_\Delta \neq \mathfrak{B}_\Delta \text{ in general,}$$
$$In\, \mathfrak{N}_\delta \supseteq \mathfrak{B}_\delta, \quad In\, \mathfrak{N}_\varepsilon \supseteq \mathfrak{B}_\varepsilon, \quad \text{and} \quad In\, \mathfrak{N}_\Delta \supseteq \mathfrak{B}_\Delta,$$

in the sense that they need not be hyperballs but only neighborhoods of any shape. The subscripts δ, ε, *and* Δ *of* $\mathfrak{N}_\delta$, $\mathfrak{N}\varepsilon$, *and* $\mathfrak{N}_\Delta$ *are the inner radii of the neighborhoods* $\mathfrak{N}_\delta$, $\mathfrak{N}\varepsilon$, *and* $\mathfrak{N}_\Delta$, *respectively. The neighborhoods* $\mathfrak{N}_\delta$, $\mathfrak{N}\varepsilon$, *and* $\mathfrak{N}_\Delta$ *can be equal to the hyperballs* $\mathfrak{B}_\delta$, $\mathfrak{B}\varepsilon$, *and* $\mathfrak{B}_\Delta$, *respectively, in special cases.*

The accepted scalar norm $\|.\| : \mathfrak{R}^K \to \mathfrak{R}_+$ on $\mathfrak{R}^K$ induces the distance function $\rho(.) : \mathfrak{R}^K \to \mathfrak{R}_+$ on $\mathfrak{R}^K$, which determines the distance $\rho(\mathbf{a}_1, \mathbf{a}_2)$ between $\mathbf{a}_1 \in \mathfrak{R}^K$ and $\mathbf{a}_2 \in \mathfrak{R}^K$ defined by

$$\rho(\mathbf{a}_1, \mathbf{a}_2) = \|\mathbf{a}_1 - \mathbf{a}_2\|, \tag{2.9}$$

and the distance function $\rho(.) : \mathfrak{R}^K \times 2^{\mathfrak{R}^K} \to \mathfrak{R}_+$ on $\mathfrak{R}^K \times 2^{\mathfrak{R}^K}$ of a point $\mathbf{a}$ from a set $\mathfrak{S}$, which is defined by

$$\rho(\mathbf{a}, \mathfrak{S}) = \inf\left[\|\mathbf{a} - \mathbf{a}^*\| : \mathbf{a}^* \in \mathrm{Cl}\mathfrak{S}\right]. \tag{2.10}$$

Definition 35 ***Neighborhood of a time-invariant connected set*** $\mathfrak{S}$
A set $\mathfrak{N}(\mathfrak{S})$ *is* ***a neighborhood of a time-invariant connected set*** $\mathfrak{S}$ *if, and only if, the closure* $Cl\mathfrak{S}$ *of* $\mathfrak{S}$ *is a subset of* $\mathfrak{N}(\mathfrak{S})$ *and the intersection* $\partial\mathfrak{S} \cap \partial\mathfrak{N}(\mathfrak{S})$ *of their boundaries* $\partial\mathfrak{S}$ *and* $\partial\mathfrak{N}(\mathfrak{S})$ *is empty set* ϕ,

$$Cl\mathfrak{S} \subset \mathfrak{N}(\mathfrak{S}), \quad \partial\mathfrak{S} \cap \partial\mathfrak{N}(\mathfrak{S}) = \phi, \tag{2.11}$$

i.e., if, and only if, the closure $Cl\, \mathfrak{S}$ *of* $\mathfrak{S}$ *is a subset of the interior* $In\, \mathfrak{N}(\mathfrak{S})$ *of* $\mathfrak{N}(\mathfrak{S})$,

$$Cl\mathfrak{S} \subset In\mathfrak{N}, \tag{2.12}$$

or equivalently, $\mathfrak{S}$ *is a subset of* $\mathfrak{N}(\mathfrak{S})$ *if, and only if,* $\mathfrak{S} \subset \mathfrak{N}(\mathfrak{S})$ *and there is* $\mu > 0$ *such that the smallest distance between the boundaries* $\partial\mathfrak{S}$ *and* $\partial\mathfrak{N}(\mathfrak{S})$ *of* $\mathfrak{S}$ *and of* $\mathfrak{N}(\mathfrak{S})$, *respectively, is bigger than* μ,

$$\mathfrak{S} \subset \mathfrak{N}(\mathfrak{S}), \quad \exists \mu > 0 \Longrightarrow$$
$$\min\left\{\inf\left[\rho(\mathbf{z}, \partial\mathfrak{S}) : \mathbf{z} \in \partial\mathfrak{N}(\mathfrak{S})\right], \inf\left[\rho\left[\mathbf{z}, \partial\mathfrak{N}(\mathfrak{S})\right] : \mathbf{z} \in \partial\mathfrak{S}\right], \right\} > \mu. \tag{2.13}$$

Definition 36 ***The*** ε***-neighborhood of a time-invariant connected set***
A set $\mathfrak{N}_\varepsilon(\mathfrak{S})$ *is* ***an*** ε***-neighborhood of a time-invariant connected set*** $\mathfrak{S}$ *if, and only if, it contains all points* $\mathbf{z} \in \mathfrak{R}^K$ *the distance of which from the set* $\mathfrak{S}$ *is smaller than* ε, $\varepsilon > 0$,

$$\mathfrak{N}_\varepsilon(\mathfrak{S}) = \left\{\mathbf{z} : \mathbf{z} \in \mathfrak{R}^K, \ \rho(\mathbf{z}, \mathfrak{S}) < \varepsilon\right\} \supset \mathfrak{S}. \tag{2.14}$$

This definition results from Definition 35. We broaden the preceding definitions to *time-varying* sets.

Definition 37 ***Distance of a point*** $\mathbf{a} \in \mathfrak{R}^K$ ***from a time-varying connected set***
The distance of a point (vector) $\mathbf{a} \in \mathfrak{R}^K$ ***from a time-varying connected set*** $\mathfrak{S}(t) \subseteq\in \mathfrak{R}^K$ ***at a moment*** $t \in \mathfrak{T}$ *is denoted by* $\rho\left[\mathbf{a}, \mathfrak{S}(t)\right]$ *and determined by*

$$\rho\left[\mathbf{a}, \mathfrak{S}(t)\right] = \inf\left[\|\mathbf{a} - \mathbf{a}^*\| : \mathbf{a}^* \in Cl\mathfrak{S}(t)\right]. \tag{2.15}$$

Definition 38 ***The fixed neighborhood of a time-varying connected set*** $\mathfrak{S}(t)$ ***on*** $\mathfrak{T}_\tau$

A time-varying set $\mathfrak{N}[\mathfrak{S}(t),\mathfrak{T}_\tau]$ *is* ***a fixed neighborhood of a time-varying connected set*** $\mathfrak{S}(t)$ ***on*** $\mathfrak{T}_\tau$ *if, and only if, the closure* $Cl\mathfrak{S}(t)$ *of* $\mathfrak{S}(t)$ *is a subset of* $\mathfrak{N}[\mathfrak{S}(t),\mathfrak{T}_\tau]$ *and the intersection* $\partial\mathfrak{S}(t)\cap\partial\mathfrak{N}[\mathfrak{S}(t),\mathfrak{T}_\tau]$ *of their boundaries* $\partial\mathfrak{S}(t)$ *and* $\partial\mathfrak{N}(\mathfrak{S})$ *is empty set* ϕ *at every* $t\in\mathfrak{T}_\tau$,

$$Cl\mathfrak{S}(t)\subset\mathfrak{N}[\mathfrak{S}(t);\mathfrak{T}_\tau],\ \ \partial\mathfrak{S}(t)\cap\partial\mathfrak{N}[\mathfrak{S}(t);\mathfrak{T}_\tau]=\phi,\ \ \forall t\in\mathfrak{T}_\tau, \tag{2.16}$$

i.e., if, and only if, the closure $Cl\mathfrak{S}(t)$ *of* $\mathfrak{S}(t)$ *is a subset of* $\mathfrak{N}[\mathfrak{S}(t);\mathfrak{T}_\tau]$ *and there is a constant* $\mu>0$ *such that the smallest distance between the boundaries* $\partial\mathfrak{S}(t)$ *and* $\partial\mathfrak{N}[\mathfrak{S}(t);\mathfrak{T}_\tau]$ *of* $\mathfrak{S}(t)$ *and* $\mathfrak{N}[\mathfrak{S}(t);\mathfrak{T}_\tau]$, *respectively, is bigger than* μ, $\forall t\in\mathfrak{T}_\tau$,

$$\begin{gathered}\forall t\in\mathfrak{T}_\tau : Cl\mathfrak{S}(t)\subset\mathfrak{N}[\mathfrak{S}(t);\mathfrak{T}_\tau],\\ \exists\mu>0\Longrightarrow\min\left\{\begin{array}{l}\inf\left[\rho\left[\mathbf{z},\partial\mathfrak{S}(t)\right]:\mathbf{z}\in\partial\mathfrak{N}[\mathfrak{S}(t);\mathfrak{T}_\tau]\right],\\ \inf\left[\rho\left[\mathbf{z},\partial\mathfrak{N}[\mathfrak{S}(t);\mathfrak{T}_\tau]\right]:\mathbf{z}\in\partial\mathfrak{S}(t)\right],\end{array}\right\}>\mu.\end{gathered} \tag{2.17}$$

The notation "$\mathfrak{T}_\tau$*", "*$\forall t\in\mathfrak{T}_\tau$*", "on* $\mathfrak{T}_\tau$*" is omitted if, and only if,* $\mathfrak{T}_\tau=\mathfrak{T}$.

Definition 39 ***The fixed*** ε***-neighborhood of a time-varying connected set*** $\mathfrak{S}(t)$ ***at*** $t\in\mathfrak{T}_\tau$

A time-varying set $\mathfrak{N}_\varepsilon(t;\mathfrak{S})$ *is* ***the*** $\varepsilon-$***neighborhood of a time-varying connected set*** $\mathfrak{S}(t)$ ***at*** $t\in\mathfrak{T}_\tau$ *if, and only if, it contains all points* $\mathbf{z}\in\mathfrak{R}^K$ *the distance of which from the set* $\mathfrak{S}(t)$ *is smaller than* ε, $\varepsilon>0$, *at* $t\in\mathfrak{T}_\tau$,

$$\mathfrak{N}_\varepsilon(t;\mathfrak{S})=\left\{\mathbf{z}:\mathbf{z}\in\mathfrak{R}^K,\ \rho\left[\mathbf{z},\mathfrak{S}(t)\right]<\varepsilon\right\}\supset\mathfrak{S}(t),\ t\in\mathfrak{T}_\tau. \tag{2.18}$$

Definition 40 ***The fixed neighborhood of the vector value*** $\mathbf{z}(t)$ ***of a vector function*** $\mathbf{z}(.)$ ***at*** $t\in\mathfrak{T}_\tau$

a) A time-varying set $\mathfrak{N}(t;\mathbf{z}(t))$ *is* ***a fixed neighborhood of the vector value*** $\mathbf{z}(t)$ ***of a vector function*** $\mathbf{z}(.)$ ***at*** $t\in\mathfrak{T}_\tau$ *if, and only if, the singleton* $\mathfrak{Z}(t)=\{\mathbf{z}(t)\}$ *is a subset of the interior* $In\,\mathfrak{N}(t,\mathbf{z}(t))$ *of* $\mathfrak{N}(t,\mathbf{z}(t))$ *at* $t\in\mathfrak{T}_\tau$,

$$\mathfrak{Z}(t)=\{\mathbf{z}(t)\}\subset In\mathfrak{N}(t,\mathbf{z}(t)),\ \ t\in\mathfrak{T}_\tau, \tag{2.19}$$

or equivalently, if, and only if, $\mathfrak{Z}(t)=\{\mathbf{z}(t)\}$ *is a subset of* $\mathfrak{N}(t,\mathbf{z}(t))$ *and there is constant* $\mu>0$ *such that the distance of* $\mathbf{z}(t)$ *from the boundary* $\partial\mathfrak{N}(t,\mathbf{z}(t))$ *of* $\mathfrak{N}(t,\mathbf{z}(t))$ *is bigger than* μ *at* $t\in\mathfrak{T}_\tau$,

$$\begin{gathered}\mathfrak{Z}(t)=\{\mathbf{z}(t)\}\subset\mathfrak{N}(t,\mathbf{z}(t)),\\ \exists\mu>0\Longrightarrow\rho\left[\mathbf{z}(t),\partial\mathfrak{N}(t,\mathbf{z}(t))\right]>\mu,\ \ t\in\mathfrak{T}_\tau,\end{gathered} \tag{2.20}$$

Figure 2.1. If, and only if, this holds for every $t\in\mathfrak{T}_\tau$ *then* $\mathfrak{N}[t;\mathbf{z}(t)]$ *is* ***a fixed neighborhood of the vector value*** $\mathbf{z}(t)$ ***of a vector function*** $\mathbf{z}(.)$ ***on*** $\mathfrak{T}_\tau$, *which is denoted by* $\mathfrak{N}[\mathbf{z}(t);\mathfrak{T}_\tau]$.

The notation "$\mathfrak{T}_\tau$*", "*$\forall t\in\mathfrak{T}_\tau$*", "on* $\mathfrak{T}_\tau$*" is omitted if, and only if,* $\mathfrak{T}_\tau=\mathfrak{T}$.

Definition 41 ***The fixed*** ε***-neighborhood of the value*** $\mathbf{z}(t)$ ***of a vector function*** $\mathbf{z}(.)$ ***at*** $t\in\mathfrak{T}_\tau$

A set $\mathfrak{N}_\varepsilon(t,\mathbf{z})$ *is* ***the fixed*** ε***-neighborhood of the value*** $\mathbf{z}(t)$ ***of a vector function*** $\mathbf{z}(.)$ ***at*** $t\in\mathfrak{T}_\tau$ *if, and only if, it contains all points* $\mathbf{w}\in\mathfrak{R}^K$ *the distance of which from the vector* $\mathbf{z}(t)$ *is smaller than* $\varepsilon>0$ *at* $t\in\mathfrak{T}_\tau$,

$$\begin{gathered}\mathfrak{N}_\varepsilon(t,\mathbf{z}(t))=\left\{\mathbf{w}:\mathbf{w}\in\mathfrak{R}^K,\ \rho\left[\mathbf{w},\mathbf{z}(t)\right]<\varepsilon\right\}\supset\mathfrak{Z}(t),\ t\in\mathfrak{T}_\tau\\ \mathfrak{Z}(t)=\{\mathbf{z}(t)\},\ \ t\in\mathfrak{T}_\tau.\end{gathered} \tag{2.21}$$

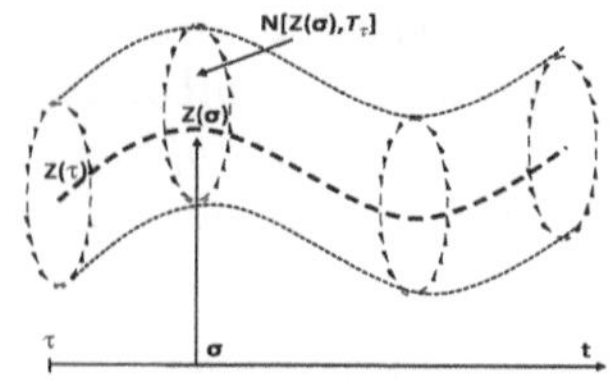

Figure 2.1: A fixed neighborhood $\mathfrak{N}\left[\mathbf{z}\left(t\right);\mathfrak{T}_{\tau}\right]$ of a vector function $\mathbf{z}\left(.\right)$ on $\mathfrak{T}_{\tau}$.

Definition 42 ***The elementwise v-neighborhood of the vector value*** $\mathbf{z}(t)$ ***of a vector function*** $\mathbf{z}\left(.\right)$ ***at*** $t\in\mathfrak{T}_{\tau}$

Let $\mathbf{v}\in\mathfrak{R}^{K}$, $\mathbf{v}=\left[v_{1}\ \ v_{2}\cdots\ v_{K}\right]^{T}$. ***The elementwise v-neighborhood of the vector value*** $\mathbf{z}(t)$ ***of a vector function*** $\mathbf{z}\left(.\right)$ ***at*** $t\in\mathfrak{T}_{\tau}$ *is the connected vector neighborhood* $\mathcal{N}_{\mathbf{v}}\left[\mathbf{z}\left(t\right)\right]$ *of* $\mathbf{z}\left(t\right)$ *at* $t\in\mathfrak{T}_{\tau}$ *defined by*

$$\begin{aligned}\mathcal{N}_{\mathbf{v}}\left[\mathbf{z}\left(t\right)\right] &= \left\{\mathbf{w}:\mathbf{w}\in\mathfrak{R}^{K},\ \left|\mathbf{w}-\mathbf{z}\left(t\right)\right|<\mathbf{v}\right\} \\ &= \left\{\mathbf{w}:\mathbf{w}\in\mathfrak{R}^{K},\ \left|w_{i}-z_{i}\left(t\right)\right|<v_{i},\ \forall i=1,2,\cdots,K\right\}. \end{aligned} \tag{2.22}$$

2.3 Continuity of sets

The set operator Δ denotes *the symmetric set difference* between two sets $\mathfrak{A}$ and $\mathfrak{B}$, which is the set of all vectors $\mathbf{Z}$ in $\mathfrak{A}$ that do not belong to $\mathfrak{B}$, and of all vectors that are in $\mathfrak{B}$ but not in $\mathfrak{A}$,

$$\mathfrak{A}\Delta\mathfrak{B}=\left(\mathfrak{A}\backslash\mathfrak{B}\right)\cup\left(\mathfrak{B}\backslash\mathfrak{A}\right). \tag{2.23}$$

It should be distinguished from the ordinary set difference $\mathfrak{A}\backslash\mathfrak{B}$, which is the set of all vectors $\mathbf{Z}$ in $\mathfrak{A}$ that do not belong to $\mathfrak{B}$,

$$\mathfrak{A}\backslash\mathfrak{B}=\left\{\mathbf{Z}:\ \mathbf{Z}\in\mathfrak{A},\ \mathbf{Z}\notin\mathfrak{B}\right\}. \tag{2.24}$$

Definition 43 ***Set continuity in time***

Let a time-varying set $\mathfrak{S}(t;t_{0})$ *be connected and bounded on* $\mathfrak{T}_{0}$ *and let* $\sigma\in\mathfrak{T}_{0}$. *Then:*

i) There exists lim $\left[\mathfrak{S}(t;t_{0}):t\longrightarrow\tau\right]$ *if, and only if, there is a connected bounded constant set* $\mathfrak{L}$ *and for every time sequence* $\mathfrak{T}_{\longrightarrow\tau}$ *converging to* τ,

$$\mathfrak{T}_{\longrightarrow\tau}=\left\{t_{k}:t_{k}\in\mathfrak{T},\ k\longrightarrow\infty\Longrightarrow t_{k}\longrightarrow\tau\right\}, \tag{2.25}$$

and for every $\varepsilon\in\mathfrak{R}^{+}$ *there exists a natural number* $\mu=\mu\left(\sigma,\varepsilon\right)$ *such that*

$$\left[\mathfrak{S}(t_{k};t_{0})\Delta\mathfrak{L}\right]\subset\mathfrak{N}\left[\partial\mathfrak{S}(t_{k};t_{0});\varepsilon\right],\quad t_{k}\in\mathfrak{T}_{\longrightarrow\tau},\ \forall k\in\left]\mu,\infty\right[,$$

which is expressed by

$$\lim\left[\mathfrak{S}(t;t_{0}):t\longrightarrow\tau\right]=\mathfrak{L}, \tag{2.26}$$

and the set $\mathfrak{L}$ *is* ***the limit set*** *of* $\mathfrak{S}(t;t_{0})$ *at* τ,

ii) The set $\mathfrak{S}(t;t_{0})$ *is continuous in* $t\in\mathfrak{T}_{\tau}$, $\mathfrak{T}_{\tau}\subseteq\mathfrak{T}_{0}$, $t_{0}\in\mathfrak{T}_{\tau}$, *if, and only if,*

$$\lim\left[\mathfrak{S}(t;t_{0}):t\longrightarrow\zeta\right]=\mathfrak{S}(\zeta;t_{0}),\quad\forall\zeta\in\mathfrak{T}_{\tau}, \tag{2.27}$$

which is denoted by $\mathfrak{S}(t;t_{0})\in\mathfrak{C}(\mathfrak{T}_{\tau})$. *This is* ***set continuity on*** $\mathfrak{T}_{\tau}$. *The expressions "****on*** $\mathfrak{T}_{\tau}$*" and "*$\mathfrak{T}_{\tau}$*" should be omitted if, and only if,* $\mathfrak{T}_{\tau}=\mathfrak{T}$.

The distance of $\mathbf{z} \in \mathfrak{R}^K$ from a *time*-varying set $\mathfrak{S}(t) \subset \mathfrak{R}^K$, (2.15), permits us to introduce *two set distances* $\rho_{(.)}\left[\mathfrak{S}_1(t), \mathfrak{S}_2(t)\right], (.) = m, M$, between *time*-varying sets $\mathfrak{S}_1(t) \subset \mathfrak{R}^K$ and $\mathfrak{S}_2(t) \subset \mathfrak{R}^K$ at a moment $t \in \mathfrak{T}$, [381].

The inner set distance $\rho_m\left[\mathfrak{S}_1(t), \mathfrak{S}_2(t)\right]$ *between the sets* $\mathfrak{S}_1(t)$ *and* $\mathfrak{S}_2(t)$ *at* $t \in \mathfrak{T}$:

$$\rho_m\left[\mathfrak{S}_1(t), \mathfrak{S}_2(t)\right] = \inf\left\{\rho\left(\mathbf{x}, \mathbf{y}\right) : \mathbf{x} \in \mathfrak{S}_1(t),\ \mathbf{y} \in \mathfrak{S}_2(t)\right\}$$
$$= \min\left\{\begin{array}{c} \inf\left\langle\rho\left[\mathbf{z}, \mathfrak{S}_1(t)\right] : \mathbf{z} \in \mathfrak{S}_2(t)\right\rangle, \\ \inf\left\langle\rho\left[\mathbf{z}, \mathfrak{S}_2(t)\right] : \mathbf{z} \in \mathfrak{S}_1(t)\right\rangle \end{array}\right\}, \tag{2.28}$$

and *the outer set distance* $\rho_M\left[\mathfrak{S}_1(t), \mathfrak{S}_2(t)\right]$ *between the sets* $\mathfrak{S}_1(t)$ *and* $\mathfrak{S}_2(t)$ *at* $t \in \mathfrak{T}$:

$$\rho_M\left[\mathfrak{S}_1(t), \mathfrak{S}_2(t)\right] = \sup\left\{\rho\left(\mathbf{x}, \mathbf{y}\right) : \mathbf{x} \in \mathfrak{S}_1(t),\ \mathbf{y} \in \mathfrak{S}_2(t)\right\}$$
$$= \max\left\{\begin{array}{c} \sup\left\langle\rho\left[\mathbf{z}, \mathfrak{S}_1(t)\right] : \mathbf{z} \in \mathfrak{S}_2(t)\right\rangle, \\ \sup\left\langle\rho\left[\mathbf{z}, \mathfrak{S}_2(t)\right] : \mathbf{z} \in \mathfrak{S}_1(t)\right\rangle \end{array}\right\}. \tag{2.29}$$

The notions of the set distances (2.28) and (2.29) enabled Grujic to define in [175, Definition 3, pp. 530, 531], [227, Definition 17, pp. 33, 34] continuity of a *time*-varying set equivalently to Definition 43.

Definition 44 ***Set continuity in time (the equivalent definition)***

i) A set $\mathfrak{L}$ *is* ***the limit set*** *of* $\mathfrak{S}(t; t_0)$ *at* $\tau \in \mathfrak{T}$, *denoted by (2.26), if, and only if,*

$$\lim\left\{\rho_M\left[\mathfrak{S}(t; t_0), \mathfrak{L}\right] : t \longrightarrow \tau\right\} = 0. \tag{2.30}$$

ii) A time-varying set is ***continuous in*** $t \in \mathfrak{T}_\tau$, $\mathfrak{T}_\tau \subseteq \mathfrak{T}$, $t_0 \in \mathfrak{T}_\tau$, *if, and only if, (2.27) holds. This is* ***set continuity on*** $\mathfrak{T}_\tau$.

The following definition links sets and system motions:

Definition 45 ***Positive invariance of time-varying sets***

A time-continuous set $\mathfrak{S}(t; t_0)$ *is positive invariant with respect to the system motions* $\mathbf{z}\left(t; t_0; \mathbf{z}_0; \mathbf{I}\right)$ *on* $\mathfrak{T}_0 \times \mathcal{J}^i$ *if, and only if* $\mathbf{z}_0 \in \mathfrak{S}(t_0; t_0)$ *implies* $\mathbf{z}\left(t; t_0; \mathbf{z}_0; \mathbf{I}\right) \in \mathfrak{S}(t; t_0)$, $\forall\left[t, \mathbf{I}\left(.\right)\right] \in \mathfrak{T}_0 \times \mathcal{J}^i$.

2.4 Set contraction

The hyperballs $\mathfrak{B}_{\delta(t_0, \varepsilon)}$ and $\mathfrak{B}_{\Delta(t_0)}$ vary in general with t_0 for *time*-varying systems. For strongly nonstationary systems their closures can contract to the singleton $\mathfrak{O}$ as t_0 goes to infinity [175]. This justified the introduction and the use (by Grujic in [175, Definition 3, p. 531], [209, Theorem 3, pp. 46-48, Theorem 4, pp. 51-53], [227, Definition 18, p. 34]) of the new family of the sets called *asymptotically contractive,* Figure *2.2. Let the set* Υ be a nonempty connected subset of $\mathfrak{R}^K$ with a nonempty boundary $\partial\Upsilon$, $\Upsilon \subset \mathfrak{R}^K$, $\partial\Upsilon \neq \phi$. Let $\rho\left(\mathbf{z}, \boldsymbol{\Upsilon}\right)$ be the distance of $\mathbf{z}$ from $\boldsymbol{\Upsilon}$.

Definition 46 ***Asymptotically contractive set relative to*** Υ

A time-varying set $\mathfrak{S}(t; t_0)$ ***asymptotically contracts to the set*** Υ, *i.e., it is* ***asymptotically contractive relative to*** Υ, *if. and only if, both*

$$In\ \mathfrak{S}(t; t_0) \supset \Upsilon, \quad \forall t \in \mathfrak{T}_0, \tag{2.31}$$

and

$$\lim\left[Cl\ \mathfrak{S}(t; t_0) : t \longrightarrow \infty\right] = Cl\ \mathfrak{u}, \tag{2.32}$$

hold. This is the asymptotic contraction relative to Υ *as* $t \to \infty$.

*The expression "****relative to*** Υ*" is to be omitted if, and only if,* $\Upsilon = \mathfrak{O} = \left\{\mathbf{0}_K\right\}$.

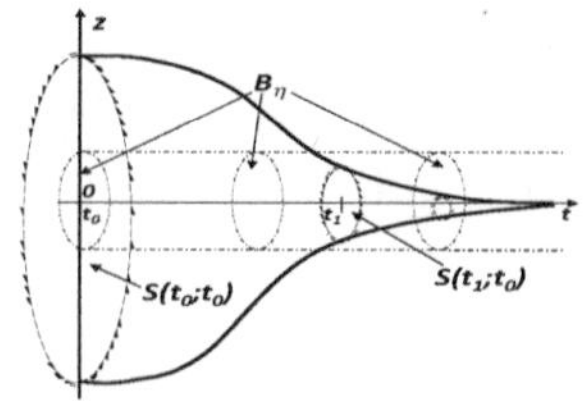

Figure 2.2: Asymptotically contractive set $S(t;t_0)$.

Example 47 *The following time varying set is asymptotically contractive:*

$$\mathfrak{S}(t)=\left\{\mathbf{z}:\|\mathbf{z}\|\left(1+\|\mathbf{z}\|\right)^{-1}<e^{-t}\right\}=\left\{\mathbf{z}:\|\mathbf{z}\|<e^{-t}\left(1-e^{-t}\right)^{-1}\right\}\Longrightarrow$$

$$In\ \mathfrak{S}(t)=\mathfrak{S}(t)=\left\{\mathbf{z}:\|\mathbf{z}\|\left(1+\|\mathbf{z}\|\right)^{-1}<e^{-t}\right\}\supset\mathfrak{O},\quad\forall t\in\mathfrak{T}_0,\quad(t_0>0)\in\mathfrak{T},$$
$$\lim\left[Cl\ \mathfrak{S}(t;t_0):t\longrightarrow\infty\right]=\lim\left[\left\{\mathbf{z}:\|\mathbf{z}\|\left(1+\|\mathbf{z}\|\right)^{-1}\leq e^{-t}\right\}:t\longrightarrow\infty\right]=\left\{\mathbf{0}_K\right\}.$$

Note 48 *The outer diameter* $\mathfrak{d}_M(t;\mathfrak{S})$ *of an asymptotically contractive set* $\mathfrak{S}(t)$ *obeys:*

$$t\longrightarrow\infty\Longrightarrow\mathfrak{d}_M(t;\mathfrak{S})\longrightarrow0,\ \ \mathfrak{d}_M(\mathfrak{T}_\tau;\mathfrak{S})=0,\quad\forall\mathfrak{T}_\tau\subseteq\mathfrak{T}.$$

Chapter 3

Systems

3.1 General dynamical systems

Michel, Hou, and Liu [427, pp. 1, 2] presented the general quadruple determination of a dynamical system the internal dynamics behavior of which, hence dynamical properties of which, are independent of the initial instant t_0. Such determination treats only the internal dynamics behavior of the system, but not its output dynamics behavior. It is appropriate for Lyapunov stability studies, but not for bounded-input bounded-output ($BIBO$) stability and tracking studies.

By following [427, pp. 1, 2] let us extend the determination of dynamical systems to the systems, on the internal dynamics and output behavior and properties of which, the initial moment t_0 can have a crucial influence. It is a characteristic of time-varying systems.

The notion *dynamical physical system is in the following sense [435, 1. Definition, p. 380], [436, 1. Definition, p. 380]:*

Definition 49 ***Dynamical physical system***

A physical system is ***dynamical on*** $\mathfrak{T}_0$ *if, and only if, it has at every moment* $\tau \in \mathfrak{T}_0$ *an internal physical situation called its* ***state at the moment*** τ *such that it and all actions on the system at the moment* τ *and after* τ *determine uniquely both its behavior (i.e., the system state and the system external response) from the moment* τ *on:* $\forall (t \geq \tau) \in \mathfrak{T}_0$.

The temporal evolution of the system state is ***the system motion.***

The temporal evolution of the system external response is ***the system output response.***

If, and only if, $\mathfrak{T}$ *replaces* $\mathfrak{T}_0$ *then "****on*** $\mathfrak{T}_0$*" is to be omitted.*

Natural sciences study the dynamical physical systems directly by investigating them as they are in reality. System science, control science, and control engineering also use mathematical descriptions, i.e., mathematical models, of the physical dynamical systems in order to study them, the dynamical physical systems. The mathematical models represent mathematical systems.

Definition 50 ***Dynamical system in general***

A mathematical dynamical system $\mathcal{S}$ ***of the order*** α, *for short* ***a dynamical system*** $\mathcal{S}$ ***of the order*** α *is a set of mappings (vector function)* $\mathcal{Z}(.)$ *and* $\mathcal{Y}(.)$ *from the Cartesian product space* $\mathfrak{T}_0 \times \mathfrak{T}_i \times \mathfrak{L}^* \times \mathfrak{S}^*(t_0) \times \mathcal{J}^{\xi}$ *into the Cartesian product space* $\mathfrak{R}^K \times \mathfrak{Y}$,

$$\mathcal{S}: \mathfrak{T}_0 \times \mathfrak{T}_i \times \mathfrak{L}^* \times \mathfrak{S}^*(t_0) \times \mathcal{J}^{\xi} \longrightarrow \mathfrak{R}^K \times \mathfrak{Y},$$

which have the following properties:

P1 ***Existence:*** *Every vector function $\mathcal{Z}(.)$ called* ***system motion*** *is defined in $\mathfrak{R}^K$ on the product set $\mathfrak{T}_0 \times \mathfrak{T}_i \times \mathfrak{L}^* \times \mathfrak{S}^*(t_0) \times \mathcal{J}^\xi$ and every vector function $\mathcal{Y}(.)$ called* ***system output response*** *is defined in $\mathfrak{Y}$ on the product set $\mathfrak{T}_0 \times \mathfrak{T}_i \times \mathfrak{L}^* \times \mathfrak{S}^*(t_0) \times \mathcal{J}^\xi$,*

$$\mathcal{Z}(t;t_0;\mathbf{Z}_0;\mathbf{I}) \in \mathfrak{R}^K, \quad \left\{ \begin{array}{c} \forall (t, t_0, \mathbf{Z}_0, \mathbf{I}) \\ \in \mathfrak{T}_0 \times \mathfrak{T}_i \times \mathfrak{L}^* \times \mathfrak{S}^*(t_0) \times \mathcal{J}^\xi \end{array} \right\}, \tag{3.1}$$

$$\mathcal{Y}(t;t_0;\mathbf{Z}_0;\mathbf{I}) \in \mathfrak{Y}, \quad \left\{ \begin{array}{c} \forall (t, t_0, \mathbf{Z}_0, \mathbf{I}) \\ \in \mathfrak{T}_0 \times \mathfrak{T}_i \times \mathfrak{L}^* \times \mathfrak{S}^*(t_0) \times \mathcal{J}^\xi \end{array} \right\}, \tag{3.2}$$

P2 ***Initial condition:*** *Every system motion $\mathcal{Z}(.)$ and every system output response $\mathcal{Y}(.)$ satisfies the initial conditions,*

$$\mathcal{Z}(t_0;t_0;\mathbf{Z}_0;\mathbf{I}) = \mathbf{Z}_0, \ \forall (t_0, \mathbf{Z}_0, \mathbf{I}) \in \mathfrak{L}^* \times \mathfrak{S}^*(t_0) \times \mathcal{J}^\xi, \tag{3.3}$$

$$\mathcal{Y}(t_0;t_0;\mathbf{Z}_0;\mathbf{I}) \in \mathbf{Y}_0, \ \forall (t_0, \mathbf{Z}_0, \mathbf{I}) \in \mathfrak{L}^* \times \mathfrak{S}^*(t_0) \times \mathcal{J}^\xi, \tag{3.4}$$

P3 ***Continuous dependence:*** *Every system motion $\mathcal{Z}(.)$ and every system output response $\mathcal{Y}(.)$ is continuous in $(t, t_0, \mathbf{Z}_0, \mathbf{I}) \in \mathfrak{T}_0 \times \mathfrak{T}_i \times \mathfrak{L}^* \times \mathfrak{S}^*(t_0) \times \mathcal{J}^\xi$,*

$$\mathcal{Z}(t;t_0;\mathbf{Z}_0;\mathbf{I}) \in \mathfrak{C}\left(\mathfrak{T}_0 \times \mathfrak{T}_i \times \mathfrak{L}^* \times \mathfrak{S}^*(t_0) \times \mathcal{J}^\xi\right), \tag{3.5}$$

$$\mathcal{Y}(t;t_0;\mathbf{Z}_0;\mathbf{I}) \in \mathfrak{C}\left(\mathfrak{T}_0 \times \mathfrak{T}_i \times \mathfrak{L}^* \times \mathfrak{S}^*(t_0) \times \mathcal{J}^\xi\right), \tag{3.6}$$

P4 ***Uniqueness:*** *For every $(t, t_0, \tau, \mathbf{Z}_0, \mathbf{I}) \in \mathfrak{T}_0 \times \mathfrak{L}^* \times \mathfrak{L}^* \times \mathfrak{S}^*(t_0) \times \mathcal{J}^\xi$ every system extended motion $\mathcal{Z}(.)$ and every system output response $\mathcal{Y}(.)$ satisfies*

$$\mathcal{Z}\left[t; t_0 + \tau; \mathcal{Z}(t_0 + \tau; t_0; \mathbf{Z}_0; \mathbf{I}); \mathbf{I}\right] = \mathcal{Z}(t;t_0;\mathbf{Z}_0;\mathbf{I}), \tag{3.7}$$

$$\mathcal{Y}\left[t; t_0 + \tau; \mathcal{Z}(t_0 + \tau; t_0; \mathbf{Z}_0; \mathbf{I}); \mathbf{I}\right] = \mathcal{Y}(t;t_0;\mathbf{Z}_0;\mathbf{I}). \tag{3.8}$$

Comment 51 ***On Property P4***

Property 4 of time-invariant dynamical systems, which is related only to the internal dynamics of the systems, reads [287, Definition 3.20, p. 81], [373, p. 28]:

$$\mathcal{Z}\left[t; \mathcal{R}(\tau; \mathbf{Z}_0; \mathbf{I}); \mathbf{I}\right] = \mathcal{Z}(t + \tau; \mathbf{Z}_0; \mathbf{I}), \tag{3.9}$$

which results from

$$\mathcal{Z}\left[t; t_0; \mathcal{Z}(t_0 + \tau; t_0; \mathbf{Z}_0; \mathbf{I}); \mathbf{I}\right] = \mathcal{Z}(t + \tau; t_0; \mathbf{Z}_0; \mathbf{I}) \tag{3.10}$$

because the initial moment does not influence motions of time-invariant systems so that it can be accepted equal to zero: $t_0 = 0$. Equation (3.10) for $t_0 = 0$ reduces to (3.9).

However, the initial moment can have a substantial influence on the character of system motions. A motion starting from $\mathbf{Z}_0$ at an initial moment t_0 can converge to the origin $\mathbf{0}_K$, but a motion starting from $\mathcal{Z}(t_0 + \tau; t_0; \mathbf{Z}_0; \mathbf{I})$ taken as the initial state at the same initial moment t_0 can diverge from the origin $\mathbf{0}_K$. Therefore, equations (3.9) and (3.10) do not hold for time-varying systems.

We use the notion of the output variables in the classical sense. To be precise:

Definition 52 ***Output variables and output vector***

A variable $Y \in \mathfrak{R}$ is an ***output variable*** *of the system if, and only if, its values result from the system behavior, they are measurable, and somebody is interested in them. The capital letter denotes the total value of the variable relative to its total zero value, if it exists, or relative to its accepted zero value.*

The number N is the maximal number of the mutually independent output variables $Y_1, Y_2, ... Y_N$ of the system. They form the ***output vector*** *$\mathbf{Y}$ of the system: $\mathbf{Y} = [Y_1 \ Y_2 \ ... \ Y_N]^T \in \mathfrak{R}^N$*

An adequate mathematical model of the dynamical physical system contains also an adequate description of the state of the physical dynamical system, which is the state of the (mathematical) dynamical system. More precisely [75, Definition 3-6, p. 83], [361, p. 105], [435, 2. Definition, p. 380], [436, 2. Definition, p. 380], [468, p. 4], [469, p. 664]:

Definition 53 ***State of a dynamical system***
The state of a dynamical system at a moment $\tau \in \mathfrak{T}$ *is the minimal amount of information about the system at the moment* τ *that, together with information about the action on the system at every moment* $(t \geq \tau) \in \mathfrak{T}$*, determines uniquely the system behavior (i.e., the system state and the output response) for all* $(t \geq \tau) \in \mathfrak{T}$*.*

The minimal number K *of mutually independent variables* Z_i, $i = 1, 2, \cdots, K$*, the values* $Z_i(\tau)$ *of which at the moment* τ *represent the state of the system at the moment* τ*, are* ***the state variables of the system.*** *They compose* ***the state vector*** $\mathbf{Z}$ ***of the system,*** $\mathbf{Z} = [Z_1 \ \ Z_2 \ \cdots \ Z_K]^T \in \mathfrak{R}^K$. *The space* $\mathfrak{R}^K$ *is* ***the state space of the system.***

Note 54 ***State and motion***
The system state vector $\mathbf{Z}(t)$ *at a moment* $t \in \mathfrak{T}$ *is the vector value of the motion* $\mathcal{Z}(t; t_0; \mathbf{Z}_0; \mathbf{I})$ *at the same moment* t*:* $\mathbf{Z}(t) \equiv \mathcal{Z}(t; t_0; \mathbf{Z}_0; \mathbf{I})$ *so that* $\mathbf{Z}(t_0) \equiv \mathcal{Z}(t_0; t_0; \mathbf{Z}_0; \mathbf{I}) \equiv \mathbf{Z}_0$*.*

3.2 Plants and control systems

3.2.1 System description

The system total input vector is $\mathbf{I} = [I_1 \ I_2 \ \cdots \ I_M]^T \in \mathfrak{R}^M$, where the value I_i is the total value of the i-th input variable. The system total output vector is $\mathbf{Y} = [Y_1 \ Y_2 \ \cdots \ Y_M]^T \in \mathfrak{R}^N$, where the value Y_j is the total value of the j-th input variable.

Let

$$\mathbf{I} = \begin{bmatrix} \mathbf{D} \\ \mathbf{U} \end{bmatrix} \in \mathfrak{R}^M, \ M = d + r \ \textit{for the plant},$$

$$\mathbf{I} = \begin{bmatrix} \mathbf{D} \\ \mathbf{Y}_d \end{bmatrix} \in \mathfrak{R}^M, \ M = d + N \ \textit{for the control system}. \tag{3.11}$$

Let α and ρ be the fixed natural numbers. The mathematical description, the mathematical model, of a large class of nonlinear *time*-varying continuous-time systems without a delay, in terms of the total coordinates has the following form.

$$\begin{gathered}
\textit{The internal dynamics, i.e., state, vector differential equation:} \\
Q\left[t,\mathbf{R}^{\alpha-1}(t)\right]\mathbf{R}^{(\alpha)}(t)+\mathbf{q}\left[t,\mathbf{R}^{\alpha-1}(t)\right]=\mathbf{h}\left[t,\mathbf{I}^{\xi}(t)\right],\quad \xi\leq\alpha, \\
\mathbf{h}\left[t,\mathbf{I}^{\xi}(t)\right]= \\
\left\{\begin{array}{c}
E(t)\,\mathbf{e}_d\left[\mathbf{D}^{\eta}(t)\right]+P(t)\,\mathbf{e}_u\left[\mathbf{U}^{\mu}(t)\right], \\
M=d+r,\ \textit{for the plant}, \\
E(t)\,\mathbf{e}_d\left[\mathbf{D}^{\eta}(t)\right]+P(t)\,\mathbf{e}_y\left[\mathbf{Y}_d^{\mu}(t)\right], \\
M=d+N,\ \textit{for the control system},
\end{array}\right\}, \\
\forall t\in\mathfrak{T},\ \eta,\mu,\xi\in\{0,1,\ \cdots\},\ \alpha,\rho,d,N,r\in\{1,2,\ \cdots\}. \\
\textit{The output vector agebraic equation:} \\
\mathbf{Y}(t)=\mathbf{s}\left[t,\mathbf{R}^{\alpha-1}(t),\mathbf{I}(t)\right]= \\
\left\{\begin{array}{c}
\mathbf{z}\left[t,\mathbf{R}^{\alpha-1}(t),\mathbf{D}(t)\right]+W(t)\,\mathbf{w}\left[\mathbf{U}(t)\right], \\
\textit{for the plant}, \\
\mathbf{z}\left[t,\mathbf{R}^{\alpha-1}(t),\mathbf{D}(t)\right]+W(t)\,\mathbf{w}\left[\mathbf{Y}_d(t)\right], \\
\textit{for the control system},
\end{array}\right\}, \\
\forall t\in\mathfrak{T},
\end{gathered}\tag{3.12}$$

where

$$\begin{aligned}
&\mathbf{R}\in\mathfrak{R}^{\rho},\ Q(.):\mathfrak{T}\times\mathfrak{R}^{\alpha\rho}\longrightarrow\mathfrak{R}^{\rho\times\rho},\ Q\left(t,\mathbf{R}^{\alpha-1}\right)\in\mathfrak{C}\left(\mathfrak{T}\times\mathfrak{R}^{\alpha\rho}\right), \\
&\mathbf{q}(.):\mathfrak{T}\times\mathfrak{R}^{\alpha\rho}\longrightarrow\mathfrak{R}^{\rho},\ \mathbf{q}\left(t,\mathbf{R}^{\alpha-1}\right)\in\mathfrak{C}\left(\mathfrak{T}\times\mathfrak{R}^{\alpha\rho}\right), \\
&\mathbf{h}(.):\mathfrak{T}\times\mathfrak{R}^{(\xi+1)M}\longrightarrow\mathfrak{R}^{\rho},\ \mathbf{h}\left(t,\mathbf{I}^{\xi}\right)\in\mathfrak{C}\left(\mathfrak{T}\times\mathfrak{R}^{(\xi+1)M}\right), \\
&E(.):\mathfrak{T}\longrightarrow\mathfrak{R}^{\rho\times d},\ E(t)\in\mathfrak{C}\left(\mathfrak{T}\right), \\
&\mathbf{e}_d(.):\mathfrak{R}^{d(\eta+1)}\longrightarrow\mathfrak{R}^{d},\ \mathbf{e}_d\left(\mathbf{D}^{\eta}\right)\in\mathfrak{C}\left(\mathfrak{R}^{d(\eta+1)}\right), \\
&P(.):\mathfrak{T}\longrightarrow\mathfrak{R}^{\rho\times r},\ P(t)\in\mathfrak{C}\left(\mathfrak{T}\right), \qquad (3.13) \\
&\mathbf{e}_u(.):\mathfrak{R}^{(\mu+1)r}\longrightarrow\mathfrak{R}^{r},\ \mathbf{e}_u\left(\mathbf{U}^{\mu}\right)\in\mathfrak{C}\left(\mathfrak{R}^{(\mu+1)r}\right), \\
&\quad \mathbf{e}_u(.)\ \textit{is invertible},\ \mathbf{e}_u^{I}(.)\ \textit{is well defined on}\ \mathfrak{R}^{r}, \\
&\quad \textit{for the plant}, \\
&\mathbf{e}_y(.):\mathfrak{R}^{N(\mu+1)}\longrightarrow\mathfrak{R}^{r},\ \mathbf{e}_y\left(\mathbf{Y}_d\right)\in\mathfrak{C}\left(\mathfrak{R}^{N(\mu+1)}\right), \\
&\quad \textit{for the control system},
\end{aligned}\tag{3.14}$$

$$\begin{aligned}
&\mathbf{s}(.):\mathfrak{T}\times\mathfrak{R}^{\alpha\rho}\times\mathfrak{R}^{M}\longrightarrow\mathfrak{R}^{N},\ \mathbf{s}(.)\in\mathfrak{C}\left(\mathfrak{T}\times\mathfrak{R}^{\rho}\times\mathfrak{R}^{M}\right), \\
&\mathbf{z}(.):\mathfrak{T}\times\mathfrak{R}^{\alpha\rho}\times\mathfrak{R}^{d}\longrightarrow\mathfrak{R}^{\rho},\ \mathbf{z}\left(t,\mathbf{R}^{\alpha-1},\mathbf{D}\right)\in\mathfrak{C}\left(\mathfrak{T}\times\mathfrak{R}^{\alpha\rho}\times\mathfrak{R}^{d}\right), \\
&W(.):\mathfrak{T}\longrightarrow\mathfrak{R}^{N\times r},\ W(t)\in\mathfrak{C}\left(\mathfrak{T}\right),
\end{aligned}\tag{3.15}$$

$$\begin{gathered}
\mathbf{w}(.):\left\{\begin{array}{c}
\mathfrak{R}^{r}\longrightarrow\mathfrak{R}^{r},\ \mathbf{w}\left(\mathbf{U}\right)\in\mathfrak{C}\left(\mathfrak{R}^{r}\right), \\
\mathbf{w}(.)\ \textit{is invertible}:\ \mathbf{w}^{I}(.)\ \textit{is well defined on}\ \mathfrak{R}^{r}, \\
\mathbf{w}^{I}\left[\mathbf{w}(\mathbf{U})\right]\equiv\mathbf{U}, \\
\textit{for the plant}, \\
\mathfrak{R}^{N}\longrightarrow\mathfrak{R}^{N},\ \mathbf{w}\left(\mathbf{Y}_d\right)\in\mathfrak{C}\left(\mathfrak{R}^{N}\right), \\
\textit{for the control system},
\end{array}\right\} \\
\alpha,\rho\in\{0,1,2,\ ...\},
\end{gathered}\tag{3.16}$$

The matrix function $Q(.)$ and the vector function $\mathbf{q}(.)$ determine the internal dynamics, i.e., the state, of the system (3.12). The matrix function $P(.)$ and the vector function $\mathbf{e}_u(.)$, $[\mathbf{e}_y(.)]$ determine how the system transmits the influence of the control vector $\mathbf{U}$ (of the desired output vector $\mathbf{Y}_d$) and of its derivatives on the state of the plant (control system), respectively. The matrix function $E(.)$ and the vector function $\mathbf{e}_d(.)$ describe how the system transfers the disturbance action on the system dynamics.

The mathematical model (3.12) is called **the plant**, ***the control system***, *or in general*, **the system** that incorporates both the plant and its control system. Complete mathematical models of mechanical plants (e.g., planes, robots, rockets. space vehicles) and of electromechanical plants (e.g., electrical machines) belong to this type of the model. The left-hand side of the first equation (3.12) describes ***the internal dynamics***, i.e., ***the state behavior, of the system***). The right-hand side of the first equation (3.12) shows how the input vector $\mathbf{I}$ acts on the state of the system. The right-hand side of the second equation (3.12) describes **the output vector dependence** on the state vector $\mathbf{R}^{\alpha-1}$, and on the input vector $\mathbf{I}$. We use the extended vector $\mathbf{R}^{\alpha-1}$ in the sense of (1.10)-(1.12):

$$\mathbf{R}^{\alpha-1}(t) = \left[\mathbf{R}^T(t) \vdots \mathbf{R}^{(1)^T}(t) \vdots ... \vdots \mathbf{R}^{(\alpha-1)^T}(t)\right]^T \in \mathfrak{R}^{\alpha\rho}. \tag{3.17}$$

The vector $\mathbf{R}^{\alpha-1} \in \mathfrak{R}^{\alpha\rho}$ is **the state vector of the system**, and $\mathbf{R}^{\alpha-1}(t) \in \mathfrak{R}^{\alpha\rho}$ is **the state vector of the system at an instant** $t \in \mathfrak{T}$, Definition 53. The space $\mathfrak{R}^{\alpha\rho}$ is **the state space of the system**, Definition 53. Formally:

Definition 55 ***The state of the system***

The vector $\mathbf{R}^{\alpha-1} \in \mathfrak{R}^{\alpha\rho}$ *is* ***the state vector of the system (3.12)*** *and* $\mathbf{R}^{\alpha-1}(\tau)$ *is* ***the state vector of the system (3.12) at a moment*** $\tau \in \mathfrak{T}$. *It, together with the extended input vector* $\mathbf{I}^{\xi}(t)$ *for all* $(t \geq \tau) \in \mathfrak{T}$, *determines completely both* $\mathbf{R}^{\alpha-1}(t)$ *itself and* $\mathbf{Y}(t)$ *at the same moment* τ *and at every moment* t *after* τ, $t > \tau$, $t \in \mathfrak{T}$.

The space $\mathfrak{R}^{\alpha\rho}$ *is* ***the state space of the system (3.12).***

The plant output vector is $\mathbf{Y}$, $\mathbf{Y} \in \mathfrak{R}^N$. When $\mathbf{Y} = \mathbf{R}$ then we also use the extended output vector $\mathbf{Y}^{\alpha-1}$ in the sense of (1.10)-(1.12), which is then the state vector of the system:

$$\mathbf{Y} = \mathbf{R} \Longrightarrow N = \rho \text{ and } \mathbf{Y}^{\alpha-1} = \left[\mathbf{Y}^T \quad \mathbf{Y}^{(1)^T} \ ... \ \mathbf{Y}^{(\alpha-1)^T}\right]^T \in \mathfrak{R}^{\alpha\rho}. \tag{3.18}$$

In the same sense we use $\mathbf{D}^{\eta}$ and $\mathbf{U}^{\mu}$.

The additional notation follows.

$\mathfrak{D}^k$ is a given, or to be determined, family of all bounded and $(k+1)$-*times* continuously differentiable on $\mathfrak{T}$ total disturbance vector functions $\mathbf{D}(.) \in \mathfrak{C}^{k+1}(\mathfrak{R}^d)$ such that they and their first $k+1$ derivatives obey *TCUP* (Principle 20),

$$\mathfrak{D}^k \subset \mathfrak{C}^{(k+1)d} = \mathfrak{C}^{k+1}(\mathfrak{R}^d). \tag{3.19}$$

$\mathfrak{D}^0 = \mathfrak{D}$ is the family of all bounded continuous and continuously differentiable total disturbance vector functions $\mathbf{D}(.) \in \mathfrak{D}$ such that they and their first derivatives obey *TCUP*,

$$\mathfrak{D} \subset \mathfrak{C}^1(\mathfrak{R}^d). \tag{3.20}$$

$\mathcal{J}^i$ is a given, or to be determined, family of all bounded i-times continuously differentiable on $\mathfrak{T}_0$ permitted input vector functions $\mathbf{I}(.)$,

$$\mathcal{J}^i \subset \mathfrak{C}^i(\mathfrak{R}^M). \tag{3.21}$$

$\mathfrak{R}_d^k$ is a given, or to be determined, family of all bounded and $(k+1)$-*times* continuously differentiable realizable total desired state vector functions $\mathbf{R}_d(.) \in \mathfrak{C}^{k+1}$ such that they and their first k derivatives obey *TCUP*,

$$\mathfrak{R}_d^k \subset \mathfrak{C}^{k+1}(\mathfrak{R}^\rho), \quad \mathbf{R}_d(.) \in \mathfrak{R}_d^k. \tag{3.22}$$

$\mathfrak{R}_d$ is a given, or to be determined, family of all bounded and continuous realizable total desired state vector functions $\mathbf{R}_d(.) \in \mathfrak{C}^1$ such that they are induced by $\mathbf{Y}_d(.) \in \mathfrak{Y}_d$ and that their first derivatives obey *TCUP*,

$$\mathfrak{R}_d \subset \mathfrak{C}^1(\mathfrak{R}^\rho), \quad \mathbf{R}_d(.) \in \mathfrak{R}_d. \tag{3.23}$$

$\mathfrak{R}_{d0}^k$ is the set of the initial values $\mathbf{R}_{d0}^{k-1} = \mathcal{R}_d^{k-1}\left(t_0; t_0; \mathbf{R}_{d0}^{\alpha-1}\right)$ of all $\mathcal{R}_d\left(.\right) \in \mathfrak{R}_d^k$,

$\mathfrak{U}^k$ is a given, or to be determined, family of all bounded and $(k+1)$-*times* continuously differentiable total control vector functions $\mathbf{U}(.) \in \mathfrak{C}^{k+1}(\mathfrak{R}^r)$ such that they and their first $k+1$ derivatives obey *TCUP*,

$$\mathfrak{U}^k \subset \mathfrak{C}^{k+1}(\mathfrak{R}^r), \quad \mathbf{U}(.) \in \mathfrak{U}^k. \tag{3.24}$$

$\mathfrak{U}^0 = \mathfrak{U}$ is the family of all bounded continuous, continuously differentiable permitted total control vector functions $\mathbf{U}(.) \in \mathfrak{U}$,

$$\mathfrak{U} \subset \mathfrak{C}^1(\mathfrak{R}^r), \quad \mathbf{U}(.) \in \mathfrak{U}. \tag{3.25}$$

$\mathfrak{Y}_d^k$ is a given, or to be determined, family of all bounded and $(k+1)$-*times* continuously differentiable realizable total desired output vector functions $\mathbf{Y}_d(.) \in \mathfrak{C}^{k+1}$ such that they and their first $k+1$ derivatives obey *TCUP*, i.e., $\mathfrak{Y}_d^k$ is a given, or to be determined, family of all bounded continuously differentiable realizable total desired extended output vector functions $\mathbf{Y}_d^k(.) \in \mathfrak{C}^1$ such that they obey *TCUP*,

$$\mathfrak{Y}_d^k \subset \mathfrak{C}^{k+1}(\mathfrak{R}^N), \quad \mathbf{Y}_d(.) \in \mathfrak{Y}_d^k, \tag{3.26}$$

$\mathfrak{Y}_d^0 = \mathfrak{Y}_d$ is *the family of all bounded continuous realizable total desired output vector functions* $\mathbf{Y}_d(.) \in \mathfrak{C}(\mathfrak{R}^N)$, such that they and their first derivatives obey *TCUP*,

$$\mathfrak{Y}_d \subset \mathfrak{C}(\mathfrak{R}^N), \quad \mathbf{Y}_d(.) \in \mathfrak{Y}_d. \tag{3.27}$$

$\mathfrak{Y}_{d0}^k$ is *the set of the initial condition* $\mathbf{Y}_{d0}^k = \mathbf{Y}_d^k(t_0)$ *of* $\mathbf{Y}_d^k(t)$ *of every* $\mathbf{Y}_d(.)\in \mathfrak{Y}_d^k$. If, and only if, $\mathbf{Y}_d(.)\in \mathfrak{Y}_d^k$ then $\mathbf{Y}_{d0}^k = \mathcal{Y}_d^k\left(t_0; t_0; \mathbf{Y}_{d0}^k\right) \in \mathfrak{Y}_{d0}^k$.

The desired system output behavior $\mathbf{Y}_d(t)$ is either determined in advance or during the system operation. It is known for the control synthesis and implementation.

3.2.2 Special form of the system

Note 56 ***The special forms of*** $\mathbf{q}\left[t, \mathbf{R}^{\alpha-1}(t)\right]$ ***and of*** $Z(t)\mathbf{z}\left(\mathbf{R}^{\alpha-1}, \mathbf{D}\right)$

The nonlinear vector terms can have the following special forms in (3.12):

$$\begin{gathered}
\mathbf{q}\left(t, \mathbf{R}^{\alpha-1}\right) = F\left(t, \mathbf{R}^{\alpha-1}\right)\mathbf{R}^{(\alpha-1)}(t), \\
F(.) : \mathfrak{T} \times \mathfrak{R}^{\alpha\rho} \longrightarrow \mathfrak{R}^{\rho\times\rho},\ F\left(t, \mathbf{R}^{\alpha-1}\right) \in \mathfrak{C}\left(\mathfrak{T} \times \mathfrak{R}^{\alpha\rho}\right), \\
\mathbf{z}\left(t, \mathbf{R}^{\alpha-1}, \mathbf{D}\right) = Z\left(t, \mathbf{R}^{\alpha-1}, \mathbf{D}\right)\mathbf{R}^{(\alpha-1)}, \\
\mathbf{Z}\left(.\right) : \mathfrak{T} \times \mathfrak{R}^{\alpha\rho} \times \mathfrak{R}^d \longrightarrow \mathfrak{R}^{N\times\rho},\ \mathbf{Z}\left(t, \mathbf{R}^{\alpha-1}, \mathbf{D}\right) \in \mathfrak{C}\left(\mathfrak{T} \times \mathfrak{R}^{\alpha\rho} \times \mathfrak{R}^d\right).
\end{gathered} \tag{3.28}$$

The plant (3.12) with these special forms of $\mathbf{q}\left(t, \mathbf{R}^{\alpha-1}\right)$ *and of* $\mathbf{z}\left(t, \mathbf{R}^{\alpha-1}, \mathbf{D}\right)$ *becomes*

$$\begin{gathered}
Q\left[t, \mathbf{R}^{\alpha-1}(t)\right]\mathbf{R}^{(\alpha)}(t) + F\left[t, \mathbf{R}^{\alpha-1}(t)\right]\mathbf{R}^{(\alpha-1)}(t) \\
= E\left(t\right)\mathbf{e}_d\left[\mathbf{D}^{\eta}(t)\right] + P\left(t\right)\mathbf{e}_u\left[\mathbf{U}^{\mu}\left(t\right)\right],\ \forall t \in \mathfrak{T}, \\
detF\left(t, \mathbf{R}^{\alpha-1}\right) \neq 0,\ \forall\left(t, \mathbf{R}^{\alpha-1}\right) \in \mathfrak{T} \times \mathfrak{R}^{\alpha\rho}, \\
\mathbf{Y}(t) = Z\left(t, \mathbf{R}^{\alpha-1}, \mathbf{D}\right)\mathbf{R}^{(\alpha-1)} + W\left(t\right)\mathbf{w}\left[\mathbf{U}(t)\right],\ \forall t \in \mathfrak{T}.
\end{gathered} \tag{3.29}$$

3.2.3 System basic properties

In order for the mathematical model of the physical system to be physically meaningful we assume the validity of the following.

Property 57 ***TCUP property of the system***
a) The system (3.12) obeys the system form (Definition 21) of TCUP (Principle 20).
b) The motion and the corresponding output response of the system (3.12) obey Principle 20 for every initial condition and for every input vector function.

We are especially interested in regular systems. They represent a special class of the systems (3.12) determined by their following property.

Property 58 ***Nonsingularity property of the matrix*** $Q\left[t,\mathbf{R}^{\alpha-1}(t)\right]$
The matrix $Q\left[t,\mathbf{R}^{\alpha-1}(t)\right]$ *of the system (3.12) is nonsingular,*

$$det\ Q\left(t,\mathbf{R}^{\alpha-1}\right)\neq 0,\ \ \forall\left(t,\mathbf{R}^{\alpha-1}\right)\in\mathfrak{T}\times\mathfrak{R}^{\alpha\rho}. \tag{3.30}$$

Definition 59 ***Regular system***
The system (3.12) that possesses Property 58 is called ***regular*** *(i.e.,* ***nonsingular****)* ***system*** *(3.12).*
Otherwise, the systems is ***singular***.

Condition (3.30) ensures solvability of (3.12) in $\mathbf{R}^{(\alpha)}(t)$,

$$\mathbf{R}^{(\alpha)}(t)=Q^{-1}\left[t,\mathbf{R}^{\alpha-1}(t)\right]\left\{\mathbf{h}\left[t,\mathbf{I}^{\xi}(t)\right]-\mathbf{q}\left[t,\mathbf{R}^{\alpha-1}(t)\right]\right\}. \tag{3.31}$$

Note 60 *The condition det* $Q\left(t,\mathbf{R}^{\alpha-1}\right)\neq 0$ *for every* $\left(t,\mathbf{R}^{\alpha-1}\right)\in\mathfrak{T}\times\mathfrak{R}^{\alpha\rho}$ *is the necessary and sufficient condition for all state variables of the plant (3.12) to have the same order* α *of their highest derivatives. This explains and justifies to call the system (3.12) that possesses Property 58 the regular system (3.12). Otherwise, the system (3.12) is* ***singular***. *We treat herein only the regular systems determined by (3.12).*

Property 61 ***Control rank property of the plant state equation***
The inverse function $\mathbf{e}_u^I(.)$ *of the vector function* $\mathbf{e}_u(.)$ *is globally defined at every moment* $t\in\mathfrak{T}$.
The matrix function $P(.)$ *has the following canonical structure:*

$$P(t)=\begin{bmatrix} P_\beta(t) & O_{\beta,r-\beta} \\ O_{\rho-\beta,\beta} & O_{\rho-\beta,r-\beta}\end{bmatrix},\ P_\beta(t)\in\mathfrak{R}^{\beta\times\beta},\ det\ P_\beta(t)\neq 0,\ \forall t\in\mathfrak{T}, \tag{3.32}$$

where β *is its rank.*

$$rank\ P(t)=\beta\leq min\ (r,\rho),\ \ \forall t\in\mathfrak{T}, \tag{3.33}$$

In view of this it is reasonable to introduce the following partitions in the state equation of (3.12):

$$\begin{gathered}
Q\left[t,\mathbf{R}^{\alpha-1}(t)\right]=\begin{bmatrix} Q_{\beta,\rho}\left[t,\mathbf{R}^{\alpha-1}(t)\right] \\ Q_{\rho-\beta,\rho}\left[t,\mathbf{R}^{\alpha-1}(t)\right]\end{bmatrix} \\
=\begin{bmatrix} Q_{\beta}\left[t,\mathbf{R}^{\alpha-1}(t)\right] & Q_{\beta,\rho-\beta}\left[t,\mathbf{R}^{\alpha-1}(t)\right] \\ Q_{\rho-\beta,\beta}\left[t,\mathbf{R}^{\alpha-1}(t)\right] & Q_{\rho-\beta,\rho-\beta}\left[t,\mathbf{R}^{\alpha-1}(t)\right]\end{bmatrix}, \\
Q_{\beta,\rho}\left[t,\mathbf{R}^{\alpha-1}(t)\right]=\begin{bmatrix} Q_{\beta}\left[t,\mathbf{R}^{\alpha-1}(t)\right] & Q_{\beta,\rho-\beta}\left[t,\mathbf{R}^{\alpha-1}(t)\right]\end{bmatrix}, \\
Q_{\rho-\beta,\rho}\left[t,\mathbf{R}^{\alpha-1}(t)\right]=\begin{bmatrix} Q_{\rho-\beta,\beta}\left[t,\mathbf{R}^{\alpha-1}(t)\right] & Q_{\rho-\beta,\rho-\beta}\left[t,\mathbf{R}^{\alpha-1}(t)\right]\end{bmatrix}, \\
Q_{\beta}\left[t,\mathbf{R}^{\alpha-1}(t)\right]\in\mathfrak{R}^{\beta\times\beta},\ det\ Q_{\beta}\left[t,\mathbf{R}^{\alpha-1}(t)\right]\neq 0,\ \forall t\in\mathfrak{T},
\end{gathered} \tag{3.34}$$

$$\mathbf{q}\left[t,\mathbf{R}^{\alpha-1}(t)\right]=\begin{bmatrix}\mathbf{q}_{\beta}\left[t,\mathbf{R}^{\alpha-1}(t)\right]\\ \mathbf{q}_{\rho-\beta}\left[t,\mathbf{R}^{\alpha-1}(t)\right]\end{bmatrix},\quad \mathbf{q}_{\beta}\left[t,\mathbf{R}^{\alpha-1}(t)\right]\in\mathfrak{R}^{\beta}, \tag{3.35}$$

$$\begin{aligned}E(t)&=\begin{bmatrix}E_{\beta,d}(t)\\ E_{\rho-\beta,d}(t)\end{bmatrix}\\ &=\begin{bmatrix}E_{\beta}(t) & E_{\beta,d-\beta}(t)\\ E_{\rho-\beta,\beta}(t) & E_{\rho-\beta,d-\beta}(t)\end{bmatrix},\quad E_{\beta}(t)\in\mathfrak{R}^{\beta\times\beta},\\ E_{\beta,d}(t)&=\begin{bmatrix}E_{\beta}(t) & E_{\beta,d-\beta}(t)\end{bmatrix},\\ E_{\rho-\beta,d}(t)&=\begin{bmatrix}E_{\rho-\beta,\beta}(t) & E_{\rho-\beta,d-\beta}(t)\end{bmatrix},\end{aligned} \tag{3.36}$$

$$\mathbf{e}_{u}\left[\mathbf{U}^{\mu}(t)\right]=\begin{bmatrix}\mathbf{e}_{u\beta}\left[\mathbf{U}^{\mu}_{\beta}(t)\right]\\ \mathbf{e}_{u,r-\beta}\left[\mathbf{U}^{\mu}_{r-\beta}(t)\right]\end{bmatrix},\quad \mathbf{e}_{u\beta}\left[\mathbf{U}^{\mu}_{\beta}(t)\right]\in\mathfrak{R}^{\beta}. \tag{3.37}$$

Lemma 62 ***Directly controllable part of the plant state equation***

If the plant (3.12) possesses Property 61 then the directly controllable part of its state equation, i.e., the subdynamics of its state behavior, on which the control acts directly, is determined by:

$$\begin{aligned}&Q_{\beta,\rho}\left[t,\mathbf{R}^{\alpha-1}(t)\right]\mathbf{R}^{(\alpha)}(t)+\mathbf{q}_{\beta}\left[t,\mathbf{R}^{\alpha-1}(t)\right]\\ &=E_{\beta,d}(t)\,\mathbf{e}_{d}\left[\mathbf{D}^{\eta}(t)\right]+P_{\beta}(t)\,\mathbf{e}_{u\beta}\left[\mathbf{U}^{\mu}_{\beta}(t)\right],\quad \forall t\in\mathfrak{T},\end{aligned} \tag{3.38}$$

Only the control β-subvector $\mathbf{U}^{\mu}_{\beta}$ influences the state of the plant. The control $(r-\beta)$-subvector $\mathbf{U}^{\mu}_{r-\beta}$ does not influence the state of the plant.

Proof. We can write the state equation of (3.12) in the following partition form due to (3.32), (3.34)-(3.37):

$$\begin{aligned}&\begin{bmatrix}Q_{\beta,\rho}\left[t,\mathbf{R}^{\alpha-1}(t)\right]\\ Q_{\rho-\beta,\rho}\left[t,\mathbf{R}^{\alpha-1}(t)\right]\end{bmatrix}\mathbf{R}^{(\alpha)}(t)+\begin{bmatrix}\mathbf{q}_{\beta}\left[t,\mathbf{R}^{\alpha-1}(t)\right]\\ \mathbf{q}_{\rho-\beta}\left[t,\mathbf{R}^{\alpha-1}(t)\right]\end{bmatrix}\\ &=\begin{bmatrix}E_{\beta,d}(t)\\ E_{\rho-\beta,d}(t)\end{bmatrix}\mathbf{e}_{d}\left[\mathbf{D}^{\eta}(t)\right]+\begin{bmatrix}P_{\beta}(t) & O_{\beta,r-\beta}\\ O_{\rho-\beta,\beta} & O_{\rho-\beta,r-\beta}\end{bmatrix}\begin{bmatrix}\mathbf{e}_{u\beta}\left[\mathbf{U}^{\mu}_{\beta}(t)\right]\\ \mathbf{e}_{u,r-\beta}\left[\mathbf{U}^{\mu}_{r-\beta}(t)\right]\end{bmatrix},\end{aligned}$$

or,

$$\begin{aligned}&\begin{bmatrix}Q_{\beta,\rho}\left[t,\mathbf{R}^{\alpha-1}(t)\right]\mathbf{R}^{(\alpha)}(t)+\mathbf{q}_{\beta}\left[t,\mathbf{R}^{\alpha-1}(t)\right]\\ Q_{\rho-\beta,\rho}\left[t,\mathbf{R}^{\alpha-1}(t)\right]\mathbf{R}^{(\alpha)}(t)+\mathbf{q}_{\rho-\beta}\left[t,\mathbf{R}^{\alpha-1}(t)\right]\end{bmatrix}\\ &=\begin{bmatrix}E_{\beta,d}(t)\,\mathbf{e}_{d}\left[\mathbf{D}^{\eta}(t)\right]+P_{\beta}(t)\,\mathbf{e}_{u\beta}\left[\mathbf{U}^{\mu}_{\beta}(t)\right]\\ E_{\rho-\beta,d}(t)\,\mathbf{e}_{d}\left[\mathbf{D}^{\eta}(t)\right]\end{bmatrix}.\end{aligned}$$

This discovers that the control acts directly only on the β-subdynamics of (3.12):

$$\begin{aligned}&Q_{\beta,\rho}\left[t,\mathbf{R}^{\alpha-1}(t)\right]\mathbf{R}^{(\alpha)}(t)+\mathbf{q}_{\beta}\left[t,\mathbf{R}^{\alpha-1}(t)\right]\\ &=E_{\beta,d}(t)\,\mathbf{e}_{d}\left[\mathbf{D}^{\eta}(t)\right]+P_{\beta}(t)\,\mathbf{e}_{u\beta}\left[\mathbf{U}^{\mu}_{\beta}(t)\right],\quad \forall t\in\mathfrak{T}.\end{aligned}$$

This is (3.38). It proves the lemma statement ■

Note 63 *The equation (3.32) implies:*

$$P^{T}(t)=\begin{bmatrix}P^{T}_{\beta}(t) & O_{\beta,\rho-\beta}\\ O_{r-\beta,\beta} & O_{r-\beta,\rho-\beta}\end{bmatrix}, \tag{3.39}$$

$$P(t)\,P^{T}(t)=\begin{bmatrix}P_{\beta}(t)\,P^{T}_{\beta}(t) & O_{\beta,\rho-\beta}\\ O_{\rho-\beta,\beta} & O_{\rho-\beta,\rho-\beta}\end{bmatrix}\in\mathfrak{R}^{\rho\times\rho}, \tag{3.40}$$

and

$$P^T(t)\,P(t) = \begin{bmatrix} P_\beta^T(t)\,P_\beta(t) & O_{\beta,r-\beta} \\ O_{r-\beta,\beta} & O_{r-\beta,r-\beta} \end{bmatrix} \in \mathfrak{R}^{r\times r}. \tag{3.41}$$

Property 64 ***Control rank property of the plant output equation***

The inverse function $\mathbf{w}^I[\mathbf{U}(t)]$ *of the vector function* $\mathbf{w}[\mathbf{U}(t)]$ *is globally defined at every moment* $t \in \mathfrak{T}$.

The matrix function $W(.)$ *has the following canonical structure:*

$$W(t) = \begin{bmatrix} W_\gamma(t) & O_{\gamma,r-\gamma} \\ O_{N-\gamma,\gamma} & O_{N-\gamma,r-\gamma} \end{bmatrix}, \quad det\, W_\gamma(t) \neq 0,\ \forall t \in \mathfrak{T}, \tag{3.42}$$

where γ *is its rank,*

$$rank\, W(t) = \gamma \leq \min(N, r), \quad \forall t \in \mathfrak{T}. \tag{3.43}$$

It is useful to introduce the following partitions in the output equation of (3.12):

$$\mathbf{z}\left[t, \mathbf{R}^{\alpha-1}(t), \mathbf{D}(t)\right] = \begin{bmatrix} \mathbf{z}_\gamma\left[t, \mathbf{R}^{\alpha-1}(t), \mathbf{D}(t)\right] \\ \mathbf{z}_{N-\gamma}\left[t, \mathbf{R}^{\alpha-1}(t), \mathbf{D}(t)\right] \end{bmatrix}, \quad \mathbf{z}_\gamma\left[t, \mathbf{R}^{\alpha-1}(t), \mathbf{D}(t)\right] \in \mathfrak{R}^\gamma, \tag{3.44}$$

$$\mathbf{w}\left[\mathbf{U}(t)\right] = \begin{bmatrix} \mathbf{w}_\gamma\left[\mathbf{U}_\gamma(t)\right] \\ \mathbf{w}_{r-\gamma}\left[\mathbf{U}_{r-\gamma}(t)\right] \end{bmatrix}, \quad \mathbf{w}_\gamma\left[\mathbf{U}_\gamma(t)\right] \in \mathfrak{R}^\gamma. \tag{3.45}$$

$$\mathbf{Y}(t) = \begin{bmatrix} \mathbf{Y}_\gamma(t) \\ \mathbf{Y}_{N-\gamma}(t) \end{bmatrix}, \quad \forall t \in \mathfrak{T}. \tag{3.46}$$

Lemma 65 ***Directly controllable part of the plant output dynamics***

If the plant (3.12) possesses Property 64 then the directly controllable part of its output dynamics, i.e., the subdynamics of its output dynamics on which the control acts directly, is determined by:

$$\mathbf{Y}_\gamma(t) = \mathbf{z}_\gamma\left[t, \mathbf{R}^{\alpha-1}(t), \mathbf{D}(t)\right] + W_\gamma(t)\,\mathbf{w}_\gamma\left[\mathbf{U}_\gamma(t)\right], \quad \forall t \in \mathfrak{T}. \tag{3.47}$$

Only the control γ*-subvector* $\mathbf{U}_\gamma$ *influences the output behavior of the plant. The control* $(r-\gamma)$*-subvector* $\mathbf{U}_{r-\gamma}$ *does not influence the internal dynamics, i.e., the state, of the plant.*

Proof. We can write the output dynamics equation of (3.12) in the following partition form due to (3.42), (3.44)-(3.46):

$$\mathbf{Y}(t) = \begin{bmatrix} \mathbf{Y}_\gamma(t) \\ \mathbf{Y}_{N-\gamma}(t) \end{bmatrix} = \begin{bmatrix} \mathbf{z}_\gamma\left[t, \mathbf{R}^{\alpha-1}(t), \mathbf{D}(t)\right] \\ \mathbf{z}_{N-\gamma}\left[t, \mathbf{R}^{\alpha-1}(t), \mathbf{D}(t)\right] \end{bmatrix}$$
$$+ \begin{bmatrix} W_\gamma(t) & O_{\gamma,r-\gamma} \\ O_{N-\gamma,\gamma} & O_{N-\gamma,r-\gamma} \end{bmatrix} \begin{bmatrix} \mathbf{w}_\gamma\left[\mathbf{U}_\gamma(t)\right] \\ \mathbf{w}_{r-\gamma}\left[\mathbf{U}_{r-\gamma}(t)\right] \end{bmatrix}, \quad \forall t \in \mathfrak{T},$$

or,

$$\begin{bmatrix} \mathbf{Y}_\gamma(t) \\ \mathbf{Y}_{N-\gamma}(t) \end{bmatrix} = \begin{bmatrix} \mathbf{z}_\gamma\left[t, \mathbf{R}^{\alpha-1}(t), \mathbf{D}(t)\right] + W_\gamma(t)\,\mathbf{w}_\gamma\left[\mathbf{U}_\gamma(t)\right] \\ \mathbf{z}_{N-\gamma}\left[t, \mathbf{R}^{\alpha-1}(t), \mathbf{D}(t)\right] \end{bmatrix}, \quad \forall t \in \mathfrak{T}.$$

This shows that the control acts directly only on the γ-subdynamics of (3.12):

$$\mathbf{Y}_\gamma(t) = \mathbf{z}_\gamma\left[t, \mathbf{R}^{\alpha-1}(t), \mathbf{D}(t)\right] + W_\gamma(t)\,\mathbf{w}_\gamma\left[\mathbf{U}_\gamma(t)\right], \ \forall t \in \mathfrak{T}.$$

This is (3.47). It proves the lemma statement ■

Condition 66 ***Plant modeling condition***

The mathematical modeling of the real plant can have the form of the system (3.12) in which the matrices multiplying control nonlinearities $\mathbf{e}_u(.)$ *and* $\mathbf{w}(.)$ *are not in their canonical forms (3.32) and (3.42), respectively. In such cases the elementary matrix transformations should be applied to the left-hand and the right-hand sides of the equations of the mathematical model so that the matrices multiplying control nonlinearities* $\mathbf{e}_u(.)$ *and* $\mathbf{w}(.)$ *have the canonical forms (3.32) and (3.42), respectively.*

Note 67 *If $\gamma > 0$ then Property 64 permits the following:*

$$\mathbf{w}\left[\mathbf{U}(t)\right] = \left\{ \begin{array}{c} W^T(t)\,\mathbf{z}_a(t),\ \mathbf{z}_a(t) \in \mathfrak{R}^N,\ \gamma = N \leq r \\ W^T(t)\,W(t)\,\mathbf{z}_b(t),\ \mathbf{z}_b(t) \in \mathfrak{R}^r,\ \gamma = r \leq N \end{array} \right\}, \tag{3.48}$$

which is solvable in $\mathbf{U}(t)$,

$$\mathbf{U}(t) = \mathbf{w}^I \left[\left\{ \begin{array}{c} W^T(t)\,\mathbf{z}_a(t),\ \mathbf{z}_a(t) \in \mathfrak{R}^N,\ \gamma = N \\ W^T(t)\,W(t)\,\mathbf{z}_b(t),\ \mathbf{z}_b(t) \in \mathfrak{R}^r,\ \gamma = r \end{array} \right\} \right] \tag{3.49}$$

for $\mathbf{z}_a(t)$ *and* $\mathbf{z}_b(t)$ *determined by*

$$\mathbf{z}_a(t) = \left\{ \begin{array}{c} \left[W(t)\,W^T(t)\right]^{-1} \\ \bullet \left[\mathbf{Y}(t) - \mathbf{z}\left[t, \mathbf{R}^{\alpha-1}(t), \mathbf{D}(t)\right]\right], \\ \gamma = N \leq r \end{array} \right\}, \tag{3.50}$$

$$\mathbf{z}_b(t) = \left\{ \begin{array}{c} \left[W^T(t)\,W(t)\right]^{-1} W^T(t) \\ \bullet \left[\mathbf{Y}(t) - \mathbf{z}\left[t, \mathbf{R}^{\alpha-1}(t), \mathbf{D}(t)\right]\right] \\ \gamma = r \leq N \end{array} \right\}. \tag{3.51}$$

If $\gamma = N \leq r$ *then* $\mathbf{U}(t)$, *(3.49), (3.50) satisfies (3.12) in the whole output space* $\mathfrak{R}^N$. *However, if* $\gamma = r \leq N$ *then* $\mathbf{U}(t)$, *(3.49), (3.51) satisfies (3.12) only in the subspace*

$$\mathfrak{R}^N_W = \left\{\mathbf{Y} : \mathbf{Y} \in \mathfrak{R}^N,\ W^T(t)\,\mathbf{Y} = \mathbf{0}_r\right\}$$

of the whole output space $\mathfrak{R}^N$.

3.2.4 Sets and systems

Definition 68 ***Inherent sets of the system*** *(3.12)*

a) The temporal interval $\mathfrak{L}^* = (l^*_m, l^*_M)$ *is the largest temporal set of the initial moments* t_0 *for which the system (3.12) has a solution. It depends, for the given system (3.12), on the initial vector* $\mathbf{R}_0^{\alpha-1} \in \mathfrak{R}^{\alpha\rho}$, *on the disturbance vector function* $\mathbf{D}(.) \in \mathfrak{D}$, *and on the control vector function* $\mathbf{U}(.) \in \mathfrak{U}^\mu$, $\mathfrak{L}^* = \mathfrak{L}^*\left[\mathbf{R}_0^{\alpha-1}, \mathbf{D}(.), \mathbf{U}(.)\right] \subseteq \mathfrak{T}$.

b) $\mathfrak{S}^*(t_0) = \mathfrak{S}_0^*$ *is the largest set of the initial vectors* $\mathbf{R}_0^{\alpha-1} \in \mathfrak{R}^{\alpha\rho}$ *at* $t_0 \in \mathfrak{T}$, *which is a nonempty subset of* $\mathfrak{R}^{\alpha\rho}$, $\mathfrak{S}^*(t_0) \neq \phi$, $\mathfrak{S}^*(t_0) \subseteq \mathfrak{R}^{\alpha\rho}$, *such that the system (3.12) has an extended solution passing through* $\mathbf{R}_0^{\alpha-1}$ *at* $t = t_0$ *for every* $\mathbf{R}_0^{\alpha-1} \in \mathfrak{R}^{\alpha\rho}$.

c) The temporal interval $\mathfrak{T}\left[t_0, \mathbf{R}_0^{\alpha-1}, \mathbf{D}(.), \mathbf{U}(.)\right]$ *is the largest temporal set over which the given system (3.12) has an extended solution passing through* $\mathbf{R}_0^{\alpha-1}$ *at* $t = t_0$. *It depends, for the given plant (3.12), on* $\left[t_0, \mathbf{R}_0^{\alpha-1}, \mathbf{D}(.), \mathbf{U}(.)\right] \in \mathfrak{L}^* \times \mathfrak{S}^*(t_0) \times \mathfrak{D} \times \mathfrak{U}^\mu$. *The interval* $\mathfrak{T}\left[t_0, \mathbf{R}_0^{\alpha-1}, \mathbf{D}(.), \mathbf{U}(.)\right] = \mathfrak{T}(t_0)$, $\mathfrak{T}(t_0) \subseteq \mathfrak{T}$, *for the fixed triplet* $\left[\mathbf{R}_0^{\alpha-1}, \mathbf{D}(.), \mathbf{U}(.)\right] \in \mathfrak{S}^*(t_0) \times \mathfrak{D} \times \mathfrak{U}^\mu$.

3.2.5 System motions

Definition 69 ***System motion (solution)***

A vector function $\mathcal{R}(.) : \mathfrak{T}_0 \times \mathfrak{T}_i \times \mathfrak{R}^{\alpha\rho} \times \mathcal{J}^\xi \longrightarrow \mathfrak{R}^\rho$ *is a* ***motion (solution)*** *of the system (3.12) through* $\mathbf{R}_0^{\alpha-1}$ *at* $t = t_0$, *if, and only if,* $\mathcal{R}(.)$, $\mathbf{R}_0^{\alpha-1} \in \mathfrak{R}^{\alpha\rho}$ *and* $t_0 \in \mathfrak{T}_i$ *obey the following for every* $[\mathbf{D}(.), \mathbf{U}(.)] \in \mathfrak{D} \times \mathfrak{U}^\mu$.

- *(P1) The vector function* $\mathcal{R}^{\alpha-1}(.)$ *is defined in* $\mathfrak{R}^{\alpha\rho}$, *on* $\mathfrak{T}(t_0)$, *where* $\mathfrak{T}(t_0) \subseteq \mathfrak{T}$,

  $$\mathcal{R}^{\alpha-1}(t; t_0, \mathbf{R}_0^{\alpha-1}; \mathbf{I}) \equiv \mathbf{R}_0^{\alpha-1}(t) \in \mathfrak{R}^{\alpha\rho},\ \forall t \in \mathfrak{T}(t_0),$$

  if, and only if, $t_0 \in \mathfrak{L}$, $\mathfrak{L} \subseteq \mathfrak{T}$, *and* $\mathbf{R}_0^{\alpha-1} \in \mathfrak{S}(t_0)$.

- *(P2) The vector function* $\mathcal{R}(.)$ *is* α*-times continuously differentiable on* $\mathfrak{T}(t_0)$ *and satisfies identically the system (3.12),*

$$\begin{gathered}\mathcal{R}(t;t_0,\mathbf{R}_0^{\alpha-1};\mathbf{D};\mathbf{U})\in\mathfrak{C}^{\alpha}\left(\mathfrak{T}^*\right),\\ Q\left[t,\mathcal{R}^{(\alpha-1)}(t;t_0;\mathbf{R}_0^{\alpha-1};\mathbf{I})\right]\mathcal{R}^{(\alpha)}(t;t_0;\mathbf{R}_0^{\alpha-1};\mathbf{I})\\ +\mathbf{q}\left[t,\mathcal{R}^{(\alpha-1)}(t;t_0;\mathbf{R}_0^{\alpha-1};\mathbf{I})\right]=\mathbf{h}\left[t,\mathbf{I}^{\xi}(t)\right],\\ t_0\in\mathfrak{L},\ \forall\left[t,\mathbf{R}_0^{\alpha-1},\mathbf{I}(.)\right]\in\mathfrak{T}(t_0)\times\mathfrak{S}(t_0)\times\mathcal{J}^{\xi}.\end{gathered}$$

- *(P3) The vector function* $\mathcal{R}(.)$ *satisfies the initial condition:*

$$\mathcal{R}^{\alpha-1}(t_0;t_0;\mathbf{R}_0^{\alpha-1};\mathbf{I})\equiv\mathbf{R}_0^{\alpha-1}.$$

- *(P4) The vector function* $\mathcal{R}(.)$ *is the unique system motion through* $\mathbf{R}_0^{\alpha-1}$ *at* $t=t_0$*; i.e., two vector functions* $\mathcal{R}_i(.):\mathfrak{T}\times\mathfrak{T}\times\mathfrak{R}^{\alpha\rho}\times\mathfrak{R}^M\longrightarrow\mathfrak{R}^\rho$, $i=1,2$, *are the system motions (solutions) through* $\mathbf{R}_0^{\alpha-1}$ *at* $t=t_0$ *if, and only if, they obey* $(P1)-(P3)$*, and they are identical,*

$$\mathcal{R}_1^{\alpha-1}(t;t_0;\mathbf{R}_0^{\alpha-1};\mathbf{I})\equiv\mathcal{R}_2^{\alpha-1}(t;t_0;\mathbf{R}_0^{\alpha-1};\mathbf{I}).\tag{3.52}$$

Comment 70 *The time interval* $\mathfrak{L}^*$ *of the initial moments* t_0 *depends on* $\mathbf{D}(.)$ *and* $\mathbf{U}(.)$ *in general. It can contract to a singleton or to the empty set* ϕ *in general. This is a characteristic of strongly nonuniform systems. The same holds for the time set* $\mathfrak{T}(t_0)$.

With this in mind we accept:

Condition 71 ***Set conditions for the system*** *(3.12)*
a) The intersection $\mathfrak{S}(\mathfrak{L})$ *or* $\mathfrak{S}$ *of* $\mathfrak{S}^*(t_0)$ *over* $\mathfrak{L}$ *or* $\mathfrak{T}$ *is a nonempty time-invariant set, respectively,*

$$\begin{gathered}\mathfrak{S}(\mathfrak{L})=\cap\left[\mathfrak{S}^*(t_0):t_0\in\mathfrak{L}\right]\neq\phi,\ \mathfrak{S}(\mathfrak{L})\subseteq\mathfrak{R}^{\alpha\rho},\\ \mathfrak{S}=\cap\left[\mathfrak{S}^*(t_0):t_0\in\mathfrak{T}\right]\neq\phi,\ \mathfrak{S}\subseteq\mathfrak{R}^{\alpha\rho}.\end{gathered}$$

b) The intersection of $\mathfrak{L}^*\left[\mathbf{R}_0^{\alpha-1},\mathbf{D}(.),\mathbf{U}(.)\right]$ *over* $\mathfrak{S}\times\mathfrak{D}\times\mathfrak{U}^\mu$ *is nonempty time-invariant interval denoted simply by* $\mathfrak{L}$,

$$\begin{gathered}\mathfrak{L}=\cap\left\{\begin{array}{c}\mathfrak{L}^*\left[\mathbf{R}_0^{\alpha-1},\mathbf{D}(.),\mathbf{U}(.)\right]:\left[\mathbf{R}_0^{\alpha-1},\mathbf{D}(.),\mathbf{U}(.)\right]\in\\ \in\mathfrak{S}\times\mathfrak{D}\times\mathfrak{U}^\mu\end{array}\right\},\\ \mathfrak{L}=(l_m,l_M)\neq\phi,\ -\infty\leq l_m<l_M\leq\infty,\ \mathfrak{L}\subseteq\mathfrak{T}.\end{gathered}$$

c) The temporal interval $\mathfrak{T}(t_0)$ *can be unbounded,* $\mathfrak{T}(t_0)\subseteq\mathfrak{T}_0$ *for every* $t_0\in\mathfrak{L}$*. The intersection* $\mathfrak{T}(\mathfrak{L})$ *of* $\mathfrak{T}(t)$ *over* $\mathfrak{L}$ *is a nonempty time-invariant interval* $\mathfrak{T}$,

$$\mathfrak{T}(\mathfrak{L})=\cap\left[\mathfrak{T}(t):t\in\mathfrak{L}\right]\neq\phi,\ \mathfrak{T}(\mathfrak{L})\subseteq\mathfrak{T}.$$

These conditions ensure that $\mathfrak{S}(t)$, $\mathfrak{L}^*\left[\mathbf{R}_0^{\alpha-1},\mathbf{D}(.),\mathbf{U}(.)\right]$ and $\mathfrak{T}(t_0)$ do not contract to singletons or to the empty set.

3.2.6 System generalized motions

Let $\mathfrak{T}^0(t_0)$ be the largest subset of $\mathfrak{T}(t_0)$ with the zero measure. "Almost always on $\mathfrak{T}(t_0)$" means "for every $t\in\left[\mathfrak{T}(t_0)\backslash\mathfrak{T}^0(t_0)\right]$".

Definition 72 ***Generalized motions (solutions) of the system*** *(3.12)*

A vector function $\mathcal{R}(.) : \mathfrak{T}\times\mathfrak{T}\times\mathfrak{R}^{\alpha\rho}\times\mathcal{J}^{\xi} \longrightarrow \mathfrak{R}^{\rho}$ *is a* ***generalized motion (generalized solution)*** *of the system (3.12) through* $\mathbf{R}_0$ *at* $t = t_0$ *if, and only if,* $\mathcal{R}(.)$, $\mathbf{R}_0^{\alpha-1} \in \mathfrak{R}^{\alpha\rho}$, *and* $t_0 \in \mathfrak{T}$ *obey the following for every* $\mathbf{I}(.) \in \mathcal{J}^{\xi}$.

- *(G1) The vector function* $\mathcal{R}(.)$ *is defined in* $\mathfrak{R}^{\rho}$, *on* $\mathfrak{T}(t_0)$, *where* $\mathfrak{T}(t_0) \subseteq \mathfrak{T}$,

$$\mathcal{R}(t; t_0, \mathbf{R}_0^{\alpha-1}; \mathbf{I}) \equiv \mathbf{R}(t) \in \mathfrak{R}^{\rho}, \ \forall t \in \mathfrak{T}(t_0),$$

if, and only if, $t_0 \in \mathfrak{L}^*$, $\mathfrak{L}^* \subseteq \mathfrak{T}$, *and* $\mathbf{R}_0^{\alpha-1} \in \mathfrak{S}^*(t_0)$.

- *(G2) The vector function* $\mathcal{R}(.)$ *is* $(\alpha - 1)$ *times continuously differentiable on* $\mathfrak{T}(t_0)$ *for* $t_0 \in \mathfrak{L}$,

$$\mathcal{R}(t; t_0, \mathbf{R}_0^{\alpha-1}; \mathbf{I}) \in \mathfrak{C}^{\alpha-1}[\mathfrak{T}(t_0)], \quad \forall t_0 \in \mathfrak{L},$$

and its $(\alpha - 1)$*st derivatives are continuously differentiable almost always on* $\mathfrak{T}(t_0)$,

$$\mathcal{R}^{(\alpha-1)}(t; t_0, \mathbf{R}_0^{\alpha-1}; \mathbf{I}) \in \mathfrak{C}^1\left[\mathfrak{T}(t_0)\backslash\mathfrak{T}^0(t_0)\right], \quad t_0 \in \mathfrak{L}.$$

At every t^* *at which* $\mathcal{R}^{(\alpha-1)}(t; t_0, \mathbf{R}_0^{\alpha-1}; \mathbf{I})$ *is not differentiable,* $t^* \in \mathfrak{T}^0(t_0)$, *the* $(\alpha - 1)$*st derivative* $\mathcal{R}^{(\alpha-1)}(t; t_0, \mathbf{R}_0^{\alpha-1}; \mathbf{I})$ *has on both sides derivatives*

$$D_l\mathcal{R}^{(\alpha-1)}(t^*; t_0, \mathbf{R}_0^{\alpha-1}; \mathbf{I}) \text{ and } D_r\mathcal{R}^{(\alpha-1)}(t^*; t_0, \mathbf{R}_0^{\alpha-1}; \mathbf{I})$$

if t^* *is an interior point of* $\mathfrak{T}(t_0)$.

If t^* *is a lower boundary point of the time interval* $\mathfrak{T}(t_0)$,

$$t^* = min\ \mathfrak{T}^*(t_0),$$

then $\mathcal{R}^{(\alpha-1)}(t; t_0, \mathbf{R}_0^{\alpha-1}; \mathbf{I})$ *has the right-hand derivative*

$$D_r\mathcal{R}^{(\alpha-1)}(t^*; t_0, \mathbf{R}_0^{\alpha-1}; \mathbf{I}).$$

If t^* *is an upper boundary point of the time interval* $\mathfrak{T}(t_0)$,

$$t^* = max\ \mathfrak{T}^*(t_0),$$

then $\mathcal{R}^{(\alpha-1)}(t; t_0, \mathbf{R}_0^{\alpha-1}; \mathbf{I})$ *has the left-hand derivative*

$$D_l\mathcal{R}^{(\alpha-1)}(t^*; t_0, \mathbf{R}_0^{\alpha-1}; \mathbf{I}).$$

$\mathcal{R}(t; t_0, \mathbf{R}_0^{\alpha-1}; \mathbf{I})$ *satisfies identically the system (3.12) almost always,*

$$Q\left[t, \mathcal{R}^{\alpha-1}(t; t_0; \mathbf{R}_0^{\alpha-1}; \mathbf{I})\right]\mathcal{R}^{(\alpha)}(t; t_0; \mathbf{R}_0^{\alpha-1}; \mathbf{I}) +$$
$$\mathbf{q}\left[t, \mathcal{R}^{(\alpha-1)}(t; t_0; \mathbf{R}_0^{\alpha-1}; \mathbf{I})\right] = \mathbf{h}\left[t, \mathbf{I}^{\xi}(t)\right],$$
$$t_0 \in \mathfrak{L}^*, \ \forall\left[t, \mathbf{R}_0^{\alpha-1}, \mathbf{I}(.)\right] \in \left[\mathfrak{T}(t_0)\backslash\mathfrak{T}^0(t_0)\right] \times \mathfrak{S}^*(t_0) \times \mathcal{J}^{\xi}.$$

- *(G3) The vector function* $\mathcal{R}(.)$ *and its derivatives up to the order* $(\alpha - 1)$ *satisfy the initial condition:*

$$\mathcal{R}^{(k)}(t_0; t_0; \mathbf{R}_0^{\alpha-1}; \mathbf{I}) \equiv \mathbf{R}_0^{(k)}, \ \ \forall k = 0, 1, ..., \ \alpha - 1.$$

- *(G4) The vector function $\mathcal{R}(.)$ is the unique system generalized motion through $\mathbf{R}_0$ at $t = t_0$,; i.e., functions $\mathcal{R}_i(.) : \mathfrak{T} \times \mathfrak{T} \times \mathfrak{R}^{\alpha\rho} \times \mathfrak{R}^M \longrightarrow \mathfrak{R}^\rho$, $i = 1, 2$, are the system-generalized motions (generalized solutions) through $\mathbf{R}_0$ at $t = t_0$ if, and only if, they obey (G1)-(G3) and they are identical almost always on $\mathfrak{T}(t_0)$,*

$$\mathcal{R}_1^{\alpha-1}(t; t_0; \mathbf{R}_0^{\alpha-1}; \mathbf{I}) = \mathcal{R}_2^{\alpha-2}(t; t_0; \mathbf{R}_0^{\alpha-1}; \mathbf{I}),$$
$$t_0 \in \mathfrak{L}^*,\ \forall \left[t, \mathbf{R}_0^{\alpha-1}, \mathbf{I}(.)\right] \in \left[\mathfrak{T}(t_0) \backslash \mathfrak{T}^0(t_0)\right] \times \mathfrak{S}^*(t_0) \times \mathcal{J}^\xi.$$

This definition extends Definition 3.8 of [287, pp. 57, 58] to *time*-varying nonlinear systems.

Definition 73 ***A (generalized) dynamical system of the order*** α
A (generalized) dynamical system $\mathcal{S}$ ***of the order*** α *is a set of mappings (vector functions) $\mathcal{R}(.)$ and $\mathcal{Y}(.)$ from the Cartesian product space $\mathfrak{T}_0 \times \mathfrak{L}^* \times \mathfrak{S}^*(t_0) \times \mathcal{J}^\xi$ into the Cartesian product space $\mathfrak{R}^\rho \times \mathfrak{R}^N$,*

$$\mathcal{S} : \mathfrak{T}_0 \times \mathfrak{L}^* \times \mathfrak{S}^*(t_0) \times \mathcal{J}^\xi \to \mathfrak{R}^\rho \times \mathfrak{R}^N,$$

which have the following properties.

P1 ***Existence:*** *Every vector function $\mathcal{R}(.)$ called* ***the system (generalized) motion*** *is defined in $\mathfrak{R}^\rho$ on $\mathfrak{T}_0 \times \mathfrak{L}^* \times \mathfrak{S}^*(t_0) \times \mathcal{J}^\xi$ and every vector function $\mathcal{Y}(.)$ called* ***the system output response*** *is defined in $\mathfrak{R}^N$ on $\mathfrak{T}_0 \times \mathfrak{L}^* \times \mathfrak{S}^*(t_0) \times \mathcal{J}^\xi$; i.e.,*

$$\mathcal{R}(t; t_0; \mathbf{R}_0^{\alpha-1}; \mathbf{I}) \in \mathfrak{R}^\rho,\ \mathcal{Y}(t; t_0; \mathbf{R}_0^{\alpha-1}; \mathbf{I}) \in \mathfrak{R}^N,$$
$$\forall (t, t_0, \mathbf{R}_0^{\alpha-1}, \mathbf{I}) \in \mathfrak{T}_0 \times \mathfrak{L}^* \times \mathfrak{S}^*(t_0) \times \mathcal{J}^\xi. \tag{3.53}$$

P2 ***Initial condition:*** *Every system (generalized) motion $\mathcal{R}(.)$, its derivatives up to the order $\alpha - 1$ and every system output response $\mathcal{Y}(.)$ satisfy the initial conditions,*

$$\mathcal{R}^{(k)}(t_0; t_0; \mathbf{R}_0^{\alpha-1}; \mathbf{I}) = \mathbf{R}_0^{(k)},\ \forall k = 0, 1, .., \alpha - 1,$$
$$\mathcal{Y}(t_0; t_0; \mathbf{R}_0^{\alpha-1}; \mathbf{I}) \in \mathbf{Y}_0,\ \forall (t_0, \mathbf{R}_0^{\alpha-1}; \mathbf{I}) \in \mathfrak{T}_0 \times \mathfrak{L}^* \times \mathfrak{S}^*(t_0) \times \mathcal{J}^\xi. \tag{3.54}$$

P3 ***Continuous dependence:*** *Every system extended (generalized) motion $\mathcal{R}^{\alpha-1}(.)$ and every output response $\mathcal{Y}(.)$ of the system are continuous in $(t, t_0, \mathbf{R}_0^{\alpha-1}, \mathbf{D}, \mathbf{U})$:*

$$\mathcal{R}^{\alpha-1}(t; t_0; \mathbf{R}_0^{\alpha-1}; \mathbf{I}) \in \mathfrak{C}\left(\mathfrak{T}_0 \times \mathfrak{L}^* \times \mathfrak{S}^*(t_0) \times \mathcal{J}^\xi\right),$$
$$\mathcal{Y}(t; t_0; \mathbf{R}_0^{\alpha-1}; \mathbf{I}) \in \mathfrak{C}\left(\mathfrak{T}_0 \times \mathfrak{L}^* \times \mathfrak{S}^*(t_0) \times \mathcal{J}^\xi\right). \tag{3.55}$$

P4 ***Uniqueness:*** *For every $(t, t_0, \tau, \mathbf{X}_0, \mathbf{D}; \mathbf{U}) \in \mathfrak{T}_0 \times \mathfrak{L}^* \times \mathfrak{S}^*(t_0) \times \mathcal{J}^\xi$ every system extended (generalized) motion $\mathcal{R}^{\alpha-1}(.)$ and every system output response $\mathcal{Y}(.)$ satisfy*

$$\mathcal{R}^{\alpha-1}\left[t; t_0 + \tau; \mathcal{R}^{\alpha-1}(t_0 + \tau; t_0; \mathbf{R}_0^{\alpha-1}; \mathbf{I}); \mathbf{I}\right]$$
$$= \mathcal{R}^{\alpha-1}(t; t_0; \mathbf{R}_0^{\alpha-1}; \mathbf{I}), \tag{3.56}$$
$$\mathcal{Y}\left[t; t_0 + \tau; \mathcal{R}^{\alpha-1}(t_0 + \tau; t_0; \mathbf{R}_0^{\alpha-1}; \mathbf{I}); \mathbf{I}\right]$$
$$= \mathcal{Y}(t; t_0; \mathbf{R}_0^{\alpha-1}; \mathbf{I}). \tag{3.57}$$

3.2.7 Existence and uniqueness of motions

Let $\mathfrak{L}^c$ and $\mathfrak{T}^c(\mathfrak{L})$ be the largest nonempty connected compact subintervals of $\mathfrak{L}$ and of $\mathfrak{T}(\mathfrak{L})$, respectively, and let $\mathfrak{S}^c(\mathfrak{L})$ be a nonempty connected compact subset of $\mathfrak{S}(\mathfrak{L})$,

$$\mathfrak{L}^c \subseteq \mathfrak{L},\ \mathfrak{T}^c(\mathfrak{L}) \subseteq \mathfrak{T}(\mathfrak{L}),\ \mathfrak{S}^c(\mathfrak{L}) \subseteq \mathfrak{S}(\mathfrak{L}),$$
$$\mathfrak{L}^c \cap \mathfrak{T}^c(\mathfrak{L}) \neq \phi,\ \ \mathfrak{S}^c(\mathfrak{L}) \neq \phi. \tag{3.58}$$

In order to discover and prove the conditions for the existence of the unique motions of the system (3.12) we recall Granwall inequality.

Theorem 74 ***Granwall inequality theorem*** *(e.g., [11, Theorem 8.1, p. 29], [427, Problem 2.14.9, p64], [433, Theorem 1.6, p. 43])*

Let r be a continuous nonnegative function on a compact interval $\mathfrak{J} = [a,b] \subset \mathfrak{T}$ and δ and k be nonnegative constants such that

$$r(t) \leq \delta + \int_a^t kr(s)\,ds, \ t \in \mathfrak{J}.$$

Then

$$r(t) \leq \delta \exp\left[k(t-a)\right], \ \forall t \in \mathfrak{J}.$$

The well-known Lipschitz condition is related to the ISO systems.

The Lipshitz condition is well known in the framework of the *input-state-output* (ISO) dynamical systems in the free regime. Its extensions to dynamical systems (3.12) in forced regimes follow.

Let

$$\mathbf{f}\left[t, \mathbf{R}^{\alpha-1}(t), \mathbf{I}^{\xi}(t)\right] = Q^{-1}\left[t, \mathbf{R}^{\alpha-1}(t)\right]\left\{\mathbf{h}\left[t, \mathbf{I}^{\xi}(t)\right] - \mathbf{q}\left[t, \mathbf{R}^{\alpha-1}(t)\right]\right\}, \tag{3.59}$$

Condition 75 ***Lipschitz condition for the system (3.12) with $\mathbf{I}(.)$ dependent only on time t***

A vector function $\mathbf{f}(.)$, $\mathbf{f}(.) : \mathfrak{T} \times \mathfrak{R}^{\alpha\rho} \times \mathcal{J}^{\xi} \longrightarrow \mathfrak{R}^{\rho}$, with $\mathbf{I}(.)$ dependent only on time t, is Lipshitz continuous on the Cartesian product set $\mathfrak{T} \times \mathfrak{S}(\mathfrak{L}) \times \mathcal{J}^{\xi}$ and obeys Lipshitz condition (3.60) in $\mathbf{R}^{\alpha-1}$ with Lipshitz constant L if, and only if,

$$\begin{aligned} &\left\|\mathbf{f}\left[t, \mathbf{R}_1^{\alpha-1}, \mathbf{I}^{\xi}(t)\right] - \mathbf{f}\left[t, \mathbf{R}_2^{\alpha-1}, \mathbf{I}^{\xi}(t)\right]\right\| \\ &\leq L\left\|\mathbf{R}_1^{\alpha-1} - \mathbf{R}_2^{\alpha-1}\right\|, \ \forall\left[t, \mathbf{R}_1^{\alpha-1}, \mathbf{R}_2^{\alpha-1}, \mathbf{I}(.)\right] \\ &\in \mathfrak{T} \times \mathfrak{S}^c(\mathfrak{L}) \times \mathfrak{S}^c(\mathfrak{L}) \times \mathcal{J}^{\xi}. \end{aligned} \tag{3.60}$$

Comment 76 ***Lipschitz condition for the feedback (closed loop) control system of the plant (3.12) with $\mathbf{D}(.)$ dependent only on time t***

*If the plant **(3.12)** is controlled by a feedback control $\mathbf{U}\left(t, \mathbf{R}^{\alpha-1}\right)$ then Lipshitz condition (3.60) becomes*

$$\begin{aligned} \left\|\mathbf{f}\left[t, \mathbf{R}_1^{\alpha-1}, \mathbf{D}^{\eta}(t), \mathbf{U}^{\mu}\left(t, \mathbf{R}_1^{\alpha-1}\right)\right] - \mathbf{f}\left[t, \mathbf{R}_2^{\alpha-1}, \mathbf{D}^{\eta}(t), \mathbf{U}^{\mu}\left(t, \mathbf{R}_2^{\alpha-1}\right)\right]\right\| \\ \leq L\left\|\mathbf{R}_1^{\alpha-1} - \mathbf{R}_2^{\alpha-1}\right\|, \\ \forall\left[t, \mathbf{R}_1^{\alpha-1}, \mathbf{R}_2^{\alpha-1}, \mathbf{D}(.), \mathbf{U}(.)\right] \in \mathfrak{T} \times \mathfrak{S}^c(\mathfrak{L}) \times \mathfrak{S}^c(\mathfrak{L}) \times \mathfrak{D}^{\eta} \times \mathfrak{U}^{\mu}. \end{aligned} \tag{3.61}$$

*With this in mind the following theorem holds also for the feedback (closed loop) control system of the system **(3.12)** with $\mathbf{D}(.)$ dependent only on time t.*

Theorem 77 ***Uniqueness of motions of the regular system***

*Let the system **(3.12)** possess Property 58. Let the function $\mathbf{f}(.)$ by (3.59) in view of (3.31), with $\mathbf{I}(.)$ dependent only on time t, be Lipshitz continuous on the time-invariant Cartesian product set*

$$\mathfrak{T}^c \times \mathfrak{S}^c(\mathfrak{L}) \times \mathcal{J}^{\xi}.$$

*Then the regular system **(3.12)** has the unique solutions $\mathcal{R}(t; t_0; \mathbf{R}^{\alpha-1}; \mathbf{I})$ on a time-invariant Cartesian product set*

$$\mathfrak{T}^c \times \mathfrak{L}^c \times \mathfrak{S}^c(\mathfrak{L}) \times \mathcal{J}^{\xi}.$$

Proof. Let the condition of the theorem hold, which implies the validity of (3.60). Equations (3.31) and (3.59) together with Property 58 yield

$$\begin{aligned}\mathbf{R}^{(\alpha)}(t) &= Q^{-1}\left[t,\mathbf{R}^{\alpha-1}(t)\right]\left\{\mathbf{h}\left[t,\mathbf{I}^{\xi}(t)\right]-\mathbf{q}\left[t,\mathbf{R}^{\alpha-1}(t)\right]\right\}\\ &= \mathbf{f}\left[t,\mathbf{R}^{\alpha-1}(t),\mathbf{I}^{\xi}(t)\right].\end{aligned}$$

Hence,

$$\mathbf{R}_1^{\alpha-1}(t)-\mathbf{R}_2^{\alpha-1}(t)=\begin{bmatrix}\mathbf{R}_1(t)-\mathbf{R}_2(t)\\ \cdots\\ \mathbf{R}_1^{(\alpha-2)}(t)-\mathbf{R}_2^{(\alpha-2)}(t)\\ \mathbf{R}_1^{(\alpha-1)}(t)-\mathbf{R}_2^{(\alpha-1)}(t)\end{bmatrix}$$

$$=\int_{t_0}^{t}\begin{bmatrix}\mathbf{R}_1^{(1)}(s)-\mathbf{R}_2^{(1)}(s)\\ \cdots\\ \mathbf{R}_1^{(\alpha-1)}(s)-\mathbf{R}_2^{(\alpha-1)}(s)\\ \mathbf{R}_1^{(\alpha)}(s)-\mathbf{R}_2^{(\alpha)}(s)\end{bmatrix}ds=\int_{t_0}^{t}\begin{bmatrix}\mathbf{R}_1^{(1)}(s)-\mathbf{R}_2^{(1)}(s)\\ \cdots\\ \mathbf{R}_1^{(\alpha-1)}(s)-\mathbf{R}_2^{(\alpha-1)}(s)\\ \mathbf{f}\left[t,\mathbf{R}_1^{\alpha-1}(s),\mathbf{I}^{\xi}(s)\right]-\\ -\mathbf{f}\left[t,\mathbf{R}_2^{\alpha-1}(s),\mathbf{I}^{\xi}(s)\right]\end{bmatrix}ds,$$

$$\Longrightarrow$$

$$\left\|\mathbf{R}_1^{\alpha-1}(t)-\mathbf{R}_2^{\alpha-1}(t)\right\|$$

$$=\left\|\int_{t_0}^{t}\begin{bmatrix}\mathbf{R}_1^{(1)}(s)-\mathbf{R}_2^{(1)}(s)\\ \cdots\\ \mathbf{R}_1^{(\alpha-1)}(s)-\mathbf{R}_2^{(\alpha-1)}(s)\\ \mathbf{f}\left[t,\mathbf{R}_1(s),\mathbf{I}^{\xi}(s)\right]-\\ -\mathbf{f}\left[t,\mathbf{R}_2^{\alpha-1}(s),\mathbf{I}^{\xi}(s)\right]\end{bmatrix}\right\|ds\le\int_{t_0}^{t}\left\|\begin{bmatrix}\mathbf{R}_1(s)-\mathbf{R}_2(s)\\ \mathbf{R}_1^{(1)}(s)-\mathbf{R}_2^{(1)}(s)\\ \cdots\\ \mathbf{R}_1^{(\alpha-1)}(s)-\mathbf{R}_2^{(\alpha-1)}(s)\\ \mathbf{f}\left[t,\mathbf{R}_1^{\alpha-1}(s),\mathbf{I}^{\xi}(s)\right]-\\ -\mathbf{f}\left[t,\mathbf{R}_2^{\alpha-1}(s),\mathbf{I}^{\xi}(s)\right]\end{bmatrix}\right\|ds$$

$$\le\int_{t_0}^{t}\left\|\begin{bmatrix}\mathbf{R}_1(s)-\mathbf{R}_2(s)\\ \mathbf{R}_1^{(1)}(s)-\mathbf{R}_2^{(1)}(s)\\ \cdots\\ \mathbf{R}_1^{(\alpha-1)}(s)-\mathbf{R}_2^{(\alpha-1)}(s)\\ \mathbf{f}\left[t,\mathbf{R}_1^{\alpha-1}(s),\mathbf{I}^{\xi}(s)\right]-\\ -\mathbf{f}\left[t,\mathbf{R}_2^{\alpha-1}(s),\mathbf{I}^{\xi}(s)\right]\end{bmatrix}\right\|ds=\int_{t_0}^{t}\left\|\begin{bmatrix}\mathbf{R}_1^{\alpha-1}(s)-\mathbf{R}_2^{\alpha-1}(s)\\ \mathbf{f}\left[t,\mathbf{R}_1^{\alpha-1}(s),\mathbf{I}^{\xi}(s)\right]-\\ -\mathbf{f}\left[t,\mathbf{R}_2^{\alpha-1}(s),\mathbf{I}^{\xi}(s)\right]\end{bmatrix}\right\|ds.$$

Lipschitz condition (3.60) applied to the right-hand side of the last inequality permits us to write

$$\left\|\mathbf{R}_1^{\alpha-1}(t)-\mathbf{R}_2^{(\alpha-1)}(t)\right\|\le\int_{t}^{t}\begin{bmatrix}\left\|\mathbf{R}_1^{\alpha-1}(s)-\mathbf{R}_2^{(\alpha-1)}(s)\right\|+\\ +L\left\|\mathbf{R}_1^{\alpha-1}(s)-\mathbf{R}_2^{(\alpha-1)}(s)\right\|\end{bmatrix}ds\Longrightarrow$$

$$\left\|\mathbf{R}_1^{\alpha-1}(t)-\mathbf{R}_2^{(\alpha-1)}(t)\right\|\le(L+1)\int_{t}^{t}\left\|\mathbf{R}_1^{\alpha-1}(s)-\mathbf{R}_2^{(\alpha-1)}(s)\right\|ds$$

This satisfies the Granwall inequality theorem 74 with $\delta=0$, $r=\|\mathbf{R}_1-\mathbf{R}_2\|$, and $k=L+1$:

$$\begin{aligned}\left\|\mathbf{R}_1^{\alpha-1}(t)-\mathbf{R}_2^{\alpha-1}(t)\right\|\le\delta\exp\left[k\left(t-a\right)\right]=0,\ \forall t\in\mathfrak{T}\Longrightarrow\\ \mathbf{R}_1^{\alpha-1}(t)=\mathbf{R}_2^{\alpha-1}(t),\ \ \forall t\in\mathfrak{T}.\end{aligned}$$

This proves the uniqueness of the system motions on the time-invariant Cartesian product set $\mathfrak{T}\times\mathfrak{L}\times\mathfrak{S}\times\mathcal{J}^{\xi}$, $(P4)$ of Definition 69 ∎

Notice that $(P4)$ of Definition 69, i.e., (3.52), implies the following for $\mathbf{R}_{01}^{\alpha-1}=\mathbf{R}_{02}^{\alpha-1}=\mathbf{R}_0^{\alpha-1}$:

$$\begin{aligned}\mathbf{R}_1^{(\alpha)}(t) &= \mathbf{R}_1^{(\alpha)}(t;t_0,\mathbf{R}_{01}^{\alpha-1};\mathbf{I})\\ &= Q^{-1}\left[t,\mathbf{R}_1^{\alpha-1}(t)\right]\left\{\mathbf{h}\left[t,\mathbf{I}^{\xi}(t)\right]-\mathbf{q}\left[t,\mathbf{R}_1^{\alpha-1}(t)\right]\right\}\\ &= Q^{-1}\left[t,\mathbf{R}_2^{\alpha-1}(t)\right]\left\{\mathbf{h}\left[t,\mathbf{I}^{\xi}(t)\right]-\mathbf{q}\left[t,\mathbf{R}_2^{\alpha-1}(t)\right]\right\}\\ &= \mathbf{R}_2^{(\alpha)}(t),\end{aligned}$$
$$\forall\left[t,\mathbf{R}_i^{\alpha-1},\mathbf{I}(.)\right]\in\mathfrak{T}^c\times\mathfrak{S}^c(\mathfrak{L})\times\mathcal{J}^{\xi},\ \forall i=1,2.$$

Altogether,

$$\mathbf{R}_1^{\alpha}(t)=\mathbf{R}_2^{\alpha}(t),\quad \forall t\in\mathfrak{T}^c,$$

i.e.,

$$\mathcal{R}_1^{\alpha}(t;t_0;\mathbf{R}_0^{\alpha-1};\mathbf{I})\equiv\mathcal{R}_2^{\alpha}(t;t_0;\mathbf{R}_0^{\alpha-1};\mathbf{I}),$$
$$\forall\left[t,t_0,\mathbf{R}_0^{\alpha-1},\mathbf{I}(.)\right]\in\mathfrak{T}^c\times\mathfrak{S}^c(\mathfrak{L})\times\mathcal{J}^{\xi}.$$

3.2.8 System advanced properties

Property 78 ***System set properties***
The sets $\mathfrak{L}^*=(l_m^*,l_M^*)$ and $\mathfrak{S}^*(t_0)$ in Definition 68 and $\mathfrak{S}$ in Condition 71 obey $\mathfrak{L}^*=\mathfrak{T}$, $\mathfrak{S}(t_0)=\mathfrak{S}(t)=\mathfrak{R}^{\alpha\rho}$.

The system ***(3.12)*** *has the unique solution $\mathcal{R}(.;t_0,\mathbf{R}_0^{\alpha-1};\mathbf{I})$ and obeys the system form (Definition 21) of TCUP (Principle 20) for every $\left[\mathbf{I}(.),\mathbf{R}_0^{\alpha-1},t_0\right]\in\mathcal{J}^{\xi}\times\mathfrak{R}^{\alpha\rho}\times\mathfrak{T}$. The corresponding output response $\mathcal{Y}(.;t_0,\mathbf{Y}_0;\mathbf{I})$ obeys TCUP (Principle 20) for every $\left[\mathbf{I}(.),\mathbf{Y}_0,t_0\right]\in\mathcal{J}^{\xi}\times\mathfrak{R}^N\times\mathfrak{T}$.*

Note 79 *Throughout this book* ***we accept the validity of Properties 57 through 78 if it is not said otherwise.***

3.2.9 Special system class: *IO* systems

Let m, N, η, μ and ξ be the fixed natural numbers. The original mathematical models of some hydraulic, pneumatic, thermal and chemical processes or plants, and their control systems, result directly in the form of the nonlinear vector *Input - Output* (*IO*) differential equation that we present in the general compact form:

$$A\left[t,\mathbf{Y}^{m-1}(t)\right]\mathbf{Y}^{(m)}(t)+\mathbf{a}\left[t,\mathbf{Y}^{m-1}(t)\right]=\mathbf{h}\left[t,\mathbf{I}^{\xi}(t)\right],\ \xi\le m,$$
$$\forall t\in\mathfrak{T},\ m\ge 1,\ \mathbf{h}(.):\mathfrak{T}\times\mathfrak{R}^{(m+1)M}\longrightarrow\mathfrak{R}^N,\ \mathbf{h}\left(t,\mathbf{I}^{\xi}\right)\in\mathfrak{C}\left(\mathfrak{T}\times\mathfrak{R}^{mM}\right),\tag{3.62}$$

where

$$\mathbf{Y}\in\mathfrak{R}^N,\ A(.):\mathfrak{T}\times\mathfrak{R}^{mN}\longrightarrow\mathfrak{R}^{N\times N},\ A\left(t,\mathbf{Y}^{m-1}\right)\in\mathfrak{C}^1\left(\mathfrak{T}\times\mathfrak{R}^{mN}\right),$$
$$\mathbf{a}(.):\mathfrak{T}\times\mathfrak{R}^{mN}\longrightarrow\mathfrak{R}^N,\ \mathbf{a}\left(t,\mathbf{Y}^{(m-1)}\right)\in\mathfrak{C}^1\left(\mathfrak{T}\times\mathfrak{R}^{mN}\right),\tag{3.63}$$

Claim 80 ***The*** *IO* ***systems are a special class of the systems*** *(3.12)*
The system (3.12) reduces to the IO system (3.62) under the following simplifying conditions:

$$\mathbf{Y}=\mathbf{R},\ \rho=N,\ \alpha=m,\ A\left[t,\mathbf{Y}^{m-1}(t)\right]\equiv Q\left[t,\mathbf{R}^{\alpha-1}(t)\right],$$
$$\mathbf{a}\left[t,\mathbf{Y}^{m-1}(t)\right]\equiv\mathbf{q}\left[t,\mathbf{R}^{\alpha-1}(t)\right],\ \mathbf{s}\left[t,\mathbf{R}(t),\mathbf{I}(t)\right]\equiv\mathbf{R}(t).\tag{3.64}$$

Conclusion 81 ***The study of the properties of the IIDO systems (3.12) is valid for the IO system (3.62)***

The systems (3.12) incorporate the IO systems (3.62).

The study of the properties of the systems (3.12) holds also for the IO systems (3.62).

3.2.10 Special system class: *ISO* systems

The qualitative theory of nonlinear dynamical systems has mostly studied the *input-state-output (ISO)* nonlinear *systems* described by *the nonlinear differential state vector equation* (3.65) and by *the nonlinear algebraic output vector equation* (3.66),

$$\frac{d\mathbf{X}(t)}{dt} = \mathbf{f}\left[t, \mathbf{X}(t), \mathbf{I}(t)\right], \ \forall t \in \mathfrak{T}, \ \mathbf{X} \in \mathfrak{R}^n, \tag{3.65}$$

$$\mathbf{Y}(t) = \mathbf{g}\left[t, \mathbf{X}(t), \mathbf{I}(t)\right], \ \forall t \in \mathfrak{T}, \tag{3.66}$$

where

$$\mathbf{f}(.) : \mathfrak{T} \times \mathfrak{R}^n \times \mathfrak{R}^M \longrightarrow \mathfrak{R}^n, \ \mathbf{f}(t, \mathbf{X}, \mathbf{I}) \in \mathfrak{C}^1\left(\mathfrak{T} \times \mathfrak{R}^n \times \mathfrak{R}^M\right), \tag{3.67}$$

$$\mathbf{g}(.) : \mathfrak{T} \times \mathfrak{R}^n \times \mathfrak{R}^M \longrightarrow \mathfrak{R}^N, \ \mathbf{g}(t, \mathbf{X}, \mathbf{I}) \in \mathfrak{C}^1\left(\mathfrak{T} \times \mathfrak{R}^n \times \mathfrak{R}^M\right). \tag{3.68}$$

Claim 82 ***The ISO systems (3.65), (3.66) are a special class of the systems (3.12)***

The system ***(3.12)*** *becomes the ISO system (3.65), (3.66) under the following special conditions:*

$$\begin{gathered}\alpha = 1, \ \ \rho = n, \ \ \xi = 0, \ \ \mathbf{R} = \mathbf{X}, \ \ Q\left[t, \mathbf{R}^{\alpha-1}(t)\right] \equiv Q\left[t, \mathbf{R}(t)\right] \equiv I_\rho, \\ \mathbf{h}\left[t, \mathbf{I}^{\xi}(t)\right] - \mathbf{q}\left[t, \mathbf{X}(t)\right] = \mathbf{h}\left[t, \mathbf{I}(t)\right] - \mathbf{q}\left[t, \mathbf{R}(t)\right] = \mathbf{f}\left[t, \mathbf{X}(t), \mathbf{I}(t)\right], \\ \mathbf{s}\left[t, \mathbf{R}(t), \mathbf{I}(t)\right] = \mathbf{g}\left[t, \mathbf{X}(t), \mathbf{I}(t)\right].\end{gathered} \tag{3.69}$$

Conclusion 83 ***The study of the properties of the systems (3.12)***

The systems (3.12) incorporate the ISO systems (3.62).

The study of the properties of the systems (3.12) holds also for the ISO systems (3.65), (3.66).

Conclusion 81 and Conclusion 82 explain that all the results of this book are directly applicable to both the *IO* systems (3.62) and the *ISO* systems (3.65), (3.66).

3.3 Existence and solvability

Let $\mathbf{Z} = \left[Z_1 \ \ Z_2 \ \cdots \ Z_K\right]^T$ and $\mathcal{Z}\left(.; t_0; \mathbf{Z}_0^{m-1}; \mathbf{w}\right)$ be a solution of (3.70) through $\mathbf{Z}_0 \in \mathfrak{R}^K$ at $t = t_0$ for a given $\mathbf{w}$, $\mathbf{Z}\left(t_0; t_0; \mathbf{Z}_0^{m-1}; \mathbf{w}\right) \equiv \mathbf{Z}_0$,

$$\begin{gathered}A(t)\,\mathbf{f}\left[t, \mathbf{Z}^m(t)\right] = M(t)\,\mathbf{g}\left[t, \mathbf{w}(t)\right], \ t \in \mathfrak{T}, \ A(.) : \mathfrak{T} \longrightarrow \mathfrak{R}^{i \times K}, \ \mathbf{Z} \in \mathfrak{R}^K, \\ \mathbf{f}(.) : \mathfrak{T} \times \mathfrak{R}^{K(m+1)} \longrightarrow \mathfrak{R}^K, \ \mathbf{w} \in \mathfrak{R}^j, \ M(.) : \mathfrak{T} \longrightarrow \mathfrak{R}^{i \times j}, \ \mathbf{g}(.) : \mathfrak{T} \times \mathfrak{R}^j \longrightarrow \mathfrak{R}^j.\end{gathered} \tag{3.70}$$

By following [426, Definition 5.3, p. 119]:

Definition 84 ***Fundamental solution***

The solution $\mathcal{Z}\left(.; t_0; \mathbf{Z}_0^{m-1}; \mathbf{w}\right)$ of (3.70) is its ***fundamental solution*** *if, and only if, all functional elements $Z_i\left(.; t_0; \mathbf{Z}_0^{m-1}; \mathbf{w}\right)$ of $\mathcal{Z}\left(.; t_0; \mathbf{Z}_0^{m-1}; \mathbf{w}\right)$ are linearly independent on $\mathfrak{T}_0$.*

Definition 85 ***Solution existence and equation solvability***

The equation (3.70) has the (unique) solution $\mathcal{Z}\left(t;t_0;\mathbf{Z}_0^{m-1}\right)$ *for (every)* $\mathbf{Z}_0 \in \mathfrak{R}^j$ *if, and only if, it is (uniquely) solvable in both* $\mathbf{f}\left[t,\mathbf{Z}^m(t)\right]$ *and then in* $\mathbf{Z}(t)$ *for (every)* $\mathbf{Z}_0 \in \mathfrak{R}^K$*, respectively.*

Theorem 86 ***Condition for the equation unique solvability***

In order for the equation (3.70) to be(uniquely) solvable in both $\mathbf{f}\left[t,\mathbf{Z}^m(t)\right]$ *and then in* $\mathbf{Z}(t)$ *for every* $\mathbf{Z}_0 \in \mathfrak{R}^K$ *(and that the solution is the fundamental solution) it is necessary and sufficient that, respectively,*

$$rank\ A(t) = \min\{i, K\} = i, \ \ (rank\ A(t) = i = K), \tag{3.71}$$

and that

$$\mathbf{f}\left[t,\mathbf{Z}^m(t)\right] = T(t)\,M(t)\,\mathbf{g}\left[t,\mathbf{w}(t)\right], \ \ T(t) = A^T(t)\left[A(t)\,A^T(t)\right]^{-1} \tag{3.72}$$

is (uniquely) solvable in $\mathbf{Z}(t)$ *for every* $\mathbf{Z}_0 \in \mathfrak{R}^K$*.*

The proof of this theorem is in Appendix C.1.

3.4 Fundamental control principle

The following *fundamental control principle* states what is possible to achieve at most with control and what is necessary for the control variables U_i, $i = 1, 2, \cdots, r$, i.e., for the control vector $\mathbf{U} \in \mathfrak{R}^r$, to satisfy in order to govern directly several mutually independent variables $Z_1, Z_2, \cdots, Z_K$, i.e., their vector $\mathbf{Z} \in \mathfrak{R}^K$, at every moment $t \in (t_1, t_2) \subseteq \mathfrak{T}$, $t_2 > t_1$.

Axiom 87 ***The fundamental control principle***

The scalar form:

a) In order for r *control variables* $U_i(.)$, $i = 1, 2, \cdots, r$, ***to control directly*** K ***mutually independent variables*** $Z_j(.)$, $j = 1, 2, \cdots, K$, ***at every moment*** $t \in (t_1, t_2) \subseteq \mathfrak{T}$, $t_2 > t_1$*, it is necessary that the number* r *of the control variables is not less than the number* K *of the variables to be controlled directly on* (t_1, t_2)*:* $r \geq K$.

b) In order for r *control variables* $U_i(.)$, $i = 1, 2, \cdots, r$, ***to control indirectly*** K ***mutually independent variables*** $Z_j(.)$, $j = 1, 2, \cdots, K$, ***at every moment*** $(t_1, t_2) \subseteq \mathfrak{T}$, $t_2 > t_1$, ***by controlling directly*** m ***mutually independent functions*** $v_k(.)$, $v_k(.) : \mathfrak{T} \times \mathfrak{R}^K \longrightarrow \mathfrak{R}$, $k = 1, 2, \cdots, m$, *depending on the variables* $Z_j(.)$ ***at every moment*** $t \in (t_1, t_2)$ *it is necessary that the number* r *of the control variables is not less than the number* m *of the functions* $v_k(.)$ *to be controlled directly on* (t_1, t_2)*:* $r \geq m$.

c) If $r < K$ *then* r ***control variables*** $U_i(.)$, $i = 1, 2, \cdots, r$, ***can control*** K ***mutually independent variables*** $Z_j(.)$, $j = 1, 2, \cdots, K$, *at every moment* $(t_1, t_2) \subseteq \mathfrak{T}$, $t_2 > t_1$, ***only indirectly by controlling directly*** m ***mutually independent functions*** $v_k(.)$, $v_k(.) : \mathfrak{T} \times \mathfrak{R}^K \longrightarrow \mathfrak{R}$, $k = 1, 2, \cdots, m$, *depending on the variables* $Z_j(.)$ *at every moment* (t_1, t_2).

The vector form:

A) In order for the control vector $\mathbf{U} \in \mathfrak{R}^r$, $\mathbf{U} = \left[U_1\ U_2 \cdots\ U_r\right]^T$, ***to control elementwise the vector variable*** $\mathbf{Z} \in \mathfrak{R}^K$, $\mathbf{Z} = \left[Z_1\ Z_2 \cdots\ Z_K\right]^T$, ***at every moment*** $(t_1, t_2) \subseteq \mathfrak{T}$, *it is necessary that the dimension* r *of the control vector* $\boldsymbol{U}$ *is not less than the dimension* K *of the vector* $\boldsymbol{Z}$ *to be controlled elementwise on* $(t_1, t_2) : r \geq K$.

B) In order for the control vector $\mathbf{U} \in \mathfrak{R}^r$, $\mathbf{U} = \left[U_1\ U_2 \cdots\ U_r\right]^T$, ***to control indirectly the vector variable*** $\mathbf{Z} \in \mathfrak{R}^K$, $\mathbf{Z} = \left[Z_1\ Z_2 \cdots\ Z_K\right]^T$, ***with mutually independent entries*** Z_j, $j = 1, 2, \cdots, K$, ***at every moment*** $(t_1, t_2) \subseteq \mathfrak{T}$, $t_2 > t_1$, ***by controlling elementwise the vector function*** $\mathbf{v} = \left[v_1\ v_2 \cdots\ v_m\right]^T$, $\mathbf{v}(.) : \mathfrak{T}x\mathfrak{R}^K \longrightarrow \mathfrak{R}^m$, ***which depends on the vector variable*** $\mathbf{Z}$, ***at every moment*** $t \in (t_1, t_2)$ *it is necessary that the* *dimension r of the control vector $\boldsymbol{U}$ is not less than the dimension m of the vector function $\mathbf{v}(.)$ to be controlled elementwise on (t_1, t_2): $r \geq m$.*

C) If $r < K$ then r control vector $\mathbf{U} \in \mathfrak{R}^r$, $\mathbf{U} = [U_1\ U_2 \cdots\ U_r]^T$, ***can control*** *$K$* ***vector variable*** *$\mathbf{Z} \in \mathfrak{R}^K$, $\mathbf{Z} = [Z_1\ Z_2 \cdots\ Z_K]^T$,* ***with mutually independent entries*** *Z_j, $j = 1, 2, \cdots, K$,* ***at every moment*** *$(t_1, t_2) \subseteq \mathfrak{T}$, $t_2 > t_1$,* ***only indirectly by controlling elementwise*** *m* ***vector function*** *$\mathbf{v} = [v_1\ v_2 \cdots\ v_m]^T$, $\mathbf{v}(.) : \mathfrak{T}x\mathfrak{R}^K \longrightarrow \mathfrak{R}^m$, $m \leq r$,* ***which depends on the vector variable*** *$\mathbf{Z}(.)$.*

Corollary 88 ***Control "perpetuum mobile" is impossible***

The scalar form: *The control "perpetuum mobile" means that r control variables $U_i(.)$, $i = 1, 2, \cdots, r$, control directly K mutually independent variables $Z_j(.)$, $j = 1, 2, \cdots, K > r$, at every moment $t \in (t_1, t_2)$. It is not possible for $K > r$.*

The vector form: *The control "perpetuum mobile" means that r control vector variable $\mathbf{U} \in \mathfrak{R}^r$, controls elementwise K vector variable $\mathbf{Z} \in \mathfrak{R}^K$, $K > r$, at every moment $t \in (t_1, t_2)$. It is not possible for $K > r$.*

Chapter 4

Desired regime

4.1 Introduction

We analyze, by referring to [252], [279], a desired regime of a system representing essentially a plant (P). The demanded plant output response $\mathbf{Y}_d(.)$ determines the plant desired regime, and vice versa.

Let $\tau \in Cl\mathfrak{T}$, $\mathfrak{T}_0 \subseteq \mathfrak{T}$, where $\tau = t_0$; i.e., $\mathfrak{T}_0 = \mathfrak{T}$, is possible.

The following coordinate transformations are called *Lyapunov coordinate transformations.* They hold for all variables,

$$\mathbf{d} = \mathbf{D} - \mathbf{D}_N, \tag{4.1}$$

$$\mathbf{i} = \mathbf{I} - \mathbf{I}_N, \tag{4.2}$$

$$\mathbf{r} = \mathbf{R} - \mathbf{R}_d = \mathbf{R} - \mathbf{R}_N, \tag{4.3}$$

$$\varepsilon_R = \mathbf{R_d} - \mathbf{R} = \mathbf{R_N} - \mathbf{R} = -\mathbf{r} \tag{4.4}$$

$$\mathbf{u} = \mathbf{U} - \mathbf{U}_N, \tag{4.5}$$

$$\mathbf{y} = \mathbf{Y} - \mathbf{Y}_N = \mathbf{Y} - \mathbf{Y}_d \tag{4.6}$$

$$\varepsilon = \mathbf{Y}_N - \mathbf{Y} = \mathbf{Y}_d - \mathbf{Y} = -\mathbf{y}. \tag{4.7}$$

These equations define them in general.

Definition 89 ***Desired regime*** *[252], [279]*

A plant is in ***a desired*** *(called also:* ***demanded*** *or* ***nominal*** *or* ***nonperturbed)*** ***regime on*** $\mathfrak{T}_0$ *if, and only if, it realizes its desired (output) response* $\mathbf{Y}_d(t)$ *all the time on* $\mathfrak{T}_0$,

$$\mathbf{Y}(t) = \mathbf{Y}_d(t), \quad \forall t \in \mathfrak{T}_0. \tag{4.8}$$

*The phrase "**on** $\mathfrak{T}_0$" is to be omitted if, and only if,* $\mathfrak{T}$ *replaces* $\mathfrak{T}_0$.

This definition implies directly a necessary (but not a sufficient) condition for a system to be in a desired (nominal, nonperturbed) regime (on $\mathfrak{T}_0$).

A change of the initial moment $t_0 \in \mathfrak{T}$ can influence the system regime. The *time*-varying system can be in a desired regime on $\mathfrak{T}_{01}$ but it does not guarantee that the system is in the desired regime on $\mathfrak{T}_{02}$, $t_{01} \neq t_{02} \Longrightarrow \mathfrak{T}_{01} \neq \mathfrak{T}_{02}$. This does not hold for *time*-invariant systems.

Proposition 90 *[252], [279] For the real output vector* $\mathbf{Y}(t)$ *to be equal to the desired output vector* $\mathbf{Y}_d(t)$ *on* $\mathfrak{T}_0$*, i.e., in order for the plant to be in a desired (desired, nonperturbed) regime on* $\mathfrak{T}_0$*, it is necessary that* $\mathbf{Y}_0 = \mathbf{Y}_{d0}$*; i.e.,* $(\mathbf{Y}(t) = \mathbf{Y}_d(t),\ \forall t \in \mathfrak{T}_0) \Longrightarrow \mathbf{Y}_0 = \mathbf{Y}_{d0}$.

The plant cannot be in a desired regime on $\mathfrak{T}_0$ if its initial real output vector $\mathbf{Y}_0$ is different from the initial desired output vector $\mathbf{Y}_{d0}$,

$$\mathbf{Y}_0 \neq \mathbf{Y}_{d0} \Longrightarrow \exists \sigma \in \mathfrak{T}_0 \Longrightarrow \mathbf{Y}(\sigma) \neq \mathbf{Y}_d(\sigma).$$

Because the real initial output vector $\mathbf{Y}_0$ is most often different from the desired initial output vector $\mathbf{Y}_{d0}$, then the system is most often in a *nondesired* (*nonnominal*, *perturbed*, *disturbed*) regime.

Definition 91 ***Nominal initial state conditions***

The initial state conditions are ***desired relative to the desired output response*** $\mathbf{Y}_d(.)$ *if, and only if, they correspond to the initial desired output vector* $\mathbf{Y}_{d0}$.

Comment 92 *Because the real initial output vector* $\mathbf{Y}(0) = \mathbf{Y}_0$ *is most often different from the desired initial output vector* $\mathbf{Y}_d(0) = \mathbf{Y}_{d0}$, *then the plant is most often in a nondesired (nonnominal, perturbed, disturbed) regime that is also called a* ***real regime***.

Definition 93 ***Nominal input*** $\mathbf{I}_N(.)$ ***relative to the plant desired response*** $\mathbf{Y}_d(.)$ ***on*** $\mathfrak{T}_0$

An input vector function $\mathbf{I}^*(.)$ *of a system is* ***nominal (desired) relative to the plant desired response*** $\mathbf{Y}_d(.)$ ***on*** $\mathfrak{T}_0$, *which is denoted by* $\mathbf{I}_N(.)$, *if, and only if,* $\mathbf{I}(.) = \mathbf{I}^*(.)$ *ensures that the corresponding real response* $\mathbf{Y}(.) = \mathbf{Y}^*(.)$ *of the plant obeys* $\mathbf{Y}^*(t) = \mathbf{Y}_d(t)$ *all the time on* $\mathfrak{T}_0$ *as soon as all the system initial (input, state and output) conditions are desired (i.e., nominal),*

$$\begin{gathered} If\ all\ initial\ conditions\ are\ desired\ then \\ \langle \mathbf{I}^*(.) = \mathbf{I}_N(.) \rangle \Longleftrightarrow \langle \mathbf{Y}^*(t) = \mathbf{Y}_d(t),\ \forall t \in \mathfrak{T}_0 \rangle . \end{gathered} \tag{4.9}$$

*The phrase "****on*** $\mathfrak{T}_0$*" is to be omitted if, and only if,* $\mathfrak{T}$ *replaces* $\mathfrak{T}_0$.

This definition is general.

Note 94 *[252], [279]An input vector function* $\mathbf{I}^*(.)$ *can be desired (i.e., nominal) relative to the desired response* $\mathbf{Y}_{d1}(.)$ *of the plant, but it need not be desired with respect to another desired response* $\mathbf{Y}_{d2}(.)$ *of the plant. This explains the relative sense of the notion "*$\mathbf{I}^*(.)$ *can be desired (i.e., nominal) relative to the desired response* $\mathbf{Y}_d(.)$*."*

Comment 95 *The desired input vector function* $\mathbf{I}_N(.)$ *of the plant incorporates both* ***the desired perturbation vector function*** *denoted by* $\mathbf{D}_N(.)$ *and* ***the desired control vector function*** $\mathbf{U}_N(.)$, $\mathbf{I}_N(.) = \left[\mathbf{D}_N^T(.) \quad \mathbf{U}_N^T(.)\right]^T$.

Definition 96 ***Nominal control*** $\mathbf{U}_N(.)$ ***of the plant relative to*** $[\mathbf{D}(.), \mathbf{Y}_d(.)]$ ***on*** $\mathfrak{T}_0 \times \mathfrak{D}^i \times \mathfrak{U}^\mu \times \mathfrak{Y}_d^k$

A control vector function $\mathbf{U}^*(.) \in \mathfrak{U}^\mu$ ***of the plant*** *is* ***nominal relative to the functional pair*** $[\mathbf{D}(.), \mathbf{Y}_d(.)]$ ***on*** $\mathfrak{T}_0 \times \mathfrak{D}^i \times \mathfrak{U}^\mu \times \mathfrak{Y}_d^k$, *which is denoted by* $\mathbf{U}_N(.)$,

$$\mathbf{U}_N(t) \equiv [\mathbf{U}_N(t; t_0; \mathbf{D}; \mathbf{Y}_d)] \in \mathfrak{U}^\mu, \tag{4.10}$$

if, and only if, $\mathbf{U}(.) = \mathbf{U}^*(.)$ *ensures that the corresponding plant real response* $\mathbf{Y}(.) = \mathbf{Y}^*(.)$ *obeys* $\mathbf{Y}^*(t) = \mathbf{Y}_d(t)$ *all the time on* $\mathfrak{T}_0$ *for every* $\mathbf{D}(.) \in \mathfrak{D}^i$ *and any chosen* $\mathbf{Y}_d(.) \in \mathfrak{Y}_d^k$ *as soon as all the system initial (input, state, and output) conditions are desired (i.e., nominal),*

$$\begin{gathered} [\mathbf{D}(.), \mathbf{Y}_d(.)] \in \mathfrak{D}^i \times \mathfrak{Y}_d^k\ and\ all\ initial\ conditions\ are\ desired\ then \\ \mathbf{U}^*(.) = \mathbf{U}_N(.)] \Longleftrightarrow \langle \mathbf{Y}^*(t) = \mathbf{Y}_d(t),\ \forall t \in \mathfrak{T}_0 \rangle . \end{gathered} \tag{4.11}$$

The time set $\mathfrak{T}_0$ *is to be omitted from the phrase "on* $\mathfrak{T}_0 \times \mathfrak{D}^i \times \mathfrak{U}^\mu \times \mathfrak{Y}_d^k$*" if, and only if,* $\mathfrak{T}$ *replaces* $\mathfrak{T}_0$.

Comment 97 *The preceding definition specifies the desired control vector function* $\mathbf{U}_N(.)$ *relative to an arbitrary* $\mathbf{D}(.) \in \mathfrak{D}^i$ *and any chosen* $\mathbf{Y}_d(.) \in \mathfrak{Y}_d^k$*, (4.10). However, the disturbance* $\mathbf{D}(.)$ *is rarely fixed. Because the disturbance vector function is unknown and unpredictable this means that for every possible* $\mathbf{D}(.)$ *belonging to the family of the permitted disturbances* $\mathfrak{D}^i$ *we should determine the corresponding nominal control. Another approach follows.*

Definition 98 ***Nominal control*** $\mathbf{U}_N(.)$ ***of the plant relative to*** $[\mathbf{D}_N(.),\ \mathbf{Y}_d(.)]$ ***on*** $\mathfrak{T}_0 \times \mathfrak{D}^i \times \mathfrak{U}^\mu \times \mathfrak{Y}_d^k$

A control vector function $\mathbf{U}^*(.) \in \mathfrak{U}^\mu$ ***of the plant*** *is* ***nominal relative to the functional pair*** $[\mathbf{D}_N(.),\ \mathbf{Y}_d(.)]$ ***on*** $\mathfrak{T}_0 \times \mathfrak{D}^i \times \mathfrak{U}^\mu \times \mathfrak{Y}_d^k$*, which is denoted by* $\mathbf{U}_N(.)$*,*

$$\mathbf{U}_N(t) \equiv [\mathbf{U}_N(t;t_0;\mathbf{D}_N;\mathbf{Y}_d)] \in \mathfrak{U}^\mu, \tag{4.12}$$

if, and only if, $\mathbf{U}(.) = \mathbf{U}^*(.)$ *ensures that the corresponding real response* $\mathbf{Y}(.) = \mathbf{Y}^*(.)$ *of the plant obeys* $\mathbf{Y}^*(t) = \mathbf{Y}_d(t)$ *all the time on* $\mathfrak{T}_0$ *for any accepted nominal disturbance-output pair* $[\mathbf{D}_N(.),\ \mathbf{Y}_d(.)] \in \mathfrak{D}^i \times \mathfrak{Y}_d^k$ *as soon as all the plant initial (input, state, and output) conditions are desired (i.e., nominal),*

$$\begin{array}{c}[\mathbf{D}_N(.), \mathbf{Y}_d(.)] \in \mathfrak{D}^i \times \mathfrak{Y}_d^k \text{ and all initial conditions are desired then} \\ \mathbf{U}^*(.) = \mathbf{U}_N(.)] \Longleftrightarrow \langle \mathbf{Y}^*(t) = \mathbf{Y}_d(t),\ \forall t \in \mathfrak{T}_0 \rangle .\end{array} \tag{4.13}$$

The time set $\mathfrak{T}_0$ *is to be omitted from the phrase "on* $\mathfrak{T}_0 \times \mathfrak{D}^i \times \mathfrak{U}^\mu \times \mathfrak{Y}_d^k$*" if, and only if,* $\mathfrak{T}$ *replaces* $\mathfrak{T}_0$*.*

Comment 99 *This definition assumes the knowledge of the nominal vector function* $\mathbf{D}_N(.)$ *relative to* $\mathbf{Y}_d(.)$*:*

$$\mathbf{D}_N(.) \equiv \mathbf{D}_N(.;\mathbf{Y}_d). \tag{4.14}$$

It is the usual design approach to determine or to accept the nominal vector function $\mathbf{D}_N(.)$ *relative to* $\mathbf{Y}_d(.)$ *for every chosen* $\mathbf{Y}_d(.) \in \mathfrak{Y}_d^k$*.*

Comment 100 *The nominal control vector function* $\mathbf{U}_N(.)$ *(on* $\mathfrak{T}_0 \times \mathfrak{D}^i \times \mathfrak{U}^\mu \times \mathfrak{Y}_d^k$*) for the plant relative to its desired output* $\mathbf{Y}_d(.)$ *is simultaneously the nominal (on* $\mathfrak{T}_0 \times \mathfrak{D}^i \times \mathfrak{U}^\mu \times \mathfrak{Y}_d^k$*) output vector function* $\mathbf{Y}_{dC}(.)$ *of the controller (C),* $\mathbf{Y}_{dC}(.) = \mathbf{U}_N(.)$ *that is to be determined from the condition to be the desired (on* $\mathfrak{T}_0 \times \mathfrak{D}^i \times \mathfrak{U}^\mu \times \mathfrak{Y}_d^k$*) control vector function of the plant relative to its desired output vector function* $\mathbf{Y}_d(.)$*.*

Comment 101 *If the controller is an output feedback controller in a closed loop control system (CS) of the plant, then its input vector function* $\mathbf{I}_C(.)$ *has two input vector functions: the desired* $\mathbf{Y}_d(.)$ *and the real* $\mathbf{Y}(.)$ *output vector functions of the plant,*

$$\mathbf{I}_C(.) = \left[\mathbf{Y}_d^T(.) \ \ \mathbf{Y}^T(.)\right]^T. \tag{4.15}$$

or equivalently, the error vector function $\boldsymbol{\varepsilon}(.)$*, or the deviation vector function* $\mathbf{y}(.)$*, of the plant output,*

$$\mathbf{I}_C(.) = \mathbf{Y}_d(.) - \mathbf{Y}(.) = \boldsymbol{\varepsilon}(.) = -\mathbf{y}(.). \tag{4.16}$$

The desired (nominal) input vector function of the controller $\mathbf{I}_{CN}(.)$ *is therefore*

$$\begin{aligned}\mathbf{I}_{CN}(.) &= \left[\mathbf{Y}_d^T(.) \ \ \mathbf{Y}_d^T(.)\right]^T, \text{ equivalently,} \\ \mathbf{I}_{CN}(.) &= \boldsymbol{\varepsilon}_N(.) = \mathbf{Y}_d(.) - \mathbf{Y}_d(.) = \mathbf{0}_N.\end{aligned}$$

This implies the desired (nominal) output error vector $\boldsymbol{\varepsilon}_N$ *of the plant as the desired input vector to the controller,* $\boldsymbol{\varepsilon}_N(t) \equiv \mathbf{Y}_d(t) - \mathbf{Y}_d(t) \equiv \mathbf{0}_N$*.*

Comment 102 *If a system represents an overall closed-loop, feedback control system (CS) then the desired input vector function* $\mathbf{I}_{CSN}(.)$ *of the whole control system incorporates both the nominal perturbation vector function* $\mathbf{D}_N(.)$ *of the controlled plant and the plant desired output vector function* $\mathbf{Y}_d(.)$*:*

$$\mathbf{I}_{CSN}(.) = \left[\mathbf{D}_N^T(.) \quad \mathbf{Y}_d^T(.)\right]^T.$$

Because the control system desired output vector function $\mathbf{Y}_{CSd}(.)$ *is the desired (output) response* $\mathbf{Y}_d(.)$ *of the plant,* $\mathbf{Y}_{CSd}(.) = \mathbf{Y}_d(.)$*, then only the desired perturbation function* $\mathbf{D}_N(.)$ *relative to* $\mathbf{Y}_d(.)$ *is to be determined.*

The control system designer's crucial interest is in a solution of the following.

Problem 103 *Under what conditions there exists a desired control vector function* $\mathbf{U}_N(.)$ *relative to the plant desired output response* $\mathbf{Y}_d(.)$ *for some* $\mathbf{D}(.)$ *in* $\mathfrak{D}^i$ *(for every* $\mathbf{D}(.)$ *in* $\mathfrak{D}^i$*); or equivalently, under what conditions the plant desired output response* $\mathbf{Y}_d(.)$ *is realizable in* $\mathfrak{D}^i$ *(is realizable on* $\mathfrak{D}^i$*), respectively?*

Some qualitative system properties (e.g., trackability properties, and tracking properties) have a sense if, and only if, there exists an affirmative solution to the preceding problem.

Definition 104 ***Realizability of*** $\mathbf{Y}_d(.)$ ***on*** $\mathfrak{T}_0 \times \mathfrak{U}^\mu \times \mathfrak{Y}_d^k$ ***under the action of the given disturbance*** $\mathbf{D}(.) \in \mathfrak{D}^i$

i) ***The given desired output behavior*** $\mathbf{Y}_d(.) \in \mathfrak{Y}_d^k$ ***of the plant*** *(3.12) is* ***realizable on*** $\mathfrak{T}_0 \times \mathfrak{U}^\mu \times \mathfrak{Y}_d^k$ ***under the perturbation of the given*** $D(.) \in \mathfrak{D}^i$ *if, and only if, there exists a control vector function* $\mathbf{U}(.) \in \mathfrak{U}^\mu$ *defined on* $\mathfrak{T}_0$ *such that* $\mathbf{Y}_d(.)$ *is the unique plant output response through* $\mathbf{Y}_d(t_0)$ *under the action of the pair* $[\mathbf{D}(.), \mathbf{U}(.)]$.

ii) If, and only if, additionally to i) the entries $Y_{di}(.)$, $\forall i = 1, 2, \cdots, n$*, are mutually independently realizable on* $\mathfrak{T}_0 \times \mathfrak{U}^\mu \times \mathfrak{Y}_d^k$ *under the action of the given disturbance* $\mathbf{D}(.) \in \mathfrak{D}^i$ *then they are* ***elementwise realizable on*** $\mathfrak{T}_0 \times \mathfrak{U}^\mu \times \mathfrak{Y}_d^k$ ***under the perturbation of the given disturbance*** $\mathbf{D}(.) \in \mathfrak{D}^i$.

iii) If, and only if, additionally to i) the dimension r *of* $\mathbf{U}$ *is the least number* j *of the entries* $U_i(.)$, $\forall i = 1, 2, \cdots, j$*, of* $\mathbf{U}$ *that satisfies i) then the realizability of* $\mathbf{Y}_d(.)$ *on* $\mathfrak{T}_0 \times \mathfrak{U}^\mu \times \mathfrak{Y}_d^k$ *under the action of the given* $\mathbf{D}(.) \in \mathfrak{D}^i$ *is* ***minimal***.

The time set $\mathfrak{T}_0$ *is to be omitted if, and only if,* $\mathfrak{T}$ *replaces* $\mathfrak{T}_0$.

This definition, the fundamental control principle 87 and Theorem 86 imply:

Lemma 105 ***Criterion for the realizability of*** $\mathbf{Y}_d(.)$ ***under the perturbation of the given disturbance*** $\mathbf{D}(.) \in \mathfrak{D}^i$

i) In order for the desired output behavior $\mathbf{Y}_d(.) \in \mathfrak{Y}_d^k$ *of the plant (3.12) to be realizable on* $\mathfrak{T}_0 \times \mathfrak{U}^\mu \times \mathfrak{Y}_d^k$ *under the perturbation of the given* $\mathbf{D}(.) \in \mathfrak{D}^i$ *it is necessary and sufficient that there exists a control vector function* $\mathbf{U}(.) \in \mathfrak{U}^\mu$ *of dimension* $r > 0$ *defined on* $\mathfrak{T}_0$ *for which* $\mathbf{Y}_d(.)$ *is the unique output solution of the plant mathematical model on* $\mathfrak{T}_0$.

ii) In order for the desired output behavior $\mathbf{Y}_d(.) \in \mathfrak{Y}_d^k$ *of the plant (3.12) to be elementwise realizable on* $\mathfrak{T}_0 \times \mathfrak{U}^\mu \times \mathfrak{Y}_d^k$ *under the perturbation of the given* $\mathbf{D}(.) \in \mathfrak{D}^i$ *it is necessary that there exists a control vector function* $\mathbf{U}(.)$ *that obeys both i) and* $r \geq N$.

iii) In order for the desired output behavior $\mathbf{Y}_d(.) \in \mathfrak{Y}_d^k$ *of the plant (3.12) to be elementwise realizable on* $\mathfrak{T}_0 \times \mathfrak{U}^\mu \times \mathfrak{Y}_d^k$ *under the action of a minimal control vector function* $\mathbf{U}(.) \in \mathfrak{U}^\mu$ *when the plant is perturbed by the given* $\mathbf{D}(.) \in \mathfrak{D}^i$ *it is necessary and sufficient that the control vector function* $\mathbf{U}(.)$ *obeys both i) and* $r = N$.

Comment 106 ***Dimension** r **of** $\mathbf{U}$ **and realizability of** $\mathbf{Y}_d(.)$ **on** $\mathfrak{T}_0 \times \mathfrak{U}^\mu \times \mathfrak{Y}_d^k$ **under the action of the given disturbance** $\mathbf{D}(.) \in \mathfrak{D}^i$*

The condition i) of Lemma 105 demands the existence of a control vector function $\mathbf{U}(.) \in \mathfrak{U}^\mu$ *of dimension* r*. This implies* $r \geq 1$ *but it does not determine its relationship to the dimension* N *of* $\mathbf{Y}_d$*.*

Lemma 105 determines realizability of a specific $\mathbf{Y}_d(.)$ for a single given $\mathbf{D}(.)$. We broaden it first to the realizability of a specific $\mathbf{Y}_d(.)$ in $\mathcal{J}^i$.

Definition 107 $\mathbf{Y}_d(.)$ ***realizable in*** $\mathcal{J}^i$ ***on*** $\mathfrak{T}_0 \times \mathfrak{Y}_d^k$

The desired response $\mathbf{Y}_d(.)$ *of the system (3.12) is* ***realizable in*** $\mathcal{J}^i$ ***on*** $\mathfrak{T}_0 \times \mathfrak{Y}_d^k$ *if, and only if, there exists* $\mathbf{I}^*(.) \in \mathcal{J}^i$ *defined on* $\mathfrak{T}_0$ *such that* $\mathbf{Y}_d(.) \in \mathfrak{Y}_d^k$ *is the unique plant output response through* $\mathbf{Y}_d(t_0)$ *under the action of* $\mathbf{I}^*(.)$*,*

$$\mathbf{I}(.) = \mathbf{I}^*(.) \implies \mathbf{Y}(.) = \mathbf{Y}_d(.).$$

The application of Definition 107 to the plant (3.12) is more significant for the tracking study.

Definition 108 ***Realizability of*** $\mathbf{Y}_d(.)$ ***in*** $\mathfrak{D}^i$ ***on*** $\mathfrak{T}_0 \times \mathfrak{U}^\mu \times \mathfrak{Y}_d^k$

i) ***The desired response*** $\mathbf{Y}_d(.)$ *of the plant (3.12) is* ***realizable in*** $\mathfrak{D}^i$ ***on the set product*** $\mathfrak{T}_0 \times \mathfrak{U}^\mu \times \mathfrak{Y}_d^k$ *if, and only if, for every* $\mathbf{Y}_d(.) \in \mathfrak{Y}_d^k$ *there exist* $\mathbf{D}^*(.) \in \mathfrak{D}^i$ *and* $\mathbf{U}^*(.) \in \mathfrak{U}^\mu$ *both defined on* $\mathfrak{T}_0$ *such that* $\mathbf{Y}_d(.)$ *is the unique plant output response through* $\mathbf{Y}_d(t_0)$ *under the action of the pair* $[\mathbf{D}^*(.), \mathbf{U}^*(.)]$*,*

$$[\mathbf{D}(.), \mathbf{U}(.)] = [\mathbf{D}^*(.), \mathbf{U}^*(.)] \implies \mathbf{Y}(.) = \mathbf{Y}_d(.).$$

ii) If, and only if, additionally to i) the control vector $\mathbf{U}$ *can act on every entry* $Y_{di}(.)$*,* $\forall i = 1, 2, .., n$*, of* $\mathbf{Y}_d$ *mutually independently when the plant (3.12) is perturbed by some disturbance* $\mathbf{D}(.) \in \mathfrak{D}^i$ *then the realizability of* $\mathbf{Y}_d(.)$ *in* $\mathfrak{D}^i$ *on* $\mathfrak{T}_0 \times \mathfrak{U}^\mu \times \mathfrak{Y}_d^k$ *is* ***elementwise****.*

iii) If, and only if, additionally to i) the dimension r *of the control vector* $\mathbf{U}$ *is the least number of the entries* $U_j(.)$*,* $\forall i = 1, 2, .., j$*, of* $\mathbf{U}$ *that satisfies i) then the control is* ***minimal*** *for the realizability of* $\mathbf{Y}_d(.)$ *in* $\mathfrak{D}^i$ *on* $\mathfrak{T}_0 \times \mathfrak{U}^\mu \times \mathfrak{Y}_d^k$*.*

The time set $\mathfrak{T}_0$ *is to be omitted if, and only if,* $\mathfrak{T}$ *replaces* $\mathfrak{T}_0$*.*

Comment 109 *This definition does not require that for every* $\mathbf{D}(.) \in \mathfrak{D}^i$ *there exists a control vector function* $\mathbf{U}(.) \in \mathfrak{U}^\mu$ *defined on* $\mathfrak{T}_0$ *such that* $\mathbf{Y}_d(.)$ *is the unique plant output response on* $\mathfrak{T}_0$ *under the action of the pair* $[\mathbf{D}(.), \mathbf{U}(.)]$*.*

Definition 108, the fundamental control principle 87 and Theorem 86 imply:

Lemma 110 ***Criterion for realizability of*** $\mathbf{Y}_d(.)$ ***in*** $\mathfrak{D}^i$ ***on*** $\mathfrak{T}_0 \times \mathfrak{U}^\mu \times \mathfrak{Y}_d^k$

i) In order for every desired output behavior $\mathbf{Y}_d(.) \in \mathfrak{Y}_d^k$ *of the plant (3.12) perturbed by some* $\mathbf{D}(.) \in \mathfrak{D}^i$ *to be realizable in* $\mathfrak{D}^i$ *on* $\mathfrak{T}_0 \times \mathfrak{U}^\mu \times \mathfrak{Y}_d^k$ *it is necessary and sufficient that for every* $\mathbf{Y}_d(.) \in \mathfrak{Y}_d^k$ *there exists a control vector function* $\mathbf{U}(.) \in \mathfrak{U}^\mu$ *defined on* $\mathfrak{T}_0$ *for which* $\mathbf{Y}_d(.)$ *is the unique output solution through* $\mathbf{Y}_d(t_0)$ *of the plant on* $\mathfrak{T}_0$*.*

ii) In order for the desired output behavior $\mathbf{Y}_d(.) \in \mathfrak{Y}_d^k$ *of the plant (3.12) perturbed by some* $\mathbf{D}(.) \in \mathfrak{D}^i$ *to be elementwise realizable in* $\mathfrak{D}^i$ *on* $\mathfrak{T}_0 \times \mathfrak{U}^\mu \times \mathfrak{Y}_d^k$ *it is necessary that there exists a control vector function* $\mathbf{U}(.)$ *that obeys both i) and* $r \geq N$*.*

iii) In order for the desired output behavior $\mathbf{Y}_d(.) \in \mathfrak{Y}_d^k$ *of the plant (3.12) to be elementwise realizable in* $\mathfrak{D}^i$ *on* $\mathfrak{T}_0 \times \mathfrak{U}^\mu \times \mathfrak{Y}_d^k$ *under the action of the minimal control* $\mathbf{U}$ *when the plant is perturbed by some* $\mathbf{D}(.) \in \mathfrak{D}^i$ *it is necessary and sufficient that there exists a control vector function* $\mathbf{U}(.) \in \mathfrak{U}^\mu$ *that obeys both i) and* $r = N$*.*

A stronger realizability property of $\mathbf{Y}_d(.)$ is its realizability on $\mathfrak{T}_0 \times \mathfrak{D}^i \times \mathfrak{U}^\mu \times \mathfrak{Y}_d^k$.

Definition 111 ***Realizability of*** $\mathbf{Y}_d(.)$ ***on*** $\mathfrak{T}_0 \times \mathfrak{D}^i \times \mathfrak{U}^\mu \times \mathfrak{Y}_d^k$

i) The desired response $\mathbf{Y}_d(.) \in \mathfrak{Y}_d^k$ *of the plant (3.12) is* ***realizable on*** $\mathfrak{T}_0\times\mathfrak{D}^i\times\mathfrak{U}^\mu\times\mathfrak{Y}_d^k$ *if, and only if, for every* $[\mathbf{D}(.), \mathbf{Y}_d(.)] \in \mathfrak{D}^i \times \mathfrak{Y}_d^k$ *there exists a control vector function* $\mathbf{U}(.) \in \mathfrak{U}^\mu$ *defined on* $\mathfrak{T}_0$ *such that* $\mathbf{Y}_d(.)$ *is the unique plant output response under the action of any* $\mathbf{D}(.) \in \mathfrak{D}^i$.

ii) If, and only if, additionally to i) the control vector $\mathbf{U}$ *can act on every entry* $Y_{di}(.)$, $\forall i = 1, 2, .., n,$ *of* $\mathbf{Y}_d$ *mutually independently when the plant is under the influence of any disturbance* $\mathbf{D}(.) \in \mathfrak{D}^i$ *then* $\mathbf{Y}_d(.)$ *is* ***elementwise realizable on*** $\mathfrak{T}_0 \times \mathfrak{D}^i \times \mathfrak{U}^\mu \times \mathfrak{Y}_d^k$.

iii) If, and only if, additionally to i) the dimension r *of the control vector* $\mathbf{U}$ *is the least number of the entries* $U_j(.)$, $\forall i = 1, 2, \cdots, j,$ *of* $\mathbf{U}$ *that satisfies i) then the control is* ***minimal*** *for the realizability of* $\mathbf{Y}_d(.)$ *on* $\mathfrak{T}_0 \times \mathfrak{D}^i \times \mathfrak{U}^\mu \times \mathfrak{Y}_d^k$.

The time set $\mathfrak{T}_0$ *is to be omitted if, and only if,* $\mathfrak{T}$ *replaces* $\mathfrak{T}_0$.

Definition 108 and the fundamental control principle 87 yield directly:

Lemma 112 ***Criterion for realizability of*** $\mathbf{Y}_d(.)$ ***on*** $\mathfrak{T}_0 \times \mathfrak{D}^i \times \mathfrak{U}^\mu \times \mathfrak{Y}_d^k$

i) In order for every desired output behavior $\mathbf{Y}_d(.) \in \mathfrak{Y}_d^k$ *of the plant (3.12) to be realizable on* $\mathfrak{T}_0 \times \mathfrak{D}^i \times \mathfrak{U}^\mu \times \mathfrak{Y}_d^k$ *it is necessary and sufficient that for every* $[\mathbf{D}(.), \mathbf{Y}_d(.)] \in \mathfrak{D}^i \times \mathfrak{Y}_d^k$ *there exists a control vector function* $\mathbf{U}(.) \in \mathfrak{U}^\mu$ *defined on* $\mathfrak{T}_0$ *for which* $\mathbf{Y}_d(.)$ *is the unique solution of the plant mathematical model on* $\mathfrak{T}_0$.

ii) In order for the desired output behavior $\mathbf{Y}_d(.) \in \mathfrak{Y}_d^k$ *of the plant (3.12) to be elementwise realizable on* $\mathfrak{T}_0 \times \mathfrak{D}^i \times \mathfrak{U}^\mu \times \mathfrak{Y}_d^k$ *it is necessary that there exists a control vector function* $\mathbf{U}(.)$ *that obeys both i) and* $r \geq N$.

iii) In order for the desired output behavior $\mathbf{Y}_d(.) \in \mathfrak{Y}_d^k$ *of the plant (3.12) to be elementwise realizable on* $\mathfrak{T}_0 \times \mathfrak{D}^i \times \mathfrak{U}^\mu \times \mathfrak{Y}_d^k$ *under the action of the minimal control* $\mathbf{U}$ *it is necessary and sufficient that there exists a control vector function* $\mathbf{U}(.) \in \mathfrak{U}^\mu$ *that obeys both i) and* $r = N$.

Comment 113 ***The realizability of*** $\mathbf{Y}_d(.)$ ***in*** $\mathcal{J}^i$ ***on*** $\mathfrak{T}_0 \times \mathfrak{Y}_d^k$ ***versus the realizability of*** $\mathbf{Y}_d(.)$ ***on*** $\mathfrak{T}_0 \times \mathcal{J}^i \times \mathfrak{Y}_d^k$

The realizability of $\mathbf{Y}_d(.)$ *in* $\mathcal{J}^i$ *on* $\mathfrak{T}_0 \times \mathfrak{Y}_d^k$, *is necessary, but not sufficient, for the realizability of* $\mathbf{Y}_d(.)$ *on* $\mathfrak{T}_0 \times \mathcal{J}^i \times \mathfrak{Y}_d^k$.

The realizability of $\mathbf{Y}_d(.)$ *on* $\mathfrak{T}_0 \times \mathcal{J}^i \times \mathfrak{Y}_d^k$, *is sufficient, but not necessary, for the realizability of* $\mathbf{Y}_d(.)$ *in* $\mathcal{J}^i$ *on* $\mathfrak{T}_0 \times \mathfrak{Y}_d^k$.

4.2 Concept and definitions

Definition 93 implies:

Definition 114 ***Nominal input-state pair*** $[\mathbf{I}_N(.)\ , \mathbf{R}_d\,(.)]$ ***on*** $\mathfrak{T}_0 \times \mathfrak{Y}_d^k$

A functional vector pair $[\mathbf{I}^*(.), \mathbf{R}^*(.)] \in \mathcal{J}^\xi \times \mathfrak{R}^\rho$ *is* ***nominal for the plant*** *(3.12)* ***on*** $\mathfrak{T}_0 \times \mathfrak{Y}_d^k$, *which is denoted by* $[\mathbf{I}_N(.)\ , \mathbf{R}_d\,(.)]$,

$$[\mathbf{I}^*(.), \mathbf{R}^*(.)] = [\mathbf{I}_N(.)\ , \mathbf{R}_d\,(.)]\,, \tag{4.17}$$

if, and only if, $[\mathbf{I}(.), \mathbf{R}(.)] = [\mathbf{I}^*(.)\mathbf{R}^*(.)]$ *ensures that for an arbitrary* $\mathbf{Y}_d(.) \in \mathfrak{Y}_d^k$ *the corresponding real response* $\mathbf{Y}(.) = \mathbf{Y}^*(.)$ *of the plant obeys* $\mathbf{Y}^*(t) = \mathbf{Y}_d(t)$ *all the time on* $\mathfrak{T}_0$ *as soon as all initial conditions are nominal:* $\mathbf{R}^{*\alpha-1}(t_0) = \mathbf{R}_d^{\alpha-1}(t_0)$ *and* $\mathbf{Y}^*(t_0) = \mathbf{Y}_d(t_0)$.

The time evolution $\mathbf{R}_d^{\alpha-1}(t; t_0; \mathbf{R}_{d0}^{\alpha-1}; \mathbf{I}_N; \mathbf{Y}_d)$, *where*

$$\mathbf{R}_d^{\alpha-1}(t_0; t_0; \mathbf{R}_{d0}^{\alpha-1}; \mathbf{I}_N; \mathbf{Y}_d) \equiv \mathbf{R}_{d0}^{\alpha-1},$$

of the desired (nominal) state vector $\mathbf{R}_d^{\alpha-1}$ *is* ***the desired (nominal) motion***

$$\mathcal{R}_d^{\alpha-1}(.; t_0; \mathbf{R}_{d0}^{\alpha-1}; \mathbf{I}_N; \mathbf{Y}_d)$$

***of the plant** (3.12) **relative to** $\mathbf{Y}_d(.) \in \mathfrak{Y}_d^k$ **on** $\mathfrak{T}_0 \times \mathcal{J}^{\xi}$,*

$$\begin{gathered}\mathbf{R}^{*\alpha-1}(t_0;t_0;\mathbf{R}_0^{*\alpha-1};\mathbf{I}_N;\mathbf{Y}_d) \equiv \mathbf{R}_0^{*\alpha-1} = \mathbf{R}_{d0}^{\alpha-1} \text{ and } \mathbf{Y}^*(t_0) = \mathbf{Y}_d(t_0) \Longrightarrow \\ \mathbf{R}^{*\alpha-1}(t;t_0;\mathbf{R}_0^{*\alpha-1};\mathbf{I}_N;\mathbf{Y}_d) \equiv \mathbf{R}_d^{\alpha-1}(t;t_0;\mathbf{R}_{d0}^{\alpha-1};\mathbf{I}_N;\mathbf{Y}_d), \\ \mathbf{Y}^*(t;t_0;\mathbf{Y}_0^*;\mathbf{I}_N) \equiv \mathbf{Y}_d(t;t_0;\mathbf{Y}_{d0}). \end{gathered} \tag{4.18}$$

$\mathfrak{T}_0$ is to be omitted if, and only if, $\mathfrak{T}$ replaces $\mathfrak{T}_0$.

Theorem 115 ***The nominal input-state pair** $[\mathbf{I}^*(.), \mathbf{R}^*(.)]$ **of the system (3.12) on** $\mathfrak{T}_0 \times \mathfrak{Y}_d^k$*

In order for a functional vector pair $[\mathbf{I}^*(.), \mathbf{R}^*(.)]$ to be nominal,

$$[\mathbf{I}^*(.), \mathbf{R}^*(.)] = [\mathbf{I}_N(.), \mathbf{R}_d(.)],$$

for the system (3.12) on $\mathfrak{T}_0 \times \mathfrak{Y}_d^k$ it is necessary and sufficient that for any $\mathbf{Y}_d(.) \in \mathfrak{Y}_d^k$:

$$\mathbf{R}_0^{*\alpha-1} = \mathbf{R}_{d0}^{\alpha-1} \text{ and } \mathbf{Y}^*(t_0) = \mathbf{Y}_d(t_0) \Longrightarrow$$

$$Q\left[t, \mathbf{R}^{*\alpha-1}(t)\right]\mathbf{R}^{*(\alpha)}(t) + \mathbf{q}\left[t, \mathbf{R}^{*\alpha-1}(t)\right] - \mathbf{h}\left[t, \mathbf{I}^{*\xi}(t)\right] = \mathbf{0}_\rho, \ \forall t \in \mathfrak{T}_0, \tag{4.19}$$

$$\mathbf{s}\left[t, \mathbf{R}^{*\alpha-1}(t), \mathbf{I}^*(t)\right] = \mathbf{Y}_d(t), \ \forall t \in \mathfrak{T}_0. \tag{4.20}$$

Proof. *Necessity.* Let $\mathbf{R}_0^{*\alpha-1} = \mathbf{R}_{d0}^{\alpha-1}$, $\mathbf{Y}^*(t_0) = \mathbf{Y}_d(t_0)$, and $[\mathbf{I}^*(.), \mathbf{R}^*(.)]$ be a nominal functional (input and state) vector pair for the system (3.12) on $\mathfrak{T}_0 \times \mathfrak{Y}_d^k$. Hence, the plant is in its nominal regime relative to any $\mathbf{Y}_d(.) \in \mathfrak{Y}_d^k$ on $\mathfrak{T}_0 \times \mathfrak{Y}_d^k$. Definition 114 shows that both $\mathbf{R}_0^{*\alpha-1} = \mathbf{R}_{d0}^{\alpha-1}$ and $[\mathbf{I}(.), \mathbf{R}(.)] = [\mathbf{I}^*(.), \mathbf{R}^*(.)]$ imply $\mathbf{Y}(.) = \mathbf{Y}_d(.)$, $\forall t \in \mathfrak{T}_0$ for every $\mathbf{Y}_d(.) \in \mathfrak{Y}_d^k$. This and the system model (3.12) yield the following equations:

$$Q\left[t, \mathbf{R}^{*\alpha-1}(t)\right]\mathbf{R}^{*(\alpha)}(t) + \mathbf{q}\left[t, \mathbf{R}^{*\alpha-1}(t)\right] = \mathbf{h}\left[t, \mathbf{I}^{*\xi}(t)\right], \ \forall t \in \mathfrak{T}_0, \tag{4.21}$$

$$\mathbf{Y}_d(t) = \mathbf{s}\left[t, \mathbf{R}^{*\alpha-1}(t), \mathbf{I}^*(t)\right], \ \forall t \in \mathfrak{T}_0. \tag{4.22}$$

(4.21), (4.22) are (4.19), (4.20), respectively.

Sufficiency. Let $\mathbf{Y}_d(.) \in \mathfrak{Y}_d^k$ be arbitrarily chosen, $\mathbf{R}_0^{*\alpha-1} = \mathbf{R}_{d0}^{\alpha-1}$ and $\mathbf{Y}^*(t_0) = \mathbf{Y}_d(t_0)$. We accept that all the conditions of the theorem hold. Let $[\mathbf{I}(.), \mathbf{R}(.)] = [\mathbf{I}^*(.), \mathbf{R}^*(.)]$. The equation (4.21) shows that the pair $[\mathbf{I}^*(.), \mathbf{R}^*(.)]$ satisfies the first equation (5.4). Equation (4.22) yields

$$\begin{gathered}\mathbf{Y}_d(t) = \mathbf{s}\left[t, \mathbf{R}^{*\alpha-1}(t), \mathbf{I}^*(t)\right] = \mathbf{Y}^*(t), \ \forall t \in \mathfrak{T}_0 \Longrightarrow \\ \mathbf{Y}^*(t) = \mathbf{Y}_d(t), \ \forall t \in \mathfrak{T}_0.\end{gathered}$$

This proves that the pair $[\mathbf{I}^*(.), \mathbf{R}^*(.)]$ satisfies the second equation (3.12). Because it also obeys the first equation (3.12) they prove that for $[\mathbf{I}(.), \mathbf{R}(.)] = [\mathbf{I}^*(.), \mathbf{R}^*(.)]$ the system (3.12) is in the nominal regime relative to $\mathfrak{T}_0 \times \mathfrak{Y}_d^k$ ∎

The application of Definition 96 to the system (3.12) leads to the following.

Definition 116 ***The desired state-control pair** $[\mathbf{R}_d(.), \mathbf{U}_N(.)]$ **of the plant (3.12) on** $\mathfrak{T}_0 \times \mathfrak{D} \times \mathfrak{Y}_d^k$*

*A functional vector pair $[\mathbf{R}^*(.), \mathbf{U}^*(.)] \in \mathfrak{R}^\rho \times \mathfrak{U}^\mu$ **is nominal for the plant** (3.12) **on** $\mathfrak{T}_0 \times \mathfrak{D} \times \mathfrak{Y}_d^k$, which is denoted by $[\mathbf{R}_d(.), \mathbf{U}_N(.)]$,*

$$[\mathbf{R}^*(.), \mathbf{U}^*(.)] = [\mathbf{R}_d(.), \mathbf{U}_N(.)], \tag{4.23}$$

if, and only if, $[\mathbf{R}(.), \mathbf{U}(.)] = [\mathbf{R}^*(.), \mathbf{U}^*(.)]$ ensures that for an arbitrary pair $[\mathbf{D}(.), \mathbf{Y}_d(.)] \in \mathfrak{D} \times \mathfrak{Y}_d^k$ the corresponding real response $\mathbf{Y}(.) = \mathbf{Y}^*(.)$ of the plant obeys $\mathbf{Y}^*(t) = \mathbf{Y}_d(t)$ all *the time on* $\mathfrak{T}_0$ *as soon as all initial conditions are nominal:* $\mathbf{R}^{*\alpha-1}(t_0) = \mathbf{R}_d^{\alpha-1}(t_0)$ *and* $\mathbf{Y}^*(t_0) = \mathbf{Y}_d(t_0)$.

The time evolution $\mathbf{R}_d^{\alpha-1}(t;t_0;\mathbf{R}_0^{\alpha-1};\mathbf{D};\mathbf{U}_N;\mathbf{Y}_d)$,

$$\mathbf{R}_d^{\alpha-1}(t_0;t_0;\mathbf{R}_{d0}^{\alpha-1};\mathbf{D};\mathbf{U}_N;\mathbf{Y}_d) \equiv \mathbf{R}_{d0}^{\alpha-1},$$

of the desired (nominal) state vector $\mathbf{R}_d^{\alpha-1}$ *is* ***the desired (nominal) motion***

$$\mathcal{R}_d^{\alpha-1}(.;t_0;\mathbf{R}_{d0}^{\alpha-1};\mathbf{D};\mathbf{U}_N;\mathbf{Y}_d)$$

of the *plant (3.12)* ***relative to*** $[\mathbf{D}(.),\mathbf{Y}_d(.)]$ *on* $\mathfrak{T}_0 \times \mathfrak{D} \times \mathfrak{U}^\mu \times \mathfrak{Y}_d^k$*, for short* ***the desired (nominal) motion,***

$$\begin{gathered}
\mathbf{R}^{*\alpha-1}(t_0;t_0;\mathbf{R}_0^{*\alpha-1};\mathbf{D};\mathbf{U}_N;\mathbf{Y}_d) \equiv \mathbf{R}_0^{*\alpha-1} = \mathbf{R}_{d0}^{\alpha-1} \quad and \quad \mathbf{Y}^*(t_0) = \mathbf{Y}_d(t_0) \Longrightarrow \\
\mathbf{R}^{*\alpha-1}(t;t_0;\mathbf{R}_0^{*\alpha-1};\mathbf{D};\mathbf{U}_N;\mathbf{Y}_d) \equiv \mathbf{R}_d^{\alpha-1}(t;t_0;\mathbf{R}_{d0}^{\alpha-1};\mathbf{D};\mathbf{U}_N;\mathbf{Y}_d), \\
\mathbf{Y}^*(t;t_0;\mathbf{Y}_0^*;\mathbf{D};\mathbf{U}_N;) \equiv \mathbf{Y}_d(t;t_0;\mathbf{Y}_{d0}).
\end{gathered} \tag{4.24}$$

$\mathfrak{T}_0$ *is to be omitted if, and only if,* $\mathfrak{T}$ *replaces* $\mathfrak{T}_0$.

Theorem 117 ***Nominal state-control pair*** $[\mathbf{R}^*(.),\mathbf{U}^*(.)]$ ***of the regular plant (3.12) on*** $\mathfrak{T}_0 \times \mathfrak{D}^\eta \times \mathfrak{Y}_d^k$

a) In order for a functional vector pair $[\mathbf{R}^*(.),\mathbf{U}^*(.)] \in \mathfrak{R}^\rho \times \mathfrak{U}^\mu$ *to be nominal for the plant (3.12) on* $\mathfrak{T}_0 \times \mathfrak{D}^\eta \times \mathfrak{Y}_d^k$, $[\mathbf{R}^*(.),\mathbf{U}^*(.)] = [\mathbf{R}_d(.),\mathbf{U}_N(.)]$, *it is necessary and sufficient that for every* $[\mathbf{D}(.),\mathbf{Y}_d(.)] \in \mathfrak{D}^\eta \times \mathfrak{Y}_d^k$:

$$\mathbf{R}_0^{*\alpha-1} = \mathbf{R}_{d0}^{\alpha-1} \quad and \quad \mathbf{Y}^*(t_0) = \mathbf{Y}_d(t_0) \Longrightarrow$$

$$Q\left[t,\mathbf{R}^{*\alpha-1}(t)\right]\mathbf{R}^{*(\alpha)}(t) + \mathbf{q}\left[t,\mathbf{R}^{*\alpha-1}(t)\right] - \left\{\begin{array}{c} E(t)\,\mathbf{e}_d\left[\mathbf{D}^\eta(t)\right] - \\ -P(t)\,\mathbf{e}_u\left[\mathbf{U}^{*\mu}(t)\right]\end{array}\right\} = \mathbf{0}_\rho,$$

$$\forall t \in \mathfrak{T}_0, \tag{4.25}$$

$$\mathbf{z}\left[t,\mathbf{R}^{*\alpha-1}(t),\mathbf{D}(t)\right] + W(t)\,\mathbf{w}\left[\mathbf{U}^*(t)\right] = \mathbf{Y}_d(t),\ \forall t \in \mathfrak{T}_0. \tag{4.26}$$

b) In order for (4.25), (4.26) to be solvable in $[\mathbf{U}^*(.),\mathbf{R}^*(.)]$:

1) It is necessary that $N \le r$.

2) It is sufficient that rank $W(t) = N \le r$ *for every* $t \in \mathfrak{T}_0$*, equivalently,*

$$det\ \left[W(t)\,W^T(t)\right] \neq 0$$

for every $t \in \mathfrak{T}_0$*, and that the following equations determine* $[\mathbf{U}^*(.),\mathbf{R}^*(.)]$:

$$\mathbf{U}^*(t) = \mathbf{w}^I\left\{W^T(t)\left[W(t)\,W^T(t)\right]^{-1}\left(\mathbf{Y}_d(t) - \mathbf{z}\left[t,\mathbf{R}^{*\alpha-1}(t),\mathbf{D}(t)\right]\right)\right\}, \tag{4.27}$$

$$\begin{gathered}
\mathbf{R}^{*(\alpha)}(t) + Q^{-1}\left[t,\mathbf{R}^{*\alpha-1}(t)\right]\mathbf{q}\left[t,\mathbf{R}^{*\alpha-1}(t)\right] \\
= \left\{\begin{array}{c} Q^{-1}\left[t,\mathbf{R}^{*\alpha-1}(t)\right]\bullet \\ \bullet E(t)\,\mathbf{e}_d\left[\mathbf{D}^\eta(t)\right] + P(t)\,\mathbf{e}_u\left[\mathbf{U}^{*\mu}(t)\right]\end{array}\right\},\quad \forall t \in \mathfrak{T}_0,
\end{gathered} \tag{4.28}$$

Proof. *Necessity.* a) Let $[\mathbf{R}^*(.),\mathbf{U}^*(.)] \in \mathfrak{R}^\rho \times \mathfrak{U}^\mu$ be a nominal functional (input and state) vector pair for the regular plant (3.12) on $\mathfrak{T}_0 \times \mathfrak{D}^\eta \times \mathfrak{Y}_d^k$. Let $\mathbf{R}_0^{*\alpha-1} = \mathbf{R}_{d0}^{\alpha-1}$ and $\mathbf{Y}^*(t_0) = \mathbf{Y}_d(t_0)$. Hence, the plant is in its nominal regime relative to $[\mathbf{D}(.),\mathbf{Y}_d(.)]$ on $\mathfrak{T}_0 \times \mathfrak{D}^\eta \times \mathfrak{Y}_d^k$. Definition 96 shows that $[\mathbf{R}(.),\mathbf{U}(.)] = [\mathbf{R}^*(.),\mathbf{U}^*(.)]$ implies $\mathbf{Y}(.) = \mathbf{Y}_d(.)$, $\forall t \in \mathfrak{T}_0$. This and the plant model (3.12) yield the following equations:

$$Q\left[t,\mathbf{R}^{*\alpha-1}(t)\right]\mathbf{R}^{*(\alpha)}(t) + \mathbf{q}\left[t,\mathbf{R}^{*\alpha-1}(t)\right] = \left\{\begin{array}{c} E(t)\,\mathbf{e}_d\left[\mathbf{D}^\eta(t)\right] - \\ -P(t)\,\mathbf{e}_u\left[\mathbf{U}^{*\mu}(t)\right]\end{array}\right\},\ \forall t \in \mathfrak{T}_0, \tag{4.29}$$

$$\mathbf{Y}_d(t) = \mathbf{z}\left[t,\mathbf{R}^{*\alpha-1}(t),\mathbf{D}(t)\right] + W(t)\,\mathbf{w}\left[\mathbf{U}^*(t)\right],\ \forall t \in \mathfrak{T}_0. \tag{4.30}$$

(4.29) multiplied on the left by $Q^{-1}\left[t,\mathbf{R}^{*\alpha-1}(t)\right]$ and (4.30) are (4.25) and (4.26), respectively.

b) Let (3.12) be solvable in $[\mathbf{U}^*(.),\mathbf{R}^*(.)]$. Solvability of (4.26) in $\mathbf{U}^*(t)$ implies that the number $N+\rho$ of the scalar equations in (4.29), (4.30) is not greater than the number $r+\rho$ of the unknown scalar variables in (4.29), (4.30): $N+\rho \leq r+\rho$. This implies necessity of $N \leq r$.

Sufficiency. Let $\mathbf{R}_0^{*\alpha-1} = \mathbf{R}_{d0}^{\alpha-1}$ and $\mathbf{Y}^*(t_0) = \mathbf{Y}_d(t_0)$. Let $[\mathbf{D}(.),\mathbf{Y}_d(.)] \in \mathfrak{D}^\eta \times \mathfrak{Y}_d^k$ be arbitrarily chosen. We accept that all the conditions of the theorem hold. Let $[\mathbf{R}(.),\mathbf{U}(.)] = [\mathbf{R}^*(.),\mathbf{U}^*(.)] \in \mathfrak{R}^\rho \times \mathfrak{U}^\mu$. The equation (4.28) multiplied on the left by $Q\left[t,\mathbf{R}^{*\alpha-1}(t)\right]$ shows that the pair $[\mathbf{R}^*(.),\mathbf{U}^*(.)]$ satisfies the first equation (3.12). Equation (4.27) yields

$$\mathbf{w}\left[\mathbf{U}^*(t)\right] = W^T(t)\left[W(t)W^T(t)\right]^{-1}\left(\mathbf{Y}_d(t) - Z(t)\,\mathbf{z}\left[\mathbf{R}^*(t),\mathbf{D}(t)\right]\right),\ \forall t \in \mathfrak{T}_0$$
$$\Longrightarrow$$
$$W(t)\,\mathbf{w}\left[\mathbf{U}^*(t)\right] = \mathbf{Y}_d(t) - Z(t)\,\mathbf{z}\left[\mathbf{R}^*(t),\mathbf{D}(t)\right],\ \forall t \in \mathfrak{T}_0 \Longrightarrow$$
$$\mathbf{Y}_d(t) = \mathbf{z}\left[t,\mathbf{R}^{*\alpha-1}(t),\mathbf{D}(t)\right] + W(t)\,\mathbf{w}\left[\mathbf{U}^*(t)\right],\ \forall t \in \mathfrak{T}_0.$$

The second equation (3.12) written for $[\mathbf{R}^*(.),\mathbf{U}^*(.)]$ reads

$$\mathbf{Y}(t) = \mathbf{z}\left[t,\mathbf{R}^{*\alpha-1}(t),\mathbf{D}(t)\right] + W(t)\,\mathbf{w}\left[\mathbf{U}^*(t)\right],\ \forall t \in \mathfrak{T}_0.$$

The preceding last two equations yield $\mathbf{Y}(t) = \mathbf{Y}_d(t)$, $\forall t \in \mathfrak{T}_0$.This proves that the pair $[\mathbf{R}^*(.),\mathbf{U}^*(.)]$ satisfies the second equation (3.12). Because it also obeys the first equation (3.12) they then prove that the plant (3.12) is in the nominal regime relative to $[\mathbf{D}(.),\mathbf{Y}_d(.)]$ on $\mathfrak{T}_0 \times \mathfrak{D}^\eta \times \mathfrak{U}^\mu \times \mathfrak{Y}_d^k$ ■

From Definition 111 and Lemma 112 we deduce the following.

Theorem 118 ***Realizability of*** $\mathbf{Y}_d(t)$ ***of the plant*** *(3.12)* ***on*** $\mathfrak{T}_0 \times \mathfrak{D}^\eta \times \mathfrak{U}^\mu \times \mathfrak{Y}_d^k$

In order for the demanded output response $\mathbf{Y}_d(t)$ *of the system (3.12) to be realizable on* $\mathfrak{T}_0 \times \mathfrak{D}^\eta \times \mathfrak{U}^\mu \times \mathfrak{Y}_d^k$ *it is sufficient that Theorem 117 holds.*

Note 119 *There are* $\rho+N$ *scalar equations and* $\rho+r$ *scalar unknown variables in (4.25), (4.26). For their solvability in* $[\mathbf{R}^*(.),\mathbf{U}^*(.)]$ *it is necessary that the number* $\rho+N$ *of the scalar equations is not bigger than the number* $\rho+r$ *of the entries of* $[\mathbf{R}^*(.),\mathbf{U}^*(.)]$*, i.e., that* $\rho+N \leq \rho+r$*, i.e., that* $N \leq r$*. This is the important guideline to the designer of the plant (see the fundamental control principle 87).*

Note 120 *The nominal pair* $[\mathbf{R}_d(.),\mathbf{U}_N(.)]$ *depends on* $[\mathbf{D}(.),\mathbf{Y}_d(.)] \in \mathfrak{D}^\eta \times \mathfrak{Y}_d^k$,

$$[\mathbf{R}_d(.),\mathbf{U}_N(.)] = [\mathbf{R}_d(.;\mathbf{D};\mathbf{Y}_d),\mathbf{U}_N(.;\mathbf{D};\mathbf{Y}_d)]. \tag{4.31}$$

The nominal pair $[\mathbf{R}_d(.),\mathbf{U}_N(.)]$ *should be determined for every* $\mathbf{D}(.) \in \mathfrak{D}^\eta$*. Because any* $\mathbf{D}(.) \in \mathfrak{D}^\eta$ *can happen, the number of possible* $\mathbf{D}(.) \in \mathfrak{D}^\eta$ *to take place is infinite. It is cumbersome to determine the nominal pair* $[\mathbf{R}_d(.),\mathbf{U}_N(.)]$ *for every possible* $\mathbf{D}(.) \in \mathfrak{D}^\eta$.

Let $\mathbf{D}_N(.) \in \mathfrak{D}^\eta$ be accepted for every $\mathbf{Y}_d(.) \in \mathfrak{Y}_d^k$, $\mathbf{D}_N(.) = \mathbf{D}_N(.;\mathbf{Y}_d)$. The extension of Definition 98 to the system (3.12) reads:

Definition 121 ***Nominal state-control pair*** $[\mathbf{R}^*(.),\mathbf{U}^*(.)]$ ***of the plant (3.12) relative to*** $\mathbf{D}_N(.) \in \mathfrak{D}^\eta$ ***on*** $\mathfrak{T}_0 \times \mathfrak{U}^\mu \times \mathfrak{Y}_d^k$

A functional vector pair $\left[\mathbf{R}^{*\alpha-1}(.),\mathbf{U}^*(.)\right] \in \mathfrak{R}^{\alpha\rho} \times \mathfrak{U}^\mu$ *is* ***nominal****, which is denoted by* $[\mathbf{R}_d(.),\mathbf{U}_N(.)]$, ***for the plant*** *(3.12)* ***relative to*** $\mathbf{D}_N(.) \in \mathfrak{D}^\eta$ ***on*** $\mathfrak{T}_0 \times \mathfrak{U}^\mu \times \mathfrak{Y}_d^k$,

$$\left[\mathbf{R}^{*\alpha-1}(.),\mathbf{U}^*(.)\right] \equiv [\mathbf{R}_d(.),\mathbf{U}_N(.)], \tag{4.32}$$

if, and only if, $\left[\mathbf{R}^{\alpha-1}(.),\mathbf{U}(.)\right]=\left[\mathbf{R}^{*\alpha-1}(.),\mathbf{U}^{*}(.)\right]\in\mathfrak{R}^{\alpha\rho}\times\mathfrak{U}^{\mu}$ *ensures that for the chosen* $\mathbf{D}_N(.)\in\mathfrak{D}^{\eta}$ *relative to any* $\mathbf{Y}_d(.)\in\mathfrak{Y}_d^k$ *the corresponding real response* $\mathbf{Y}(.)=\mathbf{Y}^{*}(.)$ *of the plant obeys* $\mathbf{Y}^{*}(t)=\mathbf{Y}_d(t)$ *all the time on* $\mathfrak{T}_0$ *as soon as all initial conditions are nominal, i.e., desired:* $\mathbf{R}_0^{*\alpha-1}=\mathbf{R}_{d0}^{\alpha-1}$ *and* $\mathbf{Y}^{*}(t_0)=\mathbf{Y}_d(t_0)$.

The time evolution $\mathbf{R}_d^{\alpha-1}(t;t_0;\mathbf{R}_{d0}^{\alpha-1};\mathbf{D};\mathbf{U}_N;\mathbf{Y}_d)$,

$$\mathbf{R}_d^{\alpha-1}(t_0;t_0;\mathbf{R}_{d0}^{\alpha-1};\mathbf{D}_N;\mathbf{U}_N;\mathbf{Y}_d)\equiv\mathbf{R}_{d0}^{\alpha-1},$$

of the desired (nominal) state vector $\mathbf{R}^{\alpha-1}$ *is* ***the desired (nominal) motion***

$$\mathcal{R}_d^{\alpha-1}(.;t_0;\mathbf{R}_{d0}^{\alpha-1};\mathbf{D}_N;\mathbf{U}_N;\mathbf{Y}_d)$$

of the *plant (3.12) relative to* $\mathbf{D}_N(.)\in\mathfrak{D}^{\eta}$ *on* $\mathfrak{T}_0\times\mathfrak{U}^{\mu}\times\mathfrak{Y}_d^k$,

$$\begin{gathered}\mathbf{R}^{*\alpha-1}(t_0;t_0;\mathbf{R}_0^{*\alpha-1};\mathbf{D}_N;\mathbf{U}_N;\mathbf{Y}_d)\equiv\mathbf{R}_0^{*\alpha-1}=\mathbf{R}_{d0}^{\alpha-1}\quad\text{and}\quad\mathbf{Y}^{*}(t_0)=\mathbf{Y}_d(t_0)\Longrightarrow\\ \mathbf{R}^{*\alpha-1}(t;t_0;\mathbf{R}_0^{*\alpha-1};\mathbf{D}_N;\mathbf{U}_N;\mathbf{Y}_d)\equiv\mathbf{R}_d^{\alpha-1}(t;t_0;\mathbf{R}_{d0}^{\alpha-1};\mathbf{D}_N;\mathbf{U}_N;\mathbf{Y}_d),\\ \mathbf{Y}^{*}(t;t_0;\mathbf{Y}_0^{*};\mathbf{D};\mathbf{U}_N;)\equiv\mathbf{Y}_d(t;t_0;\mathbf{Y}_{d0}).\end{gathered}\tag{4.33}$$

$\mathfrak{T}_0$ *is to be omitted if, and only if,* $\mathfrak{T}_0=\mathfrak{T}$.

Theorem 122 ***Nominal state-control pair*** $[\mathbf{R}^{*}(.),\mathbf{U}^{*}(.)]$ ***of the regular plant (3.12) relative to*** $\mathbf{D}_N(.)\in\mathfrak{D}^{\eta}$ ***on*** $\mathfrak{T}_0\times\mathfrak{U}^{\mu}\times\mathfrak{Y}_d^k$

a) In order for a functional vector pair $\left[\mathbf{R}^{*\alpha-1}(.),\mathbf{U}^{*}(.)\right]\in\mathfrak{R}^{\alpha\rho}\times\mathfrak{U}^{\mu}$ *to be nominal for the regular plant (3.12) relative to an arbitrary* $\mathbf{D}_N(.)\in\mathfrak{D}^{\eta}$ *on* $\mathfrak{T}_0\times\mathfrak{U}^{\mu}\times\mathfrak{Y}_d^k$, $\left[\mathbf{R}^{*\alpha-1}(.),\mathbf{U}^{*}(.)\right]=\left[\mathbf{R}_d^{\alpha-1}(.),\mathbf{U}_N(.)\right]$, *it is necessary and sufficient that for every triplet* $[t,\mathbf{D}_N(.),\mathbf{Y}_d(.)]\in\mathfrak{T}_0\times\mathfrak{D}^{\eta}\times\mathfrak{Y}_d^k$:

$$\mathbf{R}_0^{*\alpha-1}=\mathbf{R}_{d0}^{\alpha-1}\quad\text{and}\quad\mathbf{Y}^{*}(t_0)=\mathbf{Y}_d(t_0)\Longrightarrow$$

$$Q\left[t,\mathbf{R}^{*\alpha-1}(t)\right]\mathbf{R}^{*(\alpha)}(t)+\mathbf{q}\left[t,\mathbf{R}^{*\alpha-1}(t)\right]-\left\{\begin{array}{c}E(t)\,\mathbf{e}_d\left[\mathbf{D}_N^{\eta}(t)\right]\\-P(t)\,\mathbf{e}_u\left[\mathbf{U}^{*\mu}(t)\right]\end{array}\right\}=\mathbf{0}_{\rho},\tag{4.34}$$

$$\mathbf{z}\left[t,\mathbf{R}^{*\alpha-1}(t),\mathbf{D}_N(t)\right]+W(t)\,\mathbf{w}\left[\mathbf{U}^{*}(t)\right]=\mathbf{Y}_d(t).\tag{4.35}$$

b) In order for (4.34), (4.35) to be solvable in $[\mathbf{R}^{*}(.),\mathbf{U}^{*}(.)]$:

1) It is necessary that $N\leq r$.

2) It is sufficient that rank $W(t)\equiv N$ *and that the following equations determine* $[\mathbf{R}^{*}(.),\mathbf{U}^{*}(.)]$:

$$\mathbf{U}^{*}(t)=\mathbf{w}^{I}\left\{W^{T}(t)\left[W(t)\,W^{T}(t)\right]^{-1}\left(\mathbf{Y}_d(t)-\mathbf{z}\left[t,\mathbf{R}^{*\alpha-1}(t),\mathbf{D}_N(t)\right]\right)\right\},\tag{4.36}$$

$$\begin{gathered}\mathbf{R}^{*(\alpha)}(t)+Q^{-1}\left[t,\mathbf{R}^{*\alpha-1}(t)\right]\mathbf{q}\left[t,\mathbf{R}^{*\alpha-1}(t)\right]\\=\left\{\begin{array}{c}Q^{-1}\left[t,\mathbf{R}^{*\alpha-1}(t)\right]\\ \bullet E(t)\,\mathbf{e}_d\left[\mathbf{D}^{\eta}(t)\right]+P(t)\,\mathbf{e}_u\left[\mathbf{U}^{*\mu}(t)\right]\end{array}\right\},\ \forall t\in\mathfrak{T}_0,\end{gathered}\tag{4.37}$$

Proof. *Necessity.* a) Let $\mathbf{R}_0^{*\alpha-1}=\mathbf{R}_{d0}^{\alpha-1}$, $\mathbf{Y}^{*}(t_0)=\mathbf{Y}_d(t_0)$ and the functional pair $\left[\mathbf{R}^{*\alpha-1}(.),\mathbf{U}^{*}(.)\right]\in\mathfrak{R}^{\alpha\rho}\times\mathfrak{U}^{\mu}$ be a nominal functional (input-state) vector pair for the regular plant (3.12) on $\mathfrak{T}_0\times\mathfrak{D}^{\eta}\times\mathfrak{Y}_d^k$. Hence, the plant is in its nominal regime relative to any $[\mathbf{D}_N(.),\mathbf{Y}_d(.)]\in\mathfrak{D}^{\eta}\times\mathfrak{Y}_d^k$ on $\mathfrak{T}_0$. Definition 98 shows that $\left[\mathbf{R}^{\alpha-1}(.),\mathbf{U}(.)\right]=\left[\mathbf{R}^{*\alpha-1}(.),\mathbf{U}^{*}(.)\right]$ implies $\mathbf{Y}(.)=\mathbf{Y}_d(.)$, $\forall t\in\mathfrak{T}_0$. This and the plant model (3.12) yield the following equations:

$$Q\left[t,\mathbf{R}^{*\alpha-1}(t)\right]\mathbf{R}^{*(\alpha)}(t)+\mathbf{q}\left[t,\mathbf{R}^{*\alpha-1}(t)\right]=\left\{\begin{array}{c}E(t)\,\mathbf{e}_d\left[\mathbf{D}^{\eta}(t)\right]\\-P(t)\,\mathbf{e}_u\left[\mathbf{U}^{*\mu}(t)\right]\end{array}\right\},\ \forall t\in\mathfrak{T}_0,\tag{4.38}$$

$$\mathbf{Y}_d(t)=\mathbf{z}\left[t,\mathbf{R}^{*\alpha-1}(t),\mathbf{D}_N(t)\right]+W(t)\,\mathbf{w}\left[\mathbf{U}^{*}(t)\right],\ \forall t\in\mathfrak{T}_0.\tag{4.39}$$

The equations (4.38) multiplied on the left by $Q^{-1}\left[t,\mathbf{R}^{*\alpha-1}(t)\right]$ and (4.39) are (4.34) and (4.35), respectively.

b) Because $\mathbf{R}\in\mathfrak{R}^{\rho}$ and $\mathbf{U}\in\mathfrak{R}^{r}$ the number of unknown variables equals $\rho+r$. The number λ of the scalar equations in (3.12) satisfies $0<\lambda\leq N+\rho$. For (3.12) to have a solution it is necessary that $\rho+r\geq N+\rho$, i.e., that $r\geq N$.

Sufficiency. Let $\mathbf{R}_0^{*\alpha-1}=\mathbf{R}_{d0}^{\alpha-1}$ and $\mathbf{Y}^*(t_0)=\mathbf{Y}_d(t_0)$. Let $[\mathbf{D}_N(.),\mathbf{Y}_d(.)]\in\mathfrak{D}^{\eta}\times\mathfrak{Y}_d^k$ be arbitrarily chosen. We accept that all the conditions of the theorem hold. Let us accept $\left[\mathbf{R}^{\alpha-1}(.),\mathbf{U}(.)\right]=\left[\mathbf{R}^{*\alpha-1}(.),\mathbf{U}^*(.)\right]\in\mathfrak{R}^{\alpha\rho}\times\mathfrak{U}^{\mu}$. The equation (4.38) shows that $\left[\mathbf{R}^{*\alpha-1}(.),\mathbf{U}^*(.)\right]$ satisfies the first equation (3.12). The rank condition $\operatorname{rank} W(t)\equiv N$ and the equation (4.39) yield:

$$\mathbf{w}\left[\mathbf{U}^*(t)\right]=W^T(t)\left[W(t)W^T(t)\right]^{-1}\left(\mathbf{Y}_d(t)-\mathbf{z}\left[t,\mathbf{R}^{*\alpha-1}(t),\mathbf{D}_N(t)\right]\right),\ \forall t\in\mathfrak{T}_0$$

$$\Longrightarrow$$

$$W(t)\,\mathbf{w}\left[\mathbf{U}^*(t)\right]=\mathbf{Y}_d(t)-\mathbf{z}\left[t,\mathbf{R}^{*\alpha-1}(t),\mathbf{D}_N(t)\right],\ \forall t\in\mathfrak{T}_0\Longrightarrow$$
$$\mathbf{Y}_d(t)=\mathbf{z}\left[t\mathbf{R}^{*\alpha-1}(t),\mathbf{D}_N(t)\right]+W(t)\,\mathbf{w}\left[\mathbf{U}^*(t)\right],\ \forall t\in\mathfrak{T}_0.$$

The second equation (3.12) written for $\left[\mathbf{R}^{*\alpha-1}(.),\mathbf{U}^*(.)\right]$ reads

$$\mathbf{Y}(t)=\mathbf{z}\left[t,\mathbf{R}^{*\alpha-1}(t),\mathbf{D}_N(t)\right]+W(t)\,\mathbf{w}\left[\mathbf{U}^*(t)\right],\quad\forall t\in\mathfrak{T}_0$$

The preceding last two equations imply

$$\mathbf{Y}(t)=\mathbf{Y}_d(t),\quad\forall t\in\mathfrak{T}_0.$$

This proves that the pair $\left[\mathbf{R}^{*\alpha-1}(.),\mathbf{U}^*(.)\right]$ satisfies the second equation (3.12). Because it also obeys the first equation (3.12) they prove that the plant (3.12) is in the nominal regime relative to $[\mathbf{D}^*(.),\mathbf{Y}_d(.)]$ on $\mathfrak{T}_0\times\mathfrak{D}^{\eta}\times\mathfrak{U}^{\mu}\times\mathfrak{Y}_d^k$ ■

Definition 108 and Lemma 110 imply the following.

Theorem 123 ***Realizability of the desired output*** $\mathbf{Y}_d(t)$ ***of the plant*** *(3.12)* ***in*** $\mathfrak{D}^{\eta}$ ***on*** $\mathfrak{T}_0\times\mathfrak{D}^{\eta}\times\mathfrak{U}^{\mu}\times\mathfrak{Y}_d^k$

In order for the demanded output response $\mathbf{Y}_d(t)$ *of the system (3.12) to be realizable on* $\mathfrak{T}_0\times\mathfrak{D}^{\eta}\times\mathfrak{U}^{\mu}\times\mathfrak{Y}_d^k$ *it is sufficient that Theorem 122 holds.*

Chapter 5

Origins of time-varying models

5.1 Introduction

Physical phenomena (e.g., friction, matter fatigue) can vary in *time*. They cause *time*-varying properties of physical systems. The consequences are *time* variations of parameters and/or temporal variations of nonlinearities of physical systems. If the size and the speed of the variations are not small then the adequate mathematical models of the physical systems are *time*-varying. This is the physical origin of the *time*-varying nature of the mathematical models.

However, even *time*-invariant physical systems can have adequate *time*-varying mathematical models, the origin of which is purely mathematical. This holds for both *time*-invariant and *time*-varying nonlinear physical systems, the adequate mathematical models of which in terms of total coordinates are nonlinear, when their nominal motions are *time*-varying and if the mathematical models are expressed in terms of the deviations of all variables.

A nominal regime can occur only under special circumstances. Otherwise, the system is in *a nonnominal regime* also called *a perturbed regime*. Because the perturbed regime is more probable than the nominal one, we also call it *the real regime*. This rases the question of the relationship between behaviors of the system in real regimes and its nominal behavior (its behavior in a nominal regime).

We use *the output deviation (vector)* $\mathbf{y}$, (4.6), i.e., (5.1), and *the output error (vector)* ε, (4.7), i.e., (5.2), of the real behavior relative to the nominal one in order to measure their distance at every moment,

$$\textit{the output deviation vector}\text{: } \mathbf{y} = \mathbf{Y} - \mathbf{Y}_N = \mathbf{Y} - \mathbf{Y}_d, \tag{5.1}$$

$$\textit{the output error vector}\text{: } \varepsilon = \mathbf{Y}_N - \mathbf{Y} = \mathbf{Y}_d - \mathbf{Y}, \tag{5.2}$$

$$\varepsilon = -\mathbf{y}. \tag{5.3}$$

They form the coordinate transformations that we call *Lyapunov coordinate transformations* due to A. M. Lyapunov [401] who probably and possibly was the first to relate the coordinates of a real *motion* to the coordinates of the nominal (i.e., desired) *motion* of the dynamical system, Figure 5.1.

Comment 124 *The preceding equations show the equivalence among the zero error vector* $\varepsilon = \mathbf{0}_N$, *the zero deviation vector* $\mathbf{y} = \mathbf{0}_N$, *and the total desired vector* $\mathbf{Y} = \mathbf{Y}_d$, $\mathbf{Y} = \mathbf{Y}_d \Longleftrightarrow \varepsilon = \mathbf{0}_N \Longleftrightarrow \mathbf{y} = \mathbf{0}_N$. *The zero output error vector* $\varepsilon = \mathbf{0}_N$ *and the zero output deviation vector* $\mathbf{y} = \mathbf{0}_N$ *correspond to the total nominal output vector* $\mathbf{Y}_N$.

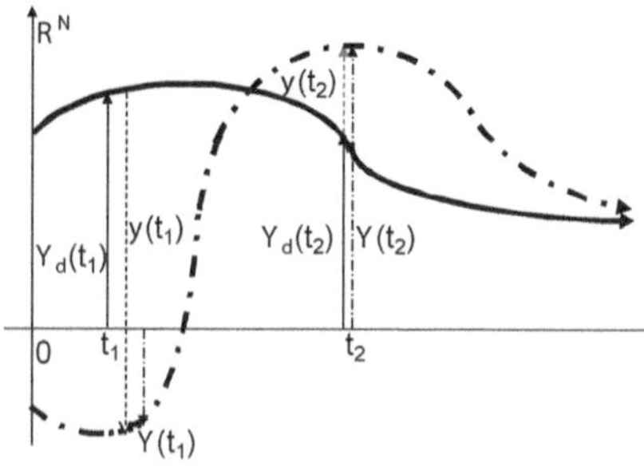

Figure 5.1: Lyapunov coordinate transformations.

Conclusion 125 *We can reduce the study of the properties of the desired (nominal) total response [of the nominal (desired) total movement] to the study of the same properties of the zero deviation vector* $\mathbf{y} = \mathbf{0}_N$*, equivalently, of the zero error vector* $\boldsymbol{\varepsilon} = \mathbf{0}_N$.

Conclusion 125 imposes the problem:

Problem 126 ***How to determine mathematical models of the systems in terms of the deviations, what are their forms, and what are their relations to the original models expressed in terms of total variables' values.***

The solution of this problem is important for the study of stability properties and for the study of tracking properties.

5.2 Deviations and mathematical models

We consider the nonlinear system (3.12):

$$Q\left[t,\mathbf{R}^{\alpha-1}(t)\right]\mathbf{R}^{(\alpha)}(t)+\mathbf{q}\left[t,\mathbf{R}^{\alpha-1}(t)\right]=\mathbf{h}\left[t,\mathbf{I}^{\xi}\left(t\right)\right],$$
$$\mathbf{Y}(t)=\mathbf{s}\left[t,\mathbf{R}^{\alpha-1}(t),\mathbf{I}(t)\right] \tag{5.4}$$

It has the *time*-varying desired output $\mathbf{Y}_d(t)$. If the system is *time*-invariant then it is described by

$$Q\left[\mathbf{R}^{\alpha-1}(t)\right]\mathbf{R}^{(\alpha)}(t)+\mathbf{q}\left[\mathbf{R}^{\alpha-1}(t)\right]=\mathbf{h}\left[\mathbf{I}^{\xi}\left(t\right)\right],$$
$$\mathbf{Y}(t)=\mathbf{s}\left[\mathbf{R}^{\alpha-1}(t),\mathbf{I}(t)\right]. \tag{5.5}$$

Let the nominal input vector $\mathbf{I}_N(t)$ and the desired state behavior $\mathbf{R}_d^{\alpha-1}(t)$ of the system (5.4) and of the system (5.5) be determined relative to the desired output $\mathbf{Y}_d(.) \in \mathfrak{Y}_d^k$.

Assumption 127 *The desired output response* $\mathbf{Y}_d(t)$ *of the system (5.4), i.e., of (5.5), is realizable in* $\mathcal{J}^i$ *on* $\mathfrak{T}_0 \times \mathcal{J}^i \times \mathfrak{Y}_d^k$.

Let the nonlinear plant (3.12) be repeated as

$$Q\left[t,\mathbf{R}^{\alpha-1}(t)\right]\mathbf{R}^{(\alpha)}(t)+\mathbf{q}\left[t,\mathbf{R}^{\alpha-1}(t)\right]=\left\{\begin{array}{c}E\left(t\right)\mathbf{e}_d\left[\mathbf{D}^{\eta}(t)\right]+\\+P\left(t\right)\mathbf{e}_u\left[\mathbf{U}^{\mu}(t)\right]\end{array}\right\},\quad\forall t\in\mathfrak{T},$$
$$\mathbf{Y}(t)=\mathbf{z}\left[t,\mathbf{R}^{\alpha-1}(t),\mathbf{D}(t)\right]+W\left(t\right)\mathbf{w}\left[\mathbf{U}(t)\right],\quad\forall t\in\mathfrak{T}. \tag{5.6}$$

The desired output $\mathbf{Y}_d(t)$ is *time*-varying, which induces the *time*-varying desired state behavior expressed by the *time*-varying state vector $\mathbf{R}_d^{\alpha-1}(t)$.

The mathematical models expressed in terms of deviations of all variables of the systems described by (3.12), i.e., of (5.4), have the following common, general, property.

Theorem 128 ***Time-varying mathematical model of the system [time-invariant (5.5) or time-varying (5.4)] in terms of the deviations (4.1), (4.3), (4.5), (4.6) with the input and the state nominal relative to $\mathbf{Y}_d(.)$ on $\mathfrak{T}_0 \times \mathcal{J}^{\xi} \times \mathfrak{Y}_d^k$***

Let Assumption 127 hold. If the mathematical model of the system [either time-invariant (5.5) or time-varying (5.4)] is nonlinear in terms of the total coordinates $\mathbf{I}$, $\mathbf{R}$, and $\mathbf{Y}$, then the system mathematical model in terms of their deviations $\mathbf{i} = \mathbf{I} - \mathbf{I}_N$, $\mathbf{r} = \mathbf{R} - \mathbf{R}_d$, and $\mathbf{y} = \mathbf{Y} - \mathbf{Y}_d$ is time-varying as soon as the system total desired output vector $\mathbf{Y}_d$ is time-varying, $\mathbf{Y}_d(t) \neq \mathbf{const.}$, so that the desired state vector $\mathbf{R}_d(t)$ is also time-varying.

Proof. What follows is fully valid for both the *time*-invariant system (5.5) and *time*-varying system (5.4). Let Assumption 127 hold. Definition 107 is valid. It and Theorem 115 imply (4.19), (4.20). The difference between (5.4) and (4.19), (4.20) reads:

$$
\begin{aligned}
Q\left[t, \mathbf{R}^{\alpha-1}(t)\right] \mathbf{R}^{(\alpha)}(t) + \mathbf{q}\left[t, \mathbf{R}^{\alpha-1}(t)\right] - Q\left[\mathbf{R}_d^{\alpha-1}(t)\right] \mathbf{R}_d^{(\alpha)}(t) - \mathbf{q}\left[\mathbf{R}_d^{\alpha-1}(t)\right] \\
= \mathbf{h}\left[t, \mathbf{I}^{\xi}(t)\right] - \mathbf{h}\left[t, \mathbf{I}_N^{\xi}(t)\right], \quad \forall t \in \mathfrak{T}, \\
\mathbf{Y}(t) - \mathbf{Y}_d(t) = \mathbf{s}\left[t, \mathbf{R}^{\alpha-1}(t), \mathbf{I}(t)\right] - \mathbf{s}\left[t, \mathbf{R}_d^{\alpha-1}(t), \mathbf{I}_N(t)\right], \quad \forall t \in \mathfrak{T}.
\end{aligned}
\tag{5.7}
$$

The application of (4.2), (4.3), and (4.6) to the preceding equations transforms them into:

$$
\begin{aligned}
& Q\left[\mathbf{R}_d^{\alpha-1}(t) + \mathbf{r}(t)^{\alpha-1}\right]\left[\mathbf{R}_d^{(\alpha)}(t) + \mathbf{r}^{(\alpha)}(t)\right] - Q\left[\mathbf{R}_d^{\alpha-1}(t)\right] \mathbf{R}_d^{(\alpha)}(t) \\
& + \mathbf{q}\left[\mathbf{R}_d^{\alpha-1}(t) + \mathbf{r}(t)^{\alpha-1}\right] - \mathbf{q}\left[\mathbf{R}_d^{\alpha-1}(t)\right] \\
& = \mathbf{h}\left[t, \mathbf{I}_N^{\xi}(t) + \mathbf{i}^{\xi}(t)\right] - \mathbf{h}\left[t, \mathbf{I}_N^{\xi}(t)\right], \quad \forall t \in \mathfrak{T}, \\
\mathbf{Y}(t) - \mathbf{Y}_d(t) & = \mathbf{s}\left[t, \mathbf{R}_d^{\alpha-1}(t) + \mathbf{r}(t), \mathbf{I}_N(t) + \mathbf{i}(t)\right] - \mathbf{s}\left[t, \mathbf{R}_d^{\alpha-1}(t), \mathbf{I}_N(t)\right], \quad \forall t \in \mathfrak{T}.
\end{aligned}
$$

These equations can be set into the following form:

$$
\begin{aligned}
& Q\left[\mathbf{R}_d^{\alpha-1}(t) + \mathbf{r}^{\alpha-1}(t)\right] \mathbf{r}^{(\alpha)}(t) \\
& + \left\langle Q\left[\mathbf{R}_d^{\alpha-1}(t) + \mathbf{r}^{\alpha-1}(t)\right] - Q\left[\mathbf{R}_d^{\alpha-1}(t)\right]\right\rangle \mathbf{R}_d^{(\alpha)}(t) \\
& + \mathbf{q}\left[\mathbf{R}_d^{\alpha-1}(t) + \mathbf{r}^{\alpha-1}(t)\right] - \mathbf{q}\left[\mathbf{R}_d^{\alpha-1}(t)\right] \\
& = \mathbf{h}\left[t, \mathbf{I}_N^{\xi}(t) + \mathbf{i}^{\xi}(t)\right] - \mathbf{h}\left[t, \mathbf{I}_N^{\xi}(t)\right], \quad \forall t \in \mathfrak{T}, \\
\mathbf{y}(t) & = \mathbf{s}\left[t, \mathbf{R}_d^{\alpha-1}(t) + \mathbf{r}(t), \mathbf{I}_N(t) + \mathbf{i}(t)\right] - \mathbf{s}\left[t, \mathbf{R}_d^{\alpha-1}(t), \mathbf{I}_N(t)\right], \quad \forall t \in \mathfrak{T}.
\end{aligned}
\tag{5.8}
$$

Let

$$
\begin{aligned}
\widetilde{Q}\left[t, \mathbf{r}^{\alpha-1}(t)\right] & = Q\left[\mathbf{R}_d^{\alpha-1}(t) + \mathbf{r}^{\alpha-1}(t)\right], \\
\widetilde{\mathbf{q}}\left[t, \mathbf{r}^{\alpha-1}(t)\right] & = \left\langle Q\left[\mathbf{R}_d^{\alpha-1}(t) + \mathbf{r}^{\alpha-1}(t)\right] - Q\left[\mathbf{R}_d^{\alpha-1}(t)\right]\right\rangle \mathbf{R}_d^{(\alpha)}(t) \\
& \quad + \mathbf{q}\left[\mathbf{R}_d^{\alpha-1}(t) + \mathbf{r}^{\alpha-1}(t)\right] - \mathbf{q}\left[\mathbf{R}_d^{\alpha-1}(t)\right], \\
\widetilde{\mathbf{h}}\left[t, \mathbf{i}^{\xi}(t)\right] & = \mathbf{h}\left[t, \mathbf{I}_N^{\xi}(t) + \mathbf{i}^{\xi}(t)\right] - \mathbf{h}\left[t, \mathbf{I}_N^{\xi}(t)\right] \\
\widetilde{\mathbf{s}}\left[t, \mathbf{r}^{\alpha-1}(t), \mathbf{i}(t)\right] & = \mathbf{s}\left[t, \mathbf{R}_d^{\alpha-1}(t) + \mathbf{r}^{\alpha-1}(t), \mathbf{I}_N(t) + \mathbf{i}(t)\right] - \mathbf{s}\left[t, \mathbf{R}_d^{\alpha-1}(t), \mathbf{I}_N(t)\right] \\
& \forall t \in \mathfrak{T}.
\end{aligned}
\tag{5.9}
$$

These definitions transform (5.8), i.e., (5.7), into

$$
\begin{aligned}
\widetilde{Q}\left[t, \mathbf{r}^{\alpha-1}(t)\right] \mathbf{r}^{(\alpha)}(t) + \widetilde{\mathbf{q}}\left[t, \mathbf{r}^{\alpha-1}(t)\right] = \widetilde{\mathbf{h}}\left[t, \mathbf{i}^{\xi}(t)\right], \quad \forall t \in \mathfrak{T}, \\
\mathbf{y}(t) = \widetilde{\mathbf{s}}\left[t, \mathbf{r}^{\alpha-1}(t), \mathbf{i}(t)\right], \quad \forall t \in \mathfrak{T}.
\end{aligned}
\tag{5.10}
$$

This is the system description in terms of deviations, which is *time*-varying ■

If the plant is *time*-invariant then the nonlinearity $\mathbf{q}(,)$ and the matrices are *time*-independent:

$$Q\left[\mathbf{R}^{\alpha-1}(t)\right]\mathbf{R}^{(\alpha)}(t)+\mathbf{q}\left[\mathbf{R}^{\alpha-1}(t)\right]=\left\{\begin{array}{c} E\mathbf{e}_d\left[\mathbf{D}^{\eta}(t)\right]+ \\ +P\mathbf{e}_u\left[\mathbf{U}^{\mu}(t)\right]\end{array}\right\},\ \forall t\in\mathfrak{T},$$
$$\mathbf{Y}(t)=\mathbf{z}\left[\mathbf{R}^{\alpha-1}(t),\mathbf{D}(t)\right]+W\mathbf{w}\left[\mathbf{U}(t)\right],\ \forall t\in\mathfrak{T}, \tag{5.11}$$

If the nominal disturbance $\mathbf{D}_N(.)\in\mathfrak{D}$ and the nominal control $\mathbf{U}_N(.)\in\mathfrak{U}^{\mu}$ are determined relative to the desired output $\mathbf{Y}_d(.)\in\mathfrak{Y}_d^k$ then the nonlinear plant (3.12) in the nominal regime reads:

$$Q\left[t,\mathbf{R}_d^{\alpha-1}(t)\right]\mathbf{R}_d^{(\alpha)}(t)+\mathbf{q}\left[t,\mathbf{R}_d^{\alpha-1}(t)\right]=\left\{\begin{array}{c} E(t)\,\mathbf{e}_d\left[\mathbf{D}_N^{\eta}(t)\right] \\ +\,P(t)\,\mathbf{e}_u\left[\mathbf{U}_N^{\mu}(t)\right]\end{array}\right\},\ \forall t\in\mathfrak{T},$$
$$\mathbf{Y}_d(t)=\mathbf{z}\left[t,\mathbf{R}_d^{\alpha-1}(t),\mathbf{D}_N(t)\right]+W(t)\,\mathbf{w}\left[\mathbf{U}_N(t)\right],\ \forall t\in\mathfrak{T}. \tag{5.12}$$

Assumption 129 *The desired output response $\mathbf{Y}_d(t)$ of the plant (3.12), i.e., of (5.6), is realizable in $\mathfrak{D}$ on $\mathfrak{T}_0\times\mathfrak{U}^{\mu}\times\mathfrak{Y}_d^k$; i.e., Theorem 122 holds, hence it is also realizable on $\mathfrak{T}_0\times\mathfrak{D}\times\mathfrak{Y}_d^k$, Theorem 123.*

Assumption 130 *The desired output response $\mathbf{Y}_d(t)$ of the plant (3.12), i.e., of (5.6), is realizable on $\mathfrak{T}_0\times\mathfrak{D}\times\mathfrak{U}^{\mu}\times\mathfrak{Y}_d^k$; i.e., Theorem 117 is valid.*

The following equations determine the plant nominal regime on $\mathfrak{T}_0\times\mathfrak{D}\times\mathfrak{U}^{\mu}\times\mathfrak{Y}_d^k$, Theorem 117, when the nominal disturbance $\mathbf{D}_N(.)\in\mathfrak{D}$ is accepted relative to any chosen $\mathbf{Y}_d(.)\in\mathfrak{Y}_d^k$:

$$Q\left[\mathbf{R}_d^{\alpha-1}(t)\right]\mathbf{R}_d^{(\alpha)}(t)+\mathbf{q}\left[\mathbf{R}_d^{\alpha-1}(t)\right]=\left\{\begin{array}{c} E(t)\,\mathbf{e}_d\left[\mathbf{D}_N^{\eta}(t)\right] \\ +\,P(t)\,\mathbf{e}_u\left[\mathbf{U}_N^{\mu}(t)\right]\end{array}\right\},\ \forall t\in\mathfrak{T},$$
$$\mathbf{Y}_d(t)=\mathbf{z}\left[t,\mathbf{R}_d^{\alpha-1}(t),\mathbf{D}_N(t)\right]+W(t)\,\mathbf{w}\left[\mathbf{U}_N(t)\right],\ \forall t\in\mathfrak{T}. \tag{5.13}$$

The control $\mathbf{U}(t)$ is nominal, $\mathbf{U}(t)\equiv\mathbf{U}_N(t)$, relative to both an accepted nominal disturbance $\mathbf{D}_N(.)\in\mathfrak{D}$ and a chosen desired output $\mathbf{Y}_d(.)\in\mathfrak{Y}_d^k$.

The mathematical models expressed in terms of deviations of all variables of the plants described by (3.12), i.e., of (5.6), have the following common general property.

Theorem 131 ***Time-varying mathematical model of the plant [time-invariant (5.11) or time-varying (5.6)] in terms of the deviations (4.1), (4.3), (4.5), (4.6) with the disturbance and control nominal relative to $\mathbf{Y}_d(.)$ on $\mathfrak{T}_0\times\mathfrak{D}^{\eta}\times\mathfrak{U}^{\mu}\times\mathfrak{Y}_d^k$***

Let Assumption 130 hold. If the mathematical model of the plant [either time-invariant (5.11) or time-varying (5.6)] is nonlinear in terms of the total coordinates **D**, **R**, **U**, *and* **Y**, *then the plant mathematical model in terms of their deviations* **d**, **r**, **u**, *and* **y** *is time-varying as soon as the plant total desired output vector $\mathbf{Y}_d$ is time-varying, $\mathbf{Y}_d(t)\neq$* **const.**, *so that the desired state vector $\mathbf{R}_d^{\alpha-1}(t)$ is also time-varying.*

Proof. What follows is fully valid for both *time*-invariant system (5.11) and *time*-varying plant (5.6). The difference between (5.6) and (5.12) reads:

$$Q\left[t,\mathbf{R}^{\alpha-1}(t)\right]\mathbf{R}^{(\alpha)}(t)+\mathbf{q}\left[t,\mathbf{R}^{\alpha-1}(t)\right]-Q\left[t,\mathbf{R}_d^{\alpha-1}(t)\right]\mathbf{R}_d^{(\alpha)}(t)-\mathbf{q}\left[t,\mathbf{R}_d^{\alpha-1}(t)\right]$$
$$=\left\{\begin{array}{c} E(t)\,\mathbf{e}_d\left[\mathbf{D}^{\eta}(t)\right] \\ +\,P(t)\,\mathbf{e}_u\left[\mathbf{U}^{\mu}(t)\right]\end{array}\right\}-\left\{\begin{array}{c} E(t)\,\mathbf{e}_d\left[\mathbf{D}_N^{\eta}(t)\right] \\ +\,P(t)\,\mathbf{e}_u\left[\mathbf{U}_N^{\mu}(t)\right]\end{array}\right\},\ \forall t\in\mathfrak{T},$$
$$\mathbf{Y}(t)-\mathbf{Y}_d(t)=\mathbf{z}\left[t,\mathbf{R}^{\alpha-1}(t),\mathbf{D}(t)\right]-\mathbf{z}\left[t,\mathbf{R}_d^{\alpha-1}(t),\mathbf{D}_N(t)\right]$$
$$+\,W(t)\,\mathbf{w}\left[\mathbf{U}(t)\right]-W(t)\,\mathbf{w}\left[\mathbf{U}_N(t)\right],\ \forall t\in\mathfrak{T}. \tag{5.14}$$

The application of **(4.1)**, **(4.3)**, **(4.5)**, and **(4.6)** to the preceding equations transforms them into:

$$Q\left[t,\mathbf{R}_d^{\alpha-1}(t)+\mathbf{r}(t)^{\alpha-1}\right]\left[\mathbf{R}_d^{(\alpha)}(t)+\mathbf{r}^{(\alpha)}(t)\right]-Q\left[t,\mathbf{R}_d^{\alpha-1}(t)\right]\mathbf{R}_d^{(\alpha)}(t)$$
$$+\mathbf{q}\left[t,\mathbf{R}_d^{\alpha-1}(t)+\mathbf{r}(t)^{\alpha-1}\right]-\mathbf{q}\left[t,\mathbf{R}_d^{\alpha-1}(t)\right]$$
$$=\left\{\begin{array}{c} E(t)\left\langle \mathbf{e}_d\left[\mathbf{D}_N^{\eta}(t)+\mathbf{d}^{\eta}(t)\right]-\mathbf{e}_d\left[\mathbf{D}_N^{\eta}(t)\right]\right\rangle \\ +P(t)\left\langle \mathbf{e}_u\left[\mathbf{U}_N^{\mu}(t)+\mathbf{u}^{\mu}(t)\right]-\mathbf{e}_u\left[\mathbf{U}_N^{\mu}(t)\right]\right\rangle \end{array}\right\},\quad \forall t\in\mathfrak{T},$$
$$\mathbf{y}(t)=\left\langle \mathbf{z}\left[t,\mathbf{R}_d^{\alpha-1}(t)+\mathbf{r}^{\alpha-1}(t),\mathbf{D}_N(t)+\mathbf{d}(t)\right]-\mathbf{z}\left[t,\mathbf{R}_d^{\alpha-1}(t),\mathbf{D}_N(t)\right]\right\rangle$$
$$+W(t)\left\langle \mathbf{w}\left[\mathbf{U}_N(t)+\mathbf{u}(t)\right]-\mathbf{w}\left[\mathbf{U}_N(t)\right]\right\rangle,\quad \forall t\in\mathfrak{T}.$$

These equations can be set into the following form:

$$\begin{array}{rcl} & & Q\left[t,\mathbf{R}_d^{\alpha-1}(t)+\mathbf{r}^{\alpha-1}(t)\right]\mathbf{r}^{(\alpha)}(t) \\ & & +\left\langle Q\left[t,\mathbf{R}_d^{\alpha-1}(t)+\mathbf{r}^{\alpha-1}(t)\right]-Q\left[t,\mathbf{R}_d^{\alpha-1}(t)\right]\right\rangle \mathbf{R}_d^{(\alpha)}(t) \\ & & +\mathbf{q}\left[t,\mathbf{R}_d^{\alpha-1}(t)+\mathbf{r}^{\alpha-1}(t)\right]-\mathbf{q}\left[t,\mathbf{R}_d^{\alpha-1}(t)\right] \\ & = & \left\{\begin{array}{c} E(t)\left\langle \mathbf{e}_d\left[\mathbf{D}_N^{\eta}(t)+\mathbf{d}^{\eta}(t)\right]-\mathbf{e}_d\left[\mathbf{D}_N^{\eta}(t)\right]\right\rangle \\ +P(t)\left\langle \mathbf{e}_u\left[\mathbf{U}_N^{\mu}(t)+\mathbf{u}^{\mu}(t)\right]-\mathbf{e}_u\left[\mathbf{U}_N^{\mu}(t)\right]\right\rangle \end{array}\right\}, \\ \mathbf{y}(t) & = & \left\langle \mathbf{z}\left[t,\mathbf{R}_d^{\alpha-1}(t)+\mathbf{r}(t),\mathbf{D}_N(t)+\mathbf{d}(t)\right]-\mathbf{z}\left[t,\mathbf{R}_d^{\alpha-1}(t),\mathbf{D}_N(t)\right]\right\rangle \\ & & +W(t)\left\langle \mathbf{w}\left[\mathbf{U}_N(t)+\mathbf{u}(t)\right]-\mathbf{w}\left[\mathbf{U}_N(t)\right]\right\rangle, \\ \forall t & \in & \mathfrak{T}. \end{array} \tag{5.15}$$

Let

$$\begin{array}{r} \widetilde{Q}\left[t,\mathbf{r}^{\alpha-1}(t)\right]=Q\left[t,\mathbf{R}_d^{\alpha-1}(t)+\mathbf{r}^{\alpha-1}(t)\right], \\ \widetilde{\mathbf{q}}\left[t,\mathbf{r}^{\alpha-1}(t)\right]=\left\langle Q\left[t,\mathbf{R}_d^{\alpha-1}(t)+\mathbf{r}^{\alpha-1}(t)\right]-Q\left[t,\mathbf{R}_d^{\alpha-1}(t)\right]\right\rangle \mathbf{R}_d^{(\alpha)}(t) \\ +\mathbf{q}\left[t,\mathbf{R}_d^{\alpha-1}(t)+\mathbf{r}^{\alpha-1}(t)\right]-\mathbf{q}\left[t,\mathbf{R}_d^{\alpha-1}(t)\right], \\ \widetilde{\mathbf{e}}_d\left[t,\mathbf{d}^{\eta}(t)\right]=\mathbf{e}_d\left[\mathbf{D}_N^{\eta}(t)+\mathbf{d}^{\eta}(t)\right]-\mathbf{e}_d\left[\mathbf{D}_N^{\eta}(t)\right], \\ \widetilde{\mathbf{e}}_u\left[t,\mathbf{u}^{\mu}(t)\right]=\mathbf{e}_u\left[\mathbf{U}_N^{\mu}(t)+\mathbf{u}^{\mu}(t)\right]-\mathbf{e}_u\left[\mathbf{U}_N^{\mu}(t)\right], \\ \widetilde{\mathbf{z}}\left[t,\mathbf{r}^{\alpha-1}(t),\mathbf{d}(t)\right]=\left\{\begin{array}{c} \mathbf{z}\left[t,\mathbf{R}_d^{\alpha-1}(t)+\mathbf{r}^{\alpha-1}(t),\mathbf{D}_N(t)+\mathbf{d}(t)\right] \\ -\mathbf{z}\left[t,\mathbf{R}_d^{\alpha-1}(t),\mathbf{D}_N(t)\right] \end{array}\right\}, \\ \widetilde{\mathbf{w}}\left[t,\mathbf{u}(t)\right]=\mathbf{w}\left[\mathbf{U}_N(t)+\mathbf{u}(t)\right]-\mathbf{w}\left[\mathbf{U}_N(t)\right],\ \forall t\in\mathfrak{T}. \end{array} \tag{5.16}$$

These definitions transform (5.15), i.e., (5.14), into

$$\begin{array}{r} \widetilde{Q}\left[t,\mathbf{r}^{\alpha-1}(t)\right]\mathbf{r}^{(\alpha)}(t)+\widetilde{\mathbf{q}}\left[t,\mathbf{r}^{\alpha-1}(t)\right]=E(t)\widetilde{\mathbf{e}}_d\left[t,\mathbf{d}^{\eta}(t)\right]+P(t)\widetilde{\mathbf{e}}_u\left[t,\mathbf{u}^{\mu}(t)\right], \\ \mathbf{y}(t)=\widetilde{\mathbf{z}}\left[t,\mathbf{r}^{\alpha-1}(t),\mathbf{d}(t)\right]+W(t)\widetilde{\mathbf{w}}\left[\mathbf{u}(t)\right],\quad \forall t\in\mathfrak{T}. \end{array} \tag{5.17}$$

This is the plant (3.12) description in terms of deviations, which is time-varying. For the special form (3.29) of the plant (3.12) and for

$$\widetilde{F}\left(t,\mathbf{r}^{\alpha-1}(t)\right)\equiv F\left(t,\mathbf{R}_d^{\alpha-1}(t)+\mathbf{r}^{\alpha-1}(t)\right) \tag{5.18}$$

the difference $\mathbf{q}\left[t,\mathbf{R}_d^{\alpha-1}(t)+\mathbf{r}^{\alpha-1}(t)\right]-\mathbf{q}\left[t,\mathbf{R}_d^{\alpha-1}(t)\right]$ in (5.16):

$$\mathbf{q}\left[t,\mathbf{R}_d^{\alpha-1}(t)+\mathbf{r}^{\alpha-1}(t)\right]-\mathbf{q}\left[t,\mathbf{R}_d^{\alpha-1}(t)\right]$$
$$=F\left(t,\mathbf{R}_d^{\alpha-1}(t)+\mathbf{r}^{\alpha-1}(t)\right)\left[\mathbf{R}_d^{\alpha-1}(t)+\mathbf{r}^{\alpha-1}(t)\right]-F\left(t,\mathbf{R}_d^{\alpha-1}(t)\right)\mathbf{R}_d^{\alpha-1}(t)$$
$$=\widetilde{F}\left(t,\mathbf{r}^{\alpha-1}(t)\right)\mathbf{r}^{\alpha-1}(t)+\left\{\begin{array}{c} F\left(t,\mathbf{R}_d^{\alpha-1}(t)+\mathbf{r}^{\alpha-1}(t)\right) \\ -F\left(t,\mathbf{R}_d^{\alpha-1}(t)\right) \end{array}\right\}\mathbf{R}_d^{\alpha-1}(t),$$

so that in (5.16) $\widetilde{\mathbf{q}}\left[t,\mathbf{r}^{\alpha-1}(t)\right]$ reads:

$$\widetilde{\mathbf{q}}\left[t,\mathbf{r}^{\alpha-1}(t)\right] = \left\langle Q\left[\mathbf{R}_d^{\alpha-1}(t)+\mathbf{r}^{\alpha-1}(t)\right] - Q\left[\mathbf{R}_d^{\alpha-1}(t)\right]\right\rangle \mathbf{R}_d^{(\alpha)}(t) \\ + \left\{F\left(t,\mathbf{R}_d^{\alpha-1}(t)+\mathbf{r}^{\alpha-1}(t)\right) - F\left(t,\mathbf{R}_d^{\alpha-1}(t)\right)\right\}\mathbf{R}_d^{\alpha-1}(t), \tag{5.19}$$

Besides, for

$$\widetilde{Z}\left[t,\mathbf{r}^{\alpha-1}(t),\mathbf{d}(t)\right] \\ \equiv Z\left[t,\mathbf{R}_d^{\alpha-1}(t)+\mathbf{r}^{\alpha-1}(t),\mathbf{D}_N(t)+\mathbf{d}(t)\right] \tag{5.20}$$

$\widetilde{\widetilde{\mathbf{z}}}\left[t,\mathbf{r}^{\alpha-1}(t),\mathbf{d}(t)\right] + \widetilde{Z}\left[t,\mathbf{r}^{\alpha-1}(t),\mathbf{d}(t)\right]\mathbf{r}^{\alpha-1}(t)$ replace $\widetilde{\mathbf{z}}\left[t,\mathbf{r}^{\alpha-1}(t),\mathbf{d}(t)\right]$ in (5.16)

$$\widetilde{\widetilde{\mathbf{z}}}\left[t,\mathbf{r}^{\alpha-1}(t),\mathbf{d}(t)\right] \\ = \left\{Z\left[t,\mathbf{R}_d^{\alpha-1}(t)+\mathbf{r}^{\alpha-1}(t),\mathbf{D}_N(t)+\mathbf{d}(t)\right] - Z\left[t,\mathbf{R}_d^{\alpha-1}(t),\mathbf{D}_N(t)\right]\right\}\mathbf{R}_d^{\alpha-1}(t)$$

The equations (5.17) become

$$\widetilde{Q}\left[t,\mathbf{r}^{\alpha-1}(t)\right]\mathbf{r}^{(\alpha)}(t) + \widetilde{\mathbf{q}}\left[t,\mathbf{r}^{\alpha-1}(t)\right] + \widetilde{F}\left[t,\mathbf{r}^{\alpha-1}(t)\right]\mathbf{r}^{\alpha-1}(t) \\ = E(t)\,\widetilde{\mathbf{e}}_d\left[t,\mathbf{d}^{\eta}(t)\right] + P(t)\,\widetilde{\mathbf{e}}_u\left[\mathbf{u}^{\mu}(t)\right], \quad \forall t\in\mathfrak{T}, \\ \mathbf{y}(t) = \widetilde{\widetilde{\mathbf{z}}}\left[t,\mathbf{r}^{\alpha-1}(t),\mathbf{d}(t)\right] + \widetilde{Z}\left[t,\mathbf{r}^{\alpha-1}(t),\mathbf{d}(t)\right]\mathbf{r}^{\alpha-1}(t) + W(t)\,\widetilde{\mathbf{w}}\left[\mathbf{u}(t)\right], \\ \forall t\in\mathfrak{T} \tag{5.21}$$

■

Conclusion 132 *The mathematical model in terms of deviations of all variables of the plant (3.12), i.e., of the plant (5.6), has the following form:*

$$\widetilde{Q}\left[t,\mathbf{r}^{\alpha-1}(t)\right]\mathbf{r}^{(\alpha)}(t) + \widetilde{\mathbf{q}}\left[t,\mathbf{r}^{\alpha-1}(t)\right] = E(t)\,\widetilde{\mathbf{e}}_d\left[t,\mathbf{d}^{\eta}(t)\right] + P(t)\,\widetilde{\mathbf{e}}_u\left[\mathbf{u}^{\mu}(t)\right], \\ \mathbf{y}(t) = \widetilde{\mathbf{z}}\left[t,\mathbf{r}^{\alpha-1}(t),\mathbf{d}(t)\right] + W(t)\,\widetilde{\mathbf{w}}\left[\mathbf{u}(t)\right], \quad \forall t\in\mathfrak{T}, \tag{5.22}$$

where all terms are determined by (5.16).

The mathematical model in terms of deviations of all variables of the special plants described by (3.29) have the following form:

$$\widetilde{Q}\left[t,\mathbf{r}^{\alpha-1}(t)\right]\mathbf{r}^{(\alpha)}(t) + \widetilde{\mathbf{q}}\left[t,\mathbf{r}^{\alpha-1}(t)\right] + \widetilde{F}\left[t,\mathbf{r}^{\alpha-1}(t)\right]\mathbf{r}^{\alpha-1}(t) \\ = E(t)\,\widetilde{\mathbf{e}}_d\left[t,\mathbf{d}^{\eta}(t)\right] + P(t)\,\widetilde{\mathbf{e}}_u\left[\mathbf{u}^{\mu}(t)\right], \quad \forall t\in\mathfrak{T}, \\ \mathbf{y}(t) = \widetilde{\widetilde{\mathbf{z}}}\left[t,\mathbf{r}^{\alpha-1}(t),\mathbf{d}(t)\right] + \widetilde{Z}\left[t,\mathbf{r}^{\alpha-1}(t),\mathbf{d}(t)\right]\mathbf{r}^{\alpha-1}(t) + W(t)\,\widetilde{\mathbf{w}}\left[\mathbf{u}(t)\right], \\ \forall t\in\mathfrak{T}, \tag{5.23}$$

where all terms are defined by (5.16) together with (5.18) and (5.19).

Comment 133 *The mathematical model determined in terms of the deviations of the [time-invariant (5.11) or time-varying (5.6)] system is time-varying as soon as the system nominal (desired) output is time-varying. This characteristic of the nonlinear systems is not possible among linear systems ([252], [279]).*

Part III

TRACKABILITY

Chapter 6

Trackability concept

6.1 On system and control concepts

Stability properties are crucial for dynamical systems in general. Among various stability concepts the most important is Lyapunov's stability concept [401]. It became the basic qualitative dynamical concept also for control systems. Its specific areas concern system stabilizability and system stabilization. They all concern the closeness of the system motions over the infinite *time* interval $[t_0, \infty[$, either to an equilibrium state or to a given specific (unperturbed, reference, desired) motion subjected to the influence of arbitrary initial conditions. They ignore the existence of the external disturbances acting on the dynamical system. In order to overcome this conceptual drawback, the concept of practical stability was introduced [84]. These concepts do not take into account the primary goal of control.

Later there appeared three typical control qualitative concepts. They are the concepts of controllability, observability, and optimality of control systems.

6.2 Controllability and observability

Kalman's *concept of the state controllability* has become a fundamental control concept, [336]-[339], [342]. It is characteristic and fully meaningful in the framework of control systems. System controllability means that for any initial state there exists control that the system state steers from the initial state to the origin in some finite *time* if the system is not subjected to actions of disturbances. The controllability concept has attracted a lot of research efforts to be further developed. E. G. Gilbert [143] generalized it to the $MIMO$ systems (plants). M. L. J. Hautus [302] established for them the simple form of the controllability criterion in the complex domain.

Kalman also introduced *the observability concept* that is fundamental for control systems as well, but, differently than controllability, it has the wider importance that also broadens to dynamical systems in general.

J. E. Bertram and P. E. Sarachik [42] broadened *the state controllability* concept to *the output controllability concept.*

Both the state controllability concept and the output controllability concept consider the system possibility of steering a state or an output from any initial state or from any initial output to another state or another output, in general, or to the zero state or to the zero output, in particular, respectively.

R. W. Brockett and M. D. Mesarović (Mesarovitch) [58] introduced *the concept of functional (output) reproducibility*, also called *the output function controllability* [11, page 313], [75, page 216], [555, pages 72 and 164], in which the target is not a particular output (e.g., the zero output) but a given function representing a reference (desired) output response. This concept, like the controllability concept, concerns systems free of any external disturbance action.

All these controllability concepts assume the nonexistence of any external perturbations acting on the system. The only external influences on the system are control actions.

6.3 Disturbance rejection or compensation

Dynamical systems, in general, and plants, hence their control systems, in particular, are really under the actions of usually unpredictable and unknown external perturbations (called usually *disturbances*).

Remark 134 ***Disturbance compensation***

In the sequel we use the terms "compensation" and "disturbance compensation", rather than the widely used term "disturbance rejection". There is not a control that can reject, i.e., eliminate, the disturbance or its action on the plant. The regulator cannot reject the wind or the exterior air temperature variations, or their consequences on the air temperature in the room. Therefore, one of the control basic goals is to compensate, i.e., to neutralize, the consequences of the disturbance actions on the plant.

S. P. Bhattacharyya [49], [50], S. P. Bhattacharyya et al. [51], E. J. Davison [102], [104], E. J. Davison et al. [105]-[109], [434], E. Fabian and W. M. Wonham [124], B. Porter et al. [486], R. Saeks and J. Murray [507], and S. Y. Zhang and C. T. Chen [573] studied largely the problem of *disturbance compensation*, which they called *disturbance rejection.*

The controllability problems and the disturbance compensation problems mainly have been studied separately. In the framework of linear systems the justification is their linearity, i.e., the validity of the superposition principle. Such an argument does not exist for nonlinear systems.

6.4 New control concepts

The above concepts do not reflect the basic control goal.

The primary and the basic control goal is that the control forces the plant real output behavior **to follow**, i.e., **to track**, its every desired behavior, *despite the plant being simultaneously subjected to disturbance actions and to arbitrary initial conditions.* The book introduces and establishes various **tracking concepts** (*perfect tracking concept* and *imperfect tracking concepts such as Lyapunov tracking, tracking with finite reachability time, high quality tracking)* and opens avenues to study other tracking concepts (e.g., *practical tracking*; for more details see Section 14.2).

Because some required kind of ***tracking*** is the very goal of the control in the real plant environment and under real operating conditions, it led to the introduction of the new control concept called ***trackability*** [279]. It is the plant property that permits the existence of control such that for any *time* $\tau \in \text{In}\,\mathfrak{T}_0$ it leads the system output behavior from an arbitrary initial output vector to become equal to the system desired output behavior at the moment τ and that they stay equal on $[\tau, \infty[$ despite unknown unpredictable disturbances act on the plant. We can consider the state behavior of the plant instead of its output behavior. In that case we treat *the state trackability* of the plant.

The trackability concept incorporates the controllability concept and the disturbance compensation concept as special cases.

The trackability concept explains whether the plant itself has a property to enable the existence of a control that can guarantee **tracking under the actions of both arbitrary initial conditions (globally or from a domain) and external disturbances**

belonging to a set $\mathfrak{D}^i$ of permitted disturbances, as well as **for every plant desired output response from a given functional family $\mathfrak{Y}_d^k$.**

Another plant property called ***natural trackability*** is a type of the plant *trackability* that permits control synthesis and control implementation without using information about the real values and forms of the disturbances and about the mathematical model of the plant state. Such control that is continuous in *time* is ***natural tracking control*** (***NTC***).

The concept of ***natural trackability*** was established and developed by discovering algorithms for synthesis of ***natural tracking control*** (*NTC*)in [165]-[169], [232]-[239], [253], [260], [263]-[265], [279], [267], [268], [269], [270], [278], [282] - [286], [445]-[453].

In what follows the tracking and trackability concepts fully established for linear *time*-invariant systems in [279] are broadened to *time*-varying nonlinear systems.

6.5 *Time* and control

PCUP (Principle 17 and 18, Chapter 1) summarizes the common general properties of physical variables. *TCUP* (*time continuity and uniqueness principle* 20, Chapter 1) jointly expresses these properties and the crucial properties of *time* [155], [255], [268], [274]-[276]. These principles enable effective *natural tracking control synthesis* for linear and nonlinear dynamical plants.

Chapter 7

Perfect trackability concepts

7.1 Perfect trackability

Let us explain the sense of ***trackability*** in general and of ***perfect trackability*** in particular.

Definition 135 ***Functionally interrelated variables***

Variables $Z_1, Z_2, \cdots, Z_K$, *are* ***functionally interrelated*** *if, and only if, there exists* $\mathbf{v}:\mathfrak{T} \times \mathfrak{R}^{(i+1)K} \longrightarrow \mathfrak{R}^n$ *such that*

$$\mathbf{v}\left(t, \mathbf{Z}^k; \sigma\right) = \mathbf{0}_m, \quad \sigma \in \mathfrak{T}_0, \ \forall\left(t \geq \sigma\right) \in \mathfrak{T}_0. \tag{7.1}$$

We specify (1.8) by

$$k \in \{0, 1, \cdots, \alpha - 1\}, \ j, l \in \{0, 1, \cdots\}, \ j \geq \eta, \ l \geq \mu. \tag{7.2}$$

Definition 136 ***The k-th-order perfect trackability of*** $\mathbf{Y}_d(.)$ ***under the perturbation of*** $\mathbf{D}(.) \in \mathfrak{D}^j$ ***on*** $\mathfrak{T}_0 \times \mathfrak{U}^l$

a) The desired output vector function $\mathbf{Y}_d(.)$ *is* ***the k-th-order perfect trackable on*** $\mathfrak{T}_0 \times \mathfrak{U}^l$ ***under the perturbation of*** $\mathbf{D}(.) \in \mathfrak{D}^j$ *if, and only if, there exists a control vector function* $\mathbf{U}(.) \in \mathfrak{U}^l$ *such that the plant real output vector* $\mathbf{Y}(t)$ *and its first k derivatives are equal to the desired plant output vector* $\mathbf{Y}_d(t)$ *and its first k derivatives for every* $t \in \mathfrak{T}_0$, *respectively, as soon as* $\mathbf{Y}^k(t_0) = \mathbf{Y}_d^k(t_0)$,

$$\begin{gathered}\mathbf{D}(.) \in \mathfrak{D}^j, \ \mathbf{Y}^k(t_0) = \mathbf{Y}_d^k(t_0) \ \text{ and } \ \exists \mathbf{U}(.) \in \mathfrak{U}^l \Longrightarrow \\ \mathbf{Y}^k(t) = \mathbf{Y}_d^k(t), \forall t \in \mathfrak{T}_0.\end{gathered} \tag{7.3}$$

b) ***The domain*** $\mathcal{D}_{PTblD}^k\left(t_0; \mathbf{D}; \mathbf{U}; \mathfrak{Y}_d^k\right)$ ***of the k-th-order perfect trackability of*** $\mathbf{Y}_d(.)$ ***under the perturbation of*** $\mathbf{D}(.) \in \mathfrak{D}^j$ ***on*** $\mathfrak{T}_0 \times \mathfrak{U}^l$ *is the largest connected set* $\mathfrak{P} = \mathfrak{P}\left(t_0; \mathbf{D}; \mathbf{U}; \mathfrak{Y}_d^k\right)$ *with the nonempty interior,* $\mathfrak{P} \subseteq \mathfrak{R}^{(k+1)N}$ *and In* $\mathfrak{P} \neq \phi$, *such that (7.3) holds for every* $\mathbf{Y}_{d0}^k = \mathbf{Y}_d^k(t_0) \in \mathfrak{P}$.

c) If, and only if, (7.3) holds for every $t_0 \in \mathfrak{T}_i$ *so that the following intersection*

$$\cap\left[\mathcal{D}_{PTblD}^k\left(t_0; \mathbf{D}; \mathbf{U}; \mathfrak{Y}_d^k\right) : t_0 \in \mathfrak{T}_i\right]$$

is connected nonempty set then the desired output vector function $\mathbf{Y}_d(.)$ *is the* ***k-th-order*** t_0***-uniform perfect trackable on*** $\mathfrak{T}_0 \times \mathfrak{T}_i \times \mathfrak{U}^l$ *under the perturbation of* $\mathbf{D}(.) \in \mathfrak{D}^j$. *If this holds when* $\mathfrak{T}$ *replaces* $\mathfrak{T}_0$ *then the expression "****on*** $\mathfrak{T}_0\times$*" is to be omitted.*

d) The preceding perfect trackability properties are ***elementwise*** *if, and only if, the control vector* $\mathbf{U}$ *can act on every entry* Y_i, $\forall i = 1, 2, \cdots, N$, *of* $\mathbf{Y}$ *mutually independently.*

e) If, and only if, additionally to some of the conditions under i)-g), the dimension r of the control vector $\mathbf{U}$ *is the least number of the entries* $U_j(.)$, $\forall i = 1, 2, .., j$, *of* $\mathbf{U}$ *that satisfies the corresponding conditions among i)-g) then the control is* ***minimal*** *for the relevant perfect trackability of* $\mathbf{Y}_d(.)$.

We denote the k-th-order right-hand side derivative of $\mathbf{Y}(t)$ at $t \in \mathfrak{T}_0$ with $D_r^k\mathbf{Y}(t)$ (Appendix B).

Lemma 137 ***The right-hand functional identity*** *[279]*
If two functions $\mathbf{Y}(.)$ *and* $\mathbf{Y}_d(.)$ *are defined, k-times continuously differentiable on* $]\sigma, \infty[$, $\sigma \in \mathfrak{T}_0$, $]\sigma, \infty[\subseteq In\, \mathfrak{T}_0$, *as well as at* $t = \sigma$ *from the right-hand side, i.e., at* $t = \sigma^+$, *and identical on* $[\sigma, \infty[$, *then all their derivatives up to the order k included are also identical on* $]\sigma, \infty[$ *and at* $t = \sigma^+$.

Note 138 *This lemma allows different vector values of the derivatives of* $\mathbf{Y}(t)$ *from the left-hand side and the right-hand side of the moment* $t = \sigma$, $\sigma \in \mathfrak{T}_0$.

Definition 104, Definition 136 and Lemma 137 directly imply the following.

Lemma 139 *If the desired output vector function* $\mathbf{Y}_d(.)$ *is differentiable at least up to the order k,* $\mathbf{Y}_d(t) \in \mathfrak{C}^k$, *then for it to be:*
i) The k-th-order perfect trackable on $\mathfrak{T}_0$ *under the perturbation of* $\mathbf{D}(.) \in \mathfrak{D}^j$ *it is necessary and sufficient to be realizable on* $\mathfrak{T}_0$ *for* $\mathbf{D}(.) \in \mathfrak{D}^j$, *equivalently, to be perfect trackable on* $\mathfrak{T}_0$ *under the action of* $\mathbf{D}(.) \in \mathfrak{D}^j$
ii) The k-th order elementwise perfect trackable on $\mathfrak{T}_0$ *under the perturbation of* $\mathbf{D}(.) \in \mathfrak{D}^j$ *it is necessary and sufficient to be elementwise realizable on* $\mathfrak{T}_0$ *for* $\mathbf{D}(.) \in \mathfrak{D}^j$, *equivalently, to be elementwise perfect trackable on* $\mathfrak{T}_0$ *under the action of* $\mathbf{D}(.) \in \mathfrak{D}^j$.

We are interested in perfect trackability of every plant desired output $\mathbf{Y}_d(.)$ from $\mathfrak{Y}_d^k$ rather than in perfect trackability of a single plant desired output.

Definition 140 ***The k-th-order perfect trackability of the plant (3.12) on*** $\mathfrak{T}_0 \times \mathfrak{D}^j \times \mathfrak{U}^l \times \mathfrak{Y}_d^k$
a) The plant (3.12) is ***the k-th-order perfect trackable on*** $\mathfrak{T}_0 \times \mathfrak{D}^j \times \mathfrak{U}^l \times \mathfrak{Y}_d^k$ *if, and only if, for every* $[\mathbf{D}(.), \mathbf{Y}_d(.)] \in \mathfrak{D}^j \times \mathfrak{Y}_d^k$ *there exists a control vector function* $\mathbf{U}(.) \in \mathfrak{U}^l$ *such that the plant real output and its first k derivatives are equal to the plant desired output and its first k derivatives on* $\mathfrak{T}_0$, *respectively, as soon as* $\mathbf{Y}^k(t_0) = \mathbf{Y}_d^k(t_0)$,

$$\mathbf{Y}^k(t_0) = \mathbf{Y}_d^k(t_0),\ \forall [\mathbf{D}(.), \mathbf{Y}_d(.)] \in \mathfrak{D}^j \times \mathfrak{Y}_d^k,\ \exists \mathbf{U}(.) \in \mathfrak{U}^l \Longrightarrow$$
$$\mathbf{Y}^k(t) = \mathbf{Y}_d^k(t),\ \forall t \in \mathfrak{T}_0. \tag{7.4}$$

b) ***The domain*** $\mathcal{D}_{PTbl}^k(t_0; \mathbf{D}; \mathbf{U}; \mathfrak{Y}_d^k)$ ***of the k-th order perfect trackability of*** $\mathbf{Y}_d(.)$ ***on*** $\mathfrak{T}_0 \times \mathfrak{D}^j \times \mathfrak{U}^l \times \mathfrak{Y}_d^k$ *is the largest connected set* $\mathfrak{P} = \mathfrak{P}(t_0; \mathbf{D}; \mathbf{U}; \mathbf{Y}_d)$ *with the nonempty interior,* $\mathfrak{P} \subseteq \mathfrak{R}^{(k+1)N}$ *and* $In\, \mathfrak{P} \neq \phi$, *such that (7.4) holds for every* $\mathbf{Y}_{d0}^k = \mathbf{Y}_d^k(t_0) \in \mathfrak{P}$.
c) If, and only if, (7.3) holds for every $\mathbf{D}(.) \in \mathfrak{D}^j$ *then the desired output vector function* $\mathbf{Y}_d(.)$ *is* ***the k-th-order perfect trackable on*** $\mathfrak{T}_0 \times \mathfrak{D}^j \times \mathfrak{U}^l \times \mathfrak{Y}_d^k$.
d) If, and only if, (7.3) holds for every $t_0 \in \mathfrak{T}_i$ *so that the following intersection*

$$\cap \left[\mathcal{D}_{PTbl}^k(t_0; \mathbf{D}; \mathbf{U}; \mathfrak{Y}_d^k) : t_0 \in \mathfrak{T}_i\right]$$

is connected nonempty set then the desired output vector function $\mathbf{Y}_d(.)$ *is the* ***k-th-order*** t_0-***uniform perfect trackable on*** $\mathfrak{T}_0 \times \mathfrak{T}_i \times \mathfrak{D}^j \times \mathfrak{U}^l \times \mathfrak{Y}_d^k$. *If this holds when* $\mathfrak{T}$ *replaces* $\mathfrak{T}_0$ *then the expression "****on*** $\mathfrak{T}_0\times$ *" is to be omitted.*

e) If, and only if, (7.3) holds for every $\mathbf{D}(.) \in \mathfrak{D}^j$ *so that the following intersection* $\cap \left[\mathcal{D}^k_{PTbl}\left(t_0;\mathbf{D};\mathbf{U};\mathfrak{Y}^k_d\right) : \mathbf{D}(.) \in \mathfrak{D}^j\right]$ *is connected nonempty set then the desired output vector function* $\mathbf{Y}_d(.)$ *is* ***the k-th-order*** $\mathbf{D}-$***uniform perfect trackable on*** $\mathfrak{T}_0 \times \mathfrak{D}^j \times \mathfrak{U}^l \times \mathfrak{Y}^k_d$.

f) If, and only if, (7.3) holds for every $[t_0, \mathbf{D}(.)] \in \mathfrak{T}_i \times \mathfrak{D}^j$ *so that*

$$\cap \left\{\mathcal{D}^k_{PTbl}\left(t_0;\mathbf{D};\mathbf{U};\mathfrak{Y}^k_d\right) : [t_0, \mathbf{D}(.)] \in \mathfrak{T}_i \times \mathfrak{D}^j\right\}$$

is connected nonempty set then the desired output vector function $\mathbf{Y}_d(.)$ *is the* ***k-th-order uniform perfect trackable on*** $\mathfrak{T}_0 \times \mathfrak{T}_i \times \mathfrak{D}^j \times \mathfrak{U}^l \times \mathfrak{Y}^k_d$. *If this holds when* $\mathfrak{T}$ *replaces* $\mathfrak{T}_0$ *then the expression "**on** $\mathfrak{T}_0$" is to be omitted.*

g) The preceding perfect trackability properties are ***elementwise*** *if, and only if, the control vector* $\mathbf{U}$ *can act on every entry* Y_i, $\forall i = 1, 2, \cdots, N$, *of* $\mathbf{Y}$ *mutually independently.*

Comment 141 *The domain* $\mathcal{D}^k_{PTbl}\left(t_0;\mathfrak{D}^j;\mathbf{U};\mathfrak{Y}^k_d\right)$ *of the perfect trackability on* $\mathfrak{T}_0 \times \mathfrak{D}^j \times \mathfrak{U}^l \times \mathfrak{Y}^k_d$ *determines the set of the initial desired output vectors* $\mathbf{Y}^{\alpha-1}_d(t_0)$ *of* $\mathbf{Y}^{\alpha-1}_d(t)$ *for which the plant can exhibit perfect tracking on* $\mathfrak{T}_0 \times \mathfrak{D}^j \times \mathfrak{U}^l \times \mathfrak{Y}^k_d$.

From Definition 111, Lemma 137, Definition 140, and Lemma 142 imply the following.

Lemma 142 ***The k-th-order perfect trackability of the plant (3.12) on*** $\mathfrak{T}_0 \times \mathfrak{D}^j \times \mathfrak{Y}^k_d$ ***and the perfect trackability on*** $\mathfrak{T}_0 \times \mathfrak{D}^j \times \mathfrak{U}^l \times \mathfrak{Y}^k_d$

i) For the plant (3.12) to be the k-th-order perfect trackable on $\mathfrak{T}_0 \times \mathfrak{D}^j \times \mathfrak{U}^l \times \mathfrak{Y}^k_d$ *it is necessary and sufficient that every* $\mathbf{Y}_d(.) \in \mathfrak{Y}^k_d$ *is realizable on* $\mathfrak{T}_0 \times \mathfrak{D}^j \times \mathfrak{U}^l \times \mathfrak{Y}^k_d$, *equivalently, to be perfect trackable on* $\mathfrak{T}_0 \times \mathfrak{D}^j \times \mathfrak{U}^l \times \mathfrak{Y}^k_d$.

ii) For the plant (3.12) to be the k-th-order elementwise perfect trackable on $\mathfrak{T}_0 \times \mathfrak{D}^j \times \mathfrak{U}^l \times \mathfrak{Y}^k_d$ *it is necessary and sufficient that every* $\mathbf{Y}_d(.) \in \mathfrak{Y}^k_d$ *is elementwise realizable on* $\mathfrak{T}_0 \times \mathfrak{D}^j \times \mathfrak{U}^l \times \mathfrak{Y}^k_d$, *equivalently, to be elementwise perfect trackable on* $\mathfrak{T}_0 \times \mathfrak{D}^j \times \mathfrak{U}^l \times \mathfrak{Y}^k_d$.

iii) In order for the desired output behavior $\mathbf{Y}_d(.) \in \mathfrak{Y}^k_d$ *of the plant (3.12) to be the k-th-order elementwise perfect trackable on* $\mathfrak{T}_0 \times \mathfrak{D}^i \times \mathfrak{U}^\mu \times \mathfrak{Y}^k_d$ *under the action of the minimal control* $\mathbf{U}(.)$ *it is necessary and sufficient that every* $\mathbf{Y}_d(.) \in \mathfrak{Y}^k_d$ *is elementwise realizable on* $\mathfrak{T}_0 \times \mathfrak{D}^j \times \mathfrak{U}^l \times \mathfrak{Y}^k_d$ *under the action of the minimal control* $\mathbf{U}(.)$, *equivalently, to be elementwise perfect trackable on* $\mathfrak{T}_0 \times \mathfrak{D}^j \times \mathfrak{U}^l \times \mathfrak{Y}^k_d$ *under the action of the minimal control* $\mathbf{U}(.)$.

These lemmas discover the equivalence between the realizability of the plant desired output and the plant perfect trackability. The type of the plant, i.e., the form of its mathematical model, governs the form of the realizability conditions.

Except for the existence requirement, the preceding definitions do not impose any other condition on the control vector function $\mathbf{U}(.)$. Its existence means that its instantaneous vector value $\mathbf{U}(t)$ is defined at every moment $t \in \mathfrak{T}_0$. This permits piecewise continuity of $\mathbf{U}(t)$; i.e., it allows $\mathbf{U}(t) \in \mathfrak{C}^-(\mathfrak{T}_0)$. A piecewise continuous variable can only be a mathematical, but not a physical variable. It is not exactly physically realizable, which is explained by *PCUP* (Principles 17 and 18). In order to be physically realizable, control variable $\mathbf{U}(.)$ should obey *PCUP*, equivalently *TCUP* (Principle 20).

The preceding definitions determine the control vector function $\mathbf{U}(.)$ in terms of the disturbance vector function $\mathbf{D}(.)$.

7.2 Perfect natural trackability

The vector form and the instantaneous value of the disturbance variable $\mathbf{D}(.)$ are most often unknown, unpredictable, and their values can only be unmeasurable. These disturbance features cause the problem of control realization if control is determined in terms of $\mathbf{D}(.)$.

Problem 143 ***Disturbance and the control synthesis problem***

Do the plant properties enable a control synthesis without using information about the real form and the value of the disturbance vector $\mathbf{D}(t)$ *at any* $t \in \mathfrak{T}_0$*? Do they enable that for every* $\mathbf{D}(.) \in \mathfrak{D}^j$*?*

Mathematical models of plants, which are the starting point for the control synthesis, are approximative both qualitatively (due to their nonlinear nature, their forms, and the dynamical complexity) and quantitatively (due to their order and parameter values).

Problem 144 ***Plant state and the control synthesis problem***

Is it possible to determine control without knowing the mathematical model of the plant state? Do the properties of the plant permit the existence of such control?

Comment 145 *Nature (e.g., the brain as a natural controller) does not use any information about the mathematical model of the plant (of any organ) in order to create very effective time-continuous control (of the organ). Moreover, nature (the brain) often does not need precise, or any, information about the forms and/or the values of disturbances. Such control exists. Nature creates such control. We call it* ***natural control*** *(NC) regardless of the controller physical nature and regardless of the creator of the controller. Natural control is the* ***natural tracking control*** *(NTC) if, and only if, it ensures a kind of tracking and its implementation does not need any information about the form and the value of any* $\mathbf{D}(.) \in \mathfrak{D}^j$ *and about the mathematical model of the plant state.*

In order to reply to the preceding questions we introduce the following definition.

Definition 146 ***The k-th-order perfect natural trackability of the plant (3.12) on*** $\mathfrak{T}_0 \times \mathfrak{D}^j \times \mathfrak{U}^l \times \mathfrak{Y}_d^k$

a) The plant (3.12) is ***the k-th-order perfect natural trackable on*** $\mathfrak{T}_0 \times \mathfrak{D}^j \times \mathfrak{U}^l \times \mathfrak{Y}_d^k$ *if, and only if, for every* $[\mathbf{D}(.), \mathbf{Y}_d(.)] \in \mathfrak{D}^j \times \mathfrak{Y}_d^k$ *there exists a control vector function* $\mathbf{U}(.) \in \mathfrak{U}^l$*, that obeys TCUP 20 on* $\mathfrak{T}_0$*, which can be synthesized without using information about the form and the value of any* $\mathbf{D}(.) \in \mathfrak{D}^j$ *and about the mathematical model of the plant state, such that the plant real output and its first k derivatives are equal to the plant desired output and its first k derivatives on* $\mathfrak{T}_0$*, respectively, as soon as* $\mathbf{Y}^k(t_0) = \mathbf{Y}_d^k(t_0)$*,*

b) ***The domain*** $\mathcal{D}_{PNTbl}^k\left(t_0; \mathfrak{D}^j; \mathbf{U}; \mathfrak{Y}_d^k\right)$ ***of the k-th-order perfect natural trackability of the plant*** *(3.12)* ***on*** $\mathfrak{T}_0 \times \mathfrak{D}^j \times \mathfrak{U}^l \times \mathfrak{Y}_d^k$ *is the largest set* $\mathfrak{P} = \mathfrak{P}\left(t_0; \mathfrak{D}^j; \mathbf{U}; \mathfrak{Y}_d^k\right)$ *with the nonempty interior,* $\mathfrak{P} \subseteq \mathfrak{R}^{(k+1)N}$ *and In* $\mathfrak{P} \neq \phi$*, such that (7.4) holds for every* $\mathbf{Y}_{d0}^k = \mathbf{Y}_d^k(t_0) \in \mathfrak{P}$*.*

c) If, and only if, (7.3) holds for every $t_0 \in \mathfrak{T}_i$ *so that the following intersection*

$$\cap \left[\mathcal{D}_{PNTbl}^k\left(t_0; \mathfrak{D}^j; \mathbf{U}; \mathfrak{Y}_d^k\right) : t_0 \in \mathfrak{T}_i\right]$$

is a connected nonempty set then the desired output vector function $\mathbf{Y}_d(.)$ *is the* ***k-th-order*** t_0***-uniform perfect natural trackable on*** $\mathfrak{T}_0 \times \mathfrak{T}_i \times \mathfrak{D}^j \times \mathfrak{U}^l \times \mathfrak{Y}_d^k$*. If this holds when* $\mathfrak{T}$ *replaces* $\mathfrak{T}_0$ *then the expression "****on*** $\mathfrak{T}_0\times$*" is to be omitted.*

d) The preceding perfect trackability properties are ***elementwise*** *if, and only if, the control vector* $\mathbf{U}$ *can act on every entry* Y_i*,* $\forall i = 1, 2, \cdots, N$*, of* $\mathbf{Y}$ *mutually independently.*

Note 147 *The k-th-order perfect natural trackability of the plant (3.12) on* $\mathfrak{T}_0 \times \mathfrak{D}^j \times \mathfrak{U}^l \times \mathfrak{Y}_d^k$ *is uniform in* $\mathbf{D}(.) \in \mathfrak{D}^j$ *by its Definition 146.*

Corollary 148 ***The k-th-order perfect trackability on*** $\mathfrak{T}_0 \times \mathfrak{D}^j \times \mathfrak{U}^l \times \mathfrak{Y}_d^k$ ***and the k-th-order perfect natural trackability on*** $\mathfrak{T}_0 \times \mathfrak{D}^j \times \mathfrak{U}^l \times \mathfrak{Y}_d^k$

Definition 140 and Definition 146 imply that the k-th-order perfect trackability on $\mathfrak{T}_0 \times \mathfrak{D}^j \times \mathfrak{U}^l \times \mathfrak{Y}_d^k$ *is necessary for the k-th-order perfect natural trackability on* $\mathfrak{T}_0 \times \mathfrak{D}^j \times \mathfrak{U}^l \times \mathfrak{Y}_d^k$*, and that the k-th-order perfect natural trackability on* $\mathfrak{T}_0 \times \mathfrak{D}^j \times \mathfrak{U}^l \times \mathfrak{Y}_d^k$ *is sufficient for the k-th-order perfect trackability on* $\mathfrak{T}_0 \times \mathfrak{D}^j \times \mathfrak{U}^l \times \mathfrak{Y}_d^k$*.*

Chapter 8

Perfect trackability criteria

8.1 Output space criteria

Let us explore which properties of the plant (3.12) ensure its perfect trackability. Let us discover those plant properties that enable the existence of a control that can force the plant to exhibit perfect tracking as soon as the initial real output vector is equal to the initial desired output vector.

If the number r of the control variables U_i is less than the number N, (ρ), of the output variables Y_i (of the state variables R_i), respectively, then the control vector function $\mathbf{U}(.)$ cannot govern mutually independently the N, (ρ), output variables Y_i (state variables R_i), respectively (Axiom 87). The solution is to accept for control to govern sets of mutually functionally interrelated output variables Y_i (state variables R_i), respectively. A suitably chosen vector function $\mathbf{v}(.)$ will express such mutual functional interrelation of the output variables Y_i (state variables R_i), respectively. We illustrate this for N output variables Y_i, i.e., for the output vector $\mathbf{U} \in \mathfrak{R}^N$.

Let $\mathbf{v}(.) : \mathfrak{T}_0 \times \mathfrak{R}^{(k+1)N} \longrightarrow \mathfrak{R}^m$ be a function of the extended error vector $\boldsymbol{\varepsilon}^k$ such that

$$\begin{gathered} 0 < m < N : \mathbf{v}\left(t, \boldsymbol{\varepsilon}^k\right) = \mathbf{0}_m \Longleftrightarrow \boldsymbol{\varepsilon}^k = \mathbf{0}_{(k+1)N}, \quad \forall t \in \mathfrak{T}, \\ m = N \;\; and \;\; k = 0 \Longrightarrow \mathbf{v}(t, \boldsymbol{\varepsilon}) = \boldsymbol{\varepsilon}, \quad \forall t \in \mathfrak{T}, \end{gathered} \tag{8.1}$$

An example of such function $\mathbf{v}(.)$ is given by

$$\mathbf{v}\left(t, \boldsymbol{\varepsilon}^k\right) \left\{ \begin{array}{l} = \Psi(t; t_0; k) \left|\boldsymbol{\varepsilon}^k(t)\right|, \quad k \in \{0, 1, \cdots, \alpha - 1\}, \; 0 < m < N, \\ = \boldsymbol{\varepsilon} = \boldsymbol{\varepsilon}^0, \; k = 0 \;\; and \;\; m = N \end{array} \right\}, \tag{8.2}$$

where

$$\boldsymbol{\varepsilon} = [\varepsilon_1 \;\; \varepsilon_2 \;\; \ldots \;\; \varepsilon_N]^T \Longrightarrow |\boldsymbol{\varepsilon}| = [|\varepsilon_1| \;\; |\varepsilon_2| \;\; \cdots \;\; |\varepsilon_N|]^T \in [\mathbf{0}_N, \infty \mathbf{1}_N],$$

$$\left|\boldsymbol{\varepsilon}^{(l)}(t)\right| = \left[\left|\varepsilon_1^{(l)}\right| \;\; \left|\varepsilon_2^{(l)}\right| \cdots \;\; \left|\varepsilon_N^{(l)}\right|\right]^T \in \mathfrak{R}^{+^N}, \tag{8.3}$$

$$\left|\boldsymbol{\varepsilon}^k(t)\right| = \left[|\boldsymbol{\varepsilon}|^T \;\; \left|\boldsymbol{\varepsilon}^{(1)}\right|^T \cdots \;\; \left|\boldsymbol{\varepsilon}^{(k)}\right|^T\right]^T \in \mathfrak{R}^{+^{(k+1)N}}, \tag{8.4}$$

and $[m \times (k+1)N]$ matrix $\Psi(t; t_0; k) \in \mathfrak{R}^{+^{m \times (k+1)N}}$:

$$\begin{gathered} \Psi(t; t_0; k) = [\Psi_1(t; t_0) \;\; \Psi_2(t; t_0) \;\; \cdots \;\; \Psi_k(t; t_0) \;\; \Psi_{k+1}(t; t_0)], \\ \Psi_l(.) = \left[\psi_{ij}^l(.)\right] : \mathfrak{T}_0 \times \mathfrak{T} \longrightarrow \mathfrak{R}^{+(m \times N)}, \end{gathered} \tag{8.5}$$

$$\left\{\begin{array}{c} N>m>0:\left\{\begin{array}{c} \psi_{ij}^{l}=\mathfrak{R}^{+},\ \forall i=1,2,\ ..,m, \\ \forall j=1,2,\cdots,N,\ \forall l\in\{1,2,\cdots,k+1\} \end{array}\right\}, \\ m=N,\ k=0:\left\{\begin{array}{c} \psi_{ij}^{1}=\delta_{ij},\ \forall i,j=1,2,\cdots,N, \\ \delta_{ij}=1 \text{ if } i=j,\ \delta_{ij}=0 \text{ if } i\neq j, \\ \forall i,j=1,2,\cdots,N, \end{array}\right\} \end{array}\right\}. \tag{8.6}$$

For example:

$$\nu\in\mathfrak{R}^{+},$$

$$\Psi_{l}\left(t;t_{0}\right)=\frac{1}{\nu^{k}}\left(1+\frac{|t-t_{0}|}{(|t|+|t_{0}|)}\right)^{k}\left[\begin{array}{ccccc} \psi_{11}^{l} & \psi_{12}^{l} & \ldots & \psi_{1.N-1}^{l} & \psi_{1.N}^{l} \\ \psi_{21}^{l} & \psi_{22}^{l} & \ldots & \psi_{2.N-1}^{l} & \psi_{2.N}^{l} \\ \ldots & \ldots & \vdots & \ldots & \ldots \\ \psi_{m1}^{l} & \psi_{m2}^{l} & \ldots & \psi_{m.N-1}^{l} & \psi_{m.N}^{l} \end{array}\right]. \tag{8.7}$$

Notice that $\Psi\left(t;t_0;0\right)$ is the $N\times N$ identity matrix I_N: $\Psi\left(t;t_0;0\right)\equiv I_N$.

Lemma 142 simplifies the study of the k-th-order perfect trackability of the plant on $\mathfrak{T}_0 \times \mathfrak{D}^j \times \mathfrak{Y}_d$ by reducing it to the study of the perfect trackability of the plant on $\mathfrak{T}_0 \times \mathfrak{D}^j \times \mathfrak{Y}_d$.

The rank of $W\left(t\right)$ is denoted by γ in (3.43); i.e.,

$$\gamma=\text{rank } W\left(t\right)\leq\min\ \left(N,r\right),\ \forall t\in\mathfrak{T}. \tag{8.8}$$

The dimension m of the vector function (8.1) cannot exceed γ,

$$0<m\leq\gamma. \tag{8.9}$$

Theorem 149 ***The conditions for the perfect trackability in the output space on the product set*** $\mathfrak{T}_0 \times \mathfrak{D}^j \times \mathfrak{Y}_d$ ***of the plant*** *(3.12)* ***with Properties*** *57, 58, 78*

In order for the plant (3.12) with Properties 57, 58, 78, to be perfect trackable via the output space on the product set $\mathfrak{T}_0 \times \mathfrak{D}^j \times \mathfrak{Y}_d$ *it is necessary and sufficient that both i) and ii) hold.*

i) The rank γ *of the control matrix* $W\left(t\right)$ *is greater than zero and cannot be greater than min* $\left(N,r\right)$ *:*

$$rank\ W\left(t\right)=\gamma\in\left\{1,2,\cdots,min\ \left(N,r\right)\right\}\ \ \forall t\in\mathfrak{T}_{0}, \tag{8.10}$$

and for the perfect elementwise trackability in the output space $m=\gamma=N\leq r$,

ii) The control vector function $\mathbf{U}\left(.\right)$ *satisfies one of the conditions 1) to 4):*

1) If γ *=rank* $W\left(t\right)=r<N$, $\forall t\in\mathfrak{T}_0$, *then the control vector function* $\mathbf{U}\left(.\right)$ *obeys (8.11) for the chosen both vector function* $\mathbf{v}\left(.\right):\mathfrak{T}_0\times\mathfrak{R}^N\longrightarrow\mathfrak{R}^m$, *(8.1) and a matrix function* $M_{r,m}\left(.\right):\mathfrak{T}_0\longrightarrow\mathfrak{R}^{r\times m}$, *where* m *is not greater than* r, $m\leq r$, *rank* $M_{r,m}\left(t\right)\equiv m$, *and* $M_{r,r}\left(t\right)\equiv I_r$*:*

$$\begin{gathered}\mathbf{U}\left(t\right)=\mathbf{w}^{I}\left\{\Gamma\left(t\right)\left[\mathbf{Y}-\mathbf{z}(t,\mathbf{R}^{\alpha-1},\mathbf{D})\right]+M_{r,m}\left(t\right)\mathbf{v}\left(t,\varepsilon\right)\right\} \\ \Gamma\left(t\right)=\left[W^{T}\left(t\right)W\left(t\right)\right]^{-1}W^{T}\left(t\right)\in\mathfrak{R}^{r\times N},\ \ \forall t\in\mathfrak{T}_{0},\end{gathered} \tag{8.11}$$

2) If γ *=rank* $W\left(t\right)=N<r$, $\forall t\in\mathfrak{T}_0$, *then the control vector function* $\mathbf{U}\left(.\right)$ *obeys (8.12) for the chosen both vector function* $\mathbf{v}\left(.\right):\mathfrak{T}_0\times\mathfrak{R}^N\longrightarrow\mathfrak{R}^m$, *(8.1) and a matrix function* $M_{r,m}\left(.\right):\mathfrak{T}_0\longrightarrow\mathfrak{R}^{r\times m}$, *rank* $M_{N,m}\left(t\right)\equiv m$ *where* m *is not greater than* N, $m\leq N$; $m=N$ *if the perfect trackability is also elementwise:*

$$\begin{gathered}\mathbf{U}\left(t\right)=\mathbf{w}^{I}\left\{\Gamma\left(t\right)\left[\mathbf{Y}-\mathbf{z}(t,\mathbf{R}^{\alpha-1},\mathbf{D})\right]+W^{T}\left(t\right)M_{N,m}\left(t\right)\mathbf{v}\left(t,\varepsilon\right)\right\} \\ \Gamma\left(t\right)=W^{T}\left(t\right)\left[W\left(t\right)W^{T}\left(t\right)\right]^{-1}\in\mathfrak{R}^{r\times N},\ \ \forall t\in\mathfrak{T}_{0},\end{gathered} \tag{8.12}$$

3) If $\gamma = $*rank* $W(t) = r = N$, $\forall t \in \mathfrak{T}_0$, *then the control vector function* $\mathbf{U}(.)$ *obeys (8.13):*

$$\mathbf{U}(t) = \mathbf{w}^I \left\langle W(t)^{-1} \left\{ \mathbf{Y}_d(t) - \mathbf{z}(\left[t, \mathbf{R}^{\alpha-1}(t), \mathbf{D}(t)\right]\right\} \right\rangle, \ \forall t \in \mathfrak{T}_0, \tag{8.13}$$

4) If the plant (3.12) also possesses Property 64 then the control vector function $\mathbf{U}(.)$ *obeys (8.14) for the chosen vector function* $\mathbf{v}(.) : \mathfrak{T}_0 \times \mathfrak{R}^N \longrightarrow \mathfrak{R}^m$, *(8.1), where* m *is not greater than* γ, $m \leq \gamma$,

$$\begin{gathered}\mathbf{w}_\gamma \left[\mathbf{U}_\gamma(t)\right] = M_{\gamma,m}(t)\, \mathbf{v}(t, \varepsilon) + W_\gamma^{-1} \left[\mathbf{Y}_\gamma - \mathbf{z}_\gamma(t, \mathbf{R}^{\alpha-1}, \mathbf{D})\right], \\ \forall \left[t, \mathbf{D}(.), \mathbf{Y}_d(.)\right] \in \mathfrak{T}_0 \times \mathfrak{D}^j \times \mathfrak{Y}_d,\end{gathered} \tag{8.14}$$

and for an accepted matrix function $M(.) : \mathfrak{T} \longrightarrow \mathfrak{R}^{r \times m}$ *that satisfies (8.15):*

$$\begin{gathered}\text{rank } M(t) = \text{rank} \begin{bmatrix} M_{\gamma,m}(t) \\ M_{r-\gamma,m}(t) \end{bmatrix} = m \leq \gamma, \ M(t) \in \mathfrak{C}(\mathfrak{T}), \text{ and} \\ \text{rank } M_{\gamma,m}(t) = m, \quad \forall t \in \mathfrak{T}.\end{gathered} \tag{8.15}$$

The domain $\mathcal{D}_{PTbl}\left(t_0; \mathfrak{D}^j; \mathbf{U}; \mathfrak{Y}_d\right)$ *of the perfect trackability on* $\mathfrak{T}_0 \times \mathfrak{D}^j \times \mathfrak{Y}_d$ *of the plant (3.12) with Properties 57, 58, 78 and with conditions under:*
ii-1) is determined by (8.16):

$$\mathcal{D}_{PTbl}\left(t_0; \mathfrak{D}^j; \mathbf{U}; \mathfrak{Y}_d\right) = \mathcal{D}_{PTbl}\left(t_0; \mathfrak{D}^j; \mathfrak{U}; \mathfrak{Y}_d\right) = \left\{ \begin{array}{c} \left(\mathbf{R}^{\alpha-1}, \mathbf{Y}\right) \in \mathfrak{R}^{\alpha\rho} \times \mathfrak{R}^N : \\ \left| \left[W^T(t_0)\, W(t_0)\right]^{-1} W^T(t_0) \left[\mathbf{Y} - \mathbf{z}\left[t_0, \mathbf{R}^{\alpha-1}, \mathbf{D}_0\right]\right] \right| \\ \leq \max\left\{ \left|\mathbf{w}(\mathbf{U})\right| : \mathbf{U} \in \mathfrak{U} \right\}, \forall \mathbf{D} \in \mathfrak{D}^j, \end{array} \right\}. \tag{8.16}$$

ii-2) is determined by (8.17):

$$\mathcal{D}_{PTbl}\left(t_0; \mathfrak{D}^j; \mathbf{U}; \mathfrak{Y}_d\right) = \mathcal{D}_{PTbl}\left(t_0; \mathfrak{D}^j; \mathfrak{U}; \mathfrak{Y}_d\right) = \left\{ \begin{array}{c} \left(\mathbf{R}^{\alpha-1}, \mathbf{Y}\right) \in \mathfrak{R}^{\alpha\rho} \times \mathfrak{R}^N : \\ \left| W^T(t_0) \left[W(t_0)\, W^T(t_0)\right]^{-1} \left[\mathbf{Y} - \mathbf{z}\left[t_0, \mathbf{R}^{\alpha-1}, \mathbf{D}_0\right]\right] \right| \\ \leq \max\left\{ \left|\mathbf{w}(\mathbf{U})\right| : \mathbf{U} \in \mathfrak{U} \right\}, \forall \mathbf{D} \in \mathfrak{D}^j, \end{array} \right\}. \tag{8.17}$$

ii-3) is determined by (8.18):

$$\mathcal{D}_{PTbl}\left(t_0; \mathfrak{D}^j; \mathbf{U}; \mathfrak{Y}_d\right) = \mathcal{D}_{PTbl}\left(t_0; \mathfrak{D}^j; \mathfrak{U}; \mathfrak{Y}_d\right) = \left\{ \begin{array}{c} \left(\mathbf{R}^{\alpha-1}, \mathbf{Y}\right) \in \mathfrak{R}^{\alpha\rho} \times \mathfrak{R}^N : \\ \left| W^{-1}(t_0) \left[\mathbf{Y}_{d0} - \mathbf{z}\left[t_0, \mathbf{R}^{\alpha-1}, \mathbf{D}_0\right]\right] \right| \\ \leq \max\left\{ \left|\mathbf{w}(\mathbf{U})\right| : \mathbf{U} \in \mathfrak{U} \right\}, \forall \mathbf{D} \in \mathfrak{D}^j, \end{array} \right\}. \tag{8.18}$$

ii-4) is determined by (8.19):

$$\mathcal{D}_{PTbl}\left(t_0; \mathfrak{D}^j; \mathbf{U}; \mathfrak{Y}_d\right) = \mathcal{D}_{PTbl}\left(t_0; \mathfrak{D}^j; \mathfrak{U}; \mathfrak{Y}_d\right) = \left\{ \begin{array}{c} \left(\mathbf{R}^{\alpha-1}, \mathbf{Y}\right) \in \mathfrak{R}^{\alpha\rho} \times \mathfrak{R}^N : \\ \left| W_\gamma^{-1}(t_0) \left[\mathbf{Y}_\gamma - \mathbf{z}_\gamma\left[t_0, \mathbf{R}^{\alpha-1}, \mathbf{D}_0\right]\right] \right| \\ \leq \max\left\{ \left|\mathbf{w}(\mathbf{U})\right| : \mathbf{U} \in \mathfrak{U} \right\}, \forall \mathbf{D} \in \mathfrak{D}^j. \end{array} \right\}. \tag{8.19}$$

The proof of this theorem is in Appendix section C.2.

Note 150 *The control* $\mathbf{U}(t)$, *(8.11)-(8.14), is* ***the perfect trackability control on the product spaces*** $\mathfrak{T}_0 \times \mathfrak{D}^j \times \mathfrak{Y}_d$ ***in the output space.***

Note 151 *This theorem permits the lesser number r of control variables than the number N of the output variables: $r < N$. This reflects the fundamental control principle 87.*

Note 152 *If $\gamma = r = N = m$ then we can set $\mathbf{v}(t,\varepsilon) \equiv \boldsymbol{\varepsilon}(t) = \mathbf{Y}_d(t) - \mathbf{Y}(t)$. In that case all output variables are controlled mutually independently, i.e., the output vector variable $\mathbf{Y}(t)$ is controlled elementwise. This reflects the fundamental control Principle 87, too.*

Comment 153 *The preceding theorem is inapplicable if rank $W(t) = 0$.*

8.2 State space criteria

If $W(t) \equiv O_{N,r}$ then the control acts on the output vector of the plant (3.12) only indirectly via $\mathbf{R}$ and its derivatives at most up to $\mathbf{R}^{\alpha-1}$. The vector $\mathbf{R}$ then assumes the control law task. In this case the control synthesis takes place in the state space. This requires the knowledge of the desired state behavior $\mathbf{R}_d^{\alpha-1}(.)$ induced by the desired output behavior $\mathbf{Y}_d(.)$ on $\mathfrak{T}_0$ for every $\mathbf{Y}_d(.) \in \mathfrak{Y}_d$.

Theorem 154 ***The conditions for the perfect trackability, via the state space, on $\mathfrak{T}_0 \times \mathfrak{D}^j \times \mathfrak{Y}_d$ of the plant** (3.12) **with Properties** 57, 58, 78*

In order for the plant (3.12) with Properties 57, 58, 78 to be perfect trackable via the state space on the product set $\mathfrak{T}_0 \times \mathfrak{D}^j \times \mathfrak{Y}_d$, it is necessary and sufficient that both i) and ii) hold:

i) The rank β of the control matrix $P(t)$ is greater than zero and cannot be greater than min (ρ, r):

$$\text{rank } P(t) = \beta \in \{1, 2, \cdots, \min(\rho, r)\}, \quad \forall t \in \mathfrak{T}_0, \tag{8.20}$$

ii) The control vector function $\mathbf{U}(.)$ satisfies one of the conditions 1) to 4):

1) If $\beta =$rank $P(t) = r < \rho$, $\forall t \in \mathfrak{T}_0$, then the control vector function $\mathbf{U}(.)$ obeys (8.21) for the chosen both vector function $\mathbf{v}(.) : \mathfrak{T}_0 \times \mathfrak{R}^N \longrightarrow \mathfrak{R}^m$, (8.1) and a matrix function $L_{r,m}(.) : \mathfrak{T}_0 \longrightarrow \mathfrak{R}^{r\times m}$, where m is not greater than r, $m \leq r$, rank $L_{r,m}(t) \equiv m$ and $L_{r,r}(t) \equiv I_r$:

$$\mathbf{e}_u\left[\mathbf{U}^{\mu}(t)\right] = \left\{ \begin{array}{c} \Gamma(t) \left[\begin{array}{c} Q\left[t, \mathbf{R}^{\alpha-1}(t)\right] \mathbf{R}^{(\alpha)}(t) \\ + \mathbf{q}\left[t, \mathbf{R}^{\alpha-1}(t)\right] - E(t)\,\mathbf{e}_d\left[\mathbf{D}^{\eta}(t)\right] \end{array} \right] \\ + L_{r,m}(t)\,\mathbf{v}(t,\varepsilon) \end{array} \right\},$$
$$\Gamma(t) = \left[P^T(t)\,P(t)\right]^{-1} P^T(t), \quad \forall t \in \mathfrak{T}_0. \tag{8.21}$$

2) If $\beta =$rank $P(t) = \rho < r$, $\forall t \in \mathfrak{T}_0$, then the control vector function $\mathbf{U}(.)$ obeys (8.22) for the chosen both vector function $\mathbf{v}(.) : \mathfrak{T}_0 \times \mathfrak{R}^N \longrightarrow \mathfrak{R}^m$, (8.1) and a matrix function $L_{\rho,m}(.) : \mathfrak{T}_0 \longrightarrow \mathfrak{R}^{\rho\times m}$, where m is not greater than ρ, $m \leq \rho$, and rank $L_{\rho,m}(t) \equiv m$:

$$\mathbf{e}_u\left[\mathbf{U}^{\mu}(t)\right] = \left\{ \begin{array}{c} \Gamma(t) \left[\begin{array}{c} Q\left[t, \mathbf{R}^{\alpha-1}(t)\right] \mathbf{R}^{(\alpha)}(t) \\ + \mathbf{q}\left[t, \mathbf{R}^{\alpha-1}(t)\right] - E(t)\,\mathbf{e}_d\left[\mathbf{D}^{\eta}(t)\right] \end{array} \right] \\ + P^T(t)\,L_{\rho,m}(t)\,\mathbf{v}(t,\varepsilon) \end{array} \right\},$$
$$\Gamma(t) = P^T(t)\left[P(t)\,P^T(t)\right]^{-1}, \quad \forall t \in \mathfrak{T}_0, \tag{8.22}$$

3) If $\beta =$ rank $P(t) = r = \rho$, $\forall t \in \mathfrak{T}_0$ then the control vector function $\mathbf{U}(.)$ obeys (8.23):

$$\mathbf{e}_u\left[\mathbf{U}^{\mu}(t)\right] = P^{-1}(t) \left\{ \begin{array}{c} Q\left[t, \mathbf{R}^{\alpha-1}(t)\right] \mathbf{R}^{(\alpha)}(t) + \mathbf{q}\left[t, \mathbf{R}^{\alpha-1}(t)\right] \\ -E(t)\,\mathbf{e}_d\left[\mathbf{D}^{\eta}(t)\right] + L_{\rho,m}(t)\,\mathbf{v}(t,\varepsilon) \end{array} \right\},$$
$$\forall t \in \mathfrak{T}_0, \tag{8.23}$$

4) If the plant (3.12) also possesses Property 61 then the control vector function $\mathbf{U}(.)$ *obeys (8.24) for a chosen vector function* $\mathbf{v}(.) : \mathfrak{T}_0 \times \mathfrak{R}^N \longrightarrow \mathfrak{R}^m$, *(8.1), where* m *is not greater than* β, $m \leq \beta$:

$$\begin{gathered} \mathbf{e}_{u\beta}\left[\mathbf{U}_{\beta}^{\mu}(t)\right] = L_{\beta,m}(t)\,\mathbf{v}(t,\boldsymbol{\varepsilon}) \\ +P_{\beta}^{-1}(t)\left\{\begin{array}{c} Q_{\beta,\rho}(t,\mathbf{R}^{\alpha-1})\mathbf{R}^{(\alpha)} \\ +\ \mathbf{q}_{\beta}(t,\mathbf{R}^{\alpha-1},\mathbf{D}^{i}) - E_{\beta,d}(t)\mathbf{e}_{d}\left[\mathbf{D}^{\eta}(t)\right] \end{array}\right\}, \\ \forall\left[t,\mathbf{D}(.),\mathbf{Y}_{d}(.)\right] \in \mathfrak{T}_0 \times \mathfrak{D}^{j} \times \mathfrak{Y}_d, \end{gathered} \tag{8.24}$$

and for an accepted matrix function $L(.)$,

$$\begin{gathered} L(.) : \mathfrak{T} \longrightarrow \mathfrak{R}^{r\times m},\ rank\ L(t) = m,\quad \forall t \in \mathfrak{T}_0, \\ Property\ 61 \Longrightarrow L(t) = \left[\begin{array}{c} L_{\beta,m}(t) \\ O_{r-\beta,m} \end{array}\right],\ rank\ L_{\beta,m}(t) = m \leq \beta,\quad \forall t \in \mathfrak{T}_0. \end{gathered} \tag{8.25}$$

The state vector function $\mathbf{R}(.)$ *is the solution of the plant (3.12) controlled by* $\mathbf{U}(t)$, *(8.24).*

The domain $\mathcal{D}_{PTbl}\left(t_0;\mathfrak{D}^j;\mathbf{U};\mathfrak{Y}_d\right)$ *of the perfect trackability on* $\mathfrak{T}_0 \times \mathfrak{D}^j \times \mathfrak{Y}_d$ *of the plant (3.12) with Properties 57, 58, 78 is determined by:*

ii-1) The equation (8.26):

$$\begin{gathered} \mathcal{D}_{PTbl}\left(t_0;\mathfrak{D}^j;\mathbf{U};\mathfrak{Y}_d\right) = \mathcal{D}_{PTbl}\left(t_0;\mathfrak{D}^j;\mathfrak{U};\mathfrak{Y}_d\right) \\ = \left\{\begin{array}{c} \mathbf{R}^{\alpha} \in \mathfrak{R}^{(\alpha+1)\rho} : \\ \left|\Gamma(t_0)\left\{\begin{array}{c} Q(t_0,\mathbf{R}^{\alpha-1})\mathbf{R}^{(\alpha)} + \mathbf{q}(t_0,\mathbf{R}^{\alpha-1},\mathbf{D}_0^{i}) \\ -\ E(t_0)\mathbf{e}_d\left(\mathbf{D}_0^{\eta}\right) \end{array}\right\}\right| \\ \leq \max\left\{\left|\mathbf{e}_{u\beta}\left(\mathbf{U}_0^{\mu}\right)\right| : \mathbf{U} \in \mathfrak{U}\right\},\ \forall \mathbf{D} \in \mathfrak{D}^j. \end{array}\right\}, \\ \Gamma(t_0) = \left[P^T(t_0)P(t_0)\right]^{-1}P^T(t_0). \end{gathered} \tag{8.26}$$

ii-2) The equation (8.27):

$$\begin{gathered} \mathcal{D}_{PTbl}\left(t_0;\mathfrak{D}^j;\mathbf{U};\mathfrak{Y}_d\right) = \mathcal{D}_{PTbl}\left(t_0;\mathfrak{D}^j;\mathfrak{U};\mathfrak{Y}_d\right) \\ = \left\{\begin{array}{c} \mathbf{R}^{\alpha} \in \mathfrak{R}^{(\alpha+1)\rho} : \\ \left|\Gamma^{-1}(t_0)\left\{\begin{array}{c} Q(t_0,\mathbf{R}^{\alpha-1})\mathbf{R}^{(\alpha)} + \mathbf{q}(t_0,\mathbf{R}^{\alpha-1},\mathbf{D}_0^{i}) \\ -E(t_0)\mathbf{e}_d\left(\mathbf{D}_0^{\eta}\right) \end{array}\right\}\right| \\ \leq \max\left\{\left|\mathbf{e}_{u\beta}\left(\mathbf{U}_0^{\mu}\right)\right| : \mathbf{U} \in \mathfrak{U}\right\},\ \forall \mathbf{D} \in \mathfrak{D}^j. \end{array}\right\}, \\ \Gamma(t_0) = P^T(t_0)\left[P(t_0)P^T(t_0)\right]^{-1}. \end{gathered} \tag{8.27}$$

ii-3) The equation (8.28):

$$\begin{gathered} \mathcal{D}_{PTbl}\left(t_0;\mathfrak{D}^j;\mathbf{U};\mathfrak{Y}_d\right) = \mathcal{D}_{PTbl}\left(t_0;\mathfrak{D}^j;\mathfrak{U};\mathfrak{Y}_d\right) \\ = \left\{\begin{array}{c} \mathbf{R}^{\alpha} \in \mathfrak{R}^{(\alpha+1)\rho} : \\ \left|P^{-1}(t_0)\left\{\begin{array}{c} Q(t_0,\mathbf{R}^{\alpha-1})\mathbf{R}^{(\alpha)} \\ +\ \mathbf{q}(t_0,\mathbf{R}^{\alpha-1},\mathbf{D}_0^{i}) \\ -\ E(t_0)\mathbf{e}_d\left(\mathbf{D}_0^{\eta}\right) \end{array}\right\}\right| \\ \leq \max\left\{\left|\mathbf{e}_{u\beta}\left(\mathbf{U}_0^{\mu}\right)\right| : \mathbf{U} \in \mathfrak{U}\right\},\ \forall \mathbf{D} \in \mathfrak{D}^j. \end{array}\right\}. \end{gathered} \tag{8.28}$$

ii-4) If the plant possesses also Property 61 then

$$\begin{gathered} \mathcal{D}_{PTbl}\left(t_0;\mathfrak{D}^j;\mathbf{U};\mathfrak{Y}_d\right) = \mathcal{D}_{PTbl}\left(t_0;\mathfrak{D}^j;\mathfrak{U};\mathfrak{Y}_d\right) \\ = \left\{\begin{array}{c} \mathbf{R}^{\alpha} \in \mathfrak{R}^{(\alpha+1)\rho} : \\ \left|P_{\beta}^{-1}(t_0)\left\{\begin{array}{c} Q_{\beta,\rho}(t_0,\mathbf{R}^{\alpha-1})\mathbf{R}^{(\alpha)} + \mathbf{q}_{\beta}(t_0,\mathbf{R}^{\alpha-1},\mathbf{D}_0^{i}) \\ -\ E_{\beta,d}(t_0)\mathbf{e}_d\left(\mathbf{D}_0^{\eta}\right) \end{array}\right\}\right| \leq \\ \leq \max\left\{\left|\mathbf{e}_{u\beta}\left(\mathbf{U}_0^{\mu}\right)\right| : \mathbf{U} \in \mathfrak{U}\right\},\ \forall \mathbf{D} \in \mathfrak{D}^j. \end{array}\right\}. \end{gathered} \tag{8.29}$$

The proof of this theorem is in Appendix C.3.

Note 155 *The control* $\mathbf{U}(t)$, *(8.24), is* ***the perfect trackability control on*** $\mathfrak{T}_0$ ***in the state space****.*

Comment 156 ***The plant trackability for*** $r < \rho$

The case ii-1) of this theorem shows that the plant is perfect trackable in state space although $r < \rho$ *as soon as rank* $P(t) > 0$, $\forall t \in \mathfrak{T}_0$, *(8.10). This broadens the results of [279].*

8.3 Both spaces criteria

Let a matrix $H(t) \in \mathfrak{R}^{(\beta+\gamma)\times m}$ be such that:

$$H(t) = \text{blockdiag}\left\{H_{\beta,m}(t) \quad H_{\gamma,m}(t)\right\}, \quad \forall t \in \mathfrak{T}_0,$$
$$\text{rank } H_{\beta,m}(t) = \text{rank } H_{\gamma,m}(t) = m \le \min\ (\ \beta, \gamma), \quad \forall t \in \mathfrak{T}_0. \tag{8.30}$$

If β =min (r, ρ) and γ =min (N, r) then control synthesis in both spaces is the direct union of the control synthesis in the output space and in the *state* space for the corresponding case. The cases when β and γ satisfy (8.31) are left open,

$$0 < \text{rank } P(t) \equiv \beta, \ 0 < \text{rank } W(t) \equiv \gamma,$$
$$\beta \le \min\ (r, \rho), \ \gamma \le \min\ (N, r), \ \beta + \gamma \le r. \tag{8.31}$$

Theorem 157 ***The conditions via both spaces for the perfect trackability on the product set*** $\mathfrak{T}_0 \times \mathfrak{D}^j \times \mathfrak{Y}_d$ ***of the plant*** *(3.12) with Properties 57, 58, 61, 64, 78*

For the plant (3.12) with Properties 57, 58, 61, 64, 78 to be perfect trackable on $\mathfrak{T}_0 \times \mathfrak{D}^j \times \mathfrak{Y}_d$ *via both spaces it is necessary and sufficient that both i) and ii) hold:*

i) The rank β *of the control matrix* $P(t)$ *obeys (8.20) holds, the rank* γ *of the control matrix* $W(t)$ *satisfies (8.10), and they jointly obey (8.31),*

and

ii) If the plant also possesses Property 61 and Property 64 then for any chosen matrix $H(t) \in \mathfrak{R}^{\theta\times m}$ *that obeys (8.30), the control vector function* $\mathbf{U}(.)$ *obeys (8.32),*

$$\left[\begin{array}{c} \mathbf{e}_{u\beta}\left[\mathbf{U}^{\mu}_{\beta}(t)\right] \\ \mathbf{w}_{\gamma}\left[\mathbf{U}_{\gamma}(t)\right] \end{array}\right] = \left[\begin{array}{c} H_{\beta,m}(t)\,\mathbf{v}(t,\varepsilon) \\ H_{\gamma,m}(t)\,\mathbf{v}(t,\varepsilon) \end{array}\right] + \left[\begin{array}{cc} P^{-1}_{\beta}(t) & O_{\beta,\gamma} \\ O_{\gamma,\beta} & W^{-1}_{\gamma}(t) \end{array}\right]$$
$$\bullet \left[\begin{array}{c} \left\{\begin{array}{c} Q_{\beta,\rho}\left[t, \mathbf{R}^{\alpha-1}(t)\right]\mathbf{R}^{(\alpha)}(t) \\ +\mathbf{q}_{\beta}\left[t, \mathbf{R}^{\alpha-1}(t)\right] - E_{\beta,d}(t)\mathbf{e}_{d}\left[\mathbf{D}^{\eta}(t)\right] \end{array}\right\} \\ \mathbf{Y}_{\gamma}(t) - \mathbf{z}_{\gamma}\left[t, \mathbf{R}^{\alpha-1}(t), \mathbf{D}(t)\right] \end{array}\right], \quad \forall t \in \mathfrak{T}_0. \tag{8.32}$$

The domain $\mathcal{D}_{PTbl}\left(t_0; \mathfrak{D}^j; \mathbf{U}; \mathfrak{Y}_d\right)$ *of the perfect trackability on* $\mathfrak{T}_0 \times \mathfrak{D}^j \times \mathfrak{Y}_d$ *of the plant (3.12) is determined by (8.33):*

$$\mathcal{D}_{PTbl}\left(t_0; \mathfrak{D}^j; \mathbf{U}; \mathfrak{Y}_d\right)$$
$$= \left\{\begin{array}{c} \left(\mathbf{R}^{(\alpha)}, \mathbf{Y}\right) : \\ \left|\begin{array}{c} \left[\begin{array}{cc} P^{-1}_{\beta}(t_0) & O_{\beta,\gamma} \\ O_{\gamma,\beta} & W^{-1}_{\gamma}(t_0) \end{array}\right] \\ \bullet \left[\begin{array}{c} \left\{\begin{array}{c} Q_{\beta,\rho}\left(t_0, \mathbf{R}^{\alpha-1}\right)\mathbf{R}^{(\alpha)} \\ +\,\mathbf{q}_{\beta}\left(t_0, \mathbf{R}^{\alpha-1}\right) - E_{\beta,d}(t_0)\mathbf{e}_{d}\left(\mathbf{D}^{\eta}_{0}\right) \end{array}\right\} \\ \mathbf{Y}_{\gamma} - \mathbf{z}_{\gamma}\left(t_0, \mathbf{R}^{\alpha-1}, \mathbf{D}_0\right) \end{array}\right] \end{array}\right| \\ \le \left[\begin{array}{c} \left|\mathbf{e}^{\mu}_{u\beta}\right|_{\max} \\ \left|\mathbf{w}_{\gamma}\right|_{\max} \end{array}\right] \end{array}\right\} \tag{8.33}$$

The proof of this theorem is in Appendix C.4.

Note 158 *The control* $\mathbf{U}(t)$, *(8.32), is* ***the perfect trackability control on*** $\mathfrak{T}_0$ ***in both spaces***.

The control vector function $\mathbf{U}(.)$ is determined in (8.32) by one differential nonlinear vector equation and one algebraic nonlinear vector equation. The nonlinearities $\mathbf{e}_{u\beta}(.)$ and $\mathbf{w}_{\gamma}(.)$ are different in general. If the equation (8.32) does not have a solution $\mathbf{U}(.)$ then the plant is not perfect trackable via both spaces. In order to reduce the problem to one differential equation and to present it in a compact clear form let:

$$\begin{aligned} \mathbf{q}^*\left[t,\mathbf{R}^{\alpha-1}(t)\right] &= Q^{-1}\left[t,\mathbf{R}^{\alpha-1}(t)\right]\mathbf{q}\left[t,\mathbf{R}^{\alpha-1}(t)\right], \\ E^*(t) &= Q^{-1}\left[t,\mathbf{R}^{\alpha-1}(t)\right]E(t), \\ P^*(t) &= Q^{-1}\left[t,\mathbf{R}^{\alpha-1}(t)\right]P(t), \\ A(t) &= \frac{\partial\mathbf{z}\left[t,\mathbf{R}^{\alpha-1}(t),\mathbf{D}(t)\right]}{\partial\mathbf{R}^{(\alpha-1)}}P^*(t), \end{aligned} \tag{8.34}$$

and

$$\Omega\left[t,\mathbf{R}^{\alpha-1}(t),\mathbf{D}(t)\right] = \left\{\begin{array}{c} \mathbf{Y}^{(1)}(t)-\frac{\partial\mathbf{z}\left[t,\mathbf{R}^{\alpha-1}(t),\mathbf{D}(t)\right]}{\partial t}-\frac{\partial\mathbf{z}\left[t,\mathbf{R}^{\alpha-1}(t),\mathbf{D}(t)\right]}{\partial\mathbf{R}^{(\alpha-1)}} \\ \bullet\left\{-\mathbf{q}^*\left[t,\mathbf{R}^{\alpha-1}(t)\right]+E^*(t)\,\mathbf{e}_d\left[\mathbf{D}^{\eta}(t)\right]\right\} \\ -\frac{\partial\mathbf{z}\left[t,\mathbf{R}^{\alpha-1}(t),\mathbf{D}(t)\right]}{\partial\mathbf{D}}\mathbf{D}^{(1)}(t) \end{array}\right\}. \tag{8.35}$$

Theorem 159 ***The conditions for the perfect trackability on*** $\mathfrak{T}_0\times\mathfrak{D}^j\times\mathfrak{Y}_d$ ***in both spaces of the regular plant (3.29) with Properties*** *57, 58, 61, 64, 78* ***and with differentiable function*** $\mathbf{z}(.)$

For the regular plant (3.12) with Properties 57, 58, 61, 64, 78 and with differentiable function $\mathbf{z}(.)$ *to be perfect trackable on* $\mathfrak{T}_0\times\mathfrak{D}^j\times\mathfrak{Y}_d$ *via both spaces it is necessary and sufficient that the equation (8.36):*

$$\begin{aligned} A(t)\,\mathbf{e}_u\left[\mathbf{U}^{\mu}(t)\right]+W(t)\,\mathbf{w}^{(1)}\left[\mathbf{U}(t)\right]+W^{(1)}(t)\,\mathbf{w}\left[\mathbf{U}(t)\right] \\ =\Omega\left[t,\mathbf{R}^{\alpha-1}(t),\mathbf{D}(t)\right]+B(t)\,\mathbf{v}(t,\varepsilon), \\ \forall\left[t,\mathbf{D}(.),\mathbf{Y}_d(.)\right]\in\mathfrak{T}_0\times\mathfrak{D}^j\times\mathfrak{Y}_d, \end{aligned} \tag{8.36}$$

has the control vector solution $\mathbf{U}(.)$ *for* $A(t)$*and* $\Omega\left[t,\mathbf{R}^{\alpha-1}(t),\mathbf{D}(t)\right]$ *determined by (8.34), (8.35), and for* $B(t)$ *(8.37),*

$$B(.):\mathfrak{T}\longrightarrow\mathfrak{R}^{N\times m},\ rank\ B(t)=m,\ \forall t\in\mathfrak{T}_0. \tag{8.37}$$

The domain $\mathcal{D}_{PTbl}\left(t_0;\mathfrak{D}^j;\mathbf{U};\mathfrak{Y}_d\right)$ *of the perfect trackability on* $\mathfrak{T}_0\times\mathfrak{D}^j\times\mathfrak{Y}_d$ *of the plant (3.12) with Properties 57, 58, 61, 64, 78 and with differentiable function* $\mathbf{z}(.)$ *is determined by:*

$$\mathcal{D}_{PTbl}\left(t_0;\mathfrak{D}^j;\mathbf{U};\mathfrak{Y}_d\right)=\mathcal{D}_{PTbl}\left(t_0;\mathfrak{D}^j;\mathfrak{U};\mathfrak{Y}_d\right)$$
$$=\left\{\begin{array}{c} \left(\mathbf{R}^{\alpha},\mathbf{Y}\right)\in\mathfrak{R}^{(\alpha+1)\rho}\times\mathfrak{R}^N: \\ \left|\Omega\left(t_0,\mathbf{R}^{\alpha-1},\mathbf{D}\right)\right| \\ \leq\max\left|\begin{array}{c} A(t_0)\,\mathbf{e}_u\left(\mathbf{U}_0^{\mu}\right)+W(t_0)\,\mathbf{w}^{(1)}\left(\mathbf{U}_0\right)+W_0^{(1)}\mathbf{w}\left(\mathbf{U}_0\right): \\ \mathbf{U}_0\in\mathfrak{U}^{\mu} \end{array}\right|, \\ \forall\mathbf{D}\in\mathfrak{D}^j \end{array}\right\}. \tag{8.38}$$

The proof of this theorem is worked out in Appendix C.5.

Note 160 *The control* $\mathbf{U}(t)$, *(8.36), is* ***the perfect trackability control on*** $\mathfrak{T}_0$ ***in both spaces when*** $\mathbf{z}(.)$ *is a differentiable function.*

If the function $\mathbf{z}(.)$ is only continuous but not differentiable, we can recall the special form (3.29) of the plant (3.12).

Let

$$\widehat{E}\left(t,\mathbf{R}^{\alpha-1},\mathbf{D}\right)=Z\left(t,\mathbf{R}^{\alpha-1},\mathbf{D}\right)F^{-1}\left[t,\mathbf{R}^{\alpha-1}(t)\right]E\left(t\right), \tag{8.39}$$

$$\widehat{Q}\left(t,\mathbf{R}^{\alpha-1},\mathbf{D}\right)=Z\left(t,\mathbf{R}^{\alpha-1},\mathbf{D}\right)F^{-1}\left[t,\mathbf{R}^{\alpha-1}(t)\right]Q\left[t,\mathbf{R}^{\alpha-1}(t)\right], \tag{8.40}$$

$$\widehat{P}\left(t,\mathbf{R}^{\alpha-1},\mathbf{D}\right)=Z\left(t,\mathbf{R}^{\alpha-1},\mathbf{D}\right)F^{-1}\left[t,\mathbf{R}^{\alpha-1}(t)\right]P\left(t\right). \tag{8.41}$$

Theorem 161 ***The conditions for the perfect trackability on*** $\mathfrak{T}_0\times\mathfrak{D}^j\times\mathfrak{Y}_d$ ***in both spaces of the plant (3.29) with Properties*** *57, 58, 78*

In order for the plant (3.29) with Properties 57, 58, 61, 64, 78, to be perfect trackable via both spaces on the product set $\mathfrak{T}_0\times\mathfrak{D}^j\times\mathfrak{Y}_d$ *, it is necessary and sufficient that the equation (8.42) has the control vector solution* $\mathbf{U}(.)$*:*

$$\begin{gathered}\widehat{P}\left(t,\mathbf{R}^{\alpha-1},\mathbf{D}\right)\mathbf{e}_u\left[\mathbf{U}^{\mu}(t)\right]+W\left(t\right)\mathbf{w}\left[\mathbf{U}(t)\right]\\ =\mathbf{Y}(t)-\widehat{E}\left(t,\mathbf{R}^{\alpha-1},\mathbf{D}\right)\mathbf{e}_d\left[\mathbf{D}^{\eta}(t)\right]\\ +\widehat{Q}\left(t,\mathbf{R}^{\alpha-1},\mathbf{D}\right)\mathbf{R}^{(\alpha)}(t)+B\left(t\right)\mathbf{v}\left(t,\varepsilon\right),\\ \forall\left[t,\mathbf{D}(.),\mathbf{Y}_d(.)\right]\in\mathfrak{T}_0\times\mathfrak{D}^j\times\mathfrak{Y}_d,\end{gathered} \tag{8.42}$$

for an accepted matrix function $B(.)$, *(8.37).*

The domain $\mathcal{D}_{PTbl}\left(t_0;\mathfrak{D}^j;\mathbf{U};\mathfrak{Y}_d\right)$ *of the perfect trackability on* $\mathfrak{T}_0\times\mathfrak{D}^j\times\mathfrak{Y}_d$ *of the plant (3.29) with Properties 57, 58, 61, 64, 78 is determined by (8.43):*

$$\mathcal{D}_{PTbl}\left(t_0;\mathfrak{D}^j;\mathbf{U};\mathfrak{Y}_d\right)=\mathcal{D}_{PTbl}\left(t_0;\mathfrak{D}^j;\mathfrak{U};\mathfrak{Y}_d\right)$$
$$=\left\{\begin{array}{c}\left(\mathbf{R}^{\alpha},\mathbf{Y}\right)\in\mathfrak{R}^{(\alpha+1)\rho}\times\mathfrak{R}^{N}:\\ \left|\begin{array}{c}\mathbf{Y}-\widehat{E}\left(t_0,\mathbf{R}^{\alpha-1},\mathbf{D}\right)\mathbf{e}_d\left(\mathbf{D}^{\eta}\right)\\ +\widehat{Q}\left(t,\mathbf{R}^{\alpha-1},\mathbf{D}\right)\mathbf{R}^{(\alpha)}(t)\end{array}\right|\\ \leq\max\left|\widehat{P}\left(t_0,\mathbf{R}_0^{\alpha-1},\mathbf{D}\right)\mathbf{e}_u\left(\mathbf{U}_0^{\mu}\right)+W\left(t_0\right)\mathbf{w}\left(\mathbf{U}_0\right)\right|,\\ \forall\mathbf{D}\in\mathfrak{D}^j\end{array}\right\}. \tag{8.43}$$

For the proof of this theorem see Appendix C.6.

Chapter 9

Perfect natural trackability criteria

9.1 Output space criteria

Definition 146 determines the perfect natural trackability of the plant (3.12) on $\mathfrak{T}_0 \times \mathfrak{D}^j \times \mathfrak{Y}_d$.

Theorem 162 ***The conditions for the perfect natural trackability on the product set*** $\mathfrak{T}_0 \times \mathfrak{D}^j \times \mathfrak{Y}_d$ ***of the plant*** *(3.12)* ***with Properties*** *57, 58, 78*

In order for the plant (3.12) with Properties 57, 58, 78 to be perfect natural trackable via the output space on $\mathfrak{T}_0 \times \mathfrak{D}^j \times \mathfrak{Y}_d$ *and for the control vector function* $\mathbf{U}(.)$ *to be perfect natural tracking control it is necessary and sufficient that both i) and ii) hold:*

i) The rank γ *of the control matrix* $W(t)$ *is greater than zero and cannot be greater than* $\min(N, r)$, *i.e., (8.10) holds, and for the perfect elementwise natural trackability in the output space* $m = \gamma = N \leq r$.

ii) The control vector function $\mathbf{U}(.)$ *satisfies one of the conditions 1) to 4):*

1) If $\gamma = \text{rank } W(t) = r < N$, $\forall t \in \mathfrak{T}_0$, *then the control vector function* $\mathbf{U}(.)$ *obeys (9.1) for the chosen both vector function* $\mathbf{v}(.) : \mathfrak{T}_0 \times \mathfrak{R}^N \longrightarrow \mathfrak{R}^m$, *(8.1) and matrix function* $M_{r,m}(.) : \mathfrak{T}_0 \longrightarrow \mathfrak{R}^{r \times m}$, *where* m *is not greater than* r, $m \leq r$, *rank* $M_{r,m}(t) \equiv m$ *and* $M_{r,r}(t) \equiv I_r$*:*

$$\mathbf{U}(t) = \mathbf{w}^I \left\{ \mathbf{w}\left[\mathbf{U}\left(t^-\right)\right] + M_{r,m}(t)\, \mathbf{v}(t, \boldsymbol{\varepsilon}) \right\}, \ \forall t \in \mathfrak{T}_0, \tag{9.1}$$

The domain $\mathcal{D}_{PNTbl}\left(t_0; \mathfrak{D}^j; \mathbf{U}; \mathfrak{Y}_d\right)$ *of the l-th-order perfect natural trackability of the plant (3.12) on* $\mathfrak{T}_0 \times \mathfrak{D}^j \times \mathfrak{Y}_d$ *is determined by (8.16).*

2) If $\gamma = \text{rank } W(t) = N < r$, $\forall t \in \mathfrak{T}_0$, *then the control vector function* $\mathbf{U}(.)$ *obeys (9.2) for the chosen both vector function* $\mathbf{v}(.) : \mathfrak{T}_0 \times \mathfrak{R}^N \longrightarrow \mathfrak{R}^m$, *(8.1) and matrix function* $M_{r,m}(.) : \mathfrak{T}_0 \longrightarrow \mathfrak{R}^{r \times m}$, *rank* $M_{N,m}(t) \equiv m$, $\forall t \in \mathfrak{T}_0$, *where* m *is not greater than* N, $m \leq N$, *and* $m = N$ *if the perfect trackability is also elementwise:*

$$\mathbf{U}(t) = \mathbf{w}^I \left\{ \Gamma(t)\, \mathbf{w}_N\left[\mathbf{U}_N\left(t^-\right)\right] + W^T(t)\, M_{N,m}(t)\, \mathbf{v}(t, \boldsymbol{\varepsilon}) \right\}$$
$$\Gamma(t) = W^T(t) \left[W(t)\, W^T(t)\right]^{-1} W(t) \in \mathfrak{R}^{r \times r}, \ \forall t \in \mathfrak{T}_0, \tag{9.2}$$

The domain $\mathcal{D}_{PNTbl}\left(t_0; \mathfrak{D}^j; \mathbf{U}; \mathfrak{Y}_d\right)$ *of the l-th-order perfect natural trackability of the plant (3.12) on* $\mathfrak{T}_0 \times \mathfrak{D}^j \times \mathfrak{Y}_d$ *is determined by (8.17).*

3) If $\gamma = \text{rank } W(t) = r = N$, $\forall t \in \mathfrak{T}_0$, *then the control vector function* $\mathbf{U}(.)$ *obeys (9.3):*

$$\mathbf{U}(t) = \mathbf{w}^I \left\langle \mathbf{w}\left[\mathbf{U}\left(t^-\right)\right] + W(t)^{-1} \boldsymbol{\varepsilon}(t) \right\rangle, \ \forall t \in \mathfrak{T}_0, \tag{9.3}$$

The domain $\mathcal{D}_{PNTbl}\left(t_0; \mathfrak{D}^j; \mathbf{U}; \mathfrak{Y}_d\right)$ *of the perfect natural trackability of the plant (3.12) on* $\mathfrak{T}_0 \times \mathfrak{D}^j \times \mathfrak{Y}_d$ *is determined by (8.18).*

4) If the plant also possesses Property 64 then the control vector function $\mathbf{U}(.)$ *obeys (9.4) for a chosen vector function* $\mathbf{v}(.) : \mathfrak{T}_0 \times \mathfrak{R}^N \longrightarrow \mathfrak{R}^N$, *(8.1),*

$$\mathbf{w}_\gamma\left[\mathbf{U}_\gamma(t)\right] = \mathbf{w}_\gamma\left[\mathbf{U}_\gamma(t^-)\right] + M_{\gamma,m}(t)\,\mathbf{v}(t,\varepsilon), \quad \forall t \in \mathfrak{T}_0, \tag{9.4}$$

and for an accepted matrix function $M(.) : \mathfrak{T} \longrightarrow \mathfrak{R}^{N\times m}$ *that satisfies (8.15).*

The domain $\mathcal{D}_{PNTbl}\left(t_0;\mathfrak{D}^j;\mathbf{U};\mathfrak{Y}_d\right)$ *of the perfect natural trackability of the plant (3.12) on* $\mathfrak{T}_0 \times \mathfrak{D}^j \times \mathfrak{Y}_d$ *is determined by (8.19).*

Appendix C.7 contains the proof of this theorem.

Comment 163 *Theorem 149 and Theorem 162 show that for the plant (3.12) to be perfect natural trackable on* $\mathfrak{T}_0 \times \mathfrak{D}^j \times \mathfrak{Y}_d$ *it is necessary and sufficient that the plant is perfect trackable on* $\mathfrak{T}_0 \times \mathfrak{D}^j \times \mathfrak{Y}_d$. *This completes Corollary 148.*

9.2 State space criteria

Theorem 164 ***The conditions for the perfect natural trackability, via the state space, on the product set*** $\mathfrak{T}_0 \times \mathfrak{D}^j \times \mathfrak{Y}_d$ ***of the plant*** *(3.12)* ***with Properties*** *57, 58, 78*

In order for the plant (3.12), which possesses Properties 57, 58, 78 to be perfect natural trackable via the state space on $\mathfrak{T}_0 \times \mathfrak{D}^j \times \mathfrak{Y}_d$ *it is necessary and sufficient that both:*

i) The rank β *of the control matrix* $P(t)$ *is greater than zero, i.e. (8.20) holds, and*

ii) The control vector function $\mathbf{U}(.)$ *satisfies one of the conditions 1)-4):*

1) If $\beta = \text{rank}\ P(t) = r < \rho$, $\forall t \in \mathfrak{T}_0$, *then the control vector function* $\mathbf{U}(.)$ *obeys (9.5) for the chosen both vector function* $\mathbf{v}(.) : \mathfrak{T}_0 \times \mathfrak{R}^N \longrightarrow \mathfrak{R}^m$, *(8.1) and matrix function* $L_{r,m}(.) : \mathfrak{T}_0 \longrightarrow \mathfrak{R}^{r\times m}$, *where* m *is not greater than* r, $m \leq r$, *rank* $L_{r,m}(t) \equiv m$ *and* $L_{r,r}(t) \equiv I_r$*:*

$$\mathbf{e}_u\left[\mathbf{U}^\mu(t)\right] = \mathbf{e}_u\left[\mathbf{U}^\mu(t^-)\right] + L_{r,m}(t)\,\mathbf{v}(t,\varepsilon), \ \forall t \in \mathfrak{T}_0. \tag{9.5}$$

The domain $\mathcal{D}_{PNTbl}\left(t_0;\mathfrak{D}^j;\mathbf{U};\mathfrak{Y}_d\right)$ *of the l-th-order perfect natural trackability of the plant (3.12) on* $\mathfrak{T}_0 \times \mathfrak{D}^j \times \mathfrak{Y}_d$ *is determined by (8.26).*

2) If $\beta = \text{rank}\ P(t) = \rho < r$, $\forall t \in \mathfrak{T}_0$, *then the control vector function* $\mathbf{U}(.)$ *obeys (9.6) for the chosen both vector function* $\mathbf{v}(.) : \mathfrak{T}_0 \times \mathfrak{R}^N \longrightarrow \mathfrak{R}^m$, *(8.1) and matrix function* $L_{r,m}(.) : \mathfrak{T}_0 \longrightarrow \mathfrak{R}^{r\times m}$, *where* m *is not greater than* r, $m \leq r$, *and rank* $L_{r,m}(t) \equiv m$*:*

$$\mathbf{e}_u\left[\mathbf{U}^\mu(t)\right] = \Gamma(t)\left[\mathbf{e}_u\left[\mathbf{U}^\mu(t^-)\right]\right] + P^T(t)\,L_{r,m}(t)\,\mathbf{v}(t,\varepsilon),$$
$$\Gamma(t) = P^T(t)\left[P(t)\,P^T(t)\right]^{-1}, \ \forall t \in \mathfrak{T}_0, \tag{9.6}$$

The domain $\mathcal{D}_{PNTbl}\left(t_0;\mathfrak{D}^j;\mathbf{U};\mathfrak{Y}_d\right)$ *of the l-th-order perfect natural trackability of the plant (3.12) on* $\mathfrak{T}_0 \times \mathfrak{D}^j \times \mathfrak{Y}_d$ *is determined by (8.27).*

3) If $\beta = \text{rank}\ P(t) = r = \rho$, $\forall t \in \mathfrak{T}_0$, *then the control vector function* $\mathbf{U}(.)$ *obeys (9.7):*

$$\mathbf{e}_u\left[\mathbf{U}^\mu(t)\right] = \ \mathbf{e}_u\left[\mathbf{U}^\mu(t^-)\right] + P(t)^{-1}\,L_{\rho,m}(t)\,\mathbf{v}(t,\varepsilon) \ , \ \forall t \in \mathfrak{T}_0, \tag{9.7}$$

The domain $\mathcal{D}_{PNTbl}\left(t_0;\mathfrak{D}^j;\mathbf{U};\mathfrak{Y}_d\right)$ *of the l-th-order perfect natural trackability of the plant (3.12) on* $\mathfrak{T}_0 \times \mathfrak{D}^j \times \mathfrak{Y}_d$ *is determined by (8.28).*

ii-4) If the plant (3.12) also possesses Property 61 then for the chosen both matrix $L_{\beta,m}(t) \in \mathfrak{R}^{\beta\times m}$*,(8.25), rank* $L_{\beta,m}(t) \equiv m$, *and vector function* $\mathbf{v}(.) : \mathfrak{T}_0 \times \mathfrak{R}^N \longrightarrow \mathfrak{R}^m$, *(8.1), the control vector function* $\mathbf{U}(.)$ *obeys (9.8),*

$$\mathbf{e}_{u\beta}\left[\mathbf{U}^\mu_\beta(t)\right] = \mathbf{e}_{u\beta}\left[\mathbf{U}^\mu_\beta(t^-)\right] + L_{\beta,m}(t)\,\mathbf{v}(t,\varepsilon), \ \forall t \in \mathfrak{T}_0, \tag{9.8}$$

for the accepted matrix function $L(.)$ such that.

The domain $\mathcal{D}_{PNTbl}\left(t_0; \mathfrak{D}^j; \mathbf{U}; \mathfrak{Y}_d\right)$ of the l-th-order perfect natural trackability of the plant (3.12) on $\mathfrak{T}_0 \times \mathfrak{D}^j \times \mathfrak{Y}_d$ is determined by (8.29).

The proof of this theorem is in Appendix C.8.

9.3 Both spaces criteria

Theorem 165 ***The conditions for the perfect natural trackability via both spaces on the product set $\mathfrak{T}_0 \times \mathfrak{D}^j \times \mathfrak{Y}_d$ of the plant** (3.12) **with Properties** 57, 58, 61, 64, 78*

For the plant (3.12) with Properties 57, 58, 61, 64, 78 to be perfect natural trackable on $\mathfrak{T}_0 \times \mathfrak{D}^j \times \mathfrak{Y}_d$ via both spaces it is necessary and sufficient that i) and ii) hold:

i) The ranks of the matrices $P(t)$ and $W(t)$ are always greater than zero and (8.31) is valid,

(ii) The control vector function $\mathbf{U}(.)$ obeys (9.9) for a chosen matrix $H_{\beta,\gamma}(t) \in \mathfrak{R}^{(\beta+\gamma)\times m}$ that obeys (8.30) and for the vector function $\mathbf{v}(.) : \mathfrak{T}_0 \times \mathfrak{R}^N \longrightarrow \mathfrak{R}^N$, (8.1):

$$\begin{bmatrix} \mathbf{e}_{u\beta}\left[\mathbf{U}_\beta^\mu(t)\right] \\ \mathbf{w}_\gamma\left[\mathbf{U}_\gamma(t)\right] \end{bmatrix} = \begin{bmatrix} \mathbf{e}_{u\beta}\left[\mathbf{U}_\beta^\mu(t^-)\right] \\ \mathbf{w}_\gamma\left[\mathbf{U}_\gamma(t^-)\right] \end{bmatrix} + H_{\beta,\gamma}(t)\,\mathbf{v}(t,\varepsilon),\ \forall t \in \mathfrak{T}_0. \tag{9.9}$$

The domain $\mathcal{D}_{PNTbl}\left(t_0; \mathfrak{D}^j; \mathbf{U}; \mathfrak{Y}_d\right)$ of the perfect natural trackability of the plant (3.12) on $\mathfrak{T}_0 \times \mathfrak{D}^j \times \mathfrak{Y}_d$ is determined by (8.33).

The proof of this theorem is the straight combination of the proof C.4 and the proof C.8.

Theorem 166 ***The conditions in both spaces for the perfect trackability on $\mathfrak{T}_0 \times \mathfrak{D}^j \times \mathfrak{Y}_d$ of the plant (3.29) with Properties** 57, 58, 61, 64, 78 **and with differentiable function $\mathbf{z}(.)$***

For the plant (3.12) with Properties 57, 58, 61, 64, 78 and with differentiable function $\mathbf{z}(.)$ to be perfect trackable on $\mathfrak{T}_0 \times \mathfrak{D}^j \times \mathfrak{Y}_d$ via both spaces it is necessary and sufficient that the equation (9.10) has the control vector solution $\mathbf{U}(.)$,

$$\begin{gathered} A(t)\,\mathbf{e}_u\left[\mathbf{U}(t)\right] + W(t)\,\mathbf{w}^{(1)}\left[\mathbf{U}(t)\right] + W^{(1)}(t)\,\mathbf{w}\left[\mathbf{U}(t)\right] = B(t)\,\mathbf{v}(t,\varepsilon) \\ +A(t^-)\,\mathbf{e}_u\left[\mathbf{U}_\beta^\mu(t^-)\right] + W(t^-)\,\mathbf{w}^{(1)}\left[\mathbf{U}(t^-)\right] + W^{(1)}(t^-)\,\mathbf{w}\left[\mathbf{U}(t)^-\right], \\ \forall\left[t, \mathbf{D}(.), \mathbf{Y}_d(.)\right] \in \mathfrak{T}_0 \times \mathfrak{D}^j \times \mathfrak{Y}_d, \end{gathered} \tag{9.10}$$

for $A(t)$ determined by (8.34) and (8.35) and for $B(t)$ (8.37).

The domain $\mathcal{D}_{PTbl}\left(t_0; \mathfrak{D}^j; \mathbf{U}; \mathfrak{Y}_d\right)$ of the perfect trackability on $\mathfrak{T}_0 \times \mathfrak{D}^j \times \mathfrak{Y}_d$ of the plant (3.12) with Properties 57, 58, 61, 64, 78 and with differentiable function $\mathbf{z}(.)$ is determined by (8.38).

The proof of this theorem is the straightforward combination of the proof C.7 and the proof C.8 in the Appendix C.

Comment 167 *The above control algorithms show that the control implementation does not need any information about the disturbance vector and about the plant state. The disturbance vector need not be measurable, which corresponds to reality.*

Chapter 10

Imperfect trackability concepts

10.1 Introduction to (imperfect) trackability

We generalize the definitions of imperfect trackability and of the imperfect natural trackability properties presented for *time*-invariant continuous-*time* linear systems in [279]. They take into account the most probable situation when the real output deviates from the desired output at the initial moment.

In order to examine completely the trackability property we explore whether the plant is trackable for every $[\mathbf{D}(.), \mathbf{Y}_d(.)] \in \mathfrak{D}^j \times \mathfrak{Y}_d^k$.

The *time* evolution of the state vector $\mathbf{R}^{\alpha-1}$ of the plant (3.12) determines completely the plant state. The output space $\mathfrak{R}^N$ is the natural environment to define the plant trackability properties and their domains.

The imperfect trackability is called, for short, ***the trackability.***

10.2 Trackability

Let

Definition 168 ***The k-th-order trackability of the plant** (3.12) **on the product set** $\mathfrak{T}_0 \times \mathfrak{D}^j \times \mathfrak{Y}_d^k$ **and its domain***

*a) The plant (3.12) is **the k-th-order trackable on** $\mathfrak{T}_0 \times \mathfrak{D}^j \times \mathfrak{Y}_d^k$ if, and only if, for every plant output desired response $\mathbf{Y}_d(.) \in \mathfrak{Y}_d^k$, for every disturbance vector function $\mathbf{D}(.) \in \mathfrak{D}^j$, and for every instant $\sigma \in In\ \mathfrak{T}_0$, there are a control vector function $\mathbf{U}(.) \in \mathfrak{U}^l$ and a connected neighborhood $\mathfrak{N}\left(t_0; \sigma; \mathbf{Y}_{d0}^k; \mathbf{D}; \mathbf{U}; \mathbf{Y}_d\right)$ of $\mathbf{Y}_{d0}^k$,*

$$\mathfrak{N}\left(t_0; \sigma; \mathbf{Y}_{d0}^k; \mathbf{D}; \mathbf{U}; \mathbf{Y}_d\right) \subseteq \mathfrak{R}^{(k+1)N},$$

such that for every initial plant output vector $\mathbf{Y}_0^k \in \mathfrak{N}\left(t_0; \sigma; \mathbf{Y}_{d0}^k; \mathbf{D}; \mathbf{U}; \mathbf{Y}_d\right)$, the difference between $\mathbf{Y}^k(t)$ and $\mathbf{Y}_d^k(t)$ becomes equal to zero at latest at the moment σ, after which it rests equal to zero on $\mathfrak{T}_0$; i.e.,

$$\begin{gathered}\forall [\mathbf{D}(.), \mathbf{Y}_d(.)] \in \mathfrak{D}^j \times \mathfrak{Y}_d^k,\ \forall \sigma \in In\ \mathfrak{T}_0,\ \exists \mathbf{U}(.) \in \mathfrak{U}^l, \\ \mathbf{U}(t) = \mathbf{U}(t; t_0; \sigma; \mathbf{D}; \mathbf{Y}_d),\ \exists \mathfrak{N}\left(t_0; \sigma; \mathbf{Y}_{d0}^k; \mathbf{D}; \mathbf{U}; \mathbf{Y}_d\right) \subseteq \mathfrak{R}^{(k+1)N} \Longrightarrow \\ \mathbf{Y}_0^k \in \mathfrak{N}\left(t_0; \sigma; \mathbf{Y}_{d0}^k; \mathbf{D}; \mathbf{U}; \mathbf{Y}_d\right) \Longrightarrow \mathbf{Y}^k(t) - \mathbf{Y}_d^k(t) = \mathbf{0}_N,\ \forall (t \geq \sigma) \in \mathfrak{T}_0, \\ k \in \{0, 1, \cdots, \alpha-1\},\ \ j,\ l \in \{0, 1, \cdots\}. \end{gathered} \tag{10.1}$$

*Such control is **the k-th-order tracking control on** $\mathfrak{T}_0 \times \mathfrak{D}^j \times \mathfrak{Y}_d^k$, for short, **the k-th-order tracking control.***

If, and only if, additionally, the output variables are ***mutually functionally interrelated*** *by m functional constraints then the k-th-order trackability on* $\mathfrak{T}_0 \times \mathfrak{D}^j \times \mathfrak{Y}_d^k$ *of* $\mathbf{Y}_d(.)$ *is* ***incomplete with*** $N-m$ ***degrees of freedom.***

If, and only if, additionally, all output variables can be controlled mutually independently $(m=0)$*, then the k-th-order trackability on* $\mathfrak{T}_0 \times \mathfrak{D}^j \times \mathfrak{Y}_d^k$ *of* $\mathbf{Y}_d(.)$ ***is complete.***

The largest connected neighborhood $\mathfrak{N}_L\left(t_0;\sigma;\mathbf{Y}_{d0}^k;\mathbf{D};\mathbf{U};\mathbf{Y}_d\right)$ *of* $\mathbf{Y}_{d0}^k$ *for any fixed* $\sigma \in In\ \mathfrak{T}_0$*, which obeys (10.1), is* ***the k-th-order trackability domain*** $\mathcal{D}_{Tbl}^k\left(t_0;\sigma;\mathbf{Y}_{d0}^k;\mathbf{D};\mathbf{U};\mathbf{Y}_d\right)$ ***of*** $\mathbf{Y}_d^k(t)$ ***for every*** $[\mathbf{D}(.),\mathbf{Y}_d(.)] \in \mathfrak{D}^j \times \mathfrak{Y}_d^k$*, i.e.,* ***on*** $\mathfrak{D}^j \times \mathfrak{Y}_d^k$*,* ***at*** $t_0 \in \mathfrak{T}_i$ ***for the fixed*** $\sigma \in In\ \mathfrak{T}_0$*,*

$$\mathcal{D}_{Tbl}^k\left(t_0;\sigma;\mathbf{Y}_{d0}^k;\mathbf{D};\mathbf{U};\mathbf{Y}_d\right) = \mathfrak{N}_L\left(t_0;\sigma;\mathbf{Y}_{d0}^k;\mathbf{D};\mathbf{U};\mathbf{Y}_d\right),\quad \sigma \in In\mathfrak{T}_0. \tag{10.2}$$

b) The intersection over $\sigma \in In\ \mathfrak{T}_0$ *of the k-th-order trackability domains*

$$\mathcal{D}_{Tbl}^k\left(t_0;\sigma;\mathbf{Y}_{d0}^k;\mathbf{D};\mathbf{U};\mathbf{Y}_d\right)$$

of $\mathbf{Y}_{d0}^k$ *(10.2) is* ***the k-th-order trackability domain*** $\mathcal{D}_{Tbl}^k\left(t_0;\mathbf{Y}_{d0}^k;\mathbf{D};\mathbf{U};\mathbf{Y}_d\right)$ ***of*** $\mathbf{Y}_d^k(t)$ ***for every*** $[\mathbf{D}(.),\mathbf{Y}_d(.)] \in \mathfrak{D}^j \times \mathfrak{Y}_d^k$*, i.e.,* ***on*** $\mathfrak{D}^j \times \mathfrak{Y}_d^k$*,* ***at*** $t_0 \in \mathfrak{T}_i$*,*

$$\mathcal{D}_{Tbl}^k\left(t_0;\mathbf{Y}_{d0}^k;\mathbf{D};\mathbf{U};\mathbf{Y}_d\right) = \cap\left[\mathcal{D}_{Tbl}^k\left(t_0;\sigma;\mathbf{Y}_{d0}^k;\mathbf{D};\mathbf{U};\mathbf{Y}_d\right) : \sigma \in In\ \mathfrak{T}_0\right], \tag{10.3}$$

if, and only if, it is a connected neighborhood of $\mathbf{Y}_{d0}^k$*.*

c) The zero $(k=0)$ *order trackability on* $\mathfrak{T}_0 \times \mathfrak{D}^j \times \mathfrak{Y}_d^k$ *is simply called* ***trackability on*** $\mathfrak{T}_0 \times \mathfrak{D}^j \times \mathfrak{Y}_d^k$*. The zero* $(k=0)$ *order tracking control on* $\mathfrak{T}_0 \times \mathfrak{D}^j \times \mathfrak{Y}_d^k$ *is simply called* ***the tracking control on*** $\mathfrak{T}_0 \times \mathfrak{D}^j \times \mathfrak{Y}_d^k$*, for short* ***the tracking control.***

d) The k-th-order trackability on $\mathfrak{T}_0 \times \mathfrak{D}^j \times \mathfrak{Y}_d^k$ *is:*

- ***Global for the fixed*** $\sigma \in In\ \mathfrak{T}_0$ ***(in the whole for the fixed*** $\sigma \in In\ \mathfrak{T}_0$*) if, and only if,* $\mathcal{D}_{Tbl}^k\left(t_0;\sigma;\mathbf{Y}_{d0}^k;\mathbf{D};\mathbf{U};\mathbf{Y}_d\right) = \mathfrak{R}^{(k+1)N}$*,* $\forall\left[\mathbf{D}(.),\mathbf{Y}_d(.)\right] \in \mathfrak{D}^j \times \mathfrak{Y}_d^k$*.*

- ***Global (in the whole)*** *if, and only if,* $\mathcal{D}_{Tbl}^k\left(t_0;\mathbf{Y}_{d0}^k;\mathbf{D};\mathbf{U};\mathbf{Y}_d\right) = \mathfrak{R}^{(k+1)N}$*, for every pair* $\left[\mathbf{D}(.),\mathbf{Y}_d(.)\right] \in \mathfrak{D}^j \times \mathfrak{Y}_d^k$*.*

e) The k-th-order trackability on $\mathfrak{T}_0 \times \mathfrak{D}^j \times \mathfrak{Y}_d^k$ *is:*

- ***Uniform on*** $\mathfrak{T}_0 \times \mathfrak{T}_i \times \mathfrak{D}^j \times \mathfrak{Y}_d^k$ ***for the fixed*** $\sigma \in In\ \mathfrak{T}_0$ *if, and only if, the intersection* $\mathcal{D}_{Tblu}^k\left(\sigma;\mathbf{Y}_{d0}^k;\mathfrak{D}^j;\mathbf{U};\mathfrak{Y}_d^k\right)$ *of all* $\mathcal{D}_{Tbl}^k\left(t_0;\sigma;\mathbf{Y}_{d0}^k;\mathbf{D};\mathbf{U};\mathbf{Y}_d\right)$ *over the product set* $\mathfrak{T}_i \times \mathfrak{D}^j \times \mathfrak{Y}_d^k$*,*

$$\mathcal{D}_{Tblu}^k\left(\sigma;\mathbf{Y}_{d0}^k;\mathfrak{D}^j;\mathbf{U};\mathfrak{Y}_d^k\right) = \cap\begin{bmatrix} \mathcal{D}_{Tbl}^k\left(t_0;\sigma;\mathbf{Y}_{d0}^k;\mathbf{D};\mathbf{U};\mathbf{Y}_d\right) : \\ \left[t_0,\mathbf{D}(.),\mathbf{Y}_d(.)\right] \in \mathfrak{T}_i \times \mathfrak{D}^j \times \mathfrak{Y}_d^k, \end{bmatrix} \tag{10.4}$$

is a connected neighborhood of $\mathbf{Y}_{d0}^k$ *of every* $\mathbf{Y}_d(.) \in \mathfrak{Y}_d^k$ *and* $\mathbf{U}(.)$ *depends on* σ *and on* $\mathfrak{D}^j \times \mathfrak{Y}_d^k$ *but not on an individual triplet* $[t_0,\mathbf{D}(.),\mathbf{Y}_d(.)\ ]$ *from* $\mathfrak{T}_i \times \mathfrak{D}^j \times \mathfrak{Y}_d^k$*,* $\mathbf{U}(t;t_0,\sigma,\mathbf{D},\mathbf{Y}_d) = \mathbf{U}(t;\sigma;\mathfrak{D}^j;\mathfrak{Y}_d^k)$*.*

Then, and only then, $\mathcal{D}_{Tblu}^k\left(\sigma;\mathbf{Y}_{d0}^k;\mathfrak{D}^j;\mathbf{U};\mathfrak{Y}_d^k\right)$ *is the* ***the k-th-order uniform trackability domain of*** $\mathbf{Y}_d^k(t)$ ***on*** $\mathfrak{T}_i \times \mathfrak{D}^j \times \mathfrak{Y}_d^k$ ***for the fixed*** $\sigma \in In\ \mathfrak{T}_0$

- ***Uniform on*** $\mathfrak{T}_0 \times \mathfrak{T}_i \times \mathfrak{D}^j \times \mathfrak{Y}_d^k$ *if, and only if, the intersection*

$$\mathcal{D}_{Tblu}^k\left(\mathbf{Y}_{d0}^k;\mathfrak{D}^j;\mathbf{U};\mathfrak{Y}_d^k\right)$$

of all $\mathcal{D}_{Tbl}^k\left(t_0;\sigma,\mathbf{Y}_{d0}^k;\mathbf{D};\mathbf{U};\mathbf{Y}_d\right)$ *over* $\left[t_0,\mathbf{D}(.),\mathbf{Y}_d(.)\right] \in \mathfrak{T}_i \times \mathfrak{D}^j \times \mathfrak{Y}_d^k$*,*

$$\mathcal{D}_{Tblu}^k\left(\mathbf{Y}_{d0}^k;\mathfrak{D}^j;\mathbf{U};\mathfrak{Y}_d^k\right) = \cap\begin{bmatrix} \mathcal{D}_{Tbl}^k\left(t_0;\mathbf{Y}_{d0}^k;\mathbf{D};\mathbf{U};\mathbf{Y}_d\right) : \\ \left[t_0,\mathbf{D}(.),\mathbf{Y}_d(.)\right] \in \mathfrak{T}_i \times \mathfrak{D}^j \times \mathfrak{Y}_d^k, \end{bmatrix} \tag{10.5}$$

is a connected neighborhood of $\mathbf{Y}_{d0}^k$ *of every* $\mathbf{Y}_d(.) \in \mathfrak{Y}_d^k$ *and* $\mathbf{U}(.)$ *depends on* $\mathfrak{D}^j \times \mathfrak{Y}_d^k$ *but not on an individual quartet* $[t_0,\sigma,\mathbf{D}(.),\mathbf{Y}_d(.)\ ]$ *from* $\mathfrak{T}_i \times \mathfrak{T}_0 \times \mathfrak{D}^j \times \mathfrak{Y}_d^k$*,* $\mathbf{U}(t;t_0,\mathbf{D},\mathbf{Y}_d) = \mathbf{U}(t;\mathfrak{D}^j;\mathfrak{Y}_d^k)$*. Then, and only then,* $\mathcal{D}_{Tblu}^k\left(\mathbf{Y}_{d0}^k;\mathfrak{D}^j;\mathbf{U};\mathfrak{Y}_d^k\right)$ *is* ***the k-th-order uniform trackability domain of*** $\mathbf{Y}_d^k(t)$ ***on the product set*** $\mathfrak{T}_i \times \mathfrak{D}^j \times \mathfrak{Y}_d^k$*.*

Theorem 169 ***The perfect versus the imperfect trackability on the product set*** $\mathfrak{T}_0 \times \mathfrak{D}^j \times \mathfrak{Y}_d^k$

If for every $\mathbf{Y}_d(.) \in \mathfrak{Y}_d^k$ *there is a connected neighborhood* $\mathfrak{N}_c\left(t_0; \mathbf{Y}_{d0}^k\right)$ *of* $\mathbf{Y}_{d0}^k$ *such that for every* $\mathbf{Y}_0^k(.) \in \mathfrak{N}_c\left(t_0; \mathbf{Y}_{d0}^k\right)$ *the plant (3.12) response* $\mathbf{Y}(t; t_0; \mathbf{Y}_0^k)$ *is continuous in* $\mathbf{Y}_0^k$ *then for the plant to be the k-th-order perfect trackable on* $\mathfrak{T}_0 \times \mathfrak{D}^j \times \mathfrak{Y}_d^k$ *it is necessary and sufficient that it is the k-th-order trackable on* $\mathfrak{T}_0 \times \mathfrak{D}^j \times \mathfrak{Y}_d^k$.

Proof. Let for every $\mathbf{Y}_d(.) \in \mathfrak{Y}_d^k$ there be a connected neighborhood $\mathfrak{N}_c\left(t_0; \mathbf{Y}_{d0}^k\right)$ of $\mathbf{Y}_{d0}^k$ such that for every $\mathbf{Y}_0^k(.) \in \mathfrak{N}_c\left(t_0; \mathbf{Y}_{d0}^k\right)$ the plant (3.12) output response $\mathbf{Y}(t; t_0; \mathbf{Y}_0^k)$ is continuous in $\mathbf{Y}_0^k$.

Necessity. We prove the statement by the contradiction. Let the plant (3.12) be the k-th-order perfect trackable on $\mathfrak{T}_0 \times \mathfrak{D}^j \times \mathfrak{Y}_d^k$ but not the k-th-order trackable $\mathfrak{T}_0 \times \mathfrak{D}^j \times \mathfrak{Y}_d^k$. Definition 140 holds. Since the plant is not the k-th-order trackable on $\mathfrak{T}_0 \times \mathfrak{D}^j \times \mathfrak{Y}_d^k$, then, due to the violation of Definition 168, for $\mathbf{Y}_0^k = \mathbf{Y}_{d0}^k$, which obeys

$$\mathbf{Y}_0^k \in \mathfrak{N}_c\left(t_0; \mathbf{Y}_{d0}^k\right) \cap \mathfrak{N}\left(t_0; \mathbf{Y}_{d0}^k\right), \forall \mathfrak{N}\left(t_0; \mathbf{Y}_{d0}^k\right) \subset \mathfrak{R}^{(k+1)N},$$

and for every $\tau \in \operatorname{In} \mathfrak{T}_0$ there exist $m \in \{0, 1, ..., k\}$ and $(\gamma \geq \tau) \in \mathfrak{T}_0$ such that

$$\|\mathbf{Y}^m(\gamma) - \mathbf{Y}_d^m(\gamma)\| > 0.$$

This and the continuity of $\mathbf{Y}(t; t_0; \mathbf{Y}_0^k)$ in $\mathbf{Y}_0^k \in \mathfrak{N}_c\left(t_0; \mathbf{Y}_{d0}^k\right)$ imply for $\mathbf{Y}_0^k = \mathbf{Y}_{d0}^k$ that

$$\left\|\mathbf{Y}^m(\gamma; t_0; \mathbf{Y}_{d0}^k) - \mathbf{Y}_d^m(\gamma)\right\| > 0.$$

This contradicts Definition 140, which is the consequence of the assumption that the plant is not the k-th-order trackable on $\mathfrak{T}_0 \times \mathfrak{D}^j \times \mathfrak{Y}_d^k$. Hence, the plant (3.12) is the k-th-order trackable on $\mathfrak{T}_0 \times \mathfrak{D}^j \times \mathfrak{Y}_d^k$.

Sufficiency. Let the plant (3.12) be the k-th-order trackable on $\mathfrak{T}_0 \times \mathfrak{D}^j \times \mathfrak{Y}_d^k$. Definition 168 holds. Let $\sigma \in \operatorname{In} \mathfrak{T}_0$ and let it be arbitrarily close to t_0 from the right-hand side, i.e., $\sigma \longrightarrow t_0^+$ in Definition 168 so that for such σ (10.1) becomes:

$$\begin{gathered}\exists \mathfrak{N}\left(t_0; \sigma; \mathbf{Y}_{d0}^k; \mathbf{D}; \mathbf{U}; \mathbf{Y}_d\right) \subseteq \mathfrak{R}^{(k+1)N}, \ \forall \mathbf{Y}_d(.) \in \mathfrak{Y}_d^k, \ \sigma \in \operatorname{In} \mathfrak{T}_0, \ \sigma \longrightarrow t_0^+, \\ \mathbf{Y}_0^k \in \mathfrak{N}\left(t_0; \sigma; \mathbf{Y}_{d0}^k; \mathbf{D}; \mathbf{U}; \mathbf{Y}_d\right) \Longrightarrow \mathbf{Y}^k(t) - \mathbf{Y}_d^k(t) = \mathbf{0}_N, \ \ \forall (t \geq \sigma) \in \mathfrak{T}_0. \qquad (10.6)\end{gathered}$$

Let $\mathbf{Y}_0^k = \mathbf{Y}_{d0}^k$ so that $\mathbf{Y}_0^k = \mathbf{Y}_{d0}^k \in \mathfrak{N}\left(t_0; \sigma; \mathbf{Y}_{d0}^k; \mathbf{D}; \mathbf{U}; \mathbf{Y}_d\right)$. Such $\mathbf{Y}_0^k$ obeys (10.6) that permits $\sigma = t_0$. They jointly imply $\mathbf{Y}^k(t) - \mathbf{Y}_d^k(t) = \mathbf{0}_N$, $\forall t \in \mathfrak{T}_0$. This satisfies Definition 140 and proves the sufficiency part of the theorem ■

Lemma 137 and Definition 168 directly imply the following.

Lemma 170 ***The k-th-order trackability and the trackability***

For the plant (3.12) to be the k-th-order trackable on $\mathfrak{T}_0 \times \mathfrak{D}^j \times \mathfrak{Y}_d^k$ *it is necessary and sufficient to be trackable on* $\mathfrak{T}_0 \times \mathfrak{D}^j \times \mathfrak{Y}_d^k$.

10.3 Natural trackability

Perfect natural trackability properties demand that $\mathbf{Y}_0^k = \mathbf{Y}_{d0}^k$, Definition 146. Let us analyze the cases when this initial condition is not satisfied, for which the perfection of the natural trackability is impossible. This leads us to introduce imperfect natural trackability properties.

Definition 171 ***The k-th-order natural trackability of the plant*** *(3.12))* ***on the product set*** $\mathfrak{T}_0 \times \mathfrak{D}^j \times \mathfrak{Y}_d^k$ ***and its domain***

a) The plant (3.12) is ***the k-th-order natural trackable on*** $\mathfrak{T}_0 \times \mathfrak{D}^j \times \mathfrak{Y}_d^k$ *if, and only if, for every plant output desired response* $\mathbf{Y}_d(.) \in \mathfrak{Y}_d^k$, *for every instant* $\sigma \in In\ \mathfrak{T}_0$, *and* *for every disturbance vector function* $\mathbf{D}(.) \in \mathfrak{D}^j$, *there is a control vector function* $\mathbf{U}(.) \in \mathfrak{U}^l$, *that can be synthesized without using information about the form and the value of* $\mathbf{D}(.) \in \mathfrak{D}^j$ *and about the mathematical model of the plant state, and there is a connected neighborhood* $\mathfrak{N}\left(t_0;\sigma;\mathbf{Y}_{d0}^k;\mathbf{D};\mathbf{U};\mathbf{Y}_d\right) \subseteq \mathfrak{R}^{(k+1)N}$ *of* $\mathbf{Y}_{d0}^k$ *such that for every plant initial output vector* $\mathbf{Y}_0^k$ *from the neighborhood* $\mathfrak{N}\left(t_0;\sigma;\mathbf{Y}_{d0}^k;\mathbf{D};\mathbf{U};\mathbf{Y}_d\right)$, *the difference between* $\mathbf{Y}^k(t)$ *and* $\mathbf{Y}_d^k(t)$ *becomes equal to zero at the latest at the moment* σ, *after which it rests equal to zero on* $\mathfrak{T}_0$; *i.e.,*

$$\begin{gathered}\forall\left[\mathbf{D}(.),\mathbf{Y}_d(.)\right] \in \mathfrak{D}^j \times \mathfrak{Y}_d^k,\ \forall \sigma \in In\ \mathfrak{T}_0,\ \exists \mathbf{U}(.) \in \mathfrak{U}^l,\\ \exists \mathbf{U}(t)=\mathbf{U}(t;t_0;\sigma;\mathfrak{D}^j;\mathbf{Y}_d),\ \exists \mathfrak{N}\left(t_0;\sigma;\mathbf{Y}_{d0}^k;\mathbf{D};\mathbf{U};\mathbf{Y}_d\right) \subseteq \mathfrak{R}^{(k+1)N} \Longrightarrow\\ \mathbf{Y}_0^k \in \mathfrak{N}\left(t_0;\sigma;\mathbf{Y}_{d0}^k;\mathbf{D};\mathbf{U};\mathbf{Y}_d\right) \Longrightarrow \mathbf{Y}^k(t)-\mathbf{Y}_d^k(t)=\mathbf{0}_N,\ \forall (t \geq \sigma) \in \mathfrak{T}_0. \end{gathered} \tag{10.7}$$

Such control is ***the k-th-order natural tracking control on*** $\mathfrak{T}_0 \times \mathfrak{D}^j \times \mathfrak{Y}_d^k$, *for short* ***the k-th-order natural tracking control.***

If, and only if, additionally, the output variables are ***mutually functionally interrelated*** *by m functional constraints then the k-th-order natural trackability on* $\mathfrak{T}_0 \times \mathfrak{D}^j \times \mathfrak{Y}_d^k$ *of* $\mathbf{Y}_d(.)$ *is* ***incomplete with*** $N-m$ ***degrees of freedom,.***

If, and only if, additionally, all output variables can be controlled mutually independently, $(m=0)$, *then the k-th-order natural trackability on* $\mathfrak{T}_0 \times \mathfrak{D}^j \times \mathfrak{Y}_d^k$ *of* $\mathbf{Y}_d(.)$ ***is complete.***

The largest connected neighborhood $\mathfrak{N}\left(t_0;\sigma;\mathbf{Y}_{d0}^k;\mathbf{D};\mathbf{U};\mathbf{Y}_d\right)$ *of* $\mathbf{Y}_{d0}^k$ *for any fixed* $\sigma \in In\ \mathfrak{T}_0$, *which obeys (10.7), is* ***the k-th-order natural trackability domain***

$$\mathcal{D}_{NTbl}^k\left(t_0;\sigma;\mathbf{Y}_{d0}^k;\mathbf{D};\mathbf{U};\mathbf{Y}_d\right)$$

of $\mathbf{Y}_d^k(t)$ *for every* $\left[\mathbf{D}(.),\mathbf{Y}_d(.)\right] \in \mathfrak{D}^j \times \mathfrak{Y}_d^k$, ***i.e., on*** $\mathfrak{D}^j \times \mathfrak{Y}_d^k$, ***at*** $t_0 \in \mathfrak{T}_i$ ***for the fixed*** $\sigma \in In\ \mathfrak{T}_0$,

$$\mathcal{D}_{NTbl}^k\left(t_0;\sigma;\mathbf{Y}_{d0}^k;\mathbf{D};\mathbf{U};\mathbf{Y}_d\right)=\mathfrak{N}\left(t_0;\sigma;\mathbf{Y}_{d0}^k;\mathbf{D};\mathbf{U};\mathbf{Y}_d\right),\quad \sigma \in In\ \mathfrak{T}_0. \tag{10.8}$$

b) The intersection over $\sigma \in In\ \mathfrak{T}_0$ *of the k-th-order natural trackability domains*

$$\mathcal{D}_{NTbl}^k\left(t_0;\sigma;\mathbf{Y}_{d0}^k;\mathbf{D};\mathbf{U};\mathbf{Y}_d\right)$$

of $\mathbf{Y}_{d0}^k$, *(10.8), is* ***the k-th-order natural trackability domain***

$$\mathcal{D}_{NTbl}^k\left(t_0;\mathbf{Y}_{d0}^k;\mathbf{D};\mathbf{U};\mathbf{Y}_d\right)$$

of $\mathbf{Y}_d^k(t)$ *for every* $\left[\mathbf{D}(.),\mathbf{Y}_d(.)\right] \in \mathfrak{D}^j \times \mathfrak{Y}_d^k$ ***, i.e., on*** $\mathfrak{D}^j \times \mathfrak{Y}_d^k$, ***at*** $t_0 \in \mathfrak{T}_i$,

$$\mathcal{D}_{NTbl}^k\left(t_0;\mathbf{Y}_{d0}^k;\mathbf{D};\mathbf{U};\mathbf{Y}_d\right)=\cap\left[\begin{array}{c}\mathcal{D}_{NTbl}^k\left(t_0;\sigma;\mathbf{Y}_{d0}^k;\mathbf{D};\mathbf{U};\mathbf{Y}_d\right):\\ \sigma \in In\ \mathfrak{T}_0\end{array}\right]. \tag{10.9}$$

c) The zero $(k=0)$ *order natural trackability on* $\mathfrak{T}_0 \times \mathfrak{D}^j \times \mathfrak{Y}_d^k$ *is called* ***natural trackability on*** $\mathfrak{T}_0 \times \mathfrak{D}^j \times \mathfrak{Y}_d^k$. *The zero,* $(k=0)$, *order natural tracking control on* $\mathfrak{T}_0 \times \mathfrak{D}^j \times \mathfrak{Y}_d^k$ *is called for short* ***natural tracking control on*** $\mathfrak{T}_0 \times \mathfrak{D}^j \times \mathfrak{Y}_d^k$, *or shorter* ***natural tracking control (NTC).***

d) The k-th-order natural trackability on $\mathfrak{T}_0 \times \mathfrak{D}^j \times \mathfrak{Y}_d^k$ *is:*

- ***Global for the fixed*** $\sigma \in In\ \mathfrak{T}_0$ ***(in the whole)*** *if, and only if,*

$$\mathcal{D}_{NTbl}^k\left(t_0;\sigma;\mathbf{Y}_{d0}^k;\mathbf{D};\mathbf{U};\mathbf{Y}_d\right)=\mathfrak{R}^{(k+1)N},\quad \forall\left[\mathbf{D}(.),\mathbf{Y}_d(.)\right] \in \mathfrak{D}^j \times \mathfrak{Y}_d^k.$$

- ***Global (in the whole)*** *if, and only if,*

$$\mathcal{D}_{NTbl}^k\left(t_0;\mathbf{Y}_{d0}^k;\mathbf{D};\mathbf{U};\mathbf{Y}_d\right)=\mathfrak{R}^{(k+1)N},\ \forall\left[\mathbf{D}(.),\mathbf{Y}_d(.)\right] \in \mathfrak{D}^j \times \mathfrak{Y}_d^k.$$

e) The k-th-order natural trackability on $\mathfrak{T}_0 \times \mathfrak{D}^j \times \mathfrak{Y}_d^k$ *is* ***uniform*** *on the product set* $\mathfrak{T}_0 \times \mathfrak{T}_i \times \mathfrak{D}^j \times \mathfrak{Y}_d^k$ *for* $\sigma \in In\,\mathfrak{T}_0$ *if, and only if, the intersection* $\mathcal{D}_{NTblu}^k\left(\mathfrak{T}_i;\sigma;\mathbf{Y}_{d0}^k;\mathfrak{D}^j;\mathbf{U};\mathfrak{Y}_d^k\right)$ *of all domains* $\mathcal{D}_{NTbl}^k\left(t_0;\sigma;\mathbf{Y}_{d0}^k;\mathbf{D};\mathbf{U};\mathbf{Y}_d\right)$ *taken over* $[t_0,\mathbf{D}(.),\ \mathbf{Y}_d(.)] \in \mathfrak{T}_i \times \mathfrak{D}^j \times \mathfrak{Y}_d^k$,

$$\mathcal{D}_{NTblu}^k\left(\mathfrak{T}_i;\sigma;\mathbf{Y}_{d0}^k;\mathfrak{D}^j;\mathbf{U};\mathfrak{Y}_d^k\right) = \cap \left[\begin{array}{c} \mathcal{D}_{NTbl}^k\left(t_0;\sigma;\mathbf{Y}_{d0}^k;\mathbf{D};\mathbf{U};\mathbf{Y}_d\right): \\ \left[t_0,\mathbf{D}(.),\mathbf{Y}_d(.)\right] \in \mathfrak{T}_i \times \mathfrak{D}^j \times \mathfrak{Y}_d^k, \end{array}\right] \tag{10.10}$$

is a connected neighborhood of $\mathbf{Y}_{d0}^k$ *of every* $\mathbf{Y}_d(.) \in \mathfrak{Y}_d^k$ *and* $\mathbf{U}(.)$ *depends on* σ *and on* $\mathfrak{D}^j \times \mathfrak{Y}_d^k$ *but not on an individual triplet* $[t_0,\mathbf{D}(.),\mathbf{Y}_d(.)\ ]$ *from* $\mathfrak{T}_i \times \mathfrak{D}^j \times \mathfrak{Y}_d^k$, $\mathbf{U}(t;t_0,\sigma,\mathbf{D},\mathbf{Y}_d) = \mathbf{U}(t;\sigma;\mathfrak{D}^j;\mathfrak{Y}_d^k)$.

Then, and only then, $\mathcal{D}_{NTblu}^k\left(\mathfrak{T}_i;\sigma;\mathbf{Y}_{d0}^k;\mathfrak{D}^j;\mathbf{U};\mathfrak{Y}_d^k\right)$ *is the* ***the k-th-order uniform trackability domain of*** $\mathbf{Y}_d^k(t)$ ***on*** $\mathfrak{T}_0 \times \mathfrak{T}_i \times \mathfrak{D}^j \times \mathfrak{Y}_d^k$ ***for the fixed*** $\sigma \in In\,\mathfrak{T}_0$.

f) The k-th-order natural trackability on the product set $\mathfrak{T}_0 \times \mathfrak{D}^j \times \mathfrak{Y}_d^k$ *is* ***uniform*** *on* $\mathfrak{T}_0 \times \mathfrak{T}_i \times \mathfrak{D}^j \times \mathfrak{Y}_d^k$ *if, and only if, interior* $In\mathcal{D}_{NTblu}^k\left(\mathfrak{T}_i;\mathbf{Y}_{d0}^k;\mathfrak{D}^j;\mathbf{U};\mathfrak{Y}_d^k\right)$ *of the intersection* $\mathcal{D}_{NTblu}^k\left(\mathfrak{T}_i;\mathbf{Y}_{d0}^k;\mathfrak{D}^j;\mathbf{U};\mathfrak{Y}_d^k\right)$ *of all k-th-order trackability domains* $\mathcal{D}_{NTbl}^k\left(t_0;\mathbf{Y}_{d0}^k;\mathbf{D};\mathbf{U};\mathbf{Y}_d\right)$ *over the product set* $\mathfrak{T}_i \times \mathfrak{D}^j \times \mathfrak{Y}_d^k$,

$$\mathcal{D}_{NTblu}^k\left(\mathfrak{T}_i;\mathbf{Y}_{d0}^k;\mathfrak{D}^j;\mathbf{U};\mathfrak{Y}_d^k\right) = \cap \left[\begin{array}{c} \mathcal{D}_{NTbl}^k\left(t_0;\mathbf{Y}_{d0}^k;\mathbf{D};\mathbf{U};\mathbf{Y}_d\right): \\ \left[t_0,\mathbf{D}(.),\mathbf{Y}_d(.)\right] \in \mathfrak{T}_i \times \mathfrak{D}^j \times \mathfrak{Y}_d^k, \end{array}\right] \tag{10.11}$$

is an open connected neighborhood of $\mathbf{Y}_{d0}^k$ *of every* $\mathbf{Y}_d(.) \in \mathfrak{Y}_d^k$ *and* $\mathbf{U}(.)$ *depends on* $\mathfrak{D}^j \times \mathfrak{Y}_d^k$ *but not on an individual triplet* $[t_0,\mathbf{D}(.),\mathbf{Y}_d(.)]$ *from* $\mathfrak{T}_i \times \mathfrak{D}^j \times \mathfrak{Y}_d^k$, $\mathbf{U}(t;t_0,\mathbf{D},\mathbf{Y}_d) = \mathbf{U}(t;\mathfrak{D}^j;\mathfrak{Y}_d^k)$.

Then, and only then, $\mathcal{D}_{Tblu}^k\left(\mathbf{Y}_{d0}^k;\mathfrak{D}^j;\mathbf{U};\mathfrak{Y}_d^k\right)$ *is the* ***the k-th-order uniform natural trackability domain of*** $\mathbf{Y}_d^k(t)$ ***on*** $\mathfrak{T} \times \mathfrak{D}^j \times \mathfrak{Y}_d^k$.

Comment 172 *Definition 168 and Definition 171 show that the k-th-order trackability on* $\mathfrak{T}_0 \times \mathfrak{D}^j \times \mathfrak{Y}_d^k$ *is necessary for the k-th-order natural trackability on* $\mathfrak{T}_0 \times \mathfrak{D}^j \times \mathfrak{Y}_d^k$*, and the natural trackability on* $\mathfrak{T}_0 \times \mathfrak{D}^j \times \mathfrak{Y}_d^k$ *is sufficient for the trackability on* $\mathfrak{T}_0 \times \mathfrak{D}^j \times \mathfrak{Y}_d^k$.

Theorem 173 ***Perfect natural trackability versus natural trackability on*** $\mathfrak{T}_0 \times \mathfrak{D}^j \times \mathfrak{Y}_d^k$

If for every $\mathbf{Y}_d(.) \in \mathfrak{Y}_d^k$ *there is a connected neighborhood* $\mathfrak{N}_c\left(t_0;\mathbf{Y}_{d0}^k\right)$ *of* $\mathbf{Y}_{d0}^k$ *such that for every* $\mathbf{Y}_0(.) \in \mathfrak{N}_c\left(t_0;\mathbf{Y}_{d0}^k\right)$ *the plant (3.12) response* $\mathbf{Y}(t;t_0;\mathbf{Y}_0^k)$ *is continuous in* $\mathbf{Y}_0^k$ *then for the plant to be the k-th-order perfect natural trackable on* $\mathfrak{T}_0 \times \mathfrak{D}^j \times \mathfrak{Y}_d^k$ *it is necessary and sufficient that it is the k-th-order natural trackable on* $\mathfrak{T}_0 \times \mathfrak{D}^j \times \mathfrak{Y}_d^k$.

The proof of this theorem is a slight modification of the proof of Theorem (169).

Lemma 137 and Definition 171 result in the following.

Lemma 174 ***The k-th-order natural trackability and the natural trackability***

For the plant (3.12) to be the k-th-order natural trackable on $\mathfrak{T}_0 \times \mathfrak{D}^j \times \mathfrak{Y}_d^k$ *it is necessary and sufficient to be the natural trackable on* $\mathfrak{T}_0 \times \mathfrak{D}^j \times \mathfrak{Y}_d^k$.

10.4 Elementwise trackability

The elementwise tracking permits us to associate different tracking requirements with different output variables of the plant. The concept of the elementwise trackability generalizes in that sense the preceding concept of the imperfect trackability.

Let a constant vector $\mathbf{w} \in \mathfrak{R}^{(k+1)N}$. It will determine the size and the shape of the corresponding neighborhood.

Definition 175 ***The k-th-order elementwise trackability of the plant (3.12) on the set product $\mathfrak{T}^{(k+1)N} \times \mathfrak{D}^j \times \mathfrak{Y}_d^k$ and its domain***

a) The plant (3.12) is ***the k-th-order elementwise trackable on*** $\mathfrak{T}_0^{(k+1)N} \times \mathfrak{D}^j \times \mathfrak{Y}_d^k$ *if, and only if, for every plant output desired response* $\mathbf{Y}_d(.) \in \mathfrak{Y}_d^k$, *for every disturbance vector function* $\mathbf{D}(.) \in \mathfrak{D}^j$, *and for every vector instant* $\boldsymbol{\sigma} \in (In\ \mathfrak{T}_0)^{(k+1)N}$, *there is an elementwise* $\mathbf{w}$*-neighborhood*

$$\mathcal{N}_{\mathbf{w}}\left(\mathbf{Y}_{d0}^k\right) = \mathcal{N}_{\mathbf{w}}\left(\mathbf{t}_0^{(k+1)N}; \boldsymbol{\sigma}; \mathbf{Y}_{d0}^k \mathbf{D}; \mathbf{U}; \mathbf{Y}_d\right)$$

of $\mathbf{Y}_{d0}^k$ *(2.22),* $\mathcal{N}_{\mathbf{w}}\left(\mathbf{Y}_{d0}^k\right) \subseteq \mathfrak{R}^{(k+1)N}$, *and there is a control vector function* $\mathbf{U}(.) \in \mathfrak{U}^l$ *such that for every plant initial output vector* $\mathbf{Y}_0^k$ *in the elementwise* $\mathbf{w}$*-neighborhood* $\mathcal{N}_{\mathbf{w}}\left(\mathbf{Y}_{d0}^k\right)$ *of* $\mathbf{Y}_{d0}^k$, *the real plant output response* $\mathbf{Y}^k(\mathbf{t}^{(k+1)N})$ *becomes elementwise equal to* $\mathbf{Y}_d^k(\mathbf{t}^{(k+1)N})$ *at the latest at the vector moment* $\boldsymbol{\sigma}$, *after which they rest equal forever, i.e.,*

$$\begin{gathered}
\forall \mathbf{Y}_d(.) \in \mathfrak{Y}_d^k,\ \forall \mathbf{D}(.) \in \mathfrak{D}^j,\ \forall \boldsymbol{\sigma} \in (In\ \mathfrak{T}_0)^{(k+1)N}, \\
\exists \mathcal{N}_{\mathbf{w}}\left(\mathbf{Y}_{d0}^k\right) = \mathcal{N}_{\mathbf{w}}\left(\mathbf{t}_0^{(k+1)N}; \boldsymbol{\sigma}; \mathbf{Y}_{d0}^k \mathbf{D}; \mathbf{U}; \mathbf{Y}_d\right) \subseteq \mathfrak{R}^{(k+1)N},\ \exists \mathbf{U}(.) \in \mathfrak{U}^l, \\
\mathbf{U}(\mathbf{t}^r) = \mathbf{U}(\mathbf{t}^r; \boldsymbol{\sigma}; \mathbf{D}; \mathbf{Y}_d),\ \mathbf{Y}_0^k \in \mathcal{N}_{\mathbf{w}}\left(\mathbf{Y}_{d0}^k\right) \Longrightarrow \\
\mathbf{Y}^k(\mathbf{t}^{(k+1)N}) = \mathbf{Y}_d^k(\mathbf{t}^{(k+1)N}),\ \forall \left(\mathbf{t}^{(k+1)N} \geq \boldsymbol{\sigma}\right) \in \mathfrak{T}_0^{(k+1)N}.
\end{gathered} \tag{10.12}$$

$\mathbf{U}(.)$ *is* ***the k-th-order elementwise tracking control on*** $\mathfrak{T}_0^{(k+1)N} \times \mathfrak{D}^j \times \mathfrak{Y}_d^k$.

The largest connected neighborhood $\mathcal{N}_{\mathbf{w}L}\left(\mathbf{t}_0^{(k+1)N}; \boldsymbol{\sigma}; \mathbf{Y}_{d0}^k \mathbf{D}; \mathbf{U}; \mathbf{Y}_d\right)$ *of* $\mathbf{Y}_{d0}^k$ *for any fixed* $\boldsymbol{\sigma} \in (In\ \mathfrak{T}_0)^{(k+1)N}$, *which obeys (10.12), is* ***the k-th-order elementwise trackability domain*** $\mathcal{D}_{ETbl}^k\left(\mathbf{t}_0^{(k+1)N}; \boldsymbol{\sigma}; \mathbf{Y}_{d0}^k \mathbf{D}; \mathbf{U}; \mathbf{Y}_d\right)$ ***of*** $\mathbf{Y}_d^k(t)$ *for every* $[\mathbf{D}(.), \mathbf{Y}_d(.)] \in \mathfrak{D}^j \times \mathfrak{Y}_d^k$, ***i.e., on*** $\mathfrak{D}^j \times \mathfrak{Y}_d^k$, ***at*** $t_0 \in \mathfrak{T}_i$ *for the fixed* $\boldsymbol{\sigma} \in (In\ \mathfrak{T}_0)^{(k+1)N}$,

$$\begin{gathered}
\mathcal{D}_{ETbl}^k\left(\mathbf{t}_0^{(k+1)N}; \boldsymbol{\sigma}; \mathbf{Y}_{d0}^k \mathbf{D}; \mathbf{U}; \mathbf{Y}_d\right) = \mathcal{N}_{\mathbf{w}L}\left(\mathbf{t}_0^{(k+1)N}; \boldsymbol{\sigma}; \mathbf{Y}_{d0}^k \mathbf{D}; \mathbf{U}; \mathbf{Y}_d\right), \\
\boldsymbol{\sigma} \in (In\ \mathfrak{T}_0)^{(k+1)N}.
\end{gathered} \tag{10.13}$$

b) The intersection over $\boldsymbol{\sigma} \in (In\ \mathfrak{T}_0)^{(k+1)N}$ *of the k-th-order elementwise trackability domains* $\mathcal{D}_{ETbl}^k\left(\mathbf{t}_0^{(k+1)N}; \boldsymbol{\sigma}; \mathbf{Y}_{d0}^k \mathbf{D}; \mathbf{U}; \mathbf{Y}_d\right)$ *of* $\mathbf{Y}_{d0}^k$, *(10.13), is* ***the k-th-order elementwise trackability domain*** $\mathcal{D}_{ETbl}^k\left(\mathbf{t}_0^{(k+1)N}; \mathbf{Y}_{d0}^k \mathbf{D}; \mathbf{U}; \mathbf{Y}_d\right)$ ***of*** $\mathbf{Y}_d^k(t)$ *for every* $[\mathbf{D}(.), \mathbf{Y}_d(.)] \in \mathfrak{D}^j \times \mathfrak{Y}_d^k$, ***i.e., on*** $\mathfrak{D}^j \times \mathfrak{Y}_d^k$, ***at*** $t_0 \in \mathfrak{T}_i$,

$$\mathcal{D}_{ETbl}^k\left(\mathbf{t}_0^{(k+1)N}; \mathbf{Y}_{d0}^k \mathbf{D}; \mathbf{U}; \mathbf{Y}_d\right) = \cap \left[\begin{array}{c} \mathcal{N}_{\mathbf{w}}\left(\mathbf{t}_0^{(k+1)N}; \boldsymbol{\sigma}; \mathbf{Y}_{d0}^k \mathbf{D}; \mathbf{U}; \mathbf{Y}_d\right): \\ \boldsymbol{\sigma} \in (In\ \mathfrak{T}_0)^{(k+1)N} \end{array}\right]. \tag{10.14}$$

c) The zero, (k = 0), order elementwise trackability on $\mathfrak{T}_0^N \times \mathfrak{D}^j \times \mathfrak{Y}_d^k$ *is called* ***elementwise trackability on*** $\mathfrak{T}_0^N \times \mathfrak{D}^j \times \mathfrak{Y}_d^k$. *The zero, (k = 0), order elementwise tracking control on* $\mathfrak{T}_0^N \times \mathfrak{D}^j \times \mathfrak{Y}_d^k$ *is called for short* ***the elementwise tracking control on*** $\mathfrak{T}^N \times \mathfrak{D}^j \times \mathfrak{Y}_d^k$, *or shorter,* ***elementwise tracking control****.*

d) The k-th-order elementwise trackability on $\mathfrak{T}_0^{(k+1)N} \times \mathfrak{D}^j \times \mathfrak{Y}_d^k$ *is:*

- ***Global for the fixed*** $\boldsymbol{\sigma} \in (In\ \mathfrak{T}_0)^{(k+1)N}$ *(equivalently,* ***in the whole for the fixed*** $\boldsymbol{\sigma} \in (In\ \mathfrak{T}_0)^{(k+1)N}$*) if, and only if,*

$$\mathcal{D}_{ETbl}^k\left(\mathbf{t}_0^{(k+1)N}; \boldsymbol{\sigma}; \mathbf{Y}_{d0}^k \mathbf{D}; \mathbf{U}; \mathbf{Y}_d\right) = \mathfrak{R}^{(k+1)N},\ \forall [\mathbf{D}(.), \mathbf{Y}_d(.)] \in \mathfrak{D}^j \times \mathfrak{Y}_d^k,$$

- ***Global (in the whole)*** *if, and only if,*

$$\mathcal{D}^k_{ETbl}\left(\mathbf{t}_0^{(k+1)N};\mathbf{Y}^k_{d0}\mathbf{D};\mathbf{U};\mathbf{Y}_d\right)=\mathfrak{R}^{(k+1)N},\ \forall\left[\mathbf{D}(.),\mathbf{Y}_d(.)\right]\in\mathfrak{D}^j\times\mathfrak{Y}^k_d.$$

e) The k-th-order elementwise trackability on $\mathfrak{T}_0^{(k+1)N}\times\mathfrak{D}^j\times\mathfrak{Y}^k_d$ *is* ***uniform for the fixed*** $\boldsymbol{\sigma}\in(In\ \mathfrak{T}_0)^{(k+1)N}$ *on* $\mathfrak{T}_0^{(k+1)N}\times\mathfrak{T}_i^{(k+1)N}\times\mathfrak{D}^j\times\mathfrak{Y}^k_d$ *if, and only if, the intersection* $\mathcal{D}^k_{ETblu}\left(\mathfrak{T}_i^{(k+1)N};\boldsymbol{\sigma};\mathbf{Y}^k_{d0};\mathfrak{D}^j;\mathbf{U};\mathfrak{Y}^k_d\right)$,

$$\mathcal{D}^k_{ETblu}\left(\mathfrak{T}_i^{(k+1)N};\boldsymbol{\sigma};\mathbf{Y}^k_{d0};\mathfrak{D}^j;\mathbf{U};\mathfrak{Y}^k_d\right)=\cap\left[\begin{array}{c}\mathcal{D}^k_{ETbl}\left(\mathbf{t}_0^{(k+1)N};\boldsymbol{\sigma};\mathbf{Y}^k_{d0}\mathbf{D};\mathbf{U};\mathbf{Y}_d\right):\\ \left[\mathbf{t}_0^{(k+1)N},\mathbf{D}(.),\mathbf{Y}_d(.)\right]\in\mathfrak{T}_i^{(k+1)N}\times\mathfrak{D}^j\times\mathfrak{Y}^k_d.\end{array}\right]$$

is an open connected neighborhood of $\mathbf{Y}^k_{d0}$ *of every* $\mathbf{Y}_d(.)\in\mathfrak{Y}^k_d$ *and* $\mathbf{U}(.)$ *depends on* $\boldsymbol{\sigma}$ *and* $\mathfrak{D}^j\times\mathfrak{Y}^k_d$ *but not on an individual triplet* $\left[\mathbf{t}_0^{(k+1)N},\mathbf{D}(.),\mathbf{Y}_d(.)\right]$ *from* $\mathfrak{T}_i^{(k+1)N}\times\mathfrak{D}^j\times\mathfrak{Y}^k_d$,

$$\mathbf{U}(\mathbf{t}^{(k+1)N};\mathbf{t}_0^{(k+1)N},\boldsymbol{\sigma},\mathbf{D},\mathbf{Y}_d)=\mathbf{U}(\mathbf{t}^{(k+1)N};\mathfrak{T}_i^{(k+1)N};\boldsymbol{\sigma};\mathfrak{D}^j;\mathfrak{Y}^k_d).$$

Then, and only then, $\mathcal{D}^k_{ETblu}\left(\mathfrak{T}_i^{(k+1)N};\boldsymbol{\sigma};\mathbf{Y}^k_{d0};\mathfrak{D}^j;\mathbf{U};\mathfrak{Y}^k_d\right)$ *is the* ***the k-th-order uniform elementwise trackability domain of*** $\mathbf{Y}^k_d(\mathbf{t}^{k)N})$ ***on*** $\mathfrak{T}_0^{(k+1)N}\times\mathfrak{T}_i^{(k+1)N}\times\mathfrak{D}^j\times\mathfrak{Y}^k_d$ ***for the fixed*** $\boldsymbol{\sigma}\in(In\ \mathfrak{T}_0)^{(k+1)N}$

f) The k-th-order elementwise trackability on $\mathfrak{T}_0^{(k+1)N}\times\mathfrak{D}^j\times\mathfrak{Y}^k_d$ *is* ***uniform*** *on the product set* $\mathfrak{T}_0^{(k+1)N}\times\mathfrak{T}_i^{(k+1)N}\times\mathfrak{D}^j\times\mathfrak{Y}^k_d$ *if, and only if, the intersection*

$$\mathcal{D}^k_{ETblu}\left(\mathfrak{T}_i^{(k+1)N};\mathbf{Y}^k_{d0};\mathfrak{D}^j;\mathbf{U};\mathfrak{Y}^k_d\right),$$

determined by

$$\mathcal{D}^k_{ETblu}\left(\mathfrak{T}_i^{(k+1)N};\mathbf{Y}^k_{d0};\mathfrak{D}^j;\mathbf{U};\mathfrak{Y}^k_d\right)=\cap\left[\begin{array}{c}\mathcal{D}^k_{ETbl}\left(\mathbf{t}_0^{(k+1)N};\mathbf{Y}^k_{d0}\mathbf{D};\mathbf{U};\mathbf{Y}_d\right):\\ \left[\mathbf{t}_0^{(k+1)N},\mathbf{D}(.),\mathbf{Y}_d(.)\right]\in\mathfrak{T}_i^{(k+1)N}\times\mathfrak{D}^j\times\mathfrak{Y}^k_d.\end{array}\right]\tag{10.15}$$

is an open connected neighborhood of $\mathbf{Y}^k_{d0}$ *of every* $\mathbf{Y}_d(.)\in\mathfrak{Y}^k_d$ *and* $\mathbf{U}(.)$ *depends on* $\mathfrak{D}^j\times\mathfrak{Y}^k_d$ *but not on an individual triplet* $\left[\mathbf{t}_0^{(k+1)N},\mathbf{D}(.),\mathbf{Y}_d(.)\right]$ *from* $\mathfrak{T}_i^{(k+1)N}\times\mathfrak{D}^j\times\mathfrak{Y}^k_d$,

$$\mathbf{U}(\mathbf{t}^{(k+1)N};\mathbf{t}_0^{(k+1)N},\mathbf{D},\mathbf{Y}_d)=\mathbf{U}(\mathbf{t}^{(k+1)N};\mathfrak{T}_i^{(k+1)N};\mathfrak{D}^j;\mathfrak{Y}^k_d).$$

Then, and only then, $\mathcal{D}^k_{ETblu}\left(\mathfrak{T}_i^{(k+1)N};\mathbf{Y}^k_{d0};\mathfrak{D}^j;\mathbf{U};\mathfrak{Y}^k_d\right)$ *is the* ***the k-th-order uniform elementwise trackability domain of*** $\mathbf{Y}^k_d(\mathbf{t}^{k)N})$ ***on (the product set)*** $\mathfrak{T}^{(k+1)N}\times\mathfrak{D}^j\times\mathfrak{Y}^k_d$.

Note 176 Complete trackability and elementwise trackability
The elementwise trackability is simultaneously the complete trackability.

The following comment in [279] also holds unchanged in this setting.

Comment 177 *Although a norm of a vector is equal to zero if, and only if, all the vector entries are zero, the equivalence between the k-th-order trackability and the k-th-order elementwise trackability does not follow from Definition 168 and Definition 175. The former specifies the same scalar reachability time $\sigma \in In\ \mathfrak{T}_0$ for all output variables and their derivatives. The latter associates different scalar reachability times with different output variables and their derivatives; i.e., the latter associates the vector reachability time $\boldsymbol{\sigma} \in (In\ \mathfrak{T}_0)^{(k+1)N}$ with the output vector and its derivatives.*

We use the following result from [279].

Lemma 178 *If two functions $\mathbf{Y}(.)$ and $\mathbf{Y}_d(.)$ are defined, k-times continuously differentiable on $]\boldsymbol{\sigma}, \infty\mathbf{1}_N[$, $\boldsymbol{\sigma} \in \mathfrak{T}_0^N$, $]\boldsymbol{\sigma}, \infty\mathbf{1}_N[\subseteq (In\mathfrak{T}_0)^N$, as well as at $\mathbf{t}^N = \boldsymbol{\sigma}$ from the right-hand side, i.e., at $\mathbf{t}^N = \boldsymbol{\sigma}^+$, and identical on $[\boldsymbol{\sigma}, \infty\mathbf{1}_N[$, $[\boldsymbol{\sigma}, \infty\mathbf{1}_N[\subseteq \mathfrak{T}_0^N$, then all their derivatives up to the order k included are also identical on $]\boldsymbol{\sigma}, \infty\mathbf{1}_N[$ and at $\mathbf{t}^N = \boldsymbol{\sigma}^+$.*

Note 179 *This lemma is the vector generalization of Lemma 137.*

Lemma 178 and Definition 175 imply the following.

Lemma 180 ***The k-th-order elementwise trackability and elementwise trackability***

For the plant (3.12) to be the k-th-order (global) elementwise trackable on $\mathfrak{T}_0^{(k+1)N} \times \mathfrak{D}^j \times \mathfrak{Y}_d^k$ it is necessary and sufficient to be (global) elementwise trackable on $\mathfrak{T}_0^{(k+1)N} \times \mathfrak{D}^j \times \mathfrak{Y}_d^k$.

10.5 Elementwise natural trackability

The natural trackability concept can also satisfy the demand for different reachability *times* to be associated with different output variables.

Definition 181 ***The k-th-order elementwise natural trackability of the plant** (3.12) **on** $\mathfrak{T}_0^{(k+1)N} \times \mathfrak{D}^j \times \mathfrak{Y}_d^k$ **and its domain***

*The plant (3.12) is **the k-th-order elementwise natural trackable on** $\mathfrak{T}_0^{(k+1)N} \times \mathfrak{D}^j \times \mathfrak{Y}_d^k$ if, and only if, for every plant output desired response $\mathbf{Y}_d(.) \in \mathfrak{Y}_d^k$, for every disturbance vector function $\mathbf{D}(.) \in \mathfrak{D}^j$, and for every vector instant $\boldsymbol{\sigma} \in (In\ \mathfrak{T}_0)^{(k+1)N}$, there is an elementwise $\boldsymbol{w}$-neighborhood $\mathcal{N}_{\mathbf{w}}\left(\mathbf{Y}_{d0}^k\right)$,*

$$\mathcal{N}_{\mathbf{w}}\left(\mathbf{Y}_{d0}^k\right) = \mathcal{N}_{\mathbf{w}}\left(\mathbf{t}_0^{(k+1)N}; \boldsymbol{\sigma}; \mathbf{Y}_{d0}^k, \mathbf{D}; \mathbf{U}; \mathbf{Y}_d\right),$$

of $\mathbf{Y}_{d0}^k$ (2.22), $\mathcal{N}_{\mathbf{w}}\left(\mathbf{Y}_{d0}^k\right) \subseteq \mathfrak{R}^{(k+1)N}$, and there is a control vector function $\mathbf{U}(.) \in \mathfrak{U}^l$ that can be synthesized without using information about the form and value of $\mathbf{D}(.) \in \mathfrak{D}^j$ and about the mathematical model of the plant state, such that for every plant initial output vector $\mathbf{Y}_0^k$ in $\mathcal{N}_{\mathbf{w}}\left(\mathbf{Y}_{d0}^k\right)$, the real plant output response $\mathbf{Y}^k(\mathbf{t}^{(l+1)N})$ becomes elementwise equal to the desired plant output response $\mathbf{Y}_d^k(\mathbf{t}^{(l+1)N})$ at the latest at the vector moment $\boldsymbol{\sigma}$, after which they rest equal forever; i.e.,

$$\begin{gathered}\forall \mathbf{Y}_d(.) \in \mathfrak{Y}_d^k,\ \forall \mathbf{D}(.) \in \mathfrak{D}^j,\ \forall \boldsymbol{\sigma} \in (In\ \mathfrak{T}_0)^{(k+1)N}, \\ \exists \mathcal{N}_{\mathbf{w}}\left(\mathbf{Y}_{d0}^k\right) = \mathcal{N}_{\mathbf{w}}\left(\mathbf{t}_0^{(k+1)N}; \boldsymbol{\sigma}; \mathbf{Y}_{d0}^k \mathbf{D}; \mathbf{U}; \mathbf{Y}_d\right) \subseteq \mathfrak{R}^{(k+1)N},\ \exists \mathbf{U}(.) \in \mathfrak{U}^l, \\ \mathbf{U}(t) \in \mathfrak{C},\ \mathbf{U}(\mathbf{t}^r) = \mathbf{U}(\mathbf{t}^r; \boldsymbol{\sigma}; \mathfrak{D}^j; \mathbf{Y}_d) \Longrightarrow \mathbf{Y}_0^k \in \mathcal{N}_{\mathbf{w}}\left(\mathbf{Y}_{d0}^k\right) \Longrightarrow \\ \mathbf{Y}^k(\mathbf{t}^{(k+1)N}) = \mathbf{Y}_d^k(\mathbf{t}^{(k+1)N}),\ \forall \left(\mathbf{t}^{(k+1)N} \geq \boldsymbol{\sigma}\right) \in \mathfrak{T}_0^{(k+1)N}. \qquad (10.16)\end{gathered}$$

Such control is **the k-th-order elementwise natural tracking control on the product set** $\mathfrak{T}_0^{(k+1)N} \times \mathfrak{D}^j \times \mathfrak{Y}_d^k$, *for short* **the k-th-order elementwise natural tracking control.**

The largest connected neighborhood $\mathcal{N}_{\mathbf{w}L}\left(\mathbf{t}_0^{(k+1)N};\boldsymbol{\sigma};\mathbf{Y}_{d0}^k\mathbf{D};\mathbf{U};\mathbf{Y}_d\right)$ *of* $\mathbf{Y}_{d0}^k$ *for any fixed* $\boldsymbol{\sigma} \in (In\ \mathfrak{T}_0)^{(k+1)N}$ *that obeys (10.16), is* **the k-th-order elementwise natural trackability domain** $\mathcal{D}_{ENTbl}^k\left(\mathbf{t}_0^{(k+1)N};\boldsymbol{\sigma};\mathbf{Y}_{d0}^k\mathbf{D};\mathbf{U};\mathbf{Y}_d\right)$ **of** $\mathbf{Y}_d^k(t)$ *for every* $[\mathbf{D}(.),\mathbf{Y}_d(.)] \in \mathfrak{D}^j \times \mathfrak{Y}_d^k$ **, i.e., on** $\mathfrak{D}^j \times \mathfrak{Y}_d^k$, **at** $t_0 \in \mathfrak{T}_i$ **for the fixed** $\boldsymbol{\sigma} \in (In\ \mathfrak{T}_0)^{(k+1)N}$,

$$\mathcal{D}_{ENTbl}^k\left(\mathbf{t}_0^{(k+1)N};\boldsymbol{\sigma};\mathbf{Y}_{d0}^k\mathbf{D};\mathbf{U};\mathbf{Y}_d\right) = \mathcal{N}_{\mathbf{w}L}\left(\mathbf{t}_0^{(k+1)N};\boldsymbol{\sigma};\mathbf{Y}_{d0}^k\mathbf{D};\mathbf{U};\mathbf{Y}_d\right), \quad \boldsymbol{\sigma} \in (In\ \mathfrak{T}_0)^{(k+1)N}. \tag{10.17}$$

b) The intersection over $\boldsymbol{\sigma} \in (In\ \mathfrak{T}_0)^{(k+1)N}$ **of the k-th-order elementwise natural trackability domains** $\mathcal{D}_{ENTbl}^k\left(\mathbf{t}_0^{(k+1)N};\boldsymbol{\sigma};\mathbf{Y}_{d0}^k\mathbf{D};\mathbf{U};\mathbf{Y}_d\right)$ *of* $\mathbf{Y}_{d0}^k$, *(10.17), is* **the k-th-order elementwise natural trackability domain**

$$\mathcal{D}_{ENTbl}^k\left(\mathbf{t}_0^{(k+1)N};\mathbf{Y}_{d0}^k\mathbf{D};\mathbf{U};\mathbf{Y}_d\right)$$

of $\mathbf{Y}_d^k(t)$ *for every* $[\mathbf{D}(.),\mathbf{Y}_d(.)] \in \mathfrak{D}^j \times \mathfrak{Y}_d^k$, **i.e., on** $\mathfrak{D}^j \times \mathfrak{Y}_d^k$, **at** $t_0 \in \mathfrak{T}_i$,

$$\mathcal{D}_{ENTbl}^k\left(\mathbf{t}_0^{(k+1)N};\mathbf{Y}_{d0}^k\mathbf{D};\mathbf{U};\mathbf{Y}_d\right) = \cap \begin{bmatrix} \mathcal{D}_{ENTbl}^k\left(\mathbf{t}_0^{(k+1)N};\boldsymbol{\sigma};\mathbf{Y}_{d0}^k\mathbf{D};\mathbf{U};\mathbf{Y}_d\right): \\ \boldsymbol{\sigma} \in (In\ \mathfrak{T}_0)^{(k+1)N}. \end{bmatrix}. \tag{10.18}$$

c) The zero $(k=0)$ *order elementwise natural trackability on* $\mathfrak{T}_0^N \times \mathfrak{D}^j \times \mathfrak{Y}_d^k$ *is called* **elementwise natural trackability on** $\mathfrak{T}_0^N \times \mathfrak{D}^j \times \mathfrak{Y}_d^k$. *The zero* $(k=0)$ *order elementwise natural tracking control on* $\mathfrak{T}_0^N \times \mathfrak{D}^j \times \mathfrak{Y}_d^k$ *is called* **elementwise natural tracking control on** $\mathfrak{T}_0^N \times \mathfrak{D}^j \times \mathfrak{Y}_d^k$, *for short,* **elementwise natural tracking control.**

d) The k-th-order elementwise natural trackability on $\mathfrak{T}_0^{(k+1)N} \times \mathfrak{D}^j \times \mathfrak{Y}_d^k$ *is:*

- **Global for the fixed** $\boldsymbol{\sigma} \in (In\ \mathfrak{T}_0)^{(k+1)N}$ *(equivalently,* **in the whole for the fixed** $\boldsymbol{\sigma} \in (In\ \mathfrak{T}_0)^{(k+1)N}$*) if, and only if,*

$$\mathcal{D}_{ENTbl}^k\left(\mathbf{t}_0^{(k+1)N};\boldsymbol{\sigma};\mathbf{Y}_{d0}^k\mathbf{D};\mathbf{U};\mathbf{Y}_d\right) = \mathfrak{R}^{(k+1)N}, \quad \forall [\mathbf{D}(.),\mathbf{Y}_d(.)] \in \mathfrak{D}^j \times \mathfrak{Y}_d^k,$$

i.e., on $\mathfrak{D}^j \times \mathfrak{Y}_d^k$, **at** $t_0 \in \mathfrak{T}_i$,

- **Global (in the whole)** *if, and only if,* $\mathcal{D}_{ENTbl}^k\left(\mathbf{t}_0^{(k+1)N};\mathbf{Y}_{d0}^k\mathbf{D};\mathbf{U};\mathbf{Y}_d\right) = \mathfrak{R}^{(k+1)N}$ *for every* $[\mathbf{D}(.),\mathbf{Y}_d(.)] \in \mathfrak{D}^j \times \mathfrak{Y}_d^k$, *i.e., on* $\mathfrak{D}^j \times \mathfrak{Y}_d^k$, **at** $t_0 \in \mathfrak{T}_i$.

e) The k-th-order elementwise natural trackability on $\mathfrak{T}_0^{(k+1)N} \times \mathfrak{D}^j \times \mathfrak{Y}_d^k$ *is:*

- **Uniform for the fixed** $\boldsymbol{\sigma} \in (In\ \mathfrak{T}_0)^{(k+1)N}$ *over* $\mathfrak{T}_0^{(k+1)N} \times \mathfrak{D}^j \times \mathfrak{Y}_d^k$ *if, and only if,* $\mathcal{D}_{ENTblu}^k\left(\boldsymbol{\sigma};\mathbf{Y}_{d0}^k;\mathfrak{D}^j;\mathbf{U};\mathfrak{Y}_d^k\right)$,

$$\mathcal{D}_{ENTblu}^k\left(\boldsymbol{\sigma};\mathbf{Y}_{d0}^k;\mathfrak{D}^j;\mathbf{U};\mathfrak{Y}_d^k\right) = \cap \begin{bmatrix} \mathcal{D}_{ENTbl}^k\left(\mathbf{t}_0^{(k+1)N};\boldsymbol{\sigma};\mathbf{Y}_{d0}^k\mathbf{D};\mathbf{U};\mathbf{Y}_d\right): \\ \left[\mathbf{t}_0^{(k+1)N},\mathbf{D}(.),\mathbf{Y}_d(.)\right] \in \mathfrak{T}^{(k+1)N} \times \mathfrak{D}^j \times \mathfrak{Y}_d^k \end{bmatrix}$$

is open connected neighborhood of $\mathbf{Y}_{d0}^k$ *of every* $\mathbf{Y}_d(.) \in \mathfrak{Y}_d^k$ *and* $\mathbf{U}(.)$ *depends on* $\boldsymbol{\sigma}$ *and on* $\mathfrak{D}^j \times \mathfrak{Y}_d^k$ *but not on an individual triplet* $\left[\mathbf{t}_0^{(k+1)N},\mathbf{D}(.),\mathbf{Y}_d(.)\right]$ *from* $\mathfrak{T}^{(k+1)N} \times \mathfrak{D}^j \times \mathfrak{Y}_d^k$, $\mathbf{U}(\mathbf{t}^r;\mathbf{t}_0^{rN},\boldsymbol{\sigma};\mathbf{D},\mathbf{Y}_d) = \mathbf{U}(\mathbf{t}^r;\boldsymbol{\sigma};\mathfrak{D}^j;\mathfrak{Y}_d^k)$.

Then, and only then, $\mathcal{D}^k_{ENTblu}\left(\boldsymbol{\sigma};\mathbf{Y}^k_{d0};\mathfrak{D}^j;\mathbf{U};\mathfrak{Y}^k_d\right)$ *is* ***the k-th-order uniform elementwise natural trackability domain of*** $\mathbf{Y}^k_d(\mathbf{t}_0^{(k+1)N})$ ***on*** $\mathfrak{T}^{(k+1)N}\times\mathfrak{D}^j\times\mathfrak{Y}^k_d$ ***for the fixed*** $\boldsymbol{\sigma}\in(In\;\mathfrak{T}_0)^{(k+1)N}$,

- ***Uniform*** *over* $\mathfrak{T}_0^{(k+1)N}\times\mathfrak{D}^j\times\mathfrak{Y}^k_d$ *if, and only if,* $\mathcal{D}^k_{ENTblu}\left(\mathbf{Y}^k_{d0};\mathfrak{D}^j;\mathbf{U};\mathfrak{Y}^k_d\right)$,

$$\mathcal{D}^k_{ENTblu}\left(\mathbf{Y}^k_{d0};\mathfrak{D}^j;\mathbf{U};\mathfrak{Y}^k_d\right)=\cap\left[\begin{array}{c}\mathcal{D}^k_{ENTbl}\left(\mathbf{t}_0^{(k+1)N};\mathbf{Y}^k_{d0}\mathbf{D};\mathbf{U};\mathbf{Y}_d\right):\\ \left[\mathbf{t}_0^{(k+1)N},\mathbf{D}(.),\mathbf{Y}_d(.)\right]\in\mathfrak{T}^{(k+1)N}\times\mathfrak{D}^j\times\mathfrak{Y}^k_d.\end{array}\right]\tag{10.19}$$

is an open connected neighborhood of $\mathbf{Y}^k_{d0}$ *of every* $\mathbf{Y}_d(.)\in\mathfrak{Y}^k_d$ *and* $\mathbf{U}(.)$ *depends on* $\mathfrak{D}^j\times\mathfrak{Y}^k_d$ *but not on an individual triplet* $\left[\mathbf{t}_0^{(k+1)N},\mathbf{D}(.),\mathbf{Y}_d(.)\right]$ *from* $\mathfrak{T}^{(k+1)N}\times\mathfrak{D}^j\times\mathfrak{Y}^k_d$, $\mathbf{U}(\mathbf{t}^r;\mathbf{t}_0^{rN},\mathbf{D},\mathbf{Y}_d)=\mathbf{U}(\mathbf{t}^r;\mathfrak{D}^j;\mathfrak{Y}^k_d)$.

Then, and only then, $\mathcal{D}^k_{ENTblu}\left(\mathbf{Y}^k_{d0};\mathfrak{D}^j;\mathbf{U};\mathfrak{Y}^k_d\right)$ *is the* ***the k-th-order uniform elementwise natural trackability domain of*** $\mathbf{Y}^k_d(\mathbf{t}_0^{(k+1)N})$ ***on*** $\mathfrak{T}^{(k+1)N}\times\mathfrak{D}^j\times\mathfrak{Y}^k_d$.

Note 182 Complete natural trackability and elementwise natural trackability
The elementwise natural trackability is simultaneously the complete natural trackability.

Comment 183 *The k-th-order elementwise trackability is necessary for the k-th-order elementwise natural trackability. The latter is sufficient for the former.*

Lemma 178 and Definition 181 induce the following.

Lemma 184 ***The k-th-order elementwise natural trackability on*** $\mathfrak{T}_0^{(k+1)N}\times\mathfrak{D}^j\times\mathfrak{Y}^k_d$ ***and the elementwise natural trackability on*** $\mathfrak{T}_0^{(k+1)N}\times\mathfrak{D}^j\times\mathfrak{Y}^k_d$

For the plant (3.12) to be the k-th-order (global) elementwise natural trackable on the product set $\mathfrak{T}_0^{(k+1)N}\times\mathfrak{D}^j\times\mathfrak{Y}^k_d$ *it is necessary and sufficient to be (global) elementwise natural trackable on the product set* $\mathfrak{T}_0^{(k+1)N}\times\mathfrak{D}^j\times\mathfrak{Y}^k_d$.

Note 185 *The trackability concept represents the important bridge between the engineer who designs the plant and the engineer who designs the controller, and the control system, for the plant.*

Trackability is the fundamental link between the manufacturer of the plant and the manufacturer of the controller, and of the control system, for the plant.

Trackability is also the link between the dynamics and mathematical modeling of the plant and the controller. It is significant for control system design.

Chapter 11

Imperfect trackability criteria

11.1 Output space criteria

The imperfect trackability, for short ***the trackability***, expresses the plant ability to permit the existence of a control that can steer the output vector from its arbitrary initial value, which belongs to a neighborhood of the initial desired output vector (Definition 168), to its desired value in a finite *time* and, after that, the control should ensure that the real output vector stays always equal to the *time*-varying desired output vector.

The nonzero initial error vector $\boldsymbol{\varepsilon}_0 \neq \mathbf{0}_N$ disables perfect tracking. It imposes the imperfect tracking that opened the problem of imperfect tracking and its quality. The low quality of tracking is when it is just ordinary tracking ensuring zero error vector only at infinite *time* $t = \infty$.

Let us accept a vector function $\mathbf{v}(.) : \mathfrak{T}_0 \times \mathrm{In}\mathfrak{T}_0 \times \mathfrak{R}^{(k+1)N} \longrightarrow \mathfrak{R}^m$ with the following properties:

$$\begin{gathered}
\mathbf{v}\left(t, \sigma, \boldsymbol{\varepsilon}^k\right) \in C\left(\mathfrak{T}_0 \times \mathrm{In}\mathfrak{T}_0 \times \mathfrak{R}^{(k+1)N}\right), \\
\mathbf{v}\left(t_0, \sigma, \boldsymbol{\varepsilon}_0^k\right) = \mathbf{0}_m, \ \ \forall \left(t_0, \sigma, \boldsymbol{\varepsilon}_0^k\right) \in \mathfrak{T} \times \mathrm{In}\mathfrak{T}_0 \times \mathfrak{R}^{(k+1)N}, \\
k = 0, \ \ m = N \Longrightarrow \mathbf{v}\left(t, \sigma, \boldsymbol{\varepsilon}\right) = \boldsymbol{\varepsilon}, \ \ \forall t \in [\sigma, \infty[, \ \forall \left(\sigma, \boldsymbol{\varepsilon}^k\right) \in \mathrm{In}\mathfrak{T}_0 \times \mathfrak{R}^{(k+1)N}, \\
\mathbf{v}\left(t, \sigma, \boldsymbol{\varepsilon}^k\right) = \mathbf{0}_m \Longleftrightarrow \boldsymbol{\varepsilon}^k = \mathbf{0}_{(k+1)N}, \ \ \forall t \in [\sigma, \infty[, \ \forall \left(\sigma, \boldsymbol{\varepsilon}^k\right) \in \mathrm{In}\mathfrak{T}_0 \times \mathfrak{R}^{(k+1)N}. \qquad (11.1)
\end{gathered}$$

If $r < N$ and rank $W(t) \equiv \gamma \leq r$ then the output vector function $\mathbf{Y}(.)$, $\mathbf{Y}(t) \in \mathfrak{R}^N$, can be controlled only indirectly by controlling elementwise the vector $\mathbf{v}(.)$, (11.1), $\mathbf{v}\left(t, \sigma, \boldsymbol{\varepsilon}^k\right) \in \mathfrak{R}^m$ with $m \leq \gamma$, Axiom 87.

An example of the function $\mathbf{v}(.)$, (11.1), is given for the *time*-invariant linear *IO* plants in [279, p. 197, the equation (9.48)]. Its generalization follows:

$$\mathbf{v}\left(t, \sigma, \boldsymbol{\varepsilon}^k\right) = \Psi\left(t; t_0; k\right)\left[\left|\boldsymbol{\varepsilon}^k(t)\right| - \psi\left(t; t_0; \sigma\right)\left|\boldsymbol{\varepsilon}^k(t_0)\right|\right] \in \mathfrak{R}^m, \qquad (11.2)$$

where $\Psi(t; t_0; k)$ can be accepted as in (8.5) for $\sigma \in \mathrm{In}\mathfrak{T}_0$ and for

$$\psi(.) : \mathfrak{T}_0 \times \mathfrak{T} \times \mathfrak{T}_0 \longrightarrow \mathfrak{R}_+, \ \psi\left(t; t_0; \sigma\right) \left\{ \begin{array}{cc} = 1, & t = t_0, \\ \in]0, 1], & \forall t \in [t_0, \sigma[, \\ = 0, & \forall t \in [\sigma, \infty[\end{array} \right\},$$

$$e.g., \ \psi\left(t; t_0; \sigma\right) = \left\{ \begin{array}{cc} \left(1 - \frac{t - t_0}{\sigma - t_0}\right)^{\nu}, & \forall t \in [t_0, \sigma], \\ 0, & \forall t \in [\sigma, \infty[\end{array} \right\}, \ \ \nu \in \mathfrak{R}^+. \qquad (11.3)$$

Notice that $\mathbf{v}\left(t,\sigma,\varepsilon^k\right)$, (11.1), becomes $\mathbf{v}\left(t,\varepsilon^k\right)$, (11.1), for $t\geq\sigma$:

$$\mathbf{v}\left(t,\sigma,\varepsilon^k\right)=\mathbf{v}\left(t,\varepsilon^k\right),\quad \forall t\geq\sigma,$$

also

$$m=N\ and\ k=0\Longrightarrow \mathbf{v}\left(t,\sigma,\varepsilon\right)=\mathbf{v}\left(t,\varepsilon\right)=\varepsilon,\ \ \forall t\geq\sigma,$$

which is important for the elementwise trackability.

Theorem 186 ***The conditions for the trackability on*** $\mathfrak{T}_0\times\mathfrak{D}^j\times\mathfrak{Y}_d^k$ ***via the output space of the plant (3.12) with Properties*** *57, 58, 64, 78*

In order for the plant (3.12) with Properties 57, 58, 78 to be trackable via the output space on the product set $\mathfrak{T}_0\times\mathfrak{D}^j\times\mathfrak{Y}_d^k$ *it is necessary and sufficient that both i) and ii) hold:*

i) The rank γ *of the control matrix* $W(t)$ *is greater than zero and not greater than min* (N,r), *i.e., (8.10) is true, and for the elementwise trackability* $m=N$ *in* $\mathbf{v}(t,\sigma,\varepsilon)$ *defined by (11.1).*

ii) The control vector function $\mathbf{U}(.)$ *satisfies one of the conditions 1) to 4):*

1) The control vector function $\mathbf{U}(.)$ *obeys (11.4) for* $\gamma=$*rank* $W(t)=r<N$, $\forall t\in\mathfrak{T}_0$, *and for the chosen vector function* $\mathbf{v}(.):\mathfrak{T}_0\times In\mathfrak{T}_0\times\mathfrak{R}^{(k+1)N}\longrightarrow\mathfrak{R}^m$, *(11.1), where* m *is not greater than* r, $m\leq r$, *rank* $M_{r,m}(t)\equiv m$ *and for* $m=r=\gamma$ *the matrix* $M_{r,m}(t)$ *is the identity matrix* I_r, $M_{r,r}(t)\equiv I_r$:

$$\begin{gathered}\mathbf{U}(t)=\mathbf{w}^I\left\{\Gamma(t)\left[\mathbf{Y}-\mathbf{z}(t,\mathbf{R}^{\alpha-1},\mathbf{D})\right]+M_{r,m}(t)\,\mathbf{v}(t,\sigma,\varepsilon)\right\}\\ \Gamma(t)=\left[W^T(t)\,W(t)\right]^{-1}W^T(t)\in\mathfrak{R}^{r\times N},\quad\forall t\in\mathfrak{T}_0,\end{gathered}\tag{11.4}$$

The domain $\mathcal{D}_{Tbl}\left(t_0;\mathfrak{D}^j;\mathbf{U};\mathfrak{Y}_d^k\right)$ *of the trackability on* $\mathfrak{T}_0\times\mathfrak{D}^j\times\mathfrak{Y}_d^k$ *of the plant (3.12) with Properties 57, 58, 78 is determined by (8.16).*

2) The control vector function $\mathbf{U}(.)$ *obeys (11.5) for* $\gamma=$*rank* $W(t)=N<r$, $\forall t\in\mathfrak{T}_0$, *and for the chosen vector function* $\mathbf{v}(.):\mathfrak{T}_0\times\mathfrak{R}^N\longrightarrow\mathfrak{R}^m$, *(11.1), rank* $M_{N,m}(t)\equiv m$ *where* m *is not greater than* N, $m\leq N$, *and* $m=N$ *if the perfect trackability is also elementwise:*

$$\begin{gathered}\mathbf{U}(t)=\mathbf{w}^I\left\{\Gamma(t)\left[\mathbf{Y}-\mathbf{z}(t,\mathbf{R}^{\alpha-1},\mathbf{D})\right]+W^T(t)\,M_{N,m}(t)\,\mathbf{v}(t,\sigma,\varepsilon)\right\}\\ \Gamma(t)=W^T(t)\left[W(t)\,W^T(t)\right]^{-1}\in\mathfrak{R}^{r\times N},\quad\forall t\in\mathfrak{T}_0,\end{gathered}\tag{11.5}$$

The domain $\mathcal{D}_{Tbl}\left(t_0;\mathfrak{D}^j;\mathbf{U};\mathfrak{Y}_d^k\right)$ *of the trackability on* $\mathfrak{T}_0\times\mathfrak{D}^j\times\mathfrak{Y}_d^k$ *of the plant (3.12) with Properties 57, 58, 78 is determined by (8.17).*

3) The control vector function $\mathbf{U}(.)$ *obeys (11.6) if* $\gamma=$*rank* $W(t)=r=N$, $\forall t\in\mathfrak{T}_0$:

$$\mathbf{U}(t)=\mathbf{w}^I\left\langle W(t)^{-1}\left\{\begin{array}{c}[1-\psi(t;t_0;\sigma)]\,\mathbf{Y}_d(t)-\\ -\mathbf{z}(\left[t,\mathbf{R}^{\alpha-1}(t),\mathbf{D}(t)\right]\end{array}\right\}\right\rangle,\ \forall t\in\mathfrak{T}_0,\tag{11.6}$$

The domain $\mathcal{D}_{Tbl}\left(t_0;\mathfrak{D}^j;\mathbf{U};\mathfrak{Y}_d^k\right)$ *of the trackability on* $\mathfrak{T}_0\times\mathfrak{D}^j\times\mathfrak{Y}_d^k$ *of the plant (3.12) with Properties 57, 58, 78 is determined by (8.18).*

4) If the plant **(3.12)** *also possesses Property 64 then the control vector function* $\mathbf{U}(.)$ *obeys (11.7),*

$$\begin{gathered}\mathbf{w}_\gamma\left[\mathbf{U}_\gamma(t)\right]=W_\gamma^{-1}(t)\left\{\begin{array}{c}\mathbf{Y}_\gamma(t)-\mathbf{z}_\gamma\left[t,\mathbf{R}^{\alpha-1}(t),\mathbf{D}(t)\right]\\ +\,M_{\gamma,m}(t)\,\mathbf{v}(t,\sigma,\varepsilon)\end{array}\right\},\\ \forall\left[t,\mathbf{D}(.),\mathbf{Y}_d(.)\right]\in\mathfrak{T}_0\times\mathfrak{D}^j\times\mathfrak{Y}_d^k,\end{gathered}\tag{11.7}$$

where the matrix $M_{\gamma,m}(t)$ *is defined in (8.15).*

The initial control satisfies

$$\mathbf{w}_{\gamma}\left(\mathbf{U}_{0}\right)=W_{\gamma}^{-1}\left(t_{0}\right)\left[\mathbf{Y}_{0}-\mathbf{z}_{\gamma}\left(t_{0},\mathbf{R}_{0}^{\alpha-1},\mathbf{D}_{0}\right)\right],\quad \forall t_{0}\in\mathfrak{T}. \tag{11.8}$$

*The trackability domain $\mathcal{D}_{Tbl}\left(t_0;\mathfrak{D}^j;\mathbf{U};\mathfrak{Y}_d^k\right)$ on $\mathfrak{T}_0\times\mathfrak{D}^j\times\mathfrak{Y}_d^k$ of the plant **(3.12)** with Properties 57, 58, 64, 78 is determined by (8.19).*

In all cases ii-1) through ii-4) the control $\mathbf{U}(t)$ becomes the perfect trackability control on $[\sigma,\infty[$ in the output space.

Proof. When $\mathbf{v}\left(.\right):\mathfrak{T}_0\times\mathfrak{R}^N\longrightarrow\mathfrak{R}^m$, (8.1), is replaced by the vector function $\mathbf{v}\left(.\right):\mathfrak{T}_0\times In\mathfrak{T}_0\times\mathfrak{R}^{(k+1)N}\longrightarrow\mathfrak{R}^m$, (11.1), in Theorem 149 then it becomes Theorem 186 ■

Comment 187 *The control (11.4) - (11.7) depends on the disturbance vector. This restricts the application of the control (11.7) only to measurable disturbances.*

11.2 State space criteria

For the tracking control synthesis only via the state space we state:

Theorem 188 ***The conditions for the trackability via the state space on $\mathfrak{T}_0\times\mathfrak{D}^j\times\mathfrak{Y}_d^{\alpha}$ of the plant (3.12) with Properties 57, 58, and 78***

For the plant (3.12), which possesses Properties 57, 58, and 78 to be trackable via the state space on $\mathfrak{T}_0\times\mathfrak{D}^j\times\mathfrak{Y}_d^{\alpha}$ it is necessary and sufficient that both i) and ii) hold:

i) the rank β of the control matrix $P\left(t\right)$ is greater than zero and not greater than $\min(\rho,r)$, i.e., (8.20) holds,

ii) The control vector function $\mathbf{U}\left(.\right)$ satisfies one of the conditions 1) to 4):

1) The control vector function $\mathbf{U}\left(.\right)$ obeys (8.21) for $\beta=\text{rank}\,P\left(t\right)=r<\rho$, $\forall t\in\mathfrak{T}_0$, and for the chosen vector function $\mathbf{v}\left(.\right):\mathfrak{T}_0\times In\mathfrak{T}_0\times\mathfrak{R}^{(k+1)N}\longrightarrow\mathfrak{R}^m$, (11.1), where m is not greater than r, $m\le r$, rank $L_{r,m}\left(t\right)\equiv m$ and $L_{r,r}\left(t\right)\equiv I_r$

The domain $\mathcal{D}_{Tbl}\left(t_0;\mathfrak{D}^j;\mathbf{U};\mathfrak{Y}_d^{\alpha}\right)$ of the trackability on $\mathfrak{T}_0\times\mathfrak{D}^j\times\mathfrak{Y}_d^{\alpha}$ of the plant (3.12) with Properties 57, 58 and 78 is determined by (8.26).

2) The control vector function $\mathbf{U}\left(.\right)$ obeys (8.22) for $\beta=\text{rank}\,P\left(t\right)=\rho<r$, $\forall t\in\mathfrak{T}_0$, and for the chosen vector function $\mathbf{v}\left(.\right):\mathfrak{T}_0\times In\mathfrak{T}_0\times\mathfrak{R}^{(k+1)N}\longrightarrow\mathfrak{R}^m$, (11.1), where m is not greater than r, $m\le r$, and rank $L_{r,m}\left(t\right)\equiv m$.

The domain $\mathcal{D}_{Tbl}\left(t_0;\mathfrak{D}^j;\mathbf{U};\mathfrak{Y}_d^{\alpha}\right)$ of the trackability on $\mathfrak{T}_0\times\mathfrak{D}^j\times\mathfrak{Y}_d^{\alpha}$ of the plant (3.12) with Properties 57, 58 and 78 is determined by (8.27).

3) For the chosen vector function $\mathbf{v}\left(.\right):\mathfrak{T}_0\times In\mathfrak{T}_0\times\mathfrak{R}^{(k+1)N}\longrightarrow\mathfrak{R}^m$, (11.1), where m is not greater than r, $m\le r$, and rank $L_{r,m}\left(t\right)\equiv m$: the control vector function $\mathbf{U}\left(.\right)$ obeys (8.23) if $\beta=\text{rank}\,P\left(t\right)=r=\rho$, $\forall t\in\mathfrak{T}_0$.

The domain $\mathcal{D}_{Tbl}\left(t_0;\mathfrak{D}^j;\mathbf{U};\mathfrak{Y}_d^{\alpha}\right)$ of the trackability on $\mathfrak{T}_0\times\mathfrak{D}^j\times\mathfrak{Y}_d^{\alpha}$ of the plant (3.12) with Properties 57, 58 and 78 is determined by (8.28).

4) If the plant also possesses Property 61 then the control vector function $\mathbf{U}\left(.\right)$ obeys (11.9),

$$\begin{gathered}\mathbf{e}_{u\beta}\left[\mathbf{U}_{\beta}^{\mu}(t)\right]\\ =P_{\beta}^{-1}\left(t\right)\left\{\begin{array}{c}Q_{\beta,\rho}(t,\mathbf{R}^{\alpha-1})\left[\begin{array}{c}\mathbf{R}^{(\alpha)}(t)\\ +\,L_{\rho,m}\left(t\right)\mathbf{v}\left(t,\sigma,\varepsilon\right)\end{array}\right]+\\ +\,\mathbf{q}_{\beta}(t,\mathbf{R}^{\alpha-1},\mathbf{D}^{i})-E_{\beta,d}(t)\mathbf{e}_{d}\left[\mathbf{D}^{\eta}(t)\right]\end{array}\right\},\\ \forall\left[t,\mathbf{D}(.),\mathbf{Y}_{d}(.)\right]\in\mathfrak{T}_{0}\times\mathfrak{D}^{j}\times\mathfrak{Y}_{d}^{\alpha}.\end{gathered} \tag{11.9}$$

where the equation (8.25) determines the matrix $L(t)$ *and* $\mathbf{v}(t,\sigma,\varepsilon)$ *is defined by (11.1). The equation (11.10) determines the initial control vector* $\mathbf{U}_0$:

$$\mathbf{e}_u\left(\mathbf{U}_0^{\mu}\right) = P_{\beta}^{-1}(t_0)\left\{ \begin{array}{c} Q_{\beta,\rho}(t_0,\mathbf{R}_0^{\alpha-1})\mathbf{R}_0^{(\alpha)} + \mathbf{q}_{\beta}(t_0,\mathbf{R}_0^{\alpha-1},\mathbf{D}_0^{i}) - \\ -E_{\beta,d}(t_0)\mathbf{e}_d\left(\mathbf{D}_0^{\eta}\right) \end{array} \right\},$$
$$\forall\,[t,\mathbf{D}(.)] \in \mathfrak{T}_0 \times \mathfrak{D}. \tag{11.10}$$

The domain $\mathcal{D}_{TBl}\left(t_0;\mathfrak{D}^j;\mathbf{U};\mathfrak{Y}_d^{\alpha}\right)$ *of the perfect trackability on* $\mathfrak{T}_0 \times \mathfrak{D}^j \times \mathfrak{Y}_d^{\alpha}$ *of the plant (3.12) with Properties 57, 58, 61, and 78 is determined by*

$$\mathcal{D}_{PTBl}\left(t_0;\mathfrak{D}^j;\mathbf{U};\mathfrak{Y}_d^{\alpha}\right) = \mathcal{D}_{PTBl}\left(t_0;\mathfrak{D}^j;\mathfrak{U};\mathfrak{Y}_d^{\alpha}\right)$$
$$= \left\{ \begin{array}{c} (\mathbf{R},\mathbf{Y}) \in \mathfrak{R}^N \times \mathfrak{R}^n : \\ \left| \begin{array}{c} P_{\beta}^{-1}(t) \\ \bullet \left\{ \begin{array}{c} Q_{\beta,\rho}(t_0,\mathbf{R}^{\alpha-1})\mathbf{R}_{d0}^{(\alpha)} \\ +\ \mathbf{q}_{\beta}(t_0,\mathbf{R}^{\alpha-1},\mathbf{D}^{i}) \\ -\ E_{\beta,d}(t_0)\mathbf{e}_d\left(\mathbf{D}^{\eta}\right) \end{array} \right\} \end{array} \right| \\ \in \left[\mathbf{e}_{umin}\left(\mathbf{U}^{\mu}\right), \mathbf{e}_{uMax}\left(\mathbf{U}^{\mu}\right)\right],\ \mathbf{U} \in \mathfrak{U}^{\mu},\ \forall \mathbf{D} \in \mathfrak{D}^j. \end{array} \right\}. \tag{11.11}$$

Proof. When $\mathbf{v}(.) : \mathfrak{T}_0 \times \mathfrak{R}^N \longrightarrow \mathfrak{R}^m$, (8.1), is replaced by the vector function $\mathbf{v}(.) : \mathfrak{T}_0 \times \text{In}\mathfrak{T}_0 \times \mathfrak{R}^{(k+1)N} \longrightarrow \mathfrak{R}^m$, (11.1), in Theorem 154 then it becomes Theorem 188 ■

The preceding theorem holds independently of the properties of the control matrix $W(t)$ of the output equation of the plant (3.12).

11.3 Both spaces criteria

Theorem 189 ***The conditions for the trackability on*** $\mathfrak{T}_0 \times \mathfrak{D}^j \times \mathfrak{Y}_d^{\alpha}$ ***of the plant (3.12) with Properties*** *57, 58, 61, 64, 78*

For the plant (3.12), which possesses Properties 57, 58, 61, 64, 78 to be trackable on $\mathfrak{T}_0 \times \mathfrak{D}^j \times \mathfrak{Y}_d^{\alpha}$ *it is necessary and sufficient that both i) and ii) hold:*

i) The ranks of the matrices $P(t)$ *and* $W(t)$ *are always greater than zero and (8.31) holds,*

and

ii) There exists control vector function $\mathbf{U}(.)$ *that obeys (11.12) for any chosen matrix* $H(t) \in \mathfrak{R}^{2r\times m}$ *that satisfies (8.30),*

$$\left[\begin{array}{c} \mathbf{e}_{u\beta}\left[\mathbf{U}_{\beta}^{\mu}(t)\right] \\ \mathbf{w}_{\gamma}\left[\mathbf{U}_{\gamma}(t)\right] \end{array}\right] = H(t)\,\mathbf{v}(t,\sigma,\varepsilon) + \left[\begin{array}{cc} P_{\beta}^{-1}(t) & O_{\beta,\gamma} \\ O_{\gamma,\beta} & W_{\gamma}^{-1}(t) \end{array}\right]$$
$$\bullet\left[\begin{array}{c} \left\{Q_{\beta,\rho}\left[t,\mathbf{R}^{\alpha-1}(t)\right]\mathbf{R}^{(\alpha)}(t) + \mathbf{q}_{\beta}\left[t,\mathbf{R}^{\alpha-1}(t)\right] - E_{\beta,d}(t)\,\mathbf{e}_d\left[\mathbf{D}^{\eta}(t)\right]\right\} \\ \mathbf{Y}_{\gamma}(t) - \mathbf{z}_{\gamma}\left[t,\mathbf{R}^{\alpha-1}(t),\mathbf{D}(t)\right] \end{array}\right],$$
$$\forall\,[t,\mathbf{D}(.),\mathbf{Y}_d(.)] \in \mathfrak{T}_0 \times \mathfrak{D}^j \times \mathfrak{Y}_d^{\alpha}. \tag{11.12}$$

This equation at $t = t_0$ *determines the initial control vector* $\mathbf{U}_0$.

The domain $\mathcal{D}_{TBl}\left(t_0;\mathfrak{D}^j;\mathbf{U};\mathfrak{Y}_d^{\alpha}\right)$ *of the perfect trackability on* $\mathfrak{T}_0 \times \mathfrak{D}^j \times \mathfrak{Y}_d^{\alpha}$ *of the plant (3.12) with Properties 57, 58, 61, 64, 78 is determined by (8.33).*

Proof. The proof of this theorem is the direct synthesis of the proofs of Theorems 157, 186, and 188 ■

Chapter 12

Imperfect natural trackability criteria

12.1 Output space criteria

Definition 171 determines the natural trackability of the plant **(3.12)** on the product set $\mathfrak{T}_0 \times \mathfrak{D}^j \times \mathfrak{Y}_d^k$.

Theorem 190 ***The conditions for the natural trackability on*** $\mathfrak{T}_0 \times \mathfrak{D}^j \times \mathfrak{Y}_d^k$ ***of the plant (3.12) with Properties*** *57, 58, 78*

In order for the plant (3.12) with Properties 57, 58, and 78 to be natural trackable via the output space on $\mathfrak{T}_0 \times \mathfrak{D}^j \times \mathfrak{Y}_d^k$ *it is necessary and sufficient that both i) and ii) hold:*

i) The rank γ *of the control matrix* $W(t)$ *is greater than zero and not greater than min* (N, r)*, i.e., (8.10) is true, and for the elementwise trackability* $m = \gamma = N$ *in* $\mathbf{v}(t, \sigma, \varepsilon)$ *defined by (11.1).*

ii) The control vector function $\mathbf{U}(.)$ *satisfies one of the conditions 1) to 4):*

1) The control vector function $\mathbf{U}(.)$ *obeys (12.1) for* γ *=rank* $W(t) = r < N$, $\forall t \in \mathfrak{T}_0$*, and for the chosen vector function* $\mathbf{v}(.) : \mathfrak{T}_0 \times In\mathfrak{T}_0 \times \mathfrak{R}^{(k+1)N} \longrightarrow \mathfrak{R}^m$*, (11.1), where* m *is not greater than* r, $m \leq r$*,rank* $M_{r,m}(t) \equiv m$ *and for* $m = r = \gamma$ *the matrix* $M_{r,m}(t)$ *is the identity matrix* I_r, $M_{r,r}(t) \equiv I_r$*:*

$$\mathbf{U}_r(t) = \mathbf{w}^I \left\{ \mathbf{w}\left[\mathbf{U}_r\left(t^-\right)\right] + M_{r,m}(t)\,\mathbf{v}(t, \sigma, \varepsilon) \right\}, \quad \forall t \in \mathfrak{T}_0, \tag{12.1}$$

The domain $\mathcal{D}_{NTbl}\left(t_0; \mathfrak{D}^j; \mathbf{U}; \mathfrak{Y}_d^k\right)$ *of the natural trackability of the plant (3.12) on the space product* $\mathfrak{T}_0 \times \mathfrak{D}^j \times \mathfrak{Y}_d^k$ *is determined by (8.16).*

2) The control vector function $\mathbf{U}(.)$ *obeys (12.2) for* γ *=rank* $W(t) = N < r$, $\forall t \in \mathfrak{T}_0$*, and for the chosen vector function* $\mathbf{v}(.) : \mathfrak{T}_0 \times \mathfrak{R}^N \longrightarrow \mathfrak{R}^m$*, (11.1), rank* $M_{N,m}(t) \equiv m$, $\forall t \in \mathfrak{T}_0$*, where* m *is not greater than* N, $m \leq N$*, and* $m = N$ *if the natural trackability is also elementwise:*

$$\mathbf{U}(t) = \mathbf{w}^I \left\{ \Gamma(t)\,\mathbf{w}\left[\mathbf{U}\left(t^-\right)\right] + W^T(t)\,M_{N,m}(t)\,\mathbf{v}(t, \sigma, \varepsilon) \right\}$$
$$\Gamma(t) = W^T(t)\left[W(t)\,W^T(t)\right]^{-1} W(t) \in \mathfrak{R}^{r \times r}, \quad \forall t \in \mathfrak{T}_0, \tag{12.2}$$

The domain $\mathcal{D}_{NTbl}\left(t_0; \mathfrak{D}^j; \mathbf{U}; \mathfrak{Y}_d^k\right)$ *of the natural trackability of the plant (3.12) on the space product* $\mathfrak{T}_0 \times \mathfrak{D}^j \times \mathfrak{Y}_d^k$ *is determined by (8.17).*

3) The control vector function $\mathbf{U}(.)$ *obeys (12.3) if* γ *=rank* $W(t) = r = N$, $\forall t \in \mathfrak{T}_0$*:*

$$\mathbf{U}(t) = \mathbf{w}^I \left\langle \mathbf{w}\left[\mathbf{U}\left(t^-\right)\right] + W(t)^{-1}\,\varepsilon(t) \right\rangle, \quad \forall t \in \mathfrak{T}_0, \tag{12.3}$$

The domain $\mathcal{D}_{NTbl}\left(t_0;\mathfrak{D}^j;\mathbf{U};\mathfrak{Y}_d^k\right)$ *of the natural trackability of the plant (3.12) on the space product* $\mathfrak{T}_0 \times \mathfrak{D}^j \times \mathfrak{Y}_d^k$ *is determined by (8.18).*

ii-4) If the plant also possesses Property 64 then the control vector function $\mathbf{U}(.)$ *obeys (12.4),*

$$\mathbf{w}_\gamma\left[\mathbf{U}_\gamma(t)\right] = \mathbf{w}_\gamma\left[\mathbf{U}_\gamma(t^-)\right] + M_{\gamma,m}(t)\,\mathbf{v}(t,\sigma,\varepsilon),\ \forall t \in \mathfrak{T}_0, \tag{12.4}$$

for the initial control vector $\mathbf{U}_0$ *determined by (11.8).*

The natural trackability domain $\mathcal{D}_{NTbl}\left(t_0;\mathfrak{D}^j;\mathbf{U};\mathfrak{Y}_d^k\right)$ *on* $\mathfrak{T}_0 \times \mathfrak{D}^j \times \mathfrak{Y}_d^k$ *for any fixed* $\sigma \in In\mathfrak{T}_0$ *of the plant (3.12) with Properties 57, 58, 64, and 78 is determined by (8.19).*

Proof. When we replace $\mathbf{v}(.) : \mathfrak{T}_0 \times \mathfrak{R}^N \longrightarrow \mathfrak{R}^m$, (8.1), by the vector function $\mathbf{v}(.) : \mathfrak{T}_0 \times In\mathfrak{T}_0 \times \mathfrak{R}^{(k+1)N} \longrightarrow \mathfrak{R}^m$, (11.1), and "perfect" by "imperfect" in Theorem 162 then it becomes Theorem 190 ■

12.2 State space criteria

Theorem 191 ***The conditions for the natural trackability on*** $\mathfrak{T}_0 \times \mathfrak{D}^j \times \mathfrak{Y}_d^\alpha$ ***of the plant (3.12) with Properties 57, 58, and 78***

In order for the plant (3.12), which possesses Properties 57, 58 and 78, to be natural trackable via the state space on $\mathfrak{T}_0 \times \mathfrak{D}^j \times \mathfrak{Y}_d^\alpha$ *it is necessary and sufficient that both:*

i) The rank β *of the control matrix* $P(t)$ *is greater than zero; i.e., (8.20) holds.*

ii) The control vector function $\mathbf{U}(.)$ *satisfies one of the conditions 1)-4):*

1) The control vector function $\mathbf{U}(.)$ *obeys (12.5) for* $\beta = \text{rank}\,P(t) = r < \rho$, $\forall t \in \mathfrak{T}_0$, *and for the chosen vector function* $\mathbf{v}(.) : \mathfrak{T}_0 \times In\mathfrak{T}_0 \times \mathfrak{R}^N \longrightarrow \mathfrak{R}^m$, *(11.1), where m is not greater than r,* $m \leq r$, *rank* $L_{r,m}(t) \equiv m$ *and* $L_{r,r}(t) \equiv I_r$*:*

$$\mathbf{e}_u\left[\mathbf{U}^\mu(t)\right] = \mathbf{e}_u\left[\mathbf{U}^\mu(t^-)\right] + L_{r,m}(t)\,\mathbf{v}(t,\sigma,\varepsilon),\ \forall t \in \mathfrak{T}_0. \tag{12.5}$$

The domain $\mathcal{D}_{NTbl}\left(t_0;\mathfrak{D}^j;\mathbf{U};\mathfrak{Y}_d^\alpha\right)$ *of the natural trackability of the plant (3.12) on the space product* $\mathfrak{T}_0 \times \mathfrak{D}^j \times \mathfrak{Y}_d^\alpha$ *is determined by (8.26).*

2) The control vector function $\mathbf{U}(.)$ *obeys (12.6) for* $\beta = \text{rank}\,P(t) = \rho < r$, $\forall t \in \mathfrak{T}_0$, *and for the chosen vector function* $\mathbf{v}(.) : \mathfrak{T}_0 \times In\mathfrak{T}_0 \times \mathfrak{R}^N \longrightarrow \mathfrak{R}^m$, *(11.1), where m is not greater than* ρ, $m \leq \rho$, *and rank* $L_{\rho,m}(t) \equiv m$*:*

$$\mathbf{e}_u\left[\mathbf{U}^\mu(t)\right] = \Gamma(t)\left\{P(t)\,\mathbf{e}_u\left[\mathbf{U}^\mu(t^-)\right]\right\} + P^T(t)\,L_{\nu,m}(t)\,\mathbf{v}(t,\sigma,\varepsilon),$$
$$\Gamma(t) = P^T(t)\left[P(t)\,P^T(t)\right]^{-1},\ \forall t \in \mathfrak{T}_0. \tag{12.6}$$

The domain $\mathcal{D}_{NTbl}\left(t_0;\mathfrak{D}^j;\mathbf{U};\mathfrak{Y}_d^\alpha\right)$ *of the natural trackability of the plant (3.12) on set product* $\mathfrak{T}_0 \times \mathfrak{D}^j \times \mathfrak{Y}_d^\alpha$ *is determined by (8.27).*

3) The control vector function $\mathbf{U}(.)$ *obeys (12.7) if* $\beta = \text{rank}\,P(t) = r = \rho$, $\forall t \in \mathfrak{T}_0$, *for the chosen vector function* $\mathbf{v}(.) : \mathfrak{T}_0 \times In\mathfrak{T}_0 \times \mathfrak{R}^N \longrightarrow \mathfrak{R}^m$, *(11.1) and rank* $L_{\rho,m}(t) \equiv m$*:*

$$\mathbf{e}_u\left[\mathbf{U}^\mu(t)\right] = \mathbf{e}_u\left[\mathbf{U}^\mu(t^-)\right] + P(t)^{-1}\,L_{\rho,m}(t)\,\mathbf{v}(t,\sigma,\varepsilon),\ \ \forall t \in \mathfrak{T}_0. \tag{12.7}$$

The domain $\mathcal{D}_{NTbl}\left(t_0;\mathfrak{D}^j;\mathbf{U};\mathfrak{Y}_d^\alpha\right)$ *of the natural trackability on the space product* $\mathfrak{T}_0 \times \mathfrak{D}^j \times \mathfrak{Y}_d^\alpha$ *of the plant (3.12) is determined by (8.28).*

4) If the plant also possesses Property 61 then for the chosen both matrix $L(t) \in \mathfrak{R}^{r\times N}$*,(8.25), and vector function* $\mathbf{v}(.) : \mathfrak{T}_0 \times In\mathfrak{T}_0 \times \mathfrak{R}^N \longrightarrow \mathfrak{R}^m$, *(11.1), the control vector function* $\mathbf{U}(.)$ *obeys (12.8),*

$$\mathbf{e}_{u\beta}\left[\mathbf{U}_\beta^\mu(t)\right] = \mathbf{e}_{u\beta}\left[\mathbf{U}_\beta^\mu(t^-)\right] + L_{\beta,m}(t)\,\mathbf{v}(t,\sigma,\varepsilon),\ \forall t \in \mathfrak{T}_0. \tag{12.8}$$

The equation (11.10) determines the initial control vector $\mathbf{U}_0$.

The domain $\mathcal{D}_{NTbl}\left(t_0;\mathfrak{D}^j;\mathbf{U};\mathfrak{Y}_d^{\alpha}\right)$ *of the natural trackability on* $\mathfrak{T}_0 \times \mathfrak{D}^j \times \mathfrak{Y}_d^{\alpha}$ *of the plant (3.12) with Properties 57, 58, 61, and 78 is determined by ((8.29).*

Proof. If we replace $\mathbf{v}(.) : \mathfrak{T}_0 \times \mathfrak{R}^N \longrightarrow \mathfrak{R}^m$, (8.1), by the vector function $\mathbf{v}(.) : \mathfrak{T}_0 \times \text{In}\mathfrak{T}_0 \times \mathfrak{R}^{(k+1)N} \longrightarrow \mathfrak{R}^m$, (11.1), and "perfect" by "imperfect" in Theorem 164 then it becomes Theorem 191 ■

12.3 Both spaces criteria

Theorem 192 ***The conditions for the trackability on*** $\mathfrak{T}_0 \times \mathfrak{D}^j \times \mathfrak{Y}_d^{\alpha}$ ***of the plant (3.12) with Properties 57, 58, 61, 64, 78***

For the plant (3.12) that possesses Properties 57, 58, 61, 64, 78 to be natural trackable on $\mathfrak{T}_0 \times \mathfrak{D}^j \times \mathfrak{Y}_d^{\alpha}$ *via both spaces it is necessary and sufficient that i) and ii) hold:*

i) The ranks of the matrices $P(t)$ *and* $W(t)$ *are always greater than zero; i.e., (8.31) is valid,*

(ii) The control vector function $\mathbf{U}(.)$ *obeys (12.9) for any chosen matrix* $H(t) \in \mathfrak{R}^{(\beta+\gamma)\times m}$ *that obeys (8.30) and for the accepted subsidiary vector function* $\mathbf{v}(.) : \mathfrak{T}_0 \times In\mathfrak{T}_0 \times \mathfrak{R}^N \longrightarrow \mathfrak{R}^N$, *(11.1),*

$$\begin{bmatrix} \mathbf{e}_{u\beta}\left[\mathbf{U}_{\beta}^{\mu}(t)\right] \\ \mathbf{w}_{\gamma}\left[\mathbf{U}_{\gamma}(t)\right] \end{bmatrix} = \begin{bmatrix} \mathbf{e}_{u\beta}\left[\mathbf{U}_{\beta}^{\mu}(t^-)\right] \\ \mathbf{w}_{\gamma}\left[\mathbf{U}_{\gamma}(t^-)\right] \end{bmatrix} + H(t)\,\mathbf{v}(t,\sigma,\varepsilon), \quad \forall t \in \mathfrak{T}_0. \tag{12.9}$$

The equations (11.8), (11.10) determine the initial control vector $\mathbf{U}_0$.

The domain $\mathcal{D}_{NTbl}\left(t_0;\mathfrak{D}^j;\mathbf{U};\mathfrak{Y}_d^{\alpha}\right)$ *of the natural trackability on* $\mathfrak{T}_0 \times \mathfrak{D}^j \times \mathfrak{Y}_d^{\alpha}$ *of the plant (3.12) with Properties 57, 58, 61, 64, 78 is determined by (8.33).*

Proof. The proof of this theorem is the direct synthesis of the proofs of Theorems 157, 186, 188, and 191 ■

Part IV

PERFECT TRACKING

Chapter 13

Tracking generally

13.1 Primary control goal

The task of control is to force a dynamical physical system representing a physical *plant to behave sufficiently close* to (i.e., *to follow* sufficiently precisely, *to track* sufficiently accurately) its *desired* (i.e., *demanded*) *output behavior* over some, usually prespecified, *time* interval and under real (usually unpredictable and unknown) both external (input) actions and initial conditions [365, pp. 121-127].

Goal 193 ***The main control goal***
The very, the primary, the essential, goal of control is to assure that the controlled plant exhibits a requested kind of ***output tracking*** *that we call, for short* ***tracking.***

The term *tracking* in the wide sense incorporates all types of the plant real output behavior *following* its desired output behavior. In the specific sense tracking signifies asymptotic tracking as *time* $t \longrightarrow \infty$.

Tracking has been mainly and largely treated in the control literature as tracking with the zero steady-state error. This simplifies the control task to the control purpose to guarantee that the controlled plant real output approaches asymptotically the plant desired output as *time* t diverges to infinity. The control theory in this sense began as the specific, simplified, tracking theory. It has been known under different names such as theory of servomechanisms/servosystems, or of regulation systems, or of control systems in general, which involves the preceding ones. The zero steady-state error study has been commonly considered as a subtopic of stability and stabilization studies. However, systematic studies of control demand the full tracking theory in its own right, which was not recognized, perhaps, until 1980 [245], [246], [487], [488] as pointed out in [279].

The notion, the sense, and the meaning of **tracking** signify in the sequel that the real plant output *tracks*, *follows*, every plant desired output that belongs to a functional family $\mathfrak{Y}_d^k$, $k \in \{0, 1, ..., \alpha - 1\}$,

- *Either the desired output is constant* (in a part of the control literature this is linked exclusively with the regulation systems) *or time-varying* (in another part of the control literature it is associated only with the servomechanisms/servosystems),

- *Under the actions of arbitrary external disturbances* that belong to a family $\mathfrak{D}^j$, $j \in \{0, 1, ...\}$

and

- *Under arbitrary (input, state, and output) initial conditions*, [279]

Tracking theory incorporates both the servomechanism/servosystem theory and the regulation theory.

Various tracking concepts exist, for which see Section 14.2.

Section 3.2 presents the properties of *the disturbance family* $\mathfrak{D}^k$, and of *the desired output family* $\mathfrak{Y}_d^k$.

Control that ensures a kind of tracking is called ***tracking control** (TC)*. The task of the tracking control synthesis is meaningful if the plant is able to exhibit tracking under an appropriate action of control, i.e., if the plant is **trackable**. The book discovers various trackability properties, defines them, and establishes conditions on the plant to possess the corresponding trackability property. This permits us to continue with ***tracking control synthesis***, which closes the main body of the book.

13.2 Tracking versus stability

We explain the differences between the *Lyapunov stability concept* and the concepts of *asymptotic tracking in the Lyapunov sense* (for short: *asymptotic tracking*) also called *Lyapunov type tracking* [279, pp. 122-124] and of tracking with a required quality (for such differences in the framework of linear systems see [279, pp. 122-124]).

The main characteristics of the Lyapunov [401] stability concept (LyS) and tracking concept(s) (TgC) are the following:

- LyS concerns the state behavior of the system.

  TgC deals directly with the output dynamical behavior, and through it, indirectly with the state behavior.

- The appropriate space to study LyS is the state space $\mathfrak{R}^{\alpha\rho}$.

  The adequate space to investigate TgC is the space product $\mathfrak{R}^N \times \mathfrak{R}^{\alpha\rho}$ in general, or its subspaces: the output space $\mathfrak{R}^N$ and the state space $\mathfrak{R}^{\alpha\rho}$.

- LyS concerns the system behavior in the nominal regime in terms of total values of variables; i.e., equivalently, in the free regime in terms of their deviations, Chapter 5,

  TgC treats the system behavior in the real, i.e., forced, regime. It takes into account both arbitrary, unpredictable, unknown disturbances from a certain functional family $\mathfrak{D}^\eta$ and every (*time*-varying or *time*-invariant) desired output vector $\mathbf{Y}_d(t)$ belonging to another functional family $\mathfrak{Y}_d^\mu$.

- LyS deals with nonzero initial conditions.

  TgC allows also arbitrary (nonzero and zero) initial conditions.

- LyS holds for the system dynamical behavior only over the unbounded and infinite *time* interval $\mathfrak{T}_0$.

  TgC treats the system dynamical behavior on any (bounded or unbounded) *time* subinterval $\mathfrak{T}_{0F}$ of $\mathfrak{T}$.

- LyS demands, except for the stability only (without the attraction), asymptotic convergence of the real motions to the desired one as *time* $t \longrightarrow \infty$.

  TgC incorporates the asymptotic convergence as *time* $t \longrightarrow \infty$ (in which case the reachability *time* $t_R = \infty$) and the finite reachability *time* $t_R < \infty$.

- *Lyapunov tracking concept* ($LyTgC$) incorporates LyS.

Besides, the practical tracking concept incorporates the practical stability concept.

The concept of tracking with a required quality suits the tracking goal better than any stability concept.

Tracking properties and stability properties are mutually independent in general [184, Example 4, p. 11], [279, Example 152, p. 123]. The zero equilibrium state can be stable, but the plant need not exhibit tracking. The plant can exhibit tracking, but its zero equilibrium state can be unstable.

Chapter 14

Tracking concepts

14.1 Tracking characterization and space

14.1.1 Output tracking and output space

The desired dynamical behavior of the plant should be well known and defined in advance, or should be fast measurable during the plant behavior. We deal in this book only with the former situation. The latter characterizes target tracking which is out of this book's scope. Y. Bar-Shalom [26], [31], Y. Bar-Shalom and T. E. Fortmann [28], Y. Bar-Shalom and X. R. Li [29], [30], Y. Bar-Shalom, X. R. Li and T. Kirubarajan [31], Y. Bar-Shalom, P. K. Willet, and X. Tian [32], S. S. Blackman [52], R. P. S. Mahler [404], D. B. Reid [496], and L. D. Stone, C. A. Barlow, and T. L. Corwin [528] developed the theory of target tracking and showed its applications.

By the very sense and by the definition, tracking (which means output tracking) concerns the relation of the real output behavior to the desired output behavior of the uncontrolled plant, of the controlled plant, and of the plant control system

Tracking takes place in *the output integral space* $\mathfrak{I}$ composed of the *time* set $\mathfrak{T}$ and the corresponding output space $\mathfrak{R}^N$ as their Cartesian product space,

$$\mathfrak{I} = \mathfrak{T} \times \mathfrak{R}^N. \tag{14.1}$$

This holds for the output tracking. It suggests studying tracking in the integral space $\mathfrak{I}$ (14.1), or equivalently in the output space $\mathfrak{R}^N$ in the course of *time* $t \in \mathfrak{T}$.

14.1.2 Output tracking and state space

We can study (output) tracking via *the state space* $\mathfrak{R}^{\alpha\rho}$. Either for theoretical reasons and/or for practical engineering reasons it is sometimes more convenient to use *the desired state behavior* that is induced by the desired output behavior $\mathbf{Y}_d(t)$, where the desired state behavior determined by $\mathbf{Y}_d(t)$ is the desired temporal evolution of the extended state vector $\mathbf{R}_d^{\alpha-1}(t)$, and the corresponding space is $\mathfrak{R}^{\alpha\rho}$.

This means to treat tracking via the state space. To do that we should calculate the desired state behavior determined by the desired output behavior. Such calculation is a complex computational task and problem in the framework of *(time*-varying) nonlinear plants. In this book we assume that the desired state behavior of the plant is well defined and known before the control implementation.

We can be interested just in the state behavior of the plant. This means that the plant state subvector $\mathbf{R}$ is simultaneously the plant output vector $\mathbf{Y} : \mathbf{R} = \mathbf{Y}$.

We can demand that the real *state behavior tracks* (sufficiently closely, with a good quality) the desired state behavior that is determined by the plant desired output behavior $\mathbf{Y}_d(t)$, and the corresponding space is $\mathfrak{R}^{\alpha\rho}$.

Comment 194 ***Tracking and state tracking***
The tracking properties reduce to the state tracking properties as soon as the state subvector $\mathbf{R}$ *is simultaneously the output vector* $\mathbf{Y} : \mathbf{R} = \mathbf{Y}$.

14.2 Various tracking concepts

Relative to the tracking target there are two main groups of tracking concepts:
- *Motion tracking*, for short *tracking*. The tracking target is a motion (or, a movement).
- *Set tracking*. The tracking target is a set that can be *time*-invariant or *time*-varying.

Tracking is ***perfect*** (***ideal***) if, and only if, the plant real output behavior is always equal to the plant desired output behavior. If the plant real output is different from the desired plant output at any moment then tracking is ***imperfect***. This book treats both perfect and imperfect tracking of *time*-varying nonlinear *IIDO* systems (3.12), hence of the *IO* systems (3.62) and of the *ISO* systems (3.65), (3.66) as the subclasses of the systems (3.12).

The definition of any tracking property should clarify the following [279].

- *The characterization of the plant behavior we are interested in*, whether we are interested in the state behavior of the plant, or in the plant output dynamical behavior, or in both
- *The space in which the demanded closeness is to be achieved*, which means that, although originally tracking concerns the output behavior, we can consider the output tracking either via *the output space* or via *the state space in general*
- *The definition of the distance between the real behavior and the desired behavior* of the plant
- *The definition of the demanded closeness of the real behavior to the desired behavior* of the plant
- *The nonempty sets of the initial conditions of all plant variables* under which the demanded closeness is to be achieved
- *The nonempty set* $\mathfrak{D}^{(\cdot)}$ *of permitted external disturbances* acting on the plant, under which the demanded closeness is to be realized
- *The nonempty set* $\mathfrak{Y}_d^{(\cdot)}$ *of realizable desired plant behaviors* that can be demanded
- *The time interval over which the demanded closeness is to be guaranteed*
- *The requested quality with which the real behavior is to follow the desired behavior* of the plant

We can differently determine the preceding imperfect tracking characteristics. Their different specifications lead to various imperfect tracking concepts. Every imperfect tracking concept comprises a number of different tracking properties. The main imperfect tracking concepts are the following.

The book [326] deals extensively with the asymptotic tracking in its basic form that considers the convergence of the real output plant behavior to its desired output behavior only as *time* t diverges to infinity, $t \longrightarrow \infty$. The references [150]-[154], [165]-[169], [173],

[177], [180], [181], [183]-[195], [196]-[197], [210], [212]-[215], [218], [221], [225], [226], [232]-[233], [234], [235]-[238], [242]-[253], [263], [267], [277], [282], [283], [279] deal with various types of *tracking in the Lyapunov sense* that is *infinite-time tracking, or* more precisely: *asymptotic tracking.* They are studied in this book.

Another tracking concept is that of *tracking with a required quality [279].*

Grujić/Gruyitch and Mounfield established the theory of control synthesis for *tracking with finite (scalar or vector) reachability time (FSRT or FVRT, respectively)* in [183], [185], [191], [196], [212], [236]-[240], [243], [269], [284], [286], [467]. The *final reachability time (FRT) tracking* satisfies better real engineering needs and customers' demands than asymptotic tracking. The *FRT* tracking concept is also a special case of the required quality tracking concept. Its further development is in [279] for the *time*-invariant continuous-*time* linear control systems. The tracking is therein treated as an infinite-*time* tracking with a finite reachability *time.* This book broadens it to *time*-varying nonlinear systems.

The concept of *tracking with a prespecified performance index [192], [219], [220],[270]* is a special case of the required quality tracking concept and incorporates *the practical tracking concept.* Grujić (Gruyitch) established their fundamentals in [183], [196], [220]. Further contributions to these concepts are in the works by Grujić/Gruyitch and Mounfield [192], [219], [220], [232]-[239], [260]-[267], [285]-[286]. A. Kökösy [351]-[355], D. Lazić (Lazitch) [376]-[377], N. Nedić (Neditch) and D. Pršić (Prshitch) [459]-[458], [492], Z. B. Ribar et al. [501], [503], M. J. Stojčić (M. Y. Stoychitch) [529], M. R. Jovanović (Yovanovitch) [568] and R. Zh. Yovanovitch [569]-[571] contributed to the further development of tracking with a prespecified performance index.

What follows contains the fundamentals and further developments of the following concepts: *Asymptotic Tracking in the Lyapunov sense, i.e., the Lyapunov type tracking* , and *Finite (Scalar or Vector) Reachability Time Tracking and Required Quality Tracking (e.g., with a given finite (scalar or vector) reachability time)* for the *time*-varying nonlinear control systems.

This book initiates the study of the set tracking.

The preceding tracking concepts open new directions in control theory to satisfy engineering needs.

Chapter 15

Perfect tracking concept

15.1 On perfect tracking generally

The concept of *perfect (ideal) tracking* enables us [279] to discover what is theoretically the best possible real dynamic total output behavior $\mathbf{Y}(t)$ of a plant *relative to its desired total output behavior* $\mathbf{Y}_d(t)$ (Definition 89). The exact meaning of perfect tracking reads as follows [279, Definition 156, p. 124].

Definition 195 ***The k-th-order perfect tracking of the plant on*** $\mathfrak{T}_0$

The plant exhibits ***the k-th-order perfect tracking on*** $\mathfrak{T}_0$, $\mathfrak{T}_0 \subseteq \mathfrak{T}$, ***of its desired k-th-order extended output vector response*** $\mathbf{Y}^k(t)$ *if, and only if, its real k-th-order output vector response* $\mathbf{Y}^k(t)$ *is always equal to its desired k-th-order output vector response* $\mathbf{Y}^k(t)$,

$$\mathbf{Y}^k(t) = \mathbf{Y}^k(t),\ \forall t \in \mathfrak{T}_0,\ \mathfrak{T}_0 \subseteq \mathfrak{T},\ k \in \{0, 1, 2, ..., \alpha - 1\}, \tag{15.1}$$

$$equivalently: \ \boldsymbol{\varepsilon}^k(t) = \mathbf{0}_{(k+1)N},\ \forall t \in \mathfrak{T}_0,\ \mathfrak{T}_0 \subseteq \mathfrak{T}, \tag{15.2}$$

$$equivalently: \ \mathbf{y}^k(t) = \mathbf{0}_{(k+1)N},\ \forall t \in \mathfrak{T}_0,\ \mathfrak{T}_0 \subseteq \mathfrak{T}. \tag{15.3}$$

If, and only if, $k = 0$*, then the zero-order perfect tracking on* $\mathfrak{T}_0$ *is simply called* ***perfect tracking on*** $\mathfrak{T}_0$.

*The expression "**on** $\mathfrak{T}_0$" is to be omitted if, and only if,* $\mathfrak{T}_0 = \mathfrak{T}$.

We now state the relationship between perfect tracking and the k-th-order perfect tracking.

Theorem 196 ***Perfect tracking on*** $\mathfrak{T}_0$ ***and the k-th-order perfect tracking on*** $\mathfrak{T}_0$ *[279, Theorem 157, p. 125]*

If the real output vector function $\mathbf{Y}(.)$ *and the desired output vector function* $\mathbf{Y}_d(.)$ *are k-times continuously differentiable on* $\mathfrak{T}_0$, $t_0 \in \mathfrak{T}$, *then for the validity of (15.1) it is necessary and sufficient that*

$$\mathbf{Y}(t) = \mathbf{Y}_d(t),\ \ \forall t \in \mathfrak{T}_0,\ t_0 \in \mathfrak{T}, \tag{15.4}$$

holds, i.e.,

$$\mathbf{Y}(t) \in \mathfrak{C}^k(\mathfrak{T}_0) \ \ and \ \ \mathbf{Y}_d(t) \in \mathfrak{C}^k(\mathfrak{T}_0) \Longrightarrow$$

$$\left\langle \mathbf{Y}^k(t) = \mathbf{Y}_d^k(t),\ \ \forall t \in \mathfrak{T}_0,\ t_0 \in \mathfrak{T} \right\rangle \Longleftrightarrow \left\langle \mathbf{Y}(t) = \mathbf{Y}_d(t),\ \forall t \in \mathfrak{T}_0,\ t_0 \in \mathfrak{T} \right\rangle. \tag{15.5}$$

Conclusion 197 *This theorem is general. Its proof given in [279, Theorem 157, p. 125] is valid regardless of the form of the mathematical model of the system, which can be either linear or nonlinear, time-invariant or time-varying. It enables us to reduce the study of the k-th-order perfect tracking on* $\mathfrak{T}_0$ *to the perfect tracking on* $\mathfrak{T}_0$.

15.2 Definitions

15.2.1 Perfect tracking and the target set of the plant

We consider the plant (3.12).

By following the general Definition 89 of the desired regime, we present Definition 195 for the plant (3.12).

Note 198 ***On system perfect tracking***

All what follows holds directly for the perfect tracking properties of the system (3.12) under the following substitutions: $\mathbf{D}$ *should be replaced by* $\mathbf{I}$, $\mathbf{D}_N$ *by* $\mathbf{I}_N$, $\mathbf{D}_N^{\eta}$ *by* $\mathbf{I}^{\xi}$, $\mathbf{U}$ *and* $\mathbf{U}_N$ *should be omitted,* $\mathbf{z}_R\left[t,\mathbf{R}^{\alpha-1}(t)\right]+\mathbf{z}_D\left[t,\mathbf{D}(t)\right]+W\left(t\right)\mathbf{w}\left[\mathbf{U}(t)\right]$ *by* $\mathbf{s}\left[t,\mathbf{R}(t),\mathbf{I}(t)\right]$, *for* $t=t_0:\mathbf{z}_R\left[t_0,\mathbf{R}^{\alpha-1}(t_0)\right]+\mathbf{z}_D\left[t_0,\mathbf{D}(t_0)\right]+W\left(t_0\right)\mathbf{w}\left[\mathbf{U}(t_0)\right]$ *by* $\mathbf{s}\left[t_0,\mathbf{R}(t_0),\mathbf{I}(t_0)\right]$.

Definition 199 ***The k-th-order perfect tracking of the plant*** *(3.12)*

The plant (3.12) exhibits ***the k-th-order perfect tracking on*** $\mathfrak{T}_0$ ***of its desired output vector response*** $\mathbf{Y}_d(t)$ *if, and only if, its real k-th-order output vector response* $\mathbf{Y}^k(t)$ *is equal to its desired k-th-order output vector response* $\mathbf{Y}_d^k(t)$ *always on* $\mathfrak{T}_0$,

$$\mathbf{Y}^k(t)=\mathbf{Y}_d^k(t),\ \ \forall t\in\mathfrak{T}_0,\ \ k\in\{0,1,\cdots,\alpha-1\}. \tag{15.6}$$

equivalently

$$\boldsymbol{\varepsilon}^k(t;t_0;\boldsymbol{\varepsilon}_0^k)=\mathbf{0}_{(k+1)N},\ \ \forall t\in\mathfrak{T}_0,\ \ k\in\{0,1,\cdots,\alpha-1\}, \tag{15.7}$$

or

$$\mathbf{y}^k(t;t_0;\mathbf{y}_0^k)=\mathbf{0}_{(k+1)N},\ \ \forall t\in\mathfrak{T}_0,\ \ k\in\{0,1,\cdots,\alpha-1\}. \tag{15.8}$$

If, and only if, $k=0$, *then the zero-order perfect tracking is simply called* ***perfect tracking.***

*The expression "**on** $\mathfrak{T}_0$" is to be omitted if, and only if,* $\mathfrak{T}_0=\mathfrak{T}$.

Note 200 ***Perfect tracking and time interval***

Let $\tau_i\in\mathfrak{T}$, $i=1,2$, $\tau_1\neq\tau_2$. *The plant (3.12) can exhibit the k-th-order perfect tracking on* $\mathfrak{T}_{\tau_1}$, *but need not on* $\mathfrak{T}_{\tau_2}$.

Definition 201 ***The k-th-order target set of the plant*** *(3.12)*

a) The set $\Upsilon^k(t;\mathbf{D};\mathbf{U};\mathbf{Y}_d)$, $\Upsilon^k(t;\mathbf{D};\mathbf{U};\mathbf{Y}_d)\subset\mathfrak{R}^{\alpha\rho}$, *of all vectors* $\mathbf{R}^{\alpha-1}\in\mathfrak{R}^{\alpha\rho}$ *such that* $\mathbf{Y}^k$ *is equal to* $\mathbf{Y}_d^k(t)$ *is* ***the k-th-order target set of the plant*** *(3.12)* ***relative to its desired k-th-order output vector response*** $\mathbf{Y}_d^k(t)$ ***at a moment*** $t\in\mathfrak{T}_0$,

$$\Upsilon^k(t;\mathbf{D};\mathbf{U};\mathbf{Y}_d)=\left\{\mathbf{R}^{\alpha-1}:\ \mathbf{Y}^k\left[\mathbf{R}^{\alpha-1}(t)\right]=\mathbf{Y}_d^k(t)\right\}$$
$$=\left\{\mathbf{R}^{\alpha-1}:\left[\begin{array}{c}\mathbf{z}\left[t,\mathbf{R}^{\alpha-1}(t),\mathbf{D}\left(t\right)\right]+W\left(t\right)\mathbf{w}\left[\mathbf{U}(t)\right]=\mathbf{Y}_d(t),\\ \left[\begin{array}{c}\mathbf{z}\left[t,\mathbf{R}^{\alpha-1}(t),\mathbf{D}\left(t\right)\right]\\+W\left(t\right)\mathbf{w}\left[\mathbf{U}(t)\right]\end{array}\right]^{(j)}=\mathbf{Y}_d^{(j)}(t),\\ \forall j=1,\cdots,k\end{array}\right]\right\}. \tag{15.9}$$

b) If, and only if, $k=0$, *then the zero-order target set* $\Upsilon_{IIDO}^0(t;\mathbf{D};\mathbf{U};\mathbf{Y}_d)$ *at a moment* t *is called* ***the target set*** $\Upsilon(t;\mathbf{D};\mathbf{U};\mathbf{Y}_d)$ ***at the moment*** t,

$$\Upsilon^0(t;\mathbf{D};\mathbf{U};\mathbf{Y}_d)\equiv\Upsilon(t;\mathbf{D};\mathbf{U};\mathbf{Y}_d)$$
$$=\left\{\mathbf{R}^{\alpha-1}:\mathbf{z}\left[t,\mathbf{R}^{\alpha-1}(t),\mathbf{D}\left(t\right)\right]+W\left(t\right)\mathbf{w}\left[\mathbf{U}(t)\right]=\mathbf{Y}_d(t)\right\}. \tag{15.10}$$

Comment 202 ***On the target set order of the plant*** *(3.12)*

The right-hand side of (15.9) shows that it is necessary to determine derivatives of the plant output nonlinearities up to the k-th-order for the k-th-order target set of the plant (3.12). This is too cumbersome for $k>0$.

Conclusion 203 ***On the tracking order of the plant*** *(3.12)*

It is reasonable to study only the zero-order ($k = 0$) tracking properties of the plant (3.12).

Proposition 204 ***The target set and the state space***

The k-th-order target set $\Upsilon^k(t;\mathbf{D};\mathbf{U};\mathbf{Y}_d)$ at a moment t of the plant (3.12) is a subset of the plant state space $\mathfrak{R}^{\alpha\rho}$,

$$\Upsilon^k(t;\mathbf{D};\mathbf{U};\mathbf{Y}_d) \subset \mathfrak{R}^{\alpha\rho},$$

rather than being a subset of the plant output space $\mathfrak{R}^N$. It depends not only on $\mathbf{Y}_d^k$ but also on $\mathbf{D}^k$, and $\mathbf{U}^k$, which means that it depends on derivatives of $\mathbf{Y}_d$, $\mathbf{D}$, and $\mathbf{U}$ in general.

Proposition 205 ***Necessary and sufficient condition for the perfect tracking and the target set***

In order for the plant (3.12) to exhibit the k-th-order perfect tracking from $\mathbf{R}_0^{\alpha-1} \in \mathfrak{R}^{\alpha\rho}$ relative to its desired k-th-order output vector response $\mathbf{Y}_d^k(t)$ on $\mathfrak{T}_0$ it is necessary and sufficient that its state vector $\mathbf{R}^{\alpha-1}(t;\mathbf{R}_0^{\alpha-1};\mathbf{D};\mathbf{U})$ is in its k-th-order target set $\Upsilon^k(t;\mathbf{D};\mathbf{U};\mathbf{Y}_d)$ always on $\mathfrak{T}_0$,

$$\mathbf{R}^{\alpha-1}(t;\mathbf{R}_0^{\alpha-1};\mathbf{D};\mathbf{U}) \in \Upsilon^k(t;\mathbf{D};\mathbf{U};\mathbf{Y}_d),\ \forall t \in \mathfrak{T}_0. \tag{15.11}$$

From Definition 195, i.e., from Definition 199, follows:

Proposition 206 ***Necessary initial condition for perfect tracking***

The necessary, but insufficient, condition for the k-th-order perfect tracking of the plant desired output response $\mathbf{Y}_d(t)$ is that the initial k-th-order real output vector $\mathbf{Y}_0^k$ of the plant is equal to its initial k-th-order desired output vector $\mathbf{Y}_{d0}^k$,

$$\mathbf{Y}_0^k = \left(\mathbf{z}\left[t_0, \mathbf{R}^{\alpha-1}(t_0), \mathbf{D}(t_0)\right] + W\left(t_0\right)\mathbf{w}\left[\mathbf{U}(t_0)\right]\right)^k = \mathbf{Y}_{d0}^k. \tag{15.12}$$

The preceding definitions imply:

Proposition 207 ***Realizability of perfect tracking***

The necessary and sufficient condition for the realizability of the perfect tracking is that the initial desired state vector $\mathbf{R}_{d0}^{\alpha-1}$ and the initial nominal control vector $\mathbf{U}_{0N}$ of (3.12) obey

$$\mathbf{Y}_0 = \mathbf{z}\left[t_0, \mathbf{R}^{\alpha-1}(t_0), \mathbf{D}\left(t_0\right)\right] + W\left(t_0\right)\mathbf{w}\left[\mathbf{U}_N(t_0)\right] = \mathbf{Y}_{d0}. \tag{15.13}$$

$\mathbf{Y}_0$ depends explicitly not only on $\mathbf{Y}_{d0}$, but also on $\mathbf{D}_0$.

Note 208 ***Imperfect (realistic) tracking***

The real initial conditions $\mathbf{R}_0^{\alpha-1}$ and $\mathbf{U}_0$ satisfy (15.13) only in very special cases. Perfect tracking is possible only in such special cases. A kind of an ***imperfect tracking****, i.e.,* ***realistic tracking****, with at least satisfactory, or with a very good, quality according to an accepted criterion is possible in the most real cases.*

15.2.2 Perfect tracking and the desired regime of the plant

We will refine Definition 89 for the plant (3.12).

Definition 209 ***The k-th-order desired (nominal) regime of the plant and perfect tracking***

The plant (3.12) is **in the k-th-order desired (nominal) regime on** $\mathfrak{T}_0$ if, and only if, it exhibits the k-th-order perfect tracking on $\mathfrak{T}_0$, (15.6),

$$\mathbf{Y}^k(t) = \mathbf{Y}_d^k(t),\ \forall t \in \mathfrak{T}_0, \tag{15.14}$$

$$\textit{equivalently}:\ \boldsymbol{\varepsilon}^k(t;t_0;\boldsymbol{\varepsilon}_0^k) = \mathbf{0}_{(k+1)N},\ \forall t \in \mathfrak{T}_0, \tag{15.15}$$

$$\textit{equivalently}:\ \mathbf{y}^k(t;t_0;\mathbf{y}_0^k) = \mathbf{0}_{(k+1)N},\ \forall t \in \mathfrak{T}_0. \tag{15.16}$$

15.2.3 Perfect tracking and the nominal control of the plant

We adjust Definition 98 to the plant (3.12).

Definition 210 ***The nominal motion and the k-th-order nominal control***

A pair $[\mathbf{R}^{*\alpha-1}(.),\ \mathbf{U}^{*}(.)]$ *is a* ***[nominal motion (nominal state) - nominal k-th-order control] pair on*** $\mathfrak{T}_{\tau}$ ***for the plant (3.12) relative to the (nominal disturbance - k-th-order desired output) pair*** $[\mathbf{D}_N(.),\ \mathbf{Y}_d^k(.)]$*, which is denoted by* $[\mathbf{R}_N^{\alpha-1}(.),\ \mathbf{U}_N(.)]$*,*

$$[\mathbf{R}^{*\alpha-1}(.), \mathbf{U}^{*}(.)] \\ = [\mathbf{R}_N^{\alpha-1}(.;\mathbf{D}_N, \mathbf{Y}_d), \mathbf{U}_N(.;\mathbf{D}_N, \mathbf{Y}_d)] = [\mathbf{R}_N^{\alpha-1}(.), \mathbf{U}_N(.)] \tag{15.17}$$

if, and only if, it guarantees (15.14):

$$[\mathbf{R}^{*\alpha-1}(.), \mathbf{U}^{*}(t)] = [\mathbf{R}_N^{\alpha-1}(t), \mathbf{U}_N(t)] \\ \Longleftrightarrow \mathbf{Y}^k(t) = \mathbf{Y}_d^k(t),\ \forall t \in \mathfrak{T}_0. \tag{15.18}$$

Theorem 211 ***The nominal (state-control) pair***

In order for a (state-control) pair $[\mathbf{R}^{*(\alpha-1)}(.), \mathbf{U}^{*}(.)]$ *to be a nominal (state-control) pair for the plant (3.12) on* $\mathfrak{T}_0$ *relative to the (nominal disturbance-desired output) pair* $[\mathbf{D}_N(.),$ $\mathbf{Y}_d(.)]$ *it is necessary and sufficient that*

$$Q\left[t, \mathbf{R}^{*(\alpha-1)}(t)\right] \mathbf{R}^{*(\alpha)}(t) + \mathbf{q}\left[t, \mathbf{R}^{*\alpha-1}(t)\right] \\ = \mathbf{e}_d\left[t, \mathbf{D}_N^{\eta}(t)\right] + P(t)\,\mathbf{e}_u\left[\mathbf{U}^{*\mu}(t)\right], \quad \forall t \in \mathfrak{T}_{\tau}, \\ \mathbf{Y}_d(t) = \mathbf{z}\left[t, \mathbf{R}^{*(\alpha-1)}(t), \mathbf{D}_N(t)\right] + W(t)\,\mathbf{w}\left[\mathbf{U}^{*}(t)\right], \quad \forall t \in \mathfrak{T}_{\tau}. \tag{15.19}$$

This theorem is a special form of Theorem 117 applied to the plant (3.12).

Note 212 ***Noncontrolled and controlled plant***

All the above related to the plant holds for both the uncontrolled and controlled plant. In the latter case $\mathbf{U}(t) \equiv \mathbf{U}[t, \mathbf{R}(t), \mathbf{Y}(t)]$*. Hence, all holds also for the control system (3.12).*

Part V

IMPERFECT TRACKING: STABLE TRACKING

Chapter 16

Output space definitions

16.1 Introduction

The extended output error vector $\boldsymbol{\varepsilon}^k$, is equal to the zero vector $\mathbf{0}_{(k+1)N}$ if, and only if, $\mathbf{Y}^k$ is equal to $\mathbf{Y}_d^k$, in view of (5.2), Comment 124. The same holds for the output error deviation vector $\mathbf{y}^k$ due to (4.6).

Notice the following relationship between the neighborhoods

$$\mathfrak{N}\left(\varepsilon; t_0; \mathbf{Y}_{d0}^k; \mathbf{D}; \mathbf{U}; \mathbf{Y}_d^k\right) \quad and \quad \mathfrak{N}_\varepsilon \left[t; \mathbf{Y}_d^k(t); \mathbf{D}; \mathbf{U}; \mathbf{Y}_d^k\right]:$$

- $\mathfrak{N}_\varepsilon \left[t; \mathbf{Y}_d^k(t); \mathbf{D}; \mathbf{U}; \mathbf{Y}_d^k\right]$ is a connected ε-neighborhood of $\mathbf{Y}_d^k(t)$ at a moment $t \in \mathfrak{T}_0$, $\mathfrak{N}_\varepsilon \left[t; \mathbf{Y}_d^k(t); \mathbf{D}; \mathbf{U}; \mathbf{Y}_d^k\right] \subseteq \mathfrak{R}^{(k+1)N}$,

$$\mathfrak{N}_\varepsilon \left[t; \mathbf{Y}_d^k(t); \mathbf{D}; \mathbf{U}; \mathbf{Y}_d^k\right] \equiv \left\{\mathbf{Y}^k\colon \mathbf{Y}^k \in \mathfrak{R}^{(k+1)n},\ \rho\left[\mathbf{Y}^k, \mathbf{Y}_d^k(t)\right] < \varepsilon\right\},$$

  and

- $\mathfrak{N}\left(\varepsilon; t_0; \mathbf{Y}_{d0}^k; \mathbf{D}; \mathbf{U}; \mathbf{Y}_d^k\right)$ is a connected neighborhood of $\mathbf{Y}_{d0}^k$ at the initial moment $t = t_0$, which is determined by $\varepsilon \in \mathfrak{R}^+, \mathbf{D}(.) \in \mathfrak{D}^j$, $\mathbf{U}(.) \in \mathfrak{U}_d^l$, and $\mathbf{Y}_d(.) \in \mathfrak{Y}_d^k$, and the outer radius of which cannot be greater than ε,

$$\begin{gathered}
\mathfrak{N}\left(\varepsilon; t_0; \mathbf{Y}_{d0}^k; \mathbf{D}; \mathbf{U}; \mathbf{Y}_d^k\right) \subseteq \mathfrak{R}^{(k+1)N}, \\
0 < \varepsilon_1 < \varepsilon_2 \Longrightarrow \mathfrak{N}\left(\varepsilon_1; t_0; \mathbf{Y}_{d0}^k; \mathbf{D}; \mathbf{U}\right) \subseteq \mathfrak{N}\left(\varepsilon_2; t_0; \mathbf{Y}_{d0}^k; \mathbf{D}; \mathbf{U}\right), \\
\varepsilon \longrightarrow 0^+ \Longrightarrow \mathfrak{N}\left(\varepsilon; t_0; \mathbf{Y}_{d0}^k; \mathbf{D}; \mathbf{U}; \mathbf{Y}_d^k\right) \longrightarrow \left\{\mathbf{Y}_{d0}^k\right\}, \\
\mathfrak{N}\left(\varepsilon; t_0; \mathbf{Y}_{d0}^k; \mathbf{D}; \mathbf{U}; \mathbf{Y}_d^k\right) \subseteq \mathfrak{N}_\varepsilon \left[t; \mathbf{Y}_d^k(t_0); \mathbf{D}; \mathbf{U}\right].
\end{gathered} \tag{16.1}$$

  The neighborhood $\mathfrak{N}\left(\varepsilon; t_0; \mathbf{Y}_{d0}^k; \mathbf{D}; \mathbf{U}; \mathbf{Y}_d^k\right)$ is usually accepted to be the δ-neighborhood $\mathfrak{N}_\delta \left[t_0; \mathbf{Y}_{d0}^k; \mathbf{D}; \mathbf{U}; \mathbf{Y}_d^k\right]$ of $\mathbf{Y}_{d0}^k$ for $0 < \delta = \delta(\varepsilon) \le \varepsilon$, $\forall \varepsilon \in \mathfrak{R}^+$. In general

$$\mathfrak{N}\left(\varepsilon; t_0; \mathbf{Y}_{d0}^k; \mathbf{D}; \mathbf{U}; \mathbf{Y}_d^k\right) \subseteq \mathfrak{N}_{\delta(\varepsilon)} \left[t_0; \mathbf{Y}_{d0}^k; \mathbf{D}; \mathbf{U}; \mathbf{Y}_d^k\right].$$

We define **in the output space** several typical **asymptotic tracking properties** in ***the Lyapunov sense***. We call them **Lyapunov tracking properties** or for short **L-tracking properties**. This means the following:

1. The plant desired output behavior $\mathbf{Y}_d(.)$ can be any from the family $\mathfrak{Y}_d^k$ of realizable desired plant output behaviors.

2. The vector disturbance function $\mathbf{D}(.)$ can be any from the family $\mathfrak{D}^k$ of permitted disturbance vector functions.

3. If, and only if, every real plant output behavior $\mathbf{Y}^k(t)$ starting initially in some connected neighborhood $\mathfrak{N}_\varepsilon\left(t_0; \mathbf{Y}_{d0}^k; \mathbf{D}; \mathbf{U}; \mathbf{Y}_d\right)$ of the desired initial plant behavior $\mathbf{Y}_{d0}^k$ at the initial moment $t_0 \in \mathfrak{T}$ converges asymptotically to the desired plant behavior $\mathbf{Y}_d^k(t)$ as time t goes to infinity, (2), then, and only then, the desired plant output behavior $\mathbf{Y}^k(.)$ is attractive and *the plant exhibits its asymptotic tracking*, for short *tracking*. If this closeness holds for every vector $\mathbf{Y}_0^k \in \mathfrak{R}^{(k+1)N}$, then, and only then, it ensures global attraction to the desired plant output behavior $\mathbf{Y}_d^k(.)$ and *the plant exhibits its global tracking.*

Additionally to the conditions under 3. we can demand the following.

4. A real behavior $\mathbf{Y}^k(t)$ of a plant should track any desired plant behavior $\mathbf{Y}_d^k(.)$, $\mathbf{Y}_d(.) \in \mathfrak{Y}_d^k$, so that Lyapunov closeness among them holds on the infinite and unbounded time interval $\mathfrak{T}_0$.

5. Lyapunov closeness means that for every ε-neighborhood

$$\mathfrak{N}_\varepsilon\left[t; \mathbf{Y}_d^k(t); \mathbf{D}; \mathbf{U}; \mathbf{Y}_d\right]$$

of $\mathbf{Y}_d^k(t)$ over the *time* interval $\mathfrak{T}_0$, there exists an initial neighborhood

$$\mathfrak{N}\left(\varepsilon; t_0; \mathbf{Y}_{d0}^k; \mathbf{D}; \mathbf{U}; \mathbf{Y}_d\right)$$

of $\mathbf{Y}_{d0}^k$ at the initial moment $t_0 \in \mathfrak{T}$ such that for every real initial plant output vector $\mathbf{Y}^k(t_0)$ in $\mathfrak{N}\left(\varepsilon; t_0; \mathbf{Y}_{d0}^k; \mathbf{D}; \mathbf{U}; \mathbf{Y}_d\right)$ the real plant behavior $\mathbf{Y}^k(t)$ rests in $\mathfrak{N}_\varepsilon\left[t; \mathbf{Y}_d^k(t); \mathbf{D}; \mathbf{U}; \mathbf{Y}_d\right]$ forever, i.e., for all $t \in \mathfrak{T}_0$. If this Lyapunov closeness holds then, and only then, it ensures Lyapunov stability to the desired output plant behavior $\mathbf{Y}_d^k(.)$. However, it does not guarantee stability of the desired motion $\mathcal{R}_d^{\alpha-1}(.)$ of the plant.

6. The initial closeness $\mathfrak{N}\left(\varepsilon; t_0; \mathbf{Y}_{d0}^k; \mathbf{D}; \mathbf{U}; \mathbf{Y}_d\right)$ of the real initial plant behavior $\mathbf{Y}_0^k = \mathbf{Y}^k(t_0)$ to the desired initial plant behavior $\mathbf{Y}_{d0}^k$ is arbitrary in its ε-closeness $\mathfrak{N}_\varepsilon\left[t; \mathbf{Y}_d^k(t); \mathbf{D}; \mathbf{U}; \mathbf{Y}_d\right]$ at the initial moment t_0, which permits arbitrariness of the initial conditions. The connected neighborhood $\mathfrak{N}\left(\varepsilon; t_0; \mathbf{Y}_{d0}^k; \mathbf{D}; \mathbf{U}; \mathbf{Y}_d\right)$ of the initial desired output $\mathbf{Y}_{d0}^k$ at the initial moment $t_0 \in \mathfrak{T}$ determines the initial closeness of $\mathbf{Y}_0^k$ to $\mathbf{Y}_{d0}^k$. The connected ϵ-neighborhood $\mathfrak{N}_\varepsilon\left[t; \mathbf{Y}_d^k(t); \mathbf{D}; \mathbf{U}; \mathbf{Y}_d\right]$ of the instantaneous desired output $\mathbf{Y}_d^k(t)$ at the moment $t \in \mathfrak{T}_0$ determines the instantaneous ε-closeness of $\mathbf{Y}^k(t)$ to $\mathbf{Y}_d^k(t)$.

7. When all above requirements under 4.-6. hold then, and only then, the desired output behavior $\mathbf{Y}_d^k(.)$ is asymptotically stable and *the plant exhibits stable tracking.*

Tracking in the Lyapunov sense demands, hence ensures, more than Lyapunov stability for the following reasons.

The requested closeness is to be realized:

i) For the k-th-order *output behavior* expressed by the temporal evolution of $\mathbf{Y}^k(t) = \mathcal{Y}^k(t; \mathbf{Y}^k; \mathbf{D}; \mathbf{U})$ for some $k \in \{0, 1, 2, \ ..., \alpha - 1\}$ rather than only for the whole state behavior $\mathbf{R}^{\alpha-1}(t) \equiv \mathbf{R}^{\alpha-1}\left(t; t_0; \mathbf{R}_0^{\alpha-1}; \mathbf{D}; \mathbf{U}\right)$ that is strictly demanded in Lyapunov stability theory (which becomes a special case of Lyapunov tracking theory)

ii) F*or every desired plant behavior from a (given, or to be determined) family* $\mathfrak{Y}_d^k$ of possibly demanded realizable plant desired behaviors $\mathbf{Y}_d(.)$, i.e., *tracking should hold over the desired output family* $\mathfrak{Y}_d^k$

iii) F*or every external disturbance* $\mathbf{D}(.)$ *from a (given, or to be determined) family* $\mathfrak{D}^j$ of permitted external disturbances, rather than only for the nominal $\mathbf{D}_N(.)$. Lyapunov stability theory does not permit nonnominal disturbances. Therefore, Lyapunov stability properties represent special cases of the corresponding asymptotic tracking properties in the Lyapunov sense.

8. The reachability *time* is infinite. This concept does not demand that the real output vector and its derivatives composing $\mathbf{Y}^k(t)$ and the desired output vector and its derivatives forming $\mathbf{Y}_d^k(t)$ become equal in finite *time*. It ensures the asymptotic convergence of the

former to the latter as *time* t escapes to infinity: $t \longrightarrow \infty$

$$t \longrightarrow \infty \Longrightarrow \mathbf{Y}^k(t) \longrightarrow \mathbf{Y}_d^k(t), \quad \textit{equivalently} \;\; \boldsymbol{\varepsilon}^k(t) \longrightarrow \mathbf{0}_{(k+1)N}.$$

It is **the infinite reachability *time* concept**.

The above consideration explains why the following definitions represent extensions and generalizations of the corresponding definitions of Lyapunov stability properties.

The precise definitions follow in the sequel.

16.2 Definitions of L-tracking properties

We recall (1.8) or (7.2):

$$k \in \{0, 1, \cdots, \alpha - 1\}, \quad j, l \in \{0, 1, \cdots\}. \tag{16.2}$$

Notice that every desired plant output vector function $\mathbf{Y}_d(.) \in \mathfrak{Y}_d^k$ and the disturbance vector function $\mathbf{D}(.) \in \mathfrak{D}^j$ are bounded, $\mathbf{Y}_d(.)$ is *k-times* continuously differentiable in *time* on $\mathfrak{T}$, and $\mathbf{D}(.)$ is *j-times* continuously differentiable in *time* on $\mathfrak{T}$. This ensures that they are adequate descriptions of the corresponding physical variables (for details see the new *Physical Continuity and Uniqueness Principle* 16-18, and *Time Continuity and Uniqueness Principle* 20).

The following definitions characterize tracking properties in the extended output space $\mathfrak{R}^{(k+1)N}$, i.e., in the extended output integral space $\mathfrak{I}_e = \mathfrak{T} \times \mathfrak{R}^{(k+1)N}$ in general, hence in the output space $\mathfrak{R}^N$, i.e., in the output integral space $\mathfrak{I} = \mathfrak{T} \times \mathfrak{R}^N$, if $k = 0$. They are basic. They generate tracking definitions in other spaces. In them the mutually independent integers k, j, and l are defined by (16.2).

Let $\mathbf{Y}_d^k(t)$ and $\mathbf{Y}_d^k(t; t_0; \mathbf{Y}_{d0}^k)$ be the abbreviations of $\mathbf{Y}_d^k(t; t_0; \mathbf{Y}_{d0}^k; \mathbf{D}; \mathbf{U})$ if the pair $[\mathbf{D}(.), \mathbf{U}(.)] \in \mathfrak{D}^j \times \mathfrak{U}^l$ is fixed,

$$\mathbf{Y}_d^k(t) \equiv \mathbf{Y}_d^k(t; t_0; \mathbf{Y}_{d0}^k) \equiv \mathbf{Y}_d^k(t; t_0; \mathbf{Y}_{d0}^k; \mathbf{D}; \mathbf{U}).$$

This implies $\mathbf{Y}_d^k(t_0) \equiv \mathbf{Y}_{d0}^k$. Hence, $\mathbf{Y}_d(.) \in \mathfrak{Y}_d^k \Longrightarrow \mathbf{Y}_{d0}^k \in \mathfrak{Y}_{d0}^k$.

We sometimes use the abbreviated notation as follows:

$$\mathbf{Y}_d^k(t) \equiv \mathbf{Y}_d^k(t; t_0; \mathbf{Y}_{d0}^k), \quad \mathbf{Y}^k(t) \equiv \mathbf{Y}^k(t; t_0; \mathbf{Y}_0^k; \mathbf{D}; \mathbf{U}). \tag{16.3}$$

We recall the set $\mathfrak{T}_i$ (1.4) of all possibly significant initial moments t_0, $t_0 \in \mathfrak{T}_i$:

$$\mathfrak{T}_i = \{t : t[T]\langle s \rangle, \;\; t \in \mathfrak{T}\} \subseteq \mathfrak{T}, \; \textit{inf} \; \mathfrak{T}_i \geq -\infty, \;\; \textit{sup} \; \mathfrak{T}_i \leq \infty. \tag{16.4}$$

We accept the following.

Assumption 213 ***The plant extended desired output vector*** $\mathbf{Y}_d^k(t)$ ***is known***

The desired output vector behavior $\mathbf{Y}_d^k(t)$ *of the plant (3.12) is well defined and known for every* $t \in \mathfrak{T}_0$.

Note 214 ***The (control) plant tracking***

The definitions are given relative to the plant (3.12). Their adaptation to the (control) plant (3.12) (of the plant), respectively, is straightforward. In fact, they hold for the control plant of the plant under the following substitutions: $\mathbf{D}$ *should be replaced by* $\mathbf{I}$, $\mathbf{D}_N$ *by* $\mathbf{I}_N$, $\mathbf{D}_N^{\eta}$ *by* $\mathbf{I}^{\xi}$, $\mathfrak{D}^j$ *by* $\mathcal{J}^j$. $\mathbf{U}$ *should be omitted.*

Definitions in the output space of various (output) tracking properties follow.

Definition 215 ***Tracking of the extended desired output behavior*** $\mathbf{Y}_d^k(t)$ ***on the set product*** $\mathfrak{T}_0 \times \mathfrak{D}^j \times \mathfrak{Y}_d^k$ ***of the plant (3.12) controlled by a control*** $\mathbf{U}(.) \in \mathfrak{U}^l$

a) The plant (3.12) exhibits ***the asymptotic output tracking of*** $\mathbf{Y}_d^k(t)$, $\mathbf{Y}_d(.) \in \mathfrak{Y}_d^k$ ***on*** $\mathfrak{T}_0 \times \mathfrak{D}^j \times \mathfrak{Y}_d^k$, *for short* ***the tracking of*** $\mathbf{Y}_d^k(t)$ ***on*** $\mathfrak{T}_0 \times \mathfrak{D}^j \times \mathfrak{Y}_d^k$ *if, and only if, for every* $[\mathbf{D}(.), \mathbf{Y}_d(.)] \in \mathfrak{D}^j \times \mathfrak{Y}_d^k$ *there exists a* t_0*-dependent connected neighborhood* $\mathfrak{N}\left(t_0; \mathbf{Y}_{d0}^k; \mathbf{D}; \mathbf{U}; \mathbf{Y}_d^k\right) \subseteq \mathfrak{R}^{(k+1)N}$ *of the plant extended desired initial output vector* $\mathbf{Y}_{d0}^k$ *at the initial moment* $t_0 \in \mathfrak{T}$ *and for every* $\varsigma > 0$ *there exists a nonnegative real number* τ, $\tau = \tau\left(t_0, \varsigma, \mathbf{Y}_{d0}^k; \mathbf{D}; \mathbf{U}; \mathbf{Y}_d^k\right) \in \mathfrak{R}_+$, *such that* $\mathbf{Y}_0^k$ *from* $\mathfrak{N}\left(t_0; \mathbf{Y}_{d0}^k; \mathbf{D}; \mathbf{U}; \mathbf{Y}_d^k\right)$ *guarantees that the extended output vector* $\mathbf{Y}^k(t; t_0; \mathbf{Y}_0^k; \mathbf{D}; \mathbf{U})$ *belongs to the* ς*-neighborhood* $\mathfrak{N}_\varsigma\left(t; \mathbf{Y}_{d0}^k; \mathbf{D}; \mathbf{U}; \mathbf{Y}_d^k\right)$ *of* $\mathbf{Y}_d^k(t; t_0; \mathbf{Y}_{d0}^k)$ *for all time* $t \in]t_0 + \tau\left(t_0, \varsigma, \mathbf{Y}_0^k; \mathbf{D}; \mathbf{U}; \mathbf{Y}_d^k\right), \infty[$, *i.e.,*

$$\begin{gathered}\forall \varsigma > 0,\ \forall [\mathbf{D}(.), \mathbf{Y}_d(.)] \in \mathfrak{D}^j \times \mathfrak{Y}_d^k, \\ \exists \mathfrak{N}\left(t_0; \mathbf{Y}_{d0}^k; \mathbf{D}; \mathbf{U}; \mathbf{Y}_d^k\right) \subseteq \mathfrak{R}^{(k+1)N},\ \mathbf{Y}_0^k \in \mathfrak{N}\left(t_0; \mathbf{Y}_{d0}^k; \mathbf{D}; \mathbf{U}; \mathbf{Y}_d^k\right) \Longrightarrow \\ \left\{\begin{array}{c} \mathbf{Y}^k(t; t_0; \mathbf{Y}_0^k; \mathbf{D}; \mathbf{U}) \in \mathfrak{N}_\varsigma\left(t; \mathbf{Y}_{d0}^k; \mathbf{D}; \mathbf{U}; \mathbf{Y}_d^k\right), \\ \forall t \in]t_0 + \tau\left(t_0, \varsigma, \mathbf{Y}_0^k; \mathbf{D}; \mathbf{U}; \mathbf{Y}_d^k\right), \infty[\end{array}\right\}. \end{gathered} \tag{16.5}$$

This is called also ***the*** *k-th-order* ***asymptotic output tracking of*** $\mathbf{Y}_d(t)$, $\mathbf{Y}_d(.) \in \mathfrak{Y}_d^k$, ***on*** $\mathfrak{T}_0 \times \mathfrak{D}^j \times \mathfrak{Y}_d^k$, *for short* ***the*** *k-th-order* ***tracking of*** $\mathbf{Y}_d(t)$ ***on*** $\mathfrak{T}_0 \times \mathfrak{D}^j \times \mathfrak{Y}_d^k$

The zero ($k = 0$) *order tracking is simply called* ***the tracking.***

b) The largest connected neighborhood $\mathfrak{N}\left(t_0; \mathbf{Y}_{d0}^k; \mathbf{D}; \mathbf{U}; \mathbf{Y}_d^k\right)$ *of* $\mathbf{Y}_{d0}^k$ *that obeys (16.5), is* ***the*** *k-th-order* ***tracking domain*** $\mathcal{D}_T^k\left(t_0; \mathbf{Y}_{d0}^k; \mathbf{D}; \mathbf{U}; \mathbf{Y}_d^k\right)$ ***of*** $\mathbf{Y}_d^k(t)$ *for every* $[\mathbf{D}(.), \mathbf{Y}_d(.)] \in \mathfrak{D}^j \times \mathfrak{Y}_d^k$ ***, i.e., on*** $\mathfrak{T}_0 \times \mathfrak{D}^j \times \mathfrak{Y}_d^k$, ***at*** $t_0 \in \mathfrak{T}$.

c) The tracking of $\mathbf{Y}_d^k(t)$ *on* $\mathfrak{T}_0 \times \mathfrak{D}^j \times \mathfrak{Y}_d^k$ *is* ***global (in the whole)*** *if, and only if,* $\mathcal{D}_T^k\left(t_0; \mathbf{Y}_{d0}^k; \mathbf{D}; \mathbf{U}; \mathbf{Y}_d^k\right) = \mathfrak{R}^{(k+1)N}$ *for every* $[\mathbf{D}(.), \mathbf{Y}_d(.)] \in \mathfrak{D}^j \times \mathfrak{Y}_d^k$.

This definition determines the tracking properties so that they depend on a particular $[t_0, \mathbf{D}(.), \mathbf{Y}_d(.)] \in \mathfrak{T} \times \mathfrak{D}^j \times \mathfrak{Y}_d^k$. Such tracking properties are nonuniform.

Definition 216 ***Uniform tracking of the extended desired output behavior*** $\mathbf{Y}_d^k(t)$, ***on*** $\mathfrak{T}_i \times \mathfrak{D}^j \times \mathfrak{Y}_d^k$ ***of the plant (3.12) controlled by a control*** $\mathbf{U}(.) \in \mathfrak{U}^l$

If, and only if, the plant (3.12) controlled by a control $\mathbf{U}(.) \in \mathfrak{U}^l$ *exhibits the tracking of* $\mathbf{Y}_d^k(t)$ *on* $\mathfrak{T}_0 \times \mathfrak{D}^j \times \mathfrak{Y}_d^k$ *for every* $t_0 \in \mathfrak{T}_i$, *and:*

a) Both the intersection $\mathcal{D}_{TU}^k\left(t_0; \mathfrak{D}^j; \mathbf{U}; \mathfrak{Y}_d^k\right)$ *of tracking domains* $\mathcal{D}_T^k\left(t_0; \mathbf{Y}_{d0}^k; \mathbf{D}; \mathbf{U}; \mathbf{Y}_d^k\right)$ *in* $\left[\mathbf{D}(.), \mathbf{Y}_d^k(.)\right] \in \mathfrak{D}^j \times \mathfrak{Y}_d^k$ *is a connected neighborhood of* $\mathbf{Y}_{d0}^k$ *of every* $\mathbf{Y}_d(.) \in \mathfrak{Y}_d^k$ *for every* $t_0 \in \mathfrak{T}_i$ *:*

$$\begin{gathered}\textit{(16.5) holds and } \exists \xi \in \mathfrak{R}^+ \Longrightarrow \mathcal{D}_{TU}^k\left(t_0; \mathfrak{D}^j; \mathbf{U}; \mathfrak{Y}_d^k\right) \\ = \cap \left[\begin{array}{c} \mathcal{D}_T^k\left(t_0; \mathbf{Y}_{d0}^k; \mathbf{D}; \mathbf{U}; \mathbf{Y}_d^k\right) : \\ \left[\mathbf{D}(.), \mathbf{Y}_d^k(.)\right] \in \mathfrak{D}^j \times \mathfrak{Y}_d^k \end{array}\right] \supset \mathfrak{N}_\xi\left(t_0; \mathfrak{D}^j; \mathbf{U}; \mathfrak{Y}_d^k\right),\ \forall t_0 \in \mathfrak{T}_i, \\ \partial \mathcal{D}_{TU}^k\left(t_0; \mathfrak{D}^j; \mathbf{U}; \mathfrak{Y}_d^k\right) \cap \partial \mathfrak{N}_\xi\left(t_0; \mathfrak{D}^j; \mathbf{U}; \mathfrak{Y}_d^k\right) = \phi,\ \forall t_0 \in \mathfrak{T}_i, \end{gathered} \tag{16.6}$$

and the minimal $\tau\left(t_0, \varsigma, \mathbf{Y}_0^k; \mathbf{D}; \mathbf{U}; \mathbf{Y}_d^k\right)$ *denoted by* $\tau_m\left(t_0, \varsigma, \mathbf{Y}_0^k; \mathbf{D}; \mathbf{U}; \mathbf{Y}_d^k\right)$, *which obeys Definition 215, satisfies (16.7):*

$$\begin{gathered}\forall \left(\zeta, t_0, \mathbf{Y}_{00}^k\right) \in \mathfrak{R}^+ \times \mathfrak{T}_i \times \mathcal{D}_{TU}^k\left(t_0; \mathfrak{D}^j; \mathbf{U}; \mathfrak{Y}_d^k\right) \Longrightarrow \tau\left(t_0, \varsigma, \mathbf{Y}_0^k; \mathfrak{D}^j; \mathbf{U}; \mathfrak{Y}_d^k\right) \\ = \sup\left[\tau_m\left(t_0, \varsigma, \mathbf{Y}_0^k; \mathbf{D}; \mathbf{U}; \mathbf{Y}_d^k\right) : \left[\mathbf{D}(.), \mathbf{Y}_d(.)\right] \in \mathfrak{D}^j \times \mathfrak{Y}_d^k\right] \in \mathfrak{R}_+ \end{gathered} \tag{16.7}$$

then the tracking of $\mathbf{Y}_d^k(t)$ *is* ***uniform in*** $[\mathbf{D}(.), \mathbf{Y}_d(.)] \in \mathfrak{D}^j \times \mathfrak{Y}_d^k$ *on* $\mathfrak{T}_i \times \mathfrak{D}^j \times \mathfrak{Y}_d^k$ *and the set* $\mathcal{D}_{TU}^k\left(t_0; \mathfrak{D}^j; \mathbf{U}; \mathfrak{Y}_d^k\right)$ *is* ***the*** $(\mathbf{D}, \mathbf{Y}_d^k)$***-uniform*** *k-th-order* ***tracking domain of*** $\mathbf{Y}_d^k(t)$ ***on*** $\mathfrak{T}_0 \times \mathfrak{D}^j \times \mathfrak{Y}_d^k$.

b) Both

$$\forall (\zeta, \mathbf{Y}_0) \in \mathfrak{R}^+ \times \mathcal{D}^k_{TU}\left(t_0; \mathfrak{D}^j; \mathbf{U}; \mathfrak{Y}^k_d\right) \Longrightarrow$$
$$\tau\left(\mathfrak{T}_i, \varsigma, \mathbf{Y}^k_0; \mathbf{D}; \mathbf{U}; \mathbf{Y}^k_d\right) = \sup\left[\tau_m\left(t_0, \varsigma, \mathbf{Y}^k_0; \mathbf{D}; \mathbf{U}; \mathbf{Y}^k_d\right) : t_0 \in \mathfrak{T}_i\right] \in \mathfrak{R}_+ \tag{16.8}$$

and the intersection $\mathcal{D}^k_{TU}\left(\mathfrak{T}_i; \mathbf{Y}^k_{d0}; \mathbf{D}; \mathbf{U}; \mathbf{Y}^k_d\right)$ *of all the tracking domains*

$$\mathcal{D}^k_T\left(t_0; \mathbf{Y}^k_{d0}; \mathbf{D}; \mathbf{U}; \mathbf{Y}^k_d\right)$$

in t_0 *over* $\mathfrak{T}_i$ *is a connected neighborhood of* $\mathbf{Y}^k_{d0}$ *for every* $\mathbf{Y}^k_d(.) \in \mathfrak{Y}^k_d$*:*

$$(16.8) \text{ holds and } \exists \xi \in \mathfrak{R}^+ \Longrightarrow \mathcal{D}^k_{TU}\left(\mathfrak{T}_i; \mathbf{Y}^k_{d0}; \mathbf{D}; \mathbf{U}; \mathbf{Y}^k_d\right)$$
$$= \cap\left[\mathcal{D}^k_T\left(t_0; \mathbf{Y}^k_{d0}; \mathbf{D}; \mathbf{U}; \mathbf{Y}^k_d\right) : t_0 \in \mathfrak{T}_i\right] \supset \mathfrak{N}_\xi\left(\mathfrak{T}_i; \mathbf{Y}^k_{d0}; \mathbf{D}; \mathbf{U}; \mathbf{Y}^k_d\right),$$
$$\partial\mathcal{D}^k_{TU}\left(\mathfrak{T}_i; \mathbf{Y}^k_{d0}; \mathbf{D}; \mathbf{U}; \mathbf{Y}^k_d\right) \cap \partial\mathfrak{N}_\xi\left(\mathfrak{T}_i; \mathbf{Y}^k_{d0}; \mathbf{D}; \mathbf{U}; \mathbf{Y}^k_d\right) = \phi. \tag{16.9}$$

then the tracking of $\mathbf{Y}^k_d(t)$ *on* $\mathfrak{T} \times \mathfrak{D}^j \times \mathfrak{Y}^k_d$ *is* ***uniform in*** t_0 ***on*** $\mathfrak{T}_i$ *and the domain*

$$\mathcal{D}^k_{TU}\left(\mathfrak{T}_i; \mathbf{Y}^k_{d0}; \mathbf{D}; \mathbf{U}; \mathbf{Y}\right)$$

is ***the domain of the*** t_0-***uniform*** *k-th-order* ***tracking of*** $\mathbf{Y}^k_d(t)$ ***on*** $\mathfrak{T}_i \times \mathfrak{T}_0 \times \mathfrak{D}^j \times \mathfrak{Y}^k_d$*,*

c) If, and only if, both the intersection $\mathcal{D}^k_T\left(\mathfrak{T}_i; \mathfrak{D}^j; \mathbf{U}; \mathfrak{Y}^k_d\right)$ *of all (asymptotic) tracking domains* $\mathcal{D}^k_T\left(t_0; \mathbf{Y}^k_{d0}; \mathbf{D}; \mathbf{U}; \mathbf{Y}^k_d\right)$ *in* $\left[t_0, \mathbf{D}(.), \mathbf{Y}^k_d(.)\right]$ *over* $\mathfrak{T}_i \times \mathfrak{D}^j \times \mathfrak{Y}^k_d$ *is a neighborhood of* $\mathbf{Y}^k_{d0}$ *of every* $\mathbf{Y}_d(.) \in \mathfrak{Y}^k_d$ *and* $\tau_m\left(t_0, \varsigma, \mathbf{Y}^k_{d0}; \mathbf{D}; \mathbf{U}; \mathbf{Y}^k_d\right)$ *obeys*

$$\forall \left(\zeta, \mathbf{Y}^k_0\right) \in \mathfrak{R}^+ \times \mathcal{D}^k_T\left(\mathfrak{T}_i; \mathfrak{D}^j; \mathbf{U}; \mathfrak{Y}^k_d\right) \Longrightarrow \tau\left(\mathfrak{T}_i, \varsigma, \mathbf{Y}^k_{d0}; \mathfrak{D}^j; \mathbf{U}; \mathfrak{Y}^k_d\right)$$
$$= \sup\left[\tau_m\left(t_0, \varsigma, \mathbf{Y}^k_0; \mathbf{D}; \mathbf{U}; \mathbf{Y}^k_d\right) : \left[t_0, \mathbf{D}(.), \mathbf{Y}_d(.)\right] \in \mathfrak{T}_i \times \mathfrak{D}^j \times \mathfrak{Y}^k_d\right] \in \mathfrak{R}_+ \tag{16.10}$$

the tracking of $\mathbf{Y}^k_d(t)$ *on* $\mathfrak{T}_0 \times \mathfrak{D}^j \times \mathfrak{Y}^k_d$ *is* ***uniform in*** $\left[t_0, \mathbf{D}(.), \mathbf{Y}_d(.)\right] \in \mathfrak{T}_i \times \mathfrak{D}^j \times \mathfrak{Y}^k_d$*; i.e.,*

$$(16.10) \text{ holds and } \exists \xi \in \mathfrak{R}^+ \Longrightarrow \mathcal{D}^k_T\left(\mathfrak{T}_i; \mathfrak{D}^j; \mathbf{U}; \mathfrak{Y}^k_d\right)$$
$$= \cap\left[\begin{array}{c} \mathcal{D}^k_T\left(t_0; \mathbf{Y}^k_{d0}; \mathbf{D}; \mathbf{U}; \mathbf{Y}^k_d\right) : \\ \left[t_0, \mathbf{D}(.), \mathbf{Y}_d(.)\right] \in \mathfrak{T}_i \times \mathfrak{D}^j \times \mathfrak{Y}^k_d \end{array}\right] \supset \mathfrak{N}_\xi\left(\mathfrak{T}_i; \mathfrak{D}^j; \mathbf{U}; \mathfrak{Y}^k_d\right),$$
$$\partial\mathcal{D}^k_T\left(\mathfrak{T}_i; \mathfrak{D}^j; \mathbf{U}; \mathfrak{Y}^k_d\right) \cap \partial\mathfrak{N}_\xi\left(\mathfrak{T}_i; \mathfrak{D}^j; \mathbf{U}; \mathfrak{Y}^k_d\right) = \phi. \tag{16.11}$$

Then, and only then, the set $\mathcal{D}^k_{TU}\left(\mathfrak{T}_i; \mathfrak{D}^j; \mathbf{U}; \mathfrak{Y}^k_d\right)$ *is the* ***the domain of the uniform k-th-order tracking of*** $\mathbf{Y}^k_d(t)$*, for every* $\mathbf{Y}_d(.) \in \mathfrak{Y}^k_d$*, on the product set* $\mathfrak{T}_0 \times \mathfrak{T}_i \times \mathfrak{D}^j \times \mathfrak{Y}^k_d$*.*

Comment 217 *The global tracking of* $\mathbf{Y}^k_d(t)$ *on* $\mathfrak{T}_0 \times \mathfrak{D}^j \times \mathfrak{Y}^k_d$ *is uniform in* $[\mathbf{D}(.), \mathbf{Y}_d(.)]$ *over* $\mathfrak{D}^j \times \mathfrak{Y}^k_d$*.*

Comment 218 *The above definitions broaden Definition 197 of [279, p. 144] to time-varying nonlinear systems.*

The tracking of $\mathbf{Y}^k_d(t)$ *and its domain* $\mathcal{D}^k_T\left(t_0; \mathbf{Y}^k_{d0}; \mathbf{D}; \mathbf{U}; \mathbf{Y}^k_d\right)$ *depend on the initial moment* t_0 *in general. If the plant is strongly temporally nonuniform then the closure* $Cl\mathcal{D}^k_T\left(t_0; \mathbf{Y}^k_{d0}; \mathbf{D}; \mathbf{U}; \mathbf{Y}^k_d\right)$ *of* $\mathcal{D}^k_T\left(t_0; \mathbf{Y}^k_{d0}; \mathbf{D}; \mathbf{U}; \mathbf{Y}^k_d\right)$ *is asymptotically contractive to the singleton* $\left\{\mathbf{Y}^k_{d0}\right\}$ *as* t_0 *escapes to infinity for a fixed* $\mathbf{Y}^k_{d0} \in \mathfrak{R}^{(k+1)N}$*,*

$$\lim\left[Cl\mathcal{D}^k_T\left(t_0; \mathbf{Y}^k_{d0}; \mathbf{D}; \mathbf{U}; \mathbf{Y}^k_d\right) : t_0 \longrightarrow \infty\right] = \left\{\mathbf{Y}^k_{d0}\right\}.$$

We do not treat such cases.

Note 219 ***Tracking and realizability of*** $\mathbf{Y}_d^k(t)$

The desired plant output $\mathbf{Y}_d^k(t)$ *can be unrealizable even though the plant can exhibit tracking. This is due to the asymptotic convergence of the real output behavior to the desired one only as* $t \longrightarrow \infty$. *This essentially means that* $\mathbf{Y}^k(t)$ *converges to* $\mathbf{Y}_d^k(t)$ *only as* $t \longrightarrow \infty$.

The preceding definitions do not require for tracking of $\mathbf{Y}_d^k(t)$ any closeness of $\mathbf{Y}^k(t)$ to $\mathbf{Y}_d^k(t)$ at any finite $t < \infty$, $t \in \mathfrak{T}_0$.

Tracking does not give any information about the real output behavior relative to the desired one at any finite moment $t \in \mathfrak{T}_0$, after the initial one, $t > t_0$. The deviation of the former from the latter can be arbitrarily large at any finite instant. The following definition eliminates this drawback.

Definition 220 ***Stable tracking of the extended desired output behavior*** $\mathbf{Y}_d^k(t)$, ***on*** $\mathfrak{T}_0 \times \mathfrak{D}^j \times \mathfrak{Y}_d^k$ ***of the plant (3.12) controlled by a control*** $\mathbf{U}(.) \in \mathfrak{U}^l$

a) The plant (3.12) exhibits **stable output tracking of** $\mathbf{Y}_d^k(t)$ **on** $\mathfrak{T}_0 \times \mathfrak{D}^j \times \mathfrak{Y}_d^k$, *for short* **the stable tracking of** $\mathbf{Y}_d^k(t)$ **on** $\mathfrak{T}_0 \times \mathfrak{D}^j \times \mathfrak{Y}_d^k$ *if, and only if, it exhibits tracking of* $\mathbf{Y}_d^k(t)$ *on* $\mathfrak{T}_0 \times \mathfrak{D}^j \times \mathfrak{Y}_d^k$, *and for every connected neighborhood* $\mathfrak{N}_\varepsilon\left[t;\mathbf{Y}_d^k(t)\right]$, $\mathfrak{N}_\varepsilon\left[t;\mathbf{Y}_d^k(t)\right] \subseteq \mathfrak{R}^{(k+1)N}$, *of* $\mathbf{Y}_d^k(t)$ *at any* $t \in \mathfrak{T}_0$, *there is a connected neighborhood* $\mathfrak{N}\left(\varepsilon;t_0;\mathbf{Y}_{d0}^k;\mathbf{D};\mathbf{U};\mathbf{Y}_d^k\right)$, *(16.1), of the plant desired initial output vector* $\mathbf{Y}_{d0}^k$ *at the initial moment* $t_0 \in \mathfrak{T}$ *such that for the initial* $\mathbf{Y}_0^k \in \mathfrak{N}\left(\varepsilon;t_0;\mathbf{Y}_{d0}^k;\mathbf{D};\mathbf{U};\mathbf{Y}_d^k\right)$ *the instantaneous* $\mathbf{Y}^k(t)$ *stays in* $\mathfrak{N}_\varepsilon\left[t;\mathbf{Y}_d^k(t)\right]$ *for all* $t \in \mathfrak{T}_0$; *i.e.,*

$$\begin{gathered}
\forall \mathfrak{N}_\varepsilon\left[t;\mathbf{Y}_d^k(t)\right] \subseteq \mathfrak{R}^{(k+1)N},\ \ \forall t \in \mathfrak{T}_0,\\
\forall\left[\mathbf{D}(.),\mathbf{Y}_d(.)\right] \in \mathfrak{D}^j \times \mathfrak{Y}_d^k,\ \exists \mathfrak{N}\left(\varepsilon;t_0;\mathbf{Y}_{d0}^k;\mathbf{D};\mathbf{U};\mathbf{Y}_d^k\right) \subseteq \mathfrak{R}^{(k+1)N},\\
\mathfrak{N}\left(\varepsilon;t_0;\mathbf{Y}_{d0}^k;\mathbf{D};\mathbf{U};\mathbf{Y}_d^k\right) \subseteq \mathcal{D}_T^k\left(t_0;\mathbf{Y}_{d0}^k;\mathbf{D};\mathbf{U};\mathbf{Y}_d^k\right) \cap \mathfrak{N}_\varepsilon\left(t_0;\mathbf{Y}_{d0}^k\right),\\
\mathbf{Y}_0^k \in \mathfrak{N}\left(\varepsilon;t_0;\mathbf{Y}_{d0}^k;\mathbf{D};\mathbf{U};\mathbf{Y}_d^k\right) \Longrightarrow\\
\mathbf{Y}^k(t;t_0;\mathbf{Y}_0^k;\mathbf{D};\mathbf{U}) \in \mathfrak{N}_\varepsilon\left[t;\mathbf{Y}_d^k(t)\right],\ \forall t \in \mathfrak{T}_0.
\end{gathered}\tag{16.12}$$

If, and only if, this holds for every $t_0 \in \mathfrak{T}_i$ *then the plant (3.12) exhibits* **stable asymptotic output tracking of** $\mathbf{Y}_d^k(t)$ **on** $\mathfrak{T}_0 \times \mathfrak{T}_i \times \mathfrak{D}^j \times \mathfrak{Y}_d^k$.

b) The largest connected neighborhood $\mathfrak{N}_L\left(\varepsilon;t_0;\mathbf{Y}_{d0}^k;\mathbf{D};\mathbf{U};\mathbf{Y}_d^k\right)$ *of the desired initial output vector* $\mathbf{Y}_{d0}^k$, *(16.12), is* **the** ε**-tracking domain** *denoted by* $\mathcal{D}_{ST}^k\left(\varepsilon;t_0;\mathbf{Y}_{d0}^k;\mathbf{D};\mathbf{U};\mathbf{Y}_d^k\right)$ **at** $t_0 \in \mathfrak{T}$ ***of the stable tracking of*** $\mathbf{Y}_d^k(t)$ **at** $t_0 \in \mathfrak{T}$ ***on*** $\mathfrak{T}_0 \times \mathfrak{D}^j \times \mathfrak{Y}_d^k$.

The domain $\mathcal{D}_{ST}^k\left(t_0;\mathbf{Y}_{d0}^k;\mathbf{D};\mathbf{U};\mathbf{Y}_d^k\right)$ **at** $t_0 \in \mathfrak{T}$ ***of the stable asymptotic output tracking of*** $\mathbf{Y}_d^k(t)$ **on** $\mathfrak{T}_0 \times \mathfrak{D}^j \times \mathfrak{Y}_d^k$, **at** $t_0 \in \mathfrak{T}$ *is the union of all* $\mathcal{D}_{ST}^k\left(\varepsilon;t_0;\mathbf{Y}_{d0}^k;\mathbf{D};\mathbf{U};\mathbf{Y}_d^k\right)$ *over* $\varepsilon \in \mathfrak{R}^+$,

$$\mathcal{D}_{ST}^k\left(t_0;\mathbf{Y}_{d0}^k;\mathbf{D};\mathbf{U};\mathbf{Y}_d^k\right) = \cup\left[\mathcal{D}_{ST}^k\left(\varepsilon;t_0;\mathbf{Y}_{d0}^k;\mathbf{D};\mathbf{U};\mathbf{Y}_d^k\right) : \varepsilon \in \mathfrak{R}^+\right].\tag{16.13}$$

If, and only if, the intersection $\mathcal{D}_{ST}^k\left(\mathfrak{T}_0,\mathfrak{T}_i;\mathbf{Y}_{d0}^k;\mathbf{D};\mathbf{U};\mathbf{Y}_d^k\right)$ *of all domains*

$$\mathcal{D}_{ST}^k\left(t_0;\mathbf{Y}_{d0}^k;\mathbf{D};\mathbf{U};\mathbf{Y}_d^k\right)$$

in $t_0 \in \mathfrak{T}_i$ *is a connected neighborhood of* $\mathbf{Y}_{d0}^k$ *for every* $t_0 \in \mathfrak{T}_i$ *then the domain*

$$\mathcal{D}_{ST}^k\left(\mathfrak{T}_0,\mathfrak{T}_i;\mathbf{Y}_{d0}^k;\mathbf{D};\mathbf{U};\mathbf{Y}_d^k\right)$$

is the domain of **the stable asymptotic output tracking of** $\mathbf{Y}_d^k(t)$ **on** $\mathfrak{T}_0 \times \mathfrak{T}_i \times \mathfrak{D}^j \times \mathfrak{Y}_d^k$.

Let $[0,\varepsilon_M)$ *be the maximal interval over which* $\mathcal{D}_{ST}^k\left(\varepsilon;t_0;\mathbf{Y}_{d0}^k;\mathbf{D};\mathbf{U};\mathbf{Y}_d^k\right)$ *is continuous in* $\varepsilon \in \mathfrak{R}_+$,

$$\mathcal{D}_{ST}^k\left(\varepsilon;t_0;\mathbf{Y}_{d0}^k;\mathbf{D};\mathbf{U};\mathbf{Y}_d^k\right) \in \mathfrak{C}\left([0,\varepsilon_M)\right),\ \left[\mathbf{D}(.),\mathbf{Y}_d(.)\right] \in \mathfrak{D}^j \times \mathfrak{Y}_d^k.$$

The strict domain $\mathcal{D}_{SST}^{k}\left(t_0;\mathbf{Y}_{d0}^{k};\mathbf{D};\mathbf{U};\mathbf{Y}_{d}^{k}\right)$ ***at*** $t_0\in\mathfrak{T}$ ***of the stable k-th-order tracking of*** $\mathbf{Y}_d(t)$ ***on*** $\mathfrak{T}_0\times\mathfrak{D}^j\times\mathfrak{Y}_d^k$ *is the union of all stable tracking domains*

$$\mathcal{D}_{ST}^{k}\left(\varepsilon;t_0;\mathbf{Y}_{d0}^{k};\mathbf{D};\mathbf{U};\mathbf{Y}_{d}^{k}\right)$$

over $\varepsilon\in[0,\varepsilon_M)$,

$$\mathcal{D}_{SST}^{k}\left(t_0;\mathbf{Y}_{d0}^{k};\mathbf{D};\mathbf{U};\mathbf{Y}_{d}^{k}\right)=\cup\left\{\mathcal{D}_{ST}^{k}\left(\varepsilon;t_0;\mathbf{Y}_{d0}^{k};\mathbf{D};\mathbf{U};\mathbf{Y}_{d}^{k}\right):\varepsilon\in[0,\varepsilon_M)\right\},$$
$$[\mathbf{D}(.),\mathbf{Y}_d(.)]\in\mathfrak{D}^j\times\mathfrak{Y}_d^k. \tag{16.14}$$

c) The stable tracking of $\mathbf{Y}_d^k(t)$ *on* $\mathfrak{T}_0\times\mathfrak{D}^j\times\mathfrak{Y}_d^k$ *is* ***global (in the whole)*** *if, and only if, it is both the global tracking of* $\mathbf{Y}_d^k(t)$ *on* $\mathfrak{T}_0\times\mathfrak{D}^j\times\mathfrak{Y}_d^k$ *and the stable tracking of* $\mathbf{Y}_d^k(t)$ *on* $\mathfrak{T}_0\times\mathfrak{D}^j\times\mathfrak{Y}_d^k$ *with* $\mathcal{D}_{ST}^{k}\left(t_0;\mathbf{Y}_{d0}^{k};\mathbf{D};\mathbf{U};\mathbf{Y}_{d}^{k}\right)=\mathfrak{R}^{(k+1)N}$ *for every* $[\mathbf{D}(.),\mathbf{Y}_d(.)]\in\mathfrak{D}^j\times\mathfrak{Y}_d^k$.

Definition 221 ***The uniform stable tracking of the desired output behavior*** $\mathbf{Y}_d^k(t)$ ***on*** $\mathfrak{T}_0\times\mathfrak{T}_i\times\mathfrak{D}^j\times\mathfrak{Y}_d^k$ ***of the plant (3.12) controlled by a control*** $\mathbf{U}(.)\in\mathfrak{U}^l$

If, and only if, the plant (3.12) controlled by a control $\mathbf{U}(.)\in\mathfrak{U}^l$ *exhibits the stable tracking of the desired output behavior* $\mathbf{Y}_d^k(t)$ *on* $\mathfrak{T}_0\times\mathfrak{D}^j\times\mathfrak{Y}_d^k$ *for every* $t_0\in\mathfrak{T}_i$ *together with:*

a) The uniform tracking in $\left[\mathbf{D}(.),\mathbf{Y}_d^k(.)\right]\in\mathfrak{D}^j\times\mathfrak{Y}_d^k$ *and the intersection*

$$\mathcal{D}_{STU}^{k}\left(t_0;\mathbf{Y}_{d0}^{k};\mathfrak{D}^j;\mathbf{U};\mathfrak{Y}_{d}^{k}\right)$$

of all $\mathcal{D}_{ST}^{k}\left(t_0;\mathbf{Y}_{d0}^{k};\mathbf{D};\mathbf{U};\mathbf{Y}_{d}^{k}\right)$,

$$\mathcal{D}_{ST}^{k}\left(t_0;\mathbf{Y}_{d0}^{k};\mathbf{D};\mathbf{U};\mathbf{Y}_{d}^{k}\right)$$

[the intersection $\mathcal{D}_{SSTU}^{k}\left(t_0;\mathbf{Y}_{d0}^{k};\mathfrak{D}^j;\mathbf{U};\mathfrak{Y}_{d}^{k}\right)$ *of all* $\mathcal{D}_{SST}^{k}\left(t_0;\mathbf{Y}_{d0}^{k};\mathbf{D};\mathbf{U};\mathbf{Y}_{d}^{k}\right)$*] over* $\mathfrak{D}^j\times\mathfrak{Y}_d^k$ *is a connected neighborhood of* $\mathbf{Y}_{d0}^k$ *for every* $t_0\in\mathfrak{T}_i$*, then, and only then, it is the* ***[strictly] stable tracking domain of*** $\mathbf{Y}_d^k(t)$ ***on*** $\mathfrak{T}_0\times\mathfrak{T}_i\times\mathfrak{D}^j\times\mathfrak{Y}_d^k$ ***uniform in the pair*** $[\mathbf{D}(.),\mathbf{Y}_d(.)]\in\mathfrak{D}^j\times\mathfrak{Y}_d^k$, *respectively,*

$$\begin{gathered}\exists\xi\in\mathfrak{R}^+\Longrightarrow\mathcal{D}_{STU}^{k}\left(t_0;\mathbf{Y}_{d0}^{k};\mathfrak{D}^j;\mathbf{U};\mathfrak{Y}_{d}^{k}\right)\\=\cap\left\{\mathcal{D}_{ST}^{k}\left(t_0;\mathbf{Y}_{d0}^{k};\mathbf{D};\mathbf{U};\mathbf{Y}_{d}^{k}\right):[\mathbf{D}(.),\mathbf{Y}_d(.)]\in\mathfrak{D}^j\times\mathfrak{Y}_d^k\right\}\\\supset\mathfrak{N}_\xi\left(t_0;\mathbf{Y}_{d0}^{k};\mathfrak{D}^j;\mathbf{U};\mathfrak{Y}_{d}^{k}\right),\quad\forall t_0\in\mathfrak{T}_i,\end{gathered}\tag{16.15}$$

$$\left[\begin{array}{c}\exists\xi\in\mathfrak{R}^+\Longrightarrow\mathcal{D}_{SSTU}^{k}\left(t_0;\mathbf{Y}_{d0}^{k};\mathfrak{D}^j;\mathbf{U};\mathfrak{Y}_{d}^{k}\right)\\=\cap\left\{\mathcal{D}_{SST}^{k}\left(t_0;\mathbf{Y}_{d0}^{k};\mathbf{D};\mathbf{U};\mathbf{Y}_{d}^{k}\right):[\mathbf{D}(.),\mathbf{Y}_d(.)]\in\mathfrak{D}^j\times\mathfrak{Y}_d^k\right\}\\\supset\mathfrak{N}_\xi\left(t_0;\mathbf{Y}_{d0}^{k};\mathfrak{D}^j;\mathbf{U};\mathfrak{Y}_{d}^{k}\right),\quad\forall t_0\in\mathfrak{T}_i,\end{array}\right].\tag{16.16}$$

b) The uniform tracking in $t_0\in\mathfrak{T}_i$ *and intersection* $\mathcal{D}_{STU}^{k}\left(\mathfrak{T}_i;\mathbf{Y}_{d0}^{k};\mathbf{D};\mathbf{U};\mathbf{Y}_{d}^{k}\right)$ *of stable tracking domains* $\mathcal{D}_{ST}^{k}\left(t_0;\mathbf{Y}_{d0}^{k};\mathbf{D};\mathbf{U};\mathbf{Y}_{d}^{k}\right)$

$$\left[\mathcal{D}_{SSTU}^{k}\left(\mathfrak{T}_i;\mathbf{Y}_{d0}^{k};\mathbf{D};\mathbf{U};\mathbf{Y}_{d}^{k}\right)\ of\mathcal{D}_{SST}^{k}\left(t_0;\mathbf{Y}_{d0}^{k};\mathbf{D};\mathbf{U};\mathbf{Y}_{d}^{k}\right)\right]$$

in t_0 *over* $\mathfrak{T}_i$ *is a connected neighborhood of* $\mathbf{Y}_{d0}^k$*, then, and only then, it is* ***the [strict] stable tracking domain of*** $\mathbf{Y}_d^k(t)$ ***on*** $\mathfrak{T}_i\times\mathfrak{D}^j\times\mathfrak{Y}_d^k$ ***uniform in*** $t_0\in\mathfrak{T}_i$, *respectively,*

$$\begin{gathered}\exists\xi\in\mathfrak{R}^+\Longrightarrow\mathcal{D}_{STU}^{k}\left(\mathfrak{T}_i;\mathbf{Y}_{d0}^{k};\mathbf{D};\mathbf{U};\mathbf{Y}_{d}^{k}\right)\\=\cap\left\{\mathcal{D}_{ST}^{k}\left(t_0;\mathbf{Y}_{d0}^{k};\mathbf{D};\mathbf{U};\mathbf{Y}_{d}^{k}\right):t_0\in\mathfrak{T}_i\right\}\supset\mathfrak{N}_\xi\left(\mathfrak{T}_i;\mathbf{Y}_{d0}^{k};\mathbf{D};\mathbf{U};\mathbf{Y}_{d}^{k}\right),\end{gathered}\tag{16.17}$$

$$\left[\begin{array}{c}\exists\xi\in\mathfrak{R}^+\Longrightarrow\mathcal{D}_{SSTU}^{k}\left(\mathfrak{T}_i;\mathbf{Y}_{d0}^{k};\mathbf{D};\mathbf{U};\mathbf{Y}_{d}^{k}\right)\\=\cap\left\{\mathcal{D}_{SST}^{k}\left(t_0;\mathbf{Y}_{d0}^{k};\mathbf{D};\mathbf{U};\mathbf{Y}_{d}^{k}\right):t_0\in\mathfrak{T}_i\right\}\supset\mathfrak{N}_\xi\left(\mathfrak{T}_i;\mathbf{Y}_{d0}^{k};\mathbf{D};\mathbf{U};\mathbf{Y}_{d}^{k}\right)\end{array}\right].\tag{16.18}$$

c) The uniform tracking in $[t_0, \mathbf{D}(.), \mathbf{Y}_d(.)] \in \mathfrak{T}_i \times \mathfrak{D}^j \times \mathfrak{Y}_d^k$ *, (16.10) holds and the intersection* $\mathcal{D}_{STU}^k(\mathfrak{T}_i;\mathfrak{D}^j;\mathbf{U};\mathfrak{Y}_d^k)$ *of all domains* $\mathcal{D}_{ST}^k(t_0;\mathbf{Y}_{d0}^k;\mathbf{D};\mathbf{U};\mathbf{Y}_d^k)$ *[the intersection* $\mathcal{D}_{SSTU}^k(\mathfrak{T};\mathfrak{D}^j;\mathbf{U};\mathfrak{Y}_d^k)$ *of all* $\mathcal{D}_{SST}^k(t_0;\mathbf{Y}_{d0}^k;\mathbf{D};\mathbf{U};\mathbf{Y}_d^k)$*] in*

$$[t_0, \mathbf{D}(.), \mathbf{Y}_d^k(.)] \in \mathfrak{T}_i \times \mathfrak{D}^j \times \mathfrak{Y}_d^k$$

is a connected neighborhood of every $\mathbf{Y}_{d0}^k \in \mathfrak{Y}_{d0}^k$ *then, and only then, it is the* ***[strictly] stable tracking domain of*** $\mathbf{Y}_d^k(t)$ *on* $\mathfrak{T}_i \times \mathfrak{D}^j \times \mathfrak{Y}_d^k$ ***uniform in*** $[t_0, \mathbf{D}(.), \mathbf{Y}_d^k(.)] \in \mathfrak{T}_i \times \mathfrak{D}^j \times \mathfrak{Y}_d^k$*, respectively,*

$$\begin{gathered}\text{(16.10) holds and } \exists \xi \in \mathfrak{R}^+ \Longrightarrow \mathcal{D}_{STU}^k(\mathfrak{T}_i;\mathfrak{D}^j;\mathbf{U};\mathfrak{Y}_d^k) \\ = \cap \left\{\mathcal{D}_{ST}^k(t_0;\mathbf{Y}_{d0}^k;\mathbf{D};\mathbf{U};\mathbf{Y}_d^k) : [t_0, \mathbf{D}(.), \mathbf{Y}_d(.)] \in \mathfrak{T}_i \times \mathfrak{D}^j \times \mathfrak{Y}_d^k\right\} \\ \supset \mathfrak{N}_\xi(\mathfrak{T}_i;\mathfrak{D}^j;\mathbf{U};\mathfrak{Y}_d^k),\end{gathered} \tag{16.19}$$

$$\left[\begin{gathered}\text{(16.10) holds and } \exists \xi \in \mathfrak{R}^+ \Longrightarrow \mathcal{D}_{SSTU}^k(\mathfrak{T}_i;\mathfrak{D}^j;\mathbf{U};\mathfrak{Y}_d^k) \\ = \cap \left\{\mathcal{D}_{SST}^k(t_0;\mathbf{Y}_{d0}^k;\mathbf{D};\mathbf{U};\mathbf{Y}_d^k) : [t_0, \mathbf{D}(.), \mathbf{Y}_d(.)] \in \mathfrak{T}_i \times \mathfrak{D}^j \times \mathfrak{Y}_d^k\right\} \\ \supset \mathfrak{N}_\xi(\mathfrak{T}_i;\mathfrak{D}^j;\mathbf{U};\mathfrak{Y}_d^k).\end{gathered}\right] \tag{16.20}$$

Theorem 222 ***Stable tracking and realizability of*** $\mathbf{Y}_d^k(t)$ ***on*** $\mathfrak{T}_0 \times \mathfrak{D}^j \times \mathfrak{Y}_d^k$

If the plant exhibits stable tracking of $\mathbf{Y}_d^k(t)$ *on* $\mathfrak{T}_0 \times \mathfrak{D}^j \times \mathfrak{Y}_d^k$*, then* $\mathbf{Y}_d^k(t)$ *is realizable on* $\mathfrak{T}_0 \times \mathfrak{D}^j \times \mathfrak{Y}_d^k$ *for the desired output initial conditions; i.e., the stable tracking on* $\mathfrak{T}_0 \times \mathfrak{D}^j \times \mathfrak{Y}_d^k$ *guarantees that*

$$\mathbf{Y}^k(0) = \mathbf{Y}_d^k(0) \Longrightarrow [\mathbf{Y}^k(t) = \mathbf{Y}_d^k(t),\ \forall t \in \mathfrak{T}_0].$$

Proof. Let $[\mathbf{D}(.), \mathbf{U}(.), \mathbf{Y}_d(.)]$ be arbitrary from $\mathfrak{D}^j \times \mathfrak{U}^l \times \mathfrak{Y}_d^k$. Let the plant (3.12) exhibits stable tracking of the desired output $\mathbf{Y}_d^k(t)$ on $\mathfrak{T}_0 \times \mathfrak{D}^j \times \mathfrak{Y}_d^k$. The conditions of Definition 220 hold and (16.12) is valid. Let us assume that $\mathbf{Y}_d^k(t)$ is not realizable. We disprove this assumption by showing that it leads to contradiction. Let $\mathbf{Y}_0^k = \mathbf{Y}_{d0}^k$. Hence, $\mathbf{Y}_0^k \in \mathfrak{N}(\varepsilon;t_0;\mathbf{Y}_{d0}^k;\mathbf{D};\mathbf{U};\mathbf{Y}_d^k)$ for every $\varepsilon \in \mathfrak{R}^+$, which implies

$$\left\|\mathbf{Y}_d^k(t;t_0;\mathbf{Y}_{d0}^k) - \mathbf{Y}^k(t;t_0;\mathbf{Y}_0^k;\mathbf{D};\mathbf{U};)\right\| < \varepsilon, \forall t \in \mathfrak{T}_0, \forall \varepsilon \in \mathfrak{R}^+,$$

due to (16.12). If $\mathbf{Y}_d^k(t)$ were unrealizable, then there would be a moment $t \in \mathfrak{T}_0$ and a number $\xi \in \mathfrak{R}^+$ such that $\left\|\mathbf{Y}_d^k(t;t_0;\mathbf{Y}_{d0}^k) - \mathbf{Y}^k(t;t_0;\mathbf{Y}_0^k;\mathbf{D};\mathbf{U})\right\| = \xi$. This would contradict that

$$\begin{gathered}\mathbf{Y}_0^k \in \mathfrak{N}(\varepsilon;t_0;\mathbf{Y}_{d0}^k;\mathbf{D};\mathbf{U};\mathbf{Y}_d^k) \text{ implies} \\ \left\|\mathbf{Y}_d^k(t;t_0;\mathbf{Y}_{d0}^k) - \mathbf{Y}^k(t;t_0;\mathbf{Y}_0^k;\mathbf{D};\mathbf{U})\right\| < \varepsilon, \forall t \in \mathfrak{T}_0, \forall \varepsilon \in \mathfrak{R}^+ \Longrightarrow \forall \varepsilon \in [0, \xi].\end{gathered}$$

Hence, the assumption that $\mathbf{Y}_d^k(t)$ is not realizable is invalid. Therefore, $\mathbf{Y}_d^k(t)$ is realizable ■

Comment 223 *The stable tracking of the desired output* $\mathbf{Y}_d^k(t)$ *on* $\mathfrak{T}_0 \times \mathfrak{D}^j \times \mathfrak{Y}_d^k$ *is sufficient for the realizability of* $\mathbf{Y}_d^k(t)$ *on* $\mathfrak{T}_0 \times \mathfrak{D}^j \times \mathfrak{Y}_d^k$*.*

However, $\mathbf{Y}_d^k(t)$ *can be realizable but the plant need not exhibit either its stable tracking or tracking. Realizability of* $\mathbf{Y}_d^k(t)$ *is not sufficient either for stable tracking or for tracking.*

The stable tracking expresses stability of the desired output behavior $\mathbf{Y}_d^k(t)$, in addition to its tracking. It does not allow arbitrarily large output error for bounded initial conditions and for the bounded input vector function. However, it does not show the rate of the convergence of the real output behavior to the desired one.

Definition 224 ***Exponential output tracking of $\mathbf{Y}_d^k(t)$ on $\mathfrak{T}_0 \times \mathfrak{D}^j \times \mathfrak{Y}_d^k$ of the plant (3.12) controlled by a control $\mathbf{U}(.) \in \mathfrak{U}^l$***

a) The plant (3.12) exhibits ***exponential asymptotic output tracking of $\mathbf{Y}_d^k(t)$, $\mathbf{Y}_d^k(.) \in \mathfrak{Y}_d^k$, on $\mathfrak{T}_0 \times \mathfrak{D}^j \times \mathfrak{Y}_d^k$****, for short* ***the exponential tracking of $\mathbf{Y}_d^k(t)$ on $\mathfrak{T}_0 \times \mathfrak{D}^j \times \mathfrak{Y}_d^k$*** *if, and only if, for every $[\mathbf{D}(.), \mathbf{Y}_d(.)] \in \mathfrak{D}^j \times \mathfrak{Y}_d^k$ there exist positive real numbers $a \geq 1$ and $b > 0$, and a connected neighborhood $\mathfrak{N}\left(t_0; \mathbf{Y}_{d0}^k; a, b; \mathbf{D}; \mathbf{U}; \mathbf{Y}_d^k\right)$ of $\mathbf{Y}_{d0}^k$ at t_0, $a = a(\mathbf{D}, \mathbf{U}, \mathbf{Y}_d^k)$ and $b = b(\mathbf{D}, \mathbf{U}, \mathbf{Y}_d^k)$, such that $\mathbf{Y}_0^k \in \mathfrak{N}\left(t_0; \mathbf{Y}_{d0}^k; a, b; \mathbf{D}; \mathbf{U}; \mathbf{Y}_d^k\right)$ guarantees that $\mathbf{Y}_d^k(t)$ approaches exponentially $\mathbf{Y}_d^k(t)$ all the time; i.e.,*

$$
\begin{gathered}
\forall [\mathbf{D}(.), \mathbf{Y}_d(.)] \in \mathfrak{D}^j \times \mathfrak{Y}_d^k,\ \exists a \in [1, \infty[,\ \exists b \in \mathfrak{R}^+,\ \exists \xi \in \mathfrak{R}^+, \\
a = a(\mathbf{D}, \mathbf{U}, \mathbf{Y}_d^k), \quad b = b(\mathbf{D}, \mathbf{U}, \mathbf{Y}_d^k), \\
\exists \mathfrak{N}\left(t_0; \mathbf{Y}_{d0}^k; a, b; \mathbf{D}; \mathbf{U}; \mathbf{Y}_d^k\right) \supset In\mathfrak{N}_\xi\left(t_0; \mathbf{Y}_{d0}^k; a, b; \mathbf{D}; \mathbf{U}; \mathbf{Y}_d^k\right) \supset \left\{\mathbf{Y}_{d0}^k\right\}, \\
\partial\mathfrak{N}\left(t_0; \mathbf{Y}_{d0}^k; a, b; \mathbf{D}; \mathbf{U}; \mathbf{Y}_d^k\right) \cap \partial\mathfrak{N}_\xi\left(t_0; \mathbf{Y}_{d0}^k; a, b; \mathbf{D}; \mathbf{U}; \mathbf{Y}_d^k\right) = \phi, \\
\mathbf{Y}_0^k \in \mathfrak{N}\left(t_0; \mathbf{Y}_{d0}^k; a, b; \mathbf{D}; \mathbf{U}; \mathbf{Y}_d^k\right) \Longrightarrow \\
\left\|\mathbf{Y}_d^k(t; t_0; \mathbf{Y}_{d0}^k) - \mathbf{Y}^k(t; t_0; \mathbf{Y}_0^k; \mathbf{D}; \mathbf{U})\right\| \\
\leq a \left\|\mathbf{Y}_{d0}^k - \mathbf{Y}_0^k\right\| exp\left[-b\left(t - t_0\right)\right], \quad \forall t \in \mathfrak{T}_0. \qquad (16.21)
\end{gathered}
$$

b) The largest connected neighborhood $\mathfrak{N}\left(t_0; \mathbf{Y}_{d0}^k; a, b; \mathbf{D}; \mathbf{U}; \mathbf{Y}_d^k\right)$ of $\mathbf{Y}_{d0}^k$ at t_0 is ***the domain $\mathcal{D}^k\left(t_0; \mathbf{Y}_{d0}^k; a, b; \mathbf{D}; \mathbf{U}; \mathbf{Y}_d^k\right)$ at $t_0 \in \mathfrak{T}$ of the exponential tracking of $\mathbf{Y}_d^k(t)$ on*** *$\mathfrak{T}_0 \times \mathfrak{D}^j \times \mathfrak{Y}_d^k$ relative to a and b. When a and b are fixed then they can be omitted,*

$$
\begin{gathered}
a \text{ and } b \text{ are fixed} \Longrightarrow \\
\mathcal{D}^k\left(t_0; \mathbf{Y}_{d0}^k; a, b; \mathbf{D}; \mathbf{U}; \mathbf{Y}_d^k\right) = \mathcal{D}^k\left(t_0; \mathbf{Y}_{d0}^k; \mathbf{D}; \mathbf{U}; \mathbf{Y}_d^k\right). \qquad (16.22)
\end{gathered}
$$

c) The exponential tracking of $\mathbf{Y}_d^k(t)$ ***on*** *$\mathfrak{T}_0 \times \mathfrak{D}^j \times \mathfrak{Y}_d^k$ is* ***global (in the whole)*** *if, and only if, the domain $\mathfrak{D}^j\left(t_0; \mathbf{Y}_{d0}^k; a, b; \mathbf{D}; \mathbf{U}; \mathbf{Y}_d^k\right) = \mathfrak{R}^{(k+1)N}$ for every $[\mathbf{D}(.), \mathbf{Y}_d(.)] \in \mathfrak{D}^j \times \mathfrak{Y}_d^k$.*

Definition 225 ***The uniform exponential output tracking of $\mathbf{Y}_d^k(t)$, on $\mathfrak{T}_0 \times \mathfrak{D}^j \times \mathfrak{Y}_d^k$ of the plant (3.12) controlled by a control $\mathbf{U}(.) \in \mathfrak{U}^l$***

a) If, and only if, the values of $a = a\left(\mathbf{D}, \mathbf{U}, \mathbf{Y}_d^k\right)$ and $b = b(\mathbf{D}, \mathbf{U}, \mathbf{Y}_d^k)$, as well as the domain $\mathcal{D}^k\left(t_0; \mathbf{Y}_{d0}^k; a, b; \mathbf{D}; \mathbf{U}; \mathbf{Y}_d^k\right)$, depend at most on the set product $\mathfrak{D}^j \times \mathfrak{Y}_d^k$,

$$
a = a\left(\mathfrak{D}^j, \mathbf{U}, \mathfrak{Y}_d^k\right), \quad b = b(\mathfrak{D}^j, \mathbf{U}, \mathfrak{Y}_d^k), \qquad (16.23)
$$

but not on a particular choice of $[\mathbf{D}(.), \mathbf{Y}_d(.)]$ from $\mathfrak{D}^j \times \mathfrak{Y}_d^k$, then the exponential tracking of $\mathbf{Y}_d^k(t)$ on $\mathfrak{T}_0 \times \mathfrak{D}^j \times \mathfrak{Y}_d^k$ is ***uniform in $[\mathbf{D}(.), \mathbf{Y}_d(.)]$ relative to (a, b)****. Its* ***domain*** *$\mathcal{D}^k\left(t_0; a, b; \mathfrak{D}^j; \mathbf{U}; \mathfrak{Y}_d^k\right)$* ***at*** *$t_0 \in \mathfrak{T}$ is the intersection of all $\mathcal{D}^k\left(t_0; \mathbf{Y}_{d0}^k; a, b; \mathbf{D}; \mathbf{U}; \mathbf{Y}_d^k\right)$ in $[\mathbf{D}(.), \mathbf{Y}_d(.)] \in \mathfrak{D}^j \times \mathfrak{Y}_d^k$ and the connected neighborhood of $\mathbf{Y}_{d0}^k$ for every $[\mathbf{D}(.), \mathbf{Y}_d(.)] \in \mathfrak{D}^j \times \mathfrak{Y}_d^k$,*

$$
\begin{gathered}
\exists \xi > 0 \Longrightarrow \mathcal{D}^k\left(t_0; a, b; \mathfrak{D}^j; \mathbf{U}; \mathfrak{Y}_d^k\right) \\
= \cap \left\{ \begin{array}{c} \mathcal{D}^k\left[t_0; \mathbf{Y}_{d0}^k; a\left(\mathbf{D}, \mathbf{U}, \mathbf{Y}_d^k\right), b\left(\mathbf{D}, \mathbf{U}, \mathbf{Y}_d^k\right); \mathbf{D}; \mathbf{U}; \mathbf{Y}_d^k\right] : \\ : [\mathbf{D}(.), \mathbf{Y}_d(.)] \in \mathfrak{D}^j \times \mathfrak{Y}_d^k \end{array} \right\} \\
\supset \mathfrak{N}_\xi\left(t_0; \mathbf{Y}_{d0}^k; \mathfrak{D}^j; \mathbf{U}; \mathfrak{Y}_d^k\right), \\
\partial\mathfrak{D}^j\left(t_0; a, b; \mathfrak{D}^j; \mathbf{U}; \mathfrak{Y}_d^k\right) \cap \mathfrak{N}_\xi\left(t_0; \mathbf{Y}_{d0}^k; \mathfrak{D}^j; \mathbf{U}; \mathfrak{Y}_d^k\right) = \phi. \qquad (16.24)
\end{gathered}
$$

b) The exponential tracking of $\mathbf{Y}_d^k(t)$ on $\mathfrak{T}_0 \times \mathfrak{D}^j \times \mathfrak{Y}_d^k$ is ***uniform in*** *$t_0 \in \mathfrak{T}$* ***relative to (a, b)*** *if, and only if, the interior of the intersection domain $\mathcal{D}^k\left(\mathfrak{T}; \mathbf{Y}_{d0}^k; a, b; \mathbf{D}; \mathbf{U}; \mathbf{Y}_d^k\right)$*

in $t_0 \in \mathfrak{T}_i$ *of all domains* $\mathcal{D}^k\left(t_0;\mathbf{Y}_{d0}^k;a,b;\mathbf{D};\mathbf{U};\mathbf{Y}_d^k\right)$ *is the connected neighborhood of* $\mathbf{Y}_{d0}^k$ *for every* $t_0 \in \mathfrak{T}_i$,

$$\begin{gathered}
\exists \xi > 0 \Longrightarrow \mathcal{D}^k\left(\mathfrak{T}_i;\mathbf{Y}_{d0}^k;a,b;\mathbf{D};\mathbf{U};\mathbf{Y}_d^k\right) \\
= \cap\left\{\mathcal{D}^k\left(t_0;\mathbf{Y}_{d0}^k;a,b;\mathbf{D};\mathbf{U};\mathbf{Y}_d^k\right) : t_0 \in \mathfrak{T}_i\right\} \\
\supset \mathfrak{N}_\xi\left(\mathfrak{T}_i;\mathbf{Y}_{d0}^k;a,b;\mathbf{D};\mathbf{U};\mathbf{Y}_d^k\right), \\
\partial\mathcal{D}^k\left(\mathfrak{T}_i;\mathbf{Y}_{d0}^k;a,b;\mathbf{D};\mathbf{U};\mathbf{Y}_d^k\right) \cap \partial\mathfrak{N}_\xi\left(\mathfrak{T}_i;\mathbf{Y}_{d0}^k;a,b;\mathbf{D};\mathbf{U};\mathbf{Y}_d^k\right) = \phi.
\end{gathered} \tag{16.25}$$

Then, and only then, the set $\mathcal{D}^k\left(\mathfrak{T}_i;\mathbf{Y}_{d0}^k;a,b;\mathbf{D};\mathbf{U};\mathbf{Y}_d^k\right)$ *is* ***the domain of the uniform exponential tracking of*** $\mathbf{Y}_d^k(t)$ ***on*** $\mathfrak{T}_0 \times \mathfrak{T}_i \times \mathfrak{D}^j \times \mathfrak{Y}_d^k$.

c) The exponential tracking of $\mathbf{Y}_d^k(t)$ *on* $\mathfrak{T}_0 \times \mathfrak{D}^j \times \mathfrak{Y}_d^k$ *is* ***uniform in*** *[*t_0,$\mathbf{D}(.)$,$\mathbf{Y}_d^k(.)$*]* ***on the space product*** $\mathfrak{T}_i \times \mathfrak{D}^j \times \mathfrak{Y}_d^k$ ***relative to*** ***(a, b)*** *if, and only if, the intersection*

$$\mathcal{D}^k\left(\mathfrak{T}_i;a,b;\mathfrak{D}^j;\mathbf{U};\mathfrak{Y}_d^k\right)$$

in $\left[t_0,\mathbf{D}(.),\mathbf{Y}_d^k(.)\right] \in \mathfrak{T}_i \times \mathfrak{D}^j \times \mathfrak{Y}_d^k$ *of all*

$$\mathcal{D}^k\left[t_0;\mathbf{Y}_{d0}^k;a\left(\mathbf{D},\mathbf{U},\mathbf{Y}_d^k\right),b\left(\mathbf{D},\mathbf{U},\mathbf{Y}_d^k\right);\mathbf{D};\mathbf{U};\mathbf{Y}_d^k\right]$$

is a connected neighborhood of $\mathbf{Y}_{d0}^k$ *for all* $\left[t_0,\mathbf{D}(.),\mathbf{Y}_d^k(.)\right] \in \mathfrak{T}_i \times \mathfrak{D}^j \times \mathfrak{Y}_d^k$,

$$\begin{gathered}
\exists \xi > 0 \Longrightarrow \mathcal{D}^k\left(\mathfrak{T}_i;a,b;\mathfrak{D}^j;\mathbf{U};\mathfrak{Y}_d^k\right) \\
= \cap\left\{\begin{array}{c}\mathcal{D}^k\left[t_0;\mathbf{Y}_{d0}^k;a\left(\mathbf{D},\mathbf{U},\mathbf{Y}_d^k\right),b\left(\mathbf{D},\mathbf{U},\mathbf{Y}_d^k\right);\mathbf{D};\mathbf{U};\mathbf{Y}_d^k\right]: \\ \left[t_0,\mathbf{D}(.),\mathbf{Y}_d^k(.)\right] \in \mathfrak{T}_i \times \mathfrak{D}^j \times \mathfrak{Y}_d^k\end{array}\right\} \\
\supset \mathfrak{N}_\xi\left(\mathfrak{T}_i;\mathfrak{D}^j;\mathbf{U};\mathfrak{Y}_d^k\right), \\
\mathcal{D}^k\left(\mathfrak{T}_i;a,b;\mathfrak{D}^j;\mathbf{U};\mathfrak{Y}_d^k\right) \cap \partial\mathfrak{N}_\xi\left(\mathfrak{T}_i;\mathfrak{D}^j;\mathbf{U};\mathfrak{Y}_d^k\right) = \phi.
\end{gathered} \tag{16.26}$$

Then, and only then, the set $\mathcal{D}^k\left(\mathfrak{T}_i;a,b;\mathfrak{D}^j;\mathbf{U};\mathfrak{Y}_d^k\right)$ *is* ***the domain of the uniform exponential tracking of*** $\mathbf{Y}_d^k(t)$ ***relative to*** $a = a\left(\mathfrak{D}^j;\mathfrak{Y}_d^k\right)$ ***and*** $b = b\left(\mathfrak{D}^j;\mathfrak{Y}_d^k\right)$ ***on*** $\mathfrak{T}_0 \times \mathfrak{T}_i \times \mathfrak{D}^j \times \mathfrak{Y}_d^k$.

*The expression "****relative to (a, b)****" can be omitted if, and only if, a and b are fixed and known.*

Theorem 226 ***Exponential tracking and stable tracking***

If the k-th-order plant exhibits exponential tracking of $\mathbf{Y}_d^k(t)$ *on the product set* $\mathfrak{T}_0 \times \mathfrak{D}^j \times \mathfrak{Y}_d^k$ *then the tracking is also stable tracking on* $\mathfrak{T}_0 \times \mathfrak{D}^j \times \mathfrak{Y}_d^k$.

Proof. Let the plant exhibit exponential tracking of $\mathbf{Y}_d^k(t)$ on $\mathfrak{T}_0 \times \mathfrak{D}^j \times \mathfrak{Y}_d^k$. The conditions of Definition 224 hold; i.e., (16.21) is valid. Let $\Delta \in \mathfrak{R}^+$ be such that the hyperball $\mathfrak{B}_\Delta(\mathbf{Y}_{d0}^k) \subset \mathcal{D}^k\left(t_0;\mathbf{Y}_{d0}^k;a,b;\mathbf{D};\mathbf{U}\right)$.This ensures that $\mathbf{Y}_0^k \in \mathfrak{B}_\Delta(\mathbf{Y}_{d0}^k)$ implies lim $\left[\left\|\mathbf{Y}^k(t) - \mathbf{Y}_d^k(t)\right\| : t \longrightarrow \infty\right] = 0$. Let $\varepsilon_M = \alpha\Delta \in \mathfrak{R}^+$. Let $\varepsilon \in]0,\varepsilon_M]$ be arbitrary and let $\delta(\varepsilon) = a^{-1}\varepsilon$ determine the hyperball $\mathfrak{B}_{\delta(\varepsilon)}(\mathbf{Y}_{d0}^k)$. Because $a \geq 1$, then $\varepsilon \leq \varepsilon_M$ implies $\delta(\varepsilon) \leq a^{-1}\alpha\Delta = \Delta$. Hence, $\delta(\varepsilon) \leq \Delta$ and $\mathfrak{B}_{\delta(\varepsilon)}(\mathbf{Y}_{d0}^k) \subset \mathfrak{B}_\Delta(\mathbf{Y}_{d0}^k)$. These results and (16.21) guarantee:

$$\begin{gathered}
\mathbf{Y}_0^k \in \mathfrak{B}_{\delta(\varepsilon)}(\mathbf{Y}_{d0}^k) \subset \mathfrak{B}_\Delta(\mathbf{Y}_{d0}^k) \Longrightarrow \left\|\mathbf{Y}_{d0}^k - \mathbf{Y}_0^k\right\| < \delta(\varepsilon) \leq \Delta \Longrightarrow \\
\left\|\mathbf{Y}_d^k(t) - \mathbf{Y}_d^k(t)\right\| \leq a\left\|\mathbf{Y}_{d0}^k - \mathbf{Y}_0^k\right\| \exp\left[-b\left(t - t_0\right)\right] < \\
< \alpha\delta(\varepsilon)\exp\left[-b\left(t - t_0\right)\right] = \alpha\alpha^{-1}\varepsilon\exp\left[-b\left(t - t_0\right)\right] \leq \varepsilon,\ \forall t \in \mathfrak{T}_0,\ \forall \varepsilon \in]0,\varepsilon_M].
\end{gathered}$$

Altogether,

$$\mathbf{Y}_0^k \in \mathfrak{B}_{\delta(\varepsilon)}(\mathbf{Y}_{d0}^k) \subset \mathfrak{B}_{\Delta}(\mathbf{Y}_{d0}^k) \Longrightarrow$$
$$\left\|\mathbf{Y}_d^k(t) - \mathbf{Y}_d^k(t)\right\| < \varepsilon, \ \forall t \in \mathfrak{T}_0, \ \forall \varepsilon \in]0, \varepsilon_M].$$

For every $\varepsilon \geq \varepsilon_M$ we accept $\delta(\varepsilon) = a^{-1}\varepsilon_M$ so that

$$\mathbf{Y}_0^k \in \mathfrak{B}_{\delta(\varepsilon)}(\mathbf{Y}_{d0}^k) \Longrightarrow \left\|\mathbf{Y}_{d0}^k - \mathbf{Y}_0^k\right\| < \delta(\varepsilon) = a^{-1}\varepsilon_M \Longrightarrow$$
$$\left\|\mathbf{Y}_d^k(t) - \mathbf{Y}_d^k(t)\right\| \leq a \left\|\mathbf{Y}_{d0}^k - \mathbf{Y}_0^k\right\| \exp\left[-b\left(t - t_0\right)\right]$$
$$< \alpha\delta(\varepsilon)\exp\left[-b\left(t - t_0\right)\right] = \alpha\alpha^{-1}\varepsilon_M \exp\left[-b\left(t - t_0\right)\right] <$$
$$< \varepsilon_M \exp\left[-b\left(t - t_0\right)\right] \leq \varepsilon \exp\left[-b\left(t - t_0\right)\right] \leq \varepsilon, \ \forall t \in \mathfrak{T}_0, \ \forall \varepsilon \in]\varepsilon_M, \infty[.$$

Altogether,

$$\mathbf{Y}_0^k \in \mathfrak{B}_{\delta(\varepsilon)}(\mathbf{Y}_{d0}^k) \subset \mathfrak{B}_{\Delta}(\mathbf{Y}_{d0}^k) \Longrightarrow$$
$$\left\|\mathbf{Y}^k(t) - \mathbf{Y}_d^k(t)\right\| < \varepsilon, \forall t \in \mathfrak{T}_0, \ \forall \varepsilon \in [\varepsilon_M, \infty[, \text{ and}$$
$$lim \left[\left\|\mathbf{Y}^k(t) - \mathbf{Y}_d^k(t)\right\| : \ t \longrightarrow \infty\right] = 0.$$

These results show that Definition 220 is satisfied, i.e., that the exponential tracking of $\mathbf{Y}_d^k(t)$ on $\mathfrak{T}_0 \times \mathfrak{D}^j \times \mathfrak{Y}_d^k$ is also the stable tracking of $\mathbf{Y}_d^k(t)$ on the product set $\mathfrak{T}_0 \times \mathfrak{D}^j \times \mathfrak{Y}_d^k$ ■

The preceding theorems imply directly the following results.

Corollary 227 ***Exponential tracking and realizability of*** $\mathbf{Y}_d^k(t)$

If the plant exhibits exponential tracking of the extended desired output $\mathbf{Y}_d^k(t)$, *then* $\mathbf{Y}_d^k(t)$ *is realizable for the extended desired output initial conditions.*

Corollary 228 ***Necessity of realizability of*** $\mathbf{Y}_d^k(t)$

Realizability of $\mathbf{Y}_d^k(t)$ *is necessary, but not sufficient, for the stable tracking, hence also for the exponential tracking.*

Note 229 *Tracking allows arbitrary big error overshoot for arbitrary small initial output error. Stable tracking eliminates this drawback. Both tracking and stable tracking permit very slow error convergence to the zero error. Exponential tracking eliminates this drawback.*

Note 230 ***Lyapunov tracking and Lyapunov stability***

The concept of Lyapunov (the infinite time) tracking (***L-tracking***) *is in the sense of Lyapunov's stability concept; i.e., in the Lyapunov sense [279]. The former broadens to disturbed and (to be) controlled plants and generalize the latter. The latter can be considered now as a special case of the former.*

The explanations under i. through iii. at the beginning of Section 16.1, together with the above definitions, show that, and why, the infinite-time tracking properties and Lyapunov stability properties are mutually different.

- *(Global) Attraction of the extended desired output behavior* $\mathbf{Y}_d^k(t)$ *of the plant is necessary, but not sufficient, for its (global) tracking of* $\mathbf{Y}_d^k(t)$ *on* $\mathfrak{T}_0 \times \mathfrak{D}^j \times \mathfrak{Y}_d^k$;
- *(Global) Tracking of the extended desired output behavior* $\mathbf{Y}_d^k(t)$ *on the product set* $\mathfrak{T}_0 \times \mathfrak{D}^j \times \mathfrak{Y}_d^k$ *is sufficient (but not necessary) for the (global) attraction of the desired state behavior* $\mathbf{R}_d^{\alpha-1}(t)$ *of the plant, i.e., the former ensures the latter;*
- *(Global) Asymptotic stability of the desired state behavior* $\mathbf{R}_d^{\alpha-1}(t)$ *of the plant is necessary, but not sufficient, for the (global) stable tracking of* $\mathbf{Y}_d^k(t)$ *on* $\mathfrak{T}_0 \times \mathfrak{D}^j \times \mathfrak{Y}_d^k$;

- *(Global) Stable tracking of the extended desired output behavior* $\mathbf{Y}_d^k(t)$ *on the product set* $\mathfrak{T}_0 \times \mathfrak{D}^j \times \mathfrak{Y}_d^k$ *is sufficient (but not necessary) for the (global) asymptotic stability of the extended desired state behavior* $\mathbf{R}_d^{\alpha-1}(t)$ *of the plant; i.e., the former ensures the latter;*
- *(Global) Exponential stability of the desired state behavior* $\mathbf{R}_d^{\alpha-1}(t)$ *of the plant is necessary, but not sufficient, for the (global) exponential tracking of* $\mathbf{Y}_d^k(t)$ *on* $\mathfrak{T}_0 \times \mathfrak{D}^j \times \mathfrak{Y}_d^k$*;*
- *(Global) Exponential tracking of the extended desired output behavior* $\mathbf{Y}_d^k(t)$ *on* $\mathfrak{T}_0 \times \mathfrak{D}^j \times \mathfrak{Y}_d^k$ *is sufficient (but not necessary) for the (global) exponential stability of the extended desired state behavior* $\mathbf{R}_d^{\alpha-1}(t)$ *of the plant; i.e., the former ensures the latter.*

Lyapunov stability properties do not guarantee the infinite time tracking properties. The former are valid only for the nominal disturbance, whereas the latter hold for any disturbance from the family $\mathfrak{D}^j$*. The former concern stability properties of a particular, single, desired state behavior* $\mathbf{R}_d^{\alpha-1}(t)$*, whereas the latter are valid for the desired output behavior* $\mathbf{Y}_d^k(t)$ *for every* $\mathbf{Y}_d(t)$ *from* $\mathfrak{Y}_d^k$*.*

The Lyapunov stability concept concerns all dynamical systems, not only plants and their control systems.

The concept of asymptotic tracking concerns the noncontrolled plant, the controlled plant, and its control system. It is the original qualitative dynamical concept of control science and of control engineering. It has full significance only in the framework of control issues.

Both the asymptotic tracking properties and Lyapunov stability properties hold on a semi-infinite time set $\mathfrak{T}_0$*. The reachability time is infinite in both of them.*

Note 231 *Tracking is necessary for all other, above-defined, tracking properties.*

Chapter 17

State space definitions

17.1 Introduction

The study of the (output) tracking properties in the state space $\mathfrak{R}^{\alpha\rho}$ is possible under the following condition.

Condition 232 ***The tracking study via the state space*** $\mathfrak{R}^{\alpha\rho}$

For every plant desired output response $\mathbf{Y}_d(.) \in \mathfrak{Y}_d^{\alpha}$ *the corresponding nominal input-state pair* $\left[\mathbf{I}_N(t), \mathbf{R}_d^{\alpha-1}(t)\right]$ *is well defined and known for all* $t \in \mathfrak{T}_0$ *and for every* $t_0 \in \mathfrak{T}_i$.

In order to satisfy this condition the plant model (3.12) should be solved in the nominal regime, (115), Theorem 115, for every $\mathbf{Y}_d(.) \in \mathfrak{Y}_d^{\alpha}$ for all $t \in \mathfrak{T}_0$ and for every $t_0 \in \mathfrak{T}_i$:

$$Q\left[t, \mathbf{R}_d^{\alpha-1}(t)\right] \mathbf{R}_d^{(\alpha)}(t) + \mathbf{q}\left[t, \mathbf{R}_d^{\alpha-1}(t)\right] - \mathbf{h}\left[t, \mathbf{I}_N^{\xi}(t)\right] = \mathbf{0}_\rho, \quad \forall t \in \mathfrak{T}_0, \tag{17.1}$$

$$\mathbf{s}\left[t, \mathbf{R}_d^{\alpha-1}(t), \mathbf{I}_N(t)\right] = \mathbf{Y}_d(t),\ \forall t \in \mathfrak{T}_0. \tag{17.2}$$

This means that the equations (17.1), (17.2) should be solved in $\mathbf{I}_N(t)$ and $\mathbf{R}_d(t)$ for every $\mathbf{Y}_d(.) \in \mathfrak{Y}_d^{\alpha}$ for all $t \in \mathfrak{T}_0$ and for every $t_0 \in \mathfrak{T}_i$. It is a very complex problem known as the problem of the inverse dynamics determination. The equations (17.1), (17.2) comprise ρ independent scalar differential equations and N independent scalar algebraic equations. The number of the scalar variables to be determined is $\rho + M$. In order for the equations (17.1), (17.2) to be solvable in $\mathbf{I}_N(t)$ and $\mathbf{R}_d(t)$ it is necessary that $N \leq M$. If $N < M$ then $(M - N)$ scalar variables can be freely chosen.

17.2 Definitions of L-tracking properties

We introduce various L-tracking properties.

Note 233 ***The (control) system tracking***

The definitions are given relative to the plant (3.12). Their adaptation to the (control) system (3.12) (of the plant), respectively, is straightforward. In fact, they hold for the control system of the plant under the following substitutions: $\mathbf{D}$ *should be replaced by* $\mathbf{I}$, $\mathbf{D}^{\eta}$ *by* $\mathbf{I}^{\xi}$, $\mathfrak{D}^j$ *by* $\mathcal{J}^j$, $\mathbf{U}, \mathbf{U}_N$ *and* $\mathbf{U}_N^{\mu}$ *should be omitted.*

What follows assumes the validity of the following.

Assumption 234 ***The plant desired motion is known***

The desired motion $\mathcal{R}_d\left(.;t_0;\mathbf{R}_{d0}^{\alpha-1};\mathbf{D}_N;\mathbf{U}_N\right)$ *of the plant (3.12) is well defined and known on* $\mathfrak{T}_0$ *for every* $\left(t_0;\mathbf{R}_{d0}^{\alpha-1};\mathbf{U_N}\right)\in\mathfrak{T}_i\times\mathfrak{R}^{\alpha\rho}\times\mathfrak{R}^r$.

This assumption means that:

- The largest temporal interval $\mathfrak{L}^*=(l_m^*,l_M^*)$ of the initial moments t_0, for which the plant (3.12) has a solution, is the whole time set $\mathfrak{T}$.

- The largest set $\mathfrak{S}^*(t_0)$ of the initial vectors $\mathbf{R}_0^{\alpha-1}\in\mathfrak{R}^{\alpha\rho}$ at any $t_0\in\mathfrak{T}_i$, for which the plant (3.12) has a solution passing through $\mathbf{R}_0^{\alpha-1}$ at $t=t_0$, is the whole space $\mathfrak{R}^{\alpha\rho}$.

- The system has a solution for every $\mathbf{U_N}(t)\in\mathfrak{R}^r$ on $\mathfrak{T}_0$ for every $t_0\in\mathfrak{T}_i$.

The desired plant motion $\mathcal{R}_d\left(.;t_0;\mathbf{R}_{d0}^{\alpha-1};\mathbf{D}_N;\mathbf{U}_N\right)$ is, together with both $\mathbf{D}_N(t)$ and $\mathbf{U}_N(t)$, a solution to

$$Q\left[t,\mathbf{R}_d^{\alpha-1}(t)\right]\mathbf{R}_d^{(\alpha)}(t)+\mathbf{q}\left[t,\mathbf{R}_d^{\alpha-1}(t)\right]-\left\{\begin{array}{c}E(t)\,\mathbf{e}_d\left[\mathbf{D}_N^{\eta}(t)\right]+\\+P(t)\,\mathbf{e}_u\left[\mathbf{U}_N^{\mu}(t)\right]\end{array}\right\}=\mathbf{0}_\rho,$$

$$\forall t\in\mathfrak{T}_0, \tag{17.3}$$

$$\mathbf{z}\left[t,\mathbf{R}_d^{\alpha-1}(t),\mathbf{D}_N(t)\right]+W(t)\,\mathbf{w}\left[\mathbf{U}_N(t)\right]=\mathbf{Y}_d(t),\ \ \forall t\in\mathfrak{T}_0. \tag{17.4}$$

due to Theorem 122. The desired plant output $\mathbf{Y}_d(t)$ determines the plant desired motion $\mathcal{R}_d\left(.;t_0;\mathbf{R}_{d0}^{\alpha-1};\mathbf{D}_N;\mathbf{U}_N\right)$. The initial nominal values $\mathbf{D}_{N0}=\mathbf{D}_N(t_0)$ and $\mathbf{U}_{N0}=\mathbf{U}_N(t_0)$, and the desired initial value $\mathbf{R}_{d0}^{\alpha-1}$ should obey (17.3) at the initial moment t_0; i.e., they should satisfy

$$\mathbf{z}\left[t_0,\mathbf{R}_d^{\alpha-1}(t_0),\mathbf{D}_N(t_0)\right]+W(t_0)\,\mathbf{w}\left[\mathbf{U}_N(t_0)\right]=\mathbf{Y}_d(t_0),\ \ t_0\in\mathfrak{T}_i. \tag{17.5}$$

Once they are determined to satisfy this equation, then they determine, in view of (3.30), the initial desired speed $\frac{d\mathbf{R}_d^{\alpha-1}(t)}{dt}\,|_{t=t_0}$ of the state vector variation,

$$\frac{d\mathbf{R}_d^{\alpha-1}(t)}{dt}\,|_{t=t_0}=Q^{-1}\left[t_0,\mathbf{R}_d^{\alpha-1}(t_0)\right]\left\{\begin{array}{c}E(t_0)\,\mathbf{e}_d\left[\mathbf{D}_N^{\eta}(t_0)\right]\\+\,P(t_0)\,\mathbf{e}_u\left[\mathbf{U}_N^{\mu}(t_0)\right]\\-\,\mathbf{q}\left[t_0,\mathbf{R}_d^{\alpha-1}(t_0)\right]\end{array}\right\}. \tag{17.6}$$

Equations (17.3), (17.4) can have several solutions, or the unique solution, or can be without a solution for a given or chosen $\mathbf{R}_{d0}$. If they do not have a solution then the desired initial state vector $\mathbf{R}_{d0}$ should be changed. It, together with $\mathbf{D}_{N0}$ and $\mathbf{U}_{N0}$, should obey (17.5).

Note 235 ***On the k-th-order tracking***

The complexity of the nonlinear output dependence on the state vector in (3.12) implies that it is reasonable to study only the zero-order tracking rather than the k-th-order tracking in the framework of the nonlinear plants (3.12).

However, the study of the k-th-order tracking is justifiable in the framework of the linear plants (3.12). It can be developed by following [252] and [279].

Definitions in the *state* space of various (output) tracking properties follow.

We repeat b) of Definition 201.

Definition 236 ***The plant target set***

The target set Υ_S ***of the plant*** *(3.12) at any moment* $t\in\mathfrak{T}$ *is the set of all state vectors* $\mathbf{R}^{\alpha-1}$ *in the state space* $\mathfrak{R}^{\alpha\rho}$ *such that the plant real output vector* $\mathbf{Y}(t)$ *is equal to the plant desired output vector* $\mathbf{Y}_d(t)$ *at the same moment* $t\in\mathfrak{T}$ *under the action of* $\mathbf{D}(t)$ *and* $\mathbf{U}(t)$, $\Upsilon_S=\Upsilon_S(t;\mathbf{D};\mathbf{U};\mathbf{Y}_d)\subset\mathfrak{R}^{\alpha\rho}$,

$$\Upsilon_S=\Upsilon_S(t;\mathbf{D};\mathbf{U};\mathbf{Y}_d)\subset\mathfrak{R}^{\alpha\rho},\ \Upsilon_S(t;\mathbf{D};\mathbf{U};\mathbf{Y}_d)$$
$$=\left\{\begin{array}{c}\mathbf{R}^{\alpha-1}\in\mathfrak{R}^{\alpha\rho}:\\:\left[\mathbf{z}\left[t,\mathbf{R}^{\alpha-1},\mathbf{D}(t)\right]+W(t)\,\mathbf{w}\left[\mathbf{U}(t)\right]\right]=\mathbf{Y}_d(t)\end{array}\right\},\ \forall t\in\mathfrak{T}. \tag{17.7}$$

This is ***the instantaneous target set of the plant*** *(3.12)* ***at the moment*** $t\in\mathfrak{T}$.

The target set is *time*-varying in general. It is a subset of $\mathfrak{R}^{\alpha\rho}$. These features hold also for *time*-invariant plants due to the *time*-variation of the desired output vector $\mathbf{Y}_d(t; t_0; \mathbf{Y}_{d0})$ [279]. Its form, size, and position in the state space $\mathfrak{R}^{\alpha\rho}$ depend on:

- *Time* $t \in \mathfrak{T}$
- The form of the output vector function $\widetilde{\mathbf{s}}(.) = Z(.)\,\mathbf{z}(.) + W(.)\,\mathbf{w}(.)$
- The instantaneous input vectors $\mathbf{D}(t)$ and $\mathbf{U}(t)$
- The instantaneous desired output vector $\mathbf{Y}_d(t)$

The definition of the perfect tracking, Definition 199, the definition of the target set, Proposition 205 and Definition 236 yield:

Proposition 237 ***The desired motion, the target set, and the perfect tracking***
In order for the plant (3.12) to exhibit the perfect tracking on $\mathfrak{T}_0$ relative to its desired output vector response $\mathbf{Y}_d(t)$ it is necessary and sufficient that its motion

$$\mathcal{R}^{\alpha-1}\left(.; t_0; \mathbf{R}_{d0}^{\alpha-1}; \mathbf{D}_N; \mathbf{U}_N; \mathbf{Y}_d\right)$$

is in $\Upsilon_S\left(t; \mathbf{D}_N; \mathbf{U}_N, \mathbf{Y}_d\right)$ all the time $t \in \mathfrak{T}_0$,

$$\mathbf{Y}(t) \equiv \mathbf{Y}_d(t) \Longleftrightarrow$$
$$\mathcal{R}^{\alpha-1}\left(t; t_0; \mathbf{R}_{d0}^{\alpha-1}; \mathbf{D}_N; \mathbf{U}_N; \mathbf{Y}_d\right) \in \Upsilon_S\left(t; \mathbf{D}_N; \mathbf{U}_N; \mathbf{Y}_d\right),\ \forall t \in \mathfrak{T}_0. \tag{17.8}$$

Corollary 238 ***The desired motion, the target, set and the perfect tracking at the initial moment***
In order for the plant (3.12) to exhibit perfect tracking on $\mathfrak{T}_0$ it is necessary (but not sufficient) that its initial desired state vector $\mathbf{R}_{d0}^{\alpha-1} = \mathbf{R}_d^{\alpha-1}(t_0)$ is in its target set

$$\Upsilon_S\left(t; \mathbf{D}_N; \mathbf{U}_N; \mathbf{Y}_d\right)$$

at the initial instant $t_0 \in \mathfrak{T}_i$,

$$\mathbf{Y}(t) \equiv \mathbf{Y}_d(t) \Longrightarrow$$
$$\mathbf{R}_{d0}^{\alpha-1} = \mathbf{R}_d^{\alpha-1}(t_0) \ \in \Upsilon_S\left(t_0; \mathbf{D}_N; \mathbf{U}_N\right). \tag{17.9}$$

We define various *(imperfect) tracking properties*, stable ones of which guarantee either asymptotic or exponential stability of the plant desired motion $\mathcal{R}_d^{\alpha-1}\left(.; t_0; \mathbf{R}_{d0}^{\alpha-1}\right)$, (4.24), the instantaneous vector value of which at a moment t is $\mathbf{R}_d^{\alpha-1}(t; t_0; \mathbf{R}_{d0}^{\alpha-1})$,

$$\mathbf{R}_d^{\alpha-1}(t) \equiv \mathbf{R}_d^{\alpha-1}(t; t_0; \mathbf{R}_{d0}^{\alpha-1}) \equiv \mathcal{R}_d^{\alpha-1}\left(t; t_0; \mathbf{R}_{d0}^{\alpha-1}\right). \tag{17.10}$$

Definition 239 ***The tracking of the desired state behavior $\mathbf{R}_d^{\alpha-1}(t)$ of the plant (3.12) controlled by control $\mathbf{U}(.) \in \mathfrak{U}$***
a) The plant (3.12) exhibits ***the asymptotic output tracking of the desired state behavior $\mathbf{R}_d^{\alpha-1}(t)$ on $\mathfrak{T}_0 \times \mathfrak{D} \times \mathfrak{R}_d^{\alpha\rho}$****, for short* ***the tracking of $\mathbf{R}_d^{\alpha-1}(t)$ on $\mathfrak{T}_0 \times \mathfrak{D} \times \mathfrak{R}_d^{\alpha\rho}$*** *if, and only if, for every $[\mathbf{D}(.), \mathbf{R}_d(.)] \in \mathfrak{D} \times \mathfrak{R}_d^{\alpha\rho}$ there exists a t_0-dependent connected neighborhood $\mathfrak{N}\left(t_0; \mathbf{R}_{d0}^{\alpha-1}; \mathbf{D}; \mathbf{R}_d; \mathbf{U}\right)$, $\mathfrak{N}\left(t_0; \mathbf{R}_{d0}^{\alpha-1}; \mathbf{D}; \mathbf{R}_d; \mathbf{U}\right) \subseteq \mathfrak{R}^{\alpha\rho}$, of the desired initial system state vector $\mathbf{R}_{d0}^{\alpha-1}$ at the initial moment $t_0 \in \mathfrak{T}$ and for every $\varsigma > 0$ there exists a nonnegative real number τ, $\tau = \tau\left(t_0, \varsigma, \mathbf{R}_{d0}^{\alpha-1}; \mathbf{D}; \mathbf{U}; \mathbf{R}_d\right) \in \mathfrak{R}_+$, such that $\mathbf{R}_0^{\alpha-1}$ from the neighborhood $\mathfrak{N}\left(t_0; \mathbf{R}_{d0}^{\alpha-1}; \mathbf{D}; \mathbf{R}_d; \mathbf{U}\right)$ of $\mathbf{R}_{d0}^{\alpha-1}$ guarantees that the plant real state vector*

$\mathbf{R}^{\alpha-1}(t;t_0;\mathbf{R}_0^{\alpha-1};\mathbf{D};\mathbf{U})$ *belongs to the instantaneous* ς*-neighborhood* $\mathfrak{N}_\varsigma\left(t;\mathbf{R}_{d0}^{\alpha-1};\mathbf{D};\mathbf{U};\mathbf{R}_d\right)$ *of* $\mathcal{R}_d^{\alpha-1}(t;t_0;\mathbf{R}_{d0}^{\alpha-1})$ *for all time* $t\in]t_0+\tau\left(t_0,\varsigma,\mathbf{R}_0^{\alpha-1};\mathbf{D};\mathbf{U};\mathbf{R}_d\right),\infty[$*; i.e.*

$$\begin{gathered}\forall\left(\varsigma,\mathbf{R}_0^{\alpha-1}\right)\in\mathfrak{R}^+\times\mathfrak{N}\left(t_0;\mathbf{R}_{d0}^{\alpha-1};\mathbf{D};\mathbf{R}_d;\mathbf{U}\right),\ \forall\left[\mathbf{D}(.),\mathbf{R}_d(.)\right]\in\mathfrak{D}\times\mathfrak{R}_d^{\alpha\rho},\\ \exists\mathfrak{N}\left(t_0;\mathbf{R}_0^{\alpha-1};\mathbf{D};\mathbf{R}_d;\mathbf{U}\right)\subseteq\mathfrak{R}^{\alpha\rho},\mathbf{R}_0^{\alpha-1}\in\mathfrak{N}\left(t_0;\mathbf{R}_{d0}^{\alpha-1};\mathbf{D};\mathbf{R}_d;\mathbf{U}\right)\Longrightarrow\\ \mathcal{R}^{\alpha-1}(t;t_0;\mathbf{R}_0^{\alpha-1};\mathbf{D};\mathbf{U})\in\mathfrak{N}_\varsigma\left(t;\mathbf{R}_{d0}^{\alpha-1};\mathbf{D};\mathbf{U};\mathbf{R}_d\right),\\ \forall t\in]t_0+\tau\left(t_0,\varsigma,\mathbf{R}_0^{\alpha-1};\mathbf{D};\mathbf{U};\mathbf{R}_d\right),\infty[.\end{gathered}\tag{17.11}$$

This is called also ***the asymptotic tracking of*** $\mathbf{R}_d^{\alpha-1}(t)$ ***on*** $\mathfrak{T}_0\times\mathfrak{D}\times\mathfrak{R}_d^{\alpha\rho}$*, for short* ***the*** α***-th-order tracking of*** $\mathbf{R}_d^{\alpha-1}(t)$ ***on*** $\mathfrak{T}_0\times\mathfrak{D}\times\mathfrak{R}_d^{\alpha\rho}$*.*

b) The largest connected neighborhood $\mathfrak{N}_L\left(t_0;\mathbf{R}_{d0}^{\alpha-1};\mathbf{D};\mathbf{R}_d;\mathbf{U}\right)$*, (17.11), is* ***the tracking domain*** $\mathcal{D}_T\left(t_0;\mathbf{R}_{d0}^{\alpha-1};\mathbf{D};\mathbf{R}_d;\mathbf{U}\right)$ ***of*** $\mathbf{R}_d^{\alpha-1}(t)$ ***on*** $\mathfrak{T}_0\times\mathfrak{D}\times\mathfrak{R}_d^{\alpha\rho}$*.*

c) The tracking of $\mathbf{R}_d^{\alpha-1}(t)$ *on* $\mathfrak{T}_0\times\mathfrak{D}\times\mathfrak{R}_d^{\alpha\rho}$ *is* ***global (in the whole)*** *if, and only if,* $\mathcal{D}_T\left(t_0;\mathbf{R}_{d0}^{\alpha-1};\mathbf{D};\mathbf{R}_d^{\alpha-1};\mathbf{U}\right)=\mathfrak{R}^{\alpha\rho}$ *for every* $[\mathbf{D}(.),\mathbf{Y}_d(.)]\in\mathfrak{D}\times\mathfrak{R}_d^{\alpha\rho}$*. It is also uniform on* $\mathfrak{D}\times\mathfrak{R}_d^{\alpha\rho}$*.*

Definition 240 ***The uniform tracking of the desired state behavior*** $\mathbf{R}_d^{\alpha-1}(t)$ ***of the plant*** *(3.12)* ***controlled by control*** $\mathbf{U}(.)\in\mathfrak{U}$

a) If, and only if, both the intersection $\mathcal{D}_{TU}\left(t_0;\mathfrak{D},\mathfrak{R}_d^{\alpha\rho};\mathbf{U}\right)$ *of the tracking domains* $\mathcal{D}_T\left(t_0;\mathbf{R}_{d0}^{\alpha-1};\mathbf{D};\mathbf{R}_d;\mathbf{U}\right)$ *of* $\mathbf{R}_d^{\alpha-1}(t)$ *in* $\left[\mathbf{D}(.),\mathbf{R}_d^{\alpha-1}(.)\right]\in\mathfrak{D}\times\mathfrak{R}_d^{\alpha\rho}$ *is also a connected neighborhood of* $\mathbf{R}_{d0}^{\alpha-1}$*,*

$$\begin{gathered}(17.11)\text{ holds and }\exists\xi\in\mathfrak{R}^+\Longrightarrow\mathcal{D}_{TU}\left(t_0;\mathfrak{D},\mathfrak{R}_d^{\alpha\rho};\mathbf{U}\right)\\ =\cap\left[\mathcal{D}_T\left(t_0;\mathbf{R}_{d0}^{\alpha-1};\mathbf{D};\mathbf{R}_d;\mathbf{U}\right):\left[\mathbf{D}(.),\mathbf{R}_d^{\alpha-1}(.)\right]\in\mathfrak{D}\times\mathfrak{R}_d^{\alpha\rho}\right]\\ \supset\mathfrak{N}_\xi\left(t_0;\mathfrak{D},\mathfrak{R}_d^{\alpha\rho};\mathbf{U}\right)\neq\phi,\\ \partial\mathcal{D}_{TU}\left(t_0;\mathfrak{D},\mathfrak{R}_d^{\alpha\rho};\mathbf{U}\right)\cap\partial\mathfrak{N}_\xi\left(t_0;\mathfrak{D},\mathfrak{R}_d^{\alpha\rho};\mathbf{U}\right)=\phi,\end{gathered}$$

and the minimal $\tau\left(t_0,\varsigma,\mathbf{R}_0^{\alpha-1};\mathbf{D};\mathbf{U};\mathbf{Y}_d^k\right)$ *denoted by* $\tau_m\left(t_0,\varsigma,\mathbf{R}_0^{\alpha-1};\mathbf{D};\mathbf{U};\mathbf{Y}_d^k\right)$*, which obeys Definition 239, satisfies*

$$\begin{gathered}\forall\left(\zeta,\mathbf{R}_0^{\alpha-1}\right)\in\mathfrak{R}^+\times\mathcal{D}_{TU}\left(t_0;\mathfrak{D},\mathfrak{R}_d^{\alpha\rho};\mathbf{U}\right)\Longrightarrow\\ \tau\left(t_0,\varsigma,\mathbf{R}_0^{\alpha-1};\mathfrak{D}^k;\mathbf{U};\mathfrak{Y}_d^{p+1}\right)=\\ =sup\left[\tau_m\left(t_0,\varsigma,\mathbf{Y}_0^p;\mathbf{D};\mathbf{U};\mathbf{Y}_d\right):\left[\mathbf{D}(.),\mathbf{Y}_d(.)\right]\in\mathfrak{D}^k\times\mathfrak{R}_d^{\alpha\rho}\right]\in\mathfrak{R}_+\end{gathered}\tag{17.12}$$

then the tracking of $\mathbf{R}_d^{\alpha-1}(t)$ *is* ***uniform in*** $\left[\mathbf{D}(.),\mathbf{R}_d^{\alpha-1}(.)\right]\in\mathfrak{D}\times\mathfrak{R}_d^{\alpha\rho}$ ***on*** $\mathfrak{T}_0\times\mathfrak{D}\times\mathfrak{R}_d^{\alpha\rho}$ *and the set* $\mathcal{D}_{TU}\left(t_0;\mathfrak{D},\mathfrak{R}_d^{\alpha\rho};\mathbf{U}\right)$ *is* ***the tracking domain of*** $\mathbf{R}_d^{\alpha-1}(t)$ ***uniform in the pair*** $\left[\mathbf{D}(.),\mathbf{R}_d^{\alpha-1}(.)\right]\in\mathfrak{D}\times\mathfrak{R}_d^{\alpha\rho}$ ***of the plant*** *(3.12)* ***on*** $\mathfrak{T}_0\times\mathfrak{D}\times\mathfrak{R}_d^{\alpha\rho}$*.*

The global tracking of $\mathbf{R}_d^{\alpha-1}(t)$ *on* $\mathfrak{T}_0\times\mathfrak{D}\times\mathfrak{R}_d^{\alpha\rho}$ *is uniform in* $\left[\mathbf{D}(.),\mathbf{R}_d^{\alpha-1}(.)\right]\in\mathfrak{D}\times\mathfrak{R}_d^{\alpha\rho}$*. The state space* $\mathfrak{R}^{\alpha\rho}$ *is* ***the tracking domain*** $\mathcal{D}_{TU}\left(t_0;\mathfrak{D},\mathfrak{R}_d^{\alpha\rho};\mathbf{U}\right)$ ***of*** $\mathbf{R}_d^{\alpha-1}(t)$ ***uniform in*** $\left[\mathbf{D}(.),\mathbf{R}_d^{\alpha-1}(.)\right]\in\mathfrak{D}\times\mathfrak{R}_d^{\alpha\rho}$ *on* $\mathfrak{T}_0\times\mathfrak{D}\times\mathfrak{R}_d^{\alpha\rho}$*.*

b) If, and only if, both the intersection $\mathcal{D}_{TU}\left(\mathfrak{T}_i;\mathbf{R}_{d0}^{\alpha-1};\mathbf{D};\mathbf{R}_d;\mathbf{U}\right)$ *of the tracking domains* $\mathcal{D}_T\left(t_0;\mathbf{R}_{d0}^{\alpha-1};\mathbf{D};\mathbf{R}_d;\mathbf{U}\right)$ *of* $\mathbf{R}_d^{\alpha-1}(t)$ *over* $t_0\in\mathfrak{T}_{ii}$ *is a connected neighborhood of* $\mathbf{R}_{d0}^{\alpha-1}$*,*

$$\begin{gathered}(17.11)\text{ holds and }\exists\eta\in\mathfrak{R}^+\Longrightarrow\mathcal{D}_{TU}\left(\mathfrak{T}_i;\mathbf{R}_{d0}^{\alpha-1};\mathbf{D};\mathbf{R}_d;\mathbf{U}\right)\\ =\cap\left[\mathcal{D}_T\left(t_0;\mathbf{R}_{d0}^{\alpha-1};\mathbf{D};\mathbf{R}_d;\mathbf{U}\right):t_0\in\mathfrak{T}_{ii}\right]\supset\mathfrak{N}_\eta\left(\mathfrak{T}_i;\mathbf{R}_{d0}^{\alpha-1};\mathbf{D};\mathbf{R}_d;\mathbf{U}\right),\\ \partial\mathcal{D}_{TU}\left(\mathfrak{T}_i;\mathbf{R}_{d0}^{\alpha-1};\mathbf{D};\mathbf{R}_d;\mathbf{U}\right)\cap\partial\mathfrak{N}_\eta\left(\mathfrak{T}_i;\mathbf{R}_{d0}^{\alpha-1};\mathbf{D};\mathbf{R}_d;\mathbf{U}\right)=\phi,\end{gathered}\tag{17.13}$$

and

$$\begin{gathered}\forall\left(\zeta,\mathbf{R}_0^{\alpha-1}\right)\in\mathfrak{R}^+\times\mathcal{D}_{TU}\left(\mathfrak{T}_i;\mathbf{R}_{d0}^{\alpha-1};\mathbf{D};\mathbf{R}_d;\mathbf{U}\right)\Longrightarrow\\ \tau\left(\mathfrak{T}_i,\varsigma,\mathbf{R}_{d0}^{\alpha-1};\mathbf{D};\mathbf{U};\mathbf{R}_d\right)=\sup\left[\tau_m\left(t_0,\varsigma,\mathbf{R}_{d0}^{\alpha-1};\mathbf{D};\mathbf{U};\mathbf{R}_d\right):t_0\in\mathfrak{T}_{ii}\right]\in\mathfrak{R}_+\end{gathered}\tag{17.14}$$

then, and only then, the tracking of $\mathbf{R}_d^{\alpha-1}(t)$ *is* ***uniform in*** t_0 ***on the product set*** $\mathfrak{T}_i \times \mathfrak{D} \times \mathfrak{R}_d^{\alpha\rho}$. *The set* $\mathcal{D}_{TU}\left(\mathfrak{T}_i;\mathbf{R}_{d0}^{\alpha-1};\mathbf{D};\mathbf{R}_d;\mathbf{U}\right)$ *is* ***the tracking domain of*** $\mathbf{R}_d^{\alpha-1}(t)$ ***uniform in*** $t_0 \in \mathfrak{T}_i$ ***on*** $\mathfrak{T}_0 \times \mathfrak{D} \times \mathfrak{R}_d^{\alpha\rho}$.

c) If, and only if, both the intersection $\mathcal{D}_{TU}\left(\mathfrak{T}_i;\mathfrak{D},\mathfrak{R}_d^{\alpha\rho};\mathbf{U}\right)$ *of the domains*

$$\mathcal{D}_T\left(t_0;\mathbf{R}_{d0}^{\alpha-1};\mathbf{D};\mathbf{R}_d;\mathbf{U}\right)$$

of $\mathbf{R}_d^{\alpha-1}(t)$ *in* $\left[t_0,\mathbf{D}(.),\mathbf{R}_d^{\alpha-1}(.)\right] \in \mathfrak{T}_i \times \mathfrak{D} \times \mathfrak{R}_d^{\alpha\rho}$ *is also a connected neighborhood of every* $\mathbf{R}_{d0}^{\alpha-1} \in \mathfrak{R}_{d0}^{\alpha-1}$,

$$\begin{gathered}
(17.11) \text{ holds and } \exists\eta \in \mathfrak{R}^+ \Longrightarrow \mathcal{D}_{TU}\left(\mathfrak{T}_i;\mathfrak{D},\mathfrak{R}_d^{\alpha\rho};\mathbf{U}\right) \\
= \cap \left[\begin{array}{c} \mathcal{D}_T\left(t_0;\mathbf{R}_{d0}^{\alpha-1};\mathbf{D};\mathbf{R}_d;\mathbf{U}\right): \\ \left[t_0,\mathbf{D}(.),\mathbf{R}_d^{\alpha-1}(.)\right] \in \mathfrak{T}_i \times \mathfrak{D} \times \mathfrak{R}_d^{\alpha\rho} \end{array}\right] \\
\supset \mathfrak{N}_\eta\left(\mathfrak{T}_i;\mathfrak{D},\mathfrak{R}_d^{\alpha\rho};\mathbf{U}\right) \neq \phi, \\
\partial\mathcal{D}_{TU}\left(\mathfrak{T}_i;\mathfrak{D},\mathfrak{R}_d^{\alpha\rho};\mathbf{U}\right) \cap \partial\mathfrak{N}_\eta\left(\mathfrak{T}_i;\mathfrak{D},\mathfrak{R}_d^{\alpha\rho};\mathbf{U}\right) = \phi,
\end{gathered} \tag{17.15}$$

and

$$\begin{gathered}
\forall\left(\varsigma,\mathbf{R}_0^{\alpha-1}\right) \in \mathfrak{R}^+ \times \mathcal{D}_{TU}\left(\mathfrak{T}_i;\mathfrak{D},\mathfrak{R}_d^{\alpha\rho};\mathbf{U}\right) \Longrightarrow \\
\tau_m\left(\mathfrak{T}_i,\varsigma,\mathbf{R}_{d0}^{\alpha-1};\mathfrak{D}^k;\mathbf{U};\mathfrak{R}_d^{\alpha\rho}\right) \\
= \sup\left[\tau\left(t_0,\varsigma,\mathbf{R}_{d0}^{\alpha-1};\mathbf{D};\mathbf{U};\mathbf{R}_d\right):\left[t_0,\mathbf{D}(.),\mathbf{R}_d(.)\right] \in \mathfrak{T}_i \times \mathfrak{D}^k \times \mathfrak{R}_d^{\alpha\rho}\right] \in \mathfrak{R}_+
\end{gathered} \tag{17.16}$$

then the tracking is ***uniform in*** $\left[t_0,\mathbf{D}(.),\mathbf{R}_d^{\alpha-1}(.)\right] \in \mathfrak{T} \times \mathfrak{D} \times \mathfrak{R}_d^{\alpha\rho}$. $\mathcal{D}_{TU}\left(\mathfrak{T}_i;\mathfrak{D},\mathfrak{R}_d^{\alpha\rho};\mathbf{U}\right)$ *is* ***the tracking domain of*** $\mathbf{R}_d^{\alpha-1}(t)$ ***uniform in*** $\left(t_0;\mathbf{D};\mathbf{R}_d^{\alpha-1}\right)$ ***on*** $\mathfrak{T} \times \mathfrak{D} \times \mathfrak{R}_d^{\alpha\rho}$.

Note 241 ***Tracking and attraction***
If the plant (3.12) exhibits (global) tracking of $\mathbf{R}_d^{\alpha-1}(t)$ *on* $\mathfrak{T}_0 \times \mathfrak{D} \times \mathfrak{R}_d^{\alpha\rho}$ *then* $\mathbf{R}_d^{\alpha-1}(t)$ *is, respectively, (globally) attractive at* t_0, *but vice versa does not hold.*

The tracking implies tracking of the desired motion $\mathcal{R}_d^{\alpha-1}\left(.;t_0;\mathbf{R}_{d0}^{\alpha-1}\right)$ and of its desired output response $\mathbf{Y}_d^k(t)$. It does not take into account the behavior either of the plant real motion $\mathbf{R}^{\alpha-1}(t;t_0;\mathbf{R}_0^{\alpha-1};\mathbf{D};\mathbf{U})$ relative to the plant extended desired output vector $\mathbf{Y}_d^k(t)$ or of the plant extended real output vector $\mathbf{Y}^k(t;t_0;\mathbf{R}_0^{\alpha-1})$ relative to the desired motion $\mathbf{R}^{\alpha-1}(t;t_0;\mathbf{R}_0^{\alpha-1};\mathbf{D};\mathbf{U})$ at any finite moment $t \in \mathfrak{T}_0$. We overcome this drawback with the following tracking property.

Definition 242 ***The stable state tracking of the desired state behavior*** $\mathbf{R}_d^{\alpha-1}(t)$ ***of the plant*** *(3.12)* ***controlled by control*** $\mathbf{U}(.) \in \mathfrak{U}$

a) The plant (3.12) exhibits ***the stable state asymptotic tracking of*** $\mathbf{R}_d^{\alpha-1}(t)$ ***on*** $\mathfrak{T}_0 \times \mathfrak{D} \times \mathfrak{R}_d^{\alpha\rho}$, *i.e.,* ***the stable state tracking of*** $\mathbf{R}_d^{\alpha-1}(t)$ ***on*** $\mathfrak{T}_0 \times \mathfrak{D} \times \mathfrak{R}_d^{\alpha\rho}$ *if, and only if, it exhibits the tracking of* $\mathbf{R}_d^{\alpha-1}(t)$ *on* $\mathfrak{T}_0 \times \mathfrak{D} \times \mathfrak{R}_d^{\alpha\rho}$, *and for every connected neighborhood* $\mathfrak{N}_\varepsilon\left[t;t_0;\mathbf{R}_d^{\alpha-1}(t)\right]$ *of the motion* $\mathbf{R}_d^{\alpha-1}(t;t_0;\mathbf{R}_{d0}^{\alpha-1})$ *at every moment* $t \in \mathfrak{T}_0$ *there is a connected neighborhood* $\mathfrak{N}\left(\varepsilon;t_0;\mathbf{R}_{d0}^{\alpha-1};\mathbf{D};\mathbf{R}_d;\mathbf{U}\right)$ *of the plant desired initial state vector* $\mathbf{R}_{d0}^{\alpha-1}$ *at the initial moment* $t_0 \in \mathfrak{T}$ *for every* $\left[\mathbf{D}(.),\mathbf{R}_d^{\alpha-1}(.)\right] \in \mathfrak{D} \times \mathfrak{R}_d^{\alpha\rho}$,

$$\begin{gathered}
\forall\mathfrak{N}_\varepsilon\left[t;t_0;\mathbf{R}_d^{\alpha-1}(t)\right] \\
\mathfrak{N}\left(\varepsilon;t_0;\mathbf{R}_{d0}^{\alpha-1};\mathbf{D};\mathbf{R}_d;\mathbf{U}\right) \subseteq \mathcal{D}_T\left(t_0;\mathbf{R}_{d0}^{\alpha-1};\mathbf{D};\mathbf{R}_d;\mathbf{U}\right) \subseteq \mathfrak{R}^{\alpha\rho}, \\
0 < \varepsilon_1 < \varepsilon_2 \Longrightarrow \mathfrak{N}\left(\varepsilon_1;t_0;\mathbf{R}_{d0}^{\alpha-1};\mathbf{D};\mathbf{R}_d;\mathbf{U}\right) \subseteq \mathfrak{N}\left(\varepsilon_2;t_0;\mathbf{R}_{d0}^{\alpha-1};\mathbf{D};\mathbf{R}_d;\mathbf{U}\right), \\
\varepsilon \longrightarrow 0^+ \Longrightarrow \mathfrak{N}\left(\varepsilon;t_0;\mathbf{R}_{d0}^{\alpha-1};\mathbf{D};\mathbf{R}_d;\mathbf{U}\right) \longrightarrow \left\{\mathbf{R}_{d0}^{\alpha-1}\right\},
\end{gathered} \tag{17.17}$$

such that $\mathbf{R}_0^{\alpha-1}$ *from* $\mathfrak{N}\left(\varepsilon; t_0; \mathbf{R}_{d0}^{\alpha-1}; \mathbf{D}; \mathbf{R}_d; \mathbf{U}\right)$, *(17.17), guarantees that the motion*

$$\mathcal{R}^{\alpha-1}(t; t_0; \mathbf{R}_0^{\alpha-1}; \mathbf{D}; \mathbf{U})$$

stays in the ε*-neighborhood* $\mathfrak{N}_\varepsilon\left(t; t_0; \mathbf{R}_d^{\alpha-1}(t)\right) \subseteq \mathfrak{R}^{\alpha\rho}$ *of* $\mathbf{R}_d^{\alpha-1}(t)$ *for all* $t \in \mathfrak{T}_0$*; i.e.,*

$$\begin{gathered}\forall \mathfrak{N}_\varepsilon\left[t; t_0; \mathbf{R}_d^{\alpha-1}(t)\right] \subseteq \mathfrak{R}^{\alpha\rho},\ \forall\left[\mathbf{D}(.), \mathbf{R}_d(.)\right] \in \mathfrak{D} \times \mathfrak{R}_d^{\alpha\rho}, \\ \exists \mathfrak{N}\left(\varepsilon; t_0; \mathbf{R}_{d0}^{\alpha-1}; \mathbf{D}; \mathbf{R}_d; \mathbf{U}\right),\ (17.17), \\ \mathbf{R}_0^{\alpha-1} \in \mathfrak{N}\left(\varepsilon; t_0; \mathbf{R}_{d0}^{\alpha-1}; \mathbf{D}; \mathbf{R}_d; \mathbf{U}\right) \Longrightarrow \\ \mathcal{R}^{\alpha-1}(t; t_0; \mathbf{R}_0^{\alpha-1}; \mathbf{D}; \mathbf{U}) \in \mathfrak{N}_\varepsilon\left[t; t_0; \mathbf{R}_d^{\alpha-1}(t)\right],\ \forall t \in \mathfrak{T}_0. \end{gathered} \quad (17.18)$$

b) The largest connected neighborhood $\mathfrak{N}_L\left(\varepsilon; t_0; \mathbf{R}_{d0}^{\alpha-1}; \mathbf{D}; \mathbf{R}_d; \mathbf{U}\right)$ *that obeys (17.18) is* ***the*** ε***-stable state tracking domain*** $\mathcal{D}_{ST}\left(\varepsilon; t_0; \mathbf{R}_{d0}^{\alpha-1}; \mathbf{D}; \mathbf{R}_d; \mathbf{U}\right)$ ***of the desired state vector*** $\mathbf{R}_d^{\alpha-1}(t)$ ***relative to*** $\left[\mathbf{D}(.), \mathbf{R}_d^{\alpha-1}(.)\right]$ ***at the initial moment*** t_0.

The stable state tracking domain $\mathcal{D}_{ST}\left(t_0; \mathbf{R}_{d0}^{\alpha-1}; \mathbf{D}; \mathbf{R}_d; \mathbf{U}\right)$ ***of the desired*** $\mathbf{R}_d^{\alpha-1}(t)$ ***on*** $\mathfrak{T}_0 \times \mathfrak{D} \times \mathfrak{R}_d^{\alpha\rho}$ ***relative to*** $\left[\mathbf{D}(.), \mathbf{R}_d^{\alpha-1}(.)\right]$ *is the union of all* ε*-stable tracking domains* $\mathcal{D}_{ST}\left(\varepsilon; t_0; \mathbf{R}_{d0}^{\alpha-1}; \mathbf{D}; \mathbf{R}_d; \mathbf{U}\right)$ *over* $\varepsilon \in \mathfrak{R}^+$ ***at the initial moment*** t_0 *,*

$$\mathcal{D}_{ST}\left(t_0; \mathbf{R}_{d0}^{\alpha-1}; \mathbf{D}; \mathbf{R}_d; \mathbf{U}\right) = \cup\left[\begin{array}{c}\mathcal{D}_{ST}\left(\varepsilon; t_0; \mathbf{R}_{d0}^{\alpha-1}; \mathbf{D}; \mathbf{R}_d; \mathbf{U}\right) \\ : \varepsilon \in \mathfrak{R}^+\end{array}\right]. \quad (17.19)$$

Let $[0,\varepsilon_M)$ *be the maximal interval over which* $\mathcal{D}_{ST}\left(\varepsilon; t_0; \mathbf{R}_{d0}^{\alpha-1}; \mathbf{D}; \mathbf{R}_d; \mathbf{U}\right)$ *is continuous in* $\varepsilon \in \mathfrak{R}_+$*,*

$$\mathcal{D}_{ST}\left(\varepsilon; t_0; \mathbf{R}_{d0}^{\alpha-1}; \mathbf{D}; \mathbf{R}_d; \mathbf{U}\right) \in \mathfrak{C}\left([0, \varepsilon_M)\right),\ \ \forall\left[\mathbf{D}(.), \mathbf{R}_d(.)\right] \in \mathfrak{D} \times \mathfrak{R}_d^{\alpha\rho}.$$

The strict stable state tracking domain $\mathcal{D}_{SST}\left(t_0; \mathbf{R}_{d0}^{\alpha-1}; \mathbf{D}; \mathbf{R}_d; \mathbf{U}\right)$ ***of*** $\mathbf{R}_d^{\alpha-1}(t)$ ***on*** $\mathfrak{T}_0 \times \mathfrak{D} \times \mathfrak{R}_d^{\alpha\rho}$ ***relative to*** $\left[\mathbf{D}(.), \mathbf{R}_d^{\alpha-1}(.)\right]$ ***at the initial moment*** t_0 *is the union over* $\varepsilon \in [0, \varepsilon_M)$ *of all* $\mathcal{D}_{ST}\left(\varepsilon; t_0; \mathbf{R}_{d0}^{\alpha-1}; \mathbf{D}; \mathbf{R}_d; \mathbf{U}\right)$,

$$\begin{gathered}\mathcal{D}_{SST}\left(t_0; \mathbf{R}_{d0}^{\alpha-1}; \mathbf{D}; \mathbf{R}_d; \mathbf{U}\right) = \cup\left\{\begin{array}{c}\mathcal{D}_{ST}\left(\varepsilon; t_0; \mathbf{R}_{d0}^{\alpha-1}; \mathbf{D}; \mathbf{R}_d; \mathbf{U}\right) \\ : \varepsilon \in [0, \varepsilon_M)\end{array}\right\}, \\ \forall\left[\mathbf{D}(.), \mathbf{R}_d(.)\right] \in \mathfrak{D} \times \mathfrak{R}_d^{\alpha\rho}. \end{gathered} \quad (17.20)$$

c) The (strict) stable state tracking of $\mathbf{R}_d^{\alpha-1}(t)$ *on* $\mathfrak{T}_0 \times \mathfrak{D} \times \mathfrak{R}_d^{\alpha\rho}$ *is* ***global*** *[****in the whole****] if, and only if, it is both the global tracking tracking of* $\mathbf{R}_d^{\alpha-1}(t)$ *on* $\mathfrak{T}_0 \times \mathfrak{D} \times \mathfrak{R}_d^{\alpha\rho}$ *and the stable tracking of* $\mathbf{R}_d^{\alpha-1}(t)$ *on* $\mathfrak{T}_0 \times \mathfrak{D} \times \mathfrak{R}_d^{\alpha\rho}$ *with the stable tracking domain* $\mathcal{D}_{ST}\left(t_0; \mathbf{R}_{d0}^{\alpha-1}; \mathbf{D}; \mathbf{R}_d; \mathbf{U}\right) = \mathfrak{R}^{(k+1)N}$ *[the stable tracking domain* $\mathcal{D}_{SST}\left(t_0; \mathbf{R}_{d0}^{\alpha-1}; \mathbf{D}; \mathbf{R}_d; \mathbf{U}\right) = \mathfrak{R}^{(k+1)N}$*], respectively, for every pair* $\left[\mathbf{D}(.), \mathbf{R}_d(.)\right] \in \mathfrak{D} \times \mathfrak{R}_d^{\alpha\rho}$.

Definition 243 ***The uniform stable state tracking of the desired state behavior*** $\mathbf{R}_d^{\alpha-1}(t)$ ***of the plant*** *(3.12)* ***controlled by control*** $\mathbf{U}(.) \in \mathfrak{U}$

If, and only if, the plant (3.12) controlled by control $\mathbf{U}(.) \in \mathfrak{U}$ *exhibits stable tracking on* $\mathfrak{T}_0 \times \mathfrak{D} \times \mathfrak{R}_d^{\alpha\rho}$ *together with:*

a) The uniform tracking in $\left[\mathbf{D}(.), \mathbf{R}_d^{\alpha-1}(.)\right]$ *on* $\mathfrak{T}_0 \times \mathfrak{D} \times \mathfrak{R}_d^{\alpha\rho}$ *and the intersection*

$$\mathcal{D}_{STU}\left(t_0; \mathfrak{D}; \mathfrak{R}_d^{\alpha\rho}; \mathbf{U}\right)$$

in $\left[\mathbf{D}(.), \mathbf{R}_d(.)\right] \in \mathfrak{D} \times \mathfrak{R}_d^{\alpha\rho}$ *of all stable tracking domains* $\mathcal{D}_{ST}\left(t_0; \mathbf{R}_{d0}^{\alpha-1}; \mathbf{D}; \mathbf{R}_d; \mathbf{U}\right)$, *[intersection* $\mathcal{D}_{SSTU}\left(t_0; \mathfrak{D}; \mathfrak{R}_d^{\alpha\rho}; \mathbf{U}\right)$ *in the pair* $\left[\mathbf{D}(.), \mathbf{R}_d(.)\right] \in \mathfrak{D} \times \mathfrak{R}_d^{\alpha\rho}$ *of all the plant strictly stable tracking domains* $\mathcal{D}_{SST}\left(t_0; \mathbf{R}_{d0}^{\alpha-1}; \mathbf{D}; \mathbf{R}_d; \mathbf{U}\right)$*] is a connected neighborhood of every*

$\mathbf{R}_{d0}^{\alpha-1} \in \mathfrak{R}_{d0}^{\alpha-1}$, *then, and only then, respectively, the (strict) stable tracking of* $\mathbf{R}_d^{\alpha-1}(t)$ *is* ***uniform in the pair*** $\left[\mathbf{D}(.), \mathbf{R}_d^{\alpha-1}(.)\right] \in \mathfrak{D} \times \mathfrak{R}_d^{\alpha\rho}$ ***on*** $\mathfrak{T}_0 \times \mathfrak{D} \times \mathfrak{R}_d^{\alpha\rho}$, *respectively,*

$$\exists \eta \in \mathfrak{R}^+, \ \mathfrak{N}_\eta\left(t_0; \mathfrak{D}; \mathfrak{R}_d^{\alpha\rho}; \mathbf{U}\right) \neq \phi \Longrightarrow$$

$$\mathcal{D}_{STU}\left(t_0; \mathfrak{D}; \mathfrak{R}_d^{\alpha\rho}; \mathbf{U}\right) = \cap \left\{ \begin{array}{c} \mathcal{D}_{ST}\left(t_0; \mathbf{R}_{d0}^{\alpha-1}; \mathbf{D}; \mathbf{R}_d; \mathbf{U}\right) \\ : \left[\mathbf{D}(.), \mathbf{R}_d(.)\right] \in \mathfrak{D} \times \ \mathfrak{R}_d^{\alpha\rho} \end{array} \right\}$$
$$\supset \mathfrak{N}_\eta\left(t_0; \mathfrak{D}; \mathfrak{R}_d^{\alpha\rho}; \mathbf{U}\right),$$

$$\left[\begin{array}{c} \mathcal{D}_{SSTU}\left(t_0; \mathfrak{D}; \mathfrak{R}_d^{\alpha\rho}; \mathbf{U}\right) = \cap \left\{ \begin{array}{c} \mathcal{D}_{SST}\left(t_0; \mathbf{R}_{d0}^{\alpha-1}; \mathbf{D}; \mathbf{R}_d; \mathbf{U}\right) \\ : \left[\mathbf{D}(.), \mathbf{R}_d(.)\right] \in \mathfrak{D} \times \ \mathfrak{R}_d^{\alpha\rho} \end{array} \right\} \supset \\ \supset \mathfrak{N}_\eta\left(t_0; \mathfrak{D}; \mathfrak{R}_d^{\alpha\rho}; \mathbf{U}\right). \end{array} \right].$$

The global (strict) stable tracking of $\mathbf{R}_d^{\alpha-1}(t)$ *on* $\mathfrak{T}_0 \times \mathfrak{D} \times \mathfrak{R}_d^{\alpha\rho}$ *is also uniform in the pair* $\left[\mathbf{D}(.); \mathbf{R}_d^{\alpha-1}(.)\right] \in \mathfrak{D} \times \mathfrak{R}_d^{\alpha\rho}$.

b) The uniform tracking in $t_0 \in \mathfrak{T}_i$ *on* $\mathfrak{T} \times \mathfrak{D} \times \mathfrak{R}_d^{\alpha\rho}$ *and the domain intersection*

$$\mathcal{D}_{STU}\left(\mathfrak{T}; \mathbf{R}_{d0}^{\alpha-1}; \mathbf{D}; \mathbf{R}_d; \mathbf{U}\right)$$

of stable tracking domains

$$\mathcal{D}_{ST}\left(t_0; \mathbf{R}_{d0}^{\alpha-1}; \mathbf{D}; \mathbf{R}_d; \mathbf{U}\right),$$

[the domain intersection $\mathcal{D}_{SSTU}\left(\mathfrak{T}_i; \mathbf{R}_{d0}^{\alpha-1}; \mathbf{D}; \mathbf{R}_d; \mathbf{U}\right)$ *of all strictly stable tracking domains* $\mathcal{D}_{SST}\left(t_0; \mathbf{R}_{d0}^{\alpha-1}; \mathbf{D}; \mathbf{R}_d; \mathbf{U}\right)$*], in* $t_0 \in \mathfrak{T}_i$ *is a connected neighborhood of* $\mathbf{R}_{d0}^{\alpha-1}$*, then, and only then, it is* ***the (strict) stable state tracking domain of*** $\mathbf{R}_d^{\alpha-1}(t)$ ***on*** $\mathfrak{T} \times \mathfrak{D} \times \mathfrak{R}_d^{\alpha\rho}$ ***uniform in*** $t_0 \in \mathfrak{T}_i$, *respectively,*

$$\exists \eta \in \mathfrak{R}^+ \Longrightarrow \mathfrak{N}_\eta\left(\mathfrak{T}_i; \mathbf{R}_{d0}^{\alpha-1}; \mathbf{D}; \mathbf{R}_d; \mathbf{U}\right) \subset \mathfrak{R}^{\alpha\rho},$$

$$\mathcal{D}_{STU}\left(\mathfrak{T}_i; \mathbf{R}_{d0}^{\alpha-1}; \mathbf{D}; \mathbf{R}_d; \mathbf{U}\right) = \cap \left\{ \begin{array}{c} \mathcal{D}_{ST}\left(t_0; \mathbf{R}_{d0}^{\alpha-1}; \mathbf{D}; \mathbf{R}_d; \mathbf{U}\right) \\ : t_0 \in \mathfrak{T}_i \end{array} \right\}$$
$$\supset \mathfrak{N}_\eta\left(\mathfrak{T}_i; \mathbf{R}_{d0}^{\alpha-1}; \mathbf{D}; \mathbf{R}_d; \mathbf{U}\right), \tag{17.21}$$

$$\left[\begin{array}{c} \mathcal{D}_{SSTU}\left(\mathfrak{T}_i; \mathbf{R}_{d0}^{\alpha-1}; \mathbf{D}; \mathbf{R}_d; \mathbf{U}\right) = \cap \left\{ \begin{array}{c} \mathcal{D}_{ST}\left(t_0; \mathbf{R}_{d0}^{\alpha-1}; \mathbf{D}; \mathbf{R}_d; \mathbf{U}\right) \\ : t_0 \in \mathfrak{T}_i \end{array} \right\} \\ \supset \mathfrak{N}_\eta\left(\mathfrak{T}_i; \mathbf{R}_{d0}^{\alpha-1}; \mathbf{D}; \mathbf{R}_d; \mathbf{U}\right) \end{array} \right]. \tag{17.22}$$

c) The uniform tracking in $\left[t_0, \mathbf{D}(.), \mathbf{R}_d^{\alpha-1}(.)\right]$ *on* $\mathfrak{T}_i \times \mathfrak{D} \times \mathfrak{R}_d^{\alpha\rho}$ *and the domain intersection* $\mathcal{D}_{STU}\left(\mathfrak{T}_i; \mathfrak{D}; \mathfrak{R}_d^{\alpha\rho}; \mathbf{U}\right)$ *of all domains* $\mathcal{D}_{ST}\left(t_0; \mathbf{R}_{d0}^{\alpha-1}; \mathbf{D}; \mathbf{R}_d; \mathbf{U}\right)$ *in* $\left[t_0, \mathbf{D}(.), \mathbf{R}_d^{\alpha-1}(.)\right] \in \mathfrak{T}_i \times \mathfrak{D} \times \mathfrak{R}_d^{\alpha\rho}$*, [the intersection* $\mathcal{D}_{SSTU}\left(\mathfrak{T}_i; \mathfrak{D}; \mathfrak{R}_d^{\alpha\rho}; \mathbf{U}\right)$ *of all* $\mathcal{D}_{SST}\left(t_0; \mathbf{R}_{d0}^{\alpha-1}; \mathbf{D}; \mathbf{R}_d; \mathbf{U}\right)$ *in the triplet* $\left[t_0, \mathbf{D}(.); \mathbf{R}_d^{\alpha-1}(.)\right] \in \mathfrak{T}_i \times \mathfrak{D} \times \mathfrak{R}_d^{\alpha\rho}$*] is a connected neighborhood of every* $\mathbf{R}_{d0}^{\alpha-1} \in \mathfrak{R}_{d0}^{\alpha-1}$*, then, and only then, it is* ***the (strict) stable state tracking domain of*** $\mathbf{R}_d^{\alpha-1}(t)$ ***on*** $\mathfrak{T}_0 \times \mathfrak{D} \times \mathfrak{R}_d^{\alpha\rho}$ ***uniform in*** $\left[t_0, \mathbf{D}(.); \mathbf{R}_d^{\alpha-1}(.)\right] \in \mathfrak{T}_i \times \mathfrak{D} \times \mathfrak{R}_d^{\alpha\rho}$, *respectively,*

$$\exists \eta \in \mathfrak{R}^+ \Longrightarrow \mathfrak{N}_\eta\left(\mathfrak{T}_i; \mathfrak{D}; \mathfrak{R}_d^{\alpha\rho}; \mathbf{U}\right) \subset \mathfrak{R}^{\alpha\rho},$$

$$\mathcal{D}_{STU}\left(\mathfrak{T}_i; \mathfrak{D}; \mathfrak{R}_d^{\alpha\rho}; \mathbf{U}\right) = \cap \left\{ \begin{array}{c} \mathcal{D}_{ST}\left(t_0; \mathbf{R}_{d0}^{\alpha-1}; \mathbf{D}; \mathbf{U}; \mathbf{R}_d\right) \\ : \left(t_0, \mathbf{R}_{d0}^{\alpha-1}, \mathbf{D}; \mathbf{R}_d\right) \\ \in \mathfrak{T}_i \times \mathfrak{R}_{d0}^{\alpha-1} \times \mathfrak{D} \times \ \mathfrak{R}_d^{\alpha\rho} \end{array} \right\}$$
$$\supset \mathfrak{N}_\eta\left(\mathfrak{T}_i; \mathfrak{D}; \mathfrak{R}_d^{\alpha\rho}; \mathbf{U}\right), \tag{17.23}$$

$$\left[\begin{array}{c} \mathcal{D}_{SSTU}\left(\mathfrak{T}_i; \mathfrak{D}; \mathfrak{R}_d^{\alpha\rho}; \mathbf{U}\right) = \cap \left\{ \begin{array}{c} \mathcal{D}_{SST}\left(t_0; \mathbf{R}_{d0}^{\alpha-1}; \mathbf{D}; \mathbf{U}; \mathbf{R}_d\right) \\ : \left(t_0, \mathbf{R}_{d0}^{\alpha-1}, \mathbf{D}; \mathbf{R}_d\right) \\ \in \mathfrak{T}_i \times \mathfrak{R}_{d0}^{\alpha-1} \times \mathfrak{D} \times \ \mathfrak{R}_d^{\alpha\rho} \end{array} \right\} \\ \supset \mathfrak{N}_\eta\left(\mathfrak{T}_i; \mathfrak{D}; \mathfrak{R}_d^{\alpha\rho}; \mathbf{U}\right) \end{array} \right]. \tag{17.24}$$

Comment 244 ***The stable state tracking in the product space***

This definition does not ensure stability of the desired output $\mathbf{Y}_d(t)$ *in spite of the fact that it guarantees asymptotic stability of* $\mathbf{R}_d^{\alpha-1}(t;t_0;\mathbf{R}_{d0}^{\alpha-1})$.

Better tracking is ***the stable state stabilizing tracking.*** *It comprises both the stable tracking of* $\mathbf{R}_d^{\alpha-1}(t)$ *of the plant (3.12) in the state space* $\mathfrak{R}^{\alpha\rho}$ *and its stable tracking in the output space* $\mathfrak{R}^N$. *This means that the plant (3.12) exhibits the stable tracking of* $[\mathbf{R}_d^{\alpha-1}(t), \mathbf{Y}_d(t)]$ *in the product space* $\mathfrak{R}^{\alpha\rho} \times \mathfrak{R}^N$.

The asymptotic convergence of the real output vector $\mathbf{Y}(t)$ to the extended desired output vector $\mathbf{Y}_d(t)$, as well as of the real motion $\mathbf{R}^{\alpha-1}\left(t;t_0;\mathbf{R}_0^{\alpha-1};\mathbf{D};\mathbf{U}\right)$ to the desired motion $\mathcal{R}_d^{\alpha-1}\left(t;t_0;\mathbf{R}_{d0}^{\alpha-1}\right)$, does not provide any information about the rate of the convergence.

Definition 245 ***The exponentially stable state tracking of the desired state behavior*** $\mathbf{R}_d^{\alpha-1}(t)$ ***of the plant (3.12) controlled by control*** $\mathbf{U}(.) \in \mathfrak{U}$

a) The plant (3.12) exhibits ***the exponentially stable state tracking of*** $\mathbf{R}_d^{\alpha-1}(t)$ ***on*** *the set product* $\mathfrak{T}_0 \times \mathfrak{D} \times \mathfrak{R}_d^{\alpha\rho}$, *for short* ***the exponentially stable state tracking of*** $\mathbf{R}_d^{\alpha-1}(t)$ ***on*** $\mathfrak{T}_0 \times \mathfrak{D} \times \mathfrak{R}_d^{\alpha\rho}$ *if, and only if for every* $[\mathbf{D}(.), \mathbf{R}_d(.)] \in \mathfrak{D} \times \mathfrak{R}_d^{\alpha\rho}$ *there exist positive real numbers* $\xi \geq 1$ *and* ψ, *and a connected neighborhood*

$$\mathfrak{N}\left(t_0;\mathbf{R}_{d0}^{\alpha-1};\mathbf{D};\mathbf{R}_d;\mathbf{U}\right) of \mathbf{R}_{d0}^{\alpha-1},$$

$\xi = \xi(t_0, \mathbf{D}, \mathbf{R}_d, \mathbf{U})$, $\psi = \psi(t_0, \mathbf{D}, \mathbf{R}_d, \mathbf{U})$, *such that*

$$\mathbf{R}_0^{\alpha-1} \in \mathfrak{N}\left(t_0;\xi;\psi;\mathbf{R}_{d0}^{\alpha-1};\mathbf{D};\mathbf{R}_d;\mathbf{U}\right)$$

implies that the plant motion $\mathbf{R}^{\alpha-1}\left(t;t_0;\mathbf{R}_0^{\alpha-1};\mathbf{D};\mathbf{U}\right)$ *approaches exponentially* $\mathbf{R}_d^{\alpha-1}(t)$ *all the time* $t \in \mathfrak{T}_0$; *i.e.,*

$$\begin{gathered}
\forall\,[\mathbf{D}(.), \mathbf{R}_d(.)] \in \mathfrak{D}\times \mathfrak{R}_d^{\alpha\rho},\ \exists \xi \in \mathfrak{R}^+,\ \xi \geq 1,\ \exists \psi \in \mathfrak{R}^+, \\
\xi = \xi(t_0, \mathbf{D}, \mathbf{R}_d, \mathbf{U}),\ \ \psi = \psi(t_0, \mathbf{D}, \mathbf{R}_d, \mathbf{U}), \\
\exists \mathfrak{N}\left(t_0;\xi;\psi;\mathbf{R}_{d0}^{\alpha-1};\mathbf{D};\mathbf{R}_d;\mathbf{U}\right) \subseteq \mathfrak{R}^{\alpha\rho}, \\
\mathbf{R}_0^{\alpha-1} \in \mathfrak{N}\left(t_0;\xi;\psi;\mathbf{R}_{d0}^{\alpha-1};\mathbf{D};\mathbf{R}_d;\mathbf{U}\right) \Longrightarrow \\
\left\|\mathbf{R}_d^{\alpha-1}\left(t;t_0;\mathbf{R}_{d0}^{\alpha-1}\right) - \mathbf{R}^{\alpha-1}\left(t;t_0;\mathbf{R}_0^{\alpha-1};\mathbf{D};\mathbf{U}\right)\right\| \\
\leq \xi \left\|\mathbf{R}_{d0}^{\alpha-1} - \mathbf{R}_0^{\alpha-1}\right\| exp\ \left[-\psi\left(t - t_0\right)\right],\ \forall t \in \mathfrak{T}_0.
\end{gathered} \tag{17.25}$$

b) The largest neighborhood $\mathfrak{N}_L\left(t_0;\mathbf{R}_{d0}^{\alpha-1};\mathbf{D};\mathbf{R}_d;\mathbf{U}\right)$ *of* $\mathbf{R}_{d0}^{\alpha-1}$ *at* $t_0 \in \mathfrak{T}_i$ *is* ***exponentially stable tracking domain*** $\mathcal{D}_{ET}\left(t_0;\xi;\psi;\mathbf{R}_{d0}^{\alpha-1};\mathbf{D};\mathbf{R}_d;\mathbf{U}\right)$ *relative to* (ξ,ψ) *of* $\mathbf{R}_d^{\alpha-1}(t)$ ***on*** $\mathfrak{T}_0 \times \mathfrak{D} \times \mathfrak{R}_d^{\alpha\rho}$.

c) The exponentially stable tracking of $\mathbf{R}_d^{\alpha-1}(t)$ *is* ***global (in the whole) on*** $\mathfrak{T}_0 \times \mathfrak{D} \times \mathfrak{R}_d^{\alpha\rho}$ *if, and only if,* $\mathcal{D}_{ET}\left(t_0;;\xi;\psi;\mathbf{R}_{d0}^{\alpha-1};\mathbf{D};\mathbf{R}_d;\mathbf{U}\right) = \mathfrak{R}^{\alpha\rho}$.

d) If, and only if, the intersection $\mathcal{D}_{ET}\left(\mathfrak{T}_i;\xi;\psi;\mathbf{R}_{d0}^{\alpha-1};\mathfrak{D};\mathfrak{R}_d^{\alpha\rho};\mathbf{U}\right)$ *of the domains*

$$\mathcal{D}_{ET}\left(t_0;;\xi;\psi;\mathbf{R}_{d0}^{\alpha-1};\mathbf{D};\mathbf{R}_d;\mathbf{U}\right)$$

in $[t_0, \mathbf{D}(.), \mathbf{R}_d(.)] \in \mathfrak{T}_i \times \mathfrak{D} \times \mathfrak{R}_d^{\alpha\rho}$ *is a nonempty connected neighborhood of* $\mathbf{R}_{d0}^{\alpha-1}$ *for every* $[t_0, \mathbf{D}(.), \mathbf{R}_d(.)] \in \mathfrak{T}_i \times \mathfrak{D} \times \mathfrak{R}_d^{\alpha\rho}$ *and the values of* ξ *and* ψ *depend at most on* $\mathfrak{T}_i \times \mathfrak{D} \times \mathfrak{R}_d^{\alpha\rho}$ *and* $\mathbf{U}$, *but not on a particular choice of* $[t_0, \mathbf{D}(.), \mathbf{R}_d(.)]$ *from* $\mathfrak{T}_i \times \mathfrak{D} \times \mathfrak{R}_d^{\alpha\rho}$, *then the exponentially stable tracking of* $\mathbf{R}_d^{\alpha-1}(t)$ *is* ***uniform on*** $\mathfrak{T}_i \times \mathfrak{D} \times \mathfrak{R}_d^{\alpha\rho}$.

e) If, and only if, additionally to d), $\mathcal{D}_{ET}\left(\mathfrak{T}_i;\xi;\psi;\mathbf{R}_{d0}^{\alpha-1};\mathfrak{D};\mathfrak{R}_d^{\alpha\rho};\mathbf{U}\right) = \mathfrak{R}^{\alpha\rho}$ *then the exponentially stable tracking of* $\mathbf{R}_d^{\alpha-1}(t)$ *is* ***globally uniform on*** $\mathfrak{T}_i \times \mathfrak{D} \times \mathfrak{R}_d^{\alpha\rho}$.

Note 246 ***The exponentially stable state tracking in the product space***
The exponentially stable tracking property ensures that the plant real motion

$$\mathbf{R}^{\alpha-1}\left(t;t_0;\mathbf{R}_0^{\alpha-1};\mathbf{D};\mathbf{U}\right)$$

converges exponentially to the plant desired motion $\mathcal{R}_d^{\alpha-1}\left(t;t_0;\mathbf{R}_{d0}^{\alpha-1}\right)$ *in the state space* $\mathfrak{R}^{\alpha\rho}$. *This does not ensure the exponential convergence of the real output vector* $\mathbf{Y}^k(t)$ *to the extended desired output vector* $\mathbf{Y}_d^k(t)$ *in the extended output space* $\mathfrak{R}^{(k+1)N}$. *In order also to ensure the exponentially stable tracking in the output space it is necessary to deal additionally with the extended output space* $\mathfrak{R}^{(k+1)N}$, *or with the extended output error space* $\mathfrak{R}^{(k+1)N}$, *rather than only with the state space. Altogether, the plant should exhibit the exponentially stable tracking in the product space* $\mathfrak{R}^{\alpha\rho}\times\mathfrak{R}^{(k+1)N}$. *This guarantees the exponential stability of the desired state motion* $\mathcal{R}_d^{\alpha-1}\left(t;t_0;\mathbf{R}_{d0}^{\alpha-1}\right)$ *and of the extended desired output behavior* $\mathbf{Y}_d^k\left(t;t_0;\mathbf{R}_{d0}^{\alpha-1}\right)$.

Note 247 *Tracking is necessary for all other above-specified tracking properties.*

Chapter 18

Set tracking

18.1 Introduction

Set tracking involves the *time*-invariant set tracking and the *time*-varying set tracking. Their meaning is that the motions $\mathcal{Z}(.;t_0;\mathbf{Z}_0;\mathbf{I})$ of the system subjected to the action of the input vector function $\mathbf{I}(.)$, or the perturbed plant motions $\mathcal{Z}(.;t_0;\mathbf{Z}_0;\mathbf{D};\mathbf{U})$ subjected to the actions of the disturbance vector function $\mathbf{D}(.)$ and of the control vector function $\mathbf{U}(.)$ reach either a given *time*-invariant set $\Upsilon \subset \mathfrak{R}^K$ or a given *time*-varying set $\Upsilon(t) \subset \mathfrak{R}^K$, $t \in \mathfrak{T}_0$, with some demanded tracking quality. The set $\Upsilon(t)$ is *the instantaneous target set at the moment* $t \in \mathfrak{T}_0$. It is induced by the set function $\Upsilon(.) : \mathfrak{T}_0 \longrightarrow 2^{\mathfrak{R}^K}$, and represents the set value of $\Upsilon(.)$ at the moment $t \in \mathfrak{T}_0$.

The *time*-invariant target set Υ can be the set $\mathfrak{Y}_d^k$ of the desired output responses of the plant. It can be an enemy military base.

A specific tracking task appears when a domestic missile should reach a set of n enemy aircrafts or warships. The set $\Upsilon(t)$ of the real positions and movements of all n enemy aircrafts or ships represents simultaneously the target set $\Upsilon(t)$ for the output behaviors of the domestic missile. The set $\Upsilon(t)$ is *time*-varying because the real positions and courses of the enemy aircrafts vary in *time*. This justifies attacking the problem of the conditions for tracking a *time*-varying set $\Upsilon(t)$ that is then the target set.

Another situation occurs when it is sufficient for the domestic aircraft or missile to enter a neighborhood of the enemy flying aircraft or to enter a neighborhood of the enemy moving submarine in order to destroy the enemy.

$\Upsilon(t)$ can be, but need not be, the set of the instantaneous desired output vectors $\mathbf{Z}_d(t) \equiv \mathbf{Y}_d(t)$ of the plant at the moment $t \in \mathfrak{T}_0$, which belong to $\mathfrak{Y}_d^k$:

$$\Upsilon(t) = \left\{\mathbf{Z}_d(t) : \mathbf{Z}_d(t) \equiv \mathbf{Y}_d(t),\ \mathbf{Y}_d(.) \in \mathfrak{Y}_d^k\right\} \subset \mathfrak{Y}_d^k,\ t \in \mathfrak{T}_0.$$

18.2 Definitions of set tracking properties

Because the definitions given for the *time*-varying set tracking hold directly for the *time*-invariant set tracking, we present only the former. They generalize the corresponding definitions of the motion tracking, i.e., the definitions of the desired behavior $\mathbf{Z}_d(t)$ tracking.

Assumption 248 ***The target set is known***
The target set $\Upsilon(t)$ is well defined and known on $\mathfrak{T}_0$.

The task of the determining the target set $\Upsilon(t)$ might be the research topic of the estimation and data acquisition theories developed by Y. Bar-Shalom and T. E. Fortmann

[28], Y. Bar-Shalom and X. R. Li [29], [30], Y. Bar-Shalom, X. R. Li, and T. Kirubarajan [31], Y. Bar-Shalom, P. K. Willet, and X. Tian [32], S. S. Blackman [52], F. E. Daum [97], F. E. Daum and R. J. Fitzgerald [98], R. P. S. Mahler [404], D. B. Reid [496] and L. D. Stone, C. A. Barlow, and T. L. Corwin [528].

Property 249 ***The target set properties***

The target set $\Upsilon(t)$ *is nonempty, connected, and continuous at every moment* $t \in \mathfrak{T}_0$:

$$\Upsilon(t) \neq \phi, \ \forall t \in \mathfrak{T}_0, \ \Upsilon(t) \in \mathfrak{C}(\mathfrak{T}_0).$$

Although these properties of $\Upsilon(t)$ permit its unboundedness, they allow simultaneously that the boundary $\partial\Upsilon(t)$ of $\Upsilon(t)$ is nonempty, $\partial\Upsilon(t) \neq \phi$, $\forall t \in \mathfrak{T}_0$. For example, the unbounded set $\Upsilon = \{x : 0 \leq x\}$ is nonempty, connected, and continuous. It has the boundary $\partial\Upsilon = \{x : x = 0\}$.

Definition 250 ***Tracking the time-varying target set*** $\Upsilon(t)$

a) The system motions $\mathcal{Z}(.;t_0;\mathbf{Z}_0;\mathbf{I})$ *exhibit* ***the tracking of the time-varying target set*** $\Upsilon(t)$ ***on*** $\mathfrak{T}_0 \times \mathcal{J}^\xi$ *if, and only if, for every* $\mathbf{I}(.) \in \mathcal{J}^\xi$ *there exists a* t_0*-dependent connected neighborhood* $\mathfrak{N}[t_0;\Upsilon(t_0);\mathbf{I}] \subseteq \mathfrak{R}^K$ *of* $\Upsilon(t_0)$ *and for every* $\varsigma \in \mathfrak{R}^+$ *there exists a nonnegative real number* τ, $\tau = \tau[t_0,\varsigma,\Upsilon(t_0);\mathbf{I}] \in \mathfrak{R}_+$, *such that* $\mathbf{Z}_0$ *from* $\mathfrak{N}[t_0;\Upsilon(t_0);\mathbf{I}]$ *guarantees that the real motion* $\mathbf{Z}(t;t_0;\mathbf{Z}_0;\mathbf{I})$ *belongs to the* ς*-neighborhood* $\mathfrak{N}_\varsigma[t;t_0;\Upsilon(t);\mathbf{I}]$ *of* $\Upsilon(t)$ *for all time* $t \in]t_0 + \tau[t_0,\varsigma,\Upsilon(t_0);\mathbf{I}], \infty[$*; i.e.*

$$\begin{gathered}\forall\varsigma \in \mathfrak{R}^+, \forall\mathbf{I}(.) \in \mathcal{J}^\xi, \exists\mathfrak{N}[t_0;\varsigma;\Upsilon(t_0);\mathbf{U}] \subseteq \mathfrak{R}^K, \\ \mathbf{Z}_0 \in \mathfrak{N}[t_0;\varsigma;\Upsilon(t_0);\mathbf{U}] \implies \mathbf{Z}(t;t_0;\mathbf{Z}_0;\mathbf{I}) \in \mathfrak{N}_\varsigma[t;t_0;\Upsilon(t);\mathbf{I}], \\ \forall t \in]t_0 + \tau[t_0,\varsigma,\Upsilon(t_0);\mathbf{I}], \infty[. \end{gathered} \tag{18.1}$$

b) ***The tracking domain*** $\mathcal{D}_{ZT}[t_0;\Upsilon(t_0);\mathbf{I}]$ ***of*** $\Upsilon(t)$ ***on*** $\mathfrak{T}_0 \times \mathcal{J}^\xi$ *is the union over* $\varsigma \in \mathfrak{R}^+$ *of the largest connected neighborhoods* $\mathfrak{N}_L(t_0;\varsigma;\Upsilon(t_0);\mathbf{I})$ *obeying (18.1):*

$$\mathcal{D}_{ZT}[t_0;\Upsilon(t_0);\mathbf{I}] = \cup\{\mathfrak{N}_L[t_0;\varsigma;\Upsilon(t_0);\mathbf{I}] : \varsigma \in \mathfrak{R}^+\} \tag{18.2}$$

c) The tracking of $\Upsilon(t)$ *on* $\mathfrak{T}_0 \times \mathcal{J}^\xi$ *is* ***global (in the whole)*** *if, and only if, the tracking domain* $\mathcal{D}_{ZT}[t_0;\Upsilon(t_0);\mathbf{I}] = \mathfrak{R}^K$ *for every* $\mathbf{I}(.) \in \mathcal{J}^\xi$.

Definition 251 ***Uniform tracking the time-varying target set*** $\Upsilon(t)$

a) If, and only if, both the intersection $\mathcal{D}_{ZTU}[t_0;\Upsilon(t_0);\mathcal{J}^\xi]$ *of the tracking domains* $\mathcal{D}_{ZT}[t_0;\Upsilon(t_0);\mathbf{I}]$ *of* $\Upsilon(t)$ *in* $\mathbf{I}(.) \in \mathcal{J}^\xi$ *is also a connected neighborhood of* $\Upsilon(t_0)$,

$$\begin{gathered}\exists\xi \in \mathfrak{R}^+ \implies \mathcal{D}_{ZTU}[t_0;\Upsilon(t_0);\mathcal{J}^\xi] \\ = \cap[\mathcal{D}_{ZT}[t_0;\Upsilon(t_0);\mathbf{I}] : \mathbf{I}(.) \in \mathcal{J}^\xi] \supset \mathfrak{N}_\xi(t_0;\Upsilon(t_0),\mathcal{J}^\xi) \neq \phi, \\ \partial\mathcal{D}_{ZTU}[t_0;\Upsilon(t_0);\mathcal{J}^\xi] \cap \partial\mathfrak{N}_\xi(t_0;\Upsilon(t_0),\mathcal{J}^\xi) = \phi, \end{gathered} \tag{18.3}$$

and the minimal $\tau(t_0,\varsigma,\Upsilon(t_0),\mathbf{I})$ *denoted by* $\tau_m(t_0,\varsigma,\Upsilon(t_0),\mathbf{I})$, *which obeys Definition 250, satisfies*

$$\begin{gathered}\forall(\zeta,\mathbf{Z}_0) \in \mathfrak{R}^+ \times \mathcal{D}_{ZTU}[t_0;\Upsilon(t_0);\mathcal{J}^\xi] \implies \\ \tau(t_0,\varsigma,\Upsilon(t_0),\mathcal{J}^\xi) = sup\ [\tau_m(t_0,\varsigma,\Upsilon(t_0),\mathbf{I}) : \mathbf{I}(.) \in \mathcal{J}^\xi] \in \mathfrak{R}_+ \end{gathered} \tag{18.4}$$

then tracking $\Upsilon(t)$ *is* ***uniform in*** $\mathbf{I}(.) \in \mathcal{J}^\xi$ ***on*** $\mathfrak{T}_0 \times \mathcal{J}^\xi$ *and* $\mathcal{D}_{ZTU}[t_0;\Upsilon(t_0);\mathcal{J}^\xi]$ *is* ***the tracking domain of*** $\Upsilon(t)$ ***uniform in*** $\mathbf{I}(.) \in \mathcal{J}^\xi$ ***on*** $\mathfrak{T}_0$.

Uniform tracking of $\Upsilon(t)$ *on* $\mathfrak{T}_0 \times \mathcal{J}^\xi$ *is* ***global (in the whole)*** *if, and only if,*

$$\mathcal{D}_{ZTU}[t_0;\Upsilon(t_0);\mathcal{J}^\xi] = \mathfrak{R}^K.$$

b) If, and only if, both the intersection $\mathcal{D}_{ZTU}(\mathfrak{T}_i;\Upsilon(t_0);\mathbf{I};)$ *of the tracking domains* $\mathcal{D}_{ZT}[t_0;\Upsilon(t_0);\mathbf{I}]$ *of* $\Upsilon(t)$ *in* $t_0\in\mathfrak{T}_i$ *is a connected neighborhood of* $\Upsilon(t_0)$,

$$\begin{gathered}\exists\eta\in\mathfrak{R}^+\Longrightarrow\mathcal{D}_{ZTU}(\mathfrak{T}_i;\Upsilon(t_0);\mathbf{I})\\=\cap[\mathcal{D}_{ZT}[t_0;\Upsilon(t_0);\mathbf{I}]:t_0\in\mathfrak{T}_i]\supset\mathfrak{N}_\eta(\mathfrak{T}_i;\Upsilon(t_0);\mathbf{I}),\\\partial\mathcal{D}_{ZTU}(\mathfrak{T}_i;\Upsilon(t_0);\mathbf{I})\cap\partial\mathfrak{N}_\eta(\mathfrak{T}_i;\Upsilon(t_0);\mathbf{I})=\phi,\end{gathered}\tag{18.5}$$

and

$$\begin{gathered}\forall(\zeta,\mathbf{Z}_0)\in\mathfrak{R}^+\times\mathcal{D}_{ZTU}(\mathfrak{T}_i;\Upsilon(t_0);\mathbf{I})\Longrightarrow\\\tau(\mathfrak{T}_i,\varsigma,\Upsilon(t_0);\mathbf{I})=sup[\tau_m(t_0,\varsigma,\Upsilon(t_0);\mathbf{I}):t_0\in\mathfrak{T}_i]\in\mathfrak{R}_+\end{gathered}\tag{18.6}$$

then, and only then, the tracking of $\Upsilon(t)$ *is* ***uniform in*** t_0 ***on*** $\mathfrak{T}_0\times\mathfrak{T}_i\times\mathcal{J}^\xi$. *The set* $\mathcal{D}_{ZTU}(\mathfrak{T}_i;\Upsilon(t_0);\mathbf{I})$ *is* ***the tracking domain of*** $\Upsilon(t)$ ***uniformly in*** $t_0\in\mathfrak{T}_i$ ***on*** $\mathfrak{T}_0\times\mathfrak{T}_i\times\mathcal{J}^\xi$.

c) If, and only if, both the intersection $\mathcal{D}_{ZTU}(\mathfrak{T}_i;\Upsilon(t_0);\mathcal{J}^\xi)$ *of the tracking domains* $\mathcal{D}_{ZTU}[t_0;\Upsilon(t_0);\mathbf{I}]$ *of* $\Upsilon(t)$ *in* $[t_0,\mathbf{I}(.),]\in\mathfrak{T}_i\times\mathcal{J}^\xi$ *is also a connected neighborhood of* $\Upsilon(t_0)$,

$$\begin{gathered}\exists\eta\in\mathfrak{R}^+\Longrightarrow\mathcal{D}_{ZTU}(\mathfrak{T}_i;\Upsilon(t_0);\mathcal{J}^\xi)\\=\cap[\mathcal{D}_{ZT}[t_0;\Upsilon(t_0);\mathbf{I}]:[t_0,\mathbf{I}(.)]\in\mathfrak{T}_i\times\mathcal{J}^\xi]\\\supset\mathfrak{N}_\eta(\mathfrak{T}_i;\Upsilon(t_0);\mathcal{J}^\xi)\neq\phi,\\\partial\mathcal{D}_{IDTU}(\mathfrak{T}_i;\Upsilon(t_0);\mathcal{J}^\xi)\cap\partial\mathfrak{N}_\eta(\mathfrak{T}_i;\Upsilon(t_0);\mathcal{J}^\xi)=\phi,\end{gathered}\tag{18.7}$$

and

$$\begin{gathered}\forall(\zeta,\mathbf{Z}_0)\in\mathfrak{R}^+\times\mathcal{D}_{ZTU}(\mathfrak{T}_i;\Upsilon(t_0);\mathcal{J}^\xi)\Longrightarrow\\\tau(\mathfrak{T}_i,\varsigma,\Upsilon(t_0);\mathcal{J}^\xi)\\=sup\ [\tau(t_0,\varsigma,\Upsilon(t_0);\mathbf{I}):[t_0,\mathbf{I}(.)]\in\mathfrak{T}_i\times\mathcal{J}^\xi]\in\mathfrak{R}_+\end{gathered}\tag{18.8}$$

then tracking $\Upsilon(t)$ *is* ***uniform in*** $[t_0,\mathbf{I}(.)]$ ***on*** $\mathfrak{T}_0\times\mathfrak{T}_i\times\mathcal{J}^\xi$. *The set* $\mathcal{D}_{ZTU}(\mathfrak{T}_i;\Upsilon(t_0);\mathcal{J}^\xi)$ *is* ***the tracking domain of*** $\Upsilon(t)$ ***uniform in*** $[t_0,\mathbf{I}(.)]$ ***on*** $\mathfrak{T}_0\times\mathfrak{T}_i\times\mathcal{J}^\xi$.

Definition 252 ***The stable tracking the time-varying set*** $\Upsilon(t)$

a) The system motions $\mathcal{Z}(.;t_0;\mathbf{Z}_0;\mathbf{I})$ *exhibit* ***the stable tracking the time-varying set*** $\Upsilon(t)$ ***on*** $\mathfrak{T}_0\times\mathcal{J}^\xi$ *if, and only if, they exhibit the tracking of* $\Upsilon(t)$ *on* $\mathfrak{T}_0\times\mathcal{J}^\xi$, *and for every connected neighborhood* $\mathfrak{N}_\varepsilon[t;t_0;\Upsilon(t)]$ *of* $\Upsilon(t)$ *at every moment* $t\in\mathfrak{T}_0$ *there is for every* $\mathbf{I}(.)\in\mathcal{J}^\xi$ *a connected neighborhood* $\mathfrak{N}[\varepsilon;t_0;\Upsilon(t_0);\mathbf{I}]$ *of* $\Upsilon(t_0)$ *at the initial moment* $t_0\in\mathfrak{T}$,

$$\begin{gathered}\forall\mathfrak{N}_\varepsilon[t;t_0;\Upsilon(t)]\subseteq\mathfrak{R}^K\\\exists\mathfrak{N}[\varepsilon;t_0;\Upsilon(t_0);\mathbf{I}]\subseteq\mathcal{D}_{ZT}[t_0;\Upsilon(t_0);\mathbf{I}],\\0<\varepsilon_1<\varepsilon_2\Longrightarrow\mathfrak{N}[\varepsilon_1;t_0;\Upsilon(t_0);\mathbf{I}]\subseteq\mathfrak{N}[\varepsilon_2;t_0;\Upsilon(t_0);\mathbf{I}],\\\varepsilon\longrightarrow0^+\Longrightarrow\mathfrak{N}(\varepsilon;t_0;\Upsilon(t_0);\mathbf{I})\longrightarrow\Upsilon(t_0),\end{gathered}\tag{18.9}$$

such that $\mathbf{Z}_0$ *from* $\mathfrak{N}[\varepsilon;t_0;\Upsilon(t_0);\mathbf{I}]$, *(18.9), guarantees that* $\mathbf{Z}(t;t_0;\mathbf{Z}_0;\mathbf{I})$ *stays in the instantaneous* ε*-neighborhood* $\mathfrak{N}_\varepsilon[t;t_0;\Upsilon(t)]\subseteq\mathfrak{R}^K$ *of* $\Upsilon(t)$ *for all* $t\in\mathfrak{T}_0$*; i.e.,*

$$\begin{gathered}\forall\mathfrak{N}_\varepsilon[t;t_0;\Upsilon(t)]\subseteq\mathfrak{R}^K,\ \forall\mathbf{I}(.)\in\mathcal{J}^\xi,\\\exists\mathfrak{N}[\varepsilon;t_0;\Upsilon(t_0);\mathbf{I}],\ (18.9),\\\mathfrak{N}[\varepsilon;t_0;\Upsilon(t_0);\mathbf{I}]\subseteq\mathcal{D}_{ZT}[t_0;\Upsilon(t_0);\mathbf{I}],\\\mathbf{Z}_0\in\mathfrak{N}[\varepsilon;t_0;\Upsilon(t_0);\mathbf{I}]\ \Longrightarrow\\\mathbf{Z}(t;t_0;\mathbf{Z}_0;\mathbf{I})\in\mathfrak{N}_\varepsilon[t;t_0;\Upsilon(t)],\ \forall t\in\mathfrak{T}_0.\end{gathered}\tag{18.10}$$

b) The largest connected neighborhood $\mathfrak{N}_l\left(\varepsilon; t_0; \mathbf{R}_{d0}^{\alpha-1}; \mathbf{D}; \Upsilon; \mathbf{U}\right)$ *that obeys (18.10) is* ***the ε-stable tracking domain***

$$\mathcal{D}_{SZT}\left[\varepsilon; t_0; \Upsilon\left(t_0\right); \mathbf{I}\right] \tag{18.11}$$

of $\Upsilon(t)$ ***at the initial moment*** t_0.

The stable tracking domain $\mathcal{D}_{SZT}\left[t_0; \Upsilon\left(t_0\right); \mathbf{I}\right]$ ***of*** $\Upsilon(t)$ ***at the initial moment*** t_0 *is the union of all* $\mathcal{D}_{SZT}\left[\varepsilon; t_0; \Upsilon\left(t_0\right); \mathbf{I}\right]$ *over* $\varepsilon \in \mathfrak{R}^+$,

$$\mathcal{D}_{SZT}\left[t_0; \Upsilon\left(t_0\right); \mathbf{I}\right] = \cup\left[\mathcal{D}_{SZT}\left[\varepsilon; t_0; \Upsilon\left(t_0\right); \mathbf{I}\right] : \varepsilon \in \mathfrak{R}^+\right]. \tag{18.12}$$

Let $[0,\varepsilon_M)$ *be the maximal interval over which* $\mathcal{D}_{SZT}\left[\varepsilon; t_0; \Upsilon\left(t_0\right); \mathbf{I}\right]$ *is continuous in* $\varepsilon \in \mathfrak{R}_+$,

$$\mathcal{D}_{SZT}\left[\varepsilon; t_0; \Upsilon\left(t_0\right); \mathbf{I}\right] \in \mathfrak{C}\left(\left[0, \varepsilon_M\right)\right), \ \forall \mathbf{I}(.) \in \mathcal{J}^{\xi}.$$

The strict stable tracking domain $\mathcal{D}_{SSZT}\left[t_0; \Upsilon\left(t_0\right); \mathbf{I}\right]$ ***of*** $\Upsilon(t)$ ***at the initial moment*** t_0 *is the union of all* $\mathcal{D}_{SZT}\left[\varepsilon; t_0; \Upsilon\left(t_0\right); \mathbf{I}\right]$ *over* $\varepsilon \in \left[0, \varepsilon_M\right)$,

$$\mathcal{D}_{SSZT}\left[t_0; \Upsilon\left(t_0\right); \mathbf{I}\right] = \cup\left\{\begin{array}{c} \mathcal{D}_{SZT}\left[\varepsilon; t_0; \Upsilon\left(t_0\right); \mathbf{I}\right] : \\ : \varepsilon \in \left[0, \varepsilon_M\right) \end{array}\right\}, \ \forall \mathbf{I}(.) \in \mathcal{J}^{\xi}. \tag{18.13}$$

c) The (strict) stable tracking of $\Upsilon(t)$ *on* $\mathfrak{T}_0 \times \mathcal{J}^{\xi}$ *is* ***global [in the whole]*** *if, and only if, it is both the global tracking of* $\Upsilon(t)$ *on* $\mathfrak{T}_0 \times \mathcal{J}^{\xi}$ *and the stable tracking of* $\Upsilon(t)$ *on* $\mathfrak{T}_0 \times \mathcal{J}^{\xi}$ *with* $\mathcal{D}_{SZT}\left[t_0; \Upsilon\left(t_0\right); \mathbf{I}\right] = \mathfrak{R}^K$ $[\mathcal{D}_{SSZT}\left[t_0; \Upsilon\left(t_0\right); \mathbf{I}\right] = \mathfrak{R}^K]$, *respectively, for every* $\mathbf{I}(.) \in \mathcal{J}^{\xi}$.

Definition 253 ***The uniform stable tracking the time-varying set*** $\Upsilon(t)$

If the system motions $\mathcal{Z}\left(.; t_0; \mathbf{Z}_0; \mathbf{I}\right)$ *exhibit stable tracking on* $\mathfrak{T}_0 \times \mathcal{J}^{\xi}$ *together with:*

a) The uniform tracking in $\mathbf{I}(.) \in \mathcal{J}^{\xi}$ *on* $\mathfrak{T}_0 \times \mathcal{J}^{\xi}$ *and the intersection domain*

$$\mathcal{D}_{SZTU}\left[t_0; \Upsilon\left(t_0\right); \mathcal{J}^{\xi}\right]$$

in $\mathbf{I}(.) \in \mathcal{J}^{\xi}$ *of stable tracking domains* $\mathcal{D}_{SZT}\left[t_0; \Upsilon\left(t_0\right); \mathbf{I}\right]$ */ the intersection domain*

$$\mathcal{D}_{SSZTU}\left[t_0; \Upsilon\left(t_0\right); \mathcal{J}^{\xi}\right]$$

in $\mathbf{I}(.) \in \mathcal{J}^{\xi}$ *of all strictly stable domains* $\mathcal{D}_{SSZT}\left[t_0; \Upsilon\left(t_0\right); \mathbf{I}\right]$*/ is a connected neighborhood of* $\Upsilon\left(t_0\right)$*, then, and only then, respectively, the (strict) stable tracking of* $\Upsilon(t)$ *is* ***uniform in*** $\mathbf{I}(.) \in \mathcal{J}^{\xi}$ ***on*** $\mathfrak{T}_0 \times \mathcal{J}^{\xi}$, *respectively,*

$$\exists \eta \in \mathfrak{R}^+, \ \mathfrak{N}_\eta\left[t_0; \Upsilon\left(t_0\right); \mathcal{J}\right] \subseteq \mathfrak{R}^K, \ \mathfrak{N}_\eta\left[t_0; \Upsilon\left(t_0\right); \mathcal{J}\right] \neq \phi \Longrightarrow$$

$$\mathcal{D}_{SZTU}\left[t_0; \Upsilon\left(t_0\right); \mathcal{J}^{\xi}\right] = \cap\left\{\mathcal{D}_{SZT}\left[t_0; \Upsilon\left(t_0\right); \mathbf{I}\right] : \mathbf{I}(.) \in \mathcal{J}^{\xi}\right\} \supset \mathfrak{N}_\eta\left[t_0; \Upsilon\left(t_0\right); \mathcal{J}\right],$$

$$\left[\begin{array}{c} \mathcal{D}_{SSZTU}\left[t_0; \Upsilon\left(t_0\right); \mathcal{J}^{\xi}\right] = \cap\left\{\begin{array}{c} \mathcal{D}_{SSZT}\left[t_0; \Upsilon\left(t_0\right); \mathbf{I}\right] : \\ : \mathbf{I}(.) \in \mathcal{J}^{\xi} \end{array}\right\} \\ \supset \mathfrak{N}_\eta\left[t_0; \Upsilon\left(t_0\right); \mathcal{J}\right]. \end{array}\right]$$

b) The uniform tracking in $t_0 \in \mathfrak{T}_i$ *on* $\mathfrak{T}_0 \times \mathcal{J}^{\xi}$ *and the domain intersection*

$$\mathcal{D}_{SZTU}\left[\mathfrak{T}_i; \Upsilon\left(t_0\right); \mathbf{I}\right]$$

of stable tracking domains $\mathcal{D}_{SZT}\left[t_0; \Upsilon\left(t_0\right); \mathbf{I}\right]$*, [the domain intersection* $\mathcal{D}_{SSZTU}\left[\mathfrak{T}_i; \Upsilon\left(t_0\right); \mathbf{I}\right]$ *of all strictly stable tracking domains* $\mathcal{D}_{SSZT}\left[t_0; \Upsilon\left(t_0\right); \mathbf{I}\right]$ *], in* $t_0 \in \mathfrak{T}_i$ *is a connected neighborhood of* $\Upsilon\left(t_0\right)$*, then, and only then, it is* ***the (strict) stable tracking domain of*** $\Upsilon(t)$

on $\mathfrak{T}_0 \times \mathfrak{T}_i \times \mathcal{J}^\xi$ ***uniform in*** $t_0 \in \mathfrak{T}_i$, *respectively,*

$$\exists \eta \in \mathfrak{R}^+ \Longrightarrow \mathfrak{N}_\eta \left[\mathfrak{T}_i; \Upsilon(t_0); \mathbf{I}\right] \subset \mathfrak{R}^K,$$
$$\mathcal{D}_{SZTU}\left[\mathfrak{T}_i; \Upsilon(t_0); \mathbf{I}\right] = \cap \left\{\mathcal{D}_{SZT}\left[t_0; \Upsilon(t_0); \mathbf{I}\right] : t_0 \in \mathfrak{T}_i\right\}$$
$$\supset \mathfrak{N}_\eta \left[\mathfrak{T}_i; \Upsilon(t_0); \mathbf{I}\right], \tag{18.14}$$

$$\left[\begin{array}{c} \mathcal{D}_{SSZTU}\left[\mathfrak{T}_i; \Upsilon(t_0); \mathbf{I}\right] = \cap \left\{\mathcal{D}_{SSZT}\left[t_0; \Upsilon(t_0); \mathbf{I}\right] : t_0 \in \mathfrak{T}_i\right\} \\ \supset \mathfrak{N}_\eta \left[\mathfrak{T}_i; \Upsilon(t_0); \mathbf{I}\right] \end{array}\right]. \tag{18.15}$$

c) The uniform tracking in $[t_0, \mathbf{I}(.)]$ *on* $\mathfrak{T}_0 \times \mathfrak{T}_i \times \mathcal{J}^\xi$ *and, respectively, the intersection* $\mathcal{D}_{SZTU}\left[\mathfrak{T}_i; \Upsilon(t_0); \mathcal{J}^\xi\right]$ *of stable tracking domains* $\mathcal{D}_{SZT}\left[t_0; \Upsilon(t_0); \mathbf{I}\right]$ *in* $[t_0, \mathbf{I}(.)] \in \mathfrak{T}_0 \times \mathcal{J}^\xi$, *[the domain intersection* $\mathcal{D}_{SZTU}\left(\mathfrak{T}_i; \Upsilon(t_0); \mathcal{J}^\xi\right)$ *of all domains* $\mathcal{D}_{SZT}\left[t_0; \Upsilon(t_0); \mathbf{I}\right]$ *in* $[t_0, \mathbf{I}(.)] \in \mathfrak{T}_0 \times \mathcal{J}^\xi$*] is a connected neighborhood of* $\Upsilon(t_0)$, *then, and only then, it is* ***the (strict) stable tracking domain of*** $\Upsilon(t)$ ***on*** $\mathfrak{T}_0 \times \mathfrak{T}_i \times \mathcal{J}^\xi$ ***uniform in*** $[t_0, \mathbf{I}(.)] \in \mathfrak{T}_i \times \mathcal{J}^\xi$,

$$\exists \eta \in \mathfrak{R}^+ \Longrightarrow \mathfrak{N}_\eta \left[\mathfrak{T}_i; \Upsilon(t_0); \mathcal{J}^\xi\right] \subset \mathfrak{R}^K,$$
$$\mathcal{D}_{SZTU}\left[\mathfrak{T}_i; \Upsilon(t_0); \mathcal{J}^\xi\right] = \cap \left\{\begin{array}{c} \mathcal{D}_{SZT}\left[t_0; \Upsilon(t_0); \mathbf{I}\right] : \\ : [t_0, \mathbf{I}(.)] \in \mathfrak{T}_i \times \mathcal{J}^\xi \end{array}\right\}$$
$$\supset \mathfrak{N}_\eta \left[\mathfrak{T}_i; \Upsilon(t_0); \mathcal{J}^\xi\right], \tag{18.16}$$

$$\left[\begin{array}{c} \mathcal{D}_{SSZTU}\left[\mathfrak{T}_i; \Upsilon(t_0); \mathcal{J}^\xi\right] = \cap \left\{\begin{array}{c} \mathcal{D}_{SSZT}\left[t_0; \Upsilon(t_0); \mathbf{I}\right] : \\ : [t_0, \mathbf{I}(.)] \in \mathfrak{T}_i \times \mathcal{J}^\xi \end{array}\right\} \\ \supset \mathfrak{N}_\eta \left[\mathfrak{T}; \Upsilon(t_0); \mathcal{J}^\xi\right] \end{array}\right]. \tag{18.17}$$

Definition 254 ***The exponential tracking of the time-varying set*** $\Upsilon(t)$

a) The system motions $\mathcal{Z}(.; t_0; \mathbf{Z}_0; \mathbf{I})$ *exhibit* ***the exponentially stable tracking of*** $\Upsilon(t)$ ***on*** $\mathfrak{T}_0 \times \mathcal{J}^\xi$ *if, and only if for every* $\mathbf{I}(.) \in \mathcal{J}^\xi$ *there exist positive real numbers* $\varsigma \geq 1$ *and* ψ, *and a connected neighborhood* $\mathfrak{N}[t_0; \Upsilon(t_0); \mathbf{I}]$ *of* $\Upsilon(t_0)$, $\varsigma = \varsigma[t_0; \Upsilon(t_0); \mathbf{I}]$, $\psi = \psi[t_0; \Upsilon(t_0); \mathbf{I}]$, *such that* $\mathbf{Z}_0 \in \mathfrak{N}[t_0; \Upsilon(t_0); \mathbf{I}]$ *guarantees that* $\mathbf{Z}(t; t_0; \mathbf{Z}_0; \mathbf{I})$ *approaches exponentially* $\Upsilon(t)$ *all the time* $t \in \mathfrak{T}_0$; *i.e.,*

$$\forall [t_0, \mathbf{I}(.)] \in \mathfrak{T}_i \times \mathcal{J}^\xi, \ \exists \varsigma \in \mathfrak{R}^+, \ \varsigma \geq 1, \ \exists \psi \in \mathfrak{R}^+,$$
$$\varsigma = \varsigma[t_0; \Upsilon(t_0); \mathbf{I}], \ \psi = \psi[t_0; \Upsilon(t_0); \mathbf{I}],$$
$$\exists \mathfrak{N}[t_0; \Upsilon(t_0); \mathbf{I}] \subseteq \mathfrak{R}^K, \ \mathbf{Z}_0 \in \mathfrak{N}[t_0; \Upsilon(t_0); \mathbf{I}] \Longrightarrow$$
$$\rho[\mathbf{Z}(t; t_0; \mathbf{Z}_0; \mathbf{I}), \Upsilon(t)]$$
$$\leq \varsigma \rho[\mathbf{Z}_0, \Upsilon(t_0)] \exp[-\psi(t - t_0)], \ \forall t \in \mathfrak{T}_0. \tag{18.18}$$

b) The largest connected neighborhood $\mathfrak{N}_L[t_0; \Upsilon(t_0); \mathbf{I}]$ *of* $\Upsilon(t_0)$ *at* $t_0 \in \mathfrak{T}_i$ *is* ***the exponentially stable tracking domain*** $\mathcal{D}_{ET}[t_0; \Upsilon(t_0); \mathbf{I}]$ ***of*** $\Upsilon(t)$ ***on*** $\mathfrak{T}_0 \times \mathcal{J}^\xi$.

c) The exponentially stable tracking of $\Upsilon(t)$ *at* $t_0 \in \mathfrak{T}_i$ *is* ***global (in the whole) on*** $\mathfrak{T}_0 \times \mathcal{J}^\xi$ *if, and only if,* $\mathcal{D}_{ET}[t_0; \Upsilon(t_0); \mathbf{I}] = \mathfrak{R}^K$.

d) The exponentially stable tracking of $\Upsilon(t)$ *on* $\mathfrak{T}_0 \times \mathcal{J}^\xi$ *is also* ***uniform in*** $(t_0, \mathbf{I})$ ***on*** $\mathfrak{T}_0 \times \mathfrak{T}_i \times \mathcal{J}^\xi$ *if, and only if, the intersection* $\mathcal{D}_{ET}\left(\mathfrak{T}_i; \Upsilon; \mathcal{J}^\xi\right)$ *of the domains* $\mathcal{D}_{ET}[t_0; \Upsilon(t_0); \mathbf{I}]$ *in* $(t_0, \mathbf{I}) \in \mathfrak{T}_i \times \mathcal{J}^\xi$ *is a nonempty connected neighborhood of* $\Upsilon(t_0)$ *for every* $(t_0, \mathbf{I}) \in \mathfrak{T}_i \times \mathcal{J}^\xi$ *and the values of* ξ *and* ψ *depend at most on* $\mathfrak{T}_i \times \mathcal{J}^\xi$, $\xi = \xi\left(\mathfrak{T}_i; \Upsilon; \mathcal{J}^\xi\right)$, $\psi = \psi\left(\mathfrak{T}_i; \Upsilon; \mathcal{J}^\xi\right)$, *but not on a particular choice of* $(t_0, \mathbf{I})$ *from* $\mathfrak{T}_i \times \mathcal{J}^\xi$, *then the exponentially stable tracking of* $\Upsilon(t)$ *is* ***uniform over*** $\mathfrak{T}_0 \times \mathfrak{T}_i \times \mathcal{J}^\xi$. *Then, and only then, the intersection* $\mathcal{D}_{ET}\left(\mathfrak{T}_i; \Upsilon; \mathcal{J}^\xi\right)$ *is* ***the domain of the uniform exponentially stable tracking of*** $\Upsilon(t)$ ***over*** $\mathfrak{T}_0 \times \mathfrak{T}_i \times \mathcal{J}^\xi$.

e) If, and only if, additionally to d) the domain $\mathcal{D}_{ET}\left(\mathfrak{T}_i; \Upsilon; \mathcal{J}^\xi\right) = \mathfrak{R}^K$ *then the uniform exponentially stable tracking of* $\Upsilon(t)$ *over* $\mathfrak{T}_0 \times \mathfrak{T}_i \times \mathcal{J}^\xi$ *is* ***global (in the whole) on*** $\mathfrak{T}_0 \times \mathfrak{T}_i \times \mathcal{J}^\xi$.

Part VI

CRITERIA FOR STABLE TRACKING

Chapter 19

Introduction

19.1 Lyapunov methods and methodologies

The famous Lyapunov method and methodologies, [401], have been developed for stability studies and stabilizing control synthesis of the *ISO* nonlinear systems. The literature on it is particularly vast, e.g. [7], [8], [9], [12], [20]-[23], [37], [38], [48], [66], [67], [71], [78], [84], [85], [86], [93], [100], [113], [248], [280], [281], [287], [292], [295], [297], [304], [305], [306], [322], [346], [349], [356], [358], [359], [363], [370], [373], [374], [383], [384], [400], [405], [406], [410], [411], [412], [413], [414], [415], [416], [417], [419]-[421], [427]-[432], [440], [454], [455], [473], [497], [504], [506], [511], [512], [531], [532]-[534], [542], [544]-[548], [552], [565], [578].

Lyapunov distinguished two general methods for studying dynamical systems. One of them, known as *the first Lyapunov method,* demands solving the system mathematical model and studying system properties by examining its solutions for every initial condition. This method can be effective for *time*-invariant linear systems, but it is generally inapplicable to the nonlinear systems. Another Lyapunov method, known as *the second* (or, *direct*) *Lyapunov method*, for short *Lyapunov method*, establishes two methodologies and conditions how to study properties of linear and nonlinear dynamical systems without solving their mathematical models, hence, without knowing their solutions.

The methodology developed for *time*-invariant linear systems starts with an arbitrary choice of a negative definite function and continues with the test of the properties of the solution of the corresponding differential equation. Its left-hand side is determined in terms of the system mathematical model, and its right-hand side contains the accepted negative definite function. The solution function is called a *Lyapunov function.* The test resolves the problem of the asymptotic stability and the exact determination of the Lyapunov function.

In the classical literature on Lyapunov stability theory and Lyapunov method, a sample of which is cited above, Lyapunov methodology for *time*-invariant linear systems was not broadened to the nonlinear systems until the publication of the paper [202], 1990.

Lyapunov developed another methodology for the application of his method to nonlinear systems. The methodology is inverse to that for the linear *time*-invariant systems. It starts with a choice of a positive definite function, without giving any hint how to select it for a given nonlinear system or for a class of nonlinear systems. If, fortunately, the total *time* derivative of the selected function is negative (semi)definite then the equilibrium state is (stable) asymptotically stable. The problem of finding a suitable function to start with was unsolved until [202], 1990. The conditions are expressed in terms of the existence of a positive definite function with negative (semi)definite total *time* derivative along the system motions.

In order to resolve the three fundamental problems of Lyapunov's method another Lyapunov methodology is established for the nonlinear system. It broadens Lyapunov's methodology for *time*-invariant linear systems to the nonlinear systems. Therefore it is called **the consistent Lyapunov methodology** (*CLM*) because it is consistent with Lyapunov's primary methodology for *time*-invariant linear systems. The paper [202], 1990, established the *consistent Lyapunov methodology* for stability studies of nonlinear systems. *CLM* was further developed in [156], [158], [160], [161], [162], [170], [174], [182], [172], [199], [200], [201], [203], [204], [205], [217], [224], [216], [255], [256], [257]. Its link with bond-graphs is in [99]. *CLM* establishes the complete conditions in the framework of both *time*-invariant and *time*-varying nonlinear systems for:

- The necessary and sufficient conditions for asymptotic and exponential stability of an equilibrium or of an invariant set of the system
- The direct determination of the Lyapunov function of the system
- The determination of the exact domain of the asymptotic stability
- The need of the single application of the methodology to get the final result

The conditions are not expressed in terms of the existence of a Lyapunov function.

Comment 255 ***Crucial developments of Lyapunov method and methodologies***

This book presents in subsequent chapters the generalization and the extension of the Lyapunov method, of Lyapunov methodology and of the Consistent Lyapunov methodology to tracking issues by establishing them for the tracking studies of the main three, large, classes of the perturbed (nonautonomous) nonlinear dynamical systems: for the IO systems (3.62), for the ISO systems (3.65), (3.66), and for the IIDO systems (3.12) that incorporate the preceding two system families.

Although Lyapunov method for stability has been established for the stability study of the motions of ISO systems (3.65) by ignoring completely the output system behavior determined by (3.66), the Lyapunov method, Lyapunov methodology and consistent Lyapunov methodology deal with both the state and output system behavior of all three classes of the systems in this book.

19.2 Tracking accuracy and definitions

The sense of the k-th-order tracking that the plant extended real output response $\mathbf{Y}^k(t)$ tracks the plant extended desired output response $\mathbf{Y}_d^k(t)$ with a demanded quality means, equivalently, that the *time* evolution of the extended output error vector $\boldsymbol{\varepsilon}^k(t) = \mathbf{Y}_d^k(t) - \mathbf{Y}^k(t)$, equivalently, of the extended output deviation vector $\mathbf{y}^k(t) = \mathbf{Y}^k(t) - \mathbf{Y}_d^k(t)$, should satisfy the required quality demand, $k \in \{0, 1, 2, ..., \alpha - 1\}$.

Lyapunov's method essentially enables us to study effectively the closeness, equivalently, *the accuracy*, denoted by $\mathcal{A}$, of $\mathbf{Y}^k(t)$ to $\mathbf{Y}_d^k(t)$. We do that by using either the extended output error vector $\boldsymbol{\varepsilon}^k$ or the extended output deviation vector $\mathbf{y}$, They are equivalent: $\mathbf{y}^k = -\boldsymbol{\varepsilon}^k$. Their norms $\|\mathbf{y}^k\|$ and $\|\boldsymbol{\varepsilon}^k\|$, $\|\mathbf{y}^k\| = \|\boldsymbol{\varepsilon}^k\|$, are the inverse measure of the closeness, i.e., the accuracy $\mathcal{A}$. The mathematical relationships among them can have different forms studied by Nedić (Neditch) in [457], for example:

$$\mathcal{A} = \frac{1}{\|\mathbf{y}^k\|} \in [0, \infty] \quad \text{or } \mathcal{A} = \ln \frac{1}{\|\mathbf{y}^k\|} \in [-\infty, \infty] \quad \text{or } \mathcal{A} = \exp\left[-\|\mathbf{y}^k\|\right] \in [0, 1].$$

This means that, in the output space, for the k-th-order tracking, the Lyapunov function and its derivative should be dependent on the extended distance of $\mathbf{Y}^k$ from $\mathbf{Y}_d^k$, i.e., on

$\mathbf{Y}^k - \mathbf{Y}_d^k = \mathbf{y}^k = -\boldsymbol{\varepsilon}^k$. In the state space the Lyapunov function and its derivative should be dependent on the state vector difference $\mathbf{R}^{\alpha-1} - \mathbf{R}_d^{\alpha-1} = \mathbf{r}^{\alpha-1} = -\boldsymbol{\varepsilon}_R^{\alpha-1}$.

The definitions of all tracking properties are presented herein in terms of the closeness of $\mathbf{Y}^k$ to $\mathbf{Y}_d^k$ or of $\mathbf{R}^{\alpha-1}$ to $\mathbf{R}_d^{\alpha-1}$. The preceding consideration shows that all tracking definitions can be equivalently presented in terms of $\mathbf{y}^k$ or of $\mathbf{r}^{\alpha-1}$, respectively. It is straightforward to do.

19.3 Suitable mathematical models

The suitable mathematical models for the tracking analysis in the state space are in terms of the deviations ($\mathbf{r}$ and $\mathbf{y}$) rather than in terms of the total valued variables ($\mathbf{R}$ and $\mathbf{Y}$). We use them in what follows.

The suitable form of the mathematical model for the tracking studies in the state space via the Lyapunov method of

- The system (3.12) is (5.10) repeated as (19.1):

$$\begin{aligned} \widetilde{Q}\left[t, \mathbf{r}^{\alpha-1}(t)\right] \mathbf{r}^{(\alpha)}(t) + \widetilde{\mathbf{q}}\left[t, \mathbf{r}^{\alpha-1}(t)\right] &= \widetilde{\mathbf{h}}\left[t, \mathbf{i}^{\xi}(t)\right], \ \forall t \in \mathfrak{T}, \\ \mathbf{y}(t) = \widetilde{\mathbf{s}}\left[t, \mathbf{r}^{\alpha-1}(t), \mathbf{i}(t)\right], \ \forall t \in \mathfrak{T}, \end{aligned} \tag{19.1}$$

- And of the plant (3.12) is (5.17) repeated as (19.2):

$$\begin{aligned} \widetilde{Q}\left[t, \mathbf{r}^{\alpha-1}(t)\right] \mathbf{r}^{(\alpha)}(t) + \widetilde{\mathbf{q}}\left[t, \mathbf{r}^{\alpha-1}(t)\right] &= E(t)\, \widetilde{\mathbf{e}}_d\left[t, \mathbf{d}^{\eta}(t)\right] + P(t)\, \widetilde{\mathbf{e}}_u\left[\mathbf{u}(t)\right], \\ &\forall t \in \mathfrak{T}, \\ \mathbf{y}(t) = \widetilde{\mathbf{z}}\left[t, \mathbf{r}^{\alpha-1}(t), \mathbf{d}(t)\right] + W(t)\, \widetilde{\mathbf{w}}\left[\mathbf{u}(t)\right], \ \forall t \in \mathfrak{T}. \end{aligned} \tag{19.2}$$

Chapter 20

Comparison and (semi)definite functions

20.1 Comparison functions

Hahn [292] introduced several classes of *comparison functions* in order to simplify the mathematical treatment of stability problems. A brief account of them follows by referring to [292].

Let $\varphi(.) : \mathfrak{R}_+ \longrightarrow \mathfrak{R}_+$ be a notation for a comparison function. Notice that the comparison functions are *time*-invariant (*time*-independent) scalar functions. This holds also for norm functions $\|.\| : \mathfrak{R}^K \longrightarrow \mathfrak{R}_+$ and for distance functions $\rho(.) : \mathfrak{R}^K \times 2^{\mathfrak{R}^K} \longrightarrow \mathfrak{R}_+$.

Definition 256 ***Classes of the comparison functions***

A function $\varphi(.) : \mathfrak{R}_+ \longrightarrow \mathfrak{R}_+$ belongs to:

(i) The class $\mathcal{K}_{[0,\alpha[}$, $0 < \alpha \leq \infty$, if, and only if, it is defined, continuous, and strictly increasing on $[0,\alpha[$, and $\varphi(0) = 0$

(ii) The class $\mathcal{K}$ if, and only if, (i) holds for $\alpha = \infty : \mathcal{K} = \mathcal{K}_{[0,\infty[}$

(iii) The class $\mathcal{KR}$ if, and only if, (ii) holds and $\zeta \longrightarrow \infty$ implies $\varphi(\zeta) \longrightarrow \infty$

(iv) The class $\mathcal{L}_{[0,\alpha[}$, $0 < \alpha \leq \infty$, if, and only if, it is defined, continuous and strictly decreasing on $[0,\alpha[$, and $\zeta \longrightarrow 0$ implies lim $[\varphi(\zeta)] = 0$

(v) The class $\mathcal{L}$ if, and only if, (iv) holds for $\alpha = \infty$: $\mathcal{L} = \mathcal{L}_{[0,\infty[}$

Example 257 *The following function $f(.) : \mathfrak{R}_+ \longrightarrow \mathfrak{R}_+$:*

1. $f(\zeta) =$sign ζ is not a comparison function. It is not continuous at $\zeta = 0$ and it is not strictly increasing or decreasing.

2. $f(\zeta) = ($sign $\zeta)$sin $\zeta = |$sin $\zeta|$ is a comparison function $\varphi(.) : \mathfrak{R}_+ \longrightarrow \mathfrak{R}_+$ of the class $\mathcal{K}_{[0,\frac{\pi}{2}[}$, $\alpha = \frac{\pi}{2}$, but not of the class $\mathcal{K}$ because it is not strictly increasing at $\zeta = \frac{\pi}{2}$ and it is decreasing on $]\frac{\pi}{2},\pi[$.

3. $f(\zeta) = 1 - e^{-\zeta}$ is a comparison function $\varphi(.) : \mathfrak{R}_+ \longrightarrow \mathfrak{R}_+$ of the class $\mathcal{K}$, but not of the class $\mathcal{KR}$.

4. $f(\zeta) = \zeta\sqrt{\zeta}$ is a comparison function $\varphi(.) : \mathfrak{R}_+ \longrightarrow \mathfrak{R}_+$ of the class $\mathcal{KR}$.

5. $f(\zeta) = 1 - \zeta$ is a comparison function $\varphi(.) : \mathfrak{R}_+ \longrightarrow \mathfrak{R}_+$ of the class $\mathcal{L}_{[0,1[}$, $\alpha = 1$.

6. $f(\zeta) = e^{-\zeta}$ is a comparison function $\varphi(.) : \mathfrak{R}_+ \longrightarrow \mathfrak{R}_+$ of the class $\mathcal{L}$.

7. $f(\zeta) =$cos ζ is not a comparison function. It is not strictly increasing or decreasing at $\zeta = 0$.

The inverse function of a comparison function $\varphi(.) : \mathfrak{R}_+ \longrightarrow \mathfrak{R}_+$ is $\varphi^I(.) : \mathfrak{R}_+ \longrightarrow \mathfrak{R}_+$, $\varphi^I[\varphi(\zeta)] \equiv \zeta$ and $\varphi[\varphi^I(\xi)] \equiv \xi$, where $\xi = \varphi(\zeta)$.

Example 258 *The following function* $f(.) : \mathfrak{R}_+ \longrightarrow \mathfrak{R}_+$*:*

1. $f(\xi) = arcsin\ \xi = |arcsin\ \xi|,\ \xi \in [0,1[$, *is the inverse to the comparison function* $\varphi(\zeta) = (sign\ \zeta) sin\ \zeta = |sin\ \zeta|$ *for* $\zeta \in [0, \frac{\pi}{2}[$ *because* $f[\varphi(\zeta)] = |arcsin\ \varphi(\zeta)| = |arcsin\ (|sin\ \zeta|)| = |arcsin\ (sin\ \zeta)| = |\zeta| = \zeta$.

2. $f(\xi) = -\ln(1-\xi)$ *is the inverse to the comparison function* $\varphi(\zeta) = 1 - e^{-\zeta}$ *because.* $f[\varphi(\zeta)] = -\ln/1 - \varphi(\zeta)/ = -\ln/1 - (1 - e^{-\zeta})/ = -\ln e^{-\zeta} = \zeta$.

3. $f(\xi) = \xi^{2/3}$ *is the inverse to the comparison function* $\varphi(\zeta) = \zeta\sqrt{\zeta} = \zeta^{3/2}$ *because* $f[\varphi(\zeta)] = \varphi^{2/3}(\zeta) = \left[\zeta^{3/2}\right]^{2/3} = \zeta$.

4. $f(\xi) = 1 - \xi$ *is the inverse to the comparison function* $\varphi(\zeta) = 1 - \zeta$ *because* $f[\varphi(\zeta)] = 1 - \varphi(\zeta) = 1 - (1 - \zeta) = \zeta$.

5. $f(V) = |\ln \xi|$ *is the inverse to the comparison function* $\varphi(\zeta) = e^{-\zeta}$ *because* $f[\varphi(\zeta)] = |\ln \varphi(\zeta)| = \left|\ln e^{-\zeta}\right| = |-\zeta| = |\zeta| = \zeta,\ \zeta \geq 0$.

Proposition 259 ***Properties of the comparison functions***

(i) *If* $\varphi(.) \in \mathcal{K}$ *and* $\psi(.) \in \mathcal{K}$ *then* $\varphi[\psi(.)] \in \mathcal{K}$.

(ii) *If* $\varphi(.) \in \mathcal{K}$ *and* $\psi(.) \in \mathcal{L}$ *then* $\varphi[\psi(.)] \in \mathcal{L}$.

(iii) *If* $\varphi(.) \in \mathcal{K}_{[0,\alpha[}$ *and* $\varphi(\alpha) = \xi$ *then* $\varphi^I(.) \in \mathcal{K}_{[0,\xi[}$.

(iv) *If* $\varphi(.) \in \mathcal{K}_{[0,\alpha[}$ *and lim* $[\varphi(\zeta) : \zeta \longrightarrow \infty] = \mu$ *then* $\varphi^I(.)$ *is not defined on* $]\mu, \infty]$.

(v) *If* $\varphi(.) \in \mathcal{K}_{[0,\alpha[}$, $\psi(.) \in \mathcal{K}_{[0,\alpha[}$, *and* $\varphi(\zeta) > \psi(\zeta)$ *on* $[0,\alpha[$ *then* $\varphi^I(\zeta) < \psi^I(\zeta)$ *on* $[0,\beta[$, *where* $\beta = \psi(\alpha)$.

Definition 260 ***Classes of the compound comparison functions***

A function $\varphi(.) : \mathfrak{R}_+ \times \mathfrak{R}_+ \longrightarrow \mathfrak{R}_+$ *belongs to:*

(i) *The class* $\mathcal{KK}_{[0;\alpha,\beta[}$, *if, and only if, both* $\varphi(*,\zeta) \in \mathcal{K}_{[0,\alpha[}$ *for every* $\zeta \in [0,\beta[$ *and* $\varphi(\zeta,*) \in \mathcal{K}_{[0,\beta[}$ *for every* $\zeta \in [0,\alpha[$

(ii) *The class* $\mathcal{KK}$ *if, and only if, (i) holds for* $\alpha = \beta = \infty$

(iii) *The class* $\mathcal{KL}_{[0;\alpha,\beta[}$ *if, and only if, both* $\varphi(*,\zeta) \in \mathcal{K}_{[0,\alpha[}$ *for every* $\zeta \in [0,\beta[$ *and* $\varphi(\zeta,*) \in \mathcal{L}_{[0,\beta[}$ *for every* $\zeta \in [0,\alpha[$

(iv) *the class* $\mathcal{KL}$ *if, and only if, (iii) holds for* $\alpha = \beta = \infty$

20.2 Semidefinite functions

We use the unified notation:

$$\mathbf{Z} \in \left\{\mathbf{Y}, \mathbf{Y}^k, \mathbf{R}^{\alpha-1}\right\},\ \mathbf{Z} = [Z_1\ Z_2\ ...Z_K]^T \in \mathfrak{R}^K,$$
$$k \in \{0,1,\ \cdots\ ,\alpha-1\}, K \in \{N, \rho, (k+1)N, \alpha\rho\},$$
$$\mathbf{z} = \mathbf{Z} - \mathbf{Z}_d,\ \mathbf{z} \in \left\{\mathbf{y}, \mathbf{y}^k, \mathbf{r}^{\alpha-1}\right\},\ \mathbf{z} \in \mathfrak{R}^K. \tag{20.1}$$

Note 261 *If*

- $\mathbf{Z} = \mathbf{Y}$ *then* $\mathbf{Z}$ *is the total output vector of the system (3.12)*
- $\mathbf{Z} = \mathbf{Y}^k$ *then* $\mathbf{Z}$ *is the extended total output vector of the system (3.12)*
- $\mathbf{Z} = \mathbf{R}^{\alpha-1}$ *then* $\mathbf{Z}$ *is the total state vector of the system (3.12)*

Lyapunov theorems determine the conditions on the function $V(.)$ called in general ***the Lyapunov function*** and on its total *time* derivative $V^{(1)}(.)$ along system motions in order for the system motions to have the requested qualitative properties. The statements of the conditions are in terms of *sign semidefinite functions, sign-definite functions,* *radially unbounded functions,* and *functions with the infinitesimally small upper bound (Russian terminology)* also called *decrescent functions (English terminology).*

Let $\tau \in \mathfrak{T}_i;$, $\tau = t_0$ is permitted.

In what follows the set Υ is a fixed *time*-invariant connected nonempty subset of $\mathfrak{R}^K$, $\Upsilon \subset \mathfrak{R}^K$, the boundary $\partial\Upsilon$ of which is nonempty, $\partial\Upsilon \neq \phi$. For example: $\Upsilon = \Upsilon_1 \cup \Upsilon_2$,

$$\Upsilon_1 = \left\{\mathbf{z} : \mathbf{z} \in \mathfrak{R}^2,\ z_1^2 + z_2^2 - 4 \leq 0\right\},\ \partial\Upsilon_1 = \left\{\mathbf{z} : \mathbf{z} \in \mathfrak{R}^2,\ z_1^2 + z_2^2 - 4 = 0\right\},$$
$$\Upsilon_2 = \left\{\mathbf{z} : \mathbf{z} \in \mathfrak{R}^2,\ |z_1| - 4 \leq 0,\ z_2 \leq 0\right\},\ \partial\Upsilon_2 = \Upsilon_2,$$
$$\partial\Upsilon = \left\{\mathbf{z} : \mathbf{z} \in \mathfrak{R}^2,\ z_1^2 + z_2^2 - 4 = 0,\ z_2 \geq 0\right\}$$
$$\cup \left\{\mathbf{z} : \mathbf{z} \in \mathfrak{R}^2,\ 2 \leq |z_1| \leq 4,\ z_2 = 0\right\}$$
$$\cup \left\{\mathbf{z} : \mathbf{z} \in \mathfrak{R}^2,\ |z_1| = 4,\ z_2 \leq 0\right\} \Longrightarrow \partial\Upsilon \neq \partial\Upsilon_1 \cup \partial\Upsilon_2.$$

By referring to Lyapunov [401, p. 57], as well as to [227, Definition 10, pp. 22, 23] and [287, Definition 4.1, pp. 94, 95] we accept:

Definition 262 ***Time-varying semidefinite function relative to*** Υ ***on*** $\mathfrak{T}_\tau$

A time-varying function $W(.;\Upsilon) : \mathfrak{T}_\tau \times \mathfrak{R}^K \longrightarrow \mathfrak{R}$ *is:*

a) ***Positive semidefinite relative to*** Υ ***on*** $\mathfrak{T}_\tau$ *if, and only if, there is a time-invariant connected neighborhood* $\mathfrak{N}(\Upsilon)$ *of* Υ *such that*

1. $W(.;\Upsilon)$ *is defined and continuous on* $\mathfrak{T}_\tau \times \mathfrak{N}(\Upsilon)$,

$$W(t,\mathbf{z};\boldsymbol{\Upsilon}) \in \mathfrak{C}[\mathfrak{T}_\tau \times \mathfrak{N}(\Upsilon)], \tag{20.2}$$

2. $W(.;\Upsilon)$ *is nonnegative on* $\mathfrak{T}_\tau \times \mathfrak{N}(\Upsilon)$,

$$W(t,\mathbf{z};\boldsymbol{\Upsilon}) \geq 0,\ \forall (t,\mathbf{z}) \in \mathfrak{T}_\tau \times \mathfrak{N}(\Upsilon), \tag{20.3}$$

3. $W(.;\Upsilon)$ *vanishes on the boundary* $\partial\Upsilon$ *of* Υ,

$$W(t,\mathbf{z};\boldsymbol{\Upsilon}) = 0,\ \ \forall (t,\mathbf{z}) \in \mathfrak{T}_\tau \times \partial\Upsilon, \tag{20.4}$$

4. *There is* $\mathbf{z}^* \in \mathfrak{N}(\Upsilon) \backslash \Upsilon$ *such that* $W(t,\mathbf{z}^*)$ *is positive on* $\mathfrak{T}_\tau$,

$$\exists \mathbf{z}^* \in \mathfrak{N}(\Upsilon) \backslash \Upsilon \Longrightarrow W(t,\mathbf{z}^*;\boldsymbol{\Upsilon}) > 0,\ \ \forall t \in \mathfrak{T}_\tau. \tag{20.5}$$

b) ***Globally positive semidefinite relative to*** Υ ***on*** $\mathfrak{T}_\tau$ *if, and only if, a) holds for* $\mathfrak{N}(\Upsilon) = \mathfrak{R}^K$,

d) ***(Globally) Negative semidefinite relative to*** Υ ***on*** $\mathfrak{T}_\tau$ *if, and only if, -* $W(.;\Upsilon)$ *is (globally) positive semidefinite relative to* Υ *on* $\mathfrak{T}_\tau$*, respectively,*

e) ***Positive semidefinite relative to*** Υ ***on*** $\mathfrak{T}_\tau$ ***and on a set*** $\mathfrak{A}$ *if, and only if,* $\mathfrak{A}$ *is a neighborhood of* Υ *such that the conditions under a) hold for* $\mathfrak{N}(\Upsilon) = \mathfrak{A}$,

f) ***Negative semidefinite relative to*** Υ ***on*** $\mathfrak{T}_\tau$ ***and on a set*** $\mathfrak{A}$ *if, and only if,* $-W(.)$ *is positive semidefinite relative to* Υ *on the set* $\mathfrak{A}$.

*The expression "**on** $\mathfrak{T}_\tau$" is to be omitted if, and only if,* $\mathfrak{T}_\tau = \mathfrak{T}$.

*The expression "**relative to** Υ" is to be omitted if, and only if,* $\Upsilon = \mathfrak{O} = \{\mathbf{0}_K\}$.

If, and only if, the function $W(.;\Upsilon)$ *is time-invariant,* $W(.;\Upsilon) : \mathfrak{R}^K \longrightarrow \mathfrak{R}$, $W(t,\mathbf{z};\boldsymbol{\Upsilon}) \equiv W(\mathbf{z};\boldsymbol{\Upsilon})$, *then "*$\mathfrak{T}_\tau \times$*", "*$\forall t \in \mathfrak{T}_\tau$*", and "on* $\mathfrak{T}_\tau$*" should be omitted.*

This definition holds directly for *time*-invariant functions $W(.) : \mathfrak{R}^K \longrightarrow \mathfrak{R}$ by omitting "$\mathfrak{T}_\tau$", "$\mathfrak{T}_\tau \times$", and "t," everywhere in the definition.

Note 263 *Continuity of a function is its crucial necessary property to be semidefinite. However, it need not be differentiable in order to be semidefinite.*

Note 264 ***Extension of Definition 262 to time-varying sets***

For the extension of the above definition to the class of time-varying semidefinite functions on time-varying neighborhoods $\mathfrak{N}(t;\Upsilon)$ of $\boldsymbol{\Upsilon}$ we apply the above definition to the set $\mathfrak{N}(\Upsilon)$ that is then the union of the sets $\mathfrak{N}(t;\Upsilon)$ for all $t \in \mathfrak{T}_{\tau}$,

$$\mathfrak{N}(\mathfrak{u}) = \cup [(t;\Upsilon) : t \in \mathfrak{T}_{\tau}] \subset \mathfrak{R}^K.$$

20.3 Definite functions

In view of Lyapunov [401, p. 57], as well as of [227, Definition 11, p. 23] and [287, Definition 4.4, pp. 96, 97], we introduce:

Definition 265 ***Time-invariant definite function relative to*** Υ

A function $W(.;\Upsilon) : \mathfrak{R}^K \longrightarrow \mathfrak{R}$ is:

a) ***Positive definite relative to*** Υ *if, and only if, there is a time-invariant connected neighborhood $\mathfrak{N}(\Upsilon)$ of $\boldsymbol{\Upsilon}$ such that*

1. $W(.;\Upsilon)$ *is defined and continuous on* $\mathfrak{N}$,

$$W(\mathbf{z};\Upsilon) \in \mathfrak{C}[\mathfrak{N}(\Upsilon)] \tag{20.6}$$

2. $W(.;\Upsilon)$ *is positive on* $\mathfrak{N}(\Upsilon)$ *out of the set* Υ,

$$W(\mathbf{z};\Upsilon) > 0, \forall \mathbf{z} \in \mathfrak{N}(\Upsilon) \setminus \Upsilon \tag{20.7}$$

3. $W(.;\Upsilon)$ *vanishes only on the boundary* $\partial\boldsymbol{\Upsilon}$ *of the set* $\boldsymbol{\Upsilon}$,

$$W(\mathbf{z};\Upsilon) = 0 \Longleftrightarrow \mathbf{z} \in \partial\boldsymbol{\Upsilon} \tag{20.8}$$

4. *There is* $\mathbf{z}^* \in \Upsilon$ *such that the minimal value of the function* $W(.;\Upsilon)$ *is at* $\mathbf{z}^*$,

$$min \left[W(\mathbf{z};\Upsilon) : \ \mathbf{z} \in \mathfrak{R}^K\right] = W(\mathbf{z}^*;\Upsilon). \tag{20.9}$$

b) ***Globally positive definite relative to*** Υ *if, and only if, a) holds for* $\mathfrak{N}(\Upsilon) = \mathfrak{R}^K$.

c) ***(Globally) Negative definite relative to*** Υ *if, and only if,* $-W(.;\Upsilon)$ *is (globally) positive definite, respectively.*

d) ***Positive definite on a set*** $\mathfrak{A}$ ***relative to*** Υ *if, and only if,* $\mathfrak{A}$ *is a neighborhood of* $\boldsymbol{\Upsilon}$ *such that the conditions under a) hold for* $\mathfrak{N}(\Upsilon) = \mathfrak{A}$,

f) ***Negative definite on a set*** $\mathfrak{A}$ ***relative to*** Υ *if, and only if,* $-W(.;\Upsilon)$ *is positive definite on the set* $\mathfrak{A}$ *relative to* Υ.

*The expression "**relative to** Υ" is to be omitted if, and only if,* $\Upsilon = \mathfrak{O} = \{\mathbf{0}_K\}$.

By following Hahn [292] we state the following.

Proposition 266 ***Comparison functions and time-invariant positive definite functions relative to*** Υ

Necessary and sufficient for a time-invariant function $W(.;\Upsilon) : \mathfrak{R}^K \longrightarrow \mathfrak{R}$ to be positive definite relative to Υ on a connected neighborhood $\mathfrak{N}(\Upsilon)$ of Υ is the existence of comparison functions $\varphi_i(.;\Upsilon) \in \mathcal{K}_{[0,\alpha[}$, $i = 1,2$, $\alpha = sup\ \{\rho(\mathbf{z},\Upsilon): \mathbf{z} \in \mathfrak{N}(\Upsilon)\}$, such that both $W(\mathbf{z};\Upsilon) \in \mathfrak{C}[\mathfrak{N}(\Upsilon)]$ and

$$\varphi_2[\rho(\mathbf{z},\Upsilon);\Upsilon] \geq W(\mathbf{z};\Upsilon) \geq \varphi_1[\rho(\mathbf{z},\Upsilon);\Upsilon], \ \forall \mathbf{z} \in \mathfrak{N}(\Upsilon) \setminus In\Upsilon. \tag{20.10}$$

Definition 265 enables us to define *time*-varying sign definite functions.

Definition 267 ***Time-varying definite function relative to*** Υ ***on*** $\mathfrak{T}_\tau \subseteq \mathfrak{T}$

A function $V(.;\Upsilon): \mathfrak{T}_\tau \times \mathfrak{R}^K \longrightarrow \mathfrak{R}$ *is:*

a) ***Positive definite relative to*** Υ ***on*** $\mathfrak{T}_\tau$ *if, and only if, there are a time-invariant connected neighborhood* $\mathfrak{N}(\Upsilon)$ *of* $\boldsymbol{\Upsilon}$ *and a time-invariant positive definite function* $W(.;\Upsilon)$ *on* $\mathfrak{N}(\Upsilon)$ *such that they obey 1) through 4):*

1) $V(.;\Upsilon)$ *is defined and continuous on* $\mathfrak{T}_\tau \times \mathfrak{N}(\Upsilon)$,

$$V(t,\mathbf{z};\Upsilon) \in \mathfrak{C}\left[\mathfrak{T}_\tau \times \mathfrak{N}(\Upsilon)\right] \tag{20.11}$$

2) $V(.;\Upsilon)$ *satisfies the following:*

$$V(t,\mathbf{z};\Upsilon) \geq W(\mathbf{z};\Upsilon), \forall (t,\mathbf{z}) \in \mathfrak{T}_\tau \times \mathfrak{N}(\Upsilon) \tag{20.12}$$

3) $V(.;\Upsilon)$ *vanishes only on the boundary* $\partial\Upsilon$ *of* Υ,

$$V(t,\mathbf{z};\Upsilon) = 0 \Longleftrightarrow \mathbf{z} \in \partial\Upsilon, \ \forall t \in \mathfrak{T}_\tau. \tag{20.13}$$

4) *There is pair* $(t^*,\mathbf{z}^*) \in \mathfrak{T}_\tau \times \Upsilon$ *such that the minimal value of the function* $V(.;\Upsilon)$ *on* $\mathfrak{T}_\tau \times \mathfrak{R}^K$ *is at* $(t^*,\mathbf{z}^*) \in \mathfrak{T}_\tau \times \Upsilon$,

$$\begin{gathered} \exists (t^*,\mathbf{z}^*) \in \mathfrak{T}_\tau \times \Upsilon \Longrightarrow \\ min \left[V(t,\mathbf{z};\Upsilon): \ (t,\mathbf{z}) \in \mathfrak{T}_\tau \times \mathfrak{R}^K\right] = V(t^*,\mathbf{z}^*;\Upsilon). \end{gathered} \tag{20.14}$$

b) ***Globally positive definite relative to*** Υ ***on*** $\mathfrak{T}_\tau$ *if, and only if, the function* $W(.;\Upsilon)$ *is globally positive definite and the conditions under a) hold for* $\mathfrak{N} = \mathfrak{R}^K$.

c) ***(Globally) Negative definite relative to*** Υ ***on*** $\mathfrak{T}_\tau$ *if, and only if,* $-V(.;\Upsilon)$ *is (globally) positive definite.*

d) ***Positive definite relative to*** Υ ***on*** $\mathfrak{T}_\tau$ ***and on a set*** $\mathfrak{A}$ *if, and only if,* $\mathfrak{A}$ *is a neighborhood of* $\mathbf{z} = \mathbf{0}_K$ *such that the conditions under a) hold for* $\mathfrak{N} = \mathfrak{A}$,

f) ***Negative definite relative to*** Υ ***on*** $\mathfrak{T}_\tau$ ***and on a set*** $\mathfrak{A}$ *if, and only if,* $-V(.;\Upsilon)$ *is positive definite on* $\mathfrak{T}_\tau$ *and on the set* $\mathfrak{A}$.

The expression " ***on*** $\mathfrak{T}_\tau$ *" is to be omitted if, and only if,* $\mathfrak{T}_\tau = \mathfrak{T}$.

*The expression "****relative to*** Υ *" is to be omitted if, and only if,* $\Upsilon = \mathfrak{O} = \{\mathbf{0}_K\}$.

Note 268 *Continuity of a function is its crucial obligatory property to be definite. However, it need not be differentiable in order to be definite.*

Proposition 266 and Definition 267 imply:

Proposition 269 ***Comparison functions and time-varying positive definite functions relative to*** Υ ***on*** $\mathfrak{T}_\tau \times \mathfrak{N}(\Upsilon)$

Necessary and sufficient for a time-varying function $V(.;\Upsilon): \mathfrak{T}_\tau \times \mathfrak{R}^K \longrightarrow \mathfrak{R}$ *to be positive definite relative to* Υ *on* $\mathfrak{T}_\tau \times \mathfrak{N}(\Upsilon)$, *where* $\mathfrak{N}(\Upsilon)$ *is a time-invariant connected neighborhood of* Υ, *is that (i) to (iv) hold:*

(i) $V(.;\Upsilon)$ *is defined and continuous on* $\mathfrak{T}_\tau \times \mathfrak{N}(\Upsilon)$,

$$V(t,\mathbf{z};\Upsilon) \in \mathfrak{C}\left[\mathfrak{T}_\tau \times \mathfrak{N}(\Upsilon)\right], \tag{20.15}$$

(ii) $V(.;\Upsilon)$ *vanishes only on the boundary* $\partial\Upsilon$ *of* Υ,

$$V(t,\mathbf{z};\Upsilon) = 0 \Longleftrightarrow \mathbf{z} \in \partial\Upsilon, \ \forall t \in \mathfrak{T}_\tau. \tag{20.16}$$

(iii) *There are* $(t^*,\mathbf{z}^*)\in\mathfrak{T}_\tau\times\Upsilon$ *such that the minimal value of the function* $V(.;\Upsilon)$ *on* $\mathfrak{T}_\tau\times\mathfrak{R}^K$ *is at* $(t^*,\mathbf{z}^*)\in\mathfrak{T}_\tau\times\Upsilon$,

$$\exists\,(t^*,\mathbf{z}^*)\in\mathfrak{T}_\tau\times\Upsilon\Longrightarrow$$
$$min\ \left[V(t,\mathbf{z};\Upsilon):\ (t,\mathbf{z})\in\mathfrak{T}_\tau\times\mathfrak{R}^K\right]=V(t^*,\mathbf{z}^*;\Upsilon). \tag{20.17}$$

(iv) *There exists a comparison function* $\varphi(.;\Upsilon)\in\mathcal{K}_{[0,\alpha[}$,

$$\alpha=sup\ \{\rho(\mathbf{z},\Upsilon):\ \mathbf{z}\in\mathfrak{N}(\Upsilon)\},$$

such that

$$V(t,\mathbf{z};\Upsilon)\geq\varphi[\rho(\mathbf{z},\Upsilon);\Upsilon],\ \forall(t,\mathbf{z})\in\mathfrak{T}_\tau\times\mathfrak{N}(\Upsilon)\setminus In\Upsilon. \tag{20.18}$$

20.4 Decrescent functions

The following functional property is indispensable for the function $V(.;\Upsilon)$ to assure that the associated set $\mathfrak{V}_\xi(t;\tau;\Upsilon)$ is not asymptotically contractive.

Definition 270 ***Time-varying decrescent function relative to*** Υ ***on*** $\mathfrak{T}_\tau$
A function $V(.;\Upsilon):\mathfrak{T}_\tau\times\mathfrak{R}^K\longrightarrow\mathfrak{R}$ *is* ***decrescent relative to*** Υ ***on*** $\mathfrak{T}_\tau$ *if, and only if,*
(i) $V(.;\Upsilon)$ *is defined and continuous on* $\mathfrak{T}_\tau\times\mathfrak{N}(\Upsilon)$,

$$V(t,\mathbf{z};\Upsilon)\in\mathfrak{C}[\mathfrak{T}_\tau\times\mathfrak{N}(\Upsilon)], \tag{20.19}$$

and
(ii) There are a time-invariant connected neighborhood $\mathfrak{N}(\Upsilon)$ *of* $\mathbf{z}=\mathbf{0}_K$ *and a time-invariant positive definite function* $W(.;\Upsilon)$ *relative to* Υ *on* $\mathfrak{N}(\Upsilon)$ *such that*

$$W(\mathbf{z};\Upsilon)\geq V(t,\mathbf{z};\Upsilon),\ \forall(t,\mathbf{z})\in\mathfrak{T}_\tau\times\mathfrak{N}(\mathrm{u}). \tag{20.20}$$

It is ***globally decrescent relative to*** Υ ***on*** $\mathfrak{T}_\tau$ *if, and only if, the function* $W(.;\Upsilon)$ *is globally positive definite relative to* Υ *and* $\mathfrak{N}=\mathfrak{R}^K$.
*The expression "**on** $\mathfrak{T}_\tau$" is to be omitted if, and only if,* $\mathfrak{T}_\tau=\mathfrak{T}$.
*The expression "**relative to** Υ" is to be omitted if, and only if,* $\Upsilon=\mathfrak{O}=\{\mathbf{0}_K\}$.

Proposition 266 and Definition 270 imply:

Proposition 271 ***Time-varying decrescent functions relative to*** Υ ***on*** $\mathfrak{T}_\tau$ ***and comparison functions***
Necessary and sufficient for a time-varying function $V(.;\Upsilon):\mathfrak{T}_\tau\times\mathfrak{R}^K\longrightarrow\mathfrak{R}$ *to be* ***decrescent relative to*** Υ ***on*** $\mathfrak{T}_\tau\times\mathfrak{N}(\Upsilon)$, *where* $\mathfrak{N}(\Upsilon)$ *is a time-invariant connected neighborhood of* Υ, *is that both (i) and (ii) hold:*
(i) $V(.;\Upsilon)$ *is defined and continuous on* $\mathfrak{T}_\tau\times\mathfrak{N}(\Upsilon)$,

$$V(t,\mathbf{z};\Upsilon)\in\mathfrak{C}[\mathfrak{T}_\tau\times\mathfrak{N}(\Upsilon)], \tag{20.21}$$

(ii) *There exists a comparison function* $\varphi(.)\in\mathcal{K}_{[0,\alpha[}$,

$$\alpha=sup\ \{\rho(\mathbf{z},\Upsilon):\ \mathbf{z}\in\mathfrak{N}(\Upsilon)\},$$

such that

$$\varphi[\rho(\mathbf{z},\Upsilon)]\geq V(t,\mathbf{z};\Upsilon),\ \forall(t,\mathbf{z})\in\mathfrak{T}_\tau\times\mathfrak{N}(\Upsilon). \tag{20.22}$$

The following functional properties are also useful. They simplify stability tests.

Definition 272 ***Time-varying radially unbounded function relative to Υ on $\mathfrak{T}_\tau$***
A function $V(.;\Upsilon):\mathfrak{T}_\tau\times\mathfrak{R}^K\longrightarrow\mathfrak{R}$ is ***radially unbounded relative to Υ on $\mathfrak{T}_\tau$*** *if, and only if, it diverges to infinity as $\rho(\mathbf{z},\Upsilon)$ diverges to infinity at every instant $t\in\mathfrak{T}_\tau$,*

$$\rho(\mathbf{z},\Upsilon)\longrightarrow\infty\Longrightarrow V(t,\mathbf{z};\Upsilon)\longrightarrow\infty,\ \ \forall t\in\mathfrak{T}_\tau.$$

*The expression "**on $\mathfrak{T}_\tau$**" is to be omitted if, and only if, $\mathfrak{T}_\tau=\mathfrak{T}$.*
*The expression "**relative to Υ**" is to be omitted if, and only if, $\Upsilon=\mathfrak{O}=\{\mathbf{0}_K\}$.*

Example 273 *The function $V(t,\mathbf{z})=\left(1+t^2\right)\left(2+t^2\right)\rho(\mathbf{z},\Upsilon)$ is radially unbounded,*

$$\rho(\mathbf{z},\Upsilon)\longrightarrow\infty\Longrightarrow V(t,\mathbf{z};\Upsilon)=\frac{1+t^2}{2+t^2}\rho(\mathbf{z},\Upsilon)\longrightarrow\infty,\ \ \forall t\in\mathfrak{T}.$$

However, globally positive definite functions defined by

$$V_1(t,\mathbf{z};\Upsilon)=\frac{1+t^2}{2+t^2}\frac{\rho(\mathbf{z},\Upsilon)}{1+\rho(\mathbf{z},\Upsilon)},\ \ V_2(t,\mathbf{z};\Upsilon)=\left(1+t^2\right)\frac{\rho(\mathbf{z},\Upsilon)}{1+\rho(\mathbf{z},\Upsilon)}$$

are not radially unbounded,

$$\rho(\mathbf{z},\Upsilon)\longrightarrow\infty\Longrightarrow V_1(t,\mathbf{z};\Upsilon)=\frac{1+t^2}{2+t^2}\frac{\rho(\mathbf{z},\Upsilon)}{1+\rho(\mathbf{z},\Upsilon)}\longrightarrow\frac{1+t^2}{2+t^2}<\infty,$$
$$\rho(\mathbf{z},\Upsilon)\longrightarrow\infty\Longrightarrow V_2(t,\mathbf{z};\Upsilon)=\left(1+t^2\right)\frac{\rho(\mathbf{z},\Upsilon)}{1+\rho(\mathbf{z},\Upsilon)}\longrightarrow\left(1+t^2\right)<\infty,$$
$$\forall t\in\mathfrak{T}.$$

Definition 274 ***Time-varying radially increasing function relative to Υ on $\mathfrak{T}_\tau$***
A function $V(.;\Upsilon):\mathfrak{T}_\tau\times\mathfrak{R}^K\longrightarrow\mathfrak{R}$ is ***radially increasing relative to Υ on $\mathfrak{T}_\tau$*** *if, and only if,*

$$\begin{array}{rcl}\forall\lambda_i & \in & \mathfrak{R}^+,\ i=1,2,\ \lambda_2>\lambda_1\Longrightarrow\\ V(t,\lambda_2\mathbf{z};\Upsilon) & > & V(t,\lambda_1\mathbf{z};\Upsilon),\ \forall(t,\mathbf{z})\in\mathfrak{T}_\tau\times\mathfrak{R}^K\backslash Cl\Upsilon.\end{array}\tag{20.23}$$

*The expression "**on $\mathfrak{T}_\tau$**" is to be omitted if, and only if, $\mathfrak{T}_\tau=\mathfrak{T}$.*
*The expression "**relative to Υ**" is to be omitted if, and only if, $\Upsilon=\mathfrak{O}=\{\mathbf{0}_K\}$.*

Example 275 *The globally positive definite functions defined by*

$$V_1(t,\mathbf{z};\Upsilon)=\frac{1+t^2}{2+t^2}\left(1-e^{-\rho(\mathbf{z},\Upsilon)}\right),$$
$$V_2(t,\mathbf{z};\Upsilon)=\left(1+t^2\right)\left(1-e^{-\rho(\mathbf{z},\Upsilon)}\right),$$
$$V_3(t,\mathbf{z};\Upsilon)=\left(1+t^2\right)\rho(\mathbf{z},\Upsilon)\left(1+e^{\rho(\mathbf{z},\Upsilon)}\right),$$

are radially increasing relative to Υ on $\mathfrak{T}_\tau$, but $V_1(t,\mathbf{z};\Upsilon)$ and $V_2(t,\mathbf{z};\Upsilon)$ are not radially unbounded, whereas $V_3(t,\mathbf{z};\Upsilon)$ is both radially increasing relative to Υ on $\mathfrak{T}_\tau$ and radially unbounded relative to Υ on $\mathfrak{T}_\tau$. However, the globally positive definite functions defined by

$$\begin{array}{rcl}V_4(t,\mathbf{z};\Upsilon) & = & \dfrac{1+t^2}{2+t^2}\dfrac{\rho(\mathbf{z},\Upsilon)}{1+\rho^2(\mathbf{z},\Upsilon)},\\ V_5(t,\mathbf{z};\Upsilon) & = & \left(1+t^2\right)\dfrac{\rho(\mathbf{z},\Upsilon)}{1+\rho^2(\mathbf{z},\Upsilon)},\\ V_6(t,\mathbf{z};\Upsilon) & = & \left(1+t^2\right)\dfrac{\left(1-e^{-\rho(\mathbf{z},\Upsilon)}\right)}{1+e^{\rho(\mathbf{z},\Upsilon)}},\end{array}$$

are neither radially increasing relative to Υ on $\mathfrak{T}_\tau$ nor radially unbounded relative to Υ on $\mathfrak{T}_\tau$.

The preceding definitions permit us to present fundamental Lyapunov results for the *ISO* nonlinear systems (3.65), (3.66). They have not been directly extended either to the *IO* nonlinear systems (3.62) or to the *IIDO* systems (3.12). The book [252] contains Lyapunov's original results for *time*-invariant *ISO* linear systems and their extensions to two other classes of *time*-invariant linear systems.

20.5 *Time*-invariant vector definite functions

We recall the unified notation ():

$$\mathbf{Z} \in \left\{\mathbf{Y}, \mathbf{Y}^k, \mathbf{R}^{\alpha-1}\right\},\ \mathbf{Z} = [Z_1\ Z_2\ ...Z_K]^T \in \mathfrak{R}^K,$$
$$k \in \{0,1,...,\alpha-1\}, K \in \{N, (k+1)N, \alpha\rho\},$$
$$\mathbf{z} = \mathbf{Z} - \mathbf{Z}_d,\ \mathbf{z} \in \left\{\mathbf{y}, \mathbf{y}^k, \mathbf{r}^{\alpha-1}\right\},\ \mathbf{z} \in \mathfrak{R}^K, \tag{20.24}$$

and note that

$$\mathbf{z}_l^1 = \left[z_l\ \ z_l^{(1)}\right]^T \in \mathfrak{R}^2,\ \ l \in \{0,1,...,K\}. \tag{20.25}$$

All vector and matrix equalities, inequalities, and powers hold elementwise. We first generalize Lyapunov's concept of scalar (semi)definite functions to *time*-invariant vector (semi)definite functions relative to a set $\mathfrak{S}$ in general. The simplest case of the set $\mathfrak{S}$ is the singleton $\mathfrak{O} = \{\mathbf{0}_K\}$. It contains only the origin $\mathbf{0}_K$ of the space $\mathfrak{R}^K$, which can be an equilibrium vector. However, if we wish to study stability properties of another equilibrium vector $\mathbf{z}_e$ then $\mathfrak{S}$ is the singleton containing only $\mathbf{z}_e$, $\mathfrak{S} = \{\mathbf{z}_e\}$.

Let ϕ be the empty set and $\mathfrak{S}$ be a *time*-invariant connected compact nonempty set,

$$\mathfrak{S} \subseteq \mathfrak{R}^K,\ \ \text{Cl}\mathfrak{S} = \mathfrak{S} \neq \phi.$$

Let $\mathfrak{N}(\mathfrak{S})$ be a connected neighborhood of the set $\mathfrak{S}$, $\mathfrak{N}(\mathfrak{S}) \subseteq \mathfrak{R}^K$.

Definition 276 ***Definition of time-invariant vector semidefinite functions***

A vector function $W(.) : \mathfrak{R}^K \to \mathfrak{R}^P$, $W(\mathbf{z}) = [W_1(\mathbf{z})\ W_2(\mathbf{z})\ ...\ W_P(\mathbf{z})]^T$, $W_i(.) : \mathfrak{R}^K \longrightarrow \mathfrak{R}$, $P \in \{1,2,3,...,K\}$, *is*

a) ***Positive (negative) semidefinite relative to the set*** $\mathfrak{S}$ *if, and only if, there is a neighborhood* $\mathfrak{N}(\mathfrak{S})$ *of* $\mathfrak{S}$ *such that (i) through (iv) hold, respectively:*

(i) $W(.)$ *is defined and continuous on* $\mathfrak{N}(\mathfrak{S})$: $W(\mathbf{z}) \in \mathfrak{C}[\mathfrak{N}(\mathfrak{S})]$

(ii) $W(\mathbf{z}) \geq \mathbf{0}_K$, $(\mathbf{W}(\mathbf{z}) \leq \mathbf{0}_K)$, $\forall \mathbf{z} \in [\mathfrak{N}(\mathfrak{S}) \backslash In\mathfrak{S}]$

(iii) $\exists \mathbf{z} \in [\mathfrak{N}(\mathfrak{S}) \backslash Cl\mathfrak{S}]$ *such that* $W(\mathbf{z}) > \mathbf{0}_K$, $(\mathbf{W}(\mathbf{z}) < \mathbf{0}_K)$

(iv) $W(\mathbf{z}) \leq \mathbf{0}_K$, $(\mathbf{W}(\mathbf{z}) \geq \mathbf{0}_K)$, $\forall \mathbf{z} \in Cl\mathfrak{S}$.

b) ***Elementwise positive (negative) semidefinite relative to the set*** $\mathfrak{S}$ *if, and only if, a) is valid, respectively, and both (b-i) and (b-ii) hold,*

(b-i) $P = K$

(b-ii) $W_l(\mathbf{z}) \equiv W_l(z_l), \forall l = 1, 2, \cdots, K$.

The preceding properties are:

d) ***Global positive (negative) semidefinite relative to the set*** $\mathfrak{S}$ *if, and only if, they hold for* $\mathfrak{N}(\mathfrak{S}) = \mathfrak{R}^K$.

e) ***On a set*** $\mathfrak{A}$ ***relative to the set*** $\mathfrak{S}$, $\mathfrak{A} \subseteq \mathfrak{R}^K$, *if, and only if, the set* $\mathfrak{A}$ *is a neighborhood of the set* $\mathfrak{S}$ *and* $\mathfrak{A} \subseteq \mathfrak{N}_L(\mathfrak{S})$, *where* $\mathfrak{N}_L(\mathfrak{S})$ *is the largest connected neighborhood* $\mathfrak{N}(\mathfrak{S})$ *of* $\mathfrak{S}$, *which obeys the corresponding conditions under a) or b).*

*The expression "**relative to the set** $\mathfrak{S}$" should be omitted if, and only if, the set $\mathfrak{S}$ is the singleton $\mathfrak{O}$, $\mathfrak{S} = \mathfrak{O} = \{\mathbf{0}_K\}$.*

Note 277 *If $P = 1$ then a) of this definition reduces to the definition of a scalar time-invariant semidefinite function.*

Example 278 *Let the set $\mathfrak{S} = \{\mathbf{z} : |\mathbf{z}| \leq \mathbf{a}, \mathbf{a} \geq \mathbf{0}_K\}$. The vector function $W(.)$,*

$$\mathbf{W}(\mathbf{z}) = \left\{ \begin{array}{c} (Z - A)(\mathbf{z} - \mathbf{a}), \ \mathbf{z} \in \mathfrak{R}^K \backslash \mathfrak{S}, \\ -(Z - A)(\mathbf{z} - \mathbf{a}), \ \mathbf{z} \in \mathfrak{S} \end{array} \right\},$$
$$Z = diag\ \{z_1\ \ z_2 \cdots\ z_K\},$$
$$\mathbf{a} = [a_1\ \ a_2 \cdots\ a_K]^T,$$
$$A = diag\ \{a_1\ \ a_2 \cdots\ a_K\},$$

is globally positive semidefinite relative to the set $\mathfrak{S} = \{\mathbf{z} : |\mathbf{z}| \leq \mathbf{a}\}$. If $\mathbf{a} = \mathbf{0}_K$ then $\mathfrak{S} = \mathfrak{O}$ and $W(\mathbf{z})$ is globally positive semidefinite.

Definition 279 ***Definition of time-invariant vector pairwise semidefinite functions***

A vector function $W(.) : \mathfrak{R}^{2K} \to \mathfrak{R}^K$, $W(\mathbf{z}^1) = [W_1(\mathbf{z}^1)\ \ W_2(\mathbf{z}^1) \cdots W_K(\mathbf{z}^1)]^T$, $W_i(.) : \mathfrak{R}^{2K} \longrightarrow \mathfrak{R}$, is ***pairwise positive (negative) semidefinite relative to the set*** *$\mathfrak{S}$ if, and only if, it is positive (negative) semidefinite relative to the set $\mathfrak{S}$ and*
$W_l(\mathbf{z}^1) \equiv W_l(\mathbf{z}_l^1)$, $\forall l = 1, 2, \cdots, K$.

Let
$$K_{ij} = j - (i-1).$$

Let $\mathbf{z}_{ij} \in \mathfrak{R}^K$,

$$\mathbf{z}_{ij} = [z_1\ \ z_2 \ldots z_{i-1}\ \widetilde{z}_i\ \ \widetilde{z}_{i+1} \cdots\ \widetilde{z}_j\ \ z_{j+1} \cdots\ z_K]^T \ \in \mathfrak{R}^K,$$

and the set $\mathfrak{S}_{ij}$,

$$\mathfrak{S}_{ij} = \left\{ \widetilde{\mathbf{z}}_{ij} : \widetilde{\mathbf{z}}_{ij} = [\widetilde{z}_i\ \ \widetilde{z}_{i+1} \cdots\ \widetilde{z}_j]^T \ \right\} \subseteq \mathfrak{R}^{K_{ij}},$$

be such that

$$\widetilde{\mathbf{z}}_{ij} \in \mathfrak{S}_{ij} \Longrightarrow \exists \mathbf{z}^1 = \left[\mathbf{z}_{ij}^T\ \ \mathbf{z}^{(1)T} \right]^T \ \in \mathfrak{S}.$$

In addition to an arbitrary form $\mathfrak{S}_a$ of the set $\mathfrak{S}$, $\mathfrak{S} = \mathfrak{S}_a$, the set $\mathfrak{S}$ can have any of the following special forms of the Cartesian product sets $\mathfrak{S}_b$, $\mathfrak{S}_c$, $\mathfrak{S}_d$,

$$\mathfrak{S}_b = \mathfrak{S}_{i_1 k_1} \times \mathfrak{S}_{i_2 k_2} \times \cdots \times \mathfrak{S}_{i_P k_P} \subset \mathfrak{R}^K,$$
$$\mathfrak{S}_c = \mathfrak{S}_{11} \times \mathfrak{S}_{22} \times \ \cdots \ \times \mathfrak{S}_{KK} \subset \mathfrak{R}^K,$$
$$\mathfrak{S} \in \{\mathfrak{S}_a, \mathfrak{S}_b, \mathfrak{S}_c\}.$$

Definition 280 ***Definition of time-invariant vector definite functions***

A vector function $V(.) : \mathfrak{R}^K \to \mathfrak{R}^P$, $V(\mathbf{z}) = [V_1(\mathbf{z})\ \ V_2(\mathbf{z})\ \ \ldots V_P(\mathbf{z})]^T$, $V_i(.) : \mathfrak{R}^K \longrightarrow \mathfrak{R}$, $\forall i = 1, 2, \cdots, P$, $P \in \{1, 2, 3, \cdots, K\}$, is

a) ***Positive (negative) definite relative to the set*** *$\mathfrak{S}$ if, and only if, there is a neighborhood $\mathfrak{N}(\mathfrak{S})$ of $\mathfrak{S}$ such that (i) through (iv) hold, respectively:*

(i) $V(.)$ is defined and continuous on $\mathfrak{N}(\mathfrak{S})$: $V(\varepsilon) \in \mathfrak{C}[\mathfrak{N}(\mathfrak{S})]$.

(ii) $V(\mathbf{z}) \geq \mathbf{0}_K$, $(\mathbf{V}(\mathbf{z}) \leq \mathbf{0}_K)$, $\forall \mathbf{z} \in [\mathfrak{N}(\mathfrak{S}) \backslash In\mathfrak{S}]$.

(iii) $V(\mathbf{z}) = \mathbf{0}_K$ for $\mathbf{z} \in [\mathfrak{N}(\mathfrak{S}) \backslash In\mathfrak{S}]$ if, and only if, $\mathbf{z} \in \partial\mathfrak{S}$.

(iv) $V(\mathbf{z}) \leq \mathbf{0}_K$, $(\mathbf{V}(\mathbf{z}) \geq \mathbf{0}_K)$, $\forall \mathbf{z} \in Cl\mathfrak{S}$.

b) ***Positive (negative) definite relative to the set*** *$\mathfrak{S}_b$ if, and only if, (i), (ii) and (iv) of (a) are valid for $\mathfrak{S} = \mathfrak{S}_b$ together with the following:*

(i) $V_l(\mathbf{z}) \equiv V_l(\mathbf{z}_{i_l k_l})$.

(ii) $V_l(\mathbf{z}_{i_l k_l}) \equiv \mathbf{0}_K$ *for* $\mathbf{z}_{i_l k_l} \in [\mathfrak{N}(\mathfrak{S}) \backslash In\mathfrak{S}]$ *if, and only if, the corresponding* $\widetilde{\mathbf{z}}_{i_l k_l} \in \partial\mathfrak{S}_{i_l k_l}$, $\forall l = 1, 2, ..., P$.

c) ***Elementwise positive (negative) definite relative to the set*** $\mathfrak{S}_d$ *if, and only if, (i), (ii), and (iv) of a) are valid for* $\mathfrak{S} = \mathfrak{S}_c$*, respectively, and (i) through (iii) hold,*

(i) $P = K$.

(ii) $V_l(\mathbf{z}) \equiv V_l(z_l)$, $\forall l = 1, 2, \cdots, K$.

(iii) $V_l(\widetilde{z}_l) \equiv 0$ *for* $\mathbf{z}_{ll} \in [\mathfrak{N}(\mathfrak{S}_d) \backslash In\mathfrak{S}_d]$ *if, and only if* $\widetilde{z}_l \in \partial\mathfrak{S}_{ll}$, $\forall l = 1, 2, \cdots, K$.

e) ***Radially strictly increasing on*** $\mathfrak{N}(\mathfrak{S}) \backslash Cl\mathfrak{S}$ *if, and only if,*

(v) $V(\lambda_1 \mathbf{z}) < V(\lambda_2 \mathbf{z})$, $0 < \lambda_1 < \lambda_2 < \infty$, $\forall \mathbf{z} \in \mathfrak{N}(\mathfrak{S}) \backslash Cl\mathfrak{S}$.

The preceding properties are:

f) ***Global*** *if, and only if, they hold for* $\mathfrak{N}(\mathfrak{S}_{(.)}) = \mathfrak{R}^K$.

g) ***On a set*** $\mathfrak{A}$ ***relative to the set*** $\mathfrak{S}_{(.)}$, $\mathfrak{A} \subseteq \mathfrak{R}^K$, *if, and only if, the set* $\mathfrak{A}$ *is a connected neighborhood of the set* $\mathfrak{S}_{(.)}$ *and* $\mathfrak{A} \subseteq \mathfrak{N}_L(\mathfrak{S}_{(.)})$, *where* $\mathfrak{N}_L(\mathfrak{S}_{(.)})$ *is the largest connected neighborhood* $\mathfrak{N}(\mathfrak{S}_{(.)})$ *of* $\mathfrak{S}_{(.)}$, *which obeys the corresponding above conditions.*

h) *A vector function* $V(.) : \mathfrak{R}^K \to \mathfrak{R}^P$, *is* ***radially unbounded*** *if, and only if, the corresponding above property is global and*

(i) $V(\lambda \mathbf{z}) \longrightarrow \infty \mathbf{1}_P$ *as* $\lambda \longrightarrow \infty$, $\forall \mathbf{z} \in \mathfrak{R}^K$.

*The expression "**relative to the set** $\mathfrak{S}_{(.)}$" should be omitted if, and only if, the set* $\mathfrak{S}_{(.)}$ *is the singleton* $\mathfrak{O}$, $\mathfrak{S}_{(.)} = \mathfrak{O} = \{\mathbf{0}_K\}$.

Note 281 *The conditions (i) through (iii) under a) do not imply positive definiteness of any entry* $V_i(.) : \mathfrak{R}^K \longrightarrow \mathfrak{R}_+$ *of* $V(.) : \mathfrak{R}^K \longrightarrow \mathfrak{R}^P$ *because* $V_i(.)$ *is defined on* $\mathfrak{R}^K$ *in general, but it is permitted to vanish only for* $z_i = 0$. *For example,* $V_i(\mathbf{z}) = z_i^2 \exp(\|\mathbf{z}\|)$. *However, they imply positive semi-definiteness on* $\mathfrak{S}$ *of every entry* $V_i(.)$ *of* $V(.)$.

Note 282 *Definition 280 is compatible with Lyapunov's original definition of scalar definite functions [401]. The former reduces to the latter for* $P = 1$.

Definition 280 is also compatible with the concept of matrix definite functions introduced in [178].

Example 283 *Vector functions* $V(.)$ *defined by:*

$$\mathbf{V}(\mathbf{z}) = \frac{1}{2} Z\mathbf{z}, \quad \mathbf{V}(\mathbf{z}) = |\mathbf{z}|,$$

are globally positive definite, radially unbounded and radially strictly increasing.

Vector functions $V(.)$ *defined by:*

$$\mathbf{V}(\mathbf{z}) = \frac{1}{2} |Z| (Z - A)(\mathbf{z} - \mathbf{a}), \quad \mathbf{V}(\mathbf{z}) = (Z - A)|\mathbf{z}|,$$

are positive definite on the set $\mathfrak{A} = \{\mathbf{z} : |\mathbf{z}| < |\mathbf{a}|\}$.

Definition 284 ***Time-invariant vector pairwise definite functions***

A vector function $V(.) : \mathfrak{R}^{2K} \to \mathfrak{R}^K$, $V(\mathbf{z}^1) = [V_1(\mathbf{z}^1) \;\; V_2(\mathbf{z}^1) \;\; ... \;\; V_K(\mathbf{z}^1)]^T$, $V_l(.) : \mathfrak{R}^{2K} \longrightarrow \mathfrak{R}$, *is* ***pairwise positive (negative) definite relative to the set*** $\mathfrak{S} \times \mathfrak{S}$ *if, and only if, it is positive (negative) definite relative to the set* $\mathfrak{S} \times \mathfrak{S}$ *and* $V_l(\mathbf{z}^1) \equiv V_l(\mathbf{z}_l^1)$, $\forall l = 1, 2, \cdots, K$.

20.6 *Time*-varying vector definite functions

Definition 285 ***Definition of time-varying vector semidefinite functions***

A vector function $W(.) : \mathfrak{T}_0 \times \mathfrak{R}^K \to \mathfrak{R}^P$, $W(t, \mathbf{z}) = [W_1(t, \mathbf{z}) \;\; W_2(t, \mathbf{z}) \;\; ... \;\; W_P(t, \mathbf{z})]^T$, $W_i(.) : \mathfrak{T}_0 \times \mathfrak{R}^K \longrightarrow \mathfrak{R}$, $\forall i = 1, 2, \cdots, P$, $P \in \{1, 2, 3, \cdots, K\}$, *is a)* ***Positive (negative) semidefinite relative to*** $\mathfrak{T}_0 \times \mathfrak{S}$ *if, and only if, there is a neighborhood* $\mathfrak{N}(\mathfrak{S})$ *of* $\mathfrak{S}$ *such that (i) through (iv) hold, respectively:*

(i) $W(.)$ *is defined and continuous on* $\mathfrak{T}_0 \times \mathfrak{N}(\mathfrak{S})$*:*

$$\mathbf{W}(t,\mathbf{z}) \in \mathfrak{C}[\mathfrak{T}_0 \times \mathfrak{N}(\mathfrak{S})].$$

(ii) $W(t,\mathbf{z}) \geq \mathbf{0}_K$, $(\mathbf{W}(t,\mathbf{z}) \leq \mathbf{0}_K)$, $\forall (t,\mathbf{z}) \in \mathfrak{T}_0 \times [\mathfrak{N}(\mathfrak{S}) \setminus In\mathfrak{S}]$.

(iii) $\forall t \in \mathfrak{T}_0$, $\exists \mathbf{z} \in [\mathfrak{N}(\mathfrak{S}) \setminus Cl\mathfrak{S}]$ *such that* $W(t,\mathbf{z}) > \mathbf{0}_K, (\mathbf{W}(t,\mathbf{z}) < \mathbf{0}_K)$.

(iv) $W(t,\mathbf{z}) \leq \mathbf{0}_K$, $[\mathbf{W}(t,\mathbf{z}) \geq \mathbf{0}_K]$, $\forall (t,\mathbf{z}) \in \mathfrak{T}_0 \times Cl\mathfrak{S}$.

b) ***Elementwise positive (negative) semidefinite relative to*** $\mathfrak{T}_0 \times \mathfrak{S}$ *if, and only if, a) is valid, respectively, and (b-i) and (b-ii) hold,*

(i) $P = K$.

(ii) $W_l(t,\mathbf{z}) \equiv W_l(t,z_l), \forall l = 1,2,...,K$.

The preceding properties are:

d) ***Global positive (negative) semidefinite relative to*** $\mathfrak{T}_0 \times \mathfrak{S}$ *if, and only if, they hold for* $\mathfrak{N}(\mathfrak{S}) = \mathfrak{R}^K$.

e) ***On a set*** $\mathfrak{A}$ ***relative to*** $\mathfrak{T}_0 \times \mathfrak{S}$, $\mathfrak{A} \subseteq \mathfrak{R}^K$, *if, and only if, the set* $\mathfrak{A}$ *is a neighborhood of the set* $\mathfrak{S}$ *and* $\mathfrak{A} \subseteq \mathfrak{N}_L(\mathfrak{S})$, *where* $\mathfrak{N}_L(\mathfrak{S})$ *is the largest neighborhood* $\mathfrak{N}(\mathfrak{S})$ *of* $\mathfrak{S}$, *which obeys the corresponding conditions under a) or b) or c).*

*The expression "**relative to*** $\mathfrak{T}_0 \times \mathfrak{S}$*" should be reduced to "**relative to*** $\mathfrak{T}_0$*" if, and only if, the set* $\mathfrak{S}$ *is the singleton* $\mathfrak{O}$, $\mathfrak{S} = \mathfrak{O} = \{\mathbf{0}_K\}$.

The notation "$\mathfrak{T}_0 \times$*" should be omitted if, and only if,* $\mathfrak{T}_0 = \mathfrak{T}$.

Example 286 *The vector function* $W(.)$,

$$\mathbf{W}(t,\mathbf{z}) = \left\{ \begin{array}{c} (1+t^2)(Z-A)(\mathbf{z}-\mathbf{a}),\ \mathbf{z} \in \mathfrak{N}(\mathfrak{S}) \setminus In\mathfrak{S} \\ -(1+t^2)(Z-A)(\mathbf{z}-\mathbf{a}),\ \mathbf{z} \in Cl\mathfrak{S} \end{array} \right\},$$

is globally positive semidefinite relative to the set $\mathfrak{S} = \{\mathbf{z} : |\mathbf{z}| \leq \mathbf{a}\}$. *If* $\mathbf{a} = \mathbf{0}_K$ *then it is globally positive semidefinite.*

Definition 287 ***Definition of time-varying vector pairwise semidefinite functions***

A vector function $W(.) : \mathfrak{T}_0 \times \mathfrak{R}^{2K} \to \mathfrak{R}^K$, $W(t,\mathbf{z}^1) = [W_1(t,\mathbf{z}^1)\ \ W_2(t,\mathbf{z}^1)\ \ ...\ \ W_K(t,\mathbf{z}^1)]^T$, $W_i(.) : \mathfrak{T}_0 \times \mathfrak{R}^{2K} \longrightarrow \mathfrak{R}$, $\forall i = 1,2,\cdots,K$, *is **pairwise positive (negative) semidefinite relative to*** $\mathfrak{T}_0 \times \mathfrak{S}$ *if, and only if, there is a connected neighborhood* $\mathfrak{N}(\mathfrak{S} \times \mathfrak{S})$ *of* $\mathfrak{S} \times \mathfrak{S}$ *such that it is positive (negative) semidefinite relative to* $\mathfrak{T}_0 \times \mathfrak{S} \times \mathfrak{S}$ *and* $W_l(t,\mathbf{z}^1) \equiv W_l(t,\mathbf{z}_l^1)$, $\forall l = 1,2,\cdots,K$.

Definition 288 ***Definition of time-varying vector definite functions***

A vector function $V(.) : \mathfrak{T}_0 \times \mathfrak{R}^K \to \mathfrak{R}^P$, $V(t,\mathbf{z}) = [V_1(t,\mathbf{z})\ \ V_2(t,\mathbf{z})\ \ ...\ \ V_P(t,\mathbf{z})]^T$, $V_i(.) : \mathfrak{T}_0 \times \mathfrak{R}^K \longrightarrow \mathfrak{R}$, $\forall i = 1,2,...,P$, $P \in \{1,2,3,\cdots,K\}$, *is*

a) ***Positive (negative) definite relative to*** $\mathfrak{T}_0 \times \mathfrak{S}$ *if, and only if, there is a neighborhood* $\mathfrak{N}(\mathfrak{S})$ *of* $\mathfrak{S}$ *such that (i) and (ii) hold, respectively:*

(i) $V(.)$ *is defined and continuous on* $\mathfrak{T}_0 \times \mathfrak{N}(\mathfrak{S})$*:* $V(t,\mathbf{z}) \in \mathfrak{C}[\mathfrak{T}_0 \times \mathfrak{N}(\mathfrak{S})]$.

(ii) There is a time-invariant vector function $W(.) : \mathfrak{R}^K \to \mathfrak{R}^P$, *which is positive (negative) definite relative to the set* $\mathfrak{S}$ *such that 1) and 2) hold:*

1) $V(t,\mathbf{z}) \geq W(\mathbf{z})$, $(\mathbf{V}(t,\mathbf{z}) \leq -\mathbf{W}(\mathbf{z}))$, $\forall (t,\mathbf{z}) \in \mathfrak{T}_0 \times [\mathfrak{N}(\mathfrak{S}) \setminus In\mathfrak{S}]$,

2) $V(t,\mathbf{z}) \leq -W(\mathbf{z})$, $(\mathbf{V}(t,\mathbf{z}) \geq \mathbf{W}(\mathbf{z}))$, $\forall (t,\mathbf{z}) \in \mathfrak{T}_0 \times Cl\mathfrak{S}$.

b) ***Positive (negative) definite relative to*** $\mathfrak{T}_0 \times \mathfrak{S}_b$ *if, and only if, a) is valid for* $\mathfrak{S} = \mathfrak{S}_b$ *together with the following:*

(i) $V_l(t,\mathbf{z}) \equiv V_l(t,\mathbf{z}_{i_l k_l})$, $\forall t \in \mathfrak{T}_0$,

(ii) $V_l(t,\mathbf{z}_{i_l k_l}) \equiv \mathbf{0}_K$ *for* $\mathbf{z}_{i_l k_l} \in [\mathfrak{N}(\mathfrak{S}) \setminus In\mathfrak{S}]$ *if, and only if, the corresponding* $\widetilde{\mathbf{z}}_{i_l k_l} \in \partial\mathfrak{S}_{i_l k_l}$, $\forall l = 1, 2,\cdots, P$, $\forall t \in \mathfrak{T}_0$.

d) ***Elementwise positive (negative) definite relative to*** $\mathfrak{T}_0 \times \mathfrak{S}_c$ *if, and only if, (a) is valid for* $\mathfrak{S} = \mathfrak{S}_c$*, and (i) through (iii) hold,*

(i) $P = K$.

(ii) $V_l(t, \mathbf{z}) \equiv V_l(t, z_l)$, $\forall l = 1, 2, ..., K$.

(iii) $V_l(t, \widetilde{z}_l) \equiv 0$ *for* $\mathbf{z}_{ll} \in [\mathfrak{N}(\mathfrak{S}_c) \backslash In\mathfrak{S}_c]$, $\forall t \in \mathfrak{T}_0$*, if, and only if, respectively,* $\widetilde{z}_l \in \partial\mathfrak{S}_{ll}$, $\forall l = 1, 2, ..., K$.

e) ***Radially strictly increasing on*** $\mathfrak{T}_0 \times [\mathfrak{N}(\mathfrak{S}) \backslash Cl\mathfrak{S}]$ *if, and only if,*

$$\mathbf{V}(t, \lambda_1 \mathbf{z}) < \mathbf{V}(t, \lambda_2 \mathbf{z}), 0 < \lambda_1 < \lambda_2 < \infty, \forall (t, \mathbf{z}) \in \mathfrak{T}_0 \times [\mathfrak{N}(\mathfrak{S}) \backslash Cl\mathfrak{S}].$$

The preceding properties are:

f) ***Global*** *if, and only if, they hold for* $\mathfrak{N}(\mathfrak{S}_{(.)}) = \mathfrak{R}^K$.

g) ***On*** $\mathfrak{T}_0 \times \mathfrak{A}$ ***relative to*** $\mathfrak{T}_0 \times \mathfrak{S}_{(.)}$, $\mathfrak{A} \subseteq \mathfrak{R}^K$*, if, and only if, the set* $\mathfrak{A}$ *is a connected neighborhood of the set* $\mathfrak{S}_{(.)}$ *and* $\mathfrak{A} \subseteq \mathfrak{N}_L(\mathfrak{S}_{(.)})$*, where* $\mathfrak{N}_L(\mathfrak{S}_{(.)})$ *is the largest connected neighborhood* $\mathfrak{N}(\mathfrak{S}_{(.)})$ *of* $\mathfrak{S}_{(.)}$*, which obeys the corresponding above conditions.*

h) *A vector function* $V(.) : \mathfrak{R}^K \to \mathfrak{R}^P$*, is* ***radially unbounded on*** $\mathfrak{T}_0$ *if, and only if, the corresponding above property is global and*

$$\mathbf{V}(t, \lambda \mathbf{z}) \longrightarrow \infty \mathbf{1}_K \text{ as } \lambda \longrightarrow \infty, \forall (t, \mathbf{z}) \in \mathfrak{T}_0 \times \mathfrak{R}^K.$$

*The expression "****relative to*** $\mathfrak{T}_0 \times \mathfrak{S}_{(.)}$*" reduces to "****relative to*** $\mathfrak{T}_0$*" if, and only if, the set* $\mathfrak{S}_{(.)}$ *is the singleton* $\mathfrak{O}$, $\mathfrak{S}_{(.)} = \mathfrak{O} = \{\mathbf{0}_K\}$.

The notation "$\mathfrak{T}_0 \times$*" should be omitted if, and only if,* $\mathfrak{T}_0 = \mathfrak{T}$.

Comment 289 *The preceding conditions 1) and 2) of (ii) under a) guarantee*

$$\mathbf{V}(t, \mathbf{z}) = \mathbf{0}_K \text{ for } (t, \mathbf{z}) \in \mathfrak{T}_0 \times [\mathfrak{N}(\mathfrak{S}) \backslash In\mathfrak{S}] \text{ if, and only if, } \mathbf{z} \in \partial\mathfrak{S}.$$

Example 290 *Vector function* $V(.)$ *defined by:*

$$\mathbf{V}(t, \mathbf{z}) = \frac{t^2 + 1}{2t^2 + 1} Z\mathbf{z},$$

is globally positive definite, radially unbounded, and radially strictly increasing. However,

$$\mathbf{V}(t, \mathbf{z}) = e^{-t} |\mathbf{z}|,$$

is not positive definite relative to $\mathfrak{T}_0$, $t_0 \in \mathfrak{T}_i$*, because* $e^{-t} \longrightarrow 0$ *as* $t \longrightarrow -\infty$*. There are not a time-invariant vector positive definite function* $W(.)$ *and a hyperball* $\mathfrak{B}$ *(centered at the origin) such that*

$$\mathbf{V}(t, \mathbf{z}) = e^{-t} |\mathbf{z}| \geq \mathbf{W}(\mathbf{z}), \ \forall (t, \mathbf{z} \neq \mathbf{0}_K) \in \mathfrak{T} \times \mathfrak{B}.$$

Let

$$\tau = -\ln\left[|Z|^{-1} \mathbf{W}(\mathbf{z})\right], \ |Z|^{-1} = diag\left\{|z_1|^{-1} \ \ |z_2|^{-1} \ \cdots \ |z_K|^{-1}\right\}, \Longrightarrow$$

$$|Z|^{-1} |\mathbf{z}| = diag\left\{|z_1|^{-1} \ \ |z_2|^{-1} \ \cdots \ |z_K|^{-1}\right\} [|z_1| \ \ |z_2| \ \cdots \ |z_K|]^T = I,$$

$$I = diag\{1 \ \ 1 \ \cdots \ 1\} \in \mathfrak{R}^{K \text{x} K},$$

then

$$e^{-t} |\mathbf{z}| < \mathbf{W}(\mathbf{z}), \ \forall t \in]\tau, \infty[.$$

Vector functions $V(.)$ *defined by:*

$$\mathbf{V}(t, \mathbf{z}) = \frac{t^2 + 1}{2} |Z| (Z - A)(\mathbf{z} - \mathbf{a}), \ \mathbf{V}(\mathbf{z}) = (e^{-t} + e^t)(A - Z) |\mathbf{z}|,$$

are positive definite on the set $\mathfrak{A} = \{\mathbf{z} : \ |\mathbf{z}| < |\mathbf{a}|\}$.

Definition 291 ***Definition of time-varying vector pairwise definite functions***
A vector function $V(.) : \mathfrak{T}_0 \times \mathfrak{R}^{2K} \rightarrow \mathfrak{R}^K$, $V(t,\mathbf{z}) = [V_1(t,\mathbf{z}) \;\; V_2(t,\mathbf{z}) \;\cdots\; V_K(t,\mathbf{z})]^T$, $V_i(.) : \mathfrak{T}_0 \times \mathfrak{R}^{2K} \longrightarrow \mathfrak{R}$, $\forall i = 1, 2, \cdots, K$, *is* ***positive (negative) pairwise definite relative to*** $\mathfrak{T}_0 \times \mathfrak{S}$ *if, and only if, it is positive (negative) definite relative to* $\mathfrak{T}_0 \times \mathfrak{S}$ *and* $V_l(t,\mathbf{z}^1) \equiv V_l(t,\mathbf{z}_l^1)$, $\forall l = 1, 2, \cdots, K$.

The decrescent property of the vector functions is crucial for uniform tracking.

Definition 292 ***Definition of time-varying decrescent vector functions***
A vector function $V(.) : \mathfrak{T}_0 \times \mathfrak{R}^K \rightarrow \mathfrak{R}^P$, $V(t,\mathbf{z}) = [V_1(t,\mathbf{z}) \;\; V_2(t,\mathbf{z}) \;\; ... \;\; V_P(t,\mathbf{z})]^T$, $V_i(.) : \mathfrak{T}_0 \times \mathfrak{R}^K \longrightarrow \mathfrak{R}$, $\forall i = 1, 2, \cdots, P$, $P \in \{1, 2, \cdots, K\}$, *is*

a) ***Decrescent relative to*** $\mathfrak{T}_0 \times \mathfrak{S}$ *if, and only if, there is a neighborhood* $\mathfrak{N}(\mathfrak{S})$ *of* $\mathfrak{S}$ *such that (i) and (ii) hold, respectively:*

(i) $V(.)$ *is defined and continuous on* $\mathfrak{T}_0 \times \mathfrak{N}(\mathfrak{S})$*:*

$$\mathbf{V}(t,\mathbf{z}) \in \mathfrak{C}\left[\mathfrak{T}_0 \times \mathfrak{N}(\mathfrak{S})\right].$$

(ii) there is a time-invariant vector function $W(.) : \mathfrak{R}^K \rightarrow \mathfrak{R}^P$ *that is positive definite relative to the set* $\mathfrak{S}$ *such that*

$$\mathbf{V}(t,\mathbf{z}) \leq \mathbf{W}(\mathbf{z}), \forall (t,\mathbf{z}) \in \mathfrak{T}_0 \times \left[\mathfrak{N}(\mathfrak{S}) \backslash In\mathfrak{S}\right].$$

b) ***Decrescent relative to*** $\mathfrak{T}_0 \times \mathfrak{S}_b$ *if, and only if, (a) is valid for* $\mathfrak{S} = \mathfrak{S}_b$ *together with the following:*

(i) $V_l(t,\mathbf{z}) \equiv V_l(t,\mathbf{z}_{i_l k_l})$, $\forall t \in \mathfrak{T}_0$.

(ii) $V_l(t,\mathbf{z}_{i_l k_l}) \equiv \mathbf{0}_K$ *for* $\mathbf{z}_{i_l k_l} \in \left[\mathfrak{N}(\mathfrak{S}) \backslash In\mathfrak{S}\right]$ *if, and only if, the corresponding* $\widetilde{\mathbf{z}}_{i_l k_l} \in \partial\mathfrak{S}_{i_l k_l}$, $\forall l = 1, 2, \cdots, P$, $\forall t \in \mathfrak{T}_0$.

d) ***Elementwise decrescent relative to*** $\mathfrak{T}_0 \times \mathfrak{S}_c$ *if, and only if, a) is valid for* $\mathfrak{S} = \mathfrak{S}_c$*, and (i) through (iii) hold,*

(i) $P = K$.

(ii) $V_l(t,\mathbf{z}) \equiv V_l(t, z_l)$, $\forall l = 1, 2, \cdots, K$.

(iii) $V_l(t,\widetilde{z}_l) \equiv 0$ *for* $\mathbf{z}_{ll} \in \left[\mathfrak{N}(\mathfrak{S}_c) \backslash In\mathfrak{S}_c\right]$, $\forall t \in \mathfrak{T}_0$, *if, and only if, respectively,* $\widetilde{z}_l \in \partial\mathfrak{S}_{ll}$, $\forall l = 1, 2, \cdots, K$.

The preceding properties are:

f) ***Global*** *if, and only if, they hold for* $\mathfrak{N}\left(\mathfrak{S}_{(.)}\right) = \mathfrak{R}^K$.

g) ***On*** $\mathfrak{T}_0$x$\mathfrak{A}$ ***relative to*** $\mathfrak{T}_0 \times \mathfrak{S}_{(.)}$, $\mathfrak{A} \subseteq \mathfrak{R}^K$, *if, and only if, the set* $\mathfrak{A}$ *is a connected neighborhood of the set* $\mathfrak{S}_{(.)}$ *and* $\mathfrak{A} \subseteq \mathfrak{N}_L\left(\mathfrak{S}_{(.)}\right)$, *where* $\mathfrak{N}_L\left(\mathfrak{S}_{(.)}\right)$ *is the largest connected neighborhood* $\mathfrak{N}\left(\mathfrak{S}_{(.)}\right)$ *of* $\mathfrak{S}_{(.)}$, *which obeys the corresponding above conditions.*

*The expression "****relative to*** $\mathfrak{T}_0 \times \mathfrak{S}_{(.)}$*" reduces to "****relative to*** $\mathfrak{T}_0$*" if, and only if, the set* $\mathfrak{S}_{(.)}$ *is the singleton* $\mathfrak{O}$, $\mathfrak{S}_{(.)} = \mathfrak{O} = \{\mathbf{0}_K\}$.

The notation "$\mathfrak{T}_0 \times$*" should be omitted if, and only if,* $\mathfrak{T}_0 = \mathfrak{T}$.

Example 293 *Vector function* $V(.)$ *defined by:*

$$\mathbf{V}(t,\mathbf{z}) = \frac{t^2+1}{2t^2+1} Z\mathbf{z} \leq \mathbf{Z}\mathbf{z},$$

is globally decrescent. However,

$$\mathbf{V}(t,\mathbf{z}) = e^t \left|\mathbf{z}\right| \;\; and \;\; \mathbf{V}(t,\mathbf{z}) = \frac{t^4+1}{2t^2+1}$$

are not decrescent relative to $\mathfrak{T}_0$, $t_0 \in \mathfrak{T}_i$, *because the term* $e^t \longrightarrow \infty$ *and the product* $\left(t^4+1\right)\left(2t^2+1\right)^{-1} \longrightarrow \infty$ *as* $t \longrightarrow \infty$.

Definition 294 ***Definition of time-varying vector pairwise decrescent functions***

A vector function $V(.) : \mathfrak{T}_0 \times \mathfrak{R}^{2K} \rightarrow \mathfrak{R}^K$, $V(t, \mathbf{z}) = [V_1(t, \mathbf{z}) \;\; V_2(t, \mathbf{z}) \; \cdots \; V_K(t, \mathbf{z})]^T$, $V_i(.) : \mathfrak{T}_0 \times \mathfrak{R}^{2K} \longrightarrow \mathfrak{R}$, $\forall i = 1, 2, \cdots, K$, *is* ***pairwise decrescent relative to*** $\mathfrak{T}_0 \times \mathfrak{S}$ *if, and only if, it is decrescent relative to* $\mathfrak{T}_0 \times \mathfrak{S}$ *and* $V_l(t, \mathbf{z}^1) \equiv V_l(t, \mathbf{z}_l^1)$, $\forall l = 1, 2, \cdots, K$.

Chapter 21

Sets and functions

21.1 Positive definite function induces sets

Every positive definite function relative to a *time*-invariant connected subset Υ of $\mathfrak{R}^K$ with a nonempty boundary $\partial\Upsilon$ on $\mathfrak{T}_\tau$ induces sets denoted by $\mathfrak{V}_\xi(t;\tau;\Upsilon)$, $\mathfrak{V}_\xi(t;\tau;\Upsilon) \subseteq \mathfrak{R}^K$, which are defined by:

Definition 295 ***The set*** $\mathfrak{V}_\xi(t;\tau;\Upsilon)$ ***induced by both the positive definite scalar function V(.;***Υ***) and its scalar value*** ξ

Let $\xi \in \mathfrak{R}^+$ *or* $\xi = \infty$. *Let the function* $V(.;\Upsilon) : \mathfrak{T}_\tau \times \mathfrak{R}^K \longrightarrow \mathfrak{R}$ *be positive definite relative to the set* Υ *on* $\mathfrak{T}_\tau$ *and let* $\mathfrak{V}_\xi(t;\tau;\Upsilon)$, $\mathfrak{V}_\xi(t;\tau;\Upsilon) \subseteq \mathfrak{R}^K$, *be the largest open connected neighborhood of the set* Υ *at* $t \in \mathfrak{T}_\tau$ *such that*

$$\xi > 0,\ \mathfrak{V}_\xi(t;\tau;\Upsilon) = \left\{\mathbf{z} : \mathbf{z} \in \mathfrak{R}^K,\ V(t,\mathbf{z};\boldsymbol{\Upsilon}) < \xi\right\},\ \ \forall t \in \mathfrak{T}_\tau, \tag{21.1}$$

and

1) $V(t,\mathbf{z};\boldsymbol{\Upsilon})$ *is strictly less than* ξ *on* $\mathfrak{V}_\xi(t;\tau;\Upsilon)$ *at every instant* $t \in \mathfrak{T}_\tau$,

$$\xi > 0,\ \ \mathbf{z} \in \mathfrak{V}_\xi(t;\tau;\Upsilon) \Longrightarrow V(t,\mathbf{z};\boldsymbol{\Upsilon}) < \xi,\ \ \forall t \in \mathfrak{T}_\tau. \tag{21.2}$$

2) If $\xi \in \mathfrak{R}^+$ *then the boundary* $\partial\mathfrak{V}_\xi(t;\tau;\Upsilon)$ *of* $\mathfrak{V}_\xi(t;\tau;\Upsilon)$ *is the hypersurface* $V(t,\mathbf{z};\boldsymbol{\Upsilon}) \equiv \xi$,

$$\xi \in \mathfrak{R}^+ \Longrightarrow V(t,\mathbf{z};\boldsymbol{\Upsilon}) = \xi \Longleftrightarrow \mathbf{z} \in \partial\mathfrak{V}_\xi(t;\tau;\Upsilon),\ \forall t \in \mathfrak{T}_\tau. \tag{21.3}$$

3) The union $\mathfrak{V}_\xi(\mathfrak{T}_\tau;\Upsilon)$ *of all* $\mathfrak{V}_\xi(t;\tau;\Upsilon)$ *over* $\mathfrak{T}_\tau$ *is the nonempty neighborhood of the set* Υ,

$$\xi > 0,\ \ \mathfrak{V}_\xi(\mathfrak{T}_\tau;\Upsilon) = \cup\left[\mathfrak{V}_\xi(t;\tau;\Upsilon) : t \in \mathfrak{T}_\tau\right] \neq \phi, \tag{21.4}$$

$$\Upsilon \subset \mathfrak{V}_\xi(\mathfrak{T}_\tau;\Upsilon). \tag{21.5}$$

*The expression "***on** $\mathfrak{T}_\tau$*" and "*τ*" are to be omitted if, and only if,* $\mathfrak{T}_\tau = \mathfrak{T}$. *Then, and only then,* $\mathfrak{V}_\xi(t;\tau;\Upsilon) = \mathfrak{V}_\xi(t;\Upsilon)$,

$$\mathfrak{T}_\tau = \mathfrak{T} \Longrightarrow \mathfrak{V}_\xi(t;\tau;\Upsilon) = \mathfrak{V}_\xi(t;\Upsilon). \tag{21.6}$$

*The expression "***relative to** Υ*" is to be omitted if, and only if,* $\Upsilon = \mathfrak{O} = \{\mathbf{0}_K\}$.

Note 296 *The condition 3) of Definition 295 is equivalent to the condition a-2) of Definition 267. They are inherent for the time-varying function* $V(.;\Upsilon)$ *to be positive definite. It reflects an important property of the set* $\mathfrak{V}_\xi(t;\tau;\Upsilon)$ *associated with the function* $V(.;\Upsilon)$ *and its values less than, or equal to,* $\xi > 0$.

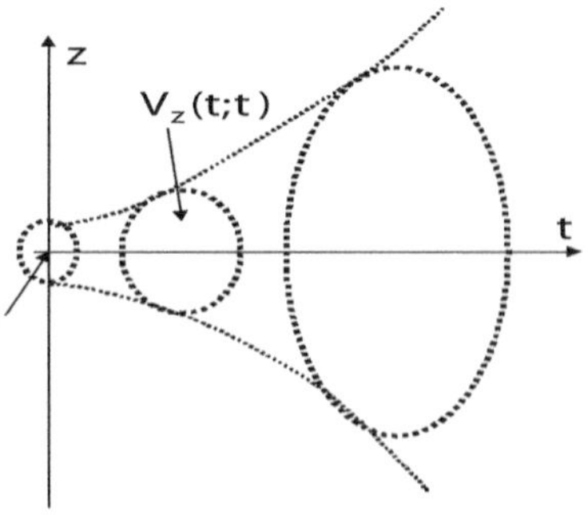

Figure 21.1: The set $\mathfrak{V}_\zeta(t;\tau;\Upsilon)$ related to the *time*-varying semidefinite function $V(t,\mathbf{z};\Upsilon) = e^{-t}\rho(\mathbf{z},\boldsymbol{\Upsilon}), \Upsilon = \mathfrak{O}$, is unbounded.

Example 297 *Let* $V(t,\mathbf{z};\boldsymbol{\Upsilon}) = e^{-t}\rho(\mathbf{z},\boldsymbol{\Upsilon})$. *It is globally positive semidefinite relative to* Υ. *It is not positive definite relative to* Υ. *There is not a time-invariant positive definite function* $W(.;\Upsilon)$ *relative to* Υ *such that* $V(.;\Upsilon)$ *obeys the condition a-2) of Definition 267 because*

$$t \longrightarrow \infty \Longrightarrow V(t,\mathbf{z};\boldsymbol{\Upsilon}) = e^{-t}\rho(\mathbf{z},\boldsymbol{\Upsilon}) \longrightarrow 0,\ \forall \mathbf{z} \in \mathfrak{R}^K \backslash \Upsilon.$$

For every $\xi \in \mathfrak{R}^+$ *the set* $\mathfrak{V}_\xi(t;\Upsilon)$ *diverges to* $\mathfrak{R}^K$ *as* $t \longrightarrow \infty$,

$$\mathfrak{V}_\xi(t;\Upsilon) = \left\{\mathbf{z} : e^{-t}\rho(\mathbf{z},\boldsymbol{\Upsilon}) < \xi\right\} = \left\{\mathbf{z} : \rho(\mathbf{z},\boldsymbol{\Upsilon}) < \xi e^t\right\} \longrightarrow \mathfrak{R}^K \text{ as } t \longrightarrow \infty$$

(Figure 21.1).

Example 298 *Let* $V(t,\mathbf{z};\boldsymbol{\Upsilon}) = \left(1+t^2\right)\rho(\mathbf{z},\boldsymbol{\Upsilon})$. *It is globally positive definite relative to* Υ. *It obeys the condition a-2). of Definition 267 for the globally positive definite time-invariant function* $W(\mathbf{z};\boldsymbol{\Upsilon}) = \rho(\mathbf{z},\boldsymbol{\Upsilon})$ *relative to* Υ. *For every* $\xi \in \mathfrak{R}^+$ *the set* $\mathfrak{V}_\xi(t;\Upsilon)$ *is a subset of the time-invariant connected neighborhood* $\mathfrak{N}_\xi(\Upsilon)$ *of* Υ *on* $\mathfrak{T}$,

$$\begin{aligned}\mathfrak{V}_\xi(t;\Upsilon) &= \left\{\mathbf{z} : \left(1+t^2\right)\rho(\mathbf{z},\boldsymbol{\Upsilon}) < \xi\right\}\\ &= \left\{\mathbf{z} : \rho(\mathbf{z},\boldsymbol{\Upsilon}) < \xi\left(1+t^2\right)^{-1}\right\} \subseteq \mathfrak{N}_\xi(\Upsilon),\ \forall t \in \mathfrak{T}.\end{aligned}$$

Notice that the set $\mathfrak{V}_\xi(t;\Upsilon)$ *is asymptotically contractive to* Υ *(Definition 46), which means that the closure* $Cl\mathfrak{V}_\xi(t;\Upsilon)$ *of the set* $\mathfrak{V}_\xi(t;\Upsilon)$ *contracts asymptotically to the closure* $Cl\Upsilon$ *of the set* Υ *as* $t \longrightarrow \infty$,

$$t \longrightarrow \infty \Longrightarrow Cl\mathfrak{V}_\xi(t;\Upsilon) = \left\{\mathbf{z} : \rho(\mathbf{z},\boldsymbol{\Upsilon}) \le \xi\left(1+t^2\right)^{-1}\right\} \longrightarrow Cl\Upsilon.$$

This means that there is not a lower open time-invariant limit of the set $\mathfrak{V}_\xi(t;\Upsilon)$ *as* $t \longrightarrow \infty$. *For arbitrarily small* $\zeta \in \mathfrak{R}^+$ *there is* $\tau \in \mathfrak{T}$ *such that* $\mathfrak{V}_\xi(t;\Upsilon)$ *is the proper subset of the open neighborhood* $\mathfrak{N}_\zeta(\Upsilon)$ *of* Υ *at every* $(t \ge \tau) \in \mathfrak{T}$,

$$\mathfrak{V}_\xi(t;\Upsilon) \subset \mathfrak{N}_\zeta(\Upsilon),\ \forall t \in \mathfrak{T}_\tau.$$

Property 299 ***The property of the boundaries of the sets*** $\mathfrak{V}_\xi(t;\tau;\Upsilon)$

The boundaries $\partial\mathfrak{V}_{\xi_i}(t;\tau;\Upsilon)$ *of the sets* $\mathfrak{V}_\xi(t;\tau;\Upsilon)$ *induced by the function* $V(.;\Upsilon)$, $i = 1,2,\ \xi_1 \neq \xi_2$, *which are subsets of the neighborhood* $\mathfrak{N}(\Upsilon)$ *introduced in Definition 267, do not ever intersect on* $\mathfrak{T}_\tau$:

$$\partial\mathfrak{V}_{\xi_1}(t;\tau;\Upsilon) \cap \partial\mathfrak{V}_{\xi_2}(t;\tau;\Upsilon) = \phi. \tag{21.7}$$

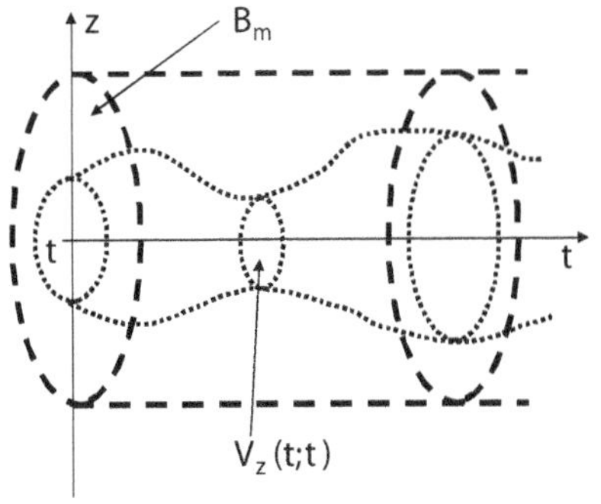

Figure 21.2: The set $\mathfrak{V}_{\varsigma}(t;\tau;\Upsilon)$ is bounded. It is all the *time* a proper subset of the hyperball $\mathfrak{N}_{\mu}(\Upsilon)$ for $\Upsilon = \mathfrak{D}$.

Proof. Let a function $V(.;\Upsilon) : \mathfrak{T}_{\tau} \times \mathfrak{R}^K \longrightarrow \mathfrak{R}$ be positive definite relative to Υ. It satisfies Definition 267. Let $\mathfrak{N}(\Upsilon)$ be the neighborhood introduced in Definition 267. Let $\xi_i \in \mathfrak{R}^+$ be such that $\mathfrak{V}_{\xi}(t;\tau;\Upsilon) \subseteq \mathfrak{N}(\mathrm{u})$, $\forall t \in \mathfrak{T}_{\tau}$, $i = 1, 2$, $\xi_1 \neq \xi_2$. Let us assume that the statement of Property 299 is wrong, i.e., that there is $\sigma \in \mathfrak{T}_{\tau}$ such that $\partial\mathfrak{V}_{\xi_1}(\sigma;\tau;\Upsilon) \cap \partial\mathfrak{V}_{\xi_2}(\sigma;\tau;\Upsilon) \neq \phi$. Let $\mathbf{z} \in \partial\mathfrak{V}_{\xi_1}(\sigma;\tau;\Upsilon) \cap \partial\mathfrak{V}_{\xi_2}(\sigma;\tau;\Upsilon)$. Because $\mathbf{z} \in \partial\mathfrak{V}_{\xi_1}(\sigma;\tau;\Upsilon)$ then $V(t, \mathbf{z}; \boldsymbol{\Upsilon}) = \xi_1$, But $\mathbf{z} \in \partial\mathfrak{V}_{\xi_2}(\sigma;\tau;\Upsilon)$ so that $V(t, \mathbf{z}; \boldsymbol{\Upsilon}) = \xi_2 \neq \xi_1$. This contradicts the continuity of the positive definite function $V(.;\Upsilon)$ relative to Υ on $\mathfrak{T}_{\tau} \times \mathfrak{N}(\Upsilon)$, (20.11). The contradiction is the consequence of the assumption that the statement of Property 299 is wrong, i.e., that (21.7) is wrong. Hence, the statement of Property 299 is correct; i.e., (21.7) is valid ■

The following proposition discovers another property of the sets $\mathfrak{V}_{\xi}(t;\tau;\Upsilon)$ induced by the positive definite function $V(.;\Upsilon)$ relative to Υ.

Proposition 300 ***Positive definiteness relative to Υ, comparison function, and the set $\mathfrak{V}_{\xi}(t;\tau;\Upsilon)$***

If the set Υ is time-invariant, connected and bounded and if a function $V(.;\Upsilon)$ is positive definite relative to Υ on $\mathfrak{T}_{\tau} \times \mathfrak{N}(\Upsilon)$, where $\mathfrak{N}(\Upsilon)$ is a time-invariant connected neighborhood of the set Υ, then there is $\eta \in \mathfrak{R}^+$ such that for every $\varsigma \in]0, \eta[$ the set $\mathfrak{V}_{\varsigma}(t;\tau;\Upsilon)$ is bounded on $\mathfrak{T}_{\tau}$ and $\mathfrak{V}_{\varsigma}(t;\tau;\Upsilon) \subseteq \mathfrak{N}(\Upsilon)$, $\forall t \in \mathfrak{T}_{\tau}$.

Proof. Let the set Υ be *time*-invariant, connected, and bounded $\mathfrak{N}(\Upsilon)$ be a bounded neighborhood of the set Υ such that $V(.;\Upsilon)$ is positive definite relative to Υ on $\mathfrak{T}_{\tau} \times \mathfrak{N}(\Upsilon)$. Let $\mu = \min\left[\rho(\mathbf{z}, \boldsymbol{\Upsilon}) : \mathbf{z} \in \partial\mathfrak{N}(\Upsilon)\right]$. Let $\varphi(.) : \mathfrak{R}_+ \longrightarrow \mathfrak{R}_+$ be a comparison function of the class $\mathcal{K}_{[0,\mu[}$, which obeys

$$V(t, \mathbf{z}; \boldsymbol{\Upsilon}) \geq \varphi\left[\rho(\mathbf{z}, \boldsymbol{\Upsilon})\right], \ \forall (t, \mathbf{z}) \in \mathfrak{T}_{\tau} \mathrm{x} \mathfrak{N}_{\mu}(\Upsilon).$$

due to (20.18). We continue the proof by contradiction. Let all the preceding conditions hold and let the set $\mathfrak{V}_{\varsigma}(t;\tau;\Upsilon)$, $\varsigma \in]0, \eta[$ for $\eta = \varphi(\mu)$, be unbounded on $\mathfrak{T}_{\tau}$. Then there is a moment $\sigma \in \mathfrak{T}_{\tau}$ such that $\mathfrak{V}_{\varsigma}(\sigma;\tau;\Upsilon) \setminus \mathfrak{N}_{\mu}(\Upsilon) \neq \phi$. Let $\mathbf{z}^* \in \mathfrak{V}_{\varsigma}(\sigma;\tau;\Upsilon) \setminus \mathfrak{N}_{\mu}(\Upsilon)$, which implies both $\rho(\mathbf{z}^*, \boldsymbol{\Upsilon}) \geq \mu$ and $V(\sigma, \mathbf{z}^*) \geq \varphi(\mu) = \eta$ due to (20.18). Hence, $\mathbf{z}^* \notin \mathfrak{V}_{\varsigma}(\sigma;\tau;\Upsilon) \setminus \mathfrak{N}_{\mu}(\Upsilon)$ (due to Definition 295), which contradicts $\mathbf{z}^* \in \mathfrak{V}_{\varsigma}(\sigma;\tau;\Upsilon) \setminus \mathfrak{N}_{\mu}(\Upsilon)$. This and $\mathfrak{V}_{\varsigma}(\sigma;\tau;\Upsilon) \cap \mathfrak{N}_{\mu}(\Upsilon) \neq \phi$, $\forall t \in \mathfrak{T}_{\tau}$ prove that the set $\mathfrak{V}_{\varsigma}(t;\tau;\Upsilon)$, $\varsigma \in]0, \eta[$, is bounded on $\mathfrak{T}_{\tau}$, and that $\mathfrak{V}_{\varsigma}(t;\tau;\Upsilon)$ is a subset of the neighborhood $\mathfrak{N}_{\mu}(\Upsilon)$ on $\mathfrak{T}_{\tau}$,i.e., $\mathfrak{V}_{\varsigma}(t;\tau;\Upsilon) \subseteq \mathfrak{N}_{\mu}(\Upsilon)$, $\forall t \in \mathfrak{T}_{\tau}$ (Figure 21.2). ■

Conclusion 301 *Example 297 and Example 298 illustrate the important and useful property of the positive definiteness relative to Υ of a function $V(.;\Upsilon)$ on $\mathfrak{T}_{\tau}$, which is the boundedness* *of the associated set* $\mathfrak{V}_{\xi}(t;\tau;\Upsilon)$ *on* $\mathfrak{T}_{\tau}$. *This need not hold for the set* $\mathfrak{V}_{\xi}(t;\tau;\Upsilon)$ *if the function* $V(.;\Upsilon)$ *is only positive semidefinite relative to* Υ *on* $\mathfrak{T}_{\tau}$.

Lyapunov classical theory concerns nonlinear systems with stability properties being valid on the *time*-invariant sets. Therefore, it deals with Lyapunov functions $V(.;\Upsilon)$ for which the associated sets $\mathfrak{V}_{\xi}(t;\tau;\Upsilon)$ are not asymptotically contractive. The works [156], [170], [175], [227, pp. 33-42], [256], [257] extend it to nonlinear systems with stability properties being valid on the sets that are asymptotically contractive, hence, they extend it to the use of Lyapunov functions $V(.;\Upsilon)$ for which the associated sets $\mathfrak{V}_{\xi}(t;\tau;\Upsilon)$ are asymptotically contractive.

Definition 302 ***Set increasing***
Let the function $V(.;\Upsilon)$, $V(t,\mathbf{z};\boldsymbol{\Upsilon})\in\mathfrak{C}\left(\mathfrak{T}_{\tau}\times\mathfrak{R}^{K}\right)$, *induce sets* $\mathfrak{V}_{\zeta_i}(t;\tau;\Upsilon)$, *(Definition 295). The sets* $\mathfrak{V}_{\zeta_i}(t;\tau;\Upsilon)$ *are* ***set increasing relative to*** Υ ***on the neighborhood*** $\mathfrak{N}(\Upsilon)$ ***of*** Υ *if, and only if,*

$$\begin{gathered}Cl\mathfrak{V}_{\xi_1}(t;\tau;\Upsilon)\subset Cl\mathfrak{V}_{\xi_2}(t;\tau;\Upsilon),\ \forall t\in\mathfrak{T}_{\tau},\\ \partial\mathfrak{V}_{\xi_1}(t;\tau;\Upsilon)\cap\partial\mathfrak{V}_{\xi_2}(t;\tau;\Upsilon)=\phi,\forall t\in\mathfrak{T}_{\tau},\\ if,\ and\ only\ if\ \xi_1<\xi_2,\ \forall t\in\mathfrak{T}_{\tau},\end{gathered}\tag{21.8}$$

for $\xi_i\in\mathfrak{R}$ *such that* $\mathfrak{V}_{\xi_i}(t;\tau;\Upsilon)\subseteq\mathfrak{N}(\Upsilon)$, $i=1,2$, $\forall t\in\mathfrak{T}_{\tau}$.

Lemma 303 *Let* $V(.;\Upsilon)$, $V(t,\mathbf{z};\boldsymbol{\Upsilon})\in\mathfrak{C}\left(\mathfrak{T}\times\mathfrak{R}^{K}\right)$, *be a positive definite function relative to* Υ. *Then the sets* $\mathfrak{V}_{\xi}(t;t_0;\Upsilon)$, $\mathfrak{V}_{\xi_j}(t;t_0)\subseteq\mathfrak{N}(\Upsilon)$, $\forall t\in\mathfrak{T}_0$, *are set increasing.*

Proof. Let the conditions of Lemma 303 statement hold. Let us assume that the sets $\mathfrak{V}_{\xi}(t;t_0;\Upsilon)$, $\mathfrak{V}_{\xi_j}(t;t_0;\Upsilon)\subseteq\mathfrak{N}(\Upsilon)$, $\forall t\in\mathfrak{T}_0$, is not set increasing. This means that there are $\sigma\in\mathfrak{T}_0$ and $\xi_i\in\mathfrak{R}^{+}$, $i=1,2$, for which $\mathfrak{V}_{\xi_j}(t;t_0;\Upsilon)\subseteq\mathfrak{N}(\Upsilon),\forall t\in\mathfrak{T}_0$, $\xi_1<\xi_2$, such that $\mathfrak{V}_{\xi_2}(\sigma;t_0;\Upsilon)\subseteq\mathfrak{V}_{\xi_1}(\sigma;t_0;\Upsilon)$. If the sets $\mathfrak{V}_{\xi_2}(\sigma;t_0;\Upsilon)$ and $\mathfrak{V}_{\xi_1}(\sigma;t_0;\Upsilon)$ were equal, $\mathfrak{V}_{\xi_2}(\sigma;t_0;\Upsilon)=\mathfrak{V}_{\xi_1}(\sigma;t_0;\Upsilon)$, then their boundaries would be equal; $\partial\mathfrak{V}_{\xi_2}(\sigma;t_0;\Upsilon)=\partial\mathfrak{V}_{\xi_1}(\sigma;t_0;\Upsilon)$ that is impossible due to Property 299. If $\mathfrak{V}_{\xi_2}(\sigma;t_0;\Upsilon)\subset\mathfrak{V}_{\xi_1}(\sigma;t_0;\Upsilon)$ were valid then there would be $\mathbf{z}\in\mathfrak{V}_{\xi_2}(\sigma;t_0;\Upsilon)$ such that both $\mathbf{z}\in\mathfrak{V}_{\xi_1}(\sigma;t_0;\Upsilon)$ and $V(\sigma,\mathbf{z};\Upsilon)>\xi_1$ were valid. However, $V(\sigma,\mathbf{z};\Upsilon)>\xi_1$ is impossible for $\mathbf{z}\in\mathfrak{V}_{\xi_1}(\sigma;t_0;\Upsilon)$ due to (21.2). The assumption that the sets $\mathfrak{V}_{\xi}(t;t_0;\Upsilon)$, $\mathfrak{V}_{\xi_j}(t;t_0;\Upsilon)\subseteq\mathfrak{N}(\Upsilon)$, $\forall t\in\mathfrak{T}_0$, are not set increasing is invalid. This implies that the sets $\mathfrak{V}_{\xi}(t;t_0;\Upsilon)$ are set increasing ∎

Proposition 304 ***Decrescent function relative to*** Υ***, comparison function and the set*** $\mathfrak{V}_{\xi}(t;\Upsilon)$.

Proposition 305 *If a function* $V(.;\Upsilon)$ *is decrescent relative to* Υ *on* $\mathfrak{T}_{\tau}\times\mathfrak{N}(\Upsilon)$, *where* $\mathfrak{N}(\Upsilon)$ *is a time-invariant connected neighborhood of the origin* $\mathbf{z}=\mathbf{0}_K$, *then there is* $\xi\in\mathfrak{R}^{+}$ *such that for every* $\varsigma\in]0,\xi[$ *there is a neighborhood* $\mathfrak{N}_{\eta}(\Upsilon)$ *of* Υ, *which is a subset of the set* $\mathfrak{V}_{\xi}(t;\tau;\Upsilon)$ *for all* $t\in\mathfrak{T}_{\tau}$,

$$\exists\xi\in\mathfrak{R}^{+}\Longrightarrow\forall\varsigma\in]0,\xi[,\ \exists\eta\in\mathfrak{R}^{+}\Longrightarrow\mathfrak{N}_{\eta}(\Upsilon)\subset\mathfrak{V}_{\xi}(t;\tau;\Upsilon),\ \forall\,t\in\mathfrak{T}_{\tau}.$$

Let $\mathfrak{N}(\Upsilon)$ be a connected neighborhood of the set Υ such that $V(.;\Upsilon)$ is decrescent relative to Υ on $\mathfrak{T}_{\tau}\times\mathfrak{N}(\Upsilon)$. Let $\mu=\min\left[\rho(\mathbf{z},\boldsymbol{\Upsilon}):\mathbf{z}\in\partial\mathfrak{N}(\Upsilon)\right]$. Let $\varphi(.):\mathfrak{R}_{+}\longrightarrow\mathfrak{R}_{+}$ be a comparison function of the class $\mathcal{K}_{[0,\mu[}$, which obeys

$$\varphi\left[\rho(\mathbf{z},\boldsymbol{\Upsilon})\right]\geq V(t,\mathbf{z};\boldsymbol{\Upsilon}),\ \forall(t,\mathbf{z})\in\mathfrak{T}_{\tau}\mathrm{x}\mathfrak{N}_{\mu}(\Upsilon)$$

due to (20.22). We continue the proof by contradiction. Let all the preceding conditions hold and let the set $\mathfrak{V}_{\zeta}(t;\tau;\Upsilon)$, $\zeta\in]0,\xi[$ for $\xi=\varphi(\mu)$, be such that for every $\eta\leq\varphi^{I}(\zeta)$, $\eta\in\mathfrak{R}^{+}$,

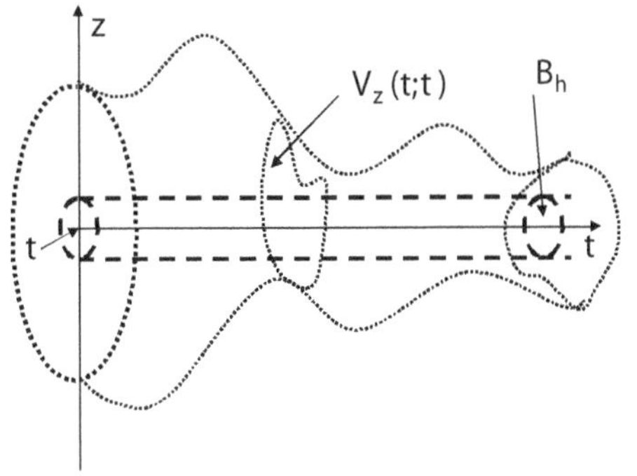

Figure 21.3: The neighborhood $\mathfrak{N}_\eta(\Upsilon)$ of $\Upsilon = \mathfrak{O}$ is the hyperball $\mathfrak{B}_\eta$ that is all the *time* the subset of the set $\mathfrak{V}_\varsigma(t;\tau;\Upsilon)$ induced by the *time*-varying decrescent function $V(.;\Upsilon)$.

there is a moment $\sigma \in \mathfrak{T}_\tau$ at which $\mathfrak{V}_\zeta(\sigma;\tau;\Upsilon)$ is a subset of $\mathfrak{N}_\eta(\Upsilon)$, $\mathfrak{V}_\zeta(\sigma;\tau;\Upsilon) \subset \mathfrak{N}_\eta(\Upsilon)$. Let $\mathbf{z}^* \in \mathfrak{N}_\eta(\Upsilon) \backslash \mathfrak{V}_\zeta(\sigma;\tau;\Upsilon)$, which implies both $\rho(\mathbf{z}^*,\mathbf{\Upsilon}) < \eta$ and $V(\sigma,\mathbf{z}^*;\Upsilon) \geq \zeta$. Hence, $\varphi[\rho(\mathbf{z}^*,\mathbf{\Upsilon})] < \varphi(\eta) \leq \zeta$. This and (20.22) imply

$$\zeta \geq \varphi(\eta) > \varphi[\rho(\mathbf{z}^*,\mathbf{\Upsilon})] \geq V(\sigma,\mathbf{z}^*;\mathbf{\Upsilon}) \geq \zeta,$$

which shows the contradiction $\zeta > \zeta$. The contradiction proves that for every $\zeta \in]0,\xi[$ there is $\eta \in \mathfrak{R}^+$ such that the neighborhood $\mathfrak{N}_\eta(\Upsilon)$ is subset of the set $\mathfrak{V}_\zeta(t;\tau;\Upsilon)$ on $\mathfrak{T}_\tau$,

$$\forall\zeta \in]0,\xi[,\ \exists\eta \in \mathfrak{R}^+ \Longrightarrow \mathfrak{V}_\zeta(t;\tau;\Upsilon) \supset \mathfrak{N}_\eta(\Upsilon),\ \forall t \in \mathfrak{T}_\tau$$

(Figure 21.3).

Example 306 *The function* $V(t,\mathbf{z};\mathbf{\Upsilon}) = (1+t^2)\,\rho(\mathbf{z},\mathbf{\Upsilon})$ *is not a decrescent function relative to* Υ. *There is not a time-invariant positive definite function* $W(.;\mathbf{\Upsilon})$ *such that* $V(.;\Upsilon)$ *obeys Definition 270, i.e., (20.20), because*

$$t \longrightarrow \infty \Longrightarrow V(t,\mathbf{z};\mathbf{\Upsilon}) = (1+t^2)\,\rho(\mathbf{z},\mathbf{\Upsilon}) \longrightarrow \infty,\ \forall(\mathbf{z} \neq \mathbf{0}_K) \in \mathfrak{R}^K.$$

For every $\xi \in \mathfrak{R}^+$ *the set* $\mathfrak{V}_\xi(t;\Upsilon)$ *is asymptotically contractive relative to* Υ *(Example 298),*

$$\mathfrak{V}_\xi(t;\Upsilon) = \left\{\mathbf{z} : \rho(\mathbf{z},\mathbf{\Upsilon}) < \xi\left(1+t^2\right)^{-1}\right\} \longrightarrow \Upsilon \text{ as } t \longrightarrow \infty.$$

For every positive definite function $W(.;\mathbf{\Upsilon})$ *there is* $\tau \in \mathfrak{T}$ *such that*

$$V(t,\mathbf{z};\mathbf{\Upsilon}) > W(\mathbf{z};\mathbf{\Upsilon}),\ \forall(t > \tau) \in \mathfrak{T}.$$

For every $\zeta \in \mathfrak{R}^+$ *there is* $\tau \in \mathfrak{T}$ *such that* $\mathfrak{V}_\xi(t;\Upsilon)$ *is the proper subset of* $\mathfrak{N}_\zeta(\Upsilon)$ *at every* $t \in \mathfrak{T}$, $t \geq \tau$,

$$\mathfrak{V}_\xi(t;\Upsilon) \subset \mathfrak{N}_\zeta(\Upsilon),\ \forall t \in \mathfrak{T},\ t \geq \tau.$$

Example 307 *The function* $V(t,\mathbf{z};\mathbf{\Upsilon}) = (1+t^2)(2+t^2)^{-1}\rho(\mathbf{z},\mathbf{\Upsilon})$ *is a decrescent function relative to* Υ *because it satisfies Definition 270 for the (globally) positive definite function* $W(\mathbf{z}) = \rho(\mathbf{z},\mathbf{\Upsilon})$ *relative to* Υ,

$$V(t,\mathbf{z};\mathbf{\Upsilon}) = \frac{1+t^2}{2+t^2}\rho(\mathbf{z},\mathbf{\Upsilon}) \leq \rho(\mathbf{z},\mathbf{\Upsilon}),\ \forall(t,\mathbf{z}) \in \mathfrak{T} \times \mathfrak{R}^K.$$

For every $\xi \in \mathfrak{R}^+$ *there is* $\zeta \in \mathfrak{R}^+$ *such that the set* $\mathfrak{W}_\zeta(\Upsilon)$,

$$\mathfrak{W}_\zeta(\Upsilon) = \{\mathbf{z} : W(\mathbf{z};\Upsilon) < \zeta\} = \{\mathbf{z} : \rho(\mathbf{z},\mathbf{\Upsilon}) < \zeta\},$$

is always a subset of the set $\mathfrak{V}_\xi(t;\Upsilon)$,

$$\mathfrak{W}_\zeta(\Upsilon) \subset \mathfrak{V}_\xi(t;\Upsilon),\ \forall t \in \mathfrak{T}.$$

21.2 Set invariance relative to a function

The concept of the set positive invariance with respect to a function was introduced in [175, Definition 8, p. 542], [209, Definition 1, p. 45], [227, Definition19, p. 34]. It enabled broadening stability criteria to hold on *time*-varying asymptotically contractive sets.

Definition 308 ***Set positive invariance relative to a function and relative to the set*** Υ

A set $\mathfrak{S}(t)$ *is:*

a) ***Positively invariant with respect to a function*** $V(.;\Upsilon)$ ***and relative to the set*** Υ *on* $\mathfrak{T}_0$ *if, and only if, respectively, for* $t_0 \in \mathfrak{T}$ *there is* $\xi_0 = \xi(t_0) \in \mathfrak{R}$ *such that both 1) and 2) hold,*

1. $\mathfrak{V}_{\xi_0}(t_0;\Upsilon) \subseteq \mathfrak{S}(t_0)$.
2. $\mathfrak{V}_{\xi_0}(t;\Upsilon) \subseteq \mathfrak{S}(t),\ \forall t \in \mathfrak{T}_0$.

b) ***Invariant with respect to a function*** $V(.;\Upsilon)$ ***and relative to the set*** Υ *if, and only if, there is* $\xi \in \mathfrak{R}$ *such that* $\mathfrak{V}_{\xi}(t;\Upsilon) \subseteq \mathfrak{S}(t),\ \forall t \in \mathfrak{T}$.

*The expression "**on** $\mathfrak{T}_0$" is to be omitted if, and only if,* $\mathfrak{T}$ *replaces* $\mathfrak{T}_0$.

*The expression "**relative to the set** Υ" is to be omitted if, and only if,* $\Upsilon = \mathfrak{O} = \{\mathbf{0}_K\}$.

A useful relationship between the function positive definiteness and sets follows.

Proposition 309 *[175, Proposition 6, p. 542], [227, Proposition 14, p. 35]* ***The (positive) invariance of a neighborhood*** $\mathfrak{N}(\Upsilon)(t)$ ***of*** Υ ***relative to a positive definite function*** $V(.;\Upsilon) : \mathfrak{T} \times \mathfrak{R}^K \longrightarrow \mathfrak{R}$, $(V(.;\Upsilon) : \mathfrak{T}_\tau \times \mathfrak{R}^K \longrightarrow \mathfrak{R})$ ***and relative to the set*** Υ

a) Every time-invariant, connected and bounded neighborhood $\mathfrak{N}(t;\Upsilon) \equiv \mathfrak{N}(\Upsilon)$ *of* Υ*, is, respectively, (positive) invariant (on* $\mathfrak{T}_\tau$*) relative to every positive definite function* $V(.;\Upsilon)$ *(on* $\mathfrak{T}_\tau$*) and relative to* Υ.

b) If $\mathfrak{N}(t;\Upsilon)$ *is a continuous time-varying neighborhood of* Υ *(on* $\mathfrak{T}_\tau$*), which is not asymptotically contractive relative to* Υ *as* $t \longrightarrow -\infty$ *and* $t \longrightarrow \infty$ *(as* $t \longrightarrow \infty$*) then it is (positive) invariant (on* $\mathfrak{T}_\tau$*) relative to every positive definite function (on* $\mathfrak{T}_\tau$*) and relative to* Υ*, respectively.*

c) If $\mathfrak{N}(t;\Upsilon)$ *is a continuous time-varying neighborhood of* Υ *(on* $\mathfrak{T}_\tau$*), which is asymptotically contractive relative to* Υ *as* $t \longrightarrow -\infty$ *and* $t \longrightarrow \infty$ *(as* $t \longrightarrow \infty$*) then it is not (positive) invariant (on* $\mathfrak{T}_\tau$*) relative to every positive definite function relative to* Υ *(on* $\mathfrak{T}_\tau$*), respectively.*

Comment 310 ***On Proposition*** *309*

In Proposition 309 the claim under:

The statement a) follows directly from Proposition 300.

The statement b) results from a) and the property of the neighborhood $\mathfrak{N}(t;\Upsilon)$ *that it is not asymptotically contractive relative to* Υ *meaning that there is a time-invariant neighborhood* $\mathfrak{N}^*(\Upsilon)$ *of* Υ *such that it is subset of* $\mathfrak{N}(t;\Upsilon)$, $\forall t \in \mathfrak{T}$, $(\forall t \in \mathfrak{T}_\tau)$*:*

$$\mathfrak{N}^*(\Upsilon) \subseteq \mathfrak{N}(t;\Upsilon), \quad \forall t \in \mathfrak{T}, \ (\forall t \in \mathfrak{T}_\tau).$$

The statement c) is the consequence of the statements a) and b).

Example 311 *The set*

$$\mathfrak{S}(t) = \left\{\mathbf{z} : \mathbf{z} \in \mathfrak{R}^2,\ (2|z_1| + |z_2|) \leq \left(1+t^2\right)^{-1}\right\}$$

is invariant relative to the function $V(.;\Upsilon)$,

$$V(t,\mathbf{z};\Upsilon) = \left(1+t^2\right)(2|z_1| + |z_2|),\ \mathbf{z} = \begin{bmatrix} z_1 \\ z_2 \end{bmatrix},$$

because the set $\mathfrak{V}_\xi(t;\mathfrak{O})$,

$$\mathfrak{V}_\xi(t;\mathfrak{O}) = \left\{\mathbf{z} : \mathbf{z} \in \mathfrak{R}^2,\ V(t,\mathbf{z};\boldsymbol{\Upsilon}) = \left(1+t^2\right)\left(2|z_1|+|z_2|\right) \leq \xi\right\}$$

obeys the conditions a) and b) of Definition 308 for every $\xi \in]0,1]$.

Example 312 *The set*

$$\mathfrak{S}(t) = \left\{\mathbf{z} : \mathbf{z} \in \mathfrak{R}^2,\ \left(2|z_1|+|z_2|\right) \leq \left(1+t^2\right)^{-1}\right\}$$

is neither invariant nor positively invariant (on any $\mathfrak{T}_\tau$*) relative to the function* $V(.;\mathfrak{O})$,

$$V(t,\mathbf{z};\mathfrak{O}) = \left(\frac{1+t^2}{2+t^2}\right)\left(2|z_1|+|z_2|\right),\ \mathbf{z} = \begin{bmatrix} z_1 \\ z_2 \end{bmatrix},$$

because the set $\mathfrak{V}_\xi(t;\mathfrak{O})$,

$$\mathfrak{V}_\xi(t;\mathfrak{O}) = \left\{\mathbf{z} : \mathbf{z} \in \mathfrak{R}^2,\ V(t,\mathbf{z};\boldsymbol{\Upsilon}) = \left(\frac{1+t^2}{2+t^2}\right)\left(2|z_1|+|z_2|\right) \leq \xi\right\}$$

does not obey either the conditions b) or a) of Definition 308 for any $\xi \in \mathfrak{R}$.

If a set $\mathfrak{S}(t)$ is bounded (on $\mathfrak{T}_\tau$) then there can exist a function $\omega(.)$ such that its constant value $\xi \in \mathfrak{R}$, $\omega(t,\mathbf{z}) = \xi$, determines a hypersurface representing the boundary $\partial\mathfrak{S}(t)$ of the set $\mathfrak{S}(t)$ (on $\mathfrak{T}_\tau$), respectively. More precisely:

Definition 313 ***The boundary function of a set***
A function $\omega(.) : \mathfrak{T} \times \mathfrak{R}^K \longrightarrow \mathfrak{R}$ *is* ***the boundary function of a connected set*** $\mathfrak{S}(t)$, $\mathfrak{S}(t) \subset \mathfrak{R}^K$, ***(on*** $\mathfrak{T}_\tau$***)*** *if, and only if, respectively, the set* $\mathfrak{S}(t)$ *has the boundary* $\partial\mathfrak{S}(t)$ *at every* $t \in \mathfrak{T}$ *(* $t \in \mathfrak{T}_\tau$*), and there is* $\xi \in \mathfrak{R}$ *such that*

$$\omega(t,\mathbf{z}) = \xi \Longleftrightarrow \mathbf{z} \in \partial\mathfrak{S}(t)\ ,\ \forall t \in \mathfrak{T}\ (\forall t \in \mathfrak{T}_\tau).$$

Example 314 *The boundary function of the bounded set:*

$$\mathfrak{S}(t) = \left\{\mathbf{z} : \mathbf{z} \in \mathfrak{R}^2,\ \left(2|z_1|+|z_2|\right) \leq \left(1+t^2\right)^{-1}\right\}$$

is

$$\omega(t,\mathbf{z}) = \left(1+t^2\right)\left(2|z_1|+|z_2|\right).$$

Its boundary is

$$\partial\mathfrak{S}(t) = \left\{\mathbf{z} : \mathbf{z} \in \mathfrak{R}^2,\ \omega(t,\mathbf{z}) = \left(1+t^2\right)\left(2|z_1|+|z_2|\right) = 1\right\}.$$

Definition 315 ***Differentiability of bounded connected sets***
A bounded connected set $\mathfrak{S}(t)$ *with the boundary function* $\omega(.) : \mathfrak{T} \times \mathfrak{R}^K \longrightarrow \mathfrak{R}$ *is* ***differentiable (on*** $\mathfrak{T}_\tau$***)*** *if, and only if, its boundary function is differentiable on* $\mathfrak{T} \times \mathfrak{R}^K$ *(on* $\mathfrak{T}_\tau \times \mathfrak{R}^K$*), respectively.*

Definition 316 ***Radially increasing boundary function on the set boundary***
The boundary function $\omega(.) : \mathfrak{T} \times \mathfrak{R}^K \longrightarrow \mathfrak{R}$ *of a bounded set* $\mathfrak{S}(t;t_\tau)$, $\mathfrak{S}(t;t_\tau) \subset \mathfrak{R}^K$, *is* ***radially increasing on its boundary*** $\partial\mathfrak{S}(t;t_\tau)$ ***on*** $\mathfrak{T}_\tau$ *if, and only if, respectively, for every* $t \in \mathfrak{T}_\tau$ *there is* $\zeta(t) \in]0,1[$ *such that* $\mathbf{z} \in \partial\mathfrak{S}(t;t_\tau)$ *implies*

$$\omega\left[t,(1-\lambda)\mathbf{z}\right] < \omega(t,\mathbf{z}) < \omega\left[t,(1+\lambda)\mathbf{z}\right],\ \forall\lambda \in]0,\zeta(t)],$$
$$\forall\mathbf{z} \in \partial\mathfrak{S}(t;t_\tau),\ \forall t \in \mathfrak{T}_\tau.$$

The expression "on $\mathfrak{T}_\tau$*" is to be omitted if, and only if,* $\mathfrak{T}_\tau = \mathfrak{T}$.

Lemma 317 ***Positive invariance of time-continuous time-varying sets relative to the set*** Υ

Let $\mathfrak{N}(t;t_0;\Upsilon)$ *be a connected nonempty continuous neighborhood of the set* Υ *at every* $t \in \mathfrak{T}_0$, *i.e., on* $\mathfrak{T}_0$. *Let both the neighborhood* $\mathfrak{N}(t;t_0;\Upsilon)$ *of* Υ *on* $\mathfrak{T}_0$ *and the system motions* $\mathbf{z}(t;t_0;\mathbf{z}_0;\mathbf{I})$ *be time-continuous on* $\mathfrak{T}_0$ *for every* $[\mathbf{z}_0, \mathbf{I}(.)] \in Cl\mathfrak{N}(t;t_0;\Upsilon) \times \mathcal{J}^i$, $t_0 \in \mathfrak{T}_i$. *If both*

1) the boundary function $\omega(.) : \mathfrak{T}_0 \times \mathfrak{R}^K \longrightarrow \mathfrak{R}$ *of* $\mathfrak{N}(t;t_0;\Upsilon)$ *is radially increasing on* $\partial\mathfrak{N}(t;t_0;\Upsilon)$ *on* $\mathfrak{T}_0$

and

2) $D^+\omega[t, \mathbf{z}(t;t_0;\mathbf{z}_0;\mathbf{I})] < 0$, $\forall(t, t_0, \mathbf{z}_0, \mathbf{I}) \in \mathfrak{T}_0 \times \mathfrak{T}_i \times Cl\mathfrak{N}(t_0;t_0;\Upsilon) \times \mathcal{J}^i$

then the closure $Cl\mathfrak{N}(t;t_0;\Upsilon)$ *of* $\mathfrak{N}(t;t_0;\Upsilon)$ *is positive invariant with respect to the system motions* $\mathbf{z}(t;t_0;\mathbf{z}_0;\mathbf{I})$ *for every* $\mathbf{I}(.) \in \mathcal{J}^i$.

Proof. Let the conditions of Lemma 317 hold. Let us assume that the closure Cl $\mathfrak{N}(t;t_0;\Upsilon)$ of $\mathfrak{N}(t;t_0;\Upsilon)$ is not positive invariant relative to the system motions $\mathbf{z}(t;t_0;\mathbf{z}_0;\mathbf{I})$ for every $\mathbf{I}(.) \in \mathcal{J}^i$. This means, due to Definition 45, that there is a quartet $[\mathbf{z}_0^*, t^*, \mathbf{z}^*, \mathbf{I}^*(.)] \in \mathrm{Cl}\mathfrak{N}(t_0;t_0;\Upsilon) \times \mathfrak{T}_0 \times \partial\mathfrak{N}(t^*;t_{0;\Upsilon}) \times \mathcal{J}^i$ and $\widetilde{t} \in \mathfrak{R}^+$ such that both $\mathbf{z}(t_0;t_0;\mathbf{z}_0;\mathbf{I}^*) = \mathbf{z}_0 \in \mathrm{Cl}\mathfrak{N}(t_0;t_0;\Upsilon)$ and $\mathbf{z}(t;t^*;\mathbf{z}^*;\mathbf{I}^*) \notin \mathrm{Cl}\mathfrak{N}(t;t_0;\Upsilon)$, $\forall t \in]t^*, \widetilde{t}[$. This, Definition 313, and Definition 316 imply that both $\omega(t^*, \mathbf{z}^*) = \xi$ and $\omega[t, \mathbf{z}(t;t^*;\mathbf{z}^*;\mathbf{I}^*)] > \xi$, $\forall t \in]t^*, \overline{t}[$, where such $\overline{t} \in]t^*, \widetilde{t}[$ exists due to the *time* continuity on $\mathfrak{T}_0$ of both $\mathfrak{N}(t;t_0;\Upsilon)$ and the motions $\mathbf{z}(t;t_0;\mathbf{z}_0;\mathbf{I})$ for every $[\mathbf{z}_0, \mathbf{I}(.)] \in \mathrm{Cl}\mathfrak{N}(t_0;t_0;\Upsilon) \times \mathcal{J}^i$. The consequence is that

$$\langle \omega[t+\theta, \mathbf{z}(t+\theta;t^*;\mathbf{z}^*;\mathbf{I}^*)] - \omega(t^*, \mathbf{z}^*) \rangle \theta^{-1} > 0, \quad \forall \theta \in]0, \overline{t} - t^*[.$$

This implies in the limit as $\theta \longrightarrow 0^+$:

$$D^+\omega[t, \mathbf{z}(t;t^*;\mathbf{z}^*;\mathbf{I}^*)]$$
$$= \lim \left\{ \langle \omega[t+\theta, \mathbf{z}(t+\theta;t^*;\mathbf{z}^*;\mathbf{I}^*)] - \omega(t^*, \mathbf{z}^*) \rangle \theta^{-1}; \theta \longrightarrow 0^+ \right\} > 0.$$

This result contradicts condition 2) of the lemma statement. The condition is the consequence of the assumption that the closure $\mathrm{Cl}\mathfrak{N}(t;t_0;\Upsilon)$ of $\mathfrak{N}(t;t_0;\Upsilon)$ is not positive invariant with respect to the system motions $\mathbf{z}(t;t_0;\mathbf{z}_0;\mathbf{I})$ for every $\mathbf{I}(.) \in \mathcal{J}^i$. The assumption fails that proves that the closure $\mathrm{Cl}\mathfrak{N}(t;t_0;\Upsilon)$ of $\mathfrak{N}(t;t_0;\Upsilon)$ is positive invariant with respect to the system motions $\mathbf{z}(t;t_0;\mathbf{z}_0;\mathbf{I})$ for every $\mathbf{I}(.) \in \mathcal{J}^i$ ■

21.3 Semidefinite functions and time-varying sets

We define semidefinite functions with respect to (w.r.t.) *time*-varying (i.e., *variable*) sets.

Let η be a positive real number or ∞, $\eta \in]0, \infty]$.

Let $\mathfrak{N}_\eta[\Upsilon(t)]$ be the η-neighborhood $\mathfrak{N}_\eta[\Upsilon(t)]$ of $\Upsilon(t)$ at $t \in \mathfrak{T}_0$, and the set $\mathfrak{N}_\eta(\Upsilon;\mathfrak{T}_0)$ be the union of all $\mathfrak{N}_\eta[\Upsilon(t)]$ in $t \in \mathfrak{T}_0$:

$$\mathfrak{N}_\eta(\Upsilon;\mathfrak{T}_0) = \cup[\mathfrak{N}_\eta[\Upsilon(t)] : t \in \mathfrak{T}_0.] \subseteq \mathfrak{R}^K.$$

Definition 318 ***Semidefinite function relative to the time-varying set*** $\Upsilon(t)$ ***on*** $\mathfrak{T}_0$

Let the set $\Upsilon(t)$ *be nonempty, connected, continuous, and with the nonempty boundary* $\partial\Upsilon(t)$, $\partial\Upsilon(t) \neq \phi$, *at every moment* $t \in \mathfrak{T}_0$, $\Upsilon(t) \in \mathfrak{C}(\mathfrak{T}_0)$.

A function $V(.;\Upsilon) : \mathfrak{T} \times \mathfrak{R}^K \longrightarrow \mathfrak{R}$ *is* ***positive*** $\langle$***negative***$\rangle$ ***semidefinite relative to the set*** $\Upsilon(t)$ ***on*** $\mathfrak{T}_0$ *if, and only if, respectively:*

i) There is constant $\eta \in]0, \infty]$ *such that the function* $V(.;\Upsilon)$ *is continuous on* $\mathfrak{T}_0 \times \mathfrak{N}_\eta(\Upsilon;\mathfrak{T}_0)$: $V(t, \mathbf{z}; \Upsilon) \in \mathfrak{C}[\mathfrak{T}_0 \times \mathfrak{N}_\eta(\Upsilon;\mathfrak{T}_0)]$.

ii) There is $\xi \in]0,\eta]$ such that the function $V(.;\Upsilon)$ has a nonnegative $\langle$nonpositive$\rangle$ value $V(t,\mathbf{z};\boldsymbol{\Upsilon}) \geq 0$ $\langle V(t,\mathbf{z};\boldsymbol{\Upsilon}) \leq 0\rangle$ for $\mathbf{z} \in \mathfrak{N}_\xi[\Upsilon(t)] \setminus Cl\Upsilon(t)$ at every $t \in \mathfrak{T}_0$: $V(t,\mathbf{z};\boldsymbol{\Upsilon}) > 0$ $\langle V(t,\mathbf{z};\boldsymbol{\Upsilon}) \leq 0\rangle$, $\forall (t,\mathbf{z}) \in \mathfrak{T}_0 \times \mathfrak{N}_\xi[\Upsilon(t)] \setminus Cl\Upsilon(t)$.

iii) There are $t^* \in \mathfrak{T}_0$ and $\mathbf{z}^* \in \mathfrak{N}_\xi[\Upsilon(t^*)] \setminus Cl\Upsilon(t^*)$ such that the value $V(t^*,\mathbf{z}^*;\boldsymbol{\Upsilon})$ of the function $V(.;\Upsilon)$ is positive $\langle$negative$\rangle$: $V(t^*,\mathbf{z}^*) > 0$ $\langle V(t^*,\mathbf{z}^*) < 0\rangle$.

iv) The value $V(t,\mathbf{z};\boldsymbol{\Upsilon})$ of the function $V(.;\Upsilon)$ is equal to zero on the boundary $\partial\Upsilon(t)$ of the set $\Upsilon(t)$: $\mathbf{z} \in \partial\Upsilon(t) \Longrightarrow V(t,\mathbf{z};\boldsymbol{\Upsilon}) = 0$, $\forall t \in \mathfrak{T}_0$.

*If, and only if, the above conditions hold when $\mathfrak{T}_0$ is replaced by $\mathfrak{T}$ then the expression "**on** $\mathfrak{T}_0$" is to be omitted.*

*If, and only if, the above conditions hold for $\eta = \infty$ then the function $V(.;\Upsilon)$ is **globally positive** $\langle$negative$\rangle$ **semidefinite with respect to the set** $\Upsilon(t)$ on $\mathfrak{T}_0$.*

If the set $\Upsilon(t) \equiv \Upsilon$ is *time*-invariant then Definition 318 reduces to the well-known definition of a positive (negative) semidefinite function relative to the set Υ.

21.4 Definite functions relative to *time*-varying sets

Let $\mathfrak{R}^-$ be the set of all negative real numbers.

Definition 319 ***Definite function relative to the time-varying set*** $\Upsilon(t)$ ***on*** $\mathfrak{T}_0$

Let the set $\Upsilon(t)$ be nonempty, connected, continuous, and with the nonempty boundary $\partial\Upsilon(t)$, $\partial\Upsilon(t) \neq \phi$, at every moment $t \in \mathfrak{T}_0$, $\Upsilon(t) \in \mathfrak{C}(\mathfrak{T}_0)$.

*A function $V(.;\Upsilon) : \mathfrak{T} \times \mathfrak{R}^K \longrightarrow \mathfrak{R}$ is **positive** $\langle$**negative**$\rangle$ **definite relative to the set** $\Upsilon(t)$ **on** $\mathfrak{T}_0$ if, and only if, respectively:*

i) There is $\eta \in \mathfrak{R}^+$ such that the function $V(.;\Upsilon)$ is continuous on $\mathfrak{T}_0 \times Cl\mathfrak{N}_\eta(\Upsilon;\mathfrak{T}_0)$: $V(t,\mathbf{z};\boldsymbol{\Upsilon}) \in \mathfrak{C}[\mathfrak{T}_0 \times Cl\mathfrak{N}_\eta(\Upsilon;\mathfrak{T}_0)]$.

ii) There is $\xi \in]0,\eta]$ such that the function $V(.;\Upsilon)$ has a positive $\langle$negative$\rangle$ value $V(t,\mathbf{z};\boldsymbol{\Upsilon}) > 0$ for every $\mathbf{z} \in \mathfrak{N}_\xi(\Upsilon;\mathfrak{T}_0) \setminus Cl\Upsilon(t)$ at every $t \in \mathfrak{T}_0$: $V(t,\mathbf{z};\boldsymbol{\Upsilon}) > 0$, $\forall (t,\mathbf{z}) \in \mathfrak{T}_0 \times \mathfrak{N}_\xi(\Upsilon;\mathfrak{T}_0) \setminus Cl\Upsilon(t)$.

iii) The value $V(t,\mathbf{z};\boldsymbol{\Upsilon})$ of the function $V(.;\Upsilon)$ is equal to zero only on the boundary $\partial\Upsilon(t)$ of the set $\Upsilon(t)$: $V(t,\mathbf{z};\boldsymbol{\Upsilon}) = 0 \Longleftrightarrow \mathbf{z} \in \partial\Upsilon(t)$, $\forall t \in \mathfrak{T}_0$.

iv) For every $\theta \in]0,\eta]$ there is $\psi \in \mathfrak{R}^+$ $\langle \psi \in \mathfrak{R}^- \rangle$ such that the smallest value $\zeta_{InfV}(\theta)$ $\langle$the biggest value $\zeta_{SupV}(\theta)\rangle$ of $V(t,\mathbf{z};\boldsymbol{\Upsilon})$ on the boundary $\partial\mathfrak{N}_\theta[\Upsilon(t)]$ of the θ-neighborhood $\mathfrak{N}_\theta[\Upsilon(t)]$ of the set $\Upsilon(t)$ for all $t \in \mathfrak{T}_0$ is a real number not less (not bigger) than ψ:

$$
\begin{gathered}
\forall\theta \in]0,\eta[,\ \exists\psi \in \mathfrak{R}^+ \ \langle \exists\psi \in \mathfrak{R}^- \rangle \Longrightarrow \\
\zeta_{InfV}(\theta) = inf\ \{V(t,\mathbf{z};\boldsymbol{\Upsilon}) : (t,\mathbf{z}) \in \mathfrak{T}_0 \times \partial\mathfrak{N}_\theta[\Upsilon(t)]\} \in \mathfrak{R}^+, \\
and\ \zeta_{InfV}(\theta) \geq \psi,\ \zeta_{InfV}(0) = 0, \\
\langle \zeta_{SupV}(\theta) = sup\ \{V(t,\mathbf{z};\boldsymbol{\Upsilon}) : (t,\mathbf{z}) \in \mathfrak{T}_0 \times \partial\mathfrak{N}_\theta[\Upsilon(t)]\} \in \mathfrak{R}^- \rangle, \\
and\ \langle \zeta_{SupV}(\theta) \leq \psi,\ \zeta_{SupV}(0) = 0 \rangle.
\end{gathered} \tag{21.9}
$$

*v) The function $V(.;\Upsilon)$ is **globally positive** $\langle$**negative**$\rangle$ **definite relative to the set** $\Upsilon(t)$ **on** $\mathfrak{T}_0$ if, and only if, the above conditions i)-iv) hold for $\xi = \eta \longrightarrow \infty$.*

*vi) The function $V(.;\Upsilon)$ is **radially unbounded relative to the set** $\Upsilon(t)$ **on** $\mathfrak{T}_0$ if, and only if, the above conditions i)-v) are valid and $\rho[\mathbf{z},\Upsilon(t)] \longrightarrow \infty$ implies $V(t,\mathbf{z};\boldsymbol{\Upsilon}) \longrightarrow \infty$ $\langle V(t,\mathbf{z};\boldsymbol{\Upsilon}) \longrightarrow -\infty \rangle$, $\forall t \in \mathfrak{T}_0$.*

*If, and only if, the above conditions hold when $\mathfrak{T}$ replaces $\mathfrak{T}_0$ then the expression "**on** $\mathfrak{T}_0$" is to be omitted.*

*If, and only if, $\Upsilon(t) \equiv \mathfrak{O} = \{\mathbf{0}_K\}$ then the expression "**relative to the set** $\Upsilon(t)$" should be omitted.*

This definition determines that a function $V(.;\Upsilon) : \mathfrak{T} \times \mathfrak{R}^K \longrightarrow \mathfrak{R}$ is **positive** $\langle$**negative**$\rangle$ **definite relative to the set** $\Upsilon(t)$ **on** $\mathfrak{T}_0$ if, and only if, respectively, $-V(.;\Upsilon)$ is **negative** $\langle$**positive**$\rangle$ **definite relative to the set** $\Upsilon(t)$ **on** $\mathfrak{T}_0$.

Property 320 ***Relationship between a positive definite function*** $V(.;\Upsilon)$ ***relative to*** $\Upsilon(t)$ ***and distance of*** $\mathbf{z}$ ***from*** $\Upsilon(t)$

If the function $V(.;\Upsilon)$ *is positive definite relative to the set* $\Upsilon(t)$ *on* $\mathfrak{T}_0$ *then from (21.9) follow:*

$$\begin{gathered}\theta \in]0,\eta[,\ \inf\{V(t,\mathbf{z};\boldsymbol{\Upsilon}) : (t,\mathbf{z}) \in \mathfrak{T}_0 \times \partial\mathfrak{N}_\theta[\Upsilon(t)]\} = \zeta = \zeta_{InfV}(\theta) \Longrightarrow \\ \mathfrak{V}_\zeta(t;\Upsilon) = \{\mathbf{z} : V(t,\mathbf{z};\boldsymbol{\Upsilon}) < \zeta\} \subseteq \mathfrak{N}_\theta[\Upsilon(t)],\ \forall t \in \mathfrak{T}_0 \\ Cl\mathfrak{V}_\zeta(t;\Upsilon) = \{\mathbf{z} : V(t,\mathbf{z};\boldsymbol{\Upsilon}) \le \zeta\} \subseteq Cl\mathfrak{N}_\theta[\Upsilon(t)],\ \forall t \in \mathfrak{T}_0 \Longrightarrow \\ \mathbf{z} \in \mathfrak{V}_\zeta(t;\Upsilon) \Longleftrightarrow \rho[\mathbf{z},\Upsilon(t)] < \theta,\ \forall t \in \mathfrak{T}_0, \\ \mathbf{z} \in Cl\mathfrak{V}_\zeta(t;\Upsilon) \Longleftrightarrow \rho[\mathbf{z},\Upsilon(t)] \le \theta,\ \forall t \in \mathfrak{T}_0,\end{gathered} \tag{21.10}$$

and for every $\zeta \in]\mu,\zeta_{InfV}(\eta)[$ *there is* $\gamma = \gamma(\zeta) \in \mathfrak{R}^+$ *such that the set* $\mathfrak{V}_\zeta(t;\Upsilon)$ *is the instantaneous subset of the* γ*-neighborhood* $\mathfrak{N}_\gamma[\Upsilon(t)]$ *of the set* $\Upsilon(t)$ *at every* $t \in \mathfrak{T}_o$ *:*

$$\begin{gathered}\eta \in \mathfrak{R}^+,\ \text{and}\ \zeta \in]0,\zeta_{InfV}(\eta)] \Longrightarrow \exists\gamma = \gamma(\zeta) \in \mathfrak{R}^+ \Longrightarrow \\ \mathfrak{V}_\zeta(t;\Upsilon) \subseteq \mathfrak{N}_\gamma[\Upsilon(t)],\ \forall t \in \mathfrak{T}_0.\end{gathered} \tag{21.11}$$

The minimal value of $\gamma(\zeta)$ *that obeys (21.11) for a given* $\zeta \in]\mu,\zeta_{InfV}(\eta)[$ *is denoted by* $\gamma_m(\zeta)$*. This prevents the unbounded spreading of the sets* $\mathfrak{V}_\zeta(t;\Upsilon)$ *to the* $\mathfrak{R}^K$ *as* $t \longrightarrow \infty$*.*

The functions $\zeta_{InfV}(.)$ *and* $\gamma_m(.)$ *are defined, continuous and strictly monotonously increasing:*

$$\begin{gathered}\zeta_{InfV}(\theta) \in \mathfrak{C}([0,\eta]),\ 0 < \theta_1 < \theta_2 \le \eta \Rightarrow \\ 0 = \zeta_{InfV}(0) < \zeta_{InfV}(\theta_1) < \zeta_{InfV}(\theta_2),\end{gathered} \tag{21.12}$$

$$\begin{gathered}\gamma_m(\zeta) \in \mathfrak{C}([0,\zeta_{InfV}(\eta)]),\ 0 < \zeta_1 < \zeta_2 \le \zeta_{InfV}(\eta) \Rightarrow \\ 0 = \gamma_m(0) < \gamma_m(\zeta_1) < \gamma_m(\zeta_2).\end{gathered} \tag{21.13}$$

They belong to the functional families $\mathcal{K}_{[0,\eta]}$ *and* $\mathcal{K}_{[0,\zeta_{InfV}(\eta)]}$*, respectively:*

$$\zeta_{InfV}(.) \in \mathcal{K}_{[0,\eta]}\ \text{and}\ \gamma_m(.) \in \mathcal{K}_{[0,\zeta_{InfV}(\eta)]}. \tag{21.14}$$

Iff additionally, the function $V(.;\Upsilon)$ *is radially unbounded relative to the set* $\Upsilon(t)$ *on* $\mathfrak{T}_0$ *then*

$$\zeta_{InfV}(.) \in \mathcal{KR}\ \text{and}\ \gamma_m(.) \in \mathcal{KR}. \tag{21.15}$$

Comment 321 ***Variation of the vector*** $\mathbf{z}$ ***in*** *(21.18)*

The vector $\mathbf{z}$ *in (21.18) need not be constant because it belongs to the boundary* $\partial\mathfrak{N}_\theta[\Upsilon(t)]$, $\mathbf{z} \in \partial\mathfrak{N}_\theta[\Upsilon(t)]$*, of the* θ*-neighborhood* $\mathfrak{N}_\theta[\Upsilon(t)]$ *of the set* $\Upsilon(t)$ *that is time-varying. In other words, the vector* $\mathbf{z}$ *in (21.18) need not be constant because* $\rho[\mathbf{z},\Upsilon(t)]$ *for* $\mathbf{z} \in \partial\mathfrak{N}_\theta[\Upsilon(t)]$ *is the constant* θ*-distance from the time-varying set* $\Upsilon(t)$*.*

If the set $\Upsilon(t) \equiv \Upsilon$ is *time*-invariant then Definition 319 reduces to the well-known definition of a positive definite function relative to the set Υ (see Definition 267). Proposition 269 establishes the link between *time*-varying positive definite functions relative to the *time*-invariant set Υ and the comparison functions $\varphi(.) : \mathfrak{R}_+ \longrightarrow \mathfrak{R}_+$ by Hahn [292], $\varphi(.) \in \mathcal{K}_{[0,\gamma[}$, $0 < \gamma \le \infty$, or $\varphi(.) \in \mathcal{K}$, or $\varphi(.) \in \mathcal{KR}$, Definition 256. An example of such a function $\varphi(.)$ in the framework of both *time*-invariant and *time*-varying sets is the function $\zeta_{\text{InfV}}(.)$ due to (21.14) and (21.15).

Theorem 322 ***Positive definite function relative to the time-varying set*** $\Upsilon(t)$ ***on*** $\mathfrak{T}_0$ ***and comparison functions***

Let the set $\Upsilon(t)$ *be nonempty, connected, continuous, and with nonempty boundary* $\partial\Upsilon(t)$ *at every moment* $t \in \mathfrak{T}_0$, $\Upsilon(t) \in \mathfrak{C}(\mathfrak{T}_0)$.

Let there be constant $\eta \in \mathfrak{R}^+$ *such that the function* $V(.;\Upsilon) : \mathfrak{T} \times \mathfrak{R}^K \longrightarrow \mathfrak{R}$ *is continuous on* $\mathfrak{T}_0 \times \mathfrak{N}_\eta(\Upsilon;\mathfrak{T}_0)$: $V(t,\mathbf{z};\boldsymbol{\Upsilon}) \in \mathfrak{C}[\mathfrak{T}_0 \times \mathfrak{N}_\eta(\Upsilon;\mathfrak{T}_0)]$. *In order for the function* $V(.;\Upsilon)$ *to be* ***positive definite relative to the set*** $\Upsilon(t)$ ***on*** $\mathfrak{T}_0$ *it is necessary and sufficient that:*

i) The value $V(t,\mathbf{z};\boldsymbol{\Upsilon})$ *of the function* $V(.;\Upsilon)$ *is equal to zero only on the boundary* $\partial\Upsilon(t)$ *of the set* $\Upsilon(t)$:

$$V(t,\mathbf{z};\boldsymbol{\Upsilon}) = 0 \Longleftrightarrow \mathbf{z} \in \partial\Upsilon(t),\ \forall t \in \mathfrak{T}_0, \tag{21.16}$$

and

there is $\varphi(.) \in \mathcal{K}_{[0,\eta[}$ *such that:*

$$V(t,\mathbf{z};\boldsymbol{\Upsilon}) \geq \varphi\{\rho[\mathbf{z},\Upsilon(t)]\},\ \forall t \in \mathfrak{T}_0. \tag{21.17}$$

ii) In order for the function $V(.;\Upsilon)$ *to be* ***globally positive definite relative to the set*** $\Upsilon(t)$ ***on*** $\mathfrak{T}_0$ *it is necessary and sufficient that the above conditions under i) hold for* $\eta = \infty$, *i.e., for* $\varphi(.) \in \mathcal{K}$.

iii) In order for the function $V(.;\Upsilon)$ *to be* ***radially unbounded relative to the set*** $\Upsilon(t)$ ***on*** $\mathfrak{T}_0$ *it is necessary and sufficient that the above conditions under i) and ii) hold for* $\varphi(.) \in \mathcal{KR}$.

If, and only if, the above conditions hold when $\mathfrak{T}$ *replaces* $\mathfrak{T}_0$ *then the expression "**on*** $\mathfrak{T}_0$*" is to be omitted.*

If, and only if, $\Upsilon(t) \equiv \mathfrak{O} = \{\mathbf{0}_K\}$ *then the expression "**relative to the set*** $\Upsilon(t)$*" should be omitted.*

Proof. *Necessity.* i) The conditions i) through iv) of Definition 319 and (21.10) through (21.14) of Property 320 prove the necessity of i) of the theorem statement.

ii) The condition v) of Definition 319 implies the necessity of ii) of the theorem statement.

iii) The condition vi) of Definition 319 implies the necessity of iii) of the theorem statement.

Sufficiency. i) The existence of a real number $\eta \in \mathfrak{R}^+$ such that the function $V(.;\Upsilon)$: $\mathfrak{T} \times \mathfrak{R}^K \longrightarrow \mathfrak{R}$ is continuous on $\mathfrak{T}_0 \times \mathfrak{N}_\eta(\Upsilon;\mathfrak{T}_0)$: $V(t,\mathbf{z};\boldsymbol{\Upsilon}) \in \mathfrak{C}[\mathfrak{T}_0 \times \mathfrak{N}_\eta(\Upsilon;\mathfrak{T}_0)]$, (21.16) and (21.17) show that the conditions i)-iii) of Definition 319 are satisfied. From (21.10) and (21.12) follow, respectively:

$$\theta = \rho[\mathbf{z},\Upsilon(t)] \Longrightarrow \mathrm{Cl}\mathfrak{V}_{\zeta_{\mathrm{Inf}V}(\theta)}(t) \subseteq \mathrm{Cl}\mathfrak{N}_\theta[\Upsilon(t)],\ \forall t \in \mathfrak{T}_0,$$
$$\theta_1 = \rho[\mathbf{z},\Upsilon(t)] > \theta \Longrightarrow \mathrm{Cl}\mathfrak{V}_{\zeta_{\mathrm{Inf}V}(\theta_1)}(t) \supset \mathrm{Cl}\mathfrak{V}_{\zeta_{\mathrm{Inf}V}(\theta)}(t),\ \forall t \in \mathfrak{T}_0,$$
$$\text{and } \mathrm{Cl}\mathfrak{V}_{\zeta_{\mathrm{Inf}V}(\theta_1)}(t) \subseteq \mathfrak{N}_{\theta_1}[\Upsilon(t)],\ \forall t \in \mathfrak{T}_0.$$

These facts and (21.14) show that

$$\zeta_{\mathrm{Inf}V}(\theta) = \inf\{V(t,\mathbf{z};\boldsymbol{\Upsilon}) : (t,\mathbf{z}) \in \mathfrak{T}_0 \times \partial\mathfrak{N}_\theta[\Upsilon(t)]\}.$$

The condition iv) of Definition 319 is also valid. The function is positive definite relative to the set $\Upsilon(t)$ on $\mathfrak{T}_0$.

ii) The condition ii) of the theorem statement proves the validity of v) of Definition 319.

iii) The condition iii) of the theorem statement proves that vi) of Definition 319 is satisfied.

The last statement of the theorem is the same as the last statement of Definition 319 ■

21.5 Decrescent functions and *time*-varying sets

Definition 323 ***Decrescent function relative to a time-varying set*** $\Upsilon(t)$ ***on*** $\mathfrak{T}_0$

Let the set $\Upsilon(t)$ *be nonempty, connected, continuous, and with nonempty boundary* $\partial\Upsilon(t)$ *at every moment* $t \in \mathfrak{T}_0$, $\Upsilon(t) \in \mathfrak{C}(\mathfrak{T}_0)$.

A function $V(.;\Upsilon) : \mathfrak{T} \times \mathfrak{R}^K \longrightarrow \mathfrak{R}_+$ *is* ***decrescent relative to the set*** $\Upsilon(t)$ ***on*** $\mathfrak{T}_0$ *if, and only if:*

i) There is $\eta \in \mathfrak{R}^+$ *such that the function* $V(.;\Upsilon)$ *is continuous on* $\mathfrak{T}_0 \times \mathfrak{N}_\eta(\Upsilon;\mathfrak{T}_0)$: $V(t,\mathbf{z};\boldsymbol{\Upsilon}) \in \mathfrak{C}[\mathfrak{T}_0 \times \mathfrak{N}_\eta(\Upsilon;\mathfrak{T}_0)]$,

ii) The value $V(t,\mathbf{z};\boldsymbol{\Upsilon})$ *of the function* $V(.;\Upsilon)$ *is equal to zero only on the boundary* $\partial\Upsilon(t)$ *of the set* $\Upsilon(t)$: $V(t,\mathbf{z};\boldsymbol{\Upsilon}) = 0 \Longleftrightarrow \mathbf{z} \in \partial\Upsilon(t)$, $\forall t \in \mathfrak{T}_0$,

iii) For every $\theta \in]0,\eta[$ *there is* $\xi \in \mathfrak{R}^+$ *such that the biggest value* $\zeta_{SupV}(\theta)$ *of* $V(t,\mathbf{z};\boldsymbol{\Upsilon})$ *on the boundary* $\partial\mathfrak{N}_\theta[\Upsilon(t)]$ *of the* θ*-neighborhood* $\mathfrak{N}_\theta[\Upsilon(t)]$ *of the set* $\Upsilon(t)$ *for all* $t \in \mathfrak{T}_0$ *is a real number not bigger than* ξ:

$$\begin{gathered}\forall\theta \in]0,\eta[,\ \exists\xi \in \mathfrak{R}^+ \Longrightarrow \\ \zeta_{SupV}(\theta) = sup\{V(t,\mathbf{z};\boldsymbol{\Upsilon}) : (t,\mathbf{z}) \in \mathfrak{T}_0 \times \partial\mathfrak{N}_\theta[\Upsilon(t)]\} \in \mathfrak{R}^+, \\ \zeta_{SupV}(\theta) \leq \xi \text{ and } \zeta_{SupV}(0) = 0. \end{gathered} \tag{21.18}$$

If, and only if, the above conditions hold for $\eta \longrightarrow \infty$ *then the function* $V(.;\Upsilon)$ *is* ***globally decrescent with respect to the set*** $\Upsilon(t)$ *on* $\mathfrak{T}_0$.

If, and only if, the above conditions hold when $\mathfrak{T}$ *replaces* $\mathfrak{T}_0$ *then the expression "**on** $\mathfrak{T}_0$" is to be omitted.*

Theorem 324 *If, and only if,* $\Upsilon(t) \equiv \mathfrak{O} = \{\mathbf{0}_K\}$ *then the expression "**relative to the set** $\Upsilon(t)$" is to be omitted.*

Property 325 ***Relationship between a decrescent function*** $V(.;\Upsilon)$ ***relative to*** $\Upsilon(t)$ ***and distance of z from*** $\Upsilon(t)$

If the function $V(.;\Upsilon)$ *is decrescent relative to the set* $\Upsilon(t)$ *on* $\mathfrak{T}_0$ *then from (21.18) follow:*

$$\begin{gathered} sup\{V(t,\mathbf{z};\boldsymbol{\Upsilon}) : (t,\mathbf{z}) \in \mathfrak{T}_0 \times \partial\mathfrak{N}_\theta[\Upsilon(t)]\} = \zeta = \zeta_{SupV}(\theta) \Longrightarrow \\ \mathfrak{V}_\zeta(t;\Upsilon) = \{\mathbf{z} : V(t,\mathbf{z};\boldsymbol{\Upsilon}) < \zeta\} \supseteq \mathfrak{N}_\theta[\Upsilon(t)],\ \forall t \in \mathfrak{T}_0 \Longrightarrow \\ \mathbf{z} \in \partial\mathfrak{V}_\zeta(t;\Upsilon) \Longleftrightarrow \rho[\mathbf{z},\Upsilon(t)] \geq \theta,\ \forall t \in \mathfrak{T}_0, \end{gathered} \tag{21.19}$$

and for every $\zeta \in]\mu,\zeta_{SupV}(\eta)[$ *there is a positive real number* γ, $\gamma = \gamma(\zeta) \in \mathfrak{R}^+$ *such that the* γ*-neighborhood* $\mathfrak{N}_\gamma[\Upsilon(t)]$ *of the set* $\Upsilon(t)$ *is the instantaneous subset of the set* $\mathfrak{V}_\zeta(t;\Upsilon)$ *at every* $t \in \mathfrak{T}_o$:

$$\begin{gathered} \mu > 0 \text{ and } \zeta \in [0,\zeta_{SupV}(\eta)[\Longrightarrow \exists\gamma = \gamma(\zeta) \in \mathfrak{R}^+ \Longrightarrow \\ \mathfrak{V}_\zeta(t;\Upsilon) \supseteq \mathfrak{N}_\gamma[\Upsilon(t)],\ \forall t \in \mathfrak{T}_0. \end{gathered} \tag{21.20}$$

This prevents the asymptotic contraction of the sets $\mathfrak{V}_\zeta(t;\Upsilon)$ *to* $\Upsilon(t)$ *as* $t \longrightarrow \infty$.

Comment 326 ***Variation of the vector z in*** *(21.18)*

The vector $\mathbf{z}$ *in (21.21) need not be constant because it belongs to the boundary* $\partial\mathfrak{N}_\theta[\Upsilon(t)]$ *of the* θ*-neighborhood* $\mathfrak{N}_\theta[\Upsilon(t)]$ *of the set* $\Upsilon(t)$ *that is time-varying. In other words, the vector* $\mathbf{z}$ *in (21.18) need not be constant because its distance* $\rho[\mathbf{z},\Upsilon(t)]$ *from the time-varying set* $\Upsilon(t)$ *is constant.*

If the set $\Upsilon(t) \equiv \Upsilon$ is *time*-invariant then Definition 323 reduces to the well-known definition of a decrescent function relative to the set Υ (see Definition 270).

Definition 323 permits us to discover the link between a *time*-varying decrescent function relative to a *time*-varying set and comparison functions $\varphi(.) : \mathfrak{R}_+ \longrightarrow \mathfrak{R}_+$, $\varphi(.) \in \mathcal{K}_{[0,\gamma[}$, $0 < \gamma \leq \infty$, or $\varphi(.) \in \mathcal{K}$, Definition 256.

Theorem 327 ***Decrescent function relative to a time-varying set*** $\Upsilon(t)$ ***on*** $\mathfrak{T}_0$ ***and comparison functions***

Let the set $\Upsilon(t)$ *be nonempty, connected, continuous, and with nonempty boundary* $\partial\Upsilon(t)$ *at every moment* $t \in \mathfrak{T}_0$, $\Upsilon(t) \in \mathfrak{C}(\mathfrak{T}_0)$.

Let there be $\eta \in \mathfrak{R}^+$ *such that the function* $V(.;\Upsilon) : \mathfrak{T} \times \mathfrak{R}^K \longrightarrow \mathfrak{R}$ *is continuous on* $\mathfrak{T}_0 \times \mathfrak{N}_\eta(\Upsilon;\mathfrak{T}_0)$: $V(t,\mathbf{z};\boldsymbol{\Upsilon}) \in \mathfrak{C}[\mathfrak{T}_0 \times \mathfrak{N}_\eta(\Upsilon;\mathfrak{T}_0)]$. *In order for the function* $V(.;\Upsilon)$ *to be* ***decrescent relative to the set*** $\Upsilon(t)$ ***on*** $\mathfrak{T}_0$ *it is necessary and sufficient that:*

i) The value $V(t,\mathbf{z};\boldsymbol{\Upsilon})$ *of the function* $V(.;\Upsilon)$ *is equal to zero only on the boundary* $\partial\Upsilon(t)$ *of the set* $\Upsilon(t)$, *i.e., (21.16) holds and there is* $\varphi(.) \in \mathcal{K}_{[0,\eta[}$ *such that:*

$$V(t,\mathbf{z};\boldsymbol{\Upsilon}) \leq \varphi\{\rho[\mathbf{z},\Upsilon(t)]\}, \ \forall t \in \mathfrak{T}_0. \tag{21.21}$$

If, and only if, the above conditions hold for $\eta = \infty$ *then the function* $V(.;\Upsilon)$ *is* ***globally decrescent with respect to the set*** $\Upsilon(t)$ ***on*** $\mathfrak{T}_0$.

If, and only if, the above conditions hold when $\mathfrak{T}_0$ *is replaced by* $\mathfrak{T}$ *then the expression "****on*** $\mathfrak{T}_0$*" is to be omitted.*

If, and only if, $\Upsilon(t) \equiv \mathfrak{O} = \{\mathbf{0}_K\}$ *then the expression "****relative to the set*** $\Upsilon(t)$*" is to be omitted.*

The proof of this theorem is fully analogous to the proof of Theorem 322 in view of Property 320 and Property 325.

21.6 Families $\mathcal{F}$ of *time*-varying sets

Definition 328 ***Family*** $\mathcal{F}$ ***of sets*** $\Upsilon(t)$

$\mathcal{F}$ *is the family of all set functions* $\Upsilon(.) : \mathfrak{T} \longrightarrow 2^{\mathfrak{R}^K}$ *with the following properties:*

1. The set $\Upsilon(t)$ *is nonempty, connected, continuous and with nonempty boundary* $\partial\Upsilon(t)$ *at every moment* $t \in \mathfrak{T}_0$, $\Upsilon(t) \in \mathfrak{C}(\mathfrak{T}_0)$.

2) There is a vector $\mathbf{z}_c$ *called* ***the center of*** $\Upsilon(t)$ ***at the moment*** $t \in \mathfrak{T}_0$, $\mathbf{z}_c = \mathbf{z}_c(t)$, *such that it always belongs to* $Cl\Upsilon(t)$ *on* $\mathfrak{T}_0$, $\mathbf{z}_c(t) \in Cl\Upsilon(t), \forall t \geq t_0$.

3) There is a function $\beta(.) : \mathfrak{T} \times \mathfrak{R}^K \longrightarrow \mathfrak{R}$, *which is* ***the boundary function of the time-varying set*** $\Upsilon(t)$, *with the following features:*

a) It is continuous on $\mathfrak{T}_0 \times \mathfrak{R}^K$: $\beta(t,\mathbf{z}) \in \mathfrak{C}(\mathfrak{T}_0 \times \mathfrak{R}^K)$,

b) There is a positive real number μ *such that:*

$$\exists \mu \in R_+ \Longrightarrow \forall t \in \mathfrak{T}_0 :$$
$$\beta(t,\mathbf{z}) > \mu \Longleftrightarrow \mathbf{z} \in \mathfrak{R}^K \backslash Cl\Upsilon(t), \ \beta(t,\mathbf{z}) = \mu \Longleftrightarrow \mathbf{z} \in \partial\Upsilon(t),$$
$$\beta(t,\mathbf{z}) < \mu \Longleftrightarrow \mathbf{z} \in In\Upsilon(t), \ \forall t \in \mathfrak{T}_0, \ \beta(\infty,\mathbf{z}) \leq \mu \Longleftrightarrow \mathbf{z} \in \Upsilon(\infty),$$
$$min \ \{\beta(t,\mathbf{z}) : \mathbf{z} \in \mathfrak{R}^K\} = \beta[t,\mathbf{z}_c(t)] = \beta_m \in \mathfrak{R}, \forall t \geq t_0.$$

c) The function $\beta(.)$ induces the sets $\mathfrak{B}_\zeta(t)$ such that:

$$
\begin{gathered}
\mathfrak{B}_\zeta(t) = \{\mathbf{z} : \beta(t,\mathbf{z}) < \varsigma\} = In\mathfrak{B}_\zeta(t), \\
\beta(t,\mathbf{z}) > \mu \Longleftrightarrow \mathbf{z} \in \mathfrak{R}^K \backslash Cl\mathfrak{B}_\mu(t), \ \forall t \geq t_0, \\
\beta(t,\mathbf{z}) = \mu \Longleftrightarrow \mathbf{z} \in \partial\mathfrak{B}_\mu(t) = \partial\Upsilon(t), \ \mathfrak{B}_\mu(t) = In\Upsilon(t), \\
\beta(t,\mathbf{z}) < \mu \Longleftrightarrow \mathbf{z} \in \mathfrak{B}_\mu(t), \ \forall t \in \mathfrak{T}_0, \\
\beta_m \leq \zeta_1 < \zeta_2 \leq \infty \Longleftrightarrow \left\{ \begin{array}{c} \mathfrak{B}_{\beta_m}(t) \subseteq \mathfrak{B}_{\zeta_1}(t) \subset \mathfrak{B}_{\zeta_2}(t), \\ \partial\mathfrak{B}_{\zeta_1}(t) \cap \partial\mathfrak{B}_{\zeta_2}(t) = \phi \end{array} \right\}, \ \forall t \in \mathfrak{T}_0, \\
\zeta \longrightarrow \infty \Longleftrightarrow \mathfrak{B}_\zeta(t) \longrightarrow \mathfrak{R}^K.
\end{gathered}
$$

4) The function $\beta(.)$ is positive definite relative to the set $\Upsilon(t)$ on $\mathfrak{T}_0$.

The following facts result from Definition 319, from 3-c, and 4) of Definition 328, from the definition of the distance function $\rho(.) : \mathfrak{R}^K \times 2^{\mathfrak{R}^K} \longrightarrow \mathfrak{R}_+$ of a vector $\mathbf{z}$ from a set $\Upsilon(t)$, and from the definition of the η-neighborhood $\mathfrak{N}_\eta[\Upsilon(t)]$ of the set $\Upsilon(t)$ at any $t \in \mathfrak{T}$, $\mathfrak{N}_\eta[\Upsilon(t)] = \{\mathbf{z} : \rho[\mathbf{z}, \Upsilon(t)] < \eta\}, \ \forall t \in \mathfrak{T}$:

$$\mathbf{z} \in \partial\mathfrak{N}_\eta[\Upsilon(t)] \Longleftrightarrow \rho[\mathbf{z}, \Upsilon(t)] = \eta = \text{const.}, \ \forall t \in \mathfrak{T}, \tag{21.22}$$

$$\eta > 0 \Longrightarrow \partial\mathfrak{N}_\eta[\Upsilon(t)] \cap \partial\Upsilon(t) = \phi, \ \forall t \in \mathfrak{T} \cup \{-\infty, \infty\}, \tag{21.23}$$

$$
\begin{gathered}
\eta > 0 \Longrightarrow \zeta_{\text{Inf}\beta}(\eta) = \\
= \inf \left\{ \begin{array}{c} \zeta : \partial\mathfrak{B}_\zeta(t) \cap \partial\mathfrak{N}_\eta[\Upsilon(t)] \neq \phi, \\ \mathfrak{B}_\zeta(t) \subset \mathfrak{N}_\eta[\Upsilon(t)] \end{array} \right\} > \mu, \ \forall t \in \mathfrak{T} \cup \{-\infty, \infty\}.
\end{gathered} \tag{21.24}
$$

Example 329

$$
\begin{aligned}
\Upsilon(t) &= \{\mathbf{z} : (1+t^2)\|\mathbf{z}\| \leq 3\} \Longrightarrow \\
\beta(t,\mathbf{z}) &= (1+t^2)\|\mathbf{z}\|, \ \mu = 3, \ \mathbf{z}_c = \mathbf{0}, \ \Upsilon(t) \in \mathcal{F}.
\end{aligned}
$$

Example 330

$$
\begin{gathered}
\Upsilon(t) = \left\{ \mathbf{z} : (1+t^2) \left\| \mathbf{z} - \left[\begin{array}{c} e^{-t} \\ (1+e^t)^{-1} \end{array} \right] \right\| \leq 4 \right\} \Longrightarrow \\
\beta(t,\mathbf{z}) = (1+t^2) \left\| \mathbf{z} - \left[\begin{array}{c} e^{-t} \\ (1+e^t)^{-1} \end{array} \right] \right\|, \ \mu = 4, \\
\mathbf{z}_c = \mathbf{z}_c(t) = \left[\begin{array}{c} e^{-t} \\ (1+e^t)^{-1} \end{array} \right], \ \Upsilon(t) \in \mathcal{F}.
\end{gathered}
$$

Example 331 *Let $v(.) : \mathfrak{R}^K \longrightarrow \mathfrak{R}_+$ be a globally positive definite function, $f(.) : \mathfrak{T} \longrightarrow \mathfrak{R}^+$ be a continuous nondecreasing function on $\mathfrak{T}$, and*

$$
\begin{gathered}
\Upsilon(t) = \left\{ \mathbf{z} : \frac{v(\mathbf{z})}{f(t)} \leq 1 \right\} \Longrightarrow \\
\beta(t,\mathbf{z}) = \frac{v(\mathbf{z})}{f(t)}, \ \ \mu = 1, \ \ \mathbf{z}_c = \mathbf{0}, \ \ \Upsilon(t) \in \mathcal{F}.
\end{gathered}
$$

Example 332 *Let $v(.) : \mathfrak{T} \times \mathfrak{R}^K \longrightarrow \mathfrak{R}_+$ be a globally positive definite function, $f(.) : \mathfrak{T} \longrightarrow \mathfrak{R}^+$ be a continuous nondecreasing function on $\mathfrak{T}$ and*

$$
\begin{gathered}
\Upsilon(t) = \left\{ \mathbf{z} : \frac{v(t,\mathbf{z})}{f(t)} \leq 7 \right\} \Longrightarrow \\
\beta(t,\mathbf{z}) = \frac{v(t,\mathbf{z})}{f(t)}, \ \ \mu = 7, \ \ \mathbf{z}_c = \mathbf{0}, \ \ \Upsilon(t) \in \mathcal{F}.
\end{gathered}
$$

Example 333 *Let* $v(.) : \mathfrak{T} \times \mathfrak{R}^K \longrightarrow \mathfrak{R}_+$ *be defined by* $v(t, \mathbf{z}) = \|H(t)\mathbf{z}\|$ *and*

$$\begin{aligned}\Upsilon(t) &= \{\mathbf{z} : v(t, \mathbf{z}) \leq 0\} = \{\mathbf{z} : v(t, \mathbf{z}) = 0\} \\ &= \{\mathbf{z} : \|H(t)\mathbf{z}\| \leq 0\} = \{\mathbf{z} : H(t)\mathbf{z} = \mathbf{0}\} \Longrightarrow \\ \beta(t, \mathbf{z}) &= v(t, \mathbf{z}) = \|H(t)\mathbf{z}\|, \ \mu = 0, \ \mathbf{z}_c = \mathbf{0}, \ \Upsilon(t) \in \mathcal{F}.\end{aligned}$$

Example 334 *Let* $\Upsilon(t)$ *be the instantaneous* μ*-neighborhood* $\mathfrak{N}_\mu[\mathbf{Y}_d(t)]$ *of the enemy real behavior that is the desired behavior* $\mathbf{Y}_d(t)$ *of the domestic aircraft or missile. Then,* $\mathbf{Y}_d(t)$ *is the center* $\mathbf{z}_c(t)$ *of* $\Upsilon(t) \equiv \mathfrak{N}_\mu[\mathbf{Y}_d(t)]$, $\mathbf{z}_c(t) \equiv \mathbf{Y}_d(t)$.

Chapter 22

Outline of the Lyapunov method

22.1 Physical origin of the Lyapunov method

Fluctuations of energy and/or of matter are among the crucial causes of *time* variations of physical variables values.

If the input vector and the initial conditions are nominal, then for the system to rest in the nominal regime it is usually necessary and sufficient to ensure the nominal energy and matter flows. If the input vector is nominal all the *time*, but some or all initial conditions are not nominal, then energy and matter flows are not nominal. The goal is then to ensure that their real flows stay close to the nominal ones, and additionally, for better system behavior, to achieve the convergence of the former to the latter.

When the values of physical variables are divided by their units then the so obtained normalized values of the variables are dimensionless. Physical variables with dimensionless values are called *dimensionless variables*. We deal only with them in this book.

Let $E(t)$ denote the (dimensionless) instantaneous accumulated energy and $M(t)$ be the (dimensionless) instantaneous accumulated matter contained in the system at a moment $t \in \mathfrak{T}$. Let $\sigma \in \mathfrak{T}$, $\sigma < t$, $t = \sigma + \varepsilon$, $\varepsilon \longrightarrow 0^+$.

Let us consider systems with the constant accumulated mass: $M(t) \equiv M$.

If the real $E(\sigma)$ is bigger than the nominal $E_N(\sigma)$, then $E(t)$ should not increase faster than $E_N(t)$, or better, $E(t)$ should increase slower than $E_N(t)$; i.e., its *time* derivative $E^{(1)}(t)$ should not be bigger than *time* derivative $E_N^{(1)}(t)$ of $E_N(t)$, or better, the former should be less than the latter,

$$E(\sigma) > E_N(\sigma) \Longrightarrow E^{(1)}(t) \leq E_N^{(1)}(t), \textit{ or better}$$
$$E^{(1)}(t) < E_N^{(1)}(t), \; t = \sigma + \varepsilon; \; \varepsilon \longrightarrow 0^+ \Longrightarrow t \longrightarrow \sigma^+. \tag{22.1}$$

If the real $E(\sigma)$ is smaller than the nominal $E_N(\sigma)$, then $E(t)$ should not decrease faster than $E_N(t)$, or better, $E(t)$ should decrease slower than $E_N(t)$; i.e., its *time* derivative $E^{(1)}(t)$ should not be smaller than $E_N^{(1)}(t)$, or better, $E^{(1)}(t)$ should be bigger than $E_N^{(1)}(t)$,

$$E(\sigma) < E_N(\sigma) \Longrightarrow E^{(1)}(t) \geq E_N^{(1)}(t), \textit{ or better}$$
$$E^{(1)}(t) > E_N^{(1)}(t), \; t = \sigma + \varepsilon; \; \varepsilon \longrightarrow 0^+ \Longrightarrow t \longrightarrow \sigma^+. \tag{22.2}$$

If the real $E(\sigma)$ is equal to the nominal $E_N(\sigma)$, then $E(t)$ should rest equal to $E_N(t)$; i.e., its *time* derivative $E^{(1)}(t)$ should be equal to $E_N^{(1)}(t)$,

$$E(\sigma) = E_N(\sigma) \Longrightarrow E^{(1)}(t) = E_N^{(1)}(t), \quad t = \sigma + \varepsilon;\ \varepsilon \longrightarrow 0^+ \Longrightarrow t \longrightarrow \sigma^+. \tag{22.3}$$

We can express this analysis in the following compact mathematical form:

$$[E(\sigma) - E_N(\sigma)]\left[E^{(1)}(t) - E_N^{(1)}(t)\right] \leq 0, \tag{22.4}$$

or better,

$$[E(\sigma) - E_N(\sigma)]\left[E^{(1)}(t) - E_N^{(1)}(t)\right]\left\{ \begin{array}{c} < 0, \text{ if } [E(\sigma) - E_N(\sigma)] \neq 0, \\ = 0, \text{ if } [E(\sigma) - E_N(\sigma)] = 0 \end{array} \right\}. \tag{22.5}$$

When we consider the limit as $\varepsilon \longrightarrow 0^+$ then $t \longrightarrow \sigma^+$ and the preceding conditions (22.1) through (22.5) yield in the limit, when $t = \sigma$:

$$[E(t) - E_N(t)]\left[E^{(1)}(t) - E_N^{(1)}(t)\right] \leq 0, \tag{22.6}$$

or better,

$$[E(t) - E_N(t)]\left[E^{(1)}(t) - E_N^{(1)}(t)\right]\left\{ \begin{array}{c} < 0, \text{ if } [E(t) - E_N(t)] \neq 0, \\ = 0, \text{ if } [E(t) - E_N(t)] = 0 \end{array} \right\}. \tag{22.7}$$

The left-hand sides of these expressions represent the *time* derivative of

$$[E(t) - E_N(t)]^2 / 2$$

so that we can them set into the following forms:

$$\frac{d}{dt}\left\{\frac{1}{2}[E(t) - E_N(t)]^2\right\} \leq 0, \tag{22.8}$$

or better, (after multiplying (22.8) with 2,

$$\frac{d}{dt}\left\{[E(t) - E_N(t)]^2\right\}\left\{ \begin{array}{c} < 0, \text{ if } [E(t) - E_N(t)] \neq 0, \\ = 0, \text{ if } [E(t) - E_N(t)] = 0 \end{array} \right\}. \tag{22.9}$$

Lyapunov generalized the function $E(.)$ in [401]. He introduced the concepts of sign semidefinite and sign definite functions denoted by $V(.)$, or by $v(.)$, which play the role of the function $E(.)$ in the general mathematical setting. Roughly speaking, just to explain the essence of the idea, we can define $V(.)$ by

$$V(t) = [E(t) - E_N(t)]^2. \tag{22.10}$$

To summarize, the appropriate function $V(.)$ and its total *time* derivative along system motions $V^{(1)}(.)$ should obey

$$V(t, ...) \geq 0, \frac{dV(t, ...)}{dt} \leq 0, \tag{22.11}$$

or better,

$$\frac{dV(t, ...)}{dt}\left\{ \begin{array}{c} < 0, \text{ if } V(t, ...) \neq 0, \\ = 0, \text{ if } V(t, ...) = 0 \end{array} \right\}. \tag{22.12}$$

Such a function $V(.)$ is a *Lyapunov function* of the system.

If the accumulated mass $M(t)$ of the system is also *time*-varying then the above consideration is directly applicable to the mass. Then the physical nature of the system induces the vector $\mathbf{V}(t)$,

$$\mathbf{V}(t) = \begin{bmatrix} [E(t) - E_N(t)]^2 \\ [M(t) - M_N(t)]^2 \end{bmatrix}. \tag{22.13}$$

This is a physical justification for *the vector Lyapunov Function* concept established coincidentally by Bellman [38] in the framework of linear systems and by Matrosov [415] established in the general setting of *time*-varying nonlinear systems. We can also use $V(t) = [E(t) - E_N(t)]^2 + [M(t) - M_N(t)]^2$ instead of $\mathbf{V}(t)$, (22.13).

22.2 Lyapunov method

22.2.1 Essence of the Lyapunov method

The Lyapunov method, which Lyapunov originally called *the second method,* is also known as *the direct Lyapunov method.* It is a mathematical method. It originally meant *the study of stability properties of the system nominal motion under the influence of arbitrary initial conditions without knowing the system motions; i.e., without solving the system mathematical model.* In order to make this general idea effective Lyapunov introduced the concepts of sign semidefinite and definite functions $V(.)$, and established stability conditions in terms of the sign properties of both the function $V(.)$ and its total *time* derivative along system motions. The derivative of the function $V(.)$ is to be determined without using any information about the motions themselves, hence without solving the system mathematical model (the system differential equation). This might seem paradoxical at first glance, but it is not. It is possible. There is not any paradox. The system mathematical model and the initial conditions contain the whole information about the system motions.

Lyapunov proposed to use them directly rather than to solve the mathematical model for all initial conditions belonging to a neighborhood of the nominal motion, which is too cumbersome a mathematical problem to solve even with the help of the most powerful future computers. This is due to the facts that the number of the initial conditions is infinite in any, arbitrarily small, neighborhood of the nominal motion, and that an infinitesimal variation of the initial conditions can abruptly change the character of the system behavior. This holds even for some linear systems. The use of computers cannot replace the use of the Lyapunov method, but their joint usage makes them both more powerful.

Method 335 ***Lyapunov method***

Information about the system motions should be extracted directly from the properties of an appropriate function $V(.)$, *and from the features of its total time derivative* $V^{(1)}(.)$ *determined via the system mathematical model itself. To achieve this effectively, Lyapunov accepted that the function* $V(.)$ *is continuous and once continuously differentiable with some additional properties on the appropriate time-invariant subset of the system state space.*

22.2.2 Application in the state space

The Lyapunov method has been well developed for the *ISO* systems (3.65), (3.66) in their state space. What follows generalizes it to the state space of the regular plant (3.12).

Comment 336 ***On application of the Lyapunov method to the regular plant*** *(3.12)* ***in the state space***

For the regular plant (3.12), i.e., for the regular plant (5.17), (Definition 59) the total time derivative $V^{(1)}(.)$ *of the differentiable function* $V(.) : \mathfrak{T} \times \mathfrak{R}^{\alpha\rho} \longrightarrow \mathfrak{R}$ *along system motions reads:*

$$V^{(1)}(t,\mathbf{r}^{\alpha-1}) = \frac{dV(t,\mathbf{r}^{\alpha-1})}{dt} = \frac{\partial V(t,\mathbf{r}^{\alpha-1})}{\partial t} + \left[gradV(t,\mathbf{r}^{\alpha-1})\right]^T \frac{d\mathbf{r}^{\alpha-1}}{dt}, \tag{22.14}$$

where

$$\left[gradV(t,\mathbf{r}^{\alpha-1})\right]^T =$$
$$= \left[\frac{\partial V(t,\mathbf{r}^{\alpha-1})}{\partial r_1}..\frac{\partial V(t,\mathbf{r}^{\alpha-1})}{\partial r_\rho}....\frac{\partial V(t,\mathbf{r}^{\alpha-1})}{\partial r_1^{(\alpha-1)}}..\frac{\partial V(t,\mathbf{r}^{\alpha-1})}{\partial r_\rho^{(\alpha-1)}}\right],$$
$$gradV(t,\mathbf{r}^{\alpha-1}) \in \mathfrak{R}^{\alpha\rho}.$$

Because

$$\frac{d\mathbf{r}^{\alpha-1}}{dt} = \left[\mathbf{r}^{(1)T} \;\; \mathbf{r}^{(2)T} \;\; ... \;\; \mathbf{r}^{(\alpha)T}\right] \in \mathfrak{R}^{\alpha\rho},$$

then

$$\left[gradV(t,\mathbf{r}^{\alpha-1})\right]^T \frac{d\mathbf{r}^{\alpha-1}}{dt} =$$
$$= \left[\frac{\partial V(t,\mathbf{r}^{\alpha-1})}{\partial r_1}..\frac{\partial V(t,\mathbf{r}^{\alpha-1})}{\partial r_\rho}....\frac{\partial V(t,\mathbf{r}^{\alpha-1})}{\partial r_1^{(\alpha-2)}}..\frac{\partial V(t,\mathbf{r}^{\alpha-1})}{\partial r_\rho^{(\alpha-2)}}\right] \frac{d\mathbf{r}^{\alpha-2}}{dt}$$
$$+ \left[\frac{\partial V(t,\mathbf{r}^{\alpha-1})}{\partial r_1^{(\alpha-1)}}....\frac{\partial V(t,\mathbf{r}^{\alpha-1})}{\partial r_\rho^{(\alpha-1)}}\right] \frac{d\mathbf{r}^{(\alpha-1)}}{dt}. \tag{22.15}$$

This requires that (5.17) is solvable in $\mathbf{r}^{(\alpha)}$ *because* $\left(d\mathbf{r}^{(\alpha-1)}/dt\right) \equiv \mathbf{r}^{(\alpha)}$*, which holds if, and only if, Property 58 holds; i.e.,* $det\widetilde{Q}\left[t,\mathbf{r}^{\alpha-1}(t)\right] \neq 0$ *for every* $t \in \mathfrak{T}_0$*. It transforms (22.14), (22.15) along motions of the regular plant (5.17) into (22.16):*

$$V^{(1)}(t,\mathbf{r}^{\alpha-1}) = \frac{\partial V(t,\mathbf{r}^{\alpha-1})}{\partial t}$$
$$+ \left[\frac{\partial V(t,\mathbf{r}^{\alpha-1})}{\partial r_1}..\frac{\partial V(t,\mathbf{r}^{\alpha-1})}{\partial r_\rho}....\frac{\partial V(t,\mathbf{r}^{\alpha-1})}{\partial r_1^{(\alpha-2)}}..\frac{\partial V(t,\mathbf{r}^{\alpha-1})}{\partial r_\rho^{(\alpha-2)}}\right] \frac{d\mathbf{r}^{\alpha-2}}{dt} +$$
$$+ \left\langle \begin{array}{c} \left[\frac{\partial V(t,\mathbf{r}^{\alpha-1})}{\partial r_1^{(\alpha-1)}}....\frac{\partial V(t,\mathbf{r}^{\alpha-1})}{\partial r_\rho^{(\alpha-1)}}\right] \\ \bullet\ \widetilde{Q}^{-1}\left[t,\mathbf{r}^{\alpha-1}(t)\right]\left\{\widetilde{\mathbf{h}}\left[t,\mathbf{i}^{\xi}(t)\right] - \widetilde{\mathbf{q}}\left[t,\mathbf{r}^{\alpha-1}(t)\right]\right\} \end{array} \right\rangle. \tag{22.16}$$

This equation takes the following form for the plant (5.17):

$$V^{(1)}(t,\mathbf{r}^{\alpha-1}) = \frac{\partial V(t,\mathbf{r}^{\alpha-1})}{\partial t}$$
$$+ \left[\frac{\partial V(t,\mathbf{r}^{\alpha-1})}{\partial r_1}..\frac{\partial V(t,\mathbf{r}^{\alpha-1})}{\partial r_\rho}....\frac{\partial V(t,\mathbf{r}^{\alpha-1})}{\partial r_1^{(\alpha-2)}}..\frac{\partial V(t,\mathbf{r}^{\alpha-1})}{\partial r_\rho^{(\alpha-2)}}\right] \frac{d\mathbf{r}^{\alpha-2}}{dt} +$$
$$+ \left\langle \begin{array}{c} \left[\frac{\partial V(t,\mathbf{r}^{\alpha-1})}{\partial r_1^{(\alpha-1)}}....\frac{\partial V(t,\mathbf{r}^{\alpha-1})}{\partial r_\rho^{(\alpha-1)}}\right] \\ \bullet\ \widetilde{Q}^{-1}\left[t,\mathbf{r}^{\alpha-1}(t)\right]\left\{E(t)\widetilde{\mathbf{e}}_d\left[t,\mathbf{d}(t)\right] + P(t)\widetilde{\mathbf{e}}_u\left[\mathbf{u}(t)\right] - \widetilde{\mathbf{q}}\left[t,\mathbf{r}^{\alpha-1}(t)\right]\right\} \end{array} \right\rangle. \tag{22.17}$$

This equation discovers that the control vector $\mathbf{u}(t)$ can influence $V^{(1)}(t,\mathbf{r}^{\alpha-1})$ only through the truncated $\mathrm{grad}_{tr}V(t,\mathbf{r}^{\alpha-1})$ of the gradient $\mathrm{grad}V(t,\mathbf{r}^{\alpha-1})$:

$$\mathrm{grad}_{tr}V(t,\mathbf{r}^{\alpha-1}) = \left[\frac{\partial V(t,\mathbf{r}^{\alpha-1})}{\partial r_1^{(\alpha-1)}}....\frac{\partial V(t,\mathbf{r}^{\alpha-1})}{\partial r_\rho^{(\alpha-1)}}\right]^T \in \mathfrak{R}^{\rho}. \tag{22.18}$$

When $\text{grad}_{tr}V(t, \mathbf{r}^{\alpha-1}) = \mathbf{0}_\rho$ then the control vector $\mathbf{u}(t)$ cannot influence $V^{(1)}(t, \mathbf{r}^{\alpha-1})$ because it cannot act on the term $\partial V(t, \mathbf{r}^{\alpha-1})/dt$. This requires that the choice of the *time-varying* function $V(.)$ for control synthesis should ensure the following compatible property of the function $V(.)$:

Definition 337 ***The function*** $V(.)$ ***state gradient compatibility***
A differentiable time-varying function $V(.) : \mathfrak{T} \times \mathfrak{R}^{\alpha\rho} \longrightarrow \mathfrak{R}$, $V(t, \mathbf{r}^{\alpha-1}) \in \mathfrak{C}^1(\mathfrak{T} \times \mathfrak{R}^{\alpha\rho})$, *is* ***state gradient compatible*** *if, and only if,*

$$\mathbf{r}^{\alpha-1} = \mathbf{0}_{\alpha\rho} \Longrightarrow V(t, \mathbf{r}^{\alpha-1}) = 0, \ \frac{\partial V(t, \mathbf{r}^{\alpha-1})}{\partial t} = 0 \ \text{and}$$
$$grad_{tr}V(t, \mathbf{r}^{\alpha-1}) = \mathbf{0}_\rho \Longleftrightarrow \mathbf{r}^{\alpha-1} = \mathbf{0}_{\alpha\rho}. \tag{22.19}$$

Definition 338 ***State Lyapunov system***
The regular plant (3.12); i.e., the regular plant (5.17), is ***the state Lyapunov system on*** $\mathfrak{T}_0$ *if, and only if, there exists a differentiable state gradient compatible positive definite function* $V(.) : \mathfrak{T}_0 \times \mathfrak{R}^{\alpha\rho} \longrightarrow \mathfrak{R}$ *on* $\mathfrak{T}_0 \times \mathfrak{R}^{\alpha\rho}$ *that obeys (22.19) such that the following*

$$grad_{tr}^T V(t, \mathbf{r}^{\alpha-1})\widetilde{Q}^{-1}\left(t, \mathbf{r}^{\alpha-1}\right) P(t) = \mathbf{0}_{1,r}, \ (t, \mathbf{r}^{\alpha-1}) \in \mathfrak{T}_0 \times \mathfrak{R}^{\alpha\rho} \tag{22.20}$$

holds if, and only if, $\mathbf{r}^{\alpha-1} = \mathbf{0}_{\alpha\rho}$*:*

$$grad_{tr}^T V(t, \mathbf{r}^{\alpha-1})\widetilde{Q}^{-1}\left(t, \mathbf{r}^{\alpha-1}\right) P(t) = \mathbf{0}_{1,r}, \ (t, \mathbf{r}^{\alpha-1}) \in \mathfrak{T}_0 \times \mathfrak{R}^{\alpha\rho} \Longleftrightarrow$$
$$\Longleftrightarrow \mathbf{r}^{\alpha-1} = \mathbf{0}_{\alpha\rho} \tag{22.21}$$

If, and only if, (22.21) is valid for $In\mathfrak{T}_0 = \mathfrak{T}$ *then the term "****on*** $\mathfrak{T}_0$*" is to be omitted; i.e., the system is* ***a state Lyapunov system****.*

Note 339 *Guide for the choice of the function* $V(.)$
Definition 337 and Definition 338 serve as a guide for the choice of the function $V(.)$ *to enable an effective synthesis of the control vector function* $\mathbf{u}(.)$ *for the regular plant (3.12); i.e., the regular plant (5.17), in the state space.*

22.2.3 Application in the output space

The output space $\mathfrak{R}^N$ is the natural space in which tracking takes place. For a higher order, for the k-th-order, tracking the natural space is the extended output space $\mathfrak{R}^{(k+1)N}$.

Comment 340 ***On application of the Lyapunov method to the system*** *(3.12)* ***in the output space***
For the plant (3.12) in terms of the deviations, i.e., for (5.17), the total time derivative $V^{(1)}(.)$ *of the function* $V(.) : \mathfrak{T} \times \mathfrak{R}^N \longrightarrow \mathfrak{R}$ *along plant output behavior reads:*

$$V^{(1)}(t, \mathbf{y}) = \frac{dV(t, \mathbf{y})}{dt} = \frac{\partial V(t, \mathbf{y})}{\partial t} + [gradV(t, \mathbf{y})]^T \frac{d\mathbf{y}}{dt}, \tag{22.22}$$

where

$$[gradV(t, \mathbf{y})] = \left[\frac{\partial V(t, \mathbf{y})}{\partial y_1} \quad \frac{\partial V(t, \mathbf{y})}{\partial y_2} .... \frac{\partial V(t, \mathbf{y})}{\partial y_N}\right]^T \in \mathfrak{R}^{\alpha N}, \tag{22.23}$$

and

$$\frac{d\mathbf{y}}{dt} = \frac{d\widetilde{\mathbf{z}}\left[t, \mathbf{r}^{\alpha-1}(t), \mathbf{d}(t)\right]}{dt} + \frac{d\left\{W(t)\,\widetilde{\mathbf{w}}\left[\mathbf{u}(t)\right]\right\}}{dt}.$$

If $\widetilde{\mathbf{z}}\left(t,\mathbf{r}^{\alpha-1},\mathbf{d}\right)\in\mathfrak{C}^{1}\left(\mathfrak{T},\mathfrak{R}^{\alpha\rho},\mathfrak{R}^{d}\right)$, $W\left(t\right)\in\mathfrak{C}^{(1)}\left(\mathfrak{T}\right)$, *and* $\widetilde{\mathbf{w}}(.)\in\mathfrak{C}^{1}\left(\mathfrak{R}^{r}\right)$ *then*

$$\begin{gathered}\frac{d\mathbf{y}}{dt}=\frac{d\left\{W\left(t\right)\widetilde{\mathbf{w}}\left[\mathbf{u}(t)\right]\right\}}{dt}+\frac{\partial\widetilde{\mathbf{z}}\left[t,\mathbf{r}^{\alpha-1}(t),\mathbf{d}(t)\right]}{\partial t}\\+\left[\frac{\partial\widetilde{\mathbf{z}}\left[t,\mathbf{r}^{\alpha-1}(t),\mathbf{d}(t)\right]}{\partial r_{1}}....\frac{\partial\widetilde{\mathbf{z}}\left[t,\mathbf{r}^{\alpha-1}(t),\mathbf{d}(t)\right]}{\partial r_{\rho}}\right]\mathbf{r}^{(1)}(t)\\+...+\left[\frac{\partial\widetilde{\mathbf{z}}\left[t,\mathbf{r}^{\alpha-1}(t),\mathbf{d}(t)\right]}{\partial r_{1}^{(\alpha-1)}}....\frac{\partial\widetilde{\mathbf{z}}\left[t,\mathbf{r}^{\alpha-1}(t),\mathbf{d}(t)\right]}{\partial r_{\rho}^{(\alpha-1)}}\right]\mathbf{r}^{(\alpha)}(t)\\+\left[\frac{\partial\widetilde{\mathbf{z}}\left[t,\mathbf{r}^{\alpha-1}(t),\mathbf{d}(t)\right]}{\partial d_{1}}....\frac{\partial\widetilde{\mathbf{z}}\left[t,\mathbf{r}^{\alpha-1}(t),\mathbf{d}(t)\right]}{\partial d_{d}}\right]\mathbf{d}^{(1)}(t).\end{gathered}\tag{22.24}$$

We can set this in a compact form by using the $N\times\rho$ *zero matrix* $O_{N,\rho}$ *:*

$$\begin{gathered}\frac{\partial\widetilde{\mathbf{z}}\left[t,\mathbf{r}^{\alpha-1}(t),\mathbf{d}(t)\right]}{\partial\mathbf{r}}=\frac{\partial\widetilde{\mathbf{z}}\left[t,\mathbf{r}^{\alpha-1}(t),\mathbf{d}(t)\right]}{\partial\mathbf{r}^{(j)}},\ j=0,\\\frac{\partial\widetilde{\mathbf{z}}\left[t,\mathbf{r}^{\alpha-1}(t),\mathbf{d}(t)\right]}{\partial\mathbf{r}^{(j)}}\\\left[\frac{\partial\widetilde{\mathbf{z}}\left[t,\mathbf{r}^{\alpha-1}(t),\mathbf{d}(t)\right]}{\partial r_{1}^{(j)}}....\frac{\partial\widetilde{\mathbf{z}}\left[t,\mathbf{r}^{\alpha-1}(t),\mathbf{d}(t)\right]}{\partial r_{\rho}^{(j)}}\right],\ j=0,1,2,\cdots,\alpha-2,\end{gathered}\tag{22.25}$$

$$\begin{gathered}\frac{\partial^{*}\widetilde{\mathbf{z}}\left[t,\mathbf{r}^{\alpha-1}(t),\mathbf{d}(t)\right]}{\partial\mathbf{r}^{\alpha-2}}\in\mathfrak{R}^{Nx\alpha\rho},\ \frac{\partial^{*}\widetilde{\mathbf{z}}\left[t,\mathbf{r}^{\alpha-1}(t),\mathbf{d}(t)\right]}{\partial\mathbf{r}^{\alpha-2}}\\=\left[O_{N,\rho}\ \ \frac{\partial\widetilde{\mathbf{z}}\left[t,\mathbf{r}^{\alpha-1}(t),\mathbf{d}(t)\right]}{\partial\mathbf{r}^{(0)}}\ \ \frac{\partial\widetilde{\mathbf{z}}\left[t,\mathbf{r}^{\alpha-1}(t),\mathbf{d}(t)\right]}{\partial\mathbf{r}^{(1)}}...\frac{\partial\widetilde{\mathbf{z}}\left[t,\mathbf{r}^{\alpha-1}(t),\mathbf{d}(t)\right]}{\partial\mathbf{r}^{(\alpha-2)}}\right],\\\frac{\partial\widetilde{\mathbf{z}}\left[t,\mathbf{r}^{\alpha-1}(t),\mathbf{d}(t)\right]}{\partial\mathbf{r}_{\rho}^{\alpha-1}}=\left[\frac{\partial\widetilde{\mathbf{z}}\left[t,\mathbf{r}^{\alpha-1}(t),\mathbf{d}(t)\right]}{\partial r_{1}^{\alpha-1}}...\frac{\partial\widetilde{\mathbf{z}}\left[t,\mathbf{r}^{\alpha-1}(t),\mathbf{d}(t)\right]}{\partial r_{\rho}^{\alpha-1}}\right],\\\frac{\partial\widetilde{\mathbf{z}}\left[t,\mathbf{r}^{\alpha-1}(t),\mathbf{d}(t)\right]}{\partial\mathbf{d}}=\left[\frac{\partial\widetilde{\mathbf{z}}\left[t,\mathbf{r}^{\alpha-1}(t),\mathbf{d}(t)\right]}{\partial d_{1}}...\frac{\partial\widetilde{\mathbf{z}}\left[t,\mathbf{r}^{\alpha-1}(t),\mathbf{d}(t)\right]}{\partial d_{d}}\right].\end{gathered}\tag{22.26}$$

The compact form of the equations (22.24) reads:

$$\begin{gathered}\frac{d\mathbf{y}}{dt}=\frac{\partial\widetilde{\mathbf{z}}\left[t,\mathbf{r}^{\alpha-1}(t),\mathbf{d}(t)\right]}{\partial t}+\frac{\partial^{*}\widetilde{\mathbf{z}}\left[t,\mathbf{r}^{\alpha-1}(t),\mathbf{d}(t)\right]}{\partial\mathbf{r}^{\alpha-2}}\mathbf{r}^{\alpha-1}\\+\frac{\partial\widetilde{\mathbf{z}}\left[t,\mathbf{r}^{\alpha-1}(t),\mathbf{d}(t)\right]}{\partial\mathbf{r}_{\rho}^{\alpha-1}}\mathbf{r}^{(\alpha)}+\frac{\partial\widetilde{\mathbf{z}}\left[t,\mathbf{r}^{\alpha-1}(t),\mathbf{d}(t)\right]}{\partial\mathbf{d}}\mathbf{d}^{(1)}(t)\\+W^{(1)}\left(t\right)\widetilde{\mathbf{w}}\left[\mathbf{u}(t)\right]+W\left(t\right)\widetilde{\mathbf{w}}^{(1)}\left[\mathbf{u}(t)\right],\ \forall t\in\mathfrak{T}.\end{gathered}\tag{22.27}$$

This demands solving

$$\widetilde{Q}\left[t,\mathbf{r}^{\alpha-1}(t)\right]\mathbf{r}^{(\alpha)}(t)+\widetilde{\mathbf{q}}\left[t,\mathbf{r}^{\alpha-1}(t)\right]=\widetilde{\mathbf{h}}\left[t,\mathbf{d}^{\xi}\left(t\right)\right],\tag{22.28}$$

either for $\mathbf{r}^{\alpha-1}(t)$ *or for* $\mathbf{r}^{(\alpha)}(t)$*, which is possible in the former case for the special plant (5.23) and in the latter case for the regular plant (5.17).*

Definition 341 ***The function*** $V(.)$ ***output gradient compatibility***

A differentiable time-varying function $V(.) : \mathfrak{T} \times \mathfrak{R}^N \longrightarrow \mathfrak{R}$ *is* ***output gradient compatible*** *if, and only if,*

$$V(t,\mathbf{y}) \in \mathfrak{C}^1\left(\mathfrak{T} \times \mathfrak{R}^N\right) \ \ and\ \mathbf{y} = \mathbf{0}_N \Longrightarrow V(t,\mathbf{y}) = 0,\ \frac{\partial V(t,\mathbf{y})}{\partial t} = 0$$
$$and\ gradV(t,\mathbf{y}) = \mathbf{0}_N \Longleftrightarrow \mathbf{y} = \mathbf{0}_N. \tag{22.29}$$

Definition 342 ***Output Lyapunov system***

The regular plant (3.12), i.e., the regular plant (5.17), is ***the output Lyapunov system on*** $\mathfrak{T}_0$ *if, and only if, there exists a differentiable output gradient compatible positive definite function* $V(.) : \mathfrak{T}_0 \times \mathfrak{R}^N \longrightarrow \mathfrak{R}$ *on* $\mathfrak{T}_0 \times \mathfrak{R}^N$*, which obeys (22.29) such that the following:*

$$grad^T V(t,\mathbf{y}) \frac{\partial \widetilde{\mathbf{z}}\left[t,\mathbf{r}^{\alpha-1}(t),\mathbf{d}(t)\right]}{\partial \mathbf{r}_\rho^{\alpha-1}} \widetilde{Q}^{-1}\left(t,\mathbf{r}^{\alpha-1}\right) P(t) = \mathbf{0}_{1,r},\ (t,\mathbf{y}) \in \mathfrak{T}_0 \times \mathfrak{R}^N \tag{22.30}$$

holds if, and only if, $\mathbf{y} = \mathbf{0}_N$*:*

$$grad^T V(t,\mathbf{y}) \widetilde{Q}^{-1}\left(t,\mathbf{r}^{\alpha-1}\right) P(t) = \mathbf{0}_{1,r},\ \left(t,\mathbf{r}^{\alpha-1},\mathbf{y}\right) \in \mathfrak{T}_0 \times \mathfrak{R}^{\alpha\rho} \times \mathfrak{R}^N$$
$$\Longleftrightarrow \mathbf{y} = \mathbf{0}_N. \tag{22.31}$$

If, and only if, (22.31) is valid for $In\mathfrak{T}_0 = \mathfrak{T}$ *then the term "****on*** $\mathfrak{T}_0$*" is to be omitted, i.e., the system is the* ***output Lyapunov system****.*

Comment 343 *The regular plant (5.17) enables:*

$$\mathbf{r}^{(\alpha)}(t) = \widetilde{Q}^{-1}\left[t,\mathbf{r}^{\alpha-1}(t)\right] \left\{ \begin{array}{c} -\widetilde{\mathbf{q}}\left[t,\mathbf{r}^{\alpha-1}(t)\right] \\ +\ E(t)\,\widetilde{\mathbf{e}}_d\left[t,\mathbf{d}(t)\right] + P(t)\,\widetilde{\mathbf{e}}_u\left[\mathbf{u}(t)\right] \end{array} \right\},\ \forall t \in \mathfrak{T},$$
$$\mathbf{y}(t) = \widetilde{\mathbf{z}}\left[t,\mathbf{r}^{\alpha-1}(t),\mathbf{d}(t)\right] + W(t)\,\widetilde{\mathbf{w}}\left[\mathbf{u}(t)\right],\ \forall t \in \mathfrak{T}, \tag{22.32}$$

and the special plant (5.23) permits:

$$\mathbf{r}^{\alpha-1}(t) = \widetilde{F}^{-1}\left[t,\mathbf{r}^{\alpha-1}(t)\right] \left\{ \begin{array}{c} -\widetilde{Q}\left[t,\mathbf{r}^{\alpha-1}(t)\right]\mathbf{r}^{(\alpha)}(t) - \widetilde{\mathbf{q}}\left[t,\mathbf{r}^{\alpha-1}(t)\right] \\ +\ E(t)\,\widetilde{\mathbf{e}}_d\left[t,\mathbf{d}(t)\right] + P(t)\,\widetilde{\mathbf{e}}_u\left[\mathbf{u}(t)\right] \\ \forall t \in \mathfrak{T}, \end{array} , \right\}$$
$$\mathbf{y}(t) = \widetilde{\mathbf{z}}\left[t,\mathbf{r}^{\alpha-1}(t),\mathbf{d}(t)\right] + \widetilde{Z}\left[t,\mathbf{r}^{\alpha-1}(t),\mathbf{d}(t)\right]\mathbf{r}^{\alpha-1}(t) + W(t)\,\widetilde{\mathbf{w}}\left[\mathbf{u}(t)\right],\ \forall t \in \mathfrak{T}, \tag{22.33}$$

Comment 344 ***On application of the Lyapunov method to the system*** *(3.12)* ***in the extended output space***

Another theoretical possibility is to use $V(.) : \mathfrak{T} \times \mathfrak{R}^{(k+1)N} \longrightarrow \mathfrak{R}$*, (7.2), and to determine its total time derivative* $V^{(1)}(.)$ *along system extended output behavior:*

$$V^{(1)}(t,\mathbf{y}^k) = \frac{dV(t,\mathbf{y}^k)}{dt}, \tag{22.34}$$

where

$$V^{(1)}(t,\mathbf{y}^k) = \frac{dV(t,\mathbf{y}^k)}{dt} = \frac{\partial V(t,\mathbf{y}^k)}{\partial t} + \left[gradV(t,\mathbf{y}^k)\right]^T \frac{d\mathbf{y}^k}{dt},$$
$$\left[gradV(t,\mathbf{y}^k)\right] \in \mathfrak{R}^{(k+1)N},$$

together with

$$\left[gradV(t,\mathbf{y}^k)\right]^T = \left[\begin{array}{c} \frac{\partial V(t,\mathbf{y}^k)}{\partial y_1} \quad \frac{\partial V(t,\mathbf{y}^k)}{\partial y_2} \cdots \frac{\partial V(t,\mathbf{y}^k)}{\partial y_N} \cdots \\ \cdots \frac{\partial V(t,\mathbf{y}^k)}{\partial y_1^k} \quad \frac{\partial V(t,\mathbf{y}^k)}{\partial y_2^k} \cdots \frac{\partial V(t,\mathbf{y}^k)}{\partial y_N^k} \end{array} \right].$$

This demands differentiating the output equation of (22.33) k times, which becomes very complex for $k > 1$. The bigger k is more complex $d\mathbf{y}^k/ddt^k$. The Lyapunov method is not effectively applicable to the study of the k-th-order tracking via the output space for $k \geq 1$.

Conclusion 345 ***Conclusion on application of the Lyapunov method to the tracking of the system*** *(3.12)*

The Lyapunov method is effectively applicable to the state tracking control synthesis in the state space by using (22.16) provided the desired motion $\mathbf{R}_d^k(t; t_0; \mathbf{R}_0^k; \mathbf{I})$ is determined for every desired output behavior $\mathbf{Y}_d(t)$.

The Lyapunov method is effectively applicable only to control synthesis to ensure tracking, but not to guarantee the k-th-order tracking if $k > 0$. In this case there is not any need for the knowledge of the desired output behavior $\mathbf{Y}_d(t)$.

22.3 Lyapunov theorems for nonlinear systems

Theorem 346 ***Original Lyapunov Theorem I on the ISO nonlinear systems in the free regime [401, pp. 59 in the Russian edition]***

If the differential equations of the perturbed motions are such that it is possible to find a sign definite function $V(.)$, the derivative $V^{(1)}(.)$ of which due to those equations is either a sign semidefinite function with the sign inverse to the sign of $V(.)$, or identically equal to zero, the unperturbed motion is stable.

Note 347 ***Original Lyapunov Note II on the ISO nonlinear systems in the free regime [401, p. 61 in the Russian edition]***

If the function $V(.)$, which satisfies the conditions of Theorem 346, simultaneously permits the infinitesimally small upper bound, and its derivative is a sign definite function, then it is possible to prove, that every perturbed motion, sufficiently close to the unperturbed one, will converge to it asymptotically.

Conclusion 348 On the original Lyapunov theorems

Lyapunov [401] developed his stability theory (definitions, method, methodologies, theorems that contain stability conditions) only in the framework of the ISO nonlinear systems ***(3.65)*** *in the free regime.*

Comment 349 *Every time-varying decrescent function has the infinitesimally small upper bound.*

22.4 Lyapunov original methodologies

Lyapunov original methodologies result from Lyapunov original theorems and they determine how to apply the Lyapunov method.

Lyapunov [401] himself determined one methodology for the *ISO time*-invariant linear systems and another one for the *ISO time*-varying linear systems and for all *ISO* nonlinear systems over *time*-invariant sets.

We meet the following question.

Question 350 *Should we start with the acceptance of a function $Q(.)$ with the demanded characteristics for the total time derivative $V^{(1)}(.)$ of the function $V(.)$, $Q(.) = V^{(1)}(.)$, and then to continue by determining the function $V(.)$ as the solution of the (nonlinear partial) differential equation*

$$\frac{dV(t,\mathbf{z})}{dt} = \frac{\partial V(t,\mathbf{z})}{\partial t} + \sum_{i=1}^{i=K} \frac{\partial V(t,\mathbf{z})}{\partial z_i}\frac{dz_i}{dt} = \frac{\partial V(t,\mathbf{z})}{\partial t} + [\operatorname{grad} V(t,\mathbf{z})]^T \frac{d\mathbf{z}}{dt} = Q(t,\mathbf{z}), \quad (22.35)$$

along system motions, or equivalently, via the system mathematical model as explained in Lyapunov method 335, and by testing the properties of the obtained solution function $V(.)$*, or, should we start with a choice of the subsidiary function* $V(.)$ *with the demanded characteristics and then continue with the test of the properties of its total time derivative* $V^{(1)}(.)$ *along system motions?*

Note 351 ***Eulerian and Dini derivatives of the function*** $V(.)$
The condition on differentiability of the function $V(.)$ *can be relaxed by requiring only its continuity. Its total Eulerian derivative* $dV(t,\mathbf{z})/dt$ *along system motions is to be then replaced by its Dini derivative* $D^{+}v(t,\mathbf{z})$ *along system motions (see Appendix B).*

Reply 352 *For the time-invariant linear systems Lyapunov's original methodology demands starting with a choice of the total time derivative* $V^{(1)}(.)$ *of the function* $V(.)$*, taken along system motions, i.e., with a choice of a function* $Q(.)$ *to be accepted for* $V^{(1)}(.)$, $V^{(1)}(.) = Q(.)$*, which should possess the demanded characteristics of* $V^{(1)}(.)$*, and then determining the function* $V(.)$ *and testing its properties. This methodology is broadened to the ISO time-invariant and time-varying nonlinear systems in [155], [156], [158], [160] - [162], [170], [172], [182], [199]-[205], [217], [255]-[257]. It is* ***consistent Lyapunov methodology*** *(CLM), for short:* ***consistent methodology*** *(CM) because it is the same as Lyapunov's original methodology for the time-invariant linear systems. It results in the stability criteria that specify the necessary and sufficient conditions on the function* $Q(.)$ *and on the solution function* $V(.)$ *to (22.35) for the linear or nonlinear ISO system to possess the requested stability property.*

If we start with a function $Q(.)$ *that has the necessary characteristics for the total time derivative* $V^{(1)}(.)$ *of the function* $V(.)$*, then we face the extremely complex mathematical problem of solving the nonlinear partial differential equation (22.35). It is solvable by the existing mathematical methods only in very special cases of the mathematical models of nonlinear systems. However, we should not forget that we use mathematics in order to cope with physical problems of physical (dynamical) systems. They use energy and/or matter for their work, for their functioning. This returns us to the physical origin of Lyapunov method 22.1. If the system accumulated mass is constant then we should determine, by using physical laws, the system accumulated energy* $E(.)$ *and power* $P(.)$*, where*

$$\frac{dE(t,\mathbf{z})}{dt} = P(t,\mathbf{z}). \tag{22.36}$$

Then we test the properties of the power $P(.)$ *and of the energy* $E(.)$*. From the control point of view, if the plant is (to be) controlled then its power also depends on control* $\mathbf{U}(.)$*:* $P(t,\mathbf{z},\mathbf{U})$*, The control vector function* $\mathbf{U}(.)$ *should be synthesized so that the plant power* $P(.)$ *possesses all requested properties. Stability criteria based on consistent Lyapunov methodology prove that the test of the properties of the energy* $E(.)$ *gives the final and complete reply to the question whether the controlled plant and its control system possess the corresponding stability property.*

Reply 353 *For the time-varying linear systems and for all nonlinear systems Lyapunov's original methodology requires us to start with a choice of the function* $V(.)$ *that has the appropriate characteristics and then to test the properties of its total time derivative* $V^{(1)}(.)$ *along system motions. This methodology is inverse to Lyapunov's original methodology for time-invariant linear systems.*

If we start with a random choice of the function $V(.)$ *with the corresponding properties, then there is not any guarantee that its selection is appropriate. Most often its derivative does not satisfy the conditions for the requested stability property. The search process for the function* $V(.)$ *with the corresponding properties such that its derivative also possesses the corresponding necessary properties can be endless, without any final result. This is the essential drawback of Lyapunov's original methodology for time-varying linear systems and for all nonlinear systems. This, inverse, Lyapunov's original methodology leaves unsolved the crucial stability problems in the framework of nonlinear systems:*

- The necessary and sufficient conditions for asymptotic stability, which are not expressed in terms of the existence of a Lyapunov function

- The direct generation of a system Lyapunov function $V(.)$

- The determination of the exact domain of the asymptotic stability

Consistent Lyapunov methodology resolves theoretically all these problems.

Chapter 23

Lyapunov method extended to tracking

23.1 Criteria: Asymptotically contractive sets

The extended k-th-order output response of the system (3.12) in terms of deviations is the output solution $\mathbf{y}^k\left(t; t_0; \mathbf{y}_0^k; \mathbf{i}\right)$ of the system (5.10):

$$\widetilde{Q}\left[t, \mathbf{r}^{\alpha-1}(t)\right] \mathbf{r}^{(\alpha)}(t) + \widetilde{\mathbf{q}}\left[t, \mathbf{r}^{\alpha-1}(t)\right] = \widetilde{\mathbf{h}}\left[t, \mathbf{i}^{\xi}(t)\right], \ \forall t \in \mathfrak{T},$$
$$\mathbf{y}(t) = \widetilde{\mathbf{s}}\left[t, \mathbf{r}(t), \mathbf{i}(t)\right], \ \forall t \in \mathfrak{T}. \tag{23.1}$$

The initial vector $\mathbf{y}_0^k$ satisfies

$$\mathbf{y}_0^k = \widetilde{\mathbf{s}}^k\left(t_0, \mathbf{r}_0^k, \mathbf{i}_0^k\right), \ t_0 \in \mathfrak{T}_i. \tag{23.2}$$

Note 354 ***If the system (3.12) is the plant (3.12)***

If the system (3.12) is the plant (3.12) then everywhere we should apply the following substitutions: $\mathbf{i}$ *to replace by* $\mathbf{d}, \mathbf{u}$*; also* $\mathbf{i}^{\xi}$ *by* $\mathbf{d}^{\eta}$, $\mathbf{u}^{\mu}$*; and* $\mathcal{J}^{\xi}$ *by* $\mathfrak{D}^{j} \times \mathfrak{U}^{l}$.

Note 355 ***The equivalence between tracking of the total desired vector $\mathbf{Z}_d(t)$ and the tracking of its deviation $\mathbf{z}_d(t) \equiv \mathbf{z}_d = \mathbf{0}_K$.***

We use the general notation in order to unify the presentation so that all that follows is valid for both the k-th-order tracking, $k \in \{0, 1, 2, \cdots, \alpha - 1\}$*, and for the state tracking. In the former case* $K = (k+1)N$, $\mathbf{Z} = \mathbf{Y}^k$, $\mathbf{Z}_d = \mathbf{Y}_d^k$, $\mathfrak{Z}_d = \mathfrak{Y}_d^{k+1}$, $\mathbf{z} = \mathbf{y}^k$, $\mathbf{z}_d = \mathbf{y}_d^k = \mathbf{0}_K$ *and* $\mathfrak{R}^K = \mathfrak{R}^{(k+1)N}$*. In the latter case* $K = \alpha N$, $\mathbf{Z} = \mathbf{R}^{\alpha-1}$, $\mathbf{Z}_d = \mathbf{R}_d^{\alpha-1}$, $\mathfrak{Z}_d = \mathfrak{R}_d^{\alpha}$, $\mathbf{z} = \mathbf{r}^{\alpha-1}$, $\mathbf{z}_d = \mathbf{r}_d^{\alpha-1} = \mathbf{0}_{\alpha N}$*, and* $\mathfrak{R}^K = \mathfrak{R}^{\alpha N}$*. The passage from the total vector* $\mathbf{Z}$ *to its deviation* $\mathbf{z}$ *reduces the study of tracking* $\mathbf{Z}_d(t)$ *to the corresponding study of tracking* $\mathbf{z}_d(t) \equiv \mathbf{z}_d = \mathbf{0}_K$.

Note 356 ***Properties of the system in terms of the deviations***

The system (23.1) in terms of the deviations possesses the same general properties as the original system (3.12) in terms of the total coordinates (values): Property 57, Property 58, and Property 78.

The study of dynamical systems behavior on asymptotically contractive sets (Definition 46) probably started 1975 in [175].

Theorem 357 ***Stable tracking of $\mathbf{Z}_d(t)$ of the system (3.12) on $\mathfrak{T}_0 \times \mathcal{J}^{\xi}$, equivalently of $\mathbf{z}_d = \mathbf{0}_K$ of the system (23.1) on $\mathfrak{T}_0 \times \mathcal{J}^{\xi}$***

In order for the system (3.12), equivalently the system (23.1), respectively, to exhibit the stable tracking of $\mathbf{Z}_d(t)$, $\mathbf{Z}_d(.) \in \mathfrak{Z}_d$, *equivalently of* $\mathbf{z}_d(t) \equiv \mathbf{0}_K$, *on* $\mathfrak{T}_0 \times \mathcal{J}^\xi$ *it is sufficient that there exist a positive definite function* $V(.) : \mathfrak{T}_0 \times \mathfrak{R}^K \times \mathcal{J}^\xi \longrightarrow \mathfrak{R}_+$ *and* $\xi_M > 0$ *such that:*

1) For every $\zeta \in]0, \xi_M]$ *and for all* $\mathbf{i}(.) \in \mathcal{J}^\xi$ *the set* $\mathfrak{V}_\zeta(t; t_0; \mathcal{J}^\xi)$ *induced by* $V(.)$ *:*

a) has the radially increasing boundary function

b) is asymptotically contractive

and

2) Dini derivative $D^+V(.)$ *of the function* $V(.)$ *along solutions* $\mathbf{z}(t; t_0; \mathbf{z}_0; \mathbf{i})$ *of the system (23.1) is negative on* $Cl\mathfrak{V}_{\xi_M}(t; t_0; \mathcal{J}^\xi) \setminus \mathfrak{O}$, $\forall [t, \mathbf{i}(.)] \in \mathfrak{T}_0 \times \mathcal{J}^\xi$ *and* $D^+V[(t, \mathbf{z}; \mathbf{i})] = 0$ *if, and only if,* $\mathbf{z} = \mathbf{0}_K$ *for every* $[t, \mathbf{i}(.)] \in \mathfrak{T}_0 \times \mathcal{J}^\xi$.

The set $Cl\mathfrak{V}_{\xi_M}(t_0; t_0; \mathcal{J}^\xi)$ *is then a subset of the domain of stable tracking on* $\mathfrak{T}_0 \times \mathcal{J}^\xi$.

Proof. Let the conditions of the theorem statement hold. All that follows holds for every $\mathbf{i}(.) \in \mathcal{J}^\xi$.

A) *Tracking.* Definition 267 and Proposition 300 guarantee the existence of both the connected neighborhood $\mathfrak{N}$ of the origin $\mathbf{z}_d = \mathbf{0}_K$ and $\xi \in \mathfrak{R}^+$ such that for every $\varsigma \in]0, \xi[$ and for all $\mathbf{i}(.) \in \mathcal{J}^\xi$ the set $\mathfrak{V}_\zeta(t; t_0; \mathcal{J}^\xi)$ is bounded on $\mathfrak{T}_\tau$ and $\mathfrak{V}_\zeta(t; t_0; \mathcal{J}^\xi) \subseteq \mathfrak{N}$, $\forall t \in \mathfrak{T}_0$. Conditions 1-a) and 2) of the theorem statement, Property 78, Note 79, and Lemma 317 imply that the closure $\mathrm{Cl}\mathfrak{V}_\zeta(t; t_0; \mathcal{J}^\xi)$ of the set $\mathfrak{V}_\zeta(t; t_0; \mathcal{J}^\xi)$ is positively invariant with respect to system motions $\mathbf{z}(t; t_0; \mathbf{z}_0; \mathbf{i})$ for every $[\mathbf{i}(.), \zeta] \in \mathcal{J}^\xi \times]0, \min\{\xi_M, \xi\}[$. The condition 1-b) of the theorem statement and Definition 46 guarantee the following:

$$\forall (\mathbf{i}, \zeta) \in \mathcal{J}^\xi \times]0, \min\{\xi_M, \xi\}[,$$
$$\mathbf{z}_0 \in \mathfrak{V}_\zeta(t_0; t_0; \mathcal{J}^\xi) \Longrightarrow \mathbf{z}(t; t_0; \mathbf{z}_0; \mathbf{i}) \longrightarrow \mathbf{z}_d = \mathbf{0}_K \text{ as } t \longrightarrow \infty. \tag{23.3}$$

The system (3.12) exhibits tracking of $\mathbf{z}_d = \mathbf{0}_K$, equivalently of $\mathbf{Z}_d(t)$, $\mathbf{Z}_d(.) \in \mathfrak{Z}_d$, on $\mathfrak{T}_0 \times \mathcal{J}^\xi$.

B) *Tracking stability.* Let for any connected neighborhood $\mathfrak{N}_\varepsilon = \mathfrak{B}_\varepsilon$ of $\mathbf{z}_d = \mathbf{0}_K$ the associated neighborhood $\mathfrak{N}(\varepsilon; t_0; \mathbf{i})$ of $\mathbf{z}_d = \mathbf{0}_K$ be set $\mathfrak{V}_{\zeta(\varepsilon)}(t; t_0; \mathcal{J}^\xi) \subset \mathfrak{B}_\varepsilon$, $\forall t \in \mathfrak{T}_0$. Such $\mathfrak{V}_{\zeta(\varepsilon)}(t; t_0; \mathcal{J}^\xi)$ exists due to the positive definiteness of $V(.)$, Definition 308, and Proposition 309. This, the condition 2) and (23.3) guarantee:

$$\forall \mathfrak{B}_\varepsilon \subseteq \mathfrak{R}^K, \ \exists \mathfrak{N}(\varepsilon; t_0; \mathbf{i}), \ \mathfrak{N}(\varepsilon; t_0; \mathbf{i}) = \mathfrak{V}_{\zeta(\varepsilon)}(t_0; t_0; \mathcal{J}^\xi),$$
$$\mathbf{z}_0 \in \mathfrak{N}(\varepsilon; t_0; \mathbf{i}) \Longrightarrow \mathbf{z}(t; t_0; \mathbf{z}_0; \mathbf{i}) \in \mathfrak{V}_{\zeta(\varepsilon)}(t; t_0; \mathcal{J}^\xi) \subset \mathfrak{B}_\varepsilon, \ \forall [t, \mathbf{i}(.)] \in \mathfrak{T}_0 \times \mathcal{J}^\xi.$$

This, Definition 220, Definition 239, the definition of $\mathbf{z}$ and a) prove that the system exhibits stable tracking of $\mathbf{z}_d = \mathbf{0}_K$; i.e., of $\mathbf{Z}_d(t)$, $\mathbf{Z}_d(.) \in \mathfrak{Z}_d$, on $\mathfrak{T}_0 \times \mathcal{J}^\xi$, and that $\mathrm{Cl}\mathfrak{V}_{\xi_M}(t_0; t_0)$ is a subset of the domain of stable tracking on $\mathfrak{T}_0 \times \mathcal{J}^\xi$ ■

Comment 358 *The asymptotic contraction of the sets* $\mathfrak{V}_\zeta(t; t_0; \mathcal{J}^\xi)$ *induced by* $V(.)$ *relaxes the condition on Dini derivative* $D^+V(.)$ *of* $V(.)$. *It should be negative out of the origin but it need not be negative definite.*

Theorem 359 ***Stable tracking of*** $\mathbf{Z}_d(t)$ ***of the system (3.12) on*** $\mathfrak{T}_0 \times \mathcal{J}^\xi$, ***equivalently of*** $\mathbf{z}_d = \mathbf{0}_K$ ***of the system (23.1) on*** $\mathfrak{T}_0 \times \mathcal{J}^\xi$

In order for the system (3.12), equivalently the system (23.1), respectively, to exhibit the stable tracking of $\mathbf{Z}_d(t)$, $\mathbf{Z}_d(.) \in \mathfrak{Z}_d$, *equivalently of* $\mathbf{z}_d(t) \equiv \mathbf{0}_K$, *on* $\mathfrak{T}_0 \times \mathcal{J}^\xi$ *it is sufficient that there are a set* $\mathfrak{S}(t; t_0)$ *and a function* $V(.) : \mathfrak{T}_0 \times \mathfrak{R}^K \times \mathcal{J}^\xi \longrightarrow \mathfrak{R}_+$ *which obey:*

1) $\mathfrak{S}(t; t_0)$ *is a bounded continuous asymptotically contractive set on* $\mathfrak{T}_0$.

2) The boundary function $\omega(.) : \mathfrak{T}_0 \times \mathfrak{R}^K \times \mathcal{J}^\xi \longrightarrow \mathfrak{R}_+$ *of the set* $\mathfrak{S}(t; t_0)$ *is radially increasing on* $\partial\mathfrak{S}(t; t_0)$, $\forall t \in \mathfrak{T}_0$.

3) The function $V(.)$ is positive definite on the set $\mathfrak{S}(\mathfrak{T}_0)$ for

$$\mathfrak{S}(\mathfrak{T}_0) = \cup\left[\mathfrak{S}(t;t_0) : t \in \mathfrak{T}_0\right].$$

4) $D^+V(.)$ along the solutions $\mathbf{z}(t;t_0;\mathbf{z}_0;\mathbf{i})$ of the system (23.1) obeys

$$D^+V\left(t,\mathbf{z};\mathcal{J}^\xi\right) \le 0,\ \forall\left[t,\mathbf{z},\mathbf{i}(.)\right] \in \mathfrak{T}_0 \times Cl\mathfrak{S}(t;t_0) \times \mathcal{J}^\xi.$$

5) $D^+\omega(.)$ along the solutions $\mathbf{z}(t;t_0;\mathbf{z}_0;\mathbf{i})$ of the system (23.1) obeys

$$D^+\omega\left(t,\mathbf{z};\mathcal{J}^\xi\right) < 0,\ \forall\left[t,\mathbf{z},\mathbf{i}(.)\right] \in \mathfrak{T}_0 \times \partial\mathfrak{S}(t;t_0) \times \mathcal{J}^\xi.$$

The set $\mathfrak{S}(t_0;t_0)$ is a subset of the domain of stable tracking on $\mathfrak{T}_0 \times \mathcal{J}^\xi$.

Proof. Let the conditions of the theorem statement be valid. All that follows holds for all $\mathbf{i}(.) \in \mathcal{J}^\xi$.

Tracking. Definition 316 together with the conditions 2) and 5) ensure positive invariance (Lemma 317) of $\mathrm{Cl}\mathfrak{S}(t;t_0)$ on $\mathfrak{T}_0 \times \mathcal{J}^\xi$ with respect to the system motions. This, Definition 46, and the condition 1) result in

$$\mathbf{z}_0 \in \mathrm{Cl}\mathfrak{S}(t_0;t_0) \Longrightarrow$$

$$\mathbf{z}(t;t_0;\mathbf{z}_0;\mathbf{i}) \in \mathrm{Cl}\mathfrak{S}(t;t_0),\ \forall\left[t,\mathbf{i}(.)\right] \in \mathfrak{T}_0 \times \mathcal{J}^\xi \Longrightarrow \tag{23.4}$$

$$t \longrightarrow \infty \Longrightarrow \mathrm{Cl}\mathfrak{S}(t;t_0) \longrightarrow \mathfrak{O} \Longrightarrow \mathbf{z}(t;t_0;\mathbf{z}_0;\mathbf{i}) \longrightarrow \mathbf{z}_d = \mathbf{0}_K,\ \forall \mathbf{i}(.) \in \mathcal{J}^\xi. \tag{23.5}$$

The system exhibits tracking of $\mathbf{z}_d = \mathbf{0}_K$, equivalently of $\mathbf{Z}_d(t)$, $\mathbf{Z}_d(.) \in \mathfrak{Z}_d$, on $\mathfrak{T}_0 \times \mathcal{J}^\xi$.

Tracking stability. Let for any connected neighborhood $\mathfrak{N}_\varepsilon = \mathfrak{B}_\varepsilon$ of $\mathbf{z}_d = \mathbf{0}_K$ the associated neighborhood $\mathfrak{N}(\varepsilon;t_0;\mathbf{i})$ of $\mathbf{z}_d = \mathbf{0}_K$ be $\mathfrak{V}_{\varsigma(\varepsilon)}\left(t_0;t_0;\mathcal{J}^\xi\right) \subset \mathfrak{B}_\varepsilon \cap \mathrm{Cl}\mathfrak{S}(t_0;t_0)$, $\forall t \in \mathfrak{T}_0$. Such $\mathfrak{V}_{\varsigma(\varepsilon)}(t;t_0)$ exists due to condition 3). This choice of $\mathfrak{N}(\varepsilon;t_0;\mathbf{i})$, $\mathfrak{N}(\varepsilon;t_0;\mathbf{i}) = \mathfrak{V}_{\varsigma(\varepsilon)}(t_0;t_0)$, (23.4), and condition 4) ensure

$$\mathbf{z}_0 \in \mathfrak{N}(\varepsilon;t_0;\mathbf{i}) \Longrightarrow \mathbf{z}(t;t_0;\mathbf{z}_0;\mathbf{i}) \in \mathfrak{V}_{\varsigma(\varepsilon)}\left(t;t_0;\mathcal{J}^\xi\right) \subset \mathfrak{B}_\varepsilon,\ \forall\left[t,\mathbf{i}(.)\right] \in \mathfrak{T}_0 \times \mathcal{J}^\xi.$$

This proves that the system exhibits tracking of $\mathbf{z}_d = \mathbf{0}_K$, equivalently of $\mathbf{Z}_d(t)$, $\mathbf{Z}_d(.) \in \mathfrak{Z}_d$, and that $\mathfrak{S}(t_0;t_0)$ is a subset of the domain of stable tracking on $\mathfrak{T}_0 \times \mathcal{J}^\xi$ in view of Definition 220, Definition 239, and the definition of $\mathbf{z}$ ■

Comment 360 *The asymptotic contraction of the set $\mathfrak{S}(t;t_0)$, the radial increasing of its boundary function on its boundary, and the negativeness of the Dini derivative of $\omega(.)$ permit the non-positiveness of $D^+V(.)$ of the positive definite function $V(.)$ and that the boundary function $\omega(.)$ is not a positive definite function.*

The use of the boundary function $\omega(.)$ can be avoided.

Theorem 361 ***Stable tracking of $\mathbf{Z}_d(t)$ of the system (3.12) on $\mathfrak{T}_0 \times \mathcal{J}^\xi$, equivalently of $\mathbf{z}_d(t) \equiv \mathbf{0}_K$, of the system (23.1) on $\mathfrak{T}_0 \times \mathcal{J}^\xi$***

In order for the system (3.12), equivalently the system (23.1), respectively, to exhibit the stable tracking of $\mathbf{Z}_d(t)$, $\mathbf{Z}_d(.) \in \mathfrak{Z}_d$, equivalently of $\mathbf{z}_d(t) \equiv \mathbf{0}_K$, on $\mathfrak{T}_0 \times \mathcal{J}^\xi$ it is sufficient that there are a set $\mathfrak{S}(t;t_0)$ and a function $V(.) : \mathfrak{T}_0 \times \mathfrak{R}^K \longrightarrow \mathfrak{R}_+$ which obey:

1) $\mathfrak{S}(t;t_0)$ is a continuous asymptotically contractive set on $\mathfrak{T}_0$.

2) The function $V(.)$ is positive definite on the set $\mathfrak{S}(\mathfrak{T}_0)$ for

$$\mathfrak{S}(\mathfrak{T}_0) = \cup\left[\mathfrak{S}(t;t_0) : t \in \mathfrak{T}_0\right].$$

3) The set $\mathfrak{S}(t;t_0)$ is positively invariant with respect to the function $V(.)$ on $\mathfrak{T}_0$,

4) $D^+V(.)$ along the solutions $\mathbf{z}(t;t_0;\mathbf{z}_0;\mathbf{i})$ of the system (23.1) obeys

$$D^+V(t,\mathbf{z},\mathbf{i}) \le 0,\ \forall\left[t,\mathbf{z},\mathbf{i}(.)\right] \in \mathfrak{T}_0 \times Cl\mathfrak{S}(t;t_0) \times \mathcal{J}^\xi.$$

The set $\mathfrak{S}(t_0;t_0)$ is a subset of the domain of stable tracking on $\mathfrak{T}_0 \times \mathcal{J}^\xi$.

Proof. Let the conditions of the theorem statement be satisfied. All that follows holds for every $\mathbf{i}(.) \in \mathcal{J}^{\xi}$.

Tracking. The conditions 1)-4) ensure positive invariance of $\mathfrak{S}(t;t_0)$ on $\mathfrak{T}_0 \times \mathcal{J}^{\xi}$ with respect to the system motions. This, Definition 46, and condition 1) result in

$$\mathbf{z}_0 \in \mathfrak{S}(t_0;t_0) \Longrightarrow$$

$$\mathbf{z}(t;t_0;\mathbf{z}_0;\mathbf{i}) \in \mathfrak{S}(t;t_0), \ \forall [t,\mathbf{i}(.)] \in \mathfrak{T}_0 \times \mathcal{J}^{\xi} \Longrightarrow \tag{23.6}$$

$$t \longrightarrow \infty \Longrightarrow \mathfrak{S}(t;t_0) \longrightarrow \mathfrak{D} \Longrightarrow \mathbf{z}(t;t_0;\mathbf{z}_0;\mathbf{i}) \longrightarrow \mathbf{z}_d = \mathbf{0}_K, \ \forall \mathbf{i}(.) \in \mathcal{J}^{\xi}, \tag{23.7}$$

Tracking stability. Let for any connected neighborhood $\mathfrak{N}_{\varepsilon} = \mathfrak{B}_{\varepsilon}$ of $\mathbf{z}_d = \mathbf{0}_K$ the associated neighborhood $\mathfrak{N}(\varepsilon;t_0;\mathbf{i})$ of $\mathbf{z}_d = \mathbf{0}_K$ be the set $\mathfrak{V}_{\zeta(\varepsilon)}(t_0;t_0;\mathcal{J}^{\xi}) \subset \mathfrak{B}_{\varepsilon} \cap \mathfrak{S}(t_0;t_0)$, $\forall t \in \mathfrak{T}_0$. Such $\mathfrak{V}_{\zeta(\varepsilon)}(t;t_0;\mathcal{J}^{\xi})$ exists due to condition 2). This choice of $\mathfrak{N}(\varepsilon;t_0;\mathbf{i})$, $\mathfrak{N}(\varepsilon;t_0;\mathbf{i}) = \mathfrak{V}_{\zeta(\varepsilon)}(t_0;t_0;\mathcal{J}^{\xi})$, (23.4), and conditions 3) and 4) ensure

$$\mathbf{z}_0 \in \mathfrak{N}(\varepsilon;t_0;\mathbf{i}) \Longrightarrow \mathbf{z}(t;t_0;\mathbf{z}_0;\mathbf{i}) \in \mathfrak{V}_{\zeta(\varepsilon)}(t;t_0;\mathcal{J}^{\xi}) \subset \mathfrak{B}_{\varepsilon},$$
$$\forall [t,\mathbf{i}(.)] \in \mathfrak{T}_0 \times \mathcal{J}^{\xi}.$$

This proves that the system exhibits tracking of $\mathbf{z}_d = \mathbf{0}_K$, equivalently of $\mathbf{Z}_d(t)$, $\mathbf{Z}_d(.) \in \mathfrak{Z}_d$, and that the set $\mathfrak{S}(t_0;t_0)$ is a subset of the domain of stable tracking on $\mathfrak{T}_0 \times \mathcal{J}^{\xi}$ in view of Definition 220, Definition 239, and the definition of $\mathbf{z}$ ∎

Note 362 *Theorem 357 through Theorem 361 reduce in the stability study of* $\mathbf{z}_d = \mathbf{0}_K$ *the system (3.12) to Theorem 12 and Theorem 13 of [227, pp. 38-40]. The illustrative examples of [227, pp. 38-41: Examples 8-10] are illustrative also for the above theorems.*

Note 363 *Theorem 357 through Theorem 361 present sufficient conditions for stable tracking that is not uniform.*

23.2 Criteria: Noncontractive *time*-varying sets

The sets $\mathfrak{V}_{\xi}(t;t_0)$, $\mathfrak{V}_{\xi}(t;t_0) \subseteq \mathfrak{R}^K$, induced by the positive definite function $V(.): \mathfrak{T}_0 \times \mathfrak{R}^K \longrightarrow \mathfrak{R}_+$, Definition 295, are crucial for the proofs of the following results. In this subsection we allow that they are not asymptotically contractive. This means that for every set $\mathfrak{V}_{\xi}(t;t_0)$ induced by the function $V(.)$ there is a connected neighborhood $\mathfrak{N}$ of the origin $\mathbf{z} = \mathbf{0}_K$, which is the hyperball $B_{\zeta(\xi)}$, such that it is always a subset of $\mathfrak{V}_{\xi}(t;t_0)$:

$$\forall \xi \in \mathfrak{R}^+, \ \exists \zeta = \zeta(\xi) \Longrightarrow \mathfrak{V}_{\xi}(t;t_0) \supseteq B_{\zeta(\xi)}, \forall t \in \mathfrak{T}_0. \tag{23.8}$$

The hyperball $B_{\zeta(\xi)}$ is a connected neighborhood of the origin $\mathbf{z} = \mathbf{0}_K$. The sets $\mathfrak{V}_{\xi}(t;t_0)$ possess this property, (23.8), if the function $V(.)$ is a decrescent function, Definition 270.

Note 354 holds also herein for what follows.

Theorem 364 ***Extended Lyapunov theorem to the uniform stable tracking of*** $\mathbf{Z}_d(t)$ ***of the system (3.12) on*** $\mathfrak{T}_0 \times \mathfrak{T}_i$ ***for every*** $\mathbf{i}(.) \in \mathcal{J}^{\xi}$***, equivalently of*** $\mathbf{z} = \mathbf{0}_K$ ***of the system (23.1) on*** $\mathfrak{T}_0 \times \mathfrak{T}_i \times \mathcal{J}^{\xi}$***, equivalently of*** $\mathbf{z} = \mathbf{0}_K$ ***of the system (23.1) on*** $\mathfrak{T}_0 \times \mathfrak{T}_i$ ***for every*** $\mathbf{i}(.) \in \mathcal{J}^{\xi}$

In order for the system (3.12), equivalently for the system (23.1), to exhibit the uniform stable tracking of $\mathbf{Z}_d(t)$, $\mathbf{Z}_d(.) \in \mathfrak{Z}_d$, *on* $\mathfrak{T}_0 \times \mathfrak{T}_i$ *for every* $\mathbf{i}(.) \in \mathcal{J}^{\xi}$ *it is sufficient that there exist scalar functions* $V(.): \mathfrak{T}_0 \times \mathfrak{R}^K \times \mathcal{J}^{\xi} \longrightarrow \mathfrak{R}_+$ *and* $G(.): \mathfrak{T}_0 \times \mathfrak{R}^K \times \mathcal{J}^{\xi} \longrightarrow \mathfrak{R}_+$ *and a connected neighborhood* $\mathfrak{N}$ *of the origin* $\mathbf{z} = \mathbf{0}_K$ *such that they and Dini derivative* $D^+V(.)$ *of* $V(.)$ *along solutions* $\mathbf{z}(t;t_0;\mathbf{z}_0;\mathbf{i})$ *of the system (23.1) obey:*

1) $V(.)$ is positive definite and decrescent on the neighborhood $\mathfrak{N}$ of the origin $\mathbf{z} = \mathbf{0}_K$ for every $\mathbf{i}(.) \in \mathcal{J}^{\xi}$.

2) $G(.)$ is positive definite on the neighborhood $\mathfrak{N}$ of the origin $\mathbf{z} = \mathbf{0}_K$ for every $\mathbf{i}(.) \in \mathcal{J}^{\xi}$.

and

3) $D^+V(.)$ along the solutions $\mathbf{z}(t; t_0; \mathbf{z}_0; \mathbf{i})$ of the system (23.1) obeys

$$D^+V(t, \mathbf{z}; \mathbf{i}) \leq -G(t, \mathbf{z}, \mathbf{i}), \ \forall [t, t_0, \mathbf{z}] \in \mathfrak{T}_0 \times \mathfrak{T}_i \times \mathfrak{N}, \ \forall \mathbf{i}(.) \in \mathcal{J}^{\xi}. \tag{23.9}$$

Proof. All that follows holds for every $\mathbf{i}(.) \in \mathcal{J}^{\xi}$. Let the conditions of the theorem statement hold. They ensure the existence of a positive definite and decrescent function $V(.) : \mathfrak{T}_0 \times \mathfrak{R}^K \longrightarrow \mathfrak{R}_+$ and a connected neighborhood $\mathfrak{N}$ of the origin $\mathbf{z} = \mathbf{0}_K$ such that they satisfy the condition 1).The function $V(.)$ obeys (20.11) through (20.13) (Definition 267 and Definition 270); i.e., $V(.)$ is defined and continuous on $\mathfrak{T}_0 \times \mathfrak{N}$,

$$V(t, \mathbf{z}; \mathbf{i}) \in \mathfrak{C}(\mathfrak{T}_0 \times \mathfrak{N}), \ \forall \mathbf{i}(.) \in \mathcal{J}^{\xi}, \tag{23.10}$$

is bounded from above and below by *time*-invariant positive definite functions $W_i(.)$, $i = 1, 2$,

$$W_1(\mathbf{z}) \geq V(t, \mathbf{z}; \mathbf{i}) \geq W_2(\mathbf{z}), \forall (t, \mathbf{z}) \in \mathfrak{T}_0 \times \mathfrak{N}, \ \forall \mathbf{i}(.) \in \mathcal{J}^{\xi} \tag{23.11}$$

and $V(.)$ vanishes at the origin,

$$V(t, \mathbf{0}_K; \mathbf{i}) = 0, \ \forall t \in \mathfrak{T}_0, \ \forall \mathbf{i}(.) \in \mathcal{J}^{\xi}. \tag{23.12}$$

Definition 267 and positive definiteness of $G(.)$ [the condition 2)] imply the existence of *time*-invariant positive definite function $W_3(.) : \mathfrak{R}^K \times \mathcal{J}^{\xi} \longrightarrow \mathfrak{R}_+$ that satisfies

$$G(t, \mathbf{z}, \mathbf{i}) \geq W_3(\mathbf{z}, \mathbf{i}), \ \forall (t, \mathbf{z}) \in \mathfrak{T}_0 \times \mathfrak{N}, \ \forall \mathbf{i}(.) \in \mathcal{J}^{\xi}. \tag{23.13}$$

Let ξ_M be the maximal $\xi \in \mathfrak{R}^+$ such that

$$\mathfrak{V}_{\xi}(t; t_0) \subseteq \mathfrak{V}_{\xi_M}(t; t_0) \subseteq \mathfrak{N}, \ \forall (t, \xi) \in \mathfrak{T}_0 \times]0, \xi_M]. \tag{23.14}$$

Such ξ_M and ξ exist due to condition 1) in view of Definition 267, Definition 295, and Proposition 300.

A) *Proof of tracking.* Let us assume that the system (3.12) does not exhibit the stable tracking of $\mathbf{Z}_d(t)$ on $\mathfrak{T}_0 \times \mathfrak{T}_i$ for every $\mathbf{i}(.) \in \mathcal{J}^{\xi}$. This means, due to the violation of Definition 215 and Definition 239, that in every hyperball $\mathfrak{B}_{\Delta}$ of the origin $\mathbf{z} = \mathbf{0}_K$, $\mathfrak{B}_{\Delta} \subseteq \mathfrak{N}$, there are $\mathbf{z}_0 \in \mathfrak{B}_{\Delta}$, $\eta \in \mathfrak{R}^+$, and $\mathbf{i}(.) \in \mathcal{J}^{\xi}$ such that

$$\inf [\mathbf{z}(t; t_0; \mathbf{z}_0; \mathbf{i}) : t \in \mathfrak{T}_0] = \eta \Longrightarrow \mathbf{z}(t; t_0; \mathbf{z}_0; \mathbf{i}) \in \mathfrak{N} \backslash \mathfrak{B}_{\eta}, \ \forall \mathbf{i}(.) \in \mathcal{J}^{\xi}. \tag{23.15}$$

This implies the existence of $\xi_f = \mathfrak{R}^+$ such that

$$\inf V \{t, [\mathbf{z}(t; t_0; \mathbf{z}_0; \mathbf{i})] : t \in \mathfrak{T}_0\} = \xi_f, \ \forall \mathbf{i}(.) \in \mathcal{J}^{\xi}. \tag{23.16}$$

Let $\gamma \in \mathfrak{R}^+$ be such that

$$\inf [W_3(\mathbf{z}, \mathbf{i}) : \mathbf{z} \in \mathfrak{N} \backslash \mathfrak{B}_{\eta}] = \gamma, \ \forall \mathbf{i}(.) \in \mathcal{J}^{\xi}.$$

This, the condition 3) and (23.13) yield:

$$D^+V(t, \mathbf{z}, \mathbf{i}) \leq -G(t, \mathbf{z}, \mathbf{i}) \leq -\gamma, \ \forall [t, \mathbf{z}] \in \mathfrak{T}_0 \times \mathfrak{N}, \ \forall \mathbf{i}(.) \in \mathcal{J}^{\xi}.$$

The integration of this inequality results in:

$$V\left[t,\mathbf{z}\left(t;t_0;\mathbf{z}_0;\mathbf{i}\right)\right]\leq V\left[t_0,\mathbf{z}\left(t_0;t_0;\mathbf{z}_0;\mathbf{i}\right)\right]-\gamma\left(t-t_0\right)$$
$$=V\left(t_0,\mathbf{z}_0\right)-\gamma\left(t-t_0\right)=V_0-\gamma\left(t-t_0\right),\ \forall t\in\mathfrak{T}_0,\ \forall\mathbf{i}\left(.\right)\in\mathcal{J}^{\xi}.$$

This and (23.16) imply

$$V\left[t,\mathbf{z}\left(t;t_0;\mathbf{z}_0;\mathbf{i}\right)\right]<\xi_f,\ \forall\left(t>\frac{V_0-\xi_f}{\gamma}+t_0\right)\in\mathfrak{T}_0,\ \forall\mathbf{i}\left(.\right)\in\mathcal{J}^{\xi},$$

which contradicts (23.16). The contradiction is the consequence of the assumption that in every hyperball $\mathfrak{B}_\Delta$ of the origin $\mathbf{z}=\mathbf{0}_K$, $\mathfrak{B}_\Delta\subseteq\mathfrak{N}$, there are $\mathbf{z}_0\in\mathfrak{B}_\Delta$, $\eta\in\mathfrak{R}^+$. and $\mathbf{i}\left(.\right)\in\mathcal{J}^{\xi}$ such that (23.15) holds. Hence, (23.15) fails; i.e., $\eta=0$ that implies the existence of $\tau>t_0$ such that

$$\inf\left[\mathbf{z}\left(\tau;t_0;\mathbf{z}_0;\mathbf{i}\right):t\in\mathfrak{T}_0\right]=0\Longrightarrow$$
$$\mathbf{z}\left(\tau;t_0;\mathbf{z}_0;\mathbf{i}\right)=\mathbf{0}_K\ \text{ and }\ V\left\{\tau,\left[\mathbf{z}\left(\tau;t_0;\mathbf{z}_0;\mathbf{i}\right)\right]\right\}=0,\ \forall\mathbf{i}\left(.\right)\in\mathcal{J}^{\xi}.$$

If $\tau<\infty$ then this and the conditions 2) and 3) guarantee

$$\mathbf{z}\left(t;t_0;\mathbf{z}_0;\mathbf{i}\right)=\mathbf{0}_K\ \text{ and }\ V\left\{t,\left[\mathbf{z}\left(t;t_0;\mathbf{z}_0;\mathbf{i}\right)\right]\right\}=0,\ \forall t\geq\tau\Longrightarrow$$
$$\lim\left[\mathbf{z}\left(t;t_0;\mathbf{z}_0;\mathbf{i}\right):t\longrightarrow\infty\right]=0,\ \forall\mathbf{z}_0\in\mathfrak{V}_{\xi_M}\left(t_0;t_0\right),\ \forall\mathbf{i}\left(.\right)\in\mathcal{J}^{\xi}.$$

If $\tau=\infty$ then again

$$\mathbf{z}\left(\infty;t_0;\mathbf{z}_0;\mathbf{i}\right)=\mathbf{0}_K\ \text{ and }\ V\left\{\infty,\left[\mathbf{z}\left(\infty;t_0;\mathbf{z}_0;\mathbf{i}\right)\right]\right\}=0\Longrightarrow$$
$$\lim\left[\mathbf{z}\left(t;t_0;\mathbf{z}_0;\mathbf{i}\right):t\longrightarrow\infty\right]=0,\ \forall\mathbf{z}_0\in\mathfrak{V}_{\xi_M}\left(t_0;t_0\right),\ \forall\mathbf{i}\left(.\right)\in\mathcal{J}^{\xi}.$$

The system exhibits tracking of $\mathbf{Z}_d\left(t\right)$ on $\mathfrak{T}_0\times\mathfrak{T}_i$ for every $\mathbf{i}\left(.\right)\in\mathcal{J}^{\xi}$. The set $\mathfrak{V}_{\xi_M}\left(t_0;t_0\right)$ is an estimate of the tracking domain on $\mathfrak{T}_0\times\mathfrak{T}_i\times\mathcal{J}^{\xi}$, $\forall\mathbf{i}\left(.\right)\in\mathcal{J}^{\xi}$.

Tracking uniformity. Let $\varsigma>0$ be arbitrarily chosen. Let $\xi\in\mathfrak{R}^+$ be such that $\mathfrak{V}_\xi\left(t;t_0\right)\subset\mathfrak{B}_\varsigma\cap\mathfrak{N}$. Such ξ exists due to condition 1), Definition 267, and Proposition 300. If $\varsigma\geq\xi_M$ then $\tau\left(t_0,\varsigma,\mathbf{Z}_{d0};\mathcal{J}^{\xi}\right)\equiv0$. If $\varsigma<\xi_M$ then

$$\tau\left(t_0,\varsigma,\mathbf{Z}_{d0};\mathcal{J}^{\xi}\right)=\left(V_0-\xi\right)\gamma_\varsigma^{-1}=\tau\left(\mathfrak{T}_{\mathrm{i}},\varsigma,\mathbf{Z}_{d0};\mathcal{J}^{\xi}\right).\tag{23.17}$$

is independent of $\left(t_0,\mathbf{i}\right)$, where $\gamma_\varsigma=\inf[W_3\left(\mathbf{z},\mathbf{i}\right):\mathbf{z}\in\mathfrak{N}\backslash\mathfrak{B}_\varsigma]$. The intersection

$$\mathfrak{V}_{\xi_M}\left(\mathfrak{T}_{\mathrm{i}}\right)=\cap\left[\mathfrak{V}_{\xi_M}\left(t_0;t_0\right):t_0\in\mathfrak{T}_{\mathrm{i}}\right]$$

is independent of $t_0\in\mathfrak{T}_{\mathrm{i}}$. It is a subset of the domain $\mathcal{D}_{TU}\left(\mathfrak{T}_{\mathrm{i}};\mathcal{J}^{\xi}\right)$ of the uniform tracking of $\mathbf{Z}_d\left(t\right)$ on $\mathfrak{T}_0\times\mathfrak{T}_i\times\mathcal{J}^{\xi}$. This and (23.17) prove that the tracking is uniform in $\left[t_0,\mathbf{i}\left(.\right)\right]\in\mathfrak{T}_i\times\mathcal{J}^{\xi}$.

Uniform tracking stability. Let $\mathbf{i}\left(.\right)\in\mathcal{J}^{\xi}$ and $t_0\in\mathfrak{T}_{\mathrm{i}}$ be arbitrary. Let $\mathfrak{N}_\varepsilon\left(t\right)$ be an arbitrary neighborhood of the origin $\mathbf{z}=\mathbf{0}_K$ at every $t\in\mathfrak{T}_0$. Let $\xi=\xi\left(\varepsilon\right)\in\mathfrak{R}^+$ be such that $\mathfrak{V}_{\xi(\varepsilon)}\left(t;t_0\right)\subset\mathfrak{N}_\varepsilon\left(t\right)\cap\mathfrak{V}_{\xi_M}\left(t;t_0\right)$ at every $t\in\mathfrak{T}_0$. The condition 1) guarantees, due to Definition 270, that there is a time invariant neighborhood $\mathfrak{M}_{\xi(\varepsilon)}$ of the origin $\mathbf{z}=\mathbf{0}_K$ such that

$$\cap\left[\mathfrak{V}_{\xi(\varepsilon)}\left(t;t_0\right):t\in\mathfrak{T}_0\right]\supseteq\mathfrak{M}_{\xi(\varepsilon)},\ \forall t\in\mathfrak{T}_0.$$

This and the conditions 1) through 3), Definition 267, Definition 295, Definition 308, and Proposition 309 guarantee

$$\mathbf{z}\left(t;t_0;\mathbf{z}_0;\mathbf{i}\right)\in\mathfrak{V}_{\xi(\varepsilon)}\left(t;t_0\right),\ \forall\left(t,\mathbf{z}_0\right)\in\mathfrak{T}_0\times\mathfrak{M}_{\xi(\varepsilon)},\ \forall\mathbf{i}\left(.\right)\in\mathcal{J}^{\xi}.$$

Because $\mathfrak{V}_{\xi(\varepsilon)}(t;t_0) \subseteq \mathfrak{N}_\varepsilon(t)$, $\forall t \in \mathfrak{T}_0$, then

$$\mathbf{z}(t;t_0;\mathbf{z}_0;\mathbf{i}) \in \mathfrak{N}_\varepsilon(t), \ \forall [t,\mathbf{z}_0;\mathbf{i}(.)] \in \mathfrak{T}_0 \times \mathfrak{M}_{\xi(\varepsilon)}, \ \forall \mathbf{i}(.) \in \mathcal{J}^\xi.$$

The tracking of the origin $\mathbf{z} = \mathbf{0}_K$, equivalently of $\mathbf{Z}_d(t)$, on $\mathfrak{T}_0 \times \mathfrak{T}_i \times \mathcal{J}^\xi$ is stable, Definition 220. The stable tracking is uniform because the neighborhood $\mathfrak{M}_{\xi(\varepsilon)}$ of the origin $\mathbf{z} = \mathbf{0}_K$ is independent of $[t_0, \mathbf{i}(.)] \in \mathfrak{T}_i \times \mathcal{J}^\xi$ for every neighborhood $\mathfrak{N}_\varepsilon(t)$ of the origin $\mathbf{z} = \mathbf{0}_K$. The intersection between the union $\mathfrak{M}_\xi$ of all $\mathfrak{M}_{\xi(\varepsilon)}$ in ε over $\mathfrak{R}^+$,

$$\mathfrak{M}_\xi = \cup \left[\mathfrak{M}_{\xi(\varepsilon)} : \ \varepsilon \in \mathfrak{R}^+\right],$$

and $\mathfrak{V}_{\xi_M}(\mathfrak{T}_\mathrm{i})$ is a subset of domain $\mathcal{D}_{STU}\left(\mathfrak{T}_i;\mathcal{J}^\xi\right)$ of the uniform stable tracking of the origin $\mathbf{z} = \mathbf{0}_K$, equivalently of $\mathbf{Z}_d(t)$, on $\mathfrak{T}_0 \times \mathfrak{T}_i$ for every $\mathbf{i}(.) \in \mathcal{J}^\xi$,

$$\mathfrak{M}_\xi \cap \mathfrak{V}_{\xi_M}(\mathfrak{T}_\mathrm{i}) \subseteq \mathcal{D}_{STU}\left(\mathfrak{T}_i;\mathcal{J}^\xi\right)$$

■

If we enhance the demands on the system then it can exhibit the exponential tracking of the origin $\mathbf{z} = \mathbf{0}_K$, equivalently of $\mathbf{Z}_d(t)$, on $\mathfrak{T}_0 \times \mathfrak{T}_i \times \mathcal{J}^\xi$:

Theorem 365 ***Extended Lyapunov theorem to the exponential tracking of*** $\mathbf{Z}_d(t)$ ***of the system (3.12) on*** $\mathfrak{T}_0 \times \mathfrak{T}_i$ ***for every*** $\mathbf{i}(.) \in \mathcal{J}^\xi$***, equivalently of*** $\mathbf{z} = \mathbf{0}_K$ ***of the system*** *(23.1)* ***on*** $\mathfrak{T}_0 \times \mathfrak{T}_i \times \mathcal{J}^\xi$***, equivalently of*** $\mathbf{z} = \mathbf{0}_K$ ***of the system*** *(23.1)* ***on*** $\mathfrak{T}_0 \times \mathfrak{T}_i$ ***for every*** $\mathbf{i}(.) \in \mathcal{J}^\xi$

In order for the system (3.12), equivalently for the system (23.1), to exhibit the uniform exponential tracking of $\mathbf{Z}_d(t)$, $\mathbf{Z}_d(.) \in \mathfrak{Z}_d$, *on* $\mathfrak{T}_0 \times \mathfrak{T}_i$ *for every* $\mathbf{i}(.) \in \mathcal{J}^\xi$ *it is sufficient that there exist scalar functions* $V(.) : \mathfrak{T}_0 \times \mathfrak{R}^K \times \mathcal{J}^\xi \longrightarrow \mathfrak{R}_+$ *and* $G(.) : \mathfrak{T}_0 \times \mathfrak{R}^K \times \mathcal{J}^\xi \longrightarrow \mathfrak{R}_+$, *the positive real numbers* η, μ_i, $\forall i \in \{1,2,3\}$, *and a connected neighborhood* $\mathfrak{N}$ *of the origin* $\mathbf{z} = \mathbf{0}_K$ *such that they and Dini derivative* $D^+V(.)$ *of* $V(.)$ *along the solutions* $\mathbf{z}(t;t_0;\mathbf{z}_0;\mathbf{i})$ *of the system (23.1) obey:*

1) $\mu_1 \|\mathbf{z}\|^\eta \leq V(t,\mathbf{z};\mathbf{i}) \leq \mu_2 \|\mathbf{z}\|^\eta$, $\forall (t,t_0,\mathbf{z}) \in \mathfrak{T}_0 \times \mathfrak{T}_i \times \mathfrak{N}$, $\forall \mathbf{i}(.) \in \mathcal{J}^\xi$,

2) $G(t,\mathbf{z};\mathbf{i}) \geq \mu_2 \|\mathbf{z}\|^\eta$, $\forall (t,t_0,\mathbf{z}) \in \mathfrak{T}_0 \times \mathfrak{T}_i \times \mathfrak{N}$, $\forall \mathbf{i}(.) \in \mathcal{J}^\xi$,

and

3) $D^+V(.)$ *along the solutions* $\mathbf{z}(t;t_0;\mathbf{z}_0;\mathbf{i})$ *of the system (23.1) obeys*

$$D^+V(t,\mathbf{z},\mathbf{i}) \leq -G(t,\mathbf{z},\mathbf{i}), \ \forall (t,t_0,\mathbf{z}) \in \mathfrak{T}_0 \times \mathfrak{T}_i \times \mathfrak{N}, \ \forall \mathbf{i}(.) \in \mathcal{J}^\xi. \tag{23.18}$$

Proof. All that follows holds for every $\mathbf{i}(.) \in \mathcal{J}^\xi$. Let all the conditions of the theorem statement be satisfied. The conditions 1) and 2) link $V(t,\mathbf{z})$ with $G(t,\mathbf{z})$:

$$G(t,\mathbf{z}) \geq \mu_2\mu_1^{-1}V(t,\mathbf{z};\mathbf{i}),$$

so that

$$D^+V(t,\mathbf{z},\mathbf{i}) \leq -\mu_2\mu_1^{-1}V(t,\mathbf{z};\mathbf{i}) \ \forall (t,t_0,\mathbf{z}) \in \mathfrak{T}_0 \times \mathfrak{T}_i \times \mathfrak{N}, \ \ \forall \mathbf{i}(.) \in \mathcal{J}^\xi.$$

The integration of this differential inequality along system motions $\mathbf{z}(t;t_0;\mathbf{z}_0;\mathbf{i})$ gives:

$$V[t,\mathbf{z}(t;t_0;\mathbf{z}_0;\mathbf{i});\mathbf{i}] \leq V(t_0;\mathbf{z}_0;\mathbf{i}) \exp\left[-\mu_2\mu_1^{-1}(t-t_0)\right],$$
$$\forall (t,t_0,\mathbf{z}) \in \mathfrak{T}_0 \times \mathfrak{T}_i \times \mathfrak{N}, \ \forall \mathbf{i}(.) \in \mathcal{J}^\xi.$$

The condition 1) permits transforming this inequality into:

$$\mu_1 \|\mathbf{z}(t;t_0;\mathbf{z}_0;\mathbf{i})\|^\eta \leq \mu_2 \|\mathbf{z}\|^\eta \exp\left[-\mu_2\mu_1^{-1}(t-t_0)\right],$$
$$\forall (t,t_0,\mathbf{z}) \in \mathfrak{T}_0 \times \mathfrak{T}_i \times \mathfrak{N}, \ \forall \mathbf{i}(.) \in \mathcal{J}^\xi,$$

or

$$\|\mathbf{z}(t;t_0;\mathbf{z}_0;\mathbf{i})\| \leq \left(\mu_1^{-1}\mu_2\right)^{1/\eta} \|\mathbf{z}\| \exp\left[-b(t-t_0)\right],$$
$$\forall (t,t_0,\mathbf{z}) \in \mathfrak{T}_0 \times \mathfrak{T}_i \times \mathfrak{N},\ \forall \mathbf{i}(.) \in \mathcal{J}^{\xi},$$

where

$$a = \left(\mu_1^{-1}\mu_2\right)^{1/\eta},\quad b = \mu_2\mu_1^{-1}\eta^{-1}.$$

Definition 225 is satisfied. The system (3.12) exhibits the uniform exponential tracking of of the origin $\mathbf{z} = \mathbf{0}_K$, equivalently of $\mathbf{Z}_d(t)$, on $\mathfrak{T}_0 \times \mathfrak{T}_i$ for every $\mathbf{i}(.) \in \mathcal{J}^{\xi}$ ■

Note 366 ***On Lyapunov theorems extended to tracking***

The characteristics of the conditions of Lyapunov theorems extended to tracking (Theorem 357 through Theorem 365) are the following:

A) The conditions are expressed in terms of the existence of the function $V(.)$ called ***the Lyapunov function****, more precisely,* ***the Lyapunov tracking function****.*

B) The conditions are sufficient but not necessary.

C) The conditions do not provide an algorithm of how to construct the Lyapunov function for the system.

D) When the conditions are satisfied the complex problem of the tracking test is resolved. Then they open a possibility to synthesize control for the given plant (3.12).

Chapter 24

CLM: Motion and set tracking

24.1 Introduction

The consistent Lyapunov methodology (CLM) was established and developed in the framework of stability studies of ISO systems [155], [156], [158], [160]-[162], [170], [172], [182], [199]-[205], [217], [255] - [257]. What follows broadens CLM to tracking studies of the general systems (3.12) that incorporate both the IO systems (3.62) and the ISO systems (3.65), (3.66).

24.2 Systems smooth properties

Consistent Lyapunov methodology (CLM) is established herein for two classes of the systems (3.12), equivalently, of the systems (23.1), in the framework of their L-tracking of sets of motions. System dynamical characteristics determine them.

Property 367 ***Target set property***

The time-invariant target set Υ, $\Upsilon \subset \mathfrak{R}^K$, is compact, connected, nonempty with the boundary $\partial\Upsilon$ being also an invariant set relative to the motions of the system ***(3.12)****.*

In the case of motion tracking then $\Upsilon = \{\boldsymbol{\varepsilon}_z : \boldsymbol{\varepsilon}_z = (\mathbf{Z}_d - \mathbf{Z}) = \mathbf{0}_K\} \subset \mathfrak{R}^K$.

Definition 368 ***Weak smoothness property***

Let Property 367 hold.

(i) There is an open, continuous, connected neighborhood $\mathfrak{S}(t;t_0;\Upsilon)$, $\mathfrak{S}(t;t_0;\Upsilon) \subseteq \mathfrak{R}^K$, of the target set Υ for every $(t,t_0) \in \mathfrak{T}_0 \times \mathfrak{T}_i$ such that

$$\mathfrak{S}(\mathfrak{T}_i;\Upsilon) = \cap\left[\mathfrak{S}(t;t_0;\Upsilon) : (t,t_0) \in \mathfrak{T}_0 \times \mathfrak{T}_i\right]$$

is also open connected neighborhood of the target set Υ, and for every

$$(t,t_0,\mathbf{Z}_0,\mathbf{I}) \in \mathfrak{T}_0 \times \mathfrak{T}_i \times \mathfrak{S}(t_0;t_0;\Upsilon) \times \mathcal{J}^{\xi}$$

the following hold.

a) The system (23.1) has a unique motion $\mathbf{Z}(t;t_0;\mathbf{Z}_0;\mathbf{I})$ through $\mathbf{Z}_0$ at t_0 on $\mathfrak{T}_0$ for every $(t_0,\mathbf{I}) \in \mathfrak{T}_i \times \mathcal{J}^{\xi}$, and

b) The system motion $\mathbf{Z}(t;t_0;\mathbf{Z}_0;\mathbf{I})$ is defined and continuous in $(t,t_0,\mathbf{Z}_0) \in \mathfrak{T}_0 \times \mathfrak{T}_i \times \mathfrak{S}(t_0;t_0;\Upsilon)$ for every $\mathbf{I}(.) \in \mathcal{J}^{\xi}$.

(ii) For every $(t_0,\mathbf{Z}_0,\mathbf{I}) \in \mathfrak{T}_i \times \left[\mathfrak{R}^K \backslash Cl\mathfrak{S}(t_0;t_0;\Upsilon)\right] \times \mathcal{J}^{\xi}$ every motion of the system is continuous in $t \in \mathfrak{T}_0$.

If, and only if, the preceding conditions hold for $\mathfrak{S}(t;t_0;\Upsilon) \equiv \mathfrak{R}^K$ then the weak smoothness property holds on $\mathfrak{R}^K$.

Definition 369 ***The first-order weak smoothness property***

(i) There is an open continuous connected neighborhood $\mathfrak{S}(t;t_0;\Upsilon)$, $\mathfrak{S}(t;t_0;\Upsilon) \subseteq \mathfrak{R}^K$, $\mathfrak{S}(t;t_0;\Upsilon) \in \mathfrak{C}^1(\mathfrak{T}_i)$, *of the target set* Υ *for every* $(t,t_0) \in \mathfrak{T}_0 \times \mathfrak{T}_i$ *such that*

$$\mathfrak{S}(\mathfrak{T}_i;\Upsilon) = \cap\left[\mathfrak{S}(t;t_0;\Upsilon) : (t,t_0) \in \mathfrak{T}_0 \times \mathfrak{T}_i\right]$$

is also an open connected neighborhood of the target set Υ, *and for every*

$$(t,t_0,\mathbf{Z}_0,\mathbf{I}) \in \mathfrak{T}_0 \times \mathfrak{T}_i \times \mathfrak{S}(t_0;t_0;\Upsilon) \times \mathcal{J}^\xi$$

the following hold.

a) The system (23.1) has a unique motion $\mathbf{Z}(t;t_0;\mathbf{Z}_0;\mathbf{I})$ *through* $\mathbf{Z}_0$ *at* t_0 *on* $\mathfrak{T}_0$ *for every* $(t_0,\mathbf{I}) \in \mathfrak{T}_i \times \mathcal{J}^\xi$.

b) $\mathbf{Z}(t;t_0;\mathbf{Z}_0;\mathbf{I})$ *is defined, continuous and continuously differentiable for all triplets* $(t,t_0,\mathbf{Z}_0) \in \mathfrak{T}_0 \times \mathfrak{T}_i \times \mathfrak{S}(t_0;t_0;\Upsilon)$ *and for every* $\mathbf{I}(.) \in \mathcal{J}^\xi$.

(ii) For every $(t_0,\mathbf{Z}_0,\mathbf{I}) \in \mathfrak{T}_i \times \left[\mathfrak{R}^K \backslash Cl\mathfrak{S}(t_0;t_0;\Upsilon)\right] \times \mathcal{J}^\xi$ *every motion of the system is continuous in* $t \in \mathfrak{T}_0$.

If, and only if, the preceding conditions hold for $\mathfrak{S}(t;t_0;\Upsilon) \equiv \mathfrak{R}^K$ *then the first-order weak smoothness property holds on* $\mathfrak{R}^K$.

Definition 370 ***Strong smoothness property***

(i) The system (23.1) possesses the weak smoothness property 368.

(ii) If the boundary $\partial\mathfrak{S}(t;t_0;\Upsilon)$ *of* $\mathfrak{S}(t;t_0;\Upsilon)$ *is nonempty at any moment* $t \in \mathfrak{T}_0$, *and for any* $t_0 \in \mathfrak{T}_i$, *then every system motion passing through* $\mathbf{Z}_0 \in \partial\mathfrak{S}(t_0;t_0;\Upsilon)$ *at every* $t_0 \in \mathfrak{T}_i$ *satisfies the following for every* $\mathbf{I}(.) \in \mathcal{J}^\xi$:

$$inf\{\rho\left[\mathbf{Z}(t;t_0;\mathbf{Z}_0;\mathbf{I}),\Upsilon\right] : t \in \mathfrak{T}_0\} > 0,\ \forall(t_0,\mathbf{Z}_0,\mathbf{I}) \in \mathfrak{T}_i \times \partial\mathfrak{S}(t_0;t_0;\Upsilon).$$

If, and only if, the preceding conditions hold for $\mathfrak{S}(t;t_0;\Upsilon) \equiv \mathfrak{R}^K$ *then the strong smoothness property holds on* $\mathfrak{R}^K$.

Definition 371 ***The first-order strong smoothness property***

(i) The system (23.1) possesses the first-order weak smoothness property 369.

(ii) If the boundary $\partial\mathfrak{S}(t;t_0;\Upsilon)$ *of* $\mathfrak{S}(t;t_0;\Upsilon)$ *is nonempty at any moment* $t \in \mathfrak{T}_0$, *and for any* $t_0 \in \mathfrak{T}_i$, *then every system motion passing through* $\mathbf{Z}_0 \in \partial\mathfrak{S}(t_0;t_0;\Upsilon)$ *at every* $t_0 \in \mathfrak{T}_i$ *satisfies the following for every* $\mathbf{I}(.) \in \mathcal{J}^\xi$:

$$inf\{\rho\left[\mathbf{Z}(t;t_0;\mathbf{Z}_0;\mathbf{I}),\Upsilon\right] : t \in \mathfrak{T}_0\} > 0,\ \forall(t_0,\mathbf{Z}_0,\mathbf{I}) \in \mathfrak{T}_i \times \partial\mathfrak{S}(t_0;t_0;\Upsilon).$$

If, and only if, the preceding conditions hold for $\mathfrak{S}(t;t_0;\Upsilon) \equiv \mathfrak{R}^K$ *then the first-order strong smoothness property holds on* $\mathfrak{R}^K$.

Comment 372 ***On smoothness properties***

Both the weak smoothness property and strong smoothness property permit nondifferentiability of the system motions $\mathbf{Z}(t;t_0;\mathbf{Z}_0;\mathbf{I})$ *with respect to* $(t,t_0,\mathbf{Z}_0) \in \mathfrak{T}_0 \times \mathfrak{T}_i \times \mathfrak{S}(t_0;t_0;\Upsilon)$, $\forall\mathbf{I}(.) \in \mathcal{J}^\xi$.

Both the first-order weak smoothness property and the first-order strong smoothness property concern systems with differentiable motions $\mathbf{Z}(t;t_0;\mathbf{Z}_0;\mathbf{I})$ *with respect to* $(t,t_0,\mathbf{Z}_0) \in \mathfrak{T}_0 \times \mathfrak{T}_i \times \mathfrak{S}(t_0;t_0;\Upsilon)$, $\forall\mathbf{I}(.) \in \mathcal{J}^\xi$, *or on the product set* $\mathfrak{T}_0 \times \mathfrak{T}_i \times \mathfrak{S}(t_0;t_0;\Upsilon) \times \mathcal{J}^\xi$.

When we treat a physical system then we can often conclude about smoothness of the system motions from the system physical properties.

24.3 Systems and generating functions

In order to determine the algorithm and conditions for the direct generation, and determination of a system Lyapunov function $V(.;\Upsilon)$ it is natural to refer to Lyapunov methodology for *time*-invariant linear systems. The methodology specifies conditions and the procedure for solving this problem, which has been solved in Lyapunov (inverse) methodology for nonlinear systems.

For *time*-invariant linear systems $\mathbf{Z}^{(1)} = A\mathbf{Z}$ the Lyapunov function $V(.)$ is the unique solution of the following differential equation for an arbitrary *time*-invariant positive definite function $W(.)$:

$$V(\mathbf{Z}) \in \mathfrak{C}^1\left(\mathfrak{T}_0 \times \mathfrak{R}^K\right) \Longrightarrow \frac{dV(\mathbf{Z})}{dt} = -W(\mathbf{Z}),\ \ \mathbf{Z} = \mathbf{Z}\left(t;\mathbf{Z}_0\right). \tag{24.1}$$

This means that Eulerian derivative $V^{(1)}(\mathbf{Z}) = dv(\mathbf{Z})/dt$ is taken along the solutions $\mathbf{Z}\left(t;\mathbf{Z}_0\right)$ of the *time*-invariant linear system.

The new *consistent Lyapunov methodology (CLM)* starts with the generalization of (24.1) to correspond to tracking of *time*-varying nonlinear system (3.12), equivalently of its form (23.1) in terms of the deviations:

$$\begin{gathered} V(t,\mathbf{Z};\mathbf{I};\boldsymbol{\Upsilon}) \in \mathfrak{C}^1\left(\mathfrak{T}_0 \times \mathfrak{R}^K\right)\ \forall \mathbf{I}\left(.\right) \in \mathcal{J}^{\xi} \Longrightarrow V^{(1)}(t,\mathbf{Z};\mathbf{I};\boldsymbol{\Upsilon}) \\ = \frac{dV(t,\mathbf{Z};\mathbf{I};\boldsymbol{\Upsilon})}{dt} = -W(t,\mathbf{Z};\mathbf{I};\boldsymbol{\Upsilon}),\ \ \mathbf{Z} = \mathbf{Z}\left(t;t_0;\mathbf{Z}_0;\mathbf{I}\right),\ \ \forall \mathbf{I}\left(.\right) \in \mathcal{J}^{\xi}, \end{gathered} \tag{24.2}$$

$$\begin{gathered} V(t,\mathbf{Z};\mathbf{I};\boldsymbol{\Upsilon}) \in \mathfrak{C}\left(\mathfrak{T}_0 \times \mathfrak{R}^K\right)\ \forall \mathbf{I}\left(.\right) \in \mathcal{J}^{\xi} \Longrightarrow \\ D^+V(t,\mathbf{Z};\mathbf{I};\boldsymbol{\Upsilon}) = -W(t,\mathbf{Z};\mathbf{I};\boldsymbol{\Upsilon}),\ \ \mathbf{Z} = \mathbf{Z}\left(t;t_0;\mathbf{Z}_0;\mathbf{I}\right),\ \ \forall \mathbf{I}\left(.\right) \in \mathcal{J}^{\xi}. \end{gathered} \tag{24.3}$$

Dini derivative $D^+V(t,\mathbf{Z};\mathbf{I};\boldsymbol{\Upsilon})$ of the function $V(.;\Upsilon)$ is taken along system motions

$$\mathbf{Z}\left(t;t_0;\mathbf{Z}_0;\mathbf{I}\right)$$

for every $\mathbf{I}\left(.\right) \in \mathcal{J}^{\xi}$. The function $W(.;\Upsilon)$ is **the generating function**. The following two classes of the generating functions enable the determination of the conditions that are both necessary and sufficient for all three fundamental tasks unsolved in original Lyapunov stability theory:

1. The establishment of the tracking conditions that are not expressed in terms of the existence of Lyapunov function $V(.;\Upsilon)$.

2. The establishment of the tracking conditions that are expressed in terms of the properties of the generating function $W(.;\Upsilon)$ and, by it induced, the solution Lyapunov function $V(.;\Upsilon)$ of (24.3).

3. The determination of both necessary and sufficient conditions for uniform stable tracking, or for exponential tracking of the *time*-varying nonlinear system (3.12), equivalently of its form (23.1) in terms of the deviations.

4. The establishment of the necessary and sufficient conditions of, and the procedure for, the direct generation of the system tracking Lyapunov function $V(.;\Upsilon)$ so that the function $V(.;\Upsilon)$ is obtained after the single application of the conditions and the procedure. The obtained result is then final.

5. The specification of the necessary and sufficient conditions for a set to be the uniform stable tracking domain.

The function $W(.;\Upsilon)$ induces the set $\mathfrak{W}_{\varsigma}\left(t,t_0;\Upsilon\right)$, Definition 295.

Definition 373 ***Family*** $L\left(\mathfrak{T}_0,\mathfrak{T}_i,\mathfrak{S},\mathcal{J}^{\xi};\Upsilon\right)$ ***of the generating functions*** $W(.;\Upsilon)$

Let $\mathfrak{S}(t;t_0;\Upsilon)$ be an open, continuous, connected neighborhood of the set Υ such that $\cup[\mathfrak{S}(t;t_0;\Upsilon):(t;t_0)\in\mathfrak{T}_0\times\mathfrak{T}_i]$ is also an open connected neighborhood of the set Υ. A function $W(.;\Upsilon):\mathfrak{T}_0\times\mathfrak{T}_i\times\mathfrak{R}^K\times\mathcal{J}^\xi\longrightarrow\mathfrak{R}_+$ belongs to the family $L\left(\mathfrak{T}_0,\mathfrak{T}_i,\mathfrak{S},\mathcal{J}^\xi;\Upsilon\right)$ if, and only if:

1) $W(.;\Upsilon)$ is continuous on $\mathfrak{T}_0\times\mathfrak{T}_i\times\mathfrak{S}(t;t_0;\Upsilon)$ for every $\mathbf{I}(.)\in\mathcal{J}^\xi$ or it is continuous on $\mathfrak{T}_0\times\mathfrak{T}_i\times\mathfrak{S}(t;t_0;\Upsilon)\times\mathcal{J}^\xi$:

$$\begin{gathered}W(t,\mathbf{Z};\mathbf{I};\mathbf{\Upsilon})\in\mathfrak{C}[\mathfrak{T}_0\times\mathfrak{T}_i\times\mathfrak{S}(t;t_0;\Upsilon)],\forall\mathbf{I}(.)\in\mathcal{J}^\xi,\ or\\ W(t,\mathbf{Z};\mathbf{I};\mathbf{\Upsilon})\in\mathfrak{C}\left[\mathfrak{T}_0\times\mathfrak{T}_i\times\mathfrak{S}(t;t_0;\Upsilon)\times\mathcal{J}^\xi\right],\end{gathered}\tag{24.4}$$

respectively.

2) The equations (24.5) and (24.6):

$$\begin{gathered}D^+V(t,\mathbf{Z};\mathbf{I};\mathbf{\Upsilon})=-W(t,\mathbf{Z};\mathbf{I};\mathbf{\Upsilon}),\\ \forall(t,t_0,\mathbf{Z})\in\mathfrak{T}_0\times\mathfrak{T}_i\times\mathfrak{S}(t;t_0;\Upsilon),\ \forall\mathbf{I}(.)\in\mathcal{J}^\xi,\quad or\\ D^+V(t,\mathbf{Z};\mathcal{J}^\xi;\Upsilon)=-W(t,\mathbf{Z};\mathcal{J}^\xi;\Upsilon),\\ \forall[t,t_0,\mathbf{Z};\mathbf{I}(.)]\in\mathfrak{T}_0\times\mathfrak{T}_i\times\mathfrak{S}(t;t_0;\Upsilon)\times\mathcal{J}^\xi,\end{gathered}\tag{24.5}$$

$$\begin{gathered}V(t,\mathbf{Z};\mathbf{I};\mathbf{\Upsilon})=0\iff\mathbf{Z}\in\partial\mathbf{\Upsilon},\ \forall(t,t_0)\in\mathfrak{T}_0\times\mathfrak{T}_i,\forall\mathbf{I}(.)\in\mathcal{J}^\xi,\quad or\\ V\left(t,\mathbf{Z};\mathcal{J}^\xi;\Upsilon\right)=0\iff\mathbf{Z}\in\partial\mathbf{\Upsilon},\ \forall[t,t_0,\mathbf{I}(.)]\in\mathfrak{T}_0\times\mathfrak{T}_i\times\mathcal{J}^\xi,\end{gathered}\tag{24.6}$$

have a solution $V(.;\Upsilon):\mathfrak{T}_0\times\mathfrak{T}_i\times\mathfrak{R}^K\longrightarrow\mathfrak{R}_+$ for every $\mathbf{I}(.)\in\mathcal{J}^\xi$, which is continuous in $(t,\mathbf{Z})\in\mathfrak{T}_i\times Cl\mathfrak{N}_\mu(\Upsilon)$ for an arbitrarily small $\mu\in\mathfrak{R}^+$, $\mu=\mu(W;\mathbf{I};\mathbf{\Upsilon})$, and which obeys (24.7) for some $\omega_\mu(\mathbf{Z};\mathbf{I};\mathbf{\Upsilon})\in\mathfrak{C}[Cl\mathfrak{N}_\mu(\Upsilon)]$ for every $\mathbf{I}(.)\in\mathcal{J}^\xi$:

$$\begin{gathered}V(t,\mathbf{Z};\mathbf{I};\mathbf{\Upsilon})\le\omega_\mu(\mathbf{Z};\mathbf{I};\mathbf{\Upsilon}),\ \forall(t,t_0,\mathbf{Z})\in\mathfrak{T}_0\times\mathfrak{T}_i\times Cl\mathfrak{N}_\mu(\Upsilon),\forall\mathbf{I}(.)\in\mathcal{J}^\xi,\ or\\ V(t,\mathbf{Z};\mathcal{J}^\xi;\Upsilon)\le\omega_\mu\left(\mathbf{Z};\mathcal{J}^\xi\right),\ \forall[t,t_0,\mathbf{Z};\mathbf{I}(.)]\in\mathfrak{T}_0\times\mathfrak{T}_i\times Cl\mathfrak{N}_\mu(\Upsilon)\times\mathcal{J}^\xi.\end{gathered}\tag{24.7}$$

3) The following holds for any $\zeta\in\mathfrak{R}^+$ satisfying $Cl\mathfrak{W}_\zeta(t,t_0;\Upsilon)\subset\mathfrak{S}(t;t_0;\Upsilon)$ for every $(t,t_0)\in\mathfrak{T}_0\times\mathfrak{T}_i$:

$$\begin{gathered}min\ \{W(t,\mathbf{Z};\mathbf{I};\mathbf{\Upsilon}):(t,t_0,\mathbf{Z})\in\mathfrak{T}_0\times\mathfrak{T}_i\times[\mathfrak{S}(t;t_0;\Upsilon)\backslash\mathfrak{W}_\zeta(t,t_0;\Upsilon)]\}=a,\\ \forall\mathbf{I}(.)\in\mathcal{J}^\xi,\quad or\\ min\ \left\{W(t,\mathbf{Z};\mathcal{J}^\xi;\Upsilon):[t,t_0,\mathbf{Z};\mathbf{I}(.)]\in\mathfrak{T}_0\times\mathfrak{T}_i\times[\mathfrak{S}(t;t_0;\Upsilon)\backslash\mathfrak{W}_\zeta(t,t_0;\Upsilon)]\times\mathcal{J}^\xi\right\}\\ =a=a(\zeta;W;\Upsilon)\in\mathfrak{R}^+.\end{gathered}\tag{24.8}$$

Definition 374 ***Family $L^1\left(\mathfrak{T}_0,\mathfrak{T}_i,\mathfrak{S},\mathcal{J}^\xi;\Upsilon\right)$ of the generating functions*** $W(.;\Upsilon)$

Let $\mathfrak{S}(t;t_0;\Upsilon)$ be an open, continuous connected neighborhood of the set Υ such that $\cup[\mathfrak{S}(t;t_0;\Upsilon):(t;t_0)\in\mathfrak{T}_0\times\mathfrak{T}_i]$ is also an open connected neighborhood of the set Υ. A function $W(.;\Upsilon):\mathfrak{T}_0\times\mathfrak{T}_i\times\mathfrak{R}^K\times\mathcal{J}^\xi\longrightarrow\mathfrak{R}_+$ belongs to the family $L^1\left(\mathfrak{T}_0,\mathfrak{T}_i,\mathfrak{S},\mathcal{J}^\xi;\Upsilon\right)$ if, and only if:

1) $W(.;\Upsilon)$ is differentiable on $\mathfrak{T}_0\times\mathfrak{T}_i\times\mathfrak{S}(t;t_0;\Upsilon)$ for every $\mathbf{I}(.)\in\mathcal{J}^\xi$ or it is differentiable on $\mathfrak{T}_0\times\mathfrak{T}_i\times\mathfrak{S}(t;t_0;\Upsilon)\times\mathcal{J}^\xi$:

$$\begin{gathered}W(t,\mathbf{Z};\mathbf{I};\mathbf{\Upsilon})\in\mathfrak{C}^1[\mathfrak{T}_0\times\mathfrak{T}_i\times\mathfrak{S}(t;t_0;\Upsilon)],\forall\mathbf{I}(.)\in\mathcal{J}^\xi,\quad or\\ W(t,\mathbf{Z};\mathbf{I};\mathbf{\Upsilon})\in\mathfrak{C}^1\left[\mathfrak{T}_0\times\mathfrak{T}_i\times\mathfrak{S}(t;t_0;\Upsilon)\times\mathcal{J}^\xi\right],\end{gathered}\tag{24.9}$$

respectively.

2) The equations (24.10) and (24.11):

$$V^{(1)}(t,\mathbf{Z};\mathbf{I};\boldsymbol{\Upsilon}) = -W(t,\mathbf{Z};\mathbf{I};\boldsymbol{\Upsilon}),$$
$$\forall (t,t_0,\mathbf{Z}) \in \mathfrak{T}_0 \times \mathfrak{T}_i \times \mathfrak{S}(t_0;t_0;\Upsilon), \forall \mathbf{I}(.) \in \mathcal{J}^\xi, \text{ or}$$
$$V^{(1)}(t,\mathbf{Z};\mathcal{J}^\xi;\Upsilon) = -W(t,\mathbf{Z};\mathcal{J}^\xi;\Upsilon),$$
$$\forall [t,t_0,\mathbf{Z};\mathbf{I}(.)] \in \mathfrak{T}_0 \times \mathfrak{T}_i \times \mathfrak{S}(t_0;t_0;\Upsilon) \times \mathcal{J}^\xi, \tag{24.10}$$

$$V(t,\mathbf{Z};\mathbf{I};\boldsymbol{\Upsilon}) = 0 \Longleftrightarrow \mathbf{Z} \in \partial\boldsymbol{\Upsilon},\ \forall (t,t_0) \in \mathfrak{T}_0 \times \mathfrak{T}_i, \forall \mathbf{I}(.) \in \mathcal{J}^\xi, \text{ or}$$
$$V(t,\mathbf{Z};\mathbf{I};\boldsymbol{\Upsilon}) = 0 \Longleftrightarrow \mathbf{Z} \in \partial\boldsymbol{\Upsilon},\ \forall [t,t_0,\mathbf{I}(.)] \in \mathfrak{T}_0 \times \mathfrak{T}_i \times \mathcal{J}^\xi, \tag{24.11}$$

have a solution $V(.;\Upsilon) : \mathfrak{T}_0 \times \mathfrak{T}_i \times \mathfrak{R}^K \longrightarrow \mathfrak{R}_+$ *for every* $\mathbf{I}(.) \in \mathcal{J}^\xi$ *that is differentiable in* $(t,\mathbf{Z}) \in \mathfrak{T}_i \times C\mathfrak{N}_\mu(\Upsilon)$ *for an arbitrarily small* $\mu \in \mathfrak{R}^+$, $\mu = \mu(W;\mathbf{I};\boldsymbol{\Upsilon})$, *and which obeys (24.7) for some* $\omega_\mu(\mathbf{Z};\mathbf{I};\boldsymbol{\Upsilon}) \in \mathfrak{C}[C\mathfrak{N}_\mu(\Upsilon)]$ *for every* $\mathbf{I}(.) \in \mathcal{J}^\xi$,

3) The condition (24.8) holds for any $\zeta \in \mathfrak{R}^+$ *satisfying* $C\mathfrak{W}_\zeta(t,t_0;\Upsilon) \subset \mathfrak{S}(t;t_0;\Upsilon)$ *for every* $(t,t_0) \in \mathfrak{T}_0 \times \mathfrak{T}_i$.

The family $E\left[\mathfrak{T}_0,\mathfrak{T}_i,\mathfrak{S}(.),\mathcal{J}^\xi;\Upsilon\right]$ of the generating functions $W(.;\Upsilon)$ is defined by:

Definition 375 ***Family*** $E\left(\mathfrak{T}_0,\mathfrak{T}_i,\mathfrak{S},\mathcal{J}^\xi;\Upsilon\right)$ ***of the generating functions*** $W(.;\Upsilon)$

Let $\mathfrak{S}(t;t_0;\Upsilon)$ *be an open, continuous connected neighborhood of the set* Υ *such that* $\cup[\mathfrak{S}(t;t_0;\Upsilon) : (t;t_0) \in \mathfrak{T}_0 \times \mathfrak{T}_i]$ *is also an open connected neighborhood of the set* Υ. *A function* $W(.;\Upsilon) : \mathfrak{T}_0 \times \mathfrak{T}_i \times \mathfrak{R}^K \times \mathcal{J}^\xi \longrightarrow \mathfrak{R}_+$ *belongs to the family* $E\left(\mathfrak{T}_0,\mathfrak{T}_i,\mathfrak{S},\mathcal{J}^\xi;\Upsilon\right)$ *if, and only if:*

1) $W(.;\Upsilon)$ *is continuous on* $\mathfrak{T}_0 \times \mathfrak{T}_i \times \mathfrak{S}(t;t_0;\Upsilon)$ *for every* $\mathbf{I}(.) \in \mathcal{J}^\xi$ *or it is continuous on* $\mathfrak{T}_0 \times \mathfrak{T}_i \times \mathfrak{S}(t;t_0;\Upsilon) \times \mathcal{J}^\xi$:

$$W(t,\mathbf{Z};\mathbf{I};\boldsymbol{\Upsilon}) \in \mathfrak{C}[\mathfrak{T}_0 \times \mathfrak{T}_i \times \mathfrak{S}(t;t_0;\Upsilon)], \forall \mathbf{I}(.) \in \mathcal{J}^\xi, \text{ or}$$
$$W(t,\mathbf{Z};\mathcal{J}^\xi;\Upsilon) \in \mathfrak{C}\left[\mathfrak{T}_0 \times \mathfrak{T}_i \times \mathfrak{S}(t;t_0;\Upsilon) \times \mathcal{J}^\xi\right], \tag{24.12}$$

respectively,

2) The equations (24.13) and (24.14):

$$D^+u(t,\mathbf{Z};\mathbf{I};\boldsymbol{\Upsilon}) = -\left[1 - u(t,\mathbf{Z};\mathbf{I};\boldsymbol{\Upsilon})\right] W(t,\mathbf{Z};\mathbf{I};\boldsymbol{\Upsilon}),$$
$$\forall (t,t_0,\mathbf{Z}) \in \mathfrak{T}_0 \times \mathfrak{T}_i \times \mathfrak{S}(t;t_0;\Upsilon),\ \forall \mathbf{I}(.) \in \mathcal{J}^\xi, \text{ or}$$
$$D^+u(t,\mathbf{Z};\mathcal{J}^\xi;\Upsilon) = -\left[1 - u(t,\mathbf{Z};\mathcal{J}^\xi;\Upsilon)\right] W(t,\mathbf{Z};\mathcal{J}^\xi;\Upsilon),$$
$$\forall [t,t_0,\mathbf{Z};\mathbf{I}(.)] \in \mathfrak{T}_0 \times \mathfrak{T}_i \times \mathfrak{S}(t;t_0;\Upsilon) \times \mathcal{J}^\xi, \tag{24.13}$$

$$u(t,\mathbf{Z};\mathbf{I};\Upsilon) = 0 \Longleftrightarrow \mathbf{Z} \in \partial\boldsymbol{\Upsilon},\ \forall (t,t_0) \in \mathfrak{T}_0 \times \mathfrak{T}_i,\ \forall \mathbf{I}(.) \in \mathcal{J}^\xi,$$
$$u(t,\mathbf{Z};\mathcal{J}^\xi;\Upsilon) = 0 \Longleftrightarrow \mathbf{Z} \in \partial\boldsymbol{\Upsilon},\ \forall [t,t_0,\mathbf{I}(.)] \in \mathfrak{T}_0 \times \mathfrak{T}_i \times \mathcal{J}^\xi, \tag{24.14}$$

respectively, have a solution $u(.;\Upsilon) : \mathfrak{T}_0 \times \mathfrak{T}_i \times \mathfrak{R}^K \times \mathcal{J}^\xi \longrightarrow \mathfrak{R}_+$ *that is continuous in* $(t,\mathbf{Z}) \in \mathfrak{T}_i \times C\mathfrak{N}_\mu(\Upsilon)$, $\forall \mathbf{I}(.) \in \mathcal{J}^\xi$, *or in* $[t,\mathbf{Z},\mathbf{I}(.)] \in \mathfrak{T}_0 \times C\mathfrak{N}_\mu(\Upsilon) \times \mathcal{J}^\xi$, *for an arbitrarily small* $\mu \in \mathfrak{R}^+$, $\mu = \mu(V;\mathbf{I};\boldsymbol{\Upsilon})$ *or* $\mu(V;\mathcal{J}^\xi;\boldsymbol{\Upsilon})$, *and which obeys (24.15) for some* $\omega_\mu(\mathbf{Z};\mathbf{I};\boldsymbol{\Upsilon}) \in \mathfrak{C}[C\mathfrak{N}_\mu(\Upsilon)]$ *for every* $\mathbf{I}(.) \in \mathcal{J}^\xi$, *or* $\omega_\mu(\mathbf{Z};\mathbf{I};\boldsymbol{\Upsilon}) \in \mathfrak{C}\left[(C\mathfrak{N}_\mu(\Upsilon)) \times \mathcal{J}^\xi\right]$:

$$u(t,\mathbf{Z};\mathbf{I};\boldsymbol{\Upsilon}) \leq \omega_\mu(\mathbf{Z};\mathbf{I};\boldsymbol{\Upsilon}),$$
$$\forall [t,t_0,\mathbf{Z};\mathbf{I}(.)] \in \mathfrak{T}_0 \times \mathfrak{T}_i \times C\mathfrak{N}_\mu(\Upsilon),\ \forall \mathbf{I}(.) \in \mathcal{J}^\xi, \text{ or}$$
$$u(t,\mathbf{Z};\mathcal{J}^\xi;\Upsilon) \leq \omega_\mu\left(\mathbf{Z};\mathcal{J}^\xi\right),\ \forall [t,t_0,\mathbf{Z};\mathbf{I}(.)] \in \mathfrak{T}_0 \times \mathfrak{T}_i \times C\mathfrak{N}_\mu(\Upsilon) \times \mathcal{J}^\xi, \tag{24.15}$$

respectively.

3) The following holds for any $\zeta \in \mathfrak{R}^+$ *satisfying* $Cl\mathfrak{W}_\zeta(t,t_0;\Upsilon) \subset \mathfrak{S}(t;t_0;\Upsilon)$ *for every* $(t,t_0) \in \mathfrak{T}_0 \times \mathfrak{T}_i$ *:*

$$min\left\{ \begin{array}{c} W(t,\mathbf{Z};\mathbf{I};\mathbf{\Upsilon}) : (t,t_0,\mathbf{Z}) \in \\ \in \mathfrak{T}_0 \times \mathfrak{T}_i \times [\mathfrak{S}(t;t_0;\Upsilon) \backslash \mathfrak{W}_\zeta(t,t_0;\mathbf{I};\mathbf{\Upsilon})] \end{array} \right\} = a, \forall \mathbf{I}(.) \in \mathcal{J}^\xi, \ or$$

$$min\left\{ \begin{array}{c} W(t,\mathbf{Z};\mathfrak{T}_i) : [t,t_0,\mathbf{Z};\mathbf{I}(.)] \in \\ \in \mathfrak{T}_0 \times \mathfrak{T}_i \times [\mathfrak{S}(t;t_0;\Upsilon) \backslash \mathfrak{W}_\zeta(t,t_0;\mathfrak{T}_i)] \times \mathcal{J}^\xi \end{array} \right\} = a,$$

$$a = a(\zeta; W; \Upsilon) \in \mathfrak{R}^+, \tag{24.16}$$

respectively.

Definition 376 ***Family*** $E^1(\mathfrak{T}_0,\mathfrak{T}_i,\mathfrak{S},\mathcal{J}^\xi;\Upsilon)$ ***of the generating functions*** $W(.;\Upsilon)$

Let $\mathfrak{S}(t;t_0;\Upsilon)$ *be an open, continuous, connected neighborhood of the set* Υ *such that* $\cup[\mathfrak{S}(t;t_0;\Upsilon) : (t;t_0) \in \mathfrak{T}_0 \times \mathfrak{T}_i]$ *is also an open connected neighborhood of the set* Υ*. A function* $W(.;\Upsilon) : \mathfrak{T}_0 \times \mathfrak{T}_i \times \mathfrak{R}^K \times \mathcal{J}^\xi \longrightarrow \mathfrak{R}_+$ *belongs to the family* $E^1(\mathfrak{T}_0,\mathfrak{T}_i,\mathfrak{S},\mathcal{J}^\xi;\Upsilon)$ *if, and only if:*

1) $W(.;\Upsilon)$ *is differentiable on* $\mathfrak{T}_0 \times \mathfrak{T}_i \times \mathfrak{S}(t;t_0;\Upsilon)$ *for every* $\mathbf{I}(.) \in \mathcal{J}^\xi$ *or it is differentiable on* $\mathfrak{T}_0 \times \mathfrak{T}_i \times \mathfrak{S}(t;t_0;\Upsilon) \times \mathcal{J}^\xi$ *:*

$$W(t,\mathbf{Z};\mathbf{I};\mathbf{\Upsilon}) \in \mathfrak{C}^1[\mathfrak{T}_0 \times \mathfrak{T}_i \times \mathfrak{S}(t;t_0;\Upsilon)], \forall \mathbf{I}(.) \in \mathcal{J}^\xi, \ or$$

$$W(t,\mathbf{Z};\mathcal{J}^\xi;\Upsilon) \in \mathfrak{C}^1\left[\mathfrak{T}_0 \times \mathfrak{T}_i \times \mathfrak{S}(t;t_0;\Upsilon) \times \mathcal{J}^\xi\right], \tag{24.17}$$

respectively

2) The equations (24.18) and (24.19):

$$u^{(1)}(t,\mathbf{Z};\mathbf{I};\mathbf{\Upsilon}) = -[1 - u(t,\mathbf{Z};\mathbf{I};\mathbf{\Upsilon})]\, W(t,\mathbf{Z};\mathbf{I};\mathbf{\Upsilon}),$$

$$\forall (t,t_0,\mathbf{Z}) \in \mathfrak{T}_0 \times \mathfrak{T}_i \times \mathfrak{S}(t;t_0;\Upsilon), \ \forall \mathbf{I}(.) \in \mathcal{J}^\xi, \ or$$

$$u^{(1)}(t,\mathbf{Z};\mathcal{J}^\xi;\mathbf{\Upsilon}) = -\left[1 - u(t,\mathbf{Z};\mathcal{J}^\xi;\Upsilon)\right] W(t,\mathbf{Z};\mathcal{J}^\xi;\Upsilon),$$

$$\forall [t,t_0,\mathbf{Z};\mathbf{I}(.)] \in \mathfrak{T}_0 \times \mathfrak{T}_i \times \mathfrak{S}(t;t_0;\Upsilon) \times \mathcal{J}^\xi, \tag{24.18}$$

$$u(t,\mathbf{Z};\mathbf{I};\Upsilon) = 0 \Longleftrightarrow \mathbf{Z} \in \partial\mathbf{\Upsilon}, \ \forall (t,t_0) \in \mathfrak{T}_0 \times \mathfrak{T}_i, \ \forall \mathbf{I}(.) \in \mathcal{J}^\xi,$$

$$u(t,\mathbf{0}_K;\mathcal{J}^\xi) = 0 \Longleftrightarrow \mathbf{Z} \in \partial\mathbf{\Upsilon}, \ \forall [t,t_0,\mathbf{I}(.)] \in \mathfrak{T}_0 \times \mathfrak{T}_i \times \mathcal{J}^\xi, \tag{24.19}$$

respectively, have a solution $u(.;\Upsilon) : \mathfrak{T}_0 \times \mathfrak{T}_i \times \mathfrak{R}^K \times \mathcal{J}^\xi \longrightarrow \mathfrak{R}_+$ *that is differentiable in* $(t,\mathbf{Z}) \in \mathfrak{T}_i \times Cl\mathfrak{N}_\mu(\Upsilon)$*, or in* $[t,\mathbf{Z},\mathbf{I}(.)] \in \mathfrak{T}_0 \times Cl\mathfrak{N}_\mu(\Upsilon) \times \mathcal{J}^\xi$*, for an arbitrarily small* $\mu \in \mathfrak{R}^+$*,* $\mu = \mu(V;\mathbf{I};\mathbf{\Upsilon})$ *or* $\mu(V;\mathcal{J}^\xi;\mathbf{\Upsilon})$*, and which obeys (24.15) for some* $\omega_\mu(\mathbf{Z};\mathbf{I};\mathbf{\Upsilon}) \in \mathfrak{C}[Cl\mathfrak{N}_\mu(\Upsilon)]$ *for every* $\mathbf{I}(.) \in \mathcal{J}^\xi$*, or* $\omega_\mu(\mathbf{Z};\mathbf{I};\mathbf{\Upsilon}) \in \mathfrak{C}\left[(Cl\mathfrak{N}_\mu(\Upsilon)) \times \mathcal{J}^\xi\right]$*, respectively.*

3) The condition (24.16) holds for any $\zeta \in \mathfrak{R}^+$ *satisfying* $Cl\mathfrak{W}_\zeta(t,t_0;\Upsilon) \subset \mathfrak{S}(t;t_0;\Upsilon)$ *for every* $(t,t_0) \in \mathfrak{T}_0 \times \mathfrak{T}_i$*, respectively.*

Note 377 ***Tracking and functional families*** $L(\mathfrak{T}_0,\mathfrak{T}_i,\mathfrak{S},\mathcal{J}^\xi;\Upsilon)$, $L^1(\mathfrak{T}_0,\mathfrak{T}_i,\mathfrak{S},\mathcal{J}^\xi;\Upsilon)$, $E(\mathfrak{T}_0,\mathfrak{T}_i,\mathfrak{S},\mathcal{J}^\xi;\Upsilon)$, ***and*** $E^1(\mathfrak{T}_0,\mathfrak{T}_i,\mathfrak{S},\mathcal{J}^\xi;\Upsilon)$

There is not any condition imposed on the system (3.12), equivalently, on its form (23.1) in terms of the deviations, in Definition 373-Definition 376. The families $L(\mathfrak{T}_0,\mathfrak{T}_i,\mathfrak{S},\mathcal{J}^\xi;\Upsilon)$*,* $L^1(\mathfrak{T}_0,\mathfrak{T}_i,\mathfrak{S},\mathcal{J}^\xi;\Upsilon)$*,* $E(\mathfrak{T}_0,\mathfrak{T}_i,\mathfrak{S},\mathcal{J}^\xi;\Upsilon)$*, and* $E^1(\mathfrak{T}_0,\mathfrak{T}_i,\mathfrak{S},\mathcal{J}^\xi;\Upsilon)$ *are independent of tracking properties of the system.*

Comment 378 ***The relationship between the functions*** $V(.;\Upsilon)$ ***and*** $u(.;\Upsilon)$ ***for the same generating function*** $W(.;\Upsilon)$

Let the generating function $W(.;\Upsilon)$ be the same in Definition 373 and Definition 375. The functions $V()$ generated by $W(.;\Upsilon) \in L\left(\mathfrak{T}_0, \mathfrak{T}_i, \mathfrak{S}, \mathcal{J}^{\xi}; \Upsilon\right)$ and the function $u()$ generated by $W(.;\Upsilon) \in E\left(\mathfrak{T}_0, \mathfrak{T}_i, \mathfrak{S}, \mathcal{J}^{\xi}; \Upsilon\right)$ are interrelated, due to (24.5) and (24.13), in the differential form, and in the explicit form [542], by:

$$\begin{gathered} D^{+}u(t,\mathbf{Z};\mathbf{I};\boldsymbol{\Upsilon}) = [1-u(t,\mathbf{Z};\mathbf{I};\boldsymbol{\Upsilon})]\, D^{+}V(t,\mathbf{Z};\mathbf{I};\boldsymbol{\Upsilon}), \text{ or} \\ u^{(1)}(t,\mathbf{Z};\mathbf{I};\boldsymbol{\Upsilon}) = [1-u(t,\mathbf{Z};\mathbf{I};\boldsymbol{\Upsilon})]\, V^{(1)}(t,\mathbf{Z};\mathbf{I};\boldsymbol{\Upsilon}), \\ \forall (t,t_0,\mathbf{Z}) \in \mathfrak{T}_0 \times \mathfrak{T}_i \times \mathfrak{S}\,(t;t_0;\Upsilon)\,, \forall \mathbf{I}\,(.) \in \mathcal{J}^{\xi}, \end{gathered} \tag{24.20}$$

$$\begin{gathered} D^{+}u(t,\mathbf{Z};\mathbf{I};\boldsymbol{\Upsilon}) = [1-u(t,\mathbf{Z};\mathbf{I};\boldsymbol{\Upsilon})]\, D^{+}V(t,\mathbf{Z};\mathbf{I};\boldsymbol{\Upsilon}), \text{ or} \\ u^{(1)}(t,\mathbf{Z};\mathbf{I};\boldsymbol{\Upsilon}) = [1-u(t,\mathbf{Z};\mathbf{I};\boldsymbol{\Upsilon})]\, V^{(1)}(t,\mathbf{Z};\mathbf{I};\boldsymbol{\Upsilon}), \\ \forall (t,t_0,\mathbf{Z}) \in \mathfrak{T}_0 \times \mathfrak{T}_i \times \mathfrak{S}\,(t;t_0;\Upsilon)\,, \forall \mathbf{I}\,(.) \in \mathcal{J}^{\xi}, \end{gathered} \tag{24.21}$$

$$\begin{gathered} D^{+}u(t,\mathbf{Z};\mathcal{J}^{\xi};\Upsilon) = \left[1-u(t,\mathbf{Z};\mathcal{J}^{\xi};\Upsilon)\right] D^{+}V(t,\mathbf{Z};\mathcal{J}^{\xi};\Upsilon), \text{ or} \\ u^{(1)}(t,\mathbf{Z};\mathcal{J}^{\xi};\boldsymbol{\Upsilon}) = [1-u(t,\mathbf{Z};\mathbf{I};\boldsymbol{\Upsilon})]\, V^{(1)}(t,\mathbf{Z};\mathcal{J}^{\xi};\Upsilon), \\ \forall\left[t,t_0,\mathbf{Z};\mathbf{I}\,(.)\right] \in \mathfrak{T}_0 \times \mathfrak{T}_i \times \mathfrak{S}\,(t;t_0;\Upsilon) \times \mathcal{J}^{\xi}, \end{gathered} \tag{24.22}$$

$$\begin{gathered} u(t,\mathbf{Z};\mathbf{I};\boldsymbol{\Upsilon}) = 1 - exp\left[-V(t,\mathbf{Z};\mathbf{I};\boldsymbol{\Upsilon})\right], \\ \forall\left[t,t_0,\mathbf{Z};\mathbf{I}\,(.)\right] \in \mathfrak{T}_0 \times \mathfrak{T}_i \times \mathfrak{S}\,(t;t_0;\Upsilon) \times \mathcal{J}^{\xi}, \text{ or} \\ u(t,\mathbf{Z};\mathcal{J}^{\xi};\Upsilon) = 1 - exp\left[-V(t,\mathbf{Z};\mathcal{J}^{\xi};\Upsilon)\right], \\ \forall\left[t,t_0,\mathbf{Z};\mathbf{I}\,(.)\right] \in \mathfrak{T}_0 \times \mathfrak{T}_i \times \mathfrak{S}\,(t;t_0;\Upsilon) \times \mathcal{J}^{\xi}, \end{gathered} \tag{24.23}$$

$$\begin{gathered} u(t,\mathbf{Z};\mathbf{I};\boldsymbol{\Upsilon}) = 0 \Longleftrightarrow V(t,\mathbf{Z};\mathbf{I};\boldsymbol{\Upsilon}) = 0, \text{ and} \\ u(t,\mathbf{Z};\mathbf{I};\boldsymbol{\Upsilon}) \longrightarrow 1 \Longleftrightarrow V(t,\mathbf{Z};\mathbf{I};\boldsymbol{\Upsilon}) \longrightarrow \infty. \end{gathered} \tag{24.24}$$

24.4 Criteria: Systems with continuous motions

The form of the uniform stable tracking conditions depend on the smoothness property of the system (Definition 368 or Definition 370) and on the family of the generating functions (Definition 373 or Definition 375).

Note 354 is valid also herein for what follows.

Theorem 379 ***Uniform stable tracking the target set Υ on $\mathfrak{T}_0 \times \mathfrak{T}_i$ by the motions of the system (3.12): strong smoothness property***

Let the system (3.12) possess strong smoothness property 370. In order for the system motions to exhibit uniform stable tracking of Υ on $\mathfrak{T}_0 \times \mathfrak{T}_i$ for every $\mathbf{I}(.) \in \mathcal{J}^{\xi}$, for a set $\mathfrak{N}(t;\mathbf{I};\boldsymbol{\Upsilon})$ to be the domain $\mathcal{D}_{ST}(t_0;\mathbf{I};\boldsymbol{\Upsilon})$ at $t = t_0 \in \mathfrak{T}_i$, $\forall \mathbf{I}(.) \in \mathcal{J}^{\xi}$, of the stable tracking of Υ on $\mathfrak{T}_0 \times \mathfrak{T}_i$ for every $\mathbf{I}(.) \in \mathcal{J}^{\xi} : \mathfrak{N}(t_0;\mathbf{I};\boldsymbol{\Upsilon}) \equiv \mathcal{D}_{ST}(t_0;\mathbf{I};\boldsymbol{\Upsilon})$, and for the set $\mathfrak{N}(\mathfrak{T}_i,\mathbf{I};\boldsymbol{\Upsilon})$, $\mathfrak{N}(\mathfrak{T}_i,\mathbf{I};\boldsymbol{\Upsilon}) = \cap\left[\mathfrak{N}(t_0;\mathbf{I};\boldsymbol{\Upsilon}) : t_0 \in \mathfrak{T}_i\right]$, $\forall \mathbf{I}(.) \in \mathcal{J}^{\xi}$, to be the domain $\mathcal{D}_{STU}(\mathfrak{T}_i;\mathbf{I};\boldsymbol{\Upsilon})$ of the uniform stable tracking of Υ on $\mathfrak{T}_0 \times \mathfrak{T}_i$ for every $\mathbf{I}(.) \in \mathcal{J}^{\xi} : \mathfrak{N}(\mathfrak{T}_i,\mathbf{I};\boldsymbol{\Upsilon}) \equiv \mathcal{D}_{ST}(\mathfrak{T}_i,\mathbf{I};\boldsymbol{\Upsilon})$ it is both necessary and sufficient that:

1) The set $\mathfrak{N}(t;\mathbf{I};\boldsymbol{\Upsilon})$ is an open continuous, connected neighborhood of Υ and $\mathfrak{N}(t;\mathbf{I};\boldsymbol{\Upsilon}) \subseteq \mathfrak{S}(t;t_0;\Upsilon)$ for every $t \in \mathfrak{T}_0$, $\forall \mathbf{I}(.) \in \mathcal{J}^{\xi}$,

2) The set $\mathfrak{N}(\mathfrak{T}_i,\mathbf{I};\boldsymbol{\Upsilon})$ is a connected neighborhood of Υ, $\forall \mathbf{I}(.) \in \mathcal{J}^{\xi}$,

3) $\boldsymbol{\Upsilon}$ is the unique invariant set in $\mathfrak{N}(t;\mathbf{I};\boldsymbol{\Upsilon})$ relative to system motions on $\mathfrak{T}_0$, $\forall \mathbf{I}(.) \in \mathcal{J}^{\xi}$,

4) For any decrescent positive definite function $W(.;\Upsilon)$ *relative to* Υ *on* $\mathfrak{T}_0 \times \mathfrak{T}_i \times \mathfrak{S}(t;t_0;\Upsilon)$, $\forall \mathbf{I}(.) \in \mathcal{J}^\xi$, *which obeys:*

a) $W(.;\Upsilon) \in L(\mathfrak{T}_0, \mathfrak{T}_i, \mathfrak{S}, \mathcal{J}^\xi; \Upsilon)$, *the equations (24.5) and (24.6) have the unique solution function* $V(.;\Upsilon)$ *with the following properties:*

(i) $V(.;\Upsilon)$ *is a decrescent positive definite function relative to* Υ *on* $\mathfrak{T}_0 \times \mathfrak{T}_i \times \mathfrak{S}(t;t_0;\Upsilon)$, *for every* $\mathbf{I}(.) \in \mathcal{J}^\xi$,

and

(ii) If the boundary $\partial\mathfrak{N}(t;\mathbf{I};\boldsymbol{\Upsilon})$ *of* $\mathfrak{N}(t;\mathbf{I};\boldsymbol{\Upsilon})$ *is nonempty then* $\mathbf{Z} \rightarrow \partial\mathfrak{N}(t;\mathbf{I};\boldsymbol{\Upsilon})$, $\mathbf{Z} \in \mathfrak{N}(t;\mathbf{I};\boldsymbol{\Upsilon})$, *implies* $V(t,\mathbf{Z};\mathbf{I}) \rightarrow \infty$, $\forall t \in \mathfrak{T}_0$, *for every* $\mathbf{I}(.) \in \mathcal{J}^\xi$;

or obeys:

b) $W(.;\Upsilon) \in E(\mathfrak{T}_0, \mathfrak{T}_i, \mathfrak{S}(.), \mathcal{J}^\xi; \Upsilon)$, *the equations (24.13) and (24.14) have the unique solution function* $u(.;\Upsilon)$ *with the following properties:*

(i) $u(.;\Upsilon)$ *is a decrescent positive definite function relative to* Υ *on* $\mathfrak{T}_0 \times \mathfrak{T}_i \times \mathfrak{S}(t;t_0;\Upsilon)$, *for every* $\mathbf{I}(.) \in \mathcal{J}^\xi$,

and

(ii) If the boundary $\partial\mathfrak{N}(t;\mathbf{I};\boldsymbol{\Upsilon})$ *of* $\mathfrak{N}(t;\mathbf{I};\boldsymbol{\Upsilon})$ *is nonempty then* $\mathbf{Z} \rightarrow \partial\mathfrak{N}(t;\mathbf{I};\boldsymbol{\Upsilon})$, $\mathbf{Z} \in \mathfrak{N}(t;\mathbf{I};\boldsymbol{\Upsilon})$, *implies* $u(t,\mathbf{Z};\mathbf{I};\boldsymbol{\Upsilon}) \rightarrow \mathbf{1}$, $\forall t \in \mathfrak{T}_0, \forall \mathbf{I}(.) \in \mathcal{J}^\xi$.

Proof of this theorem is in Appendix D.3.

Theorem 379 guarantees uniformity of the tracking with respect to the initial moment $t_0 \in \mathfrak{T}_i$, i.e., on $\mathfrak{T}_0 \times \mathfrak{T}_i$ for every $\mathbf{I}(.) \in \mathcal{J}^\xi$, but it does not ensure the tracking uniformity relative to the input total action $\mathbf{I}(.) \in \mathcal{J}^\xi$, equivalently relative to the deviation input action $\mathbf{I}(.) \in \mathcal{J}^\xi$. The tracking property, the form and size of its domain, the form of the generating function $W(.;\Upsilon)$ and of the solution function $V(.;\Upsilon)$ or $u(.;\Upsilon)$ depend on every $\mathbf{I}(.) \in \mathcal{J}^\xi$. The following theorem presents the conditions for the complete uniformity, i.e., for the uniformity in both the initial instant $t_0 \in \mathfrak{T}_i$ and in the input action $\mathbf{I}(.) \in \mathcal{J}^\xi$.

Theorem 380 ***Uniform stable tracking the target set*** Υ ***on*** $\mathfrak{T}_0 \times \mathfrak{T}_i \times \mathcal{J}^\xi$ ***by the motions of the system (3.12): strong smoothness property***

Let the system (3.12) possess strong smoothness property 370. In order for the system motions to exhibit uniform stable tracking of Υ *on* $\mathfrak{T}_0 \times \mathfrak{T}_i \times \mathcal{J}^\xi$, *for a set* $\mathfrak{N}(t;\mathcal{J}^\xi;\Upsilon)$ *to be the domain* $\mathcal{D}_{ST}(t_0;\mathcal{J}^\xi;\Upsilon)$ *at* $t = t_0 \in \mathfrak{T}_i$ *of the stable tracking of* Υ *on* $\mathfrak{T}_0 \times \mathfrak{T}_i \times \mathcal{J}^\xi$: $\mathfrak{N}(t_0;\mathbf{I};\boldsymbol{\Upsilon}) \equiv \mathcal{D}_{ST}(t_0;\mathbf{I};\boldsymbol{\Upsilon})$, *and for the set* $\mathfrak{N}(\mathfrak{T}_i,\mathcal{J}^\xi;\Upsilon)$, $\mathfrak{N}(\mathfrak{T}_i,\mathcal{J}^\xi;\Upsilon) = \cap\left[\mathfrak{N}(t_0;\mathbf{I};\boldsymbol{\Upsilon}) : [t_0,\mathbf{I}(.)] \in \mathfrak{T}_i \times \mathcal{J}^\xi\right]$ *to be the domain* $\mathcal{D}_{STU}(\mathfrak{T}_i;\times\mathcal{J}^\xi;\Upsilon)$ *of the uniform stable tracking of* Υ *on* $\mathfrak{T}_i \times \mathcal{J}^\xi$: $\mathfrak{N}(\mathfrak{T}_i,\times\mathcal{J}^\xi;\Upsilon) \equiv \mathcal{D}_{STU}(\mathfrak{T}_i,\times\mathcal{J}^\xi;\Upsilon)$ *it is both necessary and sufficient that:*

1) The set $\mathfrak{N}(t;\mathbf{I};\boldsymbol{\Upsilon})$ *is a continuous open connected neighborhood of* Υ *and* $\mathfrak{N}(t;\mathbf{I};\boldsymbol{\Upsilon}) \subseteq \mathfrak{S}(t;t_0;\Upsilon)$ *for every* $t \in \mathfrak{T}_0$, $\forall \mathbf{I}(.) \in \mathcal{J}^\xi$.

2) The set $\mathfrak{N}(\mathfrak{T}_i,\mathcal{J}^\xi;\Upsilon)$ *is a connected neighborhood of* Υ, $\forall \mathbf{I}(.) \in \mathcal{J}^\xi$.

3) $\boldsymbol{\Upsilon}$ *is the unique invariant set in* $\mathfrak{N}(t;\mathbf{I};\boldsymbol{\Upsilon})$ *relative to system motions on* $\mathfrak{T}_i$, $\forall \mathbf{I}(.) \in \mathcal{J}^\xi$.

4) For any decrescent positive definite function $W(.;\Upsilon)$ *relative to* Υ *on* $\mathfrak{T}_0 \times \mathfrak{T}_i \times \mathfrak{S}(t;t_0;\Upsilon) \times \mathcal{J}^\xi$ *that obeys:*

a) $W(.;\Upsilon) \in L(\mathfrak{T}_0, \mathfrak{T}_i, \mathfrak{S}, \mathcal{J}^\xi; \Upsilon)$, *the equations (24.5) and (24.6) have the unique solution function* $V(.;\Upsilon)$ *with the following properties:*

(i) $V(.;\Upsilon)$ *is a decrescent positive definite function relative to* Υ *on* $\mathfrak{T}_0 \times \mathfrak{T}_i \times \mathfrak{S}(t;t_0;\Upsilon) \times \mathcal{J}^\xi$

and

(ii) If the boundary $\partial\mathfrak{N}(t;\mathbf{I};\boldsymbol{\Upsilon})$ *of* $\mathfrak{N}(t;\mathbf{I};\boldsymbol{\Upsilon})$ *is nonempty then* $\mathbf{Z} \rightarrow \partial\mathfrak{N}(t;\mathbf{I};\boldsymbol{\Upsilon})$, $\mathbf{Z} \in \mathfrak{N}(t;\mathbf{I};\boldsymbol{\Upsilon})$, *implies* $V(t,\mathbf{Z};\mathbf{I}) \rightarrow \infty$, $\forall t \in \mathfrak{T}_0, \forall \mathbf{I}(.) \in \mathcal{J}^\xi$;

or obeys:

b) $W(.;\Upsilon) \in E\left(\mathfrak{T}_0, \mathfrak{T}_i, \mathfrak{S}(.), \mathcal{J}^{\xi}; \Upsilon\right)$, *the equations (24.13) and (24.14) have the unique solution function* $u(.;\Upsilon)$ *with the following properties:*

(i) $u(.;\Upsilon)$ *is a decrescent positive definite function relative to* Υ *on* $\mathfrak{T}_0 \times \mathfrak{T}_i \times \mathfrak{S}(t;t_0;\Upsilon) \times \mathcal{J}^{\xi}$

and

(ii) If the boundary $\partial\mathfrak{N}(t;\mathbf{I};\mathbf{\Upsilon})$ *of* $\mathfrak{N}(t;\mathbf{I};\mathbf{\Upsilon})$ *is nonempty then* $\mathbf{Z} \to \partial\mathfrak{N}(t;\mathbf{I};\mathbf{\Upsilon})$, $\mathbf{Z} \in \mathfrak{N}(t;\mathbf{I};\mathbf{\Upsilon})$, *implies* $u(t,\mathbf{Z};\mathbf{I};\mathbf{\Upsilon}) \to 1$, $\forall t \in \mathfrak{T}_0, \forall \mathbf{I}(.) \in \mathcal{J}^{\xi}$.

The proof of Theorem 380 is the straightforward modification of the proof D.3 of Theorem 379 as explained in D.4.

The system (3.12) possesses Property 57, Property 58, and Property 78 (Note 356). Property 58 permits us to set (3.12) into the following equivalent form by referring to (3.31):

$$\begin{gathered}\mathbf{Z}^{(1)}(t) = \mathbf{f}\left[t, \mathbf{Z}(t); \mathbf{I}; \mathbf{\Upsilon}\right], \ \forall t \in \mathfrak{T}, \ \mathbf{Z} = \mathbf{R}^{\alpha-1}, \ \mathbf{R} = \left[I_\rho \ O_\rho \ ... \ O_\rho\right]\mathbf{Z}, \\ \mathbf{f}\left[t, \mathbf{Z}(t); \mathbf{I}; \mathbf{\Upsilon}\right] = Q^{-1}\left[t, \mathbf{R}^{\alpha-1}(t)\right]\left\{\mathbf{h}\left[t, \mathbf{I}^{\xi}(t)\right] - \mathbf{q}\left[t, \mathbf{R}^{\alpha-1}(t)\right]\right\}, \\ \mathbf{Y}(t) = \mathbf{s}\left\{t, \left[I_\rho \ O_\rho \ ... \ O_\rho\right]\mathbf{Z}(t), \mathbf{I}(t)\right\}, \ \forall t \in \mathfrak{T}.\end{gathered} \tag{24.25}$$

The direct generation of Lyapunov function $V(.)$ follows.

Theorem 381 ***Uniform stable tracking the target set*** Υ ***on*** $\mathfrak{T}_0 \times \mathfrak{T}_i \times \mathcal{J}^{\xi}$ ***by the motions of the system*** *(24.25)*: ***strong smoothness property***

Let the system (24.25) possess strong smoothness property 370. Let

$$W(t,\mathbf{Z};\mathbf{I};\mathbf{\Upsilon}) = \widetilde{\mathbf{f}}^T(t,\mathbf{Z};\mathbf{I};\mathbf{\Upsilon})\,\widetilde{\mathbf{f}}(t,\mathbf{Z};\mathbf{I};\mathbf{\Upsilon}) \tag{24.26}$$

be a decrescent positive definite function relative to $\mathbf{\Upsilon}$ *on* $\mathfrak{T}_0 \times \mathfrak{T}_i \times \mathfrak{S}(t;t_0;\Upsilon) \times \mathcal{J}^{\xi}$. *Let for the function* $V(.;\Upsilon) : \mathfrak{T}_0 \times \mathfrak{T}_i \times \mathfrak{R}^K \to \mathfrak{R}_+$ *defined along the system motions* $\mathbf{Z}(.;t_0;\mathbf{Z}_0;\mathbf{I})$ *by*

$$V(t,\mathbf{Z};\mathbf{I};\mathbf{\Upsilon}) = -\int_0^{\mathbf{Z}} \widetilde{\mathbf{f}}^T(t,\mathbf{Z};\mathbf{I};\mathbf{\Upsilon})\,d\mathbf{Z}, \ \mathbf{Z} \equiv \mathbf{Z}(t) \equiv \mathbf{Z}(t;t_0;\mathbf{Z}_0;\mathbf{I}), \tag{24.27}$$

and for an arbitrarily small $\mu \in \mathfrak{R}^+$, $\mu = \mu(\mathbf{I};\mathbf{\Upsilon})$, *there exists* $\omega_\mu(\mathbf{Z};\mathbf{I};\mathbf{\Upsilon}) \in \mathfrak{C}\left[C\mathfrak{M}_\mu(\Upsilon)\right]$ *for every* $\mathbf{I}(.) \in \mathcal{J}^{\xi}$ *such that (24.7) and (24.8) hold. In order for the system motions to exhibit the uniform stable tracking of* Υ *on* $\mathfrak{T}_0 \times \mathfrak{T}_i \times \mathcal{J}^{\xi}$, *for a set* $\mathfrak{N}\left(t;\mathcal{J}^{\xi};\Upsilon\right)$ *to be the domain* $\mathcal{D}_{ST}\left(t_0;\mathcal{J}^{\xi};\Upsilon\right)$ *at* $t = t_0 \in \mathfrak{T}_i$ *of the stable tracking of* Υ *on* $\mathfrak{T}_0 \times \mathfrak{T}_i \times \mathcal{J}^{\xi}$: $\mathfrak{N}(t_0;\mathbf{I};\mathbf{\Upsilon}) \equiv \mathcal{D}_{ST}(t_0;\mathbf{I};\mathbf{\Upsilon})$, *and for the set* $\mathfrak{N}\left(\mathfrak{T}_i,\mathcal{J}^{\xi};\Upsilon\right)$,

$$\mathfrak{N}\left(\mathfrak{T}_i,\mathcal{J}^{\xi};\Upsilon\right) = \cap\left\{\mathfrak{N}(t_0;\mathbf{I};\mathbf{\Upsilon}) : \left[t_0,\mathbf{I}(.)\right] \in \mathfrak{T}_i \times \mathcal{J}^{\xi}\right\}$$

to be the domain $\mathcal{D}_{STU}\left(\mathfrak{T}_i;\mathcal{J}^{\xi};\Upsilon\right)$ *of the uniform stable tracking of* Υ *on* $\mathfrak{T}_i \times \mathcal{J}^{\xi}$: $\mathfrak{N}\left(\mathfrak{T}_i;\mathcal{J}^{\xi};\Upsilon\right) \equiv \mathcal{D}_{STU}\left(\mathfrak{T}_i;\mathcal{J}^{\xi};\Upsilon\right)$ *it is both necessary and sufficient that:*

1) The set $\mathfrak{N}(t;\mathbf{I};\mathbf{\Upsilon})$ *is a continuous open connected neighborhood of* Υ *and* $\mathfrak{N}(t;\mathbf{I};\mathbf{\Upsilon}) \subseteq \mathfrak{S}(t;t_0;\Upsilon)$ *for every* $t \in \mathfrak{T}_0$, $\forall \mathbf{I}(.) \in \mathcal{J}^{\xi}$.

2) The set $\mathfrak{N}\left(\mathfrak{T}_i;\mathcal{J}^{\xi};\Upsilon\right)$ *is a connected neighborhood of* Υ, $\forall \mathbf{I}(.) \in \mathcal{J}^{\xi}$.

3) $\mathbf{\Upsilon}$ *is the unique invariant set in* $\mathfrak{N}(t;\mathbf{I};\mathbf{\Upsilon})$ *relative to system motions on* $\mathfrak{T}_i$, $\forall \mathbf{I}(.) \in \mathcal{J}^{\xi}$.

4) The function (24.27) obeys the following:

(i) $V(.;\Upsilon)$ *is a decrescent positive definite function relative to* $\mathbf{\Upsilon}$ *on* $\mathfrak{T}_0 \times \mathfrak{T}_i \times \mathfrak{S}(t;t_0;\Upsilon) \times \mathcal{J}^{\xi}$

and

(ii) If the boundary $\partial\mathfrak{N}(t;\mathbf{I};\mathbf{\Upsilon})$ *of* $\mathfrak{N}(t;\mathbf{I};\mathbf{\Upsilon})$ *is nonempty then* $\mathbf{Z} \to \partial\mathfrak{N}(t;\mathbf{I};\mathbf{\Upsilon})$, $\mathbf{Z} \in \mathfrak{N}(t;\mathbf{I};\mathbf{\Upsilon})$, *implies* $V(t,\mathbf{Z};\mathbf{I}) \to \infty$, $\forall t \in \mathfrak{T}_0, \forall \mathbf{I}(.) \in \mathcal{J}^{\xi}$.

Proof. The function $\widetilde{\mathbf{f}}(.)$, **(24.25)**, is continuous on $\mathfrak{T} \times \mathfrak{R}^{\alpha\rho}$ due to (3.13), Property 58.

Necessity. If the system exhibits uniform stable tracking of Υ on $\mathfrak{T}_0 \times \mathfrak{T}_i \times \mathcal{J}^\xi$ then the condition 3) holds. Altogether, the function $W(t, \mathbf{Z}; \mathbf{I}; \mathbf{\Upsilon})$ defined by (24.26) is positive definite and decrescent relative to Υ on $\mathfrak{T}_0 \times \mathfrak{T}_i \times \mathfrak{S}(t; t_0; \Upsilon) \times \mathcal{J}^\xi$. The necessity of the conditions 1), 2), and 4) results from Theorem 380.

Sufficiency. Let all the conditions of the theorem statement be valid. This implies the validity of all conditions of Theorem 380, which proves that the system **(24.25)** exhibits uniform stable tracking of Υ on $\mathfrak{T}_0 \times \mathfrak{T}_i \times \mathcal{J}^\xi$ ■

Theorem 381 shows the direct link between the system Lyapunov function and the system dynamics.

Let us show how the well-known Lyapunov matrix theorem results from Theorem 381:

Theorem 382 ***Lyapunov matrix theorem for the ISO time-invariant linear systems***

In order for the unique equilibrium state $\mathbf{Z} = \mathbf{0}_\rho$ of the ISO time-invariant linear system

$$\frac{d\mathbf{Z}}{dt} = A\mathbf{Z}, \quad det\ A \neq 0, \tag{24.28}$$

to be asymptotically stable in the whole it is necessary and sufficient that for the positive definite matrix $G = G^T$ the matrix solution H of the Lyapunov matrix equation (24.29),

$$A^T H + H^T A = -G \tag{24.29}$$

is also a positive definite symmetric matrix H, $H = H^T$.

Proof. In this case $\widetilde{\mathbf{f}}(t, \mathbf{Z}; \mathbf{I}) \equiv A\mathbf{Z}$. Equation (24.27) simplifies to

$$V(t, \mathbf{Z}; \mathbf{I}) = -\int_{\mathbf{0}}^{\mathbf{Z}} \widetilde{\mathbf{f}}^T(t, \mathbf{Z}; \mathbf{I})\, d\mathbf{Z}, = -\int_{\mathbf{0}}^{\mathbf{Z}} \mathbf{Z}^T A^T d\mathbf{Z} = -\int_0^\infty \mathbf{Z}^T A^T A\mathbf{Z} dt = V(\mathbf{Z})$$

along the system solution $\mathbf{Z}(t; \mathbf{Z}) = e^{At}\mathbf{Z}$,

$$V(\mathbf{Z}) = -\int_0^\infty \mathbf{Z}^T e^{A^T t} e^{At} \mathbf{Z} dt = -\mathbf{Z}^T \left(\int_0^\infty e^{A^T t} A^T A e^{At} dt \right) \mathbf{Z},$$

or

$$\begin{aligned} V(\mathbf{Z}) &= -\frac{1}{2}\mathbf{Z}^T \left[\int_0^\infty \left[\left(Ae^{At}\right)^T Ae^{At} + \left(Ae^{At}\right)^T Ae^{At} \right] dt \right] \mathbf{Z} \\ &= -\frac{1}{2}\mathbf{Z}^T \left[\int_0^\infty \left[\left(e^{At}A\right)^T e^{At}A + \left(e^{At}A\right)^T e^{At}A \right] dt \right] \mathbf{Z} \\ &= -\left(A\mathbf{Z}^T\right)^T \underbrace{\left[\int_0^\infty \left(e^{At}\right)^T e^{At} dt \right]}_{P=P^T>O} A\mathbf{Z} = \mathbf{Z}^T \underbrace{\left(A^T P A\right)}\mathbf{Z}. \end{aligned}$$

The matrices P and H are both positive definite and symmetric, $P = P^T > O$ and $H = H^T > O$. It now follows

$$\begin{aligned} V(\times) &= \mathbf{Z}^T H \mathbf{Z}, \\ V^{(1)}(\mathbf{Z}) &= \mathbf{Z}^{(1)T}(H)\mathbf{Z} + \mathbf{Z}^T(H)\mathbf{Z}^{(1)} \\ V^{(1)}(\mathbf{Z}) &= \mathbf{Z}^T\left(A^T H\right)\mathbf{Z} + \mathbf{Z}^T(HA)\mathbf{Z} \\ V^{(1)}(\mathbf{Z}) &= \mathbf{Z}^T\left(A^T H + HA\right)\mathbf{Z}. \end{aligned}$$

For the negative definite $V^{(1)}(\mathbf{Z}) = \mathbf{Z}^T\left(A^T H + HA\right)\mathbf{Z}$ the function $V(.)$, defined by $V(\times) = \mathbf{Z}^T H\mathbf{Z}$, is positive definite if, and only if, the unique equilibrium state $\mathbf{Z} = \mathbf{0}_\rho$ of the *ISO time-* invariant linear system (24.28) is asymptotically stable (which is in the whole) (Theorem 381). Because $V^{(1)}(\mathbf{Z}) = \mathbf{Z}^T\left(A^T H + HA\right)\mathbf{Z}$ is negative definite; i.e., $G = -\left(A^T H + HA\right) = G^T$ is negative definite, and $V(\times) = \mathbf{Z}^T H\mathbf{Z}$ is positive definite; i.e., $H = H^T$ is positive definite, then the unique equilibrium state $\mathbf{Z} = \mathbf{0}_\rho$ of the *ISO time-*invariant linear system (24.28) is asymptotically stable if, and only if, for positive definite $G = G^T$ the matrix solution H of the Lyapunov matrix equation (24.29) is also positive definite and symmetric ■

Comment 383 *On application of Theorem 381*

The application of Theorem 381 demands the determination of the integral in (24.27) along motions of the system (24.25), which is possible to determine explicitly only in very simple cases.

The systems (3.12) that possess weak smoothness property 368 form another class of the systems, which is different from the class of the systems obeying strong smoothness property 370. The conditions of Theorem 379 and Theorem 380 change slightly for them, as follows.

Theorem 384 ***Uniform stable tracking the target set*** Υ ***on*** $\mathfrak{T}_0 \times \mathfrak{T}_i$ ***for every*** $\mathbf{I}(.) \in \mathcal{J}^\xi$ ***by the motions of the system*** *(24.25:* ***weak smoothness property***

Let the system (3.12) possess the weak smoothness property 368 on $\mathfrak{R}^K$. *In order for the system motions to exhibit uniform stable tracking of* Υ *on* $\mathfrak{T}_0 \times \mathfrak{T}_i$ *for every* $\mathbf{I}(.) \in \mathcal{J}^\xi$, *for a set* $\mathfrak{N}(t;\mathbf{I};\boldsymbol{\Upsilon})$ *to be the domain* $\mathcal{D}_{ST}(t_0;\mathbf{I};\boldsymbol{\Upsilon})$ *at* $t = t_0 \in \mathfrak{T}_i$, $\forall \mathbf{I}(.) \in \mathcal{J}^\xi$, *of the stable tracking on* $\mathfrak{T}_0 \times \mathfrak{T}_i$ *for every* $\mathbf{I}(.) \in \mathcal{J}^\xi : \mathfrak{N}(t_0;\mathbf{I};\boldsymbol{\Upsilon}) \equiv \mathcal{D}_{ST}(t_0;\mathbf{I};\boldsymbol{\Upsilon})$, *and for the set* $\mathfrak{N}(\mathfrak{T}_i,\mathbf{I};\boldsymbol{\Upsilon})$ *to be the domain* $\mathcal{D}_{STU}(\mathfrak{T}_i;\mathbf{I};\boldsymbol{\Upsilon})$ *of the uniform stable tracking on* $\mathfrak{T}_0 \times \mathfrak{T}_i$ *for every* $\mathbf{I}(.) \in \mathcal{J}^\xi : \mathfrak{N}(\mathfrak{T}_i,\mathbf{I};\boldsymbol{\Upsilon}) \equiv \mathcal{D}_{ST}(\mathfrak{T}_i,\mathbf{I};\boldsymbol{\Upsilon})$ *it is both necessary and sufficient that:*

1) The set $\mathfrak{N}(t;\mathbf{I};\boldsymbol{\Upsilon})$ *is a continuous open connected neighborhood of* Υ, $\forall \mathbf{I}(.) \in \mathcal{J}^\xi$.

2) the set $\mathfrak{N}(\mathfrak{T}_i,\mathbf{I};\boldsymbol{\Upsilon})$ *is a connected neighborhood of* Υ, $\forall \mathbf{I}(.) \in \mathcal{J}^\xi$.

3) $\boldsymbol{\Upsilon}$ *is the unique invariant set in* $\mathfrak{N}(t;\mathbf{I};\boldsymbol{\Upsilon})$ *relative to system motions on* $\mathfrak{T}_i$, $\forall \mathbf{I}(.) \in \mathcal{J}^\xi$.

4) For any decrescent positive definite function $W(.;\Upsilon)$ *relative to* $\boldsymbol{\Upsilon}$ *on* $\mathfrak{T}_0 \times \mathfrak{T}_i \times \mathfrak{R}^K$, $\forall \mathbf{I}(.) \in \mathcal{J}^\xi$ *that obeys:*

a) $W(.;\Upsilon) \in L\left(\mathfrak{T}_0,\mathfrak{T}_i,\mathfrak{R}^K,\mathcal{J}^\xi;\Upsilon\right)$, *the equations (24.5) and (24.6) have the unique solution function* $V(.;\Upsilon)$ *with the following properties:*

(i) $V(.;\Upsilon)$ *is a decrescent positive definite function relative to* $\boldsymbol{\Upsilon}$ *on* $\mathfrak{T}_0 \times \mathfrak{T}_i \times \mathfrak{N}(\mathfrak{T}_i,\mathbf{I};\boldsymbol{\Upsilon})$, *for every* $\mathbf{I}(.) \in \mathcal{J}^\xi$

and

(ii) If the boundary $\partial\mathfrak{N}(t;\mathbf{I};\boldsymbol{\Upsilon})$ *of* $\mathfrak{N}(t;\mathbf{I};\boldsymbol{\Upsilon})$ *is nonempty then* $\mathbf{Z} \to \partial\mathfrak{N}(t;\mathbf{I};\boldsymbol{\Upsilon})$, $\mathbf{Z} \in \mathfrak{N}(t;\mathbf{I};\boldsymbol{\Upsilon})$, *implies* $V(t,\mathbf{Z};\mathbf{I}) \to \infty$, $\forall t \in \mathfrak{T}_0$, *for every* $\mathbf{I}(.) \in \mathcal{J}^\xi$

or obeys:

b) $W(.;\Upsilon) \in E\left(\mathfrak{T}_0,\mathfrak{T}_i,\mathfrak{R}^K,\mathcal{J}^\xi;\Upsilon\right)$, *the equations (24.13) and (24.14) have the unique solution function* $u(.;\Upsilon)$ *with the following properties:*

(i) $u(.;\Upsilon)$ *is a decrescent positive definite function relative to* $\boldsymbol{\Upsilon}$ *on* $\mathfrak{T}_0 \times \mathfrak{T}_i \times \mathfrak{N}(\mathfrak{T}_i,\mathbf{I};\boldsymbol{\Upsilon})$,

for every $\mathbf{I}(.) \in \mathcal{J}^\xi$,

and

(ii) If the boundary $\partial\mathfrak{N}(t;\mathbf{I};\boldsymbol{\Upsilon})$ *of* $\mathfrak{N}(t;\mathbf{I};\boldsymbol{\Upsilon})$ *is nonempty then* $\mathbf{Z} \to \partial\mathfrak{N}(t;\mathbf{I};\boldsymbol{\Upsilon})$, $\mathbf{Z} \in \mathfrak{N}(t;\mathbf{I};\boldsymbol{\Upsilon})$, *implies* $u(t,\mathbf{Z};\mathbf{I};\boldsymbol{\Upsilon}) \to \mathbf{1}$, $\forall t \in \mathfrak{T}_0, \forall \mathbf{I}(.) \in \mathcal{J}^\xi$.

The proof of this theorem is in Appendix D.5

Comment 385 ***The comparison of Theorem 384 and Theorem 379***

The weak smoothness property 368 imposes the relaxed condition on the system properties more than the strong smoothness property 370 does. However, the conditions 1) through 4) of Theorem 384 are slightly sharper than the corresponding conditions of Theorem 379. In the former case the properties of the generating function $W(.;\Upsilon)$, and the properties of, by it induced, solution functions $V(.;\Upsilon)$ and $u(.;\Upsilon)$ should hold on the whole space $\mathfrak{R}^K$, whereas in the latter case it should hold on the set $\mathfrak{S}(t;t_0;\Upsilon)$ that is a subset of $\mathfrak{R}^K$, $\mathfrak{S}(t;t_0;\Upsilon) \subseteq \mathfrak{R}^K$.

Theorem 386 ***Uniform stable tracking the target set Υ on $\mathfrak{T}_0 \times \mathfrak{T}_i \times \mathcal{J}^\xi$ by the motions of the system (24.25): weak smoothness property***

Let the system (24.25) possess the weak smoothness property 368 on $\mathfrak{R}^K$. In order for the system motions to exhibit uniform stable tracking of Υ on $\mathfrak{T}_0 \times \mathfrak{T}_i \times \mathcal{J}^\xi$, for a set $\mathfrak{N}(t;\mathcal{J}^\xi;\Upsilon)$ to be the domain $\mathcal{D}_{ST}(t_0;\mathcal{J}^\xi;\Upsilon)$ at $t = t_0 \in \mathfrak{T}_i$ of the stable tracking on $\mathfrak{T}_0 \times \mathfrak{T}_i \times \mathcal{J}^\xi : \mathfrak{N}(t_0;\mathbf{I};\boldsymbol{\Upsilon}) \equiv \mathcal{D}_{ST}(t_0;\mathbf{I};\boldsymbol{\Upsilon})$, and for the set $\mathfrak{N}(\mathfrak{T}_i,\mathcal{J}^\xi;\Upsilon)$, $\mathfrak{N}(\mathfrak{T}_i,\mathcal{J}^\xi;\Upsilon) = \cap\left[\mathfrak{N}(t_0;\mathbf{I};\boldsymbol{\Upsilon}) : [t_0,\mathbf{I}(.)] \in \mathfrak{T}_i \times \mathcal{J}^\xi\right]$ to be the domain $\mathcal{D}_{STU}(\mathfrak{T}_i;\times\mathcal{J}^\xi;\Upsilon)$ of the uniform stable tracking on $\mathfrak{T}_i \times \mathcal{J}^\xi : \mathfrak{N}(\mathfrak{T}_i,\times\mathcal{J}^\xi;\Upsilon) \equiv \mathcal{D}_{STU}(\mathfrak{T}_i,\times\mathcal{J}^\xi;\Upsilon)$ it is both necessary and sufficient that:

1) The set $\mathfrak{N}(t;\mathbf{I};\boldsymbol{\Upsilon})$ is a continuous open connected neighborhood of Υ and $\mathfrak{N}(t;\mathbf{I};\boldsymbol{\Upsilon}) \subseteq \mathfrak{S}(t;t_0;\Upsilon)$ for every $t \in \mathfrak{T}_0$, $\forall \mathbf{I}(.) \in \mathcal{J}^\xi$.

2) The set $\mathfrak{N}(\mathfrak{T}_i,\times\mathcal{J}^\xi;\Upsilon)$ is a connected neighborhood of Υ, $\forall \mathbf{I}(.) \in \mathcal{J}^\xi$.

3) $\boldsymbol{\Upsilon}$ is the unique invariant set in $\mathfrak{N}(t;\mathbf{I};\boldsymbol{\Upsilon})$ relative to system motions on $\mathfrak{T}_i$, $\forall \mathbf{I}(.) \in \mathcal{J}^\xi$.

4) For any decrescent positive definite function $W(.;\Upsilon)$ relative to $\boldsymbol{\Upsilon}$ on $\mathfrak{T}_0 \times \mathfrak{T}_i \times \mathfrak{R}^K \times \mathcal{J}^\xi$ that obeys:

a) $W(.;\Upsilon) \in L(\mathfrak{T}_0,\mathfrak{T}_i,\mathfrak{R}^K,\mathcal{J}^\xi;\Upsilon)$, the equations (24.5) and (24.6) have the unique solution function $V(.;\Upsilon)$ with the following properties:

(i) $V(.;\Upsilon)$ is a decrescent positive definite function relative to $\boldsymbol{\Upsilon}$ on $\mathfrak{T}_0 \times \mathfrak{T}_i \times \mathfrak{N}(t;\mathbf{I};\boldsymbol{\Upsilon}) \times \mathcal{J}^\xi$

and

(ii) If the boundary $\partial\mathfrak{N}(t;\mathbf{I};\boldsymbol{\Upsilon})$ of $\mathfrak{N}(t;\mathbf{I};\boldsymbol{\Upsilon})$ is nonempty then $\mathbf{Z} \to \partial\mathfrak{N}(t;\mathbf{I};\boldsymbol{\Upsilon})$, $\mathbf{Z} \in \mathfrak{N}(t;\mathbf{I};\boldsymbol{\Upsilon})$, implies $V(t,\mathbf{Z};\mathbf{I}) \to \infty$, $\forall t \in \mathfrak{T}_0, \forall \mathbf{I}(.) \in \mathcal{J}^\xi$;

or obeys:

b) $W(.;\Upsilon) \in E(\mathfrak{T}_0,\mathfrak{T}_i,\mathfrak{S}(.),\mathcal{J}^\xi;\Upsilon)$, the equations (24.13) and (24.14) have the unique solution function $u(.;\Upsilon)$ with the following properties:

(i) $u(.;\Upsilon)$ is a decrescent positive definite function relative to $\boldsymbol{\Upsilon}$ on $\mathfrak{T}_0 \times \mathfrak{T}_i \times \mathfrak{N}(t;\mathbf{I};\boldsymbol{\Upsilon}) \times \mathcal{J}^\xi$

and

(ii) If the boundary $\partial\mathfrak{N}(t;\mathbf{I};\boldsymbol{\Upsilon})$ of $\mathfrak{N}(t;\mathbf{I};\boldsymbol{\Upsilon})$ is nonempty then $\mathbf{Z} \to \partial\mathfrak{N}(t;\mathbf{I};\boldsymbol{\Upsilon})$, $\mathbf{Z} \in \mathfrak{N}(t;\mathbf{I};\boldsymbol{\Upsilon})$, implies $u(t,\mathbf{Z};\mathbf{I};\boldsymbol{\Upsilon}) \to 1$, $\forall t \in \mathfrak{T}_0, \forall \mathbf{I}(.) \in \mathcal{J}^\xi$.

Proof. The application of the substitutions defined in the proof D.4 to the proof D.5 transforms it into the proof of Theorem 386 ■

Theorem 379 through Theorem 386 hold also for the plant (3.12) and for its form (5.10) in terms of the deviations.

24.5 Criteria: Systems with differentiable motions

If the system motions are defined, continuous, and continuously differentiable on a continuous open connected neighborhood $\mathfrak{S}(t;t_0;\Upsilon)$ of Υ for every $(t;t_0) \in \mathfrak{T}_0 \times \mathfrak{T}_i$ then it is adequate to select the generating function $W(.;\Upsilon)$ to be also defined, continuous, continuously differentiable, and positive definite on $\mathfrak{S}(t;t_0;\Upsilon)$ for every $(t;t_0) \in \mathfrak{T}_0 \times \mathfrak{T}_i$. This means that we accept:

Assumption 387 ***The first-order system smoothness***
*The system **(3.12)**, equivalently (23.1), possesses the first-order either weak or strong smoothness property, i.e., 369 or 371, rather than 368 or 371, respectively.*

Assumption 388 ***The system and the generating function***
The generating function $W(.;\Upsilon)$ is either in the family $L^1(\mathfrak{T}_0,\mathfrak{T}_i,\mathfrak{S},J^\xi;\Upsilon)$, Definition 374, or in the family $E^1(\mathfrak{T}_0,\mathfrak{T}_i,\mathfrak{S},J^\xi;\Upsilon)$, Definition 376, rather than in $L(\mathfrak{T}_0,\mathfrak{T}_i,\mathfrak{S},J^\xi;\Upsilon)$, Definition 373, or in $E(\mathfrak{T}_0,\mathfrak{T}_i,\mathfrak{S},J^\xi;\Upsilon)$, Definition 375, respectively.

The validity of both Assumption 387 and Assumption 388 transforms Theorem 379 through Theorem 386 into the following theorems: Theorem 389 through Theorem 393. This is easy to verify by making the corresponding substitutions in the proofs of Theorem 379 through Theorem 386, respectively.

Theorem 389 ***Uniform stable tracking the target set Υ by the motions of the system (3.12) on $\mathfrak{T}_0 \times \mathfrak{T}_i$ for every $\mathbf{I}(.) \in J^\xi$: The first-order strong smoothness property***
*Let the system **(3.12)** possess the first-order strong smoothness property 371. In order for the system motions to exhibit uniform stable tracking of $\mathbf{\Upsilon}$ on $\mathfrak{T}_0 \times \mathfrak{T}_i$ for every $\mathbf{I}(.) \in J^\xi$, for a set $\mathfrak{N}(t;\mathbf{I};\mathbf{\Upsilon})$ to be the domain $\mathcal{D}_{ST}(t_0;\mathbf{I};\mathbf{\Upsilon})$ at $t = t_0 \in \mathfrak{T}_i$, $\forall \mathbf{I}(.) \in J^\xi$, of the stable tracking on $\mathfrak{T}_0 \times \mathfrak{T}_i$ for every $\mathbf{I}(.) \in J^\xi : \mathfrak{N}(t_0;\mathbf{I};\mathbf{\Upsilon}) \equiv \mathcal{D}_{ST}(t_0;\mathbf{I};\mathbf{\Upsilon})$, and for the set $\mathfrak{N}(\mathfrak{T}_i;\mathbf{I};\mathbf{\Upsilon})$, $\mathfrak{N}(\mathfrak{T}_i;\mathbf{I};\mathbf{\Upsilon}) = \cap[\mathfrak{N}(t_0;\mathbf{I};\mathbf{\Upsilon}) : t_0 \in \mathfrak{T}_i]$, $\forall \mathbf{I}(.) \in J^\xi$, to be the domain $\mathcal{D}_{STU}(\mathfrak{T}_i;\mathbf{I};\mathbf{\Upsilon})$ of the uniform stable tracking on $\mathfrak{T}_0 \times \mathfrak{T}_i$ for every $\mathbf{I}(.) \in J^\xi : \mathfrak{N}(\mathfrak{T}_i;\mathbf{I};\mathbf{\Upsilon}) \equiv \mathcal{D}_{ST}(\mathfrak{T}_i;\mathbf{I};\mathbf{\Upsilon})$ it is both necessary and sufficient that:*

1) The set $\mathfrak{N}(t;\mathbf{I};\mathbf{\Upsilon})$ is a continuous open connected neighborhood of $\mathbf{\Upsilon}$ and $\mathfrak{N}(t;\mathbf{I};\mathbf{\Upsilon}) \subseteq \mathfrak{S}(t;t_0;\Upsilon)$ for every $t \in \mathfrak{T}_0$, $\forall \mathbf{I}(.) \in J^\xi$.

2) The set $\mathfrak{N}(\mathfrak{T}_i;\mathbf{I};\mathbf{\Upsilon})$ is a connected neighborhood of $\mathbf{\Upsilon}$, $\forall \mathbf{I}(.) \in J^\xi$.

3) Υ is the unique invariant set in $\mathfrak{N}(t;\mathbf{I};\mathbf{\Upsilon})$ relative to system motions on $\mathfrak{T}_0$, $\forall \mathbf{I}(.) \in J^\xi$.

4) For any decrescent positive definite function $W(.;\Upsilon)$ relative to Υ on $\mathfrak{T}_0 \times \mathfrak{T}_i \times \mathfrak{S}(t;t_0)$, $\forall \mathbf{I}(.) \in J^\xi$, that obeys:

a) $W(.;\Upsilon) \in L^1(\mathfrak{T}_0,\mathfrak{T}_i,\mathfrak{S},J^\xi;\Upsilon)$, the equations (24.10) and (24.11) have the unique solution function $V(.;\Upsilon)$ with the following properties:

(i) $V(.;\Upsilon)$ is a continuously differentiable decrescent positive definite function relative to Υ on $\mathfrak{T}_0 \times \mathfrak{T}_i \times \mathfrak{S}(t;t_0)$, for every $\mathbf{I}(.) \in J^\xi$

and

(ii) if the boundary $\partial\mathfrak{N}(t;\mathbf{I};\mathbf{\Upsilon})$ of $\mathfrak{N}(t;\mathbf{I};\mathbf{\Upsilon})$ is nonempty then $\mathbf{Z} \to \partial\mathfrak{N}(t;\mathbf{I};\mathbf{\Upsilon})$, $\mathbf{Z} \in \mathfrak{N}(t;\mathbf{I};\mathbf{\Upsilon})$, implies $V(t,\mathbf{Z};\mathbf{I};\mathbf{\Upsilon}) \to \infty$, $\forall t \in \mathfrak{T}_0$, for every $\mathbf{I}(.) \in J^\xi$

or obeys:

b) $W(.;\Upsilon) \in E^1(\mathfrak{T}_0,\mathfrak{T}_i,\mathfrak{S},J^\xi;\Upsilon)$, the equations (24.18) and (24.19) have the unique solution function $u(.)$ with the following properties:

(i) $u(.;\Upsilon)$ is a continuously differentiable decrescent positive definite function relative to Υ on $\mathfrak{T}_0 \times \mathfrak{T}_i \times \mathfrak{S}(t;t_0)$, for every $\mathbf{I}(.) \in J^\xi$

and

(ii) If the boundary $\partial\mathfrak{N}(t;\mathbf{I};\mathbf{\Upsilon})$ of $\mathfrak{N}(t;\mathbf{I};\mathbf{\Upsilon})$ is nonempty then $\mathbf{Z} \to \partial\mathfrak{N}(t;\mathbf{I};\mathbf{\Upsilon})$, $\mathbf{Z} \in \mathfrak{N}(t;\mathbf{I};\mathbf{\Upsilon})$, implies $u(t,\mathbf{Z};\mathbf{I}) \to 1$, $\forall t \in \mathfrak{T}_0, \forall \mathbf{I}(.) \in J^\xi$.

This theorem guarantees uniformity of the tracking with respect to the initial moment $t_0 \in \mathfrak{T}_i$, i.e., on $\mathfrak{T}_0 \times \mathfrak{T}_i$ for every $\mathbf{I}(.) \in J^\xi$, but it does not ensure the tracking uniformity relative to the input total action $\mathbf{I}(.) \in J^{\xi}$, equivalently relative to the deviation input action $\mathbf{I}(.) \in J^{\xi}$.

The preceding theorems establish the conditions for stable tracking uniform in the initial moment $t_0 \in \mathfrak{T}_i$, but nonuniform in the input action $\mathbf{I}(.) \in J^{\xi}$. This means that the tracking property, the form and size of its domain, the form of the generating function $W(.;\Upsilon)$ and of the solution function $V(.;\Upsilon)$ or $u(.;\Upsilon)$ depend on every $\mathbf{I}(.) \in J^{\xi}$. The following theorem establishes the conditions for the complete uniformity, i.e., for the uniformity in both the initial instant $t_0 \in \mathfrak{T}_i$ and in the input action $\mathbf{I}(.) \in J^{\xi}$.

Theorem 390 ***Uniform stable tracking the target set Υ by the motions of the system (3.12) on $\mathfrak{T}_0 \times \mathfrak{T}_i \times J^{\xi}$: The first-order strong smoothness property***

*Let the system **(3.12)** possess the first-order strong smoothness property 371. In order for the system motions to exhibit uniform stable tracking of $\mathbf{\Upsilon}$ on $\mathfrak{T}_0 \times \mathfrak{T}_i \times J^{\xi}$, for a set $\mathfrak{N}(t;J^{\xi})$ to be the domain $\mathcal{D}_{ST}(t_0;J^{\xi};\Upsilon)$ at $t = t_0 \in \mathfrak{T}_i$ of the stable tracking on $\mathfrak{T}_0 \times \mathfrak{T}_i \times J^{\xi} : \mathfrak{N}(t_0;\mathbf{I};\mathbf{\Upsilon}) \equiv \mathcal{D}_{ST}(t_0;\mathbf{I};\mathbf{\Upsilon})$, and for the set $\mathfrak{N}(\mathfrak{T}_i;J^{\xi};\Upsilon)$, $\mathfrak{N}(\mathfrak{T}_i;J^{\xi};\Upsilon) = \cap\left[\mathfrak{N}(t_0;\mathbf{I};\mathbf{\Upsilon}) : [t_0,\mathbf{I}(.)] \in \mathfrak{T}_i \times J^{\xi}\right]$ to be the domain $\mathcal{D}_{STU}(\mathfrak{T}_i;J^{\xi};\Upsilon)$ of the uniform stable tracking on $\mathfrak{T}_i \times J^{\xi} : \mathfrak{N}(\mathfrak{T}_i;J^{\xi};\Upsilon) \equiv \mathcal{D}_{STU}(\mathfrak{T}_i;J^{\xi};\Upsilon)$ it is both necessary and sufficient that:*

1) The set $\mathfrak{N}(t;\mathbf{I};\mathbf{\Upsilon})$ is a continuous open connected neighborhood of $\mathbf{\Upsilon}$ and $\mathfrak{N}(t;\mathbf{I};\mathbf{\Upsilon}) \subseteq \mathfrak{S}(t;t_0;\Upsilon)$ for every $t \in \mathfrak{T}_0$, $\forall \mathbf{I}(.) \in J^{\xi}$

2) The set $\mathfrak{N}(\mathfrak{T}_i;J^{\xi};\Upsilon)$ is a connected neighborhood of $\mathbf{\Upsilon}$, $\forall \mathbf{I}(.) \in J^{\xi}$

3) Υ is the unique invariant set in $\mathfrak{N}(t;\mathbf{I};\mathbf{\Upsilon})$ relative to system motions on $\mathfrak{T}_i$, $\forall \mathbf{I}(.) \in J^{\xi}$

4) For any decrescent positive definite function $W(.;\Upsilon)$ relative to Υ on $\mathfrak{T}_0 \times \mathfrak{T}_i \times \mathfrak{S}(t;t_0;\Upsilon) \times J^{\xi}$, that obeys:

a) $W(.;\Upsilon) \in L^1(\mathfrak{T}_0,\mathfrak{T}_i,\mathfrak{S},J^{\xi};\Upsilon)$, the equations (24.10) and (24.11) have the unique solution function $V(.;\Upsilon)$ with the following properties:

(i) $V(.;\Upsilon)$ is a continuously differentiable decrescent positive definite function relative to Υ on $\mathfrak{T}_0 \times \mathfrak{T}_i \times \mathfrak{S}(t;t_0;\Upsilon) \times J^{\xi}$

and

(ii) if the boundary $\partial\mathfrak{N}(t;\mathbf{I};\mathbf{\Upsilon})$ of $\mathfrak{N}(t;\mathbf{I};\mathbf{\Upsilon})$ is nonempty then $\mathbf{Z} \to \partial\mathfrak{N}(t;\mathbf{I};\mathbf{\Upsilon})$, $\mathbf{Z} \in \mathfrak{N}(t;\mathbf{I};\mathbf{\Upsilon})$, implies $V(t,\mathbf{Z};\mathbf{I};\mathbf{\Upsilon}) \to \infty$, $\forall t \in \mathfrak{T}_0, \forall \mathbf{I}(.) \in J^{\xi}$

or obeys:

b) $W(.;\Upsilon) \in E^1(\mathfrak{T}_0,\mathfrak{T}_i,\mathfrak{S},J^{\xi};\Upsilon)$, the equations (24.18) and (24.19) have the unique solution function $u(.;\Upsilon)$ with the following properties:

(i) $u(.;\Upsilon)$ is a continuously differentiable decrescent positive definite function relative to Υ on $\mathfrak{T}_0 \times \mathfrak{T}_i \times \mathfrak{S}(t;t_0;\Upsilon) \times J^{\xi}$

and

(ii) If the boundary $\partial\mathfrak{N}(t;\mathbf{I};\mathbf{\Upsilon})$ of $\mathfrak{N}(t;\mathbf{I};\mathbf{\Upsilon})$ is nonempty then $\mathbf{Z} \to \partial\mathfrak{N}(t;\mathbf{I};\mathbf{\Upsilon})$, $\mathbf{Z} \in \mathfrak{N}(t;\mathbf{I};\mathbf{\Upsilon})$, implies $u(t,\mathbf{Z};\mathbf{I}) \to 1$, $\forall t \in \mathfrak{T}_0, \forall \mathbf{I}(.) \in J^{\xi}$.

The tracking conditions for the systems **(3.12)** that possess the first-order weak smoothness property 369 differ from the tracking conditions for the systems obeying the first-order strong smoothness property 371. The conditions of Theorem 389 and Theorem 390 change slightly for them, as follows.

Theorem 391 ***Uniform stable tracking the target set Υ by the motions of the system (3.12) on $\mathfrak{T}_0 \times \mathfrak{T}_i$ for every $\mathbf{I}(.) \in J^{\xi}$: The first-order weak smoothness property***

*Let the system **(3.12)** possess the first-order weak smoothness property 369 on $\mathfrak{R}^K$. In order for the system motions to exhibit uniform stable tracking of $\mathbf{\Upsilon}$ on $\mathfrak{T}_0 \times \mathfrak{T}_i$ for every $\mathbf{I}(.) \in J^{\xi}$, for a set $\mathfrak{N}(t;\mathbf{I};\mathbf{\Upsilon})$ to be the domain $\mathcal{D}_{ST}(t_0;\mathbf{I};\mathbf{\Upsilon})$ at $t = t_0 \in \mathfrak{T}_i$, $\forall \mathbf{I}(.) \in J^{\xi}$, of the stable tracking on $\mathfrak{T}_0 \times \mathfrak{T}_i$ for every $\mathbf{I}(.) \in J^{\xi} : \mathfrak{N}(t_0;\mathbf{I};\mathbf{\Upsilon}) \equiv \mathcal{D}_{ST}(t_0;\mathbf{I};\mathbf{\Upsilon})$, and for* *the set $\mathfrak{N}(\mathfrak{T}_i;\mathbf{I};\mathbf{\Upsilon})$ to be the domain $\mathcal{D}_{STU}(\mathfrak{T}_i;\mathbf{I};\mathbf{\Upsilon})$ of the uniform stable tracking on $\mathfrak{T}_0 \times \mathfrak{T}_i$ for every $\mathbf{I}(.) \in J^{\xi}$: $\mathfrak{N}(\mathfrak{T}_i;\mathbf{I};\mathbf{\Upsilon}) \equiv \mathcal{D}_{ST}(\mathfrak{T}_i;\mathbf{I};\mathbf{\Upsilon})$ it is both necessary and sufficient that:*

1) The set $\mathfrak{N}(t;\mathbf{I};\mathbf{\Upsilon})$ is a continuous open connected neighborhood of $\mathbf{\Upsilon}$, $\forall \mathbf{I}(.) \in J^{\xi}$

2) The set $\mathfrak{N}(\mathfrak{T}_i;\mathbf{I};\mathbf{\Upsilon})$ is a connected neighborhood of $\mathbf{\Upsilon}$, $\forall \mathbf{I}(.) \in J^{\xi}$

3) Υ is the unique invariant set in $\mathfrak{N}(t;\mathbf{I};\mathbf{\Upsilon})$ relative to system motions on $\mathfrak{T}_0$, $\forall \mathbf{I}(.) \in J^{\xi}$

4) For any decrescent positive definite function $W(.;\Upsilon)$ relative to Υ on $\mathfrak{T}_0 \times \mathfrak{T}_i \times \mathfrak{R}^K$, $\forall \mathbf{I}(.) \in J^{\xi}$, that obeys:

a) $W(.;\Upsilon) \in L^1\left(\mathfrak{T}_0, \mathfrak{T}_i, \mathfrak{R}^K, J^{\xi};\Upsilon\right)$, the equations (24.10) and (24.11) have the unique solution function $V(.;\Upsilon)$ with the following properties:

(i) $V(.;\Upsilon)$ is a continuously differentiable decrescent positive definite function relative to Υ on $\mathfrak{T}_0 \times \mathfrak{T}_i \times \mathfrak{N}(\mathfrak{T}_i;\mathbf{I};\mathbf{\Upsilon})$, for every $\mathbf{I}(.) \in J^{\xi}$

and

(ii) If the boundary $\partial\mathfrak{N}(t;\mathbf{I};\mathbf{\Upsilon})$ of $\mathfrak{N}(t;\mathbf{I};\mathbf{\Upsilon})$ is nonempty then $\mathbf{Z} \rightarrow \partial\mathfrak{N}(t;\mathbf{I};\mathbf{\Upsilon})$, $\mathbf{Z} \in \mathfrak{N}(t;\mathbf{I};\mathbf{\Upsilon})$, implies $V(t,\mathbf{Z};\mathbf{I};\mathbf{\Upsilon}) \rightarrow \infty$, $\forall t \in \mathfrak{T}_0$, for every $\mathbf{I}(.) \in J^{\xi}a$

or obeys:

b) $W(.;\Upsilon) \in E^1\left(\mathfrak{T}_0, \mathfrak{T}_i, \mathfrak{R}^K, J^{\xi};\Upsilon\right)$, the equations (24.18) and (24.19) have the unique solution function $u(.;\Upsilon)$ with the following properties:

(i) $u(.;\Upsilon)$ is continuously differentiable decrescent positive definite function relative to Υ on $\mathfrak{T}_0 \times \mathfrak{T}_i \times \mathfrak{N}(\mathfrak{T}_i;\mathbf{I};\mathbf{\Upsilon})$

for every $\mathbf{I}(.) \in J^{\xi}$,

and

(ii) If the boundary $\partial\mathfrak{N}(t;\mathbf{I};\mathbf{\Upsilon})$ of $\mathfrak{N}(t;\mathbf{I};\mathbf{\Upsilon})$ is nonempty then $\mathbf{Z} \rightarrow \partial\mathfrak{N}(t;\mathbf{I};\mathbf{\Upsilon})$, $\mathbf{Z} \in \mathfrak{N}(t;\mathbf{I};\mathbf{\Upsilon})$, implies $u(t,\mathbf{Z};\mathbf{I}) \rightarrow \mathbf{1}$, $\forall t \in \mathfrak{T}_0, \forall \mathbf{I}(.) \in J^{\xi}$.

Comment 392 ***The comparison of Theorem** 391 **and Theorem** 389*

The first-order weak smoothness property 369 more strongly imposes the relaxed condition on the system properties than the first-order strong smoothness property 371 does. However, the conditions 1) through 4) of Theorem 391 are slightly sharper than the corresponding conditions of Theorem 389. In the former case the properties of the generating function $W(.;\Upsilon)$, and the properties of the solution functions $V(.;\Upsilon)$ and $u(.;\Upsilon)$ should hold on the whole space $\mathfrak{R}^K$, whereas in the latter case they should hold on the set $\mathfrak{S}(t;t_0;\Upsilon)$ that is a subset of $\mathfrak{R}^K$, $\mathfrak{S}(t;t_0;\Upsilon) \subseteq \mathfrak{R}^K$.

Theorem 393 ***Uniform stable tracking the target set Υ by the motions of the system (3.12) on $\mathfrak{T}_0 \times \mathfrak{T}_i \times J^{\xi}$: the first-order weak smoothness property***

*Let the system **(3.12)** possess the first-order weak smoothness property 369 on $\mathfrak{R}^K$. In order for the system motions to exhibit uniform stable tracking of $\mathbf{\Upsilon}$ on $\mathfrak{T}_0 \times \mathfrak{T}_i \times J^{\xi}$, for a set $\mathfrak{N}\left(t;J^{\xi};\Upsilon\right)$ to be the domain $\mathcal{D}_{ST}\left(t_0;J^{\xi};\Upsilon\right)$ at $t = t_0 \in \mathfrak{T}_i$ of the stable tracking on $\mathfrak{T}_0 \times \mathfrak{T}_i \times J^{\xi}$: $\mathfrak{N}(t_0;\mathbf{I};\mathbf{\Upsilon}) \equiv \mathcal{D}_{ST}(t_0;\mathbf{I};\mathbf{\Upsilon})$, and for the set $\mathfrak{N}\left(\mathfrak{T}_i,J^{\xi};\Upsilon\right)$, $\mathfrak{N}\left(\mathfrak{T}_i,J^{\xi};\Upsilon\right) = \cap\left[\mathfrak{N}(t_0;\mathbf{I};\mathbf{\Upsilon}) : [t_0,\mathbf{I}(.)] \in \mathfrak{T}_i \times J^{\xi}\right]$ to be the domain $\mathcal{D}_{STU}\left(\mathfrak{T}_i;J^{\xi};\Upsilon\right)$ of the uniform stable tracking on $\mathfrak{T}_i \times J^{\xi}$: $\mathfrak{N}\left(\mathfrak{T}_i;J^{\xi};\Upsilon\right) \equiv \mathcal{D}_{STU}\left(\mathfrak{T}_i;J^{\xi};\Upsilon\right)$ it is both necessary and sufficient that:*

1) The set $\mathfrak{N}(t;\mathbf{I};\mathbf{\Upsilon})$ is a continuous open connected neighborhood of $\mathbf{\Upsilon}$ and $\mathfrak{N}(t;\mathbf{I};\mathbf{\Upsilon}) \subseteq \mathfrak{S}(t;t_0;\Upsilon)$ for every $t \in \mathfrak{T}_0$, $\forall \mathbf{I}(.) \in J^{\xi}$.

2) The set $\mathfrak{N}\left(\mathfrak{T}_i;J^{\xi};\Upsilon\right)$ is a connected neighborhood of $\mathbf{\Upsilon}$, $\forall \mathbf{I}(.) \in J^{\xi}$

3) Υ is the unique invariant set in $\mathfrak{N}(t;\mathbf{I};\mathbf{\Upsilon})$ relative to system motions on $\mathfrak{T}_i$, $\forall \mathbf{I}(.) \in J^{\xi}$

4) For any decrescent positive definite function $W(.;\Upsilon)$ relative to Υ on $\mathfrak{T}_0 \times \mathfrak{T}_i \times \mathfrak{R}^K \times J^{\xi}$, that obeys:

a) $W(.;\Upsilon) \in L^1\left(\mathfrak{T}_0, \mathfrak{T}_i, \mathfrak{R}^K, J^{\xi};\Upsilon\right)$, the equations (24.10) and (24.11) have the unique solution function $V(.;\Upsilon)$ with the following properties:

(i) $V(.;\Upsilon)$ is a continuously differentiable decrescent positive definite function relative to Υ on $\mathfrak{T}_0 \times \mathfrak{T}_i \times \mathfrak{N}(t;\mathbf{I};\mathbf{\Upsilon}) \times J^{\xi}$

and

(ii) If the boundary $\partial\mathfrak{N}(t;\mathbf{I};\boldsymbol{\Upsilon})$ *of* $\mathfrak{N}(t;\mathbf{I};\boldsymbol{\Upsilon})$ *is nonempty then* $\mathbf{Z}\rightarrow\partial\mathfrak{N}(t;\mathbf{I};\boldsymbol{\Upsilon})$, $\mathbf{Z}\in\mathfrak{N}(t;\mathbf{I};\boldsymbol{\Upsilon})$, *implies* $V(t,\mathbf{Z};\mathbf{I};\boldsymbol{\Upsilon})\rightarrow\infty$, $\forall t\in\mathfrak{T}_0, \forall\mathbf{I}(.)\in J^{\xi}$

or obeys:

b) $W(.;\Upsilon)\in E^1\left(\mathfrak{T}_0,\mathfrak{T}_i,\mathfrak{S},J^{\xi};\Upsilon\right)$, *the equations (24.18) and (24.19) have the unique solution function* $u(.;\Upsilon)$ *with the following properties:*

(i) $u(.;\Upsilon)$ *is a continuously differentiable decrescent positive definite function relative to* Υ *on* $\mathfrak{T}_0\times\mathfrak{T}_i\times\mathfrak{N}(t;\mathbf{I};\boldsymbol{\Upsilon})\times J^{\xi}$

and

(ii) If the boundary $\partial\mathfrak{N}(t;\mathbf{I};\boldsymbol{\Upsilon})$ *of* $\mathfrak{N}(t;\mathbf{I};\boldsymbol{\Upsilon})$ *is nonempty then* $\mathbf{Z}\rightarrow\partial\mathfrak{N}(t;\mathbf{I};\boldsymbol{\Upsilon})$, $\mathbf{Z}\in\mathfrak{N}(t;\mathbf{I};\boldsymbol{\Upsilon})$, *implies* $u(t,\mathbf{Z};\mathbf{I})\rightarrow 1$, $\forall t\in\mathfrak{T}_0, \forall\mathbf{I}(.)\in J^{\xi}$.

The above theorems are directly applicable to the plant (3.12), equivalently to its form (5.10) in terms of the deviations.

Chapter 25

Time-varying set tracking

25.1 *Time*-varying set and motion tracking

Tracking a *time*-varying target set $\Upsilon(t)$ incorporates the same type of tracking the desired motion $\mathcal{Z}_d(,;t_0;\mathbf{Z}_{d0};\mathbf{I})$ of the system. The former reduces to the latter for

$$\Upsilon(t) = \{\mathbf{Z} : \mathbf{Z} = \mathbf{Z}_d(t;t_0;\mathbf{Z}_{d0};\mathbf{I})\}, \ \forall t \in \mathfrak{T}_0.$$

However, tracking the desired motion $\mathcal{Z}_d(,;t_0;\mathbf{Z}_{d0};\mathbf{I})$ does not involve the same type of tracking the target set $\Upsilon(t)$ in general.

Knowing this, by studying the tracking of the *time*-varying target set $\Upsilon(t)$ we study simultaneously the tracking of the desired motion $\mathcal{Z}_d(,;t_0;\mathbf{Z}_{d0};\mathbf{I})$ of the system.

25.2 Conditions for stable tracking

Lyapunov stability criteria rely on the concepts of positive semidefinite functions and positive definite functions relative to the origin or relative to a *time*-invariant set. They are well defined relative to *time*-invariant sets, but not with respect to *time*-varying sets. We show in what follows how the conditions for tracking a *time*-varying set can be well determined by using the novel concepts of positive semidefinite functions (Definition 318, and of definite functions relative to a *time*-varying set (Definition 319).

Definition 313 determines the boundary function of a *time*-varying set.

Theorem 394 ***Stable tracking the target set*** $\Upsilon(t)$

In order for the motions $\mathcal{Z}(,;t_0;\mathbf{Z}_0;\mathbf{I})$, $\mathbf{Z}(t) \equiv \mathcal{Z}(t;t_0;\mathbf{Z}_0;\mathbf{I})$, *of a dynamical system, which are globally defined and globally continuous for every* $\mathbf{I}(.) \in \mathcal{J}^{\xi}$, *to exhibit the stable tracking on* $\mathfrak{T}_0 \times \mathfrak{D}^i \times \mathfrak{Y}_d^k$ *of the target set* $\Upsilon(t)$ *it is sufficient that:*

i) The boundary function $\beta(.;\Upsilon)$ *of the target set* $\Upsilon(t)$ *is positive definite relative to the set* $\Upsilon(t)$ *itself on* $\mathfrak{T}_0$

ii) There is a positive definite function $w(.;\Upsilon) : \mathfrak{T} \times \mathfrak{R}^K \longrightarrow \mathfrak{R}$ *relative to the set* $\Upsilon(t)$ *on* $\mathfrak{T}_0$ *such that*

$$D^+\beta(t,\mathbf{Z};\Upsilon) \leq -w(t,\mathbf{Z};\Upsilon), \ \forall (t,\mathbf{Z}) \in \mathfrak{T}_0 \times \mathfrak{N}_\eta[\Upsilon(t)], \quad \forall \mathbf{I}(.) \in \mathcal{J}^{\xi}, \tag{25.1}$$

on some η*-neighborhood* $\mathfrak{N}_\eta[\Upsilon(t)]$ *of* $\Upsilon(t)$ *on* $\mathfrak{T}_0$.

Then the set $\mathfrak{N}_\eta[\Upsilon(t_0)]$ *is a subset of the domain of the stable tracking by the system motions the set* $\Upsilon(t)$ *on* $\mathfrak{T}_0 \times \mathcal{J}^{\xi}$.

iii) If additionally, $\eta = \infty$, the functions $\beta(.;\Upsilon)$ and $w(.;\Upsilon)$ are globally positive definite relative to the set $\Upsilon(t)$ on $\mathfrak{T}_0$ and $\beta(.;\Upsilon)$ is also radially unbounded relative to the set $\Upsilon(t)$ on $\mathfrak{T}_0$ then the stable tracking on $\mathfrak{T}_0 \times \mathcal{J}^{\xi}$ of the set $\Upsilon(t)$ is also global.

iv) If additionally to i) and ii), the function $\beta(.;\Upsilon)$ is positive definite relative to the set $\Upsilon(t)$ itself on $\mathfrak{T}_0$ for every $t_0 \in \mathfrak{T}_i$ and decrescent relative to the set $\Upsilon(t)$ on $\mathfrak{T}_0$ for every $t_0 \in \mathfrak{T}_i$, and the function $w(.;\Upsilon)$ is positive definite relative to the set $\Upsilon(t)$ on $\mathfrak{T}_0$ for every $t_0 \in \mathfrak{T}_i$ then the stable tracking of the set $\Upsilon(t)$ is uniform on $\mathfrak{T}_0 \times \mathfrak{T}_i \times \mathcal{J}^{\xi}$.

v) If all conditions i) through iv) hold then the stable tracking of the set $\Upsilon(t)$ is globally uniform on $\mathfrak{T}_0 \times \mathfrak{T}_i \times \mathcal{J}^{\xi}$.

Proof. Let the conditions i) and ii) of the theorem statement hold. The positive definiteness of the functions $\beta(.;\Upsilon)$ and $w(.;\Upsilon)$ relative to the set $\Upsilon(t)$ permits (Definition 319) the existence of some η-neighborhood $\mathfrak{N}_\eta[\Upsilon(t)]$ of $\Upsilon(t)$ on $\mathfrak{T}_0$ such that the functions $\beta(.;\Upsilon)$ and $w(.;\Upsilon)$ are positive definite on it and on $\mathfrak{T}_0$, and that (25.1) holds.

Tracking.

i) and ii) Let $\Delta \in]0,\eta]$. It determines the connected neighborhood $\mathfrak{N}_\Delta[\Upsilon(t_0)]$ of $\Upsilon(t_0)$ at the initial moment $t_0 \in \mathfrak{T}_i$, which is a subset of $\mathfrak{N}_\eta[\Upsilon(t_0)]$, $\mathfrak{N}_\Delta[\Upsilon(t_0)] \subseteq \mathfrak{N}_\eta[\Upsilon(t_0)]$. Let it be chosen and fixed. The condition (25.1) guarantees:

$$D^+\beta(t,\mathbf{Z};\mathbf{\Upsilon}) \leq -w(t,\mathbf{Z};\mathbf{\Upsilon}),\ \forall (t,\mathbf{Z}) \in \mathfrak{T}_0\times\mathfrak{R}^K \Longrightarrow$$

$$\beta[t,\mathcal{Z}(t;t_0;\mathbf{Z}_0;\mathbf{I});\Upsilon] - \beta[t_0,\mathcal{Z}(t_0;t_0;\mathbf{Z}_0;\mathbf{I});\Upsilon] \leq -\int_{t_0}^{t} w[\tau,\mathbf{Z}(\tau);\Upsilon]\,d\tau \Longrightarrow$$

$$\beta[t,\mathbf{Z}(t);\Upsilon] \leq \beta(t_0,\mathbf{Z}_0;\Upsilon) - \int_{t_0}^{t} w[\tau,\mathbf{Z}(\tau);\Upsilon]\,d\tau,\ \forall (t,\mathbf{Z}_0) \in \mathfrak{T}_0\times\mathfrak{R}^K. \tag{25.2}$$

Let

$$\gamma(\Delta) = \inf\{\min\langle w(t,\mathbf{Z};\mathbf{\Upsilon}) : \mathbf{Z} \in \partial\mathfrak{N}_\Delta[\Upsilon(t)]\rangle : t \in \mathfrak{T}_0\}.$$

The positive definiteness of $w(.;\Upsilon)$ relative to the set $\Upsilon(t)$ on $\mathfrak{T}_0$, Definition 319, and Property 320 guarantees

$$\gamma(\Delta) \in \mathfrak{R}^+,\ \forall \Delta \in]0,\eta].$$

Let us assume that there are $\mathbf{I}^*(.) \in \mathcal{J}^{\xi}$, $\Delta^* \in]0,\eta]$ and $\mathbf{Z}^* \in \mathfrak{N}_\eta[\Upsilon(t_0)]$ such that $\mathcal{Z}(t;t_0;\mathbf{Z}^*;\mathbf{I}^*) \in \mathfrak{R}^K\backslash\mathfrak{N}_{\Delta^*}[\Upsilon(t)]$, $\forall t \in \mathfrak{T}_0$. Hence, $\beta[t,\mathcal{Z}(t;t_0;\mathbf{Z}^*;\mathbf{I}^*);\Upsilon] > \beta_{\varsigma_{Inf\beta}(\Delta^*)}(t) > \mu$, $\forall t \in \mathfrak{T}_0$. Let $\gamma^* = \gamma(\Delta^*)$. However, from (25.2) follows

$$\beta[t,\mathcal{Z}(t;t_0;\mathbf{Z}^*;\mathbf{I}^*);\Upsilon] \leq \beta(t_0,\mathbf{Z}^*) - \int_{t_0}^{t} \gamma^* d\tau,\ \forall t \in \mathfrak{T}_0 \Longrightarrow$$

$$\beta[t,\mathcal{Z}(t;t_0;\mathbf{Z}^*;\mathbf{I}^*);\Upsilon] \leq \mu < \beta_{\varsigma_{Inf\beta}(\Delta^*)}(t),\ \forall t \geq t_0 + \frac{\beta(t_0,\mathbf{Z}^*;\Upsilon) - \mu}{\gamma^*} \Longrightarrow$$

$$\mathcal{Z}(t;t_0;\mathbf{Z}^*;\mathbf{I}^*) \in \mathfrak{N}_{\Delta^*}[\Upsilon(t)],\forall t \in \mathfrak{T}_0.$$

This contradicts the assumption that there are $\mathbf{I}^*(.) \in \mathcal{J}^{\xi}$, $\Delta^* \in \mathfrak{R}^+$, $\mathbf{Z}^* \in \mathfrak{R}^K\backslash\mathfrak{N}_\eta[\Upsilon(t_0)]$ such that $\mathcal{Z}(t;t_0;\mathbf{Z}^*;\mathbf{I}^*) \in \mathfrak{R}^K\backslash\mathfrak{N}_{\Delta^*}[\Upsilon(t)]$, $\forall t \in \mathfrak{T}_0$. Because $\Delta^* > 0$ can be arbitrarily small it follows that $\mathcal{Z}(t;t_0;\mathbf{Z};\mathbf{I}) \longrightarrow \Upsilon(t)$ as $t \longrightarrow \infty$ for every $[\mathbf{Z},\mathbf{I}(.)] \in \mathfrak{N}_\eta[\Upsilon(t_0)] \times \mathcal{J}^{\xi}$. This proves that the motions $\mathcal{Z}(,;t_0;\mathbf{Z}_0;\mathbf{I})$ of the dynamical system exhibit tracking of the set $\Upsilon(t)$ on $\mathfrak{T}_0 \times \mathcal{J}^{\xi}$, Definition 250.

The set $\mathfrak{N}_\eta[\Upsilon(t_0)]$ is a subset of the domain of tracking the set $\Upsilon(t)$ on $\mathfrak{T}_0 \times \mathcal{J}^{\xi}$.

iii) Let i)-iii) hold. Tracking the set $\Upsilon(t)$ on $\mathfrak{T}_0 \times \mathcal{J}^{\xi}$ is global for $\eta = \infty$ if $\beta(.;\Upsilon)$ is also radially unbounded relative to the set $\Upsilon(t)$ on $\mathfrak{T}_0$, because then $\mathfrak{N}_\eta[\Upsilon(t_0)] = \mathfrak{N}_\infty[\Upsilon(t_0)] = \mathfrak{R}^K$ and $\mathcal{Z}(t;t_0;\mathbf{Z}_0;\mathbf{I}) \longrightarrow \Upsilon(t)$ as $t \longrightarrow \infty$ for every $[\mathbf{Z}_0,\mathbf{I}(.)] \in [\mathfrak{R}^K\backslash \mathrm{Cl}\Upsilon(t_0)]\times\mathcal{J}^{\xi}$, Definition 250.

iv) If the function $\beta(.;\Upsilon)$ is positive definite and decrescent relative to the set $\Upsilon(t)$ on $\mathfrak{T}_0$ for every $t_0 \in \mathfrak{T}_i$, and the function $w(.;\Upsilon)$ is positive definite relative to the set $\Upsilon(t)$ on $\mathfrak{T}_0$ for every $t_0 \in \mathfrak{T}_i$ then (Definition 319 and Definition 323) there are $\psi \in \mathfrak{R}^+$ and $\rho \in \mathfrak{R}^+$ such that $\mathfrak{B}_\psi(t) \subseteq \mathfrak{N}_\eta[\Upsilon(t)]$ for all $(t, t_0) \in \mathfrak{T}_0 \times \mathfrak{T}_i$ and

$$\mathfrak{B}_\psi(\mathfrak{T}_0) = \cap[\mathfrak{B}_\psi(t) : t \in \mathfrak{T}_0] \supseteq \mathfrak{N}_\rho[\Upsilon(t)], \ \forall (t, t_0) \in \mathfrak{T}_0 \times \mathfrak{T}_i. \tag{25.3}$$

Hence,

$$\begin{gathered}\forall [t_0, \mathbf{Z}_0, \mathbf{I}(.)] \in \mathfrak{T}_i \times \mathfrak{B}_\psi(\mathfrak{T}_0) \times \mathcal{J}^\xi \Longrightarrow \\ \mathcal{Z}(t; t_0; \mathbf{Z}_0; \mathbf{I}) \longrightarrow \Upsilon(t) \text{ as } t \longrightarrow \infty, \ \forall t_0 \in \mathfrak{T}_i.\end{gathered} \tag{25.4}$$

Let $\zeta \in \mathfrak{R}^+$ be arbitrarily small. Let $\mathbf{Z}_0 \in \mathfrak{B}_\psi(\mathfrak{T}_0)$. Then $\beta(t_0, \mathbf{Z}_0; \Upsilon) < \psi$ so that

$$\begin{gathered}\beta[t, \mathcal{Z}(t; t_0; \mathbf{Z}_0; \mathbf{I}); \Upsilon] \le \beta(t_0, \mathbf{Z}_0; \Upsilon) - \int_{t_0}^{t} \gamma(\zeta)\, d\tau, \ \forall t \in \mathfrak{T}_0 \Longrightarrow \\ \beta[t, \mathcal{Z}(t; t_0; \mathbf{Z}_0; \mathbf{I}); \Upsilon] \le \psi - \gamma(\zeta)(t - t_0), \ \forall t \in \mathfrak{T}_0 \Longrightarrow \\ \beta[t, \mathcal{Z}(t; t_0; \mathbf{Z}_0; \mathbf{I}); \Upsilon] \le \zeta \ \forall t \ge t_0 + \frac{\psi - \zeta}{\gamma(\zeta)} = t_0 + \tau(\varsigma, \psi) \Longrightarrow \\ \mathcal{Z}(t; t_0; \mathbf{Z}_0; \mathbf{I}) \in \mathfrak{N}_\zeta[\Upsilon(t)], \forall t \in \mathfrak{T}_0, \ \forall t_0 \in \mathfrak{T}_i.\end{gathered} \tag{25.5}$$

The equation (25.3) and the results (25.4), (25.4) show that the set $\mathfrak{B}_\psi(\mathfrak{T}_0)$ and $\tau(\varsigma, \psi)$ are independent of $(t_0, \mathbf{I}) \in \mathfrak{T}_i \times \mathcal{J}^\xi$. This proves that tracking the set $\Upsilon(t)$ is uniform on $\mathfrak{T}_0 \times \mathfrak{T}_i \times \mathcal{J}^\xi$, Definition 251. The set $\mathfrak{B}_\psi(\mathfrak{T}_0)$ is a subset of the domain of the uniform tracking the set $\Upsilon(t)$ on $\mathfrak{T}_0 \times \mathfrak{T}_i \times \mathcal{J}^\xi$, Definition 251.

v) The results under i) through iv) imply that tracking the set $\Upsilon(t)$ is globally uniform on $\mathfrak{T}_0 \times \mathfrak{T}_i \times \mathcal{J}^\xi$, Definition 251.

Stable tracking.

i) and ii) Let $\varepsilon \in \mathfrak{R}^+$ be arbitrarily chosen. It determines the connected neighborhood $\mathfrak{N}_\varepsilon[t; t_0; \Upsilon(t)]$ of $\Upsilon(t)$ at every moment $t \in \mathfrak{T}_0$. Let it be fixed. Let $\zeta_0 \in]\mu, \zeta_{Inf\beta}(\varepsilon)[$ that implies $\mathfrak{B}_{\zeta_0}(t_0) \subset \mathfrak{N}_\varepsilon[t_0; t_0; \Upsilon(t_0)]$. Let $\delta \in \mathfrak{R}^+$ be such that $\mathfrak{N}_\delta[t_0; \varepsilon; \Upsilon(t_0)] \subseteq \mathfrak{B}_{\zeta_0}(t_0)$. The positive definiteness of $w(.;\Upsilon)$ and (25.2) guarantee:

$$\begin{gathered}\forall (t, \mathbf{Z}_0) \in \mathfrak{T}_0 \times \mathfrak{N}_\delta[t_0; \varepsilon; \Upsilon(t_0)] \Longrightarrow \\ \beta[t, \mathbf{Z}(t; t_0; \mathbf{Z}_0); \Upsilon] \le \beta(t_0, \mathbf{Z}_0; \Upsilon).\end{gathered}$$

This shows, in view of the positive definiteness of the function $\beta(.;\Upsilon)$ relative to the set $\Upsilon(t)$, Definition 319, that

$$\begin{gathered}\zeta(t; \mathbf{Z}_0) = \beta[t, \mathbf{Z}(t; t_0; \mathbf{Z}_0)] \le \beta(t_0, \mathbf{Z}_0; \Upsilon) \le \zeta_0 < \zeta_{\mathrm{Inf}\beta}(\varepsilon), \\ \forall (t, \mathbf{Z}_0) \in \mathfrak{T}_0 \times \mathfrak{N}_\delta[t_0; \varepsilon; \Upsilon(t_0)],\end{gathered}$$

or equivalently,

$$\mathfrak{B}_{\zeta(t;\mathbf{Z}_0)}(t) \subset \mathfrak{B}_{\zeta_{\mathrm{Inf}\beta}(\delta)}(t) \subset \mathfrak{B}_{\zeta_{\mathrm{Inf}\beta}(\varepsilon)}(t) \subset \mathfrak{N}_\varepsilon[t; t_0; \Upsilon(t_0)], \ \forall t \in \mathfrak{T}_0.$$

This and the fact that the motions $\mathcal{Z}(,; t_0; \mathbf{Z}_0; \mathbf{I})$, $\mathbf{Z}(t) \equiv \mathcal{Z}(t; t_0; \mathbf{Z}_0; \mathbf{I})$, of the dynamical system exhibit tracking of the set $\Upsilon(t)$ on $\mathfrak{T}_0 \times \mathcal{J}^\xi$ prove that the motions $\mathcal{Z}(,; t_0; \mathbf{Z}_0; \mathbf{I})$ of the dynamical system exhibit stable tracking of the set $\Upsilon(t)$ on $\mathfrak{T}_0 \times \mathcal{J}^\xi$, Definition 252. The intersection $\mathfrak{N}[t_0; \delta_{\max}(\varepsilon); \eta; \Upsilon(t_0)]$ of $\mathfrak{N}_\eta[\Upsilon(t_0)]$ and of the largest neighborhood $\mathfrak{N}_{\delta\max}[t_0; \varepsilon; \Upsilon(t_0)]$ in $\varepsilon \in \mathfrak{R}^+$,

$$\begin{gathered}\mathfrak{N}_{\delta\max}[t_0; \varepsilon; \Upsilon(t_0)] = \max\left\{\mathfrak{N}_\delta[t_0; \varepsilon; \Upsilon(t_0)] : \varepsilon \in \mathfrak{R}^+\right\}, \\ \mathfrak{N}[t_0; \delta_{\max}(\varepsilon); \eta; \Upsilon(t_0)] = \mathfrak{N}_{\delta\max}[t_0; \varepsilon; \Upsilon(t_0)] \cap \mathfrak{N}_\eta[\Upsilon(t_0)]\end{gathered}$$

is an estimate of the domain of the stable tracking of the set $\Upsilon(t)$ on $\mathfrak{T}_0 \times \mathcal{J}^{\xi}$, Definition 252.

iii) If additionally, $\eta = \infty$, the functions $\beta(.;\Upsilon)$ and $w(.;\Upsilon)$ are globally positive definite relative to the set $\Upsilon(t)$ on $\mathfrak{T}_0$ and $\beta(.;\Upsilon)$ is also radially unbounded relative to the set $\Upsilon(t)$ on $\mathfrak{T}_0$ then $\zeta_{\text{Inf}\beta}(\varepsilon) \longrightarrow \infty$ as $\varepsilon \longrightarrow \infty$, which implies that $\mathfrak{N}_{\delta\,\max}[t_0;\varepsilon;\Upsilon(t_0)] \longrightarrow \mathfrak{R}^K$ as $\varepsilon \longrightarrow \infty$. This and global tracking the set $\Upsilon(t)$ on $\mathfrak{T}_0 \times \mathcal{J}^{\xi}$ prove that then stable tracking the set $\Upsilon(t)$ on $\mathfrak{T}_0 \times \mathcal{J}^{\xi}$ is global, Definition 252.

iv) If additionally the function $\beta(.;\Upsilon)$ is positive definite and decrescent relative to the set $\Upsilon(t)$ on $\mathfrak{T}_0$ for every $t_0 \in \mathfrak{T}_i$, and the function $w(.;\Upsilon)$ is positive definite relative to the set $\Upsilon(t)$ on $\mathfrak{T}_0$ for every $t_0 \in \mathfrak{T}_i$ then (Definition 319 and Definition 323), (21.18) guarantees that

$$\forall \varepsilon \in \mathfrak{R}^+ \Longrightarrow \cap \{\mathfrak{N}[t_0;\delta_{\max}(\varepsilon);\eta;\Upsilon(t_0)] : t_0 \in \mathfrak{T}_i\}$$

is a neighborhood of $\Upsilon(t_0)$ at every $t_0 \in \mathfrak{T}_i$. This and uniform tracking the set $\Upsilon(t)$ on $\mathfrak{T}_0 \times \mathfrak{T}_i \times \mathcal{J}^{\xi}$ prove that stable tracking the set $\Upsilon(t)$ is uniform on $\mathfrak{T}_0 \times \mathfrak{T}_i \times \mathcal{J}^{\xi}$, Definition 253.

v) The results under i) through iv) imply v) ■

Conclusion 395 *Theorem 394 holds also for time-invariant sets.*

Theorem 394 is valid regardless of the time nature of the system, whether it is time-invariant or time-varying.

25.3 Conditions for exponential tracking

The characteristic of exponential tracking is the fast exponential system motions' convergence to the target set $\Upsilon(t)$, Definition 254. The following theorem shows that the slightly sharper conditions should be satisfied rather than those for stable tracking the target set $\Upsilon(t)$.

Theorem 396 ***Exponential tracking the target set*** $\Upsilon(t)$

Let the set function $\Upsilon(.)$ belong to the set family $\mathcal{F}$: $\Upsilon(.) \in \mathcal{F}$.

In order for the motions $\mathcal{Z}(.;t_0;\mathbf{Z}_0;\mathbf{I})$, $\mathbf{Z}(t) \equiv \mathcal{Z}(t;t_0;\mathbf{Z}_0;\mathbf{I})$, of a dynamical system, which are globally defined and globally continuous for every $\mathbf{I}(.) \in \mathcal{J}^{\xi}$, to exhibit the uniform exponential tracking on $\mathfrak{T}_0 \times \mathcal{J}^{\xi}$ of the target set $\Upsilon(t)$ it is sufficient that:

i) The boundary function $\beta(.;\Upsilon)$ of the target set $\Upsilon(t)$ is positive definite and decrescent relative to the set $\Upsilon(t)$ itself on $\mathfrak{T}_0$

ii) There are positive real numbers k_i, $i = 1,2,3$, and a positive definite decrescent function $w(.;\Upsilon) : \mathfrak{T} \times \mathfrak{R}^K \longrightarrow \mathfrak{R}$ relative to the set $\Upsilon(t)$ on $\mathfrak{T}_0$ such that

ii-a) The boundary function $\beta(.;\Upsilon)$ of the target set $\Upsilon(t)$ and the function $w(.;\Upsilon)$ are interrelated by (25.6) and (25.7):

$$k_1 w(t,\mathbf{Z};\boldsymbol{\Upsilon}) \leq \beta(t,\mathbf{Z};\boldsymbol{\Upsilon}) \leq k_2 w(t,\mathbf{Z};\boldsymbol{\Upsilon}), \quad \forall (t,t_0,\mathbf{Z}) \in \mathfrak{T}_0 \times \mathfrak{T}_i \times \mathfrak{N}_\eta[\Upsilon(t)], \tag{25.6}$$

$$D^+\beta(t,\mathbf{Z};\boldsymbol{\Upsilon}) \leq \frac{\partial \beta(t,\mathbf{Z};\boldsymbol{\Upsilon})}{\partial t} \leq -k_3 w(t,\mathbf{Z};\boldsymbol{\Upsilon}),$$
$$\forall (t,t_0,\mathbf{Z}) \in \mathfrak{T}_0 \times \mathfrak{T}_i \times \mathfrak{N}_\eta[\Upsilon(t)], \ \forall \mathbf{I}(.) \in \mathcal{J}^{\xi}, \tag{25.7}$$

on some η-neighborhood $\mathfrak{N}_\eta[\Upsilon(t)]$ of $\Upsilon(t)$ on $\mathfrak{T}_0$.

Then the set $\mathfrak{N}_\eta[\Upsilon(t_0)]$ is a subset of the domain of the uniform exponential tracking the set $\Upsilon(t)$ on $\mathfrak{T}_0 \times \mathfrak{T}_i \times \mathcal{J}^{\xi}$.

iii) If, additionally to i) and ii), $\eta = \infty$ and the functions $\beta(.;\Upsilon)$ and $w(.;\Upsilon)$ are also radially unbounded relative to the set $\Upsilon(t)$ on $\mathfrak{T}_0$ then uniform exponential tracking the set $\Upsilon(t)$ on $\mathfrak{T}_0 \times \mathfrak{T}_i \times \mathcal{J}^{\xi}$ is global.

Proof. Let the conditions under:

- i) and ii) hold. Then the motions $\mathcal{Z}(.;t_0;\mathbf{Z}_0;\mathbf{I})$, $\mathbf{Z}(t)\equiv\mathcal{Z}(t;t_0;\mathbf{Z}_0;\mathbf{I})$, of the dynamical system exhibit uniform stable tracking of the set $\Upsilon(t)$ on $\mathfrak{T}_0 \times \mathfrak{T}_i \times \mathcal{J}^{\xi}$, Theorem 394,
- i) through iii) hold. Then the motions $\mathcal{Z}(.;t_0;\mathbf{Z}_0;\mathbf{I})$ of the dynamical system exhibit global uniform stable tracking of the set $\Upsilon(t)$ on $\mathfrak{T}_0 \times \mathfrak{T}_i \times \mathcal{J}^{\xi}$, Theorem 394.

Let us show that the uniform stable tracking properties are also uniformly exponential. From (25.7) follows:

$$D^+\beta(t,\mathbf{Z};\boldsymbol{\Upsilon}) \leq -k_3 w(t,\mathbf{Z};\boldsymbol{\Upsilon}),\ \forall (t,t_0,\mathbf{Z}) \in \mathfrak{T}_0\times\mathfrak{T}_i\times\mathfrak{N}_\eta[\Upsilon(t)], \forall\mathbf{I}(.) \in \mathcal{J}^{\xi}.$$

This and (25.6) imply

$$D^+\beta(t,\mathbf{Z};\boldsymbol{\Upsilon}) \leq -k_2^{-1}k_3\beta(t,\mathbf{Z};\boldsymbol{\Upsilon}),\ \forall (t,t_0,\mathbf{Z}) \in \mathfrak{T}_0\times\mathfrak{T}_i\times\mathfrak{N}_\eta[\Upsilon(t)], \forall\mathbf{I}(.) \in \mathcal{J}^{\xi}.$$

The solution is

$$\begin{gathered}\beta[t,\mathcal{Z}(t;t_0;\mathbf{Z}_0;\mathbf{I});\Upsilon] \leq \beta(t_0,\mathbf{Z}_0;\Upsilon)\,exp\left[-k_2^{-1}k_3(t-t_0)\right],\\ \forall[t,t_0,\mathbf{Z}_0,\mathbf{I}(.)] \in \mathfrak{T}_0 \times \mathfrak{T}_i \times \mathfrak{N}_\eta[\Upsilon(t_0)] \times \mathcal{J}^{\xi}.\end{gathered}$$

This, Definition 319, and Definition 323 permit the following

$$\begin{gathered}\rho[\mathcal{Z}(t;t_0;\mathbf{Z}_0;\mathbf{I}),\Upsilon(t)] \leq \rho[\mathbf{Z}_0,\Upsilon(t_0)]\exp\left[-k_2^{-1}k_3(t-t_0)\right],\\ \forall[t,t_0,\mathbf{Z}_0,\mathbf{I}(.)] \in \mathfrak{T}_0 \times \mathfrak{T}_i \times \mathfrak{N}_\eta[\Upsilon(t_0)] \times \mathcal{J}^{\xi}.\end{gathered}$$

This is due to (21.10), [i.e., due to (21.17)], which in this framework has the following form:

$$\begin{gathered}inf\{\beta(t,\mathbf{Z};\boldsymbol{\Upsilon}) : (t,\mathbf{Z}) \in\mathfrak{T}_0\times\partial\mathfrak{N}_\theta[\Upsilon(t)]\} = \zeta = \zeta_{Inf\beta}(\theta) \Longrightarrow\\ \mathcal{B}_\zeta(t) = \{\mathbf{Z} : \beta(t,\mathbf{Z};\boldsymbol{\Upsilon}) < \zeta\} \subseteq \mathfrak{N}_\theta[\Upsilon(t)],\ \forall t \in \mathfrak{T}_0 \Longrightarrow\\ \mathbf{Z} \in\mathcal{B}_\zeta(t) \Longleftrightarrow \rho[\mathbf{Z},\Upsilon(t)] < \theta,\end{gathered} \tag{25.8}$$

and due to (21.19), [i.e., due to (21.21)], which in this framework has the following form:

$$\begin{gathered}\sup\{\beta(t,\mathbf{Z};\boldsymbol{\Upsilon}) : (t,\mathbf{Z}) \in \mathfrak{T}_0 \times \partial\mathfrak{N}_\theta[\Upsilon(t)]\} = \xi = \xi_{\mathrm{Sup}\beta}(\theta) \Longrightarrow\\ \mathcal{B}_\xi(t) = \{\mathbf{Z} : \beta(t,\mathbf{Z};\boldsymbol{\Upsilon}) < \xi\} \supseteq \mathfrak{N}_\theta[\Upsilon(t)],\ \forall t \in \mathfrak{T}_0 \Longrightarrow\\ \mathbf{Z} \in \mathcal{B}_\xi(t) \Longleftrightarrow \rho[\mathbf{Z},\Upsilon(t)] > \theta\end{gathered} \tag{25.9}$$

The system motions exhibit uniform exponential tracking of the target set $\Upsilon(t)$ on the set product $\mathfrak{T}_0 \times \mathfrak{T}_i \times \mathcal{J}^{\xi}$, Definition 254. The set $\mathfrak{N}_\eta[\Upsilon(t_0)]$ is an estimate of the exponential tracking domain of the target set $\Upsilon(t)$ on $\mathfrak{T}_0 \times \mathfrak{T}_i \times \mathcal{J}^{\xi}$.

If the conditions i) through iii) hold then $\mathfrak{N}_\eta[\Upsilon(t_0)] = \mathfrak{R}^K$, $\forall t_0 \in \mathfrak{T}_i$. This proves that the uniform exponential tracking of the target set $\Upsilon(t)$ is global on $\mathfrak{T}_0 \times \mathfrak{T}_i \times \mathcal{J}^{\xi}$, Definition 254 ■

Proof. Let the conditions under [illegible] and 5) hold. Then the motions [illegible] of the dynamical system exhibit uniform stable tracking of the set $\mathfrak{T}(t)$ on [illegible]. Then [illegible]

[illegible] through [illegible] hold. Then the motions [illegible] of the dynamical system exhibit global uniform stable tracking of the set $\mathfrak{T}(t)$ on [illegible] Theorem [illegible].

Let us show that the uniform stable tracking properties are also uniformly exponential. From (27.7) follows

$$[illegible]$$

[illegible] and (24.4) imply

$$[illegible]$$

The solution is [illegible]

$$[illegible]$$

[illegible] for [illegible] in the following:

$$[illegible]$$

$$[illegible]$$

The above conditions [illegible] exponential tracking [illegible] on $\mathfrak{T}(t)$ [illegible] tracking domain [illegible]

If the conditions [illegible] through [illegible] hold then [illegible] that the motions of the dynamical system exhibit [illegible] tracking of the set $\mathfrak{T}(t)$ [illegible] Definition [illegible] ■

Part VII

FINITE REACHABILITY *TIME* TRACKING

Chapter 26

Output space definitions

26.1 Finite scalar reachability *time*

Lyapunov tracking properties (in the Lyapunov sense), that is L-tracking properties, guarantee asymptotic convergence of the real output response to the desired one only for the infinite *time* [as $t \to \infty$, (2)]. They do not ensure that the real output response reaches the desired one in a finite *time* and that they stay equal from then until the final moment t_F of the plant work. In order to overcome this essential drawback from the engineering and control system customer points of view, we present definitions of some tracking properties with *finite reachability time (FRT)*. It can be *finite scalar reachability time (FSRT) (1)*, or *finite vector reachability time (FVRT)* (26.32).

Let *the final moment* t_F of the system operation obey

$$t_F \in \mathfrak{T}_0 \ or \ t_F = \infty, \quad \text{and } t_F > t_0. \tag{26.1}$$

This means that the final moment t_F can be finite ($t_F \in \mathfrak{T}_0$) or infinite ($t_F = \infty$). It determines, together with the initial instant t_0, the *time set* $\mathfrak{T}_{0F}$ over which the system should work properly,

$$\mathfrak{T}_{0F} = \left\{ t: \ t \in \mathfrak{T}, \left\langle \begin{array}{l} t_0 \leq t \leq t_F \ \text{ iff } t_F < \infty, \\ t_0 \leq t < t_F \ \text{ iff } t_F = \infty \end{array} \right\rangle \right\} \subseteq \mathfrak{T}_0, \tag{26.2}$$

so that its closure $\text{Cl}\mathfrak{T}_{0F}$ is compact if $t_F < \infty$, but

$$t_F < \infty \Longrightarrow \text{Cl}\mathfrak{T}_{0F} = \{t: \ t \in \text{Cl}\mathfrak{T}, \ t_0 \leq t \leq t_F \} \subset \mathfrak{T}_0. \tag{26.3}$$

This enables studying *the finite-time tracking* that demands $t_F < \infty$.

The concept of **finite-time tracking** is essentially different from the Lyapunov tracking concept.

A higher tracking quality than the *L*-tracking quality is *tracking with the finite scalar reachability time (FSRT)* t_R,

$$t_R \in \text{In}\mathfrak{T}_{0F}. \tag{26.4}$$

The finite scalar reachability time (FSRT) $t_R \subseteq \mathfrak{T}_0$ is the moment when the real output vector $\mathbf{Y}(t)$ becomes equal to the desired output vector $\mathbf{Y}_d(t)$, and after the moment t_R they stay equal until the final moment t_F,

$$t_F \geq t \geq t_R \Longrightarrow \mathbf{Y}(t) = \mathbf{Y}_d(t). \tag{26.5}$$

This means that the (finite scalar) reachability *time* t_R is the same for all output variables We can extend this to hold also for their derivatives,

$$t_F \geq t \geq t_R \Longrightarrow \mathbf{Y}^k(t) = \mathbf{Y}_d^k(t). \tag{26.6}$$

The finite scalar reachability *time* t_R induces the *time* sets $\mathfrak{T}_R$ and $\mathfrak{T}_{RF}$ as the subsets of $\mathfrak{T}_0$,

$$\mathfrak{T}_R = \{t \in \mathfrak{T}_0 : t_0 \leq t \leq t_R,\ t_R > t_0\} \subset \mathfrak{T}_{0F},$$
$$\mathfrak{T}_{RF} = \left\{ t \in \mathrm{Cl}\mathfrak{T}_0 : \left\langle \begin{array}{c} t_R \leq t \leq t_F < \infty, \\ t_R \leq t < t_F = \infty \end{array} \right\rangle \right\} \subset \mathfrak{T}_{0F},\ \ \mathfrak{T}_R \cup \mathfrak{T}_{RF} = \mathfrak{T}_{0F},$$
$$\mathfrak{T}_{R\infty} = \{t:\ t_R \leq t < \infty\} = [t_R, \infty[. \tag{26.7}$$

$\mathfrak{T}_R$ is *the reachability time set, and* $\mathfrak{T}_{RF}$ and $\mathfrak{T}_{R\infty}$ are *the post reachability time sets. They are the proper subsets of* $\mathfrak{T}_{0F}$.

We continue to use the abbreviated notation:

$$\mathbf{Y}_d^k(t) \equiv \mathbf{Y}_d^k(t; t_0; \mathbf{Y}_{d0}^k),\ \ \mathbf{Y}^k(t) \equiv \mathbf{Y}^k(t; t_0; \mathbf{Y}_0^k; \mathbf{D}; \mathbf{U}), \tag{26.8}$$

The following definitions determine **in the (extended) output space** $\mathfrak{R}^{(k+1)N}$ various types of ***the k-th-order tracking of*** $\mathbf{Y}_d(t)$ ***with the finite scalar reachability time (FSRT)***, i.e., ***various types of the tracking of*** $\mathbf{Y}_d^k(t)$ ***with the finite scalar reachability time (FSRT)***. The system should possess a required tracking with the final reachability *time* over the *time* set $\mathfrak{T}_{0F}$. We omit the expression "over the *time* set $\mathfrak{T}_{0F}$" from the following definitions.

The concept of **the (scalar or vector) reachability time tracking** is essentially different from the Lyapunov tracking concept.

Definition 397 ***Tracking with the finite scalar reachability time (FSRT) of the desired output*** $\mathbf{Y}_d^k(t)$ ***of the plant (3.12) controlled by a control*** $\mathbf{U}(.) \in \mathfrak{U}^l$

a) The plant (3.12) exhibits ***the output tracking of*** $\mathbf{Y}_d^k(t)$ ***with the finite scalar reachability time (FSRT)*** t_R ***on*** $\mathfrak{T}_{0F} \times \mathfrak{D}^i \times \mathfrak{Y}_d^k$, *for short* ***the tracking with the finite reachability time*** t_R ***of*** $\mathbf{Y}_d^k(t)$ ***on*** $\mathfrak{T}_{0F} \times \mathfrak{D}^i \times \mathfrak{Y}_d^k$ *if, and only if, for every* $(\mathbf{D}(.), \mathbf{Y}_d(.)) \in \mathfrak{D}^i \times \mathfrak{Y}_d^k$ *there exists a* (t_0, t_R, t_F)*-dependent connected neighborhood*

$$\mathfrak{N}\left(t_0; t_R; t_F; \mathbf{Y}_{d0}^k; \mathbf{D}; \mathbf{U}; \mathbf{Y}_d\right) \subseteq \mathfrak{R}^{(k+1)N},$$

of the plant desired initial output vector $\mathbf{Y}_{d0}^k$ *at the initial moment* $t_0 \in \mathfrak{T}_i$ *such that* $\mathbf{Y}_0^k$ *from the neighborhood* $\mathfrak{N}\left(t_0; t_R; t_F; \mathbf{Y}_{d0}^k; \mathbf{D}; \mathbf{U}; \mathbf{Y}_d\right)$ *guarantees both that* $\mathbf{Y}^k(t; t_0; \mathbf{Y}_0^k; \mathbf{D}; \mathbf{U})$ *becomes equal to* $\mathbf{Y}_d^k(t; t_0; \mathbf{Y}_{d0}^k)$ *at the moment* $t_R \in In\ \mathfrak{T}_{0F}$, *and that they stay equal afterwards on* $\mathfrak{T}_{RF}$, *i.e.,*

$$\forall\left[\mathbf{D}(.), \mathbf{Y}_d(.)\right] \in \mathfrak{D}^i \times \mathfrak{Y}_d^k,$$
$$\exists \mathfrak{N}\left(t_0; t_R; t_F; \mathbf{Y}_{d0}^k; \mathbf{D}; \mathbf{U}; \mathbf{Y}_d\right),\ \mathfrak{N}\left(t_0; t_R; t_F; \mathbf{Y}_{d0}^k; \mathbf{D}; \mathbf{U}; \mathbf{Y}_d\right) \subseteq \mathfrak{R}^{(k+1)N},$$
$$\mathbf{Y}_0^k \in \mathfrak{N}\left(t_0; t_R; t_F; \mathbf{Y}_{d0}^k; \mathbf{D}; \mathbf{U}; \mathbf{Y}_d\right) \Longrightarrow$$
$$\mathbf{Y}^k(t; t_0; \mathbf{Y}_0^k; \mathbf{D}; \mathbf{U}) = \mathbf{Y}_d^k(t; t_0; \mathbf{Y}_{d0}^k),\ \forall t \in \mathfrak{T}_{RF}. \tag{26.9}$$

b) The largest connected neighborhood $\mathfrak{N}\left(t_0; t_R; t_F; \mathbf{Y}_{d0}^k; \mathbf{D}; \mathbf{U}; \mathbf{Y}_d\right)$ *of* $\mathbf{Y}_{d0}^k$, *(26.9), is* ***the FSRT tracking domain*** *denoted by* $\mathcal{D}_T^k\left(t_0; t_R; t_F; \mathbf{Y}_{d0}^k; \mathbf{D}; \mathbf{U}; \mathbf{Y}_d\right)$ ***of*** $\mathbf{Y}_d^k(t)$ ***on*** $\mathfrak{T}_{0F} \times \mathfrak{D}^i \times \mathfrak{Y}_d^k$.

c) The FSRT tracking of $\mathbf{Y}_d^k(t)$ *on* $\mathfrak{T}_{0F} \times \mathfrak{D}^i \times \mathfrak{Y}_d^k$ *is* ***global (in the whole)*** *if, and only if,* $\mathcal{D}_T^k\left(t_0; t_R; t_F; \mathbf{Y}_{d0}^k; \mathbf{D}; \mathbf{U}; \mathbf{Y}_d\right) = \mathfrak{R}^{(k+1)N}$ *for every* $\left[\mathbf{D}(.), \mathbf{Y}_d(.)\right] \in \mathfrak{D}^i \times \mathfrak{Y}_d^k$.

The preceding definition allows nonuniformity of the tracking on $\mathfrak{D}^i \times \mathfrak{Y}_d^k$.

Definition 398 ***The uniform tracking with the finite scalar reachability time of the desired output behavior $\mathbf{Y}_d^k(t)$ on $\mathfrak{T}_{0F} \times \mathfrak{D}^i \times \mathfrak{Y}_d^k$ of the plant (3.12) controlled by a control $\mathbf{U}(.) \in \mathfrak{U}^l$***

a) If, and only if, the intersection $\mathcal{D}_{TU}^k\left(t_0;t_R;t_F;\mathfrak{D}^i;\mathbf{U};\mathfrak{Y}_d^k\right)$ of the tracking domains $\mathcal{D}_T^k\left(t_0;t_R;t_F;\mathbf{Y}_{d0}^k;\mathbf{D};\mathbf{U};\mathbf{Y}_d\right)$ in $[\mathbf{D}(.), \mathbf{Y}_d(.)] \in \mathfrak{D}^i \times \mathfrak{Y}_d^k$ is a connected neighborhood of $\mathbf{Y}_{d0}^k$ of every $\mathbf{Y}_d^{p-1}(.) \in \mathfrak{Y}_d^k$,

$$\exists \xi \in \mathfrak{R}^+ \Longrightarrow \mathcal{D}_{TU}^k\left(t_0;t_R;t_F;\mathfrak{D}^i;\mathbf{U};\mathfrak{Y}_d^k\right)$$
$$= \cap \left[\begin{array}{c} \mathcal{D}_T^k\left(t_0;t_R;t_F;\mathbf{Y}_{d0}^k;\mathbf{D};\mathbf{U};\mathbf{Y}_d\right): \\ [\mathbf{D}(.), \mathbf{Y}_d(.)] \in \mathfrak{D}^i \times \mathfrak{Y}_d^k \end{array}\right] \supset \mathfrak{N}_\xi\left(t_0;t_R;t_F;\mathfrak{D}^i;\mathbf{U};\mathfrak{Y}_d^k\right),$$
$$\partial\mathcal{D}_{TU}^k\left(t_0;t_R;t_F;\mathfrak{D}^i;\mathbf{U};\mathfrak{Y}_d^k\right) \cap \partial\mathfrak{N}_\xi\left(t_0;t_R;t_F;\mathfrak{D}^i;\mathbf{U};\mathfrak{Y}_d^k\right) = \phi, \tag{26.10}$$

then the FSRT tracking of $\mathbf{Y}_d^k(t)$ is ***uniform in*** *$[\mathbf{D}(.), \mathbf{Y}_d(.)] \in \mathfrak{D}^i \times \mathfrak{Y}_d^k$ on $\mathfrak{T}_{0F} \times \mathfrak{D}^i \times \mathfrak{Y}_d^k$ and the set $\mathcal{D}_{TU}^k\left(t_0;t_R;t_F;\mathfrak{D}^i;\mathbf{U};\mathfrak{Y}_d^k\right)$ is* ***the $(\mathbf{D}, \mathbf{Y}_d)$-uniform FSRT tracking domain of $\mathbf{Y}_d^k(t)$ on $\mathfrak{T}_{0F} \times \mathfrak{D}^i \times \mathfrak{Y}_d^k$.***

b) If, and only if, the intersection $\mathcal{D}_{TU}^k\left(\mathfrak{T};t_R;t_F;\mathbf{Y}_{d0}^k;\mathbf{D};\mathbf{U};\mathbf{Y}_d\right)$ of all the FSRT tracking domains $\mathcal{D}_T^k\left(t_0;t_R;t_F;\mathbf{Y}_{d0}^k;\mathbf{D};\mathbf{U};\mathbf{Y}_d\right)$ in t_0 over $\mathfrak{T}_i$ is a connected neighborhood of $\mathbf{Y}_{d0}^k$ then the FSRT tracking of $\mathbf{Y}_d^k(t)$ on $\mathfrak{T}_{0F} \times \mathfrak{D}^i \times \mathfrak{Y}_d^k$ is ***uniform in t_0 on $\mathfrak{T}_i$*** *and $\mathcal{D}_{TU}^k\left(\mathfrak{T}_i;t_R;t_F;\mathbf{Y}_{d0}^k;\mathbf{D};\mathbf{U};\mathbf{Y}_d\right)$ is* ***the domain of t_0-uniform FSRT tracking of $\mathbf{Y}_d^k(t)$ on $\mathfrak{T}_i \times \mathfrak{D}^i \times \mathfrak{Y}_d^k$,***

$$\exists \xi \in \mathfrak{R}^+ \Longrightarrow \mathcal{D}_{TU}^k\left(\mathfrak{T}_i;t_R;t_F;\mathbf{Y}_{d0}^k;\mathbf{D};\mathbf{U};\mathbf{Y}_d\right)$$
$$= \cap\left[\mathcal{D}_T^k\left(t_0;t_R;t_F;\mathbf{Y}_{d0}^k;\mathbf{D};\mathbf{U};\mathbf{Y}_d\right): t_0 \in \mathfrak{T}_i\right]$$
$$\supset \mathfrak{N}_\xi\left(\mathfrak{T}_i;t_R;t_F;\mathbf{Y}_{d0}^k;\mathbf{D};\mathbf{U};\mathbf{Y}_d\right),$$
$$\partial\mathcal{D}_{TU}^k\left(\mathfrak{T}_i;t_R;t_F;\mathbf{Y}_{d0}^k;\mathbf{D};\mathbf{U};\mathbf{Y}_d\right) \cap \partial\mathfrak{N}_\xi\left(\mathfrak{T}_i;t_R;t_F;\mathbf{Y}_{d0}^k;\mathbf{D};\mathbf{U};\mathbf{Y}_d\right) = \phi. \tag{26.11}$$

c) If, and only if, the intersection $\mathcal{D}_T^k\left(\mathfrak{T}_i;t_R;t_F;\mathfrak{D}^i;\mathbf{U};\mathfrak{Y}_d^k\right)$ of all FSRT tracking domains $\mathcal{D}_T^k\left(t_0;t_R;t_F;\mathbf{Y}_{d0}^k;\mathbf{D};\mathbf{U};\mathbf{Y}_d\right)$ in $[t_0, \mathbf{D}(.), \mathbf{Y}_d(.)] \in \mathfrak{T}_i \times \mathfrak{D}^i \times \mathfrak{Y}_d^k$ is a neighborhood of $\mathbf{Y}_{d0}^k$ of every $\mathbf{Y}_d(.) \in \mathfrak{Y}_d^k$ then the FSRT tracking of $\mathbf{Y}_d^k(t)$ on $\mathfrak{T}_i \times \mathfrak{D}^i \times \mathfrak{Y}_d^k$ is ***uniform in*** *$[t_0, \mathbf{D}(.), \mathbf{Y}_d(.)] \in \mathfrak{T}_i \times \mathfrak{D}^i \times \mathfrak{Y}_d^k$; i.e.,*

$$\exists \xi \in \mathfrak{R}^+ \Longrightarrow \mathcal{D}_T^k\left(\mathfrak{T}_i;t_R;t_F;\mathfrak{D}^i;\mathbf{U};\mathfrak{Y}_d^k\right)$$
$$= \cap \left[\begin{array}{c} \mathcal{D}_T^k\left(t_0;t_R;t_F;\mathbf{Y}_{d0}^k;\mathbf{D};\mathbf{U};\mathbf{Y}_d\right): \\ [t_0, \mathbf{D}(.), \mathbf{Y}_d(.)] \in \mathfrak{T}_i \times \mathfrak{D}^i \times \mathfrak{Y}_d^k \end{array}\right] \supset \mathfrak{N}_\xi\left(\mathfrak{T}_i;t_R;t_F;\mathfrak{D}^i;\mathbf{U};\mathfrak{Y}_d^k\right),$$
$$\partial\mathcal{D}_T^k\left(\mathfrak{T}_i;t_R;t_F;\mathfrak{D}^i;\mathbf{U};\mathfrak{Y}_d^k\right) \cap \partial\mathfrak{N}_\xi\left(\mathfrak{T}_i;t_R;t_F;\mathfrak{D}^i;\mathbf{U};\mathfrak{Y}_d^k\right) = \phi. \tag{26.12}$$

Then, and only then, the set $\mathcal{D}_T^k\left(\mathfrak{T}_i;t_R;t_F;\mathfrak{D}^i;\mathbf{U};\mathfrak{Y}_d^k\right)$ is ***the domain of FSRT uniform tracking of $\mathbf{Y}_d^k(t)$***, *for every $\mathbf{Y}_d(.) \in \mathfrak{Y}_d^k$, on $\mathfrak{T}_i \times \mathfrak{D}^i \times \mathfrak{Y}_d^k$.*

Comment 399 *The global FSRT tracking of $\mathbf{Y}_d^k(t)$ on $\mathfrak{T}_{0F} \times \mathfrak{D}^i \times \mathfrak{Y}_d^k$ is uniform in the functional pair $[\mathbf{D}(.), \mathbf{Y}_d(.)]$ over $\mathfrak{D}^i \times \mathfrak{Y}_d^k$.*

In order to avoid the *FSRT* tracking with a big overshoot we introduce the following.

Definition 400 ***The stable FSRT tracking of the desired output behavior $\mathbf{Y}_d^k(t)$ of the plant (3.12) controlled by a control $\mathbf{U}(.) \in \mathfrak{U}^l$***

a) The plant (3.12) exhibits ***the stable output FSRT tracking of $\mathbf{Y}_d^k(t)$ on $\mathfrak{T}_{0F} \times \mathfrak{D}^i \times \mathfrak{Y}_d^k$***, *for short* ***the stable FSRT tracking of $\mathbf{Y}_d^k(t)$ on $\mathfrak{T}_{0F} \times \mathfrak{D}^i \times \mathfrak{Y}_d^k$*** *if, and only if, it exhibits the FSRT tracking of $\mathbf{Y}_d^k(t)$ on $\mathfrak{T}_{0F} \times \mathfrak{D}^i \times \mathfrak{Y}_d^k$, and for every connected* *neighborhood* $\mathfrak{N}_{\varepsilon}\left[t;\mathbf{Y}_{d}^{k}(t)\right]$ *of* $\mathbf{Y}_{d}^{k}(t)$ *at any* $t\in\mathfrak{T}_{0F}$*, there is a connected neighborhood* $\mathfrak{N}\left(\varepsilon;t_{0};t_{R};t_{F};\mathbf{Y}_{d0}^{k};\mathbf{D};\mathbf{U};\mathbf{Y}_{d}\right)$*, (16.1), of the plant desired initial output vector* $\mathbf{Y}_{d0}^{k}$ *at the initial moment* $t_{0}\in\mathfrak{T}_{i}$ *such that for the initial* $\mathbf{Y}_{0}^{k}\in\mathfrak{N}\left(\varepsilon;t_{0};t_{R};t_{F};\mathbf{Y}_{d0}^{k};\mathbf{D};\mathbf{U};\mathbf{Y}_{d}\right)$ *the instantaneous* $\mathbf{Y}^{k}(t)$ *stays in the neighborhood* $\mathfrak{N}_{\varepsilon}\left[t;\mathbf{Y}_{d}^{k}(t)\right]$ *for all* $t\in\mathfrak{T}_{0F}$*; i.e.,*

$$\begin{gathered}\forall\mathfrak{N}_{\varepsilon}\left[t;\mathbf{Y}_{d}^{k}(t)\right]\subseteq\mathfrak{R}^{(k+1)N},\ \forall t\in\mathfrak{T}_{0F},\\ \forall\left[\mathbf{D}(.),\mathbf{Y}_{d}(.)\right]\in\mathfrak{D}^{i}\times\mathfrak{Y}_{d}^{k},\ \exists\mathfrak{N}\left(\varepsilon;t_{0};t_{R};t_{F};\mathbf{Y}_{d0}^{k};\mathbf{D};\mathbf{U};\mathbf{Y}_{d}\right)\subseteq\mathfrak{R}^{(k+1)N},\\ \mathfrak{N}\left(\varepsilon;t_{0};t_{R};t_{F};\mathbf{Y}_{d0}^{k};\mathbf{D};\mathbf{U};\mathbf{Y}_{d}\right)\subseteq\mathcal{D}_{T}^{k}\left(t_{0};t_{R};t_{F};\mathbf{Y}_{d0}^{k};\mathbf{D};\mathbf{U};\mathbf{Y}_{d}\right),\\ \mathbf{Y}_{0}^{k}\in\mathfrak{N}\left(\varepsilon;t_{0};t_{R};t_{F};\mathbf{Y}_{d0}^{k};\mathbf{D};\mathbf{U};\mathbf{Y}_{d}\right)\Longrightarrow\\ \mathbf{Y}^{k}(t;t_{0};\mathbf{Y}_{0}^{k};\mathbf{D};\mathbf{U})\in\mathfrak{N}_{\varepsilon}\left[t;\mathbf{Y}_{d}^{k}(t)\right],\ \forall t\in\mathfrak{T}_{0F}.\end{gathered}\tag{26.13}$$

b) The largest connected neighborhood $\mathfrak{N}\left(\varepsilon;t_{0};t_{R};t_{F};\mathbf{Y}_{d0}^{k};\mathbf{D};\mathbf{U};\mathbf{Y}_{d}\right)$ *of* $\mathbf{Y}_{d0}^{k}$*, (26.13), is* ***the*** ε***-tracking domain***

$$\mathcal{D}_{ST}^{k}\left(\varepsilon;t_{0};t_{R};t_{F};\mathbf{Y}_{d0}^{k};\mathbf{D};\mathbf{U};\mathbf{Y}_{d}\right)$$

at $t_{0}\in\mathfrak{T}$ ***of the stable FSRT tracking of*** $\mathbf{Y}_{d}^{k}(t)$ ***on*** $\mathfrak{T}_{0F}\times\mathfrak{D}^{i}\times\mathfrak{Y}_{d}^{k}$.

The domain $\mathcal{D}_{ST}^{k}\left(t_{0};t_{R};t_{F};\mathbf{Y}_{d0}^{k};\mathbf{D};\mathbf{U};\mathbf{Y}_{d}\right)$ ***at*** $t_{0}\in\mathfrak{T}$ ***of the stable FSRT tracking of*** $\mathbf{Y}_{d}^{k}(t)$ ***on*** $\mathfrak{T}_{0F}\times\mathfrak{D}^{i}\times\mathfrak{Y}_{d}^{k}$ *is the union of all*

$$\mathcal{D}_{ST}^{k}\left(\varepsilon;t_{0};t_{R};t_{F};\mathbf{Y}_{d0}^{k};\mathbf{D};\mathbf{U};\mathbf{Y}_{d}\right)$$

over $\varepsilon\in\mathfrak{R}^{+}$*,*

$$\mathcal{D}_{ST}^{k}\left(t_{0};t_{R};t_{F};\mathbf{Y}_{d0}^{k};\mathbf{D};\mathbf{U};\mathbf{Y}_{d}\right)=\cup\left[\begin{array}{c}\mathcal{D}_{ST}^{k}\left(\varepsilon;t_{0};t_{R};t_{F};\mathbf{Y}_{d0}^{k};\mathbf{D};\mathbf{U};\mathbf{Y}_{d}\right):\\ :\varepsilon\in\mathfrak{R}^{+}\end{array}\right].\tag{26.14}$$

Let $[0,\varepsilon_{M})$ *be the maximal interval over which*

$$\mathcal{D}_{ST}^{k}\left(\varepsilon;t_{0};t_{R};t_{F};\mathbf{Y}_{d0}^{k};\mathbf{D};\mathbf{U};\mathbf{Y}_{d}\right)$$

is continuous in $\varepsilon\in\mathfrak{R}_{+}$*,*

$$\begin{gathered}\mathcal{D}_{ST}^{k}\left(\varepsilon;t_{0};t_{R};t_{F};\mathbf{Y}_{d0}^{k};\mathbf{D};\mathbf{U};\mathbf{Y}_{d}\right)\in\mathfrak{C}\left([0,\varepsilon_{M})\right),\\ \forall\left[\mathbf{D}(.),\mathbf{Y}_{d}(.)\right]\in\mathfrak{D}^{i}\times\mathfrak{Y}_{d}^{k}.\end{gathered}$$

The domain $\mathcal{D}_{SST}^{k}\left(t_{0};t_{R};t_{F};\mathbf{Y}_{d0}^{k};\mathbf{D};\mathbf{U};\mathbf{Y}_{d}\right)$ ***at*** $t_{0}\in\mathfrak{T}$ ***of the strict stable FSRT tracking of*** $\mathbf{Y}_{d}^{k}(t)$ ***on*** $\mathfrak{T}_{0F}\times\mathfrak{D}^{i}\times\mathfrak{Y}_{d}^{k}$ *is the union of all stable FSRT tracking domains* $\mathcal{D}_{ST}^{k}\left(\varepsilon;t_{0};t_{R};t_{F};\mathbf{Y}_{d0}^{k};\mathbf{D};\mathbf{U};\mathbf{Y}_{d}\right)$ *in* $\varepsilon\in[0,\varepsilon_{M})$*,*

$$\begin{gathered}\mathcal{D}_{SST}^{k}\left(t_{0};t_{R};t_{F};\mathbf{Y}_{d0}^{k};\mathbf{D};\mathbf{U};\mathbf{Y}_{d}\right)\\ =\cup\left\{\begin{array}{c}\mathcal{D}_{ST}^{k}\left(\varepsilon;t_{0};t_{R};t_{F};\mathbf{Y}_{d0}^{k};\mathbf{D};\mathbf{U};\mathbf{Y}_{d}\right):\\ :\varepsilon\in[0,\varepsilon_{M})\end{array}\right\},\\ \forall\left[\mathbf{D}(.),\mathbf{Y}_{d}(.)\right]\in\mathfrak{D}^{i}\times\mathfrak{Y}_{d}^{k}.\end{gathered}\tag{26.15}$$

c) The stable FSRT tracking of $\mathbf{Y}_{d}^{k}(t)$ *on* $\mathfrak{T}_{0F}\times\mathfrak{D}^{i}\times\mathfrak{Y}_{d}^{k}$ *is* ***global (in the whole)*** *if, and only if, it is the global FSRT tracking of* $\mathbf{Y}_{d}^{k}(t)$ *on* $\mathfrak{T}_{0F}\times\mathfrak{D}^{i}\times\mathfrak{Y}_{d}^{k}$ *and the stable FSRT tracking of* $\mathbf{Y}_{d}^{k}(t)$ *on* $\mathfrak{T}_{0F}\times\mathfrak{D}^{i}\times\mathfrak{Y}_{d}^{k}$ *is with*

$$\mathcal{D}_{ST}^{k}\left(t_{0};t_{R};t_{F};\mathbf{Y}_{d0}^{k};\mathbf{D};\mathbf{U};\mathbf{Y}_{d}\right)=\mathfrak{R}^{(k+1)N}$$

for every $\left[\mathbf{D}(.),\mathbf{Y}_{d}(.)\right]\in\mathfrak{D}^{i}\times\mathfrak{Y}_{d}^{k}$*.*

Definition 401 ***The uniform stable FSRT tracking of the desired output behavior $\mathbf{Y}_d^k(t)$ of the plant (3.12) controlled by a control $\mathbf{U}(.) \in \mathfrak{U}^l$***

a) If, and only if, the intersection $\mathcal{D}_{STU}^k\left(t_0; t_R; t_F; \mathfrak{D}^i; \mathbf{U}; \mathfrak{Y}_d^k\right)$ of all

$$\mathcal{D}_{ST}^k\left(t_0; t_R; t_F; \mathbf{Y}_{d0}^k; \mathbf{D}; \mathbf{U}; \mathbf{Y}_d\right)$$

[$\mathcal{D}_{SSTU}^k\left(t_0; t_R; t_F; \mathfrak{D}^i; \mathbf{U}; \mathfrak{Y}_d^k\right)$ of all $\mathcal{D}_{SST}^k\left(t_0; t_R; t_F; \mathbf{Y}_{d0}^k; \mathbf{D}; \mathbf{U}; \mathbf{Y}_d\right)$] over $\mathfrak{D}^i \times \mathfrak{Y}_d^k$ is a connected neighborhood of $\mathbf{Y}_{d0}^k$ then, and only then, it is the ***(strictly) stable FSRT tracking domain of $\mathbf{Y}_d^k(t)$ on $\mathfrak{T}_{0F} \times \mathfrak{D}^i \times \mathfrak{Y}_d^k$ uniform in*** *$[\mathbf{D}(.), \mathbf{Y}_d(.)] \in \mathfrak{D}^i \times \mathfrak{Y}_d^k$, respectively,*

$$\begin{gathered} \exists \xi \in \mathfrak{R}^+ \Longrightarrow \mathcal{D}_{STU}^k\left(t_0; t_R; t_F; \mathfrak{D}^i; \mathbf{U}; \mathfrak{Y}_d^k\right) \\ = \cap\left\{\mathcal{D}_{ST}^k\left(t_0; t_R; t_F; \mathbf{Y}_{d0}^k; \mathbf{D}; \mathbf{U}; \mathbf{Y}_d\right) : [\mathbf{D}(.), \mathbf{Y}_d(.)] \in \mathfrak{D}^i \times \mathfrak{Y}_d^k\right\} \\ \supset \mathfrak{N}_\xi\left(t_0; \mathfrak{D}^i; \mathbf{U}; \mathfrak{Y}_d^k\right), \end{gathered} \tag{26.16}$$

$$\left[\begin{array}{c} \exists \xi \in \mathfrak{R}^+ \Longrightarrow \mathcal{D}_{SSTU}^k\left(t_0; t_R; t_F; \mathfrak{D}^i; \mathbf{U}; \mathfrak{Y}_d^k\right) \\ = \cap\left\{\begin{array}{c} \mathcal{D}_{SST}^k\left(t_0; t_R; t_F; \mathbf{Y}_{d0}^k; \mathbf{D}; \mathbf{U}; \mathbf{Y}_d\right) : \\ : [\mathbf{D}(.), \mathbf{Y}_d(.)] \in \mathfrak{D}^i \times \mathfrak{Y}_d^k \end{array}\right\} \\ \supset \mathfrak{N}_\xi\left(t_0; \mathfrak{D}^i; \mathbf{U}; \mathfrak{Y}_d^k\right) \end{array}\right]. \tag{26.17}$$

b) If, and only if, the intersection $\mathcal{D}_{STU}^k\left(\mathfrak{T}_i; t_R; t_F; \mathbf{Y}_{d0}^k; \mathbf{D}; \mathbf{U}; \mathbf{Y}_d\right)$ of all

$$\mathcal{D}_{ST}^k\left(t_0; t_R; t_F; \mathbf{Y}_{d0}^k; \mathbf{D}; \mathbf{U}; \mathbf{Y}_d\right)$$

[$\mathcal{D}_{SSTU}^k\left(\mathfrak{T}_i; t_R; t_F; \mathbf{Y}_{d0}^k; \mathbf{D}; \mathbf{U}; \mathbf{Y}_d\right)$ of all $\mathcal{D}_{SST}^k\left(t_0; t_R; t_F; \mathbf{Y}_{d0}^k; \mathbf{D}; \mathbf{U}; \mathbf{Y}_d\right)$] in t_0 over $\mathfrak{T}_i$ is a connected neighborhood of $\mathbf{Y}_{d0}^k$, then, and only then, it is ***the (strict) stable FSRT tracking domain of $\mathbf{Y}_d^k(t)$ on $\mathfrak{T}_{0F} \times \mathfrak{D}^i \times \mathfrak{Y}_d^k$ uniform in*** *$t_0 \in \mathfrak{T}$, respectively,*

$$\begin{gathered} \exists \xi \in \mathfrak{R}^+ \Longrightarrow \mathcal{D}_{STU}^k\left(\mathfrak{T}_i; t_R; t_F; \mathbf{Y}_{d0}^k; \mathbf{D}; \mathbf{U}; \mathbf{Y}_d\right) \\ = \cap\left\{\mathcal{D}_{ST}^k\left(t_0; t_R; t_F; \mathbf{Y}_{d0}^k; \mathbf{D}; \mathbf{U}; \mathbf{Y}_d\right) : t_0 \in \mathfrak{T}_i\right\} \\ \supset \mathfrak{N}_\xi\left(\mathfrak{T}_i; t_R; t_F; \mathbf{Y}_{d0}^k; \mathbf{D}; \mathbf{U}; \mathbf{Y}_d\right), \end{gathered} \tag{26.18}$$

$$\left[\begin{array}{c} \exists \xi \in \mathfrak{R}^+ \Longrightarrow \mathcal{D}_{SSTU}^k\left(\mathfrak{T}_i; t_R; t_F; \mathbf{Y}_{d0}^k; \mathbf{D}; \mathbf{U}; \mathbf{Y}_d\right) \\ = \cap\left\{\mathcal{D}_{SST}^k\left(t_0; t_R; t_F; \mathbf{Y}_{d0}^k; \mathbf{D}; \mathbf{U}; \mathbf{Y}_d\right) : t_0 \in \mathfrak{T}_i\right\} \\ \supset \mathfrak{N}_\xi\left(\mathfrak{T}_i; t_R; t_F; \mathbf{Y}_{d0}^k; \mathbf{D}; \mathbf{U}; \mathbf{Y}_d\right) \end{array}\right]. \tag{26.19}$$

c) If, and only if, the intersection $\mathcal{D}_{STU}^k\left(\mathfrak{T}_i; t_R; t_F; \mathfrak{D}^i; \mathbf{U}; \mathfrak{Y}_d^k\right)$ of all

$$\mathcal{D}_{ST}^k\left(t_0; t_R; t_F; \mathbf{Y}_{d0}^k; \mathbf{D}; \mathbf{U}; \mathbf{Y}_d\right)$$

[$\mathcal{D}_{SSTU}^k\left(\mathfrak{T}_i; t_R; t_F; \mathfrak{D}^i; \mathbf{U}; \mathfrak{Y}_d^k\right)$ of all $\mathcal{D}_{SST}^k\left(t_0; t_R; t_F; \mathbf{Y}_{d0}^k; \mathbf{D}; \mathbf{U}; \mathbf{Y}_d\right)$] in

$$[t_0, \mathbf{D}(.), \mathbf{Y}_d(.)] \in \mathfrak{T}_i \times \mathfrak{D}^i \times \mathfrak{Y}_d^k$$

is a connected neighborhood of every $\mathbf{Y}_{d0}^k \in \mathfrak{Y}_{d0}^{p-1}$ then, and only then, it is the ***(strictly) stable FSRT tracking domain of $\mathbf{Y}_d^k(t)$*** *on $\mathfrak{T} \times \mathfrak{D}^i \times \mathfrak{Y}_d^k$* ***uniform in*** *$[t_0, \mathbf{D}(.), \mathbf{Y}_d(.)] \in \mathfrak{T}_i \times \mathfrak{D}^i \times \mathfrak{Y}_d^k$, respectively,*

$$\begin{gathered} \exists \xi \in \mathfrak{R}^+ \Longrightarrow \mathcal{D}_{STU}^k\left(\mathfrak{T}_i; t_R; t_F; \mathfrak{D}^i; \mathbf{U}; \mathfrak{Y}_d^k\right) \\ = \cap\left\{\mathcal{D}_{ST}^k\left(t_0; t_R; t_F; \mathbf{Y}_{d0}^k; \mathbf{D}; \mathbf{U}; \mathbf{Y}_d\right) : [t_0, \mathbf{D}(.), \mathbf{Y}_d(.)] \in \mathfrak{T}_i \times \mathfrak{D}^i \times \mathfrak{Y}_d^k\right\} \\ \supset \mathfrak{N}_\xi\left(\mathfrak{T}_i; t_R; t_F; \mathfrak{D}^i; \mathbf{U}; \mathfrak{Y}_d^k\right), \end{gathered} \tag{26.20}$$

$$\left[\begin{array}{c} \exists \xi \in \mathfrak{R}^+ \Longrightarrow \mathcal{D}_{SSTU}^k\left(\mathfrak{T}_i; t_R; t_F; \mathfrak{D}^i; \mathbf{U}; \mathfrak{Y}_d^k\right) \\ = \cap\left\{\begin{array}{c} \mathcal{D}_{SST}^k\left(t_0; t_R; t_F; \mathbf{Y}_{d0}^k; \mathbf{D}; \mathbf{U}; \mathbf{Y}_d\right) : \\ : [t_0, \mathbf{D}(.), \mathbf{Y}_d(.)] \in \mathfrak{T}_i \times \mathfrak{D}^i \times \mathfrak{Y}_d^k \end{array}\right\} \\ \supset \mathfrak{N}_\xi\left(\mathfrak{T}_i; t_R; t_F; \mathfrak{D}^i; \mathbf{U}; \mathfrak{Y}_d^k\right). \end{array}\right] \tag{26.21}$$

26.2 Finite vector reachability *time*

We define **in the output space** various types of ***tracking with the finite vector reachability time (FVRT)***.

Elementwise tracking with the finite vector reachability time represents better tracking than the preceding tracking types. It allows different *FSRTs* to be associated, mutually independently, to different output variables.

We use *the elementwise unit* $(k+1)N$*-vector* $\mathbf{1}_{(k+1)N}$, all elements of which are equal to one,

$$\mathbf{1}_{(k+1)N} = \begin{bmatrix} \underbrace{1\ 1 ... 1}_{(k+1)-times} \end{bmatrix}^T \in \mathfrak{R}^{(k+1)N},\ k \in \{0, 1, 2, .., \alpha - 1\}, \tag{26.22}$$

Let us introduce *the* $(k+1)N$*-time vector* $\mathbf{t}^{(k+1)N}$ [274, p. 387], all elements of which are the same temporal variable, *time* t,

$$\mathbf{t}^{(k+1)N} = t\mathbf{1}_{(k+1)N} = [t\ t...t]^T \in \mathfrak{T}_0^{(k+1)N} \cup \{\infty\}^{(k+1)N},\ k \in \{0, 1, 2, ..., \alpha\},$$
$$\mathbf{t} = \mathbf{t}^N = t\mathbf{1}_N = [t\ t...t]^T \in \mathfrak{T}_0^N,$$
$$\mathbf{t}_0 = \mathbf{t}_0^N = t_0\mathbf{1}_N = [t_0\ t_0...t_0]^T \in \mathrm{In}\mathfrak{T}^N, \tag{26.23}$$

where

$$\mathfrak{T}_0^i = \underbrace{\mathfrak{T}_0 \times \mathfrak{T}_0 \times ... \times \mathfrak{T}_0}_{i-times}, \tag{26.24}$$

$$\mathrm{Cl}\mathfrak{T}_0^i = \underbrace{\mathrm{Cl}\mathfrak{T}_0 \times \mathrm{Cl}\mathfrak{T}_0 \times ... \times\ \mathrm{Cl}\mathfrak{T}_0}_{\text{i-times}},\ \mathrm{In}\mathfrak{T}_0^i = \underbrace{\mathrm{In}\mathfrak{T}_0 \times \mathrm{In}\mathfrak{T}_0 \times ... \times \mathrm{In}\mathfrak{T}_0}_{\text{i-times}}. \tag{26.25}$$

We can associate with every output variable Y_i its own scalar reachability *time* $t_{Ri} \in \mathrm{In}\mathfrak{T}_{0F}$, and with its derivatives $Y_i^{(1)}, Y_i^{(2)}, ..., Y_i^{(k)}$ their own scalar reachability *times* $t_{Ri(1)} \in \mathrm{In}\mathfrak{T}_{0F}$, $t_{Ri(2)} \in \mathrm{In}\mathfrak{T}_{0F}$, ..., $t_{Ri(k)} \in \mathrm{In}\mathfrak{T}_{0F}$, respectively. They compose the following *vector reachability time*:

$$\mathbf{t}_R^N = \mathbf{t}_{R(0)}^N = \begin{bmatrix} t_{R1} \\ t_{R2} \\ ... \\ t_{RN} \end{bmatrix} = \begin{bmatrix} t_{R1,(0)} \\ t_{R2,(0)} \\ ... \\ t_{RN,(0)} \end{bmatrix} \in \mathrm{In}\mathfrak{T}_{0F}^N, \tag{26.26}$$

where

$$\mathrm{Cl}\mathfrak{T}_{0F}^i = \underbrace{\mathrm{Cl}\mathfrak{T}_{0F} \times \mathrm{Cl}\mathfrak{T}_{0F} \times ... \times \mathrm{Cl}\mathfrak{T}_{0F}}_{\text{i-times}}, \tag{26.27}$$

$$\mathrm{In}\mathfrak{T}_{0F}^i = \underbrace{\mathrm{In}\mathfrak{T}_{0F} \times \mathrm{In}\mathfrak{T}_{0F} \times ... \times \mathrm{In}\mathfrak{T}_{0F}}_{\text{i-times}}. \tag{26.28}$$

The generalization of (26.26) to the j-th derivatives of the output variables reads

$$\mathbf{t}_{R(j)}^N = \begin{bmatrix} t_{R1,(j)} \\ t_{R2,(j)} \\ ... \\ t_{RN,(j)} \end{bmatrix} \in \mathrm{In}\mathfrak{T}_{0F}^N,\ j \in \{0, 1, 2, .., \alpha - 1\}. \tag{26.29}$$

In order to treat mathematically effectively and simply such cases, we define *the finite vector reachability time (FVRT)* $t_R^{(k+1)N} \in (\mathrm{In}\mathfrak{T}_{0F})^{(k+1)N}$, which is related to the output vector and its derivatives up to the order k:

$$\mathbf{t}_R^{(k+1)N} = \begin{bmatrix} \mathbf{t}_R^N \\ \mathbf{t}_{R(1)}^N \\ \mathbf{t}_{R(2)}^N \\ \cdots \\ \mathbf{t}_{R(k)}^N \end{bmatrix} = \begin{bmatrix} \mathbf{t}_{R(0)}^N \\ \mathbf{t}_{R(1)}^N \\ \mathbf{t}_{R(2)}^N \\ \cdots \\ \mathbf{t}_{R(k)}^N \end{bmatrix} \in \mathrm{In}\mathfrak{T}_{0F}^{(k+1)N}. \tag{26.30}$$

In the scalar form:

$$\mathbf{t}_R^{(k+1)N} = [t_{R1,(0)} \ldots t_{RN,(0)} \;\; t_{R1,(1)} \ldots t_{RN,(1)} \ldots t_{R1,(k)} \ldots t_{RN,(k)}]^T,$$
$$\forall k = 0,\ 1, ..., \alpha - 1, \tag{26.31}$$

where $t_{Ri} = t_{Ri(0)}$ is the reachability *time* of the i-th output variable, $i = 1, 2, \cdots, N$, and $t_{Ri,(j)}$ is the reachability *time* of the j-th derivative of the i-th output variable, $j = 0, 1, 2, \cdots, \alpha - 1$.

Let

$$\mathbf{Y}_{d0}^k(\mathbf{t}^{(k+1)N}) = \left[\mathbf{Y}^T(t) \;\; \mathbf{Y}^{(1)T}(t) \ldots \mathbf{Y}^{(k)T}(t)\right]^T \in \mathfrak{R}^{(k+1)N},$$

$$\mathbf{Y}_{d0}^k(\mathbf{t}^{(k+1)N}) = \left[\mathbf{Y}^T(\mathbf{t}_{R(0)}^N) \;\; \mathbf{Y}^{(1)T}(\mathbf{t}_{R(1)}^N) \ldots \mathbf{Y}^{(k)T}(\mathbf{t}_{R(k)}^N)\right]^T \in \mathfrak{R}^{(k+1)N},$$

$$\mathbf{Y}^{(j)T}(\mathbf{t}_{R(j)}^N) = \left[Y_1^{(j)T}(t_{R1,(j)}) \;\; Y_2^{(j)T}(t_{R2,(j)}) \ldots Y_1^{(j)T}(t_{RN,(j)})\right]^T \in \mathfrak{R}^N$$
$$j \in \{0, 1, 2, \cdots, k\}.$$

We can now summarize the above presentation about *the finite vector reachability time (FVRT)* $t^{(k+1)N} \in \mathrm{In}\mathfrak{T}_{0F}^{(k+1)N}$. It is the vector instant $t^{(k+1)N}$, since which the real output vector $\mathbf{Y}_{d0}^k(t^{(k+1)N})$ becomes elementwise equal to the desired output vector $\mathbf{Y}_d^{\alpha-1}(t^{(k+1)N})$ until the final vector instant $t^{(k+1)N}$:

$$\mathbf{Y}_{d0}^k(\mathbf{t}_R^{(k+1)N}) = \mathbf{Y}_d^k(\mathbf{t}_R^{(k+1)N}), \;\; \forall \mathbf{t}_R^{(k+1)N} \in \mathfrak{T}_{RF}^{(k+1)N} \tag{26.32}$$

We relate $t_R^{(k+1)N}$ to the tracking treated via *the extended output space* $\mathfrak{R}^{(k+1)N}$, which for $k = m - 1$ becomes *the state space* $\mathfrak{R}^{mN}$ if the plant is the *IO* plant. However, $\mathfrak{R}^{(k+1)N}$ becomes the ordinary output space $\mathfrak{R}^N$ for $k = \alpha - 1 = 0$ if the plant is the *ISO* plant because then $\alpha = 1$.

The above notation leads to

$$\mathfrak{T}^{(k+1)N} = \left\{\mathbf{t}^{(k+1)N} : \; -\infty\mathbf{1}_{(k+1)N} < \mathbf{t}_R^{(k+1)N} < \infty\mathbf{1}_{(k+1)N}\right\}, \tag{26.33}$$

$$\mathfrak{T}_R^{(k+1)N} = \left\{\mathbf{t}^{(k+1)N} : \; \mathbf{t}_0^{(k+1)N} \leq \mathbf{t}^{(k+1)N} \leq \mathbf{t}_R^{(k+1)N} < \infty\mathbf{1}_{(k+1)N}\right\},$$
$$\mathfrak{T}_{R\infty}^{(k+1)N} = \left\{\mathbf{t}^{(k+1)N} : \; \mathbf{t}_R^{(k+1)N} \leq \mathbf{t}^{(k+1)N} < \infty\mathbf{1}_{(k+1)N}\right\}, \tag{26.34}$$

and to

$$\mathfrak{T}_{RF}^{(k+1)N} = \left\{\mathbf{t}_R^{(k+1)N} : \; \left\langle \begin{array}{c} \mathbf{t}_R^{(k+1)N} \leq \mathbf{t}^{(k+1)N} \leq \mathbf{t}_F^{(k+1)N} < \infty\mathbf{1}_{(k+1)N}, \text{ or} \\ \mathbf{t}_R^{(k+1)N} \leq \mathbf{t}_R^{(k+1)N} < \mathbf{t}_F^{(k+1)N} = \infty\mathbf{1}_{(k+1)N} \end{array} \right\rangle\right\}. \tag{26.35}$$

The symbolic vector notation

$$\mathbf{Y}^k(\mathbf{t}_R^{(k+1)N}) = \mathbf{Y}_d^k(\mathbf{t}_R^{(k+1)N}), \; \forall \mathbf{t}^{(k+1)N} \in [\mathbf{t}_R^{(k+1)N}, \; \infty\mathbf{1}_{(k+1)N}[,$$
$$k \in \{0, 1, 2, ...\}$$

means in the scalar form

$$Y_i^{(j)}(t)=Y_{di}^{(j)}(t),\ \forall t\in[t_{Ri(j)},\ \infty[,\ \forall i=1,2,...,N,\ \forall j\in\{0,1,2,..,k\}.$$

Besides

$$\left|\mathbf{Y}^{(j)}(\mathbf{t}^N)-\mathbf{Y}_d^{(j)}(\mathbf{t}^N)\right|=\left|\begin{array}{c} Y_1^{(j)}(t)-Y_{d1}^{(j)}(t)\\ Y_2^{(j)}(t)-Y_{d2}^{(j)}(t)\\ \cdots\\ Y_N^{(j)}(t)-Y_{dN}^{(j)}(t)\end{array}\right|\in\mathfrak{R}_+^N,\ \forall j=0,1,2,...,k,$$

and

$$\left|\mathbf{Y}^k(\mathbf{t}_R^{(k+1)N})-\mathbf{Y}_d^k(\mathbf{t}_R^{(k+1)N})\right|=\left|\begin{array}{c}\mathbf{Y}(\mathbf{t}^N)-\mathbf{Y}_d^k(\mathbf{t}^N)\\ \mathbf{Y}^{(1)}(\mathbf{t}^N)-\mathbf{Y}_d^{(1)}(\mathbf{t}^N)\\ \cdots\\ \mathbf{Y}^{(k)}(\mathbf{t}^N)-\mathbf{Y}_d^{(k)}(\mathbf{t}^N)\end{array}\right|\in\mathfrak{R}_+^{(k+1)N},$$
$$k\in\{0,1,2,...,\alpha-1\}.$$

Let a positive real number $\varepsilon_{i(j)}$, or $\varepsilon_{i(j)}=\infty$, be associated with the j-th derivative of Y_i and of Y_{di}, and be taken for the entries of the positive N vector $\boldsymbol{\varepsilon}_{(j)}^N$, i.e., of the positive $(k+1)N-$ vector $\boldsymbol{\varepsilon}^{(k+1)N}$, respectively,

$$\boldsymbol{\varepsilon}_{(j)}^N=\left[\begin{array}{c}\varepsilon_{1,(j)}\\ \varepsilon_{2,(j)}\\ \cdots\\ \varepsilon_{N,(j)}\end{array}\right]\in\mathfrak{R}^{+^N}\cup\{\infty\}^N,\ \forall j=0,1,2,...,k,\ \varepsilon_{i,(0)}\equiv\varepsilon_i,\boldsymbol{\varepsilon}_{(0)}^N\equiv\boldsymbol{\varepsilon}^N,\tag{26.36}$$

$$\boldsymbol{\varepsilon}^{(k+1)N}=\left[\begin{array}{c}\boldsymbol{\varepsilon}_{(0)}^N\\ \boldsymbol{\varepsilon}_{(1)}^N\\ \cdots\\ \boldsymbol{\varepsilon}_{(k)}^N\end{array}\right]=\left[\begin{array}{c}\boldsymbol{\varepsilon}^N\\ \boldsymbol{\varepsilon}_{(1)}^N\\ \cdots\\ \boldsymbol{\varepsilon}_{(k)}^N\end{array}\right]\in\mathfrak{R}^{+^{(k+1)N}}\cup\{\infty\}^{(k+1)N},\ k\in\{1,2,..,\alpha-1\},\tag{26.37}$$

so that

$$\left|\mathbf{Y}_0^k-\mathbf{Y}_{d0}^k\right|<\boldsymbol{\varepsilon}^{(k+1)N},\ \forall k=0,1,2,...,\alpha-1,\tag{26.38}$$

signifies that the relationship holds element by element, i.e., **elementwise**,

$$\left|Y_{i0}^{(j)}-Y_{di0}^{(j)}\right|<\varepsilon_{i,(j)},\ \forall i=1,2,...,N,\ \forall j=0,1,2,...,\ k.\tag{26.39}$$

We use the above simplified notation in the sequel,

$$\begin{array}{c}\mathbf{Y}_d^k(\mathbf{t}_R^{(k+1)N})\equiv\mathbf{Y}_d^k(\mathbf{t}_R^{(k+1)N};\mathbf{t}_0^{(k+1)N};\mathbf{Y}_{d0}^k),\\ \mathbf{Y}^k(\mathbf{t}_R^{(k+1)N})\equiv\mathbf{Y}^k(\mathbf{t}_R^{(k+1)N};\mathbf{t}_0^{(k+1)N};\mathbf{Y}_0^k;\mathbf{D};\mathbf{U}),\end{array}\tag{26.40}$$

The following definition generalizes Definition 400.

Definition 402 ***The elementwise tracking with the finite vector reachability time $t_R^{(k+1)N}$ of the desired output $\mathbf{Y}_d^k(t_R^{(k+1)N})$ of the plant (3.12) controlled by a control $\mathbf{U}(.)\in\mathfrak{U}^l$***

a) The plant (3.12) exhibits ***the elementwise output tracking of the desired output response $\mathbf{Y}_d^k(t_R^{(k+1)N})$ with the finite vector reachability time (FVRT) $t_R^{(k+1)N}$ on $\mathfrak{T}_{0F}^{(k+1)N}\times\mathfrak{D}^i\times\mathfrak{Y}_d^k$;*** *i.e.,* ***the elementwise tracking of $\mathbf{Y}_d^k(t_R^{(k+1)N})$ with FVRT $t_R^{(k+1)N}$***

on $\mathfrak{T}_{0F}^{(k+1)N} \times \mathfrak{D}^i \times \mathfrak{Y}_d^k$ *if, and only if, for every* $\left[\mathbf{D}(.), \mathbf{Y}_d^k(.)\right] \in \mathfrak{D}^i \times \mathfrak{Y}_d^k$ *there exists a connected neighborhood* $\mathfrak{N}\left(\mathbf{t}_0^{(k+1)N}; \mathbf{t}_R^{(k+1)N}; \mathbf{t}_F^{(k+1)N}; \mathbf{Y}_{d0}^k; \mathbf{D}; \mathbf{U}; \mathbf{Y}_d^k\right)$, *which is dependent on the triplet* $\left(\mathbf{t}_0^{(k+1)N}, \mathbf{t}_R^{(k+1)N}, \mathbf{t}_F^{(k+1)N}\right)$,

$$\mathfrak{N}\left(\mathbf{t}_0^{(k+1)N}; \mathbf{t}_R^{(k+1)N}; \mathbf{t}_F^{(k+1)N}; \mathbf{Y}_{d0}^k; \mathbf{D}; \mathbf{U}; \mathbf{Y}_d^k\right) \subseteq \mathfrak{R}^{(k+1)N},$$

of the plant desired initial output vector $\mathbf{Y}_{d0}^k$ *at the initial vector moment* $t_R^{(k+1)N} \in \mathfrak{T}_{0F}^{(k+1)N}$ *such that* $\mathbf{Y}_0^k$ *from the neighborhood*

$$\mathfrak{N}\left(\mathbf{t}_0^{(k+1)N}; \mathbf{t}_R^{(k+1)N}; \mathbf{t}_F^{(k+1)N}; \mathbf{Y}_{d0}^k; \mathbf{D}; \mathbf{U}; \mathbf{Y}_d^k\right)$$

guarantees both that $\mathbf{Y}^k(t^{(k+1)N})$ *becomes equal to* $\mathbf{Y}_d^k(t^{(k+1)N})$ *at the finite vector reachability moment* $t_R^{(k+1)N} \in In\mathfrak{T}_{0F}^{(k+1)N}$, *and that they stay equal from then on* $\mathfrak{T}_{RF}^{(k+1)N}$; *i.e.,*

$$\begin{gathered}
\forall \left[\mathbf{D}(.), \mathbf{Y}_d^k(.)\right] \in \mathfrak{D}^i \times \mathfrak{Y}_d^k, \\
\exists \mathfrak{N}\left(\mathbf{t}_0^{(k+1)N}; \mathbf{t}_R^{(k+1)N}; \mathbf{t}_F^{(k+1)N}; \mathbf{Y}_{d0}^k; \mathbf{D}; \mathbf{U}; \mathbf{Y}_d^k\right), \\
\mathfrak{N}\left(\mathbf{t}_0^{(k+1)N}; \mathbf{t}_R^{(k+1)N}; \mathbf{t}_F^{(k+1)N}; \mathbf{Y}_{d0}^k; \mathbf{D}; \mathbf{U}; \mathbf{Y}_d^k\right) \subseteq \mathfrak{R}^{(k+1)N}, \\
\mathbf{Y}_0^k \in \mathfrak{N}\left(\mathbf{t}_0^{(k+1)N}; \mathbf{t}_R^{(k+1)N}; \mathbf{t}_F^{(k+1)N}; \mathbf{Y}_{d0}^k; \mathbf{D}; \mathbf{U}; \mathbf{Y}_d^k\right) \Longrightarrow \\
\mathbf{Y}^k(\mathbf{t}^{(k+1)N}) = \mathbf{Y}_d^k(\mathbf{t}^{(k+1)N}), \ \forall \mathbf{t}^{(k+1)N} \in \mathfrak{T}_{RF}^{(k+1)N}.
\end{gathered} \tag{26.41}$$

b) The largest connected neighborhood

$$\mathfrak{N}_L\left(\mathbf{t}_0^{(k+1)N}; \mathbf{t}_R^{(k+1)N}; \mathbf{t}_F^{(k+1)N}; \mathbf{Y}_{d0}^k; \mathbf{D}; \mathbf{U}; \mathbf{Y}_d^k\right)$$

of $\mathbf{Y}_{d0}^k$, *which obeys (26.41), is* ***the FVRT elementwise tracking domain***

$$\mathcal{D}_T^k\left(\mathbf{t}_0^{(k+1)N}; \mathbf{t}_R^{(k+1)N}; \mathbf{t}_F^{(k+1)N}; \mathbf{Y}_{d0}^k; \mathbf{D}; \mathbf{U}; \mathbf{Y}_d^k\right)$$

of $\mathbf{Y}_d^k(t^{(k+1)N})$ ***on*** $\mathfrak{T}_{0F}^{(k+1)N} \times \mathfrak{D}^i \times \mathfrak{Y}_d^k$.

c) FVRT elementwise tracking of $\mathbf{Y}_d^k(t^{(k+1)N})$ *on* $\mathfrak{T}_{0F}^{(k+1)N} \times \mathfrak{D}^i \times \mathfrak{Y}_d^k$ *is* ***global (in the whole)*** *if, and only if,*

$$\mathcal{D}_T^k\left(\mathbf{t}_0^{(k+1)N}; \mathbf{t}_R^{(k+1)N}; \mathbf{t}_F^{(k+1)N}; \mathbf{Y}_{d0}^k; \mathbf{D}; \mathbf{U}; \mathbf{Y}_d^k\right) = \mathfrak{R}^{(k+1)N}$$

for every $\left[\mathbf{D}(.), \mathbf{Y}_d^k(.)\right] \in \mathfrak{D}^i \times \mathfrak{Y}_d^k$. *FVRT elementwise tracking of* $\mathbf{Y}_d^k(t_R^{(k+1)N})$ *on* $\mathfrak{T}_{0F}^{(k+1)N} \times \mathfrak{D}^i \times \mathfrak{Y}_d^k$ *is then uniform in* $\left[\mathbf{D}(.), \mathbf{Y}_d^k(.)\right] \in \mathfrak{D}^i \times \mathfrak{Y}_d^k$.

We present the vector generalization of Definition 398.

Definition 403 ***The uniform elementwise tracking with the finite vector reachability time of the desired output behavior*** $\mathbf{Y}_d^k(t_R^{(k+1)N})$ ***on*** $\mathfrak{T}_{0F}^{(k+1)N} \times \mathfrak{D}^i \times \mathfrak{Y}_d^k$ ***of the plant (3.12) controlled by a control*** $\mathbf{U}(.) \in \mathfrak{U}^l$

a) If, and only if, the intersection

$$\mathcal{D}_{TU}^k\left(\mathbf{t}_0^{(k+1)N}; \mathbf{t}_R^{(k+1)N}; \mathbf{t}_F^{(k+1)N}; \mathfrak{D}^i; \mathbf{U}; \mathfrak{Y}_d^k\right)$$

of the elementwise tracking domains

$$\mathcal{D}_T^k\left(\mathbf{t}_0^{(k+1)N};\mathbf{t}_R^{(k+1)N};\mathbf{t}_F^{(k+1)N};\mathbf{Y}_{d0}^k;\mathbf{D};\mathbf{U};\mathbf{Y}_d^k\right)$$

in $\left[\mathbf{D}(.),\mathbf{Y}_d^k(.)\right]\in\mathfrak{D}^i\times\mathfrak{Y}_d^k$ *is a connected neighborhood of* $\mathbf{Y}_{d0}^k$ *of every* $\mathbf{Y}_d^k(.)\in\mathfrak{Y}_d^k$,

$$\begin{gathered}\exists\xi\in\mathfrak{R}^+\Longrightarrow\mathcal{D}_{TU}^k\left(\mathbf{t}_0^{(k+1)N};\mathbf{t}_R^{(k+1)N};\mathbf{t}_F^{(k+1)N};\mathfrak{D}^i;\mathbf{U};\mathfrak{Y}_d^k\right)\\ =\cap\left[\begin{array}{c}\mathcal{D}_T^k\left(\mathbf{t}_0^{(k+1)N};\mathbf{t}_R^{(k+1)N};\mathbf{t}_F^{(k+1)N};\mathbf{Y}_{d0}^k;\mathbf{D};\mathbf{U};\mathbf{Y}_d^k\right):\\ \left[\mathbf{D}(.),\mathbf{Y}_d^k(.)\right]\in\mathfrak{D}^i\times\mathfrak{Y}_d^k\end{array}\right]\\ \supset\mathfrak{N}_\xi\left(\mathbf{t}_0^{(k+1)N};\mathbf{t}_R^{(k+1)N};\mathbf{t}_F^{(k+1)N};\mathfrak{D}^i;\mathbf{U};\mathfrak{Y}_d^k\right),\\ \partial\mathcal{D}_{TU}^k\left(\mathbf{t}_0^{(k+1)N};\mathbf{t}_R^{(k+1)N};\mathbf{t}_F^{(k+1)N};\mathfrak{D}^i;\mathbf{U};\mathfrak{Y}_d^k\right)\\ \cap\partial\mathfrak{N}_\xi\left(\mathbf{t}_0^{(k+1)N};\mathbf{t}_R^{(k+1)N};\mathbf{t}_F^{(k+1)N};\mathfrak{D}^i;\mathbf{U};\mathfrak{Y}_d^k\right)=\phi,\end{gathered}\tag{26.42}$$

then FVRT elementwise tracking of $\mathbf{Y}_d^k(t_R^{(k+1)N})$ *is* ***uniform in*** $\left[\mathbf{D}(.),\mathbf{Y}_d^k(.)\right]\in\mathfrak{D}^i\times\mathfrak{Y}_d^k$ *on* $\mathfrak{T}_{0F}^{(k+1)N}\times\mathfrak{D}^i\times\mathfrak{Y}_d^k$ *and the set*

$$\mathcal{D}_{TU}^k\left(\mathbf{t}_0^{(k+1)N};\mathbf{t}_R^{(k+1)N};\mathbf{t}_F^{(k+1)N};\mathfrak{D}^i;\mathbf{U};\mathfrak{Y}_d^k\right)$$

is the ***$(\mathbf{D},\mathbf{Y}_d^k)$-uniform FVRT elementwise tracking domain of*** $\mathbf{Y}_d^k(\mathbf{t}_0^{(k+1)N})$ ***on*** $\mathfrak{T}_{0F}^{(k+1)N}\times\mathfrak{D}^i\times\mathfrak{Y}_d^k$.

b) If, and only if, the intersection

$$\mathcal{D}_{TU}^k\left(\mathfrak{T}_{0F}^{(k+1)N};\mathbf{t}_R^{(k+1)N};\mathbf{t}_F^{(k+1)N};\mathbf{Y}_{d0}^k;\mathbf{D};\mathbf{U};\mathbf{Y}_d^k\right)$$

of all the FVRT elementwise tracking domains

$$\mathcal{D}_T^k\left(\mathbf{t}_0^{(k+1)N};\mathbf{t}_R^{(k+1)N};\mathbf{t}_F^{(k+1)N};\mathbf{Y}_{d0}^k;\mathbf{D};\mathbf{U};\mathbf{Y}_d^k\right)$$

in $t_0^{(k+1)N}$ *over* $\mathfrak{T}_{0F}^{(k+1)N}$ *is a connected neighborhood of* $\mathbf{Y}_{d0}^k$ *then the FVRT elementwise tracking of* $\mathbf{Y}_d^k(t_R^{(k+1)N})$ *on* $\mathfrak{T}_{0F}^{(k+1)N}\times\mathfrak{D}^i\times\mathfrak{Y}_d^k$ *is* ***uniform in*** $t_0^{(k+1)N}$ ***on*** $\mathfrak{T}_0^{(k+1)N}$ *and* $\mathcal{D}_{TU}^k\left(\mathfrak{T}^{(k+1)N};\mathbf{t}_R^{(k+1)N};\mathbf{t}_F^{(k+1)N};\mathbf{Y}_{d0}^k;\mathbf{D};\mathbf{U};\mathbf{Y}_d^k\right)$ *is* ***the domain of*** $t_0^{(k+1)N}$***-uniform FVRT elementwise tracking of*** $\mathbf{Y}_d^k(t^{(k+1)N})$ ***on*** $\mathfrak{T}_0^{(k+1)N}\times\mathfrak{D}^i\times\mathfrak{Y}_d^k$,

$$\begin{gathered}\exists\xi\in\mathfrak{R}^+\Longrightarrow\mathcal{D}_{TU}^k\left(\mathfrak{T}^{(k+1)N};\mathbf{t}_R^{(k+1)N};\mathbf{t}_F^{(k+1)N};\mathbf{Y}_{d0}^k;\mathbf{D};\mathbf{U};\mathbf{Y}_d^k\right)\\ =\cap\left[\mathcal{D}_T^k\left(\mathbf{t}_0^{(k+1)N};\mathbf{t}_R^{(k+1)N};\mathbf{t}_F^{(k+1)N};\mathbf{Y}_{d0}^k;\mathbf{D};\mathbf{U};\mathbf{Y}_d^k\right):\mathbf{t}_0^{(k+1)N}\in\mathfrak{T}^{(k+1)N}\right]\\ \supset\mathfrak{N}_\xi\left(\mathfrak{T}^{(k+1)N};\mathbf{t}_R^{(k+1)N}\mathbf{t}_F^{(k+1)N};\mathbf{Y}_{d0}^k;\mathbf{D};\mathbf{U};\mathbf{Y}_d^k\right),\\ \partial\mathcal{D}_{TU}^k\left(\mathfrak{T}^{(k+1)N};\mathbf{t}_R^{(k+1)N};\mathbf{t}_F^{(k+1)N};\mathbf{Y}_{d0}^k;\mathbf{D};\mathbf{U};\mathbf{Y}_d^k\right)\\ \cap\,\partial\mathfrak{N}_\xi\left(\mathfrak{T}^{(k+1)N};\mathbf{t}_R^{(k+1)N};\mathbf{t}_F^{(k+1)N};\mathbf{Y}_{d0}^k;\mathbf{D};\mathbf{U};\mathbf{Y}_d^k\right)=\phi.\end{gathered}\tag{26.43}$$

c) If, and only if, the intersection $\mathcal{D}_T^k\left(\mathfrak{T}^{(k+1)N};\mathbf{t}_R^{(k+1)N};\mathbf{t}_F^{(k+1)N};\mathfrak{D}^i;\mathbf{U};\mathfrak{Y}_d^k\right)$ *of all FVRT elementwise tracking domains*

$$\mathcal{D}_T^k\left(\mathbf{t}_0^{(k+1)N};\mathbf{t}_R^{(k+1)N};\mathbf{t}_F^{(k+1)N};\mathbf{Y}_{d0}^k;\mathbf{D};\mathbf{U};\mathbf{Y}_d^k\right)$$

in $\left[\mathbf{t}_0^{(k+1)N},\mathbf{D}(.),\mathbf{Y}_d^k(.)\right] \in \mathfrak{T}^{(k+1)N}\times\mathfrak{D}^i \times \mathfrak{Y}_d^k$ *is a neighborhood of* $\mathbf{Y}_{d0}^k$ *of every* $\mathbf{Y}_d^k(.) \in \mathfrak{Y}_d^k$ *then the FVRT elementwise tracking of* $\mathbf{Y}_d^k(t_R^{(k+1)N})$ *on* $\mathfrak{T}_0^{(k+1)N} \times \mathfrak{D}^i \times \mathfrak{Y}_d^k$ *is* ***uniform in*** $\left[\mathbf{t}_R^{(k+1)N},\mathbf{D}(.),\mathbf{Y}_d^k(.)\right] \in \mathfrak{T}_{0F}^{(k+1)N} \times \mathfrak{D}^i \times \mathfrak{Y}_d^k$; *i.e.,*

$$\begin{gathered}
\exists \xi \in \mathfrak{R}^+ \Longrightarrow \mathcal{D}_T^k\left(\mathfrak{T}^{(k+1)N};\mathbf{t}_R^{(k+1)N};\mathbf{t}_F^{(k+1)N};\mathfrak{D}^i;\mathbf{U};\mathfrak{Y}_d^k\right) \\
= \cap \left[\begin{array}{c} \mathcal{D}_T^k\left(\mathbf{t}_0^{(k+1)N};\mathbf{t}_R^{(k+1)N};\mathbf{t}_F^{(k+1)N};\mathbf{Y}_{d0}^k;\mathbf{D};\mathbf{U};\mathbf{Y}_d^k\right): \\ \left[\mathbf{t}_0^{(k+1)N},\mathbf{D}(.),\mathbf{Y}_d^k(.)\right] \in \mathfrak{T}^{(k+1)N}\times\mathfrak{D}^i \times \mathfrak{Y}_d^k \end{array}\right] \\
\supset \mathfrak{N}_\xi\left(\mathfrak{T}_{0F}^{(k+1)N};\mathbf{t}_R^{(k+1)N};\mathbf{t}_F^{(k+1)N};\mathfrak{D}^i;\mathbf{U};\mathfrak{Y}_d^k\right), \\
\partial\mathcal{D}_T^k\left(\mathfrak{T}^{(k+1)N};\mathbf{t}_R^{(k+1)N};\mathbf{t}_F^{(k+1)N};\mathfrak{D}^i;\mathbf{U};\mathfrak{Y}_d^k\right) \\
\cap\partial\mathfrak{N}_\xi\left(\mathfrak{T}^{(k+1)N};\mathbf{t}_R^{(k+1)N};\mathbf{t}_F^{(k+1)N};\mathfrak{D}^i;\mathbf{U};\mathfrak{Y}_d^k\right) = \phi.
\end{gathered} \tag{26.44}$$

Then, and only then, the set $\mathcal{D}_T^k\left(\mathfrak{T}^{(k+1)N};\mathbf{t}_R^{(k+1)N};\mathbf{t}_F^{(k+1)N};\mathfrak{D}^i;\mathbf{U};\mathfrak{Y}_d^k\right)$ *is the* ***the domain of FVRT uniform elementwise tracking of*** $\mathbf{Y}_d^k(t^{(k+1)N})$, *for every* $\mathbf{Y}_d^k(.) \in \mathfrak{Y}_d^k$, *on* $\mathfrak{T}_{0F}^{(k+1)N} \times \mathfrak{D}^i \times \mathfrak{Y}_d^k$,

Note 404 *The global FVRT elementwise tracking of* $\mathbf{Y}_d^k(t^{(k+1)N})$ *on the product set* $\mathfrak{T}_{0F}^{(k+1)N}$ $\times\mathfrak{D}^i\times \mathfrak{Y}_d^k$ *is uniform in* $\left[\mathbf{D}(.),\mathbf{Y}_d^k(.)\right]$ *over* $\mathfrak{D}^i \times \mathfrak{Y}_d^k$.

In order to assure a stability property of the tracking with FVRT we introduce:

Definition 405 ***The stable elementwise tracking with the finite vector reachability time*** $t_R^{(k+1)N}$ ***of the desired output*** $\mathbf{Y}_d^k(t^{(k+1)N})$ ***of the plant (3.12) controlled by a control*** $\mathbf{U}(.) \in \mathfrak{U}^l$

a) The plant (3.12) exhibits ***the stable elementwise tracking of the desired output*** $\mathbf{Y}_d^k(t^{(k+1)N})$ ***with the finite vector reachability time*** $t_R^{(k+1)N}$ ***on*** $\mathfrak{T}_{0F}^{(k+1)N} \times \mathfrak{D}^i \times \mathfrak{Y}_d^k$, *i.e.,* ***the stable elementwise tracking of*** $\mathbf{Y}_d^k(t^{(k+1)N})$ ***with the finite vector reachability time*** $t_R^{(k+1)N}$ ***on*** $\mathfrak{T}_{0F}^{(k+1)N} \times \mathfrak{D}^i \times \mathfrak{Y}_d^k$ *if, and only if, it exhibits the elementwise tracking with the finite vector reachability time* $t_R^{(k+1)N}$ *on* $\mathfrak{T}_{0F}^{(k+1)N} \times \mathfrak{D}^i \times \mathfrak{Y}_d^k$, *and for every connected neighborhood* $\mathfrak{N}_\varepsilon\left[\mathbf{t}^{(k+1)N};\mathbf{Y}_d^k(\mathbf{t}_R^{(k+1)N})\right]$ *of* $\mathbf{Y}_d^k(t^{(k+1)N})$ *at any* $t^{(k+1)N} \in \mathfrak{T}_{0F}^{(k+1)N}$, *there is a connected neighborhood*

$$\mathfrak{N}\left(\varepsilon;\mathbf{t}_0^{(k+1)N};\mathbf{t}_R^{(k+1)N};\mathbf{t}_F^{(k+1)N};\mathbf{Y}_{d0}^k;\mathbf{D};\mathbf{U};\mathbf{Y}_d^k\right), \textit{(16.1)},$$

of the plant desired initial output vector $\mathbf{Y}_{d0}^k$ *at the initial vector moment* $t_0^{(k+1)N} \in \mathfrak{T}^{(k+1)N}$ *such that for the initial vector*

$$\mathbf{Y}_0^k \in \mathfrak{N}\left(\varepsilon;\mathbf{t}_0^{(k+1)N};\mathbf{t}_R^{(k+1)N};\mathbf{t}_F^{(k+1)N};\mathbf{Y}_{d0}^k;\mathbf{D};\mathbf{U};\mathbf{Y}_d^k\right)$$

the instantaneous $\mathbf{Y}^k(t_R^{(k+1)N})$ *stays in the neighborhood*

$$\mathfrak{N}_\varepsilon\left[\mathbf{t}^{(k+1)N};\mathbf{Y}_d^k(\mathbf{t}_R^{(k+1)N})\right]$$

for all $t^{(k+1)N} \in \mathfrak{T}_{0F}^{(k+1)N}$*; i.e.,*

$$\begin{gathered}\forall \mathfrak{N}_{\varepsilon}\left[\mathbf{t}^{(k+1)N}; \mathbf{Y}_d^k(\mathbf{t}_R^{(k+1)N})\right] \subseteq \mathfrak{R}^{(k+1)N},\ \forall \mathbf{t}^{(k+1)N} \in \mathfrak{T}_{0F}^{(k+1)N}, \\ \forall\left[\mathbf{D}(.), \mathbf{Y}_d^k(.)\right] \in \mathfrak{D}^i \times \mathfrak{Y}_d^k, \\ \exists \mathfrak{N}\left(\varepsilon; \mathbf{t}_0^{(k+1)N}; \mathbf{t}_R^{(k+1)N}; \mathbf{t}_F^{(k+1)N}; \mathbf{Y}_{d0}^k; \mathbf{D}; \mathbf{U}; \mathbf{Y}_d^k\right) \subseteq \mathfrak{R}^{(k+1)N}, \\ \mathfrak{N}\left(\varepsilon; \mathbf{t}_0^{(k+1)N}; \mathbf{t}_R^{(k+1)N}; \mathbf{t}_F^{(k+1)N}; \mathbf{Y}_{d0}^k; \mathbf{D}; \mathbf{U}; \mathbf{Y}_d^k\right) \\ \subseteq \mathcal{D}_T^k\left(\mathbf{t}_0^{(k+1)N}; \mathbf{t}_R^{(k+1)N}; \mathbf{t}_F^{(k+1)N}; \mathbf{Y}_{d0}^k; \mathbf{D}; \mathbf{U}; \mathbf{Y}_d^k\right), \\ \mathbf{Y}_0^k \in \mathfrak{N}\left(\varepsilon; \mathbf{t}_0^{(k+1)N}; \mathbf{t}_R^{(k+1)N}; \mathbf{t}_F^{(k+1)N}; \mathbf{Y}_{d0}^k; \mathbf{D}; \mathbf{U}; \mathbf{Y}_d^k\right) \Longrightarrow \\ \mathbf{Y}^k(\mathbf{t}^{(k+1)N}) \in \mathfrak{N}_{\varepsilon}\left[\mathbf{t}^{(k+1)N}; \mathbf{Y}_d^k(\mathbf{t}_R^{(k+1)N})\right],\ \forall \mathbf{t}^{(k+1)N} \in \mathfrak{T}_{0F}^{(k+1)N}. \end{gathered} \tag{26.45}$$

b) The largest connected neighborhood

$$\mathfrak{N}_L\left(\varepsilon; \mathbf{t}_0^{(k+1)N}; \mathbf{t}_R^{(k+1)N}; \mathbf{t}_F^{(k+1)N}; \mathbf{Y}_{d0}^k; \mathbf{D}; \mathbf{U}; \mathbf{Y}_d^k\right)$$

of $\mathbf{Y}_{d0}^k$*, (26.13), is* ***the elementwise*** ε***-tracking domain***

$$\mathcal{D}_{ST}^k\left(\varepsilon; \mathbf{t}_0^{(k+1)N}; \mathbf{t}_R^{(k+1)N}; \mathbf{t}_F^{(k+1)N}; \mathbf{Y}_{d0}^k; \mathbf{D}; \mathbf{U}; \mathbf{Y}_d^k\right)$$

at $t_0^{(k+1)N} \in \mathfrak{T}^{(k+1)N}$ ***of the stable elementwise tracking of*** $\mathbf{Y}_d^k(t_R^{(k+1)N})$ ***on*** $\mathfrak{T}_{0F}^{(k+1)N} \times \mathfrak{D}^i \times \times \mathfrak{Y}_d^k$.

The domain $\mathcal{D}_{ST}^k\left(\mathbf{t}_0^{(k+1)N}; \mathbf{t}_R^{(k+1)N}; \mathbf{t}_F^{(k+1)N}; \mathbf{Y}_{d0}^k; \mathbf{D}; \mathbf{U}; \mathbf{Y}_d^k\right)$ ***of the stable elementwise tracking of*** $\mathbf{Y}_d^k(t^{(k+1)N})$ ***on*** $\mathfrak{T}_{0F}^{(k+1)N} \times \mathfrak{D}^i \times \mathfrak{Y}_d^k$ *is the union of all*

$$\mathcal{D}_{ST}^k\left(\varepsilon; \mathbf{t}_0^{(k+1)N}; \mathbf{t}_R^{(k+1)N}; \mathbf{t}_F^{(k+1)N}; \mathbf{Y}_{d0}^k; \mathbf{D}; \mathbf{U}; \mathbf{Y}_d^k\right)$$

over $\varepsilon \in \mathfrak{R}^+$*,*

$$\begin{gathered}\mathcal{D}_{ST}^k\left(\mathbf{t}_0^{(k+1)N}; \mathbf{t}_R^{(k+1)N}; \mathbf{t}_F^{(k+1)N}; \mathbf{Y}_{d0}^k; \mathbf{D}; \mathbf{U}; \mathbf{Y}_d^k\right) \\ = \cup\left[\mathcal{D}_{ST}^k\left(\varepsilon; \mathbf{t}_0^{(k+1)N}; \mathbf{t}_R^{(k+1)N}; \mathbf{t}_F^{(k+1)N}; \mathbf{Y}_{d0}^k; \mathbf{D}; \mathbf{U}; \mathbf{Y}_d^k\right) : \varepsilon \in \mathfrak{R}^+\right]. \end{gathered} \tag{26.46}$$

Let $[0,\varepsilon_M)$ *be the maximal interval over which*

$$\mathcal{D}_{ST}^k\left(\varepsilon; \mathbf{t}_0^{(k+1)N}; \mathbf{t}_R^{(k+1)N}; \mathbf{t}_F^{(k+1)N}; \mathbf{Y}_{d0}^k; \mathbf{D}; \mathbf{U}; \mathbf{Y}_d^k\right)$$

is continuous in $\varepsilon \in \mathfrak{R}_+$*,*

$$\begin{gathered}\mathcal{D}_{ST}^k\left(\varepsilon; \mathbf{t}_0^{(k+1)N}; \mathbf{t}_R^{(k+1)N}; \mathbf{t}_F^{(k+1)N}; \mathbf{Y}_{d0}^k; \mathbf{D}; \mathbf{U}; \mathbf{Y}_d^k\right) \in \mathfrak{C}\left([0, \varepsilon_M)\right), \\ \forall\left[\mathbf{D}(.), \mathbf{Y}_d^k(.)\right] \in \mathfrak{D}^i \times \mathfrak{Y}_d^k. \end{gathered}$$

The strict domain $\mathcal{D}_{SST}^k\left(\mathbf{t}_0^{(k+1)N}; \mathbf{t}_R^{(k+1)N}; \mathbf{t}_F^{(k+1)N}; \mathbf{Y}_{d0}^k; \mathbf{D}; \mathbf{U}; \mathbf{Y}_d^k\right)$ ***of the stable tracking of*** $\mathbf{Y}_d^k(t_R^{(k+1)N})$ ***on*** $\mathfrak{T}_{0F}^{(k+1)N} \times \mathfrak{D}^i \times \mathfrak{Y}_d^k$ *is the union of all stable tracking domains*

$\mathcal{D}_{ST}^{k}\left(\varepsilon;\mathbf{t}_0^{(k+1)N};\mathbf{t}_R^{(k+1)N};\mathbf{t}_F^{(k+1)N};\mathbf{Y}_{d0}^{k};\mathbf{D};\mathbf{U};\mathbf{Y}_d^{k}\right)$ *over* $\varepsilon \in [0,\varepsilon_M)$,

$$\begin{gathered}\mathcal{D}_{SST}^{k}\left(\mathbf{t}_0^{(k+1)N};\mathbf{t}_R^{(k+1)N};\mathbf{t}_F^{(k+1)N};\mathbf{Y}_{d0}^{k};\mathbf{D};\mathbf{U};\mathbf{Y}_d^{k}\right)\\ =\cup\left\{\mathcal{D}_{ST}^{k}\left(\varepsilon;\mathbf{t}_0^{(k+1)N};\mathbf{t}_R^{(k+1)N};\mathbf{t}_F^{(k+1)N};\mathbf{Y}_{d0}^{k};\mathbf{D};\mathbf{U};\mathbf{Y}_d^{k}\right):\varepsilon\in[0,\varepsilon_M)\right\},\\ \forall\left[\mathbf{D}(.),\mathbf{Y}_d^{k}(.)\right]\in\mathfrak{D}^i\times\mathfrak{Y}_d^{k}.\end{gathered}\tag{26.47}$$

c) The stable elementwise tracking of $\mathbf{Y}_d^k(t^{(k+1)N})$ *on* $\mathfrak{T}_{0F}^{(k+1)N}\times\mathfrak{D}^i\times\mathfrak{Y}_d^k$ *is* ***global (in the whole)*** *if, and only if, it is the FVRT global tracking of* $\mathbf{Y}_d^k(t^{(k+1)N})$ *on* $\mathfrak{T}_{0F}^{(k+1)N}\times\mathfrak{D}^i\times\mathfrak{Y}_d^k$*, and the FVRT stable tracking of* $\mathbf{Y}_d^k(t^{(k+1)N})$ *on* $\mathfrak{T}_{0F}^{(k+1)N}\times\mathfrak{D}^i\times\mathfrak{Y}_d^k$ *is with*

$$\mathcal{D}_{ST}^{k}\left(\mathbf{t}_0^{(k+1)N};\mathbf{t}_R^{(k+1)N};\mathbf{t}_F^{(k+1)N};\mathbf{Y}_{d0}^{k};\mathbf{D};\mathbf{U};\mathbf{Y}_d^{k}\right)=\mathfrak{R}^{(k+1)N}$$

for every $\left[\mathbf{D}(.),\mathbf{Y}_d^k(.)\right]\in\mathfrak{D}^i\times\mathfrak{Y}_d^k$.

Definition 406 ***The uniform stable elementwise tracking of the desired output behavior*** $\mathbf{Y}_d^k(t_R^{(k+1)N})$ ***on*** $\mathfrak{T}_{0F}^{(k+1)N}\times\mathfrak{D}^i\times\mathfrak{Y}_d^k$ ***of the plant (3.12) controlled by a control*** $\mathbf{U}(.)\in\mathfrak{U}^l$

a) If, and only if, the intersection

$$\mathcal{D}_{STU}^{k}\left(\mathbf{t}_0^{(k+1)N};\mathbf{t}_R^{(k+1)N};\mathbf{t}_F^{(k+1)N};\mathfrak{D}^i;\mathbf{U};\mathfrak{Y}_d^{k}\right)$$

of all

$$\mathcal{D}_{ST}^{k}\left(\mathbf{t}_0^{(k+1)N};\mathbf{t}_R^{(k+1)N};\mathbf{t}_F^{(k+1)N};\mathbf{Y}_{d0}^{k};\mathbf{D};\mathbf{U};\mathbf{Y}_d^{k}\right)$$

[*the intersection* $\mathcal{D}_{SSTU}^{k}\left(\mathbf{t}_0^{(k+1)N};\mathbf{t}_R^{(k+1)N};\mathbf{t}_F^{(k+1)N};\mathfrak{D}^i;\mathbf{U};\mathfrak{Y}_d^{k}\right)$ *of all domains*

$$\mathcal{D}_{SST}^{k}\left(\mathbf{t}_0^{(k+1)N};\mathbf{t}_R^{(k+1)N};\mathbf{t}_F^{(k+1)N};\mathbf{Y}_{d0}^{k};\mathbf{D};\mathbf{U};\mathbf{Y}_d^{k}\right)]$$

over $\mathfrak{D}^i\times\mathfrak{Y}_d^k$ *is a connected neighborhood of* $\mathbf{Y}_{d0}^k$ *then, and only then, it is the* ***FVRT (strictly) stable elementwise tracking domain of*** $\mathbf{Y}_d^k(t^{(k+1)N})$ ***on*** $\mathfrak{T}_{0F}^{(k+1)N}\times\mathfrak{D}^i\times\mathfrak{Y}_d^k$ ***uniform in*** $\left[\mathbf{D}(.),\mathbf{Y}_d^k(.)\right]\in\mathfrak{D}^i\times\mathfrak{Y}_d^k$*, respectively,*

$$\begin{gathered}\exists\xi\in\mathfrak{R}^+\Longrightarrow\mathcal{D}_{STU}^{k}\left(\mathbf{t}_0^{(k+1)N};\mathbf{t}_R^{(k+1)N};\mathbf{t}_F^{(k+1)N};\mathfrak{D}^i;\mathbf{U};\mathfrak{Y}_d^{k}\right)\\ =\cap\left\{\begin{array}{c}\mathcal{D}_{ST}^{k}\left(\mathbf{t}_0^{(k+1)N};\mathbf{t}_R^{(k+1)N};\mathbf{t}_F^{(k+1)N};\mathbf{Y}_{d0}^{k};\mathbf{D};\mathbf{U};\mathbf{Y}_d^{k}\right):\\ :\left[\mathbf{D}(.),\mathbf{Y}_d^k(.)\right]\in\mathfrak{D}^i\times\mathfrak{Y}_d^k\end{array}\right\}\\ \supset\mathfrak{N}_\xi\left(\mathbf{t}_0^{(k+1)N};\mathfrak{D}^i;\mathbf{U};\mathfrak{Y}_d^k\right),\end{gathered}\tag{26.48}$$

$$\left[\begin{array}{c}\exists\xi\in\mathfrak{R}^+\Longrightarrow\mathcal{D}_{SSTU}^{k}\left(\mathbf{t}_0^{(k+1)N};\mathbf{t}_R^{(k+1)N};\mathbf{t}_F^{(k+1)N};\mathfrak{D}^i;\mathbf{U};\mathfrak{Y}_d^{k}\right)\\ =\cap\left\{\begin{array}{c}\mathcal{D}_{SST}^{k}\left(\mathbf{t}_0^{(k+1)N};\mathbf{t}_R^{(k+1)N};\mathbf{t}_F^{(k+1)N};\mathbf{Y}_{d0}^{k};\mathbf{D};\mathbf{U};\mathbf{Y}_d^{k}\right):\\ :\left[\mathbf{D}(.),\mathbf{Y}_d^k(.)\right]\in\mathfrak{D}^i\times\mathfrak{Y}_d^k\end{array}\right\}\\ \supset\mathfrak{N}_\xi\left(\mathbf{t}_0^{(k+1)N};\mathfrak{D}^i;\mathbf{U};\mathfrak{Y}_d^k\right)\end{array}\right].\tag{26.49}$$

b) If, and only if, the intersection

$$\mathcal{D}_{STU}^{k}\left(\mathfrak{T}^{(k+1)N};\mathbf{t}_R^{(k+1)N};\mathbf{t}_F^{(k+1)N};\mathbf{Y}_{d0}^{k};\mathbf{D};\mathbf{U};\mathbf{Y}_d^{k}\right)$$

of all domains

$$\mathcal{D}_{ST}^{k}\left(\mathbf{t}_0^{(k+1)N};\mathbf{t}_R^{(k+1)N};\mathbf{t}_F^{(k+1)N};\mathbf{Y}_{d0}^k;\mathbf{D};\mathbf{U};\mathbf{Y}_d^k\right)$$

[*the intersection* $\mathcal{D}_{SSTU}^{k}\left(\mathfrak{T}^{(k+1)N};\mathbf{t}_R^{(k+1)N};\mathbf{t}_F^{(k+1)N};\mathbf{Y}_{d0}^k;\mathbf{D};\mathbf{U};\mathbf{Y}_d^k\right)$ *of all domains*

$$\mathcal{D}_{SST}^{k}\left(\mathbf{t}_0^{(k+1)N};\mathbf{t}_R^{(k+1)N};\mathbf{t}_F^{(k+1)N};\mathbf{Y}_{d0}^k;\mathbf{D};\mathbf{U};\mathbf{Y}_d^k\right)]$$

in $t_0^{(k+1)N}$ *over* $\mathfrak{T}^{(k+1)N}$ *is a connected neighborhood of* $\mathbf{Y}_{d0}^k$*, then, and only then, it is* ***the FVRT (strict) stable elementwise tracking domain of*** $\mathbf{Y}_d^k(t^{(k+1)N})$ ***on*** $\mathfrak{T}_{0F}^{(k+1)N}\times\mathfrak{D}^i\times\mathfrak{Y}_d^k$ ***uniform in*** $t_0^{(k+1)N}\in\mathfrak{T}^{(k+1)N}$*, respectively,*

$$\begin{aligned}&\exists\xi\in\mathfrak{R}^+\Longrightarrow\mathcal{D}_{STU}^{k}\left(\mathfrak{T}^{(k+1)N};\mathbf{t}_R^{(k+1)N};\mathbf{t}_F^{(k+1)N};\mathbf{Y}_{d0}^k;\mathbf{D};\mathbf{U};\mathbf{Y}_d^k\right)\\&=\cap\left\{\mathcal{D}_{ST}^{k}\left(\mathbf{t}_0^{(k+1)N};\mathbf{t}_R^{(k+1)N};\mathbf{t}_F^{(k+1)N};\mathbf{Y}_{d0}^k;\mathbf{D};\mathbf{U};\mathbf{Y}_d^k\right):\mathbf{t}_0^{(k+1)N}\in\mathfrak{T}_{0F}^{(k+1)N}\right\}\\&\supset\mathfrak{N}_\xi\left(\mathfrak{T}^{(k+1)N};\mathbf{t}_R^{(k+1)N};\mathbf{t}_F^{(k+1)N};\mathbf{Y}_{d0}^k;\mathbf{D};\mathbf{U};\mathbf{Y}_d^k\right),\end{aligned}\tag{26.50}$$

$$\left[\begin{array}{c}\exists\xi\in\mathfrak{R}^+\Longrightarrow\mathcal{D}_{SSTU}^{k}\left(\mathfrak{T}^{(k+1)N};\mathbf{t}_R^{(k+1)N};\mathbf{t}_F^{(k+1)N};\mathbf{Y}_{d0}^k;\mathbf{D};\mathbf{U};\mathbf{Y}_d^k\right)\\=\cap\left\{\begin{array}{c}\mathcal{D}_{SST}^{k}\left(\mathbf{t}_0^{(k+1)N};\mathbf{t}_R^{(k+1)N};\mathbf{t}_F^{(k+1)N};\mathbf{Y}_{d0}^k;\mathbf{D};\mathbf{U};\mathbf{Y}_d^k\right):\\\mathbf{t}_0^{(k+1)N}\in\mathfrak{T}_{0F}^{(k+1)N}\end{array}\right\}\\\supset\mathfrak{N}_\xi\left(\mathfrak{T}^{(k+1)N};\mathbf{t}_R^{(k+1)N};\mathbf{t}_F^{(k+1)N};\mathbf{Y}_{d0}^k;\mathbf{D};\mathbf{U};\mathbf{Y}_d^k\right)\end{array}\right].\tag{26.51}$$

c) If, and only if, the intersection

$$\mathcal{D}_{STU}^{k}\left(\mathfrak{T}^{(k+1)N};\mathbf{t}_R^{(k+1)N};\mathbf{t}_F^{(k+1)N};\mathfrak{D}^i;\mathbf{U};\mathfrak{Y}_d^k\right)$$

of all domains $\mathcal{D}_{ST}^{k}\left(\mathbf{t}_0^{(k+1)N};\mathbf{t}_R^{(k+1)N};\mathbf{t}_F^{(k+1)N};\mathbf{Y}_{d0}^k;\mathbf{D};\mathbf{U};\mathbf{Y}_d^k\right)$ [*the intersection*

$$\mathcal{D}_{SSTU}^{k}\left(\mathfrak{T}^{(k+1)N};\mathbf{t}_R^{(k+1)N};\mathbf{t}_F^{(k+1)N};\mathfrak{D}^i;\mathbf{U};\mathfrak{Y}_d^k\right)$$

of all domains

$$\mathcal{D}_{SST}^{k}\left(\mathbf{t}_0^{(k+1)N};\mathbf{t}_R^{(k+1)N};\mathbf{t}_F^{(k+1)N};\mathbf{Y}_{d0}^k;\mathbf{D};\mathbf{U};\mathbf{Y}_d^k\right)]$$

in $\left[\mathbf{t}_0^{(k+1)N},\mathbf{D}(.),\mathbf{Y}_d^k(.)\right]\in\mathfrak{T}^{(k+1)N}\times\mathfrak{D}^i\times\mathfrak{Y}_d^k$ *is a connected neighborhood of every* $\mathbf{Y}_{d0}^k\in\mathfrak{Y}_{d0}^{\alpha-1}$ *then, and only then, it is the* ***FVRT (strictly) stable elementwise tracking domain of*** $\mathbf{Y}_d^k(t^{(k+1)N})$ ***on*** $\mathfrak{T}_{0F}^{(k+1)N}\times\mathfrak{D}^i\times\mathfrak{Y}_d^k$ ***uniform in*** $\left[\mathbf{t}_0^{(k+1)N},\mathbf{D}(.),\mathbf{Y}_d^k(.)\right]\in\mathfrak{T}^{(k+1)N}\times\mathfrak{D}^i\times\mathfrak{Y}_d^k$*, respectively,*

$$\begin{aligned}&\exists\xi\in\mathfrak{R}^+\Longrightarrow\mathcal{D}_{STU}^{k}\left(\mathfrak{T}^{(k+1)N};\mathbf{t}_R^{(k+1)N};\mathbf{t}_F^{(k+1)N};\mathfrak{D}^i;\mathbf{U};\mathfrak{Y}_d^k\right)\\&=\cap\left\{\begin{array}{c}\mathcal{D}_{ST}^{k}\left(\mathbf{t}_0^{(k+1)N};\mathbf{t}_R^{(k+1)N};\mathbf{t}_F^{(k+1)N};\mathbf{Y}_{d0}^k;\mathbf{D};\mathbf{U};\mathbf{Y}_d^k\right):\\:\left[\mathbf{t}_0^{(k+1)N},\mathbf{D}(.),\mathbf{Y}_d^k(.)\right]\in\mathfrak{T}^{(k+1)N}\times\mathfrak{D}^i\times\mathfrak{Y}_d^k\end{array}\right\}\\&\supset\mathfrak{N}_\xi\left(\mathfrak{T}^{(k+1)N};\mathbf{t}_R^{(k+1)N};\mathbf{t}_F^{(k+1)N};\mathfrak{D}^i;\mathbf{U};\mathfrak{Y}_d^k\right),\end{aligned}\tag{26.52}$$

$$\left[\begin{array}{c}\exists\xi\in\mathfrak{R}^+\Longrightarrow\mathcal{D}_{SSTU}^{k}\left(\mathfrak{T}^{(k+1)N};\mathbf{t}_R^{(k+1)N};\mathbf{t}_F^{(k+1)N};\mathfrak{D}^i;\mathbf{U};\mathfrak{Y}_d^k\right)\\=\cap\left\{\begin{array}{c}\mathcal{D}_{SST}^{k}\left(\mathbf{t}_0^{(k+1)N};\mathbf{t}_R^{(k+1)N};\mathbf{t}_F^{(k+1)N};\mathbf{Y}_{d0}^k;\mathbf{D};\mathbf{U};\mathbf{Y}_d^k\right):\\:\left[\mathbf{t}_0^{(k+1)N},\mathbf{D}(.),\mathbf{Y}_d^k(.)\right]\in\mathfrak{T}_{0F}^{(k+1)N}\times\mathfrak{D}^i\times\mathfrak{Y}_d^k\end{array}\right\}\\\supset\mathfrak{N}_\xi\left(\mathfrak{T}_{0F}^{(k+1)N};\mathbf{t}_R^{(k+1)N};\mathbf{t}_F^{(k+1)N};\mathfrak{D}^i;\mathbf{U};\mathfrak{Y}_d^k\right).\end{array}\right]\tag{26.53}$$

Comment 407 *Every tracking with the finite (scalar or vector) reachability time implies the perfect tracking that starts at the (scalar or vector) reachability instant* $t_R^{(k+1)N}$ *and continues until the final (scalar or vector, respectively) time* $t_F^{(k+1)N}$. *It expresses a high tracking quality.*

Chapter 27

State space definitions

27.1 Finite scalar reachability time tracking

The state finite scalar reachability time, for short in the sequel: *finite scalar reachability time (FSRT)*, t_R, $t_R \in \text{In}\, \mathfrak{T}_{0F}$, i.e., the reachability time related to the state space $\mathfrak{R}_d^{\alpha}$ means that at the moment $t = t_R$ the plant real state vector $\mathbf{R}^{\alpha-1}(t_R)$ becomes equal to the desired state vector $\mathbf{R}_d^{\alpha-1}(t_R)$, $\mathbf{R}^{\alpha-1}(t_R) = \mathbf{R}_d^{\alpha-1}(t_R)$, and that they rest equal on $\mathfrak{T}_{RF}$. This assures that the real output vector $\mathbf{Y}(t)$ becomes equal to the desired output vector $\mathbf{Y}_d(t)$ at the state reachability time t_R that is simultaneously the scalar output reachability time t_R. This comes from Definition 116 of the desired state (i.e., of the desired motion). We accept the validity of Assumption 234 in what follows.

Definition 408 ***The state tracking with the finite scalar reachability time (FSRT) of the desired behavior*** $\mathbf{R}_d^{\alpha-1}(t)$ ***of the plant (3.12) controlled by control*** $\mathbf{U}(.) \in \mathfrak{U}$

a) The plant (3.12) exhibits ***the state tracking of the desired behavior*** $\mathbf{R}_d^{\alpha-1}(t)$ ***with the finite scalar reachability time*** t_R ***on*** $\mathfrak{T}_{0F} \times \mathfrak{D}^j \times \mathfrak{R}_d^{\alpha}$, *for short* ***the state tracking of*** $\mathbf{R}_d^{\alpha-1}(t)$ ***with the finite reachability time*** t_R ***on*** $\mathfrak{T}_{0F} \times \mathfrak{D}^j \times \mathfrak{R}_d^{\alpha}$ *if, and only if, for every* $\left[\mathbf{D}(.), \mathbf{R}_{d0}^{\alpha-1}\right] \in \mathfrak{D}^j \times \mathfrak{R}_{d0}^{\alpha}$ *there exists a connected neighborhood*

$$\mathfrak{N}\left(t_0; t_R; t_F; \mathbf{R}_{d0}^{\alpha-1}; \mathbf{D}; \mathbf{U}; \mathbf{R}_d^{\alpha-1}\right), \mathfrak{N}\left(t_0; t_R; t_F; \mathbf{R}_{d0}^{\alpha-1}; \mathbf{D}; \mathbf{U}; \mathbf{R}_d^{\alpha-1}\right) \subseteq \mathfrak{R}^{\alpha\rho},$$

of the desired initial plant state vector $\mathbf{R}_{d0}^{\alpha-1}$ *at the initial moment* $t_0 \in \mathfrak{T}$ *such that* $\mathbf{R}_0^{\alpha-1}$ *from the neighborhood* $\mathfrak{N}\left(t_0; t_R; t_F; \mathbf{R}_{d0}^{\alpha-1}; \mathbf{D}; \mathbf{U}; \mathbf{R}_d^{\alpha-1}\right)$,

$$\mathbf{R}_0^{\alpha-1} \in \mathfrak{N}\left(t_0; t_R; t_F; \mathbf{R}_{d0}^{\alpha-1}; \mathbf{D}; \mathbf{U}; \mathbf{R}_d^{\alpha-1}\right),$$

guarantees that $\mathbf{R}^{\alpha-1}\left(t; t_0; \mathbf{R}_0^{\alpha-1}; \mathbf{D}; \mathbf{U}\right)$ *becomes equal to* $\mathbf{R}_d^{\alpha-1}\left(t; t_0; \mathbf{R}_{d0}^{\alpha-1}\right)$ *at the moment* $t_R \in \text{In}\, \mathfrak{T}_{0F}$, *and that they stay equal from then on* $\mathfrak{T}_{RF}$, *i.e.,*

$$\begin{gathered}\forall\left[\mathbf{D}(.), \mathbf{Y}_d(.)\right] \in \mathfrak{D}^j \times \mathfrak{R}_d^{\alpha}, \exists \mathfrak{N}\left(t_0; t_R; t_F; \mathbf{R}_{d0}^{\alpha-1}; \mathbf{D}; \mathbf{U}; \mathbf{R}_d^{\alpha-1}\right) \subseteq \mathfrak{R}^{\alpha\rho} \\ \mathbf{R}_0^{\alpha-1} \in \mathfrak{N}\left(t_0; t_R; t_F; \mathbf{R}_{d0}^{\alpha-1}; \mathbf{D}; \mathbf{U}; \mathbf{R}_d^{\alpha-1}\right) \Longrightarrow \\ \mathbf{R}^{\alpha-1}\left(t; t_0; \mathbf{R}_0^{\alpha-1}; \mathbf{D}; \mathbf{U}\right) = \mathbf{R}_d^{\alpha-1}\left(t; t_0; \mathbf{R}_{d0}^{\alpha-1}\right), \ \forall t \in \mathfrak{T}_{RF}. \end{gathered} \tag{27.1}$$

b) The largest connected neighborhood $\mathfrak{N}_L\left(t_0; t_R; t_F; \mathbf{R}_{d0}^{\alpha-1}; \mathbf{D}; \mathbf{U}; \mathbf{R}_d^{\alpha-1}\right)$ *obeying (27.1) is* ***the state FSRT tracking domain***

$$\mathcal{D}_T\left(t_0; t_R; t_F; \mathbf{R}_{d0}^{\alpha-1}; \mathbf{D}; \mathbf{U}; \mathbf{R}_d^{\alpha-1}\right)$$

of $\mathbf{R}_d^{\alpha-1}(t)$ ***on*** $\mathfrak{T}_{0F} \times \mathfrak{D}^j \times \mathfrak{R}_d^{\alpha}$.

c) The state FSRT tracking on $\mathfrak{T}_{0F} \times \mathfrak{D}^j \times \mathfrak{R}_d^{\alpha}$ *is* ***global (in the whole)*** *if, and only if,* $\mathcal{D}_T\left(t_0; t_R; t_F; \mathbf{R}_{d0}^{\alpha-1}; \mathbf{D}; \mathbf{U}; \mathbf{R}_d^{\alpha-1}\right) = \mathfrak{R}^{\alpha\rho}$ *for every* $\left[\mathbf{D}(.), \mathbf{R}_{d0}^{\alpha-1}\right] \in \mathfrak{D}^j \times \mathfrak{R}_{d0}^{\alpha}$.

Definition 409 ***The uniform state tracking with the finite scalar reachability time of the desired behavior*** $\mathbf{R}_d^{\alpha-1}(t)$ ***of the plant (3.12) controlled by control*** $\mathbf{U}(.) \in \mathfrak{U}$

a) If, and only if, the intersection $\mathcal{D}_{TU}\left(t_0; t_R; t_F; \mathfrak{D}^j; \mathbf{U}; \mathfrak{R}_d^\alpha\right)$ *of the state FSRT tracking domains* $\mathcal{D}_T\left(t_0; t_R; t_F; \mathbf{R}_{d0}^{\alpha-1}; \mathbf{D}; \mathbf{U}; \mathbf{R}_{d0}^{\alpha-1}\right)$ *over* $\left[\mathbf{D}(.), \mathbf{R}_{d0}^{\alpha-1}\right] \in \mathfrak{D}^j \times \mathfrak{R}_{d0}^\alpha$ *is also a connected neighborhood of* $\mathbf{R}_{d0}^{\alpha-1} = \mathcal{R}_d^{\alpha-1}\left(t_0; t_0; \mathbf{R}_{d0}^{\alpha-1}\right)$ *for every* $\left[\mathbf{R}_{d0}^{\alpha-1}, \mathcal{R}_d^{\alpha-1}(.)\right] \in \mathfrak{R}_{d0}^\alpha \times \mathfrak{R}_d^\alpha$:

$$\begin{gathered}\exists \xi \in \mathfrak{R}^+ \Longrightarrow \mathcal{D}_{TU}\left(t_0; t_R; t_F; \mathfrak{D}^j; \mathbf{U}; \mathfrak{R}_d^\alpha\right) \\ = \cap\left\{\mathcal{D}_T\left(t_0; t_R; t_F; \mathbf{R}_{d0}^{\alpha-1}; \mathbf{D}; \mathbf{U}; \mathbf{R}_d^{\alpha-1}\right) : \left[\mathbf{D}(.), \mathbf{R}_{d0}^{\alpha-1}\right] \in \mathfrak{D}^j \times \mathfrak{R}_{d0}^\alpha\right\} \\ \supset \mathfrak{N}_\xi\left(t_0; t_R; t_F; \mathfrak{D}^j; \mathbf{U}; \mathfrak{R}_d^\alpha\right), \\ \partial \mathcal{D}_{TU}\left(t_0; t_R; t_F; \mathfrak{D}^j; \mathbf{U}; \mathfrak{R}_d^\alpha\right) \cap \partial \mathfrak{N}_\xi\left(t_0; t_R; t_F; \mathfrak{D}^j; \mathbf{U}; \mathfrak{R}_d^\alpha\right) = \phi,\end{gathered}$$

then the state FSRT tracking is ***uniform in*** $\left[\mathbf{D}(.), \mathbf{R}_{d0}^{\alpha-1}\right] \in \mathfrak{D}^j \times \mathfrak{R}_{d0}^\alpha$ ***on*** $\mathfrak{T}_{0F} \times \mathfrak{D}^j \times \mathfrak{R}_d^\alpha$ *and the set* $\mathcal{D}_{TU}\left(t_0; t_R; t_F; \mathfrak{D}^j; \mathbf{U}; \mathfrak{R}_d^\alpha\right)$ *is* ***the uniform state FSRT tracking domain of the plant on (the product set)*** $\mathfrak{T}_{0F} \times \mathfrak{D}^j \times \mathfrak{R}_d^\alpha$.

b) If, and only if, the intersection $\mathcal{D}_{TU}\left(\mathfrak{T}_i; t_R; t_F; \mathbf{R}_{d0}^{\alpha-1}; \mathbf{D}; \mathbf{U}; \mathbf{R}_{d0}^{\alpha-1}\right)$ *of state FSRT tracking domains* $\mathcal{D}_T\left(t_0; t_R; t_F; \mathbf{R}_{d0}^{\alpha-1}; \mathbf{D}; \mathbf{U}; \mathbf{R}_{d0}^{\alpha-1}\right)$ *in* $t_0 \in \mathfrak{T}_i$ *is a connected neighborhood of* $\mathbf{R}_{d0}^{\alpha-1} = \mathcal{R}_d^{\alpha-1}\left(t_0; t_0; \mathbf{R}_{d0}^{\alpha-1}\right)$ *for every* $\mathcal{R}_d^{\alpha-1}(.) \in \mathfrak{R}_d^\alpha$,

$$\begin{gathered}\exists \eta \in \mathfrak{R}^+ \Longrightarrow \mathcal{D}_{TU}\left(\mathfrak{T}_i; t_R; t_F; \mathbf{R}_{d0}^{\alpha-1}; \mathbf{D}; \mathbf{U}; \mathbf{R}_d^{\alpha-1}\right) \\ = \cap\left[\mathcal{D}_T\left(t_0; t_R; t_F; \mathbf{R}_{d0}^{\alpha-1}; \mathbf{D}; \mathbf{U}; \mathbf{R}_d^{\alpha-1}\right) : t_0 \in \mathfrak{T}_i\right] \\ \supset \mathfrak{N}_\eta\left(\mathfrak{T}_i; t_R; t_F; \mathbf{R}_{d0}^{\alpha-1}; \mathbf{D}; \mathbf{U}; \mathbf{R}_d^{\alpha-1}\right), \\ \partial \mathcal{D}_{TU}\left(\mathfrak{T}_i; t_R; t_F; \mathbf{R}_{d0}^{\alpha-1}; \mathbf{D}; \mathbf{U}; \mathbf{R}_d^{\alpha-1}\right) \cap \\ \cap\, \partial \mathfrak{N}_\eta\left(\mathfrak{T}_i; t_R; t_F; \mathbf{R}_{d0}^{\alpha-1}; \mathbf{D}; \mathbf{U}; \mathbf{R}_d^{\alpha-1}\right) = \phi,\end{gathered}$$

then, and only then, the FSRT tracking is ***uniform in*** $t_0 \in \mathfrak{T}_i$ ***on*** $\mathfrak{T}_{0F} \times \mathfrak{D}^j \times \mathfrak{R}_d^\alpha$ ***for*** $\left[\mathbf{D}(.), \mathbf{R}_{d0}^{\alpha-1}\right] \in \mathfrak{D}^j \times \mathfrak{R}_{d0}^\alpha$. *The set* $\mathcal{D}_{TU}\left(\mathfrak{T}_i; t_R; t_F; \mathbf{R}_{d0}^{\alpha-1}; \mathbf{D}; \mathbf{U}; \mathbf{R}_d^{\alpha-1}\right)$ *is* ***the state FSRT tracking domain of the plant uniform in*** $t_0 \in \mathfrak{T}_i$ ***on*** $\mathfrak{T}_{0F} \times \mathfrak{D}^j \times \mathfrak{R}_d^\alpha$.

c) If, and only if, the intersection $\mathcal{D}_{TU}\left(\mathfrak{T}_i; t_R; t_F; \mathfrak{D}^j, \mathbf{U}; \mathfrak{R}_d^\alpha\right)$ *of the state tracking domains* $\mathcal{D}_T\left(t_0; t_R; t_F; \mathbf{R}_{d0}^{\alpha-1}; \mathbf{D}; \mathbf{U}; \mathbf{R}_d^{\alpha-1}\right)$ *in triplet* $\left[t_0, \mathbf{D}(.), \mathbf{R}_{d0}^{\alpha-1}\right] \in \mathfrak{T}_i \times \mathfrak{D}^j \times \mathfrak{R}_{d0}^\alpha$ *is a connected neighborhood of* $\mathbf{R}_{d0}^{\alpha-1} = \mathcal{R}_d^{\alpha-1}\left(t_0; t_0; \mathbf{R}_{d0}^{\alpha-1}\right)$ *for every* $\mathcal{R}_d^{\alpha-1}(.) \in \mathfrak{R}_d^\alpha$,

$$\begin{gathered}\exists \zeta \in \mathfrak{R}^+ \Longrightarrow \mathcal{D}_{TU}\left(\mathfrak{T}_i; t_R; t_F; \mathfrak{D}^j, \mathbf{U}; \mathfrak{R}_d^\alpha\right) \\ = \left\{\begin{array}{c}\cap\left[\begin{array}{c}\mathcal{D}_T\left(t_0; t_R; t_F; \mathbf{R}_{d0}^{\alpha-1}; \mathbf{D}; \mathbf{U}; \mathbf{R}_{d0}^{\alpha-1}\right) : \\ \left[t_0, \mathbf{D}(.), \mathbf{R}_{d0}^{\alpha-1}\right] \in \mathfrak{T}_i \times \mathfrak{D}^j \times \mathfrak{R}_d^\alpha\end{array}\right] \\ \supset \mathfrak{N}_\zeta\left(\mathfrak{T}_i; \mathfrak{D}^j, \mathbf{U}; \mathfrak{R}_d^\alpha\right)\end{array}\right\}, \\ \partial \mathcal{D}_{TU}\left(\mathfrak{T}_i; t_R; t_F; \mathfrak{D}^j, \mathbf{U}; \mathfrak{R}_d^\alpha\right) \cap \partial \mathfrak{N}_\zeta\left(\mathfrak{T}_i; t_R; t_F; \mathfrak{D}^j, \mathbf{U}; \mathfrak{R}_d^\alpha\right) = \phi,\end{gathered}$$

then, and only then, the tracking is ***uniform in***

$$\left[t_0, \mathbf{D}(.), \mathbf{R}_{d0}^{\alpha-1}\right] \in \mathfrak{T}_i \times \mathfrak{D}^j \times \mathfrak{R}_{d0}^\alpha$$

on $\mathfrak{T}_{0F} \times \mathfrak{D}^j \times \mathfrak{R}_d^\alpha$. *The set* $\mathcal{D}_{TU}\left(\mathfrak{T}_i; t_R; t_F; \mathfrak{D}^j, \mathbf{U}; \mathfrak{R}_d^\alpha\right)$ *is* ***the state tracking domain of the plant uniform in*** $\left[t_0, \mathbf{D}(.), \mathbf{R}_{d0}^{\alpha-1}\right]$ ***on*** $\mathfrak{T}_{0F} \times \mathfrak{D}^j \times \mathfrak{R}_d^\alpha$.

Note 410 *The global state tracking on* $\mathfrak{T}_0 \times \mathfrak{D}^j \times \mathfrak{R}_d^\alpha$ *is uniform in* $\left[\mathbf{D}(.), \mathbf{R}_{d0}^{\alpha-1}\right] \in \mathfrak{D}^j \times \mathfrak{R}_{d0}^\alpha$ *on* $\mathfrak{T}_{0F} \times \mathfrak{D}^j \times \mathfrak{R}_d^\alpha$. *The whole state space* $\mathfrak{R}_d^\alpha$ *is then the* ***the state tracking domain of the plant uniform in*** $\left(\mathbf{D}, \mathbf{R}_{d0}^{\alpha-1}\right) \in \mathfrak{D}^j \times \mathfrak{R}_d^\alpha$ ***on*** $\mathfrak{T}_{0F} \times \mathfrak{D}^j \times \mathfrak{R}_d^\alpha$.

In order to avoid the FSRT state tracking with a big overshoot we introduce the following.

Definition 411 ***The stable FSRT state tracking of the desired state behavior $\mathbf{R}_d^{\alpha-1}(t)$ on $\mathfrak{T}_{0F} \times \mathfrak{D}^j \times \mathfrak{R}_d^{\alpha}$ of the plant (3.12) controlled by a control $\mathbf{U}(.) \in \mathfrak{U}^l$***

a) The plant (3.12) exhibits ***the stable FSRT state tracking of $\mathbf{R}_d^{\alpha-1}(t)$ on $\mathfrak{T}_{0F} \times \mathfrak{D}^j \times \mathfrak{R}_d^{\alpha}$*** *if, and only if, it exhibits the FSRT state tracking of $\mathbf{R}_d^{\alpha-1}(t)$ on $\mathfrak{T}_{0F} \times \mathfrak{D}^j \times \mathfrak{R}_d^{\alpha}$, and for every connected neighborhood $\mathfrak{N}_\varepsilon\left[t;\mathbf{R}_d^{\alpha-1}(t)\right]$ of $\mathbf{R}_d^{\alpha-1}(t)$ at any $t \in \mathfrak{T}_{0F}$, there is a connected neighborhood*

$$\mathfrak{N}\left(\varepsilon;t_0;t_R;t_F;\mathbf{R}_{d0}^{\alpha-1};\mathbf{D};\mathbf{U};\mathbf{R}_d^{\alpha-1}\right),\ (16.1),$$

of the plant desired initial state vector $\mathbf{R}_{d0}^{\alpha-1}$ at the initial moment $t_0 \in \mathfrak{T}_i$ such that for the initial $\mathbf{R}_0^{\alpha-1} \in \mathfrak{N}\left(\varepsilon;t_0;t_R;t_F;\mathbf{R}_{d0}^{\alpha-1};\mathbf{D};\mathbf{U};\mathbf{R}_d^{\alpha-1}\right)$ the instantaneous plant motion $\mathbf{R}^{\alpha-1}(t;t_0;\mathbf{R}_0^{\alpha-1};\mathbf{D};\mathbf{U})$ stays in the epsilon neighborhood $\mathfrak{N}_\varepsilon\left[t;\mathbf{R}_d^{\alpha-1}(t)\right]$ for all $t \in \mathfrak{T}_{0F}$; i.e.,

$$\begin{gathered}
\forall \mathfrak{N}_\varepsilon\left[t;\mathbf{R}_d^{\alpha-1}(t)\right] \subseteq \mathfrak{R}^{\alpha\rho},\ \forall t \in \mathfrak{T}_{0F},\\
\forall\left[\mathbf{D}(.),\mathbf{R}_{d0}^{\alpha-1}\right] \in \mathfrak{D}^j \times \mathfrak{R}_{d0}^{\alpha},\ \exists \mathfrak{N}\left(\varepsilon;t_0;t_R;t_F;\mathbf{R}_{d0}^{\alpha-1};\mathbf{D};\mathbf{U};\mathbf{R}_d^{\alpha-1}\right) \subseteq \mathfrak{R}^{\alpha\rho},\\
\mathfrak{N}\left(\varepsilon;t_0;t_R;t_F;\mathbf{R}_{d0}^{\alpha-1};\mathbf{D};\mathbf{U};\mathbf{R}_d^{\alpha-1}\right) \subseteq \mathcal{D}_T\left(t_0;t_R;t_F;\mathbf{R}_{d0}^{\alpha-1};\mathbf{D};\mathbf{U};\mathbf{R}_d^{\alpha-1}\right),\\
\mathbf{R}_0^{\alpha-1} \in \mathfrak{N}\left(\varepsilon;t_0;t_R;t_F;\mathbf{R}_{d0}^{\alpha-1};\mathbf{D};\mathbf{U};\mathbf{R}_d^{\alpha-1}\right) \Longrightarrow\\
\mathbf{R}^{\alpha-1}(t;t_0;\mathbf{R}_0^{\alpha-1};\mathbf{D};\mathbf{U}) \in \mathfrak{N}_\varepsilon\left[t;\mathbf{R}_d^{\alpha-1}(t)\right],\ \forall t \in \mathfrak{T}_{0F}.
\end{gathered} \tag{27.2}$$

b) The largest connected neighborhood $\mathfrak{N}_L\left(\varepsilon;t_0;t_R;t_F;\mathbf{R}_{d0}^{\alpha-1};\mathbf{D};\mathbf{U};\mathbf{R}_d^{\alpha-1}\right)$ of $\mathbf{R}_{d0}^{\alpha-1}$, (27.2), is ***the state ε-tracking domain*** *denoted by*

$$\mathcal{D}_{ST}\left(\varepsilon;t_0;t_R;t_F;\mathbf{R}_{d0}^{\alpha-1};\mathbf{D};\mathbf{U};\mathbf{R}_d^{\alpha-1}\right)$$

at $t_0 \in \mathfrak{T}$ of the stable FSRT state tracking of $\mathbf{R}_d^{\alpha-1}(t)$ on $\mathfrak{T}_{0F} \times \mathfrak{D}^j \times \times \mathfrak{R}_d^{\alpha}$.

The domain $\mathcal{D}_{ST}\left(t_0;t_R;t_F;\mathbf{R}_{d0}^{\alpha-1};\mathbf{D};\mathbf{U};\mathbf{R}_d^{\alpha-1}\right)$ at $t_0 \in \mathfrak{T}$ of the stable FSRT state tracking of $\mathbf{R}_d^{\alpha-1}(t)$ on $\mathfrak{T}_{0F} \times \mathfrak{D}^j \times \mathfrak{R}_d^{\alpha}$ *is the union of all domains*

$$\mathcal{D}_{ST}\left(\varepsilon;t_0;t_R;t_F;\mathbf{R}_{d0}^{\alpha-1};\mathbf{D};\mathbf{U};\mathbf{R}_d^{\alpha-1}\right)$$

over $\varepsilon \in \mathfrak{R}^+$,

$$\begin{gathered}
\mathcal{D}_{ST}\left(t_0;t_R;t_F;\mathbf{R}_{d0}^{\alpha-1};\mathbf{D};\mathbf{U};\mathbf{R}_d^{\alpha-1}\right)\\
= \cup\left[\mathcal{D}_{ST}\left(\varepsilon;t_0;t_R;t_F;\mathbf{R}_{d0}^{\alpha-1};\mathbf{D};\mathbf{U};\mathbf{R}_d^{\alpha-1}\right) : \varepsilon \in \mathfrak{R}^+\right].
\end{gathered} \tag{27.3}$$

Let $[0,\varepsilon_M)$ be the maximal interval over which the ε-tracking domain

$$\mathcal{D}_{ST}\left(\varepsilon;t_0;t_R;t_F;\mathbf{R}_{d0}^{\alpha-1};\mathbf{D};\mathbf{U};\mathbf{R}_d^{\alpha-1}\right)$$

is continuous in $\varepsilon \in \mathfrak{R}_+$,

$$\begin{gathered}
\mathcal{D}_{ST}\left(\varepsilon;t_0;t_R;t_F;\mathbf{R}_{d0}^{\alpha-1};\mathbf{D};\mathbf{U};\mathbf{R}_d^{\alpha-1}\right) \in \mathfrak{C}\left([0,\varepsilon_M)\right),\\
\forall\left[\mathbf{D}(.),\mathbf{R}_{d0}^{\alpha-1}\right] \in \mathfrak{D}^j \times \mathfrak{R}_{d0}^{\alpha}.
\end{gathered}$$

The strict domain $\mathcal{D}_{SST}\left(t_0;t_R;t_F;\mathbf{R}_{d0}^{\alpha-1};\mathbf{D};\mathbf{U};\mathbf{R}_d^{\alpha-1}\right)$ at $t_0 \in \mathfrak{T}$ of the stable FSRT state tracking of $\mathbf{R}_d^{\alpha-1}(t)$ on $\mathfrak{T}_{0F} \times \mathfrak{D}^j \times \mathfrak{R}_d^{\alpha}$ *is the union of all stable FSRT state tracking domains $\mathcal{D}_{ST}\left(\varepsilon;t_0;t_R;t_F;\mathbf{R}_{d0}^{\alpha-1};\mathbf{D};\mathbf{U};\mathbf{R}_d^{\alpha-1}\right)$ over $\varepsilon \in [0,\varepsilon_M)$,*

$$\begin{gathered}
\mathcal{D}_{SST}\left(t_0;t_R;t_F;\mathbf{R}_{d0}^{\alpha-1};\mathbf{D};\mathbf{U};\mathbf{R}_d^{\alpha-1}\right)\\
= \cup\left\{\mathcal{D}_{ST}\left(\varepsilon;t_0;t_R;t_F;\mathbf{R}_{d0}^{\alpha-1};\mathbf{D};\mathbf{U};\mathbf{R}_d^{\alpha-1}\right) : \varepsilon \in [0,\varepsilon_M)\right\},\\
\forall\left[\mathbf{D}(.),\mathbf{R}_{d0}^{\alpha-1}\right] \in \mathfrak{D}^j \times \mathfrak{R}_{d0}^{\alpha}.
\end{gathered} \tag{27.4}$$

c) The stable FSRT state tracking of $\mathbf{R}_d^{\alpha-1}(t)$ *on* $\mathfrak{T}_{0F}\times\mathfrak{D}^j\times\mathfrak{R}_d^\alpha$ *is* ***global (in the whole)*** *if, and only if, it is the global FSRT state tracking of* $\mathbf{R}_d^{\alpha-1}(t)$ *on* $\mathfrak{T}_{0F}\times\mathfrak{D}^j\times\mathfrak{R}_d^\alpha$ *and the stable FSRT state tracking of* $\mathbf{R}_d^{\alpha-1}(t)$ *on* $\mathfrak{T}_{0F}\times\mathfrak{D}^j\times\mathfrak{R}_d^\alpha$ *is with* $\mathcal{D}_{ST}\left(t_0;t_R;t_F;\mathbf{R}_{d0}^{\alpha-1};\mathbf{D};\mathbf{U};\mathbf{R}_d^{\alpha-1}\right)=\mathfrak{R}^{\alpha\rho}$ *for every* $\left[\mathbf{D}(.),\mathbf{R}_{d0}^{\alpha-1}\right]\in\mathfrak{D}^j\times\mathfrak{R}_{d0}^\alpha$.

Definition 412 ***The uniform state stable FSRT tracking of the desired state behavior*** $\mathbf{R}_d^{\alpha-1}(t)$ ***on*** $\mathfrak{T}_{0F}\times\mathfrak{D}^j\times\mathfrak{R}_d^\alpha$ ***of the plant*** *(3.12)* ***controlled by a control*** $\mathbf{U}(.)\in\mathfrak{U}^l$

a) If, and only if, the intersection $\mathcal{D}_{STU}\left(t_0;t_R;t_F;\mathbf{R}_{d0}^{\alpha-1};\mathfrak{D}^j;\mathbf{U};\mathfrak{R}_d^\alpha\right)$ *of all*

$$\mathcal{D}_{ST}\left(t_0;t_R;t_F;\mathbf{R}_{d0}^{\alpha-1};\mathbf{D};\mathbf{U};\mathbf{R}_d^{\alpha-1}\right)$$

[$\mathcal{D}_{SSTU}\left(t_0;t_R;t_F;\mathbf{R}_{d0}^{\alpha-1};\mathfrak{D}^j;\mathbf{U};\mathfrak{R}_d^\alpha\right)$ *of all* $\mathcal{D}_{SST}\left(t_0;t_R;t_F;\mathbf{R}_{d0}^{\alpha-1};\mathbf{D};\mathbf{U};\mathbf{R}_d^{\alpha-1}\right)$*] over* $\mathfrak{D}^j\times\mathfrak{R}_d^\alpha$ *is a connected neighborhood of* $\mathbf{R}_{d0}^{\alpha-1}$ *then, and only then, it is the* ***(strictly) stable state FSRT tracking domain of*** $\mathbf{R}_d^{\alpha-1}(t)$ ***on (the product set)*** $\mathfrak{T}_{0F}\times\mathfrak{D}^j\times\mathfrak{R}_d^\alpha$ ***uniform in*** $\left[\mathbf{D}(.),\mathbf{R}_{d0}^{\alpha-1}\right]\in\mathfrak{D}^j\times\mathfrak{R}_{d0}^\alpha$*, respectively,*

$$\begin{array}{c}\exists\xi\in\mathfrak{R}^+\Longrightarrow\mathcal{D}_{STU}\left(t_0;t_R;t_F;\mathbf{R}_{d0}^{\alpha-1};\mathfrak{D}^j;\mathbf{U};\mathfrak{R}_d^\alpha\right)\\ =\cap\left\{\begin{array}{c}\mathcal{D}_{ST}\left(t_0;t_R;t_F;\mathbf{R}_{d0}^{\alpha-1};\mathbf{D};\mathbf{U};\mathbf{R}_d^{\alpha-1}\right)\\ :\left[\mathbf{D}(.),\mathbf{R}_{d0}^{\alpha-1}\right]\in\mathfrak{D}^j\times\mathfrak{R}_{d0}^\alpha\end{array}\right\}\\ \supset\mathfrak{N}_\xi\left(t_0;\mathfrak{D}^j;\mathbf{U};\mathfrak{R}_d^\alpha\right),\end{array}\tag{27.5}$$

$$\left[\begin{array}{c}\exists\xi\in\mathfrak{R}^+\Longrightarrow\mathcal{D}_{SSTU}\left(t_0;t_R;t_F;\mathfrak{D}^j;\mathbf{U};\mathfrak{R}_d^\alpha\right)\\ =\cap\left\{\begin{array}{c}\mathcal{D}_{SST}\left(t_0;t_R;t_F;\mathbf{R}_{d0}^{\alpha-1};\mathbf{D};\mathbf{U};\mathbf{R}_d^{\alpha-1}\right)\\ :\left[\mathbf{D}(.),\mathbf{R}_{d0}^{\alpha-1}\right]\in\mathfrak{D}^j\times\mathfrak{R}_{d0}^\alpha\end{array}\right\}\\ \supset\mathfrak{N}_\xi\left(t_0;\mathfrak{D}^j;\mathbf{U};\mathfrak{R}_d^\alpha\right)\end{array}\right].\tag{27.6}$$

b) If, and only if, the intersection $\mathcal{D}_{STU}\left(\mathfrak{T}_i;t_R;t_F;\mathbf{R}_{d0}^{\alpha-1};\mathbf{D};\mathbf{U};\mathbf{R}_d^{\alpha-1}\right)$ *of all*

$$\mathcal{D}_{ST}\left(t_0;t_R;t_F;\mathbf{R}_{d0}^{\alpha-1};\mathbf{D};\mathbf{U};\mathbf{R}_d^{\alpha-1}\right)$$

[$\mathcal{D}_{SSTU}\left(\mathfrak{T}_i;t_R;t_F;\mathbf{R}_{d0}^{\alpha-1};\mathbf{D};\mathbf{U};\mathbf{R}_d^{\alpha-1}\right)$ *of* $\mathcal{D}_{SST}\left(t_0;t_R;t_F;\mathbf{R}_{d0}^{\alpha-1};\mathbf{D};\mathbf{U};\mathbf{R}_d^{\alpha-1}\right)$*] in* t_0 *over* $\mathfrak{T}_i$ *is a connected neighborhood of* $\mathbf{R}_{d0}^{\alpha-1}$*, then, and only then, it is* ***the (strict) stable state FSRT tracking domain of*** $\mathbf{R}_d^{\alpha-1}(t)$ ***on*** $\mathfrak{T}_{0F}\times\mathfrak{D}^j\times\mathfrak{R}_d^\alpha$ ***uniform in*** $t_0\in\mathfrak{T}_i$*, respectively,*

$$\begin{array}{c}\exists\xi\in\mathfrak{R}^+\Longrightarrow\mathcal{D}_{STU}\left(\mathfrak{T}_i;t_R;t_F;\mathbf{R}_{d0}^{\alpha-1};\mathbf{D};\mathbf{U};\mathbf{R}_d^{\alpha-1}\right)\\ =\cap\left\{\mathcal{D}_{ST}\left(t_0;t_R;t_F;\mathbf{R}_{d0}^{\alpha-1};\mathbf{D};\mathbf{U};\mathbf{R}_d^{\alpha-1}\right):t_0\in\mathfrak{T}_i\right\}\\ \mathfrak{N}_\xi\left(\mathfrak{T}_i;t_R;t_F;\mathbf{R}_{d0}^{\alpha-1};\mathbf{D};\mathbf{U};\mathbf{R}_d^{\alpha-1}\right),\end{array}\tag{27.7}$$

$$\left[\begin{array}{c}\exists\xi\in\mathfrak{R}^+\Longrightarrow\mathcal{D}_{SSTU}\left(\mathfrak{T}_i;t_R;t_F;\mathbf{R}_{d0}^{\alpha-1};\mathbf{D};\mathbf{U};\mathbf{R}_{d0}^{\alpha-1}\right)\\ =\cap\left\{\mathcal{D}_{SST}\left(t_0;t_R;t_F;\mathbf{R}_{d0}^{\alpha-1};\mathbf{D};\mathbf{U};\mathbf{R}_{d0}^{\alpha-1}\right):t_0\in\mathfrak{T}_i\right\}\\ \supset\mathfrak{N}_\xi\left(\mathfrak{T}_i;t_R;t_F;\mathbf{R}_{d0}^{\alpha-1};\mathbf{D};\mathbf{U};\mathbf{R}_{d0}^{\alpha-1}\right)\end{array}\right].\tag{27.8}$$

c) If, and only if, the intersection $\mathcal{D}_{STU}\left(\mathfrak{T}_i;t_R;t_F;\mathfrak{D}^j;\mathbf{U};\mathfrak{R}_d^\alpha\right)$ *of all*

$$\mathcal{D}_{ST}\left(t_0;t_R;t_F;\mathbf{R}_{d0}^{\alpha-1};\mathbf{D};\mathbf{U};\mathbf{R}_d^{\alpha-1}\right)$$

[$\mathcal{D}_{SSTU}\left(\mathfrak{T}_i;t_R;t_F;\mathfrak{D}^j;\mathbf{U};\mathfrak{R}_d^\alpha\right)$ *of all* $\mathcal{D}_{SST}\left(t_0;t_R;t_F;\mathbf{R}_{d0}^{\alpha-1};\mathbf{D};\mathbf{U};\mathbf{R}_d^{\alpha-1}\right)$*] in*

$$\left[t_0,\mathbf{D}(.),\mathbf{R}_{d0}^{\alpha-1}\right]\in\mathfrak{T}_i\times\mathfrak{D}^j\times\mathfrak{R}_{d0}^\alpha$$

is a connected neighborhood of every $\mathbf{R}_{d0}^{\alpha-1} \in \mathfrak{R}_{d0}^{\alpha}$ *then, and only then, it is the* ***(strictly) stable state FSRT tracking domain of*** $\mathbf{R}_d^{\alpha-1}(t)$ *on* $\mathfrak{T}_{0F} \times \mathfrak{D}^j \times \mathfrak{R}_d^{\alpha}$ ***uniform in*** $\left[t_0, \mathbf{D}(.), \mathbf{R}_{d0}^{\alpha-1}\right] \in \mathfrak{T}_i \times \mathfrak{D}^j \times \mathfrak{R}_{d0}^{\alpha}$*, respectively,*

$$\begin{gathered}\exists \xi \in \mathfrak{R}^+ \Longrightarrow \mathcal{D}_{STU}\left(\mathfrak{T}_i; t_R; t_F; \mathfrak{D}^j; \mathbf{U}; \mathfrak{R}_d^{\alpha}\right) \\ = \cap \left\{ \begin{array}{l} \mathcal{D}_{ST}\left(t_0; t_R; t_F; \mathbf{R}_{d0}^{\alpha-1}; \mathbf{D}; \mathbf{U}; \mathbf{R}_{d0}^{\alpha-1}\right) : \\ : \left[t_0, \mathbf{D}(.), \mathbf{R}_{d0}^{\alpha-1}\right] \in \mathfrak{T} \times \mathfrak{D}^j \times \mathfrak{R}_{d0}^{\alpha} \end{array} \right\} \\ \supset \mathfrak{N}_{\xi}\left(\mathfrak{T}_i; t_R; t_F; \mathfrak{D}^j; \mathbf{U}; \mathfrak{R}_d^{\alpha}\right), \end{gathered} \tag{27.9}$$

$$\left[\begin{gathered}\exists \xi \in \mathfrak{R}^+ \Longrightarrow \mathcal{D}_{SSTU}\left(\mathfrak{T}_i; t_R; t_F; \mathfrak{D}^j; \mathbf{U}; \mathfrak{R}_d^{\alpha}\right) \\ = \cap \left\{ \begin{array}{l} \mathcal{D}_{SST}\left(t_0; t_R; t_F; \mathbf{R}_{d0}^{\alpha-1}; \mathbf{D}; \mathbf{U}; \mathbf{R}_{d}^{\alpha-1}\right) : \\ : \left[t_0, \mathbf{D}(.), \mathbf{R}_{d0}^{\alpha-1}\right] \in \mathfrak{T} \times \mathfrak{D}^j \times \mathfrak{R}_{d0}^{\alpha} \end{array} \right\} \\ \supset \mathfrak{N}_{\xi}\left(\mathfrak{T}_i; t_R; t_F; \mathfrak{D}^j; \mathbf{U}; \mathfrak{R}_d^{\alpha}\right). \end{gathered}\right] \tag{27.10}$$

Because the exponential convergence of the real state $\mathbf{R}^{\alpha-1}(t)$ to the desired state $\mathbf{R}_d^{\alpha-1}(t)$ does not imply the exponential convergence of the real output vector $\mathbf{Y}(t)$ to the desired output vector $\mathbf{R}_d^{\alpha-1}(t)$, we do not present the definition of the state exponential tracking with the finite scalar reachability time (for which see [191]).

27.2 Elementwise state *FVRT* tracking

Let $\mathbf{t}_R^{\alpha\rho}$ be *the state finite vector reachability time*, for short in the sequel: *finite vector reachability time (FVRT)*,

$$\mathbf{t}_R^{\alpha\rho} = \left[t_{R1}\; t_{R2}\; ...\; t_{R,\alpha\rho}\right]^T \in (\text{In}\;\mathfrak{T}_{0F})^{\alpha\rho}, \; \mathfrak{T}_R^{\alpha\rho} = \{\mathbf{t}^{\alpha\rho} : \; \mathbf{t}_0^{\alpha\rho} \leq \; \mathbf{t}^{\alpha\rho} \leq \mathbf{t}_R^{\alpha\rho}\}. \tag{27.11}$$

For the plant (3.12) the state vector reachability *time* $\mathbf{t}_R^{\alpha\rho}$ is different from its output vector reachability *time* $\mathbf{t}_R^N$. This comes from the fact that the state variables R_i, $i = 1, 2, ..., \alpha\rho$, are different from the output variables Y_i, $i = 1, 2, ..., N$, and that $\alpha\rho \neq N$ in general. The same holds for the final state vector moment $\mathbf{t}_F^{\alpha\rho}$, $\mathbf{t}_F^{\alpha\rho} \in \mathfrak{T}^{\alpha\rho}$, and the final output vector moment $\mathbf{t}_F^N \in \mathfrak{T}^N$. Although the initial output scalar moment t_0 and the initial state scalar moment t_{0ID} are equal, $t_0 = t_{0ID} \in \mathfrak{T}$, the corresponding initial vector moments $\mathbf{t}_0^N$ and $\mathbf{t}_0^{\alpha\rho}$ are different due to $\alpha\rho \neq N$ in general,

$$\mathbf{t}_0^N = \left[t_0\;\; t_0\; ...\; t_0\right]^T \in \mathfrak{T}^N, \;\; \mathbf{t}_0^{\alpha\rho} = \left[t_0\;\; t_0\; ...\; t_0\right]^T \; \in \mathfrak{T}^{\alpha\rho}, \;\; \mathbf{t}_0^N \neq \mathbf{t}_0^{\alpha\rho}.$$

The notation

$$\mathbf{R}^{\alpha-1}(\mathbf{t}_R^{\alpha\rho}) = \mathbf{R}_d^{\alpha-1}(\mathbf{t}_R^{\alpha\rho}), \; \forall \mathbf{t}_R^{\alpha\rho} \in \mathfrak{T}_{RF}^{\alpha\rho} = \left[\mathbf{t}_R^{\alpha\rho}, \; \mathbf{t}_F^{\alpha\rho}\right] \subset \mathfrak{T}_{0F}^{\alpha\rho} \tag{27.12}$$

means that the real state vector $\mathbf{R}^{\alpha-1}(\mathbf{t}_R^{\alpha\rho})$ becomes equal element by element; i.e., elementwise, to the desired state vector $\mathbf{R}_d^{\alpha-1}(\mathbf{t}_R^{\alpha\rho})$ at the latest at the state vector reachability time $\mathbf{t}_R^{\alpha\rho}$ and thereafter they rest equal until the final vector moment $\mathbf{t}_F^{\alpha\rho} \in \mathfrak{T}_{0F}^{\alpha\rho}$,

$$R_i^{(j)}(t) = R_{di}^{(j)}(t), \; \forall t \in \left[t_{Ri(j)}, \; t_{Fi(j)}\right], \; \forall i = 1, 2, ..., \rho, \; \forall j = 0, 1, .., \alpha - 1;$$

i.e.,

$$\mathbf{R}^{\alpha-1}(\mathbf{t}_R^{\alpha\rho}) = \mathbf{R}_d^{\alpha-1}(\mathbf{t}_R^{\alpha\rho}), \; \forall \mathbf{t}_R^{\alpha\rho} \in \mathfrak{T}_{RF}^{\alpha\rho} = \left[\mathbf{t}_R^{\alpha\rho}, \; \mathbf{t}_F^{\alpha\rho}\right].$$

The state vector reachability time $\mathbf{t}_R^{\alpha\rho}$ (27.12) ensures that the real output vector $\mathbf{Y}(t)$ becomes elementwise equal to its desired vector value $\mathbf{Y}_d(t)$ at the latest at the scalar output reachability time t_{RM},

$$\begin{gathered}\mathbf{Y}(t) = \mathbf{Y}_d(t), \; \forall t \in \left[t_{RM}, \; t_{Fm}\right], \\ t_{RM} = \max\left\{t_{R1}\;, \; t_{R2}\;, ..\; t_{Rn}\;\right\}, \;\; t_{Fm} = \min\left\{t_{F1}\;, \; t_{F2}\;, ..\; t_{Fn}\;\right\}. \end{gathered} \tag{27.13}$$

Definition 413 ***The state elementwise tracking with the state finite vector reachability time $\mathbf{t}_R^{\alpha\rho}$ of the desired state of the plant (3.12) controlled by a control $\mathbf{U}(.) \in \mathfrak{U}^l$***

a) The plant (3.12) exhibits ***the state elementwise tracking with the state finite vector reachability time $\mathbf{t}_R^{\alpha\rho}$ on $\mathfrak{T}_{0F}^{\alpha\rho} \times \mathfrak{D}^j \times \mathfrak{R}_d^{\alpha}$*** *if, and only if, for every $\left[\mathbf{D}(.), \mathbf{R}_{d0}^{\alpha-1}\right] \in \mathfrak{D}^j \times \mathfrak{R}_{d0}^{\alpha}$ there exists a connected neighborhood $\mathfrak{N}\left(\mathbf{t}_0^{\alpha\rho}; \mathbf{t}_R^{\alpha\rho}; \mathbf{t}_F^{\alpha\rho}; \mathbf{R}_{d0}^{\alpha-1}; \mathbf{D}; \mathbf{U}; \mathbf{R}_d^{\alpha-1}\right) \subseteq \mathfrak{R}^{\alpha\rho}$ dependent on $(\mathbf{t}_0^{\alpha\rho}, \mathbf{t}_R^{\alpha\rho}, \mathbf{t}_F^{\alpha\rho})$ of the plant desired initial state vector $\mathbf{R}_{d0}^{\alpha-1}$ at the initial vector moment $\mathbf{t}_0^{\alpha\rho} \in \mathfrak{T}_i^{\alpha\rho}$ such that $\mathbf{R}_0^{\alpha-1}$ from the neighborhood*

$$\mathfrak{N}\left(\mathbf{t}_0^{\alpha\rho}; \mathbf{t}_R^{\alpha\rho}; \mathbf{t}_F^{\alpha\rho}; \mathbf{R}_{d0}^{\alpha-1}; \mathbf{D}; \mathbf{U}; \mathbf{R}_d^{\alpha-1}\right)$$

guarantees both that $\mathbf{R}^{\alpha-1}(\mathbf{t}_R^{\alpha\rho}; \mathbf{t}_0^{\alpha\rho}; \mathbf{R}_0^{\alpha-1}; \mathbf{D}; \mathbf{U})$ becomes equal to $\mathbf{R}_d^{\alpha-1}(\mathbf{t}_R^{\alpha\rho})$ at the finite vector reachability moment $\mathbf{t}_R^{\alpha\rho} \in In\mathfrak{T}_{0F}^{\alpha\rho}$, and that they stay equal from then on $\mathfrak{T}_{RF}^{\alpha\rho}$, i.e.,

$$\begin{gathered}
\forall \left[\mathbf{D}(.), \mathbf{R}_{d0}^{\alpha-1}\right] \in \mathfrak{D}^j \times \mathfrak{R}_{d0}^{\alpha}, \\
\exists \mathfrak{N}\left(\mathbf{t}_0^{\alpha\rho}; \mathbf{t}_R^{\alpha\rho}; \mathbf{t}_F^{\alpha\rho}; \mathbf{R}_{d0}^{\alpha-1}; \mathbf{D}; \mathbf{U}; \mathbf{R}_d^{\alpha-1}\right) \subseteq \mathfrak{R}^{\alpha\rho}, \\
\mathbf{R}_0^{\alpha-1} \in \mathfrak{N}\left(\mathbf{t}_0^{\alpha\rho}; \mathbf{t}_R^{\alpha\rho}; \mathbf{t}_F^{\alpha\rho}; \mathbf{R}_{d0}^{\alpha-1}; \mathbf{D}; \mathbf{U}; \mathbf{R}_d^{\alpha-1}\right) \Longrightarrow \\
\mathbf{R}^{\alpha-1}(\mathbf{t}_R^{\alpha\rho}; \mathbf{t}_0^{\alpha\rho}; \mathbf{R}_0^{\alpha-1}; \mathbf{D}; \mathbf{U}) = \mathbf{R}_d^{\alpha-1}(\mathbf{t}_R^{\alpha\rho}),\ \forall \mathbf{t}_R^{\alpha\rho} \in \mathfrak{T}_{RF}^{\alpha\rho}. \qquad (27.14)
\end{gathered}$$

b) The largest connected neighborhood $\mathfrak{N}_L\left(\mathbf{t}_0^{\alpha\rho}; \mathbf{t}_R^{\alpha\rho}; \mathbf{t}_F^{\alpha\rho}; \mathbf{R}_{d0}^{\alpha-1}; \mathbf{D}; \mathbf{U}; \mathbf{R}_d^{\alpha-1}\right)$ of $\mathbf{R}_{d0}^{\alpha-1}$, which obeys (27.14), is ***the FVRT state tracking domain***

$$\mathcal{D}_T\left(\mathbf{t}_0^{\alpha\rho}; \mathbf{t}_R^{\alpha\rho}; \mathbf{t}_F^{\alpha\rho}; \mathbf{R}_{d0}^{\alpha-1}; \mathbf{D}; \mathbf{U}; \mathbf{R}_d^{\alpha-1}\right)$$

of $\mathbf{R}_d^{\alpha-1}(\mathbf{t}^{\alpha\rho})$ on $\mathfrak{T}_{0F}^{\alpha\rho} \times \mathfrak{D}^j \times \mathfrak{R}_d^{\alpha}$.

c) The FVRT state tracking of $\mathbf{R}_d^{\alpha-1}(\mathbf{t}^{\alpha\rho})$ on $\mathfrak{T}_{0F}^{\alpha\rho} \times \mathfrak{D}^j \times \mathfrak{R}_d^{\alpha}$ is ***global (in the whole)*** *if, and only if, $\mathcal{D}_T\left(\mathbf{t}_0^{\alpha\rho}; \mathbf{t}_R^{\alpha\rho}; \mathbf{t}_F^{\alpha\rho}; \mathbf{R}_{d0}^{\alpha-1}; \mathbf{D}; \mathbf{U}; \mathbf{R}_d^{\alpha-1}\right) = \mathfrak{R}^{\alpha\rho}$ for every $\left[\mathbf{D}(.), \mathbf{R}_{d0}^{\alpha-1}\right] \in \mathfrak{D}^j \times \mathfrak{R}_{d0}^{\alpha}$. The FVRT state tracking of $\mathbf{R}_d^{\alpha-1}(\mathbf{t}^{\alpha\rho})$ on $\mathfrak{T}_{0F}^{\alpha\rho} \times \mathfrak{D}^j \times \mathfrak{R}_d^{\alpha}$ is then uniform in $\left[\mathbf{D}(.), \mathbf{R}_{d0}^{\alpha-1}\right] \in \mathfrak{D}^j \times \mathfrak{R}_{d0}^{\alpha}$.*

This tracking property implies attraction of the desired motion $\mathbf{R}_d^{\alpha-1}(t)$ with the state finite vector reachability time $\mathbf{t}_R^{\alpha\rho}$, but attraction of $\mathbf{R}_d^{\alpha-1}(t)$ does not guarantee the elementwise tracking with the finite state vector reachability time $\mathbf{t}_R^{\alpha\rho}$.

We present the vector generalization of Definition 409:

Definition 414 ***The uniform state tracking with the finite vector reachability time of the desired behavior $\mathbf{R}_d^{\alpha-1}(\mathbf{t}^{\alpha\rho})$ on $\mathfrak{T}_{0F}^{\alpha\rho} \times \mathfrak{D}^j \times \times \mathfrak{R}_d^{\alpha}$ of the plant (3.12) controlled by a control $\mathbf{U}(.) \in \mathfrak{U}^l$***

a) If, and only if, the intersection $\mathcal{D}_{TU}\left(\mathbf{t}_0^{\alpha\rho}; \mathbf{t}_R^{\alpha\rho}; \mathbf{t}_F^{\alpha\rho}; \mathfrak{D}^j; \mathbf{U}; \mathfrak{R}_d^{\alpha}\right)$ of the state tracking domains $\mathcal{D}_T\left(\mathbf{t}_0^{\alpha\rho}; \mathbf{t}_R^{\alpha\rho}; \mathbf{t}_F^{\alpha\rho}; \mathbf{R}_{d0}^{\alpha-1}; \mathbf{D}; \mathbf{U}; \mathbf{R}_d^{\alpha-1}\right)$ in

$$\left[\mathbf{D}(.), \mathbf{R}_{d0}^{\alpha-1}\right] \in \mathfrak{D}^j \times \mathfrak{R}_{d0}^{\alpha}$$

is a connected neighborhood of $\mathbf{R}_{d0}^{\alpha-1}$ of every $\mathbf{R}_d^{\alpha-1}(.) \in \mathfrak{R}_d^{\alpha}$,

$$\begin{gathered}
\exists \xi \in \mathfrak{R}^+ \Longrightarrow \mathcal{D}_{TU}\left(\mathbf{t}_0^{\alpha\rho}; \mathbf{t}_R^{\alpha\rho}; \mathbf{t}_F^{\alpha\rho}; \mathfrak{D}^j; \mathbf{U}; \mathfrak{R}_d^{\alpha}\right) \\
= \cap \left[\begin{array}{c} \mathcal{D}_T\left(\mathbf{t}_0^{\alpha\rho}; \mathbf{t}_R^{\alpha\rho}; \mathbf{t}_F^{\alpha\rho}; \mathbf{R}_{d0}^{\alpha-1}; \mathbf{D}; \mathbf{U}; \mathbf{R}_d^{\alpha-1}\right) : \\ \left[\mathbf{D}(.), \mathbf{R}_{d0}^{\alpha-1}\right] \in \mathfrak{D}^j \times \mathfrak{R}_{d0}^{\alpha} \end{array}\right] \supset \mathfrak{N}_\xi\left(\mathbf{t}_0^{\alpha\rho}; \mathbf{t}_R^{\alpha\rho}; \mathbf{t}_F^{\alpha\rho}; \mathfrak{D}^j; \mathbf{U}; \mathfrak{R}_d^{\alpha}\right), \\
\partial \mathcal{D}_{TU}\left(\mathbf{t}_0^{\alpha\rho}; \mathbf{t}_R^{\alpha\rho}; \mathbf{t}_F^{\alpha\rho}; \mathfrak{D}^j; \mathbf{U}; \mathfrak{R}_d^{\alpha}\right) \cap \partial \mathfrak{N}_\xi\left(\mathbf{t}_0^{\alpha\rho}; \mathbf{t}_R^{\alpha\rho}; \mathbf{t}_F^{\alpha\rho}; \mathfrak{D}^j; \mathbf{U}; \mathfrak{R}_d^{\alpha}\right) = \phi, \qquad (27.15)
\end{gathered}$$

then the FVRT state tracking of $\mathbf{R}_d^{\alpha-1}(\mathbf{t}^{\alpha\rho})$ is ***uniform in*** *$\left[\mathbf{D}(.), \mathbf{R}_{d0}^{\alpha-1}\right] \in \mathfrak{D}^j \times \mathfrak{R}_{d0}^{\alpha}$ on the product set $\mathfrak{T}_{0F}^{\alpha\rho} \times \mathfrak{D}^j \times \mathfrak{R}_d^{\alpha}$ and $\mathcal{D}_{TU}\left(\mathbf{t}_0^{\alpha\rho}; \mathbf{t}_R^{\alpha\rho}; \mathbf{t}_F^{\alpha\rho}; \mathfrak{D}^j; \mathbf{U}; \mathfrak{R}_d^{\alpha}\right)$ is* ***the $(\mathbf{D}, \mathbf{R}_{d0}^{\alpha-1})$-uniform FVRT state tracking domain of $\mathbf{R}_d^{\alpha-1}(\mathbf{t}^{\alpha\rho})$ on $\mathfrak{T}_{0F}^{\alpha\rho} \times \mathfrak{D}^j \times \mathfrak{R}_d^{\alpha}$.***

b) If, and only if, the intersection $\mathcal{D}_{TU}\left(\mathfrak{T}_i^{\alpha\rho};\mathbf{t}_R^{\alpha\rho};\mathbf{t}_F^{\alpha\rho};\mathbf{R}_{d0}^{\alpha-1};\mathbf{D};\mathbf{U};\mathbf{R}_d^{\alpha-1}\right)$ *of all the FVRT state tracking domains*

$$\mathcal{D}_T\left(\mathbf{t}_0^{\alpha\rho};\mathbf{t}_R^{\alpha\rho};\mathbf{t}_F^{\alpha\rho};\mathbf{R}_{d0}^{\alpha-1};\mathbf{D};\mathbf{U};\mathbf{R}_d^{\alpha-1}\right)$$

in $\mathbf{t}_0^{\alpha\rho}$ *over* $\mathfrak{T}_i^{\alpha\rho}$ *is a connected neighborhood of* $\mathbf{R}_{d0}^{\alpha-1}$ *then the FVRT state tracking of* $\mathbf{R}_d^{\alpha-1}(\mathbf{t}^{\alpha\rho})$ *on the set product* $\mathfrak{T}_{0F}^{\alpha\rho}\times\mathfrak{D}^j\times\mathfrak{R}_d^{\alpha}$ *is* ***uniform in*** $\mathbf{t}_0^{\alpha\rho}$ ***on*** $\mathfrak{T}_i^{\alpha\rho}$ *and*

$$\mathcal{D}_{TU}\left(\mathfrak{T}_i^{\alpha\rho};\mathbf{t}_R^{\alpha\rho};\mathbf{t}_F^{\alpha\rho};\mathbf{R}_{d0}^{\alpha-1};\mathbf{D};\mathbf{U};\mathbf{R}_d^{\alpha-1}\right)$$

is ***the domain of*** $\mathbf{t}_0^{\alpha\rho}$***-uniform FVRT state tracking of the desired state vector*** $\mathbf{R}_d^{\alpha-1}(\mathbf{t}^{\alpha\rho})$ ***on*** $\mathfrak{T}_{0F}^{\alpha\rho}\times\mathfrak{T}_i^{\alpha\rho}\times\mathfrak{D}^j\times\mathfrak{R}_d^{\alpha}$,

$$\begin{gathered}\exists\xi\in\mathfrak{R}^+\Longrightarrow\mathcal{D}_{TU}\left(\mathfrak{T}_i^{\alpha\rho};\mathbf{t}_R^{\alpha\rho};\mathbf{t}_F^{\alpha\rho};\mathbf{R}_{d0}^{\alpha-1};\mathbf{D};\mathbf{U};\mathbf{R}_d^{\alpha-1}\right)\\ =\cap\left[\mathcal{D}_T\left(\mathbf{t}_0^{\alpha\rho};\mathbf{t}_R^{\alpha\rho};\mathbf{t}_F^{\alpha\rho};\mathbf{R}_{d0}^{\alpha-1};\mathbf{D};\mathbf{U};\mathbf{R}_d^{\alpha-1}\right):\mathbf{t}_0^{\alpha\rho}\in\mathfrak{T}_i^{\alpha\rho}\right]\\ \supset\mathfrak{N}_\xi\left(\mathfrak{T}_i^{\alpha\rho};\mathbf{t}_R^{\alpha\rho}\mathbf{t}_F^{\alpha\rho};\mathbf{R}_{d0}^{\alpha-1};\mathbf{D};\mathbf{U};\mathbf{R}_d^{\alpha-1}\right),\\ \partial\mathcal{D}_{TU}\left(\mathfrak{T}_i^{\alpha\rho};\mathbf{t}_R^{\alpha\rho};\mathbf{t}_F^{\alpha\rho};\mathbf{R}_{d0}^{\alpha-1};\mathbf{D};\mathbf{U};\mathbf{R}_d^{\alpha-1}\right)\\ \cap\,\partial\mathfrak{N}_\xi\left(\mathfrak{T}_i^{\alpha\rho};\mathbf{t}_R^{\alpha\rho};\mathbf{t}_F^{\alpha\rho};\mathbf{R}_{d0}^{\alpha-1};\mathbf{D};\mathbf{U};\mathbf{R}_d^{\alpha-1}\right)=\phi.\end{gathered}\tag{27.16}$$

c) If, and only if, the intersection $\mathcal{D}_T\left(\mathfrak{T}_i^{\alpha\rho};\mathbf{t}_R^{\alpha\rho};\mathbf{t}_F^{\alpha\rho};\mathfrak{D}^j;\mathbf{U};\mathfrak{R}_d^{\alpha}\right)$ *of all FVRT state tracking domains* $\mathcal{D}_T\left(\mathbf{t}_0^{\alpha\rho};\mathbf{t}_R^{\alpha\rho};\mathbf{t}_F^{\alpha\rho};\mathbf{R}_{d0}^{\alpha-1};\mathbf{D};\mathbf{U};\mathbf{R}_d^{\alpha-1}\right)$ *in* $\left[\mathbf{t}_0^{\alpha\rho},\mathbf{D}(.),\mathbf{R}_{d0}^{\alpha-1}\right]\in\mathfrak{T}_i^{\alpha\rho}\times\mathfrak{D}^j\times\mathfrak{R}_{d0}^{\alpha}$ *is a neighborhood of* $\mathbf{R}_{d0}^{\alpha-1}$ *for every* $\mathbf{R}_{d0}^{\alpha-1}(.)\in\mathfrak{R}_{d0}^{\alpha}$ *then the FVRT state tracking of* $\mathbf{R}_d^{\alpha-1}(\mathbf{t}^{\alpha\rho})$ *on* $\mathfrak{T}_{0F}^{\alpha\rho}\times\mathfrak{D}^j\times\mathfrak{R}_d^{\alpha}$ *is* ***uniform in*** $\left[\mathbf{t}_0^{\alpha\rho},\mathbf{D}(.),\mathbf{R}_{d0}^{\alpha-1}\right]\in\mathfrak{T}_i^{\alpha\rho}\times\mathfrak{D}^j\times\mathfrak{R}_{d0}^{\alpha}$ ***on*** $\mathfrak{T}_{0F}^{\alpha\rho}\times\mathfrak{D}^j\times\mathfrak{R}_d^{\alpha}$*; i.e.,*

$$\begin{gathered}\exists\xi\in\mathfrak{R}^+\Longrightarrow\mathcal{D}_{TU}\left(\mathfrak{T}_i^{\alpha\rho};\mathbf{t}_R^{\alpha\rho};\mathbf{t}_F^{\alpha\rho};\mathfrak{D}^j;\mathbf{U};\mathfrak{R}_d^{\alpha}\right)\\ =\cap\left[\begin{array}{c}\mathcal{D}_T\left(\mathbf{t}_0^{\alpha\rho};\mathbf{t}_R^{\alpha\rho};\mathbf{t}_F^{\alpha\rho};\mathbf{R}_{d0}^{\alpha-1};\mathbf{D};\mathbf{U};\mathbf{R}_{d0}^{\alpha-1}\right):\\ :\left[\mathbf{t}_0^{\alpha\rho},\mathbf{D}(.),\mathbf{R}_{d0}^{\alpha-1}\right]\in\mathfrak{T}_i^{\alpha\rho}\times\mathfrak{D}^j\times\mathfrak{R}_{d0}^{\alpha}\end{array}\right]\\ \supset\mathfrak{N}_\xi\left(\mathfrak{T}_i^{\alpha\rho};\mathbf{t}_R^{\alpha\rho};\mathbf{t}_F^{\alpha\rho};\mathfrak{D}^j;\mathbf{U};\mathfrak{R}_d^{\alpha}\right),\\ \partial\mathcal{D}_{TU}\left(\mathfrak{T}_i^{\alpha\rho};\mathbf{t}_R^{\alpha\rho};\mathbf{t}_F^{\alpha\rho};\mathfrak{D}^j;\mathbf{U};\mathfrak{R}_d^{\alpha}\right)\cap\partial\mathfrak{N}_\xi\left(\mathfrak{T}_i^{\alpha\rho};\mathbf{t}_R^{\alpha\rho};\mathbf{t}_F^{\alpha\rho};\mathfrak{D}^j;\mathbf{U};\mathfrak{R}_d^{\alpha}\right)=\phi.\end{gathered}\tag{27.17}$$

Then, and only then, the set $\mathcal{D}_{TU}\left(\mathfrak{T}_i^{\alpha\rho};\mathbf{t}_R^{\alpha\rho};\mathbf{t}_F^{\alpha\rho};\mathfrak{D}^j;\mathbf{U};\mathfrak{R}_d^{\alpha}\right)$ *is* ***the domain of FVRT uniform state tracking of*** $\mathbf{R}_d^{\alpha-1}(\mathbf{t}^{\alpha\rho})$*, for every* $\mathbf{R}_{d0}^{\alpha-1}\in\mathfrak{R}_{d0}^{\alpha}$*, on* $\mathfrak{T}_{0F}^{\alpha\rho}\times\mathfrak{T}_i^{\alpha\rho}\times\mathfrak{D}^j\times\mathfrak{R}_d^{\alpha}$,

Note 415 *The global FVRT state tracking of* $\mathbf{R}_d^{\alpha-1}(\mathbf{t}^{\alpha\rho})$ *on the set product* $\mathfrak{T}_{0F}^{\alpha\rho}\times\mathfrak{D}^j\times\mathfrak{R}_d^{\alpha}$ *is uniform in* $\left[\mathbf{D}(.),\mathbf{R}_{d0}^{\alpha-1}\right]$ *over* $\mathfrak{D}^j\times\mathfrak{R}_d^{\alpha}$.

In order to assure a stability property of the state tracking with FVRT we introduce:

Definition 416 ***The stable elementwise state tracking of*** $\mathbf{R}_d^{\alpha-1}(\mathbf{t}^{\alpha\rho})$ ***with the finite vector reachability time*** $\mathbf{t}_R^{\alpha\rho}$ ***of the plant*** *(3.12)* ***controlled by a control*** $\mathbf{U}(.)\in\mathfrak{U}^l$

a) The plant (3.12) exhibits ***the stable elementwise state tracking of*** $\mathbf{R}_d^{\alpha-1}(\mathbf{t}^{\alpha\rho})$ ***with the finite vector reachability time*** $\mathbf{t}_R^{\alpha\rho}$ ***on*** $\mathfrak{T}_{0F}^{\alpha\rho}\times\mathfrak{D}^j\times\mathfrak{R}_d^{\alpha}$ *if, and only if, it exhibits the elementwise state tracking with the finite vector reachability time* $\mathbf{t}_R^{\alpha\rho}$ *on* $\mathfrak{T}_{0F}^{\alpha\rho}\times\mathfrak{D}^j\times\mathfrak{R}_d^{\alpha}$*, and for every connected neighborhood* $\mathfrak{N}_\varepsilon\left[\mathbf{t}_R^{\alpha\rho};\mathbf{R}_d^{\alpha-1}(\mathbf{t}^{\alpha\rho})\right]$ *of* $\mathbf{R}_d^{\alpha-1}(\mathbf{t}^{\alpha\rho})$ *at any* $\mathbf{t}^{\alpha\rho}\in\mathfrak{T}_{0F}^{\alpha\rho}$*, there is a connected neighborhood* $\mathfrak{N}\left(\varepsilon;\mathbf{t}_0^{\alpha\rho};\mathbf{t}_R^{\alpha\rho};\mathbf{t}_F^{\alpha\rho};\mathbf{R}_{d0}^{\alpha-1};\mathbf{D};\mathbf{U};\mathbf{R}_d^{\alpha-1}\right)\subseteq\mathfrak{R}^{\alpha\rho}$ *of the plant desired initial output vector* $\mathbf{R}_{d0}^{\alpha-1}$ *at the initial vector moment* $\mathbf{t}_0^{\alpha\rho}\in\mathfrak{T}^{\alpha\rho}$ *such that for the initial vector*

$$\mathbf{R}_0^{\alpha-1}\in\mathfrak{N}\left(\varepsilon;\mathbf{t}_0^{\alpha\rho};\mathbf{t}_R^{\alpha\rho};\mathbf{t}_F^{\alpha\rho};\mathbf{R}_{d0}^{\alpha-1};\mathbf{D};\mathbf{U};\mathbf{R}_d^{\alpha-1}\right)$$

the instantaneous $\mathbf{R}^{\alpha-1}(\mathbf{t}^{\alpha\rho})$ *stays in the neighborhood* $\mathfrak{N}_{\varepsilon}\left[\mathbf{t}_{R}^{\alpha\rho};\mathbf{R}_{d}^{\alpha-1}(\mathbf{t}^{\alpha\rho})\right]$ *for all* $\mathbf{t}^{\alpha\rho}\in\mathfrak{T}_{0F}^{\alpha\rho}$*; i.e.,*

$$\begin{gathered}\forall\mathfrak{N}_{\varepsilon}\left[\mathbf{t}_{R}^{\alpha\rho};\mathbf{R}_{d}^{\alpha-1}(\mathbf{t}^{\alpha\rho})\right]\subseteq\mathfrak{R}^{\alpha\rho},\ \forall\mathbf{t}^{\alpha\rho}\in\mathfrak{T}_{0F}^{\alpha\rho},\\ \forall\left[\mathbf{D}\left(.\right),\mathbf{R}_{d0}^{\alpha-1}\right]\in\mathfrak{D}^{j}\times\mathfrak{R}_{d0}^{\alpha},\\ \exists\mathfrak{N}\left(\varepsilon;\mathbf{t}_{0}^{\alpha\rho};\mathbf{t}_{R}^{\alpha\rho};\mathbf{t}_{F}^{\alpha\rho};\mathbf{R}_{d0}^{\alpha-1};\mathbf{D};\mathbf{U};\mathbf{R}_{d}^{\alpha-1}\right)\subseteq\mathfrak{R}^{\alpha\rho},\\ \mathfrak{N}\left(\varepsilon;\mathbf{t}_{0}^{\alpha\rho};\mathbf{t}_{R}^{\alpha\rho};\mathbf{t}_{F}^{\alpha\rho};\mathbf{R}_{d0}^{\alpha-1};\mathbf{D};\mathbf{U};\mathbf{R}_{d}^{\alpha-1}\right)\\ \subseteq\ \mathcal{D}_{T}\left(\mathbf{t}_{0}^{\alpha\rho};\mathbf{t}_{R}^{\alpha\rho};\mathbf{t}_{F}^{\alpha\rho};\mathbf{R}_{d0}^{\alpha-1};\mathbf{D};\mathbf{U};\mathbf{R}_{d}^{\alpha-1}\right),\\ \mathbf{R}_{0}^{\alpha-1}\in\mathfrak{N}\left(\varepsilon;\mathbf{t}_{0}^{\alpha\rho};\mathbf{t}_{R}^{\alpha\rho};\mathbf{t}_{F}^{\alpha\rho};\mathbf{R}_{d0}^{\alpha-1};\mathbf{D};\mathbf{U};\mathbf{R}_{d}^{\alpha-1}\right)\implies\\ \mathbf{R}^{\alpha-1}(\mathbf{t}^{\alpha\rho};\mathbf{t}_{0}^{\alpha\rho};\mathbf{R}_{0}^{\alpha-1};\mathbf{D};\mathbf{U})\in\mathfrak{N}_{\varepsilon}\left[\mathbf{t}_{R}^{\alpha\rho};\mathbf{R}_{d}^{\alpha-1}(\mathbf{t}^{\alpha\rho})\right],\ \forall\mathbf{t}^{\alpha\rho}\in\mathfrak{T}_{0F}^{\alpha\rho}.\end{gathered}\tag{27.18}$$

b) The largest connected neighborhood $\mathfrak{N}_{L}\left(\varepsilon;\mathbf{t}_{0}^{\alpha\rho};\mathbf{t}_{R}^{\alpha\rho};\mathbf{t}_{F}^{\alpha\rho};\mathbf{R}_{d0}^{\alpha-1};\mathbf{D};\mathbf{U};\mathbf{R}_{d}^{\alpha-1}\right)$ *of* $\mathbf{R}_{d0}^{\alpha-1}$*, (27.18), is* ***the state*** ε***-tracking domain***

$$\mathcal{D}_{ST}\left(\varepsilon;\mathbf{t}_{0}^{\alpha\rho};\mathbf{t}_{R}^{\alpha\rho};\mathbf{t}_{F}^{\alpha\rho};\mathbf{R}_{d0}^{\alpha-1};\mathbf{D};\mathbf{U};\mathbf{R}_{d}^{\alpha-1}\right)$$

at $\mathbf{t}_{0}^{\alpha\rho}\in\mathfrak{T}_{i}^{\alpha\rho}$ ***of the stable state tracking of*** $\mathbf{R}_{d}^{\alpha-1}(\mathbf{t}^{\alpha\rho})$ ***on*** $\mathfrak{T}_{0F}^{\alpha\rho}\times\mathfrak{D}^{j}\times\times\mathfrak{R}_{d}^{\alpha}$.

The domain $\mathcal{D}_{ST}\left(\mathbf{t}_{0}^{\alpha\rho};\mathbf{t}_{R}^{\alpha\rho};\mathbf{t}_{F}^{\alpha\rho};\mathbf{R}_{d0}^{\alpha-1};\mathbf{D};\mathbf{U};\mathbf{R}_{d}^{\alpha-1}\right)$ ***of the stable state tracking of*** $\mathbf{R}_{d}^{\alpha-1}(\mathbf{t}^{\alpha\rho})$ ***on*** $\mathfrak{T}_{0F}^{\alpha\rho}\times\mathfrak{D}^{j}\times\mathfrak{R}_{d}^{\alpha}$ *is the union of all*

$$\mathcal{D}_{ST}\left(\varepsilon;\mathbf{t}_{0}^{\alpha\rho};\mathbf{t}_{R}^{\alpha\rho};\mathbf{t}_{F}^{\alpha\rho};\mathbf{R}_{d0}^{\alpha-1};\mathbf{D};\mathbf{U};\mathbf{R}_{d}^{\alpha-1}\right)$$

over $\varepsilon\in\mathfrak{R}^{+}$*,*

$$\begin{gathered}\mathcal{D}_{ST}\left(\mathbf{t}_{0}^{\alpha\rho};\mathbf{t}_{R}^{\alpha\rho};\mathbf{t}_{F}^{\alpha\rho};\mathbf{R}_{d0}^{\alpha-1};\mathbf{D};\mathbf{U};\mathbf{R}_{d}^{\alpha-1}\right)\\ =\cup\left[\mathcal{D}_{ST}\left(\varepsilon;\mathbf{t}_{0}^{\alpha\rho};\mathbf{t}_{R}^{\alpha\rho};\mathbf{t}_{F}^{\alpha\rho};\mathbf{R}_{d0}^{\alpha-1};\mathbf{D};\mathbf{U};\mathbf{R}_{d}^{\alpha-1}\right):\varepsilon\in\mathfrak{R}^{+}\right].\end{gathered}\tag{27.19}$$

Let [0,ε_{M}*) be the maximal interval over which*

$$\mathcal{D}_{ST}\left(\varepsilon;\mathbf{t}_{0}^{\alpha\rho};\mathbf{t}_{R}^{\alpha\rho};\mathbf{t}_{F}^{\alpha\rho};\mathbf{R}_{d0}^{\alpha-1};\mathbf{D};\mathbf{U};\mathbf{R}_{d}^{\alpha-1}\right)$$

is continuous in $\varepsilon\in\mathfrak{R}_{+}$*,*

$$\begin{gathered}\mathcal{D}_{ST}\left(\varepsilon;\mathbf{t}_{0}^{\alpha\rho};\mathbf{t}_{R}^{\alpha\rho};\mathbf{t}_{F}^{\alpha\rho};\mathbf{R}_{d0}^{\alpha-1};\mathbf{D};\mathbf{U};\mathbf{R}_{d}^{\alpha-1}\right)\in\mathfrak{C}\left([0,\varepsilon_{M})\right),\\ \forall\left[\mathbf{D}\left(.\right),\mathbf{R}_{d0}^{\alpha-1}\right]\in\mathfrak{D}^{j}\times\mathfrak{R}_{d0}^{\alpha}.\end{gathered}$$

The strict domain $\mathcal{D}_{SST}\left(\mathbf{t}_{0}^{\alpha\rho};\mathbf{t}_{R}^{\alpha\rho};\mathbf{t}_{F}^{\alpha\rho};\mathbf{R}_{d0}^{\alpha-1};\mathbf{D};\mathbf{U};\mathbf{R}_{d}^{\alpha-1}\right)$ ***of the stable state tracking of*** $\mathbf{R}_{d}^{\alpha-1}(\mathbf{t}^{\alpha\rho})$ ***on*** $\mathfrak{T}_{0F}^{\alpha\rho}\times\mathfrak{D}^{j}\times\mathfrak{R}_{d}^{\alpha}$ *is the union of all stable state tracking domains*

$$\mathcal{D}_{ST}\left(\varepsilon;\mathbf{t}_{0}^{\alpha\rho};\mathbf{t}_{R}^{\alpha\rho};\mathbf{t}_{F}^{\alpha\rho};\mathbf{R}_{d0}^{\alpha-1};\mathbf{D};\mathbf{U};\mathbf{R}_{d}^{\alpha-1}\right)$$

over $\varepsilon\in[0,\varepsilon_{M})$*,*

$$\begin{gathered}\mathcal{D}_{SST}\left(\mathbf{t}_{0}^{\alpha\rho};\mathbf{t}_{R}^{\alpha\rho};\mathbf{t}_{F}^{\alpha\rho};\mathbf{R}_{d0}^{\alpha-1};\mathbf{D};\mathbf{U};\mathbf{R}_{d}^{\alpha-1}\right)\\ =\cup\left\{\mathcal{D}_{ST}\left(\varepsilon;\mathbf{t}_{0}^{\alpha\rho};\mathbf{t}_{R}^{\alpha\rho};\mathbf{t}_{F}^{\alpha\rho};\mathbf{R}_{d0}^{\alpha-1};\mathbf{D};\mathbf{U};\mathbf{R}_{d}^{\alpha-1}\right):\varepsilon\in[0,\varepsilon_{M})\right\},\\ \forall\left[\mathbf{D}\left(.\right),\mathbf{R}_{d0}^{\alpha-1}\right]\in\mathfrak{D}^{j}\times\mathfrak{R}_{d0}^{\alpha}.\end{gathered}\tag{27.20}$$

c) The stable state tracking of $\mathbf{R}_{d}^{\alpha-1}(\mathbf{t}^{\alpha\rho})$ *on* $\mathfrak{T}_{0F}^{\alpha\rho}\times\mathfrak{D}^{j}\times\mathfrak{R}_{d}^{\alpha}$ *is* ***global (in the whole)*** *if, and only if, it is the FVRT global state tracking of* $\mathbf{R}_{d}^{\alpha-1}(\mathbf{t}^{\alpha\rho})$ *on* $\mathfrak{T}_{0F}^{\alpha\rho}\times\mathfrak{D}^{j}\times\mathfrak{R}_{d}^{\alpha}$ *and the FVRT stable state tracking of* $\mathbf{R}_{d}^{\alpha-1}(\mathbf{t}^{\alpha\rho})$ *on*

$$\mathfrak{T}_{0F}^{\alpha\rho}\times\mathfrak{D}^{j}\times\mathfrak{R}_{d0}^{\alpha}$$

with $\mathcal{D}_{ST}\left(\mathbf{t}_{0}^{\alpha\rho};\mathbf{t}_{R}^{\alpha\rho};\mathbf{t}_{F}^{\alpha\rho};\mathbf{R}_{d0}^{\alpha-1};\mathbf{D};\mathbf{U};\mathbf{R}_{d}^{\alpha-1}\right)=\mathfrak{R}^{\alpha\rho},\ \forall\left[\mathbf{D}\left(.\right),\mathbf{R}_{d0}^{\alpha-1}\right]\in\mathfrak{D}^{j}\times\mathfrak{R}_{d0}^{\alpha}$.

Definition 417 ***The uniform stable elementwise state tracking of the desired state behavior*** $\mathbf{R}_d^{\alpha-1}(\mathbf{t}^{\alpha\rho})$ ***on*** $\mathfrak{T}_{0F}^{\alpha\rho}\times\mathfrak{D}^j\times\mathfrak{R}_d^{\alpha}$ ***of the plant*** *(3.12)* ***controlled by a control*** $\mathbf{U}(.)\in\mathfrak{U}^l$

a) If, and only if, the intersection $\mathcal{D}_{STU}\left(\mathbf{t}_0^{\alpha\rho};\mathbf{t}_R^{\alpha\rho};\mathbf{t}_F^{\alpha\rho};\mathfrak{D}^j;\mathbf{U};\mathfrak{R}_d^{\alpha}\right)$ *of all*

$$\mathcal{D}_{ST}\left(\mathbf{t}_0^{\alpha\rho};\mathbf{t}_R^{\alpha\rho};\mathbf{t}_F^{\alpha\rho};\mathbf{R}_{d0}^{\alpha-1};\mathbf{D};\mathbf{U};\mathbf{R}_d^{\alpha-1}\right)$$

[the intersection $\mathcal{D}_{SSTU}\left(\mathbf{t}_0^{\alpha\rho};\mathbf{t}_R^{\alpha\rho};\mathbf{t}_F^{\alpha\rho};\mathfrak{D}^j;\mathbf{U};\mathfrak{R}_d^{\alpha}\right)$ *of all*

$$\mathcal{D}_{SST}\left(\mathbf{t}_0^{\alpha\rho};\mathbf{t}_R^{\alpha\rho};\mathbf{t}_F^{\alpha\rho};\mathbf{R}_{d0}^{\alpha-1};\mathbf{D};\mathbf{U};\mathbf{R}_d^{\alpha-1}\right)]$$

over $\mathfrak{D}^j\times\mathfrak{R}_d^{\alpha}$ *is a connected neighborhood of* $\mathbf{R}_{d0}^{\alpha-1}$ *then, and only then, it is the* ***FVRT (strictly) stable state tracking domain of*** $\mathbf{R}_d^{\alpha-1}(\mathbf{t}^{\alpha\rho})$ ***on (the product set)*** $\mathfrak{T}_{0F}^{\alpha\rho}\times\mathfrak{D}^j\times\mathfrak{R}_{d0}^{\alpha}$ ***uniform in*** $\left[\mathbf{D}(.),\mathbf{R}_{d0}^{\alpha-1}\right]\in\mathfrak{D}^j\times\mathfrak{R}_{d0}^{\alpha}$ *, respectively,*

$$\begin{gathered}\exists\xi\in\mathfrak{R}^+\Longrightarrow\mathcal{D}_{STU}\left(\mathbf{t}_0^{\alpha\rho};\mathbf{t}_R^{\alpha\rho};\mathbf{t}_F^{\alpha\rho};\mathfrak{D}^j;\mathbf{U};\mathfrak{R}_d^{\alpha}\right)\\=\cap\left\{\begin{array}{c}\mathcal{D}_{ST}\left(\mathbf{t}_0^{\alpha\rho};\mathbf{t}_R^{\alpha\rho};\mathbf{t}_F^{\alpha\rho};\mathbf{R}_{d0}^{\alpha-1};\mathbf{D};\mathbf{U};\mathbf{R}_d^{\alpha-1}\right):\\:\left[\mathbf{D}(.),\mathbf{R}_{d0}^{\alpha-1}\right]\in\mathfrak{D}^j\times\mathfrak{R}_{d0}^{\alpha}\end{array}\right\}\\\supset\mathfrak{N}_{\xi}\left(\mathbf{t}_0^{\alpha\rho};\mathfrak{D}^j;\mathbf{U};\mathfrak{R}_d^{\alpha}\right),\end{gathered}\tag{27.21}$$

$$\left[\begin{gathered}\exists\xi\in\mathfrak{R}^+\Longrightarrow\mathcal{D}_{SSTU}\left(\mathbf{t}_0^{\alpha\rho};\mathbf{t}_R^{\alpha\rho};\mathbf{t}_F^{\alpha\rho};\mathfrak{D}^j;\mathbf{U};\mathfrak{R}_d^{\alpha}\right)\\=\cap\left\{\begin{array}{c}\mathcal{D}_{SST}\left(\mathbf{t}_0^{\alpha\rho};\mathbf{t}_R^{\alpha\rho};\mathbf{t}_F^{\alpha\rho};\mathbf{R}_{d0}^{\alpha-1};\mathbf{D};\mathbf{U};\mathbf{R}_d^{\alpha-1}\right):\\:\left[\mathbf{D}(.),\mathbf{R}_{d0}^{\alpha-1}\right]\in\mathfrak{D}^j\times\mathfrak{R}_{d0}^{\alpha}\end{array}\right\}\\\supset\mathfrak{N}_{\xi}\left(\mathbf{t}_0^{\alpha\rho};\mathfrak{D}^j;\mathbf{U};\mathfrak{R}_d^{\alpha}\right)\end{gathered}\right].\tag{27.22}$$

b) If, and only if, the intersection $\mathcal{D}_{STU}\left(\mathfrak{T}_i^{\alpha\rho};\mathbf{t}_R^{\alpha\rho};\mathbf{t}_F^{\alpha\rho};\mathbf{R}_{d0}^{\alpha-1};\mathbf{D};\mathbf{U};\mathbf{R}_d^{\alpha-1}\right)$ *of all*

$$\mathcal{D}_{ST}\left(\mathbf{t}_0^{\alpha\rho};\mathbf{t}_R^{\alpha\rho};\mathbf{t}_F^{\alpha\rho};\mathbf{R}_{d0}^{\alpha-1};\mathbf{D};\mathbf{U};\mathbf{R}_d^{\alpha-1}\right)$$

[the intersection $\mathcal{D}_{SSTU}\left(\mathfrak{T}_i^{\alpha\rho};\mathbf{t}_R^{\alpha\rho};\mathbf{t}_F^{\alpha\rho};\mathbf{R}_{d0}^{\alpha-1};\mathbf{D};\mathbf{U};\mathbf{R}_d^{\alpha-1}\right)$ *of all*

$$\mathcal{D}_{SST}\left(\mathbf{t}_0^{\alpha\rho};\mathbf{t}_R^{\alpha\rho};\mathbf{t}_F^{\alpha\rho};\mathbf{R}_{d0}^{\alpha-1};\mathbf{D};\mathbf{U};\mathbf{R}_d^{\alpha-1}\right)]$$

in $\mathbf{t}_0^{\alpha\rho}$ *over* $\mathfrak{T}_i^{\alpha\rho}$ *is a connected neighborhood of* $\mathbf{R}_{d0}^{\alpha-1}$*, then, and only then, it is* ***the FVRT (strict) stable state tracking domain of*** $\mathbf{R}_{d0}^{\alpha-1}(\mathbf{t}_R^{\alpha\rho})$ ***on*** $\mathfrak{T}_{0F}^{\alpha\rho}\times\mathfrak{D}^j\times\mathfrak{R}_d^{\alpha}$ ***uniform in*** $\mathbf{t}_0^{\alpha\rho}\in\mathfrak{T}_i^{\alpha\rho}$*, respectively,*

$$\begin{gathered}\exists\xi\in\mathfrak{R}^+\Longrightarrow\mathcal{D}_{STU}\left(\mathfrak{T}_i^{\alpha\rho};\mathbf{t}_R^{\alpha\rho};\mathbf{t}_F^{\alpha\rho};\mathbf{R}_{d0}^{\alpha-1};\mathbf{D};\mathbf{U};\mathbf{R}_d^{\alpha-1}\right)\\=\cap\left\{\mathcal{D}_{ST}\left(\mathbf{t}_0^{\alpha\rho};\mathbf{t}_R^{\alpha\rho};\mathbf{t}_F^{\alpha\rho};\mathbf{R}_{d0}^{\alpha-1};\mathbf{D};\mathbf{U};\mathbf{R}_d^{\alpha-1}\right):t_0\in\mathfrak{T}_i^{\alpha\rho}\right\}\\\supset\mathfrak{N}_{\xi}\left(\mathfrak{T}_i^{\alpha\rho};\mathbf{t}_R^{\alpha\rho};\mathbf{t}_F^{\alpha\rho};\mathbf{R}_{d0}^{\alpha-1};\mathbf{D};\mathbf{U};\mathbf{R}_d^{\alpha-1}\right),\end{gathered}\tag{27.23}$$

$$\left[\begin{gathered}\exists\xi\in\mathfrak{R}^+\Longrightarrow\mathcal{D}_{SSTU}\left(\mathfrak{T}_i^{\alpha\rho};\mathbf{t}_R^{\alpha\rho};\mathbf{t}_F^{\alpha\rho};\mathbf{R}_{d0}^{\alpha-1};\mathbf{D};\mathbf{U};\mathbf{R}_d^{\alpha-1}\right)\\=\cap\left\{\mathcal{D}_{SST}\left(\mathbf{t}_0^{\alpha\rho};\mathbf{t}_R^{\alpha\rho};\mathbf{t}_F^{\alpha\rho};\mathbf{R}_{d0}^{\alpha-1};\mathbf{D};\mathbf{U};\mathbf{R}_d^{\alpha-1}\right):t_0\in\mathfrak{T}_i^{\alpha\rho}\right\}\\\supset\mathfrak{N}_{\xi}\left(\mathfrak{T}_i^{\alpha\rho};\mathbf{t}_R^{\alpha\rho};\mathbf{t}_F^{\alpha\rho};\mathbf{R}_{d0}^{\alpha-1};\mathbf{D};\mathbf{U};\mathbf{R}_d^{\alpha-1}\right)\end{gathered}\right].\tag{27.24}$$

c) If, and only if, the intersection $\mathcal{D}_{STU}\left(\mathfrak{T}_i^{\alpha\rho};\mathbf{t}_R^{\alpha\rho};\mathbf{t}_F^{\alpha\rho};\mathfrak{D}^j;\mathbf{U};\mathfrak{R}_d^{\alpha}\right)$ *of all*

$$\mathcal{D}_{ST}\left(\mathbf{t}_0^{\alpha\rho};\mathbf{t}_R^{\alpha\rho};\mathbf{t}_F^{\alpha\rho};\mathbf{R}_{d0}^{\alpha-1};\mathbf{D};\mathbf{U};\mathbf{R}_d^{\alpha-1}\right)$$

[the intersection $\mathcal{D}_{SSTU}\left(\mathfrak{T}_i^{\alpha\rho};\mathbf{t}_R^{\alpha\rho};\mathbf{t}_F^{\alpha\rho};\mathfrak{D}^j;\mathbf{U};\mathfrak{R}_d^{\alpha}\right)$ *of all*

$$\mathcal{D}_{SST}\left(\mathbf{t}_0^{\alpha\rho};\mathbf{t}_R^{\alpha\rho};\mathbf{t}_F^{\alpha\rho};\mathbf{R}_{d0}^{\alpha-1};\mathbf{D};\mathbf{U};\mathbf{R}_d^{\alpha-1}\right)]$$

in $\left[\mathbf{t}_0^{\alpha\rho}, \mathbf{D}(.), \mathbf{R}_{d0}^{\alpha-1}\right] \in \mathfrak{T}_i^{\alpha\rho} \times \mathfrak{D}^j \times \mathfrak{R}_{d0}^{\alpha}$ *is a connected neighborhood of every* $\mathbf{R}_{d0}^{\alpha-1} \in \mathfrak{R}_{d0}^{\alpha}$ *then, and only then, it is the* ***FVRT (strictly) stable state tracking domain of*** $\mathbf{R}_d^{\alpha-1}(\mathbf{t}^{\alpha\rho})$ *on*

$$\mathfrak{T}_{0F}^{\alpha\rho} \times \mathfrak{D}^j \times \mathfrak{R}_{d0}^{\alpha}$$

uniform in $\left[\mathbf{t}_0^{\alpha\rho}, \mathbf{D}(.), \mathbf{R}_{d0}^{\alpha-1}\right] \in \mathfrak{T}_i^{\alpha\rho} \times \mathfrak{D}^j \times \mathfrak{R}_{d0}^{\alpha}$*, respectively,*

$$\begin{array}{c} \exists \xi \in \mathfrak{R}^{+} \Longrightarrow \mathcal{D}_{STU}\left(\mathfrak{T}_i^{\alpha\rho}; \mathbf{t}_R^{\alpha\rho}; \mathbf{t}_F^{\alpha\rho}; \mathfrak{D}^j; \mathbf{U}; \mathfrak{R}_d^{\alpha}\right) \\ = \cap \left\{ \begin{array}{c} \mathcal{D}_{ST}\left(\mathbf{t}_0^{\alpha\rho}; \mathbf{t}_R^{\alpha\rho}; \mathbf{t}_F^{\alpha\rho}; \mathbf{R}_{d0}^{\alpha-1}; \mathbf{D}; \mathbf{U}; \mathbf{R}_d^{\alpha-1}\right) : \\ : \left[\mathbf{t}_0^{\alpha\rho}, \mathbf{D}(.), \mathbf{R}_{d0}^{\alpha-1}\right] \in \mathfrak{T}_i^{\alpha\rho} \times \mathfrak{D}^j \times \mathfrak{R}_{d0}^{\alpha} \end{array} \right\} \\ \supset \mathfrak{N}_{\xi}\left(\mathfrak{T}_i^{\alpha\rho}; \mathbf{t}_R^{\alpha\rho}; \mathbf{t}_F^{\alpha\rho}; \mathfrak{D}^j; \mathbf{U}; \mathfrak{R}_d^{\alpha}\right), \end{array} \tag{27.25}$$

$$\left[\begin{array}{c} \exists \xi \in \mathfrak{R}^{+} \Longrightarrow \mathcal{D}_{SSTU}\left(\mathfrak{T}_i^{\alpha\rho}; \mathbf{t}_R^{\alpha\rho}; \mathbf{t}_F^{\alpha\rho}; \mathfrak{D}^j; \mathbf{U}; \mathfrak{R}_d^{\alpha}\right) \\ = \cap \left\{ \begin{array}{c} \mathcal{D}_{SST}\left(\mathbf{t}_0^{\alpha\rho}; \mathbf{t}_R^{\alpha\rho}; \mathbf{t}_F^{\alpha\rho}; \mathbf{R}_{d0}^{\alpha-1}; \mathbf{D}; \mathbf{U}; \mathbf{R}_{d0}^{\alpha-1}\right) : \\ : \left[\mathbf{t}_0^{\alpha\rho}, \mathbf{D}(.), \mathbf{R}_{d0}^{\alpha-1}\right] \in \mathfrak{T}_i^{\alpha\rho} \times \mathfrak{D}^j \times \mathfrak{R}_{d0}^{\alpha} \end{array} \right\} \\ \supset \mathfrak{N}_{\xi}\left(\mathfrak{T}_i^{\alpha\rho}; \mathbf{t}_R^{\alpha\rho}; \mathbf{t}_F^{\alpha\rho}; \mathfrak{D}^j; \mathbf{U}; \mathfrak{R}_d^{\alpha}\right). \end{array} \right] \tag{27.26}$$

Comment 418 *Every state tracking with the finite (scalar or vector) reachability time implies the perfect state tracking that starts at the (scalar or vector) reachability instant and continues until the final (scalar or vector) time, respectively. It expresses very high state tracking quality.*

Chapter 28

Criteria on contractive sets

28.1 Introduction

We continue with the unified study of tracking conditions. They incorporate the tracking conditions in the output space and those in the state space.

Note 354 through Note 356 are applicable to what follows. They enable us to reduce the study of the tracking properties of the system (3.12) described in terms of the total values of all variables to the tracking properties of the system (23.1) in terms of the deviations of all variables, which are induced by the former.

Definition 419 ***Definition of the family $\mathcal{F}_s$ of scalar functions***

A scalar function $f(.) : \mathfrak{T} \times \mathfrak{R}_+ \times In\mathfrak{T}_0 \longrightarrow \mathfrak{R}_+$ belongs to the family $\mathcal{F}_s$ of scalar functions, $f(.) \in \mathcal{F}_s$, if and only if, for the given scalar reachability time $t_R \in In\mathfrak{T}_0$ there is $\zeta > 0$, $\zeta = \zeta(t_R)$, such that:

a) $f(t, V; t_R) \geq 0 \ \forall (t, V) \in \mathfrak{T}_0 \times \mathcal{F}_\zeta$, $\mathcal{F}_\zeta = \{V : \ 0 \leq V < \zeta\}$.

b) $f(t, V; t_R) = 0$ if, and only if, $V = 0$, $\forall t \in \mathfrak{T}$.

c) The maximal solution $V(t, t_0; V_0; t_R)$ of

$$D^+V(t) \leq -f[t, V(t); t_R], \ \forall (t, V) \in \mathfrak{T}_0 \times \mathfrak{V}_\zeta(t) \tag{28.1}$$

obeys

$$V(t, t_0; V_0; t_R) \in \mathfrak{V}_\zeta(t), \ \forall (t, t_0; V_0) \in \mathfrak{T}_0 \times \mathfrak{T}_i \times \mathfrak{V}_\zeta(t_0), \tag{28.2}$$

$$V(t, t_0; V_0; t_R) = 0, \ \forall (t, t_0; V_0) \in [t_R, \infty[\times \mathfrak{T}_i \times \mathfrak{V}_\zeta(t_0), \tag{28.3}$$

$$V(t, t_0; 0; t_R) = 0, \ \forall (t, t_0) \in \mathfrak{T}_0 \times \mathfrak{T}_i. \tag{28.4}$$

Example 420 ***Possible forms of $f(.) \in \mathcal{F}_s$***

1) For $f(t, V; t_R) \equiv k \, signV$; i.e. $f(t, V; t_R) = k$ if, and only if, $V > 0$, and $f(t, V; t_R) = 0$ if, and only if, $V = 0$, $k = k(V_0) = V_0 (t_R - t_0)^{-1}$, the solution $V(t, t_0; V_0; t_R)$ of (28.1) reads:

$$V(t, t_0; V_0; t_R) = \left\{ \begin{array}{c} V_0 \left[1 - (t_R - t_0)^{-1} (t - t_0)\right], \ t \in [t_0, t_R], \\ 0, \ t \in [t_R, \infty[. \end{array} \right\} \tag{28.5}$$

2) For $f(t, V; t_R) \equiv kV^\gamma$, $k = k(V_0) = V_0^{1-\gamma} [(1-\gamma)(t_R - t_0)]^{-1}$, $\gamma \in]0, 1[$, the solution $V(t, t_0; V_0; t_R)$ of (28.1) reads:

$$V(t, t_0; V_0; t_R) = \left\{ \begin{array}{c} V_0 \left[1 - (t_R - t_0)^{-1} (t - t_0)\right]^{1/(1-\gamma)}, \ t \in [t_0, t_R], \\ 0, \ t \in [t_R, \infty[. \end{array} \right\}.$$

3) For $f(t,V;t_R) = kexp(-V)$, $k = k(V_0) = k = [exp(V_0) - 1](t_R - t_0)^{-1\cdot}$, *the solution* $V(t,t_0;V_0;t_R)$ *of (28.1) reads:*

$$V(t,t_0;V_0;t_R)$$
$$= \left\{ \left\langle \begin{array}{c} \ln\left\{\exp(V_0) - [\exp(V_0) - 1](t_R - t_0)^{-1\cdot}(t - t_0)\right\}, \\ t \in [t_0, t_R], \\ 0,\ t \in [t_R, \infty[. \end{array} \right\rangle \right\}$$

Definition 421 ***Definition of the family*** $\mathcal{F}_V$ ***of vector functions***

A vector function $\mathbf{f}(.) : \mathfrak{T}_0^K \times \mathfrak{R}_+^K \times In\mathfrak{T}_0^K \longrightarrow \mathfrak{R}_+^K$ *belongs to the family* $\mathcal{F}_V$ *of vector functions,* $\mathbf{f}(.) \in \mathcal{F}_V$, *if and only if, for the given vector reachability time* $\mathbf{t}_R \in In\mathfrak{T}_0^K$ *there is a vector* $\boldsymbol{\zeta} > \mathbf{0}_K$, $\boldsymbol{\zeta} = \boldsymbol{\zeta}(\mathbf{t}_R)$, *such that:*

a) $\mathbf{f}(\mathbf{t}^K, \mathbf{V}; \mathbf{t}_R) \geq \mathbf{0}_K$, $\forall (\mathbf{t}^K, \mathbf{V}) \in \mathfrak{T}_0^K \times \mathfrak{F}_\zeta$, $\mathfrak{F}_\zeta = \{\mathbf{V} : \mathbf{0}_K \leq \mathbf{V} < \zeta\}.$,

b) $\mathbf{f}(\mathbf{t}^K, \mathbf{V}; \mathbf{t}_R) = \mathbf{0}_K$ *if, and only if,* $V = \mathbf{0}_K$, $\forall \mathbf{t}^K \in \mathfrak{T}_0^K$.

c) The maximal solution $V(\mathbf{t}^K, \mathbf{t}_0^K; \mathbf{V}_0; \mathbf{t}_R)$ *of*

$$D^+\mathbf{V} \leq -\mathbf{f}(\mathbf{t}^K, \mathbf{V}; \mathbf{t}_R),\ \forall (\mathbf{t}^K, \mathbf{V}) \in \mathfrak{T}_0^K \times \mathfrak{F}_\zeta \tag{28.6}$$

obeys

$$\mathbf{V}(\mathbf{t}^K, \mathbf{t}_0^K; \mathbf{V}_0; t_R) \in \mathfrak{V}_\zeta(\mathbf{t}^K),\ \forall (\mathbf{t}^K, \mathbf{t}_0^K; \mathbf{V}_0) \in \mathfrak{T}_0^K \times \mathfrak{T}_i^K \times \mathfrak{F}_\zeta, \tag{28.7}$$
$$\mathbf{V}(\mathbf{t}^K, \mathbf{t}_0^K; \mathbf{V}_0; t_R) = \mathbf{0}_K,\ \forall (\mathbf{t}^K, \mathbf{t}_0^K; \mathbf{V}_0) \in [t_R, \infty\mathbf{1}_K[\times \mathfrak{T}_i^K \times \mathfrak{F}_\zeta, \tag{28.8}$$
$$\mathbf{V}(\mathbf{t}^K, \mathbf{t}_0^K; \mathbf{0}_K; t_R) = \mathbf{0}_K,\ \forall (\mathbf{t}^K, \mathbf{t}_0^K) \in \mathfrak{T}_0^K \times \mathfrak{T}_i^K. \tag{28.9}$$

Example 422 ***Possible forms of*** $\mathbf{f}(.) \in \mathcal{F}_V$

Let

$$I_K = diag\{1\ \ 1\ ...\ 1\} \in \mathfrak{R}^{K\times K}, L = diag\{l_1\ \ l_2\ ...\ l_K\} \in \mathfrak{R}^{K\times K},$$
$$t_R = [\tau_{R1}\ \ \tau_{R2}\ ....\ \ \tau_{RK}]^T, T = diag\{t\ \ t\ ....\ \ t\} \in \mathfrak{R}^{K\times K},$$
$$T_R = diag\{\tau_{R1}\ \ \tau_{R2}\ ....\ \ \tau_{RK}\} \in \mathfrak{R}^{K\times K}, T_0 = diag\{t_0\ \ t_0\ ....\ \ t_0\} \in \mathfrak{R}^{K\times K},$$

$$\mathbf{V} = [V_1\ V_2\ ...\ V_K]^T \in \mathfrak{R}_+^K,\ sign\mathbf{V} = \left\{ \begin{array}{c} \mathbf{1}_K,\ \mathbf{V} > \mathbf{0}_K, \\ = \mathbf{0}_K,\ \mathbf{V} = \mathbf{0}_K. \end{array} \right\},$$
$$\mathbf{V}^\gamma = [V_1^\gamma\ V_2^\gamma\ ...\ V_K^\gamma]^T,$$
$$exp(-\mathbf{V}) = [exp(-V_1)\ \ exp(-V_2)\ \ \cdots\ \ exp(-V_K)]^T,$$
$$\ln(\mathbf{V}) = [\ln(V_1)\ \ \ln(V_2)\ \ \cdots\ \ \ln(V_K)]^T,$$

and $\infty 0 = 0$.

1) For $\mathbf{f}(\mathbf{t}^K, \mathbf{V}; t_R) \equiv LsignV$, $L = L(\mathbf{V}_0) = (T_R - T_0)^{-1}V_0$, *the solution vector function* $V(\mathbf{t}^K, \mathbf{t}_0^K; \mathbf{V}_0; t_R)$ *of (28.1) reads:*

$$\mathbf{V}(\mathbf{t}^K, \mathbf{t}_0^K; \mathbf{V}_0; t_R) = \left\{ \begin{array}{c} \left[I_K - (T_R - T_0)^{-1}(T - T_0)\right]\mathbf{V}_0,\ T \in [T_0, T_R], \\ \mathbf{0}_K,\ T \in [T_R, \infty I_K[. \end{array} \right\} \tag{28.10}$$

2) For $\mathbf{f}(\mathbf{t}^K, \mathbf{V}; t_R) \equiv LV^\gamma$, $L = L(\mathbf{V}_0) = [(1-\gamma)(T_R - T_0)]^{-1}V_0^{1-\gamma}$, $\gamma \in]0,1[$, *the solution* $V(\mathbf{t}^K, \mathbf{t}_0^K; \mathbf{V}_0; t_R)$ *of (28.1) reads:*

$$\mathbf{V}(\mathbf{t}^K, \mathbf{t}_0^K; \mathbf{V}_0; t_R) = \left\{ \left\langle \begin{array}{c} \left[I_K - (T_R - T_0)^{-1}(T - T_0)\right]^{1/(1-\gamma)}\mathbf{V}_0, \\ T \in [T_0, T_R], \\ \mathbf{0}_K,\ T \in [T_R, \infty I_K[. \end{array} \right\rangle \right\}. \tag{28.11}$$

3) For $\mathbf{f}\left(\mathbf{t}^K,\mathbf{V};t_R\right) \equiv L exp(-\mathbf{V})$, $L = (T_R - T_0)^{-1\cdot}\left[exp(\mathbf{V}_0) - \mathbf{1}_K\right]$, $L = L(\mathbf{V}_0)$, *the solution* $V\left(\mathbf{t}^K,\mathbf{t}_0^K;\mathbf{V}_0;t_R\right)$ *of (28.1) reads:*

$$\mathbf{V}\left(\mathbf{t}^K,\mathbf{t}_0^K;\mathbf{V}_0;t_R\right)$$
$$= \left\{ \begin{array}{ll} \left\langle \begin{array}{c} \ln\left\{\exp(\mathbf{V}_0) - (T_R - T_0)^{-1\cdot}(T - T_0)\left[\exp(\mathbf{V}_0) - \mathbf{1}_K\right]\right\}, \\ T \in [T_0, T_R], \end{array} \right\rangle & \\ \mathbf{0}_K, & T \in [T_R, \infty I_K[. \end{array} \right\} \tag{28.12}$$

28.2 Stable tracking with *FSRT*

The following theorem represents the form of Theorem 357 adjusted to the stable tracking with the finite scalar reachability *time* (*FSRT*).

Theorem 423 ***Stable tracking with FSRT*** t_R ***of*** $\mathbf{Z}_d(t)$ ***of the system (3.12) on*** $\mathfrak{T}_0 \times \mathcal{J}^\xi$, ***equivalently of*** $\mathbf{z}_d(t) \equiv \mathbf{0}_K$ ***of the system*** *(23.1)* ***on*** $\mathfrak{T}_0 \times \mathcal{J}^\xi$

In order for the system (3.12), equivalently for the system (23.1), to exhibit, respectively, the stable tracking with FSRT t_R *of* $\mathbf{Z}_d(t)$, $\mathbf{Z}_d(.) \in \mathfrak{Z}_d$, *equivalently of* $\mathbf{z}_d(t) \equiv \mathbf{0}_K$, *on* $\mathfrak{T}_0 \times \mathcal{J}^\xi$ *it is sufficient that there exist a positive definite function* $V(.) : \mathfrak{T}_0 \times \mathfrak{R}^K \longrightarrow \mathfrak{R}_+$, *a function* $f(.) : \mathfrak{T}_0 \times \mathfrak{R}_+ \times In\mathfrak{T}_0 \longrightarrow \mathfrak{R}_+$, *and* $\xi_M > 0$ *such that:*

1) For every $\xi \in]0, \xi_M]$ *the set* $\mathfrak{V}_\xi(t;t_0)$ *induced by the function* $V(.)$:

a) Has the radially increasing boundary function.

b) Is asymptotically contractive.

2) The function $f(.)$ *belongs to the family* $\mathcal{F}_s$ *of scalar functions:* $f(.) \in \mathcal{F}_s$, *and*

3) Dini derivative $D^+V(.)$ *of the function* $V(.)$ *along solutions* $\mathbf{z}(t;t_0;\mathbf{z}_0;\mathbf{I})$ *of the system (23.1) obeys*

$$D^+V\left(t;t_0;V_0;t_R;\mathcal{J}^\xi\right) \leq -f\left[V\left(t,t_0;V_0;t_R;\mathcal{J}^\xi\right);t_R\right],$$
$$\forall [t, V_0, \mathbf{I}(.)] \in \mathfrak{T}_0 \times Cl\mathfrak{V}_{\xi_M}(t;t_0) \times \mathcal{J}^\xi. \tag{28.13}$$

The set $Cl\mathfrak{V}_{\xi_M}(t_0;t_0)$ *is a subset of the domain of the stable tracking with FSRT* t_R *on* $\mathfrak{T}_0 \times \mathcal{J}^\xi$.

Proof. The positive definiteness of $V(.)$ and the properties a) to c) of $f(.) \in \mathcal{F}_s$ together with the conditions 1) through 3) of the theorem statement show that all the conditions of Theorem 357 hold. Therefore, the system (3.12) exhibits the stable tracking of $\mathbf{Z}_d(t)$, $\mathbf{Z}_d(.) \in \mathfrak{Z}_d$, equivalently of $\mathbf{z}_d(t) \equiv \mathbf{0}_K$, on $\mathfrak{T}_0 \times \mathcal{J}^\xi$. From $f(.) \in \mathcal{F}_s$ follows the validity of Definition 419. Its conditions a) to c), i.e., (28.1) through (28.4), and (28.13) imply:

$$V\left(t,t_0;V_0;t_R;\mathcal{J}^\xi\right) \in \mathfrak{V}_\eta(t), \ \forall [t, t_0; V_0, \mathbf{I}(.)] \in \mathfrak{T}_0 \times \mathfrak{T}_i \times \mathfrak{V}_\eta(t_0) \times \mathcal{J}^\xi, \tag{28.14}$$

$$V\left(t,t_0;V_0;t_R;\mathcal{J}^\xi\right) = 0, \ \forall [t, t_0; V_0, \mathbf{I}(.)] \in [t_R, \infty[\times \mathfrak{T}_i \times \mathfrak{V}_\eta(t_0) \times \mathcal{J}^\xi, \tag{28.15}$$

$$V\left(t,t_0;0;t_R;\mathcal{J}^\xi\right) = 0, \ \forall (t,t_0) \in \mathfrak{T}_0 \times \mathfrak{T}_i, \tag{28.16}$$

for $\eta = \min(\zeta, \xi_M)$. The positive definiteness of $V(.)$ and (28.14) through (28.16) prove:

$$\mathbf{z}(t;t_0;\mathbf{z}_0;\mathbf{I}) \in \mathfrak{V}_\eta(t), \ \forall [t, t_0, \mathbf{z}_0, \mathbf{I}(.)] \in \mathfrak{T}_0 \times \mathfrak{T}_i \times \mathfrak{V}_\eta(t_0) \times \mathcal{J}^\xi, \tag{28.17}$$

$$\mathbf{z}(t;t_0;\mathbf{z}_0;\mathbf{I}) = \mathbf{0}_K, \ \forall [t, t_0, \mathbf{z}_0, \mathbf{I}(.)] \in [t_R, \infty[\times \mathfrak{T}_i \times \mathfrak{V}_\eta(t_0) \times \mathcal{J}^\xi, \tag{28.18}$$

$$\mathbf{z}(t;t_0;\mathbf{0}_K;\mathbf{I}) = \mathbf{0}_K, \ \forall [t, t_0, \mathbf{I}(.)] \in \mathfrak{T}_0 \times \mathfrak{T}_i \times \mathcal{J}^\xi. \tag{28.19}$$

The stable tracking is with FSRT t_R, Definition 400 ∎

Note 424 *By following Theorem 357 and Theorem 423 we can easily accommodate Theorem 359 and Theorem 361 to determine the conditions for the stable tracking with FSRT.*

28.3 Stable tracking with *FVRT*

The vector generalization of Theorem 423 reads:

Theorem 425 ***Elementwise stable tracking with FVRT*** t_R ***of*** $\mathbf{Z}_d(t)$ ***of the system (3.12) on*** $\mathfrak{T}_0 \times \mathcal{J}^\xi$***, equivalently of*** $\mathbf{z}_d(t) \equiv \mathbf{0}_K$ ***of the system (23.1) on*** $\mathfrak{T}_0 \times \mathcal{J}^\xi$

In order for the system (3.12), equivalently for the system (23.1), to exhibit, respectively, the elementwise stable tracking with FVRT t_R *of* $\mathbf{Z}_d(t)$, $\mathbf{Z}_d(.) \in \mathfrak{Z}_d$, *equivalently of* $\mathbf{z}_d(t) \equiv \mathbf{0}_K$, *on* $\mathfrak{T}_0 \times \mathcal{J}^\xi$ *it is sufficient that there exist a positive definite vector function* $V(.) : \mathfrak{T}_0^K \times \mathfrak{R}^K \longrightarrow \mathfrak{R}_+$, *a vector function* $\mathbf{f}(.) : \mathfrak{T}_0^K \times \mathfrak{R}^K \times In\mathfrak{T}_0^K \longrightarrow \mathfrak{R}_+^K$, *and a vector* $\boldsymbol{\xi}_M > \mathbf{0}_K$ *such that:*

1) For every vector $\boldsymbol{\xi} \in]\mathbf{0}_K, \boldsymbol{\xi}_M]$ *the set* $\mathfrak{V}_{\boldsymbol{\xi}}(\mathbf{t}^K)$ *induced by the vector function* $V(.)$ *:*

a) Has the elementwise radially increasing boundary function.

b) Is asymptotically contractive.

2) The vector function $\mathbf{f}(.)$ *belongs to the family* $\mathcal{F}_V$ *of vector functions:* $\mathbf{f}(.) \in \mathcal{F}_V$, *and*

3) Dini derivative $D^+V(.)$ *of the vector function* $V(.)$ *along the solutions* $\mathbf{z}(\mathbf{t}^K; \mathbf{t}_0^K; \mathbf{z}_0; \mathbf{I})$ *of the system (23.1) obeys*

$$D^+\mathbf{V}\left(\mathbf{t}^K; \mathbf{t}_0^K; \mathbf{V}_0; t_R; \mathcal{J}^\xi\right) \leq -\mathbf{f}\left[\mathbf{t}^K, \mathbf{V}\left(\mathbf{t}^K; \mathbf{t}_0^K; \mathbf{V}_0; t_R; \mathcal{J}^\xi\right); t_R\right],$$
$$\forall \left[\mathbf{t}^K, \mathbf{V}_0, \mathbf{I}(.)\right] \in \mathfrak{T}_0^K \times Cl\mathfrak{V}_{\boldsymbol{\xi}_M}\left(\mathbf{t}^K\right) \times \mathcal{J}^\xi. \tag{28.20}$$

The set $Cl\mathfrak{V}_{\boldsymbol{\xi}_M}(t_0; t_0)$ *is a subset of the domain of the elementwise stable tracking with FVRT* t_R *on* $\mathfrak{T}_0 \times \mathcal{J}^\xi$.

Proof. The proof is the straightforward vector generalization of the proof of Theorem 423. The positive definiteness of $V(.)$ and the properties a) to c) of $\mathbf{f}(.) \in \mathcal{F}_V$ together with the conditions 1) through 3) of the theorem statement show that all the conditions of Theorem 357 hold elementwise. Therefore, the system (3.12) exhibits the elementwise stable tracking of $\mathbf{Z}_d(t)$, $\mathbf{Z}_d(.) \in \mathfrak{Z}_d$, equivalently of $\mathbf{z}_d(t) \equiv \mathbf{0}_K$, on $\mathfrak{T}_0 \times \mathcal{J}^\xi$. From $\mathbf{f}(.) \in \mathcal{F}_V$ follows the validity of Definition 421. Its conditions a) to c), i.e., (28.6) through (28.9), and (28.20) imply:

$$\mathbf{V}\left(\mathbf{t}^K; \mathbf{t}_0^K; \mathbf{V}_0; t_R; \mathcal{J}^\xi\right) \in \mathfrak{V}_{\boldsymbol{\eta}}\left(\mathbf{t}^K\right),$$
$$\forall \left[\mathbf{t}^K, \mathbf{t}_0^K; \mathbf{V}_0, \mathbf{I}(.)\right] \in \mathfrak{T}_0^K \times \mathfrak{T}_i^K \times \mathfrak{V}_{\boldsymbol{\eta}}\left(\mathbf{t}^K\right) \times \mathcal{J}^\xi, \tag{28.21}$$
$$\mathbf{V}\left(\mathbf{t}^K; \mathbf{t}_0^K; \mathbf{V}_0; t_R; \mathcal{J}^\xi\right) = \mathbf{0}_K,$$
$$\forall \left[\mathbf{t}^K, \mathbf{t}_0^K; \mathbf{V}_0, \mathbf{I}(.)\right] \in [t_R, \infty \mathbf{1}_K[\times \mathfrak{T}_i^K \times \mathfrak{V}_{\boldsymbol{\eta}}\left(\mathbf{t}_0^K\right) \times \mathcal{J}^\xi, \tag{28.22}$$
$$\mathbf{V}\left(\mathbf{t}^K; \mathbf{t}_0^K; \mathbf{0}_K; t_R; \mathcal{J}^\xi\right) = \mathbf{0}_K, \ \forall \left(\mathbf{t}^K, \mathbf{t}_0^K\right) \in \mathfrak{T}_0^K \times \mathfrak{T}_i^K, \tag{28.23}$$

for $\boldsymbol{\eta} = \min\left(\boldsymbol{\zeta}, \boldsymbol{\xi}_M\right)$. The positive definiteness of $V(.)$ and (28.21) through (28.23) prove:

$$\mathbf{z}\left(\mathbf{t}^K; \mathbf{t}_0^K; \mathbf{z}_0; \mathbf{I}\right) \in \mathfrak{V}_\eta(t), \ \forall \left[\mathbf{t}^K, \mathbf{t}_0^K, \mathbf{z}_0, \mathbf{I}(.)\right] \in \mathfrak{T}_0^K \times \mathfrak{T}_i^K \times \mathfrak{V}_{\boldsymbol{\eta}}\left(\mathbf{t}^K\right) \times \mathcal{J}^\xi,$$
$$\mathbf{z}\left(\mathbf{t}^K; \mathbf{t}_0^K; \mathbf{z}_0; \mathbf{I}\right) = \mathbf{0}_K, \ \forall \left[\mathbf{t}^K, \mathbf{t}_0^K, \mathbf{z}_0, \mathbf{I}(.)\right] \in [t_R, \infty \mathbf{1}_K[\times \mathfrak{T}_i^K \times \mathfrak{V}_{\boldsymbol{\eta}}\left(\mathbf{t}^K\right) \times \mathcal{J}^\xi,$$
$$\mathbf{z}\left(\mathbf{t}^K; \mathbf{t}_0^K; \mathbf{0}_K; \mathbf{I}\right) = \mathbf{0}_K, \ \forall \left[\mathbf{t}^K, \mathbf{t}_0^K, \mathbf{I}(.)\right] \in \mathfrak{T}_0^K \times \mathfrak{T}_i^K \times \mathcal{J}^\xi.$$

The stable tracking is with FVRT t_R, Definition 405 ∎

Note 426 *By following Theorem 357, Theorem 423, and Theorem 425 we can easily accommodate Theorem 359 and Theorem 361 to determine the conditions for the stable tracking with FVRT.*

Chapter 29

Criteria on noncontractive sets

29.1 Stable tracking with *FSRT*

The following theorem shows how to adjust Theorem 364 to the uniform stable tracking with *FSRT* of $\mathbf{z} = \mathbf{0}_K$ of the system (23.1) on $\mathfrak{T}_0 \times \mathfrak{T}_i \times \mathcal{J}^{\xi}$.

Theorem 427 ***Extended Lyapunov theorem to the uniform stable tracking with FSRT of*** $\mathbf{Z}_d(t)$ ***of the system (3.12) on*** $\mathfrak{T}_0 \times \mathfrak{T}_i$ ***for every*** $\mathbf{i}(.) \in J^j$***, equivalently of*** $\mathbf{z} = \mathbf{0}_K$ ***of the system (23.1) on*** $\mathfrak{T}_0 \times \mathfrak{T}_i \times \mathcal{J}^{\xi}$***, equivalently of*** $\mathbf{z} = \mathbf{0}_K$ ***of the system (23.1) on*** $\mathfrak{T}_0 \times \mathfrak{T}_i$ ***for every*** $\mathbf{i}(.) \in J^j$

In order for the system (3.12), equivalently for the system (23.1), to exhibit the uniform stable tracking with FSRT of $\mathbf{Z}_d(t)$, $\mathbf{Z}_d(.) \in \mathfrak{Z}_d$, *equivalently of* $\mathbf{z}_d(t) \equiv \mathbf{0}_K$, *on* $\mathfrak{T}_0 \times \mathfrak{T}_i$ *for every* $\mathbf{i}(.) \in J^j$ *it is sufficient that there exist scalar functions* $V(.) : \mathfrak{T}_0 \times \mathfrak{R}^K \times \mathcal{J}^{\xi} \longrightarrow \mathfrak{R}_+$ *and* $f(.) : \mathfrak{T}_0 \times \mathfrak{R}^K \times In\mathfrak{T}_0 \longrightarrow \mathfrak{R}_+$, *and a connected neighborhood* $\mathfrak{N}$ *of the origin* $\mathbf{z} = \mathbf{0}_K$ *such that:*

1) The function $V(.)$ *is positive definite and decrescent on the neighborhood* $\mathfrak{N}$ *of the origin* $\mathbf{z} = \mathbf{0}_K$ *for every* $\mathbf{i}(.) \in \mathcal{J}^{\xi}$.

2) The function $f(.)$ *belongs to the family* $\mathcal{F}_s$ *of scalar functions:* $f(.) \in \mathcal{F}_s$.

3) Dini derivative $D^+V(.)$ *of* $V(.)$ *along solutions* $\mathbf{z}(t; t_0; \mathbf{z}_0; \mathbf{i})$ *of the system (23.1) obeys (28.13).*

Proof. Let all the conditions of the theorem statement hold. The conditions 1) through 3) and the properties a) to c) of $f(.) \in \mathcal{F}_s$, Definition 419, show that all the conditions of Theorem 364 are satisfied. This guarantees that the system (3.12), equivalently the system (23.1), exhibits the uniform stable tracking with *FSRT* of $\mathbf{Z}_d(t)$, $\mathbf{Z}_d(.) \in \mathfrak{Z}_d$, equivalently of $\mathbf{z}_d(t) \equiv \mathbf{0}_K$, on $\mathfrak{T}_0 \times \mathfrak{T}_i$ for every $\mathbf{i}(.) \in J^j$. The condition c) of Definition 419; i.e., (28.2)-(28.2), proves (28.17)-(28.19). The stable tracking is with *FSRT* t_R, Definition 400 ∎

29.2 Stable tracking with *FVRT*

In the case where the finite reachability *time* is vector $\mathbf{t}_R$ rather than scalar t_R then Theorem 427 becomes

Theorem 428 ***Elementwise uniform stable tracking with FVRT*** $\mathbf{t}_R$ ***of*** $\mathbf{Z}_d(t)$ ***of the system (3.12) on*** $\mathfrak{T}_0 \times \mathcal{J}^{\xi}$***, equivalently of*** $\mathbf{z}_d(t) \equiv \mathbf{0}_K$ ***of the system (23.1) on*** $\mathfrak{T}_0 \times \mathcal{J}^{\xi}$

In order for the system (3.12), equivalently for the system (23.1), to exhibit, respectively, the elementwise stable tracking with FVRT $\mathbf{t}_R$ *of* $\mathbf{Z}_d(t)$, $\mathbf{Z}_d(.) \in \mathfrak{Z}_d$, *equivalently of* $\mathbf{z}_d(t) \equiv \mathbf{0}_K$, *on* $\mathfrak{T}_0 \times \mathcal{J}^{\xi}$ *it is sufficient that there exist a vector function* $V(.) : \mathfrak{T}_0^K \times \mathfrak{R}^K \longrightarrow \mathfrak{R}_+$ *and a vector function* $\mathbf{f}(.) : \mathfrak{T}_0^K \times \mathfrak{R}^K \times In\mathfrak{T}_0^K \longrightarrow \mathfrak{R}_+^K$ *such that:*

1) The vector function $V(.)$ *is positive definite and decrescent on the neighborhood* $\mathfrak{N}$ *of the origin* $\mathbf{z} = \mathbf{0}_K$ *for every* $\mathbf{i}(.) \in \mathcal{J}^{\xi}$.

2) The vector function $\mathbf{f}(.)$ *belongs to the family* $\mathcal{F}_V$ *of vector functions:* $\mathbf{f}(.) \in \mathcal{F}_V$.

And

3) Dini derivative $D^+V(.)$ *of the vector function* $V(.)$ *along the solutions* $\mathbf{z}(.;\mathbf{t}_0^K;\mathbf{z}_0;\mathbf{i})$ *of the system (23.1) obeys*

$$D^+\mathbf{V}\left(\mathbf{t}^K;\mathbf{t}_0^K;\mathbf{V}_0;\mathbf{t}_R;\mathcal{J}^{\xi}\right) \leq -\mathbf{f}\left[\mathbf{t}^K,\mathbf{V}\left(\mathbf{t}^K;\mathbf{t}_0^K;\mathbf{V}_0;\mathbf{t}_R;\mathcal{J}^{\xi}\right);\mathbf{t}_R\right],$$
$$\forall\left[\mathbf{t}^K,\mathbf{V}_0,\mathbf{i}(.)\right] \in \mathfrak{T}_0^K \times C\mathfrak{V}_{\xi_M}\left(\mathbf{t}^K\right) \times \mathcal{J}^{\xi}. \tag{29.1}$$

The set $C\mathfrak{V}_{\xi_M}(t_0;t_0)$ *is a subset of the domain of the elementwise stable tracking with FVRT* $\mathbf{t}_R$ *on* $\mathfrak{T}_0 \times \mathcal{J}^{\xi}$.

Proof. In order to prove the sufficiency of the conditions of the theorem statement we accept that they are all fulfilled. The conditions 1)-3) and Definition 421, which is satisfied due to $\mathbf{f}(.) \in \mathcal{F}_V$, i.e., (28.7) - (28.9), together with Definition 288 and Definition 292, which are satisfied due to the condition 1), imply that all the conditions of Definition 402 through Definition 406 are fulfilled ■

Chapter 30

FRT tracking control synthesis

30.1 Internal dynamics space

30.1.1 Derivative of *V(.)* along motions

The following is necessary to present the results in a compact form:

$$\psi(t,\mathbf{r}^{\alpha-1}) = \frac{\partial V(t,\mathbf{r}^{\alpha-1})}{\partial t}$$
$$+\left[\frac{\partial V(t,\mathbf{r}^{\alpha-1})}{\partial r_1}..\frac{\partial V(t,\mathbf{r}^{\alpha-1})}{\partial r_\rho}....\frac{\partial V(t,\mathbf{r}^{\alpha-1})}{\partial r_1^{(\alpha-2)}}..\frac{\partial V(t,\mathbf{r}^{\alpha-1})}{\partial r_\rho^{(\alpha-2)}}\right]\frac{d\mathbf{r}^{\alpha-2}}{dt}, \tag{30.1}$$

$$\mathrm{grad}_{tr}V(t,\mathbf{r}^{\alpha-1}) = \begin{bmatrix} \frac{\partial V(t,\mathbf{r}^{\alpha-1})}{\partial r_1^{(\alpha-1)}} \\ . \\ . \\ . \\ \frac{\partial V(t,\mathbf{r}^{\alpha-1})}{\partial r_\rho^{(\alpha-1)}} \end{bmatrix} \in \mathfrak{R}^\rho, \tag{30.2}$$

where $\mathrm{grad}_{tr}V(t,\mathbf{r}^{\alpha-1})$ is introduced in (22.18).

Assumption 429 ***Choice of the function V(.)***

The functions $V(.)$ and $W(.)$ satisfy:

1) $V(.)$ is an ID gradient compatible, (Definition 337), continuously differentiable function on $\mathfrak{T}_0 \times \mathfrak{R}^{\alpha\rho}$, $V(t,\mathbf{r}^{\alpha-1}) \in C^1(\mathfrak{T}_0 \times \mathfrak{R}^{\alpha\rho})$, and positive definite.

2) The following rank condition holds:

$$rank\left\{\left[grad_{tr}V(t,\mathbf{r}^{\alpha-1})\right]^T \widetilde{Q}^{-1}(t,\mathbf{r}^{\alpha-1})P(t)\right\} = 0 \Longleftrightarrow \mathbf{r}^{\alpha-1} = \mathbf{0}_{\alpha\rho},\ \forall t \in \mathfrak{T}_0. \tag{30.3}$$

3) There exists $\xi_M > 0$ such that for every $\zeta \in]0,\xi_M]$ and for all $\mathbf{d}(.) \in \mathfrak{D}^j$ the set $\mathfrak{V}_\zeta(t;t_0;\mathfrak{D}^j)$ induced by $V(.)$:

a) Has the radially increasing boundary function.

b) Is asymptotically contractive.

4) $W(.;\mathbf{d})$ is positive on $\mathfrak{T}_0 \times (\mathfrak{R}^{\alpha\rho}\backslash\mathfrak{O})$ for every $\mathbf{d} \in \mathfrak{D}^j$ and $W(t,\mathbf{r}^{\alpha-1};\mathbf{d}) = 0$ if, and only if, $\mathbf{r}^{\alpha-1} = \mathbf{0}_{\alpha\rho}$, for every $(t,\mathbf{d}) \in \mathfrak{T}_0 \times \mathfrak{D}^j$.

Comment 430 ***The guide for the choice of the function $V(.)$***

The function $V(.)$ should be chosen to be ID gradient compatible positive definite and to satisfy the rank condition (30.3) relative to the plant characteristics $\widetilde{Q}^{-1}(t,\mathbf{r}^{\alpha-1})$ and $P(t)$. This helps to relate the choice of the function $V(.)$ to the plant characteristics.

Lemma 431 ***Relationship among*** $V(.)$, $W(.)$, ***and*** $\mathbf{u}(.)$

Let Assumption 429 hold. Let $r \geq \rho$ *and* $rankP(t) = \rho$ *for all* $t \in \mathfrak{T}_0$.

In order for the function $-W(.)$ *to be the derivative* $V^{(1)}(.)$ *of the function* $V(.)$,

$$-W(t, \mathbf{r}^{\alpha-1}; \mathbf{d}) \equiv V^{(1)}(t, \mathbf{r}^{\alpha-1}; \mathbf{d}), \tag{30.4}$$

such that

$$\begin{gathered} -W(\tau, \mathbf{r}^{\alpha-1}; \mathbf{d}) \equiv V^{(1)}(\tau, \mathbf{r}^{\alpha-1}; \mathbf{d}) \equiv 0 \Longrightarrow \\ -W(t, \mathbf{r}^{\alpha-1}; \mathbf{d}) \equiv V^{(1)}(t, \mathbf{r}^{\alpha-1}; \mathbf{d}) = 0, \ \forall t \geq \tau, \end{gathered} \tag{30.5}$$

along the motions of the plant (23.1) controlled by control $\mathbf{u}(.)$ *it is necessary and sufficient that control* $\mathbf{u}(.)$ *obeys*

$$\begin{gathered} \mathbf{r}^{\alpha-1}(t) \neq \mathbf{0}_{\alpha\rho} \Longrightarrow \widetilde{\mathbf{e}}_u\left[\mathbf{u}^{\mu}(t)\right] = \left\{\left[grad_{tr}V(t, \mathbf{r}^{\alpha-1})\right]^T \widetilde{Q}^{-1}\left[t, \mathbf{r}^{\alpha-1}(t)\right] P(t)\right\}^T \\ \bullet \left\langle \begin{array}{c} \left\{\left[grad_{tr}V(t, \mathbf{r}^{\alpha-1})\right]^T \widetilde{Q}^{-1}\left[t, \mathbf{r}^{\alpha-1}(t)\right] P(t)\right\} \bullet \\ \bullet \left\{\left[grad_{tr}V(t, \mathbf{r}^{\alpha-1})\right]^T \widetilde{Q}^{-1}\left[t, \mathbf{r}^{\alpha-1}(t)\right] P(t)\right\}^T \end{array} \right\rangle^{-1} \\ \bullet \left\{ \begin{array}{c} \psi(t, \mathbf{r}^{\alpha-1}) - W(t, \mathbf{r}^{\alpha-1}; \mathbf{d}) + \left[grad_{tr}V(t, \mathbf{r}^{\alpha-1})\right]^T \widetilde{Q}^{-1}\left[t, \mathbf{r}^{\alpha-1}(t)\right] \\ \bullet \left\{\widetilde{\mathbf{q}}\left[t, \mathbf{r}^{\alpha-1}(t)\right] - E(t)\widetilde{\mathbf{e}}_d\left[\mathbf{d}^{\eta}(t)\right]\right\} \end{array} \right\}, \\ \forall t \in [t_0, \tau[, \\ \mathbf{r}^{\alpha-1}(\tau) = \mathbf{0}_{\alpha\rho} \Longrightarrow \\ \widetilde{\mathbf{e}}_u\left[\mathbf{u}^{\mu}(t)\right] = P^T(t)\left[P(t)P^T(t)\right]^{-1}\left\{\widetilde{\mathbf{q}}\left[t, \mathbf{r}^{\alpha-1}(t)\right] - E(t)\widetilde{\mathbf{e}}_d\left[\mathbf{d}^{\eta}(t)\right]\right\}, \\ \forall t \geq \tau \end{gathered} \tag{30.6}$$

The proof of this Lemma is in Appendix D.6.

This lemma is applicable to perfect trackable plants due to the conditions: $r \geq \rho$ and $\text{rank}P(t) = \rho$, Theorem 154.

30.1.2 Asymptotically contractive sets

Lemma 431 enables the simple proof of the following theorem:

Theorem 432 ***Stable tracking of*** $\mathbf{r}_d^{\alpha-1}(t) \equiv \mathbf{0}_{\alpha\rho}$ ***of the plant*** *(23.1)* ***on*** $\mathfrak{T}_0 \times \mathfrak{D}^j$

Let Lemma 431 be valid.

1) In order for the plant (23.1) controlled by the control $\mathbf{u}(t)$, *to exhibit the stable tracking of* $\mathbf{r}_d^{\alpha-1}(t) \equiv \mathbf{0}_{\alpha\rho}$, *on* $\mathfrak{T}_0 \times \mathfrak{D}^j$, *it is sufficient that the control* $\mathbf{u}(.)$ *obeys Lemma 431, i.e., (30.6).*

2) If, additionally to 1), $W(.;\mathbf{d})$ *depends on* $V(.)$ *so that* $W(t, V; \mathbf{d};; t_R) \in \mathcal{F}_s$ *for every* $\mathbf{d} \in \mathfrak{D}^j$ *then the plant (23.1) exhibits the stable tracking with FSRT* t_R *of* $\mathbf{r}_d^{\alpha-1}(t) \equiv \mathbf{0}_{\alpha\rho}$, *on* $\mathfrak{T}_0 \times \mathfrak{D}^j$.

Proof. Let Lemma 431 be valid. Assumption 429 is therefore also valid.

1) The validity of Assumption 429 and Lemma 431, i.e., (30.6) imply the validity of all the conditions of Theorem 357 for $\mathbf{Z} = \mathbf{R}^{\alpha-1}$, i.e., for $\mathbf{z} = \mathbf{r}^{\alpha-1}$. The plant (3.12), equivalently the plant (23.1), respectively, controlled by the control $\mathbf{u}(t)$ (30.6) exhibits the stable tracking of $\mathbf{r}_d^{\alpha-1}(t) \equiv \mathbf{0}_{\alpha\rho}$, on $\mathfrak{T}_0 \times \mathfrak{D}^j$.

2) The conditions under 1) and $W(t, V; \mathbf{d};; t_R) \in \mathcal{F}_s$ for every $\mathbf{d} \in \mathfrak{D}^j$ imply that the conditions of Theorem 423 are satisfied. The plant (3.12), equivalently the plant (23.1), respectively, controlled by the control $\mathbf{u}(t)$ (30.6) exhibits the stable tracking with *FSRT* t_R of $\mathbf{r}_d^{\alpha-1}(t) \equiv \mathbf{0}_{\alpha\rho}$, on $\mathfrak{T}_0 \times \mathfrak{D}^j$. ∎

Note 433 *This theorem is the basis for synthesis of a robust control that is independent of* $\mathbf{d} \in \mathfrak{D}^j$. *Various approaches to design the robust stabilizing control proposed in [1], [5], [72], [88], [89], [91], [100], [106], [114], [119], [121], [136], [270], [291], [408], [423], [437], [495], [527], [575], and [576] can be explored to appropriately adjust Theorem 432 to robust stable tracking.*

Comment 434 *The control (30.6) depends on the real form and the instantaneous value of* $E(t)\widetilde{\mathbf{e}}_d[\mathbf{d}^\eta(t)]$. *The instantaneous value of* $E(t)\widetilde{\mathbf{e}}_d[\mathbf{d}^\eta(t)]$, *hence of the extended disturbance deviation* $\mathbf{d}^\eta(t)$, *should be measurable and measured at every moment* $t \in \mathfrak{T}_0$ *for control realization and implementation.*

Note 435 ***Relaxation of the conditions on control and robustness***
Theorem 423 permits replacing both $W(.)$ *by* $W(.)$ *and the condition (30.4) by (30.7):*

$$-W(t,\mathbf{r}^{\alpha-1};\mathfrak{D}^j) \geq V^{(1)}(t,\mathbf{r}^{\alpha-1};\mathbf{d}),\ \forall [t,\mathbf{d}(.)] \in \mathfrak{T}_0 \times \mathfrak{D}^j, \tag{30.7}$$

so that the condition (30.5) rests unchanged. This permits expressing the control in terms of the boundaries $\widetilde{\mathbf{e}}_{dA}$ *of* $E(t)\widetilde{\mathbf{e}}_d[\mathbf{d}^\eta(t)]$,

$$|E(t)\widetilde{\mathbf{e}}_d[\mathbf{d}^\eta(t)]| \leq \widetilde{\mathbf{e}}_{dA},\ \forall [t,\mathbf{d}(.)] \in \mathfrak{T}_0 \times \mathfrak{D}^j. \tag{30.8a}$$

For the following control

$$\begin{gathered}
\mathbf{r}^{\alpha-1}(t) \neq \mathbf{0}_{\alpha\rho} \Longrightarrow \widetilde{\mathbf{e}}_u[\mathbf{u}^\mu(t)] = \left\{\left[grad_{t_r}V(t,\mathbf{r}^{\alpha-1})\right]^T \widetilde{Q}^{-1}\left[t,\mathbf{r}^{\alpha-1}(t)\right]P(t)\right\}^T \\
\bullet\left\langle \begin{array}{c} \left\{\left[grad_{t_r}V(t,\mathbf{r}^{\alpha-1})\right]^T \widetilde{Q}^{-1}\left[t,\mathbf{r}^{\alpha-1}(t)\right]P(t)\right\} \\ \bullet\left\{\left[grad_{t_r}V(t,\mathbf{r}^{\alpha-1})\right]^T \widetilde{Q}^{-1}\left[t,\mathbf{r}^{\alpha-1}(t)\right]P(t)\right\}^T \end{array}\right\rangle^{-1} \\
\bullet\left\{ \begin{array}{c} \psi(t,\mathbf{r}^{\alpha-1}) - W(t,\mathbf{r}^{\alpha-1};\mathbf{d}) - \left|\left[grad_{t_r}V(t,\mathbf{r}^{\alpha-1})\right]^T \widetilde{Q}^{-1}\left[t,\mathbf{r}^{\alpha-1}(t)\right]\right|\widetilde{\mathbf{e}}_{dA} \\ +\left[grad_{t_r}V(t,\mathbf{r}^{\alpha-1})\right]^T \widetilde{Q}^{-1}\left[t,\mathbf{r}^{\alpha-1}(t)\right]\widetilde{\mathbf{q}}\left[t,\mathbf{r}^{\alpha-1}(t)\right], \end{array}\right\}, \\
t \in [t_0,\tau], \\
\mathbf{r}^{\alpha-1}(\tau) = \mathbf{0}_{\alpha\rho} \Longrightarrow \\
\widetilde{\mathbf{e}}_u[\mathbf{u}^\mu(t)] = P^T(t)\left[P(t)P^T(t)\right]^{-1}\left\{\widetilde{\mathbf{q}}\left[t,\mathbf{r}^{\alpha-1}(t)\right] - E(t)\widetilde{\mathbf{e}}_d[\mathbf{d}^\eta(t)]\right\}, \\
\forall t \in [\tau,\infty[
\end{gathered} \tag{30.9}$$

$V^{(1)}(t,\mathbf{r}^{\alpha-1})$ *satisfies (30.7); i.e.,*

$$V^{(1)}(t,\mathbf{r}^{\alpha-1}) \leq -W(t,\mathbf{r}^{\alpha-1};\mathbf{D}),\ \forall [t,\mathbf{d}(.)] \in \mathfrak{T}_0 \times \mathfrak{D}^j, \tag{30.10}$$

because

$$\begin{gathered}
\mathbf{r}^{\alpha-1}(t) \neq \mathbf{0}_{\alpha\rho} \Longrightarrow V^{(1)}(t,\mathbf{r}^{\alpha-1}) = \psi(t,\mathbf{r}^{\alpha-1}) \\
+\left[\nabla V(t,\mathbf{r}^{\alpha-1})\right]^T \widetilde{Q}^{-1}\left[t,\mathbf{r}^{\alpha-1}(t)\right]\left\{ \begin{array}{c} E(t)\widetilde{\mathbf{e}}_d[\mathbf{d}^\eta(t)] \\ +P(t)\widetilde{\mathbf{e}}_u[\mathbf{u}^\mu(t)] - \mathbf{q}\left[t,\mathbf{r}^{\alpha-1}(t)\right] \end{array}\right\}
\end{gathered}$$

$$\begin{gathered}
= \psi(t,\mathbf{r}^{\alpha-1}) + \left[grad_{t_r}V(t,\mathbf{r}^{\alpha-1})\right]^T \widetilde{Q}^{-1}\left[t,\mathbf{r}^{\alpha-1}(t)\right]E(t)\widetilde{\mathbf{e}}_d[\mathbf{d}^\eta(t)] \\
+\left[grad_{t_r}V(t,\mathbf{r}^{\alpha-1})\right]^T \widetilde{Q}^{-1}\left[t,\mathbf{r}^{\alpha-1}(t)\right]P(t)\widetilde{\mathbf{e}}_u[\mathbf{u}^\mu(t)] \\
-\left[grad_{t_r}V(t,\mathbf{r}^{\alpha-1})\right]^T \widetilde{Q}^{-1}\left[t,\mathbf{r}^{\alpha-1}(t)\right]\mathbf{q}\left[t,\mathbf{r}^{\alpha-1}(t)\right]
\end{gathered}$$

$$= \left[grad_{tr}V(t,\mathbf{r}^{\alpha-1})\right]^T \widetilde{Q}^{-1}\left[t,\mathbf{r}^{\alpha-1}(t)\right] E(t)\widetilde{\mathbf{e}}_d\left[\mathbf{d}^{\eta}(t)\right]$$
$$-\left|\left[grad_{tr}V(t,\mathbf{r}^{\alpha-1})\right]^T \widetilde{Q}^{-1}\left[t,\mathbf{r}^{\alpha-1}(t)\right]\right| \widetilde{\mathbf{e}}_{dA} - W(t,\mathbf{r}^{\alpha-1};\mathbf{D}) \leq -W(t,\mathbf{r}^{\alpha-1};\mathbf{D}),$$
$$\forall\left[t,\mathbf{d}(.)\right] \in \mathfrak{T}_0 \times \mathfrak{D}^j.$$

The control (30.9) does not depend on the real form and the instantaneous value of the product $E(t)\widetilde{\mathbf{e}}_d\left[\mathbf{d}^{\eta}(t)\right]$ *because it is replaced by its absolute upper bound* $\widetilde{\mathbf{e}}_{dA}$, *(30.8a), in (30.6).*

Note 436 *It is easy to apply Theorem 359 and Theorem 361 to the stable tracking control synthesis by following Theorem 432, its proof, and Note 435.*

30.1.3 Noncontractive *time*-varying sets

We will illustrate how Theorem 364 and Theorem 365 can be used for Lyapunov synthesis of the stable tracking control. We will do this by using Theorem 364.

Assumption 437 ***Choice of the function V(.)***
The functions $V(.)$ *and* $W(.)$ *satisfy:*
1) $V(.)$ *is positive definite and decrescent.*
2) The rank condition (30.3) holds.
3) $W(.;\mathbf{d})$ *is positive definite on the neighborhood* $\mathfrak{N}$ *of* $\mathbf{r}_d^{\alpha-1}(t) \equiv \mathbf{0}_{\alpha\rho}$ *for every* $\mathbf{d} \in \mathfrak{D}^j$.

Theorem 438 ***Stable tracking of*** $\mathbf{r}_d^{\alpha-1}(t) \equiv \mathbf{0}_{\alpha\rho}$ ***of the plant*** *(23.1)* ***on*** $\mathfrak{T}_0 \times \mathfrak{D}^j$
Let Lemma 431 be valid.
1) In order for the plant (3.12), equivalently the plant (23.1), respectively, controlled by the control $\mathbf{u}(t)$, *to exhibit the stable tracking of* $\mathbf{r}_d^{\alpha-1}(t) \equiv \mathbf{0}_{\alpha\rho}$, *on* $\mathfrak{T}_0 \times \mathfrak{D}^j$ *it is sufficient that control* $\mathbf{u}(.)$ *obeys Lemma 431, i.e., (30.6).*
2) If, additionally to 1), $W(.;\mathbf{d})$ *depends on* $V(.)$ *so that* $W(t,V;\mathbf{d};;t_R) \in \mathcal{F}_s$ *for every* $\mathbf{d} \in \mathfrak{D}^j$ *then the plant (23.1) exhibits the stable tracking with FSRT* t_R *of* $\mathbf{r}_d^{\alpha-1}(t) \equiv \mathbf{0}_{\alpha\rho}$, *on* $\mathfrak{T}_0 \times \mathfrak{D}^j$.

Proof. Let Lemma 431 be valid. Assumption 429 is therefore also valid.
1) Assumption 437 and Lemma 431, i.e., (30.6), prove the validity of all the conditions of Theorem 364. The plant (3.12), equivalently the plant (23.1), respectively, controlled by the control $\mathbf{u}(t)$ (30.6) exhibits the stable tracking of $\mathbf{r}_d^{\alpha-1}(t) \equiv \mathbf{0}_{\alpha\rho}$, on $\mathfrak{T}_0 \times \mathfrak{D}^j$.
2) The conditions under 1) and $W(t,V;\mathbf{d};;t_R) \in \mathcal{F}_s$ for every $\mathbf{d} \in \mathfrak{D}^j$ imply that the conditions of Theorem 427 are fulfilled. The plant (3.12), equivalently the plant (23.1), respectively, controlled by the control $\mathbf{u}(t)$ (30.6) exhibits the stable tracking with FSRT t_R of $\mathbf{r}_d^{\alpha-1}(t) \equiv \mathbf{0}_{\alpha\rho}$, on $\mathfrak{T}_0 \times \mathfrak{D}^j$. ∎
Note 435 is applicable also to Theorem 438.

30.2 Output space

30.2.1 Derivative of $V(.)$ along the output behavior

We refer to Conclusion 345.

There are two possibilities for the application of the Lyapunov method to the output Lyapunov plant (5.17), Definition 342, in order to study tracking in the output space:
A) To eliminate $\mathbf{r}^{\alpha}(t)$ from $V^{(1)}(t,\mathbf{y})$ by using (22.32)
B) To eliminate $\mathbf{r}^{\alpha-1}(t)$ from $V^{(1)}(t,\mathbf{y})$ by using (22.33)

A) Let the function $V(.) : \mathfrak{T} \times \mathfrak{R}^N \longrightarrow \mathfrak{R}$ be a decrescent, positive definite, gradient compatible, differentiable function. We use jointly (22.22), (22.23), (22.27):

$$
\begin{gathered}
V^{(1)}(t, \mathbf{y}) \\
= [\mathrm{grad}V(t, \mathbf{y})]^T \left\{ \begin{array}{c}
\frac{\partial \widetilde{\mathbf{z}}\left[t, \mathbf{r}^{\alpha-1}(t), \mathbf{d}(t)\right]}{\partial t} + \frac{\partial^* \widetilde{\mathbf{z}}\left[t, \mathbf{r}^{\alpha-1}(t), \mathbf{d}(t)\right]}{\partial \mathbf{r}^{\alpha-2}} \mathbf{r}^{\alpha-1} \\
+ \frac{\partial \widetilde{\mathbf{z}}\left[t, \mathbf{r}^{\alpha-1}(t), \mathbf{d}(t)\right]}{\partial \mathbf{r}_\rho^{\alpha-1}} \mathbf{r}^{(\alpha)} + \frac{\partial \widetilde{\mathbf{z}}\left[t, \mathbf{r}^{\alpha-1}(t), \mathbf{d}(t)\right]}{\partial \mathbf{d}} \mathbf{d}^{(1)} \\
+ W^{(1)}(t)\, \widetilde{\mathbf{w}}(\mathbf{u}) + W(t)\, \widetilde{\mathbf{w}}^{(1)}(\mathbf{u}),
\end{array} \right\}, \\
\forall t \in \mathfrak{T}.
\end{gathered} \tag{30.11}
$$

The equation (22.32) permits us to eliminate $\mathbf{r}^{(\alpha)}(t)$ from (30.11):

$$
\begin{gathered}
V^{(1)}(t, \mathbf{y}) \\
= [\mathrm{grad}V(t, \mathbf{y})]^T \left\{ \begin{array}{c}
\frac{\partial \widetilde{\mathbf{z}}\left[t, \mathbf{r}^{\alpha-1}(t), \mathbf{d}(t)\right]}{\partial t} + \frac{\partial^* \widetilde{\mathbf{z}}\left[t, \mathbf{r}^{\alpha-1}(t), \mathbf{d}(t)\right]}{\partial \mathbf{r}^{\alpha-2}} \mathbf{r}^{\alpha-1} \\
+ \frac{\partial \widetilde{\mathbf{z}}\left[t, \mathbf{r}^{\alpha-1}(t), \mathbf{d}(t)\right]}{\partial \mathbf{d}} \mathbf{d}^{(1)} \\
+ W^{(1)}(t)\, \widetilde{\mathbf{w}}(\mathbf{u}) + W(t)\, \widetilde{\mathbf{w}}^{(1)}(\mathbf{u}),
\end{array} \right\} \\
+ [\mathrm{grad}V(t, \mathbf{y})]^T \left\{ \begin{array}{c}
\frac{\partial \widetilde{\mathbf{z}}\left[t, \mathbf{r}^{\alpha-1}(t), \mathbf{d}(t)\right]}{\partial \mathbf{r}_\rho^{\alpha-1}} \widetilde{Q}^{-1}\left[t, \mathbf{r}^{\alpha-1}(t)\right] \\
\bullet \left\{ \begin{array}{c} -\widetilde{\mathbf{q}}\left[t, \mathbf{r}^{\alpha-1}(t)\right] \\ + E(t)\, \widetilde{\mathbf{e}}_d\left[t, \mathbf{d}(t)\right] + P(t)\, \widetilde{\mathbf{e}}_u\left[\mathbf{u}^\mu(t)\right] \end{array} \right\}
\end{array} \right\}.
\end{gathered}
$$

The following compact terms:

$$
\begin{gathered}
\Theta\left[t, \mathbf{r}^{\alpha-1}(t), \mathbf{y}(t), \mathbf{d}(t)\right] \\
= [\mathrm{grad}V(t, \mathbf{y})]^T \left\{ \begin{array}{c}
\frac{\partial \widetilde{\mathbf{z}}\left[t, \mathbf{r}^{\alpha-1}(t), \mathbf{d}(t)\right]}{\partial t} + \frac{\partial^* \widetilde{\mathbf{z}}\left[t, \mathbf{r}^{\alpha-1}(t), \mathbf{d}(t)\right]}{\partial \mathbf{r}^{\alpha-2}} \mathbf{r}^{\alpha-1} \\
+ \frac{\partial \widetilde{\mathbf{z}}\left[t, \mathbf{r}^{\alpha-1}(t), \mathbf{d}(t)\right]}{\partial \mathbf{r}_\rho^{\alpha-1}} \mathbf{r}^{(\alpha)} + \frac{\partial \widetilde{\mathbf{z}}\left[t, \mathbf{r}^{\alpha-1}(t), \mathbf{d}(t)\right]}{\partial \mathbf{d}} \mathbf{d}^{(1)} \\
+ \left\{ \begin{array}{c} \frac{\partial \widetilde{\mathbf{z}}\left[t, \mathbf{r}^{\alpha-1}(t), \mathbf{d}(t)\right]}{\partial \mathbf{r}_\rho^{\alpha-1}} \widetilde{Q}^{-1}\left[t, \mathbf{r}^{\alpha-1}(t)\right] \\ \bullet \left\{ -\widetilde{\mathbf{q}}\left[t, \mathbf{r}^{\alpha-1}(t)\right] + E(t)\, \widetilde{\mathbf{e}}_d\left[t, \mathbf{d}(t)\right] \right\} \end{array} \right\}
\end{array} \right\},
\end{gathered} \tag{30.12}
$$

$$
\widehat{P}(t) = \frac{\partial \widetilde{\mathbf{z}}\left[t, \mathbf{r}^{\alpha-1}(t), \mathbf{d}(t)\right]}{\partial \mathbf{r}_\rho^{\alpha-1}} \widetilde{Q}^{-1}\left[t, \mathbf{r}^{\alpha-1}(t)\right] P(t), \tag{30.13}
$$

$$
W^{(1)}(t)\, \widetilde{\mathbf{w}}(\mathbf{u}) + W(t)\, \widetilde{\mathbf{w}}^{(1)}(\mathbf{u}) + \widehat{P}(t)\, \widetilde{\mathbf{e}}_u\left[\mathbf{u}^\mu(t)\right] = \mathbf{h}(t), \tag{30.14}
$$

transform the above equation for $V^{(1)}(t, \mathbf{y})$ into:

$$
V^{(1)}(t, \mathbf{y}) = \Theta\left[t, \mathbf{r}^{\alpha-1}(t), \mathbf{y}(t), \mathbf{d}(t)\right] + \mathrm{grad}^T V(t, \mathbf{y}) \mathbf{h}(t), \tag{30.15}
$$

The control vector function $\mathbf{u}(.)$ is the solution of (30.13), (30.14) for $\mathbf{h}(t)$ determined by

$$
\begin{gathered}
\mathbf{h}(t) = -\mathrm{grad}V(t, \mathbf{y}) \left(\mathrm{grad}^T V(t, \mathbf{y}) \mathrm{grad}V(t, \mathbf{y})\right)^{-1} \\
\bullet \left[\Theta\left[t, \mathbf{r}^{\alpha-1}(t), \mathbf{y}(t), \mathbf{d}(t)\right] + \Phi(t, \mathbf{y})\right].
\end{gathered} \tag{30.16}
$$

The function $\Phi(t, \mathbf{y}) : \mathfrak{T}_0 \times \mathfrak{R}^N \longrightarrow \mathfrak{R}$ is chosen with the appropriate properties to ensure the required tracking property. Finally, (30.15) and (30.16) jointly become:

$$
V^{(1)}(t, \mathbf{y}) = -\Phi(t, \mathbf{y}),\ \forall (t, \mathbf{y}) \in \mathfrak{T}_0 \times \mathfrak{R}^N. \tag{30.17}
$$

B) After replacing $\mathbf{r}^{\alpha-1}(t)$ by the right-hand side of (22.33) into the output equation of the state Lyapunov plant (23.1), Definition 338, then it becomes:

$$
\mathbf{y}(t) = \widehat{\widetilde{\mathbf{z}}}\left[t, \mathbf{r}^{\alpha-1}(t), \mathbf{d}(t)\right] + \widetilde{P}(t)\, \widetilde{\mathbf{e}}_u\left[\mathbf{u}^\mu(t)\right] + W(t)\, \widetilde{\mathbf{w}}\left[\mathbf{u}(t)\right],\ \forall t \in \mathfrak{T}, \tag{30.18}
$$

where

$$\widehat{\widetilde{\mathbf{z}}}\left[t,\mathbf{r}^{\alpha-1}(t),\mathbf{d}(t)\right]=\widetilde{\mathbf{z}}\left[t,\mathbf{r}^{\alpha-1}(t),\mathbf{d}(t)\right]+\widetilde{Z}\left[t,\mathbf{r}^{\alpha-1}(t),\mathbf{d}(t)\right]\widetilde{F}^{-1}\left[t,\mathbf{r}^{\alpha-1}(t)\right]$$
$$\bullet\left\{-\widetilde{Q}\left[t,\mathbf{r}^{\alpha-1}(t)\right]\mathbf{r}^{(\alpha)}(t)-\widetilde{\mathbf{q}}\left[t,\mathbf{r}^{\alpha-1}(t)\right]+E\left(t\right)\widetilde{\mathbf{e}}_{d}\left[t,\mathbf{d}(t)\right],\right\},\ \forall t\in\mathfrak{T},$$
$$\widetilde{P}\left(t\right)=\widetilde{Z}\left[t,\mathbf{r}^{\alpha-1}(t),\mathbf{d}(t)\right]\widetilde{F}^{-1}\left[t,\mathbf{r}^{\alpha-1}(t)\right]P\left(t\right),\ \forall t\in\mathfrak{T}. \tag{30.19}$$

The first derivative of (30.18) reads:

$$\mathbf{y}^{(1)}(t)=\widehat{\widetilde{\mathbf{z}}}^{(1)}\left[t,\mathbf{r}^{\alpha-1}(t),\mathbf{d}(t)\right]+\widetilde{P}^{(1)}\left(t\right)\widetilde{\mathbf{e}}_{u}\left[\mathbf{u}^{\mu}(t)\right]+\widetilde{P}\left(t\right)\widetilde{\mathbf{e}}_{u}^{(1)}\left[\mathbf{u}^{\mu}(t)\right]$$
$$+W^{(1)}\left(t\right)\widetilde{\mathbf{w}}\left[\mathbf{u}(t)\right]+W\left(t\right)\widetilde{\mathbf{w}}^{(1)}\left[\mathbf{u}(t)\right],\ \forall t\in\mathfrak{T}. \tag{30.20}$$

This permits us to determine the first derivative of the accepted decrescent positive definite compatible differentiable function $V(.):\mathfrak{T}\times\mathfrak{R}^{N}\longrightarrow\mathfrak{R}$ by applying (30.20) to (22.22):

$$V^{(1)}(t,\mathbf{y})=\frac{\partial V(t,\mathbf{y})}{\partial t}+\left[\mathrm{grad}V(t,\mathbf{y})\right]^{T}\frac{d\mathbf{y}}{dt}$$
$$=\frac{\partial V(t,\mathbf{y})}{\partial t}+\left[\mathrm{grad}V(t,\mathbf{y})\right]^{T}\widehat{\widetilde{\mathbf{z}}}^{(1)}\left[t,\mathbf{r}^{\alpha-1}(t),\mathbf{d}(t)\right]$$
$$+\left[\mathrm{grad}V(t,\mathbf{y})\right]^{T}\left\{\begin{array}{c}\widetilde{P}^{(1)}\left(t\right)\widetilde{\mathbf{e}}_{u}\left[\mathbf{u}^{\mu}(t)\right]+\widetilde{P}\left(t\right)\widetilde{\mathbf{e}}_{u}^{(1)}\left[\mathbf{u}^{\mu}(t)\right]\\+W^{(1)}\left(t\right)\widetilde{\mathbf{w}}\left[\mathbf{u}(t)\right]+W\left(t\right)\widetilde{\mathbf{w}}^{(1)}\left[\mathbf{u}(t)\right]\end{array}\right\},\ \mathbf{y}\neq\mathbf{0}_{N},$$

or

$$V^{(1)}(t,\mathbf{y})=\omega\left[t,\mathbf{r}^{\alpha-1}(t),\mathbf{d}(t)\right]+\left[\mathrm{grad}V(t,\mathbf{y})\right]^{T}\mathbf{g}(t),$$
$$\omega\left[t,\mathbf{r}^{\alpha-1}(t),\mathbf{d}(t),\mathbf{y}\right]=\frac{\partial V(t,\mathbf{y})}{\partial t}+\left[\mathrm{grad}V(t,\mathbf{y})\right]^{T}\widehat{\widetilde{\mathbf{z}}}^{(1)}\left[t,\mathbf{r}^{\alpha-1}(t),\mathbf{d}(t)\right],\ \mathbf{y}\neq\mathbf{0}_{N},$$

and

$$\widetilde{P}^{(1)}\left(t\right)\widetilde{\mathbf{e}}_{u}\left[\mathbf{u}^{\mu}(t)\right]+\widetilde{P}\left(t\right)\widetilde{\mathbf{e}}_{u}^{(1)}\left[\mathbf{u}^{\mu}(t)\right]+W^{(1)}\left(t\right)\widetilde{\mathbf{w}}\left[\mathbf{u}(t)\right]+W\left(t\right)\widetilde{\mathbf{w}}^{(1)}\left[\mathbf{u}(t)\right]=\mathbf{g}(t)$$
$$\mathbf{g}(t)=-\left[\mathrm{grad}V(t,\mathbf{y})\right]\left\{\left[\mathrm{grad}V(t,\mathbf{y})\right]^{T}\left[\mathrm{grad}V(t,\mathbf{y})\right]\right\}^{-1}$$
$$\bullet\left\{\omega\left[t,\mathbf{r}^{\alpha-1}(t),\mathbf{d}(t),\mathbf{y}\right]+\chi\left(t,\mathbf{y}\right)\right\},\ \mathbf{y}\neq\mathbf{0}_{N}, \tag{30.21}$$

where $\chi(.):\mathfrak{T}\times\mathfrak{R}^{N}\longrightarrow\mathfrak{R}_{+}$ is a chosen positive definite function. The final form of $V^{(1)}(t,\mathbf{y})$ is:

$$V^{(1)}(t,\mathbf{y})=-\chi\left(t,\mathbf{y}\right),\ \mathbf{y}\neq\mathbf{0}_{N}. \tag{30.22}$$

The requested tracking property and the corresponding Lyapunov tracking conditions govern the choice of the function $\chi(.)$.

The control $\mathbf{u}(t)$ is the solution of

$$\widetilde{P}^{(1)}\left(t\right)\widetilde{\mathbf{e}}_{u}\left[\mathbf{u}^{\mu}(t)\right]+\widetilde{P}\left(t\right)\widetilde{\mathbf{e}}_{u}^{(1)}\left[\mathbf{u}^{\mu}(t)\right]+W^{(1)}\left(t\right)\widetilde{\mathbf{w}}\left[\mathbf{u}(t)\right]+W\left(t\right)\widetilde{\mathbf{w}}^{(1)}\left[\mathbf{u}(t)\right]$$
$$=-\left[\mathrm{grad}V(t,\mathbf{y})\right]\left\{\left[\mathrm{grad}V(t,\mathbf{y})\right]^{T}\left[\mathrm{grad}V(t,\mathbf{y})\right]\right\}^{-1}$$
$$\bullet\left\{\omega\left[t,\mathbf{r}^{\alpha-1}(t),\mathbf{d}(t),\mathbf{y}\right]+\chi\left(t,\mathbf{y}\right)\right\},\ \mathbf{y}\neq\mathbf{0}_{N}. \tag{30.23}$$

If $\mathbf{y}=\mathbf{0}_{N}$ then the control is the perfect trackable control.

30.2.2 Asymptotically contractive sets

The above consideration simplifies the proofs of the following theorems.

Assumption 439 ***Choice of the function V(.)***

The functions $V(.)$ and $\chi(.)$ satisfy:

1) $V(.)$ is a gradient compatible continuously differentiable function on $\mathfrak{T} \times \mathfrak{R}^N$ and positive definite.

2) There exists $\xi_M > 0$ such that for every $\zeta \in]0, \xi_M]$ and for all $\mathbf{d}(.) \in \mathfrak{D}^j$ the set $\mathfrak{V}_\zeta(t; t_0; \mathfrak{D}^j)$ induced by $V(.)$:

a) Has the radially increasing boundary function.,

b) Is asymptotically contractive.

3) $\Phi(t, \mathbf{y})$ is positive on $Cl\mathfrak{V}_{\xi_M}(t; t_0; \mathfrak{D}^j) \backslash \mathfrak{O}$ and $\Phi(t, \mathbf{y}) = 0$ if, and only if, $\mathbf{y} = \mathbf{0}_N$.

Comment 440 ***The guide for the choice of the function*** $V(.)$

The function $V(.)$ should be chosen to be compatible continuously differentiable positive definite and to satisfy the rank condition (30.3) relative to the plant characteristics $\widetilde{Q}^{-1}(t, \mathbf{r}^{\alpha-1})$ and $P(t)$. This helps to relate the choice of the function $V(.)$ to the plant characteristics.

Theorem 441 ***Stable tracking of $\mathbf{Y}_d(t)$ of the plant (3.12) on $\mathfrak{T}_0 \times \mathfrak{D}^j \times \mathfrak{Y}_d$, equivalently of $\mathbf{y}_d(t) \equiv \mathbf{0}_N$ of the plant (23.1) on $\mathfrak{T}_0 \times \mathfrak{D}^j \times \mathfrak{Y}_d$***

Let Assumption 439 be satisfied.

1) In order for the plant (23.1) controlled by the control $\mathbf{u}(t)$, to exhibit the stable tracking of $\mathbf{y}_d(t) \equiv \mathbf{0}_N$, on $\mathfrak{T}_0 \times \mathfrak{D}^j$, it is sufficient that control $\mathbf{u}(.)$ obeys (30.14) together with (30.12), (30.13), and (30.16) for $\mathbf{y} \neq \mathbf{0}_N$ and that it is the perfect trackable control for $\mathbf{y} = \mathbf{0}_N$.

2) If, additionally to 1), $\Phi(.)$ depends on $V(.)$ so that $\Phi(t, V; t_R) \in \mathcal{F}_s$ then the plant (23.1) exhibits the stable tracking with FSRT t_R of $\mathbf{y}_d(t) \equiv \mathbf{0}_N$, on $\mathfrak{T}_0 \times \mathfrak{D}^j \times \mathfrak{Y}_d$.

Proof. The stable tracking of $\mathbf{y}_d(t) \equiv \mathbf{0}_N$ means the stable tracking of $\mathbf{Y}_d(t)$ because $\mathbf{y}(t) = \mathbf{Y}_d(t) - \mathbf{Y}(t) \equiv \mathbf{0}_N \equiv \mathbf{y}_d(t)$ holds if, and only if, $\mathbf{Y}(t) \equiv \mathbf{Y}_d(t)$.

Let Assumption 439 be valid and let the control $\mathbf{u}(.)$ obey (30.14) together with (30.13), (30.12), and (30.16).

1) The validity of Assumption 439, (30.14), (30.13), (30.12), and (30.16) implies the validity of all the conditions of Theorem 357 for $\mathbf{Z} = \mathbf{Y}$, i.e., for $\mathbf{z} = \mathbf{y}$. The plant (23.1), controlled by the control $\mathbf{u}(t)$ determined by (30.14) together with (30.13), (30.12), and (30.16) for $\mathbf{y} \neq \mathbf{0}_N$ and for $\mathbf{y} = \mathbf{0}_N$ determined to be the perfect trackable control, exhibits the stable tracking of $\mathbf{y}_d(t) \equiv \mathbf{0}_N$, on $\mathfrak{T}_0 \times \mathfrak{D}^j \times \mathfrak{Y}_d$.

2) The conditions under 1) and $\Phi(t, V; t_R) \in \mathcal{F}_s$ imply that the conditions of Theorem 423 are satisfied. The plant (23.1), controlled by the control $\mathbf{u}(t)$ determined by (30.14) together with (30.13), (30.12) and (30.16) for $\mathbf{y} \neq \mathbf{0}_N$ and for $\mathbf{y} = \mathbf{0}_N$ determined to be the perfect trackable control, exhibits the stable tracking with *FSRT* t_R of $\mathbf{y}_d(t) \equiv \mathbf{0}_N$, on $\mathfrak{T}_0 \times \mathfrak{D}^j \times \mathfrak{Y}_d$ ■

The conditions change for the stable tracking of $\mathbf{y}_d(t) \equiv \mathbf{0}_N$ of the plant (23.1).

Assumption 442 ***Choice of the function V(.)***

The functions $V(.)$ and $\chi(.)$ satisfy:

1) $V(.)$ is a gradient compatible, continuously differentiable function on $\mathfrak{T} \times \mathfrak{R}^N$ and positive definite.

2) There exists $\xi_M > 0$ such that for every $\zeta \in]0, \xi_M]$ and for all $\mathbf{d}(.) \in \mathfrak{D}^j$ the set $\mathfrak{V}_\zeta(t; t_0; \mathfrak{D}^j)$ induced by $V(.)$:

a) Has the radially increasing boundary function

b) Is asymptotically contractive

3) $\chi(t, \mathbf{y})$ is positive on $Cl\mathfrak{V}_{\xi_M}(t; t_0; \mathfrak{D}^j) \backslash \mathfrak{O}$ and $\chi(t, \mathbf{y}) = 0$ if, and only if, $\mathbf{y} = \mathbf{0}_N$.

Theorem 443 ***Stable tracking of*** $\mathbf{Y}_d(t)$ ***of the plant*** *(3.29)* ***on*** $\mathfrak{T}_0 \times \mathfrak{D}^j \times \mathfrak{Y}_d$***, equivalently of*** $\mathbf{y}_d(t) \equiv \mathbf{0}_N$ ***of the plant*** *(23.1)* ***on*** $\mathfrak{T}_0 \times \mathfrak{D}^j \times \mathfrak{Y}_d$

Let Assumption 442 be satisfied.

1) In order for the plant (23.1) controlled by the control $\mathbf{u}(t)$*, to exhibit the stable tracking of* $\mathbf{y}_d(t) \equiv \mathbf{0}_N$*, on* $\mathfrak{T}_0 \times \mathfrak{D}^j$*, it is sufficient that control* $\mathbf{u}(.)$ *obeys (30.23) together with (30.19) for* $\mathbf{y} \neq \mathbf{0}_N$ *and that it is the perfect trackable control for* $\mathbf{y} = \mathbf{0}_N$*.*

2) If, additionally to 1), $\chi(.)$ *depends on* $V(.)$ *so that* $\chi(t, V; t_R) \in \mathcal{F}_s$ *then the plant (23.1) exhibits the stable tracking with FSRT* t_R *of* $\mathbf{y}_d(t) \equiv \mathbf{0}_N$*, on* $\mathfrak{T}_0 \times \mathfrak{D}^j \times \mathfrak{Y}_d$*.*

Proof. The stable tracking of $\mathbf{y}_d(t) \equiv \mathbf{0}_N$ means the stable tracking of $\mathbf{Y}_d(t)$ because $\mathbf{y}(t) = \mathbf{Y}_d(t) - \mathbf{Y}(t) \equiv \mathbf{0}_N \equiv \mathbf{y}_d(t)$ holds if, and only if, $\mathbf{Y}(t) \equiv \mathbf{Y}_d(t)$.

Let Assumption 442 be valid and let the control $\mathbf{u}(.)$ obey (30.23) together with (30.19). If $\mathbf{y} = \mathbf{0}_N$ then the control is the perfect trackable control.

1) The validity of both Assumption 442 and (30.23) together with (30.19) implies the validity of all the conditions of Theorem 357 for $\mathbf{Z} = \mathbf{Y}$, i.e., for $\mathbf{z} = \mathbf{y}$. The plant (23.1), controlled by the control $\mathbf{u}(t)$ determined by (30.23) together with (30.19) for $\mathbf{y} \neq \mathbf{0}_N$ and for $\mathbf{y} = \mathbf{0}_N$ determined to be the perfect trackable control, exhibits the stable tracking of $\mathbf{y}_d(t) \equiv \mathbf{0}_N$, on $\mathfrak{T}_0 \times \mathfrak{D}^j \times \mathfrak{Y}_d$.

2) The conditions under 1) and $\chi(t, V; t_R) \in \mathcal{F}_s$ imply that the conditions of Theorem 423 are satisfied. The plant (23.1), controlled by the control $\mathbf{u}(t)$ determined by (30.23) together with (30.19) for $\mathbf{y} \neq \mathbf{0}_N$ and for $\mathbf{y} = \mathbf{0}_N$ determined to be the perfect trackable control, exhibits the stable tracking with *FSRT* t_R of $\mathbf{y}_d(t) \equiv \mathbf{0}_N$, on $\mathfrak{T}_0 \times \mathfrak{D}^j \times \mathfrak{Y}_d$ ■

30.2.3 Noncontractive *time*-varying sets

Theorem 364 and Theorem 365 are used for Lyapunov synthesis of the stable tracking control.

Assumption 444 ***Choice of the function V(.)***

The functions $V(.)$ *and* $\Phi(.)$ *satisfy:*

1) $V(.)$ *is a gradient compatible continuously differentiable positive definite and decrescent function.*

2) $\Phi(.; \mathbf{d})$ *is positive definite on the neighborhood* $\mathfrak{N}$ *of the origin* $\mathbf{y}_d(t) \equiv \mathbf{0}_N$ *for every* $\mathbf{d} \in \mathfrak{D}^j$*.*

Theorem 445 ***Stable tracking of the desired output deviation*** $\mathbf{y}_d(t) \equiv \mathbf{0}_N$ ***of the plant*** *(23.1)* ***on*** $\mathfrak{T}_0 \times \mathfrak{D}^j \times \mathfrak{Y}_d$

Let Assumption 444 be valid.

1) In order for the plant (23.1), controlled by the control $\mathbf{u}(t)$*, to exhibit the stable tracking of* $\mathbf{y}_d(t) \equiv \mathbf{0}_N$ *on* $\mathfrak{T}_0 \times \mathfrak{D}^j \times \mathfrak{Y}_d$ *it is sufficient that the control* $\mathbf{u}(.)$ *obeys Assumption 444 and (30.14) together with (30.12), (30.13), and (30.16) for* $\mathbf{y} \neq \mathbf{0}_N$ *and that it is the perfect trackable control for* $\mathbf{y} = \mathbf{0}_N$*.*

2) If, additionally to 1), $\Phi(.; \mathbf{d})$ *depends on* $V(.)$ *so that* $\Phi(t, V; \mathbf{d};; t_R) \in \mathcal{F}_s$ *for every* $\mathbf{d} \in \mathfrak{D}^j$ *then the plant (23.1) exhibits the stable tracking with FSRT* t_R *of* $\mathbf{y}_d(t) \equiv \mathbf{0}_N$*, on* $\mathfrak{T}_0 \times \mathfrak{D}^j \times \mathfrak{Y}_d$*.*

Proof. 1) Let all the conditions of the theorem statement under 1) be valid. All the conditions of Theorem 364 are satisfied. The plant (23.1), controlled by the control $\mathbf{u}(t)$, which is determined by Assumption 444 and (30.14) together with (30.13), (30.12), and (30.16) for $\mathbf{y} \neq \mathbf{0}_N$ and which is the perfect trackable control for $\mathbf{y} = \mathbf{0}_N$, exhibits the stable tracking of $\mathbf{y}_d(t) \equiv \mathbf{0}_N$, on $\mathfrak{T}_0 \times \mathfrak{D}^j \times \mathfrak{Y}_d$.

2) The conditions under 1) and $\Phi(t, V; \mathbf{d};; t_R) \in \mathcal{F}_s$ for every $\mathbf{d} \in \mathfrak{D}^j$ imply that the conditions of Theorem 427 are fulfilled. The plant (23.1), controlled by the control $\mathbf{u}(t)$

determined by Assumption 444 and (30.14) together with (30.12), (30.13), and (30.16) for $\mathbf{y} \neq \mathbf{0}_N$ and which is the perfect trackable control for $\mathbf{y} = \mathbf{0}_N$, exhibits the stable tracking with $FSRT$ t_R of $\mathbf{y}_d(t) \equiv \mathbf{0}_N$, on $\mathfrak{T}_0 \times \mathfrak{D}^j \times \mathfrak{Y}_d$ ■

The tracking conditions on *time*-invariant sets for the plant (23.1) read:

Assumption 446 ***Choice of the function V(.)***

The functions $V(.)$ and $\chi(.)$ satisfy:

1) $V(.)$ is a compatible continuously differentiable positive definite and decrescent function.

2) $\chi(.;\mathbf{d})$ is positive definite on the neighborhood $\mathfrak{N}$ of the origin $\mathbf{y}_d(t) \equiv \mathbf{0}_N$ for every $\mathbf{d} \in \mathfrak{D}^j$.

Theorem 447 ***Stable tracking of $\mathbf{y}_d(t) \equiv \mathbf{0}_N$ of the plant (23.1) on the product set $\mathfrak{T}_0 \times \mathfrak{D}^j \times \mathfrak{Y}_d$***

Let Assumption 446 be valid.

1) In order for the plant (23.1), controlled by the control $\mathbf{u}(t)$, to exhibit the stable tracking of $\mathbf{y}_d(t) \equiv \mathbf{0}_N$ on $\mathfrak{T}_0 \times \mathfrak{D}^j \times \mathfrak{Y}_d$ it is sufficient that the control $\mathbf{u}(.)$ obeys Assumption 446 and (30.23) together with (30.19) for $\mathbf{y} \neq \mathbf{0}_N$ and that it is the perfect trackable control for $\mathbf{y} = \mathbf{0}_N$

2) If, additionally to 1), $\chi(.;\mathbf{d})$ depends on $V(.)$ so that $\chi(t,V;\mathbf{d};;t_R) \in \mathcal{F}_s$ for every $\mathbf{d} \in \mathfrak{D}^j$ then the plant (23.1) exhibits the stable tracking with FSRT t_R of $\mathbf{y}_d(t) \equiv \mathbf{0}_N$, on $\mathfrak{T}_0 \times \mathfrak{D}^j \times \mathfrak{Y}_d$.

Proof. 1) Let all the conditions of the theorem statement under 1) be valid. All the conditions of Theorem 364 are satisfied. The plant (23.1), controlled by the control $\mathbf{u}(t)$, which is determined by Assumption 446 and (30.23) together with (30.19) for $\mathbf{y} \neq \mathbf{0}_N$, and which is the perfect trackable control for $\mathbf{y} = \mathbf{0}_N$, exhibits the stable tracking of $\mathbf{y}_d(t) \equiv \mathbf{0}_N$, on $\mathfrak{T}_0 \times \mathfrak{D}^j \times \mathfrak{Y}_d$.

2) The conditions under 1) and $\chi(t,V;\mathbf{d};;t_R) \in \mathcal{F}_s$ for every $\mathbf{d} \in \mathfrak{D}^j$ imply that the conditions of Theorem 427 are fulfilled. The plant (23.1), controlled by the control $\mathbf{u}(t)$, which is determined by Assumption 446 and (30.23) together with (30.19) for $\mathbf{y} \neq \mathbf{0}_N$, and which is the perfect trackable control for $\mathbf{y} = \mathbf{0}_N$, exhibits the stable tracking with $FSRT$ t_R of $\mathbf{y}_d(t) \equiv \mathbf{0}_N$, on $\mathfrak{T}_0 \times \mathfrak{D}^j \times \mathfrak{Y}_d$ ■

[illegible] by Assumption [illegible] and [illegible] together with [illegible] is the perfect [illegible] [illegible] ■

The [illegible] for the plant [illegible]

Proposition 4.10 *[illegible] of the function* [illegible]

[illegible]

2. [illegible]

Theorem 4.17 *Stable tracking of* [illegible] *of the plant* [illegible] *on the product set* [illegible]

Let Assumption 4.10 [illegible]

[illegible]

[illegible]

Part VIII

REQUIRED TRACKING QUALITY AND CONTROL SYNTHESIS

Part VIII

Chapter 31

Natural tracking control concept

What does nature, i.e., the brain as part of nature, use to create control of any organ? It evidently uses information about the error ε of the real organ behavior $\mathbf{Y}(.)$ relative to its desired behavior $\mathbf{Y}_d(.)$. But this is not the only information that the brain uses to create the control. For example, in order to control the position of a hand, of a finger, of a leg, the brain uses information about the difference between their desired and real positions, which is information about their position errors. However, the brain simultaneously uses information about the forces of the muscles acting on the organs. The muscle force is a control variable. The brain, as the central part of *the natural controller*, uses information about the (realized) control itself. This is one essential characteristic of the control created by the brain, i.e., by nature.

The brain, in general nature, does not have any information about a mathematical model of the controlled organ. This is another crucial characteristic of the control created by the brain, i.e., by nature. It leads to the new control concept (Section 6.4, Comment 145, Definition 171):

Definition 448 ***Natural Control (NC)***

A control $\mathbf{U}$ *is* ***natural control (NC)*** *if, and only if:*

1. It obeys the ***time continuity and uniqueness principle*** *(TCUP, Principle 20).*

2. Its synthesis and effective implementation use information about both the output error vector ε *(and possibly its derivatives and/or its integral) and the control action* $\mathbf{U}$ *itself.*

3. Its synthesis and effective implementation do not use information either about the plant mathematical model or about the mathematical description of the plant state or about the real instantaneous values of disturbances,

$$\mathbf{U} = \mathbf{U}(\varepsilon, \mathbf{U}), \quad \mathbf{U}(t) \in \mathfrak{C}(\mathfrak{T}_0). \tag{31.1}$$

The controller should possess an internal local feedback from its output to its input in order to generate *natural control.* A mathematical rather than a physical consideration determines clearly and precisely the sign, the character and the strength of such local feedback. We refer to [166]-[169], [232]-[241], [253], [263]-[265], [282]-[286], [444]-[453] for the following definition.

Definition 449 ***Natural Tracking Control (NTC)***

Natural control is ***natural tracking control (NTC)*** *if, and only if, it ensures a (demanded) type of tracking determined by a tracking algorithm described by an operator* $\boldsymbol{T}(.)$,

$$\mathbf{U} = \mathbf{U}(\varepsilon, \mathbf{U}; \mathbf{T}), \; \mathbf{U}(t) \in \mathfrak{C}(\mathfrak{T}_0). \tag{31.2}$$

We present and further develop the fundamentals of the *NTC* theory, the mathematical root of which is in the papers [190, Note 11, p. 19],[220, Note 11, p. S-38]. These papers showed the mathematical possibility to replace the internal object dynamics and the external disturbance action by the control used to compensate completely their influence on the object behavior. The mathematics showed that such control demands *the unit positive local feedback without delay in the controller.* The unit positive feedback without delay is forbidden in control theory because such an isolated feedback system is totally unstable and will blow immediately in reality. Z. B. Ribar and this author simulated effectively on an analog computer the *NTC* of a second-order linear plant in the Laboratory of Automatic Control, Faculty of Mechanical Engineering, Belgrade University, Serbia (Spring 1988). The feedback *NT* controller is in the closed loop of the overall control system. Its local unit positive feedback operates in full harmony with the global negative feedback of the control system. This is the control principle that is the basis of the life of every human cell and of the whole organism. Such control is *self-adaptive control.* The further development of it showed that its more adequate name is the ***natural tracking control*** (*NTC*) [166]-[169], [232]-[239], [253], [263]-[265], [282]-[286], [444]-[453]. In the papers [232]-[239], [282]-[286], [444]-[453] William Pratt Mounfield, J*r.* introduced the concept of *high-gain NTC* and worked out all the examples by solving the difficult problem of digital simulations of the plant behavior controlled by time-continuous *NTC* that incorporates the local unit positive feedback. He was the first to do such simulations successfully and to show effective applications to technical plants. Other developments of the *NTC* and of its various applications to control of continuous-time technical plants can be found in the Ph.D./D.Sci dissertations by A. Kökösy [351] and D. V. Lazitch [376], in the papers by Kökösy [353]-[355], D. V. Lazitch [377]-[380], N. Nedić (Neditch) and D. Pršić (Prshitch) [458]-[460], [492], Z. B. Ribar [500], Z. B. Ribar et al. [501], [503], and in the M.Sci thesis by M. R. Jovanović (Yovanovitch) [568].

What follows shows how the *NTC* can assure the demanded high quality tracking defined by various tracking algorithms.

Chapter 32

Tracking quality: Output space

32.1 Output space tracking operator

If Property 64 characterizes the plant (3.12) then Lemma 65 determines that the dimension of the controllable part (3.47) of the plant output equation is equal to $\gamma \leq \min(N, r)$, where r is the dimension of the control vector $\mathbf{U} \in \mathfrak{R}^r$. Only the output γ-subvector $\mathbf{Y}_\gamma = [Y_1 \ \ Y_2...Y_\gamma]^T \in \mathfrak{R}^\gamma$ can be then controlled directly, i.e. elementwise, Axiom 87, Property 64, and Lemma 65,

$$\mathbf{Y} = [Y_1 \ Y_2..Y_N]^T \in \mathfrak{R}^N, \ \ \mathbf{Y}_\gamma = [Y_1 \ Y_2..Y_\gamma]^T \ \in \mathfrak{R}^\gamma, \ 1 \leq \gamma \leq \min(N, r). \tag{32.1}$$

Nevertheless, the whole output vector $\mathbf{Y} \in \mathfrak{R}^N$, hence the whole output error vector $\boldsymbol{\varepsilon} \in \mathfrak{R}^N$, should be controlled at least indirectly. The γ-subvector $\mathbf{Y}_\gamma$ of $\mathbf{Y}$ does not contain the entries $Y_{\gamma+1}, Y_{\gamma+2}, ..Y_N$ of $\mathbf{Y}$.

Comment 450 ***Controllable output variables and their errors***

When $\gamma < N$, in order to solve the control problem, by following Axiom 87, we used subsidiary vector functions $\mathbf{v}(,)$ with dimension m, $m \leq \gamma$, to establish the perfect trackability conditions in which $\mathbf{v}(,)$ is determined by (8.1), and the trackability conditions in which $\mathbf{v}(,)$ obeys (11.1), for the control synthesis in the output space. This suggests for us to introduce the following subsidiary, modified, output, and error variables and by them induced subsidiary, modified, output and error vectors:

$$\left\{ \begin{array}{c} Y_\gamma^* = Y_j, \\ \varepsilon_\gamma^* = \varepsilon_j \end{array} \right\} \ \textit{iff} \ |\varepsilon_j| = max(|\varepsilon_i| : \forall i = \gamma, \gamma+1, ..., N),$$

$$j \in \{\gamma, \gamma+1, ..., N\}, \ \gamma \in \{1, 2, ..., N\},$$

$$\mathbf{Y}_\gamma^* = \left[Y_1 \ \ Y_2 \ ... \ Y_{\gamma-1} \ Y_\gamma^* \ \right]^T \in \mathfrak{R}^\gamma, \tag{32.2}$$

$$\boldsymbol{\varepsilon}_\gamma^* = \left[\varepsilon_1 \ \ \varepsilon_2 \ ... \ \varepsilon_{\gamma-1} \ \varepsilon_\gamma^* \ \right]^T \in \mathfrak{R}^\gamma. \tag{32.3}$$

The γ-subvector $\mathbf{Y}_\gamma^*$ of $\mathbf{Y}$ does contain explicitly the entries $Y_{\gamma+1}, Y_{\gamma+2}, ..Y_N$ of $\mathbf{Y}$. The γ-subvector $\boldsymbol{\varepsilon}_\gamma^*$ of $\boldsymbol{\varepsilon}$ contains only implicitly the entries $\varepsilon_{\gamma+1}, \varepsilon_{\gamma+2}, ..\varepsilon_N$ of $\boldsymbol{\varepsilon}$. The analogy holds for ε_γ^* that has the following characteristics:

$$\begin{aligned} \varepsilon_\gamma^* &= 0 \Longleftrightarrow \varepsilon_j = 0, \ \forall j = \gamma, \gamma+1, ..., N, \ \forall \xi > 0, \ \textit{and} \\ |\varepsilon_\gamma^*| &< \xi \Longleftrightarrow |\varepsilon_j| < \xi, \forall j = \gamma, \gamma+1, ..., N, \ \forall \xi > 0, \end{aligned} \tag{32.4}$$

are important for the indirect control of the output variables $Y_{\gamma+1}, Y_{\gamma+2}, ..Y_N$ that are not directly controllable. The use of the vectors $\mathbf{Y}_\gamma^*$ and $\boldsymbol{\varepsilon}_\gamma^*$ enables us to synthesize the control

$\mathbf{U} \in \mathfrak{R}^r$ that guarantees stable tracking. This permits $r < N$ that is the characteristic of many plants. If $\gamma = N \leq r$ then the control can guarantee elementwise stable tracking.

If we wish to ensure both stable tracking and state stable tracking then we introduce a differentiable radially unbounded, decrescent positive definite function $v(.) : \mathfrak{T} \times \mathfrak{R}^{\alpha\rho} \longrightarrow \mathfrak{R}_+$. Its desired value $v_d(t, \varepsilon)$ is zero: $v_d\left(t, \varepsilon_R^{\alpha-1}\right) \equiv 0$, and its error $\varepsilon_v = -v\left(t, \varepsilon_R^{\alpha-1}\right)$, where $\varepsilon_R = \mathbf{R}_d - \mathbf{R}$. The error ε_v changes the modified error vector ε_γ^* to the jointly modified error vector ε_v^*:

$$\begin{gathered}\varepsilon_R = \mathbf{R}_d - \mathbf{R}, \quad \varepsilon_v = -v\left(t, \varepsilon_R^{\alpha-1}\right), \\ \varepsilon_v^* = \left[\varepsilon_1 \ \varepsilon_2 \ \ldots \ \varepsilon_{\gamma-2} \ \varepsilon_{\gamma-1}^* \ \varepsilon_v\right]^T = \left[\varepsilon_{\gamma-1}^{*T} \ \ \varepsilon_v\right]^T \in \mathfrak{R}^\gamma, \\ \varepsilon_v^* = \mathbf{0}_\gamma \Longleftrightarrow both\ \varepsilon_{\gamma-1}^* = \mathbf{0}_{\gamma-1} \text{ and } \varepsilon_v = 0. \end{gathered} \tag{32.5}$$

Note 451 ***On the use of the vector*** $\varepsilon_v^*(t)$

The use of the vector $\varepsilon_v^*(t)$, (32.5), demands the usage of its derivative $\varepsilon_v^{*(1)}(t)$ along the plant output behavior and along its motions because of the necessity to use the derivative $\varepsilon_v^{(1)}(t) = v^{(1)}\left(t, \mathbf{r}^{\alpha-1}\right)$. This shows that the derivative $\varepsilon_v^{*(1)}(t)$ of $\varepsilon_v^*(t)$ is very complex, which is the reason to continue with the modified error vector $\varepsilon_\gamma^*(t)$; (32.3) rather than with $\varepsilon_v^*(t)$. The control synthesis by applying the vector $\varepsilon_v^*(t)$, (32.5), might be the topic of some future research.

Conclusion 452 ***The advantages of using*** ε_γ^* ***rather than*** ε

By using ε_γ^* instead of ε the control synthesized only in the output space can assure at least stable tracking.

The number r of control variables can be less than the number N of the output variables.

We define the criterion of the tracking quality to be a solution of the following differential equation of the output error ε_γ^*, (32.3), of its derivatives

$$\varepsilon_\gamma^*(t), \varepsilon_\gamma^{*(1)}(t), \ldots, \varepsilon_\gamma^{*(k)}(t), \ k \in \{0, 1, \ldots, \alpha - 1\}, \tag{32.6}$$

and/or of its integral [279],

$$\begin{gathered}\mathbf{T}\left(t, \varepsilon_\gamma^*(t), \varepsilon_\gamma^{*(1)}(t), \ldots, \varepsilon_\gamma^{*(k)}(t), \int_{t_0}^t \varepsilon_\gamma^*(t)\, dt\right) \\ = \mathbf{T}\left(t, \varepsilon_\gamma^{*k}(t), \int_{t_0}^t \varepsilon_\gamma^*(t)\, dt\right) = \mathbf{0}_m, \ \forall t \in \mathfrak{T}_{0F}, \ m \leq \gamma. \end{gathered} \tag{32.7}$$

The vector function $T(.)$ represents *the vector tracking operator*,

$$\mathbf{T}(.) : \mathfrak{T} \times \mathfrak{R}^{(k+1)\gamma} \times \mathfrak{R}^\gamma \longrightarrow \mathfrak{R}^m, \ \mathbf{T}(.) \in \mathfrak{C}^-. \tag{32.8}$$

It determines the class of *the tracking algorithms* in the output space, which express the demanded tracking qualities. It is the basis for the synthesis of the high-quality tracking.

Task 453 ***Tracking quality and the control task***

The control should force the plant output behavior $\mathbf{Y}(.)$ to satisfy the demanded tracking quality specified by (32.7).

The basic task of the quality tracking control synthesis is to determine such control.

The last equation in (32.7) holds in the cases for which the reachability moment t_R is the same for all entries of the modified error vector ε_γ^*, (32.3), and for their derivatives.

Analogously, the final moment t_F is the same for all entries of the modified error vector ε^*_γ and for their derivatives.

However, if different reachability *times* $t_{Ri,(j)}$ and/or different final *times* $t_{Fi,(j)}$ are associated with the elements of the modified error vector ε^*_γ and/or with the elements of the j-th derivative $\varepsilon^{*(j)}_\gamma$ of ε^*_γ, then we use the vectors

$$\mathbf{t}^{(k+1)\gamma} = [t \;\; t \; ... \; t]^T \in \mathfrak{T}^{(k+1)\gamma} \quad and \quad \varepsilon^{*k}_\gamma = \left[\varepsilon^{*T}_\gamma \;\; \varepsilon^{*(1)T}_\gamma \; ... \; \varepsilon^{*(k)T}_\gamma\right]^T \in \mathfrak{R}^{(k+1)\gamma}.$$

The use of the *time vector* $\mathbf{t}^{(k+1)\gamma}$ permits us to simplify, formally mathematically, the treatment of the vector relationships. This has the full advantage in the framework of the elementwise tracking with the finite vector reachability time $\mathbf{t}_R^{(k+1)\gamma}$ iff $\gamma = N$. For the same reason we introduce the output error diagonal matrices $E^*_\gamma(t) \equiv E^*_\gamma(\mathbf{t}^\gamma)$ and $E(t) \equiv E(\mathbf{t}^N)$,

$$E^*_\gamma(t) = \text{diag}\left\{\varepsilon_1(t) \;\; \varepsilon_2(t) \; \ldots \; \varepsilon^*_\gamma(t) \;\right\} = E^*_\gamma(\mathbf{t}^\gamma), \;\; E\left(\mathbf{t}^N\right) = E^*_N\left(\mathbf{t}^N\right). \tag{32.9}$$

Let

$$\begin{aligned}\varepsilon^{*k}_\gamma\left(\mathbf{t}^{(k+1)\gamma}\right) &\equiv \varepsilon^{*k}_\gamma\left(\mathbf{t}^{(k+1)\gamma}; \mathbf{t}_0^{(k+1)\gamma}; \varepsilon_0^k\right) \\ &\equiv \varepsilon^{*k}_\gamma\left(\mathbf{t}^{(k+1)\gamma}; \mathbf{t}_0^{(k+1)\gamma}; \varepsilon_0^k; \mathbf{D}; \mathbf{U}; \mathbf{Y}_d\right),\end{aligned} \tag{32.10}$$

so that

$$\varepsilon^*_\gamma(\mathbf{t}^\gamma) \equiv \varepsilon^*_\gamma(\mathbf{t}^\gamma; \mathbf{t}_0^\gamma; \varepsilon_0) \equiv \varepsilon^*_\gamma\left(\mathbf{t}^\gamma; \mathbf{t}_0^\gamma; \varepsilon_0^k; \mathbf{D}; \mathbf{U}; \mathbf{Y}_d\right). \tag{32.11}$$

This notation simplifies formally (32.7) to:

$$\mathbf{T}\left(\mathbf{t}^m, \varepsilon^{*k}_\gamma\left(\mathbf{t}^{(k+1)\gamma}\right), \int_{\mathbf{t}_0^N}^{\mathbf{t}^N} E^*_\gamma(\mathbf{t}^\gamma)\, d\mathbf{t}^\gamma\right) = \mathbf{0}_m, \;\; \forall \mathbf{t}^{(k+1)\gamma} \in \mathfrak{T}_{0F}^{(k+1)\gamma}. \tag{32.12}$$

The control task 453 determines the tracking control goal:

Goal 454 ***The basic tracking control goal***

The tracking control should force the plant output error behavior $\varepsilon^*_\gamma(.)$ to obey the tracking quality determined by (32.12).

32.2 Tracking operator properties

The following properties, [279], of *the vector tracking operator* $T(.)$, (32.12),

$$\mathbf{T}(.) : \mathfrak{T}_{0F}^m \times \mathfrak{R}^{(k+1)\gamma} \times \mathfrak{R}^\gamma \longrightarrow \mathfrak{R}^m, \;\; m \leq \gamma \leq \min(N, r), \tag{32.13}$$

determine the class of *the tracking algorithms* that express the required tracking qualities, and herein are the basis for the control synthesis to fulfill the high-quality demand.

Property 455 ***Continuity of the error solution***

The solution $\varepsilon^{*k}_\gamma\left(\mathbf{t}^{(k+1)\gamma}\right) = \varepsilon^{*k}_\gamma\left(\mathbf{t}^\gamma; \mathbf{t}_0^{(k+1)\gamma}; \varepsilon_0^k\right)$ *of (32.12) is continuous in time on* $\mathfrak{T}_{0F}^{(k+1)\gamma}$,

$$\begin{aligned}&\mathbf{T}\left(\mathbf{t}^m, \varepsilon^{*k}_\gamma\left(\mathbf{t}^{(k+1)\gamma}\right), \int_{\mathbf{t}_0^\gamma}^{\mathbf{t}^\gamma} E^*_\gamma(\mathbf{t}^\gamma)\, d\mathbf{t}^\gamma\right) = \mathbf{0}_m, \;\; \forall \mathbf{t}^{(k+1)\gamma} \in \mathfrak{T}_{0F}^{(k+1)\gamma} \\ &\qquad \Longrightarrow \varepsilon^*_\gamma\left(\mathbf{t}^\gamma; \mathbf{t}_0^\gamma; \varepsilon_0^k\right) \in \mathfrak{C}(\mathfrak{T}_{0F}^\gamma).\end{aligned} \tag{32.14}$$

Property 456 ***The operator*** $T(.)$ ***vanishes at the origin***
The operator $\mathbf{T}(.)$ *vanishes at the origin at every moment,*

$$\boldsymbol{\varepsilon}_{\gamma}^{*k} = \mathbf{0}_{(k+1)\gamma} \Longrightarrow \mathbf{T}\left(\mathbf{t}^{m}, \mathbf{0}_{(k+1)\gamma}, \int_{\mathbf{t}_0^{\gamma}}^{\mathbf{t}^{\gamma}} O_{\gamma} d\mathbf{t}^{\gamma}\right) = \mathbf{0}_{m}, \ \forall \mathbf{t}^{\gamma} \in \mathfrak{T}_{0F}^{\gamma}. \tag{32.15}$$

Property 457 ***The solution of*** *(32.12)* ***for all zero initial conditions***
The solution of (32.12) for all zero initial conditions is identically equal to the zero vector,

$$\boldsymbol{\varepsilon}_{\gamma 0}^{*k} = \mathbf{0}_{(k+1)\gamma} \text{ and}$$
$$\mathbf{T}\left(\mathbf{t}^{m}, \boldsymbol{\varepsilon}_{\gamma}^{*k}\left(\mathbf{t}^{(k+1)\gamma}\right), \int_{\mathbf{t}_0^{\gamma}}^{\mathbf{t}^{\gamma}} E_{\gamma}^{*}\left(\mathbf{t}^{\gamma}\right) d\mathbf{t}^{\gamma}\right) = \mathbf{0}_{m}, \ \forall \mathbf{t}^{(k+1)\gamma} \in \mathfrak{T}_{0F}^{(k+1)\gamma} \Longrightarrow$$
$$\boldsymbol{\varepsilon}_{\gamma}^{*}\left(\mathbf{t}^{\gamma}; \mathbf{t}_0^{\gamma}; \mathbf{0}_{(k+1)\gamma}\right) = \mathbf{0}_{\gamma}, \ \forall \mathbf{t}^{\gamma} \in \mathfrak{T}_{0F}^{\gamma}. \tag{32.16}$$

Note 458 *If (32.12) holds, then the condition* $\forall \mathbf{t}^{(k+1)\gamma} \in \mathfrak{T}_{0F}^{(k+1)\gamma}$ *demands that the initial output error vector, its derivatives, and its integral also obey (32.12) at the initial moment* $\mathbf{t}_0^{m}$.

Property 459 ***The operator*** $T(.)$ ***vanishes at the initial moment***
The initial output error vector, its initial derivatives, and its integral obey (32.12) at the initial moment t_0*; i.e., at* $\mathbf{t}_0^{m}$ *:*

$$\mathbf{T}\left(\mathbf{t}_0^{m}, \boldsymbol{\varepsilon}_{\gamma}^{*k}\left(\mathbf{t}_0^{(k+1)\gamma}\right), \int_{\mathbf{t}_0^{\gamma}}^{\mathbf{t}_0^{\gamma}} E_{\gamma}^{*}\left(\mathbf{t}^{\gamma}\right) d\mathbf{t}^{\gamma}\right) = \mathbf{T}\left(\mathbf{t}_0^{m}, \boldsymbol{\varepsilon}_{\gamma 0}^{*k}, \mathbf{0}_{\gamma}\right) = \mathbf{0}_{m},$$
$$\forall \boldsymbol{\varepsilon}_{\gamma 0}^{*k} \in \mathfrak{R}^{(k+1)\gamma}. \tag{32.17}$$

Note 460 *The real initial modified output error vector* $\boldsymbol{\varepsilon}_{\gamma}^{*}$ *can be unpredictable, uncontrollable, and arbitrary. It results from the past behavior of the plant, which is untouchable, from the initial actions of both the disturbances and the control variables. The consequence is that the real output error* γ*-subvector* $\boldsymbol{\varepsilon}_{\gamma 0}^{*}$ *is different from the zero vector:* $\boldsymbol{\varepsilon}_{\gamma 0}^{*} \neq \mathbf{0}_{N}$*. Therefore, it sometimes does not satisfy the condition (32.17), which then implies the violation of (32.12) and its nonrealizability at every* $\mathbf{t}^{(k+1)\gamma} \in \mathfrak{T}_{0F}^{(k+1)\gamma}$ *for such initial conditions.*

There are mathematical conditions under which the initial control vector $\mathbf{U}_0$ *can be chosen so to assure that the real output error vector* $\boldsymbol{\varepsilon}_0$ *is equal to the zero vector. The mathematical conditions are that* $N \leq r$ *and*

$$\mathbf{U}_0 = \mathbf{w}^{I}\left\{W_0^{T}\left(W_0 W_0^{T}\right)^{-1}\left[\mathbf{Y}_{dy0} - \mathbf{z}\left(t_0, \mathbf{R}_0^{\alpha-1}, \mathbf{D}_0\right)\right]\right\}.$$

The proof is immediate. For such initial control vector $\mathbf{U}_0$ *the output equation of (3.12) reads:*

$$\mathbf{Y}_0 = \mathbf{z}\left(t_0, \mathbf{R}_0^{\alpha-1}, \mathbf{D}_0\right) + W_0\left\{\begin{array}{c} W_0^{T}\left(W_0 W_0^{T}\right)^{-1} \\ \bullet\left[\begin{array}{c} \mathbf{Y}_{dy0} \\ -\mathbf{z}\left(t_0, \mathbf{R}_0^{\alpha-1}, \mathbf{D}_0\right) \end{array}\right] \end{array}\right\} = \mathbf{Y}_{dy0};$$

i.e. $\boldsymbol{\varepsilon}_0 = \mathbf{0}_{N}$*. This is mathematically possible, but it is questionable whether such* $\mathbf{U}_0$ *is physically realizable.*

Property 461 ***Tracking operator stability***
The tracking operator $\mathbf{T}(.)$ *(32.12) is such that its solution*

$$\boldsymbol{\varepsilon}_{\gamma}^{*}\left(\mathbf{t}^{\gamma}; \mathbf{t}_0^{(k+1)\gamma}; \mathbf{0}_{(k+1)\gamma}\right) \equiv \mathbf{0}_{\gamma}$$

is asymptotically (or exponentially) stable [in the whole].

Comment 462 *If the tracking operator* $\mathbf{T}(.)$ *in (32.12) possesses Property 461 and its solution along the output behavior of the plant (3.12) obeys the equation (32.12) then the plant exhibits (exponentially) stable tracking [in the whole] if* $\varepsilon_\gamma^* = \varepsilon_v^*$. *However, if* ε_γ^* *is replaced by* ε_v^* *then the plant exhibits both stable state tracking and (exponentially) stable tracking [in the whole].*

Problem 463 ***Matching the error vector, its derivatives, and integral with the tracking algorithm on*** $\mathfrak{T}_{0F}^{(k+1)\gamma}$

How can we guarantee for arbitrary error vector, error vector derivatives and the error vector integral to satisfy the tracking algorithm (32.12) on $\mathfrak{T}_{0F}^{(k+1)\gamma}$*?*

The solution to this problem was proposed in [274] and further developed in [279] for *time*-invariant linear systems. We generalize it to *time*-varying nonlinear systems.

Note 464 *The value* $t_R^{(k+1)\gamma}$ *is infinite if, and only if, tracking should be only asymptotic,* $t_R^{(k+1)\gamma} = \infty\mathbf{1}_{(k+1)\gamma}$, *which implies* $\mathfrak{T}_{R\infty}^{(k+1)\gamma} = \left\{\infty\mathbf{1}_{(k+1)\gamma}\right\}$. *Otherwise* $t_R^{(k+1)\gamma} \in \mathfrak{T}_{0F}^{(k+1)\gamma}$; *hence,* $\mathfrak{T}_{R\infty}^{(k+1)\gamma} \subset \mathfrak{T}_{0F}^{(k+1)\gamma}$, *and* $\mathbf{t}_0^{(k+1)\gamma} < t_R^{(k+1)\gamma} \leq \mathbf{t}_F^{(k+1)\gamma} < \infty\mathbf{1}_{(k+1)\gamma}$ *so that*

$$\mathbf{t}_R^{(k+1)\gamma} < \infty\mathbf{1}_{(k+1)\gamma} \Longrightarrow$$
$$\mathfrak{T}_{RF}^{(k+1)\gamma} = \left\{\mathbf{t}^{(k+1)\gamma} : \ \mathbf{t}_0^{(k+1)\gamma} < \mathbf{t}_R^{(k+1)\gamma} \leq \mathbf{t}^{(k+1)\gamma} \leq \mathbf{t}_F^{(k+1)\gamma}\right\} \subset \mathfrak{T}_{0F}^{(k+1)\gamma}. \tag{32.18}$$

In what follows we accept the finite vector reachability time $t_R^{(k+1)\gamma}$, *(32.18).*

Solution 465 ***The relaxed demand in general***

The root of the problem 463 is not in a particular initial error, but in the possible incompatibility of some initial errors with the tracking algorithm (32.12) at the initial moment $\mathbf{t}_0^{(k+1)\gamma}$ *due to the arbitrariness of the initial errors; i.e., it is possible that*

$$\exists \varepsilon_{\gamma 0}^{*k} \in \mathfrak{R}^{(k+1)\gamma} \Longrightarrow \mathbf{T}\left(\mathbf{t}_0^m; \varepsilon_{\gamma 0}^{*k}, \int_{\mathbf{t}_0^\gamma}^{\mathbf{t}^\gamma} E_\gamma^*\left(\mathbf{t}^\gamma\right) d\mathbf{t}^\gamma\right) = \mathbf{T}\left(\mathbf{t}_0^m; \varepsilon_{\gamma 0}^{*k}, \mathbf{0}_\gamma\right) \neq \mathbf{0}_m, \tag{32.19}$$

It is a fact that the control cannot influence such $\varepsilon_{\gamma 0}^{*k}$. *Hence, let the demand be relaxed so that (32.12) holds only on* $\mathfrak{T}_{RF}^{(k+1)\gamma}$ *rather than on the whole* $\mathfrak{T}_{0F}^{(k+1)\gamma}$ *for such* $\varepsilon_{\gamma 0}^{*k}$,

$$\mathbf{T}\left(\mathbf{t}^m, \varepsilon_\gamma^{*k}\left(\mathbf{t}^{(k+1)\gamma}; \mathbf{t}_0^{(k+1)\gamma}; \varepsilon_0^k\right), \int_{\mathbf{t}_0^\gamma}^{\mathbf{t}^\gamma} E_\gamma^*\left(\mathbf{t}^\gamma\right) d\mathbf{t}^\gamma\right) = \mathbf{0}_m, \ \forall \mathbf{t}^{(k+1)\gamma} \in \mathfrak{T}_{RF}^{(k+1)\gamma}, \tag{32.20}$$

rather than on the whole $\mathfrak{T}_{0F}^{(k+1)\gamma}$. *This relaxed demand should be satisfied for every initial output error vector* $\varepsilon_{\gamma 0}^{*k}$ *including those that violate (32.17), (Property 459), i.e., for which (32.19) holds.*

The preceding analysis opens the need to modify the control goal [279].

Goal 466 ***The modified control goal***

The control should force the plant to behave so that the following tracking algorithm holds [instead of (32.12)]:

$$\left\{\begin{array}{c} \mathbf{T}\left(\mathbf{t}^m, \varepsilon_\gamma^{*k}\left(\mathbf{t}^{(k+1)\gamma}\right), \int_{t_0^\gamma}^{\mathbf{t}^\gamma} E d\mathbf{t}^\gamma\right) = \mathbf{0}_m, \\ \forall \mathbf{t}^{(k+1)\gamma} \left\{\begin{array}{l} \in \mathfrak{T}_{0F}^{(k+1)\gamma} \ iff \ \mathbf{T}\left(\mathbf{t}_0^m; \varepsilon_{\gamma 0}^{*k}, \mathbf{0}_\gamma\right) = \mathbf{0}_m \\ \in \mathfrak{T}_{RF}^{(k+1)\gamma} \ iff \ \mathbf{T}\left(\mathbf{t}_0^m; \varepsilon_{\gamma 0}^{*k}, \mathbf{0}_\gamma\right) \neq \mathbf{0}_m \end{array}\right\} \end{array}\right\}, \tag{32.21}$$

$$
\text{and} \left\{ \begin{array}{c} \varepsilon_{\gamma}^{*k}\left(\mathbf{t}^{(k+1)\gamma};\mathbf{t}_0^{(k+1)\gamma};\varepsilon_0^k\right) \\ = \left\{ \begin{array}{c} -\mathbf{f}_{\gamma}^k\left(\mathbf{t}^{(k+1)\gamma};\mathbf{f}_0^{*k}\right),\ \forall \mathbf{t}^{(k+1)\gamma} \in \mathfrak{T}_{0F}^{(k+1)\gamma}, \\ iff\ \mathbf{T}\left(\mathbf{t}_0^m;\varepsilon_{\gamma 0}^{*k},\mathbf{0}_{\gamma}\right) \neq \mathbf{0}_m,\ k \in \{0,1,2,...\} \end{array} \right\} \end{array} \right\}
$$

$$
\text{where } \mathbf{f}_{\gamma 0}^k = -\varepsilon_{\gamma 0}^{*k} \text{ and } \mathbf{f}_{\gamma}^k\left(\mathbf{t}^{(k+1)\gamma};\mathbf{f}_{\gamma 0}^k\right) = \mathbf{0}_{(k+1)\gamma},\ \forall \mathbf{t}^{(k+1)\gamma} \in \mathfrak{T}_{RF}^{\gamma}, \tag{32.22}
$$

for a given or to be determined both $\mathbf{t}_R^{(k+1)\gamma} \in In\mathfrak{T}_0^{(k+1)\gamma}$ *and the vector function* $\mathbf{f}(.) : \mathfrak{T}^N \times \mathfrak{R}^N \longrightarrow \mathfrak{R}^N$ *, which implies its* γ*-subfunction* $\mathbf{f}_{\gamma} : \mathfrak{T}^{\gamma} \times \mathfrak{R}^{\gamma} \longrightarrow \mathfrak{R}^{\gamma}$.

Note 467 *We can use (32.21) instead of (32.12) also in the cases when the initial error vector* $\varepsilon_{\gamma 0}^{*k}$ *is different from the zero vector,* $\varepsilon_{\gamma 0}^{*k} \neq \mathbf{0}_{(k+1)\gamma}$*, but the tracking algorithm is satisfied at the initial moment; i.e.,*

$$
\mathbf{T}\left(\mathbf{t}_0^m, \varepsilon_{\gamma 0}^{*k}, \int_{t_0}^{t_0} E_{\gamma}^*\left(\mathbf{t}^{\gamma}\right) d\mathbf{t}^{\gamma}\right) = \mathbf{0}_m.
$$

We present several characteristic simple forms of the tracking algorithm $\mathbf{T}(.)$. They satisfy (32.14) through (32.17), i.e., they obey Properties 455-459.

Comment 468 *If we allow the parameters of the tracking algorithm to depend on the initial error vector* $\varepsilon_{\gamma}^{*k}$*, then we can define them in terms of* $\varepsilon_{\gamma}^{*k}$ *so that the tracking algorithm* $T(.)$ *(32.12) obeys (32.17). It is possible only if there are* $i \in \{0,1,...,k\}$ *and* $j \in \{0,1,...,k\}$*,* $i \neq j$*, such that* $E^{(i)}(t_0)E^{(j)}(t_0) < O$.

32.3 Reference output

The reality of the plant operations is often such that the real output vector is not desired: $\mathbf{Y}_0 \neq \mathbf{Y}_{d0}$, i.e., that the initial output error is not zero vector: $\varepsilon_0 \neq \mathbf{0}_N$.

Solution 469 ***The reference output vectors*** $\mathbf{Y}_R$ ***and*** $\mathbf{Y}_{R\gamma}^*$

The wise demand for the desired output behavior takes into account the real initial output vector $\mathbf{Y}_0$ *and defines* ***the realizable desired output vector*** $\mathbf{Y}_R$ *that starts from* $\mathbf{Y}_0$ *at the initial moment* t_0*. It is also called* ***the realizable output reference vector*** $\mathbf{Y}_R$*, for short:* ***the reference output vector*** $\mathbf{Y}_R$ *:*

$$
\mathbf{Y}_R = \left[Y_{R1}\ Y_{R2}\ \ ...\ Y_{RN}\ \right]^T \in \mathfrak{R}^N, \tag{32.23}
$$

It should be such that at the initial instant it is equal to the initial real output vector $\mathbf{Y}_0$*, and because the finite vector reachability time* $\mathbf{t}_R^N$ *has elapsed it is always equal to the desired output vector* $\mathbf{Y}_d$*,*

$$
\begin{gathered}
\mathbf{Y}_{R0} = \mathbf{Y}_R\left(\mathbf{t}_0^N;\mathbf{t}_0^N;\mathbf{Y}_{R0}\right) = \mathbf{Y}\left(\mathbf{t}_0^N;\mathbf{t}_0^N;\mathbf{Y}_0\right) = \mathbf{Y}_0, \\
\mathbf{Y}_R\left(\mathbf{t}^N;\mathbf{t}_0^N;\mathbf{Y}_{R0}\right) = \mathbf{Y}_d\left(\mathbf{t}^N\right),\ \forall \mathbf{t}^N \in \mathfrak{T}_{RF}^N.
\end{gathered} \tag{32.24}
$$

The fact that only the γ*-subvector* $\mathbf{Y}_{\gamma}$*,*

$$
\mathbf{Y}_{\gamma} = \left[Y_1\ Y_2\ \ ...\ Y_{\gamma}\ \right]^T \in \mathfrak{R}^{\gamma}, \tag{32.25}
$$

is directly controllable, Axiom 87, Property 64, Lemma 65, and the control goal demand is that the whole output vector $\mathbf{Y}$ *should be controlled at least indirectly, which is possible in principle, Axiom 87, lead to the use of the* γ*-subvector* $\mathbf{Y}_{R\gamma}^*$ *of the reference vector* $\mathbf{Y}_R$ *:*

$$
\begin{gathered}
\mathbf{Y}_{R\gamma}^* = \left[Y_{R1}\ Y_{R2}\ \ ...Y_{R\gamma-1}\ Y_{R\gamma}^*\ \right]^T \in \mathfrak{R}^{\gamma}, \\
Y_{R\gamma}^* = Y_{Rj} \Longleftrightarrow \left|Y_{Rj} - Y_j\right| = max\left(\left|Y_{Ri} - Y_i\right| : \forall i = \gamma, \gamma+1, ..., N\right), \\
j \in \{\gamma, \gamma+1,\ \cdots, N\},\ \gamma \in \{1, 2,\ \cdots, N\}.
\end{gathered} \tag{32.26}
$$

The vector $\mathbf{Y}_{R\gamma}^{*}$ *represents* ***the directly (i.e., elementwise) controllable reference output*** *γ-subvector of the plant, such that it is continuous, possibly continuously differentiable i-times, at the initial moment it and its derivatives are equal to the real initial output γ-subvector $\mathbf{Y}_{\gamma 0}$ and its initial derivatives, that the former become and stay equal to the desired output γ-subvector $\mathbf{Y}_{\gamma d}$ and its derivatives as soon as the vector reachability time $\mathbf{t}_{R}^{\gamma}$ has elapsed:*

$$\begin{gathered}\mathbf{Y}_{R\gamma}^{*}(\mathbf{t}^{\gamma};\mathbf{t}_{0}^{\gamma};\mathbf{Y}_{R0})\in\mathfrak{C}^{i}(\mathfrak{T}_{0F}^{\gamma}),\ i\in\{0,1,2,...\},\\ \mathbf{Y}_{R\gamma 0}^{*}=\mathbf{Y}_{R\gamma}^{*}(\mathbf{t}_{0}^{\gamma};\mathbf{t}_{0}^{\gamma};\mathbf{Y}_{R0})=\mathbf{Y}_{\gamma}(\mathbf{t}_{0}^{\gamma};\mathbf{t}_{0}^{\gamma};\mathbf{Y}_{\gamma 0})=\mathbf{Y}_{\gamma 0},\\ \mathbf{Y}_{R\gamma}^{*}(\mathbf{t}^{\gamma};\mathbf{t}_{0}^{\gamma};\mathbf{Y}_{R0})=\mathbf{Y}_{d\gamma}^{k}(\mathbf{t}^{\gamma}),\ \forall\mathbf{t}^{\gamma}\in\mathfrak{T}_{RF}^{\gamma}.\end{gathered}\tag{32.27}$$

The initial $\mathbf{Y}_{R\gamma 0}^{*k}=\mathbf{Y}_{R\gamma}^{*k}(\mathbf{t}_{0}^{(k+1)\gamma})$ *is also well defined due to (32.27):* $\mathbf{Y}_{R\gamma 0}^{*k}=\mathbf{Y}_{\gamma 0}^{k}$. *However, (32.27) does not define* $\mathbf{Y}_{R\gamma}^{*k}(\mathbf{t}^{(k+1)\gamma})$ *on* $In\mathfrak{T}_{0R}^{(k+1)\gamma}$. *The definition of* $\mathbf{Y}_{R\gamma}^{*k}(\mathbf{t}^{(k+1)\gamma})$ *on* $In\mathfrak{T}_{0R}^{(k+1)N}$ *is left open for the free appropriate choice.*

The equations (32.23), (32.24) lead to **the subsidiary error vector** $\boldsymbol{\epsilon}$,

$$\begin{gathered}\epsilon_{j}=Y_{Rj}-Y_{j},\ \ \forall j=1,2,..,N,\\ \boldsymbol{\epsilon}=\mathbf{Y}_{R}-\mathbf{Y}=\boldsymbol{\epsilon}(\mathbf{Y})=\boldsymbol{\epsilon}\left[\boldsymbol{\varepsilon}\left(\mathbf{t}_{0}^{N}\right)\right],\ \boldsymbol{\epsilon}=\left[\epsilon_{1}\ \ \epsilon_{2}\ \ ...\ \epsilon_{N}\right]^{T}\in\mathfrak{R}^{N}.\end{gathered}\tag{32.28}$$

The initial vector $\boldsymbol{\epsilon}_{0}=\boldsymbol{\epsilon}\left(\mathbf{t}_{0}^{N}\right)=\boldsymbol{\epsilon}\left[\mathbf{Y}\left(\mathbf{t}_{0}^{N}\right)\right]$ is equal to the zero vector for every initial real output vector $\mathbf{Y}_{0}=\mathbf{Y}\left(\mathbf{t}_{0}^{N}\right)$; i.e., for every $\boldsymbol{\varepsilon}_{0}=\boldsymbol{\varepsilon}\left(\mathbf{t}_{0}^{N}\right)$ due to (32.27):

$$\begin{gathered}\mathbf{Y}_{R}=\mathbf{Y}_{d}\Longrightarrow\boldsymbol{\epsilon}=\mathbf{Y}_{R}-\mathbf{Y}=\mathbf{Y}_{d}-\mathbf{Y}=\boldsymbol{\varepsilon},\\ \mathbf{Y}_{R}\neq\mathbf{Y}_{d}\Longrightarrow\boldsymbol{\epsilon}\left(\mathbf{t}_{0}^{N}\right)=\boldsymbol{\epsilon}\left[\mathbf{Y}\left(\mathbf{t}_{0}^{N}\right)\right]=\boldsymbol{\epsilon}\left[\boldsymbol{\varepsilon}\left(\mathbf{t}_{0}^{N}\right)\right]=\mathbf{Y}_{R}\left(\mathbf{t}_{0}^{N}\right)-\mathbf{Y}\left(\mathbf{t}_{0}^{N}\right)\\ =\mathbf{Y}\left(\mathbf{t}_{0}^{N}\right)-\mathbf{Y}\left(\mathbf{t}_{0}^{N}\right)\equiv\mathbf{0}_{N},\ \forall\mathbf{Y}\left(\mathbf{t}_{0}^{N}\right)\in\mathfrak{R}^{N},\end{gathered}\tag{32.29}$$

and

$$\begin{gathered}\mathbf{Y}_{R}\neq\mathbf{Y}_{d}\Longrightarrow\boldsymbol{\epsilon}_{\gamma}^{*}\left(\mathbf{t}_{0}^{\gamma}\right)=\boldsymbol{\epsilon}_{\gamma}^{*}\left[\mathbf{Y}_{\gamma}^{*}\left(\mathbf{t}_{0}^{\gamma}\right)\right]=\mathbf{Y}_{R\gamma}^{*}\left(\mathbf{t}_{0}^{\gamma}\right)-\mathbf{Y}_{\gamma}^{*}\left(\mathbf{t}_{0}^{\gamma}\right)\\ =\mathbf{Y}_{\gamma}^{*}\left(\mathbf{t}_{0}^{\gamma}\right)-\mathbf{Y}_{\gamma}^{*}\left(\mathbf{t}_{0}^{\gamma}\right)\equiv\mathbf{0}_{\gamma},\ \forall\mathbf{Y}_{\gamma}^{*}\left(\mathbf{t}_{0}^{\gamma}\right)\in\mathfrak{R}^{v},\end{gathered}\tag{32.30}$$

This is the crucial property of the vector $\boldsymbol{\epsilon}$. The γ-subvector $\boldsymbol{\epsilon}_{\gamma}^{*}$ of $\boldsymbol{\epsilon}$ contains, explicitly or implicitly, all entries of $\boldsymbol{\epsilon}$:

$$\begin{gathered}\epsilon_{\gamma}^{*}=\epsilon_{j}\text{ iff }\left|\epsilon_{j}\right|\geq\max\left(\left|\epsilon_{i}\right|:\forall i=\gamma,\gamma+1,...,N\right),\ j\in\{k,k+1,...,N\},\\ \gamma\in\{1,2,...,N\}\\ \boldsymbol{\epsilon}_{\gamma}^{*}=\left[\epsilon_{1}\ \ \epsilon_{2}\ ...\ \epsilon_{\gamma-1}\ \epsilon_{\gamma}^{*}\ \right]\in\mathfrak{R}^{\gamma};\ \gamma=N\Longleftrightarrow\boldsymbol{\epsilon}_{\gamma}^{*}=\boldsymbol{\epsilon},\ N\leq r.\end{gathered}\tag{32.31}$$

It has the same properties, which are exposed in (32.4), as $\boldsymbol{\varepsilon}_{\gamma}^{*}$, (32.5):

$$\begin{gathered}\epsilon_{\gamma}^{*}=0\Longleftrightarrow\epsilon_{j}=0,\ \forall j=\gamma,\gamma+1,...,N,\ \text{and}\\ \left|\epsilon_{\gamma}^{*}\right|<\xi\Longleftrightarrow\left|\epsilon_{j}\right|<\xi,\forall j=\gamma,\gamma+1,...,N,\forall\xi>0.\end{gathered}\tag{32.32}$$

This is another inherent feature for control. The γ-subvector ϵ_{γ}^{*} induces

$$\ominus_{\gamma}^{*}\left(\boldsymbol{\epsilon}_{\gamma}^{*}\right)=\operatorname{diag}\left\{\epsilon_{1}\ \ \epsilon_{2}\ ...\ \epsilon_{\gamma-1}\ \epsilon_{\gamma}^{*}\right\}\in\mathfrak{R}^{\gamma\times\gamma}.\tag{32.33}$$

Comment 470 *The reference output vector* $\mathbf{Y}_{R}$ *equals the desired output vector* $\mathbf{Y}_{d}$ *if, and only if,* $\mathbf{T}\left(\mathbf{t}_{0}^{m};\boldsymbol{\varepsilon}_{\gamma}^{*k},\mathbf{0}_{\gamma}\right)=\mathbf{0}_{m}$. *The reference output vector* $\mathbf{Y}_{R}$ *replaces the desired output vector* $\mathbf{Y}_{d}$ *if, and only if,* $\mathbf{T}\left(\mathbf{t}_{0}^{m};\boldsymbol{\varepsilon}_{\gamma}^{*k},\mathbf{0}_{\gamma}\right)\neq\mathbf{0}_{m}$. *Therefore, the reference output γ-subvector* $\mathbf{Y}_{R\gamma}^{*}$ *replaces the desired output γ-subvector* $\mathbf{Y}_{\gamma d}$ *if, and only if,* $\mathbf{T}\left(\mathbf{t}_{0}^{m};\boldsymbol{\varepsilon}_{\gamma}^{*k},\mathbf{0}_{\gamma}\right)\neq\mathbf{0}_{m}$. *Then, and only then, the subsidiary error γ-subvector* $\boldsymbol{\epsilon}_{\gamma}^{*}$ *replaces the γ-subvector* $\boldsymbol{\varepsilon}_{\gamma}$. *This leads to the structure of the feedback control system shown in Figure 32.2 for* $\gamma=N$.

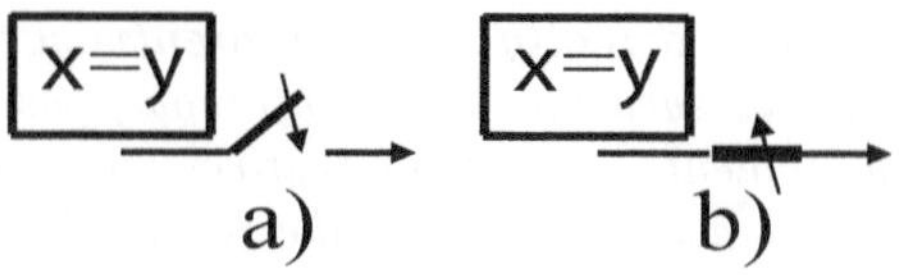

Figure 32.1: a) Switch *closes* if, and only if, $\mathbf{x} = \mathbf{y}$. b) Switch *opens* if, and only if, $\mathbf{x} = \mathbf{y}$.

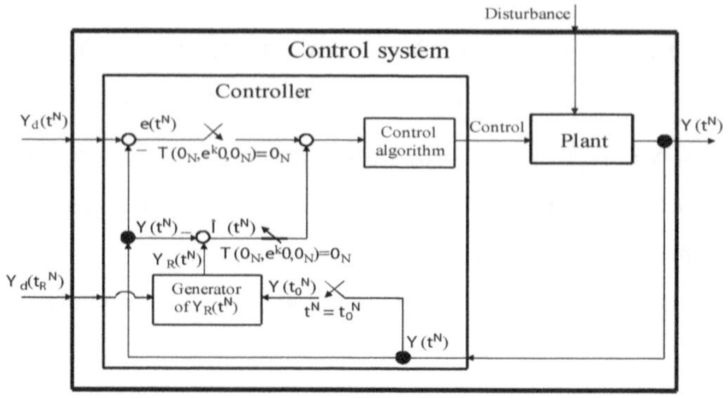

Figure 32.2: The structural diagram of the control system. The induced error vector $\boldsymbol{\epsilon}$ is denoted by $\in$ in this figure for printing reasons.

Solution 471 ***On the determination of the reference output vector function*** $\mathbf{Y}_R(.)$
One approach to construct the reference output vector function $\mathbf{Y}_R(.)$, *hence* $\mathbf{Y}^*_{R\gamma}(.)$, *is to apply the solution obtained in the theory of optimal control systems relative to the accepted optimality criterion [4], [15], [36], [40], [41], [44], [55], [62]-[64], [73], [129], [299], [319], [347], [381], [481], [482], [526], [522], [543]. If the optimality criterion is the maximal deviation of* $\mathbf{Y}_R(\mathbf{t}^N)$ *from* $\mathbf{Y}_d(\mathbf{t}^N)$ *over the time interval* $[\mathbf{t}_0^N, \mathbf{t}_R^N]$, *where* $\mathbf{t}_0^N$ *and* $\mathbf{t}_R^N$ *are known (prespecified), then its optimal value is the minimal maximal deviation over the time interval* $[\mathbf{t}_0^N, \mathbf{t}_R^N]$. *If the optimality criterion is the energy consumption over the time interval* $[\mathbf{t}_0^N, \mathbf{t}_R^N]$ *then its optimal value is the minimal energy consumption over the time interval* $[\mathbf{t}_0^N, \mathbf{t}_R^N]$. *The optimality criterion can be the reachability time. The corresponding optimal solution is the minimal reachability time.*

Another reasonable choice is the following:

$$\mathbf{Y}_R(\mathbf{t}^N) = \left\{ \begin{array}{c} \mathbf{Y}_d(\mathbf{t}^N) \text{ iff } \mathbf{T}\left(\mathbf{t}_0^m, \boldsymbol{\varepsilon}^k, \mathbf{0}_N\right) = \mathbf{0}_m, \\ \mathbf{Y}_d(\mathbf{t}^N) + \mathbf{f}(\mathbf{t}^N) \text{ iff } \mathbf{T}\left(\mathbf{t}_0^m, \boldsymbol{\varepsilon}^k, \mathbf{0}_N\right) \neq \mathbf{0}_m \end{array} \right\}, \ \forall \mathbf{t}^N \in \mathfrak{T}_{0F}^N. \tag{32.34}$$

This implies, due to Solution 465,

$$\mathbf{Y}^*_{R\gamma}(\mathbf{t}^N) = \left\{ \begin{array}{c} \mathbf{Y}^*_{d\gamma}(\mathbf{t}^\gamma) \text{ iff } \mathbf{T}\left(\mathbf{t}_0^m, \boldsymbol{\varepsilon}_\gamma^{*k}, \mathbf{0}_\gamma\right) = \mathbf{0}_m, \\ \mathbf{Y}^*_{d\gamma}(\mathbf{t}^\gamma) + \mathbf{f}_\gamma(\mathbf{t}^\gamma) \text{ iff } \mathbf{T}\left(\mathbf{t}_0^m, \boldsymbol{\varepsilon}_\gamma^{*k}, \mathbf{0}_\gamma\right) \neq \mathbf{0}_m \end{array} \right\}, \ \forall \mathbf{t}^\gamma \in \mathfrak{T}_{0F}^\gamma. \tag{32.35}$$

The equation (32.34) determines the block diagram of the generator of $\mathbf{Y}_R(\mathbf{t}^N)$, *Figure 32.3.The subsidiary vectors* $\mathbf{f}(\mathbf{t}^N)$ *and its* γ*-subvector* $\mathbf{f}_\gamma(\mathbf{t}^\gamma)$,

$$\begin{gathered} \mathbf{f}(\mathbf{t}^N) = [f_1(t) \quad f_2(t) \ .. \ f_N(t)]^T \in \mathfrak{R}^N, \\ \mathbf{f}_\gamma(\mathbf{t}^\gamma) = [f_1(t) \quad f_2(t) \ .. \ f_\gamma(t)]^T \in \mathfrak{R}^\gamma, \end{gathered} \tag{32.36}$$

should obey the following conditions that result from (32.22), (32.23) and (32.24):

$$\begin{gathered} \mathbf{f}(\mathbf{t}_0^N) = -\boldsymbol{\varepsilon}(\mathbf{t}_0^N), \ \forall \boldsymbol{\varepsilon}(\mathbf{t}_0^N) \in \mathfrak{R}^N; \ \ \mathbf{f}(\mathbf{t}^N) \ = \mathbf{0}_N, \ \forall \mathbf{t}^N \in \mathfrak{T}_{RF}^N, \\ \mathbf{f}_\gamma(\mathbf{t}_0^\gamma) = -\boldsymbol{\varepsilon}_\gamma^*(\mathbf{t}_0^\gamma), \ \forall \boldsymbol{\varepsilon}_\gamma^*(\mathbf{t}_0^\gamma) \in \mathfrak{R}^\gamma; \ \mathbf{f}_\gamma(\mathbf{t}^\gamma) = \mathbf{0}_\gamma, \ \forall \mathbf{t}^\gamma \in \mathfrak{T}_{RF}^\gamma. \end{gathered} \tag{32.37}$$

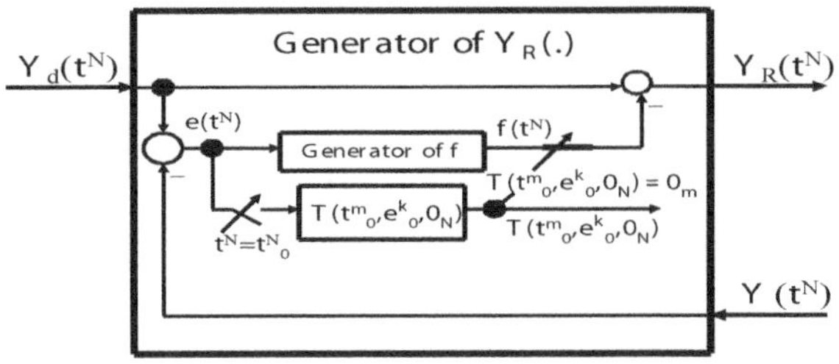

Figure 32.3: The block diagram of the generator of the subsidiary reference output vector function $\mathbf{Y}_R(.)$.

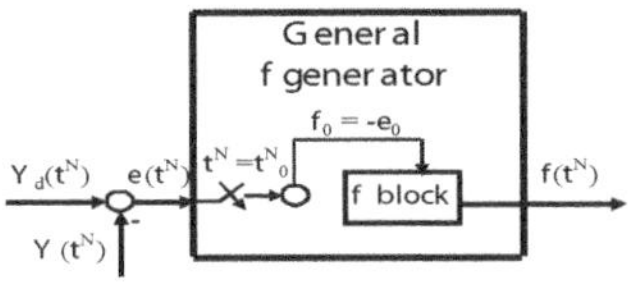

Figure 32.4: The block diagram of the general generator of the subsidiary function $\mathbf{f}(.)$.

These equations determine the general block diagram of the generator of $\mathbf{f}(.)$, *Figure 32.4, and they link* $\mathbf{Y}^*_{R\gamma}(\mathbf{t}^\gamma)$ *with* $\mathbf{Y}^*_{d\gamma}(\mathbf{t}^\gamma)$ *via* $\mathbf{f}_\gamma(\mathbf{t}^\gamma)$, *(32.35). Besides (32.31), (32.34), and (32.35) furnish*

$$\forall \mathbf{t}^N \in \mathfrak{T}^N_{0F} \Longrightarrow \boldsymbol{\epsilon}(\mathbf{t}^N) = \mathbf{Y}_R(\mathbf{t}^N) - \mathbf{Y}(\mathbf{t}^N)$$
$$= \left\{ \begin{array}{c} \mathbf{Y}_d(\mathbf{t}^N) \text{ iff } \mathbf{T}\left(\mathbf{t}^m_0, \boldsymbol{\varepsilon}^k, \mathbf{0}_N\right) = \mathbf{0}_m, \\ \mathbf{Y}_d(\mathbf{t}^N) + \mathbf{f}(\mathbf{t}^N) \text{ iff } \mathbf{T}\left(\mathbf{t}^m_0, \boldsymbol{\varepsilon}^k, \mathbf{0}_N\right) \neq \mathbf{0}_m \end{array} \right\} - \mathbf{Y}(\mathbf{t}^N),$$

and

$$\forall \mathbf{t}^\gamma \in \mathfrak{T}^\gamma_{0F} \Longrightarrow \boldsymbol{\epsilon}^*_\gamma(\mathbf{t}^\gamma) = \mathbf{Y}^*_{R\gamma}(\mathbf{t}^\gamma) - \mathbf{Y}^*_\gamma(\mathbf{t}^\gamma)$$
$$= \left\{ \begin{array}{c} \mathbf{Y}^*_{d\gamma}(\mathbf{t}^\gamma) \text{ iff } \mathbf{T}\left(\mathbf{t}^m_0, \boldsymbol{\varepsilon}^{*k}_\gamma, \mathbf{0}_\gamma\right) = \mathbf{0}_m, \\ \mathbf{Y}^*_{d\gamma}(\mathbf{t}^\gamma) + \mathbf{f}_\gamma(\mathbf{t}^\gamma) \text{ iff } \mathbf{T}\left(\mathbf{t}^m_0, \boldsymbol{\varepsilon}^{*k}_\gamma, \mathbf{0}_\gamma\right) \neq \mathbf{0}_m \end{array} \right\} - \mathbf{Y}^*_{R\gamma}(\mathbf{t}^\gamma).$$

The beginning and the end of these equations, together with (4.7), yield

$$\boldsymbol{\epsilon}(\mathbf{t}^N) = \boldsymbol{\varepsilon}(\mathbf{t}^N) + \left\{ \begin{array}{c} \mathbf{0}_N \text{ iff } \mathbf{T}\left(\mathbf{t}^m_0, \boldsymbol{\varepsilon}^k, \mathbf{0}_N\right) = \mathbf{0}_m, \\ \mathbf{f}(\mathbf{t}^N) \text{ iff } \mathbf{T}\left(\mathbf{t}^m_0, \boldsymbol{\varepsilon}^k, \mathbf{0}_N\right) \neq \mathbf{0}_m \end{array} \right\}, \ \forall \mathbf{t}^N \in \mathfrak{T}^\gamma_{0F}, \tag{32.38}$$

and

$$\boldsymbol{\epsilon}^*_\gamma(\mathbf{t}^\gamma) = \left\{ \begin{array}{c} \boldsymbol{\varepsilon}^*_\gamma(\mathbf{t}^\gamma) \text{ iff } \mathbf{T}\left(\mathbf{t}^m_0, \boldsymbol{\varepsilon}^{*k}_\gamma, \mathbf{0}_\gamma\right) = \mathbf{0}_m, \\ \boldsymbol{\varepsilon}^*_\gamma(\mathbf{t}^\gamma) + \mathbf{f}_\gamma(\mathbf{t}^\gamma) \text{ iff } \mathbf{T}\left(\mathbf{t}^m_0, \boldsymbol{\varepsilon}^{*k}_\gamma, \mathbf{0}_\gamma\right) \neq \mathbf{0}_m \end{array} \right\}, \ \forall \mathbf{t}^\gamma \in \mathfrak{T}^\gamma_{0F}. \tag{32.39}$$

At the initial moment $\mathbf{f}(\mathbf{t}^N_0) = -\boldsymbol{\varepsilon}(\mathbf{t}^N_0)$, *(32.37), which reduces (32.38) to*

$$\boldsymbol{\epsilon}(\mathbf{t}^N_0) = \boldsymbol{\epsilon}\left[\boldsymbol{\varepsilon}(\mathbf{t}^N_0)\right] = \mathbf{0}_N, \ \forall \boldsymbol{\varepsilon}(\mathbf{t}^N_0) \in \mathfrak{R}^N, \tag{32.40}$$

and (32.39) to:

$$\boldsymbol{\epsilon}^*_\gamma(\mathbf{t}^\gamma_0) = \boldsymbol{\epsilon}^*_\gamma\left[\boldsymbol{\varepsilon}^*_\gamma(\mathbf{t}^\gamma_0)\right] = \mathbf{0}_\gamma, \ \forall \boldsymbol{\varepsilon}^*_\gamma(\mathbf{t}^\gamma_0) \in \mathfrak{R}^\gamma. \tag{32.41}$$

These equations agree with (32.29) and (32.30), respectively. In addition if the control guarantees $\boldsymbol{\epsilon}(\mathbf{t}^N) = \mathbf{0}_N$, $\forall \mathbf{t}^N \in \mathfrak{T}^N_{0F}$, *then*

$$\boldsymbol{\varepsilon}(\mathbf{t}^N) = -\mathbf{f}(\mathbf{t}^N), \ \forall \mathbf{t}^N \in \mathfrak{T}^N_{0F}, \ \boldsymbol{\varepsilon}^*_\gamma(\mathbf{t}^\gamma) = -\mathbf{f}_\gamma(\mathbf{t}^\gamma), \ \forall \mathbf{t}^\gamma \in \mathfrak{T}^\gamma_{0F}, \tag{32.42}$$

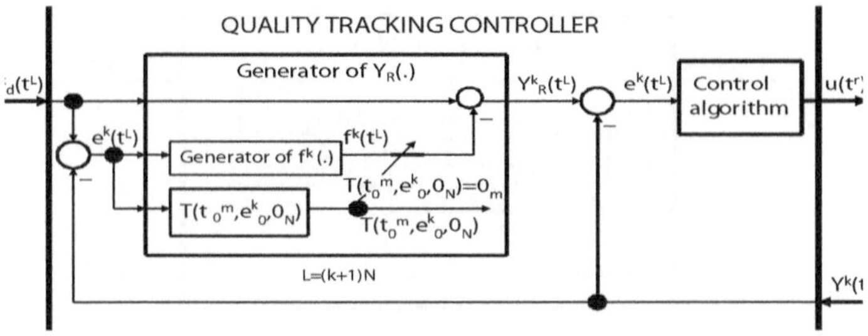

Figure 32.5: The block diagram of the quality tracking controller.

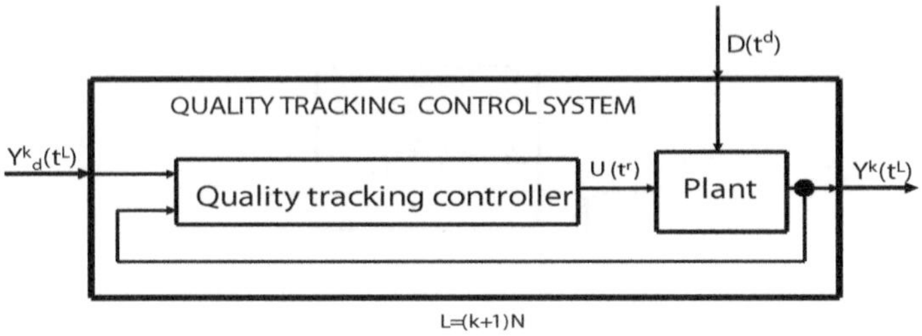

Figure 32.6: The structural diagram of the quality tracking control system of the plant perturbed by the action of the disturbance **D**.

due to (32.38). This and (32.37) lead to the following conclusion:

$$\epsilon(\mathbf{t}^N) = \mathbf{0}_N, \forall \mathbf{t}^N \in \mathfrak{T}_{RF}^N \Longrightarrow \varepsilon(\mathbf{t}^N) = \mathbf{0}_N,\ \forall \mathbf{t}^N \in \mathfrak{T}_{RF}^N,$$
$$\epsilon_\gamma^*(\mathbf{t}^\gamma) = \mathbf{0}_\gamma, \forall \mathbf{t}^\gamma \in \mathfrak{T}_{RF}^\gamma \Longrightarrow \varepsilon_\gamma^*(\mathbf{t}^\gamma) = \mathbf{0}_\gamma,\ \forall \mathbf{t}^\gamma \in \mathfrak{T}_{RF}^\gamma, \tag{32.43}$$

The control vector acts only on the output γ-subvector $\mathbf{Y}_\gamma$, Comment 450, i.e., on the error γ-subvector ε_γ. Because the whole real output vector $\mathbf{Y}\left(\mathbf{t}^N\right)$ should reach the whole desired output vector $\mathbf{Y}_d\left(\mathbf{t}^N\right)$ at latest at the final reachability time, then we should use only the $\gamma-$subequations of (32.29)-(32.43) in the sequel if $\gamma < N$.

The subsidiary output vector $\mathbf{Y}_R^k(\mathbf{t}^{(k+1)N})$ replaces the desired output vector $\mathbf{Y}_d^k(\mathbf{t}^{(k+1)N})$ in the quality tracking controller, Figure 32.5. Figure 32.6 shows the structural diagram of the quality tracking control system of the plant.

A possible specific form of the subsidiary function $\mathbf{f}_\gamma^*(.)$ was proposed in [274], Figure 32.7: We determine $\mathbf{f}_\gamma^*(.)$ completely by:

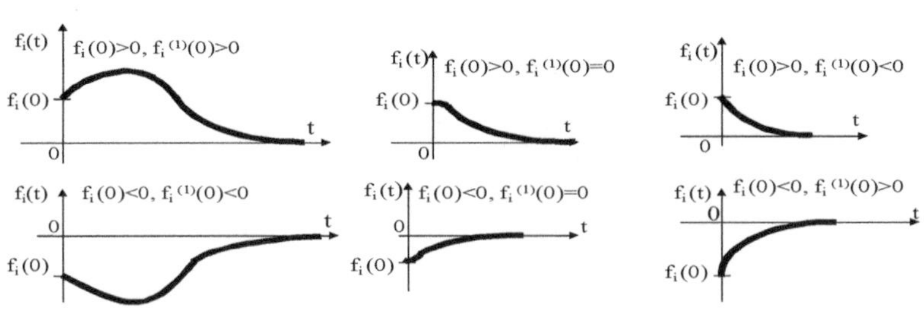

Figure 32.7: The i-th entry $f_i(.) = f(.)$ of the subsidiary vector function $\mathbf{f}(.)$.

$$f_i(t)$$
$$\left\{\begin{array}{c}\left\{\begin{array}{c}k_i + A_i \sin\left[w_i\left(t-t_0\right)+h_i\right],\ \forall t \in [t_0, t_{Ri}],\\ 0,\ \forall t \in [t_{Ri},\ t_{Fi}],\ t_{Fi}<\infty,\\ \text{if}: \left(\varepsilon_{i0}\varepsilon_{i0}^{(1)}>0\right) \text{ or } \left(\varepsilon_{i0}\neq 0, \varepsilon_{i0}^{(1)}=0\right)\\ \text{or } \left(\varepsilon_{i0}=0, \varepsilon_{i0}^{(1)}\neq 0\right),\end{array}\right\},\\ 0,\ \forall t \in [t_0,\ t_{Fi}],\ t_{Fi}<\infty,\ \text{if } \varepsilon_{i0}=\varepsilon_{i0}^{(1)}=0,\\ \left\{\begin{array}{c}a_i\left(t-t_0\right)^2+b_i\left(t-t_0\right)+c_i,\ \forall t\in[t_0, t_{Ri}],\\ 0,\ \forall t \in [t_{Ri},\ t_{Fi}],\ t_{Fi}<\infty,\ \text{if } \varepsilon_{i0}\varepsilon_{i0}^{(1)}<0\end{array}\right\},\\ \forall i=1,2,\ \cdots,\gamma\end{array}\right\} \tag{32.44}$$

where

$$f_i(t_0)=\varepsilon_{i0},\ \forall i=1,2,\ \cdots,\gamma,\ f_i^{(1)}(t_0)=\varepsilon_{i0}^{(1)},\ \ \forall i=1,2,\ \cdots,\gamma, \tag{32.45}$$

$$f_i(t_{Ri})=0,\ f_i^{(1)}(t_{Ri})=0,\ \ \forall i=1,2,\ \cdots,\gamma. \tag{32.46}$$

The equations (32.44)-(32.46) imply the complete definition of the continuously differentiable subsidiary function $\mathbf{f}_\gamma^*(.)$ by (32.47)-(32.51):

$$f_i(t)=\left\{\begin{array}{c}\left\langle\begin{array}{c}\varepsilon_{i0}\left\{1+\sin\left[\frac{\varepsilon_{i0}^{(1)}}{\varepsilon_{i0}}\left(t-t_0\right)\right]\right\},\ \forall t\in[t_0, t_{Ri}],\\ t_{Ri}=t_0+\frac{3\pi\varepsilon_{i0}}{2\varepsilon_{i0}^{(1)}}\end{array}\right\rangle\\ 0,\ \forall t\in[t_{Ri}, t_{Fi}),\ \text{iff } \varepsilon_{i0}\varepsilon_{i0}^{(1)}>0,\end{array}\right\}, \tag{32.47}$$

$$f_i(t)=\left\{\begin{array}{c}\left\langle\begin{array}{c}\frac{\varepsilon_{i0}}{2}\left\{1+\sin\left[\frac{\pi(t-t_0)}{t_{Ri}-t_0}+\frac{\pi}{2}\right]\right\},\ \forall t\in[t_0, t_{Ri}],\\ t_{Ri} \text{ can be given or chosen freely, } t_{Ri}\in]t_0, t_{Fi}[,\end{array}\right\rangle\\ 0,\ \forall t\in[t_{Ri}, t_{Fi}),\ \text{iff } \varepsilon_{i0}\neq 0 \text{ and } \varepsilon_{i0}^{(1)}=0,\end{array}\right\} \tag{32.48}$$

$$f_i(t)=\left\{\begin{array}{c}\left\{\begin{array}{c}\frac{\varepsilon_{i0}^{(1)}(t_{Ri}-t_0)}{\pi}\sin\left[\frac{\pi}{(t_{Ri}-t_0)}\left(t-t_0\right)\right],\\ \forall t\in[t_0, \frac{1}{2}t_{Ri}],\\ t_{Ri} \text{ can be given or chosen freely, } t_{Ri}\in]t_0, t_{Fi}[,\end{array}\right\},\\ \left\{\begin{array}{c}\frac{\varepsilon_{i0}^{(1)}(t_{Ri}-t_0)}{\pi}\left\langle 1+\cos\left[\frac{\pi}{t_{Ri}-t_0}\left(t-t_0\right)\right]\right\rangle,\\ \forall t\in[\frac{1}{2}t_{Ri}, t_{Fi}],\end{array}\right\},\\ \text{iff } \varepsilon_{i0}=0 \textit{ and } \varepsilon_{i0}^{(1)}\neq 0,\end{array}\right\} \tag{32.49}$$

$$f_i(t)=0,\ \forall t\in[t_0,\ t_{Fi}],\ \text{iff } \varepsilon_{i0}=0 \textit{ and } \varepsilon_{i0}^{(1)}=0, \tag{32.50}$$

$$f_i(t)=\left\{\begin{array}{c}\left\{\begin{array}{c}\frac{\left(\varepsilon_{i0}^{(1)}\right)^2}{4\varepsilon_{i0}}\left(t-t_0\right)^2+\varepsilon_{i0}^{(1)}\left(t-t_0\right)+\varepsilon_{i0},\\ \forall t\in[t_0,\ t_{Ri}],\ t_{Ri}=t_0-\frac{\varepsilon_{i0}}{\varepsilon_{i0}^{(1)}},\end{array}\right\}\\ 0,\ \forall t\in[t_{Ri},\ t_{Fi}],\ \text{iff } \varepsilon_{i0}\varepsilon_{i0}^{(1)}<0.\end{array}\right\} \tag{32.51}$$

The equations (32.47)-(32.51) determine the combined i-th entry $f_i(.)$ of $\mathbf{f}_\gamma^*(.)$, Figure 32.8. It is the combination of the sinusoidal function and the parabolic function. Figure 32.9 represents the block of the combined subsidiary vector function $\mathbf{f}^1(.)$.

The formulae (32.47)-(32.51) show the dependence of the finite vector reachability *time* $\mathbf{t}_R^\gamma$ on the initial conditions. Bounded initial conditions bounded the finite vector reachability *time* $\mathbf{t}_R^\gamma$. This is reasonable and rational.

As the summary of the above consideration we state:

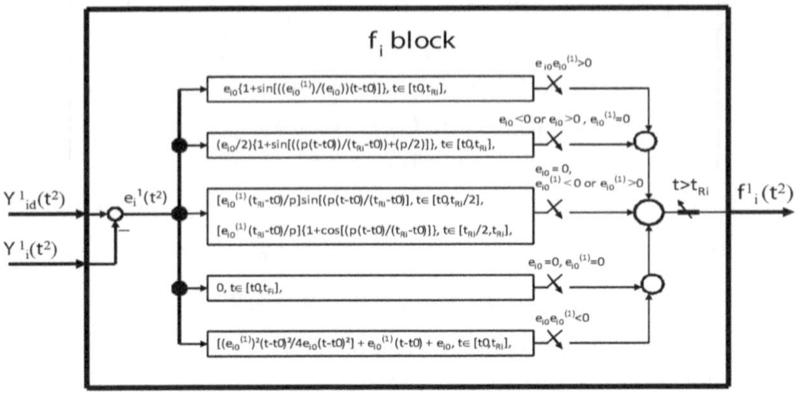

Figure 32.8: The block diagram of the combined *i-th* $f_i(.)$ entry of $\mathbf{f}(.)$.

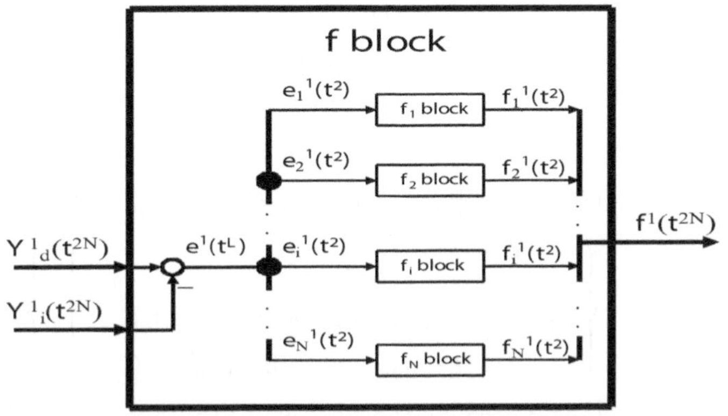

Figure 32.9: The block diagram of the generator of the combined subsidiary vector function $\mathbf{f}(.)$.

Goal 472 ***The general modified control goal***

The control should force the plant to behave so that the following tracking algorithm holds [instead of (32.12)]:

$$\mathbf{T}\left(\mathbf{t}^m, \boldsymbol{\epsilon}_\gamma^{*k}\left(\mathbf{t}^{(k+1)\gamma}\right), \int_{\mathbf{t}_0^\gamma}^{\mathbf{t}^\gamma} \ominus_\gamma^*\left(\mathbf{t}^\gamma\right) d\mathbf{t}^\gamma\right) = \mathbf{0}_m, \ \forall \mathbf{t}^{(k+1)\gamma} \in \mathfrak{T}_{0F}^{(k+1)\gamma}. \tag{32.52}$$

Property 473 ***The properties of the tracking algorithm*** *(32.13)*

The algorithm (32.13) expressed in terms of $\boldsymbol{\epsilon}_\gamma^*\left(\mathbf{t}^\gamma\right)$ and its derivative as in (32.52) possesses, in addition to Properties 455 through 457, also Property 459 due to (32.30).

32.4 Tracking algorithm and initial conditions

The following theorem enables us to establish the unified tracking control for both cases $\boldsymbol{\varepsilon}^{*(k)} = \mathbf{0}_{(k+1)\gamma}$ and $\boldsymbol{\varepsilon}^{*(k)} \neq \mathbf{0}_{(k+1)\gamma}$.

Theorem 474 ***The main theorem on the tracking algorithm and initial conditions***

Let (32.23) through (32.43) be valid. In order for the tracking algorithm determined by (32.21), (32.22) in terms of the real error vector $\boldsymbol{\varepsilon}_\gamma^*$, its derivatives and integral, to hold, it is necessary and sufficient that the tracking algorithm $\mathbf{T}(.)$ (32.52) holds.

Proof. Notice that the tracking algorithms (32.12), (32.21), and (32.52) possess Properties 455 through 457. The tracking algorithm (32.52) possesses also Property 459 as explained in Property 473.

Let (32.38) through (32.43) be valid.

Necessity. Let (32.21), (32.22) hold. We separate the case $\mathbf{T}\left(\mathbf{t}^m, \boldsymbol{\varepsilon}_\gamma^{*k}, \mathbf{0}_\gamma\right) = \mathbf{0}_m$ from the case $\mathbf{T}\left(\mathbf{t}^m, \boldsymbol{\varepsilon}_\gamma^{*k}, \mathbf{0}_\gamma\right) \neq \mathbf{0}_m$.

If $\boldsymbol{\varepsilon}_\gamma^{*k}$ is such that $\mathbf{T}\left(\mathbf{t}^m, \boldsymbol{\varepsilon}_\gamma^{*k}, \mathbf{0}_\gamma\right) = \mathbf{0}_m$, then $\boldsymbol{\epsilon}^k(\mathbf{t}^{(k+1)\gamma}) = \boldsymbol{\varepsilon}_\gamma^{*k}(\mathbf{t}^{(k+1)\gamma})$ for every $\mathbf{t}^{(k+1)\gamma} \in \mathfrak{T}_0^{(k+1)\gamma}$ due to (32.38) so that

$$\mathbf{T}\left(\mathbf{t}^m, \boldsymbol{\varepsilon}_\gamma^{*k}(\mathbf{t}_R^{(k+1)\gamma}), \int_{\mathbf{t}_0^\gamma}^{\mathbf{t}^\gamma} E_\gamma^*\left(\mathbf{t}^\gamma\right) d\mathbf{t}^\gamma\right)$$
$$= \mathbf{T}\left(\mathbf{t}^m, \boldsymbol{\epsilon}^{*k}(\mathbf{t}_R^{(k+1)\gamma}), \int_{\mathbf{t}_0^\gamma}^{\mathbf{t}^\gamma} \ominus_\gamma^*\left(\mathbf{t}^\gamma\right) d\mathbf{t}^\gamma\right) = \mathbf{0}_m, \ \forall \mathbf{t}_R^{(k+1)\gamma} \in \mathfrak{T}_{0F}^{(k+1)\gamma}.$$

This proves necessity of (32.52) if $\mathbf{T}\left(\mathbf{t}^m, \boldsymbol{\varepsilon}_\gamma^{*k}, \mathbf{0}_\gamma\right) = \mathbf{0}_m$.

If $\boldsymbol{\varepsilon}_\gamma^{*k}$ violates (32.17), i.e., $\mathbf{T}\left(\mathbf{t}^m, \boldsymbol{\varepsilon}_\gamma^{*k}, \mathbf{0}_\gamma\right) \neq \mathbf{0}_m$ holds, then the conditions (32.22) and (32.39) imply $\boldsymbol{\epsilon}^k(\mathbf{t}^{(k+1)\gamma}) = \mathbf{0}_{(k+1)\gamma}$, $\forall \mathbf{t}^{(k+1)\gamma} \in \mathfrak{T}_{0F}^{(k+1)\gamma}$. This proves that $\boldsymbol{\epsilon}^k(\mathbf{t}^{(k+1)\gamma})$ satisfies (32.52), i.e., that (32.52) holds.

Sufficiency. Let (32.52) be valid.

If $\boldsymbol{\varepsilon}_{\gamma 0}^{*k}$ is such that $\mathbf{T}\left(\mathbf{t}_0^m, \boldsymbol{\varepsilon}_{\gamma 0}^{*k}, \mathbf{0}_\gamma\right) = \mathbf{0}_m$, then $\boldsymbol{\epsilon}^k(\mathbf{t}^{(k+1)\gamma}) = \boldsymbol{\varepsilon}_\gamma^{*k}(\mathbf{t}^{(k+1)\gamma})$ for every $\mathbf{t}^{(k+1)\gamma} \in \mathfrak{T}_{0F}^{(k+1)\gamma}$ due to (32.39) so that, in view of (32.52),

$$\mathbf{T}\left(\mathbf{t}^m, \boldsymbol{\epsilon}^k(\mathbf{t}^{(k+1)\gamma}), \int_{\mathbf{t}_0^\gamma}^{\mathbf{t}^\gamma} \ominus_\gamma^*\left(\mathbf{t}^\gamma\right) d\mathbf{t}^\gamma\right)$$
$$= \mathbf{T}\left(\mathbf{t}^m, \boldsymbol{\varepsilon}_\gamma^{*k}(\mathbf{t}^{(k+1)\gamma}), \int_{\mathbf{t}_0^\gamma}^{\mathbf{t}^\gamma} E_\gamma^*\left(\mathbf{t}^\gamma\right) d\mathbf{t}^\gamma\right) = \mathbf{0}_m,$$
$$\forall \mathbf{t}^{(k+1)\gamma} \in \mathfrak{T}_{0F}^{(k+1)\gamma}. \tag{32.53}$$

This proves validity of (32.21) for the case $\mathbf{T}\left(\mathbf{t}^m, \boldsymbol{\varepsilon}_\gamma^{*k}, \mathbf{0}_\gamma\right) = \mathbf{0}_m$.

If $\boldsymbol{\varepsilon}_{\gamma 0}^{*k}$ such that $\mathbf{T}\left(\mathbf{t}_0^m, \boldsymbol{\varepsilon}_{\gamma 0}^{*k}, \mathbf{0}_\gamma\right) \neq \mathbf{0}_m$ holds, then (32.52), (32.41), and (32.16) imply

$$\boldsymbol{\epsilon}_\gamma^{*k}(\mathbf{t}^{(k+1)\gamma}; \boldsymbol{\epsilon}_{\gamma 0}^{*k}) = \mathbf{0}_{(k+1)\gamma}, \ \forall \mathbf{t}^{(k+1)\gamma} \in \mathfrak{T}_{0F}^{(k+1)\gamma};$$

hence, $\boldsymbol{\varepsilon}_\gamma^{*k}(\mathbf{t}^{(k+1)\gamma}; \boldsymbol{\varepsilon}_\gamma^{*k}) + \mathbf{f}_\gamma^k(\mathbf{t}^{(k+1)\gamma}; \mathbf{f}_{\gamma 0}^k) = \mathbf{0}_{(k+1)\gamma}$ for every $\mathbf{t}^{(k+1)\gamma} \in \mathfrak{T}_{0F}^{(k+1)\gamma}$ due to (32.39). This proves (32.22) and, together with (32.37), proves $\boldsymbol{\varepsilon}_\gamma^{*k}(\mathbf{t}^{(k+1)\gamma}) = \mathbf{0}_{(k+1)\gamma}$ for every $\mathbf{t}^{(k+1)\gamma} \in \mathfrak{T}_{RF}^{(k+1)\gamma}$. Therefore, (32.21) also holds ∎

We present several characteristic simple forms of the tracking algorithm $\mathbf{T}(.)$ in the output space. They satisfy (32.14) through (32.17); i.e., they obey Properties 455-459. They also satisfy Solution 465 and Solution 469.

Chapter 33

Tracking algorithms: Output space

33.1 Matrix notation meaning

The examples to follow show *time*-invariant tracking operators $\mathbf{T}(.)$. This permits us to accept the zero moment for the initial moment: $t_0 = 0$. This simplifies the formulae.

The definition of the integer k is in (1.8). The diagonal elements of the constant diagonal $\gamma \times \gamma$ matrix $K_{0\gamma}$ are positive real numbers:

$$K_{0\gamma} = \text{diag}\{k_{01}\ k_{02}\ \cdots k_{0\gamma}\} \in \mathfrak{R}_+^{\gamma\times\gamma},\ \ k_{0i} \in \mathfrak{R}^+,\ \ \forall i = 1, 2,\ \cdots, \gamma,\ K_{0N} = K_0.$$

The $\gamma \times \gamma$ diagonal matrix T is the independent *time* matrix the elements of which are the same: *time* t:

$$T_\gamma = \text{diag}\{t\ t\ \cdots\ t\} \in \mathfrak{T}^{\gamma\times\gamma},\ \ T_N = T,\ \ \mathfrak{T}^{\gamma\times\gamma} = \{T_\gamma : T_\gamma \geq O_\gamma\},$$

$$T_{1\gamma} = \text{diag}\{t_1\ t_2\ ...t_\gamma\},\ \ t_i > 0,\ \forall i = 1, 2, ..., \gamma,\ \ T_{1N} = T_1.$$

The matrix exponential function $e^{-K_{0\gamma}T_{1\gamma}^{-1}T_\gamma}$ is a $\gamma \times \gamma$ diagonal exponential matrix function defined by:

$$e^{-K_{0\gamma}T_{1\gamma}^{-1}T_\gamma} = \exp\left[-K_{0\gamma}T_{1\gamma}^{-1}T_\gamma\right] =$$
$$= \text{diag}\left\{e^{-k_{01}t_1^{-1}t}\ e^{-k_{02}t_2^{-1}t}\ \ ...\ e^{-k_{0N}t_\gamma^{-1}t}\right\} \in \mathfrak{R}^{\gamma\times\gamma}.$$

33.2 Examples of tracking algorithms

Example 475 ***The first-order linear elementwise exponential tracking algorithm***

In this case $\mathbf{t}_R^\gamma = \mathbf{t}_F^\gamma = \infty\mathbf{1}_\gamma$; *i.e.,* $\mathfrak{T}_{RF}^\gamma = \mathfrak{T}_{F\infty}^\gamma = \{\infty\mathbf{1}_\gamma\}$. *The following tracking algorithm*

$$\mathbf{T}\left(\boldsymbol{\epsilon}_\gamma^{*\,1}\right) = \mathbf{T}\left(\boldsymbol{\epsilon}_\gamma^*, \boldsymbol{\epsilon}_\gamma^{*\,(1)}\right) = T_{1\gamma}\boldsymbol{\epsilon}_\gamma^{*\,(1)}(\mathbf{t}^\gamma) + K_{0\gamma}\boldsymbol{\epsilon}_\gamma^*(\mathbf{t}^\gamma) = \mathbf{0}_\gamma, \forall \mathbf{t}^\gamma \in \mathfrak{T}_{0F}^\gamma,$$
$$T_{1\gamma} = diag\{t_1\ t_2\ \ \cdots\ t_\gamma\},\ t_i > 0, \forall i = 1, 2, ..., \gamma \leq N, \tag{33.1}$$

determines the global exponential tracking of $\mathbf{Y}_d\left(\mathbf{t}^N\right)$ *on* $\mathfrak{T}_{0F}^N$ *or the perfect tracking of*

$\mathbf{Y}_R(\mathbf{t}^N)$ *on* $\mathfrak{T}_{0F}^N$*, which is illustrated by its solution*

$$\mathbf{T}(\boldsymbol{\varepsilon}_{\gamma 0}^{*1}) = \mathbf{0}_\gamma \Longrightarrow$$
$$\boldsymbol{\epsilon}_\gamma^*(\mathbf{t}^\gamma;\boldsymbol{\epsilon}_{\gamma 0}^*) = \boldsymbol{\varepsilon}_\gamma^*(\mathbf{t}^\gamma;\boldsymbol{\varepsilon}_{\gamma 0}^*) = \epsilon_\gamma^{*-K_0 T_1^{-1} T}\boldsymbol{\varepsilon}_{\gamma 0}^*, \forall(\mathbf{t}^\gamma,\boldsymbol{\varepsilon}_{\gamma 0}^*) \in \mathfrak{T}_{0F}^\gamma \times \mathfrak{R}^\gamma,$$
$$\mathbf{T}(\boldsymbol{\varepsilon}_{\gamma 0}^{*1}) \neq \mathbf{0}_\gamma \Longrightarrow \boldsymbol{\epsilon}_\gamma^*(\mathbf{t}^\gamma;\boldsymbol{\epsilon}_{\gamma 0}^*) = \boldsymbol{\varepsilon}_\gamma^*(\mathbf{t}^\gamma;\boldsymbol{\varepsilon}_{\gamma 0}) = \mathbf{0}_\gamma,\ \forall \mathbf{t}^\gamma \in \mathfrak{T}_{RF}^\gamma.$$

The reachability time is infinite iff $\mathbf{T}(\boldsymbol{\varepsilon}_{\gamma 0}^{*1}) = \mathbf{0}_\gamma$*. The convergence to the zero error vector is then elementwise and exponentially asymptotic. Such tracking of* $\mathbf{Y}_d(\mathbf{t}^N)$ *is also stable. However, if* $\mathbf{T}(\boldsymbol{\varepsilon}_{\gamma 0}^{*1}) \neq \mathbf{0}_\gamma$ *then the tracking of* $\mathbf{Y}_d(\mathbf{t}^N)$ *is perfect on* $\mathfrak{T}_{RF}^N$*.*

The tracking operator $\mathbf{T}(\mathbf{t}^\gamma,\boldsymbol{\varepsilon}_\gamma^{*1})$ *(33.1) possesses Properties 455-461 and* $\mathbf{T}(.) \in \mathfrak{C}^-$*. The tracking is exponentially stable in the whole if* $\mathbf{T}(\boldsymbol{\varepsilon}_{\gamma 0}^{*1}) = \mathbf{0}_\gamma$*.*

We can apply Comment 468 by choosing $T_{1\gamma} \in \mathfrak{R}^{+\gamma\times\gamma}$ *and/or* $K_{0\gamma} \in \mathfrak{R}^{+\gamma\times\gamma}$ *to obey*

$$T_1(\boldsymbol{\varepsilon}_{\gamma 0}^{*1})\boldsymbol{\varepsilon}_{\gamma 0}^{*(1)} + K_0(\boldsymbol{\varepsilon}_{\gamma 0}^{*1})\boldsymbol{\varepsilon}_{\gamma 0}^* = \mathbf{0}_\gamma \ if,\ and\ only\ if,\ E_{\gamma 0}^* E_{\gamma 0}^{*(1)} < 0,$$

e.g.,

$$T_{1\gamma}(\boldsymbol{\varepsilon}_{\gamma 0}^{*1}) = -K_{0\gamma}E_{\gamma 0}^*\left(E_{\gamma 0}^{*(1)}\right)^{-1} \in \mathfrak{R}^{+^{\gamma\times\gamma}},$$

or

$$K_{0\gamma}(\boldsymbol{\varepsilon}_{\gamma 0}^{*1}) = -T_{1\gamma}\left(E_{\gamma 0}^*\right)^{-1}E_{\gamma 0}^{*(1)} \in \mathfrak{R}^{+^{\gamma\times\gamma}}.$$

Such a choice of $T_1 \in \mathfrak{R}^+$ *and/or* $K_0 \in \mathfrak{R}^+$ *assures that the tracking algorithm* $T(.)$ *obeys (32.17) for the given* $\boldsymbol{\varepsilon}_{\gamma 0}^{*1}$*.*

Let

$$E_\gamma^{(\eta)} = [E_{\gamma 0}\quad E_{\gamma 1}\ \ldots\ E_{\gamma\eta}] \in \mathfrak{R}^{\gamma\times(\eta+1)\gamma},\ \eta \leq \alpha - 1,$$

be given, or to be determined, constant extended (block) matrix, the entries of which are constant submatrices $E_{\gamma k} \in \mathfrak{R}^{\gamma\times\gamma}$, $k = 0, 1, .., \eta \leq \alpha - 1$.

Example 476 ***The higher-order linear elementwise exponential tracking algorithm***

In this case the tracking should be a kind of asymptotic tracking so that the vector reachability time $\mathbf{t}_R^\gamma$ *is infinite,* $\mathbf{t}_R^\gamma = \infty\mathbf{1}_\gamma$*. This implies* $\mathfrak{T}_{R\infty} = \{\infty\mathbf{1}_\gamma\}$*. We define the higher-order linear elementwise exponential tracking algorithm by*

$$\mathbf{T}\left(\mathbf{t}^\gamma,\boldsymbol{\epsilon}_\gamma^*,\boldsymbol{\epsilon}_\gamma^{*(1)},..,\boldsymbol{\epsilon}_\gamma^{*(\eta)},\int_{\mathbf{t}_0^\gamma=\mathbf{0}_\gamma}^{\mathbf{t}^\gamma}\ominus_\gamma^*(\mathbf{t}^\gamma)\,d\mathbf{t}^\gamma\right) = \sum_{k=0}^{k=\eta\leq\alpha-1}E_{\gamma k}\boldsymbol{\epsilon}_\gamma^{*(k)}(\mathbf{t}^\gamma)$$
$$= E_\gamma^{(\eta)}\boldsymbol{\epsilon}_\gamma^{*\,\eta}(\mathbf{t}^{(\eta+1)\gamma}) = \mathbf{0}_\gamma,\ \forall\mathbf{t}^{(\eta+1)\gamma} \in \mathfrak{T}_{0F}^{(\eta+1)\gamma}, \tag{33.2}$$

with the matrices $E_{\gamma k}$ *such that the real parts of the roots of the characteristic polynomial* $f(s)$ *of* $\mathbf{T}(s)$*,*

$$f(s) = det\left(\sum_{k=0}^{k=\eta\leq\alpha-1}E_{\gamma k}s^k\right),$$

are negative.

The linear differential equation (33.2) is with the constant matrix coefficients. It has the unique solution for every initial condition $\boldsymbol{\epsilon}_\gamma^{*\,\eta-1}(\mathbf{t}_0^{\eta\gamma}) \in \mathfrak{R}^{\eta\gamma}$*.*

The reachability time is infinite. The error vector $\boldsymbol{\epsilon}_\gamma^{*\,\eta-1}(\mathbf{t}^{\eta\gamma}) = \boldsymbol{\varepsilon}_\gamma^{*\eta-1}(\mathbf{t}^{\eta\gamma})$ *converges to the zero error vector elementwise, exponentially and asymptotically iff* $\mathbf{T}(\boldsymbol{\varepsilon}_{\gamma 0}^{*1}) = \mathbf{0}_\gamma$*. Such tracking of* $\mathbf{Y}_d(\mathbf{t}^N)$ *is stable. However, iff* $\mathbf{T}(\boldsymbol{\varepsilon}_{\gamma 0}^{*1}) \neq \mathbf{0}_\gamma$ *then the tracking of* $\mathbf{Y}_R(\mathbf{t}^N)$ *is perfect on* $\mathfrak{T}_{RF}^N$*, but not of that* $\mathbf{Y}_d(\mathbf{t}^N)$*.*

The tracking operator (33.2) possesses Properties 455-461 and $\mathbf{T}(.) \in \mathfrak{C}^-$*. The tracking is exponentially stable in the whole if* $\mathbf{T}(\boldsymbol{\varepsilon}_{\gamma 0}^{*1}) = \mathbf{0}_\gamma$ *in the output space.*

Example 477 ***The sharp elementwise stable tracking with the finite vector reachability time*** $\mathbf{t}_R^\gamma$

We accept the finite vector reachability time t_R^γ, $t_R^\gamma = [t_{R1}\ t_{R2}\ ..\ t_{R\gamma}]^T$ *, so that*

$$\mathfrak{T}_{0R}^\gamma = [\mathbf{0}_\gamma, \mathbf{t}_R^\gamma] \subset \mathfrak{T}_{0F}^\gamma,\ \ \mathfrak{T}_{RF}^\gamma = [\mathbf{t}_R^\gamma,\ \mathbf{t}_F^\gamma],\ \ \mathbf{t}_F^\gamma < \infty \mathbf{1}_\gamma. \tag{33.3}$$

The algorithm for the elementwise stable tracking with the finite vector reachability time t_R^γ,

$$\mathbf{T}\left(\boldsymbol{\varepsilon}_{\gamma 0}^{*1}\right) = \mathbf{0}_\gamma \Longrightarrow \mathbf{t}_R^\gamma = T_1 K_0^{-1}\left|\boldsymbol{\varepsilon}_{\gamma 0}^*\right|,$$

$$\left|\boldsymbol{\varepsilon}_{\gamma 0}^*\right| = \left[|\varepsilon_{10}|\quad |\varepsilon_{20}|\ \ldots |\varepsilon_{2,\gamma-1,0}|\quad |\varepsilon_{\gamma 0}^*|\right]^T \in \mathfrak{R}^\gamma,$$

reads

$$\mathbf{T}\left(\boldsymbol{\epsilon}_\gamma^*, \boldsymbol{\epsilon}_\gamma^{*(1)}\right) = T_{1\gamma}\boldsymbol{\epsilon}_\gamma^{*(1)}(\mathbf{t}^\gamma) + K_{0\gamma} sign\boldsymbol{\epsilon}_\gamma^*(\mathbf{t}_0^\gamma) = \mathbf{0}_\gamma,\ \forall \mathbf{t}^\gamma \in \mathfrak{T}_{0F}^\gamma\ , \tag{33.4}$$

where

$$sign\epsilon_k = \left\{\begin{array}{ll} |\epsilon_k|^{-1}\epsilon_k, & \epsilon_k \neq 0, \\ 0, & \epsilon_k = 0 \end{array}\right\},\ sign\boldsymbol{\epsilon}_\gamma^* = \left[sign\epsilon_1\quad sign\epsilon_2\ \cdots\ sign\epsilon_\gamma^*\right]^T.$$

Let

$$S\left(\boldsymbol{\varepsilon}_{\gamma 0}^*\right) = diag\left\{sign\varepsilon_{10}\quad sign\varepsilon_{20}\ \cdots\ sign\varepsilon_{\gamma 0}^*\right\},$$

The solution $\boldsymbol{\epsilon}_\gamma^*(\mathbf{t}^\gamma;\boldsymbol{\epsilon}_{\gamma 0}^*)$,

$$\begin{aligned} \mathbf{T}\left(\boldsymbol{\varepsilon}_{\gamma 0}^{*1}\right) = \mathbf{0}_\gamma &\Longrightarrow \boldsymbol{\epsilon}_\gamma^*(\mathbf{t}^\gamma; \boldsymbol{\epsilon}_{\gamma 0}^*) = \boldsymbol{\varepsilon}_\gamma^*(\mathbf{t}^\gamma; \boldsymbol{\varepsilon}_{\gamma 0}^*) \\ &= \boldsymbol{\varepsilon}_{\gamma 0}^* - T_{1\gamma}^{-1} K_{0\gamma} S\left(\boldsymbol{\varepsilon}_{\gamma 0}^*\right)\mathbf{t}^\gamma,\ \forall \mathbf{t}^\gamma \in \mathfrak{T}_{0R}^\gamma, \\ \boldsymbol{\epsilon}_\gamma^*(\mathbf{t}^\gamma; \boldsymbol{\epsilon}_{\gamma 0}^*) &= \boldsymbol{\varepsilon}_\gamma^*(\mathbf{t}^\gamma; \boldsymbol{\varepsilon}_{\gamma 0}^*) = \mathbf{0}_\gamma,\ \forall \mathbf{t}^\gamma \in \mathfrak{T}_{RF}^\gamma, \\ \mathbf{T}\left(\boldsymbol{\varepsilon}_{\gamma 0}^{*1}\right) \neq \mathbf{0}_\gamma &\Longrightarrow \boldsymbol{\epsilon}_\gamma^*(\mathbf{t}^\gamma; \boldsymbol{\epsilon}_{\gamma 0}^*) = \mathbf{0}_\gamma,\ \forall \mathbf{t}^\gamma \in \mathfrak{T}_{RF}^\gamma, \end{aligned}$$

to (33.4) determines the output error behavior $\boldsymbol{\varepsilon}_\gamma^*(\mathbf{t}^\gamma; \boldsymbol{\varepsilon}_{\gamma 0}^*)$ *that approaches sharply the zero error vector* $\boldsymbol{\varepsilon}_\gamma^* = \mathbf{0}_\gamma$ *in the linear form (along a straight line) with the nonzero constant velocity* $T_{1\gamma}^{-1}K_{0\gamma}S\left(\boldsymbol{\varepsilon}_{\gamma 0}^*\right)\mathbf{1}_\gamma$ *iff* $\mathbf{T}\left(\boldsymbol{\varepsilon}_{\gamma 0}^{*1}\right) = \mathbf{0}_\gamma$. *Otherwise,* $\mathbf{T}\left(\boldsymbol{\varepsilon}_{\gamma 0}^{*1}\right) \neq \mathbf{0}_\gamma$ *and the tracking of* $\mathbf{Y}_d\left(\mathbf{t}^N\right)$ *is perfect on* $\mathfrak{T}_{RF}^N$. *The tracking of* $\mathbf{Y}_d\left(\mathbf{t}^N\right)$ *becomes perfect at the finite vector reachability time* $\mathbf{t}_R^\gamma$ *and rests perfect on* $\mathfrak{T}_{RF}^N$.

Iff $\gamma = N$ *and* $\mathbf{T}\left(\boldsymbol{\varepsilon}_0^1\right) = \mathbf{0}_N$ *then the convergence to the zero error vector; i.e., the convergence of* $\mathbf{Y}\left(\mathbf{t}^N\right)$ *to* $\mathbf{Y}_d\left(\mathbf{t}^N\right)$ *is elementwise, strictly monotonous, continuous, and*

$$\begin{gathered} \mathbf{T}\left(\boldsymbol{\varepsilon}_0^1\right) = \mathbf{0}_N \Longrightarrow \left|\boldsymbol{\varepsilon}(t; \boldsymbol{\varepsilon}_0^1)\right| \leq |\boldsymbol{\varepsilon}_0|\,,\ \forall t \in \mathfrak{T}_{0F} \Longrightarrow \\ \left\|\boldsymbol{\varepsilon}(t; \boldsymbol{\varepsilon}_0^1)\right\| \leq \|\boldsymbol{\varepsilon}_0\|\,,\ \forall t \in \mathfrak{T}_{0F} \Longrightarrow \\ \forall \varepsilon \in \mathfrak{R}^+,\ \exists \delta \in \mathfrak{R}^+,\ \delta = \delta\left(\varepsilon\right) = \varepsilon \Longrightarrow \left\|\boldsymbol{\varepsilon}_0^1\right\| < \delta \Longrightarrow \left\|\boldsymbol{\varepsilon}(t; \boldsymbol{\varepsilon}_0^1)\right\| \leq \varepsilon,\ \forall t \in \mathfrak{T}_{0F}. \end{gathered}$$

The tracking of $\mathbf{Y}_d\left(\mathbf{t}^N\right)$ *is stable on* $\mathfrak{T}_{0F}^N$ *if* $\mathbf{T}\left(\boldsymbol{\varepsilon}_{\gamma 0}^{*1}\right) = \mathbf{0}_\gamma$ *and* $\gamma = N$.

Iff $\mathbf{T}\left(\boldsymbol{\varepsilon}_{\gamma 0}^{*1}\right) = \mathbf{0}_\gamma$ *then the bigger* $K_{0\gamma}$, *the smaller* t_R^γ *is for the fixed* $T_{1\gamma}$ *and* $\boldsymbol{\varepsilon}_{\gamma 0}^{*1}$, *and vice versa. The smaller* $T_{1\gamma}$, *the smaller* t_R^γ *for the fixed* $K_{0\gamma}$ *and* $\boldsymbol{\varepsilon}_{\gamma 0}^{*1}$, *and vice versa. The bigger* $\left|\boldsymbol{\varepsilon}_{\gamma 0}^{*1}\right|$, *the bigger* t_R^γ *is for the fixed* $T_{1\gamma}$ *and* $K_{0\gamma}$, *and vice versa. These relationships hold elementwise.*

The tracking operator (33.4) possesses Properties 455-461 and $\mathbf{T}(.) \in \mathfrak{C}^-$. *The tracking is stable in the whole with FVRT* t_R^γ *if* $\mathbf{T}\left(\boldsymbol{\varepsilon}_{\gamma 0}^{*1}\right) = \mathbf{0}_\gamma$.

In the case $E_{\gamma 0}^* E_{\gamma 0}^{*(1)} < O_\gamma$ *we can adjust (Comment 468) the matrix parameters* $T_{1\gamma}$ *and* $K_{0\gamma}$ *of the tracking algorithm:*

$$T_{1\gamma}E_{\gamma 0}^{*(1)}(0) + K_{0\gamma}S\left(E_{\gamma 0}^{*(1)}\right) = O_\gamma,\ if,\ and\ only\ if,\ E_{\gamma 0}^* E_{\gamma 0}^{*(1)} < O_\gamma,$$

where the matrix E_γ^* *is defined by (32.9), i.e.,*

$$E_\gamma^* = diag\{\varepsilon_1 \;\; \varepsilon_2 \;\; \cdots \;\; \varepsilon_\gamma^*\}, \quad S\left(E_{\gamma 0}^*\right) = S\left(\varepsilon_{\gamma 0}^*\right).$$

Let

$$\left|E_\gamma^*\right|^{1/2} = \mathrm{diag}\left\{\left|\varepsilon_1(t)\right|^{1/2} \;\; \left|\varepsilon_2(t)\right|^{1/2} \;\; \cdots \;\; \left|\varepsilon_\gamma^*(t)\right|^{1/2}\right\}, \tag{33.5}$$

$$\left|\ominus_\gamma^*(\mathbf{t}^\gamma)\right|^{1/2} = \mathrm{diag}\left\{\left|\epsilon_1(t)\right|^{1/2} \;\; \left|\epsilon_2(t)\right|^{1/2} \;\; \cdots \;\; \left|\epsilon_\gamma^*(t)\right|^{1/2}\right\}. \tag{33.6}$$

Example 478 ***The first-power smooth elementwise stable tracking with the finite vector reachability time*** $\mathbf{t}_R^\gamma$

If the control acting on the plant ensures

$$\mathbf{T}\left(\boldsymbol{\epsilon}_\gamma^*, \boldsymbol{\epsilon}_\gamma^{*(1)}\right) = T_{1\gamma}\boldsymbol{\epsilon}_\gamma^{*(1)}(\mathbf{t}^\gamma) + 2K_{0\gamma}\left|\ominus_\gamma^*(\mathbf{t}^\gamma)\right|^{1/2} sign\boldsymbol{\epsilon}_{\gamma 0}^* = \mathbf{0}_\gamma, \; \forall \mathbf{t}^\gamma \in \mathfrak{T}_{0F}^\gamma, \tag{33.7}$$

then the plant exhibits the elementwise stable tracking with the finite vector reachability time t_R^γ*,*

$$\mathbf{T}\left(\boldsymbol{\varepsilon}_{\gamma 0}^{*1}\right) = \mathbf{0}_\gamma \Longrightarrow \mathbf{t}_R^\gamma = T_{1\gamma}K_{0\gamma}^{-1}\left|E_{\gamma 0}^*\right|^{1/2}\mathbf{1}_\gamma = T_{1\gamma}K_{0\gamma}^{-1}\left|\boldsymbol{\varepsilon}_{\gamma 0}^*\right|^{1/2},$$
$$\left|\boldsymbol{\varepsilon}_{\gamma 0}^*\right|^{1/2} = \left[\left|\varepsilon_1(0)\right|^{1/2} \;\; \left|\varepsilon_2(0)\right|^{1/2} \;\; \cdots \;\; \left|\varepsilon_\gamma^*(0)\right|^{1/2}\right]^T,$$

which is determined by the output error behavior,

$$\mathbf{T}\left(\boldsymbol{\varepsilon}_{\gamma 0}^{*1}\right) = \mathbf{0}_\gamma \Longrightarrow \boldsymbol{\epsilon}_\gamma^*(\mathbf{t}^\gamma; \boldsymbol{\epsilon}_{\gamma 0}^*)$$
$$= \boldsymbol{\varepsilon}_\gamma^*(\mathbf{t}^\gamma; \boldsymbol{\varepsilon}_{\gamma 0}^*)\left\{\begin{array}{l} = \left\{\begin{array}{c} \left[\left|E_{\gamma 0}^{*(1)}\right|^{1/2} - T_{1\gamma}^{-1}K_{0\gamma}T_\gamma\right]^2 sign\boldsymbol{\varepsilon}_{\gamma 0}^*, \\ \forall \mathbf{t}^\gamma \in \mathfrak{T}_{0R}^\gamma, \end{array}\right\} \\ = \mathbf{0}_\gamma, \qquad \forall \mathbf{t}^\gamma \in \mathfrak{T}_{RF}^\gamma, \end{array}\right\}$$
$$\mathbf{T}\left(\boldsymbol{\varepsilon}_{\gamma 0}^{*1}\right) \neq \mathbf{0}_\gamma \Longrightarrow \boldsymbol{\epsilon}_\gamma^*(\mathbf{t}^\gamma; \boldsymbol{\epsilon}_{\gamma 0}^*) = \boldsymbol{\varepsilon}_\gamma^*(\mathbf{t}^\gamma; \boldsymbol{\varepsilon}_{\gamma 0}^*) = \mathbf{0}_\gamma, \; \forall \mathbf{t}^\gamma \in \mathfrak{T}_{RF}^\gamma. \tag{33.8}$$

This implies

$$sign\boldsymbol{\epsilon}_\gamma^*(\mathbf{t}^\gamma) = sign\boldsymbol{\epsilon}_{\gamma 0}^*, \; \forall \mathbf{t}^\gamma \in In\mathfrak{T}_{0R}^\gamma.$$

The tracking operator (33.7) possesses Properties 455-461 and $\mathbf{T}(.) \in \mathfrak{C}^-$*. The tracking is stable in the whole with FVRT* t_R^γ *if* $\mathbf{T}\left(\boldsymbol{\varepsilon}_{\gamma 0}^{*1}\right) = \mathbf{0}_\gamma$*.*

If $\gamma = N$ *then the output error vector* $\boldsymbol{\varepsilon}(\mathbf{t}^N)$ *approaches smoothly elementwise the zero output vector in the finite vector reachability time* t_R^N *iff* $\mathbf{T}\left(\boldsymbol{\varepsilon}_0^1\right) = \mathbf{0}_N$*. The convergence is strictly monotonous and continuous. It is also without any oscillation, overshoot, or undershoot iff* $\mathbf{T}\left(\boldsymbol{\varepsilon}_0^1\right) = \mathbf{0}_N$*. Then,* $\boldsymbol{\varepsilon}(\mathbf{t}^N) = \mathbf{0}_N, \; \forall \mathbf{t}^N \in \mathfrak{T}_{RF}^N$*, but also if* $\mathbf{T}\left(\boldsymbol{\varepsilon}_0^1\right) \neq \mathbf{0}_N$ *then* $\boldsymbol{\varepsilon}(\mathbf{t}^\gamma; \boldsymbol{\varepsilon}_0) = \mathbf{0}_N, \; \forall \mathbf{t}^\gamma \in \mathfrak{T}_{RF}^\gamma$*. The tracking of* $\mathbf{Y}_d\left(\mathbf{t}^N\right)$ *is stable on* $\mathfrak{T}_{0F}^N$*.*

Iff $\mathbf{T}\left(\boldsymbol{\varepsilon}_{\gamma 0}^{*1}\right) = \mathbf{0}_\gamma$ *then the solution (33.8) obeys the following:*

$$\mathbf{T}\left(\boldsymbol{\varepsilon}_{\gamma 0}^{*1}\right) = \mathbf{0}_\gamma \Longrightarrow \left\|\boldsymbol{\varepsilon}_\gamma^*(\mathbf{t}^\gamma; \boldsymbol{\varepsilon}_{\gamma 0}^*)\right\| \leq \left\|\boldsymbol{\varepsilon}_{\gamma 0}^*\right\|, \; \forall \mathbf{t}^\gamma \in \mathfrak{T}_{0F}^\gamma \Longrightarrow \delta(\varepsilon) = \varepsilon$$
$$\Longrightarrow \forall \varepsilon \in \mathfrak{R}^+, \; \exists \delta \in \mathfrak{R}^+, \; \delta = \delta(\varepsilon) = \varepsilon \Longrightarrow$$
$$\left\|\boldsymbol{\varepsilon}_{\gamma 0}^*\right\| < \delta \Longrightarrow \left\|\boldsymbol{\varepsilon}_\gamma^*(\mathbf{t}^\gamma; \boldsymbol{\varepsilon}_{\gamma 0}^*)\right\| \leq \varepsilon, \; \forall \mathbf{t}^\gamma \in \mathfrak{T}_{0F}^\gamma.$$

Therefore, such tracking of $\mathbf{Y}_d\left(\mathbf{t}^N\right)$ *is stable on* $\mathfrak{T}_{0F}^N$*. Then the bigger* $K_{0\gamma}$*, the smaller* t_R^γ *is for the fixed* $T_{1\gamma}$ *and* $\boldsymbol{\varepsilon}_{\gamma 0}^{*1}$*, and vice versa. The smaller* $T_{1\gamma}$*, the smaller* t_R^γ *is for the fixed* $K_{0\gamma}$ *and* $\boldsymbol{\varepsilon}_{\gamma 0}^{*1}$*, and vice versa. The bigger* $\left|\boldsymbol{\varepsilon}_{\gamma 0}^{*1}\right|$*, the bigger* t_R^γ *is for the fixed* $T_{1\gamma}$ *and* $K_{0\gamma}$*, and vice versa. These claims are in the elementwise sense iff* $\gamma = N$*. We can adjust (Comment 468) the matrix parameters* $T_{1\gamma}$ *and* $K_{0\gamma}$ *of the tracking algorithm so that elementwise*

$$T_{1\gamma}E_\gamma^{*(1)}(0) + 2K_{0\gamma}\left|E_\gamma^*(0)\right|^{1/2} S\left(E_{\gamma 0}^*\right) = O_\gamma, \; if, \text{ and only } if, \; E_{\gamma 0}^*E_{\gamma 0}^{*(1)} < O_\gamma.$$

Example 479 ***The higher-power smooth elementwise stable tracking with the finite vector reachability time*** t_R^γ

Let the tracking algorithm be

$$\mathbf{T}\left(\epsilon_\gamma^{*1}\right) = T_{1\gamma}\epsilon_\gamma^{*(1)}(\mathbf{t}^\gamma) + K_{0\gamma}\left|\ominus_\gamma^*(\mathbf{t}^\gamma)\right|^{I-K_\gamma^{-1}} sign\epsilon_{\gamma 0}^* = \mathbf{0}_\gamma,\ \forall \mathbf{t}^\gamma \in \mathfrak{T}_{0F}^\gamma,$$
$$K_\gamma = diag\{k_1\ k_2\ ...k_\gamma\},\ k_i \in \{2,\ 3,\ ...\},\ \forall i = 1,2,\ \cdots,\gamma,$$
$$\left|\ominus_\gamma^*(\mathbf{t}^\gamma)\right|^{I-K_\gamma^{-1}} = diag\left\{\left|\epsilon_1(t)\right|^{1-k_1^{-1}}\ \left|\epsilon_2(t)\right|^{1-k_2^{-1}}\ \cdots\ \left|\epsilon_\gamma^*(t)\right|^{1-k_\gamma^{-1}}\right\}. \tag{33.9}$$

Iff $\mathbf{T}\left(\varepsilon_{\gamma 0}^{*1}\right) = \mathbf{0}_\gamma$ *then (33.10) determines the following solution* $\epsilon_\gamma^*(\mathbf{t}^\gamma;\epsilon_{\gamma 0}^*)$ *to* $T\left(\mathbf{t}^\gamma, \epsilon_\gamma^*, \epsilon_\gamma^{*(1)}\right) = \mathbf{0}_\gamma,\ \forall \mathbf{t}^\gamma \in \mathfrak{T}_{0F}^\gamma$:

$$\mathbf{T}\left(\varepsilon_{\gamma 0}^{*1}\right) = \mathbf{0}_\gamma \Longrightarrow \epsilon_\gamma^*(\mathbf{t}^\gamma;\epsilon_{\gamma 0}^*) = \varepsilon_\gamma^*(\mathbf{t}^\gamma;\varepsilon_{\gamma 0}^*)$$
$$= \frac{1}{2}S\left(\varepsilon_{\gamma 0}^*\right)\left\{\begin{array}{c}\left\{\begin{array}{c} I_\gamma + \\ +S\left[\left|\varepsilon_{\gamma 0}^*\right|^{K^{-1}} - T_{1\gamma}^{-1}K_{0\gamma}\mathbf{t}^\gamma\right]\end{array}\right\} \\ \bullet\left[\left|\varepsilon_{\gamma 0}^*\right|^{K_\gamma^{-1}} - T_{1\gamma}^{-1}K_\gamma^{-1}K_{0\gamma}\mathbf{t}^\gamma\right]^K\end{array}\right\}$$
$$= \left\{\begin{array}{ll}\left[\left|E_{\gamma 0}^{*(1)}\right|^{K^{-1}} - T_{1\gamma}^{-1}K_\gamma^{-1}K_{0\gamma}T_\gamma\right]^K sign\varepsilon_{\gamma 0}^*, & \forall \mathbf{t}^\gamma \in \mathfrak{T}_{0R}^\gamma, \\ \mathbf{0}_\gamma, & \forall \mathbf{t}^\gamma \in \mathfrak{T}_{RF}^\gamma\end{array}\right\} \Longrightarrow$$
$$\mathbf{T}\left(\varepsilon_{\gamma 0}^{*1}\right) = \mathbf{0}_\gamma \Longrightarrow \mathbf{t}_R^\gamma = T_{1\gamma}K_\gamma K_{0\gamma}^{-1}\left|\varepsilon_{\gamma 0}^*\right|^{K^{-1}}, \tag{33.10}$$

but iff $\mathbf{T}\left(\varepsilon_{\gamma 0}^{*1}\right) \neq \mathbf{0}_\gamma$, *then the solution* $\epsilon_\gamma^*(\mathbf{t}^\gamma;\epsilon_{\gamma 0}^*)$ *to* $\mathbf{T}\left(\epsilon_\gamma^*, \epsilon_\gamma^{*(1)}\right) = \mathbf{0}_\gamma,\ \forall \mathbf{t}^\gamma \in \mathfrak{T}_{RF}^\gamma$, *satisfies the following:*

$$\mathbf{T}\left(\varepsilon_{\gamma 0}^{*1}\right) \neq \mathbf{0}_\gamma \Longrightarrow \epsilon_\gamma^*(\mathbf{t}^\gamma;\epsilon_{\gamma 0}^*) = \varepsilon_\gamma^*(\mathbf{t}^\gamma;\varepsilon_{\gamma 0}^*) = \mathbf{0}_\gamma,\ \forall \mathbf{t}^\gamma \in \mathfrak{T}_{RF}^\gamma.$$

Notice that

$$\left[\left|\varepsilon_{\gamma 0}^*\right|^{K_\gamma^{-1}} - T_{1\gamma}^{-1}K_{0\gamma}\mathbf{t}^\gamma\right]^K == \left[\begin{array}{c}\left[\left|\varepsilon_{10}\right|^{k_1^{-1}} - t_1^{-1}k_{01}t\right]^{k_1} \\ \left[\left|\varepsilon_{20}\right|^{k_2^{-1}} - t_2^{-1}k_{02}t\right]^{k_2} \\ \cdots\cdots\cdots \\ \left[\left|\varepsilon_{\gamma 0}^*\right|^{k_\gamma^{-1}} - t_\gamma^{-1}k_{0\gamma}t\right]^{k_\gamma}\end{array}\right] \in \mathfrak{R}^\gamma. \tag{33.11}$$

The tracking operator (33.9) possesses Properties 455-461 and $\mathbf{T}(.) \in \mathfrak{C}^-$. *The tracking is stable in the whole with FVRT* t_R^γ *if* $\mathbf{T}\left(\varepsilon_{\gamma 0}^{*1}\right) = \mathbf{0}_\gamma$.

The equations (33.10) express the elementwise nonlinear convergence to the zero error vector iff $\mathbf{T}\left(\varepsilon_{\gamma 0}^{*1}\right) = \mathbf{0}_\gamma$. *Then the convergence is strictly monotonous and continuous, without any oscillation, overshoot, or undershoot. The errors enter the zero values smoothly. Therefore, such tracking of* $\mathbf{Y}_d\left(\mathbf{t}^N\right)$ *is stable on* $\mathfrak{T}_{0F}^N$ *due to (33.10) and (33.11):*

$$\mathbf{T}\left(\varepsilon_{\gamma 0}^{*1}\right) = \mathbf{0}_\gamma \Longrightarrow \left\|\varepsilon_\gamma^*(\mathbf{t}^\gamma;\varepsilon_{\gamma 0}^*)\right\| \leq \left\|\varepsilon_{\gamma 0}^*\right\|,\ \forall \mathbf{t}^\gamma \in \mathfrak{T}_{0F}^\gamma \Longrightarrow \delta(\varepsilon) = \varepsilon$$
$$\Longrightarrow \forall \varepsilon \in \mathfrak{R}^+,\ \exists \delta \in \mathfrak{R}^+,\ \delta = \delta(\varepsilon) = \varepsilon \Longrightarrow$$
$$\left\|\varepsilon_{\gamma 0}^*\right\| < \delta \Longrightarrow \left\|\varepsilon_\gamma^*(\mathbf{t}^\gamma;\varepsilon_{\gamma 0}^*)\right\| \leq \varepsilon,\ \forall \mathbf{t}^\gamma \in \mathfrak{T}_{0F}^\gamma.$$

Then the bigger $K_{0\gamma}$, *the smaller* t_R^γ *is for the fixed* $T_{1\gamma}$ *and* $\varepsilon_{\gamma 0}^*$, *and vice versa. The smaller* $T_{1\gamma}$, *the smaller* t_R^γ *is for the fixed* $K_{0\gamma}$ *and* $\varepsilon_{\gamma 0}^*$, *and vice versa. The bigger* $\left|\varepsilon_{\gamma 0}^*\right|$, *the bigger* t_R^γ *is for the fixed* $T_{1\gamma}$ *and* $K_{0\gamma}$, *and vice versa.*

Iff $\mathbf{T}\left(\boldsymbol{\varepsilon}_{\gamma 0}^{*1}\right) \neq \mathbf{0}_{\gamma}$ *then the tracking of* $\mathbf{Y}_{d\gamma}\left(\mathbf{t}^{\gamma}\right)$ *is perfect only on* $\mathfrak{T}_{RF}^{\gamma}$.

If we wish to adjust (Comment 468) the matrix parameters $T_{1\gamma}$ *and* $K_{0\gamma}$ *of the tracking algorithm, then they should satisfy*

$$T_{1\gamma}E_{\gamma 0}^{*(1)} + K_{0\gamma}\left|E_{\gamma 0}\right|^{I-K^{-1}} S\left(E_{\gamma 0}^{*}\right) = O_{\gamma}, \text{ if, and only if, } E_{\gamma 0}^{*}E_{\gamma 0}^{*(1)} < 0.$$

Let

$$\sigma\left(\varepsilon_i^{(k)}, \varepsilon_i^{(k+1)}\right) = \left\{\begin{array}{c} -1, \left\{\begin{array}{c} \text{if } \varepsilon_i^{(k)} < 0, \forall \varepsilon_i^{(k+1)} \in \mathfrak{R}; \\ \text{or if } \varepsilon_i^{(k)} = 0 \text{ and } \varepsilon_i^{(k+1)} < 0, \end{array}\right\} \\ 0, \text{ iff } \varepsilon_i^{(k)} = 0 \text{ and } \varepsilon_i^{(k+1)} = 0, \\ 1, \left\{\begin{array}{c} \text{if } \varepsilon_i^{(k)} > 0, \forall \varepsilon_i^{(k+1)} \in \mathfrak{R}; \\ \text{or if } \varepsilon_i^{(k)} = 0 \text{ and } \varepsilon_i^{(k+1)} > 0 \end{array}\right\}, \\ \forall i = 1, 2, \cdots, \gamma - 1, \quad i = \gamma \Longrightarrow \varepsilon_{i0} = \varepsilon_{\gamma 0}^{*} \end{array}\right\}, \tag{33.12}$$

$$\Sigma\left(\boldsymbol{\varepsilon}_{\gamma}^{*(k)}, \boldsymbol{\varepsilon}_{\gamma}^{*(k+1)}\right) = diag\left\{\sigma\left(\varepsilon_1^{(k)}, \varepsilon_1^{(k+1)}\right) \sigma\left(\varepsilon_2^{(k)}, \varepsilon_2^{(k+1)}\right) .\sigma\left(\varepsilon_{\gamma}^{*(k)}, \varepsilon_{\gamma}^{*(k+1)}\right)\right\},$$
$$\forall k = 1, 2, ..., \tag{33.13}$$

and

$$\Sigma\left(\boldsymbol{\varepsilon}_{\gamma}^{*1}\right) = diag\left\{\sigma\left(\varepsilon_1, \varepsilon_1^{(1)}\right) \sigma\left(\varepsilon_2, \varepsilon_2^{(1)}\right) \dots \sigma\left(\varepsilon_{\gamma}^{*}, \varepsilon_{\gamma}^{*(1)}\right)\right\}, \tag{33.14}$$

The equations (33.12) - (33.14) enable us to determine Dini derivative $D^{+}\left|\boldsymbol{\varepsilon}_{\gamma}^{*}(t)\right|$ *of* $\left|\boldsymbol{\varepsilon}_{\gamma}^{*}(t)\right|$ *(for details see Appendix B) :*

$$D^{+}\left|\boldsymbol{\varepsilon}_{\gamma}^{*}(t)\right| = \Sigma\left(\boldsymbol{\varepsilon}_{\gamma}^{*}, \boldsymbol{\varepsilon}_{\gamma}^{*(1)}\right) \boldsymbol{\varepsilon}_{\gamma}^{*(1)}(t). \tag{33.15}$$

Example 480 ***Sharp absolute error vector tracking elementwise and stable with the finite vector reachability time*** t_R^{γ}

The solution to the following tracking algorithm [in which we use (33.12), (33.13), and (33.15)]:

$$\mathbf{T}\left(\boldsymbol{\epsilon}_{\gamma}^{*}, \boldsymbol{\epsilon}_{\gamma}^{*(1)}\right) = T_{1\gamma}D^{+}\left|\boldsymbol{\epsilon}_{\gamma}^{*}\right| + K_{0\gamma} sign\left|\boldsymbol{\epsilon}_{\gamma 0}^{*}\right|$$
$$= T_{1\gamma}\Sigma\left(\boldsymbol{\epsilon}_{\gamma}^{*1}\right)\boldsymbol{\epsilon}_{\gamma}^{*(1)} + K_{0\gamma} sign\left|\boldsymbol{\epsilon}_{\gamma 0}^{*}\right| = \mathbf{0}_{\gamma}, \ \forall \mathbf{t}^{\gamma} \in \mathfrak{T}_{0F}^{\gamma}, \tag{33.16}$$

implies for $\mathbf{T}\left(\boldsymbol{\varepsilon}_{\gamma 0}^{*1}\right) = \mathbf{0}_{\gamma}$:

$$\mathbf{T}\left(\boldsymbol{\varepsilon}_{\gamma 0}^{*1}\right) = \mathbf{0}_{\gamma} \Longrightarrow \boldsymbol{\epsilon}_{\gamma}^{*}(\mathbf{t}^{\gamma}; \boldsymbol{\epsilon}_{\gamma 0}^{*}) = \boldsymbol{\varepsilon}_{\gamma}^{*}(\mathbf{t}^{\gamma}; \boldsymbol{\varepsilon}_{\gamma 0}^{*})$$
$$= \left\{\begin{array}{ll} \left[\left|\boldsymbol{\varepsilon}_{\gamma 0}^{*}\right| - T_{1\gamma}^{-1}K_{0\gamma}T_{\gamma}\right] sign\left|\boldsymbol{\varepsilon}_{\gamma 0}^{*}\right|, & \forall \mathbf{t}^{\gamma} \in \mathfrak{T}_{0R}^{\gamma}, \\ \mathbf{0}_{\gamma}, & \forall \mathbf{t}^{\gamma} \in \mathfrak{T}_{RF}^{\gamma}. \end{array}\right\},$$
$$\mathbf{t}_R^{\gamma} = T_{1\gamma}K_{0\gamma}^{-1}\left|\boldsymbol{\varepsilon}_{\gamma 0}^{*}\right|, \tag{33.17}$$

but for $\mathbf{T}\left(\boldsymbol{\varepsilon}_{\gamma 0}^{*1}\right) \neq \mathbf{0}_{\gamma}$ *it implies:*

$$\mathbf{T}\left(\boldsymbol{\varepsilon}_{\gamma 0}^{*1}\right) \neq \mathbf{0}_{\gamma} \Longrightarrow \boldsymbol{\epsilon}_{\gamma}^{*}(\mathbf{t}^{\gamma}; \boldsymbol{\epsilon}_{\gamma 0}^{*}) = \boldsymbol{\varepsilon}_{\gamma}^{*}(\mathbf{t}^{\gamma}; \boldsymbol{\varepsilon}_{\gamma 0}^{*}) = \mathbf{0}_{\gamma}, \ \forall \mathbf{t}^{\gamma} \in \mathfrak{T}_{RF}^{\gamma}.$$

The tracking operator (33.16) possesses Properties 455-461 and $\mathbf{T}(.) \in \mathfrak{C}^{-}$. *The tracking is stable in the whole with FVRT* t_R^{γ} *if* $\mathbf{T}\left(\boldsymbol{\varepsilon}_{\gamma 0}^{*1}\right) = \mathbf{0}_{\gamma}$.

Equations (33.17) permit

$$\mathbf{T}\left(\varepsilon_{\gamma 0}^{*1}\right)=\mathbf{0}_{\gamma} \Longrightarrow \left|\varepsilon_{\gamma}^{*}(\mathbf{t}^{\gamma};\varepsilon_{\gamma 0}^{*})\right| \leq \left|\varepsilon_{\gamma 0}^{*}\right|,\ \forall \mathbf{t}^{\gamma} \in \mathfrak{T}_{0F}^{\gamma} \Longrightarrow$$
$$\left\|\varepsilon_{\gamma}^{*}(\mathbf{t}^{\gamma};\varepsilon_{\gamma 0}^{*})\right\| \leq \left\|\varepsilon_{\gamma 0}^{*}\right\|,\ \forall \mathbf{t}^{\gamma} \in \mathfrak{T}_{0F}^{\gamma} \Longrightarrow$$
$$\forall \varepsilon \in \mathfrak{R}^{+},\ \exists \delta \in \mathfrak{R}^{+},\ \delta=\delta(\varepsilon)=\varepsilon \Longrightarrow$$
$$\left\|\varepsilon_{\gamma 0}^{*}\right\| < \delta \Longrightarrow \left\|\varepsilon_{\gamma}^{*}(\mathbf{t}^{\gamma};\varepsilon_{\gamma 0}^{*})\right\| \leq \varepsilon,\ \forall \mathbf{t}^{\gamma} \in \mathfrak{T}_{0F}^{\gamma}.$$

The tracking is elementwise stable with the finite vector reachability time $t_{R}^{\gamma}=T_{1\gamma}K_{0\gamma}^{-1}\left|\varepsilon_{\gamma 0}^{*}\right|$. *It is strictly monotonous and continuous without oscillation, overshoot, and undershoot.*

In order to follow Comment 468 we adjust the matrix parameters $T_{1\gamma}$ *and* $K_{0\gamma}$ *of the tracking algorithm to satisfy*

$$T_{1\gamma}\Sigma\left(\varepsilon_{\gamma 0}^{*1},\varepsilon_{\gamma 0}^{*(1)}\right)E_{\gamma 0}^{*(1)}+K_{0\gamma}S\left(E_{\gamma 0}^{*(1)}\right)=O_{\gamma},\ if,\ and\ only\ if,\ E_{\gamma 0}^{*(1)}E_{\gamma 0}^{*(1)}<0.$$

Example 481 ***The exponential absolute error vector tracking elementwise and stable with the finite vector reachability time*** t_{R}^{γ}

The tracking algorithm is in terms of the elementwise absolute value of the error vector,

$$\forall \mathbf{t}^{\gamma} \in \mathfrak{T}_{0F}^{\gamma} \Longrightarrow \mathbf{T}\left(\epsilon_{\gamma}^{*},\epsilon_{\gamma}^{*(1)}\right)=T_{1\gamma}D^{+}\left|\epsilon_{\gamma}^{*}\right|+K_{\gamma}\left(\left|\epsilon_{\gamma}^{*}\right|+K_{0\gamma}sign\left|\epsilon_{\gamma 0}^{*}\right|\right)=$$
$$=T_{1\gamma}\Sigma\left(\epsilon_{\gamma}^{*},\epsilon_{\gamma}^{*(1)}\right)\epsilon_{\gamma}^{*(1)}+K_{\gamma}\left(\left|\epsilon_{\gamma}^{*}\right|+K_{0\gamma}sign\left|\epsilon_{\gamma 0}^{*}\right|\right)=\mathbf{0}_{\gamma}. \tag{33.18}$$

The solution of the differential equation (33.18) written in the matrix diagonal form

$$D^{+}\left[\left|\ominus_{\gamma}^{*}\right|+K_{0\gamma}S\left(\left|\epsilon_{\gamma 0}^{*}\right|\right)\right]=-T_{1\gamma}^{-1}K_{\gamma}\left[\left|\ominus_{\gamma}^{*}\right|+K_{0\gamma}S\left(\left|\epsilon_{\gamma 0}^{*}\right|\right)\right] \Longrightarrow$$
$$\left[\left|\ominus_{\gamma}^{*}\right|+K_{0\gamma}S\left(\left|\epsilon_{\gamma 0}^{*}\right|\right)\right]^{-1}D^{+}\left[\left|\ominus_{\gamma}^{*}\right|+K_{0\gamma}S\left(\left|\epsilon_{\gamma 0}^{*}\right|\right)\right]=-T_{1\gamma}^{-1}K_{\gamma} \Longrightarrow$$
$$D^{+}\left\{ln\left[\left|\ominus_{\gamma}^{*}\right|+K_{0\gamma}S\left(\left|\epsilon_{\gamma 0}^{*}\right|\right)\right]\right\}=-T_{1\gamma}^{-1}K_{\gamma}, \tag{33.19}$$

reads in the matrix form

$$ln\left\{\left[\left|\ominus_{\gamma}^{*}\right|+K_{0\gamma}S\left(\left|\epsilon_{\gamma 0}^{*}\right|\right)\right]\left[\left|\ominus_{\gamma 0}^{*}\right|+K_{0\gamma}S\left(\left|\epsilon_{\gamma 0}^{*}\right|\right)\right]^{-1}\right\}=-T_{1\gamma}^{-1}K_{\gamma}T_{\gamma},$$
$$\forall T_{\gamma} \in \mathfrak{T}_{0}^{\gamma \times \gamma}.$$

The final form of the solution is

$$\ominus_{\gamma}^{*}(t;\epsilon_{\gamma 0}^{*})$$
$$=\left\{\begin{array}{c}\left\{\begin{array}{c}e^{*-T_{1\gamma}^{-1}K_{\gamma}T_{\gamma}}\left(\left|\ominus_{\gamma 0}^{*}\right|+K_{0\gamma}\right)S\left(\left|\epsilon_{\gamma 0}^{*}\right|\right)-K_{0\gamma}S\left(\left|\epsilon_{\gamma 0}^{*}\right|\right), \\ \forall T_{\gamma} \in \left[O_{\gamma},T_{\gamma R}\right], \\ O_{\gamma}, \qquad \forall T_{\gamma} \in \left[O_{\gamma},T_{\gamma R}\right],\end{array}\right\} \\ where\ 0\infty=0,\end{array}\right\},$$

$$T_{R\gamma}=\left\{\begin{array}{ll}K_{\gamma}^{-1}T_{1\gamma}ln\left\{K_{0\gamma}^{-1}\left[\left|\Psi_{\gamma 0}^{*}\right|+K_{0\gamma}\right]\right\}, & \epsilon_{\gamma 0}^{*}\neq\mathbf{0}_{\gamma}, \\ O_{\gamma}, & \epsilon_{\gamma 0}^{*}=\mathbf{0}_{\gamma}\end{array}\right\}, \tag{33.20}$$

$$\mathbf{t}_{R}^{\gamma}=\left[t_{R1}\ t_{R2}\ \cdots\ t_{R\gamma}\right]^{T} \Longleftrightarrow T_{\gamma R}=diag\left\{t_{R1}\ t_{R2}\ \cdots\ t_{R\gamma}\right\}. \tag{33.21}$$

If $\mathbf{T}\left(\varepsilon_{\gamma 0}^{*1}\right)=\mathbf{0}_{\gamma}$ *then we can set the solution (33.20) in the equivalent vector form*

$$\epsilon_{\gamma}^{*}(\mathbf{t}^{\gamma};\epsilon_{\gamma 0}^{*})=\left\{\begin{array}{c}\left[e^{-T_{1\gamma}^{-1}K_{\gamma}T_{\gamma}}\left(\left|\epsilon_{\gamma 0}^{*}\right|+K_{0\gamma}\right)-K_{0\gamma}\right]sign\left(\left|\epsilon_{\gamma 0}^{*}\right|\right), \\ \forall \mathbf{t}^{\gamma} \in \left[\mathbf{0}_{\gamma},\mathbf{t}_{R}^{\gamma}\right],\ i.e.,\ \forall T_{\gamma} \in \left[O_{\gamma},T_{\gamma R}\right], \\ \mathbf{0}_{\gamma},\ \forall \mathbf{t}^{\gamma} \in \left[\mathbf{t}_{R}^{\gamma},\ \mathbf{t}_{F}^{\gamma}\right],\ i.e.,\ \forall T_{\gamma} \in \left[T_{\gamma R},T_{\gamma F}\right].\end{array}\right\}. \tag{33.22}$$

The tracking operator (36.16) possesses Properties 455-461 and $\mathbf{T}(.) \in \mathfrak{C}^-$. *The tracking is stable in the whole with FVRT* t_R^γ *if* $\mathbf{T}\left(\boldsymbol{\varepsilon}_{\gamma 0}^{*1}\right) = \mathbf{0}_\gamma$.

Iff $\mathbf{T}\left(\boldsymbol{\varepsilon}_{\gamma 0}^{*1}\right) = \mathbf{0}_\gamma$ *then the solution is continuous and monotonous without oscillation, overshoot, and undershoot, and obeys*

$$\mathbf{T}\left(\boldsymbol{\varepsilon}_{\gamma 0}^{*1}\right) = \mathbf{0}_\gamma \Longrightarrow \boldsymbol{\epsilon}_\gamma^*(\mathbf{t}^\gamma; \boldsymbol{\epsilon}_{\gamma 0}^*) = \boldsymbol{\varepsilon}_\gamma^*(\mathbf{t}^\gamma; \boldsymbol{\varepsilon}_{\gamma 0}^*) \Longrightarrow$$
$$\left|\boldsymbol{\varepsilon}_\gamma^*(\mathbf{t}^\gamma; \boldsymbol{\varepsilon}_{\gamma 0}^*)\right| \leq \left|\boldsymbol{\varepsilon}_{\gamma 0}^*\right|, \ \forall \mathbf{t}^\gamma \in \mathfrak{T}_{0F}^\gamma \Longrightarrow$$
$$\forall \varepsilon \in \mathfrak{R}^+, \ \exists \delta \in \mathfrak{R}^+, \ \delta = \delta(\varepsilon) = \varepsilon \Longrightarrow$$
$$\left\|\boldsymbol{\varepsilon}_{\gamma 0}^*\right\| < \delta \Longrightarrow \left\|\boldsymbol{\varepsilon}_\gamma^*(\mathbf{t}^\gamma; \boldsymbol{\varepsilon}_{\gamma 0}^*)\right\| \leq \varepsilon, \ \forall \mathbf{t}^\gamma \in \mathfrak{T}_{0F}^\gamma,$$

The tracking of $\mathbf{Y}_d\left(\mathbf{t}^N\right)$ *is stable on* $\mathfrak{T}_{0F}^N$. *It converges with the exponential rate to the zero error vector and reaches (elementwise iff* $\gamma = N$*) the origin in finite vector reachability time* t_R^γ *(33.20), (33.21).*

Iff $\mathbf{T}\left(\boldsymbol{\varepsilon}_{\gamma 0}^{*1}\right) \neq \mathbf{0}_\gamma$ *then the solution reads*

$$\mathbf{T}\left(\boldsymbol{\varepsilon}_{\gamma 0}^{*1}\right) \neq \mathbf{0}_\gamma \Longrightarrow$$
$$\boldsymbol{\epsilon}_\gamma^*(\mathbf{t}^\gamma; \boldsymbol{\epsilon}_{\gamma 0}^*) = \boldsymbol{\varepsilon}_\gamma^*(\mathbf{t}^\gamma; \boldsymbol{\varepsilon}_{\gamma 0}^*) = \mathbf{0}_\gamma, \quad \forall \mathbf{t}^\gamma \in [\mathbf{t}_R^\gamma, \ \mathbf{t}_F^\gamma].$$

Comment 468 leads to the following adjustment of the matrix parameters $T_{1\gamma}$ *and* $K_{0\gamma}$ *of the tracking algorithm:*

$$T_{1\gamma}\Sigma\left(\boldsymbol{\varepsilon}_{\gamma 0}^*, \boldsymbol{\varepsilon}_{\gamma 0}^{*(1)}\right) E_{\gamma 0}^{*(1)} + K_\gamma\left[\left|E_{\gamma 0}^{*(1)}\right| + K_0 S\left(\left|E_{\gamma 0}^{*(1)}\right|\right)\right] = O_\gamma,$$
$$if, \text{ and only } if, \ E_{\gamma 0}^* E_{\gamma 0}^{*(1)} < 0.$$

Note 482 *We can effectively use the tracking algorithms proposed in the preceding examples, Example 475 through Example 481, also when we synthesize tracking control by applying various design methods; e.g., Lyapunov-like, adaptive control, sliding mode, and natural tracking control method.*

The system behavior results from the actions of control, of disturbances, and of initial conditions. The control cannot influence disturbances and initial conditions, but it can take into account their consequences that are the plant state error and the output error.

Chapter 34

NTC synthesis: Output space

34.1 General *NTC* theorem: Output space

Property 455 through Property 459 constitute the basis for the following general theorem.

Theorem 483 ***General NTC synthesis in the output space***
Let (32.23) through (32.43) be valid.

In order for the natural trackable plant (3.12), with Properties 57, 58, 78, controlled by the natural tracking control $\mathbf{U}$ *to exhibit tracking on* $\mathfrak{T}_{0F}^{(k+1)\gamma} \times \mathfrak{D}^j \times \mathfrak{Y}_d^k$ *determined by the tracking algorithm* $\mathbf{T}(.)$*, (32.21), (32.22), it is necessary and sufficient that both i) and ii) hold:*

i) the rank γ *of the control matrix* $W(t)$ *is greater than zero and not greater than* $min(\gamma, r)$*, i.e., (8.10) is true, and for the elementwise trackability* $m = \gamma = N \leq r$ *in* $\mathbf{T}(.)$ *defined by (32.21).*

ii) The control vector function $\mathbf{U}(.)$ *satisfies one of the conditions 1) to 4):1) The control vector function* $\mathbf{U}(.)$ *obeys (34.1) for* $\gamma = rankW(t) = r < N$, $\forall t \in \mathfrak{T}_0$, *and for the chosen matrix function* $M_{r,m}(.) : \mathfrak{T}_0 \longrightarrow \mathfrak{R}^{r\times m}$, *where* m *is not greater than* r, $m \leq r$, *rank* $M_{r,m}(t) \equiv m$, *and* $M_{r,r}(t) \equiv I_r$:

$$\mathbf{U}_r(t) = \mathbf{w}_r^I \left\{ \begin{array}{c} \mathbf{w}_r\left[\mathbf{U}_r(t^-)\right] \\ + M_{r,m}(t)\,\mathbf{T}\left[\mathbf{t}^m, \boldsymbol{\epsilon}_r^{*k}\left(\mathbf{t}^{(k+1)r}\right), \int_{t_0^\gamma}^{\mathbf{t}^\gamma} \ominus_r^*\left(\mathbf{t}^r\right) d\mathbf{t}^r\right] \end{array} \right\}, \\ \forall \mathbf{t}^{(k+1)r} \in \mathfrak{T}_0^{(k+1)r}, \tag{34.1}$$

The domain $\mathcal{D}_{T\mathcal{B}l}\left(t_0; \mathfrak{D}^j; \mathbf{U}; \mathfrak{Y}_d^k\right)$ *of tracking on* $\mathfrak{T}_0 \times \mathfrak{D}^j \times \mathfrak{Y}_d^k$ *equals the perfect trackability domain* $\mathcal{D}_{PT\mathcal{B}l}\left(t_0; \mathfrak{D}^j; \mathbf{U}; \mathfrak{Y}_d\right)$ *on* $\mathfrak{T}_0 \times \mathfrak{D}^j \times \mathfrak{Y}_d^k$ *determined by (8.16).*

2) The control vector function $\mathbf{U}(.)$ *obeys (34.2) for* $\gamma = rankW(t) = N < r$, $\forall t \in \mathfrak{T}_0$, *and for the chosen matrix function* $M_{N,m}(.) : \mathfrak{T}_0 \longrightarrow \mathfrak{R}^{N\times m}$, *where rank* $M_{N,m}(t) \equiv m$, $\forall t \in \mathfrak{T}_0$, m *is not greater than* N, $m \leq N$, *and* $m = N$ *if the perfect trackability is also elementwise:*

$$\mathbf{U}_N(t) = \mathbf{w}_N^I \left\{ \begin{array}{c} \Gamma(t)\,\mathbf{w}_N\left[\mathbf{U}_N(t^-)\right] \\ + W^T(t)\,M_{N,m}(t)\,\mathbf{T}\left[\mathbf{t}^m, \boldsymbol{\epsilon}^k\left(\mathbf{t}^{(k+1)N}\right), \int_{t_0^\gamma}^{\mathbf{t}^\gamma} \ominus\left(\mathbf{t}^N\right) d\mathbf{t}^N\right] \end{array} \right\} \\ \forall \mathbf{t}^{(k+1)N} \in \mathfrak{T}_0^{(k+1)N}, \\ \Gamma(t) = W^T(t)\left[W(t)\,W^T(t)\right]^{-1} W(t) \in \mathfrak{R}^{r\times r},\ \forall t \in \mathfrak{T}_0, \tag{34.2}$$

The domain $\mathcal{D}_{T\mathcal{B}l}\left(t_0; \mathfrak{D}^j; \mathbf{U}; \mathfrak{Y}_d^k\right)$ *of tracking on* $\mathfrak{T}_0 \times \mathfrak{D}^j \times \mathfrak{Y}_d^k$ *equals the perfect trackability domain* $\mathcal{D}_{PT\mathcal{B}l}\left(t_0; \mathfrak{D}^j; \mathbf{U}; \mathfrak{Y}_d\right)$ *on* $\mathfrak{T}_0 \times \mathfrak{D}^j \times \mathfrak{Y}_d^k$ *determined by (8.17).*

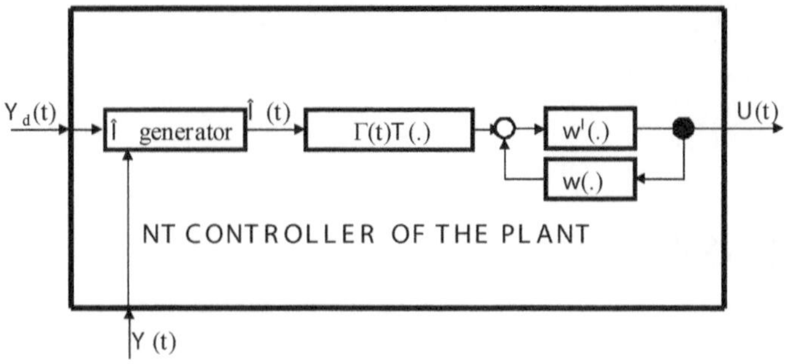

Figure 34.1: The full block diagram of the natural tracking (NT) controller of the plant (3.12) for $\gamma = N$.

3) The control vector function $\mathbf{U}(.)$ *obeys (34.3) if* γ =*rank*$W(t) = r = N$, $\forall t \in \mathfrak{T}_0$*:*

$$\mathbf{U}\left(\mathbf{t}^N\right) = \mathbf{w}^I \left\langle \begin{array}{c} \mathbf{w}\left[\mathbf{U}\left(\mathbf{t}^{N-}\right)\right] \\ + W\left(\mathbf{t}^N\right)^{-1}\mathbf{T}\left[\mathbf{t}^N, \epsilon^k\left(\mathbf{t}^{(k+1)N}\right), \int_{t_0^\gamma}^{\mathbf{t}^\gamma} \ominus\left(\mathbf{t}^N\right) d\mathbf{t}^N\right] \end{array} \right\rangle, \quad \forall \mathbf{t}^N \in \mathfrak{T}_0^N, \tag{34.3}$$

The domain $\mathcal{D}_{Tbl}\left(t_0; \mathfrak{D}^j; \mathbf{U}; \mathfrak{Y}_d^k\right)$ *of tracking on* $\mathfrak{T}_0 \times \mathfrak{D}^j \times \mathfrak{Y}_d^k$ *equals the perfect trackability domain* $\mathcal{D}_{PTbl}\left(t_0; \mathfrak{D}^j; \mathbf{U}; \mathfrak{Y}_d\right)$ *on* $\mathfrak{T}_0 \times \mathfrak{D}^j \times \mathfrak{Y}_d^k$ *determined by (8.18).*

4) If the plant also possesses Property 64 then the control vector function $\mathbf{U}(.)$ *obeys (34.4)*

$$\mathbf{U}_\gamma(t) = \mathbf{w}_\gamma^I \left\{ \mathbf{w}_\gamma\left[\mathbf{U}_\gamma(t^-)\right] + \mathbf{T}\left[\mathbf{t}^\gamma, \epsilon_\gamma^{*k}\left(\mathbf{t}^{(k+1)\gamma}\right), \int_{t_0^\gamma}^{\mathbf{t}^\gamma} \Psi_\gamma^*\left(\mathbf{t}^\gamma\right) d\mathbf{t}^\gamma\right]\right\}, \quad \forall t \in \mathfrak{T}_0. \tag{34.4}$$

The domain $\mathcal{D}_{Tbl}\left(t_0; \mathfrak{D}^j; \mathbf{U}; \mathfrak{Y}_d^k\right)$ *of tracking on* $\mathfrak{T}_0 \times \mathfrak{D}^j \times \mathfrak{Y}_d^k$ *equals the perfect trackability domain* $\mathcal{D}_{PTbl}\left(t_0; \mathfrak{D}^j; \mathbf{U}; \mathfrak{Y}_d\right)$ *on* $\mathfrak{T}_0 \times \mathfrak{D}^j \times \mathfrak{Y}_d^k$ *determined by (8.19).*

Proof. When we replace $\mathbf{v}(t, \sigma, \varepsilon)$ by $\mathbf{T}\left[\mathbf{t}^m, \epsilon_r^{*k}\left(\mathbf{t}^{(k+1)r}\right), \int_{t_0^\gamma}^{\mathbf{t}^\gamma} \ominus_\gamma^*\left(\mathbf{t}^r\right) d\mathbf{t}^r\right]$ then this theorem becomes Theorem 190 ■

Figure 34.1 shows the full block diagram of the *natural tracking* (NT) *controller* of the plant (3.12).

Comment 484 *The implementation of the NTC, (34.1)-(34.4), does not need any information about the state of the plant, about the plant mathematical model, or about the disturbance vector value. This expresses the superiority of the NTC over other tracking control approaches.*

34.2 NTC synthesis: Output space

In what follows $m = \gamma$ and $t_0 = 0$.

Condition 485 *The plant (3.12) possesses Properties 57, 58 and 78.*

All following theorems result directly from Theorem 483.

Theorem 486 ***NTC for the first-order linear elementwise exponential tracking, Example 475***

Let (32.23) through (32.43) be valid. Let the reachability time $\tau_R = \tau_F = \infty$*; i.e.,* $\mathfrak{T}_{R\infty} = \mathfrak{T}_{RF} = \{\infty\}$.

For the plant (3.12) obeying Condition (485) to be controlled by the natural tracking control $\mathbf{U}$ *in order to exhibit tracking on* $\mathfrak{T}_{0F}^{(k+1)\gamma} \times \mathfrak{D}^j \times \mathfrak{Y}_d^k$ *determined by the following tracking algorithm*

$$\mathbf{T}\left(\varepsilon_\gamma^*, \varepsilon_\gamma^{*(1)}\right) = T_{1\gamma}\varepsilon_\gamma^{*(1)}(t) + K_{0\gamma}\varepsilon_\gamma^*(t) = \mathbf{0}_\gamma,\ \forall t \in \mathfrak{T}_{0F},\ if\ \mathbf{T}\left(0, \varepsilon_{\gamma 0}^{*1}\right) = \mathbf{0}_\gamma,$$
$$\varepsilon_\gamma^{*1}\left(t; t_0; \varepsilon_{\gamma 0}^{*1}\right) = -\mathbf{f}_\gamma^1(t),\ \forall t \in \mathfrak{T}_{0F},\ if\ \mathbf{T}\left(0, \varepsilon_{\gamma 0}^{*1}\right) \neq \mathbf{0}_\gamma,$$
$$T_{1\gamma} = diag\{\tau_1\ \tau_2\ \cdots\ \tau_\gamma\},\ T_{1N} = T_1,\ K_{0\gamma} = diag\{k_{01}\ k_{02}\ \cdots\ k_{0\gamma}\},\quad K_{0N} = K_0,$$
$$\tau_i > 0\ \ and\ k_{0i} > 0,\ \forall i = 1, 2, \cdots, \gamma, \tag{34.5}$$

it is necessary and sufficient that both i) and ii) hold:

i) The rank γ *of the control matrix* $W(t)$ *is greater than zero and not greater than* $min(\gamma, r)$*; i.e., (8.10) is true, and for the elementwise trackability* $m = \gamma = N$ *in* $\mathbf{T}(.)$ *defined by (34.5),*

ii) The control vector function $\mathbf{U}(.)$ *satisfies one of the conditions 1) to 4):*

1) The control vector function $\mathbf{U}(.)$ *obeys (34.6) for* $m = \gamma = rankW(t) = r < N$, $\forall t \in \mathfrak{T}_0$,

$$\mathbf{U}(t) = \mathbf{w}^I\left\{\mathbf{w}\left[\mathbf{U}\left(t^-\right)\right] + T_{1r}\boldsymbol{\epsilon}_r^{*(1)}(t) + K_{0r}\boldsymbol{\epsilon}_r^*(t)\right\},\ \forall t \in \mathfrak{T}_{0F}, \tag{34.6}$$

2) The control vector function $\mathbf{U}(.)$ *obeys (34.7) for* $m = \gamma = rankW(t) = N < r$, $\forall t \in \mathfrak{T}_0$*, so that the trackability is also elementwise:*

$$\mathbf{U}(t) = \mathbf{w}^I\left\{\Gamma(t)\,\mathbf{w}\left[\mathbf{U}\left(t^-\right)\right] + W^T(t)\,\mathbf{T}\left[T_1^{(1)}\boldsymbol{\epsilon}(t) + K_0\boldsymbol{\epsilon}(t)\right]\right\},$$
$$\Gamma(t) = W^T(t)\left[W(t)\,W^T(t)\right]^{-1}W(t) \in \mathfrak{R}^{r\times r},\ \forall t \in \mathfrak{T}_{0F}, \tag{34.7}$$

3) The control vector function $\mathbf{U}(.)$ *obeys (34.8) if* $m = \gamma = rankW(t) = r = N$, $\forall t \in \mathfrak{T}_0$*, so that the trackability is also elementwise:*

$$\mathbf{U}(t) = \mathbf{w}^I\left\langle \begin{array}{c} \mathbf{w}\left[\mathbf{U}\left(t^-\right)\right] \\ +\,W(t)^{-1}\left[T_1\boldsymbol{\epsilon}^{(1)}(t) + K_0\boldsymbol{\epsilon}(t)\right] \end{array}\right\rangle,\ \forall t \in \mathfrak{T}_{0F}, \tag{34.8}$$

4) If the plant also possesses Property 64 and $r \geq \gamma = m$ *then the control vector function* $\mathbf{U}(.)$ *satisfies (34.9):*

$$\mathbf{U}_\gamma(t) = \mathbf{w}_\gamma^I\left\{\mathbf{w}_\gamma\left[\mathbf{U}_\gamma(t^-)\right] + T_{1\gamma}\boldsymbol{\epsilon}_\gamma^{*(1)}(t) + K_{0\gamma}\boldsymbol{\epsilon}_\gamma^*(t)\right\},\ \forall t \in \mathfrak{T}_{0F}. \tag{34.9}$$

We use the complex variable s in:

$$S_\gamma^{(\eta)}(s) = \left[s^0 I_\gamma \vdots s^1 I_i \vdots s^2 I_\gamma \vdots \ldots \vdots s^\eta I_\gamma\right]^{\mathbf{T}} \in \mathfrak{C}^{(\eta+1)\gamma\times\gamma}.$$

Theorem 487 ***NTC for the higher-order linear elementwise exponential tracking, Example 476***

Let (32.23) through (32.43) be valid. Let the reachability time $\tau_R = \tau_F = \infty$*; i.e.,* $\mathfrak{T}_{R\infty} = \mathfrak{T}_{RF} = \{\infty\}$.

For the plant (3.12) obeying Condition (485) to be controlled by the natural tracking control $\mathbf{U}$ *in order to exhibit tracking on* $\mathfrak{T}_{0F}^{(k+1)\gamma} \times \mathfrak{D}^j \times \mathfrak{Y}_d^k$ *determined by the tracking algorithm (34.10),*

$$\mathbf{T}\left(\varepsilon_{\gamma}^{*}, \varepsilon_{\gamma}^{*(1)}, ..., \varepsilon_{\gamma}^{*(\eta)}\right) = \sum_{k=0}^{k=\eta\leq\nu-1} E_{\gamma k}\varepsilon_{\gamma}^{*(k)}(t) = E_{\gamma}^{(\eta)}\varepsilon_{\gamma}^{*\eta}(t) = \mathbf{0}_{\gamma},$$

$$\forall t \in \mathfrak{T}_{0F} \ if \ \mathbf{T}\left(\varepsilon_{\gamma 0}^{*\eta}\right) = \mathbf{0}_{\gamma},$$

$$\varepsilon_{\gamma}^{*\eta}\left(t; \varepsilon_{\gamma 0}^{*\eta}\right) = -\mathbf{f}_{\gamma}^{\eta}(t), \ \forall t \in \mathfrak{T}_{0F} \ if \ \mathbf{T}\left(\varepsilon_{\gamma 0}^{*\eta}\right) \neq \mathbf{0}_{\gamma}, \tag{34.10}$$

with the matrices $E_{\gamma k} \in \mathfrak{R}^{\gamma\times\gamma}$ *such that the real parts* Res_i *of the roots* s_i *of its characteristic polynomial* $f(s)$, $f(s) = det\left(E_{\gamma}^{(\eta)} S_{\gamma}^{(\eta)}(s)\right)$,

$$f(s_i) = det\left(E_{\gamma}^{(\eta)} S_{\gamma}^{(\eta)}(s_i)\right) = 0 \Longrightarrow Res_i < 0, \ \forall i = 1, 2, ..., \eta\gamma, \tag{34.11}$$

it is necessary and sufficient that both i) and ii) hold:

i) the rank γ *of the control matrix* $W(t)$ *is greater than zero and not greater than* $min(N, r)$, *i.e., (8.10) is true, and for the elementwise trackability* $\gamma = N \leq r$ *in* $\mathbf{T}(.)$ *defined by (34.10),*

ii) The control vector function $\mathbf{U}(.)$ *satisfies one of the conditions 1) to 4):*

1) The control vector function $\mathbf{U}(.)$ *obeys (34.12) for* $\gamma = rankW(t) = r < N$, $\forall t \in \mathfrak{T}_0$:

$$\mathbf{U}(t) = \mathbf{w}^{I}\left\{\mathbf{w}\left[\mathbf{U}\left(t^{-}\right)\right] + E_{r}^{(\eta)}\boldsymbol{\epsilon}_{r}^{*\eta}(t)\right\}, \ \forall t \in \mathfrak{T}_{0F}, \tag{34.12}$$

2) The control vector function $\mathbf{U}(.)$ *obeys (34.13) for* $\gamma = rankW(t) = N < r$, $\forall t \in \mathfrak{T}_0$, *so that the trackability is also elementwise:*

$$\mathbf{U}(t) = \mathbf{w}^{I}\left\{\Gamma(t)\mathbf{w}\left[\mathbf{U}\left(t^{-}\right)\right] + W^{T}(t)E^{(\eta)}\boldsymbol{\epsilon}^{\eta}(t)\right\}$$

$$\Gamma(t) = W^{T}(t)\left[W(t)W^{T}(t)\right]^{-1}W(t) \in \mathfrak{R}^{r\times r}, \ \forall t \in \mathfrak{T}_{0F}, \tag{34.13}$$

3) The control vector function $\mathbf{U}(.)$ *obeys (34.14) if* $\gamma = rankW(t) = r = N$, $\forall t \in \mathfrak{T}_0$:

$$\mathbf{U}(t) = \mathbf{w}^{I}\left\langle\mathbf{w}\left[\mathbf{U}\left(t^{-}\right)\right] + W(t)^{-1}E^{(\eta)}\boldsymbol{\epsilon}^{\eta}(t)\right\rangle, \forall t \in \mathfrak{T}_{0F}, \tag{34.14}$$

4) If the plant also possesses Property 64 then the control vector function $\mathbf{U}(.)$ *satisfies (34.15):*

$$\mathbf{U}_{\gamma}(t) = \mathbf{w}_{\gamma}^{I}\left\{\mathbf{w}_{\gamma}\left[\mathbf{U}_{\gamma}(t^{-})\right] + E_{\gamma}^{(\eta)}\boldsymbol{\epsilon}_{\gamma}^{*\eta}(t)\right\}, \ \forall t \in \mathfrak{T}_{0F}. \tag{34.15}$$

The solution $\varepsilon_{\gamma}^{*}(.)$ *shows that the tracking is exponentially stable in the whole if* $\mathbf{T}\left(\varepsilon_{\gamma 0}^{*\eta}\right) = \mathbf{0}_{\gamma}$.

Theorem 488 ***NTC for the sharp elementwise stablewise tracking with the finite vector reachability time*** $\mathbf{t}_R^{\gamma}$***, Example 477***

Let (32.23) through (32.43) be valid.

For the plant (3.12) obeying Condition (485) to be controlled by the natural tracking control $\mathbf{U}$ *in order to exhibit tracking on* $\mathfrak{T}_{0F}^{(k+1)\gamma} \times \mathfrak{D}^{j} \times \mathfrak{Y}_{d}^{k}$ *determined by the algorithm*

$$\mathbf{T}\left(\varepsilon_{\gamma}^{*}, \varepsilon_{\gamma}^{*(1)}\right) = T_{1\gamma}\varepsilon_{\gamma}^{*(1)}(\mathbf{t}^{\gamma}) + K_{0\gamma}\,sign\varepsilon_{\gamma 0}^{*} = \mathbf{0}_{\gamma},$$

$$\forall t \in \mathfrak{T}_{0F} \ if \ \mathbf{T}\left(\varepsilon_{\gamma 0}^{*1}\right) = \mathbf{0}_{\gamma},$$

$$\varepsilon_{\gamma}^{*1}\left(\mathbf{t}^{2N}; \varepsilon_{\gamma 0}^{*1}\right) = -\mathbf{f}_{\gamma}^{1}\left(\mathbf{t}^{2N}\right), \ \forall \mathbf{t}^{2N} \in \mathfrak{T}_{0F}^{2N}, \ if \ \mathbf{T}\left(\varepsilon_{\gamma 0}^{*1}\right) \neq \mathbf{0}_{\gamma}, \tag{34.16}$$

for the sharp elementwise stablewise tracking with the finite vector reachability time $\mathbf{t}_R$,

$$\mathbf{T}\left(\varepsilon_{\gamma 0}^{*1}\right) = \mathbf{0}_\gamma \Longrightarrow \mathbf{t}_R^\gamma = T_{1\gamma} K_{0\gamma}^{-1} \left|\varepsilon_{\gamma 0}^*\right|, \ \varepsilon_{\gamma 0}^* \in \mathfrak{R}^\gamma, \tag{34.17}$$

it is necessary and sufficient that both i) and ii) hold:

i) the rank γ *of the control matrix* $W(t)$ *is greater than zero and not greater than* $\min(N, r)$, *i.e., (8.10) is true, and for the elementwise trackability* $\gamma = N \leq r$ *in* $\mathbf{T}(.)$ *defined by (34.16),*

ii) The control vector function $\mathbf{U}(.)$ *satisfies one of the conditions 1) to 4):*

1) The control vector function $\mathbf{U}(.)$ *obeys (34.18) for* $\gamma = rankW(t) = r < N$, $\forall t \in \mathfrak{T}_0$:

$$\mathbf{U}(t) = \mathbf{w}^I \left\{ \mathbf{w}\left[\mathbf{U}\left(t^-\right)\right] + T_{1r}\boldsymbol{\epsilon}_r^{*(1)}(\mathbf{t}^r) + K_{0r}\, sign\boldsymbol{\epsilon}_{r0}^* \right\}, \ \forall \mathbf{t}^r \in \mathfrak{T}_{0F}^r, \tag{34.18}$$

then the error vector $\boldsymbol{\epsilon}_r^*(\mathbf{t}^r)$ *is controlled elementwise with the finite vector reachability time* $\mathbf{t}_R^r$ *is*

$$\mathbf{T}\left(\varepsilon_{\gamma 0}^{*1}\right) = \mathbf{0}_\gamma \Longrightarrow \mathbf{t}_R^r = T_{1r} K_{0r}^{-1} \left|\varepsilon_{r0}^*\right|, \ \varepsilon_{r0}^* \in \mathfrak{R}^r. \tag{34.19}$$

2) The control vector function $\mathbf{U}(.)$ *obeys (34.20) for* $\gamma = rankW(t) = N < r$, $\forall t \in \mathfrak{T}_0$, *so that the trackability is also elementwise:*

$$\mathbf{U}(t) = \mathbf{w}^I \left\{ \begin{array}{c} \Gamma(t)\, \mathbf{w}\left[\mathbf{U}\left(t^-\right)\right] \\ + W^T(t) \left[T_1 \boldsymbol{\epsilon}^{(1)}(\mathbf{t}^N) + K_0 sign\boldsymbol{\epsilon}_0\right] \end{array} \right\}$$

$$\Gamma(t) = W^T(t)\left[W(t)\, W^T(t)\right]^{-1} W(t) \in \mathfrak{R}^{r \times r}, \ \forall t \in \mathfrak{T}_{0F}, \tag{34.20}$$

then the error vector $\boldsymbol{\epsilon}(\mathbf{t}^N)$ *is controlled elementwise with the finite vector reachability time* $\mathbf{t}_R$,

$$\mathbf{T}\left(\mathbf{0}_N, \varepsilon_0^1\right) = \mathbf{0}_N \Longrightarrow \mathbf{t}_R = \mathbf{t}_R^N = T_1 K_0^{-1} \left|\varepsilon_0\right|, \ \varepsilon_0 \in \mathfrak{R}^N. \tag{34.21}$$

3) The control vector function $\mathbf{U}(.)$ *obeys (34.22) if* $\gamma = rankW(t) = r = N$, $\forall t \in \mathfrak{T}_0$:

$$\mathbf{U}(t) = \mathbf{w}^I \left\langle \begin{array}{c} \mathbf{w}\left[\mathbf{U}\left(t^-\right)\right] \\ + W(t)^{-1} \left[\begin{array}{c} T_1 \boldsymbol{\epsilon}^{(1)}(\mathbf{t}^N) \\ + K_0 sign\boldsymbol{\epsilon}(\mathbf{t}_0^N) \end{array} \right] \end{array} \right\rangle, \ \forall t \in \mathfrak{T}_{0F}, \tag{34.22}$$

then the error vector $\boldsymbol{\epsilon}(\mathbf{t}^N)$ *is controlled elementwise with the finite vector reachability time* $\mathbf{t}_R$ *determined by (34.21).*

4) If the plant also possesses Property 64 then the control vector function $\mathbf{U}(.)$ *satisfies (34.23):*

$$\mathbf{U}_\gamma(t) = \mathbf{w}_\gamma^I \left\langle \mathbf{w}_\gamma \left[\mathbf{U}_\gamma(t^-)\right] + T_{1\gamma}\boldsymbol{\epsilon}_\gamma^{*(1)}(\mathbf{t}^\gamma) + K_{0\gamma} sign\boldsymbol{\epsilon}_{\gamma 0}^* \right\rangle,$$
$$\forall t \in \mathfrak{T}_{0F}, \tag{34.23}$$

then the error vector $\boldsymbol{\epsilon}_\gamma^*(\mathbf{t}^\gamma)$ *is controlled elementwise with the finite vector reachability time* $\mathbf{t}_R^\gamma$ *determined by (34.17).*

The solution $\varepsilon_\gamma^*(.)$ *shows that the tracking is stable in the whole with FVRT* t_R^γ *if* $\mathbf{T}\left(0, \varepsilon_{\gamma 0}^{*1}\right) = \mathbf{0}_\gamma$.

We refer to (33.9) for:

$$\left|\ominus_\gamma^*(t)\right|^{1/2} = \text{diag}\left\{ \left|\epsilon_1(t)\right|^{1/2} \ \left|\epsilon_2(t)\right|^{1/2} \ \ldots \ \left|\epsilon_\gamma^*(t)\right|^{1/2} \right\},$$
$$\text{with } \boldsymbol{\epsilon}_\gamma^* = \left[\epsilon_1 \ \epsilon_2 \ \cdots \ \epsilon_\gamma^*\right]^{\mathbf{T}},$$

Theorem 489 ***The NTC synthesis for the first power smooth elementwise stablewise tracking with the finite vector reachability time $\mathbf{t}_R^{\gamma}$, Example 478***

Let (32.23) through (32.43) be valid.

For the plant (3.12) obeying Condition (485) to be controlled by the natural tracking control $\mathbf{U}$ *in order to exhibit tracking on* $\mathfrak{T}_{0F}^{(k+1)\gamma} \times \mathfrak{D}^j \times \mathfrak{Y}_d^k$ *determined by*

$$\mathbf{T}\left(\boldsymbol{\varepsilon}_{\gamma}^{*}, \boldsymbol{\varepsilon}_{\gamma}^{*(1)}\right) = T_{1\gamma}\boldsymbol{\varepsilon}_{\gamma}^{*(1)}(t) + 2K_{0\gamma}\left|E_{\gamma}^{*}(t)\right|^{1/2} sign\boldsymbol{\varepsilon}_{\gamma 0}^{*} = \mathbf{0}_{\gamma},$$
$$\forall t \in \mathfrak{T}_{0F} \ if \ \mathbf{T}\left(\boldsymbol{\varepsilon}_{\gamma 0}^{*1}\right) = \mathbf{0}_{\gamma},$$
$$\boldsymbol{\varepsilon}_{\gamma}^{*1}\left(\mathbf{t}^{2\gamma}; \boldsymbol{\varepsilon}_{\gamma 0}^{*1}\right) = -\mathbf{f}_{\gamma}^{1}\left(\mathbf{t}^{2\gamma}\right), \ \forall \mathbf{t}^{2\gamma} \in \mathfrak{T}_{0F}^{2\gamma}, \ if \ \mathbf{T}\left(\boldsymbol{\varepsilon}_{\gamma 0}^{*1}\right) \neq \mathbf{0}_{\gamma},$$
$$\left|E_{\gamma}^{*}(t)\right|^{1/2} = diag\left\{\left|\varepsilon_1(t)\right|^{1/2} \ \left|\varepsilon_2(t)\right|^{1/2} \ \ldots \left|\varepsilon_{\gamma-1}(t)\right|^{1/2} \ \left|\varepsilon_{\gamma}^{*}(t)\right|^{1/2}\right\}, \quad (34.24)$$

for the first-power smooth elementwise stablewise tracking with the finite vector reachability time $\mathbf{t}_R^{\gamma}$,

$$\mathbf{T}\left(\boldsymbol{\varepsilon}_{\gamma 0}^{*1}\right) = \mathbf{0}_{\gamma} \Longrightarrow \mathbf{t}_R^{\gamma} = \mathbf{t}_0^{\gamma} + T_{1\gamma}K_{0\gamma}^{-1}\left|E_{\gamma 0}^{*}\right|^{1/2}\mathbf{1}_{\gamma} = \mathbf{t}_0^{\gamma} + T_{1\gamma}K_{0\gamma}^{-1}\left|\boldsymbol{\varepsilon}_{\gamma 0}^{*}\right|^{1/2},$$
$$\left|\boldsymbol{\varepsilon}_{\gamma 0}^{*}\right|^{1/2} = \left[\left|\varepsilon_1(0)\right|^{1/2} \ \left|\varepsilon_2(0)\right|^{1/2} \ \ldots \ \left|\varepsilon_{\gamma-1}(0)\right|^{1/2} \ \left|\varepsilon_{\gamma}^{*}(0)\right|^{1/2}\right]^T, \quad (34.25)$$

it is necessary and sufficient that both i) and ii) hold:

i) the rank γ of the control matrix $W(t)$ is greater than zero and not greater than min(N, r), i.e., (8.10) is true, and for the elementwise trackability $\gamma = N \leq r$ in $\mathbf{T}(.)$ *defined by (34.24),*

ii) The control vector function $\mathbf{U}(.)$ *satisfies one of the conditions 1) to 4):*

1) The control vector function $\mathbf{U}(.)$ *obeys (34.26) for γ =rank$W(t) = r < N$, $\forall t \in \mathfrak{T}_0$:*

$$\mathbf{U}(t) = \mathbf{w}^I\left\{\begin{array}{c}\mathbf{w}\left[\mathbf{U}(t^-)\right] + T_1\boldsymbol{\epsilon}_r^{*(1)}(\mathbf{t}^r) + \\ +2K_{0r}\left|\Theta_r^{*}(\mathbf{t}^r)\right|^{1/2} sign\boldsymbol{\epsilon}_{r0}^{*}\end{array}\right\}, \ \forall t \in \mathfrak{T}_{0F}, \quad (34.26)$$

then the finite vector reachability time $\mathbf{t}_R^r$ *is*

$$\mathbf{T}\left(\boldsymbol{\varepsilon}_{\gamma 0}^{*1}\right) = \mathbf{0}_{\gamma} \Longrightarrow \mathbf{t}_R^r = T_{1r}K_{0r}^{-1}\left|\boldsymbol{\varepsilon}_{r0}^{*}\right|^{1/2}, \ \boldsymbol{\varepsilon}_{r0}^{*} \in \mathfrak{R}^r, \quad (34.27)$$

2) The control vector function $\mathbf{U}(.)$ *obeys (34.28) for γ =rank$W(t) = N < r$, $\forall t \in \mathfrak{T}_0$, so that the trackability is also elementwise:*

$$\mathbf{U}(t) = \mathbf{w}^I\left\{\Gamma(t)\mathbf{w}\left[\mathbf{U}(t^-)\right] + W^T(t)\left[\begin{array}{c}T_1\boldsymbol{\epsilon}^{(1)}(\mathbf{t}^N) \\ +\ 2K_0\left|\Theta\left(\mathbf{t}^N\right)\right|^{1/2} sign\boldsymbol{\epsilon}_0\end{array}\right]\right\}$$
$$\Gamma(t) = W^T(t)\left[W(t)W^T(t)\right]^{-1}W(t) \in \mathfrak{R}^{r\times r}, \ \forall t \in \mathfrak{T}_{0F}, \quad (34.28)$$

then the finite vector reachability time $\mathbf{t}_R$ *is*

$$\mathbf{T}\left(\mathbf{0}_N, \boldsymbol{\varepsilon}_0^1\right) = \mathbf{0}_N \Longrightarrow \mathbf{t}_R = \mathbf{t}_R^N = T_1K_0^{-1}\left|\boldsymbol{\varepsilon}_0\right|^{1/2}, \ \boldsymbol{\varepsilon}_0 \in \mathfrak{R}^N. \quad (34.29)$$

3) The control vector function $\mathbf{U}(.)$ *obeys (34.30) if γ =rank$W(t) = r = N$, $\forall t \in \mathfrak{T}_0$:*

$$\mathbf{U}(t) = \mathbf{w}^I\left\langle\begin{array}{c}\mathbf{w}\left[\mathbf{U}(t^-)\right] \\ +W(t)^{-1}\left[\begin{array}{c}T_1\boldsymbol{\epsilon}^{(1)}(\mathbf{t}^N) \\ +\ 2K_0\left|\Theta\left(\mathbf{t}^N\right)\right|^{1/2} sign\boldsymbol{\epsilon}_0\end{array}\right]\end{array}\right\rangle, \ \forall t \in \mathfrak{T}_{0F}. \quad (34.30)$$

Then the finite vector reachability time $\mathbf{t}_R$ *is determined by (34.29).*

4) If the plant also possesses Property 64 then the control vector function $\mathbf{U}(.)$ *satisfies (34.31):*

$$\mathbf{U}_{\gamma}(t)=\mathbf{w}_{\gamma}^{I}\left\langle \mathbf{w}_{\gamma}\left[\mathbf{U}_{\gamma}(t^{-})\right]+T_{1\gamma}\boldsymbol{\epsilon}_{\gamma}^{*(1)}(\mathbf{t}^{\gamma})+2K_{0\gamma}\left|\ominus_{\gamma}^{*}(\mathbf{t}^{\gamma})\right|^{1/2} sign\boldsymbol{\epsilon}_{\gamma 0}^{*}\right\rangle, \quad \forall t\in\mathfrak{T}_{0F}, \tag{34.31}$$

then the finite vector reachability time $\mathbf{t}_{R}^{\gamma}$ *is determined by (34.25).*

The solution $\varepsilon_{\gamma}^{*}(.)$ *shows that the tracking is stable in the whole with FVRT* t_{R}^{γ} *if* $\mathbf{T}\left(0,\varepsilon_{\gamma 0}^{*1}\right)=\mathbf{0}_{\gamma}$.

We defined in (33.9):

$$K_{\gamma}=\operatorname{diag}\left\{k_{1}\, k_{2}\, \cdots\, k_{\gamma}\right\},\ k_{i}\in\{2,\ 3,\ \cdots\},\ \forall i=1,2,\ \cdots,\gamma,\ K_{N}=K,$$
$$\left|\ominus_{\gamma}^{*}(\mathbf{t}^{\gamma})\right|^{I-K^{-1}}=\operatorname{diag}\left\{\left|\epsilon_{1}(t)\right|^{1-k_{1}^{-1}}\ \left|\epsilon_{2}(t)\right|^{1-k_{2}^{-1}}\ \cdots\ \left|\epsilon_{\gamma-1}(t)\right|^{1-k_{\gamma}^{-1}}\ \left|\epsilon_{\gamma}^{*}(t)\right|^{1-k_{\gamma}^{-1}}\right\}.$$

Theorem 490 ***The NTC synthesis for the higher power smooth elementwise stablewise tracking with the finite vector reachability time*** $\mathbf{t}_{R}^{\gamma}$***, Example 479***

Let (32.23) through (32.43) be valid.

For the plant (3.12) obeying Condition (485) to be controlled by the natural tracking control $\mathbf{U}$ *in order to exhibit tracking on* $\mathfrak{T}_{0F}^{(k+1)\gamma}\times\mathfrak{D}^{j}\times\mathfrak{Y}_{d}^{k}$ *determined by the tracking algorithm*

$$\mathbf{T}\left(\varepsilon_{\gamma}^{*},\varepsilon_{\gamma}^{*(1)}\right)=T_{1\gamma}\varepsilon_{\gamma}^{*(1)}(t)+K_{0\gamma}\left|E_{\gamma}^{*}(t)\right|^{I-K_{\gamma}^{-1}} sign\varepsilon_{\gamma 0}^{*}=\mathbf{0}_{\gamma},$$
$$\forall t\in\mathfrak{T}_{0F}\ if\ \mathbf{T}\left(\varepsilon_{\gamma 0}^{*1}\right)=\mathbf{0}_{\gamma},$$
$$\varepsilon_{\gamma}^{*1}\left(\mathbf{t}^{2\gamma};\mathbf{t}_{0}^{2\gamma};\varepsilon_{0}^{1}\right)=-\mathbf{f}_{\gamma}^{1}\left(\mathbf{t}^{2\gamma}\right),\ \forall\mathbf{t}^{2\gamma}\in\mathfrak{T}_{0F}^{2\gamma},\ if\ \mathbf{T}\left(\varepsilon_{\gamma 0}^{*1}\right)\neq\mathbf{0}_{\gamma}, \tag{34.32}$$

for the higher-power smooth elementwise stablewise tracking with the finite vector reachability time $\mathbf{t}_{R}^{\gamma}$,

$$\mathbf{T}\left(\varepsilon_{\gamma 0}^{*1}\right)=\mathbf{0}_{\gamma}\Longrightarrow\mathbf{t}_{R}^{\gamma}=T_{1\gamma}K_{\gamma}K_{0\gamma}^{-1}\left|\varepsilon_{\gamma 0}^{*}\right|^{K_{\gamma}^{-1}},\ \varepsilon_{\gamma 0}^{*}\in\mathfrak{R}^{\gamma}, \tag{34.33}$$

it is necessary and sufficient that both i) and ii) hold:

i) the rank γ *of the control matrix* $W(t)$ *is greater than zero and not greater than* $min(N,r)$*, i.e., (8.10) is true, and for the elementwise trackability* $\gamma=N\leq r$ *in* $\mathbf{T}(.)$ *defined by (34.32),*

ii) The control vector function $\mathbf{U}(.)$ *satisfies one of the conditions 1) to 4):*

1) The control vector function $\mathbf{U}(.)$ *obeys (34.34) for* $\gamma=rankW(t)=r<N$, $\forall t\in\mathfrak{T}_{0}$*:*

$$\mathbf{U}(t)=\mathbf{w}^{I}\left\{\begin{array}{c}\mathbf{w}\left[\mathbf{U}(t^{-})\right]\\ +T_{1r}\boldsymbol{\epsilon}_{r}^{*(1)}(t)+K_{0r}\left|\ominus_{r}^{*}(t)\right|^{I-K_{\gamma}^{-1}} sign\boldsymbol{\epsilon}_{r0}^{*}\end{array}\right\},\ \forall t\in\mathfrak{T}_{0F}, \tag{34.34}$$

then the finite vector reachability time $\mathbf{t}_{R}^{r}$ *is determined by:*

$$\mathbf{T}\left(\varepsilon_{\gamma 0}^{*1}\right)=\mathbf{0}_{\gamma}\Longrightarrow\mathbf{t}_{R}^{r}=\mathbf{t}_{0}^{r}+T_{1r}K_{r}K_{0r}^{-1}\left|\varepsilon_{r0}^{*}\right|^{K_{r}^{-1}},\ \varepsilon_{r0}^{*}\in\mathfrak{R}^{r}. \tag{34.35}$$

2) The control vector function $\mathbf{U}(.)$ *obeys (34.36) for* $\gamma=rankW(t)=N<r$, $\forall t\in\mathfrak{T}_{0}$*, so that the trackability is also elementwise:*

$$\mathbf{U}(t)=\mathbf{w}^{I}\left\{\Gamma(t)\,\mathbf{w}\left[\mathbf{U}(t^{-})\right]+W^{T}(t)\left[\begin{array}{c}T_{1}\boldsymbol{\epsilon}^{(1)}(t)\\ +K_{0}\left|\ominus(\mathbf{t}^{N})\right|^{I-K^{-1}} sign\boldsymbol{\epsilon}_{0}\end{array}\right]\right\}$$
$$\Gamma(t)=W^{T}(t)\left[W(t)\,W^{T}(t)\right]^{-1}W(t)\in\mathfrak{R}^{r\times r},\ \forall t\in\mathfrak{T}_{0F}, \tag{34.36}$$

then the finite vector reachability time $\mathbf{t}_R$ *is determined by:*

$$\mathbf{T}\left(\varepsilon_0^1\right) = \mathbf{0}_N \Longrightarrow \mathbf{t}_R = \mathbf{t}_R^N = T_1 K K_0^{-1} \left|\varepsilon_0\right|^{K_r^{-1}}, \ \varepsilon_0 \in \mathfrak{R}^N. \tag{34.37}$$

3) The control vector function $\mathbf{U}(.)$ *obeys (34.38) if* $\gamma = rankW(t) = r = N$, $\forall t \in \mathfrak{T}_0$:

$$\mathbf{U}(t) = \mathbf{w}^I \left\langle \begin{array}{c} \mathbf{w}\left[\mathbf{U}\left(t^-\right)\right] \\ +W(t)^{-1} \left[\begin{array}{c} T_1 \boldsymbol{\epsilon}^{(1)}(\mathbf{t}^N) \\ +K_0 \left|\ominus\left(\mathbf{t}^N\right)\right|^{I-K^{-1}} sign\boldsymbol{\epsilon}_0 \end{array} \right] \end{array} \right\rangle, \ \forall t \in \mathfrak{T}_{0F}, \tag{34.38}$$

then the finite vector reachability time $\mathbf{t}_R$ *is determined by (34.37).*

4) If the plant also possesses Property 64 then the control vector function $\mathbf{U}(.)$ *satisfies (34.39):*

$$\mathbf{U}_\gamma(t) = \mathbf{w}_\gamma^I \left\langle \mathbf{w}_\gamma \left[\mathbf{U}_\gamma(t^-)\right] + T_{1\gamma} \boldsymbol{\epsilon}_\gamma^{*(1)}(t) + K_{0\gamma} \left|\ominus_\gamma^*(t)\right|^{I-K_\gamma^{-1}} sign\boldsymbol{\epsilon}_{\gamma 0}^* \right\rangle, \ \forall t \in \mathfrak{T}_{0F}, \tag{34.39}$$

then the finite vector reachability time $\mathbf{t}_R$ *is determined by (34.33).*

The solution $\varepsilon_\gamma^*(.)$ *shows that the tracking is stable in the whole with FVRT* t_R^γ *if* $\mathbf{T}\left(\varepsilon_{\gamma 0}^{*1}\right) = \mathbf{0}_\gamma$.

Theorem 491 ***The NTC synthesis for the sharp absolute error vector value tracking elementwise and stablewise with the finite vector reachability time*** $\mathbf{t}_R^\gamma$, ***Example 480***

Let (32.23) through (32.43) be valid.

For the plant (3.12) obeying Condition (485) to be controlled by the natural tracking control $\mathbf{U}$ *in order to exhibit tracking on* $\mathfrak{T}_{0F}^{2N} \times \mathfrak{D}^j \times \mathfrak{Y}_d^k$ *determined by the following tracking algorithm*

$$\begin{gathered} \mathbf{T}\left(\varepsilon_\gamma^*, \varepsilon_\gamma^{*(1)}\right) = T_{1\gamma}\Sigma\left(\varepsilon_\gamma^*, \varepsilon_\gamma^{*(1)}\right)\varepsilon_\gamma^{*(1)} + K_{0\gamma} sign\left|\varepsilon_{\gamma 0}^*\right| = \mathbf{0}_\gamma, \\ \forall \mathbf{t}^{2N} \in \mathfrak{T}_{0F}^{2N} \ if \ \mathbf{T}\left(\varepsilon_{\gamma 0}^{*1}\right) = \mathbf{0}_\gamma, \\ \varepsilon_\gamma^{*1}\left(\mathbf{t}^{2\gamma}; \mathbf{t}_0^{2\gamma}; \varepsilon_0^1\right) = -\mathbf{f}_\gamma^1\left(\mathbf{t}^{2\gamma}\right), \ \forall \mathbf{t}^{2\gamma} \in \mathfrak{T}_{0F}^{2\gamma}, \ if \ \mathbf{T}\left(\varepsilon_{\gamma 0}^{*1}\right) \neq \mathbf{0}_\gamma, \end{gathered} \tag{34.40}$$

for the sharp absolute error vector value tracking elementwise stablewise with the finite vector reachability time $\mathbf{t}_R^\gamma$,

$$\mathbf{T}\left(\varepsilon_{\gamma 0}^{*1}\right) = \mathbf{0} \Longrightarrow \mathbf{t}_R^\gamma = T_{1\gamma} K_{0\gamma}^{-1} \left|\varepsilon_{\gamma 0}^*\right|, \ \varepsilon_{\gamma 0}^* \in \mathfrak{R}^\gamma, \tag{34.41}$$

it is necessary and sufficient that both i) and ii) hold:

i) the rank γ *of the control matrix* $W(t)$ *is greater than zero and not greater than* $min(N, r)$, *i.e., (8.10) is true, and for the elementwise trackability* $\gamma = N \leq r$ *in* $\mathbf{T}(.)$ *defined by (34.32),*

ii) The control vector function $\mathbf{U}(.)$ *satisfies one of the conditions 1) to 4):*

1) The control vector function $\mathbf{U}(.)$ *obeys (34.42) for* $\gamma = rankW(t) = r < N$, $\forall t \in \mathfrak{T}_0$:

$$\mathbf{U}(t) = \mathbf{w}^I \left\{ \begin{array}{c} \mathbf{w}\left[\mathbf{U}\left(t^-\right)\right] + T_{1r}\boldsymbol{\epsilon}_r^{*(1)}(t) \\ +T_{1r}\Sigma\left(\boldsymbol{\epsilon}_r^*, \boldsymbol{\epsilon}_r^{*(1)}\right)\boldsymbol{\epsilon}_r^{*(1)} + K_{0r} sign\left|\boldsymbol{\epsilon}_{r0}^*\right| \end{array} \right\}, \ \forall t \in \mathfrak{T}_{0F}, \tag{34.42}$$

then the finite vector reachability time $\mathbf{t}_R^r$ *is determined by:*

$$\mathbf{T}\left(\varepsilon_{\gamma 0}^{*1}\right) = \mathbf{0} \Longrightarrow \mathbf{t}_R^r = T_{1r} K_{0r}^{-1} \left|\varepsilon_{r0}^*\right|, \ \varepsilon_{r0}^* \in \mathfrak{R}^r, \tag{34.43}$$

2) The control vector function $\mathbf{U}(.)$ *obeys (34.44) for* $\gamma = rankW(t) = N < r$, $\forall t \in \mathfrak{T}_0$, *so that the trackability is also elementwise:*

$$\mathbf{U}(t) = \mathbf{w}^I \left\{ \begin{array}{c} \mathbf{w}\left[\mathbf{U}(t^-)\right] + \Gamma(t)\,\mathbf{w}\left[\mathbf{U}(t^-)\right] \\ +\, W^T(t)\left[T_1\Sigma\left(\boldsymbol{\epsilon}, \boldsymbol{\epsilon}^{(1)}\right)\boldsymbol{\epsilon}^{(1)} + K_0 sign\left|\boldsymbol{\epsilon}_0\right|\right] \end{array} \right\}$$
$$\Gamma(t) = W^T(t)\left[W(t)W^T(t)\right]^{-1}W(t) \in \mathfrak{R}^{r\times r},\ \forall t \in \mathfrak{T}_{0F}, \tag{34.44}$$

then the finite vector reachability time $\mathbf{t}_R$ *is determined by:*

$$\mathbf{T}\left(\varepsilon_0^1\right) = \mathbf{0} \Longrightarrow \mathbf{t}_R = \mathbf{t}_R^N = T_1K_0^{-1}\left|\varepsilon_0\right|,\ \varepsilon_0 \in \mathfrak{R}^N. \tag{34.45}$$

3) The control vector function $\mathbf{U}(.)$ *obeys (34.46) if* $\gamma = rankW(t) = r = N$, $\forall t \in \mathfrak{T}_0$*:*

$$\mathbf{U}(t) = \mathbf{w}^I \left\langle \begin{array}{c} \mathbf{w}\left[\mathbf{U}(t^-)\right] \\ +\, W(t)^{-1}\left[\begin{array}{c} T_1\Sigma\left(\boldsymbol{\epsilon}, \boldsymbol{\epsilon}^{(1)}\right)\boldsymbol{\epsilon}^{(1)} \\ +\, K_0 sign\left|\boldsymbol{\epsilon}_0\right| \end{array}\right] \end{array} \right\rangle,\ \forall t \in \mathfrak{T}_{0F}. \tag{34.46}$$

Then the finite vector reachability time $\mathbf{t}_R$ *is determined by (34.45).*

4) If the plant also possesses Property 64 then the control vector function $\mathbf{U}(.)$ *satisfies (34.47):*

$$\mathbf{U}_\gamma(t) = \mathbf{w}_\gamma^I\left\{\mathbf{w}_\gamma\left[\mathbf{U}_\gamma(t^-)\right] + T_{1\gamma}\Sigma\left(\boldsymbol{\epsilon}_\gamma^*, \boldsymbol{\epsilon}_\gamma^{*(1)}\right)\boldsymbol{\epsilon}_\gamma^{*(1)} + K_{0\gamma} sign\left|\boldsymbol{\epsilon}_{\gamma 0}^*\right|\right\},$$
$$\forall \mathbf{t}^{2\gamma} \in \mathfrak{T}_{0F}^{2\gamma}, \tag{34.47}$$

then the finite vector reachability time $\mathbf{t}_R$ *is determined by (34.41).*

The solution $\varepsilon_\gamma^*(.)$ *shows that the tracking is stable in the whole with FVRT* t_R^γ *if* $\mathbf{T}\left(\varepsilon_{\gamma 0}^{*1}\right) = \mathbf{0}_\gamma$.

Theorem 492 ***NTC synthesis for the exponential absolute error vector value tracking elementwise and stablewise with the finite vector reachability time*** t_R^γ**, *Example 481***

Let (32.23) through (32.43) be valid.

For the plant (3.12) obeying Condition (485) to be controlled by the natural tracking control $\mathbf{U}$ *in order to exhibit tracking on* $\mathfrak{T}_{0F}^{2N} \times \mathfrak{D}^j \times \mathfrak{Y}_d^k$ *determined by*

$$\mathbf{T}\left(\varepsilon_\gamma^*, \varepsilon_\gamma^{*(1)}\right) = T_{1\gamma}D^+\left|\varepsilon_\gamma^*\right| + K_\gamma\left(\left|\varepsilon_\gamma^*\right| + K_{0\gamma} sign\left|\varepsilon_{\gamma 0}^*\right|\right)$$
$$= T_{1\gamma}\Sigma\left(\varepsilon_\gamma^*, \varepsilon_\gamma^{*(1)}\right)\varepsilon_\gamma^{*(1)} + K_\gamma\left(\left|\varepsilon_\gamma^*\right| + K_{0\gamma} sign\left|\varepsilon_{\gamma 0}^*\right|\right) = \mathbf{0}_\gamma,$$
$$\forall \mathbf{t}^{2\gamma} \in \mathfrak{T}_{0F}^{2\gamma}\ if\ \mathbf{T}\left(\varepsilon_{\gamma 0}^{*1}\right) = \mathbf{0}_\gamma,$$
$$\boldsymbol{\varepsilon}_\gamma^{*1}\left(\mathbf{t}^{2\gamma}; \mathbf{t}_0^{2\gamma}; \varepsilon_0^1\right) = \mathbf{f}_\gamma^1\left(\mathbf{t}^{2\gamma}\right),\ \forall \mathbf{t}^{2\gamma} \in \mathfrak{T}_{0F}^{2\gamma},\ if\ \mathbf{T}\left(\varepsilon_{\gamma 0}^{*1}\right) \neq \mathbf{0}_\gamma, \tag{34.48}$$

for the exponential absolute error vector value tracking elementwise and stablewise with the finite vector reachability time t_R^γ*, (33.20), (33.20):*

$$\mathbf{T}\left(\varepsilon_{\gamma 0}^{*1}\right) = \mathbf{0}_\gamma \Longrightarrow$$
$$T_{R\gamma} = \left\{ \begin{array}{ll} K_\gamma^{-1}T_{1\gamma}ln\left\{K_{0\gamma}^{-1}\left[\left|\ominus_{\gamma 0}^*\right| + K_{0\gamma}\right]\right\}, & \boldsymbol{\epsilon}_{\gamma 0}^* \neq \mathbf{0}_\gamma, \\ O_\gamma, & \boldsymbol{\epsilon}_{\gamma 0}^* = \mathbf{0}_\gamma \end{array} \right\},$$
$$T_{R\gamma} = diag\{t_{R1}\ t_{R2}\ \cdots\ t_{R\gamma}\} \Longleftrightarrow \mathbf{t}_R^\gamma = \left[t_{R1}\ t_{R2}\ \cdots\ t_{R\gamma}\right]^T,$$
$$T_{RN} = T_R,\ \mathbf{t}_R^N = \mathbf{t}_R. \tag{34.49}$$

it is necessary and sufficient that both i) and ii) hold:

i) The rank γ of the control matrix $W(t)$ is greater than zero and not greater than min(N,r), i.e., (8.10) is true, and for the elementwise trackability $\gamma = N \leq r$ in $\mathbf{T}(.)$ defined by (34.32),

ii) The control vector function $\mathbf{U}(.)$ satisfies one of the conditions 1) to 4):

1) The control vector function $\mathbf{U}(.)$ obeys (34.50) for γ =rank$W(t) = r < N$, $\forall t \in \mathfrak{T}_0$:

$$\mathbf{U}(t) = \mathbf{w}^I \left\{ \begin{array}{c} \mathbf{w}\left[\mathbf{U}(t^-)\right] + T_{1r}D^+ \left|\epsilon_r^*\right| \\ + K_r\left(\left|\epsilon_r^*\right| + K_{0r}\,sign\left|\epsilon_{r0}^*\right|\right) \end{array} \right\},\ \forall t \in \mathfrak{T}_{0F}. \tag{34.50}$$

Then the finite vector reachability time $\mathbf{t}_R^r$ is determined by (34.49) and by:

$$\mathbf{T}\left(\mathbf{0}_r, \boldsymbol{\varepsilon}_{r0}^{*1}\right) = \mathbf{0}_r \Longrightarrow \boldsymbol{\varepsilon}_{r0}^{*1} = \boldsymbol{\epsilon}_{r0}^* \Longrightarrow$$

$$T_{Rr} = \left\{ \begin{array}{ll} K_r^{-1}T_{1r}\ln\left\{K_{0r}^{-1}\left[\left|\ominus_{r0}^*\right| + K_{0r}\right]\right\}, & \boldsymbol{\epsilon}_{r0}^* \neq \mathbf{0}_r, \\ O_r, & \boldsymbol{\epsilon}_{r0}^* = \mathbf{0}_r \end{array} \right\}. \tag{34.51}$$

2) The control vector function $\mathbf{U}(.)$ obeys (34.52) for γ =rank$W(t) = N < r$, $\forall t \in \mathfrak{T}_0$, so that the trackability is also elementwise:

$$\mathbf{U}(t) = \mathbf{w}^I \left\{ \begin{array}{c} \Gamma(t)\,\mathbf{w}\left[\mathbf{U}(t^-)\right] \\ + W^T(t)\left[T_1D^+\left|\boldsymbol{\epsilon}(t)\right| + K\left(\left|\boldsymbol{\epsilon}(t)\right| + K_0\,sign\left|\boldsymbol{\epsilon}_0\right|\right)\right] \end{array} \right\}$$

$$\Gamma(t) = W^T(t)\left[W(t)W^T(t)\right]^{-1}W(t) \in \mathfrak{R}^{r\times r},\ \forall t \in \mathfrak{T}_{0F}. \tag{34.52}$$

Then the finite vector reachability time $\mathbf{t}_R$ is determined by (34.49) and by:

$$\mathbf{T}\left(\boldsymbol{\varepsilon}_0^1\right) = \mathbf{0}_\gamma \Longrightarrow$$

$$T_R = \left\{ \begin{array}{ll} K^{-1}T_1 ln\left\{K_0^{-1}\left[\left|E_0\right| + K_0\right]\right\}, & \boldsymbol{\epsilon}_0 \neq \mathbf{0}_N, \\ O_N, & \boldsymbol{\epsilon}_0 = \mathbf{0}_N \end{array} \right\}. \tag{34.53}$$

3) The control vector function $\mathbf{U}(.)$ obeys (34.54) if γ =rank$W(t) = r = N$, $\forall t \in \mathfrak{T}_0$:

$$\mathbf{U}(t) = \mathbf{w}^I \left\langle \begin{array}{c} \mathbf{w}\left[\mathbf{U}(t^-)\right] \\ + W(t)^{-1}\left[\begin{array}{c} T_1D^+\left|\boldsymbol{\epsilon}(t)\right| \\ + K\left(\left|\boldsymbol{\epsilon}(t)\right| + K_0\,sign\left|\boldsymbol{\epsilon}_0\right|\right) \end{array} \right] \end{array} \right\rangle,\ \forall t \in \mathfrak{T}_{0F}. \tag{34.54}$$

Then the finite vector reachability time $\mathbf{t}_R$ is determined by (34.49) and by (34.53).

4) If the plant also possesses Property 64 then the control vector function $\mathbf{U}(.)$ satisfies (34.55):

$$\mathbf{U}_\gamma(t) = \mathbf{w}_\gamma^I\left\{\mathbf{w}_\gamma\left[\mathbf{U}_\gamma(t^-)\right] + T_{1\gamma}D^+\left|\boldsymbol{\epsilon}_\gamma^*\right| + K_\gamma\left(\left|\boldsymbol{\epsilon}_\gamma^*\right| + K_{0\gamma}\,sign\left|\boldsymbol{\epsilon}_{\gamma0}^*\right|\right)\right\},$$
$$\forall \mathbf{t}^{2\gamma} \in \mathfrak{T}_{0F}^{2\gamma}, \tag{34.55}$$

then the finite vector reachability time $\mathbf{t}_R$ is determined by (34.49).

The solution $\boldsymbol{\varepsilon}_\gamma^*(.)$ shows that the tracking is stable in the whole with FVRT t_R^γ if $\mathbf{T}\left(\boldsymbol{\varepsilon}_{\gamma0}^{*1}\right) = \mathbf{0}_\gamma$.

The following comments generalize the corresponding comments in [279, Comments 384-389, pp. 278,279].

Comment 493 *The above-synthesized NTC control $\mathbf{U}$ does not need knowledge of the state of the plant.*

Comment 494 ***Implementation of the NTC algorithm***

The controller should comprise the corresponding component that generates the subsidiary vector function $\mathbf{f}_\gamma(.)$ *defined by (32.36), (32.37), Figure 32.4.*

The controller uses information about the subsidiary vector function $\mathbf{f}_\gamma(.)$ *and about the real output error vector* $\boldsymbol{\varepsilon}_\gamma^*$*, determines the subsidiary reference output vector* $\mathbf{Y}_R$*, (32.23)-(32.27), and the subsidiary error vector* $\boldsymbol{\epsilon}_\gamma^*$*, (32.28)-(32.32), and uses them to generate the NTC according to the accepted algorithm.*

The natural tracking controller does not use information about the plant mathematical model. It does not use any information about the disturbance vector or about high derivatives of the real output vector $\mathbf{Y}(t)$ *at the moment* t*. The NTC is fully robust. The preceding control algorithms verify these claims.*

The usage of $\mathbf{U}(t^-)$ *in order to generate* $\mathbf{U}(t)$ *at the moment* t *expresses a kind of memory and expresses the adaptability of the NTC.*

The transmission of the signal from the controller output to its input without delay, or with a negligible delay, is the crucial demand on the controller. It is linked with the continuity of the time flow. The time continuity of NTC is crucial for its effective implementation.

Comment 495 *The control gain* $\Gamma(t)$ *resolves the complex problem of channeling the control action on the MIMO plant (3.12).*

Comment 496 *The higher-order linear or nonlinear NTC algorithms need the measurement of the higher output error vector derivatives that are not often accessible for measurement. For this reason we presented herein the nonlinear NTC algorithms only of the first order. They ensure high tracking qualities including the finite reachability time with the full robustness relative to the fluctuations of the plant state and relative to the external disturbances.*

Chapter 35

Tracking quality: State space

35.1 State space tracking operator

The description of the tracking quality in the state space is essentially the same as in the output space. Comparing with Chapter 32 the difference is only in the notation as shown in the sequel.

For the plant (3.12) with Property 61 holds Lemma 62, which determines that the dimension of the controllable part (3.38) of the plant state equation is equal to $\beta \leq \min(\rho, r)$, where r is the dimension of the control vector $\mathbf{U} \in \mathfrak{R}^r$. Only the state β-subvector $\mathbf{R}_\beta = [R_1 \ \ R_2 ... R_\beta]^T \in \mathfrak{R}^\beta$ can be then controlled directly; i.e., elementwise, Axiom 87 and Lemma 62,

$$\mathbf{R} = [R_1 \ R_2 .. R_\rho]^T \in \mathfrak{R}^\rho, \ \mathbf{R}_\beta = [R_1 \ R_2 .. R_\beta]^T \in \mathfrak{R}^\beta, \ 1 \leq \beta \leq \min(\rho, r). \tag{35.1}$$

Regardless of this fact, the whole state vector $\mathbf{R} \in \mathfrak{R}^\rho$, hence the whole state error vector $\varepsilon_R \in \mathfrak{R}^\rho$, (4.4),

$$\boldsymbol{\varepsilon}_R = \mathbf{R_d} - \mathbf{R} = [\varepsilon_{R1} \ \varepsilon_{R2} .. \varepsilon_{R\rho}]^T \in \mathfrak{R}^\rho, \tag{35.2}$$

should be controlled at least indirectly. The β-subvector $\mathbf{R}_\beta$ of $\mathbf{R}$ does not contain the entries $R_{\beta+1}, R_{\beta+2}, .. R_\rho$ of $\mathbf{R}$.

Comment 497 ***Controllable state variables and their errors***

When $\beta < \rho$, in order to solve the control problem, by following Axiom 87, we used subsidiary vector functions $\mathbf{v}(,)$ with dimension m, $m \leq \beta$, to establish the perfect trackability conditions in which $\mathbf{v}(,)$ is determined by (8.1), and the trackability conditions in which $\mathbf{v}(,)$ obeys (11.1), for the control synthesis in the state space. This suggests to us to introduce the following subsidiary, modified, state error variables and by them induced subsidiary, modified, state error vectors:

$$\left\{ \begin{array}{c} R_\beta^* = R_j, \\ \varepsilon_\beta^* = \varepsilon_{Rj} \end{array} \right\} \ iff \ |\varepsilon_{Rj}| = max\left(|\varepsilon_{Ri}| : \forall i = \beta, \beta+1, \ \cdots \ , \rho\right),$$

$$j \in \{\beta, \beta+1, \ \cdots \ , \rho\}, \ \beta \in \{1, 2, \ \cdots \ , \rho\},$$

$$\mathbf{R}_\beta^* = \left[R_1 \ \ R_2 \ \cdots \ R_{\beta-1} \ R_\beta^* \ \right]^T \in \mathfrak{R}^\beta, \tag{35.3}$$

$$\boldsymbol{\varepsilon}_\beta^* = \left[\varepsilon_{R1} \ \ \varepsilon_{R2} \ \cdots \ \varepsilon_{R,\beta-1} \ \varepsilon_\beta^* \ \right]^T \in \mathfrak{R}^\beta. \tag{35.4}$$

The β-subvector $\mathbf{R}_\beta^*$ of $\mathbf{R}$ does contain implicitly the entries $R_{\beta+1}, R_{\beta+2}, .. R_\rho$ of $\mathbf{R}$ and the β-subvector ε_β^* of ε_R does contain implicitly the entries $\varepsilon_{R,\beta+1}, \varepsilon_{R,\beta+2}, .. \varepsilon_{R,\rho}$ of ε_R. The *following characteristics of* ε_{β}^{*}*:*

$$\begin{aligned} \varepsilon_{\beta}^{*} = 0 &\Longleftrightarrow \varepsilon_{Rj} = 0, \ \forall j = \beta, \beta+1, ..., \rho, \ \forall \xi > 0, \quad \textit{and} \\ \left|\varepsilon_{\beta}^{*}\right| < \xi &\Longleftrightarrow \left|\varepsilon_{Rj}\right| < \xi, \forall j = \beta, \beta+1, ..., \rho, \ \forall \xi > 0, \end{aligned} \tag{35.5}$$

are important for indirect control of the state variables $R_{\beta+1}$, $R_{\beta+2}$, ..R_{ρ} *that are not directly controllable. The use of the vectors* $\mathbf{R}_{\beta}^{*}$ *and* ε_{β}^{*} *enables us to synthesize the control* $\mathbf{U} \in \mathfrak{R}^{r}$ *that can guarantee both the stable tracking and the stable state tracking. This permits* $r < \rho$ *that is the characteristic of many plants. If* $\beta = \rho \leq r$ *then the control can guarantee the elementwise stable state tracking.*

Conclusion 498 ***The advantages of using*** ε_{β}^{*} ***rather than*** ε

By using ε_{β}^{*} *instead of* ε *the control synthesized only in the state space can guarantee both stable state tracking and tracking.*

The number r *of control variables can be less than the number* ρ *of the state variables.*

We define the criterion of the tracking quality to be a solution of the following differential equation expressed in terms of the error ε_{β}^{*}, (35.4), its derivatives

$$\varepsilon_{\beta}^{*}(t), \boldsymbol{\varepsilon}_{\beta}^{*(1)}(t), ..., \boldsymbol{\varepsilon}_{\beta}^{*(\alpha-1)}(t), \tag{35.6}$$

and/or in its integral [279],

$$\begin{aligned} &\mathbf{T}\left(t, \varepsilon_{\beta}^{*}(t), \boldsymbol{\varepsilon}_{\beta}^{*(1)}(t), ..., \boldsymbol{\varepsilon}_{\beta}^{*(\alpha-1)}(t), \int_{t_0}^{t} \varepsilon_{\beta}^{*}(t)\, dt\right) \\ &= \mathbf{T}\left(t, \boldsymbol{\varepsilon}_{\beta}^{*\alpha-1}(t), \int_{t_0}^{t} \varepsilon_{\beta}^{*}(t)\, dt\right) = \mathbf{0}_{m}, \ \forall t \in \mathfrak{T}_{0F}, \ m \leq \beta. \end{aligned} \tag{35.7}$$

The vector function $T(.)$ represents *the vector tracking operator*,

$$\mathbf{T}(.) : \mathfrak{T} \times \mathfrak{R}^{\alpha\beta} \times \mathfrak{R}^{\beta} \longrightarrow \mathfrak{R}^{m}, \ \mathbf{T}(.) \in \mathfrak{C}^{-}. \tag{35.8}$$

It determines the class of *the tracking algorithms* in the state space, which express the demanded tracking qualities. It is the basis for the synthesis of the high-quality tracking.

Task 499 ***Tracking quality and the control task in the state space***

The control should force the plant state behavior $\mathbf{R}(.)$ *to satisfy the demanded tracking quality specified by (35.7).*

The basic task of the state quality tracking control synthesis is to determine such control.

The last equation in (35.7) holds in the cases for which the reachability moment t_R is the same for all entries of the modified error vector ε_{β}^{*}, (35.4), and for their derivatives. Analogously, the final moment t_F is the same for all entries of the modified error vector ε_{β}^{*} and for their derivatives.

However, if different reachability *times* $t_{Ri,(j)}$ and/or different final *times* $t_{Fi,(j)}$ are associated with the elements of the modified error vector ε_{β}^{*} and/or with the elements of the j-th derivative $\varepsilon_{\beta}^{*(j)}$ of ε_{β}^{*}, then we use:

$$\mathbf{t}^{\alpha\beta} = [t \ t \ ... \ t]^{T} \in \mathfrak{T}^{\alpha\beta} \text{ related to } \boldsymbol{\varepsilon}_{\beta}^{*\alpha-1}.$$

The use of the *time vector* $\mathbf{t}^{\alpha\beta}$ permits us to simplify, formally mathematically, the treatment of the vector relationships. This has the full advantage in the framework of the elementwise tracking with the finite vector reachability time $\mathbf{t}_R^{\alpha\beta}$ iff $\beta = \rho$, $\mathbf{t}_R^{\alpha\beta} = \mathbf{t}_R^{\alpha\rho}$. For the same reason we introduce the state error diagonal matrices $E_\beta^*(t) \equiv E_\beta^*(\mathbf{t}^\beta)$ and $E_R(t) \equiv E_R(\mathbf{t}^\rho)$,

$$E_\beta^*(t) = \operatorname{diag}\left\{\varepsilon_{r1}(t) \quad \varepsilon_{r2}(t) \quad \cdots \quad \varepsilon_\beta^*(t)\ \right\} = E_\beta^*(\mathbf{t}^\beta),\ E_R(\mathbf{t}^\rho) = E_\rho^*(\mathbf{t}^\rho). \tag{35.9}$$

Let

$$\boldsymbol{\varepsilon}_\beta^{*\alpha-1}(\mathbf{t}^{\alpha\beta}) \equiv \boldsymbol{\varepsilon}_\beta^{*\alpha-1}\left(\mathbf{t}^{\alpha\beta}; \mathbf{t}_0^{\alpha\beta}; \boldsymbol{\varepsilon}_{R0}^{\alpha-1}\right) \equiv \boldsymbol{\varepsilon}_\beta^{*\alpha-1}\left(\mathbf{t}^{\alpha\beta}; \mathbf{t}_0^{\alpha\beta}; \boldsymbol{\varepsilon}_{R0}^{\alpha-1}; \mathbf{D}; \mathbf{U}; \mathbf{R}_d\right), \tag{35.10}$$

so that

$$\varepsilon_\beta^*(\mathbf{t}^\beta) \equiv \boldsymbol{\varepsilon}_\beta^*\left(\mathbf{t}^\beta; \mathbf{t}_0^\beta; \boldsymbol{\varepsilon}_{R0}^{\alpha-1}\right) \equiv \boldsymbol{\varepsilon}_\beta^*\left(\mathbf{t}^\beta; \mathbf{t}_0^\beta; \boldsymbol{\varepsilon}_{R0}^{\alpha-1}; \mathbf{D}; \mathbf{U}; \mathbf{R}_d\right). \tag{35.11}$$

This notation simplifies formally (35.7) to:

$$\mathbf{T}\left(\mathbf{t}^m, \boldsymbol{\varepsilon}_\beta^{*\alpha-1}(\mathbf{t}^{\alpha\beta}), \int_{\mathbf{t}_0^\rho}^{\mathbf{t}^\rho} E_\beta^*(\mathbf{t}^\beta)\, d\mathbf{t}^\beta\right) = \mathbf{0}_m,\ \forall \mathbf{t}^{\alpha\beta} \in \mathfrak{T}_{0F}^{\alpha\beta}. \tag{35.12}$$

The control task 499 determines the tracking control goal:

Goal 500 ***The basic tracking control goal***
The tracking control should force the plant state error behavior $\boldsymbol{\varepsilon}_\beta^*(.)$ to obey the tracking quality determined by (35.12).

35.2 Tracking operator properties

The following properties of *the vector tracking operator* $T(.)$, (35.12),

$$\mathbf{T}(.) : \mathfrak{T}_{0F}^m \times \mathfrak{R}^{\alpha\beta} \times \mathfrak{R}^\beta \longrightarrow \mathfrak{R}^m, m \le \beta \le \min(\rho, r), \tag{35.13}$$

determine the class of *the tracking algorithms* that express the required tracking qualities, and which are herein the basis for the control synthesis to fulfill the high quality demand.

Property 501 ***Continuity of the error solution***
The solution $\boldsymbol{\varepsilon}_\beta^{*\alpha-1}(\mathbf{t}^{\alpha\beta}) = \boldsymbol{\varepsilon}_\beta^*\left(\mathbf{t}^\beta; \mathbf{t}_0^\beta; \boldsymbol{\varepsilon}_0^{\alpha-1}\right)$ of (35.12) is continuous in time on $\mathfrak{T}_{0F}^{\alpha\beta}$,

$$\mathbf{T}\left(\mathbf{t}^m, \boldsymbol{\varepsilon}_\beta^{*\alpha-1}(\mathbf{t}^{\alpha\beta}), \int_{\mathbf{t}_0^\beta}^{\mathbf{t}^\beta} E_\beta^*(\mathbf{t}^\beta)\, d\mathbf{t}^\beta\right) = \mathbf{0}_m,\ \forall \mathbf{t}^{\alpha\beta} \in \mathfrak{T}_{0F}^{\alpha\beta}$$
$$\Longrightarrow \boldsymbol{\varepsilon}_\beta^*\left(\mathbf{t}^\beta; \mathbf{t}_0^\beta; \boldsymbol{\varepsilon}_0^{\alpha-1}\right) \in \mathfrak{C}(\mathfrak{T}_{0F}^\beta). \tag{35.14}$$

Property 502 ***The operator*** $T(.)$ ***vanishes at the origin***
The operator $T(.)$ vanishes at the origin at every moment,

$$\boldsymbol{\varepsilon}_\beta^{*\alpha-1} = \mathbf{0}_{\alpha\beta} \Longrightarrow \mathbf{T}\left(\mathbf{t}^m, \mathbf{0}_{\alpha\beta}, \int_{\mathbf{t}_0^\beta}^{\mathbf{t}^\beta} O_\beta d\mathbf{t}^\beta\right) = \mathbf{0}_m,\ \forall \mathbf{t}^\beta \in \mathfrak{T}_{0F}^\beta. \tag{35.15}$$

Property 503 ***The solution of*** *(35.12)* ***for all zero initial conditions***
The solution of (35.12) for all zero initial conditions is identically equal to the zero vector,

$$\boldsymbol{\varepsilon}_{\beta 0}^{*\alpha-1} = \mathbf{0}_{\alpha\beta}\ \textit{and}$$
$$\mathbf{T}\left(\mathbf{t}^m, \boldsymbol{\varepsilon}_\beta^{*\alpha-1}(\mathbf{t}^{\alpha\beta}), \int_{\mathbf{t}_0^\beta}^{\mathbf{t}^\beta} E_\beta^*(\mathbf{t}^\beta)\, d\mathbf{t}^\beta\right) = \mathbf{0}_m,\ \forall \mathbf{t}^{\alpha\beta} \in \mathfrak{T}_{0F}^{\alpha\beta} \Longrightarrow$$
$$\varepsilon_\beta^*\left(\mathbf{t}^\beta; \mathbf{t}_0^\beta; \mathbf{0}_{\alpha\beta}\right) = \mathbf{0}_\beta,\ \forall \mathbf{t}^\beta \in \mathfrak{T}_{0F}^\beta. \tag{35.16}$$

Note 504 *If (35.12) holds, then the condition* $\forall \mathbf{t}^{\alpha\beta} \in \mathfrak{T}_{0F}^{\alpha\beta}$ *demands that the initial state error vector, its derivatives, and its integral also obey (35.12) at the initial moment* $\mathbf{t}_0^m$.

Property 505 ***The operator*** $T(.)$ ***vanishes at the initial moment***

The initial state error vector, its initial derivatives, and its integral obey (35.12) at the initial moment t_0*; i.e., at* $\mathbf{t}_0^m$:

$$\mathbf{T}\left(\mathbf{t}_0^m, \boldsymbol{\varepsilon}_\beta^{*\alpha-1}\left(\mathbf{t}^{\alpha\beta}\right), \int_{\mathbf{t}_0^\beta}^{\mathbf{t}_0^\beta} E_\beta^*\left(\mathbf{t}^\beta\right) d\mathbf{t}^\beta\right) = \mathbf{T}\left(\mathbf{t}_0^m, \boldsymbol{\varepsilon}_\beta^{*\alpha-1}, \mathbf{0}_\beta\right) = \mathbf{0}_m,$$
$$\forall \boldsymbol{\varepsilon}_\beta^{*\alpha-1} \in \mathfrak{R}^{\alpha\beta}. \tag{35.17}$$

Note 506 *The real initial modified state error vector* $\varepsilon_{\beta 0}^*$ *is unpredictable, uncontrollable, and arbitrary. It results from the past behavior of the plant, which is untouchable. It determines the initial control vector* $\mathbf{U}_{\beta 0}$ *in the state space, rather than depends on* $\mathbf{U}_{\beta 0}$*. It sometimes does not satisfy the condition (35.17), which then implies the violation of (35.12) and its nonrealizability at every* $\mathbf{t}^{\alpha\beta} \in \mathfrak{T}_{0F}^{\alpha\beta}$ *for such initial conditions.*

Problem 507 ***Matching the error vector, its derivatives, and integral with the tracking algorithm on*** $\mathfrak{T}_{0F}^{\alpha\beta}$

How can we guarantee the arbitrary error vector, error vector derivatives, and the error vector integral to satisfy the tracking algorithm (35.12) on $\mathfrak{T}_{0F}^{\alpha\beta}$*?*

We generalize the solution to this problem proposed in [274] and further developed in [279] for the *time*-invariant linear systems.

Note 508 *The vector value* $t_R^{\alpha\beta}$ *is infinite,* $t_R^{\alpha\beta} = \infty \mathbf{1}_{\alpha\beta}$*, if and only if, the tracking should be only asymptotic, which implies* $\mathfrak{T}_{R\infty}^{\alpha\beta} = \{\infty \mathbf{1}_{\alpha\beta}\}$*. Otherwise* $t_R^{\alpha\beta} \in In\mathfrak{T}_{0F}^{\alpha\beta}$*; hence,* $\mathfrak{T}_{R\infty}^{\alpha\beta} \subset \mathfrak{T}_{0F}^{\alpha\beta}$*, and* $\mathbf{t}_0^{\alpha\beta} < t_R^{\alpha\beta} \leq \mathbf{t}_F^{\alpha\beta} < \infty \mathbf{1}_{\alpha\beta}$ *so that*

$$\mathbf{t}_R^{\alpha\beta} < \infty \mathbf{1}_{\alpha\beta} \Longrightarrow$$
$$\mathfrak{T}_{RF}^{\alpha\beta} = \left\{\mathbf{t}^{\alpha\beta} : \mathbf{t}_0^{\alpha\beta} < \mathbf{t}_R^{\alpha\beta} \leq \mathbf{t}^{\alpha\beta} \leq \mathbf{t}_F^{\alpha\beta}\right\} \subset \mathfrak{T}_{0F}^{\alpha\beta}. \tag{35.18}$$

In what follows we accept the finite reachability time $\mathbf{t}_R^{\alpha\beta}$*, (35.18).*

Solution 509 ***The relaxed demand in general***

The root of Problem 507 is not in a particular initial error, but in the possible incompatibility of some initial errors with the tracking algorithm (35.12) at the initial moment $\mathbf{t}_0^{\alpha\beta}$ *due to the arbitrariness of the initial errors, i.e., it is possible that*

$$\mathbf{T}\left(\mathbf{t}_0^m; \boldsymbol{\varepsilon}_{\beta 0}^{*\alpha-1}, \int_{\mathbf{t}_0^\beta}^{\mathbf{t}^\beta} E_\beta^*\left(\mathbf{t}^\beta\right) d\mathbf{t}^\beta\right) = \mathbf{T}\left(\mathbf{t}_0^m; \boldsymbol{\varepsilon}_{\beta 0}^{*\alpha-1}, \mathbf{0}_\beta\right) \neq \mathbf{0}_m,$$
$$for\ some\ \boldsymbol{\varepsilon}_{\beta 0}^{*\alpha-1} \in \mathfrak{R}^{\alpha\beta}. \tag{35.19}$$

It is a fact that the control cannot influence such $\boldsymbol{\varepsilon}_{\beta 0}^{*\alpha-1}$*. Hence, let the demand be relaxed so that (35.12) holds only on* $\mathfrak{T}_{RF}^{\alpha\beta}$ *rather than on the whole* $\mathfrak{T}_{0F}^{\alpha\beta}$ *for such* $\boldsymbol{\varepsilon}_{\beta 0}^{*\alpha-1}$,

$$\mathbf{T}\left(\mathbf{t}^m, \boldsymbol{\varepsilon}_\beta^{*\alpha-1}\left(\mathbf{t}^{\alpha\beta}; \mathbf{t}_0^{\alpha\beta}; \boldsymbol{\varepsilon}_{R0}^{\alpha-1}\right), \int_{\mathbf{t}_0^\beta}^{\mathbf{t}^\beta} E_\beta^*\left(\mathbf{t}^\beta\right) d\mathbf{t}^\beta\right) = \mathbf{0}_m,\ \forall \mathbf{t}^{\alpha\beta} \in \mathfrak{T}_{RF}^{\alpha\beta}, \tag{35.20}$$

rather than on the whole $\mathfrak{T}_{0F}^{\alpha\beta}$*. This relaxed demand should be satisfied for every initial state error vector* $\boldsymbol{\varepsilon}_{\beta 0}^{*\alpha-1}$ *that violates (35.17), (Property 505), i.e., for which (35.19) holds.*

The preceding analysis opens the need to modify the control goal [279].

Goal 510 ***The modified control goal***

The control should force the plant to behave so that the following tracking algorithm holds [instead of (35.12)]:

$$\left\{\begin{array}{c}\mathbf{T}\left(\mathbf{t}^{m},\boldsymbol{\varepsilon}_{\beta}^{*\alpha-1}\left(\mathbf{t}_{F}^{\alpha\beta}\right),\int_{t_{0}^{\beta}}^{\mathbf{t}^{\beta}}E_{\beta}^{*}\left(\mathbf{t}^{\beta}\right)d\mathbf{t}^{\beta}\right)=\mathbf{0}_{m},\\ \forall\mathbf{t}_{F}^{\alpha\beta}\left\{\begin{array}{c}\in\mathfrak{T}_{0F}^{\alpha\beta}\ iff\ \mathbf{T}\left(\mathbf{t}_{0}^{m};\boldsymbol{\varepsilon}_{\beta0}^{*\alpha-1},\mathbf{0}_{\beta}\right)=\mathbf{0}_{m}\\ \in\mathfrak{T}_{RF}^{\alpha\beta}\ iff\ \mathbf{T}\left(\mathbf{t}_{0}^{m};\boldsymbol{\varepsilon}_{\beta0}^{*\alpha-1},\mathbf{0}_{\beta}\right)\neq\mathbf{0}_{m}\end{array}\right\}\end{array}\right\}, \tag{35.21}$$

$$and\left\{\begin{array}{c}\boldsymbol{\varepsilon}_{\beta}^{*\alpha-1}\left(\mathbf{t}_{F}^{\alpha\beta};\mathbf{t}_{0}^{\alpha\beta};\boldsymbol{\varepsilon}_{0}^{\alpha-1}\right)\\ =\left\{\begin{array}{c}-\mathbf{f}_{\beta}^{\alpha-1}\left(\mathbf{t}_{F}^{\alpha\beta};\mathbf{f}_{\beta0}^{\alpha-1}\right),\ \forall\mathbf{t}^{\alpha\beta}\in\mathfrak{T}_{0F}^{\alpha\beta},\\ iff\ \mathbf{T}\left(\mathbf{t}_{0}^{m};\boldsymbol{\varepsilon}_{\beta0}^{*\alpha-1},\mathbf{0}_{\beta}\right)\neq\mathbf{0}_{m},\end{array}\right\}\end{array}\right\}$$

$$where\ \mathbf{f}_{\beta0}^{\alpha-1}=-\boldsymbol{\varepsilon}_{\beta0}^{*\alpha-1}\ and\ \mathbf{f}_{\beta}^{\alpha-1}\left(\mathbf{t}_{F}^{\alpha\beta};\mathbf{f}_{\beta0}^{\alpha-1}\right)=\mathbf{0}_{\alpha\beta},\ \forall\mathbf{t}^{\alpha\beta}\in\mathfrak{T}_{RF}^{\alpha\beta}, \tag{35.22}$$

for a given or to be determined both $\mathbf{t}_R^{\alpha\beta}\in In\mathfrak{T}_0^{\alpha\beta}$ *and the vector function* $\mathbf{f}(.) : \mathfrak{T}^{\rho}\times\mathfrak{R}^{\rho}\longrightarrow\mathfrak{R}^{\rho}$ *, which implies its* β*-subfunction* $\mathbf{f}_{\beta}:\mathfrak{T}^{\beta}\times\mathfrak{R}^{\beta}\longrightarrow\mathfrak{R}^{\beta}$.

Note 511 *We can use (35.21) instead of (35.12) also in the cases when the initial error vector* $\varepsilon_{\beta0}^{*\alpha-1}$ *is different from the zero vector,* $\varepsilon_{\beta0}^{*\alpha-1}\neq\mathbf{0}_{\alpha\beta}$*, but the tracking algorithm is satisfied at the initial moment, i.e.,*

$$\mathbf{T}\left(\mathbf{t}_{0}^{m},\boldsymbol{\varepsilon}_{\beta0}^{*\alpha-1},\int_{t_{0}}^{t_{0}}E_{\beta}^{*}\left(\mathbf{t}^{\beta}\right)d\mathbf{t}^{\beta}\right)=\mathbf{0}_{m}.$$

35.3 Reference state vector $\mathbf{R}_R$

The reality of the plant operations is often such that the real *state* vector is not desired: $\mathbf{R}_0\neq\mathbf{R}_{d0}$, i.e., that the initial *state* error is not zero vector: $\varepsilon_{R0}\neq\mathbf{0}_{\rho}$.

Solution 512 ***The reference state vectors*** $\mathbf{R}_R$ ***and*** $\mathbf{R}_{R\beta}^*$

The wise demand for the desired state behavior takes into account the real initial state vector $\mathbf{R}_0$ *and defines* ***the realizable desired state vector*** $\mathbf{R}_R$ *that starts from* $\mathbf{R}_0$ *at the initial moment* t_0*. It is called also* ***the realizable state reference vector*** $\mathbf{R}_R$*, for short:* ***the reference state vector*** $\mathbf{R}_R$*:*

$$\mathbf{R}_R=\left[R_{R1}\ R_{R2}\ \cdots\ R_{R\rho}\ \right]^T\in\mathfrak{R}^{\rho}, \tag{35.23}$$

It should be such that at the initial instant it is equal to the initial real state vector $\mathbf{R}_0$*, and because the finite vector reachability time* $\mathbf{t}_R^{\rho}$ *has elapsed it is always equal to the desired state vector* $\mathbf{R}_d$*,*

$$\begin{array}{c}\mathbf{R}_{R0}=\mathbf{R}_R\left(\mathbf{t}_0^{\rho};\mathbf{t}_0^{\rho};\mathbf{R}_{R0}\right)=\mathbf{R}\left(\mathbf{t}_0^{\rho};\mathbf{t}_0^{\rho};\mathbf{R}_0\right)=\mathbf{R}_0,\\ \mathbf{R}_R\left(\mathbf{t}^{\rho};\mathbf{t}_0^{\rho};\mathbf{R}_{R0}\right)=\mathbf{R}_d\left(\mathbf{t}^{\rho}\right),\ \forall\mathbf{t}^{\rho}\in\mathfrak{T}_{RF}^{\rho}.\end{array} \tag{35.24}$$

The facts that only the β*-subvector* $\mathbf{R}_{\beta}$*,*

$$\mathbf{R}_{\beta}=\left[R_1\ R_2\ \ldots\ R_{\beta}\ \right]^T\in\mathfrak{R}^{\beta}, \tag{35.25}$$

is directly controllable, Axiom 87, Property 61, Lemma 62, and that the control goal demands for the whole state vector $\mathbf{R}$ *to be controlled, at least indirectly, which is possible in principle, Axiom 87, lead to the use of the* β*-subvector* $\mathbf{R}^*_{R\beta}$ *of the reference vector* $\mathbf{R}_R$:

$$\mathbf{R}^*_{R\beta} = \left[R_{R1}\ R_{R2}\ ...R_{R\gamma-1}\ R^*_{R\beta}\ \right]^T \in \mathfrak{R}^\beta,$$
$$R^*_{R\beta} = R_{Rj} \Longleftrightarrow |R_{Rj} - R_j| = max\left(|R_{Ri} - R_i| : \forall i = \beta, \beta+1, ..., \rho\right),$$
$$j \in \{\beta, \beta+1, ..., \rho\},\ \beta \in \{1, 2, ..., \rho\}, \tag{35.26}$$

The vector $\mathbf{R}^*_{R\beta}$ *represents* ***the directly (i.e., elementwise) controllable reference state*** β*-subvector of the plant, such that it is continuous, possibly continuously differentiable i-times, at the initial moment it and its derivatives are equal to the real initial state* β*-subvector* $\mathbf{R}_{\beta0}$ *and its initial derivatives, that the former become and stay equal to the desired state* β*-subvector* $\mathbf{R}_{\beta d}$ *and its derivatives because the vector reachability time* $\mathbf{t}^\beta_R$ *has elapsed:*

$$\mathbf{R}^*_{R\beta}(\mathbf{t}^\beta; \mathbf{t}^\beta_0; \mathbf{R}^*_{R\beta0}) \in \mathfrak{C}^i(\mathfrak{T}^\beta_{0F}),\ i \in \{0, 1, 2, ...\},$$
$$\mathbf{R}^*_{R\beta0} = \mathbf{R}^*_{R\beta}\left(\mathbf{t}^\beta_0; \mathbf{t}^\beta_0; \mathbf{R}^*_{R\beta0}\right) = \mathbf{R}_\beta\left(\mathbf{t}^\beta_0; \mathbf{t}^\beta_0; \mathbf{R}_{\beta0}\right) = \mathbf{R}_{\beta0},$$
$$\mathbf{R}^*_{R\beta}\left(\mathbf{t}^\beta; \mathbf{t}^\beta_0; \mathbf{R}^*_{R\beta0}\right) = \mathbf{R}_{d\beta}\left(\mathbf{t}^\beta\right),\ \forall \mathbf{t}^\beta \in \mathfrak{T}^\beta_{RF}. \tag{35.27}$$

The initial $\mathbf{R}^{*\alpha-1}_{R\beta0} = \mathbf{R}^{*\alpha-1}_{R\beta}(\mathbf{t}^{\alpha\beta}_0)$ *is also well defined due to (35.27):* $\mathbf{R}^{*\alpha-1}_{R\beta0} = \mathbf{R}^{\alpha-1}_{\beta0}$*.However, (35.27) does not define* $\mathbf{R}^{*\alpha-1}_{R\beta}(\ \mathbf{t}^{\alpha\beta})$ *on* $In\mathfrak{T}^{\alpha\beta}_{0R}$*. The definition of* $\mathbf{R}^{*\alpha-1}_{R\beta}(\ \mathbf{t}^{\alpha\beta})$ *on* $In\mathfrak{T}^{\alpha\rho}_{0R}$ *is left open for the free appropriate choice.*

The equations (35.23), (35.24) lead to **the subsidiary error vector** ϵ_R,

$$\epsilon_{Rj} = R_{Rj} - R_j,\ \ \forall j = 1, 2, \cdots, \rho,$$
$$\boldsymbol{\epsilon}_R = \mathbf{R}_R - \mathbf{R} = \boldsymbol{\epsilon}_R\left(\mathbf{R}\right),\ \boldsymbol{\epsilon}_R = \left[\epsilon_{R1}\ \ \epsilon_{R2}\ \ \cdots\ \ \epsilon_{R\rho}\right]^T \in \mathfrak{R}^\rho. \tag{35.28}$$

The initial vector $\boldsymbol{\epsilon}_{R0} = \boldsymbol{\epsilon}_R\left(\mathbf{t}^\rho_0\right) = \boldsymbol{\epsilon}_R\left[\mathbf{R}\left(\mathbf{t}^\rho_0\right)\right]$ is equal to the zero vector for every initial real *state* vector $\mathbf{R}_0 = \mathbf{R}\left(\mathbf{t}^\rho_0\right)$ due to (35.27):

$$\mathbf{R}_R = \mathbf{R}_d \Longrightarrow \boldsymbol{\epsilon}_R = \mathbf{R}_R - \mathbf{R} = \mathbf{R}_d - \mathbf{R} = \boldsymbol{\varepsilon},$$
$$\mathbf{R}_R \neq \mathbf{R}_d \Longrightarrow \boldsymbol{\epsilon}_R\left(\mathbf{t}^\rho_0\right) = \boldsymbol{\epsilon}_R\left[\mathbf{R}\left(\mathbf{t}^\rho_0\right)\right] = \mathbf{R}_R\left(\mathbf{t}^\rho_0\right) - \mathbf{R}\left(\mathbf{t}^\rho_0\right)$$
$$= \mathbf{R}\left(\mathbf{t}^\rho_0\right) - \mathbf{R}\left(\mathbf{t}^\rho_0\right) \equiv \mathbf{0}_\rho,\ \forall \mathbf{R}\left(\mathbf{t}^\rho_0\right) \in \mathfrak{R}^\rho, \tag{35.29}$$

and

$$\mathbf{R}_R = \mathbf{R}_d \Longrightarrow \boldsymbol{\epsilon}^*_\beta = \mathbf{R}_{R\beta} - \mathbf{R}_\beta = \mathbf{R}_{d\beta} - \mathbf{R}_\beta = \boldsymbol{\varepsilon}^*_\beta,$$
$$\mathbf{R}_R \neq \mathbf{R}_d \Longrightarrow \boldsymbol{\epsilon}^*_\beta\left(\mathbf{t}^\beta_0\right) = \boldsymbol{\epsilon}^*_\beta\left[\mathbf{R}^*_\beta\left(\mathbf{t}^\beta_0\right)\right] = \mathbf{R}^*_{R\beta}\left(\mathbf{t}^\beta_0\right) - \mathbf{R}^*_\beta\left(\mathbf{t}^\beta_0\right)$$
$$= \mathbf{R}^*_\beta\left(\mathbf{t}^\beta_0\right) - \mathbf{R}^*_\beta\left(\mathbf{t}^\beta_0\right) \equiv \mathbf{0}_\beta,\ \forall \mathbf{R}^*_\beta\left(\mathbf{t}^\beta_0\right) \in \mathfrak{R}^\beta, \tag{35.30}$$

This is the crucial property of the vectors $\boldsymbol{\epsilon}_R$ and $\boldsymbol{\epsilon}^*_\beta$. The β-subvector $\boldsymbol{\epsilon}^*_\beta$ of $\boldsymbol{\epsilon}_R$ contains, explicitly or implicitly, all entries of $\boldsymbol{\epsilon}_R$:

$$\boldsymbol{\epsilon}^*_\beta = \epsilon_{Rj}\ iff\ \ |\epsilon_{Rj}| \geq \max\left(|\epsilon_{Ri}| : \forall i = \beta, \beta+1,\ \cdots, \rho\right),\ \beta \in \{1, 2,\ \cdots, \rho\},$$
$$\boldsymbol{\epsilon}^*_\beta = \left[\epsilon_{R1}\ \ \epsilon_{R2} \ldots \epsilon_{R,\beta-1}\ \epsilon^*_\beta\ \right] \in \mathfrak{R}^\beta;\ \beta = \rho \Longleftrightarrow \boldsymbol{\epsilon}^*_\beta = \boldsymbol{\epsilon}_R,\ \rho \leq r. \tag{35.31}$$

It has the same properties as ε^*_β, which are exposed in (35.5):

$$\boldsymbol{\epsilon}^*_\beta = 0 \Longleftrightarrow \epsilon_{Rj} = 0,\ \ \forall j = \beta, \beta+1, ..., \rho,\ \text{and}$$
$$\left|\boldsymbol{\epsilon}^*_\beta\right| < \xi \Longleftrightarrow |\epsilon_{Rj}| < \xi, \forall j = \beta, \beta+1, ..., \rho, \forall \xi > 0. \tag{35.32}$$

This is another inherent feature of $\boldsymbol{\epsilon}_\beta^*$ for control. The β-subvector $\boldsymbol{\epsilon}_\beta^*$ induces

$$\ominus_\beta^*\left(\boldsymbol{\epsilon}_\beta^*\right) = \operatorname{diag}\left\{\epsilon_{R1} \;\; \epsilon_{R2} \; ... \; \epsilon_{R,\beta-1} \;\; \epsilon_\beta^*\right\} \in \mathfrak{R}^{\beta\times\beta}. \tag{35.33}$$

Comment 513 *The reference state vector* $\mathbf{R}_R$ *equals the desired state vector* $\mathbf{R}_d$ *if, and only if,* $\mathbf{T}\left(\mathbf{t}_0^m;\boldsymbol{\varepsilon}_\beta^{*\alpha-1},\mathbf{0}_\beta\right) = \mathbf{0}_m$. *The reference state vector* $\mathbf{R}_R$ *replaces the desired state vector* $\mathbf{R}_d$ *if, and only if,* $\mathbf{T}\left(\mathbf{t}_0^m;\boldsymbol{\varepsilon}_\beta^{*\alpha-1},\mathbf{0}_\beta\right) \neq \mathbf{0}_m$. *Therefore, the reference state β-subvector* $\mathbf{R}_{R\beta}^*$ *replaces the desired state β-subvector* $\mathbf{R}_{\beta d}$ *if, and only if,* $\mathbf{T}\left(\mathbf{t}_0^m;\boldsymbol{\varepsilon}_\beta^{*\alpha-1},\mathbf{0}_\beta\right) \neq \mathbf{0}_m$. *Then, and only then, the subsidiary error β-subvector* $\boldsymbol{\epsilon}_\beta^*$ *replaces the β-subvector* $\boldsymbol{\varepsilon}_\beta^*$. *This leads to the structure of the feedback control system shown in Figure 471 when* $\mathbf{R}$, $\mathbf{R}_d$, $\mathbf{R}_R$, $\alpha-1$, ρ, $\boldsymbol{\varepsilon}_R$, $\boldsymbol{\epsilon}_\rho$, *replace, respectively,* $\mathbf{Y}$, $\mathbf{Y}_d$, $\mathbf{Y}_R$, k, N, $\boldsymbol{\varepsilon}$, $\boldsymbol{\epsilon}$.

Solution 514 ***On the determination of the reference state vector function*** $\mathbf{R}_R(.)$
The choice of $\mathbf{R}_R(.)$ *is the following:*

$$\mathbf{R}_R(\mathbf{t}^\rho) = \left\{\begin{array}{c} \mathbf{R}_d(\mathbf{t}^\rho) \; iff \; \mathbf{T}\left(\mathbf{t}_0^m,\boldsymbol{\varepsilon}^{\alpha-1},\mathbf{0}_\rho\right) = \mathbf{0}_m, \\ \mathbf{R}_d(\mathbf{t}^\rho) + \mathbf{f}(\mathbf{t}^\rho) \; iff \; \mathbf{T}\left(\mathbf{t}_0^m,\boldsymbol{\varepsilon}^{\alpha-1},\mathbf{0}_\rho\right) \neq \mathbf{0}_m \end{array}\right\}, \; \forall \mathbf{t}^\rho \in \mathfrak{T}_{0F}^\rho. \tag{35.34}$$

so that

$$\mathbf{R}_{R\beta}^*(\mathbf{t}^\rho) = \left\{\begin{array}{c} \mathbf{R}_{d\beta}^*(\mathbf{t}^\beta) \; iff \; \mathbf{T}\left(\mathbf{t}_0^m,\boldsymbol{\varepsilon}_\beta^{*\alpha-1},\mathbf{0}_\beta\right) = \mathbf{0}_m, \\ \mathbf{R}_{d\beta}^*(\mathbf{t}^\beta) + \mathbf{f}_\beta(\mathbf{t}^\beta) \; iff \; \mathbf{T}\left(\mathbf{t}_0^m,\boldsymbol{\varepsilon}_\beta^{*\alpha-1},\mathbf{0}\right) \neq \mathbf{0}_m \end{array}\right\}, \; \forall \mathbf{t}^\beta \in \mathfrak{T}_{0F}^\beta. \tag{35.35}$$

The equation (35.34) determines the block diagram of the generator of $\mathbf{R}_R^{\alpha-1}(\mathbf{t}^{\alpha\rho})$, *which has the same structure as the generator of* $\mathbf{Y}_R(\mathbf{t}^N)$ *Figure 32.3.*

The subsidiary vectors $\mathbf{f}(\mathbf{t}^\rho)$ *and its β-subvector* $\mathbf{f}_\beta(\mathbf{t}^\beta)$,

$$\mathbf{f}(\mathbf{t}^\rho) = \left[f_1(t) \;\; f_2(t) \; .. \; f_\rho(t)\right]^T \in \mathfrak{R}^\rho, \; \mathbf{f}_\beta(\mathbf{t}^\beta) = \left[f_1(t) \;\; f_2(t) \; .. \; f_\beta(t)\right]^T \in \mathfrak{R}^\beta, \tag{35.36}$$

should obey the following conditions that result from (35.22), (35.34) and (35.35):

$$\begin{array}{c} \mathbf{f}(\mathbf{t}_0^\rho) = -\boldsymbol{\varepsilon}_R(\mathbf{t}_0^\rho), \; \forall \boldsymbol{\varepsilon}_R(\mathbf{t}_0^\rho) \in \mathfrak{R}^\rho; \; \mathbf{f}(\mathbf{t}^\rho) = \mathbf{0}_\rho, \; \forall \mathbf{t}^\rho \in \mathfrak{T}_{RF}^\rho, \\ \mathbf{f}_\beta(\mathbf{t}_0^\beta) = -\boldsymbol{\varepsilon}_\beta^*(\mathbf{t}_0^\beta), \; \forall \boldsymbol{\varepsilon}_\beta^*(\mathbf{t}_0^\beta) \in \mathfrak{R}^\beta; \; \mathbf{f}_\beta(\mathbf{t}^\beta) = \mathbf{0}_\beta, \; \forall \mathbf{t}^\beta \in \mathfrak{T}_{RF}^\beta. \end{array} \tag{35.37}$$

These equations determine the general block diagram of the generator of $\mathbf{f}(.)$, *which is analogous to the **f**-generator shown in Figure 32.4. They link* $\mathbf{R}_{R\beta}^*(\mathbf{t}^\beta)$ *with* $\mathbf{R}_{d\beta}^*(\mathbf{t}^\beta)$ *via* $\mathbf{f}_\beta(\mathbf{t}^\beta)$, *(35.35). Also (35.31), (35.34), and (35.35) furnish*

$$\forall \mathbf{t}^\rho \in \mathfrak{T}_{0F}^\rho \Longrightarrow \boldsymbol{\epsilon}_R(\mathbf{t}^\rho) = \mathbf{R}_R(\mathbf{t}^\rho) - \mathbf{R}(\mathbf{t}^\rho)$$
$$= \left\{\begin{array}{c} \mathbf{R}_d(\mathbf{t}^\rho) \; iff \; \mathbf{T}\left(\mathbf{t}_0^m,\boldsymbol{\varepsilon}_R^{\alpha-1},\mathbf{0}_\rho\right) = \mathbf{0}_m, \\ \mathbf{R}_d(\mathbf{t}^\rho) + \mathbf{f}(\mathbf{t}^\rho) \; iff \; \mathbf{T}\left(\mathbf{t}_0^m,\boldsymbol{\varepsilon}_R^{\alpha-1},\mathbf{0}_\rho\right) \neq \mathbf{0}_m \end{array}\right\} - \mathbf{R}(\mathbf{t}^\rho),$$

and

$$\forall \mathbf{t}^\beta \in \mathfrak{T}_{0F}^\beta \Longrightarrow \boldsymbol{\epsilon}_\beta^*(\mathbf{t}^\beta) = \mathbf{R}_{R\beta}^*(\mathbf{t}^\beta) - \mathbf{R}_\beta^*(\mathbf{t}^\beta)$$
$$= \left\{\begin{array}{c} \mathbf{R}_{d\beta}^*(\mathbf{t}^\beta) \; iff \; \mathbf{T}\left(\mathbf{t}_0^m,\boldsymbol{\varepsilon}_{r\beta}^{*(\alpha-1)\beta},\mathbf{0}_\beta\right) = \mathbf{0}_m, \\ \mathbf{R}_{d\beta}^*(\mathbf{t}^\beta) + \mathbf{f}_\beta(\mathbf{t}^\beta) \; iff \; \mathbf{T}\left(\mathbf{t}_0^m,\boldsymbol{\varepsilon}_{r\beta}^{*(\alpha-1)\beta},\mathbf{0}\right) \neq \mathbf{0}_m \end{array}\right\} - \mathbf{R}_\beta^*(\mathbf{t}^\beta).$$

The beginning and the end of these equations, together with (4.4), yield

$$\boldsymbol{\epsilon}_R(\mathbf{t}^\rho) = \boldsymbol{\varepsilon}_R(\mathbf{t}^\rho) + \left\{\begin{array}{c} \mathbf{0}_\rho \; iff \; \mathbf{T}\left(\mathbf{t}_0^m,\boldsymbol{\varepsilon}_R,\mathbf{0}_\rho\right) = \mathbf{0}_m, \\ \mathbf{f}(\mathbf{t}^\rho) \; iff \; \mathbf{T}\left(\mathbf{t}_0^m,\boldsymbol{\varepsilon}_R,\mathbf{0}_\rho\right) \neq \mathbf{0}_m \end{array}\right\}, \; \forall \mathbf{t}^\rho \in \mathfrak{T}_{0F}^\beta, \tag{35.38}$$

and

$$\epsilon_\beta^*(\mathbf{t}^\beta) = \left\{ \begin{array}{c} \varepsilon_\beta^*(\mathbf{t}^\beta) \text{ iff } \mathbf{T}\left(\mathbf{t}_0^m, \boldsymbol{\varepsilon}_{r\beta}^{*(\alpha-1)\beta}, \mathbf{0}_\beta\right) = \mathbf{0}_m, \\ \varepsilon_\beta^*(\mathbf{t}^\beta) + \mathbf{f}_\beta(\mathbf{t}^\beta) \text{ iff } \mathbf{T}\left(\mathbf{t}_0^m, \boldsymbol{\varepsilon}_{r\beta}^{*(\alpha-1)\beta}, \mathbf{0}_\beta\right) \neq \mathbf{0}_m \end{array} \right\}$$
$$\forall \mathbf{t}^\beta \in \mathfrak{T}_{0F}^\beta. \tag{35.39}$$

At the initial moment $\mathbf{f}(\mathbf{t}_0^\rho) = -\varepsilon_R(\mathbf{t}_0^\rho)$, *(35.37), which reduces (35.38) to*

$$\epsilon_R(\mathbf{t}_0^\rho) = \epsilon_R\left[\varepsilon_R(\mathbf{t}_0^\rho)\right] = \mathbf{0}_\rho, \ \forall \varepsilon_R(\mathbf{t}_0^\rho) \in \mathfrak{R}^\rho, \tag{35.40}$$

and (35.39) to:

$$\epsilon_\beta^*(\mathbf{t}_0^\beta) = \epsilon_\beta^*\left[\varepsilon_\beta^*(\mathbf{t}_0^\beta)\right] = \mathbf{0}_\beta, \ \forall \varepsilon_\beta^*(\mathbf{t}_0^\beta) \in \mathfrak{R}^\beta. \tag{35.41}$$

These equations agree with (35.29) and (35.30), respectively. In addition if the control guarantees $\epsilon_R(\mathbf{t}^\rho) = \mathbf{0}_\rho$, $\forall \mathbf{t}^\rho \in \mathfrak{T}_{0F}^\beta$, *then*

$$\varepsilon_R(\mathbf{t}^\rho) = -\mathbf{f}(\mathbf{t}^\rho), \ \forall \mathbf{t}^\rho \in \mathfrak{T}_{0F}^\beta. \tag{35.42}$$

due to (35.38). This and (35.37) lead to the following conclusion:

$$\epsilon_R(\mathbf{t}^\rho) = \mathbf{0}_\rho, \ \forall \mathbf{t}^\rho \in \mathfrak{T}_{RF}^\beta \Longrightarrow \varepsilon_R(\mathbf{t}^\rho) = \mathbf{0}_\rho, \ \forall \mathbf{t}^\rho \in \mathfrak{T}_{RF}^\beta. \tag{35.43}$$

Comment 497 shows that the control vector acts only on the *state* β-subvector $\mathbf{R}_\beta$, i.e., on the error β-subvector ε_β^*. Because the whole real *state* vector $\mathbf{R}(\mathbf{t}^\rho)$ should reach the whole desired *state* vector $\mathbf{R}_d(\mathbf{t}^\rho)$ at the latest at the final reachability time, then we should use only the β-subequations of (35.29)-(35.43) in the sequel.

The subsidiary reference *state* vector $\mathbf{R}_R^{\alpha\text{-}1}(\mathbf{t}^{\alpha\rho})$ replaces the desired *state* vector $\mathbf{R}_d^{\alpha\text{-}1}(\mathbf{t}^{\alpha\rho})$ in the quality tracking controller, the block diagram of which is analogous to that shown in Figure 32.5. The structural diagram of the *state* quality tracking control system of the plant is analogous to that of Figure 32.6.

A possible specific form of the subsidiary function $\mathbf{f}_\beta^*(.)$ was proposed in [274], Figure 32.7. We determine completely $\mathbf{f}_\beta^*(.)$ by (32.44) through (32.51) when we do in them the following replacements:

$$\mathbf{Y} \longrightarrow \mathbf{R}, \ \varepsilon_{i0} \longrightarrow \varepsilon_{Ri0}, \ \varepsilon_{i0}^{(1)} \longrightarrow \varepsilon_{Ri0}^{(1)}, \ \gamma \longrightarrow \beta. \tag{35.44}$$

Then the equations (32.47)-(32.51) determine the combined i-th entry $f_i(.)$ of $\mathbf{f}_\beta^*(.)$, Figure 32.8 with the replacements (35.44). It is the combination of the sinusoidal function and the parabolic function. Figure 32.9 with the replacements (35.44) represents the block of the combined subsidiary vector function $\mathbf{f}_\beta^*(.)$.

The formulae (32.47)-(32.51) with the replacements (35.44) show the dependence of the finite vector reachability *time* $\mathbf{t}^\beta$ on the initial conditions: bounded initial conditions, bounded the finite vector reachability *time* $\mathbf{t}^\beta$. This is reasonable and rational.

As the summary of the above consideration we state:

Goal 515 ***The general modified control goal***

The control should force the plant to behave so that the following tracking algorithm holds [instead of (35.12)]:

$$\mathbf{T}\left(\mathbf{t}^m, \epsilon_\beta^{*\alpha-1}(\mathbf{t}^{\alpha\beta}), \int_{t_0^\beta}^{\mathbf{t}^\beta} \ominus_\beta^*(\mathbf{t}^\beta)\, d\mathbf{t}^\beta\right) = \mathbf{0}_m, \ \forall \mathbf{t}^{\alpha\beta} \in \mathfrak{T}_{0F}^{\alpha\beta}. \tag{35.45}$$

Property 516 ***The properties of the tracking algorithm*** *(35.13)*

The algorithm (35.13) expressed in terms of $\epsilon_\beta^*(\mathbf{t}^\beta)$ *and its derivative as in (35.45) possesses, in addition to Properties 501 through 503, also Property 505 due to (35.30).*

35.4 Tracking algorithm and initial conditions

The following theorem enables us to establish the unified tracking control for both cases $\varepsilon_R = \mathbf{0}_{\alpha\beta}$ and $\varepsilon_R \neq \mathbf{0}_{\alpha\beta}$.

Theorem 517 ***The main theorem on the tracking algorithm and initial conditions*** *Let (35.23) through (35.43) be valid. In order for the tracking algorithm determined by (35.21), (35.22) in terms of the real error vector ε^*_β, its derivatives, and integral to hold, it is necessary and sufficient that the tracking algorithm $\mathbf{T}(.)$ (35.45) holds.*

Proof. Notice that the tracking algorithms (35.12), (35.21), and (*35.45*) possess Properties 501 through 503. The tracking algorithm (*35.45*) possesses also Property 505 as explained in Property *516*

Let (35.38) through (35.43) be valid.

Necessity. Let (35.21), (35.22) hold. We separate the case $\mathbf{T}\left(\mathbf{t}^m, \boldsymbol{\varepsilon}_\beta^{*\alpha-1}, \mathbf{0}_\beta\right) = \mathbf{0}_m$ from the case $\mathbf{T}\left(\mathbf{t}^m, \boldsymbol{\varepsilon}_\beta^{*\alpha-1}, \mathbf{0}_\beta\right) \neq \mathbf{0}_m$.

If $\boldsymbol{\varepsilon}_\beta^{*\alpha-1}$ is such that $\mathbf{T}\left(\mathbf{t}^m, \boldsymbol{\varepsilon}_\beta^{*\alpha-1}, \mathbf{0}_\beta\right) = \mathbf{0}_m$, then $\epsilon_\beta^{*\alpha-1}(\mathbf{t}^{\alpha\beta}) = \varepsilon_\beta^{*\alpha-1}(\mathbf{t}^{\alpha\beta})$ for every $\mathbf{t}^{\alpha\beta} \in \mathfrak{T}_0^{\alpha\beta}$ due to (35.38) so that

$$\mathbf{T}\left(\mathbf{t}^m, \boldsymbol{\varepsilon}_\beta^{*\alpha-1}(\mathbf{t}_R^{\alpha\beta}), \int_{\mathbf{t}_0^\beta}^{\mathbf{t}^\beta} E_\beta^*\left(\mathbf{t}^\beta\right)\left(\mathbf{t}^\beta\right) d\mathbf{t}^\beta\right)$$

$$= \mathbf{T}\left(\mathbf{t}^m, \boldsymbol{\epsilon}_\beta^{*\alpha-1}(\mathbf{t}_R^{\alpha\beta}), \int_{\mathbf{t}_0^\beta}^{\mathbf{t}^\beta} \ominus_\beta^*\left(\mathbf{t}^\beta\right) d\mathbf{t}^\beta\right) = \mathbf{0}_m,\ \forall \mathbf{t}_R^{\alpha\beta} \in \mathfrak{T}_{0F}^{\alpha\beta}.$$

This proves necessity of (*35.45*) if $\mathbf{T}\left(\mathbf{t}^m, \boldsymbol{\varepsilon}_\beta^{*\alpha-1}, \mathbf{0}_\beta\right) = \mathbf{0}_m$.

If $\varepsilon_\beta^{*\alpha-1}$ violates (35.17), i.e., $\mathbf{T}\left(\mathbf{t}^m, \boldsymbol{\varepsilon}_\beta^{*\alpha-1}, \mathbf{0}_\beta\right) \neq \mathbf{0}_m$ holds, then the conditions (35.22) and (35.38) imply $\epsilon_\beta^{*\alpha-1}(\mathbf{t}^{\alpha\beta}) = \mathbf{0}_{\alpha\beta},\ \forall \mathbf{t}^{\alpha\beta} \in \mathfrak{T}_{0F}^{\alpha\beta}$. This proves that $\epsilon_\beta^{*\alpha-1}(\mathbf{t}^{\alpha\beta})$ satisfies (*35.45*), i.e., proves that (*35.45*) holds.

Sufficiency. Let (*35.45*) be valid.

If $\boldsymbol{\varepsilon}_{\beta 0}^{*\alpha-1}$ is such that $\mathbf{T}\left(\mathbf{t}_0^m, \boldsymbol{\varepsilon}_{\beta 0}^{*\alpha-1}, \mathbf{0}_\beta\right) = \mathbf{0}_m$, then $\epsilon_\beta^{*\alpha-1}(\mathbf{t}^{\alpha\beta}) = \varepsilon_\beta^{*\alpha-1}(\mathbf{t}^{\alpha\beta})$ for every $\mathbf{t}^{\alpha\beta} \in \mathfrak{T}_{0F}^{\alpha\beta}$ due to (35.39) so that, in view of (*35.45*),

$$\mathbf{T}\left(\mathbf{t}^m, \boldsymbol{\epsilon}_R(\mathbf{t}_R^{\alpha\beta}), \int_{\mathbf{t}_0^\beta}^{\mathbf{t}^\beta} \ominus_\beta^*\left(\mathbf{t}^\beta\right) d\mathbf{t}^\beta\right)$$

$$= \mathbf{T}\left(\mathbf{t}^m, \boldsymbol{\varepsilon}_\beta^{*\alpha-1}(\mathbf{t}_R^{\alpha\beta}), \int_{\mathbf{t}_0^\beta}^{\mathbf{t}^\beta} E_\beta^*\left(\mathbf{t}^\beta\right)\left(\mathbf{t}^\beta\right) d\mathbf{t}^\beta\right) = \mathbf{0}_m,\ \forall \mathbf{t}_R^{\alpha\beta} \in \mathfrak{T}_{0F}^{\alpha\beta}. \tag{35.46}$$

This proves validity of (35.21) for the case $\mathbf{T}\left(\mathbf{t}^m, \boldsymbol{\varepsilon}_\beta^{*\alpha-1}, \mathbf{0}_\beta\right) = \mathbf{0}_m$.

If $\boldsymbol{\varepsilon}_{\beta 0}^{*\alpha-1}$ such that $\mathbf{T}\left(\mathbf{t}_0^m, \boldsymbol{\varepsilon}_{\beta 0}^{*\alpha-1}, \mathbf{0}_\beta\right) \neq \mathbf{0}_m$ holds, then (35.41), (*35.45*), and Property 503 imply $\boldsymbol{\epsilon}_\beta^{*\alpha-1}(\mathbf{t}^{\alpha\beta};\boldsymbol{\epsilon}_{\beta 0}^{*\alpha-1}) = \mathbf{0}_{\alpha\beta}$ for every $\mathbf{t}^{\alpha\beta} \in \mathfrak{T}_{0F}^{\alpha\beta}$; hence,

$$\boldsymbol{\varepsilon}_\beta^{*\alpha-1}(\mathbf{t}^{\alpha\beta}; \boldsymbol{\varepsilon}_\beta^{*\alpha-1}) + \mathbf{f}_\beta^{\alpha-1}(\mathbf{t}^{\alpha\beta}; \mathbf{f}_{\beta 0}^{\alpha-1}) = \mathbf{0}_{\alpha\beta}$$

for every $\mathbf{t}^{\alpha\beta} \in \mathfrak{T}_{0F}^{\alpha\beta}$ due to (35.39). This proves (35.22) and, together with (35.37), proves $\varepsilon_\beta^{*\alpha-1}(\mathbf{t}^{\alpha\beta}) = \mathbf{0}_{\alpha\beta}$ for every $\mathbf{t}^{\alpha\beta} \in \mathfrak{T}_{RF}^{\alpha\beta}$. Therefore, (35.21) also holds ∎

We present several characteristic simple forms of the tracking algorithm $T(.)$ in the *state* space. They satisfy (35.14) through (35.17); i.e., they obey Properties 501-505. They also satisfy Solution 509 and Solution 512.

Chapter 36

Tracking algorithms: State space

36.1 Matrix notation meaning

Let $t_0 = 0$.

The definition of the integer k is in (1.8). The diagonal elements of the constant diagonal $\beta \times \beta$ matrix $K_{0\beta}$ are positive real numbers:

$$K_{0\beta} = \operatorname{diag}\{k_{0\beta,1}\ k_{0\beta,2}\ \cdots\ k_{0\beta,\beta}\} \in \mathfrak{R}_{+}^{\beta\times\beta},\ k_{0\beta,i} \in \mathfrak{R}^{+},\ \forall i = 1, 2,\ \cdots, \beta.$$

The $\beta \times \beta$ diagonal matrix T_β is the independent *time* matrix the elements of which are the same: *time* t:

$$T_\beta = \operatorname{diag}\{t\ t\ \cdots\ t\} \in \mathfrak{T}^{\beta\times\beta}.$$

The matrix exponential function $e^{-K_{0\beta}T_{1\beta}^{-1}T_\beta}$ is a diagonal matrix function defined by

$$e^{-K_{0\beta}T_{1\beta}^{-1}T_\beta} = \exp\left[-K_{0\beta}T_1^{-1}T_\beta\right] = \operatorname{diag}\left\{e^{-k_{01}t_1^{-1}t}\ e^{-k_{02}t_2^{-1}t}\ \cdots\ e^{-k_{0\beta}t_\beta^{-1}t}\right\}.$$

36.2 Examples of tracking algorithms

Example 518 ***The first-order linear elementwise exponential tracking algorithm***
In this case $\mathbf{t}_R^\beta = \mathbf{t}_F^\beta = \infty\mathbf{1}_\beta$, *i.e.,* $\mathfrak{T}_{RF}^\beta = \mathfrak{T}_{F\infty}^\beta = \{\infty\mathbf{1}_\beta\}$. *The following tracking algorithm*

$$\mathbf{T}\left(\boldsymbol{\epsilon}_\beta^{*1}\right) = T_{1\beta}\boldsymbol{\epsilon}_{\boldsymbol{\beta}}^{*(1)}(\mathbf{t}^\beta) + K_{0\beta}\boldsymbol{\epsilon}_{\boldsymbol{\beta}}^{*}(\mathbf{t}^\beta) = \mathbf{0}_\beta, \forall \mathbf{t}^\beta \in \mathfrak{T}_{0F}^\beta,$$
$$T_{1\beta} = diag\{t_{\beta,1}\ t_{\beta,2}\ \cdots\ t_{\beta,\beta}\},\ t_{\beta,i} > 0, \forall i = 1, 2, ..., \beta, \tag{36.1}$$

determines the global exponential tracking of $\mathbf{Y}_d\left(\mathbf{t}^N\right)$ *on* $\mathfrak{T}_{0F}^N$ *or the perfect tracking of* $\mathbf{Y}_R\left(\mathbf{t}^N\right)$ *on* $\mathfrak{T}_{0F}^N$, *which is illustrated by its solution*

$$\mathbf{T}\left(\varepsilon_{\beta 0}^{*1}\right) = \mathbf{0}_\beta \Longrightarrow$$
$$\boldsymbol{\epsilon}_{\boldsymbol{\beta}}^*(\mathbf{t}^\beta; \boldsymbol{\epsilon}_{\beta 0}^*) = \boldsymbol{\varepsilon}_\beta^*(\mathbf{t}^\beta; \boldsymbol{\varepsilon}_{\beta 0}^*) = \epsilon_\beta^{*-K_{0\beta}T_{1\beta}T_\beta}\boldsymbol{\varepsilon}_{\beta 0}^*, \forall\left(\mathbf{t}^\beta, \boldsymbol{\varepsilon}_{\beta 0}^*\right) \in \mathfrak{T}_{0F}^\beta \times \mathfrak{R}^\beta,$$
$$\mathbf{T}\left(\varepsilon_{\beta 0}^{*1}\right) \neq \mathbf{0}_\beta \Longrightarrow \boldsymbol{\epsilon}_{\boldsymbol{\beta}}^*(\mathbf{t}^\beta; \boldsymbol{\epsilon}_{\beta 0}^*) = \boldsymbol{\varepsilon}_\beta^*(\mathbf{t}^\beta; \boldsymbol{\varepsilon}_{\beta 0}^*) = \mathbf{0}_\beta,\ \forall \mathbf{t}^\beta \in \mathfrak{T}_{RF}^\beta.$$

The reachability time is infinite iff $\mathbf{T}\left(\varepsilon_{\beta 0}^{*1}\right) = \mathbf{0}_\beta$. *The convergence to the zero error vector is then elementwise and exponentially asymptotic. Such tracking of* $\mathbf{Y}_d\left(\mathbf{t}^N\right)$ *is also stable. However, if* $\mathbf{T}\left(\varepsilon_{\beta 0}^{*1}\right) \neq \mathbf{0}_\beta$ *then the tracking of* $\mathbf{Y}_d\left(\mathbf{t}^N\right)$ *is perfect on* $\mathfrak{T}_{RF}^N$.

The tracking operator $\mathbf{T}\left(\varepsilon_{\beta 0}^{*1}\right)$ *(36.1) possesses Properties 501-505 and* $\mathbf{T}(.) \in \mathfrak{C}^-$. *The tracking is exponentially stable in the whole if* $\mathbf{T}\left(\varepsilon_{\beta 0}^{*1}\right) = \mathbf{0}_\beta$.

Let

$$E_\beta^{(\eta)} = [E_{\beta,0} \quad E_{\beta,1} \ \ldots \ E_{\beta,\eta}] \in \mathfrak{R}^{\beta\times(\eta+1)\beta},\ \eta \leq \alpha - 1,$$

be given, or to be determined, constant extended (block) matrix, the entries of which are constant submatrices $E_{\beta,k} \in \mathfrak{R}^{\beta\times\ \beta}$, $k = 0, 1, \cdots, \eta \leq \alpha - 1$.

Example 519 ***The higher-order linear elementwise exponential tracking algorithm***

In this case the tracking should be a kind of asymptotic tracking so that the vector reachability time $\mathbf{t}_R^\beta$ *is infinite,* $\mathbf{t}_R^\beta = \infty\mathbf{1}_\beta$. *This implies* $\mathfrak{T}_{R\infty} = \{\infty\mathbf{1}_\beta\}$. *We define the higher-order linear elementwise exponential tracking algorithm by*

$$\begin{aligned}\mathbf{T}\left(\boldsymbol{\epsilon}_{\boldsymbol{\beta}}^*, \boldsymbol{\epsilon}_{\boldsymbol{\beta}}^{*\,(1)}, ..., \boldsymbol{\epsilon}_{\boldsymbol{\beta}}^{*\,(\eta)}\right) &= \sum_{k=0}^{k=\eta\leq\alpha-1} E_{\beta k}\boldsymbol{\epsilon}_{\boldsymbol{\beta}}^{*\,(k)}(\mathbf{t}^\beta) \\ &= E_\beta^{(\eta)}\boldsymbol{\epsilon}_{\boldsymbol{\beta}}^{*\,\eta}(\mathbf{t}^{(\eta+1)\beta}) = \mathbf{0}_\beta,\ \forall \mathbf{t}^{(\eta+1)\beta} \in \mathfrak{T}_{0F}^{(\eta+1)\beta}\ ,\end{aligned} \tag{36.2}$$

with the matrices $E_{\beta k}$ *such that the real parts of the roots of the characteristic polynomial* $f(s)$ *of* $\mathbf{T}(s)$,

$$f(s) = \det\left(\sum_{k=0}^{k=\eta\leq\alpha-1} E_{\beta k}s^k\right),$$

are negative.

The linear differential equation (36.2) has constant matrix coefficients. It has the unique solution for every initial condition $\epsilon_\beta^{*\eta-1}(\mathbf{t}_0^{\eta\beta}) \in \mathfrak{R}^{\eta\beta}$.

The reachability time is infinite. The error vector $\epsilon_\beta^{*\eta-1}(\mathbf{t}^{\eta\beta}) = \varepsilon_\beta^{*\eta-1}(\mathbf{t}^{\eta\beta})$ *converges to the zero error vector elementwise, exponentially, and asymptotically iff* $\mathbf{T}\left(\varepsilon_{\beta 0}^{*1}\right) = \mathbf{0}_\beta$. *Such tracking of* $\mathbf{Y}_d\left(\mathbf{t}^N\right)$ *is stable. However, iff* $\mathbf{T}\left(\varepsilon_{\beta 0}^{*1}\right) \neq \mathbf{0}_\beta$ *then the tracking of* $\mathbf{Y}_d\left(\mathbf{t}^N\right)$ *is perfect on* $\mathfrak{T}_{RF}^N$.

The tracking operator (36.2) possesses Properties 501-505 and $\mathbf{T}(.) \in \mathfrak{C}^-$. *The tracking is exponentially stable in the whole if* $\mathbf{T}\left(\varepsilon_{\beta 0}^{*1}\right) = \mathbf{0}_\beta$.

Example 520 ***The sharp elementwise stable tracking with the finite vector state reachability time (FVIDRT)*** $\mathbf{t}_R^\beta$

We accept the finite vector state reachability time t_R^β, $t_R^\beta = [t_{R1}\ t_{R2}\ ..\ t_{R\beta}]^T$, *so that*

$$\mathfrak{T}_{0R}^\beta = [\mathbf{0}_\beta, \mathbf{t}_R^\beta] \subset \mathfrak{T}_{0F}^\beta,\ \mathfrak{T}_{RF}^\beta = [\mathbf{t}_R^\beta,\ \mathbf{t}_F^\beta],\ \mathbf{t}_F^\beta < \infty\mathbf{1}_\beta. \tag{36.3}$$

The algorithm for the elementwise stable tracking with the finite vector state reachability time t_R^β,

$$\mathbf{T}\left(\varepsilon_{\beta 0}^{*1}\right) = \mathbf{0}_\beta \Longrightarrow \mathbf{t}_R^\beta = T_{1\beta}K_{0\beta}\left|\varepsilon_{\beta 0}^*\right|,$$

$$\left|\epsilon_{\beta 0}^*\right| = \left[\left|\epsilon_{\beta 10}\right| \quad \left|\epsilon_{\beta 20}\right| \ \ldots \left|\epsilon_{\beta,\beta-1,0}\right| \quad \left|\epsilon_{\beta 0}^*\right|\right]^T \in \mathfrak{R}^\beta,$$

reads

$$\mathbf{T}\left(\boldsymbol{\epsilon}_{\beta 0}^{*}, \boldsymbol{\epsilon}_{\beta 0}^{*(1)}\right) = T_{1\beta}\boldsymbol{\epsilon}_{\boldsymbol{\beta}}^{*(1)}(\mathbf{0}_{\beta}) + K_{0\beta}sign\boldsymbol{\epsilon}_{\boldsymbol{\beta}}^{*}(\mathbf{0}_{\beta}) = \mathbf{0}_{\beta}. \tag{36.4}$$

Let

$$S\left(\varepsilon_{\beta 0}^{*}\right) = diag\left\{sign\varepsilon_{10} \quad sign\varepsilon_{20} \quad \cdots \quad sign\varepsilon_{\beta-1,0} \quad sign\varepsilon_{\beta 0}^{*}\right\},$$

The solution $\epsilon_{\beta}^{*}(\mathbf{0}_{\beta};\epsilon_{\beta 0}^{*})$,

$$\begin{gathered}\mathbf{T}\left(\varepsilon_{\beta 0}^{*1}\right) = \mathbf{0}_{\beta} \Longrightarrow \boldsymbol{\epsilon}_{\boldsymbol{\beta}}^{*}(\mathbf{t}^{\beta};\boldsymbol{\epsilon}_{\beta 0}^{*}) = \boldsymbol{\varepsilon}_{\beta}^{*}(\mathbf{t}^{\beta};\boldsymbol{\varepsilon}_{\beta 0}^{*}) \\ = \boldsymbol{\varepsilon}_{\beta 0}^{*} - T_{1\beta}K_{0\beta}S\left(\boldsymbol{\varepsilon}_{\beta 0}^{*}\right)\mathbf{t}^{\beta},\ \forall \mathbf{t}^{\beta} \in \mathfrak{T}_{0R}^{\beta}, \\ \boldsymbol{\epsilon}_{\boldsymbol{\beta}}^{*}(\mathbf{t}^{\beta};\boldsymbol{\epsilon}_{\beta 0}^{*}) = \boldsymbol{\varepsilon}_{\beta}^{*}(\mathbf{t}^{\beta};\boldsymbol{\varepsilon}_{\beta 0}^{*}) = \mathbf{0}_{\beta}, \forall \mathbf{t}^{\beta} \in \mathfrak{T}_{RF}^{\beta},\end{gathered}$$

$$\mathbf{T}\left(\boldsymbol{\varepsilon}_{\beta 0}^{*1}\right) \neq \mathbf{0}_{\beta} \Longrightarrow \boldsymbol{\epsilon}_{\boldsymbol{\beta}}^{*}(\mathbf{t}^{\beta};\boldsymbol{\epsilon}_{\beta 0}^{*}) = \mathbf{0}_{\beta}, \forall \mathbf{t}^{\beta} \in \mathfrak{T}_{RF}^{\beta},$$

to (36.4) determines the output error behavior $\boldsymbol{\varepsilon}_{\beta}^{*}(\mathbf{t}^{\beta};\boldsymbol{\varepsilon}_{\beta 0}^{*})$ *that approaches sharply the zero error vector* $\boldsymbol{\varepsilon}_{\beta}^{*} = \mathbf{0}_{\beta}$ *in the linear form (along a straight line) with the nonzero constant velocity* $T_{1\beta}K_{0\beta}S\left(\boldsymbol{\varepsilon}_{\beta 0}^{*}\right)\mathbf{1}_{\beta}$ *iff* $\mathbf{T}\left(\boldsymbol{\varepsilon}_{\beta 0}^{*1}\right) = \mathbf{0}_{\beta}$. *Otherwise,* $\mathbf{T}\left(\boldsymbol{\varepsilon}_{\beta 0}^{*1}\right) \neq \mathbf{0}_{\beta}$ *and the tracking of* $\mathbf{Y}_{d}\left(\mathbf{t}^{N}\right)$ *is perfect on* $\mathfrak{T}_{RF}^{N}$. *The tracking of* $\mathbf{Y}_{d}\left(\mathbf{t}^{N}\right)$ *becomes perfect at the finite vector state reachability time* $\mathbf{t}_{R}^{\beta}$ *and rests perfect on* $\mathfrak{T}_{RF}^{N}$.

Iff $\beta = \rho$ *and* $\mathbf{T}\left(\boldsymbol{\varepsilon}_{R0}^{1}\right) = \mathbf{0}_{\rho}$ *then*

$$\begin{gathered}\mathbf{T}\left(\boldsymbol{\varepsilon}_{R0}^{1}\right) = \mathbf{0}_{\rho} \Longrightarrow \left|\boldsymbol{\varepsilon}_{R}(\mathbf{t}^{\rho};\boldsymbol{\varepsilon}_{R0}^{1})\right| \leq \left|\boldsymbol{\varepsilon}_{R0}\right|,\ \forall \mathbf{t}^{\rho} \in \mathfrak{T}_{0F}^{\rho} \Longrightarrow \\ \left\|\boldsymbol{\varepsilon}_{R}(\mathbf{t}^{\rho};\boldsymbol{\varepsilon}_{R0}^{1})\right\| \leq \left\|\boldsymbol{\varepsilon}_{R0}\right\|,\ \forall \mathbf{t}^{\rho} \in \mathfrak{T}_{0F}^{\rho} \Longrightarrow \\ \forall \varepsilon \in \mathfrak{R}^{+},\ \exists \delta \in \mathfrak{R}^{+},\ \delta = \delta\left(\varepsilon\right) = \varepsilon \Longrightarrow \\ \left\|\boldsymbol{\varepsilon}_{R0}^{1}\right\| < \delta \Longrightarrow \left\|\boldsymbol{\varepsilon}_{R}(\mathbf{t}^{\rho};\boldsymbol{\varepsilon}_{R0}^{1})\right\| \leq \varepsilon,\ \forall \mathbf{t}^{\rho} \in \mathfrak{T}_{0F}^{\rho}.\end{gathered}$$

The tracking of $\mathbf{Y}_{d}\left(\mathbf{t}^{N}\right)$ *is stable state on* $\mathfrak{T}_{0F}^{N}$ *if* $\mathbf{T}\left(\boldsymbol{\varepsilon}_{R0}^{1}\right) = \mathbf{0}_{\rho}$.

Iff $\mathbf{T}\left(\boldsymbol{\varepsilon}_{\beta 0}^{*1}\right) = \mathbf{0}_{\beta}$ *then the bigger* $K_{0\beta}$, *the smaller* t_{R}^{β} *is for fixed* $T_{1\beta}$ *and* $\boldsymbol{\varepsilon}_{\beta 0}^{*1}$, *and vice versa. The smaller* $T_{1\beta}$, *the smaller* t_{R}^{β} *is for fixed* $K_{0\beta}$ *and* $\boldsymbol{\varepsilon}_{\beta 0}^{*1}$, *and vice versa. The bigger* $\left|\boldsymbol{\varepsilon}_{\beta 0}^{*1}\right|$, *the bigger* t_{R}^{β} *is for fixed* $T_{1\beta}$ *and* $K_{0\beta}$, *and vice versa. These relationships hold elementwise.*

The tracking operator (33.4) possesses Properties 501-505 and $\mathbf{T}(.) \in \mathfrak{C}^{-}$. *The tracking is stable in the whole with FVIDRT* t_{R}^{β} *if* $\mathbf{T}\left(\boldsymbol{\varepsilon}_{\beta 0}^{*1}\right) = \mathbf{0}_{\beta}$.

Let

$$\sigma\left(\varepsilon_{Ri}^{(k)}, \varepsilon_{Ri}^{(k+1)}\right) = \left\{\begin{array}{c} -1, \left\{\begin{array}{c} if\ \varepsilon_{Ri}^{(k)} < 0, \forall \varepsilon_{Ri}^{(k+1)} \in \mathfrak{R}; \\ or\ if\ \varepsilon_{Ri}^{(k)} = 0\ and\ \varepsilon_{Ri}^{(k+1)} < 0, \end{array}\right\} \\ 0,\ iff\ \varepsilon_{Ri}^{(k)} = 0\ and\ \varepsilon_{Ri}^{(k+1)} = 0, \\ 1, \left\{\begin{array}{c} if\ \varepsilon_{Ri}^{(k)} > 0, \forall \varepsilon_{Ri}^{(k+1)} \in \mathfrak{R}; \\ or\ if\ \varepsilon_{Ri}^{(k)} = 0\ and\ \varepsilon_{Ri}^{(k+1)} > 0 \end{array}\right\}, \\ \forall i = 1, 2, ..., \beta - 1,\ i = \beta \Longrightarrow \varepsilon_{\beta 0} = \varepsilon_{\beta 0}^{*} \end{array}\right\}, \tag{36.5}$$

$$\Sigma\left(\boldsymbol{\varepsilon}_{R\beta}^{*(k)}, \boldsymbol{\varepsilon}_{R\beta}^{*(k+1)}\right) = \mathrm{diag}\left\{\sigma\left(\varepsilon_{R1}^{(k)}, \varepsilon_{R1}^{(k+1)}\right)\ \sigma\left(\varepsilon_{R2}^{(k)}, \varepsilon_{R2}^{(k+1)}\right) ... \sigma\left(\varepsilon_{\beta}^{*(k)}, \varepsilon_{\beta}^{*(k+1)}\right)\right\}$$
$$\forall k = 1, 2, ..., \tag{36.6}$$

$$\Sigma\left(\boldsymbol{\varepsilon}_{R\beta}^{*1}\right) = \mathrm{diag}\left\{\sigma\left(\varepsilon_{R1}, \varepsilon_{R1}^{(1)}\right)\ \sigma\left(\varepsilon_{R2}, \varepsilon_{R2}^{(1)}\right)\ ...\ \sigma\left(\varepsilon_{R\beta}^{*}, \varepsilon_{R\beta}^{*(1)}\right)\right\}, \tag{36.7}$$

The equations (36.5)-(36.7) enable us to determine Dini derivative $D^+\left|\varepsilon_{R\beta}^*(t)\right|$ of $\left|\varepsilon_{R\beta}^*(t)\right|$ (for details see Appendix B) :

$$D^+\left|\varepsilon_{R\beta}^*(t)\right| = \Sigma\left(\varepsilon_{R\beta}^*, \varepsilon_{R\beta}^{*(1)}\right)\varepsilon_{R\beta}^{*(1)}(t). \tag{36.8}$$

Example 521 ***The first-power smooth elementwise stable tracking with the finite vector state reachability time*** $\mathbf{t}_R^\beta$

If the control acting on the plant ensures

$$\mathbf{T}\left(\epsilon_\beta^*, \epsilon_\beta^{*(1)}\right) = T_{1\beta}\epsilon_\beta^{*(1)}(\mathbf{0}_\beta) + 2K_{0\beta}\left|\ominus_\beta^*(\mathbf{0}_\beta)\right|^{1/2} sign\epsilon_{\beta 0}^* = \mathbf{0}_\beta,\ \forall \mathbf{0}_\beta \in \mathfrak{T}_{0F}^\beta, \tag{36.9}$$

where

$$\left|\ominus_\beta^*\left[\epsilon_\beta^*(\mathbf{t}^\beta)\right]\right|^{1/2} = diag\left\{|\epsilon_1(t)|^{1/2}\ \ |\epsilon_2(t)|^{1/2}\ \ \ldots\ \ |\varepsilon_\beta^*(t)|^{1/2}\right\},$$

then the plant exhibits the elementwise stable tracking with the finite vector state reachability time t_R^β,

$$\mathbf{T}\left(\varepsilon_{\beta 0}^{*1}\right) = \mathbf{0}_\beta \Longrightarrow \mathbf{t}_R^\beta = T_{1\beta}K_{0\beta}\left|\ominus_{\beta 0}^*\right|^{1/2}\mathbf{1}_\beta = T_{1\beta}K_{0\beta}\left|\epsilon_{\beta 0}^*\right|^{1/2},$$
$$\left|\epsilon_{\beta 0}^*\right|^{1/2} = \left[|\epsilon_1(0)|^{1/2}\ \ |\epsilon_2(0)|^{1/2}\ \ \ldots\ \ |\epsilon_\beta^*(0)|^{1/2}\right]^T,$$

which is determined by the output error behavior,

$$\mathbf{T}\left(\varepsilon_{\beta 0}^{*1}\right) = \mathbf{0}_\beta \Longrightarrow \epsilon_\beta^*(\mathbf{t}^\beta;\epsilon_{\beta 0}^*) = \varepsilon_\beta^*(\mathbf{t}^\beta;\varepsilon_{\beta 0}^*)$$
$$\left\{\begin{array}{ll} = \left[\left|E_{\beta 0}^{*(1)}\right|^{1/2} - T_{1\beta}K_{0\beta}T_\beta\right]^2 sign\epsilon_{\beta 0}^*, & \forall \mathbf{t}^\beta \in \mathfrak{T}_{0R}^\beta, \\ = \mathbf{0}_\beta, & \forall \mathbf{t}^\beta \in \mathfrak{T}_{RF}^\beta, \end{array}\right\}$$
$$\mathbf{T}\left(\varepsilon_{\beta 0}^{*1}\right) \neq \mathbf{0}_\beta \Longrightarrow \epsilon_\beta^*(\mathbf{t}^\beta;\epsilon_{\beta 0}^*) = \varepsilon_\beta^*(\mathbf{t}^\beta;\varepsilon_{\beta 0}^*) = \mathbf{0}_\beta,\ \forall \mathbf{t}^\beta \in \mathfrak{T}_{RF}^\beta. \tag{36.10}$$

This implies

$$sign\epsilon_\beta^*(\mathbf{t}^\beta) = sign\epsilon_{\beta 0}^*,\ \forall \mathbf{t}^\beta \in In\mathfrak{T}_{RF}^\beta.$$

The tracking operator (36.9) possesses Properties 501-505 and $\mathbf{T}(.) \in \mathfrak{C}^-$. *The tracking is stable in the whole with FVIDRT* t_R^β *if* $\mathbf{T}\left(\varepsilon_{\beta 0}^{*1}\right) = \mathbf{0}_\beta$.

Iff $\mathbf{T}\left(\varepsilon_{\beta 0}^{*1}\right) = \mathbf{0}_\beta$ *then the solution (36.10) obeys the following:*

$$\mathbf{T}\left(\varepsilon_{\beta 0}^{*1}\right) = \mathbf{0}_\beta \Longrightarrow \left\|\varepsilon_\beta^*(\mathbf{t}^\beta;\varepsilon_{\beta 0}^*)\right\| \le \left\|\varepsilon_{\beta 0}^*\right\|,\ \forall \mathbf{t}^\beta \in \mathfrak{T}_{0F}^\beta \Longrightarrow \delta(\varepsilon) = \varepsilon$$
$$\Longrightarrow \forall \varepsilon \in \mathfrak{R}^+,\ \exists \delta \in \mathfrak{R}^+,\ \delta = \delta(\varepsilon) = \varepsilon \Longrightarrow$$
$$\left\|\varepsilon_{\beta 0}^*\right\| < \delta \Longrightarrow \left\|\varepsilon_\beta^*(\mathbf{t}^\beta;\varepsilon_{\beta 0}^*)\right\| \le \varepsilon,\ \forall \mathbf{t}^\beta \in \mathfrak{T}_{0F}^\beta.$$

Therefore, such tracking of $\mathbf{Y}_d(\mathbf{t}^N)$ *is stable state on* $\mathfrak{T}_{0F}^N$. *Then the bigger* $K_{0\beta}$, *the smaller* t_R^β *is for fixed* $T_{1\beta}$ *and* $\varepsilon_{\beta 0}^{*1}$, *and vice versa. The smaller* $T_{1\beta}$, *the smaller* t_R^β *is for fixed* $K_{0\beta}$ *and* $\varepsilon_{\beta 0}^{*1}$, *and vice versa. The bigger* $\left|\varepsilon_{\beta 0}^{*1}\right|$, *the bigger* t_R^β *is for fixed* $T_{1\beta}$ *and* $K_{0\beta}$, *and vice versa. These claims are in the elementwise sense iff* $\beta = \rho$.

Example 522 ***The higher-power smooth elementwise stable tracking with the finite vector state reachability time*** t_R^β

Let the tracking algorithm be

$$\mathbf{T}\left(\boldsymbol{\epsilon}_{\boldsymbol{\beta}}^{*1}\right) = T_{1\beta}\boldsymbol{\epsilon}_{\boldsymbol{\beta}}^{*(1)}(\mathbf{t}^{\beta}) + K_{0\beta}\left|\ominus_{\beta}^{*}(\mathbf{t}^{\beta})\right|^{I-K^{-1}} sign\boldsymbol{\epsilon}_{\beta 0}^{*} = \mathbf{0}_{\beta},\ \forall \mathbf{t}^{\beta} \in \mathfrak{T}_{0F}^{\beta},$$
$$K = diag\{k_1\ k_2\ \cdots\ k_{\beta}\},\ k_i \in \{2,\ 3,\ \cdots\},\ \forall i = 1, 2,\ \cdots,\ \beta,$$
$$\left|\ominus_{\beta}^{*}(\mathbf{t}^{\beta})\right|^{I-K^{-1}} = diag\left\{ |\epsilon_1(t)|^{1-k_1^{-1}}\ \ |\epsilon_2(t)|^{1-k_2^{-1}}\ \cdots\ |\epsilon_{\beta}^{*}(t)|^{1-k_{\beta}^{-1}} \right\}. \tag{36.11}$$

Iff $\mathbf{T}\left(\boldsymbol{\varepsilon}_{\beta 0}^{*1}\right) = \mathbf{0}_{\beta}$ *then (36.12) determines the following solution* $\epsilon_{\beta}^{*}(\mathbf{t}^{\beta};\epsilon_{\beta 0}^{*})$ *to* $T\left(\boldsymbol{\epsilon}_{\boldsymbol{\beta}}^{*1}\right) = \mathbf{0}_{\beta}$, $\forall \mathbf{t}^{\beta} \in \mathfrak{T}_{0F}^{\beta}$:

$$\mathbf{T}\left(\varepsilon_{\beta 0}^{*1}\right) = \mathbf{0}_{\beta} \Longrightarrow \boldsymbol{\epsilon}_{\boldsymbol{\beta}}^{*}(\mathbf{t}^{\beta};\boldsymbol{\epsilon}_{\beta 0}^{*}) = \boldsymbol{\varepsilon}_{\beta}^{*}(\mathbf{t}^{\beta};\boldsymbol{\varepsilon}_{\beta 0}^{*})$$
$$= \frac{1}{2}S\left(\varepsilon_{\beta 0}^{*}\right)\left\{ \begin{array}{c} \left\{ \begin{array}{c} I_{\beta} \\ +\ S\left[\left|\boldsymbol{\varepsilon}_{\beta 0}^{*}\right|^{K^{-1}} - T_{1\beta}^{-1}K^{-1}K_{0\beta}T_{\beta}\right] \end{array} \right\} \\ \bullet \left[\left|\boldsymbol{\varepsilon}_{\beta 0}^{*}\right|^{K^{-1}} - T_{1\beta}^{-1}K^{-1}K_{0\beta}T_{\beta}\right]^{K} \end{array} \right\}$$
$$= \left\{ \begin{array}{ll} \left[\left|E_{\beta 0}^{*(1)}\right|^{K^{-1}} - T_{1\beta}K^{-1}K_{0\beta}T_{\beta}\right]^{K} sign\boldsymbol{\varepsilon}_{\beta 0}^{*}, & \forall \mathbf{t}^{\beta} \in \mathfrak{T}_{0R}^{\beta}, \\ \mathbf{0}_{\beta}, & \forall \mathbf{t}^{\beta} \in \mathfrak{T}_{RF}^{\beta} \end{array} \right\} \Longrightarrow$$
$$\mathbf{t}_{R}^{\beta} = T_{1\beta}KK_{0\beta}^{-1}\left|\boldsymbol{\varepsilon}_{\beta 0}^{*}\right|^{K^{-1}}, \tag{36.12}$$

but iff $\mathbf{T}\left(\varepsilon_{\beta 0}^{*1}\right) \neq \mathbf{0}_{\beta}$, *then the solution* $\epsilon_{\beta}^{*}(\mathbf{t}^{\beta};\epsilon_{\beta 0}^{*})$ *to* $T\left(\boldsymbol{\epsilon}_{\boldsymbol{\beta}}^{*}, \boldsymbol{\epsilon}_{\boldsymbol{\beta}}^{*(1)}\right) = \mathbf{0}_{\beta}$, $\forall \mathbf{t}^{\beta} \in \mathfrak{T}_{RF}^{\beta}$, *satisfies the following:*

$$\mathbf{T}\left(\varepsilon_{\beta 0}^{*1}\right) \neq \mathbf{0}_{\beta} \Longrightarrow \boldsymbol{\epsilon}_{\boldsymbol{\beta}}^{*}(\mathbf{t}^{\beta};\boldsymbol{\epsilon}_{\beta 0}^{*}) = \boldsymbol{\varepsilon}_{\boldsymbol{\beta}}^{*}(\mathbf{t}^{\beta};\boldsymbol{\varepsilon}_{\beta 0}^{*}) = \mathbf{0}_{\beta},\ \forall \mathbf{t}^{\beta} \in \mathfrak{T}_{RF}^{\beta}.$$

Notice that

$$\left[\left|\boldsymbol{\varepsilon}_{\beta 0}^{*}\right|^{K^{-1}} - T_{1\beta}K_{0\beta}\mathbf{t}^{\beta}\right]^{K}$$
$$= \begin{bmatrix} \left[|\varepsilon_{10}|^{k_1^{-1}} - t_1^{-1}k_{01}t\right]^{k_1} \\ \left[|\varepsilon_{20}|^{k_2^{-1}} - t_2^{-1}k_{02}t\right]^{k_2} \\ \cdots\cdots\cdots \\ \left[\left|\varepsilon_{\beta 0}^{*}\right|^{k_{\beta}^{-1}} - t_{\beta}^{-1}k_{0\gamma}t\right]^{k_{\beta}} \end{bmatrix} \in \mathfrak{R}^{\beta}. \tag{36.13}$$

The tracking operator (36.11) possesses Properties 501-505 and $\mathbf{T}(.) \in \mathfrak{C}^{-}$. *The tracking is stable in the whole with FVIDRT* t_{R}^{β} *if* $\mathbf{T}\left(\boldsymbol{\varepsilon}_{\beta 0}^{*1}\right) = \mathbf{0}_{\beta}$.

The equations (36.12) express the elementwise nonlinear convergence to the zero error vector $\mathbf{0}_{\beta}$ *iff* $\mathbf{T}\left(\varepsilon_{\beta 0}^{*1}\right) = \mathbf{0}_{\beta}$. *Then the convergence is strictly monotonous and continuous, without any oscillation, overshoot, or undershoot. Therefore, such tracking of* $\mathbf{Y}_d\left(\mathbf{t}^{N}\right)$ *is stable state on* $\mathfrak{T}_{0F}^{N}$. *The errors enter the zero values smoothly. Also, (36.12) and (36.13) imply*

$$\mathbf{T}\left(\varepsilon_{\beta 0}^{*1}\right) = \mathbf{0}_{\beta} \Longrightarrow \left\|\boldsymbol{\varepsilon}_{\beta}^{*}(\mathbf{t}^{\beta};\boldsymbol{\varepsilon}_{\beta 0}^{*})\right\| \leq \left\|\boldsymbol{\varepsilon}_{\beta 0}^{*}\right\|,\ \forall \mathbf{t}^{\beta} \in \mathfrak{T}_{0F}^{\beta} \Longrightarrow \delta(\varepsilon) = \varepsilon$$
$$\Longrightarrow \forall \varepsilon \in \mathfrak{R}^{+},\ \exists \delta \in \mathfrak{R}^{+},\ \delta = \delta(\varepsilon) = \varepsilon \Longrightarrow$$
$$\left\|\boldsymbol{\varepsilon}_{\beta 0}^{*}\right\| < \delta \Longrightarrow \left\|\boldsymbol{\varepsilon}_{\beta}^{*}(\mathbf{t}^{\beta};\boldsymbol{\varepsilon}_{\beta 0}^{*})\right\| \leq \varepsilon,\ \forall \mathbf{t}^{\beta} \in \mathfrak{T}_{0F}^{\beta}.$$

Then the bigger $K_{0\beta}$, *the smaller* t_R^β *is for fixed* $T_{1\beta}$ *and* $\varepsilon_{\beta 0}^*$, *and vice versa. The smaller* $T_{1\beta}$, *the smaller* t_R^β *is for fixed* $K_{0\beta}$ *and* $\varepsilon_{\beta 0}^*$, *and vice versa. The bigger* $\left|\varepsilon_{\beta 0}^*\right|$, *the bigger* t_R^β *is for fixed* $T_{1\beta}$ *and* $K_{0\beta}$, *and vice versa.*

Iff $\mathbf{T}\left(\varepsilon_{\beta 0}^{*1}\right) \neq \mathbf{0}_\beta$ *then the tracking of* $\mathbf{Y}_R\left(\mathbf{t}^N\right)$ *is perfect on* $\mathfrak{T}_{0F}^N$, *but not of* $\mathbf{Y}_d\left(\mathbf{t}^N\right)$ *on* $\mathfrak{T}_{0F}^N$.

Example 523 ***Sharp absolute error vector tracking elementwise and stable with the finite vector state reachability time*** t_R^β

The solution to the following tracking algorithm [in which we use (36.5), (36.6), and (36.8)]:

$$\mathbf{T}\left(\epsilon_\beta^*, \epsilon_\beta^{*(1)}\right) = T_{1\beta}D^+\left|\epsilon_\beta^*\right| + K_{0\beta}sign\left|\epsilon_{\beta 0}^*\right|$$
$$= T_{1\beta}\Sigma\left(\epsilon_{\boldsymbol{\beta}}^{*\,1}\right)\epsilon_{\boldsymbol{\beta}}^{*\,(1)} + K_{0\beta}sign\left|\epsilon_{\beta 0}^*\right| = \mathbf{0}_\beta,\ \forall \mathbf{t}^\beta \in \mathfrak{T}_{0F}^\beta\ , \tag{36.14}$$

yields for $\mathbf{T}\left(\varepsilon_{\beta 0}^{*1}\right) = \mathbf{0}_\beta$:

$$\mathbf{T}\left(\varepsilon_{\beta 0}^{*1}\right) = \mathbf{0}_\beta \Longrightarrow \epsilon_{\boldsymbol{\beta}}^*(\mathbf{t}^\beta;\epsilon_{\beta 0}^*) = \varepsilon_\beta^*(\mathbf{t}^\beta;\varepsilon_{\beta 0}^*)$$
$$= \left\{\begin{array}{ll} \left[\left|\varepsilon_{\beta 0}^*\right| - T_{1\beta}^{-1}K_{0\beta}T_\beta\right] sign\left|\varepsilon_{\beta 0}^*\right|, & \forall \mathbf{t}^\beta \in \mathfrak{T}_{0R}^\beta, \\ \mathbf{0}_\beta, & \forall \mathbf{t}^\beta \in \mathfrak{T}_{RF}^\beta. \end{array}\right\},$$
$$\mathbf{t}_R^\beta = T_{1\beta}K_{0\beta}^{-1}\left|\varepsilon_{\beta 0}^*\right|, \tag{36.15}$$

but for $\mathbf{T}\left(\varepsilon_{\beta 0}^{*1}\right) \neq \mathbf{0}_\beta$ *it yields:*

$$\mathbf{T}\left(\varepsilon_{\beta 0}^{*1}\right) \neq \mathbf{0}_\beta \Longrightarrow \epsilon_{\boldsymbol{\beta}}^*(\mathbf{t}^\beta;\epsilon_{\beta 0}^*) = \varepsilon_\beta^*(\mathbf{t}^\beta;\varepsilon_{\beta 0}^*) = \mathbf{0}_\beta,\ \forall \mathbf{t}^\beta \in \mathfrak{T}_{RF}^\beta.$$

The tracking operator (36.14) possesses Properties 501-505 and $\mathbf{T}(.) \in \mathfrak{C}^-$. *The tracking is stable in the whole with FVIDRT* t_R^β *if* $\mathbf{T}\left(\varepsilon_{\beta 0}^{*1}\right) = \mathbf{0}_\beta$.

Equations (36.15) permit

$$\mathbf{T}\left(\varepsilon_{\beta 0}^{*1}\right) = \mathbf{0}_\beta \Longrightarrow \left|\varepsilon_{\boldsymbol{\beta}}^*(\mathbf{t}^\beta;\varepsilon_{\beta 0}^*)\right| \leq \left|\varepsilon_{\beta 0}^*\right|,\ \forall \mathbf{t}^\beta \in \mathfrak{T}_{0F}^\beta, \Longrightarrow$$
$$\left\|\varepsilon_{\boldsymbol{\beta}}^*(\mathbf{t}^\beta;\varepsilon_{\beta 0}^*)\right\| \leq \left\|\varepsilon_{\beta 0}^*\right\|,\ \forall \mathbf{t}^\beta \in \mathfrak{T}_{0F}^\beta \Longrightarrow \forall \varepsilon \in \mathfrak{R}^+,\ \exists \delta \in \mathfrak{R}^+,\ \delta = \delta(\varepsilon) = \varepsilon \Longrightarrow$$
$$\left\|\varepsilon_{\beta 0}^*\right\| < \delta \Longrightarrow \left\|\varepsilon_{\boldsymbol{\beta}}^*(\mathbf{t}^\beta;\varepsilon_{\beta 0}^*)\right\| \leq \varepsilon,\ \forall \mathbf{t}^\beta \in \mathfrak{T}_{0F}^\beta.$$

Then the tracking is stable state with the finite vector state reachability time

$$\mathbf{t}_R^\beta = T_{1\beta}K_{0\beta}^{-1}\left|\varepsilon_{\beta 0}^*\right|.$$

It is strictly monotonous and continuous without oscillation, overshoot, and undershoot.

Example 524 ***The exponential absolute error vector tracking elementwise and stable with the finite vector state reachability time*** t_R^β

The tracking algorithm is in terms of the elementwise absolute value of the error vector,

$$\mathbf{T}\left(\epsilon_\beta^*, \epsilon_\beta^{*(1)}\right) = T_{1\beta}D^+\left|\epsilon_\beta^*\right| + K\left(\left|\epsilon_\beta^*\right| + K_{0\beta}sign\left|\epsilon_{\beta 0}^*\right|\right)$$
$$= T_{1\beta}\Sigma\left(\epsilon_{\boldsymbol{\beta}}^*, \epsilon_{\boldsymbol{\beta}}^{*\,(1)}\right)\epsilon_{\boldsymbol{\beta}}^{*\,(1)} + K\left(\left|\epsilon_\beta^*\right| + K_{0\beta}sign\left|\epsilon_{\beta 0}^*\right|\right) = \mathbf{0}_\beta,\ \forall \mathbf{t}^\beta \in \mathfrak{T}_{0F}^\beta. \tag{36.16}$$

The solution of the differential equation (36.16) written in the matrix diagonal form

$$D^+\left[\left|\ominus_\beta^*\right|+K_{0\beta}S\left(\left|\boldsymbol{\epsilon}_{\beta 0}^*\right|\right)\right]=-T_{1\beta}K\left[\left|\ominus_\beta^*\right|+K_{0\beta}S\left(\left|\boldsymbol{\epsilon}_{\beta 0}^*\right|\right)\right]\Longrightarrow$$
$$\left[\left|\ominus_\beta^*\right|+K_{0\beta}S\left(\left|\boldsymbol{\varepsilon}_{\beta 0}^*\right|\right)\right]^{-1}D^+\left[\left|\ominus_\beta^*\right|+K_{0\beta}S\left(\left|\boldsymbol{\varepsilon}_{\beta 0}^*\right|\right)\right]=-T_{1\beta}K\Longrightarrow$$
$$D^+\left\{ln\left[\left|\ominus_\beta^*\right|+K_{0\beta}S\left(\left|\boldsymbol{\varepsilon}_{\beta 0}^*\right|\right)\right]\right\}=-T_{1\beta}K, \tag{36.17}$$

reads in the matrix form

$$ln\left\{\left[\left|\ominus_\beta^*\right|+K_{0\beta}S\left(\left|\boldsymbol{\varepsilon}_{\beta 0}^*\right|\right)\right]\left[\left|\ominus_{\beta 0}^*\right|+K_{0\beta}S\left(\left|\boldsymbol{\varepsilon}_{\beta 0}^*\right|\right)\right]^{-1}\right\}=-T_{1\beta}KT_\beta,$$
$$\forall T\in T_{0\beta}.$$

The final form of the solution is

$$\ominus_\beta^*(t;\boldsymbol{\epsilon}_{\beta 0}^*)$$
$$=\left\{\begin{array}{c}\left\{\begin{array}{c}e^{-T_{1\beta}K_\beta T_\beta}\left(\left|\ominus_{\beta 0}^*\right|+K_{0\beta}\right)S\left(\left|\boldsymbol{\epsilon}_{\beta 0}^*\right|\right)-K_{0\beta}S\left(\left|\boldsymbol{\epsilon}_{\beta 0}^*\right|\right),\\ \forall T_\beta\in\left[O_\beta,T_{\beta R}\right],\end{array}\right\}\\ O_\beta,\qquad \forall T_\beta\in\left[O_\beta,T_{\beta R}\right],\\ where\ 0\infty=0,\end{array}\right\},$$

$$\mathbf{T}\left(\boldsymbol{\varepsilon}_{\beta 0}^{*1}\right)=\mathbf{0}_\beta\Longrightarrow T_{\beta R}=\left\{\begin{array}{ll}K^{-1}T_{1\beta}ln\left\{K_{0\beta}\left[\left|\ominus_{\beta 0}^*\right|+K_{0\beta}\right]\right\}, & \boldsymbol{\epsilon}_{\beta 0}^*\neq\mathbf{0}_\beta,\\ O_\beta, & \boldsymbol{\epsilon}_{\beta 0}^*=\mathbf{0}_\beta\end{array}\right\}, \tag{36.18}$$

$$\mathbf{t}_R^\beta=\left[t_{R1}\ \ t_{R2}\ \ldots\ t_{R\beta}\right]^T\Longleftrightarrow T_{\beta R}=diag\left\{t_{R1}\ \ t_{R2}\ \ldots\ t_{R\beta}\right\}. \tag{36.19}$$

If $\mathbf{T}\left(\boldsymbol{\varepsilon}_{\beta 0}^{*1}\right)=\mathbf{0}_\beta$ *then we can set the solution (36.18) in the equivalent vector form*

$$\boldsymbol{\epsilon}_{\boldsymbol{\beta}}^*(\mathbf{0}_\beta;\boldsymbol{\varepsilon}_{\beta 0}^*)=\left\{\begin{array}{c}\left[e^{-T_{1\beta}K_\beta T_\beta}\left(\left|\boldsymbol{\varepsilon}_{\beta 0}^*\right|+K_{0\beta}\right)-K_{0\beta}\right]sign\left(\left|\boldsymbol{\varepsilon}_{\beta 0}^*\right|\right),\\ \forall\mathbf{t}^\beta\in\left[\mathbf{0}_\beta,\ \mathbf{t}_R^\beta\right];\ i.e.,\ \forall T_\beta\in\left[O_\beta,T_{\beta R}\right],\\ \mathbf{0}_\beta,\ \ \forall\mathbf{t}^\beta\in\left[\mathbf{t}_R^\beta,\ \mathbf{t}_F^\beta\right];\ i.e.,\ \forall T_\beta\in\left[T_{\beta R},T_{\beta F}\right].\end{array}\right\}. \tag{36.20}$$

The tracking operator (36.16) possesses Properties 501-505 and $\mathbf{T}(.)\in\mathfrak{C}^-$. *The tracking is stable in the whole with FVIDRT* t_R^β *if* $\mathbf{T}\left(\boldsymbol{\varepsilon}_{\beta 0}^{*1}\right)=\mathbf{0}_\beta$.

Iff $\mathbf{T}\left(\boldsymbol{\varepsilon}_{\beta 0}^{*1}\right)=\mathbf{0}_\beta$ *then the solution is continuous and monotonous without oscillation, overshoot, and undershoot, and obeys*

$$\mathbf{T}\left(\boldsymbol{\varepsilon}_{\beta 0}^{*1}\right)=\mathbf{0}_\beta\Longrightarrow\boldsymbol{\epsilon}_{\boldsymbol{\beta}}^*(\mathbf{t}^\beta;\boldsymbol{\varepsilon}_{\beta 0}^*)=\boldsymbol{\varepsilon}_\beta^*(\mathbf{t}^\beta;\boldsymbol{\varepsilon}_{\beta 0}^*)\Longrightarrow$$
$$\left|\boldsymbol{\varepsilon}_\beta^*(\mathbf{t}^\beta;\boldsymbol{\varepsilon}_{\beta 0}^*)\right|\leq\left|\boldsymbol{\varepsilon}_{\beta 0}^*\right|,\ \forall\mathbf{t}^\beta\in\mathfrak{T}_{0F}^\beta\Longrightarrow$$
$$\forall\varepsilon\in\mathfrak{R}^+,\ \exists\delta\in\mathfrak{R}^+,\ \delta=\delta\left(\varepsilon\right)=\varepsilon\Longrightarrow$$
$$\left\|\boldsymbol{\varepsilon}_{R0}\right\|<\delta\Longrightarrow\left\|\boldsymbol{\varepsilon}_\beta^*(\mathbf{t}^\beta;\boldsymbol{\varepsilon}_{\beta 0}^*)\right\|\leq\varepsilon,\ \forall\mathbf{t}^\beta\in\mathfrak{T}_{0F}^\beta,$$

Then the tracking of $\mathbf{Y}_d\left(\mathbf{t}^N\right)$ *is stable on* $\mathfrak{T}_{0F}^N$. *It converges with the exponential rate to the zero error vector and reaches (elementwise iff* $\beta=\rho$*) the origin in finite vector state reachability time* t_R^β *(36.18), (36.19).*

Iff $\mathbf{T}\left(\boldsymbol{\varepsilon}_{\beta 0}^{*1}\right)\neq\mathbf{0}_\beta$ *then the solution reads*

$$\mathbf{T}\left(\boldsymbol{\varepsilon}_{\beta 0}^{*1}\right)\neq\mathbf{0}_\beta\Longrightarrow\boldsymbol{\epsilon}_{\boldsymbol{\beta}}^*(\mathbf{t}^\beta;\boldsymbol{\epsilon}_{\beta 0}^*)=\boldsymbol{\varepsilon}_\beta^*(\mathbf{t}^\beta;\boldsymbol{\varepsilon}_{\beta 0}^*)=\mathbf{0}_\beta,\quad\forall\mathbf{0}_\beta\in\left[\mathbf{t}_R^\beta,\ \mathbf{t}_F^\beta\right].$$

Note 525 *We can effectively use the tracking algorithms proposed in the preceding examples, Example 518 through Example 524, also when we synthesize tracking control by applying various design methods; e.g., Lyapunov-like, adaptive control, sliding mode, and natural tracking control method.*

The system behavior results from the actions of control, of disturbances, and of initial conditions. The control cannot influence disturbances and initial conditions, but can take into account their consequences that are the plant state error and the output error.

Chapter 37

NTC synthesis: State space

37.1 General NTC theorem: State space

In what follows $t_0 = 0$.

Property 501 through Property 505 form the basis for the following general theorems.

Theorem 526 ***General NTC synthesis in the state space***

Let (35.23) through (35.43) be valid.

In order for the plant (3.12), with Properties 57, 58, 78, controlled by the natural tracking control $\mathbf{U}$ *to exhibit tracking on* $\mathfrak{T}_{0F}^{\alpha\rho} \times \mathfrak{D}^j \times \mathfrak{Y}_d^{\alpha}$ *determined by the tracking algorithm* $\mathbf{T}(.)$*, (35.21), (35.22), it is necessary and sufficient that both:*

i) the rank β *of the control matrix* $P(t)$ *is greater than zero and not greater than* $\min(\rho, r)$*, i.e. (8.20) holds,*

and

ii) The control vector function $\mathbf{U}(.)$ *satisfies one of the conditions 1) - 4):*

1) The control vector function $\mathbf{U}(.)$ *obeys (37.1) for* $\beta = rankP(t) = r < \rho$, $\forall t \in \mathfrak{T}_0$*, and for the chosen matrix function* $L_{r,m}(.) : \mathfrak{T}_0 \longrightarrow \mathfrak{R}^{r\times m}$*, where* m *is not greater than* r*,* $m \leq r$*, and rank* $L_{r,m}(t) \equiv m$*:*

$$\mathbf{e}_{ur}\left[\mathbf{U}_r^{\mu}(t)\right] = \left\{ \begin{array}{c} \mathbf{e}_{ur}\left[\mathbf{U}_r^{\mu}(t^-)\right] \\ + L_{r,m}(t)\,\mathbf{T}\left(\mathbf{t}^m, \boldsymbol{\epsilon}_r^{*\alpha-1}(\mathbf{t}^{\alpha r}), \int_{t_0^{\beta}}^{\mathbf{t}^{\beta}} \ominus(\mathbf{t}^r)\, d\mathbf{t}^r\right) \end{array} \right\}, \forall \mathbf{t}^{\alpha r} \in \mathfrak{T}_0^{\alpha r}. \quad (37.1)$$

The domain $\mathcal{D}_T\left(t_0; \mathfrak{D}^j; \mathbf{U}; \mathfrak{Y}_d^{\alpha}\right)$ *of tracking on* $\mathfrak{T}_{0F}^{\alpha\rho} \times \mathfrak{D}^j \times \mathfrak{Y}_d^{\alpha}$ *is equal to the domain* $\mathcal{D}_{PTbl}\left(t_0; \mathfrak{D}^j; \mathbf{U}; \mathfrak{Y}_d^{\alpha}\right)$ *of the perfect trackability of the plant (3.12) on* $\mathfrak{T}_0 \times \mathfrak{D}^j \times \mathfrak{Y}_d^{\alpha}$*, which is determined by (8.26).*

2) The control vector function $\mathbf{U}(.)$ *obeys (37.2) for* $\beta = rankP(t) = \rho < r$, $\forall t \in \mathfrak{T}_0$*, and for the chosen matrix function* $L_{\rho,m}(.) : \mathfrak{T}_0 \longrightarrow \mathfrak{R}^{\rho\times m}$*, where* m *is not greater than* ρ*,* $m \leq \rho$*, and rank* $L_{\rho,m}(t) \equiv m$*:*

$$\mathbf{e}_{u}\left[\mathbf{U}^{\mu}(t)\right] = \left\{ \begin{array}{c} \Gamma(t)\left[P(t)\,\mathbf{e}_u\left[\mathbf{U}^{\mu}(t^-)\right]\right] \\ + P^T(t)\,L_{\rho,m}(t)\,\mathbf{T}\left(\mathbf{t}^m, \boldsymbol{\epsilon}^{\alpha-1}(\mathbf{t}^{\alpha\rho}), \int_{t_0^{\beta}}^{\mathbf{t}^{\beta}} \ominus(\mathbf{t}^{\rho})\, d\mathbf{t}^{\rho}\right) \end{array} \right\},$$

$$\Gamma(t) = P^T(t)\left[P(t)\,P^T(t)\right]^{-1}, \ \forall t \in \mathfrak{T}_0, \quad (37.2)$$

The domain $\mathcal{D}_T\left(t_0; \mathfrak{D}^j; \mathbf{U}; \mathfrak{Y}_d^{\alpha}\right)$ *of tracking on* $\mathfrak{T}_{0F}^{\alpha\rho} \times \mathfrak{D}^j \times \mathfrak{Y}_d^{\alpha}$ *is equal to the domain*

$$\mathcal{D}_{PTbl}\left(t_0; \mathfrak{D}^j; \mathbf{U}; \mathfrak{Y}_d^{\alpha}\right)$$

of the perfect trackability of the plant (3.12) on $\mathfrak{T}_0 \times \mathfrak{D}^j \times \mathfrak{Y}_d^{\alpha}$*, which is determined by (8.27).*

3) The control vector function $\mathbf{U}(.)$ obeys (37.3) if $\beta = \mathrm{rank}P(t) = r = \rho$, $\forall t \in \mathfrak{T}_0$, and for the chosen matrix function $L_{r,m}(.) : \mathfrak{T}_0 \longrightarrow \mathfrak{R}^{r\times m}$, where m is not greater than r, $m \leq r$, and rank $L_{r,m}(t) \equiv m$::

$$\mathbf{e}_u\left[\mathbf{U}^{\mu}(t)\right] = \left\{ \begin{array}{c} \mathbf{e}_u\left[\mathbf{U}^{\mu}(t^-)\right] \\ +\ P(t)^{-1} L_{r,m}(t)\,\mathbf{T}\left(\mathbf{t}^m, \boldsymbol{\epsilon}^{\alpha-1}(\mathbf{t}^{\alpha r}), \int_{t_0^{\beta}}^{\mathbf{t}^{\beta}} \ominus(\mathbf{t}^r)\, d\mathbf{t}^r\right) \end{array} \right\}, \quad \forall t \in \mathfrak{T}_0, \tag{37.3}$$

The domain $\mathcal{D}_T(t_0; \mathfrak{D}^j; \mathbf{U}; \mathfrak{Y}_d^{\alpha})$ of tracking on $\mathfrak{T}_{0F}^{\alpha\rho} \times \mathfrak{D}^j \times \mathfrak{Y}_d^{\alpha}$ is equal to the domain $\mathcal{D}_{PTbl}(t_0; \mathfrak{D}^j; \mathbf{U}; \mathfrak{Y}_d^{\alpha})$ of the perfect trackability of the plant (3.12) on $\mathfrak{T}_0 \times \mathfrak{D}^j \times \mathfrak{Y}_d^{\alpha}$, which is determined by (8.28).

4) If the plant also possesses Property 61 then for the chosen matrix function $L_{\beta,m}(.) : \mathfrak{T}_0 \longrightarrow \mathfrak{R}^{\beta\times m}$, the control vector function $\mathbf{U}(.)$ obeys (8.29),

$$\mathbf{e}_{u\beta}\left[\mathbf{U}_{\beta}^{\mu}(t)\right] = \mathbf{e}_{u\beta}\left[\mathbf{U}_{\beta}^{\mu}(t^-)\right]$$
$$+\ L_{\beta,m}(t)\,\mathbf{T}\left(\mathbf{t}^m, \boldsymbol{\epsilon}_{\beta}^{\alpha-1}(\mathbf{t}^{\alpha\beta}), \int_{t_0^{\beta}}^{\mathbf{t}^{\beta}} \ominus(\mathbf{t}^{\beta})\, d\mathbf{t}^{\beta}\right), \quad \forall t \in \mathfrak{T}_0. \tag{37.4}$$

The equation (11.10) determines the initial control vector $\mathbf{U}_0$.

The domain $\mathcal{D}_T(t_0; \mathfrak{D}^j; \mathbf{U}; \mathfrak{Y}_d^{\alpha})$ of tracking on $\mathfrak{T}_{0F}^{\alpha\rho} \times \mathfrak{D}^j \times \mathfrak{Y}_d^{\alpha}$ is equal to the domain $\mathcal{D}_{PTbl}(t_0; \mathfrak{D}^j; \mathbf{U}; \mathfrak{Y}_d^{\alpha})$ of the perfect trackability of the plant (3.12) with Properties 57, 58, 61, and 78, which is determined by (11.11).

Proof. If we replace the vector function $\mathbf{v}(.) : \mathfrak{T}_0\times \mathrm{In}\mathfrak{T}_0 \times \mathfrak{R}^{(k+1)N} \longrightarrow \mathfrak{R}^m$, (11.1) by $\mathbf{T}\left(\mathbf{t}^m, \boldsymbol{\epsilon}^{\alpha-1}(\mathbf{t}^{\alpha\rho}), \int_{t_0^{\beta}}^{\mathbf{t}^{\beta}} \ominus(\mathbf{t}^{\rho})\, d\mathbf{t}^{\rho}\right)$ in Theorem 191 then it becomes Theorem 526 ■

37.2 NTC synthesis in the state space

In what follows $m = \beta$.

The vectors $\boldsymbol{\varepsilon}_{\beta}^*$, (35.3) and (35.4), and $\boldsymbol{\epsilon}_{\beta}^*$, (35.31), become the vectors $\boldsymbol{\varepsilon}_{Rr}^*$ and $\boldsymbol{\epsilon}_{Rr}^*$ when the subscript "β" is replaced by the subscript "Rr":

$$\boldsymbol{\varepsilon}_{Rr}^* = \left[\varepsilon_{R1}\ \varepsilon_{R2}\ ...\ \varepsilon_{R,r-1}\ \ \varepsilon_{R,r1}^*\right]^T, \quad \boldsymbol{\epsilon}_{Rr}^* = \left[\epsilon_{R1}\ \epsilon_{R2}\ ...\ \epsilon_{R,r-1}\ \ \epsilon_{R,r1}^*\right]^T. \tag{37.5}$$

Condition 527 *The plant **(3.12)** possesses Properties 57, 58, and 78.*

All following theorems result directly from Theorem *526*.

Theorem 528 ***NTC for the first-order linear elementwise exponential tracking, Example 518***

Let (35.23) through (35.29) be valid. Let the reachability time $\tau_{Rstate} = \tau_{Fstate} = \infty$, i.e., $\mathfrak{T}_{R\infty state} = \mathfrak{T}_{RFstate} = \{\infty\}$.

For the plant (3.12) satisfying Condition (527) to be controlled by the natural tracking control $\mathbf{U}$ in order to exhibit tracking on $\mathfrak{T}_{0F}^{2\rho} \times \mathfrak{D}^j \times \mathfrak{Y}_d^{\alpha}$ determined by the following *tracking algorithm*

$$\mathbf{T}\left(\boldsymbol{\varepsilon}_{\beta}^{*},\boldsymbol{\varepsilon}_{\beta}^{*(1)}\right)=T_{1\beta}\boldsymbol{\varepsilon}_{\boldsymbol{\beta}}^{*(\mathbf{1})}(\mathbf{t}^{\beta})+K_{0\beta}\boldsymbol{\varepsilon}_{\beta}^{*}(\mathbf{t}^{\beta})=\mathbf{0}_{\beta},\ \forall\mathbf{t}^{\beta}\in\mathfrak{T}_{0F}^{\beta},$$
$$if\ \mathbf{T}\left(\boldsymbol{\varepsilon}_{\beta0}^{*1}\right)=\mathbf{0}_{\beta},$$
$$\boldsymbol{\varepsilon}_{\beta}^{*}\left(t;\boldsymbol{\varepsilon}_{\beta0}^{*}\right)=-\mathbf{f}_{\beta}^{1}\left(t\right),\ \forall t\in\mathfrak{T}_{0F},\ if\ \mathbf{T}\left(\boldsymbol{\varepsilon}_{\beta0}^{*1}\right)\neq\mathbf{0}_{\beta},$$
$$T_{1\beta}=diag\{\tau_{1R1}\ \tau_{1R2}\ \cdots\ \tau_{1R\beta}\},\ K_{0\beta}=diag\{k_{0R1}\ k_{0R2}\ \cdots\ k_{0R\beta}\},$$
$$T_{1Rr}=diag\{\tau_{1R1}\ \tau_{1R2}\ \cdots\ \tau_{1Rr}\},\ K_{0Rr}=diag\{k_{0R1}\ k_{0R2}\ \cdots\ k_{0Rr}\},$$
$$T_{1R}=diag\{\tau_{1R1}\ \tau_{1R2}\ \cdots\ \tau_{1R\rho}\},\ K_{0R}=diag\{k_{0R1}\ k_{0R2}\ \cdots\ k_{0R\rho}\},$$
$$\tau_{1Ri}>0,\ k_{0Ri}>0,\ \forall i=1,2,...,(\beta\ or\ r\ or\ R),\tag{37.6}$$

it is necessary and sufficient that both:

i) The rank β of the control matrix $P(t)$ is greater than zero and not greater than $min(\rho,r)$; i.e., (8.20) holds,

and

ii) The control vector function $\mathbf{U}(.)$ satisfies one of the conditions 1)-4):

1) The control vector function $\mathbf{U}(.)$ obeys (37.7) for $\beta = rankP(t) = r < \rho$, $\forall t\in\mathfrak{T}_0$,

$$\mathbf{e}_u\left[\mathbf{U}^{\mu}(t)\right]=\mathbf{e}_u\left[\mathbf{U}^{\mu}(t^-)\right]+T_{1r}\boldsymbol{\epsilon}_r^{*(1)}(t)+K_{0r}\boldsymbol{\epsilon}_r^{*}(t),\ \forall t\in\mathfrak{T}_0.\tag{37.7}$$

2) The control vector function $\mathbf{U}(.)$ obeys (37.8) for $\beta = rankP(t) = \rho < r$, $\forall t\in\mathfrak{T}_0$,

$$\mathbf{e}_u\left[\mathbf{U}^{\mu}(t)\right]=\left\{\begin{array}{c}\Gamma(t)\left[P(t)\,\mathbf{e}_u\left[\mathbf{U}^{\mu}(t^-)\right]\right]\\+P^T(t)\left[T_1\boldsymbol{\epsilon}_\rho^{(1)}(t)+K_0\boldsymbol{\epsilon}_\rho(t)\right]\end{array}\right\},$$
$$\Gamma(t)=P^T(t)\left[P(t)P^T(t)\right]^{-1},\ \forall t\in\mathfrak{T}_0,\tag{37.8}$$

3) The control vector function $\mathbf{U}(.)$ obeys (37.9) if $\beta = rankP(t) = r = \rho$, $\forall t\in\mathfrak{T}_0$:

$$\mathbf{e}_u\left[\mathbf{U}^{\mu}(t)\right]=\mathbf{e}_u\left[\mathbf{U}^{\mu}(t^-)\right]+P(t)^{-1}\left[T_1\boldsymbol{\epsilon}_\rho^{(1)}(t)+K_0\boldsymbol{\epsilon}_\rho(t)\right],$$
$$\forall t\in\mathfrak{T}_0.\tag{37.9}$$

4) If the plant also possesses Property 61 then the control vector function $\mathbf{U}(.)$ obeys (37.10),

$$\mathbf{e}_{u\beta}\left[\mathbf{U}_{\beta}^{\mu}(t)\right]=\mathbf{e}_{u\beta}\left[\mathbf{U}_{\beta}^{\mu}(t^-)\right]+T_{1\beta}\boldsymbol{\epsilon}_{\beta}^{*(1)}(t)+K_{0\beta}\boldsymbol{\epsilon}_{\beta}^{*}(t),\ \forall t\in\mathfrak{T}_0.\tag{37.10}$$

The solution $\boldsymbol{\varepsilon}_{\beta}^{*}(.)$ shows that the tracking is exponentially stable in the whole with if $\mathbf{T}\left(\boldsymbol{\varepsilon}_{\beta0}^{*1}\right)=\mathbf{0}_{\beta}$.

We use the complex variable s in:

$$S_{\rho}^{(\eta)}(s)=\left[s^0I_\rho\ \vdots\ s^1I_\rho\ \vdots\ s^2I_\rho\ \vdots\ ...\ \vdots\ s^{\eta}I_\rho\right]^{\mathbf{T}}\in\mathfrak{C}^{(\eta+1)\rho\times\rho}.$$

Theorem 529 ***NTC for the higher-order linear elementwise exponential tracking, Example 519***

Let (35.23) through (35.29) be valid. Let the reachability time $\tau_R=\tau_F=\infty$; i.e., $\mathfrak{T}_{R\infty}=\mathfrak{T}_{RF}=\{\infty\}$.

For the plant (3.12) satisfying Condition (527) to be controlled by the natural tracking control $\mathbf{U}$ *in order to exhibit tracking on* $\mathfrak{T}_{0F}^{2\rho} \times \mathfrak{D}^{j} \times \mathfrak{Y}_{d}^{\alpha}$ *determined by the tracking algorithm (37.11),*

$$\mathbf{T}\left(\boldsymbol{\varepsilon}_{\beta}^{*\eta}\right) = \sum_{k=0}^{k=\eta\leq\alpha-1} E_{\beta k}\boldsymbol{\varepsilon}_{\beta}^{*}(\mathbf{t}^{\beta}) = E_{R\beta}^{(\eta)}\boldsymbol{\varepsilon}_{\beta}^{*}(\mathbf{t}^{\beta}) = \mathbf{0}_{\beta},$$
$$\forall\mathbf{t}^{(\eta+1)\beta} \in \mathfrak{T}_{0F}^{(\eta+1)\beta} \ if\ \mathbf{T}\left(\boldsymbol{\varepsilon}_{\beta0}^{*\eta}\right) = \mathbf{0}_{\beta},$$
$$\boldsymbol{\varepsilon}_{\beta}^{*\eta}\left(\mathbf{t}^{(\eta+1)\beta};\boldsymbol{\epsilon}_{\beta0}^{*\eta}\right) = -\mathbf{f}_{R\beta}^{\eta}\left(\mathbf{t}^{(\eta+1)\beta}\right),\ \forall\mathbf{t}^{(\eta+1)\beta} \in \mathfrak{T}_{0F}^{(\eta+1)\beta} \ if\ \mathbf{T}\left(\boldsymbol{\varepsilon}_{\beta0}^{*\eta}\right) \neq \mathbf{0}_{\beta}, \tag{37.11}$$

with the matrices $E_{\beta k} \in \mathfrak{R}^{\beta\times\ \beta}$ *such that the real parts* $Res_{\beta i}$ *of the roots* $s_{\beta i}$ *of its characteristic polynomial* $f_{\beta}(s)$, $f_{\beta}(s) = det\left(E_{\beta}^{(\eta)}S_{\beta}^{(\eta)}(s)\right)$, *are negative,*

$$f(s_{\beta i}) = det\left(E_{\beta}^{(\eta)}S_{\beta}^{(\eta)}(s_{\beta i})\right) = 0 \Longrightarrow Res_{\beta i} < 0,\ \forall i = 1,2,\ \cdots,\eta\beta, \tag{37.12}$$

it is necessary and sufficient that both:

i) The rank β *of the control matrix* $P(t)$ *is greater than zero and not greater than* $min(\rho, r)$*; i.e., (8.20) holds,*

and

ii) The control vector function $\mathbf{U}(.)$ *satisfies one of the conditions 1)-4):*

1) The control vector function $\mathbf{U}(.)$ *obeys (37.13) for* $\beta = rankP(t) = r < \rho$, $\forall t \in \mathfrak{T}_0$,

$$\mathbf{e}_u\left[\mathbf{U}^{\mu}(t)\right] = \mathbf{e}_u\left[\mathbf{U}^{\mu}\left(t^{-}\right)\right] + E_r^{(\eta)}\boldsymbol{\epsilon}_r^{*\eta}(t),\ \forall t \in \mathfrak{T}_0. \tag{37.13}$$

2) The control vector function $\mathbf{U}(.)$ *obeys (37.14) for* $\beta = rankP(t) = \rho < r$, $\forall t \in \mathfrak{T}_0$:

$$\mathbf{e}_u\left[\mathbf{U}^{\mu}(t)\right] = \Gamma(t)\left[P(t)\mathbf{e}_u\left[\mathbf{U}^{\mu}\left(t^{-}\right)\right]\right] + P^T(t)\,E_{\rho}^{(\eta)}\boldsymbol{\epsilon}_{\rho}^{\eta}(t),$$
$$\Gamma(t) = P^T(t)\left[P(t)P^T(t)\right]^{-1},\ \forall t \in \mathfrak{T}_0. \tag{37.14}$$

3) The control vector function $\mathbf{U}(.)$ *obeys (37.15) if* $\beta = rankP(t) = r = \rho$, $\forall t \in \mathfrak{T}_0$:

$$\mathbf{e}_u\left[\mathbf{U}^{\mu}(t)\right] = \mathbf{e}_u\left[\mathbf{U}^{\mu}\left(t^{-}\right)\right] + P(t)^{-1}E_{\rho}^{(\eta)}\boldsymbol{\epsilon}_{\rho}^{\eta}(t),\ \forall t \in \mathfrak{T}_0, \tag{37.15}$$

4) If the plant also possesses Property 61 then the control vector function $\mathbf{U}(.)$ *obeys (37.16),*

$$\mathbf{e}_{u\beta}\left[\mathbf{U}_{\beta}^{\mu}(t)\right] = \mathbf{e}_{u\beta}\left[\mathbf{U}_{\beta}^{\mu}(t^{-})\right] + E_{\beta}^{(\eta)}\boldsymbol{\epsilon}_{\beta}^{*\eta}(t),\ \forall t \in \mathfrak{T}_0. \tag{37.16}$$

The solution $\boldsymbol{\varepsilon}_{\beta}^{*}(.)$ *shows that the tracking is exponentially stable in the whole if* $\mathbf{T}\left(\boldsymbol{\epsilon}_{\beta0}^{*\eta}\right) = \mathbf{0}_{\rho}$.

Theorem 530 ***NTC for the sharp elementwise stablewise tracking with the finite vector state reachability time*** $\mathbf{t}_R^{\beta}$***, Example 520***

Let (35.23) through (35.29) be valid.

For the plant (3.12) satisfying Condition (527) to be controlled by the natural tracking control $\mathbf{U}$ *in order to exhibit tracking on* $\mathfrak{T}_{0F}^{2\rho} \times \mathfrak{D}^{j} \times \mathfrak{Y}_{d}^{\alpha}$ *determined by the algorithm*

$$\mathbf{T}\left(\boldsymbol{\varepsilon}_{\beta}^{*}, \boldsymbol{\varepsilon}_{\beta}^{*(1)}\right) = T_{1\beta}\boldsymbol{\varepsilon}_{\beta}^{*(1)}(\mathbf{t}^{\beta}) + K_{0\beta}sign\boldsymbol{\varepsilon}_{\beta0}^{*} = \mathbf{0}_{\beta},$$
$$\forall\mathbf{t}^{\beta} \in \mathfrak{T}_{0F}^{\beta} \ if\ \mathbf{T}\left(\boldsymbol{\varepsilon}_{\beta0}^{*1}\right) = \mathbf{0}_{\beta},$$
$$\boldsymbol{\varepsilon}_{\beta}^{*1}\left(\mathbf{t}^{2\beta};\mathbf{t}_0^{2\beta};\boldsymbol{\epsilon}_{\beta0}^{*}\right) = -\mathbf{f}_R^{1}\left(\mathbf{t}^{2\beta}\right),\ \forall\mathbf{t}^{2\beta} \in \mathfrak{T}_{0F}^{2\beta},\ if\ \mathbf{T}\left(\boldsymbol{\varepsilon}_{\beta0}^{*1}\right) \neq \mathbf{0}_{\beta}, \tag{37.17}$$

for the sharp elementwise stablewise tracking with the finite vector state reachability time $\mathbf{t}_R^{\beta}$,

$$\mathbf{T}\left(\varepsilon_{\beta 0}^{*1}\right)=\mathbf{0}_{\beta}\Longrightarrow \mathbf{t}_R^{\beta}=T_{1\beta}K_{0\beta}^{-1}\left|\varepsilon_{\beta 0}^{*}\right|,\ \varepsilon_{\beta 0}^{*}\in\mathfrak{R}^{\beta}, \tag{37.18}$$

it is necessary and sufficient that both:

i) The rank β *of the control matrix* $P(t)$ *is greater than zero and not be greater than* $min(\rho, r)$*; i.e., (8.20) holds,*

and

ii) The control vector function $\mathbf{U}(.)$ *satisfies one of the conditions 1)-4):*

1) the control vector function $\mathbf{U}(.)$ *obeys (37.19) for* $\beta = rankP(t) = r < \rho$, $\forall t \in \mathfrak{T}_0$,

$$\mathbf{e}_u\left[\mathbf{U}^{\mu}(t)\right]=\mathbf{e}_u\left[\mathbf{U}^{\mu}\left(t^{-}\right)\right]+T_{1r}\boldsymbol{\epsilon}_r^{*(1)}(\mathbf{t}^r)+K_{0r}sign\boldsymbol{\epsilon}_{Rr0}^{*},\forall t\in\mathfrak{T}_0. \tag{37.19}$$

Then the error vector $\boldsymbol{\epsilon}_r^*(\mathbf{t}^r)$ *is controlled elementwise with the finite vector state reachability time* $\mathbf{t}_R^r$,

$$\mathbf{T}\left(\varepsilon_{r0}^{*1}\right)=\mathbf{0}_r\Longrightarrow \mathbf{t}_R^{r}=T_{1r}K_{0r}^{-1}\left|\varepsilon_{r0}^{*}\right|,\ \varepsilon_{r0}^{*}\in\mathfrak{R}^{r}, \tag{37.20}$$

2) The control vector function $\mathbf{U}(.)$ *obeys (37.21) for* $\beta = rankP(t) = \rho < r$, $\forall t \in \mathfrak{T}_0$,

$$\mathbf{e}_u\left[\mathbf{U}^{\mu}(t)\right]=\Gamma(t)\left[P(t)\,\mathbf{e}_u\left[\mathbf{U}^{\mu}\left(t^{-}\right)\right]\right]+P^T(t)\left[T_{1R}\boldsymbol{\epsilon}_{\rho}^{(1)}(\mathbf{t}^{\rho})+K_{0R}sign\boldsymbol{\epsilon}_{\rho 0}\right],$$
$$\Gamma(t)=P^T(t)\left[P(t)\,P^T(t)\right]^{-1},\ \forall t\in\mathfrak{T}_0, \tag{37.21}$$

Then the error vector $\boldsymbol{\epsilon}_{\rho}(\mathbf{t}^{\rho})$ *is controlled elementwise with the finite vector state reachability time* $\mathbf{t}_R^{\rho}$,

$$\mathbf{T}\left(\varepsilon_{\rho 0}^{1}\right)=\mathbf{0}_{\rho}\Longrightarrow \mathbf{t}_R^{\rho}=T_{1\rho}K_{0\rho}^{-1}\left|\varepsilon_{\rho 0}\right|,\ \varepsilon_{\rho 0}\in\mathfrak{R}^{\rho}, \tag{37.22}$$

3) The control vector function $\mathbf{U}(.)$ *obeys (37.23) if* $\beta = rankP(t) = r = \rho$, $\forall t \in \mathfrak{T}_0$:

$$\mathbf{e}_u\left[\mathbf{U}^{\mu}(t)\right]=\mathbf{e}_u\left[\mathbf{U}^{\mu}\left(t^{-}\right)\right]+P(t)^{-1}\left[T_{1\rho}\boldsymbol{\epsilon}_{\rho}^{(1)}(\mathbf{t}^{\rho})+K_{0\rho}sign\boldsymbol{\epsilon}_{\rho 0}\right],\ \forall \mathbf{t}^{\rho}\in\mathfrak{T}_0^{\rho}, \tag{37.23}$$

Then the error vector $\boldsymbol{\epsilon}_{\rho}(\mathbf{t}^{\rho})$ *is controlled elementwise with the finite vector state reachability time* $\mathbf{t}_R^{\rho}$ *determined by (37.22).*

4) If the plant also possesses Property 61 then the control vector function $\mathbf{U}(.)$ *obeys (37.24),*

$$\mathbf{e}_{u\beta}\left[\mathbf{U}_{\beta}^{\mu}(t)\right]=\mathbf{e}_{u\beta}\left[\mathbf{U}_{\beta}^{\mu}(t^{-})\right]+T_{1\beta}\boldsymbol{\epsilon}_{\beta}^{*(1)}(\mathbf{t}^{\beta})+K_{0\beta}sign\boldsymbol{\epsilon}_{\beta 0}^{*},\ \forall t\in\mathfrak{T}_0. \tag{37.24}$$

Then the error vector $\boldsymbol{\epsilon}_{\beta}^*(\mathbf{t}^{\beta})$ *is controlled elementwise with the finite vector state reachability time* $\mathbf{t}_R^{\beta}$ *determined by (37.18).*

The solution $\varepsilon_{\beta}^*(.)$ *shows that the tracking is stable in the whole with FVIDRT* t_R^{β} *if* $\mathbf{T}\left(\varepsilon_{\beta 0}^{*1}\right)=\mathbf{0}_{\beta}$.

We define $\left|\Psi_R(t)\right|^{1/2}$:

$$\left|\Psi_R(t)\right|^{1/2}=\operatorname{diag}\left\{\left|\epsilon_{R1}(t)\right|^{1/2}\ \ \left|\epsilon_{R2}(t)\right|^{1/2}\ \ \cdots\ \ \left|\epsilon_{R\rho}(t)\right|^{1/2}\right\},$$
$$\text{for } \boldsymbol{\epsilon}_R=\left[\epsilon_{R1}\ \ \epsilon_{R2}\ \ \cdots\ \ \epsilon_{R\rho}\right]^{\mathbf{T}},$$

so that

$$\left|\Psi_{\beta}^*(t)\right|^{1/2}=\operatorname{diag}\left\{\left|\epsilon_{R1}(t)\right|^{1/2}\ \ \left|\epsilon_{R2}(t)\right|^{1/2}\ \ \cdots\ \ \left|\epsilon_{R,\beta-1}(t)\right|^{1/2}\ \ \left|\epsilon_{\beta}^*(t)\right|^{1/2}\right\},$$
$$for\ \boldsymbol{\epsilon}_{\boldsymbol{\beta}}^*=\left[\epsilon_{R1}\ \ \epsilon_{R2}\ \ \cdots\ \ \epsilon_{R,\beta-1}\ \ \epsilon_{\beta}^*\right]^{\mathbf{T}},$$

$$\left|\Psi_{Rr}^{*}(t)\right|^{1/2}=\operatorname{diag}\left\{\left|\epsilon_{R1}(t)\right|^{1/2}\quad\left|\epsilon_{R2}(t)\right|^{1/2}\quad\cdots\quad\left|\epsilon_{R,r-1}(t)\right|^{1/2}\quad\left|\epsilon_{Rr}^{*}(t)\right|^{1/2}\right\},$$
$$for\ \epsilon_{r}^{*}=\left[\epsilon_{R1}\ \ \epsilon_{R2}\ \ \cdots\ \ \epsilon_{R,r-1}\ \ \epsilon_{Rr}^{*}\right]^{\mathbf{T}}.$$

Theorem 531 ***The NTC synthesis for the first-power smooth elementwise stablewise tracking with the finite vector state reachability time*** $\mathbf{t}_R^{\rho}$, ***Example 521***

Let (35.23) through (35.29) be valid.

For the plant (3.12) satisfying Condition (527) to be controlled by the natural tracking control $\mathbf{U}$ *in order to exhibit tracking on* $\mathfrak{T}_{0F}^{2\rho}\times\mathfrak{D}^{j}\times\mathfrak{Y}_{d}^{\alpha}$ *determined by*

$$\begin{gathered}\mathbf{T}\left(\varepsilon_{\beta}^{*},\varepsilon_{\beta}^{*(1)}\right)=T_{1\beta}\varepsilon_{\beta}^{*(1)}(\mathbf{t}^{\beta})+2K_{0\beta}\left|E_{\beta}\left(\mathbf{t}^{\beta}\right)\right|^{1/2}sign\varepsilon_{\beta0}^{*}=\mathbf{0}_{\beta},\\ \forall\mathbf{t}^{\beta}\in\mathfrak{T}_{0F}^{\beta}\ if\ \mathbf{T}\left(\varepsilon_{\beta0}^{*1}\right)=\mathbf{0}_{\beta},\\ \varepsilon_{\beta}^{*1}\left(\mathbf{t}^{2\beta};\epsilon_{\beta0}^{*1}\right)=-\mathbf{f}_{R}^{1}\left(\mathbf{t}^{2\beta}\right),\ \forall\mathbf{t}^{2\beta}\in\mathfrak{T}_{0F}^{2\beta},\ if\ \mathbf{T}\left(\varepsilon_{\beta0}^{*1}\right)\neq\mathbf{0}_{\beta},\\ \left|E_{\beta}^{*}(t)\right|^{1/2}=diag\left\{\left|\varepsilon_{R1}(t)\right|^{1/2}\quad\left|\varepsilon_{R2}(t)\right|^{1/2}\quad\ldots\quad\left|\varepsilon_{R\beta}^{*}(t)\right|^{1/2}\right\},\end{gathered}\tag{37.25}$$

for the first-power smooth elementwise stablewise tracking with the finite vector state reachability time $\mathbf{t}_R^{\beta}$ *if* $\mathbf{T}\left(\varepsilon_{\beta0}^{*1}\right)=\mathbf{0}_{\beta}$ *:*

$$\begin{gathered}\mathbf{t}_{R}^{\beta}=T_{1\beta}K_{0\beta}^{-1}\left|E_{\beta0}^{*}\right|^{1/2}\mathbf{1}_{\beta}=T_{1\beta}K_{0\beta}^{-1}\left|\varepsilon_{\beta0}^{*}\right|^{1/2},\ \varepsilon_{\beta0}^{*}\in\mathfrak{R}^{\beta},\\ \left|\varepsilon_{\beta0}^{*}\right|^{1/2}=\left[\left|\varepsilon_{R1}(0)\right|^{1/2}\quad\left|\varepsilon_{R2}(0)\right|^{1/2}\quad\ldots\quad\left|\varepsilon_{R\beta}^{*}(0)\right|^{1/2}\right]^{T},\end{gathered}\tag{37.26}$$

it is necessary and sufficient that both:

i) The rank β *of the control matrix* $P(t)$ *is greater than zero and not greater than* $min(\rho,r)$*; i.e., (8.20) holds,*

and

ii) The control vector function $\mathbf{U}(.)$ *satisfies one of the conditions 1)-4):*

1) The control vector function $\mathbf{U}(.)$ *obeys (37.27) for* $\beta=rankP(t)=r<\rho,\forall t\in\mathfrak{T}_0$,

$$\mathbf{e}_{u}\left[\mathbf{U}^{\mu}(t)\right]=\mathbf{e}_{u}\left[\mathbf{U}^{\mu}\left(t^{-}\right)\right]+T_{1r}\epsilon_{r}^{*(1)}(\mathbf{t}^{r})+2K_{0r}\left|\Psi_{r}^{*}\left(\mathbf{t}^{r}\right)\right|^{1/2}sign\epsilon_{r0}^{*},\forall t\in\mathfrak{T}_{0}.\tag{37.27}$$

Then the error vector $\epsilon_r^{*(1)}(\mathbf{t}^r)$ *is controlled elementwise with the finite vector state reachability time* $\mathbf{t}_R^r$,

$$\mathbf{T}\left(\varepsilon_{r0}^{*1}\right)=\mathbf{0}_{r}\Longrightarrow\mathbf{t}_{R}^{r}=T_{1r}K_{0r}^{-1}\left|E_{r0}^{*}\right|^{1/2}\mathbf{1}_{r}=T_{1r}K_{0r}^{-1}\left|\varepsilon_{r0}^{*}\right|^{1/2},\ \varepsilon_{r0}^{*}\in\mathfrak{R}^{r}.\tag{37.28}$$

2) The control vector function $\mathbf{U}(.)$ *obeys (37.29) for* $\beta=rankP(t)=\rho<r,\forall t\in\mathfrak{T}_0$,

$$\begin{gathered}\mathbf{e}_{u}\left[\mathbf{U}^{\mu}(t)\right]=\Gamma(t)\left[P(t)\,\mathbf{e}_{u}\left[\mathbf{U}^{\mu}\left(t^{-}\right)\right]\right]+P^{T}(t)\left[\begin{array}{c}T_{1\rho}\epsilon_{\rho}^{(1)}(\mathbf{t}^{\rho})\\ +\ 2K_{0\rho}\left|\Psi_{\rho}\left(\mathbf{t}^{\rho}\right)\right|^{1/2}sign\epsilon_{\rho0}\end{array}\right]\\ \Gamma(t)=P^{T}(t)\left[P(t)\,P^{T}(t)\right]^{-1},\ \forall t\in\mathfrak{T}_{0},\end{gathered}\tag{37.29}$$

Then the error vector $\epsilon_\rho(\mathbf{t}^\rho)$ *is controlled elementwise with the finite vector state reachability time* $\mathbf{t}_R^{\rho}$,

$$\mathbf{T}\left(\varepsilon_{\rho0}^{1}\right)=\mathbf{0}_{\rho}\Longrightarrow\mathbf{t}_{R}^{\rho}=T_{1\rho}K_{0\rho}^{-1}\left|E_{\rho0}\right|^{1/2}\mathbf{1}_{\rho}=\mathbf{t}_{0}^{\rho}+T_{1\rho}K_{0\rho}^{-1}\left|\epsilon_{\rho0}\right|^{1/2},\ \varepsilon_{\rho0}\in\mathfrak{R}^{\rho}.\tag{37.30}$$

3) The control vector function $\mathbf{U}(.)$ *obeys (37.31) if* $\beta=rankP(t)=r=\rho,\forall t\in\mathfrak{T}_0$:

$$\mathbf{e}_{u}\left[\mathbf{U}^{\mu}(t)\right]=\mathbf{e}_{u}\left[\mathbf{U}^{\mu}\left(t^{-}\right)\right]+P(t)^{-1}\left[\begin{array}{c}T_{1\rho}\epsilon_{\rho}^{(1)}(\mathbf{t}^{\rho})\\ +\ 2K_{0R}\left|\Psi_{\rho}\left(\mathbf{t}^{\rho}\right)\right|^{1/2}sign\epsilon_{\rho0}\end{array}\right],\ \forall\mathbf{t}^{\rho}\in\mathfrak{T}_{0}^{\rho},\tag{37.31}$$

Then the error vector $\boldsymbol{\epsilon}_R(\mathbf{t}^\rho)$ *is controlled elementwise with the finite vector state reachability time* $\mathbf{t}_R^\rho$ *determined in (37.30).*

4) If the plant possesses also Property 61 then the control vector function $\mathbf{U}(.)$ *obeys (37.32),*

$$\mathbf{e}_{u\beta}\left[\mathbf{U}_\beta^\mu(t)\right] = \mathbf{e}_{u\beta}\left[\mathbf{U}_\beta^\mu(t^-)\right] + T_{1\beta}\boldsymbol{\epsilon}_\beta^{*(1)}(\mathbf{t}^\beta) + 2K_{0\beta}\left|\Psi_\beta^*\left(\mathbf{t}^\beta\right)\right|^{1/2} sign\boldsymbol{\epsilon}_{\beta 0}^*,\ \forall t \in \mathfrak{T}_0. \quad (37.32)$$

Then the error vector $\boldsymbol{\epsilon}_\beta(\mathbf{t}^\beta)$ *is controlled elementwise with the finite vector state reachability time* $\mathbf{t}_R^\beta$ *is determined in (37.26).*

The solution $\varepsilon_\beta^*(.)$ *shows that the tracking is stable in the whole with FVIDRT* t_R^β *if* $\mathbf{T}\left(\varepsilon_{\beta 0}^{*1}\right) = \mathbf{0}_\beta$.

We define

$$K_\rho = \operatorname{diag}\{k_1\ k_2\ \cdots\ k_\rho\},\ k_i \in \{2,\ 3,\ \cdots\},\ \forall i = 1, 2,\ \cdots,\rho,$$
$$K_\beta = \operatorname{diag}\{k_1\ k_2\ \cdots\ k_\beta\},\ k_i \in \{2,\ 3,\ \cdots\},$$
$$\left|\Psi_\beta^*\left(\mathbf{t}^\beta\right)\right|^{I-K_\beta^{-1}} = \operatorname{diag}\left\{\left|\epsilon_1(t)\right|^{1-k_1^{-1}}\ \ \left|\epsilon_2(t)\right|^{1-k_2^{-1}}\ \cdots\ \left|\epsilon_\beta^*(t)\right|^{1-k_\beta^{-1}}\right\},$$
$$\left|\Psi_\rho\left(\mathbf{t}^\rho\right)\right|^{I-K^{-1}} = \operatorname{diag}\left\{\left|\epsilon_1(t)\right|^{1-k_1^{-1}}\ \ \left|\epsilon_2(t)\right|^{1-k_2^{-1}}\ \cdots\ \left|\epsilon_\rho(t)\right|^{1-k_\rho^{-1}}\right\}.$$

Theorem 532 ***The NTC synthesis for the higher-power smooth elementwise stablewise tracking with the finite vector state reachability time*** $\mathbf{t}_R^\rho$***, Example 522***

Let (35.23) through (35.29) be valid.

For the plant (3.12) satisfying Condition (527) to be controlled by the natural tracking control $\mathbf{U}$ *in order to exhibit tracking on* $\mathfrak{T}_{0F}^{2\rho} \times \mathfrak{D}^j \times \mathfrak{Y}_d^\alpha$ *determined by the tracking algorithm*

$$\mathbf{T}\left(\varepsilon_\beta^*, \varepsilon_\beta^{*(1)}\right) = T_{1\beta}\boldsymbol{\varepsilon}_\beta^{*(1)}(\mathbf{t}^\beta) + K_{0\beta}\left|E_\beta\left(\mathbf{t}^\beta\right)\right|^{I-K_\beta^{-1}} sign\boldsymbol{\varepsilon}_{\beta 0}^* = \mathbf{0}_\beta,$$
$$\forall \mathbf{t}^\beta \in \mathfrak{T}_{0F}^\beta\ if\ \mathbf{T}\left(\boldsymbol{\varepsilon}_{\beta 0}^{*1}\right) = \mathbf{0}_\beta,$$
$$\varepsilon_\beta^{*1}\left(\mathbf{t}^{2\beta}; \boldsymbol{\epsilon}_{\beta 0}^{*1}\right) = -\mathbf{f}_\beta^1\left(\mathbf{t}^{2\beta}\right),\ \forall \mathbf{t}^{2\beta} \in \mathfrak{T}_{0F}^{2\beta},\ if\ \mathbf{T}\left(\boldsymbol{\varepsilon}_{\beta 0}^{*1}\right) \neq \mathbf{0}_\beta, \quad (37.33)$$

for the higher-power smooth elementwise stablewise tracking with the finite vector state reachability time $\mathbf{t}_R^\beta$,

$$\mathbf{T}\left(\varepsilon_{\beta 0}^{*1}\right) = \mathbf{0}_\beta \Longrightarrow \mathbf{t}_R^\beta = T_{1\beta}K_\beta K_{0\beta}^{-1}\left|\boldsymbol{\varepsilon}_{\beta 0}^*\right|^{K_\beta^{-1}},\ \boldsymbol{\varepsilon}_{\beta 0}^* \in \mathfrak{R}^\beta, \quad (37.34)$$

it is necessary and sufficient that both:

i) The rank β *of the control matrix* $P(t)$ *is greater than zero and not greater than* $min(\rho, r)$*, i.e., (8.20) holds,*

and

ii) The control vector function $\mathbf{U}(.)$ *satisfies one of the conditions 1) - 4):*

1) The control vector function $\mathbf{U}(.)$ *obeys (37.35) for* $\beta = rankP(t) = r < \rho$, $\forall t \in \mathfrak{T}_0$,

$$\mathbf{e}_u\left[\mathbf{U}^\mu(t)\right] = \mathbf{e}_u\left[\mathbf{U}^\mu\left(t^-\right)\right] + T_{1r}\boldsymbol{\epsilon}_r^{*(1)}(\mathbf{t}^r) + K_{0r}\left|\Psi_r^*\left(\mathbf{t}^r\right)\right|^{I-K_r^{-1}} sign\boldsymbol{\epsilon}_{r0}^*,$$
$$\forall t \in \mathfrak{T}_0. \quad (37.35)$$

Then the error vector $\boldsymbol{\epsilon}_r^{*(1)}(\mathbf{t}^r)$ *is controlled elementwise with the finite vector state reachability time* $\mathbf{t}_R^r$,

$$\mathbf{T}\left(\varepsilon_{r0}^{*1}\right) = \mathbf{0}_r \Longrightarrow \mathbf{t}_R^r = T_{1r}K_r K_{0r}^{-1}\left|\varepsilon_{r0}^*\right|^{K^{-1}},\ \varepsilon_{r0}^* \in \mathfrak{R}^r.$$

2) The control vector function $\mathbf{U}(.)$ *obeys (37.36) for* $\beta = rankP(t) = \rho < r, \forall t \in \mathfrak{T}_0$,

$$\mathbf{e}_u\left[\mathbf{U}^{\mu}(t)\right] = \left\{ \begin{array}{c} \Gamma(t)\left[P(t)\,\mathbf{e}_u\left[\mathbf{U}^{\mu}(t^-)\right]\right] \\ +\,P^T(t)\left[\begin{array}{c} T_{1\rho}\boldsymbol{\epsilon}_{\rho}^{(1)}(\mathbf{t}^{\rho}) \\ +\,K_{0\rho}\left|\Psi_{\rho}(\mathbf{t}^{\rho})\right|^{I-K_{\rho}^{-1}} sign\boldsymbol{\epsilon}_{\rho 0} \end{array}\right] \end{array} \right\},$$

$$\Gamma(t) = P^T(t)\left[P(t)\,P^T(t)\right]^{-1}, \ \forall t \in \mathfrak{T}_0. \tag{37.36}$$

Then the error vector $\boldsymbol{\epsilon}_{\rho}^{(1)}(\mathbf{t}^{\rho})$ *is controlled elementwise with the finite vector state reachability time* $\mathbf{t}_R^{\rho}$,

$$\mathbf{T}\left(\varepsilon_{\rho 0}^{1}\right) = \mathbf{0}_{\rho} \Longrightarrow \mathbf{t}_R^{\rho} = T_{1\rho}K_{\rho}K_{0\rho}^{-1}\left|\varepsilon_{\rho 0}\right|^{K_R^{-1}}, \ \varepsilon_{\rho 0} \in \mathfrak{R}^{\rho}.$$

3)The control vector function $\mathbf{U}(.)$ *obeys (37.37) if* $\beta = rankP(t) = r = \rho, \forall t \in \mathfrak{T}_0$:

$$\mathbf{e}_u\left[\mathbf{U}^{\mu}(t)\right] = \mathbf{e}_u\left[\mathbf{U}^{\mu}(t^-)\right] + P(t)^{-1}\left[\begin{array}{c} T_{1\rho}\boldsymbol{\epsilon}_{\rho}^{(1)}(\mathbf{t}^{\rho}) \\ +\,K_{0\rho}\left|\Psi_{\rho}(\mathbf{t}^{\rho})\right|^{I-K_{\rho}^{-1}} sign\boldsymbol{\epsilon}_{\rho 0} \end{array}\right],$$

$$\forall \mathbf{t}^{\rho} \in \mathfrak{T}_0^{\rho}. \tag{37.37}$$

4) If the plant also possesses Property 61 then the control vector function $\mathbf{U}(.)$ *obeys (37.38),*

$$\mathbf{e}_{u\beta}\left[\mathbf{U}_{\beta}^{\mu}(t)\right] = \mathbf{e}_{u\beta}\left[\mathbf{U}_{\beta}^{\mu}(t^-)\right] + T_{1\beta}\boldsymbol{\epsilon}_{\beta}^{*(1)}(\mathbf{t}^{\beta}) + K_{0\beta}\left|\Psi_{\beta}^{*}(\mathbf{t}^{\beta})\right|^{I-K_{\rho}^{-1}} sign\boldsymbol{\epsilon}_{\beta 0}^{*},$$

$$\forall t \in \mathfrak{T}_0. \tag{37.38}$$

The solution $\varepsilon_{\beta}^{*}(.)$ *shows that the tracking is stable in the whole with FVIDRT* t_R^{β} *if* $\mathbf{T}\left(\varepsilon_{\beta 0}^{*1}\right) = \mathbf{0}_{\beta}$.

Theorem 533 ***The NTC synthesis for the sharp absolute error vector value tracking elementwise and stablewise with the finite vector state reachability time*** $\mathbf{t}_R^{\rho}$***, Example 523***

Let (35.23) through (35.29) be valid.

For the plant (3.12) satisfying Condition (527) to be controlled by the natural tracking control $\mathbf{U}$ *in order to exhibit tracking on* $\mathfrak{T}_{0F}^{2\rho} \times \mathfrak{D}^j \times \mathfrak{Y}_d^{\alpha}$ *determined by the following tracking algorithm*

$$\mathbf{T}\left(\varepsilon_{\beta}^{*}, \varepsilon_{\beta}^{*(1)}\right) = T_{1\beta}\Sigma\left(\varepsilon_{\beta}^{*}, \varepsilon_{\beta}^{*(1)}\right)\varepsilon_{\beta}^{*(1)} + K_{0\beta}sign\left|\varepsilon_{\beta 0}^{*}\right| = \mathbf{0}_{\beta},$$

$$\forall \mathbf{t}^{\beta} \in \mathfrak{T}_{0F}^{\beta} \ if \ \mathbf{T}\left(\mathbf{0}_{\beta}, \boldsymbol{\epsilon}_{\beta 0}^{*1}\right) = \mathbf{0}_{\beta},$$

$$\varepsilon_{\beta}^{*1}\left(\mathbf{t}^{2\beta}; \mathbf{t}_0^{2\beta}; \boldsymbol{\epsilon}_{\beta 0}^{*\eta}\right) = -\mathbf{f}_{\beta}^{1}\left(\mathbf{t}^{2\beta}\right), \ \forall \mathbf{t}^{2\beta} \in \mathfrak{T}_{0F}^{2\beta}, \ if \ \mathbf{T}\left(\mathbf{0}_{\beta}, \boldsymbol{\epsilon}_{\beta 0}^{*1}\right) \neq \mathbf{0}_{\beta}, \tag{37.39}$$

for the sharp absolute error vector value tracking elementwise stablewise with the finite vector state reachability time $\mathbf{t}_R^{\beta}$,

$$\mathbf{T}\left(\mathbf{0}_{\beta}, \boldsymbol{\epsilon}_{\beta 0}^{*1}\right) = \mathbf{0}_{\beta} \Longrightarrow \mathbf{t}_R^{\beta} = T_1K_0^{-1}\left|\boldsymbol{\epsilon}_{\beta 0}^{*}\right|, \ \varepsilon_{\beta 0}^{*} \in \mathfrak{R}^{\beta},$$

it is necessary and sufficient that both:

i) The rank β *of the control matrix* $P(t)$ *is greater than zero and not greater than* $min(\rho, r)$*, i.e., (8.20) holds,*

and

ii) The control vector function $\mathbf{U}(.)$ *satisfies one of the conditions 1)-4):*

1) The control vector function $\mathbf{U}(.)$ *obeys (37.40) for* $\beta = rankP(t) = r < \rho$, $\forall t \in \mathfrak{T}_0$,

$$\mathbf{e}_u\left[\mathbf{U}^{\mu}(t)\right] = \mathbf{e}_u\left[\mathbf{U}^{\mu}(t)\right] + T_{1r}\Sigma\left(\boldsymbol{\epsilon}_r^*, \boldsymbol{\epsilon}_r^{*(1)}\right)\boldsymbol{\epsilon}_r^{*(1)} + K_{0r}\,sign\left|\boldsymbol{\epsilon}_{r0}^*\right|, \forall t \in \mathfrak{T}_0. \quad (37.40)$$

2) The control vector function $\mathbf{U}(.)$ *obeys (37.41) for* $\beta = rankP(t) = \rho < r$, $\forall t \in \mathfrak{T}_0$,

$$\mathbf{e}_u\left[\mathbf{U}^{\mu}(t)\right] = \Gamma(t)\left[P(t)\,\mathbf{e}_u\left[\mathbf{U}^{\mu}(t)\right]\right] + P^T(t)\left[\begin{array}{c} T_{1R}\Sigma\left(\boldsymbol{\epsilon}_R, \boldsymbol{\epsilon}_R^{(1)}\right)\boldsymbol{\epsilon}_R^{(1)} \\ + K_{0R}\,sign\left|\boldsymbol{\epsilon}_{R0}\right| \end{array}\right],$$

$$\Gamma(t) = P^T(t)\left[P(t)\,P^T(t)\right]^{-1}, \ \forall t \in \mathfrak{T}_0, \quad (37.41)$$

3) The control vector function $\mathbf{U}(.)$ *obeys (37.42) if* $\beta = rankP(t) = r = \rho$, $\forall t \in \mathfrak{T}_0$:

$$\mathbf{e}_u\left[\mathbf{U}^{\mu}(t)\right] = \mathbf{e}_u\left[\mathbf{U}^{\mu}(t)\right] + P(t)^{-1}\left[\begin{array}{c} T_{1R}\Sigma\left(\boldsymbol{\epsilon}_R, \boldsymbol{\epsilon}_R^{(1)}\right)\boldsymbol{\epsilon}_R^{(1)} \\ + K_{0R}\,sign\left|\boldsymbol{\epsilon}_{R0}\right| \end{array}\right], \ \forall t \in \mathfrak{T}_0, \quad (37.42)$$

4) If the plant also possesses Property 61 then the control vector function $\mathbf{U}(.)$ *obeys (37.43),*

$$\mathbf{e}_{u\beta}\left[\mathbf{U}_{\beta}^{\mu}(t)\right] = \mathbf{e}_{u\beta}\left[\mathbf{U}_{\beta}^{\mu}(t^-)\right] + T_{1\beta}\Sigma\left(\boldsymbol{\epsilon}_{\beta}^*, \boldsymbol{\epsilon}_{\beta}^{*(1)}\right)\boldsymbol{\epsilon}_{\beta}^{*(1)} + K_{0\beta}\,sign\left|\boldsymbol{\epsilon}_{\beta 0}^*\right|,$$
$$\forall t \in \mathfrak{T}_0. \quad (37.43)$$

The solution $\boldsymbol{\varepsilon}_{\beta}^*(.)$ *shows that the tracking is stable in the whole with FVIDRT* t_R^{ρ} *if* $\mathbf{T}\left(\mathbf{0}_{\beta}, \boldsymbol{\epsilon}_{\beta 0}^{*1}\right) = \mathbf{0}_{\beta}$.

Theorem 534 ***NTC synthesis for the exponential absolute error vector value tracking elementwise and stablewise with the finite vector state reachability time*** t_R^{ρ}, ***Example 524***

Let (35.23) through (35.29) be valid.

For the plant (3.12) satisfying Condition (527) to be controlled by the natural tracking control $\mathbf{U}$ *in order to exhibit tracking on* $\mathfrak{T}_{0F}^{2\rho} \times \mathfrak{D}^j \times \mathfrak{Y}_d^{\alpha}$ *determined by*

$$\mathbf{T}\left(\boldsymbol{\varepsilon}_{\beta}^*, \boldsymbol{\varepsilon}_{\beta}^{*(1)}\right) = T_{1\beta}D^+\left|\boldsymbol{\varepsilon}_{\beta}^*\right| + K_{\beta}\left(\left|\boldsymbol{\varepsilon}_{\beta}^*\right| + K_{0\beta}\,sign\left|\boldsymbol{\varepsilon}_{\beta 0}^*\right|\right)$$
$$= T_{1\beta}\Sigma\left(\boldsymbol{\varepsilon}_{\beta,\beta}^{*\ *(1)}\right)\boldsymbol{\varepsilon}_{\beta}^{*(1)} + K_{\beta}\left(\left|\boldsymbol{\varepsilon}_{\beta}^*\right| + K_{0\beta}\,sign\left|\boldsymbol{\varepsilon}_{\beta 0}^1\right|\right) = \mathbf{0}_{\beta},$$
$$\forall \mathbf{t}^{\beta} \in \mathfrak{T}_{0F}^{\beta}\ if\ \mathbf{T}\left(\boldsymbol{\varepsilon}_{\beta 0}^{*1}\right) = \mathbf{0}_{\beta},$$
$$\boldsymbol{\varepsilon}_{\beta}^*\left(\mathbf{t}^{2\beta}; \mathbf{t}_0^{2\beta}; \boldsymbol{\varepsilon}_{\beta 0}^*\right) = \mathbf{f}_{\beta}^1\left(\mathbf{t}^{2\beta}\right), \ \forall \mathbf{t}^{2\beta} \in \mathfrak{T}_{0F}^{2\beta},\ if\ \mathbf{T}\left(\boldsymbol{\varepsilon}_{\beta 0}^{*1}\right) \neq \mathbf{0}_{\beta}, \quad (37.44)$$

for the exponential absolute error vector value tracking elementwise and stablewise with the finite vector state reachability time t_R^{ρ} *(37.45), (37.46):*

$$T_{R\beta} = \left\{\begin{array}{ll} K_{\beta}^{-1}T_{1\beta}ln\left\{K_{0\beta}^{-1}\left[\left|\Psi_{\beta 0}^*\right| + K_{0\beta}\right]\right\}, & \boldsymbol{\epsilon}_{\beta 0}^* \neq \mathbf{0}_{\beta}, \\ O_{\beta}, & \boldsymbol{\epsilon}_{\beta 0}^* = \mathbf{0}_{\beta} \end{array}\right\}, \quad (37.45)$$

$$T_{R\beta} = diag\left\{t_{R1}\ t_{R2}\ \cdots\ t_{R\beta}\right\} \Longleftrightarrow \mathbf{t}_R^{\beta} = \left[t_{R1}\ t_{R2}\ \cdots\ t_{R\beta}\right]^T. \quad (37.46)$$

it is necessary and sufficient that both:

i) The rank β *of the control matrix* $P(t)$ *is greater than zero and not greater than* $min(\rho, r)$*; i.e., (8.20) holds,*

and

ii) The control vector function $\mathbf{U}(.)$ *satisfies one of the conditions 1)-4):*

1) the control vector function $\mathbf{U}(.)$ *obeys (37.47) for* $\beta = rankP(t) = r < \rho$, $\forall t \in \mathfrak{T}_0$,

$$\mathbf{e}_u\left[\mathbf{U}^{\mu}(t)\right] = \mathbf{e}_u\left[\mathbf{U}^{\mu}\left(t^-\right)\right] + T_{1r}D^+\left|\boldsymbol{\epsilon}_r^*\right| + K_r\left(\left|\boldsymbol{\epsilon}_r^*\right| + K_{0r}sign\left|\boldsymbol{\epsilon}_{r0}^*\right|\right), \forall t \in \mathfrak{T}_0. \tag{37.47}$$

2) The control vector function $\mathbf{U}(.)$ *obeys (37.48) for* $\beta = rankP(t) = \rho < r$, $\forall t \in \mathfrak{T}_0$,

$$\mathbf{e}_u\left[\mathbf{U}^{\mu}(t)\right] = \Gamma(t)\left[P(t)\,\mathbf{e}_u\left[\mathbf{U}\left(t^-\right)\right]\right] + P^T(t)\begin{bmatrix} T_{1R}D^+\left|\boldsymbol{\epsilon}_R\right| \\ +\,K_R\left(\left|\boldsymbol{\epsilon}_R\right| + K_{0R}sign\left|\boldsymbol{\epsilon}_{R0}\right|\right)\end{bmatrix}$$

$$\Gamma(t) = P^T(t)\left[P(t)\,P^T(t)\right]^{-1}, \ \forall t \in \mathfrak{T}_0, \tag{37.48}$$

3) The control vector function $\mathbf{U}(.)$ *obeys (37.49) if* $\beta = rankP(t) = r = \rho$, $\forall t \in \mathfrak{T}_0$:

$$\mathbf{e}_u\left[\mathbf{U}^{\mu}(t)\right] = \mathbf{e}_u\left[\mathbf{U}\left(t^-\right)\right] + P(t)^{-1}\begin{bmatrix} T_{1R}D^+\left|\boldsymbol{\epsilon}_R\right| \\ +\,K_R\left(\left|\boldsymbol{\epsilon}_R\right| + K_{0R}sign\left|\boldsymbol{\epsilon}_{R0}\right|\right)\end{bmatrix}, \quad \forall t \in \mathfrak{T}_0. \tag{37.49}$$

4) If the plant also Possesses Property 61 then the control vector function $\mathbf{U}(.)$ *obeys (37.50),*

$$\mathbf{e}_{u\beta}\left[\mathbf{U}_{\beta}^{\mu}(t)\right] = \mathbf{e}_{u\beta}\left[\mathbf{U}_{\beta}^{\mu}(t^-)\right] + T_{1\beta}D^+\left|\boldsymbol{\epsilon}_{\beta}^*\right| + K_{\beta}\left(\left|\boldsymbol{\epsilon}_{\beta}^*\right| + K_{0\beta}sign\left|\boldsymbol{\epsilon}_{\beta 0}^*\right|\right), \quad \forall t \in \mathfrak{T}_0. \tag{37.50}$$

The solution $\boldsymbol{\varepsilon}_{\beta}^*(.)$ *shows that the tracking is stable in the whole with FVIDRT* t_R^{ρ} *if* $\mathbf{T}\left(\mathbf{0}_{\rho}, \boldsymbol{\epsilon}_{\beta 0}^{*1}\right) = \mathbf{0}_{\beta}$.

The following comments generalize the corresponding comments in [279, Comments 384-389, pp. 278, 279].

Comment 535 *The above-synthesized NTC control* $\mathbf{U}$ *does not demand knowledge of the state of the plant.*

Comment 536 ***Implementation of the NTC algorithm***

The controller should comprise the corresponding component that generates the subsidiary vector function $\mathbf{f}_R^1(.)$ *defined by (35.36),(35.37).*

The controller uses information about the subsidiary vector function $\mathbf{f}_R^1(.)$ *and about the real output error vector* $\boldsymbol{\varepsilon}_{\beta}^*$*, determines the subsidiary reference output vector* $\mathbf{R}_R$*, (35.34), and the subsidiary error vector* $\boldsymbol{\epsilon}_R$*, (35.28), and uses them to generate the NTC according to the accepted algorithm.*

The natural tracking controller does not use information about the plant mathematical model. It does not use any information about the disturbance vector function $\mathbf{D}(.)$ *or about high derivatives of the real state vector* $\mathbf{R}^{\alpha-1}(t)$ *at the moment* t*. The NTC is fully robust. The preceding control algorithms verify these claims.*

The usage of $\mathbf{U}(t^-)$ *in order to generate* $\mathbf{U}(t)$ *at the moment* t *expresses a kind of memory and expresses the adaptability of the NTC.*

The transmission of the signal from the controller output to its input without delay, or with negligible delay, is the crucial demand on the controller.

Chapter 38

Tracking quality: Both spaces

38.1 Both spaces (BS) tracking operator

Let $\boldsymbol{\epsilon}^*_{\boldsymbol{\theta}}$,

$$\boldsymbol{\varepsilon}^*_{\theta} = \begin{bmatrix} \boldsymbol{\varepsilon}^*_{\boldsymbol{\beta}} \\ \boldsymbol{\varepsilon}^*_{\gamma} \end{bmatrix} \in \mathfrak{R}^{\theta}, \ \ \boldsymbol{\epsilon}^*_{\boldsymbol{\theta}} = \begin{bmatrix} \boldsymbol{\epsilon}^*_{\boldsymbol{\beta}} \\ \boldsymbol{\epsilon}^*_{\gamma} \end{bmatrix} \in \mathfrak{R}^{\theta}, \ \theta = \beta + \gamma, \tag{38.1}$$

and

$$\boldsymbol{\varepsilon}^*_{2r} = \begin{bmatrix} \boldsymbol{\varepsilon}^*_{Rr} \\ \boldsymbol{\varepsilon}^*_{r} \end{bmatrix} \in \mathfrak{R}^{2r}, \ \ \boldsymbol{\epsilon}^*_{2r} = \begin{bmatrix} \boldsymbol{\epsilon}^*_{Rr} \\ \boldsymbol{\epsilon}^*_{r} \end{bmatrix} \in \mathfrak{R}^{2r}, \tag{38.2}$$

be the subvectors of the joint error vectors $\boldsymbol{\varepsilon}_{\boldsymbol{\mu}}$ and $\boldsymbol{\epsilon}_{\boldsymbol{\mu}}$,

$$\boldsymbol{\varepsilon}_{\mu} = \begin{bmatrix} \boldsymbol{\varepsilon}_{R} \\ \boldsymbol{\varepsilon} \end{bmatrix} \in \mathfrak{R}^{\rho+N}, \ \ \boldsymbol{\epsilon}_{\mu} = \begin{bmatrix} \boldsymbol{\epsilon}_{R} \\ \boldsymbol{\epsilon} \end{bmatrix} \in \mathfrak{R}^{\rho+N}, \ \mu = \rho + N, \tag{38.3}$$

where $\boldsymbol{\varepsilon}^*_{\boldsymbol{\beta}}$ is defined by (35.3), (35.4), $\boldsymbol{\epsilon}^*_{\boldsymbol{\beta}}$ by (35.31), $\boldsymbol{\varepsilon}^*_{\gamma}$ by (32.2), and $\boldsymbol{\epsilon}^*_{\gamma}$ by (32.31). The vectors $\boldsymbol{\varepsilon}^*_{\boldsymbol{\beta}}$ and $\boldsymbol{\epsilon}^*_{\boldsymbol{\beta}}$ become the vectors $\boldsymbol{\varepsilon}^*_{Rr}$ and $\boldsymbol{\epsilon}^*_{Rr}$ when β is equal to r, respectively. Analogously, the vectors $\boldsymbol{\varepsilon}^*_{\gamma}$ and $\boldsymbol{\epsilon}^*_{\gamma}$ become the vectors $\boldsymbol{\varepsilon}^*_{r}$ and $\boldsymbol{\epsilon}^*_{r}$ when γ is equal to r, respectively.

The description of the tracking quality in both spaces is essentially the same as in the output space, Chapter 32, and as in the *state* space, Chapter 35. Comparing with Chapter 32 and Chapter 35 the difference is only in the joint notation defined by (38.1)-(38.3).

If the plant (3.12) possesses both Property 61 and Property 64 then Lemma 62 and Lemma 65 determine that the dimension of the controllable part (3.38), (3.47) of the plant is equal to $\theta = \beta + \gamma$, with $\beta \leq$min(ρ, r) and $\gamma \leq$min(N, r), Only the $(\beta + \gamma)$-subvector $\left[\mathbf{R}^T_{\beta} \ \mathbf{Y}^T_{\gamma}\right]^T \in \mathfrak{R}^{\theta}$, (35.1) and (32.1), can be then controlled directly; i.e., elementwise, Axiom 87, Property 64, Lemma 62, Property 64, and Lemma 65. The whole joint vector $\left[\mathbf{R}^T \ \mathbf{Y}^T\right]^T \in \mathfrak{R}^{\rho+N} = \mathfrak{R}^{\theta}$, hence the whole joint error vectors $\boldsymbol{\varepsilon}_{\mu} \in \mathfrak{R}^{\mu}$ and $\boldsymbol{\epsilon}_{\mu}$, (38.3),

$$\boldsymbol{\varepsilon}_{\mu} = \begin{bmatrix} \boldsymbol{\varepsilon}_{\mu} \\ \boldsymbol{\varepsilon} \end{bmatrix} = \begin{bmatrix} \mathbf{R}_d - \mathbf{R} \\ \mathbf{Y}_d - \mathbf{Y} \end{bmatrix} = \left[\varepsilon_{\mu 1} \ \varepsilon_{\mu 2} .. \varepsilon_{\mu,\mu}\right]^T \in \mathfrak{R}^{\mu}, \tag{38.4}$$

$$\boldsymbol{\epsilon}_{\mu} = \begin{bmatrix} \boldsymbol{\epsilon}_{R} \\ \boldsymbol{\epsilon} \end{bmatrix} = \begin{bmatrix} \mathbf{R}_R - \mathbf{R} \\ \mathbf{Y}_R - \mathbf{Y} \end{bmatrix} = \left[\epsilon_{\mu 1} \ \epsilon_{\mu 2} .. \epsilon_{\mu,\mu}\right]^T \in \mathfrak{R}^{\mu}, \tag{38.5}$$

should be controlled at least indirectly. The θ-subvector $\left[\mathbf{R}^T_{\beta} \ \mathbf{Y}^T_{\gamma}\right]^T$ of $\left[\mathbf{R}^T \ \mathbf{Y}^T\right]^T$ does not contain either the entries $R_{\beta+1}, R_{\beta+2}, ..R_{\rho}$ of $\mathbf{R}$ or the entries $Y_{\gamma+1}, Y_{\gamma+2}, ..Y_N$ of $\mathbf{Y}$. This implies that the θ-subvector $\left[\boldsymbol{\varepsilon}^T_{\beta} \ \boldsymbol{\varepsilon}^T_{\gamma}\right]^T$ of $\left[\boldsymbol{\varepsilon}^T_{R} \ \boldsymbol{\varepsilon}^T\right]^T$ does not contain either the errors

$\varepsilon_{R,\beta+1}$, $\varepsilon_{R,\beta+2}$, $..\varepsilon_{R\rho}$ of ε_R or the errors $\varepsilon_{\gamma+1}$, $\varepsilon_{\gamma+2}$, $..\varepsilon_N$ of ε. The same holds for the θ-subvector $\left[\epsilon_\beta^T \;\; \epsilon_\gamma^T\right]^T$ of $\left[\epsilon_R^T \;\; \epsilon^T\right]^T$.

Let

$$E_\theta^*(t) = \text{blockdiag}\left\{E_\beta^*(t) \quad E_\gamma^*(t)\right\}, \; \forall t \in \mathfrak{T}_0^{\alpha\theta}. \tag{38.6}$$

Task 537 ***Tracking quality and the control task in both spaces***
The control should force the plant joint behavior $\left[\mathbf{R}^T(.) \; \mathbf{Y}^T(.)\right]^T$ *to satisfy the demanded tracking quality specified by (38.7):*

$$\mathbf{T}\left(t, \varepsilon_\theta^{*\alpha-1}(t), \int_{t_0}^{t} E_\theta^*(t)\, dt\right) = \mathbf{0}_m, \; \forall t \in \mathfrak{T}_{0F}^{\alpha\theta}. \tag{38.7}$$

The basic task of the quality tracking control synthesis is to determine such control.

The equation (38.7) holds in the cases for which the reachability moment t_R is the same for all entries of the modified error vector ε_θ^*, (38.1), and for their derivatives. Analogously, the final moment t_F is the same for all entries of the modified error vector ε_θ^* and for their derivatives.

However, if different reachability *times* $t_{Ri,(j)}$ and/or different final *times* $t_{Fi,(j)}$ are associated with the elements of the modified error vector ε_θ^* and/or with the elements of the j-th derivative $\varepsilon_\theta^{*(j)}$ of ε_θ^*, $j = 1, 2, \cdots, \alpha - 1$, then we use:

$$\mathbf{t}^{\alpha\theta} = [t \;\; t \; ... \; t]^T \in \mathfrak{T}^{\alpha\theta} \text{ related to } \varepsilon_\theta^{*\alpha-1}.$$

The use of the *time vector* $\mathbf{t}^{\alpha\theta}$ permits us to simplify, formally mathematically, the treatment of the vector relationships. This has the full advantage in the framework of elementwise tracking with the finite vector reachability time $\mathbf{t}_R^{\alpha\theta}$. For the same reason we introduce both spaces (*BS*) error diagonal matrices $E_\theta^*(\mathbf{t}^\theta)$ and $E(\mathbf{t}^\rho)$,

$$E_\theta^*(\mathbf{t}^\theta) = \text{blockdiag}\left\{E_\beta^*(\mathbf{t}^\beta) \quad E_\gamma^*(\mathbf{t}^\gamma)\right\}, \; E_\theta(\mathbf{t}^\theta) = E_{\rho+N}(\mathbf{t}^{\rho+N}), \tag{38.8}$$

where $E_\beta^*(t)$ is determined by (35.9) and (32.9) defines $E_\gamma^*(t)$. Let

$$\begin{aligned}\varepsilon_\theta^{*\alpha-1}(\mathbf{t}^{\alpha\theta}) &\equiv \varepsilon_\theta^{*\alpha-1}(\mathbf{t}^{\alpha\theta}; \mathbf{t}_0^{\alpha\theta}; \varepsilon_{\theta0}^{*\alpha-1}) \\ &\equiv \varepsilon_\theta^{*\alpha-1}(\mathbf{t}^{\alpha\theta}; \mathbf{t}_0^{\alpha\theta}; \varepsilon_{\theta0}^{*\alpha-1}; \mathbf{D}; \mathbf{U}; \mathbf{R}_d),\end{aligned} \tag{38.9}$$

so that

$$\varepsilon_\theta^*(\mathbf{t}^\theta) \equiv \varepsilon_\theta^*(\mathbf{t}^\theta; \mathbf{t}_0^\theta; \varepsilon_{\theta0}) \equiv \varepsilon_\theta^*(\mathbf{t}^\theta; \mathbf{t}_0^\theta; \varepsilon_{\theta0}; \mathbf{D}; \mathbf{U}; \mathbf{R}_d). \tag{38.10}$$

This notation simplifies formally (38.7) to:

$$\mathbf{T}\left(\mathbf{t}^m, \varepsilon_\theta^{*\alpha-1}(\mathbf{t}^{\alpha\theta}), \int_{\mathbf{t}_0^\rho}^{\mathbf{t}^\rho} E_\theta^*(\mathbf{t}^\theta)\, d\mathbf{t}^\theta\right) = \mathbf{0}_m, \;\; \forall \mathbf{t}^{\alpha\theta} \in \mathfrak{T}_{0F}^{\alpha\theta}. \tag{38.11}$$

The control task 537 determines the tracking control goal:

Goal 538 ***The basic tracking control goal***
The tracking control should force the plant BS error behavior $\varepsilon_\theta^*(.)$ *to obey the tracking quality determined by (38.11).*

38.2 Tracking operator properties

The following properties of *the vector tracking operator* $T(.)$, (38.11),

$$\mathbf{T}(.): \mathfrak{T}_{0F}^{m} \times \mathfrak{R}^{\alpha\theta} \times \mathfrak{R}^{\theta} \longrightarrow \mathfrak{R}^{m}, m \leq \theta \leq [\min(\rho, r) + \min(N, r)], \tag{38.12}$$

determine the class of *the tracking algorithms* that express the required tracking qualities. They are the basis for the control synthesis to fulfill the high quality demand.

Property 539 ***Continuity of the error solution***
The solution $\varepsilon_{\theta}^{*\alpha-1}(\mathbf{t}^{\alpha\theta}) = \varepsilon_{\theta}^{*}(\mathbf{t}^{\theta}; \mathbf{t}_{0}^{\theta}; \varepsilon_{\theta 0}^{\alpha-1})$ *of (38.11) is continuous in time on* $\mathfrak{T}_{0F}^{\alpha\theta}$,

$$\mathbf{T}\left(\mathbf{t}^{m}, \varepsilon_{\theta}^{*\alpha-1}(\mathbf{t}^{\alpha\theta}), \int_{\mathbf{t}_{0}^{\theta}}^{\mathbf{t}^{\theta}} E_{\theta}^{*}(\mathbf{t}^{\theta})\, d\mathbf{t}^{\theta}\right) = \mathbf{0}_{m}, \quad \forall \mathbf{t}^{\alpha\theta} \in \mathfrak{T}_{0F}^{\alpha\theta}$$
$$\Longrightarrow \varepsilon_{\theta}^{*}(\mathbf{t}^{\theta}; \mathbf{t}_{0}^{\theta}; \varepsilon_{\theta 0}^{*\alpha-1}) \in \mathfrak{C}(\mathfrak{T}_{0F}^{\theta}). \tag{38.13}$$

Property 540 ***The operator*** $T(.)$ ***vanishes at the origin***
The operator $T(.)$ *vanishes at the origin at every moment,*

$$\varepsilon_{\theta}^{*\alpha-1} = \mathbf{0}_{\alpha\theta} \Longrightarrow \mathbf{T}\left(\mathbf{t}^{m}, \mathbf{0}_{\alpha\theta}, \int_{\mathbf{t}_{0}^{\theta}}^{\mathbf{t}_{0}^{\theta}} O_{\theta} d\mathbf{t}^{\theta}\right) = \mathbf{0}_{m}, \quad \forall \mathbf{t}^{\theta} \in \mathfrak{T}_{0F}^{\theta}. \tag{38.14}$$

Property 541 ***The solution of*** *(38.11)* ***for all zero initial conditions***
The solution of (38.11) for all zero initial conditions is identically equal to the zero vector,

$$\varepsilon_{\theta 0}^{*\alpha-1} = \mathbf{0}_{\alpha\theta} \quad \text{and}$$
$$\mathbf{T}\left(\mathbf{t}^{m}, \varepsilon_{\theta}^{*\alpha-1}(\mathbf{t}^{\alpha\theta}), \int_{\mathbf{t}_{0}^{\theta}}^{\mathbf{t}^{\theta}} E_{\theta}^{*}(\mathbf{t}^{\theta})\, d\mathbf{t}^{\theta}\right) = \mathbf{0}_{m}, \quad \forall \mathbf{t}^{\alpha\theta} \in \mathfrak{T}_{0F}^{\alpha\theta} \Longrightarrow$$
$$\varepsilon_{\theta}^{*}(\mathbf{t}^{\theta}; \mathbf{t}_{0}^{\theta}; \mathbf{0}_{\alpha\theta}) = \mathbf{0}_{\theta}, \quad \forall \mathbf{t}^{\theta} \in \mathfrak{T}_{0F}^{\theta}. \tag{38.15}$$

Note 542 *If (38.11) holds, then the condition* $\forall \mathbf{t}^{\alpha\theta} \in \mathfrak{T}_{0F}^{\alpha\theta}$ *demands that the initial BS error vector, its derivatives, and its integral also obey (38.11) at the initial moment* $\mathbf{t}_{0}^{m}$.

Property 543 ***The operator*** $T(.)$ ***vanishes at the initial moment***
The initial BS error vector, its initial derivatives, and its integral obey (38.11) at the initial moment t_0, *i.e., at* $\mathbf{t}_{0}^{m}$:

$$\mathbf{T}\left(\mathbf{t}_{0}^{m}, \varepsilon_{\theta}^{*\alpha-1}(\mathbf{t}^{\alpha\theta}), \int_{\mathbf{t}_{0}^{\theta}}^{\mathbf{t}^{\theta}} E_{\theta}^{*}(\mathbf{t}^{\theta})\, d\mathbf{t}^{\theta}\right) = \mathbf{T}(\mathbf{t}_{0}^{m}, \varepsilon_{\theta}^{*\alpha-1}, \mathbf{0}_{\theta}) = \mathbf{0}_{m},$$
$$\forall \varepsilon_{\theta}^{*\alpha-1} \in \mathfrak{R}^{\alpha\beta}. \tag{38.16}$$

Note 544 *The real initial modified BS error vector* $\varepsilon_{\theta 0}^{*}$ *is unpredictable, uncontrollable, and arbitrary. It results from the past behavior of the plant, which is untouchable. It determines the initial control vector* $\mathbf{U}_{\beta 0}$ *in both spaces, rather than to depend on* $\mathbf{U}_{\beta 0}$. *It sometimes does not satisfy the condition (38.16), which then implies the violation of (38.11) and its nonrealizability at every* $\mathbf{t}^{\alpha\theta} \in \mathfrak{T}_{0F}^{\alpha\theta}$ *for such initial conditions.*

Problem 545 ***Matching the error vector, its derivatives, and integral with the tracking algorithm on*** $\mathfrak{T}_{0F}^{\alpha\theta}$
How can we guarantee for arbitrary error vector, error vector derivatives and the error vector integral to satisfy the tracking algorithm (38.11) on $\mathfrak{T}_{0F}^{\alpha\theta}$?

We follow the solution to this problem proposed in [274] and further extended in [279] for the *time*-invariant linear systems.

Note 546 *The value* $t_R^{\alpha\theta}$ *is infinite,* $t_R^{\alpha\theta} = \infty \mathbf{1}_{\alpha\theta}$*, if and only if, the tracking should be only asymptotic, which implies* $\mathfrak{T}_{R\infty}^{\alpha\theta} = \{\infty \mathbf{1}_{\alpha\theta}\}$. *Otherwise* $t_R^{\alpha\theta} \in In\mathfrak{T}_{0F}^{\alpha\theta}$*; hence,* $\mathfrak{T}_{R\infty}^{\alpha\theta} \subset \mathfrak{T}_{0F}^{\alpha\theta}$*, and* $\mathbf{t}_0^{\alpha\theta} < t_R^{\alpha\theta} \leq \mathbf{t}_F^{\alpha\theta} < \infty \mathbf{1}_{\alpha\theta}$ *so that*

$$\mathbf{t}_R^{\alpha\theta} < \infty \mathbf{1}_{\alpha\theta} \Longrightarrow$$
$$\mathfrak{T}_{RF}^{\alpha\theta} = \left\{ \mathbf{t}^{\alpha\theta} : \mathbf{t}_0^{\alpha\theta} < \mathbf{t}_R^{\alpha\theta} \leq \mathbf{t}^{\alpha\theta} \leq \mathbf{t}_F^{\alpha\theta} \right\} \subset \mathfrak{T}_{0F}^{\alpha\theta}. \tag{38.17}$$

In what follows we accept the finite reachability time $t_R^{\alpha\theta}$, (38.17).

Solution 547 ***The relaxed demand in general***

The root of the problem 545 is not in a particular initial error, but in the possible incompatibility of some initial errors with the tracking algorithm (38.11) at the initial moment $\mathbf{t}_0^{\alpha\theta}$ *due to the arbitrariness of the initial errors; i.e., it is possible that*

$$\mathbf{T}\left(\mathbf{t}_0^m; \boldsymbol{\varepsilon}_{\theta 0}^{*\alpha-1}, \int_{\mathbf{t}_0^\theta}^{\mathbf{t}^\theta} E_\theta^*\left(\mathbf{t}^\theta\right) d\mathbf{t}^\theta\right) = \mathbf{T}\left(\mathbf{t}_0^m; \boldsymbol{\varepsilon}_{\theta 0}^{*\alpha-1}, \mathbf{0}_\theta\right) \neq \mathbf{0}_m,$$
$$\text{for some } \boldsymbol{\varepsilon}_{\theta 0}^{*\alpha-1} \in \mathfrak{R}^{\alpha\theta}. \tag{38.18}$$

The control cannot influence such $\boldsymbol{\varepsilon}_{\theta 0}^{*\alpha-1}$. *Let the demand be relaxed so that (38.11) holds only on* $\mathfrak{T}_{RF}^{\alpha\theta}$ *rather than on the whole* $\mathfrak{T}_{0F}^{\alpha\theta}$ *for such* $\boldsymbol{\varepsilon}_{\theta 0}^{*\alpha-1}$,

$$\mathbf{T}\left(\mathbf{t}^m, \boldsymbol{\varepsilon}_\theta^{*\alpha-1}\left(\mathbf{t}^{\alpha\theta}; \mathbf{t}_0^{\alpha\theta}; \boldsymbol{\varepsilon}_{\theta 0}^{*\alpha-1}\right), \int_{\mathbf{t}_0^\theta}^{\mathbf{t}^\theta} E_\theta^*\left(\mathbf{t}^\theta\right) d\mathbf{t}^\theta\right) = \mathbf{0}_m, \ \forall \mathbf{t}^{\alpha\theta} \in \mathfrak{T}_{RF}^{\alpha\theta}, \tag{38.19}$$

rather than on the whole $\mathfrak{T}_{0F}^{\alpha\theta}$. *This relaxed demand should be satisfied for every initial BS error vector* $\boldsymbol{\varepsilon}_{\theta 0}^{*\alpha-1}$ *that violates (38.16), (Property 543), i.e., for which (38.18) holds.*

In view of the preceding analysis we modify the control goal [279].

Goal 548 ***The modified control goal***

The control should force the plant to behave so that the following tracking algorithm holds [instead of (38.11)]:

$$\left\{\begin{array}{c} \mathbf{T}\left(\mathbf{t}^m, \boldsymbol{\varepsilon}_\theta^{*\alpha-1}\left(\mathbf{t}^{\alpha\theta}\right), \int_{t_0^\theta}^{\mathbf{t}^\theta} E_\theta^*\left(\mathbf{t}^\theta\right) d\mathbf{t}^\theta\right) = \mathbf{0}_m, \\ \forall \mathbf{t}^{\alpha\theta} \left\{\begin{array}{l} \in \mathfrak{T}_{0F}^{\alpha\theta} \text{ iff } \mathbf{T}\left(\mathbf{t}_0^m; \boldsymbol{\varepsilon}_{\theta 0}^{*\alpha-1}, \mathbf{0}_\theta\right) = \mathbf{0}_m \\ \in \mathfrak{T}_{RF}^{\alpha\theta} \text{ iff } \mathbf{T}\left(\mathbf{t}_0^m; \boldsymbol{\varepsilon}_{\theta 0}^{*\alpha-1}, \mathbf{0}_\theta\right) \neq \mathbf{0}_m \end{array}\right\} \end{array}\right\}, \tag{38.20}$$

$$\text{and} \left\{\begin{array}{c} \boldsymbol{\varepsilon}_\theta^{*\alpha-1}\left(\mathbf{t}^{\alpha\theta}; \mathbf{t}^{\alpha\theta}; \boldsymbol{\varepsilon}_0^{\alpha-1}\right) = \\ = \left\{\begin{array}{c} -\mathbf{f}_\theta^{\alpha-1}\left(\mathbf{t}^{\alpha\theta}; \mathbf{t}_0^{\alpha\theta}; \mathbf{f}_{\theta 0}^{\alpha-1}\right), \ \forall \mathbf{t}^{\alpha\theta} \in \mathfrak{T}_{0F}^{\alpha\theta}, \\ \text{iff } \mathbf{T}\left(\mathbf{t}_0^m; \boldsymbol{\varepsilon}_{\theta 0}^{*\alpha-1}, \mathbf{0}_\theta\right) \neq \mathbf{0}_m, \end{array}\right\} \end{array}\right\}$$
$$\text{where } \mathbf{f}_{\theta 0}^{\alpha-1} = -\boldsymbol{\varepsilon}_{\theta 0}^{*\alpha-1} \text{ and } \mathbf{f}_\theta^{\alpha-1}\left(\mathbf{t}^{\alpha\theta}; \mathbf{t}_0^{\alpha\theta}; \mathbf{f}_{\theta 0}^{\alpha-1}\right) = \mathbf{0}_{\alpha\theta}, \ \forall \mathbf{t}^{\alpha\theta} \in \mathfrak{T}_{RF}^{\alpha\theta}, \tag{38.21}$$

for a given or to be determined both $t_R^{\alpha\theta} \in In\mathfrak{T}_0^{\alpha\theta}$ *and the vector function* $\mathbf{f}(.) : \mathfrak{T}^\theta \times \mathfrak{R}^\theta \longrightarrow \mathfrak{R}^\theta$, *which implies its* θ*-subfunction* $\mathbf{f}_\theta : \mathfrak{T}^\theta \times \mathfrak{R}^\theta \longrightarrow \mathfrak{R}^\theta$.

Note 549 *We can use (38.20) instead of (38.11) also in the cases when the initial error vector* $\varepsilon_{\theta 0}^{*\alpha-1}$ *is different from the zero vector,* $\varepsilon_{\theta 0}^{*\alpha-1} \neq \mathbf{0}_{\alpha\theta}$*, but the tracking algorithm is satisfied at the initial moment; i.e.,*

$$\mathbf{T}\left(\mathbf{t}_0^m, \boldsymbol{\varepsilon}_{\theta 0}^{*\alpha-1}, \int_{t_0^\theta}^{t_0^\theta} E_\theta^*\left(\mathbf{t}^\theta\right) d\mathbf{t}^\theta\right) = \mathbf{0}_m.$$

38.3 The reference *BS* vector

The reality of the plant operations is often such that the real *BS* vector is not desired: $\left[\mathbf{R}_0^T \;\; \mathbf{Y}_0^T\right]^T \neq \left[\mathbf{R}_{d0}^T \;\; \mathbf{Y}_{d0}^T\right]^T$, i.e., that the initial *BS* error vector is not the zero vector, $\varepsilon_{\theta 0} \neq \mathbf{0}_\theta$.

Solution 550 ***The reference BS vectors*** $\left[\mathbf{R}_R^T \;\; \mathbf{Y}_R^T\right]^T$ ***and*** $\left[\mathbf{R}_{R\beta}^{*T} \;\; \mathbf{Y}_{R\gamma}^{*T}\right]^T$

The wise demand for the desired BS behavior takes into account the real initial BS vector $\left[\mathbf{R}_0^T \;\; \mathbf{Y}_0^T\right]^T$ *and defines* ***the realizable desired BS vector*** $\left[\mathbf{R}_R^T \;\; \mathbf{Y}_R^T\right]^T$ *that starts from* $\left[\mathbf{R}_0^T \;\; \mathbf{Y}_0^T\right]^T$ *at the initial moment* t_0*. It is called also* ***the realizable BS reference vector*** $\left[\mathbf{R}_R^T \;\; \mathbf{Y}_R^T\right]^T$*, for short* ***the reference BS vector*** $\left[\mathbf{R}_R^T \;\; \mathbf{Y}_R^T\right]^T$ *:*

$$\left[\mathbf{R}_R^T \;\; \mathbf{Y}_R^T\right]^T = \left[R_{R1} \; R_{R2} \; ... \; R_{R\rho} \;\; Y_{R1} \; Y_{R2} \; ... \; Y_{RN}\;\right]^T \in \mathfrak{R}^\theta, \tag{38.22}$$

It should be such that at the initial instant it is equal to the initial real BS vector $\left[\mathbf{R}_0^T \;\; \mathbf{Y}_0^T\right]^T$*, and because the finite vector reachability time* $\mathbf{t}_R^\theta$ *has elapsed it is always equal to the desired BS vector* $\left[\mathbf{R}_d^T \;\; \mathbf{Y}_d^T\right]^T$*,*

$$\begin{gathered}\left[\begin{array}{c}\mathbf{R}_{R0}\\ \mathbf{Y}_{R0}\end{array}\right]=\left[\begin{array}{c}\mathbf{R}_R\left(\mathbf{t}_0^\rho;\mathbf{t}_0^\rho;\mathbf{R}_{R0}\right)\\ \mathbf{Y}_R\left(\mathbf{t}_0^N;\mathbf{t}_0^N;\mathbf{Y}_{R0}\right)\end{array}\right]=\left[\begin{array}{c}\mathbf{R}\left(\mathbf{t}_0^\rho;\mathbf{t}_0^\rho;\mathbf{R}_0\right)\\ \mathbf{Y}\left(\mathbf{t}_0^N;\mathbf{t}_0^N;\mathbf{Y}_0\right)\end{array}\right]=\left[\begin{array}{c}\mathbf{R}_0\\ \mathbf{Y}_0\end{array}\right],\\ \left[\begin{array}{c}\mathbf{R}_R\left(\mathbf{t}^\rho\right)\\ \mathbf{Y}_R\left(\mathbf{t}^N\right)\end{array}\right]\equiv\left[\begin{array}{c}\mathbf{R}_R\left(\mathbf{t}^\rho;\mathbf{t}_0^\rho;\mathbf{R}_{R0}\right)\\ \mathbf{Y}_R\left(\mathbf{t}^N;\mathbf{t}_0^N;\mathbf{Y}_{R0}\right)\end{array}\right]=\left[\begin{array}{c}\mathbf{R}_d\left(\mathbf{t}^\rho\right)\\ \mathbf{Y}_d\left(\mathbf{t}^N\right)\end{array}\right],\ \forall\mathbf{t}^\mu\in\mathfrak{T}_{RF}^\mu,\ \mu=\rho+N.\end{gathered} \tag{38.23}$$

The facts that only the θ*-subvector* $\left[\mathbf{R}_\beta^T \;\; \mathbf{Y}_\gamma^T\right]^T$*,*

$$\left[\mathbf{R}_\beta^T \;\; \mathbf{Y}_\gamma^T\right]^T = \left[R_1 \; R_2 \; \cdots \; R_\beta \;\; Y_1 \; Y_2 \; \cdots \; Y_\gamma\right]^T \in \mathfrak{R}^\theta, \tag{38.24}$$

is directly controllable, Axiom 87, Property 61, Lemma 62, Property 64, Lemma 65, and that the control goal demands for the whole BS vector $\left[\mathbf{R}^T \;\; \mathbf{Y}^T\right]^T$ *to be controlled, at least indirectly, which is possible in principle, Axiom 87, lead to the use of the* θ*-subvector* $\left[\mathbf{R}_{R\beta}^{*T} \;\; \mathbf{Y}_{R\gamma}^{*T}\right]^T$ *of the reference vector* $\left[\mathbf{R}_R^T \;\; \mathbf{Y}_R^T\right]^T$*, where* $\mathbf{R}_{R\beta}^*$ *and* $\mathbf{Y}_{R\gamma}^*$ *are determined by (35.26), (35.27) and by (32.26), (32.27), respectively.*

The equations (38.22), (38.23) lead to **the subsidiary error vector** ϵ_μ,

$$\begin{gathered}\epsilon_{\mu j}=R_{Rj}-R_j,\ \forall j=1,2,\ \cdots\ ,\rho,\\ \epsilon_{\mu j}=Y_{R,j-\rho}-Y_{j-\rho},\ \forall j=\rho+1,\rho+2,\ \cdots\ ,N,\\ \boldsymbol{\epsilon}_\mu=\left[\begin{array}{c}\mathbf{R}_R-\mathbf{R}\\ \mathbf{Y}_R-\mathbf{Y}\end{array}\right]=\boldsymbol{\epsilon}_\mu\left(\mathbf{R},\mathbf{Y}\right)\in\mathfrak{R}^\mu,\ \mu=\rho+N.\end{gathered} \tag{38.25}$$

The initial vector $\boldsymbol{\epsilon}_{\mu 0} = \epsilon_\mu\left(\mathbf{t}_0^\mu\right) = \epsilon_\mu\left[\mathbf{R}\left(\mathbf{t}_0^\rho\right), \mathbf{Y}\left(\mathbf{t}_0^N\right)\right]$ is equal to the zero vector for every initial real *BS* vector $\left[\mathbf{R}_0^T \;\; \mathbf{Y}_0^T\right]^T = \left[\mathbf{R}^T\left(\mathbf{t}_0^\rho\right) \;\; \mathbf{Y}^T\left(\mathbf{t}_0^N\right)\right]^T$ due to (35.27) and (32.27):

$$\begin{gathered}\left[\mathbf{R}_R^T \;\; \mathbf{Y}_R^T\right]^T\neq\left[\mathbf{R}_d^T \;\; \mathbf{Y}_d^T\right]^T\Longrightarrow\boldsymbol{\epsilon}_\mu=\boldsymbol{\varepsilon}_\mu,\\ \left[\mathbf{R}_R^T \;\; \mathbf{Y}_R^T\right]^T\neq\left[\mathbf{R}_d^T \;\; \mathbf{Y}_d^T\right]^T\Longrightarrow\\ \boldsymbol{\epsilon}_\mu\left(\mathbf{t}_0^\mu\right)=\boldsymbol{\epsilon}_\mu\left[\mathbf{R}\left(\mathbf{t}_0^\rho\right),\mathbf{Y}\left(\mathbf{t}_0^N\right)\right]=\left[\begin{array}{c}\mathbf{R}_R\left(\mathbf{t}_0^\rho\right)-\mathbf{R}\left(\mathbf{t}_0^\rho\right)\\ \mathbf{Y}_R\left(\mathbf{t}_0^N\right)-\mathbf{Y}\left(\mathbf{t}_0^N\right)\end{array}\right]\\ =\left[\begin{array}{c}\mathbf{R}\left(\mathbf{t}_0^\rho\right)-\mathbf{R}\left(\mathbf{t}_0^\rho\right)\\ \mathbf{Y}\left(\mathbf{t}_0^N\right)-\mathbf{Y}\left(\mathbf{t}_0^N\right)\end{array}\right]\equiv\mathbf{0}_\mu,\ \forall\left[\mathbf{R}^T\left(\mathbf{t}_0^\rho\right) \;\; \mathbf{Y}^T\left(\mathbf{t}_0^N\right)\right]^T\in\mathfrak{R}^\mu,\end{gathered} \tag{38.26}$$

and

$$\boldsymbol{\epsilon}_\theta^*\left(\mathbf{t}_0^\theta\right)=\boldsymbol{\epsilon}_\theta^*\left[\mathbf{R}_\beta^*\left(\mathbf{t}_0^\beta\right),\mathbf{Y}_\gamma^*\left(\mathbf{t}_0^\gamma\right)\right]=\begin{bmatrix}\mathbf{R}_{R\beta}^*\left(\mathbf{t}_0^\beta\right)-\mathbf{R}_\beta^*\left(\mathbf{t}_0^\beta\right)\\ \mathbf{Y}_{R\gamma}^*\left(\mathbf{t}_0^\gamma\right)-\mathbf{Y}_\gamma^*\left(\mathbf{t}_0^\gamma\right)\end{bmatrix}$$
$$=\begin{bmatrix}\mathbf{R}_\beta^*\left(\mathbf{t}_0^\beta\right)-\mathbf{R}_\beta^*\left(\mathbf{t}_0^\beta\right)\\ \mathbf{Y}_\gamma^*\left(\mathbf{t}_0^\gamma\right)-\mathbf{Y}_\gamma^*\left(\mathbf{t}_0^\gamma\right)\end{bmatrix}\equiv\mathbf{0}_\theta,\ \forall\left[\mathbf{R}_\beta^*\left(\mathbf{t}_0^\beta\right),\mathbf{Y}_\gamma^*\left(\mathbf{t}_0^\gamma\right)\right]\in\mathfrak{R}^\theta, \tag{38.27}$$

This is the crucial property of the vector ϵ_μ. The θ-subvector $\boldsymbol{\epsilon}_\theta^*$ of ϵ_μ contains, explicitly or implicitly, all entries of ϵ_μ. It has the same properties as ε_β^*, (35.5), and ε_γ^*, (32.4). This is their another inherent feature for control. The θ-subvector $\boldsymbol{\epsilon}_\theta^*$,

$$\boldsymbol{\epsilon}_\theta^*=\left[\epsilon_{\theta,1}\ \ \epsilon_{\theta,2}\ \ \cdots\ \ \epsilon_{\theta,\theta-1}\ \epsilon_{\theta,\theta}^*\right]^T\in\mathfrak{R}^\theta, \tag{38.28}$$

induces

$$\ominus_\theta^*\left(\boldsymbol{\epsilon}_\theta^*\right)=\operatorname{diag}\left\{\epsilon_{\theta,1}\ \ \epsilon_{\theta,2}\ \ \cdots\ \ \epsilon_{\theta,\theta-1}\ \epsilon_{\theta,\theta}^*\right\}\in\mathfrak{R}^{\theta\times\theta}. \tag{38.29}$$

Solution 551 ***On the determination of the reference vector function*** $\left[\mathbf{R}_R^T(.)\ \mathbf{Y}_R^T(.)\right]^T$
The equations (32.34), (32.35), (35.34), (35.35) define the joint reference vector

$$\left[\mathbf{R}_R^T(.)\ \ \mathbf{Y}_R^T(.)\right]^T.$$

The subsidiary vectors $\mathbf{f}_\mu(\mathbf{t}^\mu)$ *and its* θ*-subvector* $\mathbf{f}_\theta(\mathbf{t}^\theta)$,

$$\mathbf{f}_\mu(\mathbf{t}^\mu)=\left[f_1(t)\ \ f_2(t)\ ..\ f_\mu(t)\right]^T\in\mathfrak{R}^\mu,\ \mathbf{f}_\theta(\mathbf{t}^\theta)=\left[f_1(t)\ \ f_2(t)\ ..\ f_\theta(t)\right]^T\in\mathfrak{R}^\theta, \tag{38.30}$$

should obey the conditions (32.37) and (35.37). The equations (32.38), (32.39), (35.38), and (35.39) yield

$$\boldsymbol{\epsilon}_\mu(\mathbf{t}^\mu)=\varepsilon_\mu(\mathbf{t}^\mu)+\left\{\begin{array}{c}\mathbf{0}_\mu\ iff\ \mathbf{T}\left(\mathbf{t}_0^m,\varepsilon_\mu^{\alpha-1},\mathbf{0}_\mu\right)=\mathbf{0}_m,\\ \mathbf{f}_\mu(\mathbf{t}^\mu)\ iff\ \mathbf{T}\left(\mathbf{t}_0^m,\boldsymbol{\varepsilon}_\mu^{\alpha-1},\mathbf{0}_\mu\right)\neq\mathbf{0}_m\end{array}\right\},$$
$$\forall\mathbf{t}^\mu\in\mathfrak{T}_{0F}^\mu, \tag{38.31}$$

and

$$\boldsymbol{\epsilon}_\theta^*(\mathbf{t}^\theta)=\left\{\begin{array}{c}\varepsilon_\theta^*(\mathbf{t}^\theta)\ iff\ \mathbf{T}\left(\mathbf{t}_0^m,\varepsilon_\beta^{*\alpha-1},\mathbf{0}_\theta\right)=\mathbf{0}_m,\\ \varepsilon_\theta^*(\mathbf{t}^\theta)+\mathbf{f}_\theta(\mathbf{t}^\theta)\ iff\ \mathbf{T}\left(\mathbf{t}_0^m,\varepsilon_{r\beta}^{*\alpha-1},\mathbf{0}_\theta\right)\neq\mathbf{0}_m\end{array}\right\}$$
$$\forall\mathbf{t}^\theta\in\mathfrak{T}_{0F}^\theta. \tag{38.32}$$

At the initial moment $\mathbf{f}_\mu(\mathbf{t}_0^\mu)=-\varepsilon_\mu(\mathbf{t}_0^\mu)$ *due to (32.37) and (35.37), which reduces (38.31) to*

$$\boldsymbol{\epsilon}_\mu(\mathbf{t}_0^\mu)=\boldsymbol{\epsilon}_\mu\left[\varepsilon_\mu(\mathbf{t}_0^\mu)\right]=\mathbf{0}_\mu,\ \forall\varepsilon_\mu(\mathbf{t}_0^\mu)\in\mathfrak{R}^\mu, \tag{38.33}$$

and (38.32) to:

$$\boldsymbol{\epsilon}_\theta^*(\mathbf{t}_0^\theta)=\boldsymbol{\epsilon}_\theta^*\left[\varepsilon_\theta^*(\mathbf{t}_0^\theta)\right]=\mathbf{0}_\theta,\ \forall\varepsilon_\theta^*(\mathbf{t}_0^\theta)\in\mathfrak{R}^\theta. \tag{38.34}$$

These equations agree with (38.26) and (38.27), respectively. In addition if the control guarantees $\epsilon_\mu(\mathbf{t}^\mu)=\mathbf{0}_\mu,\ \forall\mathbf{t}^\mu\in\mathfrak{T}_{0F}^\mu$, *then*

$$\varepsilon_\mu(\mathbf{t}^\mu)=-\mathbf{f}_\mu(\mathbf{t}^\mu),\ \ \forall\mathbf{t}^\mu\in\mathfrak{T}_{0F}^\mu. \tag{38.35}$$

due to (38.31). This, (32.37), (35.37) and (38.30) lead to the following conclusion:

$$\boldsymbol{\epsilon}_\mu(\mathbf{t}^\mu)=\mathbf{0}_\mu,\forall\mathbf{t}^\mu\in\mathfrak{T}_{RF}^\mu\Longrightarrow\varepsilon_\mu(\mathbf{t}^\mu)=\mathbf{0}_\mu,\ \ \forall\mathbf{t}^\mu\in\mathfrak{T}_{RF}^\mu. \tag{38.36}$$

The summary of the above consideration reads:

Goal 552 ***The general modified control goal***
The control should force the plant to behave so that the following tracking algorithm holds [instead of (38.11)]:

$$\mathbf{T}\left(\mathbf{t}^{m}, \boldsymbol{\epsilon}_{\theta}^{*\alpha-1}\left(\mathbf{t}^{\alpha\theta}\right), \int_{t_0^{\theta}}^{\mathbf{t}^{\theta}} \ominus_{\theta}^{*}\left(\mathbf{t}^{\theta}\right) d\mathbf{t}^{\theta}\right) = \mathbf{0}_{m}, \ \forall \mathbf{t}^{\alpha\theta} \in \mathfrak{T}_{0F}^{\alpha\theta}. \tag{38.37}$$

Property 553 ***The properties of the tracking algorithm*** *(38.12)*
The algorithm (38.12) expressed in terms of $\boldsymbol{\epsilon}_{\theta}^{*}\left(\mathbf{t}^{\theta}\right)$ and its derivative as in (38.37) possesses, in addition to Properties 539 through 541, also Property 543 due to (38.27).

38.4 Tracking algorithm and initial conditions

The following theorem enables us to establish the unified tracking control for both cases $\varepsilon_R = \mathbf{0}_{\alpha\theta}$ and $\varepsilon_R \neq \mathbf{0}_{\alpha\theta}$.

Theorem 554 ***The main theorem on the tracking algorithm and initial conditions***
Let (38.22) through (38.36) be valid. In order for the tracking algorithm determined by (38.20), (38.21) in terms of the real error vector ε_{θ}^{*}, its derivatives, and integral, to hold, it is necessary and sufficient that the tracking algorithm $\mathbf{T}(.)$ (38.37) holds.

Proof. The proof of this theorem is the straightforward unification of the proofs of Theorem 474 and Theorem 517 ■

Chapter 39

Tracking algorithms: Both spaces

39.1 Matrix notation meaning

Let $t_0 = 0$.

The definition of the integer k is in (1.8). The diagonal elements of the constant diagonal $\theta \times \theta$ matrix K_0 are positive real numbers:

$$K_{0\nu} = \text{diag}\{k_{0\nu 1}\ k_{0\nu 2}\ \cdots\ k_{0\nu,\nu}\} \in \mathfrak{R}_+^{\nu \times \nu},\ k_{0\nu i} \in \mathfrak{R}^+,\ \forall i = 1, 2, .., \nu,$$
$$\nu \in \{\theta, 2r, \mu\}. \tag{39.1}$$

The $\nu \times \nu$ diagonal matrix T_ν is the independent *time* matrix the elements of which are the same: *time* t:

$$T_\nu = \text{diag}\{t\ t\ \cdots\ t\} \in \mathfrak{T}^{\nu \times \nu},\ \mathfrak{T}^{\nu \times \nu} = \{T_\nu : O_\nu \leq T_\nu\}, \tag{39.2}$$

$$T_{1\nu} = diag\{t_1\ t_2\ \cdots\ t_\nu\},\ t_i \in \mathfrak{R}^+,\ \forall i = 1, 2, ..., \nu. \tag{39.3}$$

The matrix exponential function $e^{-K_{0\nu} T_{1\nu}^{-1} T_\nu}$ is a diagonal matrix function defined by

$$e^{-K_{0\nu} T_{1\nu}^{-1} T_\nu} = \exp\left[-K_{0\nu} T_{1\nu}^{-1} T_\nu\right]$$
$$= diag\left\{e^{-k_{01} t_1^{-1} t}\ e^{-k_{02} t_2^{-1} t}\ \ldots\ e^{-k_{0\nu} t_\nu^{-1} t}\right\}. \tag{39.4}$$

39.2 Examples of tracking algorithms

Example 555 ***The first-order linear elementwise exponential tracking algorithm***

In this case $\mathbf{t}_R^\theta = \mathbf{t}_F^\theta = \infty \mathbf{1}_\theta$; *i.e.,* $\mathfrak{T}_{RF}^\theta = \mathfrak{T}_{F\infty}^\theta = \{\infty \mathbf{1}_\theta\}$. *The following tracking algorithm*

$$\mathbf{T}\left(\boldsymbol{\epsilon}_\theta^*, \boldsymbol{\epsilon}_\theta^{*(1)}\right) = T_{1\theta} \boldsymbol{\epsilon}_\theta^{*(1)}(\mathbf{t}^\theta) + K_{0\theta} \boldsymbol{\epsilon}_\theta^*(\mathbf{t}^\theta) = \mathbf{0}_\theta,\ \forall \mathbf{t}^\theta \in \mathfrak{T}_{0F}^\theta,$$
$$T_{1\theta} = diag\{t_1\ t_2\ \cdots\ t_\theta\},\ t_i > 0, \forall i = 1, 2, ..., \theta, \tag{39.5}$$

determines the global exponential tracking of $\mathbf{Y}_d\left(\mathbf{t}^N\right)$ *on* $\mathfrak{T}_{0F}^N$ *or the perfect tracking of* $\mathbf{Y}_R\left(\mathbf{t}^N\right)$ *on* $\mathfrak{T}_{RF}^\theta$*, which is illustrated by its solution*

$$\mathbf{T}\left(\boldsymbol{\varepsilon}_\theta^{*1}\right) = \mathbf{0}_\theta \Longrightarrow$$
$$\boldsymbol{\epsilon}_\theta^*(\mathbf{t}^\theta; \boldsymbol{\epsilon}_{\theta 0}^*) = \boldsymbol{\varepsilon}_\theta^*(\mathbf{t}^\theta; \boldsymbol{\varepsilon}_{\theta 0}^*) = e^{-K_{0\theta} T_{1\theta}^{-1} T_\theta} \boldsymbol{\varepsilon}_{\theta 0}^*, \forall \left(\mathbf{t}^\theta, \boldsymbol{\varepsilon}_{\theta 0}^*\right) \in \mathfrak{T}_{0F}^\theta \times \mathfrak{R}^\theta,$$
$$\mathbf{T}\left(\boldsymbol{\varepsilon}_\theta^{*1}\right) \neq \mathbf{0}_\theta \Longrightarrow \boldsymbol{\epsilon}_\theta^*(\mathbf{t}^\theta; \boldsymbol{\epsilon}_{\theta 0}^*) = \boldsymbol{\varepsilon}_\theta^*(\mathbf{t}^\theta; \boldsymbol{\varepsilon}_{\theta 0}^*) = \mathbf{0}_\theta,\ \forall \mathbf{t}^\theta \in \mathfrak{T}_{RF}^\theta.$$

The reachability time is infinite iff $\mathbf{T}\left(\boldsymbol{\varepsilon}_{\theta}^{*1}\right)=\mathbf{0}_{\theta}$*. The convergence to the zero error vector is then elementwise and exponentially asymptotic. Such tracking of* $\mathbf{Y}_d\left(\mathbf{t}^N\right)$ *is also stable. If* $\mathbf{T}\left(\boldsymbol{\varepsilon}_{\theta0}^{*1}\right)\neq\mathbf{0}_{\theta}$ *then the tracking of* $\mathbf{Y}_d\left(\mathbf{t}^N\right)$ *is perfect on* $\mathfrak{T}_{RF}^{N}$*.*

The tracking operator $\mathbf{T}\left(\boldsymbol{\varepsilon}_{\theta}^{*1}\right)$ *(39.5) possesses Properties 539-541 and* $\mathbf{T}(.)\in\mathfrak{C}^{-}$*. The tracking is exponentially stable in the whole if* $\mathbf{T}\left(\boldsymbol{\varepsilon}_{\theta0}^{*1}\right)=\mathbf{0}_{\theta}$*.*

Let

$$E^{(\eta)}=[E_0\quad E_1\quad\cdots\quad E_\eta]\in\mathfrak{R}^{\theta\times(\eta+1)\theta},\quad \eta\leq\alpha-1,$$

be given, or to be determined, constant extended (block) matrix, the entries of which are constant submatrices $E_k\in\mathfrak{R}^{\theta\times\,\theta}$, $k=0,1,\cdots,\eta\leq\alpha-1$.

Example 556 ***The higher-order linear elementwise exponential tracking algorithm***

In this case the tracking should be a kind of asymptotic tracking so that the vector reachability time $\mathbf{t}_R^{\theta}$ *is infinite,* $\mathbf{t}_R^{\theta}=\infty\mathbf{1}_{\theta}$*. This implies* $\mathfrak{T}_{R\infty}=\{\infty\mathbf{1}_{\theta}\}$*. We define the higher-order linear elementwise exponential tracking algorithm by*

$$\mathbf{T}\left(\boldsymbol{\epsilon}_{\theta}^{*},\boldsymbol{\epsilon}_{\theta}^{*(1)},\ \cdots\ ,\boldsymbol{\epsilon}_{\theta}^{*(\eta)}\right)=\sum_{k=0}^{k=\eta\leq\alpha-1}E_k\boldsymbol{\epsilon}_{\theta}^{*(k)}(\mathbf{t}^{\theta})$$
$$=E^{(\eta)}\boldsymbol{\epsilon}_{\theta}^{*\eta}(\mathbf{t}^{(\eta+1)\theta})=\mathbf{0}_{\theta},\ \forall\mathbf{t}^{(\eta+1)\theta}\in\mathfrak{T}_{0F}^{(\eta+1)\theta}\ , \tag{39.6}$$

with the matrices E_k *such that the real parts of the roots of the characteristic polynomial* $f(s)$ *of* $\mathbf{T}(s)$*,*

$$f(s)=det\left(\sum_{k=0}^{k=\eta\leq m-1}E_ks^k\right),$$

are negative.

The linear differential equation (39.6) is with the constant matrix coefficients. It has the unique solution for every initial condition $\boldsymbol{\epsilon}_{\theta}^{*\eta-1}(\mathbf{t}_0^{\eta\theta})\in\mathfrak{R}^{\eta\theta}$*.*

The reachability time is infinite. The error vector $\boldsymbol{\epsilon}_{\theta}^{*\eta-1}(\mathbf{t}^{\eta\theta})=\boldsymbol{\varepsilon}_{\theta}^{*\eta-1}(\mathbf{t}^{\eta\theta})$ *converges to the zero error vector elementwise, exponentially, and asymptotically in the BS iff* $\mathbf{T}\left(\boldsymbol{\varepsilon}_{\theta}^{*1}\right)=\mathbf{0}_{\theta}$*. Such tracking of* $\mathbf{Y}_d\left(\mathbf{t}^N\right)$ *is stable. Iff* $\mathbf{T}\left(\boldsymbol{\varepsilon}_{\theta}^{*1}\right)\neq\mathbf{0}_{\theta}$ *then the tracking of* $\mathbf{Y}_d\left(\mathbf{t}^N\right)$ *is perfect on* $\mathfrak{T}_{RF}^{N}$*.*

The tracking operator (39.6) possesses Properties 539-541 and $\mathbf{T}(.)\in\mathfrak{C}^{-}$*. The tracking is exponentially stable in the whole in the BS if* $\mathbf{T}\left(\boldsymbol{\varepsilon}_{\theta0}^{*1}\right)=\mathbf{0}_{\theta}$*.*

Example 557 ***The sharp elementwise stable tracking with the finite vector reachability time*** $\mathbf{t}_R^{\theta}$

We accept the finite vector reachability time t_R^{θ}*,* $t_R^{\theta}=[t_{R1}\ t_{R2}\ ..\ t_{R\theta}]^T$*, so that*

$$\mathfrak{T}_{0R}^{\theta}=[\mathbf{0}_{\theta},\mathbf{t}_R^{\theta}]\subset\mathfrak{T}_{0F}^{\theta},\ \ \mathfrak{T}_{RF}^{\theta}=[\mathbf{t}_R^{\theta},\ \mathbf{t}_F^{\theta}],\ \ \mathbf{t}_F^{\theta}<\infty\mathbf{1}_{\theta}. \tag{39.7}$$

The algorithm for the elementwise stable tracking with the finite vector reachability time t_R^{θ}*,*

$$\mathbf{T}\left(\boldsymbol{\varepsilon}_{\theta0}^{*1}\right)=\mathbf{0}_{\theta}\Longrightarrow\mathbf{t}_R^{\theta}=T_{1\theta}K_{0\theta}^{-1}\left|\boldsymbol{\varepsilon}_{\theta0}^{*}\right|,$$
$$\left|\boldsymbol{\varepsilon}_{\theta0}^{*}\right|=[\left|\varepsilon_{\theta10}\right|\quad\left|\varepsilon_{\theta20}\right|\quad\cdots\quad\left|\varepsilon_{\theta,\theta-1,0}\right|\quad\left|\varepsilon_{\theta0}^{*}\right|]^T\in\mathfrak{R}^{\theta},$$
$$\left|\boldsymbol{\epsilon}_{\theta0}^{*}\right|=[\left|\epsilon_{\theta10}\right|\quad\left|\epsilon_{\theta20}\right|\quad\cdots\quad\left|\epsilon_{\theta,\theta-1,0}\right|\quad\left|\epsilon_{\theta0}^{*}\right|]^T\in\mathfrak{R}^{\theta},$$

reads

$$\mathbf{T}\left(\boldsymbol{\epsilon}_{\theta}^{*},\boldsymbol{\epsilon}_{\theta}^{*(1)}\right)=T_{1\theta}\boldsymbol{\epsilon}_{\theta}^{*(1)}(\mathbf{t}^{\theta})+K_{0\theta}sign\boldsymbol{\epsilon}_{\theta}^{*}(\mathbf{t}_0^{\theta})=\mathbf{0}_{\theta},\ \forall\mathbf{t}^{\theta}\in\mathfrak{T}_{0F}^{\theta}\ . \tag{39.8}$$

Let

$$S(\varepsilon_{\theta 0}^{*}) = diag\{sign\varepsilon_{10} \quad sign\varepsilon_{20} \quad \cdots \quad sign\varepsilon_{\theta-1,0} \ sign\varepsilon_{\theta 0}^{*}\},$$

The solution $\boldsymbol{\epsilon}_{\theta}^{*}(\mathbf{t}^{\theta};\boldsymbol{\epsilon}_{\theta 0}^{*})$,

$$\begin{aligned}\mathbf{T}(\varepsilon_{\theta 0}^{*1}) = \mathbf{0}_{\theta} &\Longrightarrow \boldsymbol{\epsilon}_{\theta}^{*}(\mathbf{t}^{\theta};\boldsymbol{\epsilon}_{\theta 0}^{*}) = \boldsymbol{\varepsilon}_{\theta}^{*}(\mathbf{t}^{\theta};\boldsymbol{\varepsilon}_{\theta 0}^{*}) \\ &= \boldsymbol{\varepsilon}_{\theta 0}^{*} - T_{1\theta}^{-1}K_{0\theta}S(\boldsymbol{\varepsilon}_{\theta 0}^{*})\,\mathbf{t}^{\theta}, \ \forall \mathbf{t}^{\theta}\in\mathfrak{T}_{0R}^{\theta}, \\ \boldsymbol{\epsilon}_{\theta}^{*}(\mathbf{t}^{\theta};\boldsymbol{\epsilon}_{\theta 0}^{*}) &= \boldsymbol{\varepsilon}_{\theta}^{*}(\mathbf{t}^{\theta};\boldsymbol{\varepsilon}_{\theta 0}^{*}) = \mathbf{0}_{\theta}, \ \forall \mathbf{t}^{\theta}\in\mathfrak{T}_{RF}^{\theta}, \\ \mathbf{T}(\varepsilon_{\theta 0}^{*1}) \neq \mathbf{0}_{\theta} &\Longrightarrow \boldsymbol{\epsilon}_{\theta}^{*}(\mathbf{t}^{\theta};\boldsymbol{\epsilon}_{\theta 0}^{*}) = \mathbf{0}_{\theta}, \ \forall \mathbf{t}^{\theta}\in\mathfrak{T}_{RF}^{\theta},\end{aligned}$$

to (39.8) determines the BS error behavior $\boldsymbol{\varepsilon}_{\theta}^{*}(\mathbf{t}^{\theta};\boldsymbol{\varepsilon}_{\theta 0}^{*})$ *that approaches sharply the zero error vector* $\boldsymbol{\varepsilon}_{\theta}^{*} = \mathbf{0}_{\theta}$ *in the linear form (along a straight line) with the nonzero constant velocity* $T_{1\theta}^{-1}K_{0\theta}S(\boldsymbol{\varepsilon}_{\theta 0}^{*})\,\mathbf{1}_{\theta}$ *iff* $\mathbf{T}(\varepsilon_{\theta 0}^{*1}) = \mathbf{0}_{\theta}$. *Otherwise,* $\mathbf{T}(\varepsilon_{\theta 0}^{*1}) \neq \mathbf{0}_{\theta}$ *and the tracking of* $\mathbf{Y}_d(\mathbf{t}^N)$ *is perfect on* $\mathfrak{T}_{RF}^{\theta}$. *The tracking of* $\mathbf{Y}_d(\mathbf{t}^N)$ *becomes perfect at the finite vector reachability time* $\mathbf{t}_R^{\theta}$ *and rests perfect on* $\mathfrak{T}_{RF}^{\theta}$.

Iff $\mathbf{T}(\varepsilon_{\mu 0}^{*1}) = \mathbf{0}_{\mu}$ *then the convergence to the zero error vector; i.e., the convergence of* $\mathbf{Y}(\mathbf{t}^N)$ *to* $\mathbf{Y}_d(\mathbf{t}^N)$ *is elementwise, strictly monotonous, continuous, and*

$$\begin{aligned}\left|\varepsilon(t;\varepsilon_0^1)\right| \leq \left|\varepsilon_0\right|, \ \forall t\in\mathfrak{T}_{0F} &\Longrightarrow \left\|\varepsilon(t;\varepsilon_0^1)\right\| \leq \left\|\varepsilon_0\right\|, \ \forall t\in\mathfrak{T}_{0F} \Longrightarrow \\ \forall\varepsilon\in\mathfrak{R}^{+}, \ \exists\delta\in\mathfrak{R}^{+}, \ &\delta=\delta(\varepsilon)=\varepsilon \Longrightarrow \\ \left\|\varepsilon_0^1\right\| < \delta &\Longrightarrow \left\|\varepsilon(t;\varepsilon_0^1)\right\| \leq \varepsilon, \ \forall t\in\mathfrak{T}_{0F}.\end{aligned}$$

The tracking of $\mathbf{Y}_d(\mathbf{t}^N)$ *is stable on* $\mathfrak{T}_{0F}^{\mu}$ *if* $\mathbf{T}(\varepsilon_{\mu 0}^{*1}) = \mathbf{0}_{\mu}$.

The tracking operator (39.8) possesses Properties 539-541 and $\mathbf{T}(.)\in\mathfrak{C}^{-}$. *The tracking is stable in the whole with FVRT* t_R^{θ} *if* $\mathbf{T}(\varepsilon_{\theta 0}^{*1}) = \mathbf{0}_{\theta}$.

Let

$$\left|\ominus_{\theta}^{*}(\mathbf{t}^{\theta})\right|^{1/2} = \mathrm{diag}\left\{\left|\epsilon_{\theta 1}(t)\right|^{1/2} \quad \left|\epsilon_{\theta 2}(t)\right|^{1/2} \quad \cdots \quad \left|\epsilon_{\theta,\theta-1}(t)\right|^{1/2} \quad \left|\epsilon_{\theta,\theta}^{*}(t)\right|^{1/2}\right\}.$$

Example 558 ***The first-power smooth elementwise stable tracking with the finite vector reachability time*** $\mathbf{t}_R^{\theta}$

If the control acting on the plant ensures

$$\mathbf{T}\left(\boldsymbol{\epsilon}_{\theta}^{*},\boldsymbol{\epsilon}_{\theta}^{*(1)}\right) = T_{1\theta}\boldsymbol{\epsilon}_{\theta}^{*(1)}(\mathbf{t}^{\theta}) + 2K_{0\theta}\left|\ominus_{\theta}^{*}(\mathbf{t}^{\theta})\right|^{1/2} sign\boldsymbol{\epsilon}_{\theta 0}^{*} = \mathbf{0}_{\theta}, \ \forall \mathbf{t}^{\theta}\in\mathfrak{T}_{0F}^{\theta}, \tag{39.9}$$

then the plant exhibits the elementwise stable tracking with the finite vector reachability time t_R^{θ},

$$\begin{aligned}\mathbf{T}(\varepsilon_{\theta 0}^{*1}) = \mathbf{0}_{\theta} \Longrightarrow \mathbf{t}_R^{\theta} = T_{1\theta}K_{0\theta}^{-1}\left|E_{\theta 0}^{*}\right|^{1/2}\mathbf{1}_{\theta} = T_{1\theta}K_0^{-1}\left|\varepsilon_{\theta 0}^{*}\right|^{1/2}, \\ \left|\varepsilon_{\theta 0}^{*}\right|^{1/2} = \left[\left|\varepsilon_{\theta 1}(0)\right|^{1/2} \quad \left|\varepsilon_{\theta 2}(0)\right|^{1/2} \quad \ldots \quad \left|\varepsilon_{\theta-1}(0)\right|^{1/2} \quad \left|\varepsilon_{\theta}^{*}(0)\right|^{1/2}\right]^{T},\end{aligned}$$

which is determined by the output error behavior,

$$\begin{aligned}&\mathbf{T}(\varepsilon_{\theta 0}^{*1}) = \mathbf{0}_{\theta} \Longrightarrow \boldsymbol{\epsilon}_{\theta}^{*}(\mathbf{t}^{\theta};\boldsymbol{\epsilon}_{\theta 0}^{*}) \\ &= \boldsymbol{\varepsilon}_{\theta}^{*}(\mathbf{t}^{\theta};\boldsymbol{\varepsilon}_{\theta 0}^{*})\left\{\begin{array}{c} = \left\{\begin{array}{c}\left[\left|E_{\theta 0}^{*(1)}\right|^{1/2} - T_{1\theta}^{-1}K_{0\theta}T_{\theta}\right]^2 sign\varepsilon_{\theta 0}^{*}, \\ \forall \mathbf{t}^{\theta}\in\mathfrak{T}_{0R}^{\theta},\end{array}\right\} \\ = \mathbf{0}_{\theta}, \qquad \forall \mathbf{t}^{\theta}\in\mathfrak{T}_{RF}^{\theta},\end{array}\right\} \\ &\mathbf{T}(\varepsilon_{\theta 0}^{*1}) \neq \mathbf{0}_{\theta} \Longrightarrow \boldsymbol{\epsilon}_{\theta}^{*}(\mathbf{t}^{\theta};\boldsymbol{\epsilon}_{\theta 0}^{*}) = \boldsymbol{\varepsilon}_{\theta}^{*}(\mathbf{t}^{\theta};\boldsymbol{\varepsilon}_{\theta 0}^{*}) = \mathbf{0}_{\theta}, \ \forall \mathbf{t}^{\theta}\in\mathfrak{T}_{RF}^{\theta}.\end{aligned} \tag{39.10}$$

This implies

$$sign\boldsymbol{\epsilon}_\theta^*(\mathbf{t}^\theta) = sign\boldsymbol{\epsilon}_{\theta 0}^*,\ \forall \mathbf{t}^\theta \in In\mathfrak{T}_{0R}^\theta.$$

The tracking operator (39.9) possesses Properties 539-541 and $\mathbf{T}(.) \in \mathfrak{C}^-$. *The tracking is stable in the whole with FVRT* t_R^θ *if* $\mathbf{T}\left(\boldsymbol{\varepsilon}_{\theta 0}^{*1}\right) = \mathbf{0}_\theta$.

Iff $\mathbf{T}\left(\boldsymbol{\varepsilon}_{\theta 0}^{*1}\right) = \mathbf{0}_\theta$ *then the solution (39.10) obeys the following:*

$$\begin{gathered}
\|\boldsymbol{\varepsilon}_\theta^*(t;\boldsymbol{\varepsilon}_{\theta 0}^*)\| \leq \|\boldsymbol{\varepsilon}_{\theta 0}^*\|,\ \forall t \in \mathfrak{T}_{0F} \Longrightarrow \delta(\varepsilon) = \varepsilon \Longrightarrow \\
\forall \varepsilon \in \mathfrak{R}^+,\ \exists \delta \in \mathfrak{R}^+,\ \delta = \delta(\varepsilon) = \varepsilon \Longrightarrow \\
\|\boldsymbol{\varepsilon}_{\theta 0}^*\| < \delta \Longrightarrow \|\boldsymbol{\varepsilon}_\theta^*(t;\boldsymbol{\varepsilon}_{\theta 0}^*)\| \leq \varepsilon,\ \forall t \in \mathfrak{T}_{0F}.
\end{gathered}$$

Therefore, such tracking of $\mathbf{Y}_d\left(\mathbf{t}^N\right)$ *is stable on* $\mathfrak{T}_{0F}^N$. *Then the bigger* $K_{0\theta}$, *the smaller* t_R^θ *is for fixed* $T_{1\theta}$ *and* $\boldsymbol{\varepsilon}_{\theta 0}^{*1}$, *and vice versa. The smaller* $T_{1\theta}$, *the smaller* t_R^θ *is for fixed* K_0 *and* $\boldsymbol{\varepsilon}_{\theta 0}^{*1}$, *and vice versa. The bigger* $\left|\boldsymbol{\varepsilon}_{\theta 0}^{*1}\right|$, *the bigger* t_R^θ *is for fixed* T_1 *and* K_0, *and vice versa.*

Example 559 ***The higher-power smooth elementwise stable tracking with the finite vector reachability time*** t_R^θ

Let the tracking algorithm be

$$\begin{gathered}
\mathbf{T}\left(\boldsymbol{\epsilon}_\theta^{*1}\right) = T_{1\theta}\boldsymbol{\epsilon}_\theta^{*(1)}(\mathbf{t}^\theta) + K_{0\theta}\left|\ominus\left(\mathbf{t}^\theta\right)\right|^{I-K_\theta^{-1}} sign\boldsymbol{\epsilon}_{\theta 0}^* = \mathbf{0}_\theta,\ \forall \mathbf{t}^\theta \in \mathfrak{T}_{0F}^\theta, \\
K_\theta = diag\{k_1\ k_2\ \cdots\ k_\theta\},\ k_i \in \{2,\ 3,\ \cdots\},\ \forall i = 1,2,\ \cdots\ ,\theta, \\
\left|\ominus_\theta^*\left(\mathbf{t}^\theta\right)\right|^{I-K_\theta^{-1}} = diag\left\{\left|\epsilon_{\theta 1}(t)\right|^{1-k_1^{-1}}\ \left|\epsilon_{\theta 2}(t)\right|^{1-k_2^{-1}}\ \cdots\ \left|\epsilon_{\theta,\theta}^*(t)\right|^{1-k_\theta^{-1}}\right\}.
\end{gathered} \tag{39.11}$$

Iff $\mathbf{T}\left(\boldsymbol{\varepsilon}_{\theta 0}^{*1}\right) = \mathbf{0}_\theta$ *then (39.12) determines the following solution* $\boldsymbol{\epsilon}_\theta^*(\mathbf{t}^\theta;\boldsymbol{\epsilon}_{\theta 0}^*)$ *to the tracking algorithm* $T\left(\boldsymbol{\epsilon}_\theta^*, \boldsymbol{\epsilon}_\theta^{*(1)}\right) = \mathbf{0}_\theta,\ \forall \mathbf{t}^\theta \in \mathfrak{T}_{0F}^\theta$:

$$\begin{gathered}
\mathbf{T}\left(\boldsymbol{\varepsilon}_{\theta 0}^{*1}\right) = \mathbf{0}_\theta \Longrightarrow \boldsymbol{\epsilon}_\theta^*(\mathbf{t}^\theta;\boldsymbol{\epsilon}_{\theta 0}^*) = \boldsymbol{\varepsilon}_\theta^*(\mathbf{t}^\theta;\boldsymbol{\varepsilon}_{\theta 0}^*) \\
= \frac{1}{2}S\left(\boldsymbol{\varepsilon}_{\theta 0}^*\right)\left\{\begin{array}{c}\left\{\begin{array}{c} I_\theta + \\ +S\left[\left|\boldsymbol{\varepsilon}_{\theta 0}^*\right|^{K_\theta^{-1}} - T_{1\theta}K_{0\theta}\mathbf{t}^\theta\right]\end{array}\right\} \\ \bullet\left[\left|\boldsymbol{\varepsilon}_{\theta 0}^*\right|^{K_\theta^{-1}} - T_{1\theta}K_\theta^{-1}K_{0\theta}\mathbf{t}^\theta\right]^{K_\theta}\end{array}\right\} = \\
= \left\{\begin{array}{ll}\left[\left|E_{\theta 0}^{*(1)}\right|^{K_\theta^{-1}} - T_{1\theta}K_\theta^{-1}K_{0\theta}\left(T_\theta - T_{\theta 0}\right)\right]^{K_\theta} sign\boldsymbol{\varepsilon}_{\theta 0}^*, & \forall \mathbf{t}^\theta \in \mathfrak{T}_{0R}^\theta, \\ \mathbf{0}_\theta, & \forall \mathbf{t}^\theta \in \mathfrak{T}_{RF}^\theta\end{array}\right\} \Longrightarrow \\
\mathbf{t}_R^\theta = \mathbf{t}_0^\theta + T_{1\theta}KK_0^{-1}\left|\boldsymbol{\varepsilon}_{\theta 0}^*\right|^{K_\theta^{-1}},
\end{gathered} \tag{39.12}$$

but iff $\mathbf{T}\left(\boldsymbol{\varepsilon}_{\theta 0}^{*1}\right) \neq \mathbf{0}_\theta$, *then the solution* $\boldsymbol{\epsilon}_\theta^*(\mathbf{t}^\theta;\boldsymbol{\epsilon}_{\theta 0}^*)$ *to* $T\left(\boldsymbol{\epsilon}_\theta^*, \boldsymbol{\epsilon}_\theta^{*(1)}\right) = \mathbf{0}_\theta,\ \forall \mathbf{t}^\theta \in \mathfrak{T}_{RF}^\theta$, *satisfies the following:*

$$\mathbf{T}\left(\boldsymbol{\varepsilon}_{\theta 0}^{*1}\right) \neq \mathbf{0}_\theta \Longrightarrow \boldsymbol{\epsilon}_\theta^*(\mathbf{t}^\theta;\boldsymbol{\epsilon}_{\theta 0}^*) = \boldsymbol{\varepsilon}_\theta^*(\mathbf{t}^\theta;\boldsymbol{\varepsilon}_{\theta 0}^*) = \mathbf{0}_\theta,\ \forall \mathbf{t}^\theta \in \mathfrak{T}_{RF}^\theta.$$

We use the following:

$$\begin{gathered}
\left[\left|\boldsymbol{\varepsilon}_{\theta 0}^*\right|^{K_\theta^{-1}} - T_{1\theta}K_{0\theta}\mathbf{t}^\theta\right]^{K_\theta} \\
= \left[\begin{array}{c}\left[\left|\varepsilon_{10}\right|^{k_1^{-1}} - t_1^{-1}k_{01}t\right]^{k_1} \\ \left[\left|\varepsilon_{20}\right|^{k_2^{-1}} - t_2^{-1}k_{02}t\right]^{k_2} \\ \cdots\cdots\cdots \\ \left[\left|\boldsymbol{\varepsilon}_{\theta 0}^*\right|^{k_\theta^{-1}} - t_\theta^{-1}k_{0\gamma}t\right]^{k_\theta}\end{array}\right] \in \mathfrak{R}^\theta.
\end{gathered} \tag{39.13}$$

The tracking operator (39.11) possesses Properties 539- 541 and $\mathbf{T}(.) \in \mathfrak{C}^-$. *The tracking is stable in the whole with FVRT* t_R^θ *if* $\mathbf{T}\left(\varepsilon_{\theta 0}^{*1}\right) = \mathbf{0}_\theta$.

The equations (39.12) and (39.13) imply

$$\|\varepsilon_\theta^*(t;\varepsilon_{\theta 0}^*)\| \le \|\varepsilon_{\theta 0}^*\|,\ \forall t \in \mathfrak{T}_{0F} \Longrightarrow \delta(\varepsilon) = \varepsilon \Longrightarrow$$
$$\forall \varepsilon \in \mathfrak{R}^+,\ \exists \delta \in \mathfrak{R}^+,\ \delta = \delta(\varepsilon) = \varepsilon \Longrightarrow$$
$$\|\varepsilon_{\theta 0}^*\| < \delta \Longrightarrow \|\varepsilon_\theta^*(t;\varepsilon_{\theta 0}^*)\| \le \varepsilon,\ \forall t \in \mathfrak{T}_{0F}.$$

Iff $\mathbf{T}\left(\varepsilon_{\theta 0}^{*1}\right) \neq \mathbf{0}_\theta$ *then the tracking of* $\mathbf{Y}_d\left(\mathbf{t}^N\right)$ *is perfect on* $\mathfrak{T}_{RF}^\theta$.

Example 560 ***Sharp absolute error vector tracking elementwise and stable with the finite vector reachability time*** t_R^θ

The solution to the following tracking algorithm:

$$\mathbf{T}\left(\epsilon_\theta^*, \epsilon_\theta^{*(1)}\right) = T_{1\theta} D^+ |\epsilon_\theta^*| + K_{0\theta} sign|\epsilon_{\theta 0}^*| =$$
$$= T_{1\theta}\Sigma\left(\epsilon_\theta^{*1}\right)\epsilon_\theta^{*(1)} + K_{0\theta} sign|\epsilon_{\theta 0}^*| = \mathbf{0}_\theta,\ \forall \mathbf{t}^\theta \in \mathfrak{T}_{0F}^\theta\ , \tag{39.14}$$

reads for $\mathbf{T}\left(\varepsilon_{\theta 0}^{*1}\right) = \mathbf{0}_\theta$:

$$\mathbf{T}\left(\varepsilon_{\theta 0}^{*1}\right) = \mathbf{0}_\theta \Longrightarrow \epsilon_\theta^*(\mathbf{t}^\theta;\epsilon_{\theta 0}^*) = \varepsilon_\theta^*(\mathbf{t}^\theta;\varepsilon_{\theta 0}^*)$$
$$= \left\{\begin{array}{cc} \left[|\varepsilon_{\theta 0}^*| - T_{1\theta}K_{0\theta}\left(T_\theta - T_{\theta 0}\right)\right] sign|\varepsilon_{\theta 0}^*|, & \forall \mathbf{t}^\theta \in \mathfrak{T}_{0R}^\theta, \\ \mathbf{0}_\theta, & \forall \mathbf{t}^\theta \in \mathfrak{T}_{RF}^\theta. \end{array}\right\},$$
$$\mathbf{t}_R^\theta = \mathbf{t}_0^\theta + T_{1\theta}K_{0\theta}|\varepsilon_{\theta 0}^*|, \tag{39.15}$$

but for $\mathbf{T}\left(\varepsilon_{\theta 0}^{*1}\right) \neq \mathbf{0}_\theta$ *it reads:*

$$\mathbf{T}\left(\varepsilon_{\theta 0}^{*1}\right) \neq \mathbf{0}_\theta \Longrightarrow \epsilon_\theta^*(\mathbf{t}^\theta;\epsilon_{\theta 0}^*) = \varepsilon_\theta^*(\mathbf{t}^\theta;\varepsilon_{\theta 0}^*) = \mathbf{0}_\theta,\ \forall \mathbf{t}^\theta \in \mathfrak{T}_{RF}^\theta.$$

The tracking operator (39.14) possesses Properties 539-541 and $\mathbf{T}(.) \in \mathfrak{C}^-$. *The tracking is stable in the whole with FVRT* t_R^θ *if* $\mathbf{T}\left(\varepsilon_{\theta 0}^{*1}\right) = \mathbf{0}_\theta$.

Equations (39.15) permit

$$\mathbf{T}\left(\varepsilon_{\theta 0}^{*1}\right) = \mathbf{0}_\theta \Longrightarrow \left|\varepsilon_\theta^*(t;\varepsilon_{\theta 0}^{*1})\right| \le |\varepsilon_{\theta 0}^*|,\ \forall t \in \mathfrak{T}_{0F} \Longrightarrow$$
$$\left\|\varepsilon_\theta^*(t;\varepsilon_{\theta 0}^{*1})\right\| \le \|\varepsilon_{\theta 0}^*\|,\ \forall t \in \mathfrak{T}_{0F} \Longrightarrow$$
$$\forall \varepsilon \in \mathfrak{R}^+,\ \exists \delta \in \mathfrak{R}^+,\ \delta = \delta(\varepsilon) = \varepsilon \Longrightarrow$$
$$\|\varepsilon_{\theta 0}^*\| < \delta \Longrightarrow \left\|\varepsilon_\theta^*(t;\varepsilon_{\theta 0}^{*1})\right\| \le \varepsilon,\ \forall t \in \mathfrak{T}_{0F}.$$

The tracking is stable state with the finite vector reachability time $t_R^\theta = \mathbf{t}_0^\theta + T_{1\theta}K_{0\theta}|\varepsilon_{\theta 0}^*|$. *It is strictly monotonous and continuous without oscillation, overshoot, and undershoot.*

Example 561 ***The exponential absolute error vector tracking elementwise and stable with the finite vector reachability time*** t_R^θ

The tracking algorithm is in terms of the elementwise absolute value of the error vector,

$$\mathbf{T}\left(\epsilon_\theta^*, \epsilon_\theta^{*(1)}\right) = T_{1\theta} D^+ |\epsilon_\theta^*| + K_\theta\left(|\epsilon_\theta^*| + K_{0\theta} sign|\epsilon_{\theta 0}^*|\right) =$$
$$= T_{1\theta}\Sigma\left(\epsilon_\theta^*, \epsilon_\theta^{*(1)}\right)\epsilon_\theta^{*(1)} + K_\theta\left(|\epsilon_\theta^*| + K_{0\theta} sign|\epsilon_{\theta 0}^*|\right) = \mathbf{0}_\theta,\ \forall \mathbf{t}^\theta \in \mathfrak{T}_{0F}^\theta. \tag{39.16}$$

The solution of the differential equation (39.16), which is written in the matrix diagonal form (39.17):

$$D^+\left[|\ominus_\theta^*| + K_{0\theta}S\left(|\epsilon_{\theta 0}^*|\right)\right] = -T_{1\theta}K_\theta\left[|\ominus_\theta^*| + K_{0\theta}S\left(|\epsilon_{\theta 0}^*|\right)\right] \Longrightarrow$$
$$\left[|\ominus_\theta^*| + K_{0\theta}S\left(|\epsilon_{\theta 0}^*|\right)\right]^{-1} D^+\left[|\ominus_\theta^*| + K_{0\theta}S\left(|\epsilon_{\theta 0}^*|\right)\right] = -T_{1\theta}K_\theta \Longrightarrow$$
$$D^+\left\{ln\left[|\ominus_\theta^*| + K_{0\theta}S\left(|\epsilon_{\theta 0}^*|\right)\right]\right\} = -T_{1\theta}K_\theta, \tag{39.17}$$

reads in the matrix form

$$ln\left\{\left[\left|\ominus_{\theta}^{*}\right| + K_{0\theta}S\left(\left|\boldsymbol{\epsilon}_{\theta 0}^{*}\right|\right)\right]\left[\left|\ominus_{\theta 0}^{*}\right| + K_{0\theta}S\left(\left|\boldsymbol{\epsilon}_{\theta 0}^{*}\right|\right)\right]^{-1}\right\} = -T_{1\theta}K_{\theta}\left(T_{\theta} - T_{\theta 0}\right),$$
$$\forall T_{\theta} \in \mathfrak{T}_{0}^{\theta \times \theta}.$$

The final form of the solution is

$$\ominus_{\theta}^{*}(t;\boldsymbol{\epsilon}_{\theta 0}^{*})$$
$$= \left\{ \begin{array}{c} \left\{ \begin{array}{c} e^{-T_{1\theta}K_{\theta}(T_{\theta}-T_{\theta 0})}\left(\left|\ominus_{\theta 0}^{*}\right| + K_{0\theta}\right)S\left(\left|\boldsymbol{\epsilon}_{\theta 0}^{*}\right|\right) - K_{0\theta}S\left(\left|\boldsymbol{\epsilon}_{\theta 0}^{*}\right|\right), \\ \forall T_{\theta} \in \left[T_{\theta 0}, T_{\theta R}\right], \\ O_{\theta}, \qquad \forall T_{\theta} \in \left[T_{\theta 0}, T_{\theta R}\right], \end{array} \right\} \\ where\ 0\infty = 0, \end{array} \right\},$$

$$\mathbf{T}\left(\boldsymbol{\varepsilon}_{\theta 0}^{*1}\right) = \mathbf{0}_{\theta} \Longrightarrow T_{R\theta} = \left\{ \begin{array}{ll} K_{\theta}^{-1}T_{1\theta}ln\left\{K_{0\theta}\left[\left|\ominus_{\theta 0}^{*}\right| + K_{0\theta}\right]\right\}, & \boldsymbol{\epsilon}_{\theta 0}^{*} \neq \mathbf{0}_{\theta}, \\ O_{\theta}, & \boldsymbol{\epsilon}_{\theta 0}^{*} = \mathbf{0}_{\theta} \end{array} \right\}, \tag{39.18}$$

$$\mathbf{t}_{R}^{\theta} = \left[t_{R1}\ \ t_{R2}\ \ldots\ t_{R\theta}\right]^{T} \Longleftrightarrow T_{\theta R} = diag\left\{t_{R1}\ \ t_{R2}\ \cdots\ t_{R\theta}\right\}. \tag{39.19}$$

We can set the solution (39.18) in the equivalent vector form

$$\boldsymbol{\epsilon}_{\theta}^{*}(\mathbf{t}^{\theta};\boldsymbol{\epsilon}_{\theta 0}^{*}) = \left\{ \begin{array}{c} \left[\epsilon_{\theta}^{-T_{1\theta}K_{\theta}T}\left(\left|\boldsymbol{\epsilon}_{\theta 0}^{*}\right| + K_{0\theta}\right) - K_{0\theta}\right] sign\left(\left|\boldsymbol{\epsilon}_{\theta 0}^{*}\right|\right), \\ \forall \mathbf{t}^{\theta} \in \left[\mathbf{0}_{\theta}, \mathbf{t}_{R}^{\theta}\right],\ i.e.,\ \forall T_{\theta} \in \left[O_{\theta}, T_{\theta R}\right], \\ \mathbf{0}_{\theta},\ \ \forall \mathbf{t}^{\theta} \in \left[\mathbf{t}_{R}^{\theta},\ \mathbf{t}_{F}^{\theta}\right],\ i.e.,\ \forall T_{\theta} \in \left[T_{\theta R}, T_{\theta F}\right]. \end{array} \right\}. \tag{39.20}$$

The tracking operator (39.16) possesses Properties 539-541 and $\mathbf{T}(.) \in \mathfrak{C}^{-}$. *The tracking is stable in the whole with FVRT* t_{R}^{θ} *if* $\mathbf{T}\left(\boldsymbol{\varepsilon}_{\theta 0}^{*1}\right) = \mathbf{0}_{\theta}$.

Iff $\mathbf{T}\left(\boldsymbol{\varepsilon}_{\theta 0}^{*1}\right) = \mathbf{0}_{\theta}$ *then the solution is continuous and monotonous without oscillation, overshoot, and undershoot, and obeys*

$$\mathbf{T}\left(\boldsymbol{\varepsilon}_{\theta 0}^{*1}\right) = \mathbf{0}_{\theta} \Longrightarrow \boldsymbol{\epsilon}_{\theta}^{*}(\mathbf{t}^{\theta};\boldsymbol{\epsilon}_{\theta 0}^{*}) = \boldsymbol{\varepsilon}_{\theta}^{*}(\mathbf{t}^{\theta};\boldsymbol{\varepsilon}_{\theta 0}^{*}) \Longrightarrow$$
$$\left|\boldsymbol{\varepsilon}_{\theta}^{*}(\mathbf{t}^{\theta};\boldsymbol{\varepsilon}_{\theta 0}^{*})\right| \leq \left|\boldsymbol{\varepsilon}_{\theta 0}^{*}\right|,\ \forall \mathbf{t}^{\theta} \in \mathfrak{T}_{0F}^{\theta} \Longrightarrow$$
$$\forall \varepsilon \in \mathfrak{R}^{+},\ \exists \delta \in \mathfrak{R}^{+},\ \delta = \delta\left(\varepsilon\right) = \varepsilon \Longrightarrow$$
$$\left\|\boldsymbol{\varepsilon}_{\theta 0}\right\| < \delta \Longrightarrow \left\|\boldsymbol{\varepsilon}_{\theta}^{*}(\mathbf{t}^{\theta};\boldsymbol{\varepsilon}_{\theta 0}^{*})\right\| \leq \varepsilon,\ \forall \mathbf{t}^{\theta} \in \mathfrak{T}_{0F}^{\theta},$$

Then the tracking of $\mathbf{Y}_{d}\left(\mathbf{t}^{N}\right)$ *is stable on* $\mathfrak{T}_{0F}^{\theta}$. *It converges with the exponential rate to the zero error vector and reaches (elementwise iff* $\theta = N$*) the origin in the finite vector reachability time* t_{R}^{θ} *(39.18), (39.19).*

Iff $\mathbf{T}\left(\boldsymbol{\varepsilon}_{\theta 0}^{*1}\right) \neq \mathbf{0}_{\theta}$ *then the solution reads*

$$\mathbf{T}\left(\boldsymbol{\varepsilon}_{\theta 0}^{*1}\right) \neq \mathbf{0}_{\theta} \Longrightarrow \boldsymbol{\epsilon}_{\theta}^{*}(\mathbf{t}^{\theta};\boldsymbol{\epsilon}_{\theta 0}^{*}) = \boldsymbol{\epsilon}_{\theta}^{*}(\mathbf{t}^{\theta};\boldsymbol{\epsilon}_{\theta 0}^{*}) = \mathbf{0}_{\theta},\ \ \forall \mathbf{t}^{\theta} \in \left[\mathbf{t}_{R}^{\theta},\ \mathbf{t}_{F}^{\theta}\right].$$

Note 562 *We can effectively use the tracking algorithms proposed in the preceding examples, Example 555 through Example 561, also when we synthesize tracking control by applying various design methods; e.g., Lyapunov-like, adaptive control, sliding mode, and natural tracking control method.*

Chapter 40

NTC synthesis: Both spaces

40.1 General NTC theorem: Both spaces (BS)

Property 539 through Property 543 form the basis for the following general theorem.

Theorem 563 ***General NTC synthesis in both spaces (BS)***
Let (38.22) through (38.36) be valid.

In order for the plant (3.12), with Properties 57, 58 through 78, controlled by the natural tracking control $\mathbf{U}$ *to exhibit tracking on* $\mathfrak{T}_{0F}^{\alpha\rho} \times \mathfrak{D}^j \times \mathfrak{Y}_d^{\alpha}$ *determined by the tracking algorithm* $\mathbf{T}(.)$*, (38.20), (38.21), it is necessary and sufficient that both:*

i) The rank β *of the control matrix* $P(t)$ *is greater than zero; i.e., (8.20) holds, and the rank* γ *of the control matrix* $W(t)$ *is greater than zero and not greater than* $min(N, r)$*, i.e., (8.10) is true, i.e., jointly: (8.31) is valid,*

and

ii) If the plant also possesses Property 61 and Property 64 then for any chosen matrix $H(t) \in \mathfrak{R}^{\theta \times m}$ *that obeys (8.30), the control vector function* $\mathbf{U}(.)$ *obeys (40.1),*

$$\begin{bmatrix} \mathbf{e}_{u\beta}\left[\mathbf{U}_{\beta}^{\mu}(t)\right] \\ \mathbf{w}_{\gamma}\left[\mathbf{U}_{\gamma}(t)\right] \end{bmatrix} = \begin{bmatrix} \mathbf{e}_{u\beta}\left[\mathbf{U}_{\beta}^{\mu}(t^{-})\right] \\ \mathbf{w}_{\gamma}\left[\mathbf{U}_{\gamma}(t^{-})\right] \end{bmatrix}$$
$$+ H(t)\,\mathbf{T}\left(\mathbf{t}^{m}, \boldsymbol{\epsilon}_{\theta}^{*\alpha-1}\left(\mathbf{t}^{\alpha\theta}\right), \int_{t_0^{\theta}}^{\mathbf{t}^{\theta}} \ominus_{\theta}^{*}\left(\mathbf{t}^{\theta}\right) d\mathbf{t}^{\theta}\right),\ \forall \mathbf{t}^{\alpha\theta} \in \mathfrak{T}_0^{\alpha\theta}. \tag{40.1}$$

The domain $\mathcal{D}_T\left(t_0; \mathfrak{D}^j; \mathbf{U}; \mathfrak{Y}_d\right)$ *of tracking on* $\mathfrak{T}_0 \times \mathfrak{D}^j \times \mathfrak{Y}_d^{\alpha}$ *is equal to the domain* $\mathcal{D}_{PTbl}\left(t_0; \mathfrak{D}^j; \mathbf{U}; \mathfrak{Y}_d\right)$ *of the perfect trackability of the plant (3.12) on* $\mathfrak{T}_0 \times \mathfrak{D}^j \times \mathfrak{Y}_d^{\alpha}$ *determined by (8.33).*

Proof. If we replace the vector function $\mathbf{v}(.) : \mathfrak{T}_0 \times \mathrm{In}\mathfrak{T}_0 \times \mathfrak{R}^{(k+1)N} \longrightarrow \mathfrak{R}^m$, (11.1) by $\mathbf{T}\left(\mathbf{t}^{m}, \boldsymbol{\epsilon}^{\alpha-1}\left(\mathbf{t}^{\alpha\rho}\right), \int_{t_0^{\beta}}^{\mathbf{t}^{\beta}} \ominus\left(\mathbf{t}^{\rho}\right) d\mathbf{t}^{\rho}\right)$ in Theorem 192 then it becomes Theorem 563 ∎

40.2 NTC synthesis in both spaces

The vectors $\boldsymbol{\varepsilon}_{\beta}^{*}$, (35.4), and $\boldsymbol{\epsilon}_{\beta}^{*}$, (35.31), become the vectors $\boldsymbol{\varepsilon}_{\theta}^{*}$ and $\boldsymbol{\epsilon}_{\theta}^{*}$ when the subscript "β" is replaced by the subscript "θ":

$$\boldsymbol{\varepsilon}_{\theta}^{*} = \left[\varepsilon_{\theta 1}\ \varepsilon_{\theta 2}\ \dots\ \varepsilon_{\theta,\theta-1}\ \ \varepsilon_{\theta,\theta}^{*}\right]^{T},\quad \boldsymbol{\epsilon}_{\theta}^{*} = \left[\epsilon_{\theta 1}\ \epsilon_{\theta 2}\ \dots\ \epsilon_{\theta,\theta-1}\ \ \epsilon_{\theta,\theta}^{*}\right]^{T}. \tag{40.2}$$

Condition 564 *The plant (3.12) possesses Properties 57 through 78.*

All following theorems result directly from Theorem *563.*
Let $t_0 = 0$.

Theorem 565 ***NTC for the first-order linear elementwise exponential tracking, Example 555***

Let (38.22) through (38.36) be valid. Let the reachability time $\tau_{RBS} = \tau_{FBS} = \infty$; *i.e.,* $\mathfrak{T}_{R\infty BS} = \mathfrak{T}_{RFBS} = \{\infty\}$.

For the plant (3.12) obeying Condition (564) to be controlled by the natural tracking control $\mathbf{U}$ *in order to exhibit tracking on* $\mathfrak{T}_{0F}^{\theta} \times \mathfrak{D}^j \times \mathfrak{Y}_d^{\alpha}$ *determined by the following tracking algorithm*

$$\mathbf{T}\left(\varepsilon_{\theta}^{*}, \varepsilon_{\theta}^{*(1)}\right) = T_{1\theta}\varepsilon_{\theta}^{*(1)}(t) + K_{0\theta}\varepsilon_{\theta}^{*}(t) = \mathbf{0}_{\theta},\ \forall t \in \mathfrak{T}_{0F},$$
$$if\ \mathbf{T}\left(\varepsilon_{\theta 0}^{*1}\right) = \mathbf{0}_{\theta},$$
$$\varepsilon_{\theta}^{*}\left(t; t_0; \varepsilon_{\theta 0}^{*1}\right) = -\mathbf{f}_{R\theta}^{1}(t),\ \forall t \in \mathfrak{T}_{0F},\ if\ \mathbf{T}\left(\varepsilon_{\theta 0}^{*1}\right) \neq \mathbf{0}_{\theta},$$
$$T_{1\theta} = blockdiag\{T_{1\beta}\ \ T_{1\gamma}\},\ K_{0\theta} = blocdiag\{K_{0\beta}\ \ K_{0\gamma}\},$$
$$T_{1R\theta} = diag\{\tau_{1R1}\ \tau_{1R2}\ \cdots\ \tau_{1R\theta}\},\ K_{0R\theta} = diag\{k_{0R1}\ k_{0R2}\ \cdots\ k_{0R\theta}\},$$
$$T_{1R} = diag\{\tau_{1R1}\ \tau_{1R2}\ \cdots\ \tau_{1R\rho}\},\ K_{0R} = diag\{k_{0R1}\ k_{0R2}\ \cdots\ k_{0R\rho}\}$$
$$\tau_{1Ri} > 0,\ k_{0Ri} > 0,\ \forall i = 1, 2, \cdots, (\theta\ or\ 2r\ or\ \mu), \tag{40.3}$$

it is necessary and sufficient that both:

i) The rank β *of the control matrix* $P(t)$ *is greater than zero; i.e., (8.20) holds, and the rank* γ *of the control matrix* $W(t)$ *is greater than zero and not greater than* $min(\rho, r)$, *i.e., (8.10) is true; or jointly: (8.31) is valid,*

and

ii) If the plant possesses also Property 61 and Property 64 then for any chosen matrix $H(t) \in \mathfrak{R}^{\theta \times m}$ *that obeys (8.30), the control vector function* $\mathbf{U}(.)$ *obeys (40.4),*

$$\begin{bmatrix} \mathbf{e}_{u\beta}\left[\mathbf{U}_{\beta}(t)\right] \\ \mathbf{w}_{\gamma}\left[\mathbf{U}_{\gamma}(t)\right] \end{bmatrix} = \begin{bmatrix} \mathbf{e}_{u\beta}\left[\mathbf{U}_{\beta}(t^{-})\right] \\ \mathbf{w}_{\gamma}\left[\mathbf{U}_{\gamma}(t^{-})\right] \end{bmatrix}$$
$$+ H(t)\left[T_{1\theta}\epsilon_{\theta}^{*(1)}(t) + K_{0,\theta}\epsilon_{\theta}^{*}(t)\right],\ \forall t \in \mathfrak{T}_0. \tag{40.4}$$

The domain $\mathcal{D}_T\left(t_0; \mathfrak{D}^j; \mathbf{U}; \mathfrak{Y}_d\right)$ *of tracking on* $\mathfrak{T}_0 \times \mathfrak{D}^j \times \mathfrak{Y}_d^{\alpha}$ *is equal to the domain* $\mathcal{D}_{PTbl}\left(t_0; \mathfrak{D}^j; \mathbf{U}; \mathfrak{Y}_d\right)$ *of the perfect trackability of the plant (3.12) on* $\mathfrak{T}_0 \times \mathfrak{D}^j \times \mathfrak{Y}_d^{\alpha}$ *determined by (8.33).*

The solution $\varepsilon_{\theta}^{*}(.)$ *shows that the tracking is exponentially stable in the whole if* $\mathbf{T}\left(\varepsilon_{\theta 0}^{*1}\right) = \mathbf{0}_{\theta}$.

Theorem 566 ***NTC for the higher-order linear elementwise exponential tracking, Example 556***

Let (38.22) through (38.36) be valid. Let the reachability time $\tau_R = \tau_F = \infty$; *i.e.,* $\mathfrak{T}_{R\infty} = \mathfrak{T}_{RF} = \{\infty\}$.

For the plant (3.12) obeying Condition (564) to be controlled by the natural tracking control $\mathbf{U}$ *in order to exhibit tracking on* $\mathfrak{T}_{0F}^{\theta} \times \mathfrak{D}^j \times \mathfrak{Y}_d^{\alpha}$ *determined by the tracking algorithm (40.5),*

$$\mathbf{T}\left(t, \varepsilon_{\theta}^{*\eta}\right) = \sum_{k=0}^{k=\eta \leq \alpha-1} E_{\theta k}\varepsilon_{\theta}^{*}(t) = E_{\theta}^{(\eta)}\varepsilon_{\theta}^{*\eta}(\mathbf{t}^{(\eta+1)\theta}) = \mathbf{0}_{\theta},$$
$$\forall \mathbf{t}^{(\eta+1)\theta} \in \mathfrak{T}_{0F}^{(\eta+1)\theta}\ if\ \mathbf{T}\left(0, \varepsilon_{\theta 0}^{*\eta}\right) = \mathbf{0}_{\theta},$$
$$\varepsilon_{\theta}^{*\eta}\left(\mathbf{t}^{(\eta+1)\theta}; \varepsilon_{\theta 0}^{*\eta}\right) = -\mathbf{f}_{\beta}^{\eta}\left(\mathbf{t}^{(\eta+1)\theta}\right),\ \forall \mathbf{t}^{(\eta+1)\theta} \in \mathfrak{T}_{0F}^{(\eta+1)\theta}$$
$$if\ \mathbf{T}\left(0, \varepsilon_{\theta 0}^{*\eta}\right) \neq \mathbf{0}_{\theta}, \tag{40.5}$$

with the matrices $E_{\theta k} \in \mathfrak{R}^{\theta \times \theta}$ *such that the real parts* Res_{ri} *of the roots* s_i *of its characteristic polynomial* $f_{R\theta}(s)$, $f_{R\theta}(s) = det\left(E_{\beta}^{(\eta)} S_{\theta}^{(\eta)}(s)\right)$, *are negative,*

$$f(s_i) = det\left(E_{\theta}^{(\eta)} S_{\theta}^{(\eta)}(s_i)\right) = 0 \Longrightarrow Res_i < 0, \ \forall i = 1, 2, ..., \eta\theta, \tag{40.6}$$

it is necessary and sufficient that both:

i) The rank β *of the control matrix* $P(t)$ *is greater than zero and not greater than* $min(\rho, r)$*, i.e., (8.20) holds, and the rank* γ *of the control matrix* $W(t)$ *is greater than zero and not greater than* $min(\gamma, r)$*, i.e., (8.10) is true; or jointly: (8.31) is valid,*

and

ii) if the plant also possesses Property 61 and Property 64 then for any chosen matrix $H(t) \in \mathfrak{R}^{\theta \times m}$ *that obeys (8.30), the control vector function* $\mathbf{U}(.)$ *obeys (40.7),*

$$\begin{gathered} \begin{bmatrix} \mathbf{e}_{u\beta}\left[\mathbf{U}_{\beta}^{\mu}(t)\right] \\ \mathbf{w}_{\gamma}\left[\mathbf{U}_{\gamma}(t)\right] \end{bmatrix} = \begin{bmatrix} \mathbf{e}_{u\beta}\left[\mathbf{U}_{\beta}^{\mu}(t^{-})\right] \\ \mathbf{w}_{\gamma}\left[\mathbf{U}_{\gamma}(t^{-})\right] \end{bmatrix} \\ + H(t)\left[E_{\theta}^{(\eta)} \boldsymbol{\varepsilon}_{\theta}^{*}(\mathbf{t}^{(\eta+1)\theta})\right], \ \forall t \in \mathfrak{T}_0. \end{gathered} \tag{40.7}$$

The domain $\mathcal{D}_T(t_0; \mathfrak{D}^j; \mathbf{U}; \mathfrak{Y}_d)$ *of tracking on* $\mathfrak{T}_0 \times \mathfrak{D}^j \times \mathfrak{Y}_d^{\alpha}$ *is equal to the domain* $\mathcal{D}_{PTbl}(t_0; \mathfrak{D}^j; \mathbf{U}; \mathfrak{Y}_d)$ *of the perfect trackability of the plant (3.12) on* $\mathfrak{T}_0 \times \mathfrak{D}^j \times \mathfrak{Y}_d^{\alpha}$ *determined by (8.33).*

The solution $\boldsymbol{\varepsilon}_{\theta}^{*}(.)$ *shows that the tracking is exponentially stable in the whole if* $\mathbf{T}(\boldsymbol{\varepsilon}_{\theta 0}^{*1}) = \mathbf{0}_{\theta}$.

Theorem 567 ***NTC for the sharp elementwise stablewise tracking with the finite vector state reachability time*** $\mathbf{t}_R^{\theta}$***, Example 557***

Let (38.22) through (38.36) be valid.

For the plant (3.12) obeying Condition (564) to be controlled by the natural tracking control $\mathbf{U}$ *in order to exhibit tracking on* $\mathfrak{T}_{0F}^{\theta} \times \mathfrak{D}^j \times \mathfrak{Y}_d^{\alpha}$ *determined by the algorithm*

$$\begin{gathered} \mathbf{T}\left(\boldsymbol{\varepsilon}_{\theta}^{*}, \boldsymbol{\varepsilon}_{\theta}^{*(1)}\right) = T_{1\theta}\boldsymbol{\varepsilon}_{\theta}^{*(1)}(\mathbf{t}^{\theta}) + K_{0\theta} sign \boldsymbol{\varepsilon}_{\theta 0}^{*} = \mathbf{0}_{\theta}, \\ \forall \mathbf{t}^{\theta} \in \mathfrak{T}_{0F}^{\theta} \ if \ \mathbf{T}(\boldsymbol{\varepsilon}_{\theta o}^{*1}) = \mathbf{0}_{\theta}, \\ \boldsymbol{\varepsilon}_{\theta}^{*1}(\mathbf{t}^{2\theta}; \boldsymbol{\varepsilon}_{\theta 0}^{*1}) = -\mathbf{f}_{\theta}^{1}(\mathbf{t}^{2\theta}), \ \forall \mathbf{t}^{2\theta} \in \mathfrak{T}_{0F}^{2\theta}, \ if \ \mathbf{T}(\boldsymbol{\varepsilon}_{\theta o}^{*1}) \neq \mathbf{0}_{\theta}, \end{gathered} \tag{40.8}$$

for the sharp elementwise stablewise tracking with the finite vector state reachability time $\mathbf{t}_R^{\theta}$,

$$\mathbf{t}_R^{\theta} = T_{1\theta} K_{0\theta} \left|\boldsymbol{\varepsilon}_{\theta 0}^{*}\right|, \ \boldsymbol{\varepsilon}_{\theta 0}^{*} \in \mathfrak{R}^{\theta}, \tag{40.9}$$

it is necessary and sufficient that both:

i) The rank β *of the control matrix* $P(t)$ *is greater than zero and not greater than* $min(\rho, r)$*, i.e., (8.20) holds, and the rank* γ *of the control matrix* $W(t)$ *is greater than zero and not greater than* $min(\gamma, r)$*; i.e., (8.10) is true, i.e., jointly: (8.31) is valid,*

and

ii) if the plant also possesses Property 61 and Property 64 then for any chosen matrix $H(t) \in \mathfrak{R}^{\theta \times m}$ *that obeys (8.30), the control vector function* $\mathbf{U}(.)$ *obeys (40.10),*

$$\begin{gathered} \begin{bmatrix} \mathbf{e}_{u\beta}\left[\mathbf{U}_{\beta}^{\mu}(t)\right] \\ \mathbf{w}_{\gamma}\left[\mathbf{U}_{\gamma}(t)\right] \end{bmatrix} = \begin{bmatrix} \mathbf{e}_{u\beta}\left[\mathbf{U}_{\beta}^{\mu}(t^{-})\right] \\ \mathbf{w}_{\gamma}\left[\mathbf{U}_{\gamma}(t^{-})\right] \end{bmatrix} \\ + H(t)\left[T_{1\theta}\boldsymbol{\epsilon}_{\theta}^{*(1)}(t) + K_{0\theta} sign \boldsymbol{\epsilon}_{\theta 0}^{*}\right], \ \forall t \in \mathfrak{T}_0. \end{gathered} \tag{40.10}$$

The domain $\mathcal{D}_T(\mathrm{t}_0;\mathfrak{D}^j;\mathbf{U};\mathfrak{Y}_d)$ *of tracking on* $\mathfrak{T}_0 \times \mathfrak{D}^j \times \mathfrak{Y}_d^\alpha$ *is equal to the domain* $\mathcal{D}_{PTbl}(\mathrm{t}_0;\mathfrak{D}^j;\mathbf{U};\mathfrak{Y}_d)$ *of the perfect trackability of the plant (3.12) on* $\mathfrak{T}_0 \times \mathfrak{D}^j \times \mathfrak{Y}_d^\alpha$ *determined by (8.33).*

The solution $\varepsilon_\theta^*(.)$ *shows that the tracking is stable in the whole with FVIDRT* t_R^ρ *if* $\mathbf{T}(\varepsilon_{\theta 0}^{*1}) = \mathbf{0}_\theta$.

We defined $\ominus_\theta^*(\epsilon_\theta^*) \equiv \ominus_\theta^*[\epsilon_\theta^*(\mathbf{t}^\theta)] \equiv \ominus_\theta^*(\mathbf{t}^\theta)$ in (38.29) so that

$$|\ominus_\theta^*(\epsilon_\theta^*)|^{1/2} = \operatorname{diag}\left\{|\epsilon_{\theta,1}|^{1/2} \quad |\ \epsilon_{\theta,2}|^{1/2} \quad \cdots \quad |\epsilon_{\theta,\theta-1}|^{1/2} \quad |\epsilon_{\theta,\theta}^*|^{1/2}\right\},$$
$$\text{for } \epsilon_\theta^* = \begin{bmatrix}\epsilon_{\mu,1} & \epsilon_{\mu,2} & \cdots & \epsilon_{\mu,\theta-1} & \epsilon_{\mu,\theta}^*\end{bmatrix}^T \in \mathfrak{R}^\theta,$$

$$|\Psi_{2r}^*(\epsilon_{2r}^*)|^{1/2}$$
$$= \operatorname{diag}\left\{|\epsilon_{2r,1}(t)|^{1/2} \quad |\epsilon_{2r,2}(t)|^{1/2} \quad \cdots \quad |\epsilon_{2r,2r-1}(t)|^{1/2} \quad |\epsilon_{2r,2r}^*(t)|^{1/2}\right\},$$
$$for\ \epsilon_{2r}^* = \begin{bmatrix}\epsilon_{2r,1} & \epsilon_{2r,2} & \cdots & \epsilon_{2r,2r-1} & \epsilon_{2r,2r}^*\end{bmatrix}^T.$$

Theorem 568 ***The NTC synthesis for the first-power smooth elementwise stablewise tracking with the finite vector BS reachability time*** $\mathbf{t}_R^\rho$, ***Example 558***

Let (38.22) through (38.36) be valid.

For the plant (3.12) obeying Condition (564) to be controlled by the natural tracking control $\mathbf{U}$ *in order to exhibit tracking on* $\mathfrak{T}_{0F}^\theta \times \mathfrak{D}^j \times \mathfrak{Y}_d^\alpha$ *determined by*

$$\mathbf{T}\left(\varepsilon_\theta^*, \varepsilon_\theta^{*(1)}\right) = T_{1\theta}\varepsilon_\theta^{*(1)}(\mathbf{t}^\theta) + 2K_{0\theta}\left|E_\theta^*(\mathbf{t}^\theta)\right|^{1/2} sign\varepsilon_{\theta 0}^* = \mathbf{0}_\theta,$$
$$\forall \mathbf{t}^\theta \in \mathfrak{T}_{0F}^\theta\ if\ \mathbf{T}(\varepsilon_{\theta o}^{*1}) = \mathbf{0}_\theta,$$
$$\varepsilon_\theta^{*1}(\mathbf{t}^{2\theta};\varepsilon_{\theta 0}^{*1}) = -\mathbf{f}_\theta^1(\mathbf{t}^{2\theta}),\ \forall \mathbf{t}^{2\theta} \in \mathfrak{T}_{0F}^{2\theta},\ if\ \mathbf{T}(\varepsilon_{\theta o}^{*1}) \neq \mathbf{0}_\theta,$$
$$|E_\theta^*(t)|^{1/2} = diag\left\{|\varepsilon_{\theta,1}(t)|^{1/2} \quad |\varepsilon_{\theta,2}(t)|^{1/2} \quad \cdots \quad |\varepsilon_{\theta,\theta-1}(t)|^{1/2} \quad |\varepsilon_\theta^*(t)|^{1/2}\right\}, \quad (40.11)$$

for the first-power smooth elementwise stablewise tracking with the finite vector BS reachability time $\mathbf{t}_R^\theta$,

$$\mathbf{t}_R^\theta = T_{1\theta}K_{0\theta}\left|E_{\theta 0}^*\right|^{1/2}\mathbf{1}_\theta = T_{1\theta}K_{0\theta}\left|\varepsilon_{\theta 0}^*\right|^{1/2},\ \varepsilon_{\theta 0}^* \in \mathfrak{R}^\theta,$$
$$\left|\varepsilon_{\theta 0}^*\right|^{1/2} = \left[|\varepsilon_{\theta,1}(0)|^{1/2} \quad |\varepsilon_{\theta,2}(0)|^{1/2} \quad .. \quad |\varepsilon_{\theta,\theta-1}(0)|^{1/2} \quad |\varepsilon_\theta^*(0)|^{1/2}\right]^T, \quad (40.12)$$

it is necessary and sufficient that both:

i) The rank β *of the control matrix* $P(t)$ *is greater than zero and not greater than* $\min(\rho, r)$, *i.e., (8.20) holds, and the rank* γ *of the control matrix* $W(t)$ *is greater than zero and not greater than* $\min(\gamma, r)$, *i.e., (8.10) is true, i.e., jointly: (8.31) is valid,*

and

ii) If the plant also possesses Property 61 and Property 64 then for any chosen matrix $H(t) \in \mathfrak{R}^{\theta \times m}$ *that obeys (8.30), the control vector function* $\mathbf{U}(.)$ *obeys (40.13),*

$$\begin{bmatrix}\mathbf{e}_{u\beta}\left[\mathbf{U}_\beta^\mu(t)\right] \\ \mathbf{w}_\gamma\left[\mathbf{U}_\gamma(t)\right]\end{bmatrix} = \begin{bmatrix}\mathbf{e}_{u\beta}\left[\mathbf{U}_\beta^\mu(t^-)\right] \\ \mathbf{w}_\gamma\left[\mathbf{U}_\gamma(t^-)\right]\end{bmatrix}$$
$$+ H(t)\left[T_{1\theta}\epsilon_\theta^{*(1)}(t) + K_{0\theta}sign\epsilon_{\theta 0}^*\right],\ \forall t \in \mathfrak{T}_0. \quad (40.13)$$

The domain $\mathcal{D}_T(\mathrm{t}_0;\mathfrak{D}^j;\mathbf{U};\mathfrak{Y}_d)$ *of tracking on* $\mathfrak{T}_0 \times \mathfrak{D}^j \times \mathfrak{Y}_d^\alpha$ *is equal to the domain* $\mathcal{D}_{PTbl}(\mathrm{t}_0;\mathfrak{D}^j;\mathbf{U};\mathfrak{Y}_d)$ *of the perfect trackability of the plant (3.12) on* $\mathfrak{T}_0 \times \mathfrak{D}^j \times \mathfrak{Y}_d^\alpha$ *determined by (8.33).*

The solution $\varepsilon_\theta^*(.)$ *shows that the tracking is stable in the whole with FVIDRT* t_R^ρ *if* $\mathbf{T}(\varepsilon_{\theta 0}^{*1}) = \mathbf{0}_\theta$.

We define

$$K_\theta = diag\{k_1\ k_2\ \cdots\ k_\theta\},\ k_i \in \{2,\ 3,\ \cdots\},\ \forall i = 1,2,\ \cdots,\theta,$$

$$\left|\Psi_\theta^*\left(\mathbf{t}^\theta\right)\right|^{I-K_\theta^{-1}}$$
$$= diag\left\{\left|\epsilon_{\theta 1}(t)\right|^{1-k_1^{-1}}\ \left|\epsilon_{\theta 2}(t)\right|^{1-k_2^{-1}}\ \cdots\ \left|\epsilon_{\theta,\theta-1}(t)\right|^{1-k_{\theta-1}^{-1}}\ \left|\epsilon_\theta^*(t)\right|^{1-k_\theta^{-1}}\right\}.$$

Theorem 569 ***The NTC synthesis for the higher power smooth elementwise stablewise tracking with the finite vector BS reachability time $\mathbf{t}_R^\rho$, Example 559***

Let (38.22) through (38.36) be valid.

For the plant (3.12) obeying Condition (564) to be controlled by the natural tracking control $\mathbf{U}$ *in order to exhibit tracking on* $\mathfrak{T}_{0F}^\theta \times \mathfrak{D}^j \times \mathfrak{Y}_d^\alpha$ *determined by the tracking algorithm*

$$\mathbf{T}\left(\varepsilon_\theta^*, \varepsilon_\theta^{*(1)}\right) = T_{1\theta}\varepsilon_\theta^{*(1)}(\mathbf{t}^\theta) + K_{0\theta}\left|E_\theta\left(\mathbf{t}^\theta\right)\right|^{I-K_\theta^{-1}} sign\varepsilon_{\theta 0}^* = \mathbf{0}_\theta,$$
$$\forall \mathbf{t}^\theta \in \mathfrak{T}_{0F}^\theta\ if\ \mathbf{T}\left(\varepsilon_{\theta 0}^{*1}\right) = \mathbf{0}_\theta,$$
$$\varepsilon_\theta^{*1}\left(\mathbf{t}^{2\theta};\varepsilon_{\theta 0}^{*1}\right) = -\mathbf{f}_\theta^1\left(\mathbf{t}^{2\theta}\right),\ \forall \mathbf{t}^{2\theta} \in \mathfrak{T}_{0F}^{2\theta},\ if\ \mathbf{T}\left(\varepsilon_{\theta 0}^{*1}\right) \neq \mathbf{0}_\theta, \tag{40.14}$$

for the higher-power smooth elementwise stablewise tracking with the finite vector BS reachability time $\mathbf{t}_R^\theta$,

$$\mathbf{t}_R^\theta = T_{1\theta}K_\theta K_{0\theta}^{-1}\left|\varepsilon_{\theta 0}^*\right|^{K_\theta^{-1}},\ \varepsilon_{\theta 0}^* \in \mathfrak{R}^\theta, \tag{40.15}$$

it is necessary and sufficient that both:

i) The rank β of the control matrix $P(t)$ is greater than zero and not greater than $min(\rho, r)$, i.e., (8.20) holds, and the rank γ of the control matrix $W(t)$ is greater than zero and not greater than $min(\gamma, r)$, i.e., (8.10) is true, i.e., jointly: (8.31) is valid,

and

ii) If the plant also possesses Property 61 and Property 64 then for any chosen matrix $H(t) \in \mathfrak{R}^{\theta\times m}$ that obeys (8.30), the control vector function $\mathbf{U}(.)$ *obeys (40.16),*

$$\begin{bmatrix} \mathbf{e}_{u\beta}\left[\mathbf{U}_\beta^\mu(t)\right] \\ \mathbf{w}_\gamma\left[\mathbf{U}_\gamma(t)\right] \end{bmatrix} = \begin{bmatrix} \mathbf{e}_{u\beta}\left[\mathbf{U}_\beta^\mu(t^-)\right] \\ \mathbf{w}_\gamma\left[\mathbf{U}_\gamma(t^-)\right] \end{bmatrix}$$
$$+ H(t)\left[T_{1\theta}\boldsymbol{\epsilon}_\theta^{*(1)}(\mathbf{t}^\theta) + K_{0\theta}\left|\Psi_\theta^*\left(\mathbf{t}^\theta\right)\right|^{I-K^{-1}} sign\boldsymbol{\epsilon}_{\theta 0}^*\right],\ \forall t \in \mathfrak{T}_0. \tag{40.16}$$

The domain $\mathcal{D}_T\left(t_0;\mathfrak{D}^j;\mathbf{U};\mathfrak{Y}_d\right)$ of tracking on $\mathfrak{T}_0 \times \mathfrak{D}^j \times \mathfrak{Y}_d^\alpha$ is equal to the domain $\mathcal{D}_{PTbl}\left(t_0;\mathfrak{D}^j;\mathbf{U};\mathfrak{Y}_d\right)$ of the perfect trackability of the plant (3.12) on $\mathfrak{T}_0 \times \mathfrak{D}^j \times \mathfrak{Y}_d^\alpha$ determined by (8.33).

The solution $\varepsilon_\theta^*(.)$ shows that the tracking is stable in the whole with FVIDRT t_R^ρ if $\mathbf{T}\left(\varepsilon_{\theta 0}^{*1}\right) = \mathbf{0}_\theta$.

Theorem 570 ***The NTC synthesis for the sharp absolute error vector value tracking elementwise and stablewise with the finite vector BS reachability time $\mathbf{t}_R^\rho$, Example 560***

Let (38.22) through (38.36) be valid.

For the plant (3.12) obeying Condition (564) to be controlled by the natural tracking control $\mathbf{U}$ *in order to exhibit tracking on* $\mathfrak{T}_{0F}^\theta \times \mathfrak{D}^j \times \mathfrak{Y}_d^\alpha$ *determined by the following tracking algorithm*

$$\mathbf{T}\left(\varepsilon_\theta^*, \varepsilon_\theta^{*(1)}\right) = T_{1\theta}\Sigma\left(\varepsilon_\theta^*, \varepsilon_\theta^{*(1)}\right)\varepsilon_\theta^{*(1)} + K_{0\theta}sign\left|\varepsilon_{\theta 0}^*\right| = \mathbf{0}_\theta,$$
$$\forall \mathbf{t}^\theta \in \mathfrak{T}_{0F}^\theta\ if\ \mathbf{T}\left(\varepsilon_{\theta 0}^{*1}\right) = \mathbf{0}_\theta,$$
$$\varepsilon_\theta^{*1}\left(\mathbf{t}^{2\theta};\varepsilon_{\theta 0}^{*1}\right) = -\mathbf{f}_\theta^1\left(\mathbf{t}^{2\beta}\right),\ \forall \mathbf{t}^{2\beta} \in \mathfrak{T}_{0F}^{2\beta},\ if\ \mathbf{T}\left(\varepsilon_{\theta 0}^{*1}\right) \neq \mathbf{0}_\theta, \tag{40.17}$$

for the sharp absolute error vector value tracking elementwise stablewise with the finite vector BS reachability time $\mathbf{t}_R^\theta$,

$$\mathbf{t}_R^\theta = T_1 K_0^{-1} \left|\varepsilon_{\theta 0}^*\right|, \quad \varepsilon_{\theta 0}^* \in \mathfrak{R}^\theta,$$

it is necessary and sufficient that both:

i) The rank β *of the control matrix* $P(t)$ *is greater than zero and not greater than* $min(\rho, r)$, *i.e., (8.20) holds, and the rank* γ *of the control matrix* $W(t)$ *is greater than zero and not greater than* $min(\gamma, r)$, *i.e., (8.10) is true, i.e., jointly: (8.31) is valid,*

and

ii) if the plant also possesses Property 61 and Property 64 then for any chosen matrix $H(t) \in \mathfrak{R}^{\theta \times m}$ *that obeys (8.30), the control vector function* $\mathbf{U}(.)$ *obeys (40.18),*

$$\begin{bmatrix} \mathbf{e}_{u\beta}\left[\mathbf{U}_\beta^\mu(t)\right] \\ \mathbf{w}_\gamma\left[\mathbf{U}_\gamma(t)\right] \end{bmatrix} = \begin{bmatrix} \mathbf{e}_{u\beta}\left[\mathbf{U}_\beta^\mu(t^-)\right] \\ \mathbf{w}_\gamma\left[\mathbf{U}_\gamma(t^-)\right] \end{bmatrix}$$
$$+ H(t)\left[T_{1\theta}\Sigma\left(\boldsymbol{\epsilon}_\theta^*, \boldsymbol{\epsilon}_\theta^{*(1)}\right)\boldsymbol{\epsilon}_\theta^{*(1)} + K_{0\theta} sign\left|\boldsymbol{\epsilon}_{\theta 0}^*\right|\right], \ \forall t \in \mathfrak{T}_0. \tag{40.18}$$

The domain $\mathcal{D}_T\left(t_0; \mathfrak{D}^j; \mathbf{U}; \mathfrak{Y}_d\right)$ *of tracking on* $\mathfrak{T}_0 \times \mathfrak{D}^j \times \mathfrak{Y}_d^\alpha$ *is equal to the domain* $\mathcal{D}_{PTbl}\left(t_0; \mathfrak{D}^j; \mathbf{U}; \mathfrak{Y}_d\right)$ *of the perfect trackability of the plant (3.12) on* $\mathfrak{T}_0 \times \mathfrak{D}^j \times \mathfrak{Y}_d^\alpha$ *determined by (8.33).*

The solution $\varepsilon_\theta^*(.)$ *shows that the tracking is stable in the whole with FVIDRT* t_R^ρ *if* $\mathbf{T}\left(\varepsilon_{\theta 0}^{*1}\right) = \mathbf{0}_\theta$.

Theorem 571 ***NTC synthesis for the exponential absolute error vector value tracking elementwise and stablewise with the finite vector BS reachability time*** t_R^ρ, ***Example 561***

Let (38.22) through (38.36) be valid.

For the plant (3.12) obeying Condition (564) to be controlled by the natural tracking control $\mathbf{U}$ *in order to exhibit tracking on* $\mathfrak{T}_{0F}^\theta \times \mathfrak{D}^j \times \mathfrak{Y}_d^\alpha$ *determined by*

$$\mathbf{T}\left(\varepsilon_\theta^*, \varepsilon_\theta^{*(1)}\right) = T_{1\theta} D^+ \left|\varepsilon_\theta^*\right| + K_\theta\left(\left|\varepsilon_\theta^*\right| + K_{0\theta} sign\left|\varepsilon_{\theta 0}^*\right|\right)$$
$$= T_{1\theta}\Sigma\left(\varepsilon_{\theta,\theta}^{*\ *(1)}\right)\varepsilon_\theta^{*(1)} + K_\theta\left(\left|\varepsilon_\theta^*\right| + K_{0\theta} sign\left|\varepsilon_{\theta 0}^*\right|\right) = \mathbf{0}_\theta,$$
$$\forall \mathbf{t}^\theta \in \mathfrak{T}_{0F}^\theta \ if \ \mathbf{T}\left(\varepsilon_{\theta 0}^{*1}\right) = \mathbf{0}_\theta,$$
$$\varepsilon_\theta^*\left(\mathbf{t}^{2\theta}; \varepsilon_{\theta 0}^*\right) = \mathbf{f}_\theta^1\left(\mathbf{t}^{2\theta}\right), \ \forall \mathbf{t}^{2\theta} \in \mathfrak{T}_{0F}^{2\theta}, \ if \ \mathbf{T}\left(\varepsilon_{\theta 0}^{*1}\right) \neq \mathbf{0}_\theta, \tag{40.19}$$

for the exponential absolute error vector value tracking elementwise and stablewise with the finite vector BS reachability time t_R^θ *(40.20), (40.21):*

$$T_{R\theta} = T_{\theta 0} + \left\{ \begin{array}{ll} K_\theta^{-1} T_{1\theta} \ln\left\{K_{0\theta}\left[\left|\Psi_{\theta 0}^*\right| + K_{0\theta}\right]\right\}, & \boldsymbol{\epsilon}_{\theta 0}^* \neq \mathbf{0}_\theta, \\ O_\theta, & \boldsymbol{\epsilon}_{\theta 0}^* = \mathbf{0}_\theta \end{array} \right\}, \tag{40.20}$$

$$T_{R\theta} = diag\left\{t_{R1} \ t_{R2} \ \cdots \ t_{R\theta}\right\} \Longleftrightarrow \mathbf{t}_R^\theta = \left[t_{R1} \ t_{R2} \ \cdots \ t_{R\theta}\right]^T, \tag{40.21}$$

it is necessary and sufficient that both:

i) The rank β *of the control matrix* $P(t)$ *is greater than zero and not greater than* $min(\rho, r)$, *i.e., (8.20) holds, and the rank* γ *of the control matrix* $W(t)$ *is greater than zero and not greater than* $min(\gamma, r)$, *i.e., (8.10) is true, i.e., jointly: (8.31) is valid,*

and

ii) if the plant also possesses Property 61 and Property 64 then for any chosen matrix $H(t) \in \mathfrak{R}^{\theta \times m}$ *that obeys (8.30), the control vector function* $\mathbf{U}(.)$ *obeys (40.22),*

$$\begin{bmatrix} \mathbf{e}_{u\beta}\left[\mathbf{U}_\beta^\mu(t)\right] \\ \mathbf{w}_\gamma\left[\mathbf{U}_\gamma(t)\right] \end{bmatrix} = \begin{bmatrix} \mathbf{e}_{u\beta}\left[\mathbf{U}_\beta^\mu(t^-)\right] \\ \mathbf{w}_\gamma\left[\mathbf{U}_\gamma(t^-)\right] \end{bmatrix}$$
$$+ H(t)\left[T_{1\theta} D^+ \left|\boldsymbol{\epsilon}_\theta^*\right| + K_\theta\left(\left|\boldsymbol{\epsilon}_\theta^*\right| + K_{0\theta} sign\left|\boldsymbol{\epsilon}_{\theta 0}^*\right|\right)\right], \ \forall t \in \mathfrak{T}_0. \tag{40.22}$$

The domain $\mathcal{D}_T\left(t_0;\mathfrak{D}^j;\mathbf{U};\mathfrak{Y}_d\right)$ of tracking on $\mathfrak{T}_0 \times \mathfrak{D}^j \times \mathfrak{Y}_d^\alpha$ is equal to the domain $\mathcal{D}_{PTbl}\left(t_0;\mathfrak{D}^j;\mathbf{U};\mathfrak{Y}_d\right)$ of the perfect trackability of the plant (3.12) on $\mathfrak{T}_0 \times \mathfrak{D}^j \times \mathfrak{Y}_d^\alpha$ determined by (8.33).

The solution $\varepsilon_\theta^*\left(\mathbf{t}^\theta;\varepsilon_{\theta o}^*\right)$ shows that the tracking is stable in the whole with FVIDRT t_R^θ if $\mathbf{T}\left(\varepsilon_{\theta 0}^{*1}\right) = \mathbf{0}_\theta$.

The following comments generalize the corresponding comments in [279, Comments 384-389, pp. 278, 279].

Comment 572 *The above-synthesized NTC control* **U** *does not demand knowledge of the internal dynamics of the plant.*

Comment 573 ***Implementation of the NTC algorithm***

The controller determines the subsidiary reference BS vector $\left[\mathbf{R}_R^T \;\; \mathbf{Y}_R^T\right]^T$, (38.22), the subsidiary vector function $\mathbf{f}_\theta^1(.)$, and the subsidiary error vector $\boldsymbol{\epsilon}_\theta^*$, (38.28), and uses them to generate the NTC according to the accepted algorithm.

The natural tracking controller does not use information about the plant mathematical model. It does not use any information about the disturbance vector function $\mathbf{D}(.)$ or about high derivatives of the real BS vector $\left[\mathbf{R}^T(t) \;\; \mathbf{Y}^T(t)\right]^T$ at the moment t. The NTC is fully robust. The preceding control algorithms verify these claims.

The usage of $\mathbf{U}(t^-)$ in order to generate $\mathbf{U}(t)$ at the moment t expresses a kind of memory and expresses the adaptability of the NTC.

The transmission of the signal from the controller output to its input without delay, or with negligible delay, is the crucial demand on the controller.

Part IX

CONCLUSION

Chapter 41

Systems, control, tracking, trackability

41.1 Perturbed systems

Every dynamical physical system is under the influence of the initial conditions and external (e.g., disturbance and/or control) actions. Their adequate mathematical models reflect this fact. This posed the problem of discovering the conditions for the existence and uniqueness of system solutions.

The book treats the systems **(3.12)** described by *time*-varying nonlinear vector both differential equation of any order and algebraic equation. The former is the system state equation. The latter is the system output equation. One subclass of these systems is described by a *time*-varying nonlinear vector differential equation of any order in terms of the output vector (3.12). Another subclass of the systems is described by a *time*-varying nonlinear vector differential equation of the first-order (3.65) and by a *time*-varying nonlinear vector algebraic equation (3.66) (known as the state-space system description).

All new concepts, definitions, and results holding for the system **(3.12)** hold simultaneously for the systems (3.12) and (3.65), (3.66).

The solution for the problem of the conditions for the existence and uniqueness of system solutions in the perturbed regimes has the classical form of the Lipshitz condition.

41.2 Control goal: Tracking

The fundamental control principle, Principle 87, has the same importance for control science and for control engineering as the law of matter and energy preservation or as the second law of thermodynamics on the system entropy variation for physics. It determines what is at most possible to achieve with r control variables in order to control N output variables.

The tracking issue has existed as long as the control one due to its sense to be *the very control goal, the primary control task.* Its complexity was the inherent obstacle to be deeply and completely treated. It was considered in the stability framework in its simplest form to show the asymptotic convergence of the plant real behavior to its desired behavior as *time* escapes to infinity. It was reduced to the demand for zero steady-state error.

The control goal and the control task are much more profound.

Since 1980/1981 [245], [246], [487], [488] the study of the tracking phenomena, of the tracking properties, and of the tracking control synthesis has been creating the basis for the new control scientific fundamental, for ***the tracking theory*** of linear and nonlinear dynamical systems.

The tracking theory systematically satisfies the primary goal and purpose of control science and of control engineering. The book establishes the general tracking theory in the framework of *time*-varying nonlinear systems described by the any order *time*-varying nonlinear vector differential equation and by the *time*-varying nonlinear algebraic equation. The former describes the system internal dynamics; i.e., the system state behavior. The latter links the system output with its state behavior and with the external action on the system.

The tracking theory incorporates stability theory. The former might be considered as the generalization of the latter by treating the systems behavior under the simultaneous actions of both arbitrary initial conditions and unknown and unpredictable external actions in order to achieve the control goal with a demanded quality. All solutions of the tracking problems established in this book are directly applicable to solving the corresponding stability problems.

The tracking signifies that the plant behavior is under the influence of both initial conditions and external disturbances. These two crucial actions on the plant cannot be separately treated in the framework of the nonlinear plants regardless of their *time* nature.

41.3 Tracking demands trackability

The new control concept that is ***the trackability concept*** is the prerequisite to study the tracking. The tracking control synthesis task has a solution only for the trackable plants. The trackability can be perfect or imperfect. They are precisely defined. The trackability concept comprises definitions of various tracking properties and of their domains. Both the necessary and sufficient conditions for them are proved to have simple forms.

The natural trackability, either perfect or imperfect, clarifies whether the control can be synthesized without using any information about the plant internal dynamics and about the real form and values of the external disturbances. The criteria for the natural trackability and its domain have simple forms.

Chapter 42

Lyapunov theory and tracking

42.1 Lyapunov theory extended to tracking

Lyapunov method and Lyapunov methodology for nonlinear systems have been established for stability studies by following Lyapunov [401]. They are originally effectively applicable only in the state space; i.e., to the systems described by the first-order state differential equation and by the algebraic equation. (Conclusion 345). This demands determining the desired motion $\mathbf{R}_d^{\alpha-1}(t; t_0; \mathbf{R}_0^{\alpha-1}; \mathbf{I})$ for every desired output behavior $\mathbf{Y}_d(t) \in \mathfrak{Y}_d$. The direct application of the Lyapunov method and Lyapunov methodology to tracking studies concerned mainly asymptotic tracking; i.e., the zero steady-state error at infinite moment $t = \infty$.

The Lyapunov method [401] permits us to study the dynamical behavior of the system subjected to actions of arbitrary initial conditions without solving the system mathematical model. It transforms the problem of solving, in general, the system of higher-order nonlinear scalar ordinary differential equations to the problem of solving the first-order nonlinear partial differential equation for one boundary condition.

The tracking properties in a Lyapunov sense compose ***the Lyapunov tracking concept***. They incorporate Lyapunov stability properties and the Lyapunov stability concept as special cases. The Lyapunov tracking concept introduces and defines a number of various tracking types.

The book broadens the Lyapunov method and Lyapunov methodology for nonlinear systems to the stable tracking of the desired plant behavior on asymptotically contractive sets and to the uniform stable tracking on *time*-varying sets that are not asymptotically contractive and on *time*-invariant sets. It is further broadened to tracking a *time*-invariant or a *time*-varying set, which is crucial for defense tasks. The established conditions are sufficient, but not necessary. Their application is effective when a Lyapunov function can be chosen or when it is given. Then, they are useful for tracking control synthesis. However, if a Lyapunov function is neither given nor can be chosen, then a fundamental problem appears: how to determine, how to construct, and how to generate system Lyapunov function for the tracking study, for the tracking control synthesis. Another unsolved fundamental problem of both Lyapunov methods and their preceding extensions to tracking is the problem of the determination of the conditions for tracking, which are both necessary and sufficient and which are not expressed in terms of the existence of a Lyapunov function, as well as the problem of the determination of the exact tracking domains. If the conditions are satisfied then it means that the system exhibits the tested tracking property. However, if the conditions are not satisfied, then we cannot conclude anything. We should look for another tentative Lyapunov function. Such a search can be endless.

42.2 Consistent Lyapunov methodology: Tracking

Lyapunov himself established the inverse methodology for the application of his method [401]. The inverse methodology (inverse to Lyapunov's methodology for the linear systems) starts with a choice of a positive definite function $v(.)$ and continues with the negative definiteness test of its derivative $v^{(1)}$ along system motions. This methodology does not give any guide on how to select the positive definite function $v(.)$. Also, it establishes only sufficient conditions. Even if all the sufficient conditions are satisfied the result does not determine the exact tracking (stability) domain. Assumption 429 and Assumption 437 determine a guideline on how to select the function $v(.)$ relative to the given system, Comment 430.

The new, *consistent Lyapunov methodology* established in the framework of the stability of motion is extended herein to tracking. It enabled us to resolve all of the preceding unsolved problems in Lyapunov stability theory.

Consistent Lyapunov methodology (*CLM*) resolves effectively theoretically the following fundamental problems:

- The determination of the conditions that are both necessary and sufficient for uniform stable tracking.
- The conditions are not expressed in terms of the existence of a Lyapunov function.
- The single application of the conditions results in the final clear response whether the system exhibits the tested tracking or not.
- The generation of the system Lyapunov function after the single application of the conditions.
- The determination of the exact tracking domains of the system.

Consistent Lyapunov methodology establishes the general solutions for the general class of the systems (3.12) and for its two subclasses (3.62) and (3.65), (3.66). Weak smoothness property 368 and strong smoothness property 370 determine their properties that enabled us to prove the solutions in the form of necessary and sufficient conditions.

Theorems 379, 380, 381, 389, 390, 391, and 393 express consistent Lyapunov methodology. It provides the solutions to all three fundamental stability problems: the determination of the system Lyapunov function $v(.)$, the discovery of the necessary and sufficient conditions for asymptotic or exponential tracking (hence, also stability) that are not in terms of the existence of a Lyapunov function, and the determination of the exact tracking (hence, stability) domain. Unfortunately, even for the regular system (Definition 59); i.e., if

$$\det\widetilde{Q}\left(t,\mathbf{r}^{\alpha-1}\right) \neq 0,\ \forall\left(t,\mathbf{r}^{\alpha-1}\right) \in \mathfrak{T} \times \mathfrak{R}^{\alpha\rho},$$

the mathematical problem of determining a solution $v\left[t,\mathbf{r}^{\alpha-1}(t)\right]$ of the nonlinear partial differential equation along system behavior:

$$v^{(1)}\left[t,\mathbf{r}^{\alpha-1}(t)\right] = -w\left[t,\mathbf{r}^{\alpha-1}(t);\mathbf{i}\right],$$

where $w(.)$ is an arbitrary definite function and

$$v^{(1)}\left[t,\mathbf{r}^{\alpha-1}(t)\right] = \frac{\partial v\left[t,\mathbf{r}^{\alpha-1}(t)\right]}{\partial t}$$
$$+\left\{\text{grad}V\left[t,\mathbf{r}^{\alpha-1}(t)\right]\right\}^T \widetilde{Q}^{-1}\left[t,\mathbf{r}^{\alpha-1}(t)\right]\left\{\widetilde{\mathbf{h}}\left[t,\mathbf{i}^{\xi}(t)\right]-\widetilde{\mathbf{q}}\left[t,\mathbf{r}^{\alpha-1}(t)\right]\right\},$$

is solvable only in very simple cases. Fortunately, the physical engineering approach overcomes this problem in the framework of physical, rather than purely mathematical, dynamical systems. Lyapunov function $v(.)$ represents the accumulated energy and mass in the physical system. The function $-w(.)$ represents the system generalized power; i.e., its power and the mass flow through the system. After determining the accumulated system power and mass, the system power and the mass flow, we have determined the functions $v(.)$ and $-w(.)$. If $w(.)$ is positive definite then the system exhibits stable tracking of the desired state behavior if, and only if, $v(.)$ is also positive definite. If $w(.)$ and/or $v(.)$ is not positive definite then the control should be changed and/or the system itself should be improved.

If the generating function $w(.)$ multiplied by -1, i.e., $-w(.)$, is the system power $p(.)$, $-w(.) = p(.)$, then the solution Lyapunov function $v(.)$ is the system energy $e(.)$, $v(.) = e(.)$. The essential sense of the conditions is the following: if the system power $p(.)$ is negative definite then for the system to exhibit the requested tracking it is both necessary and sufficient that the system energy is a positive definite function.

This physical interpretation is significant for control engineering applications of *CLM* to tracking. Understanding that mass (flow) and energy (flow) create the system processes, movements, and dynamics, it is clear that *CLM* suggests starting with the determination of the system power $p(.)$ and continuing with its negative definiteness test. If the result is affirmative then the system energy should be determined and tested. If the system power is not negative definite then changes should be made in the system to ensure negative definiteness of its power.

42.3 Lyapunov control synthesis

Lyapunov tracking conditions are the basis for the synthesis of tracking control that guarantees a Lyapunov tracking property.

The Lyapunov method originally demands knowledge of the system mathematical model in the form of the classical state-space description; i.e., in Cauchy (normal) form, the measurability of all system parameters, nonlinearities, as well as the measurability of all disturbances. This is unavoidable for the synthesis of the control to preserve the plant desired behavior in spite of nonnominal disturbances acting on the plant, which is expressed in (30.6) and (30.9) for the case $\mathbf{r}^{\alpha-1}(\tau) = \mathbf{0}_{\alpha\rho}$.

The Lyapunov method has been extended to synthesis of a robust control by majorizing disturbances. This is explained in Note 435. It also shows how we can majorize the system state vector so to avoid knowledge of its exact mathematical model. The result is the control that always acts on the plant as the plant is subjected to the most sever disturbance actions and variations of the system state. Therefore, the control algorithm then can be too conservative, too stringent.

The most crucial disadvantage of the Lyapunov method for control synthesis is that the control tends to infinity as the gradient, grad$V(.)$, of the function $v(.)$, or the whole term

$$\left\{\left[\Delta V(t,\mathbf{r}^{\alpha-1})\right]^T \widetilde{Q}^{-1}\left[t,\mathbf{r}^{\alpha-1}(t)\right] P(t)\right\}$$

in the case of the plant (3.12), approaches zero vector; i.e., the origin. If the control is bounded, then it can assure only a kind of practical tracking (hence, stability) [183], [184], [186], [187], [196], [197], [376]-[380].

The book shows how the Lyapunov method can be directly applied to the systems (3.12) despite they're not being in the classical state-space form.

Chapter 43

High quality tracking: Control synthesis

43.1 Tracking with finite reachability time

The fact that any human-made technical system has a finite life *time* imposed the need for introducing the concept of ***the tracking with the finite, either scalar or vector, reachability time (FSRT*** or ***FVRT)***. For various *FSRT* and *FVRT* tracking properties well defined herein, the proved tracking conditions constitute the basis for the control synthesis to ensure the appropriate tracking with the (possibly demanded) *FSRT* or *FVRT*.

The herein established Lyapunov conditions to various tracking properties with the finite reachability time are the basis for control synthesis to ensure any demanded of such properties.

43.2 Demanded tracking quality

A higher tracking quality can be specified in the form of the zero vector value of the tracking operator $\mathbf{T}(.)$ at every instant belonging to a finite or infinite *time* set. The tracking operator $\mathbf{T}(.)$ depends on the plant (output, or internal dynamics or joint) error vector, its derivatives, and possibly on its integral. Because the initial error vector is unpredictable and uncontrollable, there exist an infinite number of them for which the tracking operator $\mathbf{T}(.)$ is different from the zero vector at the initial moment. This led to recall the wise engineering strategy: to replace the desired plant behavior with the reference plant behavior such that it is equal to the real initial plant behavior and to became equal to the plant desired behavior at the reachability time, after which they coincide all the *time*. This induced the new error vector relative to the plant reference behavior. It replaced the plant error vector relative to the plant desired behavior. Seven typical high-quality tracking operators determine high tracking quality. They are used to synthesize ***the natural tracking control*** (*NTC*) that ensures the high quality of the plant tracking behavior.

The fact that all concepts, definitions, conditions, criteria, control algorithms, and control laws are established in the framework of the large class of *time*-varying nonlinear systems shows that the established tracking theory is general. It can be further developed to other classes of the plants as well as to other tracking concepts and can be linked with various tracking quality criteria (e.g., with optimal tracking control criteria).

43.3 Trackability theory and tracking theory importance

As stability theory is fundamental for dynamical systems in general, so tracking theory is fundamental for control systems.

The establishment of general tracking theory, and of its special subtheory for linear systems in [279], permits and requests setting the tracking issue as the central topic of the undergraduate, postgraduate, and doctoral control courses as well as to become the key topic of control research.

The book contributes to the general qualitative theory of dynamical systems, to control science, and control engineering by covering their fundamental lacunae.

Part X

APPENDICES

Appendix A

Notation

The meaning of the notation is explained in the text at its first use. What follows is the complete list of the notation.

The notation "$\mathfrak{T}_\tau$", "$\forall t \in \mathfrak{T}_\tau$", and "on $\mathfrak{T}_\tau$" is to be omitted if, and only if, $\mathfrak{T}_\tau = \mathfrak{T}$. The instant τ can be equal to the initial instant t_0; $\tau = t_0$ is permitted, but not required.

A.1 Abbreviations

We present abbreviations used most frequently in the book.

BS *both spaces*

Cl$\mathfrak{S}$ *the closure of the set* $\mathfrak{S}$

CLM *consistent Lyapunov methodology*

FRT *finite reachability time that can be either scalar or vector finite reachability time*

FSRT *finite scalar reachability time*

FVRT *finite vector reachability time*

iff *if, and only if*

In$\mathfrak{S}$ *the interior of the set* $\mathfrak{S}$

ID *internal dynamics*

IO *input-output*

IO system *input-output system*

IIDO system *input- high-order internal dynamics (state)-output system* that is simply called: *the system*

ISO *input-the first-order state-output*

L-tracking *Lyapunov tracking*

LTC *Lyapunov tracking control*

LyS *Lyapunov stability concept*

LyTgC *Lyapunov tracking concept*

NTC *natural tracking control*

P *plant*

PCUP *physical continuity and uniqueness principle*

TC *tracking control*

TCUP *time continuity and uniqueness principle*

TgC *tracking concept(s)*

w.r.t. *with respect to*

A.2 Indexes

A.2.1 Subscripts

The typical subscripts follow.

CS control system

d the subscript d denotes "desired"

e equilibrium,

i the subscript i denotes "the i-th"

j the subscript j denotes "the j-th"

P for plant

$zero$ the subscript zero denotes "the zero value"

0 the subscript 0 (zero) associated with a variable x denotes its initial value x_0; however, if it is associated with a subset of the set $\mathfrak{T}$ then the subscript 0 (zero) denotes the corresponding *time* set; e.g., $\mathfrak{T}_0$.

A.2.2 Superscript

I the superscript I denotes "inverse", $\mathbf{f}_3^I\left[\mathbf{f}_3(\mathbf{U})\right] = \mathbf{U}, \forall \mathbf{U} \in \mathfrak{R}^r$, if $\mathbf{f}_3(\mathbf{U}) = \mathbf{W}$ then $\mathbf{f}_3\left[\mathbf{f}_3^I(\mathbf{W})\right] = \mathbf{W}, \forall \mathbf{W} \in \mathfrak{R}^r$.

A.3 Letters

Lower-case block or italic letters are used for scalars. Bold (lower case and capital, Greek and Roman) block letters denote vectors. Upper-case block letters denote matrices, or points. Upper-case calligraphic and $\mathfrak{fraktur}$ letters designate sets or spaces.

The notation "$; t_{(.)0}$" is omitted as an argument of a variable if, and only if, a choice of the initial moment $t_{(.)0}$ does not have any influence on the value of the variable.

A.3.1 Calligraphic letters

With a few exceptions, calligraphic letters are used for sets.

$\mathcal{A}$ *accuracy*

$\mathcal{C}$ *controller*

$\mathcal{CS}$ *control system*

$\mathcal{D}$ *domain*

$\mathfrak{D}_E(..;\alpha,\beta,\Delta;)$ *the exponential tracking domain relative to* α, β, and Δ

$\mathfrak{D}_E^k(.)$ *the k-th-order exponential tracking domain*

$\mathcal{D}_{ENTblu}^l(\cdot)$ *the l-th-order uniform elementwise natural trackability domain*

$\mathcal{D}_{ETbl}^l(\cdot)$ *the l-th-order elementwise trackability domain*

$\mathfrak{D}_{EU}^k(\cdot)$ *the k-th-order uniform exponential tracking domain*

$\mathcal{D}_{NTbl}^l(\cdot)$ *the domain of the l-th-order natural trackability of the plant*

$\mathcal{D}_{PNTbl}^l(\cdot)$ *the domain of the l-th-order perfect natural trackability of the plant*

$\mathcal{D}_{PTbl}^l(\cdot)$ *the domain of the l-th-order perfect trackability of the plant*

$\mathcal{D}_{PTblD}^l(\cdot)$ *the domain of the l-th-order perfect trackability of* $\mathbf{Y}_d(.)$ *for a fixed* $\mathbf{D}(.) \in \mathfrak{D}^k$

$\mathcal{D}_{SStT}(..;\varepsilon;.)$ *the* ε*-stable state tracking domain*

$\mathcal{D}_{SStT}(.)$ *the stable state tracking domain*

$\mathcal{D}_{StTU}(.)$ *the uniform state tracking domain*

$\mathcal{D}_{SSStT}(.)$ *the strict stable state tracking domain*

$\mathcal{D}_T^k(.)$ *the k-th-*order *tracking domain*

$\mathcal{D}_{Tbl}^l(.;\sigma;.)$ *the l-th-order trackability domain for fixed* $\sigma \in \mathrm{In}\mathfrak{T}_0$

$\mathcal{D}^l_{Tbl}(.)$ *the l-th-order trackability domain*

$\mathcal{D}^l_{TblU}(.)$ *the l-th-order uniform trackability domain*

$\mathcal{D}^k_{TU}(.)$ *the uniform k-th-order tracking domain*

$\mathcal{F}$ *the family of all set functions* $\Upsilon(.) : \mathfrak{T} \longrightarrow 2^{\mathfrak{R}^K}$

$\mathcal{J}^i$ *a given, or to be determined, family of all bounded i-times continuously differentiable on* $\mathfrak{T}_0$ *permitted input vector functions* $\mathbf{I}(.)$, $\mathcal{J}^i \subset \mathfrak{C}^i$

$\mathcal{K}_{[0,\alpha[}$ *the class of functions* $\varphi(.) : \mathfrak{R}_+ \longrightarrow \mathfrak{R}_+$ *defined, continuous, and strictly increasing on* $[0, \alpha[$ *such that vanish at the origin,* $\varphi(0) = 0, 0 < \alpha \leq \infty$

$\mathcal{K} = \mathcal{K}_{[0,\infty[}$

$\mathcal{P}$ *plant*

$\mathcal{R}^{\alpha-1}(.) : \mathfrak{T} \times \mathfrak{T} \times \mathfrak{R}^{\alpha\rho} \times \mathfrak{R}^{d\beta} \times \mathfrak{R}^{r\gamma} \longrightarrow \mathfrak{R}^{\alpha\rho}$ *a motion (solution) of the system,*

$$\mathcal{R}^{\alpha-1}(t_0; t_0, \mathbf{R}_0^{\alpha-1}; \mathbf{D}; \mathbf{U}) \equiv \mathbf{R}_0^{\alpha-1}$$

$\mathcal{S}$ *a dynamical system*

$\mathcal{Y}(.)$ a mapping called *output response* that describes the output system behavior,

$$\mathcal{Y}(.) : \mathfrak{T}_0 \times \mathfrak{T}_i \times \mathfrak{Y} \times \mathcal{J}^i \rightarrow \mathfrak{Y},\ \mathcal{Y}(t_0; t_0; \mathbf{Y}_0; \mathbf{I}) \equiv \mathbf{Y}_0$$

$\mathcal{Z}(.)$ a mapping called *motion (solution)* that describes the state behavior of any system,

$$\mathcal{Z}(.) : \mathfrak{T}_0 \times \mathfrak{T}_i \times \mathfrak{Z}_0 \times \mathcal{J}^i \rightarrow \mathfrak{Z},\ \mathcal{Z}(t_0; t_0; \mathbf{Z}_0; \mathbf{I}) \equiv \mathbf{Z}_0$$

A.3.2 Fraktur letters

Capital fraktur letters are used for spaces or for sets.

$\mathfrak{B} \subseteq \mathfrak{R}^n$ *a nonempty subset of* $\mathfrak{R}^n$

$\mathfrak{B}_\xi(\mathbf{a})$ an *open hyperball with the radius* ξ *centered at the point* $\mathbf{a} \in \mathfrak{R}^K$ *in the space* $\mathfrak{R}^K$, $\mathfrak{B}_\xi(\mathbf{a}) = \{\mathbf{w} : \|\mathbf{w} - \mathbf{a}\| < \xi\}$

$\mathfrak{B}_\xi$ *an open hyperball with the radius* ξ *centered at the origin of the corresponding space,* $\mathfrak{B}_\xi = \mathfrak{B}_\xi(\mathbf{0}_K)$

$\mathfrak{C}$ *the family of all continuous functions on* $\mathfrak{T}$

$\mathfrak{C}^{ki}(\mathfrak{S})$ *the family of all functions defined, continuous and k-times continuously differentiable on the set* $\mathfrak{S} \subseteq \mathfrak{R}^i$, $\mathfrak{C}^{ki}(\mathfrak{R}^i) = \mathfrak{C}^k(\mathfrak{R}^i) = \mathfrak{C}^{ki}$

$\mathfrak{C}^k = \mathfrak{C}^k(\mathfrak{T}_0)$ *the family of all functions defined, continuous and k-times continuously differentiable on* $\mathfrak{T}_0$, $\mathfrak{C}^k = \mathfrak{C}^k(\mathfrak{T})$ and $\mathfrak{C}^0 = \mathfrak{C}$

$\mathfrak{C}^0(\mathfrak{S}) = \mathfrak{C}(\mathfrak{S})$ *the family of all functions defined and continuous on the set* $\mathfrak{S}$, $\mathfrak{C}^0(\mathfrak{R}^i) = \mathfrak{C}^{0,i}$

$\mathfrak{C}^{k-}(\mathfrak{R}^i)$ *the family of all functions defined and continuous everywhere on* $\mathfrak{R}^i$ and *k-times continuously differentiable on* $\mathfrak{R}^i \backslash \{\mathbf{0}_i\}$, which have defined and continuous derivatives at the origin $\mathbf{0}_i$ of $\mathfrak{R}^i$ up to the order $(k-1)$, which have defined the left and the right k-th order derivative at the origin $\mathbf{0}_i$

$\mathfrak{C}^-(\mathfrak{R}^N)$ *the family of all functions defined everywhere on* $\mathfrak{R}^N$

$\mathfrak{d}_m = \mathfrak{d}_m(t; \mathfrak{S}) \in \mathfrak{R}_+ \cup \{\infty\}$ *the inner diameter of a time-varying set* $\mathfrak{S}(t)$ *at a moment* $t \in \mathfrak{T}$

$\mathfrak{d}_m(\mathfrak{T}_\tau; \mathfrak{S})$ *the inner diameter of a time-varying set* $\mathfrak{S}(t)$ *on* $\mathfrak{T}_\tau$,

$$\mathfrak{d}_m(\mathfrak{T}_\tau; \mathfrak{S}) = \inf[\mathfrak{d}_m(t; \mathfrak{S}) : t \in \mathfrak{T}_\tau]$$

$\mathfrak{d}_M = \mathfrak{d}_M(t; \mathfrak{S}) \in \mathfrak{R}_+ \cup \{\infty\}$ *the outer diameter of a time-varying set* $\mathfrak{S}(t)$ *at a moment* $t \in \mathfrak{T}$

$\mathfrak{d}_M = \mathfrak{d}_M(\mathfrak{T}_\tau; \mathfrak{S})$ *the outer diameter of a time-varying set* $\mathfrak{S}(t)$ *on* $\mathfrak{T}_\tau$,

$$\mathfrak{d}_M(\mathfrak{T}_\tau; \mathfrak{S}) = \sup[\mathfrak{d}_M(t; \mathfrak{S}) : t \in \mathfrak{T}_\tau]$$

$\mathfrak{d}_{(.)}(\mathfrak{T};\mathfrak{S}) = \mathfrak{d}_{(.)}(\mathfrak{S})$

$\mathfrak{D}^k$ *a given, or to be determined, family of all bounded $(k+1)$-times continuously differentiable on $\mathfrak{T}$ permitted total disturbance vector functions $\mathbf{D}(.)$, $\mathfrak{D}^k \subset \mathfrak{C}^{k+1}(\mathfrak{R}^d)$*

$\mathfrak{D}^0 = \mathfrak{D}$ *the family of all bounded continuous and continuously differentiable on $\mathfrak{T}$ permitted total disturbance vector functions $\mathbf{D}(.) \in \mathfrak{D}$ such that they obey TCUP, $\mathfrak{D} \subset \mathfrak{C}^1(\mathfrak{R}^d)$*

$\mathfrak{I}$ *the output integral space, $\mathfrak{I} = \mathfrak{T} \times \mathfrak{R}^K$*

$\mathfrak{I}_e$ *the extended output integral space, $\mathfrak{I}_e = \mathfrak{T} \times \mathfrak{R}^{(k+1)N}$*

$\mathfrak{J}$ *a compact time interval on which Granwall inequality theorem 74 holds,*

$$\mathfrak{J} = [a, b] \subset \mathfrak{T}$$

$\mathfrak{L}$ *the largest temporal set of the initial moments t_0 for which the given system has a solution, it is the intersection of $\mathfrak{L}^*\left[\mathbf{R}_0^{\alpha-1}, \mathbf{D}(.), \mathbf{U}(.)\right]$ over the product set $\mathfrak{S} \times \mathfrak{D}^\beta \times \mathfrak{U}^\gamma$,*

$$\mathfrak{L} = \cap \left\{ \begin{array}{c} \mathfrak{L}^*\left[\mathbf{R}_0^{\alpha-1}, \mathbf{D}(.), \mathbf{U}(.)\right] : \left[\mathbf{R}_0^{\alpha-1}, \mathbf{D}(.), \mathbf{U}(.)\right] \\ \in \mathfrak{S} \times \mathfrak{D}^\beta \times \mathfrak{U}^\gamma \end{array} \right\}$$

$$\mathfrak{L} = (l_m, l_M) \neq \phi, \ -\infty \le l_m < l_M \le \infty, \ \mathfrak{L} \subseteq \mathfrak{T}$$

$\mathfrak{L}^c$ *an arbitrary nonempty connected compact subinterval of $\mathfrak{L}$*

$\mathfrak{L}^*$ *is the largest temporal set of the initial moments t_0 for which the system has a solution, $\mathfrak{L}^* = (l_m^*, l_M^*)$*

$\mathfrak{N}_{(.)} \subseteq \mathfrak{R}^K$ *a neighborhood of the origin $\mathbf{z} = \mathbf{0}_K$*

$\mathfrak{N}_\varepsilon(\mathfrak{S})$ *the ε-neighborhood of the set $\mathfrak{S}$,*

$$\mathfrak{N}_\varepsilon(\mathfrak{S}) = \left\{\mathbf{z} : \mathbf{z} \in \mathfrak{R}^K, \ \rho(\mathbf{z}, \mathfrak{S}) < \varepsilon\right\} \supset \mathfrak{S}$$

$\mathfrak{N}(\varepsilon; \mathfrak{S})$ *a connected neighborhood of the set $\mathfrak{S}$,* which is determined by ε and the outer radius of which cannot be greater than ε, such that

$$0 < \varepsilon_1 < \varepsilon_2 \Longrightarrow \mathfrak{N}(\varepsilon_1; \mathfrak{S}) \subseteq \mathfrak{N}(\varepsilon_2; \mathfrak{S}),$$
$$\varepsilon \longrightarrow 0^+ \Longrightarrow \mathfrak{N}(\varepsilon; \mathfrak{S}) \longrightarrow \mathfrak{S}, \ \mathfrak{N}(\varepsilon; \mathfrak{S}) \subseteq \mathfrak{N}_\varepsilon(\mathfrak{S})$$

$\mathfrak{N}(t; \mathfrak{S})$ *a fixed neighborhood of a time-varying bounded set $\mathfrak{S}(t)$* at $t \in \mathfrak{T}_\tau$, $\mathrm{Cl}\mathfrak{S}(t) \subset \mathfrak{N}(t; \mathfrak{S}), \forall t \in \mathfrak{T}_\tau$

$\mathfrak{N}(t; \varepsilon; \mathfrak{S})$ *the connected neighborhood of a time-varying bounded set $\mathfrak{S}(t)$* determined by ε at $t \in \mathfrak{T}_\tau$

$\mathfrak{N}_\varepsilon(t; \mathfrak{S})$ *the connected ε-neighborhood of a time-varying bounded set $\mathfrak{S}(t)$* at $t \in \mathfrak{T}_\tau$, $\mathfrak{N}_\varepsilon(t; \mathfrak{S}) = \left\{\mathbf{z} : \mathbf{z} \in \mathfrak{R}^K, \ \rho[\mathbf{z}, \mathfrak{S}(t)] < \varepsilon\right\} \supset \mathfrak{S}(t)$, $\forall t \in \mathfrak{T}_\tau$

$\mathfrak{N}[t; \mathbf{z}(t)]$ *a fixed neighborhood of the vector value $\mathbf{z}(t)$ of a vector function $\mathbf{z}(.)$ at $t \in \mathfrak{T}_\tau$*

$\mathfrak{N}_\varepsilon\left[t; \mathbf{Y}_d^k(t)\right]$ *a connected ε- neighborhood of $\mathbf{Y}_d^k(t)$ at a moment $t \in \mathfrak{T}$, $\mathfrak{N}_\varepsilon\left[t; \mathbf{Y}_d^k(t)\right] \subseteq \mathfrak{R}^{(k+1)N}$*

$\mathfrak{N}\left(t_0; \mathbf{Y}_{d0}^k; \mathbf{D}; \mathbf{U}\right)$ *a connected neighborhood of the desired initial system behavior $\mathbf{Y}_{d0}^k$ at the initial moment $t_0 \in \mathfrak{T}$, $\mathfrak{N}\left(t_0; \mathbf{Y}_{d0}^k; \mathbf{D}; \mathbf{U}\right) \subseteq \mathfrak{R}^{(k+1)N}$*

$\mathfrak{N}\left(t_0; \varepsilon; \mathbf{Y}_{d0}^k; \mathbf{D}; \mathbf{U}\right)$ *a connected neighborhood of $\mathbf{Y}_{d0}^k$ in $\mathfrak{R}^{(k+1)N}$ space,* which is determined by $(t_0; \varepsilon; \mathbf{D}; \mathbf{Y}_d) \in \mathfrak{T} \times]0, \infty] \times \mathfrak{D} \times \mathfrak{Y}_d^k$, $\mathfrak{N}\left(t_0; \varepsilon; \mathbf{Y}_{d0}^k; \mathbf{D}; \mathbf{U}\right) \subseteq \mathfrak{R}^{(k+1)N}$, and the outer radius of which cannot be greater than ε

$\mathfrak{O} \subset \mathfrak{R}^K$ *the singleton containing only the origin $\mathbf{0}_K$* of $\mathfrak{R}^K$, $\mathfrak{O} = \{\mathbf{0}_K\}$, $K = 1, 2, \cdots$

$\mathfrak{P}$ *a time-invariant connected open nonempty set $\mathfrak{P}$, $\mathfrak{P} \subseteq \mathfrak{R}^{(p+1)N}$, such that $\mathfrak{P} \subseteq \mathfrak{S}$*

$\mathfrak{R}$ *the set of all real numbers*

$\mathfrak{R}^+$ *the set of all positive real numbers*
$\mathfrak{R}_+$ *the set of all non-negative real numbers*
$\mathfrak{R}^-$ *the set of all negative real numbers*
$\mathfrak{R}^K$ *the K−dimensional real vector space,*

$$\mathfrak{R}^K \in \left\{\mathfrak{R}^N,\ \mathfrak{R}^\rho, \mathfrak{R}^{(k+1)N},\ \mathfrak{R}^{\alpha\rho}\right\}$$

$\mathfrak{R}^{M\times N}$ *the $M \times N$ matrix space, the zero matrix in which is $O_{M\times N}$. If $M = N$ then for short $O_{N\times N} = O_N$*

$\mathfrak{R}_d^k$ *a given, or to be determined, family of all bounded and $(k+1)$-times continuously differentiable realizable total desired output vector functions $\mathbf{R}_d(.) \in \mathfrak{C}^{k+1}$ such that they and their first $k+1$ derivatives obey TCUP,*

$$\mathfrak{R}_d^k \subset \mathfrak{C}^{k+1}(\mathfrak{R}^\rho),$$

i.e., *it is a family of all bounded continuously differentiable realizable total desired state vector functions $\mathcal{R}_d^{\alpha-1}(.) \in \mathfrak{C}^1$ such that they and their derivatives obey TCUP*

$\mathfrak{R}_{d0}^{\alpha-1}$ *the set of the initial vector values $\mathbf{R}_{d0}^{\alpha-1}$ of $\mathbf{R}_d^{\alpha-1}(t)$ of every $\mathbf{R}_d(.) \in \mathfrak{R}_d^{\alpha\rho}$. If, and only if, $\mathbf{R}_d(.) \in \mathfrak{R}_d^{\alpha}$ then $\mathbf{R}_{d0}^{\alpha-1} = \mathcal{R}_d^{\alpha-1}\left(t_0; t_0; \mathbf{R}_{d0}^{\alpha-1}\right) \in \mathfrak{R}_{d0}^{\alpha\rho}$*

$\mathfrak{S}$ *time-invariant connected set, $\mathfrak{S} \subseteq \mathfrak{R}^K$, the intersection of $\mathfrak{S}(t)$ over $\mathfrak{T}$,*

$$\mathfrak{S} = \cap[\mathfrak{S}(t) : t \in \mathfrak{T}] \subseteq \mathfrak{R}^{\alpha\rho},\ \mathfrak{S} \neq \phi$$

$\mathfrak{S}(t)$ *time-varying connected set, $\mathfrak{S}(t) \subseteq \mathfrak{R}^K, \forall t \in \mathfrak{T}$,*

$\mathfrak{S}(t; t_0)$ *time-varying connected set, $\mathfrak{S}(t) \subseteq \mathfrak{R}^K, \forall t \in \mathfrak{T}_0$, $t_0 \in \mathfrak{T}_i$,* if, and only if, $\mathfrak{T}_i = \mathfrak{T}$ then $\mathfrak{S}(t; t_0) \equiv \mathfrak{S}(t)$

$\mathfrak{S}(t_0)$ *the largest connected connected set $\mathfrak{S}(t)$ of the system such that the system has a unique solution passing through $\mathbf{R}_0^{\alpha-1} \in \mathfrak{S}(t_0)$ at $t = t_0$, $\mathfrak{S}(t_0) \subseteq \mathfrak{R}^{\alpha\rho}$, $t_0 \in \mathfrak{T}$*

$\mathfrak{S}(\mathfrak{L})$ *the intersection of $\mathfrak{S}(t)$ over $\mathfrak{L}$,*

$$\mathfrak{S}(\mathfrak{L}) = \cap[\mathfrak{S}(t) : t \in \mathfrak{L}] \subseteq \mathfrak{R}^{\alpha\rho},\ \mathfrak{S}(\mathfrak{L}) \neq \phi$$

$\mathfrak{S}^c(\mathfrak{L})$ *the largest nonempty connected compact subset of the set $\mathfrak{S}(\mathfrak{L})$;* if, and only if, $\mathfrak{L} = \mathfrak{T}$ *then* $\mathfrak{S}^c(\mathfrak{L}) = \mathfrak{S}^c(\mathfrak{T}) = \mathfrak{S}^c$

$\mathfrak{T}$ *the accepted reference time set,* the arbitrary element of which is an arbitrary moment t and the time unit of which is second s, $1_t = s$, $t\langle s\rangle$,

$$\mathfrak{T} = \{t : t[T]\langle s\rangle,\ \text{num}\ t \in \mathfrak{R},\ dt > 0\},\ inf\ \mathfrak{T} = -\infty,\ \ sup\ \mathfrak{T} = \infty$$

$\mathfrak{T}_i$ *the set of the initial instants t_0,* which is an unbounded on the right subset of $\mathfrak{T}$, $\mathfrak{T}_i = \{t : t[T]\langle s\rangle,\ t \in \mathfrak{T}\} \subseteq \mathfrak{T}$, *inf* $\mathfrak{T}_i \geq -\infty$, *sup* $\mathfrak{T}_i = \infty$

$\mathfrak{T}^c$ *a connected nonempty compact subset of $\mathfrak{T}$*

$\mathfrak{T}^0$ *a subset of $\mathfrak{T}$,* the measure of which is zero, $\mathfrak{T}^0 \subset \mathfrak{T}$

$\mathfrak{T}_{0\mp}$ *the subset of $\mathfrak{T}$,* which has the minimal element min $\mathfrak{T}_{0\mp}$ that is the initial instant $t_{0\mp}$, num $t_{0\mp} = 0^{\mp}$,

$$\mathfrak{T}_{0\mp} = \{t : t \in \mathfrak{T},\ t \geq t_{0\mp},\ \text{num}\ t_{0\mp} \in \mathfrak{R}\}, \mathfrak{T}_{0\mp} \subset \mathfrak{T},$$
$$min\mathfrak{T}_{0\mp} = t_{0\mp} \in \mathfrak{T},\ \ sup\ \mathfrak{T}_{0\mp} = \infty$$

$\mathfrak{T}_{0F}$ a subinterval of $\mathfrak{T}_0$ determined by both $t_0 \in \mathfrak{T}$ and $t_F \in \mathfrak{T}_0$, $t_F > t_0$,

$$\mathfrak{T}_{0F} = \left\{t :\ t \in \mathfrak{T}, \left\langle \begin{array}{l} t_0 \leq t \leq t_F \ \text{ iff } t_F < \infty, \\ t_0 \leq t < t_F \ \text{ iff } t_F = \infty \end{array} \right\rangle\right\} \subseteq \mathfrak{T}_0,$$

so that its closure $\mathrm{Cl}\mathfrak{T}_{0F}$ is compact if

$$t_F < \infty, t_F < \infty \Longrightarrow Cl\mathfrak{T}_{0F} = \{t : \ t \in \mathfrak{T}_0, \ t_0 \leq t \leq t_F < \infty\} \subseteq \mathfrak{T}_{0F}$$

$\mathfrak{T}(t_0)$ *the largest connected temporal set on which the system has a unique solution passing through* $\mathbf{R}_0^{\alpha-1}$ *at* $t = t_0$ for fixed $\left[\mathbf{R}_0^{\alpha-1}, \mathbf{D}(.), \mathbf{U}(.)\right]$, $\mathfrak{T}(t_0) \subseteq \mathfrak{T}_0$ for every $t_0 \in \mathfrak{L}_{()}$

$\mathfrak{T}^c(t_0)$ *a nonempty connected compact subinterval of* $\mathfrak{T}(t_0)$

$\mathfrak{T}^0(t_0)$ *a subset of* $\mathfrak{T}(t_0)$ *with the zero measure*

$\mathfrak{T}(\mathfrak{L})$ *the intersection of* $\mathfrak{T}(t_0)$ *over* $\mathfrak{L}$, $\mathfrak{T}(\mathfrak{L}) = \cap\left[\mathfrak{T}(t_0) : t_0 \in \mathfrak{L}\right]$

$\mathfrak{T}_\tau$ *a nonempty connected subset of* $\mathfrak{T}$, $\mathfrak{T}_\tau = \{t : \ t \in \mathfrak{T}, \ \tau \leq t \ < \infty\}$

$\mathfrak{T}_{t^*}^-$ *a set composed of a sequence* t_k *converging to* t^* *from the left as* $k \longrightarrow \infty$ if $t^* > -\infty$, ($t^* = \infty$ is permitted)

$$\mathfrak{T}_{t^*}^- = \{t_k : t_k \in \mathfrak{T}, \ t_k < t^*, \ k = 0, 1, 2, \ \cdots \ , \ k \longrightarrow \infty \Longrightarrow \ t_k \longrightarrow t^*\}, \\ t^* \in]-\infty, \infty]$$

$\mathfrak{T}_{t^*}^+$ *a set composed of a sequence* t_k *converging to* t^* *from the right as* $k \longrightarrow \infty$ if $t^* < \infty$, ($t^* = -\infty$ is permitted),

$$\mathfrak{T}_{t^*}^+ = \{t_k : t_k \in \mathfrak{T}, \ t_k > t^*, \ k = 0, 1, 2, \ \cdots \ , \ k \longrightarrow \infty \Longrightarrow \ t_k \longrightarrow t^*\}, \\ t^* \in [-\infty, \infty[$$

$\mathfrak{T}_{t^*}$ *a set composed of a sequence* t_k *converging to* t^* *as the integer* $k \longrightarrow \infty$,

$$\mathfrak{T}_{t^*} = \{t_k : t_k \in \mathfrak{T}, \ k = 0, 1, 2, \ \cdots \ , \ k \longrightarrow \infty \Longrightarrow \ t_k \longrightarrow t^*\}, \ t^* \in [-\infty, \infty]$$

$\mathfrak{T}_R$ *the reachability time set*, $\mathfrak{T}_R = \{t \in \mathfrak{T}_0 : t_0 \leq t \leq t_R < \infty\} \subset \mathfrak{T}_{0F}$

$\mathfrak{T}_{RF}$ *the post reachability time subset of* $\mathfrak{T}_{0F}$,

$$\mathfrak{T}_{RF} = \left\{ t \in Cl\mathfrak{T}_0 : \left\langle \begin{array}{c} t_R \leq t \leq t_F < \infty, \\ t_R \leq t < t_F = \infty \end{array} \right\rangle \right\} \subset \mathfrak{T}_{0F}$$

$\mathfrak{T}_{F\infty}$ *the infinite post final time set*,

$$\mathfrak{T}_{F\infty} = \{t : \ t \in [t_F, \infty[\}$$

$\mathfrak{T}_0^i$ *the time set product*,

$$\mathfrak{T}_0^i = \underbrace{\mathfrak{T}_0 \times \mathfrak{T}_0 \times ... \times \mathfrak{T}_0}_{i-times} = \left\{\mathbf{t}^i : \mathbf{t}_0^i \leq \mathbf{t}^i < \infty \mathbf{1}_i\right\},$$

$$\mathrm{Cl}\mathfrak{T}_0^i = \underbrace{\mathrm{Cl}\mathfrak{T}_0 \times \mathrm{Cl}\mathfrak{T}_0 \times \ \cdots \ \times \mathrm{Cl}\mathfrak{T}_0}_{\text{i-times}},$$

$$\mathrm{In}\mathfrak{T}_0^i = \underbrace{\mathrm{In}\mathfrak{T}_0 \times \mathrm{In}\mathfrak{T}_0 \times \ \cdots \ \times \mathrm{In}\mathfrak{T}_0}_{\text{i-times}}$$

$\mathfrak{T}_{0F}^{(k+1)N}$ *a subinterval of* $\mathfrak{T}_0^{(k+1)N}$ determined by

$$\mathfrak{T}_{0F}^{(k+1)N} = \left\{ \begin{array}{c} \mathbf{t}_R^{(k+1)N} : \mathbf{t}_R^{(k+1)N} \in \mathfrak{T}^{(k+1)N}, \\ \mathbf{t}_0^{(k+1)N} \leq \mathbf{t}_R^{(k+1)N} < \mathbf{t}_F^{(k+1)N} \ \ iff \ \mathbf{t}_F^{(k+1)N} = \infty \mathbf{1}_{(k+1)N}, \\ \mathbf{t}_0^{(k+1)N} \leq \mathbf{t}_R^{(k+1)N} \leq \mathbf{t}_F^{(k+1)N} \ \ iff \ \mathbf{t}_F^{(k+1)N} < \infty \mathbf{1}_{(k+1)N}, \end{array} \right\}$$

$\mathfrak{T}_{R\infty}^{(k+1)N}$ *the vector infinite post reachability time set,*

$$\mathfrak{T}_{R\infty}^{(k+1)N} = \underbrace{\mathfrak{T}_{R\infty} \times \mathfrak{T}_{R\infty} \times \quad \cdots \quad \times \mathfrak{T}_{R\infty}}_{(k+1)N-times} =$$
$$= \left\{ \mathbf{t}^{(k+1)N} : \ \mathbf{t}^{(k+1)N} \in [\mathbf{t}_R^{(k+1)N}, \infty \mathbf{1}_{(k+1)N}[\right\}$$

$\mathfrak{T}_R^N$ *the vector pre-reachability time set,*

$$\mathfrak{T}_R^N = \underbrace{\mathfrak{T}_R \times \mathfrak{T}_R \times \quad \cdots \quad \times \mathfrak{T}_R}_{\text{N-times}} = \left\{ \mathbf{t}^N : \ \mathbf{t}^N \in [\mathbf{t}_0^N, \mathbf{t}_R^N] \right\}$$

$\mathfrak{T}_R^{(k+1)N}$ *the general vector pre-reachability time set,*

$$\mathfrak{T}_R^{(k+1)N} = \underbrace{\mathfrak{T}_R \times \mathfrak{T}_R \times \quad \cdots \quad \times \mathfrak{T}_R}_{(k+1)N-times} =$$
$$= \left\{ \mathbf{t}^{(k+1)N} : \ \mathbf{t}^{(k+1)N} \in [\mathbf{t}_0^{(k+1)N}, \mathbf{t}_R^{N(k+1)N}] \right\}$$

$\mathfrak{T}_{F\infty}^{(k+1)N}$ *the vector infinite post final time set iff* $\mathbf{t}_F^{(k+1)N} < \infty \mathbf{1}_{(k+1)N}$,

$$\mathfrak{T}_{F\infty}^{(k+1)N} = \underbrace{\mathfrak{T}_{F\infty} \times \mathfrak{T}_{\mathsf{F}\infty} \times \quad \cdots \quad \times \mathfrak{T}_{F\infty}}_{(k+1)N-times} =$$
$$= \left\{ \mathbf{t}^{(k+1)N} : \ \mathbf{t}^{(k+1)N} \in [\mathbf{t}_F^{(k+1)N}, \infty \mathbf{1}_{(k+1)N}[\right\}$$

$\mathfrak{T}_{R\infty}^{(k+1)N}$ *the vector infinite post reachability time set iff* $\mathbf{t}_R^{(k+1)N} < \infty \mathbf{1}_{(k+1)N}$,

$$\mathfrak{T}_{R\infty}^{(k+1)N} = \underbrace{\mathfrak{T}_{R\infty} \times \mathfrak{T}_{\mathsf{R}\infty} \times ... \times \mathfrak{T}_{R\infty}}_{(k+1)\text{N-times}} =$$
$$= \left\{ \mathbf{t}^{(k+1)N} : \ \mathbf{t}^{(k+1)N} \in [\mathbf{t}_R^{N(k+1)N}, \infty \mathbf{1}_{(k+1)N}[\right\}$$

$\mathfrak{T}_{RF}^{(k+1)N}$ *the vector post reachability time set iff* $\mathbf{t}_R^{(k+1)N} < \infty \mathbf{1}_{(k+1)N}$,

$$\mathfrak{T}_{RF}^{(k+1)N} = \underbrace{\mathfrak{T}_{RF} \times \mathfrak{T}_{RF} \times \quad \cdots \quad \times \mathfrak{T}_{RF}}_{(k+1)\text{N-times}} =$$
$$\left\{ \mathbf{t}^{(k+1)N} : \ \mathbf{t}_R^{(k+1)N} \leq \mathbf{t}^{(k+1)N} \leq \mathbf{t}_F^{(k+1)N} < \infty \mathbf{1}_{(k+1)N} \right\}$$

$\mathfrak{T}_{\longrightarrow\sigma}$ a *time* sequence converging to σ,

$$\mathfrak{T}_{\longrightarrow\sigma} = \{ t_k : t_k \in \mathfrak{T}, \ k \longrightarrow \infty \Longrightarrow t_k \longrightarrow \sigma \}$$

$\mathfrak{U}^l$ *a given, or to be determined, family of all bounded and* $(l+1)$*-times continuously differentiable permitted total control vector functions* $\mathbf{U}(.)$,

$$\mathfrak{U}^l \subset \mathfrak{C}^{l+1}(\mathfrak{R}^r), \ \mathbf{U}(.) \in \mathfrak{C}^{l+1}(\mathfrak{R}^r),$$

such that every $\mathbf{U}(.) \in \mathfrak{U}^l$ is permitted for the system, $l \in \{0, 1, 2, \quad \cdots \quad, \alpha\}$

$\mathfrak{U}^0 = \mathfrak{U}$ *the family of all bounded continuous and continuously differentiable permitted total control vector functions* $\mathbf{U}(.) \in \mathfrak{U}$, $\mathfrak{U}^0 = \mathfrak{U} \subset \mathfrak{C}^1(\mathfrak{R}^r)$

$\mathfrak{U}(t) \subset \mathfrak{R}^l$ *time-varying uniquely bounded set*

$\partial\mathfrak{V}_\xi(t;\tau)$ *the boundary of the set* $\mathfrak{V}_\xi(t;\tau)$

$\mathfrak{V}_\xi(t;\tau) \subseteq \mathfrak{R}^K$ *the set associated with the function* $V(.) : \mathfrak{T}_\tau \times \mathfrak{R}^K \longrightarrow \mathfrak{R}$ *and with its value* $\xi \in \mathfrak{R}^+$, or $\xi = \infty$, at $t \in \mathfrak{T}_\tau$ so that it is the largest open connected neighborhood $\mathfrak{V}_\xi(t;\tau)$ of the origin $\mathbf{0}_K$ on $\mathfrak{T}_\tau$ such that

1. If $\xi \in \mathfrak{R}^+$ then the boundary $\partial\mathfrak{V}_\xi(t;\tau)$ of $\mathfrak{V}_\xi(t;\tau)$ is the hypersurface $V(t,\mathbf{z}) = \xi$:

$$\xi \in \mathfrak{R}^+ \Longrightarrow \partial\mathfrak{V}_\xi(t;\tau) = \{\mathbf{z} : V(t,\mathbf{z}) = \xi\}, \ \forall t \in \mathfrak{T}_\tau,$$
$$V(t,\mathbf{z}) = \xi \Longleftrightarrow \mathbf{z} \in \partial\mathfrak{V}_\xi(t;\tau), \ \forall t \in \mathfrak{T}_\tau,$$

2. The intersection $\mathfrak{V}_\xi(\mathfrak{T}_\tau)$ of $\mathfrak{V}_\xi(t;\tau)$ over $\mathfrak{T}_\tau$,

$$\mathfrak{V}_\xi(\mathfrak{T}_\tau) = \cap[\mathfrak{V}_\xi(t;\tau) : t \in \mathfrak{T}_\tau],$$

is also a neighborhood of the origin $\mathbf{0}_K$:

$$\exists\zeta \in \mathfrak{R}^+ \Longrightarrow \mathfrak{B}_\zeta \subset \mathfrak{V}_\xi(\mathfrak{T}_\tau),$$

and

3. $V(t,\mathbf{z})$ is strictly less than ξ on $\mathfrak{V}_\xi(t;\tau)$ at every instant $t \in \mathfrak{T}_\tau$:

$$V(t,\mathbf{z}) < \xi, \ \forall(\mathbf{z}, \ t) \in \mathfrak{V}_\xi(t;\tau) \times \mathfrak{T}_\tau,$$

where the expression "*on* $\mathfrak{T}_\tau$" and the notation "$;\tau$" are to be omitted if, and only if, $\mathfrak{T}_\tau = \mathfrak{T}$

$\mathfrak{V}_\mathbf{c}(t;\tau) \subseteq \mathfrak{R}^K$ *the largest open connected neighborhood of* $\mathbf{z} = \mathbf{0}_K$ at *an arbitrary instant* $t \in \mathfrak{T}_\tau$, which is associated with the vector function $\mathbf{v}(.)$ and its vector value $\mathbf{c} \in \mathfrak{R}^K$ so that they obey

$$\mathbf{v}(t,\mathbf{z}) < \mathbf{c}, \ \forall(t,\mathbf{z}) \in \mathfrak{T}_0 \times \mathfrak{V}_\mathbf{c}(t;\tau),$$
$$\mathbf{v}(t,\mathbf{z}) = \mathbf{c} \Longleftrightarrow \mathbf{z} \in \partial\mathfrak{V}_\mathbf{c}(t;\tau), \ \forall t \in \mathfrak{T}_\tau,$$
$$\exists\zeta \in \mathfrak{R}^+ \Longrightarrow \mathfrak{B}_\zeta \subseteq \mathfrak{V}_\mathbf{c}(\mathfrak{T}_\tau) = \cap[\mathfrak{V}_\mathbf{c}(t;\tau) : t \in \mathfrak{T}_\tau]$$

$\mathrm{Cl}\mathfrak{V}_\mathbf{c}(t;\tau)$ is *the closure of the set* $\mathfrak{V}_\mathbf{c}(t;\tau)$ *at* $t \in \mathfrak{T}_\tau$, and $\partial\mathfrak{V}_\mathbf{c}(t;\tau)$ is *its boundary at* $t \in \mathfrak{T}_\tau$ if the boundary exists

$\mathfrak{Y} \subseteq \mathfrak{R}^N$ *the total output vector space of any system*

$\mathfrak{Y}_d^k$ *a given, or to be determined, family of all bounded and* $(k+1)$*-times continuously differentiable realizable total desired output vector functions* $\mathbf{Y}_d(.) \in \mathfrak{C}^{k+1}$ such that they and their first $k+1$ derivatives obey *TCUP*,

$$\mathfrak{Y}_d^k \subset \mathfrak{C}^{k+1}(\mathfrak{R}^N),$$

i.e., it is a family of all bounded continuously differentiable realizable total desired extended output vector functions $\mathbf{Y}_d^k(.) \in \mathfrak{C}^1$ such that they and their derivatives obey *TCUP*

$\mathfrak{Y}_{d0}^k$ *the set of the initial vector values* $\mathbf{Y}_{d0}^k$ *of* $\mathbf{Y}_d^k(t)$ *of every* $\mathbf{Y}_d(.) \in \mathfrak{Y}_d^k$. If, and only if, $\mathbf{Y}_d(.) \in \mathfrak{Y}_d^k$ then $\mathbf{Y}_{d0}^k = \mathcal{Y}_d^k(t_0;t_0;\mathbf{Y}_{d0}^k) \in \mathfrak{Y}_{d0}^k$

$\mathfrak{Y}_d = \mathfrak{Y}_d^0$ *the family of all bounded continuous and continuously differentiable realizable desired total output vector functions* $\mathbf{Y}_d(.)$, such that they and their first derivatives obey *TCUP*,

$$\mathfrak{Y}_d \subset \mathfrak{C}^1(\mathfrak{R}^N)$$

$\mathfrak{Z} \subseteq \mathfrak{R}^K$ *the general total state vector space*

$\mathfrak{Z}_0$ *the set of initial conditions* $\mathbf{Z}_0$ or $\mathbf{z}_0$ for which the system has a solution

$\mathfrak{Z}(t)$ the singleton containing only $\mathbf{z}(t)$, $\mathfrak{Z}(t) = \{\mathbf{z}(t)\}$.

A.3.3 Greek letters

Lower-case Greek letters denote scalars.

α *a natural number, the order of the () system*

β is *the rank of the matrix* $P(t) \in \mathfrak{R}^{\rho \times r}$: $\beta \leq \min(\rho, r)$

γ is *the rank of the matrix* $W(t) \in \mathfrak{R}^{N \times r}$: $\gamma \leq \min(N, r)$

δ_{ij} *the Kronecker delta*, $\delta_{ij} = 1$ for $i = j$, and $\delta_{ij} = 0$ for $i \neq j$

$\boldsymbol{\varepsilon}$ *the output error vector* $\boldsymbol{\varepsilon} \in R^N$,

$$\boldsymbol{\varepsilon} = \mathbf{Y_d} - \mathbf{Y} = -\mathbf{y},\ \boldsymbol{\varepsilon} = [\varepsilon_1\ \varepsilon_2\ \cdots\ \varepsilon_N\ ]^T,\ \varepsilon_i \equiv Y_{di} - Y_i,$$

or a positive real vector in general

$\boldsymbol{\varepsilon}^k$ *the extended output error vector*,

$$\boldsymbol{\varepsilon}^k = \left[\boldsymbol{\varepsilon}^T\ \boldsymbol{\varepsilon}^{(1)^T}\ \cdots\ \boldsymbol{\varepsilon}^{(k)^T}\ \right]^T \in \mathfrak{R}^{(k+1)N}$$

$\boldsymbol{\varepsilon}_R$ the *state error vector* $\boldsymbol{\varepsilon}_R \in R^\rho$,

$$\boldsymbol{\varepsilon}_R = \mathbf{R_d} - \mathbf{R} = -\mathbf{r},\ \boldsymbol{\varepsilon}_R = [\varepsilon_{R1}\ \varepsilon_{R2}\ \cdots\ \varepsilon_{R\rho}\ ]^T,\ \varepsilon_{Ri} \equiv R_{di} - R_i,\ \varepsilon_\rho = \varepsilon_{R\rho}$$

ε^*_β *the modified state error variable*,

$$\varepsilon^*_\beta = \varepsilon_{Rj}\ iff\ |\varepsilon_{Rj}| \geq \max\left(|\varepsilon_{Ri}| : \forall i = \beta, v+1, .., \rho\right),\ j \in \{\beta, \beta+1, .., \rho\},$$
$$\beta \in \{1, 2,\ \cdots\ , \rho\}$$
$$\varepsilon^*_\beta = 0\ iff\ \varepsilon_{Rj} = 0,\ and\ \left|\varepsilon^*_\beta\right| < \xi\ iff\ |\varepsilon_{Rj}| < \xi,\ \forall j = \beta, \beta+1, .., \rho,\ \forall \xi > 0$$

$\boldsymbol{\varepsilon}^*_\beta$ *the β-subvector vector of $\boldsymbol{\varepsilon}_R$ that implicitly contains the entries* $\varepsilon_{\beta+1}$, $\varepsilon_{\beta+2}$, ... ε_ρ *of* $\boldsymbol{\varepsilon}$,

$$\boldsymbol{\varepsilon}^*_\beta = \left[\varepsilon_{R1}\ \varepsilon_{R2}\ \cdots\ \varepsilon_{R,\beta-1}\ \varepsilon^*_\beta\ \right] \in \mathfrak{R}^\beta,\ \beta = \rho \Longleftrightarrow \boldsymbol{\varepsilon}^*_\beta = \boldsymbol{\varepsilon}_R = \boldsymbol{\varepsilon}_\rho$$

ε^*_γ *the modified output error variable*,

$$\varepsilon^*_\gamma = \varepsilon_j\ iff\ |\varepsilon_j| \geq \max\left(|\varepsilon_i| : \forall i = \gamma, \gamma+1,\ \cdots\ , N\right),\ j \in \{k, k+1,\ \cdots\ , N\},$$
$$\gamma \in \{1, 2, ..., N\}$$
$$\varepsilon^*_\gamma = 0\ iff\ \varepsilon_j = 0,\ \text{and}\ \left|\varepsilon^*_\gamma\right| < \xi\ iff\ |\varepsilon_j| < \xi,\ \forall j = \gamma, \gamma+1,\ \cdots\ , N,\ \forall \xi > 0$$

$\boldsymbol{\varepsilon}^*_\gamma$ *the γ-subvector vector of $\boldsymbol{\varepsilon}$ that implicitly contains the entries* $\varepsilon_{\gamma+1}$, $\varepsilon_{\gamma+2}$, ... ε_N *of* $\boldsymbol{\varepsilon}$,

$$\boldsymbol{\varepsilon}^*_\gamma = \left[\varepsilon_1\ \ \varepsilon_2\ \ \cdots\ \ \varepsilon_{\gamma-1}\ \varepsilon^*_\gamma\ \right] \in \mathfrak{R}^\gamma,\ \gamma = N \Longleftrightarrow \boldsymbol{\varepsilon}^*_\gamma = \boldsymbol{\varepsilon},\ N \leq r$$

ε_v *subsidiary error variable in the output error space*, $\varepsilon_v = -v\left(t, \boldsymbol{\varepsilon}_R^{\alpha-1}\right)$, where $v(.) : \mathfrak{T} \times \mathfrak{R}^{\alpha\rho} \longrightarrow \mathfrak{R}_+$ is a differentiable radially unbounded decrescent positive definite function in the *state* space $\mathfrak{R}^{\alpha\rho}$

$\boldsymbol{\varepsilon}^N_{(j)}$ a positive real number $\varepsilon_{i(j)}$, or $\varepsilon_{i(j)} = \infty$, associated with the j-th derivative of Y_i and of Y_{di}, and taken for the entries of the positive N vector $\boldsymbol{\varepsilon}_{(j)}$, i.e., of the positive $(k+1)N-$ vector $\boldsymbol{\varepsilon}^{kN}$,

$$\boldsymbol{\varepsilon}^N_{(j)} = \begin{bmatrix} \varepsilon_{1,(j)} \\ \varepsilon_{2,(j)} \\ \cdots \\ \varepsilon_{N,(j)} \end{bmatrix} \in \mathfrak{R}^{+^N} \cup \{\infty\}^N,\ \forall j = 0, 1, 2, \cdots, k,\ \varepsilon_{i,(0)} \equiv \varepsilon_i, \boldsymbol{\varepsilon}^N_{(0)} \equiv \boldsymbol{\varepsilon}^N$$

$\varepsilon^{(k+1)N}$ a positive $(k+1)N$- vector ε^{kN} the entries of which are vectors $\varepsilon_{(j)}^{N}$, $j=0,1,2,\cdots,k$,

$$\varepsilon^{(k+1)N}=\begin{bmatrix}\varepsilon_{(0)}^{N}\\ \varepsilon_{(1)}^{N}\\ \cdots\\ \varepsilon_{(k)}^{N}\end{bmatrix}=\begin{bmatrix}\varepsilon^{N}\\ \varepsilon_{(1)}^{N}\\ \cdots\\ \varepsilon_{(k)}^{N}\end{bmatrix}\in\mathfrak{R}^{+^{(k+1)N}}\cup\{\infty\}^{(k+1)N},\ k\in\{1,2,\cdots,p-1\},$$

so that

$$\left|\mathbf{Y}_0^k-\mathbf{Y}_{d0}^k\right|<\varepsilon^{(k+1)N},\ \forall k=0,1,2,\cdots,p-1,$$

signifies

$$\left|Y_{i0}^{(j)}-Y_{di0}^{(j)}\right|<\varepsilon_{i,(j)},\ \forall i=1,2,\cdots,N,\ \forall j=0,1,2,\cdots,\ k$$

ϵ the subsidiary output error vector ϵ relative to the reference output vector Y_R,

$$\epsilon=\mathbf{Y}_R-\mathbf{Y},\ \epsilon=[\epsilon_1\ \ \epsilon_2\ \ ...\ \epsilon_N\ ]^T$$

Ψ *the diagonal matrix associated with* ϵ,

$$\Psi=\operatorname{diag}\{\epsilon_1\ \ \epsilon_2\ \ ...\ \epsilon_N\}$$

θ a nonnegative integer
η *a nonnegative integer*
μ *a nonnegative integer*
ρ *a nonnegative integer, the dimension of the state subvector* $\mathbf{R}$ *of the system*
$\sigma\left(\varepsilon_i^{(k)},\varepsilon_i^{(k+1)}\right)$ *a subsidiary scalar functional operator defined by*

$$\sigma\left(\varepsilon_i^{(k)},\varepsilon_i^{(k+1)}\right)=\left\{\begin{array}{c}-1,\ \varepsilon_i^{(k)}<0,\forall\varepsilon_i^{(k+1)}\in\mathfrak{R};\ \varepsilon_i^{(k)}=0\ \mathit{and}\ \varepsilon_i^{(k+1)}<0,\\ 0,\ \varepsilon_i^{(k)}=0\ \mathit{and}\ \varepsilon_i^{(k+1)}=0,\\ 1,\ \varepsilon_i^{(k)}>0,\forall\varepsilon_i^{(k+1)}\in\mathfrak{R};\ \varepsilon_i^{(k)}=0\ \mathit{and}\ \varepsilon_i^{(k+1)}>0\end{array}\right\}$$

$\Sigma\left(\varepsilon^{(k)},\varepsilon^{(k+1)}\right)$ *a subsidiary matrix functional operator defined by*

$$\Sigma\left(\varepsilon^{(k)},\varepsilon^{(k+1)}\right)=\operatorname{diag}\left\{\sigma\left(\varepsilon_1^{(k)},\varepsilon_1^{(k+1)}\right)\ \ \sigma\left(\varepsilon_2^{(k)},\varepsilon_2^{(k+1)}\right)\ \ \cdots\ \ \sigma\left(\varepsilon_N^{(k)},\varepsilon_N^{(k+1)}\right)\right\}$$

$\Sigma\left(\varepsilon^{v}\right)$

$$\Sigma\left(\varepsilon^{v}\right)=\operatorname{blockdiag}\left\{\Sigma\left(\varepsilon,\varepsilon^{(1)}\right)\ \ \Sigma\left(\varepsilon^{(1)},\varepsilon^{(2)}\right)\ \ \cdots\ \ \Sigma\left(\varepsilon^{(\nu-1)},\varepsilon^{(\nu)}\right)\right\}$$

τ *a subsidiary notation for time t*
Υ^k *the* k-th *order target set*
Υ *the (zero-order) target set called simply: the target set*
ρ *a natural number, the dimension of the vector* $\mathbf{R}$
ϕ *the empty set*
$\varphi(.):\mathfrak{R}_+\longrightarrow\mathfrak{R}_+$ *a comparison function*
$\psi(.):\mathfrak{T}_0\times\mathfrak{T}\longrightarrow\mathfrak{R}^+$ *a subsidiary function,*

$$\psi(.):\mathfrak{T}_0\times\mathfrak{T}\longrightarrow\mathfrak{R}^+,\ e.g.,\ \psi(t;t_0)=(1+t-t_0)$$

$\psi(.):\mathfrak{T}_0\times\mathfrak{T}\times\mathfrak{T}_0\longrightarrow\mathfrak{R}$ *a subsidiary function,*

$$\psi(t;t_0;\sigma)\left\{\begin{array}{l} =1,\ t=t_0 \\ \neq 0,\ t\in]t_0,\sigma[, \\ 0,\ \forall t\in[\sigma,\infty[\end{array}\right\},\ e.g.,$$

$$\psi(t;t_0;\sigma)=\left\{\begin{array}{ll} \left(1-\frac{t-t_0}{\sigma-t_0}\right)^{\nu}, & \forall t\in[t_0,\sigma], \\ 0, & \forall t\in[\sigma,\infty[\end{array}\right\},\ \nu\in\mathfrak{R}^+$$

$\ominus_\beta^*(.) : \mathfrak{R}^\beta \longrightarrow \mathfrak{R}^{\beta\times\beta}$ *a subsidiary diagonal matrix function in the state error space,*

$$\ominus_\beta^*\left(\boldsymbol{\epsilon}_\beta^*\right)=\mathrm{diag}\left\{\epsilon_{R1}\ \ \epsilon_{R2}\ \ \cdots\ \ \epsilon_{R,\beta-1}\ \ \epsilon_\beta^*\right\}\in\mathfrak{R}^{\beta\times\beta}$$

$\ominus_\gamma^*(.) : \mathfrak{R}^\gamma \longrightarrow \mathfrak{R}^{\gamma\times\gamma}$ *a subsidiary diagonal matrix function in the output error space,*

$$\ominus_\gamma^*\left(\boldsymbol{\epsilon}_\gamma^*\right)=\mathrm{diag}\left\{\epsilon_1\ \ \epsilon_2\ \ \cdots\ \ \epsilon_{\gamma-1}\ \ \epsilon_\gamma^*\right\}\in\mathfrak{R}^{\gamma\times\gamma}$$

$\omega_i\in\mathfrak{R}^+$ *the angular speed of the sinusoidal oscillation*

A.3.4 Roman letters

Lower-case Roman letters designate scalars, but if they are bold then they represent vectors. Capital Roman letters are used for matrices, but if they are bold then they represent vectors.

$d\in R$ a natural number, $d\in\{1,2,...\}$, *the dimension of the disturbance vector* $\mathbf{D}$

$\mathbf{D}\in\mathfrak{R}^d$ *the real total disturbance vector*

$\mathbf{d}\in\mathfrak{R}^d$ *the deviation of the disturbance vector,* $\mathbf{d}=\mathbf{D}-\mathbf{D}_N$

$\mathbf{e}_d(.) : \mathfrak{R}^d \longrightarrow \mathfrak{R}^d$ *the vector function that transmits the disturbance action on the dynamics of the system,* $\mathbf{e}_d(\mathbf{D})\in\mathfrak{C}^1\left(\mathfrak{R}^d\right)$

$\mathbf{e}_u(.) : \mathfrak{R}^r \longrightarrow \mathfrak{R}^r$ *the vector function that transmits the control action on the dynamics of the plant,* $\mathbf{e}_u(\mathbf{U})\in\mathfrak{C}^1\left(\mathfrak{R}^r\right)$

$\mathbf{e}_y(.) : \mathfrak{R}^N \longrightarrow \mathfrak{R}^r$ *the vector function that transmits the desired output action on the dynamics of the control system,* $\mathbf{e}_y(\mathbf{Y}_d)\in\mathfrak{C}\left(\mathfrak{R}^N\right)$

$E(.) : \mathfrak{T} \longrightarrow \mathfrak{R}$ *the (dimensionless) instantaneous accumulated energy of the system at a moment* $t\in\mathfrak{T}$

$E(.) : \mathfrak{T} \longrightarrow \mathfrak{R}^{N\times N}$ *the output error diagonal matrix function,*

$$E(t)=\mathrm{diag}\left\{\varepsilon_1(t)\ \ \varepsilon_2(t)\ \ \cdots\ \ \varepsilon_N(t)\right\}=E\left(\mathbf{t}^N\right)$$

$E(.) : \mathfrak{T} \longrightarrow \mathfrak{R}^{\rho\times d}$ *the matrix function that transmits the disturbance action on the dynamics of the* control *system,* $E(t)\in\mathfrak{C}^1(\mathfrak{T})$

f *a natural number,* $f\in\{1,2,\ \cdots\ \}$

$$\mathbf{f}\left[t,\mathbf{y}^{m-1}(t)\right]=\left(\begin{array}{c} \left\{A\left[\mathbf{Y}_d^{m-1}(t)+\mathbf{y}^{m-1}(t)\right]-A\left[\mathbf{Y}_d^{m-1}(t)\right]\right\}\mathbf{Y}_d^{(m)}(t) \\ +\,\mathbf{a}\left[\mathbf{Y}_d^{m-1}(t)+\mathbf{y}^{m-1}(t)\right]-\mathbf{a}\left[\mathbf{Y}_d^{m-1}(t)\right] \end{array}\right)$$

$\mathrm{grad}V(t,\mathbf{r}^{\alpha-1})\in\mathfrak{R}^{\alpha\rho}$ *the gradient of the function* $V(.)$,

$$\left[\mathrm{grad}V(t,\mathbf{r}^{\alpha-1})\right]^T=$$

$$=\left[\frac{\partial V(t,\mathbf{r}^{\alpha-1})}{\partial r_1}..\frac{\partial V(t,\mathbf{r}^{\alpha-1})}{\partial r_\rho}\ \ \cdots\ \ \frac{\partial V(t,\mathbf{r}^{\alpha-1})}{\partial r_1^{(\alpha-1)}}\ \ \cdots\ \ \frac{\partial V(t,\mathbf{r}^{\alpha-1})}{\partial r_\rho^{(\alpha-1)}}\right]$$

$\mathrm{grad}_{tr}V(t,\mathbf{r}^{\alpha-1})\in\mathfrak{R}^\rho$ *the truncated gradient of the function* $V(.)$,

$$\mathrm{grad}_{tr}V(t,\mathbf{r}^{\alpha-1})=\left[\frac{\partial V(t,\mathbf{r}^{\alpha-1})}{\partial r_1^{(\alpha-1)}}....\frac{\partial V(t,\mathbf{r}^{\alpha-1})}{\partial r_\rho^{(\alpha-1)}}\right]^T,$$

$\mathbf{h}(.) : \mathfrak{T} \times \mathfrak{R}^{(\xi+1)M} \longrightarrow \mathfrak{R}^{\rho}$ *the vector function describing the influence of the input vector on the state of the system,* $\mathbf{h}(t, \mathbf{I}^{\xi}) \in \mathfrak{C}(\mathfrak{T} \times \mathfrak{R}^{(\xi+1)M})$

$\mathbf{i} \in \mathfrak{R}^M$ *the input deviation vector,* $\mathbf{i} = \mathbf{I} - \mathbf{I}_N$,

$I_K \in \mathfrak{R}^{K \times K}$ *the* $K \times K$ *unity matrix,* $I_K = \text{diag}\{1 \ 1 \ \cdots \ 1\}$

$I_i \in R$ *the total value of the* $i-th$ *input variable of the system*

$\mathbf{I} \in R^M$ *the total real input vector of the system,*

$$\mathbf{I} = [I_1 \ I_2 \ ... \ I_M]^T \in \mathfrak{R}^M,$$

$$\mathbf{I} = \begin{bmatrix} \mathbf{D} \\ \mathbf{U} \end{bmatrix} \text{ for the plant, } M = d + r,$$

$$\mathbf{I} = \begin{bmatrix} \mathbf{D} \\ \mathbf{Y}_d \end{bmatrix} \text{ for the control system, } M = d + N$$

$\mathbf{I}_N \in \mathfrak{R}^M$ *the total nominal input vector of the system*

$\mathbf{I}^{\mu} \in \mathfrak{R}^{(\mu+1)M}$ *the extended total input vector*

$$\mathbf{I}^{\mu} = \left[\mathbf{I}^T \vdots \mathbf{I}^{(1)T} \vdots ... \vdots \mathbf{I}^{(\mu)T}\right]^T \in \mathfrak{R}^{(\mu+1)M}, \mathbf{I}^{-1}(t) \equiv \mathbf{0}_M$$

k, j, l *mutually independent nonnegative integers,*

$$k \in \{0, 1, \ \cdots \ , \alpha - 1\}, \ j, l \in \{0, 1, \ \cdots \ ,\}$$

K *nonnegative integer,* $K \in \{N, \rho, (k+1)N, \alpha\rho\}$

$M(.) : \mathfrak{T} \longrightarrow \mathfrak{R}$ *the (dimensionless) instantaneous accumulated mass of the system at a moment* $t \in \mathfrak{T}$

N *the dimension of the output vector* Y

$O_{M \times N}$ *the* $M \times N$ *zero matrix.* If $M = N$ then for short $O_{N \times N} = O_N$

$P(.) : \mathfrak{T} \longrightarrow \mathfrak{R}^{\rho \times r}$ *the matrix function of the system (3.12) that transmits the input (control or desired output) action on the system state*

$\mathbf{q}(.) : \mathfrak{T} \times \mathfrak{R}^{\alpha\rho} \longrightarrow \mathfrak{R}^{\rho}$ *the vector nonlinearity of the system* (3.12),

$$\mathbf{q}(t, \mathbf{R}^{\alpha-1}) \in \mathfrak{C}(\mathfrak{T} \times \mathfrak{R}^{\alpha\rho})$$

r *the dimension of the control vector* $\mathbf{U}$

$\mathbf{R} \in \mathfrak{R}^{\rho}$ *the total real state subvector of the system,*

$$\mathbf{R} = [R_1 \ R_2 \ \cdots \ R_{\rho}]^T \in \mathfrak{R}^{\rho}$$

$\mathbf{R}^{\alpha-1} \in \mathfrak{R}^{\alpha\rho}$ *the total real state vector of the system,*

$$\mathbf{R}^{\alpha-1} = \left[\mathbf{R}^T \ \mathbf{R}^{(1)T} \ \cdots \ \mathbf{R}^{(\alpha-1)T}\right]^T \in \mathfrak{R}^{\alpha\rho}$$

$\mathbf{R}_N^{\alpha-1} \in \mathfrak{R}^{\alpha\rho}$ *the total nominal state vector of the system*

s *the complex number or the complex variable*

$\mathbf{s}(.) : \mathfrak{T} \times \mathfrak{R}^{\rho} \times \mathfrak{R}^M \longrightarrow \mathfrak{R}^N$ *the output vector nonlinearity of the system,*

$$\mathbf{s}(.) \in \mathfrak{C}^1(\mathfrak{T} \times \mathfrak{R}^{\rho} \times \mathfrak{R}^M)$$

$t \in \mathfrak{T}$ *time, an arbitrary instant, an arbitrary moment*

$t_0 \in \mathfrak{T}$ *an accepted initial moment*

t^- *the left-hand side of the moment* t, $t^- = \lim(t - \varepsilon : \varepsilon \longrightarrow 0, \ \varepsilon > 0)$

t^+ *the right-hand side of the moment* t, $t^+ = \lim(t + \varepsilon : \varepsilon \longrightarrow 0, \ \varepsilon > 0)$

t_{zero} *the relative zero moment,*

$\overline{t} \in \mathfrak{R}$ *the nondimensional mathematical temporal variable (the mathematical time),*

$$\overline{t} = \frac{t}{1_t}[-], \ \text{num}\ \overline{t} = \text{num}\ t$$

$T \in \mathfrak{R}^+$ *the period of a motion or of a response*
$T \in \mathfrak{R}^{K\times K}$ *the temporal matrix,* $T = \text{diag}\{t \ \ t \ \ \cdots \ \ t\}$
$T_0 \in \mathfrak{R}^{K\times K}$, $T_0 = \text{diag}\{t_0 \ \ t_0 \ \ \cdots \ \ t_0\}$
$T_R \in \mathfrak{R}^{K\times K}$, $T_R = \text{diag}\{\tau_{R1} \ \ \tau_{R2} \ \ \cdots \ \ \tau_{RK}\}$
T_t *a chosen time scale with the corresponding time unit* 1_t
T *the temporal (the time) dimension*
$T(.) : \mathfrak{R} \times \mathfrak{R}^{(k+1)N} \times \mathfrak{R}^N \longrightarrow \mathfrak{R}^N$ *the vector tracking operator, in general:*

$$\mathbf{T}(.) : \mathfrak{T}^N \times \mathfrak{R}^{(k+1)N} \times \mathfrak{R}^N \longrightarrow \mathfrak{R}^N$$

$t_F \in \text{Cl}\mathfrak{T}_0$ *the final moment*
$\mathbf{t}_F^N = [t_{F1} \ \ t_{F2} \ \ \cdots \ \ t_{FN}]^T \in \mathfrak{T}_0^N \cup \{\infty\}^N$ *the vector final moment related to the system output behavior*
$t_R \in \text{In}\mathfrak{T}_{0F}$ *the scalar reachability time*
$t_{Ri} \in \text{In}\mathfrak{T}_{0F}$ *the scalar reachability time associated with the i-th output variable*
$\mathbf{t}^{(k+1)N}$ *the* $(k+1)N$*-time vector,*

$$\mathbf{t}^{(k+1)N} = t\mathbf{1}_{(k+1)N} = [t \ t...t]^T \in \mathfrak{T}_0^{(k+1)N}, \ k \in \{0, 1, 2, \ \cdots \ , \alpha - 1\},$$
$$\mathbf{t} = \mathbf{t}^N = t\mathbf{1}_N = [t \ t...t]^T \in \mathfrak{T}_0^N,$$
$$\mathbf{t}_0 = \mathbf{t}_0^N = t_0\mathbf{1}_N = [t_0 \ t_0...t_0]^T \in \text{In}\mathfrak{T}^N$$

$t_{Ri,(j)} \in \text{In}\mathfrak{T}_{0F}$ *the scalar reachability time associated with the j-th derivative of the i-th output variable*
$t_R^N \in \text{In}\mathfrak{T}_{0F}^N$ *the vector reachability time, the i-th entry of which is* t_{Ri}, which is related to the system output behavior,

$$\mathbf{t}_R^N = \mathbf{t}_{R(0)}^N = \begin{bmatrix} t_{R1} \\ t_{R2} \\ \cdots \\ t_{RN} \end{bmatrix} = \begin{bmatrix} t_{R1,(0)} \\ t_{R2,(0)} \\ \cdots \\ t_{RN,(0)} \end{bmatrix} \in (\text{In}\mathfrak{T}_{0F})^N$$

$\mathbf{t}_{R(j)}^N \in \text{In}\mathfrak{T}_{0F}^N$ *the vector reachability time associated with the j-th derivative* $\mathbf{Y}^{(j)}(t)$ *of the output vector* $\mathbf{Y}(t)$,

$$\mathbf{t}_{R(j)}^N = \begin{bmatrix} t_{R1,(j)} \\ t_{R2,(j)} \\ \cdots \\ t_{RN,(j)} \end{bmatrix} \in (\text{In}\mathfrak{T}_{0F})^N, \ \ \mathbf{t}_{R(0)}^N = \mathbf{t}_R^N$$

$\mathbf{t}_R^{(k+1)N} \in \text{In}\mathfrak{T}_{0F}^{(k+1)N}$ *the vector reachability time associated with the output vector* $\mathbf{Y}(t)$ *and its first k derivatives* $\mathbf{Y}^{(1)}(t), \ \cdots, \ \mathbf{Y}^{(k)}(t)$ *in the extended output space* $\mathfrak{R}^{(k+1)N}$,

$$\mathbf{t}_R^{(k+1)N} = \begin{bmatrix} \mathbf{t}_{R(0)}^N \\ \mathbf{t}_{R(1)}^N \\ \mathbf{t}_{R(2)}^N \\ \cdots \\ \mathbf{t}_{R(k)}^N \end{bmatrix} = \begin{bmatrix} \mathbf{t}_R^N \\ \mathbf{t}_{R(1)}^N \\ \mathbf{t}_{R(2)}^N \\ \cdots \\ \mathbf{t}_{R(k)}^N \end{bmatrix} \in (\text{In}\mathfrak{T}_0)^{(k+1)N},$$
$$k \in \{0, 1, 2, \cdots, \ p-1\}, \ \mathbf{t}_{R[0]}^N = \mathbf{t}_{R(0)}^N = \mathbf{t}_R^N$$

$\mathbf{U} \in \mathfrak{R}^r$ *the real total control vector*

v_t (v_τ) *the speed of the evolution (of the flow) of the time value and of its numerical value*

$V(.) : \mathfrak{R}^K \rightarrow \mathfrak{R}$ a *time-invariant semidefinite or definite function*

$V(.) : \mathfrak{T}_0 \times \mathfrak{R}^K \rightarrow \mathfrak{R}$ *a time-varying semidefinite or definite vector function*

$\mathbf{v} \in \mathfrak{R}^K$ $\mathbf{v} = [v_1\ v_2\ \cdots\ v_K]^T$ implies:

$$\mathbf{v}^\gamma = [v_1^\gamma\ v_2^\gamma\ \cdots\ v_K^\gamma]^T,$$
$$\exp(-\mathbf{v}) = [\exp(-v_1)\ \exp(-v_2)\ \cdots\ \exp(-v_K)]^T,$$
$$\ln(\mathbf{v}) = [\ln(v_1)\ \ln(v_2)\ \cdots\ \ln(v_K)]^T$$

$\mathbf{v}(.) : \mathfrak{T}_0 \times \mathfrak{R}^{mN} \longrightarrow \mathfrak{R}^N$ a subsidiary function

$$\mathbf{v}\left(t, \boldsymbol{\varepsilon}^{m-1}\right) = \mathbf{0}_N \Longleftrightarrow \boldsymbol{\varepsilon}^{m-1} = \mathbf{0}_{mN}$$

$\mathbf{v}(.) : \mathfrak{T}_0 \times \text{In}\mathfrak{T}_0 \times \mathfrak{R}^{mN} \longrightarrow \mathfrak{R}^N$ a subsidiary function

$$\mathbf{v}(t, \sigma, \boldsymbol{\varepsilon}^p) \in C\left(\mathfrak{T}_0 \times \text{In}\mathfrak{T}_0 \times \mathfrak{R}^{(p+1)N}\right),$$
$$\mathbf{v}(t, \sigma, \boldsymbol{\varepsilon}^p) \neq \mathbf{0}_N,\ t \in]t_0, \sigma[,\ \forall (\sigma, \boldsymbol{\varepsilon}^p) \in \text{In}\mathfrak{T}_0 \times \mathfrak{R}^{(p+1)N},$$
$$\mathbf{v}(t_0, \sigma, \boldsymbol{\varepsilon}_0^p) = \mathbf{0}_N,\ \forall (t_0, \sigma, \boldsymbol{\varepsilon}_0^p) \in \mathfrak{T}_i \times \text{In}\mathfrak{T}_0 \times \mathfrak{R}^{(p+1)N},$$
$$\mathbf{v}(t, \sigma, \boldsymbol{\varepsilon}^p) = \mathbf{0}_N \Longleftrightarrow \boldsymbol{\varepsilon}^p(t) = \mathbf{0}_{(p+1)N},\ \forall (t \geq \sigma, \sigma) \in \mathfrak{T}_0 \times \text{In}\mathfrak{T}_0$$

$V(.) : \mathfrak{R}^K \rightarrow \mathfrak{R}^P$ a *time*-invariant semidefinite or definite vector function,

$$\mathbf{V}(\mathbf{z}) = [V_1(\mathbf{z})\ V_2(\mathbf{z})\ \cdots\ V_P(\mathbf{z})]^T,\ V_i(.) : \mathfrak{R}^K \longrightarrow \mathfrak{R},$$
$$P \in \{1, 2, 3, \cdots, K\}$$

$V(.) : \mathfrak{T}_0 \times \mathfrak{R}^K \rightarrow \mathfrak{R}^P$ a *time*-varying semidefinite or definite vector function,

$$\mathbf{V}(t, \mathbf{z}) = [V_1(t, \mathbf{z})\ V_2(t, \mathbf{z})\ \cdots\ V_P(t, \mathbf{z})]^T, V_i(.) : \mathfrak{T}_0 \times \mathfrak{R}^K \longrightarrow \mathfrak{R},$$
$$\forall i = 1, 2, \cdots, P, P \in \{1, 2, 3,\ \cdots, K\}$$

$W(.) : \mathfrak{R}^K \rightarrow \mathfrak{R}$ a *time-invariant semidefinite or definite function*

$W(.) : \mathfrak{T}_0 \times \mathfrak{R}^K \rightarrow \mathfrak{R}^P$ a *time-varying semidefinite or definite function*

$\mathbf{w} \in \mathfrak{R}^K$ $\mathbf{w} = [w_1\ w_2\ \cdots\ w_K]^T$,

$\mathbf{w}(.) : \mathfrak{R}^r \longrightarrow \mathfrak{R}^r$ *the output vector nonlinearity of the plant,* $\mathbf{w}(\mathbf{U}) \in \mathfrak{C}^1(\mathfrak{R}^r)$

$\mathbf{w}(.) : \mathfrak{R}^N \longrightarrow \mathfrak{R}^N$ *the output vector nonlinearity of the control system,*

$$\mathbf{w}(\mathbf{Y}_d) \in \mathfrak{C}^1\left(\mathfrak{R}^N\right)$$

Y_j *the total value of the* j-th *output variable of the plant*

$\mathbf{y} \in \mathfrak{R}^N$ *the output deviation vector of the plant,*

$$\mathbf{y} = \mathbf{Y} - \mathbf{Y}_d,\ \mathbf{y} = [y_1\ y_2\ \cdots\ y_N]^T$$

$\mathbf{y}^1 = \left[\mathbf{y}^T\ \mathbf{y}^{(1)T}\right]^T \in \mathfrak{R}^{2N}$

$\mathbf{y}^{\alpha-1} = \left[\mathbf{y}^T\ \mathbf{y}^{(1)T}\ \cdots\ \mathbf{y}^{(\alpha-1)T}\right]^T \in \mathfrak{R}^{\alpha N}$

$\mathbf{Y} \in \mathfrak{R}^N$ *the real total output vector of the system,*

$$\mathbf{Y} = [Y_1\ Y_2\ \cdots\ Y_N]^T \in \mathfrak{R}^N$$

$\mathbf{Y}_{\gamma}$ *the* $\gamma-$*subvector of* $\mathbf{Y}$ *that does not contain the entries* $Y_{\gamma+1}$, $Y_{\gamma+2}$,..., Y_N *of* $\mathbf{Y}$,

$$\mathbf{Y}_{\gamma} = [Y_1 \ Y_2..Y_{\gamma}]^T \in \mathfrak{R}^{\gamma}, 1 \leq \gamma \leq min\,(N, r)$$

$\mathbf{Y}_{\gamma}^{*}$ *the* γ*-subvector of* $\mathbf{Y}$ *that contains the entries* $Y_{\gamma+1}$, $Y_{\gamma+2}$, $..Y_N$ of $\mathbf{Y}$,

$$Y_{\gamma}^{*} = Y_j \ iff \left\{ \begin{array}{c} |\varepsilon_j| \geq \\ \geq \max\left(|\varepsilon_i| : \forall i = \gamma, \gamma+1, \cdots, N\right) \end{array} \right\},$$

$$j \in \{\gamma, \gamma+1, \cdots, N\}, \ \gamma \in \{1, 2, \cdots, N\},$$

$$\mathbf{Y}_{\gamma}^{*} = \left[Y_1 \ \ Y_2 \ \ \cdots \ \ Y_{\gamma-1} \ Y_{\gamma}^{*}\ \right]^T \in \mathfrak{R}^{\gamma}$$

$\mathbf{Y}_d \in \mathfrak{R}^N$ *the desired total output vector of the plant*
$\mathbf{Y}^p \in \mathfrak{R}^{(p+1)N}$ *the extended total output vector,*

$$\mathbf{Y}^m = \left[\mathbf{Y}^T \vdots \mathbf{Y}^{(1)T} \vdots \ldots \vdots \mathbf{Y}^{(p)T}\right]^T$$

$\mathbf{z} \in \mathfrak{R}^K$ $\quad \mathbf{z} = [z_1 \ z_2 \ \cdots \ z_K]^T$
$\mathbf{z}$ deviation of the subsidiary vector $\mathbf{Z}$,

$$\mathbf{z} = \mathbf{Z} - \mathbf{Z}_d, \ \mathbf{z} = [z_1 \ \ z_2 \ \ \cdots \ \ z_K]^T \in \mathfrak{R}^K, \ \mathbf{z} \in \left\{\mathbf{r}^{\alpha-1}, \mathbf{y}^k\right\}$$

$\mathbf{z}(.) : \mathfrak{T} \times \mathfrak{R}^{\alpha\rho} \times \mathfrak{R}^d \longrightarrow \mathfrak{R}^{\rho}$ *the vector function describing the influences of the state and of the disturbance vector on the output of the system,*

$$\mathbf{z}\left(t, \mathbf{R}^{\alpha-1}, \mathbf{D}\right) \in \mathfrak{C}^1\left(\mathfrak{T} \times \mathfrak{R}^{\alpha\rho} \times \mathfrak{R}^d\right)$$

A.4 Names

Elementwise means *element by element*

Lyapunov's method and methodology *the original method and methodologies established by Lyapunov on time-invariant sets*

Lyapunov method *the extensions of Lyapunov's method to both time-invariant sets and time-varying sets*

Lyapunov Tracking Control (LTC) *the tracking concept in the Lyapunov sense, which broadens Lyapunov's stability concept to tracking*

NTC *the natural tracking control*

A.5 Symbols

∞ *infinite value,* $\infty 0 = 0$

$\{.\}$ *large brackets,* or *a set*

$(.)$ *an arbitrary variable,* or *an index*

(a, b) *an interval* that can be $]a, b[$ or $]a, b]$ or $[a, b[$ or $[a, b]$:

$$(a, b) \in \{]a, b[, \quad]a, b], \quad [a, b[, \quad [a, b]\}$$

$].[$ *an open interval*

$].]$ *a semiopen semiclosed interval* that is open on the left and closed on the right

$[.[$ *a semiclosed semiopen interval* that is closed on the left and open on the right

$[.]$ *a closed interval,* or *the notation for the physical dimension*

$\exists$ *there exist(s); there is (there are),*

$\nexists$ *there does not exist (do not exist); there is not (there are not),*
$\exists!$ *there exists exactly one,*
$\langle 1_t \rangle$ the notation for *the units*; e.g., if the *time* unit is second s then $t\,\langle 1_t \rangle = t\,\langle s \rangle$,
$||.|| : \mathfrak{R}^K \to \mathfrak{R}_+$ *the norm on the vector space* $\mathfrak{R}^K$, it can be any norm on $\mathfrak{R}^K$, if it is the Euclidean norm on $\mathfrak{R}^K$ then

$$\mathbf{z} = [z_1\ z_2\ ...\ z_K]^T \in \mathfrak{R}^K \Longrightarrow ||\mathbf{z}|| = \sqrt{\mathbf{z}^T\mathbf{z}} = \sqrt{\sum_{i=1}^{i=k} z_i^2}$$

$\exp(\alpha t)$ $\exp(\alpha t) \equiv e^{\alpha t}$
$x\,[.]$ the notation for *the physical dimension of the variable* x, if $x = t$ then $x\,[.] = t\,[\mathrm{T}]$, $x\,[\text{-}]$ means that the variable x is nondimensional (dimensionless)
$\det(.)$ *the determinant of the square matrix* (.)
$\dim Y$ *the mathematical dimension,* $\dim Y$ is the mathematical dimension of the vector Y, $\dim Y = N$,
$\inf$ *infimum,*
$\min(n, r)$ *the smaller number between* n *and* r,

$$\min(n, r) = \left\{ \begin{array}{c} n,\ n \le r, \\ r,\ r \le n \end{array} \right\}$$

$\mathrm{num}\ x$ *the numerical value of the variable* $x \in \mathfrak{R}$
$\mathrm{sign}(.) : \mathfrak{R} \to \{-1, 0, 1\}$ *the scalar signum function,*

$$\mathrm{sign}\ x = |x|^{-1} x\ if\ x \neq 0,\ and\ \mathrm{sign}\ 0 = 0$$

$\sup$ *supremum*

A.6 Matrix and vector notation

$\mathbf{0}_k \in \mathfrak{R}^k$ *the zero vector in* R^k, $\mathbf{0}_k = [0\ \ 0\ \ \cdots\ \ 0]^T \in \mathfrak{R}^k$
$O_{M \times N}$ *the zero matrix in the* $M \times N$ *dimensional matrix space* $R^{M \times N}$
O_N *the zero matrix in* $\mathfrak{R}^{N \times N}$, $O_N = O_{N \times N}$
$\mathbf{1}_k \in \mathfrak{R}^k$ *the unit vector in* R^k, $\mathbf{1}_k = [1\ \ 1\ \ \cdots\ \ 1]^T \in \mathfrak{R}^k$
$\mathrm{diag}\ A \in \mathfrak{R}^{k \times k}$ *the diagonal matrix determined by the diagonal elements of the square matrix* $A = [a_{ij}]$,

$$A = [a_{ij}] \in \mathfrak{R}^{k \times k} \Longrightarrow \mathrm{diag}\ A = \mathrm{diag}\,\{a_{11}\ \ a_{22}\ \ \cdots\ \ a_{kk}\} \in \mathfrak{R}^{k \times k}$$

$$\mathbf{r}^{\alpha-1} = \left[\mathbf{r}^T\ \ \mathbf{r}^{(1)T}\ \ \cdots\ \ \mathbf{r}^{(\alpha-1)T}\right]^T$$

$$\mathbf{r}^{\alpha-1,1} = \left[\mathbf{r}^{\alpha-1^T}\ \ \left(\mathbf{r}^{\alpha-1}\right)^{(1)T}\right]^T$$

$S\,(.) : \mathfrak{R}^N \longrightarrow \mathfrak{R}^{N \times N}$ *the sign diagonal matrix function,*

$$\boldsymbol{\varepsilon} = [\varepsilon_1\ \ \varepsilon_2\ \ \cdots\ \ \varepsilon_N]^T \Longrightarrow$$
$$S\,(\boldsymbol{\varepsilon}) = \mathrm{diag}\,\{\mathrm{sign}\varepsilon_1\ \ \mathrm{sign}\varepsilon_2\ \ \cdots\ \ \mathrm{sign}\varepsilon_N\} \in \{-I_N, O_N, I_N\},$$
$$S\,(|\boldsymbol{\varepsilon}|) = \mathrm{diag}\,\{\mathrm{sign}\,|\varepsilon_1|\ \ \mathrm{sign}\,|\varepsilon_2|\ \ \cdots\ \ \mathrm{sign}\,|\varepsilon_N|\} \in \{O_N, I_N\}$$

$\mathrm{sign}\,(.) : \mathfrak{R}^k \longrightarrow \mathfrak{R}^k$

$$\mathrm{sign}\ \mathbf{v} = \left\{ \begin{array}{c} \mathbf{1}_K,\ \mathbf{v} > \mathbf{0}_K, \\ = \mathbf{0}_K,\ \mathbf{v} = \mathbf{0}_K \\ -\mathbf{1}_K,\ \mathbf{v} > \mathbf{0}_K. \end{array} \right\}$$

$|\mathbf{v}| \in \mathfrak{R}_+^k$ the vector absolute value,

$$|\mathbf{v}| = [|v_1| \quad |v_2| \quad \cdots \quad |v_N|]^T$$

$\mathbf{v}(.) : \mathfrak{T}_0 \times \mathrm{In}\mathfrak{T}_0 \times \mathfrak{R}^{mN} \longrightarrow \mathfrak{R}^N$ *a subsidiary function* defined by

$$\mathbf{v}\left(t, \sigma, \boldsymbol{\varepsilon}^{m-1}\right) \in C\left(\mathfrak{T}_0 \times \mathrm{In}\mathfrak{T}_0 \times \mathfrak{R}^{mN}\right); \; \mathbf{v}\left(t, \sigma, \boldsymbol{\varepsilon}^{m-1}\right) \neq \mathbf{0}_N, \; t \in]t_0, \sigma[,$$
$$\mathbf{v}\left(t_0, \sigma, \boldsymbol{\varepsilon}_0^{m-1}\right) = \mathbf{0}_N, \; \forall \sigma \in \mathrm{In}\mathfrak{T}_0,$$
$$\mathbf{v}\left(t, \sigma, \boldsymbol{\varepsilon}^{m-1}\right) = \mathbf{0}_N \Longleftrightarrow \boldsymbol{\varepsilon}^{m-1} = \mathbf{0}_{mN}, \; \forall (t \geq \sigma, \sigma) \in \mathfrak{T}_0 \times \mathrm{In}\mathfrak{T}_0,$$

e.g.,

$$\mathbf{v}\left(t, \sigma, \boldsymbol{\varepsilon}^{m-1}\right) = \sum_{i=0}^{i=m-1} \left|\boldsymbol{\varepsilon}^{(i)}(t)\right| - \left\{ \begin{array}{c} \sum_{i=0}^{i=m-1} \left\{ \begin{array}{c} \left|\boldsymbol{\varepsilon}^{(i)}(t_0)\right| \; \psi(t; t_0; \sigma), \\ t \in [t_0, \sigma] \end{array} \right\}, \\ \mathbf{0}_N, \; \forall t \in [\sigma, \infty[\end{array} \right\}$$

$|V| \in \mathfrak{R}_+^{k \times k}$ the matrix absolute value,

$$|V| = \mathrm{diag}\{|v_1| \quad |v_2| \quad \cdots \quad |v_N|\}$$

$\mathbf{Y} = [Y_1 \; Y_2 \; \cdots \; Y_N]^T \in \mathfrak{R}^{N \times 1}, \; Y_i \in \mathfrak{R}, \; Y_i \neq 0, \; \forall i = 1, 2, \cdots, N$

$\mathbf{y}^1 = \left[\mathbf{y}^T \; \mathbf{y}^{(1)T}\right]^T$

$\mathbf{y}^{p-1} = \left[\mathbf{y}^T \; \mathbf{y}^{(1)T} \; \cdots \; \mathbf{y}^{(p-1)T}\right]^T$

A.7 Sets

$\mathfrak{A} \backslash \mathfrak{B}$ *the set difference* between the set $\mathfrak{A}$ and the set $\mathfrak{B}$ is the set of all vectors $\mathbf{Z}$ in $\mathfrak{A}$ that do not belong to $\mathfrak{B}$,

$$\mathfrak{A} \backslash \mathfrak{B} = \{\mathbf{Z} : \; \mathbf{Z} \in \mathfrak{A}, \; \mathbf{Z} \notin \mathfrak{B}\}$$

$\mathfrak{A} \Delta \mathfrak{B}$ *the symmetric set difference* between the set $\mathfrak{A}$ and the set $\mathfrak{B}$ is the set of all vectors $\mathbf{Z}$ in $\mathfrak{A}$, which do not belong to $\mathfrak{B}$, and of all vectors that are in $\mathfrak{B}$ but not in $\mathfrak{A}$,

$$\mathfrak{A} \Delta \mathfrak{B} = (\mathfrak{A} \backslash \mathfrak{B}) \cup (\mathfrak{B} \backslash \mathfrak{A})$$

$\mathfrak{B} \subseteq \mathfrak{R}^n$ *a nonempty subset of* $\mathfrak{R}^n$

$\mathfrak{B}_\xi(z)$ an *open hyperball with the radius* ξ *centered at the point* z *in the corresponding space,*

$$\mathfrak{B}_\xi(\mathbf{z}) = \{\mathbf{w} : \; \|\mathbf{w} - \mathbf{z}\| < \xi\}$$

$\mathfrak{B}_\xi$ *an open hyperball with the radius* ξ *centered at the origin of the corresponding space,*

$$\mathfrak{B}_\xi = \mathfrak{B}_\xi(\mathbf{0})$$

$\mathrm{Cl}\mathfrak{S}$ *the closure of the set* $\mathfrak{S}$

$\partial\mathfrak{S}$ *the boundary of the set* $\mathfrak{S}$

$\mathrm{In}\mathfrak{S}$ *the interior of the set* $\mathfrak{S}$

$\mathfrak{N}(\mathfrak{S})$ *a neighborhood of the set* $\mathfrak{S}$, $\mathfrak{S} \subset \mathfrak{N}(\mathfrak{S})$

$\mathfrak{N}(\mathfrak{S}; \varepsilon)$ *the* ε*-neighborhood of the set* $\mathfrak{S}$,

$$\mathfrak{N}(\mathfrak{S}; \varepsilon) = \{\mathbf{z} : d(\mathbf{z}, \mathfrak{S}) < \varepsilon\} \supset \mathfrak{S}$$

$\mathfrak{P}(\mathfrak{S})$ *the power set of the set* $\mathfrak{S}$, which is denoted also by $2^{\mathfrak{S}}$, is the set of all subsets of a set $\mathfrak{S}$

$\mathfrak{S}(.): \mathfrak{T} \rightarrow \mathfrak{P}(\mathfrak{S})$ a *set-valued function*

$\mathfrak{S}(t)$ *the set value of the set-valued function* $\mathfrak{S}(.)$ at a moment $t \in \mathfrak{T}$, which is a *time-varying set* (at the moment $t \in \mathfrak{T}$)

$\mathfrak{S}_l(\mathfrak{T}_\tau)$ *the lower set limit* on $\mathfrak{T}_\tau$ of the set $\mathfrak{S}(t;t_0)$,

$$\mathfrak{S}_l(\mathfrak{T}_\tau) = \cap\left[\mathfrak{S}(t;t_0) : (t;t_0) \in \mathfrak{T}_0 \times \mathfrak{T}_\tau\right]$$

$\mathfrak{S}^u(\mathfrak{T}_\tau)$ *the upper set limit* on $\mathfrak{T}_\tau$ of the set $\mathfrak{S}(t;t_0)$,

$$\mathfrak{S}^u(\mathfrak{T}_\tau) = \cup\left[\mathfrak{S}(t;t_0) : (t;t_0) \in \mathfrak{T}_0 \times \mathfrak{T}_\tau\right]$$

$\rho\left[\mathbf{z}, \mathfrak{S}(t)\right]$ *the distance* of $\mathbf{z}$ from $\mathfrak{S}(t)$ at the moment $t \in \mathfrak{T}$,

$$d\left[\mathbf{z}, \mathfrak{S}(t)\right] = \inf\left[\|\mathbf{z} - \mathbf{w}\| : \mathbf{w} \in \mathfrak{S}(t)\right]$$

$\rho_m\left[\mathfrak{S}_1(t), \mathfrak{S}_2(t)\right]$ *the inner set distance between the sets* $\mathfrak{S}_1(t)$ *and* $\mathfrak{S}_2(t)$ *at the moment* $t \in \mathfrak{T}$,

$$\begin{gathered}\rho_m\left[\mathfrak{S}_1(t), \mathfrak{S}_2(t)\right] \\ = \inf\left\{\rho\left(\mathbf{x}, \mathbf{y}\right) : \mathbf{x} \in \mathfrak{S}_1(t),\ \mathbf{y} \in \mathfrak{S}_2(t)\right\} = \\ = \min\left\{\begin{array}{c}\inf\left\langle\rho\left[\mathbf{z}, \mathfrak{S}_1(t)\right] : \mathbf{z} \in \mathfrak{S}_2(t)\right\rangle, \\ \inf\left\langle\rho\left[\mathbf{z}, \mathfrak{S}_2(t)\right] : \mathbf{z} \in \mathfrak{S}_1(t)\right\rangle\end{array}\right\}\end{gathered}$$

$\rho_M\left[\mathfrak{S}_1(t), \mathfrak{S}_2(t)\right]$ *the outer set distance between the sets* $\mathfrak{S}_1(t)$ *and* $\mathfrak{S}_2(t)$ *at the moment* $t \in \mathfrak{T}$,

$$\begin{gathered}\rho_M\left[\mathfrak{S}_1(t), \mathfrak{S}_2(t)\right] = \\ = \sup\left\{\rho\left(\mathbf{x}, \mathbf{y}\right) : \mathbf{x} \in \mathfrak{S}_1(t),\ \mathbf{y} \in \mathfrak{S}_2(t)\right\} = \\ = \max\left\{\begin{array}{c}\sup\left\langle\rho\left[\mathbf{z}, \mathfrak{S}_1(t)\right] : \mathbf{z} \in \mathfrak{S}_2(t)\right\rangle, \\ \sup\left\langle\rho\left[\mathbf{z}, \mathfrak{S}_2(t)\right] : \mathbf{z} \in \mathfrak{S}_1(t)\right\rangle\end{array}\right\}\end{gathered}$$

$\partial\mathfrak{V}_\xi\left(t;\tau\right)$ *the boundary of the set* $\mathfrak{V}_\xi\left(t;\tau\right)$

$\mathfrak{V}_\xi\left(t;\tau\right) \subseteq \mathfrak{R}^K$ *the set associated with the function* $V\left(.\right) : \mathfrak{T}_\tau \times \mathfrak{R}^K \longrightarrow \mathfrak{R}$ *and with its value* $\xi \in \mathfrak{R}^+$, or $\xi = \infty$, at $t \in \mathfrak{T}_\tau$ so that it is the largest open connected neighborhood $\mathfrak{V}_\xi\left(t;\tau\right)$ of the origin $\mathbf{0}_K$ on $\mathfrak{T}_\tau$ such that

1. If $\xi \in \mathfrak{R}^+$ then the boundary $\partial\mathfrak{V}_\xi\left(t;\tau\right)$ of $\mathfrak{V}_\xi\left(t;\tau\right)$ is the hypersurface $V\left(t, \mathbf{z}\right) = \xi$,

$$\begin{gathered}\xi \in \mathfrak{R}^+ \Longrightarrow \partial\mathfrak{V}_\xi\left(t;\tau\right) = \left\{\mathbf{z} : V\left(t, \mathbf{z}\right) = \xi\right\},\ \forall t \in \mathfrak{T}_\tau, \\ V\left(t, \mathbf{z}\right) = \xi \Longleftrightarrow \mathbf{z} \in \partial\mathfrak{V}_\xi\left(t;\tau\right)\end{gathered}$$

2. The intersection $\mathfrak{V}_\xi\left(\mathfrak{T}_\tau\right)$ of $\mathfrak{V}_\xi\left(t;\tau\right)$ over $\mathfrak{T}_\tau$,

$$\mathfrak{V}_\xi\left(\mathfrak{T}_\tau\right) = \cap\left[\mathfrak{V}_\xi\left(t;\tau\right) : t \in \mathfrak{T}_\tau\right],$$

is also a neighborhood of the origin $\mathbf{0}_K$,

$$\exists\zeta \in \mathfrak{R}^+ \Longrightarrow \mathfrak{B}_\zeta \subset \mathfrak{V}_\xi\left(\mathfrak{T}_\tau\right)$$

and

3. $V\left(t, \mathbf{z}\right)$ is strictly less than ξ on $\mathfrak{V}_\xi\left(t;\tau\right)$ at every instant $t \in \mathfrak{T}_\tau$,

$$V\left(t, \mathbf{z}\right) < \xi,\ \forall\left(\mathbf{z},\ t\right) \in \mathfrak{V}_\xi\left(t;\tau\right) \times \mathfrak{T}_\tau$$

where the expression "*on* $\mathfrak{T}_\tau$" and the notation "$;\tau$" are to be omitted if, and only if, $\mathfrak{T}_\tau = \mathfrak{T}$

A.8 Units

$1_{(.)}$ *the unit of a physical variable* (.)

1_t *the time unit of the reference time axis* T, if second s is the *time* unit then $1_t = s$

Appendix B

Dini derivatives

B.1 Definitions of derivatives

Note 574 *The notation t^- designates mathematically the left-hand side of the moment t,*

$$t^- = lim(t - \varepsilon : \varepsilon \longrightarrow 0, \varepsilon > 0). \tag{B.1}$$

The notation t^+ denotes mathematically the right-hand side of the moment t,

$$t^+ = lim(t + \varepsilon : \varepsilon \longrightarrow 0, \varepsilon > 0). \tag{B.2}$$

Note 575 *From the physical point of view, the left-hand side of the moment t is the beginning of the (infinitesimal) duration of the moment t. The right-hand side of the moment t represents the end of the (infinitesimal) duration of the moment t.*

Let $t^* \in \mathfrak{R} \cup \{-\infty\} \cup \{\infty\}$. Let $\mathfrak{T}^-_{\longrightarrow t^*}$ be a set composed of a sequence t_k converging to t^* from the left as $k \longrightarrow \infty$ if $t^* > -\infty$, ($t^* = \infty$ is permitted),

$$\mathfrak{T}^-_{\longrightarrow t^*} = \left\{t_k : t_k \in \mathfrak{T},\ t_k < t^*,\ k = 0, 1, 2, \cdots,\ k \longrightarrow \infty \Longrightarrow t_k \longrightarrow t^{*-}\right\},$$
$$t^* \in]-\infty, \infty],$$

and, analogously, let $\mathfrak{T}^+_{\longrightarrow t^*}$ be a set composed of a sequence t_k converging to t^* from the right as $k \longrightarrow \infty$ if $t^* < \infty$, ($t^* = -\infty$ is permitted),

$$\mathfrak{T}^+_{\longrightarrow t^*} = \left\{t_k : t_k \in \mathfrak{T},\ t_k > t^*,\ k = 0, 1, 2, \cdots,\ k \longrightarrow \infty \Longrightarrow t_k \longrightarrow t^{*+}\right\},$$
$$t^* \in [-\infty, \infty[.$$

We accept the following definition from [287, Definition 3.5, p. 54].

Definition 576 *a) A number $\alpha \in \mathfrak{R}$ is* ***the left, [right], partial limit*** *of a function $f(.)$, $f(.) : \mathfrak{R} \longrightarrow \mathfrak{R}$, over a sequence set $\mathfrak{T}^-_{\longrightarrow t^*}$, $\left[\mathfrak{T}^+_{\longrightarrow t^*}\right]$, if, and only if, for every $\varepsilon \in \mathfrak{R}^+$ there is an integer N such that $k > N$ implies $|f(t_k) - \alpha| < \varepsilon$,*

b) The symbol $\alpha = -\infty$, $[\alpha = \infty]$, is the left [right] partial limit of the function $f(.)$ over a sequence set $\mathfrak{T}^-_{\longrightarrow t^*}$, $\left(\mathfrak{T}^+_{\longrightarrow t^*}\right)$, if, and only if, for every $\varepsilon \in \mathfrak{R}^+$ there is an integer N such that, respectively, $k > N$ implies $f(t_k) < -\varepsilon^{-1}$, $\left[f(t_k) > \varepsilon^{-1}\right]$,

c) The greatest [the smallest] partial limit of the function $f(.)$ over all the sequence sets $\mathfrak{T}^-_{\longrightarrow t^*}$ is its left upper [left lower] limit at $t = t^*$, which is denoted, respectively, by

$$lim\ sup\ \left[f(t) : t \longrightarrow t^{*-}\right] = \overline{lim}\ \left[f(t) : t \longrightarrow t^{*-}\right],$$

$$lim\ inf\left[f(t) : t \longrightarrow t^{*-}\right] = \underline{lim}\ \left[f(t) : t \longrightarrow t^{*-}\right],$$

d) the greatest [the smallest] partial limit of the function $f(.)$ *over all the sequence sets* $\mathfrak{T}^{+}_{\longrightarrow t^{*}}$ *is its right upper [right lower] limit at* $t = t^{*}$, *which is denoted, respectively, by*

$$lim\ sup\ \left[f(t) : t \longrightarrow t^{*+}\right] = \overline{lim}\ \left[f(t) : t \longrightarrow t^{*+}\right],$$

$$lim\ inf\ \left[f(t) : t \longrightarrow t^{*+}\right] = \underline{lim}\ \left[f(t) : t \longrightarrow t^{*+}\right].$$

Example 577 *Let* $f(0) = 0$ *and* $f(t) = 5sign\left[7sin\left(t^{-4}\right)\right]$ *for* $t \neq 0$. *Then,*

$$\overline{lim}\ \left[f(t) : t \longrightarrow 0^{-}\right] = \overline{lim}\ \left[f(t) : t \longrightarrow 0^{+}\right] = 5,$$
$$\underline{lim}\ \left[f(t) : t \longrightarrow 0^{-}\right] = \underline{lim}\ \left[f(t) : t \longrightarrow 0^{+}\right] = -5.$$

The definition of Dini derivatives follows by referring to [287], [425], [565].

Definition 578 ***Forward-time derivatives***
Let $V(.)$ *be a continuous scalar function of* $(t, \mathbf{z}) \in \mathfrak{T} \times \mathfrak{R}^{K}$, $V(t, \mathbf{z}) \in \mathfrak{C}\left(\mathfrak{T} \times \mathfrak{R}^{K}\right)$. *Let* $\mathcal{Z}(.; t_0; \mathbf{z}; \mathbf{i})$ *be a motion of a dynamical system through* $\mathbf{z}$ *at* $t = t_0$, $\mathcal{Z}(t; t_0; \mathbf{z}; \mathbf{i}) \equiv \mathbf{z}(t)$, $\mathcal{Z}(t_0; t_0; \mathbf{z}; \mathbf{i}) \equiv \mathbf{z}$. *Then,*

- The forward-time upper left Dini derivative $D^{-}V(t, \mathbf{z})$ *of* $V(.)$ *along* $\mathcal{Z}(.; t_0; \mathbf{z}; \mathbf{i})$ *at* $(t, \mathbf{z})$ *is*

$$D^{-}V(t, \mathbf{z}) = lim\ sup\left\{\frac{V\left[t+\theta, \mathcal{Z}(t+\theta; t; \mathbf{z}; \mathbf{i})\right] - V(t, \mathbf{z})}{\theta} : \theta \longrightarrow 0^{-}\right\}.$$

- The forward-time lower left Dini derivative $D_{-}V(t, \mathbf{z})$ *of* $V(.)$ *along* $\mathcal{Z}(.; t_0; \mathbf{z}; \mathbf{i})$ *at* $(t, \mathbf{z})$ *is*

$$D_{-}V(t, \mathbf{z}) = lim\ inf\left\{\frac{V\left[t+\theta, \mathcal{Z}(t+\theta; t; \mathbf{z}; \mathbf{i})\right] - V(t, \mathbf{z})}{\theta} : \theta \longrightarrow 0^{-}\right\}.$$

- The forward-time upper right Dini derivative $D^{+}V(t, \mathbf{z})$ *of* $V(.)$ *along* $\mathcal{Z}(.; t_0; \mathbf{z}; \mathbf{i})$ *at* $(t, \mathbf{z})$ *is*

$$D^{+}V(t, \mathbf{z}) = lim\ sup\left\{\frac{V\left[t+\theta, \mathcal{Z}(t+\theta; t; \mathbf{z}; \mathbf{i})\right] - V(t, \mathbf{z})}{\theta} : \theta \longrightarrow 0^{+}\right\}.$$

- The forward-time lower right Dini derivative $D_{+}V(t, \mathbf{z})$ *of* $V(.)$ *along* $\mathcal{Z}(.; t_0; \mathbf{z}; \mathbf{i})$ *at* $(t, \mathbf{z})$ *is*

$$D_{+}V(t, \mathbf{z}) = lim\ inf\left\{\frac{V\left[t+\theta, \mathcal{Z}(t+\theta; t; \mathbf{z}; \mathbf{i})\right] - V(t, \mathbf{z})}{\theta} : \theta \longrightarrow 0^{+}\right\}.$$

- The function $V(.)$ *has the forward-time left Dini derivative along the motion* $\mathcal{Z}(.; t_0; \mathbf{z}; \mathbf{i})$ *at* $(t, \mathbf{z})$, *which is denoted by* $D_{l}V(t, \mathbf{z})$ *if, and only if,* $D^{-}V(t, \mathbf{z}) = D_{-}V(t, \mathbf{z})$ *and then*

$$D_{l}V(t, \mathbf{z}) = D^{-}V(t, \mathbf{z}) = D_{-}V(t, \mathbf{z}).$$

- The function $V(.)$ *has the forward-time right Dini derivative along the motion* $\mathcal{Z}(.; t_0; \mathbf{z}; \mathbf{i})$ *at* $(t, \mathbf{z})$, *which is denoted by* $D_{r}V(t, \mathbf{z})$ *if, and only if,* $D^{+}V(t, \mathbf{z}) = D_{+}V(t, \mathbf{z})$ *and then*

$$D_{r}V(t, \mathbf{z}) = D^{+}V(t, \mathbf{z}) = D_{+}V(t, \mathbf{z}).$$

- The function $V(.)$ *has the forward-time Eulerian derivative along the motion* $\mathcal{Z}(.; t_0; \mathbf{z}; \mathbf{i})$ *at* $(t, \mathbf{z})$, *which is denoted by* $d_{f}V(t, \mathbf{z})/dt$ *or by* $V_{f}^{(1)}(t, \mathbf{z})$ *or simpler by* $dV(t, \mathbf{z})/dt$ *or by*

$V^{(1)}(t,\mathbf{z})$ *if, and only if, it has both* $D_l V(t,\mathbf{z})$ *and* $D_r V(t,\mathbf{z})$ *such that* $D_l V(t,\mathbf{z}) = D_r V(t,\mathbf{z})$ *and then*

$$\frac{dV(t,\mathbf{z})}{dt} = V^{(1)}(t,\mathbf{z}) = \frac{d_f V(t,\mathbf{z})}{dt} = V_f^{(1)}(t,\mathbf{z}) = D_l V(t,\mathbf{z}) = D_r V(t,\mathbf{z}).$$

$D^* V(t,\mathbf{z})$ *means that either* $D^- V(t,\mathbf{z})$ *or* $D^+ V(t,\mathbf{z})$ *can be equally used. Analogously,* $D_* V(t,\mathbf{z})$ *signifies that either* $D_- V(t,\mathbf{z})$ *or* $D_+ V(t,\mathbf{z})$ *can be equally used.*

Note 579 *From the physical point of view, only the forward-time derivatives have the sense [273], [274] so that* $V_f^{(1)}(t,\mathbf{z})$ *is simply denoted by* $V^{(1)}(t,\mathbf{z})$,

$$\frac{dv(t,\mathbf{z})}{dt} = V^{(1)}(t,\mathbf{z}) = \frac{d_f V(t,\mathbf{z})}{dt} = V_f^{(1)}(t,\mathbf{z}) = D_l V(t,\mathbf{z}) = D_r V(t,\mathbf{z}). \tag{B.3}$$

From the mathematical point of view we can also introduce the backward-time derivatives as follows.

Definition 580 ***Backward-time derivatives***

Let $V(.)$ *be a continuous scalar function of* $(t,\mathbf{z}) \in \mathfrak{T} \times \mathfrak{R}^K$, $V(t,\mathbf{z}) \in \mathfrak{C}\left(\mathfrak{T} \times \mathfrak{R}^K\right)$. *Let* $\mathcal{Z}(.;t_0;\mathbf{z};\mathbf{i})$ *be a motion of a dynamical system through* $\mathbf{z}$ *at* $t = t_0$, $\mathcal{Z}(t;t_0;\mathbf{z};\mathbf{i}) \equiv \mathbf{z}(t)$, $\mathcal{Z}(t_0;t_0;\mathbf{z};\mathbf{i}) \equiv \mathbf{z}$. *Then,*

- The backward-time upper left Dini derivative ${}^-Dv(t,\mathbf{z})$ *of* $V(.)$ *along* $\mathcal{Z}(.;t_0;\mathbf{z};\mathbf{i})$ *at* $(t,\mathbf{z})$ *is*

$${}^-Dv(t,\mathbf{z}) = \lim\sup\left\{\frac{V\left[t-\theta, \mathcal{Z}(t-\theta;t_0;\mathbf{z};\mathbf{i})\right] - V(t,\mathbf{z})}{\theta} : \theta \longrightarrow 0^-\right\}.$$

- The backward-time lower left Dini derivative ${}_-Dv(t,\mathbf{z})$ *of* $V(.)$ *along* $\mathcal{Z}(.;t_0;\mathbf{z};\mathbf{i})$ *at* $(t,\mathbf{z})$ *is*

$${}_-Dv(t,\mathbf{z}) = \lim\inf\left\{\frac{V\left[t-\theta, \mathcal{Z}(t-\theta;t_0;\mathbf{z};\mathbf{i})\right] - V(t,\mathbf{z})}{\theta} : \theta \longrightarrow 0^-\right\}.$$

- The backward-time upper right Dini derivative ${}^+Dv(t,\mathbf{z})$ *of* $V(.)$ *along* $\mathcal{Z}(.;t_0;\mathbf{z};\mathbf{i})$ *at* $(t,\mathbf{z})$ *is*

$${}^+Dv(t,\mathbf{z}) = \lim\sup\left\{\frac{V\left[t-\theta, \mathcal{Z}(t-\theta;t_0;\mathbf{z};\mathbf{i})\right] - V(t,\mathbf{z})}{\theta} : \theta \longrightarrow 0^+\right\}.$$

- The backward-time lower right Dini derivative ${}_+Dv(t,\mathbf{z})$ *of* $V(.)$ *along the motion* $\mathcal{Z}(.;t_0;\mathbf{z};\mathbf{i})$ *at* $(t,\mathbf{z})$ *is*

$${}_+Dv(t,\mathbf{z}) = \lim\inf\left\{\frac{V\left[t-\theta, \mathcal{Z}(t-\theta;t_0;\mathbf{z};\mathbf{i})\right] - V(t,\mathbf{z})}{\theta} : \theta \longrightarrow 0^+\right\}.$$

- The function $V(.)$ *has the backward-time left Dini derivative along* $\mathcal{Z}(.;t_0;\mathbf{z};\mathbf{i})$ *at* $(t,\mathbf{z})$, *which is denoted by* ${}_lDv(t,\mathbf{z})$ *if, and only if,* ${}^-Dv(t,\mathbf{z}) = {}_-Dv(t,\mathbf{z})$ *and then*

$${}_lDv(t,\mathbf{z}) = {}^-Dv(t,\mathbf{z}) = {}_-Dv(t,\mathbf{z}).$$

- The function $V(.)$ *has the backward-time right Dini derivative along* $\mathcal{Z}(.;t_0;\mathbf{z};\mathbf{i})$ *at* $(t,\mathbf{z})$, *which is denoted by* ${}_rDv(t,\mathbf{z})$ *if, and only if,* ${}^+Dv(t,\mathbf{z}) = {}_+Dv(t,\mathbf{z})$ *and then*

$${}_rDv(t,\mathbf{z}) = {}^+Dv(t,\mathbf{z}) = {}_+Dv(t,\mathbf{z}).$$

- *The function* $V(.)$ *has the backward-time Eulerian derivative along* $\mathcal{Z}(.;t_0;\mathbf{z};\mathbf{i})$ *at* $(t,\mathbf{z})$, *which is denoted by* $d_bV(t,\mathbf{z})/dt$ *or by* $V_b^{(1)}(t,\mathbf{z})$ *if, and only if, it has both* ${}_lDv(t,\mathbf{z})$ *and* ${}_rDv(t,\mathbf{z})$ *such that* ${}_lDv(t,\mathbf{z}) =_r Dv(t,\mathbf{z})$ *and then*

$$\frac{d_bV(t,\mathbf{z})}{dt} = V_b^{(1)}(t,\mathbf{z}) = {}_lDv(t,\mathbf{z}) = {}_rDv(t,\mathbf{z}).$$

${}^*Dv(t,\mathbf{z})$ *means that either* ${}^-Dv(t,\mathbf{z})$ *or* ${}^+Dv(t,\mathbf{z})$ *can be equally used. Analogously,* ${}_*Dv(t,\mathbf{z})$ *signifies that either* ${}_-Dv(t,\mathbf{z})$ *or* ${}_+Dv(t,\mathbf{z})$ *can be equally used.*

B.2 Properties

Definition 578, Definition 580 and the following relationships:

$$\sup f(.) = -\inf\left[-f(.)\right], \ \inf f(.) = -\sup\left[-f(.)\right]$$

yield:

$$\begin{gathered} {}^-Dv(t,\mathbf{z}) = -D_+V(t,\mathbf{z}), \ {}_-Dv(t,\mathbf{z}) = -D^+V(t,\mathbf{z}), \\ {}^+Dv(t,\mathbf{z}) = -\left[D_-V(t,\mathbf{z})\right], \ {}_+Dv(t,\mathbf{z}) = -D^-V(t,\mathbf{z}), \\ {}_lDv(t,\mathbf{z}) = -D_rV(t,\mathbf{z}), \ {}_rDv(t,\mathbf{z}) = -D_lV(t,\mathbf{z}), \\ V_b^{(1)}(t,\mathbf{z}) = -V_f^{(1)}(t,\mathbf{z}). \end{gathered}$$

Let $\delta_{ij} = 1$ if $i = j$, $\delta_{ij} = 0$ if $i \neq j$, $\Delta_j\mathbf{z} = \Delta s\left[\delta_{1j} \ \ \delta_{2j}....\delta_{kj}\right]^T$, and, by referring to [209, p. 84],

$$D_t^+V(t,\mathbf{z}) = \lim\sup\left\{\frac{V(t+\theta,\mathbf{z}) - V(t,\mathbf{z})}{\theta} : \theta \longrightarrow 0^+\right\},$$

$$D_s^+V(t,\mathbf{z}) = \sum_{j=1}^{j=k}\lim\sup\left\{\frac{V(t,\mathbf{z}+\Delta_j\mathbf{z}) - V(t,\mathbf{z})}{\Delta s}\frac{\Delta s}{\theta} : \theta \longrightarrow 0^+, \ \Delta s \longrightarrow 0\right\}.$$

These definitions and the fact that

$$\sup\left[f_1(t) + f_2(t)\right] \leq \sup f_1(t) + \sup f_2(t)$$

imply, [209, p. 84],

$$D_t^+V(t,\mathbf{z}) \leq D_t^+V(t,\mathbf{z}) + D_s^+V(t,\mathbf{z}). \tag{B.4}$$

Let

$\mathfrak{I}^c$ any connected nonempty compact subset of the output integral space $\mathfrak{I} \subseteq \mathfrak{T} \times \mathfrak{R}^K$

$\mathfrak{L}^c$ any connected nonempty compact subset of $\mathfrak{T}$

$\mathfrak{T}^c$ any connected nonempty compact subset of $\mathfrak{T}$

$\mathfrak{C}_Z(\mathfrak{T}^c \times \mathfrak{L}^c \times \mathfrak{I}^c)$ be the family of the motions $\mathcal{Z}(t;t_0;\mathbf{z};\mathbf{i})$ continuous in $\mathbf{z} \in \mathfrak{R}^K, \forall (t;t_0) \in \mathfrak{T}^c \times \mathfrak{L}^c$

$\mathfrak{C}_t(\mathfrak{T}^c \times \mathfrak{L}^c \times \mathfrak{I}^c)$ be the family of the motions $\mathcal{Z}(t;t_0;\mathbf{z};\mathbf{i})$ continuous in time $t \in \mathfrak{T}^c$, $\forall (t_0;\mathbf{z}_0) \in \mathfrak{L}^c \times \mathfrak{I}^c$

$In\mathfrak{I}$ be the interior of $\mathfrak{I}$.

The following result is presented in [287, Lemma 4.15, p. 114] by noting that there is a typographical error in the proof. In the fourth line of the proof in [287, Lemma 4.15, p. 114] there should have been written Lemma 4.14 instead of wrongly written Lemma 4.15. The complete proof is in [373, 6.2 Lemma, p.30].

Lemma 581 ***The fundamental lemma on Dini derivatives***

Let $\mathcal{Z}(.;t_0;\mathbf{z}_0;\mathbf{i})$ *be a motion of a dynamical system through* $\mathbf{z}_0$ *at* $t=t_0$, $\mathcal{Z}(t;t_0;\mathbf{z}_0;\mathbf{i})\equiv\mathbf{z}(t)$, $\mathcal{Z}(t_0;t_0;\mathbf{z}_0;\mathbf{i})\equiv\mathbf{z}(0)\equiv\mathbf{z}_0$. *Let* $\mathcal{Z}(t;t_0;\mathbf{z}_0;\mathbf{i})$ *be continuous in time* $t\in\mathfrak{T}^c$, $\forall(t_0;\mathbf{z}_0)\in\mathfrak{L}^c\times\mathfrak{I}^c$, *i.e.* $\mathcal{Z}(t;t_0;\mathbf{z}_0;\mathbf{i})\in\mathfrak{C}_t(\mathfrak{T}^c\times\mathfrak{L}^c\times\mathfrak{I}^c\times\mathcal{J}^i)$, *i.e. and be continuous in* $\mathbf{z}_0\in\mathfrak{I}^c,\forall(t;t_0,\mathbf{i})\in\mathfrak{T}^c\times\mathfrak{L}^c\times\mathcal{J}^i$; *i.e.,* $\mathcal{Z}(t;t_0;\mathbf{z}_0;\mathbf{i})\in\mathfrak{C}_s(\mathfrak{T}^c\times\mathfrak{L}^c\times\mathfrak{I}^c\times\mathcal{J}^i)$. *Let* $V(t,\mathbf{z})\in\mathfrak{C}(\mathfrak{I})$. *If i) or ii) holds,*

i) $D_+V(t,\mathbf{z})\leq 0$, $\forall(t;\mathbf{z})\in\mathfrak{T}^c\times In\mathfrak{I}$,

ii) $D^+V(t,\mathbf{z})\leq 0$, $\forall(t;\mathbf{z})\in\mathfrak{T}^c\times In\mathfrak{I}$,

then

a) $V(.)$ *is non-increasing on* $[t_0,t_f[$, *where* t_f *is the first moment when* $\mathcal{Z}(t;t_0;\mathbf{z}_0;\mathbf{i})$ *is not in* $In\mathfrak{I}$,

$$\mathcal{Z}(t;t_0;\mathbf{z}_0;\mathbf{i})\in In\mathfrak{I},\ \forall t\in[t_0,t_f[,\ \text{and}\ \mathcal{Z}(t_f;t_0;\mathbf{z}_0)\notin In\mathfrak{I},$$

b) $V(.)$ *is differentiable almost everywhere on* $[t_0,t_f[$,

and

c) either

1. $$\int_{t_0}^{t} D_+V[\sigma,\mathcal{Z}(\sigma;t_0;\mathbf{z}_0)]\geq V[t,\mathcal{Z}(t;t_0;\mathbf{z}_0;\mathbf{i})]-V(t_0;\mathbf{z}_0),\ \forall t\in[t_0,t_f[,$$

2. $$\int_{t_0}^{t} D^+V[\sigma,\mathcal{Z}(\sigma;t_0;\mathbf{z}_0)]\geq V[t,\mathcal{Z}(t;t_0;\mathbf{z}_0;\mathbf{i})]-V(t_0;\mathbf{z}_0),\ \forall t\in[t_0,t_f[,$$

respectively.

Appendix C

Proofs for Part III

C.1 Proof of Theorem 89

Proof. The matrices $A(t) \in \mathfrak{R}^{i\times K}$ and $M(t) \in \mathfrak{R}^{i\times j}$, and the vectors $\mathbf{f}[t, \mathbf{Z}^m(t)] \in \mathfrak{R}^K$ and $\mathbf{g}[t, \mathbf{w}(t)] \in \mathfrak{R}^j$ can be set in the equivalent forms:

$$A(t) = \begin{bmatrix} A_\rho(t) & O_{\rho,(K-\rho)} \\ O_{(i-\rho),\rho} & O_{(i-\rho),(K-\rho)} \end{bmatrix}, \ \det A_\rho(t) \neq 0, \ \forall t \in \mathfrak{T}, \tag{C.1}$$

$$M(t) = \begin{bmatrix} M_\rho(t) & M_{\rho,(j-\rho)}(t) \\ M_{(i-\rho),\rho}(t) & M_{(i-\rho),(j-\rho)}(t) \end{bmatrix}, \ \forall t \in \mathfrak{T}, \tag{C.2}$$

$$\mathbf{f}[t, \mathbf{Z}^m(t)] = \begin{bmatrix} \mathbf{f}_\rho[t, \mathbf{Z}^m(t)] \\ \mathbf{f}_{K-\rho}[t, \mathbf{Z}^m(t)] \end{bmatrix}, \ \forall [t, \mathbf{Z}^m(t)] \in \mathfrak{T}\times\mathfrak{R}^K, \tag{C.3}$$

$$\mathbf{g}[t, \mathbf{w}(t)] = \begin{bmatrix} \mathbf{g}_\rho[t, \mathbf{w}(t)] \\ \mathbf{g}_{j-\rho}[t, \mathbf{w}(t)] \end{bmatrix}, \ \forall [t, \mathbf{w}(t)] \in \mathfrak{T}\times\mathfrak{R}^j, \tag{C.4}$$

so that (3.70) becomes, after applying the elementary matrix transformations to it,

$$A(t)\mathbf{f}[t, \mathbf{Z}^m(t)] = \begin{bmatrix} A_\rho(t)\mathbf{f}_\rho[t, \mathbf{Z}^m(t)] \\ \mathbf{0}_{i-\rho} \end{bmatrix} = M(t)\mathbf{g}[t, \mathbf{w}(t)]$$

$$= \begin{bmatrix} M_\rho(t)\mathbf{g}_\rho[t, \mathbf{w}(t)] + M_{\rho,(j-\rho)}(t)\mathbf{g}_{j-\rho}[t, \mathbf{w}(t)] \\ M_{(i-\rho),\rho}(t)\mathbf{g}_\rho[t, \mathbf{w}(t)] + M_{i-\rho,j-\rho}(t)\mathbf{g}_{j-\rho}[t, \mathbf{w}(t)] \end{bmatrix}, \ \forall t \in \mathfrak{T}. \tag{C.5}$$

Case 1: $i > K$ and $\text{rank}A(t) = \rho \leq K < i$. This implies the existence of only ρ unique scalar solutions $z_j(t; t_0; \mathbf{Z}_0)$, $j = 1, 2, .., \rho$, to only ρ scalar equations of (3.70), which are determined as solutions to

$$\mathbf{f}_\rho[t, \mathbf{Z}^m(t)] = A_\rho^{-1}(t)\left\{M_\rho(t)\mathbf{g}_\rho[t, \mathbf{w}(t)] + M_{\rho,(j-\rho)}(t)\mathbf{g}_{j-\rho}[t, \mathbf{w}(t)]\right\}.$$

Other $i - \rho$ scalar equations of (3.70) cannot be satisfied provided all i scalar equations of (3.70) are linearly independent. The unique solution $\mathbf{Z}(t; t_0; \mathbf{Z}_0)$ to (3.70) does not exist for $i > K$. This is clear because there are more, i.e., i, scalar equations than the number, i.e., K, of the unknown scalar variables to be determined. This implies $i \leq K$ in (3.71).

Case 2: $i \leq K$ and $\text{rank}A(t) \equiv \rho = i \leq K$. Then, $\text{rank}\left[A(t)A^T(t)\right] \equiv i$, $\det\left[A(t)A^T(t)\right]$

$\neq 0$, and (C.1)-(C.4) yield

$$A(t)A^T(t) = \begin{bmatrix} A_\rho(t) & O_{\rho,(K-\rho)} \\ O_{(i-\rho),\rho} & O_{(i-\rho),(K-\rho)} \end{bmatrix} \begin{bmatrix} A_\rho^T(t) & O_{\rho,(i-\rho)} \\ O_{(K-\rho),\rho} & O_{(K-\rho),(i-\rho)} \end{bmatrix}$$

$$\begin{bmatrix} A_\rho(t)A_\rho^T(t) & O_{\rho,(i-\rho)} \\ O_{(i-\rho),\rho} & O_{(i-\rho),(i-\rho)} \end{bmatrix}, \det\left[A_\rho(t)A_\rho^T(t)\right] \neq 0,\ \forall t \in \mathfrak{T},\ \rho = i,$$

$$\mathbf{f}\left[t, \mathbf{Z}^m(t)\right] = \begin{bmatrix} \mathbf{f}_\rho\left[t, \mathbf{Z}^m(t)\right] \\ \mathbf{f}_{K-\rho}\left[t, \mathbf{Z}^m(t)\right] \end{bmatrix},\ \forall t \in \mathfrak{T},$$

$$M(t)\mathbf{g}\left[t, \mathbf{w}(t)\right] = \begin{bmatrix} M_\rho(t)\mathbf{g}_\rho\left[t, \mathbf{w}(t)\right] + M_{\rho,(j-\rho)}(t)\mathbf{g}_{j-\rho}\left[t, \mathbf{w}(t)\right] \\ M_{(i-\rho),\rho}(t)\mathbf{g}_\rho\left[t, \mathbf{w}(t)\right] + M_{i-\rho,j-\rho}(t)\mathbf{g}_{j-\rho}\left[t, \mathbf{w}(t)\right] \\ \forall t \in \mathfrak{T} \end{bmatrix}. \tag{C.6}$$

This determines

$$\mathbf{f}_\rho\left[t, \mathbf{Z}^m(t)\right] = A_\rho^T(t)\left[A_\rho(t)A_\rho^T(t)\right]^{-1}$$
$$\bullet \left[M_\rho(t)\mathbf{g}_\rho\left[t, \mathbf{w}(t)\right] + M_{\rho,(j-\rho)}(t)\mathbf{g}_{j-\rho}\left[t, \mathbf{w}(t)\right]\right],\ \forall t \in \mathfrak{T}. \tag{C.7}$$

The function $\mathbf{f}_\rho\left[., \mathbf{Z}^m(t)\right]$ determined by (C.7) satisfies the first ρ equations (3.70) as shown now:

$$A_\rho(t)\mathbf{f}_\rho\left[t, \mathbf{Z}^m(t)\right] = A_\rho(t)A_\rho^T(t)\left[A_\rho(t)A_\rho^T(t)\right]^{-1}$$
$$\bullet \left[M_\rho(t)\mathbf{g}_\rho\left[t, \mathbf{w}(t)\right] + M_{\rho,(j-\rho)}(t)\mathbf{g}_{j-\rho}\left[t, \mathbf{w}(t)\right]\right]$$
$$= \left[M_\rho(t)\mathbf{g}_\rho\left[t, \mathbf{w}(t)\right] + M_{\rho,(j-\rho)}(t)\mathbf{g}_{j-\rho}\left[t, \mathbf{w}(t)\right]\right],\ \forall t \in \mathfrak{T},$$

Other $i-\rho$ equations (3.70) have arbitrary solutions. This implies the non-uniqueness of the solutions of all i equations (3.70).

If $\rho = i$ then the solution of (3.70) is the solution of (C.7), i.e., of (3.72) for

$$T(t) = A^T(t)\left[A(t)A^T(t)\right]^{-1} \in \mathfrak{R}^{i \times K},\ \forall t \in \mathfrak{T}, \tag{C.8}$$

because $A(t)A^T(t) \in \mathfrak{R}^{i \times i}$, $i \leq K$ and $\rho = \text{rank}A(t) \equiv i$ imply $det\left[A(t)A^T(t)\right] \neq 0, \forall t \in \mathfrak{T}$. The equation (3.72) is solvable in $\mathbf{Z}(t)$ for every $\mathbf{Z}_0 \in \mathfrak{R}^K$. It has the unique solution that is also the fundamental solution for

$$\rho = rankA(t) \equiv i = K.$$

This is the second part of (3.71). The number K of unknown variables is equal to the number i of the equations. This explains why the system has the unique solutions. The choice of the matrix $T(t)$ determined by (C.8) enables the uniqueness of the solution for $\rho = \text{rank}A(t) \equiv i = K$, which is then also the fundamental solution ■

C.2 Proof of Theorem 152

Proof. Let the plant (3.12) have Properties 57, 58, 78.

Necessity. Let the plant (3.12) with Properties 57, 58, 78 be perfect trackable on $\mathfrak{T}_0 \times \mathfrak{D}^j \times \mathfrak{Y}_d$. Definition 140 is satisfied. There exists a control vector function $\mathbf{U}(.)$ for which $\mathbf{Y}(t) = \mathbf{Y}_d(t)$, i.e., $\varepsilon(t) = \mathbf{0}_N$, $\forall t \in \mathfrak{T}_0$, which validates (8.1) for $m \leq \gamma < \min(N, r)$, and for the perfect elementwise trackability $m = \gamma = N = \min(N, r)$ due to Fundamental control principle 87. This shows that at least one control variable acts on the plant, which is through the matrix $W(t)$. If rank of $W(t)$ were equal to zero, then $W(t)$ would be the zero matrix due to Property 64. No one control variable would act on the plant output behavior. This proves $\gamma \geq 1$. If the plant is elementwise perfect trackable then $\gamma = \text{rank}W(t) = N < r$, $\forall t \in \mathfrak{T}_0$, due to Axiom 87. This completes the validity of the condition i).

ii-1) If $\gamma = \text{rank}W(t) = r < N$, $\forall t \in \mathfrak{T}_0$, then $\det\left[W^T(t)W(t)\right] \neq 0$, $\forall t \in \mathfrak{T}_0$, so that the output equation of (3.12) multiplied on the left by the matrix product

$$\left[W^T(t)W(t)\right]^{-1}W^T(t)$$

becomes

$$\left[W^T(t)W(t)\right]^{-1}W^T(t)\left[\mathbf{Y}-\mathbf{z}(t,\mathbf{R}^{\alpha-1},\mathbf{D})\right] = \mathbf{w}\left[\mathbf{U}(t)\right], \forall t \in \mathfrak{T}_0.$$

Because $\gamma = \text{rank}W(t) = r < N$ then the plant output vector can be controlled only indirectly through an arbitrarily accepted vector function $\mathbf{v}(.) : \mathfrak{T}_0 \times \mathfrak{R}^N \longrightarrow \mathfrak{R}^m$, (8.1), Axiom 87. It is necessary to extend the output equation of (3.12). The validity of (8.1) permits us to subtract $M_{r,m}(t)\mathbf{v}(t,\varepsilon)$ from the right-hand side of the preceding equation:

$$\left[W^T(t)W(t)\right]^{-1}W^T(t)\left[\mathbf{Y}-\mathbf{z}(t,\mathbf{R}^{\alpha-1},\mathbf{D})\right] = \mathbf{w}\left[\mathbf{U}(t)\right] - M_{r,m}(t)\mathbf{v}(t,\varepsilon)$$
$$\forall t \in \mathfrak{T}_0,$$

or

$$\mathbf{w}\left[\mathbf{U}(t)\right] = \Gamma(t)\left[\mathbf{Y}-\mathbf{z}(t,\mathbf{R}^{\alpha-1},\mathbf{D})\right] + M_{r,m}(t)\mathbf{v}(t,\varepsilon)$$
$$\Gamma(t) = \left[W^T(t)W(t)\right]^{-1}W^T(t) \in \mathfrak{R}^{r\times N}, \ \forall t \in \mathfrak{T}_0,$$

i.e.,

$$\mathbf{U}(t) = \mathbf{w}^I\left\{\Gamma(t)\left[\mathbf{Y}-\mathbf{z}(t,\mathbf{R}^{\alpha-1},\mathbf{D})\right] + M_{r,m}(t)\mathbf{v}(t,\varepsilon)\right\}$$
$$\Gamma(t) = \left[W^T(t)W(t)\right]^{-1}W^T(t) \in \mathfrak{R}^{r\times N}, \ \forall t \in \mathfrak{T}_0.$$

This equation is (8.11) that at $t = t_0$ determines (8.16) since $\mathbf{v}(t_0,\varepsilon) \equiv \mathbf{0}_m$, (8.1).

ii-2) If the plant is perfect trackable then $\gamma = \text{rank}W(t) = N < r$, $\forall t \in \mathfrak{T}_0$, and then $\det\left[W(t)W^T(t)\right] \neq 0$, $\forall t \in \mathfrak{T}_0$, so that the output equation of (3.12) yields

$$W^T(t)\left[W(t)W^T(t)\right]^{-1}\left[\mathbf{Y}-\mathbf{z}(t,\mathbf{R}^{\alpha-1},\mathbf{D})\right] = \mathbf{w}\left[\mathbf{U}(t)\right], \ \forall t \in \mathfrak{T}_0.$$

The validity of (8.1) permits us to subtract $W^T(t)M_{N,m}(t)\mathbf{v}(t,\varepsilon)$ from the right-hand side of the preceding equation:

$$W^T(t)\left[W(t)W^T(t)\right]^{-1}\left[\mathbf{Y}-\mathbf{z}(t,\mathbf{R}^{\alpha-1},\mathbf{D})\right]$$
$$= \mathbf{w}\left[\mathbf{U}(t)\right] - W^T(t)M_{N,m}(t)\mathbf{v}(t,\varepsilon), \ \forall t \in \mathfrak{T}_0,$$

or

$$\mathbf{w}\left[\mathbf{U}(t)\right] = \Gamma(t)\left[\mathbf{Y}-\mathbf{z}(t,\mathbf{R}^{\alpha-1},\mathbf{D})\right] + W^T(t)M_{r,m}(t)\mathbf{v}(t,\varepsilon)$$
$$\Gamma(t) = W^T(t)\left[W(t)W^T(t)\right]^{-1} \in \mathfrak{R}^{r\times N}, \ \forall t \in \mathfrak{T}_0,$$

i.e.,

$$\mathbf{U}(t) = \mathbf{w}^I\left\{\Gamma(t)\left[\mathbf{Y}-\mathbf{z}(t,\mathbf{R}^{\alpha-1},\mathbf{D})\right] + W^T(t)M_{r,m}(t)\mathbf{v}(t,\varepsilon)\right\}$$
$$\Gamma(t) = W^T(t)\left[W(t)W^T(t)\right]^{-1} \in \mathfrak{R}^{r\times N}, \ \forall t \in \mathfrak{T}_0,$$

This equation is (8.12) that at $t = t_0$ determines (8.17) because $\mathbf{v}(t,\varepsilon) \equiv \mathbf{0}_m$, (8.1).

ii-3) If $\gamma = rankW(t) = r = N$, $\forall t \in \mathfrak{T}_0$, then $detW(t) \neq 0$, $\forall t \in \mathfrak{T}_0$, so that the output equation of (3.12) multiplied on the left by $W(t)^{-1}$ becomes

$$W(t)^{-1}\left[\mathbf{Y} - \mathbf{z}(t, \mathbf{R}^{\alpha-1}, \mathbf{D})\right] = \mathbf{w}\left[\mathbf{U}(t)\right], \forall t \in \mathfrak{T}_0.$$

Because $\boldsymbol{\varepsilon}(t) = \mathbf{0}_N$, $\forall t \in \mathfrak{T}_0$, we may then add it to $\mathbf{Y}$, and because $\mathbf{Y}+\boldsymbol{\varepsilon} = \mathbf{Y}+\mathbf{Y}_d-\mathbf{Y} = \mathbf{Y}_d$, the result is

$$\mathbf{w}\left[\mathbf{U}(t)\right] = W(t)^{-1}\left\{\mathbf{Y}_d(t) - \mathbf{z}(\left[t, \mathbf{R}^{\alpha-1}(t), \mathbf{D}(t)\right]\right\}, \forall t \in \mathfrak{T}_0,$$

i.e.

$$\mathbf{U}(t) = \mathbf{w}^I\left\langle W(t)^{-1}\left\{\mathbf{Y}_d(t) - \mathbf{z}(\left[t, \mathbf{R}^{\alpha-1}(t), \mathbf{D}(t)\right]\right\}\right\rangle, \forall t \in \mathfrak{T}_0,$$

This equation is (8.13) that at $t = t_0$ determines (8.18) inasmuch $\mathbf{v}(t_0, \boldsymbol{\varepsilon}) \equiv \mathbf{0}_m$,

ii-4) Let the plant (3.12) also possess Property 64. The complete control action on the plant output behavior is $W_\gamma(t)\mathbf{w}_\gamma\left[\mathbf{U}_\gamma(t)\right] \in \mathfrak{R}^\gamma$ due to Property 64. If the rank γ of the control matrix $W(t)$ is less than N and $N \leq r$, $\gamma < \min(N, r)$, then the output control vector can be controlled only indirectly through an arbitrarily accepted vector function $\mathbf{v}(.) : \mathfrak{T}_0 \times \mathfrak{R}^N \longrightarrow \mathfrak{R}^m$, (8.1), Axiom 87. It is necessary to extend the output equation of (3.12),

$$\mathbf{Y} = \mathbf{z}(t, \mathbf{R}^{\alpha-1}, \mathbf{D}) + W(t)\mathbf{w}\left[\mathbf{U}(t)\right], \ \mathbf{Y} \in \mathfrak{R}^N, \ \mathbf{U} \in \mathfrak{R}^r,$$

by using (8.1) and (8.15) in it as follows:

$$W(t)\left[\mathbf{w}\left[\mathbf{U}(t)\right] - M(t)\mathbf{v}(t, \boldsymbol{\varepsilon})\right] = \mathbf{Y} - \mathbf{z}(t, \mathbf{R}^{\alpha-1}, \mathbf{D}),$$

or, to apply Lemma 65 and (3.47):

$$\mathbf{Y}_\gamma(t) = \mathbf{z}_\gamma\left[t, \mathbf{R}^{\alpha-1}(t), \mathbf{D}(t)\right] + W_\gamma(t)\mathbf{w}_\gamma\left[\mathbf{U}(t)\right], \ \forall t \in \mathfrak{T},$$

and then to use (8.1) and (8.15). Its solution for $\mathbf{w}_\gamma\left[\mathbf{U}_\gamma(t)\right]$ reads

$$\mathbf{w}_\gamma\left[\mathbf{U}_\gamma(t)\right] = M_{\gamma,m}(t)\mathbf{v}(t, \boldsymbol{\varepsilon}) + W_\gamma^{-1}(t)\left\{\mathbf{Y}_\gamma(t) - \mathbf{z}_\gamma(\mathbf{t}, \mathbf{R}^{\alpha-1}, \mathbf{D})\right\},$$

due to the nonsingularity of the square matrix $W_\gamma(t)$, (3.42). Hence,

$$\mathbf{U}_\gamma(t) = \mathbf{w}_\gamma^I\left\langle M_{\gamma,m}(t)\mathbf{v}(t, \boldsymbol{\varepsilon}) + W_\gamma^{-1}(t)\left\{\mathbf{Y}_\gamma(t) - \mathbf{z}_\gamma(\mathbf{t}, \mathbf{R}^{\alpha-1}, \mathbf{D})\right\}\right\rangle,$$

This equation is (8.14) that determines (8.19) because $\mathbf{v}(t_0, \boldsymbol{\varepsilon}_0) = \mathbf{0}_m$, (8.1).

Sufficiency. Let the conditions of the theorem statement hold.

ii-1) The equation (8.11) can be set in the following form:

$$W(t)\left[W^T(t)W(t)\right]^{-1}W^T(t)\left[\mathbf{Y} - \mathbf{z}(t, \mathbf{R}^{\alpha-1}, \mathbf{D})\right]$$
$$= W(t)\left\{\mathbf{w}\left[\mathbf{U}(t)\right] - M_{r,m}(t)\mathbf{v}(t, \boldsymbol{\varepsilon})\right\}, \ \forall t \in \mathfrak{T}_0,$$

or, after multiplying this equation on the left by $W^T(t)$:

$$W^T(t)\left[\mathbf{Y} - \mathbf{z}(t, \mathbf{R}^{\alpha-1}, \mathbf{D})\right]$$
$$= W^T(t)W(t)\left\{\mathbf{w}\left[\mathbf{U}(t)\right] - M_{r,m}(t)\mathbf{v}(t, \boldsymbol{\varepsilon})\right\}, \ \forall t \in \mathfrak{T}_0,$$

Let us multiply on the left the output equation of (3.12) by $W^T(t)$:

$$W^T(t)\left[\mathbf{Y} - \mathbf{z}(t, \mathbf{R}^{\alpha-1}, \mathbf{D})\right] = W^T(t)W(t)\left\{\mathbf{w}\left[\mathbf{U}(t)\right]\right\}, \ \forall t \in \mathfrak{T}_0.$$

The last two equations imply

$$W^T(t)W(t)M_{r,m}(t)\mathbf{v}(t, \boldsymbol{\varepsilon}) = \mathbf{0}_r, \ \forall t \in \mathfrak{T}_0.$$

This reduces to

$$M_{r,m}(t)\,\mathbf{v}(t,\boldsymbol{\varepsilon}) = \mathbf{0}_r,\ \forall t \in \mathfrak{T}_0,$$

due to $\det\left[W^T(t)\,W(t)\right] \neq 0$, $\forall t \in \mathfrak{T}_0$. The rank of $M_{r,m}(t)$, $\operatorname{rank}M_{r,m}(t) \equiv m \leq r$, $\forall t \in \mathfrak{T}_0$, ensures $\det\left[M_{r,m}^T(t)\,M_{r,m}(t)\right] \neq 0$, $\forall t \in \mathfrak{T}_0$, so that

$$\mathbf{v}(t,\boldsymbol{\varepsilon}) = \mathbf{0}_m,\ \forall t \in \mathfrak{T}_0.$$

This and (8.1) imply

$$\boldsymbol{\varepsilon}(t) = \mathbf{Y}_d(t) - \mathbf{Y}(t) = \mathbf{0}_m,\ \forall t \in \mathfrak{T}_0.$$

The plant is the perfect trackable on $\mathfrak{T}_0 \times \mathfrak{D}^j \times \mathfrak{Y}_d$, Definition 140. The control is defined for $\left(\mathbf{R}_0^{\alpha-1},\mathbf{Y}_0\right) \in \mathcal{D}_{PTbl}\left(t_0;\mathfrak{D}^j;\mathbf{U};\mathfrak{Y}_d\right)$ determined by (8.16).

ii-2) The equation (8.12) can be set in the following form:

$$W(t)\,\mathbf{w}\left[\mathbf{U}(t)\right] = W(t)\,W^T(t)\,M_{r,m}(t)\,\mathbf{v}(t,\boldsymbol{\varepsilon})$$
$$+\,W(t)\,W^T(t)\left[W(t)\,W^T(t)\right]^{-1}\left[\mathbf{Y}(t) - \mathbf{z}(t,\mathbf{R}^{\alpha-1},\mathbf{D})\right],\ \forall t \in \mathfrak{T}_0,$$

i.e.,

$$W(t)\,\mathbf{w}\left[\mathbf{U}(t)\right] = W(t)\,W^T(t)\,M_{r,m}(t)\,\mathbf{v}(t,\boldsymbol{\varepsilon})$$
$$+\,\left[\mathbf{Y}(t) - \mathbf{z}(t,\mathbf{R}^{\alpha-1},\mathbf{D})\right] = \mathbf{Y}(t) - \mathbf{z}(t,\mathbf{R}^{\alpha-1},\mathbf{D}),\ \forall t \in \mathfrak{T}_0,$$

due to the output equation of (3.12). The result is:

$$W(t)\,W^T(t)\,M_{r,m}(t)\,\mathbf{v}(t,\boldsymbol{\varepsilon}) = \mathbf{0}_r,\ \forall t \in \mathfrak{T}_0.$$

This reduces to

$$\mathbf{v}(t,\boldsymbol{\varepsilon}) = \mathbf{0}_r,\ \forall t \in \mathfrak{T}_0,$$

due to $\det\left[W(t)\,W^T(t)\right] \neq 0$, $\forall t \in \mathfrak{T}_0$. This and (8.1) imply

$$\boldsymbol{\varepsilon}(t) = \mathbf{Y}_d(t) - \mathbf{Y}(t) = \mathbf{0}_m,\ \forall t \in \mathfrak{T}_0.$$

The plant is the perfect trackable on $\mathfrak{T}_0 \times \mathfrak{D}^j \times \mathfrak{Y}_d$, Definition 140. The control is defined for $\left(\mathbf{R}_0^{\alpha-1},\mathbf{Y}_0\right) \in \mathcal{D}_{PTbl}\left(t_0;\mathfrak{D}^j;\mathbf{U};\mathfrak{Y}_d\right)$ determined by (8.17).

ii-3) Another form of the equation (8.13) reads:

$$W(t)\,\mathbf{w}\left[\mathbf{U}(t)\right] = \mathbf{Y}_d(t) - \mathbf{z}(\left[t,\mathbf{R}^{\alpha-1}(t),\mathbf{D}(t)\right],\ \forall t \in \mathfrak{T}_0.$$

This and the output equation of (3.12) imply directly

$$\mathbf{Y}_d(t) = \mathbf{Y}(t),\ \forall t \in \mathfrak{T}_0,\ \text{i.e., } \boldsymbol{\varepsilon}(t) = \mathbf{0}_m,\ \forall t \in \mathfrak{T}_0.$$

The plant is the perfect trackable on $\mathfrak{T}_0 \times \mathfrak{D}^j \times \mathfrak{Y}_d$, Definition 140. The control is defined for $\left(\mathbf{R}_0^{\alpha-1},\mathbf{Y}_0\right) \in \mathcal{D}_{PTbl}\left(t_0;\mathfrak{D}^j;\mathbf{U};\mathfrak{Y}_d\right)$ determined by (8.18).

ii-4) The equation (8.14):

$$\mathbf{w}_\gamma\left[\mathbf{U}_\gamma(t)\right] = M_{\gamma,m}(t)\,\mathbf{v}(t,\boldsymbol{\varepsilon}) + W_\gamma^{-1}\left[\mathbf{Y}_\gamma - \mathbf{z}_\gamma(t,\mathbf{R}^{\alpha-1},\mathbf{D})\right],$$
$$\forall\left[t,\mathbf{D}(.),\mathbf{Y}_d(.)\right] \in \mathfrak{T}_0 \times \mathfrak{D}^j \times \mathfrak{Y}_d, \tag{C.9}$$

and the output equation of (3.12) for the first γ subequations, which reads (3.47), i.e.,

$$\mathbf{w}_\gamma\left[\mathbf{U}_\gamma(t)\right] = W_\gamma^{-1}(t)\left[\mathbf{Y}_\gamma - \mathbf{z}_\gamma(t,\mathbf{R}^{\alpha-1},\mathbf{D}\right],\ \forall t \in \mathfrak{T}_0,$$

imply

$$M_{\gamma,m}(t)\,\mathbf{v}(t,\boldsymbol{\varepsilon}) = \mathbf{0}_\gamma,\ \forall t \in \mathfrak{T}_0,$$

After multiplying this equation on the left by $M_{\gamma,m}^{T}(t)$, the result is

$$M_{\gamma,m}^{T}(t)\,M_{\gamma,m}(t)\,\mathbf{v}(t,\varepsilon)=\mathbf{0}_m,\ \forall t\in\mathfrak{T}_0.$$

The matrix on the left-hand side of this equation is square and nonsingular due to (8.15), so that the equation reduces to:

$$\mathbf{v}(t,\varepsilon)=\mathbf{0}_m,\ \forall t\in\mathfrak{T}_0.$$

This equation implies $\varepsilon(t)=\mathbf{0}_N$, $\forall t\in\mathfrak{T}_0$, in view of (8.1). The initial control vector $\mathbf{U}_0\in\mathcal{D}_{PTbl}\left(t_0;\mathfrak{D}^j;\mathbf{U};\mathfrak{Y}_d^{\alpha}\right)$ is well defined for the perfect trackable domain $\mathrm{D}_{PTbl}\left(t_0;\mathfrak{D}^j;\mathbf{U};\mathfrak{Y}_d^{\alpha}\right)$ determined by (8.19) ■

C.3 Proof of Theorem 157

Proof. *Necessity.* Let the plant (3.12) with Properties 57, 58, 78 be perfect trackable via the ID space on the product set $\mathfrak{T}_0\times\mathfrak{D}^j\times\mathfrak{Y}_d$. Definition 140 holds so that $\varepsilon(t)\equiv\mathbf{0}_N$, and $\mathbf{v}(t,\varepsilon)\equiv\mathbf{0}_m$ due to (8.1). Because the control ensures $\varepsilon(t)\equiv\mathbf{Y}_d(t)-\mathbf{Y}(t)\equiv\mathbf{0}_N$ then rank $P(t)=\beta\geq 1$.This proves the condition i).

ii-1) If $\beta=$rank $P(t)=r<\rho$, $\forall t\in\mathfrak{T}_0$, then $\det\left[P^T(t)\,P(t)\right]\neq 0$, $\forall t\in\mathfrak{T}_0$, so that the state equation of (3.12) multiplied on the left by the matrix product $\left[P^T(t)\,P(t)\right]^{-1}P^T(t)$ becomes

$$\left[P^T(t)\,P(t)\right]^{-1}P^T(t)\begin{bmatrix}Q\left[t,\mathbf{R}^{\alpha-1}(t)\right]\mathbf{R}^{(\alpha)}(t)\\+\,\mathbf{q}\left[t,\mathbf{R}^{\alpha-1}(t)\right]-E(t)\,\mathbf{e}_d\left[\mathbf{D}^{\eta}(t)\right]\end{bmatrix}$$
$$=\mathbf{e}_u\left[\mathbf{U}^{\mu}(t)\right],\forall t\in\mathfrak{T}_0.$$

Because $\beta=$rank $P(t)=r<\rho$, $\forall t\in\mathfrak{T}_0$, the plant state vector can be controlled only indirectly through an arbitrarily accepted vector function $\mathbf{v}(.):\mathfrak{T}_0\times\mathfrak{R}^N\longrightarrow\mathfrak{R}^m$, (8.1), Axiom 87. It is necessary to extend the state equation of (3.12). The validity of (8.1) permits us to subtract $L_{r,m}(t)\,\mathbf{v}(t,\varepsilon)$ from the right-hand side of the preceding equation:

$$\left[P^T(t)\,P(t)\right]^{-1}P^T(t)\begin{bmatrix}Q\left[t,\mathbf{R}^{\alpha-1}(t)\right]\mathbf{R}^{(\alpha)}(t)\\+\,\mathbf{q}\left[t,\mathbf{R}^{\alpha-1}(t)\right]-E(t)\,\mathbf{e}_d\left[\mathbf{D}^{\eta}(t)\right]\end{bmatrix}$$
$$=\mathbf{e}_u\left[\mathbf{U}^{\mu}(t)\right]-L_{r,m}(t)\,\mathbf{v}(t,\varepsilon),\ \forall t\in\mathfrak{T}_0,$$

or

$$\mathbf{e}_u\left[\mathbf{U}^{\mu}(t)\right]=\Gamma(t)\begin{bmatrix}Q\left[t,\mathbf{R}^{\alpha-1}(t)\right]\mathbf{R}^{(\alpha)}(t)\\+\,\mathbf{q}\left[t,\mathbf{R}^{\alpha-1}(t)\right]-E(t)\,\mathbf{e}_d\left[\mathbf{D}^{\eta}(t)\right]\end{bmatrix}+L_{r,m}(t)\,\mathbf{v}(t,\varepsilon),$$
$$\Gamma(t)=\left[P^T(t)\,P(t)\right]^{-1}P^T(t),\ \forall t\in\mathfrak{T}_0.$$

This equation is (8.21) that at $t=t_0$ determines (8.26) because $\mathbf{v}(t,\varepsilon)\equiv\mathbf{0}_m$, (8.1).

ii-2) If $\beta=$rank $P(t)=\rho<r$, $\forall t\in\mathfrak{T}_0$, then $\det\left[P(t)\,P^T(t)\right]\neq 0$, $\forall t\in\mathfrak{T}_0$, so that the ID equation of (3.12) yields

$$P^T(t)\left[P(t)\,P^T(t)\right]^{-1}\begin{bmatrix}Q\left[t,\mathbf{R}^{\alpha-1}(t)\right]\mathbf{R}^{(\alpha)}(t)\\+\,\mathbf{q}\left[t,\mathbf{R}^{\alpha-1}(t)\right]-E(t)\,\mathbf{e}_d\left[\mathbf{D}^{\eta}(t)\right]\end{bmatrix}$$
$$=\mathbf{e}_u\left[\mathbf{U}^{\mu}(t)\right],\forall t\in\mathfrak{T}_0.$$

The validity of (8.1) permits us to subtract $P^T(t)\,L_{\rho,m}(t)\,\mathbf{v}(t,\varepsilon)$ from the right-hand side of the preceding equation and by setting $\Gamma(t)=P^T(t)\left[P(t)\,P^T(t)\right]^{-1}$:

$$\Gamma(t)\begin{bmatrix}Q\left[t,\mathbf{R}^{\alpha-1}(t)\right]\mathbf{R}^{(\alpha)}(t)\\+\,\mathbf{q}\left[t,\mathbf{R}^{\alpha-1}(t)\right]-E(t)\,\mathbf{e}_d\left[\mathbf{D}^{\eta}(t)\right]\end{bmatrix}=\mathbf{e}_u\left[\mathbf{U}^{\mu}(t)\right]$$
$$-\,P^T(t)\,L_{\rho,m}(t)\,\mathbf{v}(t,\varepsilon),\ \Gamma(t)=P^T(t)\left[P(t)\,P^T(t)\right]^{-1},\ \forall t\in\mathfrak{T}_0.$$

i.e.,

$$\mathbf{e}_u\left[\mathbf{U}^{\mu}(t)\right]=\left\{\begin{array}{c}\Gamma(t)\left[\begin{array}{c}Q\left[t,\mathbf{R}^{\alpha-1}(t)\right]\mathbf{R}^{(\alpha)}(t)\\+\mathbf{q}\left[t,\mathbf{R}^{\alpha-1}(t)\right]-E(t)\,\mathbf{e}_d\left[\mathbf{D}^{\eta}(t)\right]\end{array}\right]\\+P^T(t)\,L_{\rho,m}\mathbf{v}(t,\varepsilon)\end{array}\right\},$$
$$\Gamma(t)=P^T(t)\left[P(t)\,P^T(t)\right]^{-1},\ \forall t\in\mathfrak{T}_0,$$

This equation is (8.22) that at $t=t_0$ determines (8.27) because $\mathbf{v}(t,\varepsilon)\equiv\mathbf{0}_m$, (8.1).

ii-3) If β =rank $P(t)=r=\rho$, $\forall t\in\mathfrak{T}_0$, then $\det P(t)\neq 0$, $\forall t\in\mathfrak{T}_0$, so that the state equation of (3.12) multiplied on the left by $P(t)^{-1}$ becomes

$$P(t)^{-1}\left[\begin{array}{c}Q\left[t,\mathbf{R}^{\alpha-1}(t)\right]\mathbf{R}^{(\alpha)}(t)\\+\mathbf{q}\left[t,\mathbf{R}^{\alpha-1}(t)\right]-E(t)\,\mathbf{e}_d\left[\mathbf{D}^{\eta}(t)\right]\end{array}\right]=\mathbf{e}_u\left[\mathbf{U}^{\mu}(t)\right],\ \forall t\in\mathfrak{T}_0.$$

Inasmuchas (8.1) is valid we may add $P(t)^{-1}L_{\rho,m}(t)\,\mathbf{v}(t,\varepsilon)$ to the left-hand side of this equation:

$$\mathbf{e}_u\left[\mathbf{U}^{\mu}(t)\right]=P(t)^{-1}\left\{\begin{array}{c}Q\left[t,\mathbf{R}^{\alpha-1}(t)\right]\mathbf{R}^{(\alpha)}(t)\\+\mathbf{q}\left[t,\mathbf{R}^{\alpha-1}(t)\right]-E(t)\,\mathbf{e}_d\left[\mathbf{D}^{\eta}(t)\right]+\\+L_{\rho,m}(t)\,\mathbf{v}(t,\varepsilon)\end{array}\right\},$$
$$\forall t\in\mathfrak{T}_0.$$

This equation is (8.23) that at $t=t_0$ determines (8.28) because $\mathbf{v}(t,\varepsilon)\equiv\mathbf{0}_m$.

ii-4) If the plant (3.12) possesses Property 61 then in view of (8.1), Lemma 62, (3.38), and Axiom 87 we can write the controllable part of the ID equation of the plant (3.12) as

$$P_\beta(t)\left\{\mathbf{e}_{u\beta}\left[\mathbf{U}_\beta^{\mu}(t)\right]-L_{\beta,m}(t)\,\mathbf{v}(t,\varepsilon)\right\}=Q_{\beta,\rho}(t,\mathbf{R}^{\alpha-1})\mathbf{R}^{(\alpha)}$$
$$+\mathbf{q}_\beta(t,\mathbf{R}^{\alpha-1},\mathbf{D}^{i})-E_{\beta,d}(t)\mathbf{e}_d\left[\mathbf{D}^{\eta}(t)\right],\ \forall t\in\mathfrak{T}. \tag{C.10}$$

The equations (3.32) and (C.10) permit us to find:

$$\mathbf{e}_{u\beta}\left[\mathbf{U}_\beta^{\mu}(t)\right]=L_{\beta,m}(t)\,\mathbf{v}(t,\varepsilon)+$$
$$+P_\beta^{-1}(t)\left\{\begin{array}{c}Q_{\beta,\rho}(t,\mathbf{R}^{\alpha-1})\mathbf{R}^{(\alpha)}+\\+\mathbf{q}_\beta(t,\mathbf{R}^{\alpha-1},\mathbf{D}^{i})-E_{\beta,d}(t)\mathbf{e}_d\left[\mathbf{D}^{\eta}(t)\right]\end{array}\right\}.$$

This is (8.24). This equation and (8.1) prove (8.29).

Sufficiency. Let all the conditions hold.

ii-1) The equation (8.21) can be set in the following form:

$$P(t)\left[P^T(t)\,P(t)\right]^{-1}P^T(t)\left[\begin{array}{c}Q\left[t,\mathbf{R}^{\alpha-1}(t)\right]\mathbf{R}^{(\alpha)}(t)\\+\mathbf{q}\left[t,\mathbf{R}^{\alpha-1}(t)\right]-E(t)\,\mathbf{e}_d\left[\mathbf{D}^{\eta}(t)\right]\end{array}\right]$$
$$=P(t)\left\{\mathbf{e}_u\left[\mathbf{U}^{\mu}(t)\right]-L_{r,m}(t)\,\mathbf{v}(t,\varepsilon)\right\},\ \forall t\in\mathfrak{T}_0,$$

or, after multiplying this equation on the left by $P^T(t)$:

$$P^T(t)\left[\begin{array}{c}Q\left[t,\mathbf{R}^{\alpha-1}(t)\right]\mathbf{R}^{(\alpha)}(t)\\+\mathbf{q}\left[t,\mathbf{R}^{\alpha-1}(t)\right]-E(t)\,\mathbf{e}_d\left[\mathbf{D}^{\eta}(t)\right]\end{array}\right]$$
$$=P^T(t)\,P(t)\left\{\mathbf{e}_u\left[\mathbf{U}^{\mu}(t)\right]-L_{r,m}(t)\,\mathbf{v}(t,\varepsilon)\right\},\ \forall t\in\mathfrak{T}_0,$$

Let us multiply on the left the ID equation of (3.12) by $P^T(t)$:

$$P^T(t)\left[\begin{array}{c}Q\left[t,\mathbf{R}^{\alpha-1}(t)\right]\mathbf{R}^{(\alpha)}(t)\\+\mathbf{q}\left[t,\mathbf{R}^{\alpha-1}(t)\right]-E(t)\,\mathbf{e}_d\left[\mathbf{D}^{\eta}(t)\right]\end{array}\right]$$
$$=P^T(t)\,P(t)\,\mathbf{e}_u\left[\mathbf{U}^{\mu}(t)\right],\ \forall t\in\mathfrak{T}_0,$$

The last two equations imply

$$P^T(t)\,P(t)\,L_{r,m}(t)\,\mathbf{v}(t,\varepsilon) = \mathbf{0}_r,\ \forall t \in \mathfrak{T}_0.$$

This reduces to

$$L_{r,m}(t)\,\mathbf{v}(t,\varepsilon) = \mathbf{0}_r,\ \forall t \in \mathfrak{T}_0,$$

due to $\det\left[P^T(t)\,P(t)\right] \neq 0$, $\forall t \in \mathfrak{T}_0$. The rank of $L_{r,m}(t)$, $\mathrm{rank}L_{r,m}(t) \equiv m \leq r$, $\forall t \in \mathfrak{T}_0$, ensures $\det\left[L_{r,m}^T(t)\,L_{r,m}(t)\right] \neq 0$, $\forall t \in \mathfrak{T}_0$, so that

$$\mathbf{v}(t,\varepsilon) = \mathbf{0}_m,\ \forall t \in \mathfrak{T}_0.$$

This and (8.1) imply

$$\varepsilon(t) = \mathbf{Y}_d(t) - \mathbf{Y}(t) = \mathbf{0}_m,\ \forall t \in \mathfrak{T}_0.$$

The plant is the perfect trackable on $\mathfrak{T}_0 \times \mathfrak{D}^j \times \mathfrak{Y}_d$, Definition 140. The control is defined for $\left(\mathbf{R}_0^{\alpha-1}, \mathbf{Y}_0\right) \in \mathcal{D}_{PTbl}\left(t_0; \mathfrak{D}^j; \mathbf{U}; \mathfrak{Y}_d\right)$ determined by (8.26).

ii-2) The equation (8.22) can be set in the following form:

$$\begin{gathered} P(t)\,\mathbf{e}_u\left[\mathbf{U}^\mu(t)\right] - P(t)\,P^T(t)\,L_{\rho,m}(t)\,\mathbf{v}(t,\varepsilon) \\ = P(t)\,P^T(t)\left[P(t)\,P^T(t)\right]^{-1} \\ \bullet \begin{bmatrix} Q\left[t,\mathbf{R}^{\alpha-1}(t)\right]\mathbf{R}^{(\alpha)}(t) \\ +\,\mathbf{q}\left[t,\mathbf{R}^{\alpha-1}(t)\right] - E(t)\,\mathbf{e}_d\left[\mathbf{D}^\eta(t)\right] \end{bmatrix},\ \forall t \in \mathfrak{T}_0, \end{gathered}$$

i.e.,

$$\begin{gathered} P(t)\,\mathbf{e}_u\left[\mathbf{U}^\mu(t)\right] = P(t)\,P^T(t)\,L_{\rho,m}(t)\,\mathbf{v}(t,\varepsilon) \\ + \begin{bmatrix} Q\left[t,\mathbf{R}^{\alpha-1}(t)\right]\mathbf{R}^{(\alpha)}(t) \\ +\,\mathbf{q}\left[t,\mathbf{R}^{\alpha-1}(t)\right] - E(t)\,\mathbf{e}_d\left[\mathbf{D}^\eta(t)\right] \end{bmatrix},\ \forall t \in \mathfrak{T}_0. \end{gathered}$$

This and the state equation of (3.12) result in:

$$P(t)\,P^T(t)\,L_{\rho,m}(t)\,\mathbf{v}(t,\varepsilon) = \mathbf{0}_r,\ \forall t \in \mathfrak{T}_0.$$

This reduces to

$$L_{r,m}(t)\,\mathbf{v}(t,\varepsilon) = \mathbf{0}_r,\ \forall t \in \mathfrak{T}_0,$$

due to $\det\left[P(t)\,P^T(t)\right] \neq 0$, $\forall t \in \mathfrak{T}_0$. The rank of $L_{r,m}(t)$, $\mathrm{rank}L_{r,m}(t) \equiv m \leq r$, $\forall t \in \mathfrak{T}_0$, ensures $\det\left[L_{r,m}^T(t)\,L_{r,m}(t)\right] \neq 0$, $\forall t \in \mathfrak{T}_0$, so that

$$\mathbf{v}(t,\varepsilon) = \mathbf{0}_m,\ \forall t \in \mathfrak{T}_0.$$

This and (8.1) imply

$$\varepsilon(t) = \mathbf{Y}_d(t) - \mathbf{Y}(t) = \mathbf{0}_m,\ \forall t \in \mathfrak{T}_0.$$

The plant is the perfect trackable on $\mathfrak{T}_0 \times \mathfrak{D}^j \times \mathfrak{Y}_d$, Definition 140. The control is defined for $\left(\mathbf{R}_0^{\alpha-1}, \mathbf{Y}_0\right) \in \mathcal{D}_{PTbl}\left(t_0; \mathfrak{D}^j; \mathbf{U}; \mathfrak{Y}_d\right)$ determined by (8.27).

ii-3) Another form of the equation (8.23) reads:

$$P(t)\,\mathbf{e}_u\left[\mathbf{U}^\mu(t)\right] = \left\{ \begin{array}{c} Q\left[t,\mathbf{R}^{\alpha-1}(t)\right]\mathbf{R}^{(\alpha)}(t) \\ +\,\mathbf{q}\left[t,\mathbf{R}^{\alpha-1}(t)\right] - E(t)\,\mathbf{e}_d\left[\mathbf{D}^\eta(t)\right] + L_{\rho,m}(t)\,\mathbf{v}(t,\varepsilon) \end{array} \right\},$$
$$\forall t \in \mathfrak{T}_0.$$

This and the ID equation of (3.12) imply directly

$$L_{\rho,m}(t)\,\mathbf{v}(t,\varepsilon) = \mathbf{0}_\rho,\ \forall t \in \mathfrak{T}_0.$$

The rank of $L_{\rho,m}(t)$, rank$L_{\rho,m}(t) \equiv m \leq \rho$, $\forall t \in \mathfrak{T}_0$, ensures

$$\det\left[L^T_{\rho,m}(t)\,L_{\rho,m}(t)\right] \neq 0,\ \forall t \in \mathfrak{T}_0,$$

so that

$$\mathbf{v}(t,\varepsilon) = \mathbf{0}_m,\quad \forall t \in \mathfrak{T}_0.$$

This and (8.1) imply

$$\varepsilon(t) = \mathbf{Y}_d(t) - \mathbf{Y}(t) = \mathbf{0}_m,\ \forall t \in \mathfrak{T}_0.$$

The plant is the perfect trackable on $\mathfrak{T}_0 \times \mathfrak{D}^j \times \mathfrak{Y}_d$, Definition 140. The control is defined for $\left(\mathbf{R}_0^{\alpha-1}, \mathbf{Y}_0\right) \in \mathcal{D}_{PTbl}\left(t_0;\mathfrak{D}^j;\mathbf{U};\mathfrak{Y}_d\right)$ determined by (8.28).

ii-4) Let the plant (3.12) also possess Property 61. We multiply (8.24) on the left by $P_\beta(t)$ so that:

$$\begin{gathered} P_\beta(t)\,\mathbf{e}_{u\beta}\left[\mathbf{U}^\mu_\beta(t)\right] = P_\beta(t)\,L_{\beta,m}(t)\,\mathbf{v}(t,\varepsilon) \\ +\left[\begin{array}{c} Q_{\beta,\rho}(t,\mathbf{R}^{\alpha-1})\mathbf{R}^{(\alpha)} \\ +\ \mathbf{q}_\beta(t,\mathbf{R}^{\alpha-1},\mathbf{D}^i) - E_{\beta,d}(t)\mathbf{e}_d\left[\mathbf{D}^\eta(t)\right] \end{array}\right], \\ \forall\left[t,\mathbf{D}(.),\mathbf{Y}_d(.)\right] \in \mathfrak{T}_0 \times \mathfrak{D}^j \times \mathfrak{Y}_d, \end{gathered} \tag{C.11}$$

The first β equations of the ID equations of (3.12) read, due to Lemma 62, (3.38):

$$\begin{gathered} P_\beta(t)\,\mathbf{e}_{u\beta}\left[\mathbf{U}^\mu_\beta(t)\right] = Q_{\beta,\rho}(t,\mathbf{R}^{\alpha-1})\mathbf{R}^{(\alpha)} \\ +\ \mathbf{q}_\beta(t,\mathbf{R}^{\alpha-1},\mathbf{D}^i) - E_{\beta,d}(t)\mathbf{e}_d\left[\mathbf{D}^\eta(t)\right], \\ \forall\left[t,\mathbf{D}(.),\mathbf{Y}_d(.)\right] \in \mathfrak{T}_0 \times \mathfrak{D}^j \times \mathfrak{Y}_d. \end{gathered}$$

The preceding two equations confirm:

$$P_\beta(t)\,L_{\beta,m}(t)\,\mathbf{v}(t,\varepsilon) = \mathbf{0}_\beta,\ \forall t \in \mathfrak{T}_0,$$

or equivalently

$$L_{\beta,m}(t)\,\mathbf{v}(t,\varepsilon) = \mathbf{0}_\beta,\ \forall t \in \mathfrak{T}_0, \tag{C.12}$$

due to rank$P_\beta(t) = m \leq \beta$, $\forall t \in \mathfrak{T}_0$, (8.20). Because $\beta \geq m$, (8.25), and rank$M_{\beta,m}(t) \equiv m$, (8.25), $L^T_{\beta,m}(t)\,L_{\beta,m}(t)$ is a square nonsingular matrix on $\mathfrak{T}_0$. We multiply (C.12) on the left by $\left[L^T_{\beta,m}(t)\,L_{\beta,m}(t)\right]^{-1} L^T_{\beta,m}(t)$. The result is

$$\mathbf{v}(t,\varepsilon) = \mathbf{0}_m,\ \forall t \in \mathfrak{T}_0.$$

This equation and (8.1) prove

$$\varepsilon(t) = \varepsilon(t;\mathbf{D};\mathbf{Y}_d) = \mathbf{0}_N,\ \forall\left[t,\mathbf{D}(.),\mathbf{Y}_d(.)\right] \in \mathfrak{T}_0 \times \mathfrak{D}^j \times \mathfrak{Y}_d.$$

The plant exhibits perfect tracking on $\mathfrak{T}_0 \times \mathfrak{D}^j \times \mathfrak{Y}_d$, Definition 140. The control $\mathbf{U}_0$ is well defined for every $(\mathbf{R}_0, \mathbf{Y}_0) \in \mathcal{D}_{PTbl}\left(t_0;\mathfrak{D}^j;\mathbf{U};\mathfrak{Y}^\alpha_d\right)$, (8.29) ■

C.4 Proof of Theorem 160

Proof. *Necessity.* Let the plant (3.12) with Properties 57, 58, 61, 64, 78 and with (8.31) be perfect trackable on $\mathfrak{T}_0 \times \mathfrak{D}^j \times \mathfrak{Y}_d$ via both spaces. There exists a control vector function

$\mathbf{U}(.)$ that satisfies (3.12) and for which $\mathbf{Y}(t) = \mathbf{Y}_d(t)$, i.e., $\boldsymbol{\varepsilon}(t) = \mathbf{0}_N$, $\forall t \in \mathfrak{T}_0$, Definition 136:

$$Q\left[t, \mathbf{R}^{\alpha-1}(t)\right] \mathbf{R}^{(\alpha)}(t) + \mathbf{q}\left[t, \mathbf{R}^{\alpha-1}(t)\right] = E(t)\, \mathbf{e}_d\left[\mathbf{D}^{\eta}(t)\right] + P(t)\, \mathbf{e}_u\left[\mathbf{U}^{\mu}(t)\right]$$
$$\mathbf{Y}(t) = \mathbf{z}\left[t, \mathbf{R}^{\alpha-1}(t), \mathbf{D}(t)\right] + W(t)\, \mathbf{w}\left[\mathbf{U}(t)\right],\ \forall t \in \mathfrak{T}_0,$$

i.e., due to the perfect trackability on $\mathfrak{T}_0 \times \mathfrak{D}^j \times \mathfrak{Y}_d$, so that $\mathbf{v}(t, \varepsilon) = \mathbf{0}_m$, $\forall t \in \mathfrak{T}_0$, in view of $\boldsymbol{\varepsilon}(t) = \mathbf{0}_N$, $\forall t \in \mathfrak{T}_0$, (8.1). This means that the control acts on both the internal dynamics behavior and the output behavior of the plant, which is possible only if (8.31) is true. The condition i) is valid. We can set the preceding equations in the following form:

$$P(t)\, \mathbf{e}_u\left[\mathbf{U}^{\mu}(t)\right] = Q\left[t, \mathbf{R}^{\alpha-1}(t)\right] \mathbf{R}^{(\alpha)}(t) + \mathbf{q}\left[t, \mathbf{R}^{\alpha-1}(t)\right] - E(t)\, \mathbf{e}_d\left[\mathbf{D}^{\eta}(t)\right]$$
$$W(t)\, \mathbf{w}\left[\mathbf{U}(t)\right] = \mathbf{Y}(t) - \mathbf{z}\left[t, \mathbf{R}^{\alpha-1}(t), \mathbf{D}(t)\right],\ \forall t \in \mathfrak{T}_0,$$

and in the joint form:

$$\begin{bmatrix} P(t) & O_{\rho,r} \\ O_{N,r} & W(t) \end{bmatrix} \begin{bmatrix} \mathbf{e}_u\left[\mathbf{U}^{\mu}(t)\right] \\ \mathbf{w}\left[\mathbf{U}(t)\right] \end{bmatrix}$$
$$= \begin{bmatrix} Q\left[t, \mathbf{R}^{\alpha-1}(t)\right] & O_{\rho,N} \\ O_{N,\rho} & O_{N,N} \end{bmatrix} \begin{bmatrix} \mathbf{R}^{(\alpha)}(t) \\ \mathbf{0}_N \end{bmatrix}$$
$$+ \begin{bmatrix} I_\rho & O_{\rho,N} \\ O_{N.\rho} & I_N \end{bmatrix} \begin{bmatrix} \mathbf{q}\left[t, \mathbf{R}^{\alpha-1}(t)\right] - E(t)\, \mathbf{e}_d\left[\mathbf{D}^{\eta}(t)\right] \\ \mathbf{Y}(t) - \mathbf{z}\left[t, \mathbf{R}^{\alpha-1}(t), \mathbf{D}(t)\right] \end{bmatrix}$$
$$\forall t \in \mathfrak{T}_0,$$

or, by applying (3.32) and (3.42) from Properties 61 and 64, respectively:

$$\begin{bmatrix} \begin{matrix} P_\beta(t) & O_{\beta,r-\beta} \\ O_{\rho-\beta,\beta} & O_{\rho-\beta,r-\beta} \end{matrix} & O_{\rho,r} \\ O_{N,r} & \begin{matrix} W_\gamma(t) & O_{\gamma,r-\gamma} \\ O_{N-\gamma,\gamma} & O_{N-\gamma,r-\gamma} \end{matrix} \end{bmatrix} \begin{bmatrix} \begin{bmatrix} \mathbf{e}_{u\beta}\left[\mathbf{U}^{\mu}_{\beta}(t)\right] \\ \mathbf{e}_{u,r-\beta}\left[\mathbf{U}^{\mu}_{r-\beta}(t)\right] \end{bmatrix} \\ \begin{bmatrix} \mathbf{w}_\gamma\left[\mathbf{U}_\gamma(t)\right] \\ \mathbf{w}_{r-\gamma}\left[\mathbf{U}_{r-\gamma}(t)\right] \end{bmatrix} \end{bmatrix},$$

i.e.,

$$= \begin{bmatrix} \begin{matrix} Q_{\beta,\rho}\left[t, \mathbf{R}^{\alpha-1}(t)\right] \\ Q_{\rho-\beta,\rho}\left[t, \mathbf{R}^{\alpha-1}(t)\right] \end{matrix} & O_{\rho,N} \\ O_{N,\rho} & O_N \end{bmatrix} \begin{bmatrix} \mathbf{R}^{(\alpha)}(t) \\ \mathbf{0}_N \end{bmatrix}$$

$$+ \begin{bmatrix} \begin{matrix} I_\beta & O_{\beta,\rho-\beta} \\ O_{\rho-\beta,\beta} & I_{\rho-\beta} \end{matrix} & O_{\rho,N} \\ O_{N.\rho} & \begin{matrix} I_\gamma & O_{\gamma,N-\gamma} \\ O_{N-\gamma,\gamma} & I_{N-\gamma} \end{matrix} \end{bmatrix}$$
$$\bullet \begin{bmatrix} \begin{bmatrix} \mathbf{q}_\beta\left[t, \mathbf{R}^{\alpha-1}(t)\right] - E_{\beta,d}(t)\mathbf{e}_d\left[\mathbf{D}^{\eta}(t)\right] \\ \mathbf{q}_{\rho-\beta}\left[t, \mathbf{R}^{\alpha-1}(t)\right] - \left[E_{\rho-\beta,d}(t)\right] \mathbf{e}_d\left[\mathbf{D}^{\eta}(t)\right] \end{bmatrix} \\ \begin{bmatrix} \mathbf{Y}_\gamma(t) - \mathbf{z}_\gamma\left[t, \mathbf{R}^{\alpha-1}(t), \mathbf{D}(t)\right] \\ \mathbf{Y}_{N-\gamma}(t) - \mathbf{z}_{N-\gamma}\left[t, \mathbf{R}^{\alpha-1}(t), \mathbf{D}(t)\right] \end{bmatrix} \end{bmatrix}$$
$$\forall t \in \mathfrak{T}_0 \Longrightarrow$$

$$\begin{bmatrix} \begin{bmatrix} P_\beta(t)\, \mathbf{e}_{u\beta}\left[\mathbf{U}^{\mu}_{\beta}(t)\right] \\ \mathbf{0}_{\rho-\beta} \end{bmatrix} \\ \begin{bmatrix} W_\gamma(t)\, \mathbf{w}_\gamma\left[\mathbf{U}_\gamma(t)\right] \\ \mathbf{0}_{N-\gamma} \end{bmatrix} \end{bmatrix}$$

$$= \left[\begin{array}{c} \left[\begin{array}{c} Q_{\beta,\rho}\left[t,\mathbf{R}^{\alpha-1}(t)\right]\mathbf{R}^{(\alpha)}(t) \\ Q_{\rho-\beta,\rho}\left[t,\mathbf{R}^{\alpha-1}(t)\right]\mathbf{R}^{(\alpha)}(t) \end{array} \right] \\ \mathbf{0}_N \end{array} \right]$$

$$+ \left[\begin{array}{c} \left[\begin{array}{c} \mathbf{q}_{\beta}\left[t,\mathbf{R}^{\alpha-1}(t)\right] - E_{\beta,d}(t)\mathbf{e}_d\left[\mathbf{D}^{\eta}(t)\right] \\ \mathbf{q}_{\rho-\beta}\left[t,\mathbf{R}^{\alpha-1}(t)\right] - \left[E_{\rho-\beta,d}(t)\right]\mathbf{e}_d\left[\mathbf{D}^{\eta}(t)\right] \end{array} \right] \\ \left[\begin{array}{c} \mathbf{Y}_{\gamma}(t) - \mathbf{z}_{\gamma}\left[t,\mathbf{R}^{\alpha-1}(t),\mathbf{D}(t)\right] \\ \mathbf{Y}_{N-\gamma}(t) - \mathbf{z}_{N-\gamma}\left[t,\mathbf{R}^{\alpha-1}(t),\mathbf{D}(t)\right] \end{array} \right] \end{array} \right]$$

$$\forall t \in \mathfrak{T}_0 \Longrightarrow$$

$$\left[\begin{array}{cc} P_{\beta}(t) & O_{\beta,\gamma} \\ O_{\gamma,\beta} & W_{\gamma}(t) \end{array} \right] \left[\begin{array}{c} \mathbf{e}_{u\beta}\left[\mathbf{U}^{\mu}_{\beta}(t)\right] \\ \mathbf{w}_{\gamma}\left[\mathbf{U}_{\gamma}(t)\right] \end{array} \right]$$
$$= \left[\begin{array}{c} \left\{ \begin{array}{c} Q_{\beta,\rho}\left[t,\mathbf{R}^{\alpha-1}(t)\right]\mathbf{R}^{(\alpha)}(t) + \\ \mathbf{q}_{\beta}\left[t,\mathbf{R}^{\alpha-1}(t)\right] - E_{\beta,d}(t)\mathbf{e}_d\left[\mathbf{D}^{\eta}(t)\right] \end{array} \right\} \\ \mathbf{Y}_{\gamma}(t) - \mathbf{z}_{\gamma}\left[t,\mathbf{R}^{\alpha-1}(t),\mathbf{D}(t)\right] \end{array} \right], \tag{C.13}$$

and

$$\left[\begin{array}{c} \mathbf{0}_{\rho-\beta} \\ \mathbf{0}_{N-\gamma} \end{array} \right]$$
$$= \left[\begin{array}{c} \left\{ \begin{array}{c} Q_{\rho-\beta,\rho}\left[t,\mathbf{R}^{\alpha-1}(t)\right]\mathbf{R}^{(\alpha)}(t) + \\ \mathbf{q}_{\rho-\beta}\left[t,\mathbf{R}^{\alpha-1}(t)\right] - E_{\rho-\beta,d}(t)\mathbf{e}_d\left[\mathbf{D}^{\eta}(t)\right] \end{array} \right\} \\ \mathbf{Y}_{N-\gamma}(t) - \mathbf{z}_{N-\gamma}\left[t,\mathbf{R}^{\alpha-1}(t),\mathbf{D}(t)\right] \end{array} \right].$$

Only the first equation of the last two equations contains the control vector $\mathbf{U}(t)$. We may write it in the following form due to (8.1), (8.30) and Axiom 87:

$$\left[\begin{array}{cc} P_{\beta}(t) & O_{\beta,\gamma} \\ O_{\gamma,\beta} & W_{\gamma}(t) \end{array} \right] \left\{ \left[\begin{array}{c} \mathbf{e}_{u\beta}\left[\mathbf{U}^{\mu}_{\beta}(t)\right] \\ \mathbf{w}_{\gamma}\left[\mathbf{U}_{\gamma}(t)\right] \end{array} \right] + H(t)\,\mathbf{v}(t,\varepsilon) \right\}$$
$$= \left[\begin{array}{c} \left\{ \begin{array}{c} Q_{\beta,\rho}\left[t,\mathbf{R}^{\alpha-1}(t)\right]\mathbf{R}^{(\alpha)}(t) \\ + \mathbf{q}_{\beta}\left[t,\mathbf{R}^{\alpha-1}(t)\right] - E_{\beta,d}(t)\mathbf{e}_d\left[\mathbf{D}^{\eta}(t)\right] \end{array} \right\} \\ \mathbf{Y}_{\gamma}(t) - \mathbf{z}_{\gamma}\left[t,\mathbf{R}^{\alpha-1}(t),\mathbf{D}(t)\right] \end{array} \right],$$

or

$$\left[\begin{array}{c} \mathbf{e}_{u\beta}\left[\mathbf{U}^{\mu}_{\beta}(t)\right] \\ \mathbf{w}_{\gamma}\left[\mathbf{U}_{\gamma}(t)\right] \end{array} \right]$$
$$= \left[\begin{array}{c} H_{\beta,m}(t)\,\mathbf{v}(t,\varepsilon) \\ H_{\gamma,m}(t)\,\mathbf{v}(t,\varepsilon) \end{array} \right] + \left[\begin{array}{cc} P^{-1}_{\beta}(t) & O_{\beta,\gamma} \\ O_{\gamma,\beta} & W^{-1}_{\gamma}(t) \end{array} \right]$$
$$\bullet \left[\begin{array}{c} \left\{ \begin{array}{c} Q_{\beta,\rho}\left[t,\mathbf{R}^{\alpha-1}(t)\right]\mathbf{R}^{(\alpha)}(t) \\ + \mathbf{q}_{\beta}\left[t,\mathbf{R}^{\alpha-1}(t)\right] - E_{\beta,d}(t)\mathbf{e}_d\left[\mathbf{D}^{\eta}(t)\right] \end{array} \right\} \\ \mathbf{Y}_{\gamma}(t) - \mathbf{z}_{\gamma}\left[t,\mathbf{R}^{\alpha-1}(t),\mathbf{D}(t)\right] \end{array} \right],$$

This is (8.32). It implies (8.33) due to $\mathbf{v}(t_0,\varepsilon_0) = \mathbf{0}_m$, (8.1), which comes from $\varepsilon_0 = \mathbf{0}_N$.

Sufficiency. Let all the conditions of the theorem statement be valid. The equivalent form of the equation (8.32) is:

$$\left[\begin{array}{cc} P_\beta(t) & O_{\beta,\gamma} \\ O_{\gamma,\beta} & W_\gamma(t) \end{array}\right]\left(\left[\begin{array}{c} \mathbf{e}_{u\beta}\left[\mathbf{U}^{\mu}_{\beta}(t)\right] \\ \mathbf{w}_\gamma\left[\mathbf{U}_\gamma(t)\right] \end{array}\right]+\left[\begin{array}{c} H_{\beta,m}(t)\,\mathbf{v}(t,\varepsilon) \\ H_{\gamma,m}(t)\,\mathbf{v}(t,\varepsilon) \end{array}\right]\right)$$
$$=\left[\begin{array}{c} \left\{\begin{array}{c} Q_{\beta,\rho}\left[t,\mathbf{R}^{\alpha-1}(t)\right]\mathbf{R}^{(\alpha)}(t) \\ +\,\mathbf{q}_\beta\left[t,\mathbf{R}^{\alpha-1}(t)\right]-E_{\beta,d}(t)\mathbf{e}_d\left[\mathbf{D}^{\eta}(t)\right] \end{array}\right\} \\ \mathbf{Y}_\gamma(t)-\mathbf{z}_\gamma\left[t,\mathbf{R}^{\alpha-1}(t),\mathbf{D}(t)\right] \end{array}\right],$$
$$\forall\left[t,\mathbf{D}(.),\mathbf{Y}_d(.)\right]\in\mathfrak{T}_0\times\mathfrak{D}^j\times\mathfrak{Y}_d, \tag{C.14}$$

Only the equations (C.13) of (3.12) contain the control vector $\mathbf{U}(t)$, thus we subtract them from (C.14). The result is

$$\left[\begin{array}{cc} P_\beta(t) & O_{\beta,\gamma} \\ O_{\gamma,\beta} & W_\gamma(t) \end{array}\right]\left[\begin{array}{c} H_{\beta,m}(t)\,\mathbf{v}(t,\varepsilon) \\ H_{\gamma,m}(t)\,\mathbf{v}(t,\varepsilon) \end{array}\right]\equiv\left[\begin{array}{c} \mathbf{0}_\beta \\ \mathbf{0}_\gamma \end{array}\right].$$

The multiplication of this equation on the left by

$$\left[\begin{array}{cc} P_\beta(t) & O_{\beta,\gamma} \\ O_{\gamma,\beta} & W_\gamma(t) \end{array}\right]^{-1}$$

reduces it to

$$\left[\begin{array}{c} H_{\beta,m}(t) \\ H_{\gamma,m}(t) \end{array}\right]\mathbf{v}(t,\varepsilon)\equiv\left[\begin{array}{c} \mathbf{0}_\beta \\ \mathbf{0}_\gamma \end{array}\right].$$

We multiply this on the left by

$$\left[\left\{\left[H^T_{\beta,m}(t)\quad H^T_{\gamma,m}(t)\right]\left[\begin{array}{c} H_{\beta,m}(t) \\ H_{\gamma,m}(t) \end{array}\right]\right\}\right]^{-1}\left[H^T_{\beta,m}(t)\quad H^T_{\gamma,m}(t)\right],$$

which is permitted due to (8.30) and get $\mathbf{v}(t,\varepsilon)\equiv\mathbf{0}_m$. This and (8.1) prove $\varepsilon(t)\equiv\mathbf{0}_N$. The plant (3.12) with Properties 57, 58, 61, 64, 78 is perfect trackable on $\mathfrak{T}_0\times\mathfrak{D}^j\times\mathfrak{Y}_d$. Definition 136 is satisfied. It and and (8.32) show that the domain $\mathcal{D}_{PTbl}\left(t_0;\mathfrak{D}^j;\mathbf{U};\mathfrak{Y}_d\right)$ of the perfect trackability on $\mathfrak{T}_0\times\mathfrak{D}^j\times\mathfrak{Y}_d$ of the plant (3.12) with Properties 57, 58, 61, 64, 78 is determined by (8.33) ■

C.5 Proof of Theorem 162

Proof. *Necessity.* Let the plant (3.12) with Properties 57, 58, 61, 64, 78, with differentiable $\mathbf{z}\left[t,\mathbf{R}^{\alpha-1}(t),\mathbf{D}(t)\right]$ and with (8.31), be perfect trackable on $\mathfrak{T}_0\times\mathfrak{D}^j\times\mathfrak{Y}_d$ via both spaces. There exists a control vector function $\mathbf{U}(.)$ that satisfies (3.12) and for which $\mathbf{Y}(t)=\mathbf{Y}_d(t)$, i.e., $\varepsilon(t)=\mathbf{0}_N$, $\forall t\in\mathfrak{T}_0$, Definition 136. We recall (3.12).

$$Q\left[t,\mathbf{R}^{\alpha-1}(t)\right]\mathbf{R}^{(\alpha)}(t)+\mathbf{q}\left[t,\mathbf{R}^{\alpha-1}(t)\right]=E(t)\,\mathbf{e}_d\left[\mathbf{D}^{\eta}(t)\right]+P(t)\,\mathbf{e}_u\left[\mathbf{U}^{\mu}(t)\right],$$
$$\mathbf{Y}(t)=\mathbf{z}\left[t,\mathbf{R}^{\alpha-1}(t),\mathbf{D}(t)\right]+W(t)\,\mathbf{w}\left[\mathbf{U}(t)\right].$$

Let (8.34) be valid:

$$\mathbf{q}^*\left[t,\mathbf{R}^{\alpha-1}(t)\right]=Q^{-1}\left[t,\mathbf{R}^{\alpha-1}(t)\right]\mathbf{q}\left[t,\mathbf{R}^{\alpha-1}(t)\right],$$
$$E^*(t)=Q^{-1}\left[t,\mathbf{R}^{\alpha-1}(t)\right]E(t),\;\; P^*(t)=Q^{-1}\left[t,\mathbf{R}^{\alpha-1}(t)\right]P(t).$$

These equations permit to set (3.12) into

$$\mathbf{R}^{(\alpha)}(t)=-\mathbf{q}^*\left[t,\mathbf{R}^{\alpha-1}(t)\right]+E^*(t)\,\mathbf{e}_d\left[\mathbf{D}^{\eta}(t)\right]+P^*(t)\,\mathbf{e}_u\left[\mathbf{U}^{\mu}(t)\right],$$
$$W(t)\,\mathbf{w}\left[\mathbf{U}(t)\right]=\mathbf{Y}(t)-\mathbf{z}\left[t,\mathbf{R}^{\alpha-1}(t),\mathbf{D}(t)\right].$$

The derivative of the last equation reads:

$$W(t)\,\mathbf{w}^{(1)}[\mathbf{U}(t)] + W^{(1)}(t)\,\mathbf{w}[\mathbf{U}(t)] = \mathbf{Y}^{(1)}(t) - \mathbf{z}^{(1)}\left[t, \mathbf{R}^{\alpha-1}(t), \mathbf{D}(t)\right],$$

where

$$\begin{gathered}\mathbf{z}^{(1)}\left[t, \mathbf{R}^{\alpha-1}(t), \mathbf{D}(t)\right] = \frac{\partial \mathbf{z}\left[t, \mathbf{R}^{\alpha-1}(t), \mathbf{D}(t)\right]}{\partial t} \\ + \frac{\partial \mathbf{z}\left[t, \mathbf{R}^{\alpha-1}(t), \mathbf{D}(t)\right]}{\partial \mathbf{R}^{\alpha-2}} \mathbf{R}^{\alpha-1} \\ + \frac{\partial \mathbf{z}\left[t, \mathbf{R}^{\alpha-1}(t), \mathbf{D}(t)\right]}{\partial \mathbf{R}^{(\alpha-1)}} \left\{-\mathbf{q}^*\left[t, \mathbf{R}^{\alpha-1}(t)\right] + E^*(t)\,\mathbf{e}_d\left[\mathbf{D}^{\eta}(t)\right]\right\} \\ + \frac{\partial \mathbf{z}\left[t, \mathbf{R}^{\alpha-1}(t), \mathbf{D}(t)\right]}{\partial \mathbf{R}^{(\alpha-1)}} P^*(t)\,\mathbf{e}_u\left[\mathbf{U}^{\mu}(t)\right] + \frac{\partial \mathbf{z}\left[t, \mathbf{R}^{\alpha-1}(t), \mathbf{D}(t)\right]}{\partial \mathbf{D}} \mathbf{D}^{(1)}(t) \Longrightarrow\end{gathered}$$

$$\begin{gathered}\frac{\partial \mathbf{z}\left[t, \mathbf{R}^{\alpha-1}(t), \mathbf{D}(t)\right]}{\partial \mathbf{R}^{(\alpha-1)}} P^*(t)\,\mathbf{e}_u\left[\mathbf{U}^{\mu}(t)\right] + W(t)\,\mathbf{w}^{(1)}[\mathbf{U}(t)] + W^{(1)}(t)\,\mathbf{w}[\mathbf{U}(t)] \\ = \mathbf{Y}^{(1)}(t) - \frac{\partial \mathbf{z}\left[t, \mathbf{R}^{\alpha-1}(t), \mathbf{D}(t)\right]}{\partial t} \\ - \frac{\partial \mathbf{z}\left[t, \mathbf{R}^{\alpha-1}(t), \mathbf{D}(t)\right]}{\partial \mathbf{R}^{(\alpha-1)}} \left\{-\mathbf{q}^*\left[t, \mathbf{R}^{\alpha-1}(t)\right] + E^*(t)\,\mathbf{e}_d\left[\mathbf{D}^{\eta}(t)\right]\right\} \\ - \frac{\partial \mathbf{z}\left[t, \mathbf{R}^{\alpha-1}(t), \mathbf{D}(t)\right]}{\partial \mathbf{D}} \mathbf{D}^{(1)}(t) = \Omega\left[t, \mathbf{R}^{\alpha-1}(t), \mathbf{D}(t)\right],\end{gathered}$$

where $A(t)$ and $\Omega\left[t, \mathbf{R}^{\alpha-1}(t), \mathbf{D}(t)\right]$ are determined by (8.34) and (8.35), which read, respectively:

$$A(t) = \frac{\partial \mathbf{z}\left[t, \mathbf{R}^{\alpha-1}(t), \mathbf{D}(t)\right]}{\partial \mathbf{R}^{(\alpha-1)}} P^*(t)$$

$$\Omega\left[t, \mathbf{R}^{\alpha-1}(t), \mathbf{D}(t)\right] = \left\{ \begin{array}{c} \mathbf{Y}^{(1)}(t) - \frac{\partial \mathbf{z}\left[t, \mathbf{R}^{\alpha-1}(t), \mathbf{D}(t)\right]}{\partial t} - \frac{\partial \mathbf{z}\left[t, \mathbf{R}^{\alpha-1}(t), \mathbf{D}(t)\right]}{\partial \mathbf{R}^{(\alpha-1)}} \\ \bullet \left\{-\mathbf{q}^*\left[t, \mathbf{R}^{\alpha-1}(t)\right] + E^*(t)\,\mathbf{e}_d\left[\mathbf{D}^{\eta}(t)\right]\right\} \\ - \frac{\partial \mathbf{z}\left[t, \mathbf{R}^{\alpha-1}(t), \mathbf{D}(t)\right]}{\partial \mathbf{D}} \mathbf{D}^{(1)}(t) \end{array} \right\},$$

so that

$$\begin{gathered}A(t)\,\mathbf{e}_u\left[\mathbf{U}^{\mu}(t)\right] + W(t)\,\mathbf{w}^{(1)}[\mathbf{U}(t)] + W^{(1)}(t)\,\mathbf{w}[\mathbf{U}(t)] \\ = \Omega\left[t, \mathbf{R}^{\alpha-1}(t), \mathbf{D}(t)\right]\end{gathered} \tag{C.15}$$

and, due to (8.37), $\varepsilon(t) = \mathbf{0}_N$, $\forall t \in \mathfrak{T}_0$, (8.1), and Axiom 87:

$$\begin{gathered}A(t)\,\mathbf{e}_u\left[\mathbf{U}^{\mu}(t)\right] + W(t)\,\mathbf{w}^{(1)}[\mathbf{U}(t)] + W^{(1)}(t)\,\mathbf{w}[\mathbf{U}(t)] \\ = \Omega\left[t, \mathbf{R}^{\alpha-1}(t), \mathbf{D}(t)\right] + B(t)\,\mathbf{v}(t, \varepsilon).\end{gathered} \tag{C.16}$$

This is (8.36).

Sufficiency. The equation (3.12) with differentiable $\mathbf{z}(.)$ can be set in the form (C.15). It and (8.36) imply $B(t)\,\mathbf{v}(t, \varepsilon) \equiv \mathbf{0}_N$. We multiply this equation on the left by

$$\left[B^T(t)\,B(t)\right]^{-1} B^T(t),$$

which is possible due to (8.37). The result is $\mathbf{v}(t, \varepsilon) \equiv \mathbf{0}_m$, which, together with (8.1), implies $\varepsilon(t) \equiv \mathbf{0}_N$. Definition 140 is satisfied. The plant (3.12) with Properties 57, 58, 78 and with differentiable $\mathbf{z}(.)$ is perfect trackable on $\mathfrak{T}_0 \times \mathfrak{D}^j \times \mathfrak{Y}_d$ The equation (8.36) and the constraints on the nonlinearities $\mathbf{e}_{u\beta}(.)$ and $\mathbf{w}_{\gamma}(.)$ determine the domain $\mathcal{D}_{PTbl}\left(t_0; \mathfrak{D}^j; \mathbf{U}; \mathfrak{Y}_d\right)$ by (8.38) ■

C.6 Proof of Theorem 164

Proof. The special form of the plant (3.12) reads (3.29):

$$Q\left[t,\mathbf{R}^{\alpha-1}(t)\right]\mathbf{R}^{(\alpha)}(t)+F\left[t,\mathbf{R}^{\alpha-1}(t)\right]\mathbf{R}^{(\alpha-1)}(t)$$
$$=E\left(t\right)\mathbf{e}_{d}\left[\mathbf{D}^{\eta}(t)\right]+P\left(t\right)\mathbf{e}_{u}\left[\mathbf{U}(t)\right],\ \forall t\in\mathfrak{T},$$
$$\det F\left(t,\mathbf{R}^{\alpha-1}\right)\neq 0,\ \forall\left(t,\mathbf{R}^{\alpha-1}\right)\in\mathfrak{T}\times\mathfrak{R}^{\alpha\rho},$$
$$\mathbf{Y}(t)=Z\left(t,\mathbf{R}^{\alpha-1},\mathbf{D}\right)\mathbf{R}^{(\alpha-1)}+W\left(t\right)\mathbf{w}\left[\mathbf{U}(t)\right],\ \forall t\in\mathfrak{T},\qquad(C.17)$$

The nonsingularity of $F\left[t,\mathbf{R}^{\alpha-1}(t)\right]$ permits us to solve the ID equation of (C.17) for $\mathbf{R}^{(\alpha-1)}(t)$:

$$\mathbf{R}^{(\alpha-1)}(t)=F^{-1}\left[t,\mathbf{R}^{\alpha-1}(t)\right]\left\{\begin{array}{c}E\left(t\right)\mathbf{e}_{d}\left[\mathbf{D}^{\eta}(t)\right]+P\left(t\right)\mathbf{e}_{u}\left[\mathbf{U}(t)\right]\\-Q\left[t,\mathbf{R}^{\alpha-1}(t)\right]\mathbf{R}^{(\alpha)}(t)\end{array}\right\}.$$

which transforms the output equation of (C.17) into:

$$\mathbf{Y}(t)=Z\left(t,\mathbf{R}^{\alpha-1},\mathbf{D}\right)F^{-1}\left[t,\mathbf{R}^{\alpha-1}(t)\right]$$
$$\bullet\left\{\begin{array}{c}E\left(t\right)\mathbf{e}_{d}\left[\mathbf{D}^{\eta}(t)\right]+P\left(t\right)\mathbf{e}_{u}\left[\mathbf{U}(t)\right]\\-Q\left[t,\mathbf{R}^{\alpha-1}(t)\right]\mathbf{R}^{(\alpha)}(t)\end{array}\right\}+W\left(t\right)\mathbf{w}\left[\mathbf{U}(t)\right],\ \forall t\in\mathfrak{T},\qquad(C.18)$$

or, equivalently,

$$\mathbf{Y}(t)=Z\left(t,\mathbf{R}^{\alpha-1},\mathbf{D}\right)F^{-1}\left[t,\mathbf{R}^{\alpha-1}(t)\right]\left\{\begin{array}{c}E\left(t\right)\mathbf{e}_{d}\left[\mathbf{D}^{\eta}(t)\right]\\-Q\left[t,\mathbf{R}^{\alpha-1}(t)\right]\mathbf{R}^{(\alpha)}(t)\end{array}\right\}$$
$$+Z\left(t,\mathbf{R}^{\alpha-1},\mathbf{D}\right)F^{-1}\left[t,\mathbf{R}^{\alpha-1}(t)\right]P\left(t\right)\mathbf{e}_{u}\left[\mathbf{U}(t)\right]+W\left(t\right)\mathbf{w}\left[\mathbf{U}(t)\right],\ \forall t\in\mathfrak{T}.\qquad(C.19)$$

The equations (8.39) through (8.41) simplify the notation in (C.19):

$$\widehat{P}\left(t,\mathbf{R}^{\alpha-1},\mathbf{D}\right)\mathbf{e}_{u}\left[\mathbf{U}(t)\right]+W\left(t\right)\mathbf{w}\left[\mathbf{U}(t)\right]$$
$$=\mathbf{Y}(t)-\widehat{E}\left(t,\mathbf{R}^{\alpha-1},\mathbf{D}\right)\mathbf{e}_{d}\left[\mathbf{D}^{\eta}(t)\right]+\widehat{Q}\left(t,\mathbf{R}^{\alpha-1},\mathbf{D}\right)\mathbf{R}^{(\alpha)}(t).\qquad(C.20)$$

Necessity. Let the plant (3.29) with Properties 57, 58, 61, 64, 78 be perfect trackable on $\mathfrak{T}_0\times\mathfrak{D}^j\times\mathfrak{Y}_d^\alpha$ via both spaces. Definition 140 is satisfied: $\boldsymbol{\varepsilon}\left(t\right)\equiv\mathbf{0}_N$. The equation (8.1) holds. Hence, $B\left(t\right)\mathbf{v}\left(t,\boldsymbol{\varepsilon}\right)\equiv\mathbf{0}_N$, where $B\left(t\right)$ is determined by (8.37). This product can be added (Axiom 87) to the right–hand side of (C.20) that becomes (8.42). The condition (8.43) results from the constraints on the control vector $\mathbf{U}$.

Sufficiency. The equations (8.42) and (C.20) together with (8.1) and (8.37) imply $\boldsymbol{\varepsilon}\left(t\right)\equiv\mathbf{0}_N$. Definition 140 is satisfied. The plant (3.29) with Properties 57, 58, 61, 64, 78 is perfect trackable on $\mathfrak{T}_0\times\mathfrak{D}^j\times\mathfrak{Y}_d^\alpha$ via both spaces ■

C.7 Proof of Theorem 165

Proof. Let the plant (3.12) possess Properties 57, 58, and 78.

Necessity. Let the plant (3.12) be perfect natural trackable on $\mathfrak{T}_0\times\mathfrak{D}^j\times\mathfrak{Y}_d^\alpha$. Definition 146 holds. Definition 146 and Definition 140 imply that the plant (3.12) is perfect trackable on $\mathfrak{T}_0\times\mathfrak{D}^j\times\mathfrak{Y}_d^\alpha$. This verifies that there is a control $\mathbf{U}(t)$ that acts via the matrix $W\left(t\right)$ on the output vector. This and (3.42) of Property 64 imply $\gamma\equiv\operatorname{rank}W\left(t\right)\geq 1,\ \forall t\in\mathfrak{T}_0$. From $W\left(t\right)\in\mathfrak{R}^{N\times r}$ follows $\gamma\leq\min(N,r)$. This and Axiom 87 determine $\gamma=\min(N,r)=N$ for the elementwise trackability, which completes the condition i). The error vector $\boldsymbol{\varepsilon}\left(t\right)\equiv\mathbf{0}_N$

. This implies $\mathbf{v}(t,\varepsilon) \equiv \mathbf{0}_N$ due to (8.1). Theorem 149 is valid, i.e., the equations (8.11) through (8.19) hold. Property 57 permits us to set (8.11) into the following form:

$$\begin{aligned} W\left(t^{-}\right)\mathbf{w}\left[\mathbf{U}(t^{-})\right] &= \mathbf{Y}(t^{-}) - \mathbf{z}\left[t^{-},\mathbf{R}^{\alpha-1}(t^{-}),\mathbf{D}(t^{-})\right] = W\left(t\right)\mathbf{w}\left[\mathbf{U}(t^{-})\right] \\ &= \mathbf{Y}(t) - \mathbf{z}\left[t,\mathbf{R}^{\alpha-1}(t),\mathbf{D}(t)\right] = W\left(t\right)\mathbf{w}\left[\mathbf{U}(t)\right], \ \forall t \in \mathfrak{T}_0. \end{aligned} \tag{C.21}$$

ii-1) If $\gamma = \text{rank}W\left(t\right) = r < N$, $\forall t \in \mathfrak{T}_0$, then (8.11) and (8.16) hold, Theorem 149:

$$\begin{aligned} \mathbf{U}\left(t\right) &= \mathbf{w}^I\left\{\Gamma\left(t\right)\left[\mathbf{Y} - \mathbf{z}(t,\mathbf{R}^{\alpha-1},\mathbf{D})\right] + M_{r,m}\left(t\right)\mathbf{v}\left(t,\varepsilon\right)\right\} \\ \Gamma\left(t\right) &= \left[W^T\left(t\right)W\left(t\right)\right]^{-1}W^T\left(t\right) \in \mathfrak{R}^{r\times N}, \ \forall t \in \mathfrak{T}_0. \end{aligned}$$

This and (C.21) yield:

$$\begin{aligned} \mathbf{U}\left(t\right) &= \mathbf{w}^I\left\{\Gamma\left(t\right)W\left(t\right)\mathbf{w}\left[\mathbf{U}(t^{-})\right] + M_{r,m}\left(t\right)\mathbf{v}\left(t,\varepsilon\right)\right\} \\ &= \mathbf{w}^I\left\{\left[W^T\left(t\right)W\left(t\right)\right]^{-1}W^T\left(t\right)W\left(t\right)\mathbf{w}\left[\mathbf{U}(t^{-})\right] + M_{r,m}\left(t\right)\mathbf{v}\left(t,\varepsilon\right)\right\}, \ \forall t \in \mathfrak{T}_0 \Longrightarrow \\ \mathbf{U}\left(t\right) &= \mathbf{w}^I\left\{\mathbf{w}\left[\mathbf{U}(t^{-})\right] + M_{r,m}\left(t\right)\mathbf{v}\left(t,\varepsilon\right)\right\}, \ \forall t \in \mathfrak{T}_0. \end{aligned}$$

This is (9.1).

ii-2) If $\gamma = \text{rank}W\left(t\right) = N < r$, $\forall t \in \mathfrak{T}_0$, then (8.12) and (8.17) hold, Theorem 149:

$$\mathbf{U}\left(t\right) = \mathbf{w}^I\left\{\begin{array}{c} W^T\left(t\right)\left[W\left(t\right)W^T\left(t\right)\right]^{-1}\left[\mathbf{Y} - \mathbf{z}(t,\mathbf{R}^{\alpha-1},\mathbf{D})\right] \\ + W^T\left(t\right)M_{r,m}\left(t\right)\mathbf{v}\left(t,\varepsilon\right) \end{array}\right\}, \ \forall t \in \mathfrak{T}_0.$$

This and (C.21) imply:

$$\mathbf{U}\left(t\right) = \mathbf{w}^I\left\{\begin{array}{c} W^T\left(t\right)\left[W\left(t\right)W^T\left(t\right)\right]W\left(t\right)\mathbf{w}\left[\mathbf{U}(t^{-})\right] \\ + W^T\left(t\right)M_{r,m}\left(t\right)\mathbf{v}\left(t,\varepsilon\right) \end{array}\right\}, \forall t \in \mathfrak{T}_0,$$

i.e.,

$$\begin{aligned} \mathbf{U}\left(t\right) &= \mathbf{w}^I\left\{\Gamma\left(t\right)\mathbf{w}\left[\mathbf{U}\left(t^{-}\right)\right] + W^T\left(t\right)M_{r,m}\left(t\right)\mathbf{v}\left(t,\varepsilon\right)\right\} \\ \Gamma\left(t\right) &= W^T\left(t\right)\left[W\left(t\right)W^T\left(t\right)\right]^{-1}W\left(t\right) \in \mathfrak{R}^{r\times r}, \ \forall t \in \mathfrak{T}_0. \end{aligned}$$

This is (9.2).

ii-3) The equation (8.13):

$$\mathbf{U}\left(t\right) = \mathbf{w}^I\left\langle W\left(t\right)^{-1}\left\{\mathbf{Y}_d\left(t\right) - \mathbf{z}(\left[t,\mathbf{R}^{\alpha-1}\left(t\right),\mathbf{D}\left(t\right)\right]\right\}\right\rangle, \ \forall t \in \mathfrak{T}_0,$$

can be modified by using $\mathbf{Y}_d\left(t\right) = \varepsilon\left(t\right) + \mathbf{Y}\left(t\right), \forall t \in \mathfrak{T}_0$:

$$\mathbf{U}\left(t\right) = \mathbf{w}^I\left\langle W\left(t\right)^{-1}\left\{\varepsilon\left(t\right) + \mathbf{Y}\left(t\right) - \mathbf{z}(\left[t,\mathbf{R}^{\alpha-1}\left(t\right),\mathbf{D}\left(t\right)\right]\right\}\right\rangle, \ \forall t \in \mathfrak{T}_0.$$

The output equation of (3.12) determines

$$\mathbf{Y}\left(t\right) - \mathbf{z}(\left[t,\mathbf{R}^{\alpha-1}\left(t\right),\mathbf{D}\left(t\right)\right] = W\left(t\right)\mathbf{w}\left[\mathbf{U}\left(t\right)\right],$$

which together with (C.21) transforms the preceding equation into

$$\mathbf{U}\left(t\right) = \mathbf{w}^I\left\langle\mathbf{w}\left[\mathbf{U}\left(t^{-}\right)\right] + W\left(t\right)^{-1}\varepsilon\left(t\right)\right\rangle, \ \forall t \in \mathfrak{T}_0.$$

This is (9.3). It implies (8.18).

ii-4) Let the plant also possess Property 64. The equations (C.21), Lemma 65, (3.47), (8.15), Axiom 87, and $\mathbf{v}(t,\boldsymbol{\varepsilon}) \equiv \mathbf{0}_N$ permit

$$W_\gamma(t)\,\mathbf{w}_\gamma\left[\mathbf{U}_\gamma(t)\right] = W_\gamma(t)\,\mathbf{w}_\gamma\left[\mathbf{U}_\gamma(t^-)\right] + W_\gamma(t)\,M_{\gamma,m}(t)\,\mathbf{v}(t,\boldsymbol{\varepsilon}),\ \forall t \in \mathfrak{T}_0.$$

This equation multiplied on the left by $W_\gamma^{-1}(t)$ becomes the equation (9.4). It determines the domain $\mathcal{D}_{PNTbl}\left(t_0;\mathfrak{D}^j;\mathbf{U};\mathfrak{Y}_d\right)$, (8.19) The condition ii-4) is valid.

Sufficiency. Let all the conditions of the theorem statement hold.

ii-1) The rank $\gamma = \operatorname{rank} W(t) = r < N$, $\forall t \in \mathfrak{T}_0$, and the chosen vector function $\mathbf{v}(.) : \mathfrak{T}_0 \times \mathfrak{R}^N \longrightarrow \mathfrak{R}^m$ obeys (8.1), where m is not greater than r, $m \le r$, and rank $M_{r,m}(t) \equiv m$. The rank $W(t) = r < N$ ensures nonsingularity of $W^T(t)\,W(t)$. We multiply on the left the output equation of (3.12) by $\left[W^T(t)\,W(t)\right]^{-1} W^T(t)$. The result reads

$$\begin{gathered}\mathbf{w}\left[\mathbf{U}(t)\right] = \left[W^T(t)\,W(t)\right]^{-1} W^T(t)\left\{\mathbf{Y}(t) - \mathbf{z}\left[t,\mathbf{R}^{\alpha-1}(t),\mathbf{D}(t)\right]\right\},\\ \mathbf{w}\left[\mathbf{U}(t^-)\right] = \left[W^T(t^-)\,W(t^-)\right]^{-1} W^T(t^-)\left\{\mathbf{Y}(t^-) - \mathbf{z}\left[t^-,\mathbf{R}^{\alpha-1}(t^-),\mathbf{D}(t^-)\right]\right\},\\ \forall t \in \mathfrak{T}_0.\end{gathered}$$

The equation (C.21) permits:

$$\begin{aligned}\mathbf{w}\left[\mathbf{U}(t^-)\right] &= \left[W^T(t^-)\,W(t^-)\right]^{-1} W^T(t^-)\left\{\mathbf{Y}(t^-) - \mathbf{z}\left[t^-,\mathbf{R}^{\alpha-1}(t^-),\mathbf{D}(t^-)\right]\right\}\\ &= \left[W^T(t)\,W(t)\right]^{-1} W^T(t)\left\{\mathbf{Y}(t) - \mathbf{z}\left[t,\mathbf{R}^{\alpha-1}(t),\mathbf{D}(t)\right]\right\},\\ &\forall t \in \mathfrak{T}_0,\end{aligned}$$

which transforms the control algorithm (9.1):

$$\mathbf{U}(t) = \mathbf{w}^I\left\{\begin{array}{c}\left[W^T(t)\,W(t)\right]^{-1} W^T(t)\left\{\mathbf{Y}(t) - \mathbf{z}\left[t,\mathbf{R}^{\alpha-1}(t),\mathbf{D}(t)\right]\right\}\\ +\,M_{r,m}(t)\,\mathbf{v}(t,\boldsymbol{\varepsilon})\end{array}\right\},\quad \forall t \in \mathfrak{T}_0,$$

i.e.,

$$W^T(t)\,W(t)\,\mathbf{w}\left[\mathbf{U}(t)\right] = \left\{\begin{array}{c}W^T(t)\left\{\mathbf{Y}(t) - \mathbf{z}\left[t,\mathbf{R}^{\alpha-1}(t),\mathbf{D}(t)\right]\right\}\\ +\,W^T(t)\,W(t)\,M_{r,m}(t)\,\mathbf{v}(t,\boldsymbol{\varepsilon})\end{array}\right\},\quad \forall t \in \mathfrak{T}_0.$$

This and the output equation of (3.12) multiplied on the left by $W^T(t)$ imply:

$$W^T(t)\,W(t)\,M_{r,m}(t)\,\mathbf{v}(t,\boldsymbol{\varepsilon}) = \mathbf{0}_r,\ \forall t \in \mathfrak{T}_0,$$

i.e.,

$$M_{r,m}(t)\,\mathbf{v}(t,\boldsymbol{\varepsilon}) = \mathbf{0}_r,\ \forall t \in \mathfrak{T}_0,$$

due to the nonsingularity of $W^T(t)\,W(t)$. The rank $M_{r,m}(t) \equiv m \le r$ ensures the existence of $\left[M_{r,m}^T(t)\,M_{r,m}(t)\right]^{-1}$. We multiply the preceding equation on the left by

$$\left[M_{r,m}^T(t)\,M_{r,m}(t)\right]^{-1} M_{r,m}^T(t)$$

to obtain:

$$\mathbf{v}(t,\boldsymbol{\varepsilon}) = \mathbf{0}_m,\ \forall t \in \mathfrak{T}_0.$$

This and (8.1) imply

$$\boldsymbol{\varepsilon}(t) = \mathbf{0}_N,\ \forall t \in \mathfrak{T}_0.$$

The plant (3.12) is perfect natural trackable on $\mathfrak{T}_0 \times \mathfrak{D}^j \times \mathfrak{Y}_d^\alpha$, Definition 146 for every vector pair $\left(\mathbf{R}_0^{\alpha-1}, \mathbf{Y}_0\right) \in \mathcal{D}_{PNTbl}\left(t_0; \mathfrak{D}^j; \mathbf{U}; \mathfrak{Y}_d\right)$, (8.16).

ii-2) This case holds for $\gamma = \text{rank } W(t) = N < r$, $\forall t \in \mathfrak{T}_0$, and for the chosen vector function $\mathbf{v}(.) : \mathfrak{T}_0 \times \mathfrak{R}^N \longrightarrow \mathfrak{R}^m$, (8.1), where m is not greater than r, $m \leq r$, (9.2) is valid. The rank $W(t) = N < r$, $\forall t \in \mathfrak{T}_0$, ensures the nonsingularity of $W(t) W^T(t)$. The output equation of (3.12) at t^- reads:

$$W\left(t^-\right) \mathbf{w}\left[\mathbf{U}\left(t^-\right)\right] = \left\{\mathbf{Y}(t^-) - \mathbf{z}\left[t^-, \mathbf{R}^{\alpha-1}(t^-), \mathbf{D}(t^-)\right]\right\}, \ \forall t \in \mathfrak{T}_0,$$

or

$$W(t) \mathbf{w}\left[\mathbf{U}\left(t^-\right)\right] = \left\{\mathbf{Y}(t) - \mathbf{z}\left[t, \mathbf{R}^{\alpha-1}(t), \mathbf{D}(t)\right]\right\}, \ \forall t \in \mathfrak{T}_0. \tag{C.22}$$

This transforms (9.2) into:

$$\mathbf{w}\left[\mathbf{U}(t)\right] = W^T(t)\left[W(t) W^T(t)\right]^{-1}\left\{\mathbf{Y}(t) - \mathbf{z}\left[t, \mathbf{R}^{\alpha-1}(t), \mathbf{D}(t)\right]\right\}$$
$$+ W^T(t) M_{N,m}(t) \mathbf{v}(t, \boldsymbol{\varepsilon}), \ \forall t \in \mathfrak{T}_0,$$

or

$$W(t) \mathbf{w}\left[\mathbf{U}(t)\right] = \left\{\mathbf{Y}(t) - \mathbf{z}\left[t, \mathbf{R}^{\alpha-1}(t), \mathbf{D}(t)\right]\right\}$$
$$+ W(t) W^T(t) M_{r,m}(t) \mathbf{v}(t, \boldsymbol{\varepsilon}), \ \forall t \in \mathfrak{T}_0.$$

This, the output equation of (3.12), and the nonsingularity of $\left[W(t) W^T(t)\right]$ due to rank $W(t) = N < r$ imply

$$M_{N,m}(t) \mathbf{v}(t, \boldsymbol{\varepsilon}) = \mathbf{0}_N, \ \forall t \in \mathfrak{T}_0,$$

Because rank $M_{N,m}(t) \equiv m \leq N$, $\forall t \in \mathfrak{T}_0$, this equation together with (8.1) reduces to:

$$\boldsymbol{\varepsilon}(t) = \mathbf{0}_N, \ \forall t \in \mathfrak{T}_0.$$

The plant (3.12) is perfect natural trackable on $\mathfrak{T}_0 \times \mathfrak{D}^j \times \mathfrak{Y}_d^\alpha$, Definition 146 for every vector pair $\left(\mathbf{R}_0^{\alpha-1}, \mathbf{Y}_0\right) \in \mathcal{D}_{PNTbl}\left(t_0; \mathfrak{D}^j; \mathbf{U}; \mathfrak{Y}_d\right)$, (8.17).

ii-3) Because $\gamma = \text{rank } W(t) = r = N$, $\forall t \in \mathfrak{T}_0$ then (9.3) determines the control. Its another form is

$$W(t) \mathbf{w}\left[\mathbf{U}(t)\right] = W(t) \mathbf{w}\left[\mathbf{U}\left(t^-\right)\right] + \boldsymbol{\varepsilon}(t), \ \forall t \in \mathfrak{T}_0.$$

This and (C.22) result in:

$$W(t) \mathbf{w}\left[\mathbf{U}(t)\right] = \mathbf{Y}(t) - \mathbf{z}\left[t, \mathbf{R}^{\alpha-1}(t), \mathbf{D}(t)\right] + \boldsymbol{\varepsilon}(t), \ \forall t \in \mathfrak{T}_0.$$

The output equation of (3.12) transforms this into

$$W(t) \mathbf{w}\left[\mathbf{U}(t)\right] = W(t) \mathbf{w}\left[\mathbf{U}(t)\right] + \boldsymbol{\varepsilon}(t), \ \forall t \in \mathfrak{T}_0,$$

that proves

$$\boldsymbol{\varepsilon}(t) = \mathbf{0}_N, \ \forall t \in \mathfrak{T}_0.$$

The plant (3.12) is perfect natural trackable on $\mathfrak{T}_0 \times \mathfrak{D}^j \times \mathfrak{Y}_d^\alpha$, Definition 146 for every vector pair $\left(\mathbf{R}_0^{\alpha-1}, \mathbf{Y}_0\right) \in \mathcal{D}_{PNTbl}\left(t_0; \mathfrak{D}^j; \mathbf{U}; \mathfrak{Y}_d\right)$, (8.18).

ii-4) Let the plant also possess Property 64. Lemma 65, i.e., (3.47), permits to set the γ vector subequation of the output equation of (3.12) into the following form:

$$\mathbf{w}_\gamma\left[\mathbf{U}_\gamma(t)\right] = W_\gamma^{-1}(t)\left\{\mathbf{Y}_\gamma(t) - \mathbf{z}_\gamma\left[t, \mathbf{R}^{\alpha-1}(t), \mathbf{D}(t)\right]\right\}, \ \forall t \in \mathfrak{T}_0, \tag{C.23}$$

i.e.,

$$\mathbf{w}_{\gamma}\left[\mathbf{U}_{\gamma}(t^{-})\right]=W_{\gamma}^{-1}\left(t^{-}\right)\left\{\mathbf{Y}_{\gamma}(t^{-})-\mathbf{z}_{\gamma}\left[t^{-},\mathbf{R}^{\alpha-1}(t^{-}),\mathbf{D}(t^{-})\right]\right\}$$
$$\forall t\in\mathfrak{T}_0,$$

or

$$\mathbf{w}_{\gamma}\left[\mathbf{U}_{\gamma}(t^{-})\right]=W_{\gamma}^{-1}\left(t\right)\left\{\mathbf{Y}_{\gamma}(t)-\mathbf{z}_{\gamma}\left[t,\mathbf{R}^{\alpha-1}(t),\mathbf{D}(t)\right]\right\},\ \forall t\in\mathfrak{T}_0, \tag{C.24}$$

due to Property 64. The equations (9.4) and (C.24) imply

$$M_{\gamma,m}\left(t\right)\mathbf{v}\left(t,\boldsymbol{\varepsilon}\right)=\mathbf{0}_{\gamma},\ \forall t\in\mathfrak{T}_0, \tag{C.25}$$

equivalently, due to rank $M_{\gamma,m}\left(t\right)\equiv m\leq\gamma$ and (8.1),

$$\boldsymbol{\varepsilon}\left(t\right)=\mathbf{0}_N,\ \forall t\in\mathfrak{T}_0, \tag{C.26}$$

due to (8.1). The plant **(3.12)** is perfect trackable on $\mathfrak{T}_0\times\mathfrak{D}^j\times\mathfrak{Y}_d^{\alpha}$ (Definition 140). If $m=\gamma=N\leq r$ then $\mathbf{v}\left(t,\boldsymbol{\varepsilon}\right)\equiv\boldsymbol{\varepsilon}$ so that (C.25) is immediately (C.26). The perfect natural trackability is elementwise. The control (9.4) is independent of both the plant internal dynamics and disturbances. It is natural control. Altogether, the plant **(3.12)** is perfect natural trackable on $\mathfrak{T}_0\times\mathfrak{D}^j\times\mathfrak{Y}_d^{\alpha}$ (Definition 146) for every $\mathbf{Y}_{d0}\in\mathcal{D}_{PNTbl}\left(t_0;\mathfrak{D}^j;\mathbf{U};\mathfrak{Y}_d^{\alpha}\right)$, (8.19) ■

C.8 Proof of Theorem 167

Proof. Let the plant (3.12) possess Properties 57, 58, 78,

Necessity. Let the plant (3.12) be perfect natural trackable on $\mathfrak{T}_0\times\mathfrak{D}^j\times\mathfrak{Y}_d^{\alpha}$. Definition 146 and Definition 140 imply that the plant (3.12) is also perfect trackable on $\mathfrak{T}_0\times\mathfrak{D}^j\times\mathfrak{Y}_d^{\alpha}$. Definition 140 holds so that $\boldsymbol{\varepsilon}\left(t\right)\equiv\mathbf{0}_N$, and $\mathbf{v}\left(t,\boldsymbol{\varepsilon}\right)\equiv\mathbf{0}_m$ due to (8.1). Theorem 154 holds, which implies the validity of the condition i) and the validity of (8.21) through (8.29).

ii-1) Let β =rank $P\left(t\right)=r<\rho$, $\forall t\in\mathfrak{T}_0$, and the chosen vector function $\mathbf{v}\left(.\right):\mathfrak{T}_0\times\mathfrak{R}^N\longrightarrow\mathfrak{R}^m$ obey (8.1), where m is not greater than r, $m\leq r$, and rank $L_{r,m}\left(t\right)\equiv m$. The equation (8.21) originally reads:

$$\mathbf{e}_u\left[\mathbf{U}^{\mu}\left(t\right)\right]=\left\{\begin{array}{c}\Gamma\left(t\right)\left[\begin{array}{c}Q\left[t,\mathbf{R}^{\alpha-1}(t)\right]\mathbf{R}^{(\alpha)}(t)\\+\mathbf{q}\left[t,\mathbf{R}^{\alpha-1}(t)\right]-E\left(t\right)\mathbf{e}_d\left[\mathbf{D}^{\eta}(t)\right]\end{array}\right]\\+L_{r,m}\left(t\right)\mathbf{v}\left(t,\boldsymbol{\varepsilon}\right)\end{array}\right\},$$
$$\Gamma\left(t\right)=\left[P^T\left(t\right)P\left(t\right)\right]^{-1}P^T\left(t\right),\ \forall t\in\mathfrak{T}_0,$$

Property 57 confirms:

$$\begin{aligned}&P\left(t\right)\mathbf{e}_u\left[\mathbf{U}^{\mu}\left(t\right)\right]\\&=Q\left[t,\mathbf{R}^{\alpha-1}(t)\right]\mathbf{R}^{(\alpha)}(t)+\mathbf{q}\left[t,\mathbf{R}^{\alpha-1}(t)\right]-E\left(t\right)\mathbf{e}_d\left[\mathbf{D}^{\eta}(t)\right]\\&=Q\left[t,\mathbf{R}^{\alpha-1}(t^{-})\right]\mathbf{R}^{(\alpha)}(t^{-})+\mathbf{q}\left[t^{-},\mathbf{R}^{\alpha-1}(t^{-})\right]-E\left(t^{-}\right)\mathbf{e}_d\left[\mathbf{D}^{\eta}(t^{-})\right]\\&=P\left(t^{-}\right)\mathbf{e}_u\left[\mathbf{U}^{\mu}\left(t^{-}\right)\right]=P\left(t\right)\mathbf{e}_u\left[\mathbf{U}^{\mu}\left(t^{-}\right)\right].\end{aligned} \tag{C.27}$$

This transforms the equation (8.21) into:

$$\mathbf{e}_u\left[\mathbf{U}^{\mu}\left(t\right)\right]=\left\{\begin{array}{c}\left[P^T\left(t\right)P\left(t\right)\right]^{-1}P^T\left(t\right)P\left(t\right)\mathbf{e}_u\left[\mathbf{U}^{\mu}\left(t^{-}\right)\right]\\+L_{r,m}\left(t\right)\mathbf{v}\left(t,\boldsymbol{\varepsilon}\right)\end{array}\right\}\Longrightarrow$$
$$\mathbf{e}_u\left[\mathbf{U}^{\mu}\left(t\right)\right]=\mathbf{e}_u\left[\mathbf{U}^{\mu}\left(t^{-}\right)\right]+L_{r,m}\left(t\right)\mathbf{v}\left(t,\boldsymbol{\varepsilon}\right),\forall t\in\mathfrak{T}_0.$$

The last equation is (9.5).

ii-2) Let $\beta = \text{rank } P(t) = \rho < r$, $\forall t \in \mathfrak{T}_0$. Let the chosen vector function $\mathbf{v}(.) : \mathfrak{T}_0 \times \mathfrak{R}^N \longrightarrow \mathfrak{R}^m$ fulfill (8.1), where m is not greater than r, $m \leq r$, and rank $L_{r,m}(t) \equiv m$. Theorem 154 confirms (8.22):

$$\mathbf{e}_u[\mathbf{U}^\mu(t)] = \left\{ \begin{array}{c} \Gamma(t) \left[\begin{array}{c} Q\left[t, \mathbf{R}^{\alpha-1}(t)\right] \mathbf{R}^{(\alpha)}(t) \\ + \mathbf{q}\left[t, \mathbf{R}^{\alpha-1}(t)\right] - E(t)\, \mathbf{e}_d[\mathbf{D}^\eta(t)] \end{array} \right] \\ + P^T(t)\, L_{\rho,m}(t)\, \mathbf{v}(t, \boldsymbol{\varepsilon}) \end{array} \right\},$$

$$\Gamma(t) = P^T(t) \left[P(t) P^T(t)\right]^{-1}, \ \forall t \in \mathfrak{T}_0,$$

This can take the following form by applying (C.27) :

$$\mathbf{e}_u[\mathbf{U}^\mu(t)] = \Gamma(t)\left[P(t)\, \mathbf{e}_u\left[\mathbf{U}^\mu(t^-)\right]\right] + P^T(t)\, L_{\rho,m}(t)\, \mathbf{v}(t, \boldsymbol{\varepsilon}),$$
$$\Gamma(t) = P^T(t)\left[P(t) P^T(t)\right]^{-1}, \ \forall t \in \mathfrak{T}_0.$$

This is (9.6).

ii-3) For $\beta = \text{rank } P(t) = r = \rho$, $\forall t \in \mathfrak{T}_0$, (8.23) is valid due to Theorem 154:

$$\mathbf{e}_u[\mathbf{U}^\mu(t)] = \left\langle P(t)^{-1} \left\{ \begin{array}{c} Q\left[t, \mathbf{R}^{\alpha-1}(t)\right] \mathbf{R}^{(\alpha)}(t) \\ + \mathbf{q}\left[t, \mathbf{R}^{\alpha-1}(t)\right] - E(t)\, \mathbf{e}_d[\mathbf{D}^\eta(t)] \\ + L_{\rho,m}(t)\, \mathbf{v}(t, \boldsymbol{\varepsilon}) \end{array} \right\} \right\rangle,$$
$$\forall t \in \mathfrak{T}_0,$$

or, in view of (C.27),

$$\mathbf{e}_u[\mathbf{U}^\mu(t)] = P(t)^{-1} \left\{ \ P(t)\, \mathbf{e}_u[\mathbf{U}^\mu(t^-)] + L_{\rho,m}(t)\, \mathbf{v}(t, \boldsymbol{\varepsilon}) \ \right\}, \ \forall t \in \mathfrak{T}_0,$$

i.e.,

$$\mathbf{e}_u[\mathbf{U}^\mu(t)] = \ \mathbf{e}_u[\mathbf{U}^\mu(t^-)] + P(t)^{-1} L_{\rho,m}(t)\, \mathbf{v}(t, \boldsymbol{\varepsilon}) \ , \ \forall t \in \mathfrak{T}_0,$$

which is (9.7).

ii-4) Let the plant (3.12) also possesses Property 61. Theorem 154 implies the validity of the equations (8.24) and (8.29). Property 57 ensures that

$$\begin{array}{c} P_\beta(t)\, \mathbf{e}_{u\beta}\left[\mathbf{U}^\mu_\beta(t)\right] + E_{\beta,d}(t)\mathbf{e}_d[\mathbf{D}^\eta(t)] \\ = P_\beta(t^-)\, \mathbf{e}_{u\beta}\left[\mathbf{U}^\mu_\beta(t^-)\right] + E_{\beta,d}(t^-)\mathbf{e}_d\left[\mathbf{D}^\eta(t^-)\right] \\ = Q_{\beta,\rho}(t, \mathbf{R}^{\alpha-1})\mathbf{R}^{(\alpha)} + \mathbf{q}_\beta(t, \mathbf{R}^{\alpha-1}, \mathbf{D}^i) \\ = Q_{\beta,\rho}\left[t^-, \mathbf{R}^{\alpha-1}(t^-)\right] \mathbf{R}^{(\alpha)}(t^-) + \mathbf{q}_\beta\left[t^-, \mathbf{R}^{\alpha-1}(t^-)\right] \\ \forall [t, \mathbf{D}(.), \mathbf{Y}_d(.)] \in \mathfrak{T}_0 \times \mathfrak{D}^j \times \mathfrak{Y}^\alpha_d. \end{array} \tag{C.28}$$

This can be set, due to (8.1) and Axiom 87, into the following form:

$$\begin{array}{c} P_\beta(t)\, \mathbf{e}_{u\beta}\left[\mathbf{U}^\mu_\beta(t)\right] = P_\beta(t)\left\{\mathbf{e}_{u\beta}\left[\mathbf{U}^\mu_\beta(t)\right] - L_{\beta,m}(t)\, \mathbf{v}(t, \boldsymbol{\varepsilon})\right\} \\ = Q_{\beta,\rho}(t, \mathbf{R}^{\alpha-1})\mathbf{R}^{(\alpha)} + \mathbf{q}_\beta(t, \mathbf{R}^{\alpha-1}, \mathbf{D}^i) - E_{\beta,d}(t)\mathbf{e}_d[\mathbf{D}^\eta(t)] \\ = Q_{\beta,\rho}\left[t^-, \mathbf{R}^{\alpha-1}(t^-)\right] \mathbf{R}^{(\alpha)}(t^-) + \mathbf{q}_\beta(t^-, \mathbf{R}^{\alpha-1}, \mathbf{D}^i) - E_{\beta,d}(t^-)\mathbf{e}_d\left[\mathbf{D}^\eta(t^-)\right] \\ = P_\beta(t^-)\, \mathbf{e}_{u\beta}\left[\mathbf{U}(t^-)\right], \ \forall [t, \mathbf{D}(.), \mathbf{Y}_d(.)] \in \mathfrak{T}_0 \times \mathfrak{D}^j \times \mathfrak{Y}^\alpha_d, \end{array} \tag{C.29}$$

i.e.,

$$P_\beta(t)\left\{\mathbf{e}_{u\beta}\left[\mathbf{U}^\mu_\beta(t)\right] - L_{\beta,m}(t)\, \mathbf{v}(t, \boldsymbol{\varepsilon})\right\} = P_\beta(t^-)\, \mathbf{e}_{u\beta}\left[\mathbf{U}^\mu_\beta(t^-)\right],$$
$$\forall [t, \mathbf{D}(.), \mathbf{Y}_d(.)] \in \mathfrak{T}_0 \times \mathfrak{D}^j \times \mathfrak{Y}^\alpha_d.$$

Multiplying this equation on the left by $P_{\beta}^{-1}(t)$ that is possible due to condition i) and using $P_{\beta}(t) \equiv P_{\beta}(t^{-})$ due to Property 57 the result is (9.8).

The domain $\mathcal{D}_{PNT\mathfrak{b}l}(t_0; \mathfrak{D}^j; \mathbf{U}; \mathfrak{Y}_d)$ of the l-th order perfect natural trackability of the plant (3.12) on $\mathfrak{T}_0 \times \mathfrak{D}^j \times \mathfrak{Y}_d^{\alpha}$ under ii-1) through ii-4) is determined by (8.26) through (8.29), respectively, due to Definition 140.

Sufficiency. Let all the conditions of the theorem statement be satisfied.

ii-1) Let $\beta = \text{rank } P(t) = r < \rho$, $\forall t \in \mathfrak{T}_0$, and the chosen vector function $\mathbf{v}(.) : \mathfrak{T}_0 \times \mathfrak{R}^N \longrightarrow \mathfrak{R}^m$ obey (8.1), where m is not greater than r, $m \leq r$, and rank $L_{r,m}(t) \equiv m$. The equation (9.5),

$$\mathbf{e}_u\left[\mathbf{U}^{\mu}(t)\right] = \mathbf{e}_u\left[\mathbf{U}^{\mu}(t^{-})\right] + L_{r,m}(t)\,\mathbf{v}(t,\boldsymbol{\varepsilon}),\ \forall t \in \mathfrak{T}_0.$$

and (C.27) lead to:

$$P(t)\,\mathbf{e}_u\left[\mathbf{U}(t)\right] = Q\left[t, \mathbf{R}^{\alpha-1}(t)\right]\mathbf{R}^{(\alpha)}(t) + \mathbf{q}\left[t, \mathbf{R}^{\alpha-1}(t)\right] - E(t)\,\mathbf{e}_d\left[\mathbf{D}^{\eta}(t)\right] + P(t)\,L_{r,m}(t)\,\mathbf{v}(t,\boldsymbol{\varepsilon}),\ \forall t \in \mathfrak{T}_0,$$

i.e.,

$$Q\left[t, \mathbf{R}^{\alpha-1}(t)\right]\mathbf{R}^{(\alpha)}(t) + \mathbf{q}\left[t, \mathbf{R}^{\alpha-1}(t)\right] = E(t)\,\mathbf{e}_d\left[\mathbf{D}^{\eta}(t)\right] + P(t)\left[\mathbf{e}_u\left[\mathbf{U}(t)\right] - L_{r,m}(t)\,\mathbf{v}(t,\boldsymbol{\varepsilon})\right],\ \forall t \in \mathfrak{T}_0.$$

The *ID* equation of (3.12) reduces the preceding equation to:

$$P(t)\,L_{r,m}(t)\,\mathbf{v}(t,\boldsymbol{\varepsilon}) = \mathbf{0}_{\rho},\ \forall t \in \mathfrak{T}_0.$$

The rank $P(t) = r < \rho$, $\forall t \in \mathfrak{T}_0$, guarantees the existence of $\left[P^T(t)\,P(t)\right]^{-1}$. After the multiplication on the left of the preceding equation by the matrix product $\left[P^T(t)\,P(t)\right]^{-1} P^T(t)$ it becomes

$$L_{r,m}(t)\,\mathbf{v}(t,\boldsymbol{\varepsilon}) = \mathbf{0}_r,\ \forall t \in \mathfrak{T}_0.$$

The rank of $L_{r,m}(t)$, rank $L_{r,m}(t) \equiv m \leq r$, $\forall t \in \mathfrak{T}_0$, ensures the nonsingularity of $\left[L_{r,m}^T(t)\,L_{r,m}(t)\right]$, $\forall t \in \mathfrak{T}_0$, so that

$$\mathbf{v}(t,\boldsymbol{\varepsilon}) = \mathbf{0}_m,\ \forall t \in \mathfrak{T}_0.$$

This and (8.1) imply

$$\boldsymbol{\varepsilon}(t) = \mathbf{Y}_d(t) - \mathbf{Y}(t) = \mathbf{0}_m,\ \forall t \in \mathfrak{T}_0.$$

The plant is the perfect natural trackable on $\mathfrak{T}_0 \times \mathfrak{D}^j \times \mathfrak{Y}_d^{\alpha}$, Definition 146 because the control is independent of both the plant dynamics and the disturbances. The control is well defined for $\left(\mathbf{R}_0^{\alpha-1}, \mathbf{Y}_0\right) \in \mathcal{D}_{PNT\mathfrak{b}l}(t_0; \mathfrak{D}^j; \mathbf{U}; \mathfrak{Y}_d)$ determined by (8.26).

ii-2) The control law (9.6) holds for $\beta = \text{rank } P(t) = \rho < r$, $\forall t \in \mathfrak{T}_0$, and for the chosen vector function $\mathbf{v}(.) : \mathfrak{T}_0 \times \mathfrak{R}^N \longrightarrow \mathfrak{R}^m$, (8.1), where m is not greater than r, $m \leq r$, and rank $L_{r,m}(t) \equiv m$:

$$\mathbf{e}_u\left[\mathbf{U}^{\mu}(t)\right] = \Gamma(t)\left[P(t)\,\mathbf{e}_u\left[\mathbf{U}^{\mu}(t^{-})\right]\right] + P^T(t)\,L_{\rho,m}(t)\,\mathbf{v}(t,\boldsymbol{\varepsilon}),$$
$$\Gamma(t) = P^T(t)\left[P(t)\,P^T(t)\right]^{-1},\ \forall t \in \mathfrak{T}_0,$$

or

$$P(t)\,\mathbf{e}_u\left[\mathbf{U}(t)\right] = P(t)\,\mathbf{e}_u\left[\mathbf{U}^{\mu}(t^{-})\right] + P(t)\,P^T(t)\,L_{\rho,m}(t)\,\mathbf{v}(t,\boldsymbol{\varepsilon}),\ \forall t \in \mathfrak{T}_0.$$

This and (C.27) imply

$$P(t)\,P^T(t)\,L_{\rho,m}(t)\,\mathbf{v}(t,\boldsymbol{\varepsilon}) = \mathbf{0}_{\rho},\ \forall t \in \mathfrak{T}_0.$$

This and $\det\left[P(t)\,P^T(t)\right] \neq 0,\ \forall t \in \mathfrak{T}_0$ give

$$L_{r,m}(t)\,\mathbf{v}(t,\boldsymbol{\varepsilon}) = \mathbf{0}_r,\ \forall t \in \mathfrak{T}_0.$$

The rank of $L_{r,m}(t)$, $\text{rank}L_{r,m}(t) \equiv m \leq r$, $\forall t \in \mathfrak{T}_0$, ensures the nonsingularity of $\left[L^T_{r,m}(t)\,L_{r,m}(t)\right]$, $\forall t \in \mathfrak{T}_0$, so that

$$\mathbf{v}(t,\boldsymbol{\varepsilon}) = \mathbf{0}_m,\ \forall t \in \mathfrak{T}_0.$$

This and (8.1) imply

$$\boldsymbol{\varepsilon}(t) = \mathbf{Y}_d(t) - \mathbf{Y}(t) = \mathbf{0}_m,\ \forall t \in \mathfrak{T}_0.$$

The plant is the perfect natural trackable on $\mathfrak{T}_0 \times \mathfrak{D}^j \times \mathfrak{Y}^\alpha_d$, Definition 146 because the control is independent of both the plant dynamics and the disturbances. The control is well defined for $\left(\mathbf{R}^{\alpha-1}_0, \mathbf{Y}_0\right) \in \mathcal{D}_{PNTbl}\left(t_0; \mathfrak{D}^j; \mathbf{U}; \mathfrak{Y}_d\right)$ determined by (8.27).

ii-3) In this case the control vector function $\mathbf{U}(.)$ obeys (9.7) if $\beta = \text{rank}\,P(t) = r = \rho$, $\forall t \in \mathfrak{T}_0$:

$$\mathbf{e}_u\left[\mathbf{U}^\mu(t)\right] = \ \mathbf{e}_u\left[\mathbf{U}^\mu(t^-)\right] + P(t)^{-1}\,L_{\rho,m}(t)\,\mathbf{v}(t,\boldsymbol{\varepsilon})\ ,\ \forall t \in \mathfrak{T}_0,$$

or

$$P(t)\,\mathbf{e}_u\left[\mathbf{U}(t)\right] = P(t)\,\mathbf{e}_u\left[\mathbf{U}^\mu\left(t^-\right)\right] + L_{\rho,m}(t)\,\mathbf{v}(t,\boldsymbol{\varepsilon}),\ \forall t \in \mathfrak{T}_0,$$

i.e.,

$$\begin{aligned} P(t)\,\mathbf{e}_u\left[\mathbf{U}(t)\right] = Q\left[t, \mathbf{R}^{\alpha-1}(t)\right]\mathbf{R}^{(\alpha)}(t) + \mathbf{q}\left[t, \mathbf{R}^{\alpha-1}(t)\right] \\ - E(t)\,\mathbf{e}_d\left[\mathbf{D}^\eta(t)\right] + L_{\rho,m}(t)\,\mathbf{v}(t,\boldsymbol{\varepsilon}), \\ \forall t \in \mathfrak{T}_0. \end{aligned}$$

The replacement of $P(t)\,\mathbf{e}_u\left[\mathbf{U}(t)\right]$ from the *ID* equation of (3.12) into the preceding equation results in

$$L_{\rho,m}(t)\,\mathbf{v}(t,\boldsymbol{\varepsilon}) = \mathbf{0}_r,\ \forall t \in \mathfrak{T}_0.$$

The rank of $L_{\rho,m}(t)$, rank $L_{r,m}(t) \equiv m \leq r = \rho$, $\forall t \in \mathfrak{T}_0$, ensures the nonsingularity of $\left[L^T_{\rho r,m}(t)\,L_{\rho,m}(t)\right]$, $\forall t \in \mathfrak{T}_0$, so that

$$\mathbf{v}(t,\boldsymbol{\varepsilon}) = \mathbf{0}_m,\ \forall t \in \mathfrak{T}_0.$$

This and (8.1) imply

$$\boldsymbol{\varepsilon}(t) = \mathbf{Y}_d(t) - \mathbf{Y}(t) = \mathbf{0}_m,\ \forall t \in \mathfrak{T}_0.$$

The plant is the perfect natural trackable on $\mathfrak{T}_0 \times \mathfrak{D}^j \times \mathfrak{Y}^\alpha_d$, Definition 146 because the control is independent of both the plant dynamics and the disturbances. The control is well defined for $\left(\mathbf{R}^{\alpha-1}_0, \mathbf{Y}_0\right) \in \mathcal{D}_{PNTbl}\left(t_0; \mathfrak{D}^j; \mathbf{U}; \mathfrak{Y}_d\right)$ determined by (8.28).

ii-4) Let the plant (3.12) also possesses Property 61. The equations (C.28) multiplied on the left by $P^{-1}_\beta(t) \equiv P^{-1}_\beta(t^-)$, which is true in view of Lemma 65, (3.47), become

$$\begin{aligned} \mathbf{e}_{u\beta}\left[\mathbf{U}^\mu_\beta(t)\right] = P^{-1}_\beta(t)\left\{\begin{array}{c} Q_{\beta,\rho}(t, \mathbf{R}^{\alpha-1})\mathbf{R}^{(\alpha)}(t) \\ + \mathbf{q}_\beta(t, \mathbf{R}^{\alpha-1}, \mathbf{D}^i) - E_{\beta,d}(t)\mathbf{e}_d\left[\mathbf{D}^\eta(t)\right] \end{array}\right\} \\ = P^{-1}_\beta(t^-)\left\{\begin{array}{c} Q_{\beta,\rho}(t^-, \mathbf{R}^{\alpha-1})\mathbf{R}^{(\alpha)}(t^-) \\ + \mathbf{q}_\beta(t^-, \mathbf{R}^{\alpha-1}, \mathbf{D}^i) - E_{\beta,d}(t^-)\mathbf{e}_d\left[\mathbf{D}^\eta(t^-)\right] \end{array}\right\} = \mathbf{e}_{u\beta}\left[\mathbf{U}^\mu_\beta(t^-)\right]. \end{aligned}$$

These equations imply $\mathbf{e}_u\left[\mathbf{U}_\beta(t)\right] = \mathbf{e}_u\left[\mathbf{U}_\beta(t^-)\right]$, which together with (9.8), furnishes:

$$L_{\beta,m}\mathbf{v}(t,\boldsymbol{\varepsilon}) = \mathbf{0}_\rho,\ \forall t \in \mathfrak{T}_0.$$

Because rank $L_{\beta,m} = m \leq \beta$ then $\det\left(L_{\beta,m}^T L_{\beta,m}\right) \neq 0$. We multiply $L_{\beta,m}\mathbf{v}(t,\boldsymbol{\varepsilon}) = \mathbf{0}_\rho$ on the left by $\left(L_{\beta,m}^T L_{\beta,m}\right)^{-1} L_{\beta,m}^T$. The result is $\mathbf{v}(t,\boldsymbol{\varepsilon}) = \mathbf{0}_m$, $\forall t \in \mathfrak{T}_0$; i.e., $\boldsymbol{\varepsilon}(t) = \mathbf{0}_N$, $\forall t \in \mathfrak{T}_0$. The plant (3.12) is perfect trackable on $\mathfrak{T}_0 \times \mathfrak{D}^j \times \mathfrak{Y}_d^\alpha$, Definition 140. The control $\mathbf{U}(t)$, (9.8), is well defined for every $(\mathbf{R},\mathbf{Y}) \in \mathcal{D}_{PNT\mathfrak{bl}}\left(t_0;\mathfrak{D}^j;\mathbf{U};\mathfrak{Y}_d^\alpha\right)$, ((8.29). It is independent of the plant internal dynamics and of disturbances. It is natural control and the plant (3.12) is perfect natural trackable on $\mathfrak{T}_0 \times \mathfrak{D}^j \times \mathfrak{Y}_d^\alpha$, Definition 146 ■

Appendix D

Proofs for Part VII

D.1 Lemma 1

Lemma 582 ***Statement of Lemma 1***

Let the system (3.12) possess weak smoothness property, Definition 368. Let the system exhibit uniform tracking of Υ *on* $\mathfrak{T}_i \times \mathcal{J}^j$ *with the instantaneous tracking domain* $\mathcal{D}_T(t_0;\mathbf{I};\boldsymbol{\Upsilon})$ *on* $\mathfrak{T}_i \times \mathcal{J}^j$ *obeying* $\mathcal{D}_T(t_0;\mathbf{I};\boldsymbol{\Upsilon}) \subseteq \mathfrak{S}(t;t_0;\Upsilon)$ *for all* $t \in \mathfrak{T}_0$ *and for every* $\mathbf{I}(.) \in \mathcal{J}^j$, $t_0 \in \mathfrak{T}_i$, *and with the domain* $\mathcal{D}_{TU}(\mathfrak{T}_i;\mathbf{I};\boldsymbol{\Upsilon})$ *of uniform tracking on* $\mathfrak{T}_i \times \mathcal{J}^j$.

a) If $\mathfrak{T}_i \subset \mathfrak{T}$ *then:*

1) $(t_0, \mathbf{Z}_0) \in \mathfrak{T}_i \times \mathcal{D}_T(t_0;\mathbf{I};\boldsymbol{\Upsilon})$ *implies* $\mathbf{Z}(t;t_0;\mathbf{Z}_0;\mathbf{I}) \in \mathcal{D}_T(\mathfrak{T}_i;\mathbf{I};\boldsymbol{\Upsilon})$ *for all* $(t,t_0) \in \mathfrak{T}_0 \times \mathfrak{T}_i$ *and for every* $\mathbf{I}(.) \in \mathcal{J}^j$, *which means that* $\mathcal{D}_T(\mathfrak{T}_i;\mathbf{I};\boldsymbol{\Upsilon})$ *is invariant set of the system on* $\mathfrak{T}_i$ *for every* $\mathbf{I}(.) \in \mathcal{J}^j$,

2) $\mathcal{D}_T(t_0;\mathbf{I};\boldsymbol{\Upsilon})$ *is an open continuous connected neighborhood of* $\boldsymbol{\Upsilon}$ *at any* $t_0 \in \mathfrak{T}_i$ *for every* $\mathbf{I}(.) \in \mathcal{J}^j$ *:*

$$\mathcal{D}_T(t_0;\mathbf{I};\boldsymbol{\Upsilon}) \equiv In\, \mathcal{D}_T(t_0;\mathbf{I};\boldsymbol{\Upsilon}),$$

$\mathcal{D}_T(t_0;\mathbf{I};\boldsymbol{\Upsilon}) \in \mathfrak{C}(\mathfrak{T}_i)$ *for every* $\mathbf{I}(.) \in \mathcal{J}^j$,

3) $\mathcal{D}_{TU}(\mathfrak{T}_i;\mathbf{I};\boldsymbol{\Upsilon})$ *is a connected neighborhood of* $\boldsymbol{\Upsilon}$ *on* $\mathfrak{T}_i$ *for every* $\mathbf{I}(.) \in \mathcal{J}^j$. *If* $\mathcal{D}_T(t_0;\mathbf{I};\boldsymbol{\Upsilon}) \equiv \mathcal{D}_{TU}(\mathfrak{T}_i;\mathbf{I};\boldsymbol{\Upsilon})$ *on* $\mathfrak{T}_i$ *for every* $\mathbf{I}(.) \in \mathcal{J}^j$ *then* $\mathcal{D}_{TU}(\mathfrak{T}_i;\mathbf{I};\boldsymbol{\Upsilon})$ *is also an open neighborhood of* Υ *and invariant set of the system on* $\mathfrak{T}_i$ *for every* $\mathbf{I}(.) \in \mathcal{J}^j$,

b) If $\mathfrak{T}_i = \mathfrak{T}$ *then:*

1) $\mathcal{D}_T(t_0;\mathbf{I};\boldsymbol{\Upsilon})$ *is an invariant set of the system on* $\mathfrak{T}$ *for every* $\mathbf{I}(.) \in \mathcal{J}^j$ *that is,* $(t_0, \mathbf{Z}_0) \in \mathfrak{T} \times \mathcal{D}_T(t_0;\mathbf{I};\boldsymbol{\Upsilon})$ *implies* $\mathbf{Z}(t;t_0;\mathbf{Z}_0;\mathbf{I}) \in \mathcal{D}_T(t_0;\mathbf{I};\boldsymbol{\Upsilon})$ *for all* $(t,t_0) \in \mathfrak{T}_0 \times \mathfrak{T}$ *and for every* $\mathbf{I}(.) \in \mathcal{J}^j$,

2) $\mathcal{D}_T(t_0;\mathbf{I};\boldsymbol{\Upsilon})$ *is an open continuous connected neighborhood of* $\boldsymbol{\Upsilon}$ *at any* $t_0 \in \mathfrak{T}$ *for every* $\mathbf{I}(.) \in \mathcal{J}^j$ *:*

$$\mathcal{D}_T(t_0;\mathbf{I};\boldsymbol{\Upsilon}) \equiv In\mathcal{D}_T(t_0;\mathbf{I};\boldsymbol{\Upsilon}),$$

$\mathcal{D}_T(t_0;\mathbf{I};\boldsymbol{\Upsilon}) \in \mathfrak{C}(\mathfrak{T})$ *for every* $\mathbf{I}(.) \in \mathcal{J}^j$,

3) $\mathcal{D}_{TU}(\mathfrak{T};\mathbf{Z}_d;\mathbf{I})$ *is a connected neighborhood of* $\boldsymbol{\Upsilon}$ *on* $\mathfrak{T}$ *for every* $\mathbf{I}(.) \in \mathcal{J}^j$. *If* $\mathcal{D}_T(t_0;\mathbf{I};\boldsymbol{\Upsilon}) \equiv \mathcal{D}_T(\mathfrak{T};\mathbf{I};\boldsymbol{\Upsilon}) = \mathcal{D}_T(\mathbf{Z}_d;\mathbf{I})$ *on* $\mathfrak{T}$ *for every* $\mathbf{I}(.) \in \mathcal{J}^j$ *then* $\mathcal{D}_T(\mathbf{Z}_d;\mathbf{I})$ *is also an open neighborhood of* $\boldsymbol{\Upsilon}$ *and invariant set of the system on* $\mathfrak{T}$ *for every* $\mathbf{I}(.) \in \mathcal{J}^j$,

Proof. Proof of Lemma 1

Let the system (3.12) possess weak smoothness property, Definition 368). Let the system exhibit uniform tracking of Υ on $\mathfrak{T}_i \times \mathcal{J}^j$ with the instantaneous tracking domain $\mathcal{D}_T(t_0;\mathbf{I};\boldsymbol{\Upsilon})$ on $\mathfrak{T}_i \times \mathcal{J}^j$ obeying $\mathcal{D}_T(t_0;\mathbf{I};\boldsymbol{\Upsilon}) \subseteq \mathfrak{S}(t;t_0;\Upsilon)$ for all $t \in \mathfrak{T}_0$ and for every $\mathbf{I}(.) \in \mathcal{J}^j$, and with the domain $\mathcal{D}_{TU}(\mathfrak{T}_i;\mathbf{I};\boldsymbol{\Upsilon})$ of uniform tracking on $\mathfrak{T}_i \times \mathcal{J}^j$, $\mathcal{D}_{TU}(\mathfrak{T}_i;\mathbf{I};\boldsymbol{\Upsilon}) = \cap\left[\mathcal{D}_T(t_0;\mathbf{I};\boldsymbol{\Upsilon}) : t_0 \in \mathfrak{T}_i\right]$, (Definition 216).

a) Let $t_0 \in \mathfrak{T}_i$ and $t^* \in \mathfrak{T}_i$, $t^* \neq t_0$. Let $\mathbf{Z}^* = \mathbf{Z}(t^*; t_0; \mathbf{Z}_0; \mathbf{I})$ for any $[\mathbf{Z}_0, \mathbf{I}(.)] \in \mathcal{D}_T(t_0; \mathbf{I}; \mathbf{\Upsilon}) \times \mathcal{J}^j$. Then, $\mathbf{Z}(t; t_0; \mathbf{Z}_0; \mathbf{I}) \longrightarrow \mp$ as $t \longrightarrow \infty$. Because

$$\mathbf{Z}(t; t^*; \mathbf{Z}^*; \mathbf{I}) \equiv \mathbf{Z}[t; t^*; \mathbf{Z}(t^*; t_0; \mathbf{Z}_0; \mathbf{I}); \mathbf{I}],$$

which is true due to (i) of weak smoothness property and $\mathcal{D}_T(t_0; \mathbf{I}; \mathbf{\Upsilon}) \subseteq \mathfrak{S}(t; t_0; \Upsilon)$ for all $t \in \mathfrak{T}_0$ and for every $\mathbf{I}(.) \in \mathcal{J}^j$, then $\mathbf{Z}(t; t^*; \mathbf{Z}^*; \mathbf{I}) \longrightarrow \mp$ as $t \longrightarrow \infty$. Hence, $\mathbf{Z}^* = \mathbf{Z}(t^*; t_0; \mathbf{Z}_0; \mathbf{I}) \in \mathcal{D}_T(t^*; \mathbf{Z}_d; \mathbf{I})$ that proves the statement under a-1). Let $\zeta \in \mathfrak{R}^+$ be such that $\mathfrak{N}_{2\zeta}(\Upsilon) \subset \mathcal{D}_{TU}(\mathfrak{T}_i; \mathbf{I}; \mathbf{\Upsilon})$. It exists (Definition 216). Let it be assumed that $\mathcal{D}_{TU}(\mathfrak{T}_i; \mathbf{I}; \mathbf{\Upsilon})$ is not open for all $t_0 \in \mathfrak{T}_i$ and for some $\mathbf{I}(.) \in \mathcal{J}^j$. Let there exist $t_0' \in \mathfrak{T}_i$ and $\mathbf{Z}_0' \in \partial\mathcal{D}_T(t_0'; \mathbf{I}; \mathbf{\Upsilon}) \cap \mathcal{D}_T(t_0'; \mathbf{I}; \mathbf{\Upsilon})$. Let $\varepsilon \in]0, \zeta/2[$. Then, (i) of weak smoothness property and $\mathcal{D}_T(t_0; \mathbf{I}; \mathbf{\Upsilon}) \subseteq \mathfrak{S}(t; t_0; \Upsilon)$ for all $t \in \mathfrak{T}_0$ and for every $\mathbf{I}(.) \in \mathcal{J}^j$, $t_0 \in \mathfrak{T}_i$, imply existence of $\theta \in \mathfrak{R}^+$, $\theta = \theta(t_0', \mathbf{Z}_0', \varepsilon, \mathbf{I})$, such that $\|\mathbf{Z}_0 - \mathbf{Z}_0'\| < \theta$ ensures $\|\mathbf{Z}(t_0' + 2\sigma'; t_0'; \mathbf{Z}_0; \mathbf{I}) - \mathbf{Z}(t_0' + 2\sigma'; t_0'; \mathbf{Z}_0'; \mathbf{I})\| < \varepsilon$, where $\sigma' = \tau(t_0, \varsigma, \mathbf{Z}_0; \mathbf{I})$, Definition 216. Because $\varepsilon < \zeta/2$ and

$$\rho[\mathbf{Z}(t_0' + 2\sigma'; t_0'; \mathbf{Z}_0'; \mathbf{I}), \Upsilon] < \zeta$$

then $\mathbf{Z}(t_0' + 2\sigma'; t_0'; \mathbf{Z}_0'; \mathbf{I}) \in \mathfrak{N}_{2\zeta}(\Upsilon) \subset \mathcal{D}_{TU}(\mathfrak{T}_i; \mathbf{I}; \mathbf{\Upsilon})$. Hence, $\mathbf{Z}_0 \in \mathcal{D}_T(t_0'; \mathbf{I}; \mathbf{\Upsilon})$. Any $\mathbf{Z}_0$ obeying $\|\mathbf{Z}_0 - \mathbf{Z}_0'\| < \theta$ may be selected in a θ-neighborhood of $\mathbf{Z}_0'$ out of $\mathcal{D}_T(t_0'; \mathbf{I}; \mathbf{\Upsilon})$, which is contradicted by the obtained $\mathbf{Z}_0 \in \mathcal{D}_T(t_0'; \mathbf{I}; \mathbf{\Upsilon})$. The former is true and the latter is wrong showing that there are not $t_0' \in \mathfrak{T}_i$ and $\mathbf{Z}_0' \in \partial\mathcal{D}_T(t_0'; \mathbf{I}; \mathbf{\Upsilon}) \cap \mathcal{D}_T(t_0'; \mathbf{I}; \mathbf{\Upsilon})$. If $\mathbf{Z}_0' \in \partial\mathcal{D}_T(t_0'; \mathbf{I}; \mathbf{\Upsilon})$ then $\mathbf{Z}_0' \notin \mathcal{D}_T(t_0'; \mathbf{I}; \mathbf{\Upsilon})$. The set $\mathcal{D}_T(t_0; \mathbf{I}; \mathbf{\Upsilon})$ is open for all $t_0 \in \mathfrak{T}_i$, for every $\mathbf{I}(.) \in \mathcal{J}^j$, and it is a neighborhood of Υ due to Definition 216 on $\mathfrak{T}_i$, for every $\mathbf{I}(.) \in \mathcal{J}^j$. Therefore, $\mathcal{D}_T(t; \mathbf{I}; \mathbf{\Upsilon}) \equiv \text{In}\mathcal{D}_T(t; \mathbf{I}; \mathbf{\Upsilon})$ and it is a neighborhood of Υ on $\mathfrak{T}_i$. Altogether, $\mathcal{D}_T(t; \mathbf{I}; \mathbf{\Upsilon})$ is neighborhood of Υ on $\mathfrak{T}_i$. In order to prove $\mathcal{D}_T(t_0; \mathbf{I}; \mathbf{\Upsilon}) \in \mathfrak{C}(\mathfrak{T}_i)$ for every $\mathbf{I}(.) \in \mathcal{J}^j$ we use a contradiction. Let there exist $[t_0^*, \mathbf{I}^*(.)] \in \mathfrak{T}_i \times \mathcal{J}^j$ such that $\mathcal{D}_T(t_0; \mathbf{I}; \mathbf{\Upsilon})$ is discontinuous at $[t_0^*, \mathbf{I}^*(.)]$. As a consequence, there are $\varepsilon^* \in \mathfrak{R}^+$ and a sequence $K^* \subseteq \{1, 2, ..., n, ...\}$ such that $t_k \longrightarrow t_0^*$, $k \longrightarrow \infty$, $k \in K^*$, and that there is $\mathbf{Z}^* \in \mathcal{D}_T(t_0^*; \mathbf{Z}_d; \mathbf{I})$ for which $\rho[\mathbf{Z}^*, \mathcal{D}_T(t_k; \mathbf{Z}_d; \mathbf{I})] \geq \varepsilon^*$, $\forall k \in K^*$, and/or there is $w^* \in \mathcal{D}_T(t_k; \mathbf{Z}_d; \mathbf{I})$, $\forall k \in K^*$, for which $\rho[\mathbf{w}^*, \mathcal{D}_T(t_0^*; \mathbf{Z}_d; \mathbf{I})] \geq \varepsilon^*$. Let $0 < \xi < \varepsilon/2$ and $\mathfrak{N}_\xi(\Upsilon) \subseteq \mathcal{D}_T(\mathfrak{T}_i; \mathbf{I}; \mathbf{\Upsilon})$, $\forall [t, t_0, \mathbf{I}(.)] \in \mathfrak{T}_0 \times \mathfrak{T}_i \times \mathcal{J}^j$, which is possible due to uniform tracking of Υ on $\mathfrak{T}_i \times \mathcal{J}^j$, (Definition 216). Let $m \in K^*$ be such that $t_m > t_0^* + \tau(t_0^*, \xi/2, \mathbf{Z}^*; \mathbf{I})$, $t_m \in \mathfrak{T}_i$. This guarantees (Definition 216): $\mathbf{Z}(t_m; t_0; \mathbf{Z}^*; \mathbf{I}) \in \mathfrak{N}_{\xi/2}(\Upsilon)$. Let $\delta = \delta(t_0^*, \mathbf{Z}^*, m, \xi/2, \mathbf{I}) \in \mathfrak{R}^+$, $\delta < \xi/2$ and $\psi = \psi(t_0^*, \mathbf{Z}^*, m, \xi/2, \mathbf{I}) \in \mathfrak{R}^+$ obey that

$$|t_j - t_0^*| < \psi \text{ and } \|\mathbf{Z}_0 - \mathbf{Z}^*\| < \delta, \; j \in K^*$$
$$\text{imply } \|\mathbf{Z}(t_m; t_j; \mathbf{Z}_0; \mathbf{I}) - \mathbf{Z}(t_m; t_0^*; \mathbf{Z}^*; \mathbf{I})\| < \xi/2,$$

which is possible due to continuity of the system motions (weak smoothness property induced by strong smoothness property). Hence, $\mathbf{Z}(t_m; t_0^*; \mathbf{Z}^*; \mathbf{I}) \in \mathfrak{N}_{\xi/2}(\Upsilon)$ implies

$$\mathbf{Z}(t_m; t_j; \mathbf{Z}_0; \mathbf{I}) \in \mathfrak{N}_\xi(\Upsilon).$$

This further yields

$$\mathbf{Z}(t_m; t_j; \mathbf{Z}_0; \mathbf{I}) \in \mathcal{D}_T(t_m; \mathbf{Z}_d; \mathbf{I})$$

and $\mathbf{Z}_0 \in \mathcal{D}_T(t_j; \mathbf{Z}_d; \mathbf{I})$. Also, $\|\mathbf{Z}_0 - \mathbf{Z}^*\| < \delta < \xi/2 < \varepsilon/4$ and $\mathbf{Z}_0 \in \mathcal{D}_T(t_j; \mathbf{I}; \mathbf{\Upsilon})$ imply $\rho[\mathbf{Z}^*, \mathcal{D}_T(t_j; \mathbf{I}; \mathbf{\Upsilon})] < \varepsilon^*$ that contradicts

$$\rho[\mathbf{Z}^*, \mathcal{D}_T(t_k; \mathbf{I}; \mathbf{\Upsilon})] \geq \varepsilon^*, \forall k \in K^*,$$

and disproves the existence of $[t_0^*, \mathbf{Z}^*, \mathbf{I}^*(.)] \in \mathfrak{T}_i \times \mathcal{D}_T(t_0^*; \mathbf{I}; \mathbf{\Upsilon}) \times \mathcal{J}^j$ for which

$$\rho[\mathbf{Z}^*, \mathcal{D}_T(t_k; \mathbf{I}; \mathbf{\Upsilon})] \geq \varepsilon^*, \forall k \in K^*.$$

In the analogous way we show that there are not w^* and t_0^* as defined above. This proves continuity of $\mathcal{D}_T(\mathfrak{T}_i; \mathbf{I}; \boldsymbol{\Upsilon})$ on $\mathfrak{T}_i$. The statement under a-2) is correct. Furthermore, $\mathcal{D}_{TU}(\mathfrak{T}_i; \mathbf{I}; \boldsymbol{\Upsilon})$ is a neighborhood of Υ by definition, (Definition 216). Its connectedness is proved by contradiction as follows. Let us assume that it is not connected. Then, there are disjoint sets $\mathcal{D}_{TUk}$, $k = 1, 2, \cdots, N$, such that $\mathcal{D}_{TU}(\mathfrak{T}_i; \mathbf{I}; \boldsymbol{\Upsilon}) = \cup[\mathcal{D}_{TUk} : k = 1, 2, ..., N]$. At least one of $\mathcal{D}_{TUk}$ is not a neighborhood of $\boldsymbol{\Upsilon}$, e.g., $\mathcal{D}_{TU1}$ and let $\mathcal{D}_{TUm}$, $\mathcal{D}_{TUm} \subset \mathcal{D}_{TU}(\mathfrak{T}_i; \mathbf{I}; \boldsymbol{\Upsilon})$, $m \in \{2, 3, \cdots, N\}$, be a connected neighborhood of Υ, which is possible because $\boldsymbol{\Upsilon}$ is a compact connected set. Then $\mathbf{Z}_0 \in \mathcal{D}_{TU1}$ implies $\mathbf{Z}(t; t_0; \mathbf{Z}_0; \mathbf{I}) \longrightarrow \mp$ as $t \longrightarrow \infty$, $\forall [t_0, \mathbf{I}(.)] \in \mathfrak{T}_i \times \mathcal{J}^j$.There is $t_1 \in \mathfrak{T}_0$ such that $\mathbf{Z}(t_1; t_0; \mathbf{Z}_0; \mathbf{I}) \notin \mathcal{D}_{TU}(\mathfrak{T}_i; \mathbf{I}; \boldsymbol{\Upsilon})$ because of continuity of $\mathbf{Z}(t; t_0; \mathbf{Z}_0; \mathbf{I})$ in $t \in \mathfrak{T}_0$, $\forall t_0 \in \mathfrak{T}_i$, and because $\mathcal{D}_{TU1}$ is a disjoint subset of $\mathcal{D}_{TU}(\mathfrak{T}_i; \mathbf{I}; \boldsymbol{\Upsilon})$, which is not a neighborhood of $\boldsymbol{\Upsilon}$. However, this is impossible due to the fact that $\mathbf{Z}[t; t_1; \mathbf{Z}(t_1; t_0; \mathbf{Z}_0; \mathbf{I}); \mathbf{I}] \equiv \mathbf{Z}(t; t_0; \mathbf{Z}_0; \mathbf{I}) \longrightarrow \mp$ as $t \longrightarrow \infty$, $\forall [t_0, \mathbf{I}(.)] \in \mathfrak{T}_i \times \mathcal{J}^j$. Hence, the assumption on disconnectedness of $\mathcal{D}_{TU}(\mathfrak{T}_i; \mathbf{I}; \boldsymbol{\Upsilon})$ is incorrect, which completes the proof of all statements under a) by noting that the invariance of $\mathcal{D}_{TU}(\mathfrak{T}_i; \mathbf{I}; \boldsymbol{\Upsilon})$ under a-3) results directly from 1) in the case $\mathcal{D}_T(\mathfrak{T}_i; \mathbf{I}; \boldsymbol{\Upsilon}) = \mathcal{D}_{TU}(\mathfrak{T}_i; \mathbf{I}; \boldsymbol{\Upsilon})$ for all $t \in \mathfrak{T}_i$.

b) The assertions under b) follow directly from those under a) in the case $\mathfrak{T}_i = \mathfrak{T}$ ∎

D.2 Lemma 2

Lemma 583 ***Statement of Lemma 2***

a) If the system (23.1) possessing weak smoothness property (Definition 368) exhibits stable tracking of $\boldsymbol{\Upsilon}$ *on* $\mathfrak{T}_0 \times \mathcal{J}^j$, $\forall t_0 \in \mathfrak{T}_i$, *and its domains* $\mathcal{D}_T(t_0; \mathbf{I}; \boldsymbol{\Upsilon})$ *of tracking and* $\mathcal{D}_{ST}(t_0; \mathbf{I}; \boldsymbol{\Upsilon})$ *of stable tracking at* $t_0 \in \mathfrak{T}_i$ *obey, respectively,* $\mathcal{D}_T(t_0; \mathbf{I}; \boldsymbol{\Upsilon}) \subseteq \mathfrak{S}(t_0; t_0; \Upsilon)$ *and* $\mathcal{D}_{ST}(t_0; \mathbf{I}; \boldsymbol{\Upsilon}) \subseteq \mathfrak{S}(t_0; t_0; \Upsilon)$ *then the domains* $\mathcal{D}_T(t_0; \mathbf{I}; \boldsymbol{\Upsilon})$, $\mathcal{D}_{ST}(t_0; \mathbf{I}; \boldsymbol{\Upsilon})$, $\mathcal{D}_T(\mathfrak{T}_i; \mathbf{I}; \boldsymbol{\Upsilon})$, *and* $\mathcal{D}_{ST}(\mathfrak{T}_i; \mathbf{I}; \boldsymbol{\Upsilon})$ *are interrelated by*

$$\mathcal{D}_{ST}(t_0; \mathbf{I}; \boldsymbol{\Upsilon}) \subseteq \mathcal{D}_T(t_0; \mathbf{I}; \boldsymbol{\Upsilon}), \ \forall [t_0, \mathbf{I}(.)] \in \mathfrak{T}_i \times \mathcal{J}^j,$$
$$\mathcal{D}_{ST}(\mathfrak{T}_i; \mathbf{I}; \boldsymbol{\Upsilon}) \subseteq \mathcal{D}_T(\mathfrak{T}_i; \mathbf{I}; \boldsymbol{\Upsilon}), \ \forall \mathbf{I}(.) \in \mathcal{J}^j.$$

b) If the system (23.1) possessing weak smoothness property (Definition 368) exhibits uniform stable tracking of $\boldsymbol{\Upsilon}$ *on* $\mathfrak{T}_0 \times \mathcal{J}^j$, $\forall t_0 \in \mathfrak{T}_i$, *and its domains* $\mathcal{D}_{TU}(\mathfrak{T}_i; \mathbf{I}; \boldsymbol{\Upsilon})$ *of uniform tracking and* $\mathcal{D}_{STU}(\mathfrak{T}_i; \mathbf{I}; \boldsymbol{\Upsilon})$ *of stable uniform tracking on* $\mathfrak{T}_i \times \mathcal{J}^j$ *obey, respectively,* $\mathcal{D}_{TU}(\mathfrak{T}_i; \mathbf{I}; \boldsymbol{\Upsilon}) \subseteq \mathfrak{S}(\mathfrak{T}_i; \Upsilon)$ *and* $\mathcal{D}_{STU}(\mathfrak{T}_i; \mathbf{I}; \boldsymbol{\Upsilon}) \subseteq \mathfrak{S}(\mathfrak{T}_i; \Upsilon)$, *where* $\mathfrak{S}(\mathfrak{T}_i; \Upsilon) = \cap[\mathfrak{S}(t; t_0; \Upsilon) : (t; t_0) \in \mathfrak{T}_0 \times \mathfrak{T}_i]$ *is a neighborhood of* Υ, *then the domains* $\mathcal{D}_{TU}(\mathfrak{T}_i; \mathbf{I}; \boldsymbol{\Upsilon})$, *and* $\mathcal{D}_{STU}(\mathfrak{T}_i; \mathbf{I}; \boldsymbol{\Upsilon})$ *are interrelated by*

$$\mathcal{D}_{STU}(\mathfrak{T}_i; \mathbf{I}; \boldsymbol{\Upsilon}) \subseteq \mathcal{D}_{TU}(\mathfrak{T}_i; \mathbf{I}; \boldsymbol{\Upsilon}), \ \forall \mathbf{I}(.) \in \mathcal{J}^j.$$

Proof. Proof of Lemma 2

a) Let the conditions under a) hold. From Definition 216 and Definition 215 it follows that $\mathbf{Z}_0 \in \mathcal{D}_{ST}(t_0; \mathbf{I}; \boldsymbol{\Upsilon})$ implies $\mathbf{Z}_0 \in \mathcal{D}_T(t_0; \mathbf{I}; \boldsymbol{\Upsilon})$ and that $\mathbf{Z}_0^* \in \mathcal{D}_T(t_0; \mathbf{I}; \boldsymbol{\Upsilon})$ need not imply $\mathbf{Z}_0^* \in \mathcal{D}_{ST}(t_0; \mathbf{I}; \boldsymbol{\Upsilon})$. This proves $\mathcal{D}_{ST}(t_0; \mathbf{I}; \boldsymbol{\Upsilon}) \subseteq \mathcal{D}_T(t_0; \mathbf{I}; \boldsymbol{\Upsilon})$, $\forall [t_0, \mathbf{I}(.)] \in \mathfrak{T}_i \times \mathcal{J}^j$. This,

$$\mathcal{D}_{ST}(\mathfrak{T}_i; \mathbf{I}; \boldsymbol{\Upsilon}) = \cap[\mathcal{D}_{ST}(t_0; \mathbf{I}; \boldsymbol{\Upsilon}) : t_0 \in \mathfrak{T}_i],$$

and $\mathcal{D}_T(\mathfrak{T}_i; \mathbf{I}; \boldsymbol{\Upsilon}) = \cap[\mathcal{D}_T(t_0; \mathbf{I}; \boldsymbol{\Upsilon}) : t_0 \in \mathfrak{T}_i]$ imply

$$\mathcal{D}_{ST}(\mathfrak{T}_i; \mathbf{I}; \boldsymbol{\Upsilon}) \subseteq \mathcal{D}_T(\mathfrak{T}_i; \mathbf{I}; \boldsymbol{\Upsilon}), \forall \mathbf{I}(.) \in \mathcal{J}^j.$$

The proof of the statement under a) is complete.

b) The proof of the statement under b) is analogous to the proof under a) due to the analogy between Definition 215 linked with Definition 216 and Definition 220 linked with Definition 221 ∎

D.3 Proof of Theorem 363

Proof. Proof of Theorem 379.

The function $W(;\Upsilon)$ induces the set $\mathfrak{W}_{\varsigma}(t,t_0)$, Definition 295.

Necessity. Let the system (3.12) possess Strong smoothness property 370. Let its motions exhibit uniform stable tracking of Υ on $\mathfrak{T}_0 \times \mathfrak{T}_i$, $\forall \mathbf{I}(.) \in \mathcal{J}^{\xi}$. Definition 221 and Definition 220 imply that the system has the domain $\mathcal{D}_{STU}(\mathfrak{T}_i;\mathbf{I};\mathbf{\Upsilon})$ of the uniform stable tracking on $\mathfrak{T}_0 \times \mathfrak{T}_i$, $\forall \mathbf{I}(.) \in \mathcal{J}^{\xi}$, and domain $\mathcal{D}_{ST}(t_0;\mathbf{I};\mathbf{\Upsilon})$ at every $t_0 \in \mathfrak{T}_i$ of the stable tracking on $\mathfrak{T}_0 \times \mathfrak{T}_i$, $\forall \mathbf{I}(.) \in \mathcal{J}^{\xi}$. The preceding definitions, Definition 215 and Definition 216, show that the system has also the uniform tracking domain $\mathcal{D}_{TU}(\mathfrak{T}_i;\mathbf{I};\mathbf{\Upsilon})$ on $\mathcal{J}^{\xi}$ and the instantaneous tracking domain $\mathcal{D}_T(t_0;\mathbf{I};\mathbf{\Upsilon})$ at every $t_0 \in \mathfrak{T}_i$ on $\mathfrak{T}_0 \times \mathfrak{T}_i$, $\forall \mathbf{I}(.) \in \mathcal{J}^{\xi}$, so that (D.1) holds:

$$\mathcal{D}_{ST}(t_0;\mathbf{I};\mathbf{\Upsilon}) \subseteq \mathcal{D}_T(t_0;\mathbf{I};\mathbf{\Upsilon}), \ \forall t_0 \in \mathfrak{T}_i, \ \forall \mathbf{I}(.) \in \mathcal{J}^{\xi}, \text{ and}$$
$$\mathcal{D}_{STU}(\mathfrak{T}_i;\mathbf{I};\mathbf{\Upsilon}) \subseteq \mathcal{D}_{TU}(\mathfrak{T}_i;\mathbf{I};\mathbf{\Upsilon}), \ \forall \mathbf{I}(.) \in \mathcal{J}^{\xi}. \tag{D.1}$$

Also, $\mathcal{D}_T(t_0;\mathbf{I};\mathbf{\Upsilon})$ is a neighborhood of $\mathbf{\Upsilon}$ for every $t_0 \in \mathfrak{T}_i$, $\forall \mathbf{I}(.) \in \mathcal{J}^{\xi}$, and $\mathcal{D}_{TU}(\mathfrak{T}_i;\mathbf{I};\mathbf{\Upsilon})$ is a neighborhood of $\mathbf{\Upsilon}$ for every $\mathbf{I}(.) \in \mathcal{J}^{\xi}$. The set $\mathfrak{S}(t,t_0;\Upsilon)$ is a neighborhood of $\mathbf{\Upsilon}$ at every $t_0 \in \mathfrak{T}_i$ and

$$\mathfrak{S}(\mathfrak{T}_i;\Upsilon) = \cap \left[\mathfrak{S}(t,t_0;\Upsilon) : (t;t_0) \in \mathfrak{T}_0 \times \mathfrak{T}_i\right]$$

is also a neighborhood of $\mathbf{\Upsilon}$ (weak smoothness property 368 induced by strong smoothness property 370). Hence, $\mathcal{D}_T(t_0;\mathbf{I};\mathbf{\Upsilon}) \cap \mathfrak{S}(t,t_0;\Upsilon) \neq \phi$ for all $t_0 \in \mathfrak{T}_i$ and $\mathcal{D}_{TU}(\mathfrak{T}_i;\mathbf{I};\mathbf{\Upsilon}) \cap \mathfrak{S}(\mathfrak{T}_i;\Upsilon) \neq \phi$.

Let us prove

$$\mathcal{D}_T(t_0;\mathbf{I};\mathbf{\Upsilon}) \subseteq \mathfrak{S}(t,t_0;\Upsilon), \forall (t,t_0) \in \mathfrak{T}_0 \times \mathfrak{T}_i, \forall \mathbf{I}(.) \in \mathcal{J}^{\xi}. \tag{D.2}$$

If this were not true then there would exist

$$[t_0, \mathbf{Z}, \mathbf{I}(.)] \in \mathfrak{T}_i \times \left[\mathcal{D}_T(t_0;\mathbf{I};\mathbf{\Upsilon}) \backslash \mathfrak{S}(t,t_0;\Upsilon)\right], \forall \mathbf{I}(.) \in \mathcal{J}^{\xi},$$

which would mean

$$\mathbf{Z} \in \mathcal{D}_T(t_0;\mathbf{I};\mathbf{\Upsilon}) \cap \partial\mathfrak{S}(t,t_0;\Upsilon) \ \ or \ \mathbf{Z} \in \mathcal{D}_T(t_0;\mathbf{I};\mathbf{\Upsilon}) \backslash Cl\mathfrak{S}(t,t_0;\Upsilon)$$

due to $\mathcal{D}_T(t_0;\mathbf{I};\mathbf{\Upsilon}) \cap \mathfrak{S}(t,t_0;\Upsilon) \neq \phi$ and the fact that $\mathfrak{S}(t,t_0;\Upsilon)$ is open (weak smoothness property 368 induced by strong smoothness property 370). If $\mathbf{Z} \in \mathcal{D}_T(t_0;\mathbf{I};\mathbf{\Upsilon}) \cap \partial\mathfrak{S}(t,t_0;\Upsilon)$ then $\inf[\rho[\mathbf{Z}(t;t_0;\mathbf{Z};\mathbf{I}),\Upsilon] : t \in \mathfrak{T}_0] > 0$ due to (ii) of strong smoothness property 370), which would mean $\mathbf{Z} \notin \mathcal{D}_T(t_0;\mathbf{I};\mathbf{\Upsilon})$ and would contradict $\mathbf{Z} \in \mathcal{D}_T(t_0;\mathbf{I};\mathbf{\Upsilon}) \cap \partial\mathfrak{S}(t,t_0;\Upsilon)$. Hence, $\mathbf{Z} \notin \mathcal{D}_T(t_0;\mathbf{I};\mathbf{\Upsilon}) \cap \partial\mathfrak{S}(t,t_0;\Upsilon)$ and $\mathcal{D}_T(t_0;\mathbf{I};\mathbf{\Upsilon}) \cap \partial\mathfrak{S}(t,t_0;\Upsilon) = \phi$. If

$$\mathbf{Z} \in \mathcal{D}_T(t_0;\mathbf{I};\mathbf{\Upsilon}) \backslash \mathrm{Cl}\mathfrak{S}(t,t_0;\Upsilon)$$

then $\lim[\rho[\mathbf{Z}(t;t_0;\mathbf{Z};\mathbf{I}),\Upsilon] : t \longrightarrow \infty] = 0$, which together with (i) of strong smoothness property 370, (ii) of weak smoothness property 368 and $\mathfrak{S}(t,t_0;\Upsilon) \in \mathfrak{C}(\mathfrak{T}_i)$ (due to (i) of weak smoothness property 368) would imply existence of $(t^*,t_0^*) \in (\mathrm{In}\mathfrak{T}_0) \times \mathfrak{T}_i$, $\forall \mathbf{I}^*(.) \in \mathcal{J}^{\xi}$, such that $\mathbf{Z}(t^*;t_0^*;\mathbf{Z};\mathbf{I}) \in \partial\mathfrak{S}(t^*,t_0^*)$. This is impossible as shown above. Assume triplet $(t^*,t_0^*) \in (\mathrm{In}\mathfrak{T}_0) \times \mathfrak{T}_i$, $\forall \mathbf{I}^*(.) \in \mathcal{J}^{\xi}$, does not exist. Hence, $\mathcal{D}_T(t_0;\mathbf{I};\mathbf{\Upsilon}) \backslash \mathrm{Cl}\mathfrak{S}(t,t_0;\Upsilon) = \phi$, which, together with $\mathcal{D}_T(t_0;\mathbf{I};\mathbf{\Upsilon}) \cap \partial\mathfrak{S}(t,t_0;\Upsilon) = \phi$ and $\mathcal{D}_T(t_0;\mathbf{I};\mathbf{\Upsilon}) \cap \mathfrak{S}(t,t_0;\Upsilon) \neq \phi$, implies $\mathfrak{S}(t,t_0;\Upsilon) \supseteq \mathcal{D}_T(t_0;\mathbf{I};\mathbf{\Upsilon}) \supseteq \mathcal{D}_{ST}(t_0;\mathbf{I};\mathbf{\Upsilon})$ by having in mind that both $\mathcal{D}_T(t_0;\mathbf{I};\mathbf{\Upsilon})$ and $\mathfrak{S}(t,t_0;\Upsilon)$ are open neighborhoods of $\mathbf{\Upsilon}$ on $\mathfrak{T}_0 \times \mathfrak{T}_i$, $\forall \mathbf{I}(.) \in \mathcal{J}^{\xi}$, in view of Lemma 1, (i) of weak smoothness property 368 and (i) of strong smoothness property 370, and that $\mathcal{D}_T(t_0;\mathbf{I};\mathbf{\Upsilon}) \supseteq \mathcal{D}_{ST}(t_0;\mathbf{I};\mathbf{\Upsilon})$ due to Lemma 2). This completes the proof of (D.2).

Let

$$\mathfrak{N}\left(t_0;\mathbf{I};\boldsymbol{\Upsilon}\right) \equiv \mathcal{D}_T\left(t_0;\mathbf{I};\boldsymbol{\Upsilon}\right) \tag{D.3}$$

so that $\mathfrak{S}\left(t_0,t_0\right) \supseteq \mathfrak{N}\left(t_0;\mathbf{I};\boldsymbol{\Upsilon}\right)$. Hence, $\mathfrak{N}\left(t_0;\mathbf{I};\boldsymbol{\Upsilon}\right)$ is an open continuous connected neighborhood of $\boldsymbol{\Upsilon}$ on $\mathfrak{T}_0 \times \mathfrak{T}_i$, $\forall \mathbf{I}(.) \in \mathcal{J}^{\xi}$, (a-2 of Lemma 1), and $\mathfrak{N}\left(\mathfrak{T}_i;\mathbf{I}\right) = \mathcal{D}_{TU}\left(\mathfrak{T}_i;\mathbf{I};\boldsymbol{\Upsilon}\right)$ is a connected neighborhood of $\boldsymbol{\Upsilon}$. Also,

$$\mathfrak{N}\left(\mathfrak{T}_i;\mathbf{I}\right) = \cap\left[\mathfrak{N}\left(t_0;\mathbf{I};\boldsymbol{\Upsilon}\right) : t_0 \in \mathfrak{T}_i\right]$$

because of $\mathfrak{N}\left(t_0;\mathbf{I};\boldsymbol{\Upsilon}\right) \equiv \mathcal{D}_T\left(t_0;\mathbf{I};\boldsymbol{\Upsilon}\right)$ and

$$\begin{aligned}\mathfrak{N}\left(\mathfrak{T}_i;\mathbf{I}\right) = \cap\left[\mathfrak{N}\left(t_0;\mathbf{I};\boldsymbol{\Upsilon}\right) : t_0 \in \mathfrak{T}_i\right] &= \cap\left[\mathcal{D}_T\left(t_0;\mathbf{I};\boldsymbol{\Upsilon}\right) : t_0 \in \mathfrak{T}_i, \forall \mathbf{I}(.) \in \mathcal{J}^{\xi}\right] \\ &= \mathcal{D}_{TU}\left(\mathfrak{T}_i;\mathbf{I};\boldsymbol{\Upsilon}\right), \forall \mathbf{I}(.) \in \mathcal{J}^{\xi}.\end{aligned}$$

These results prove the necessity of the conditions 1) and 2). From $\mathcal{D}_T\left(t_0;\mathbf{I};\boldsymbol{\Upsilon}\right) = \mathfrak{N}\left(t_0;\mathbf{I};\boldsymbol{\Upsilon}\right)$, Definition 215, it follows that Υ is a unique invariant set in $\mathfrak{N}\left(t_0;\mathbf{I};\boldsymbol{\Upsilon}\right)$; i.e., that there is not an invariant set in $\left[\mathfrak{N}\left(t_0;\mathbf{I};\boldsymbol{\Upsilon}\right)\backslash\Upsilon\right]$, $\forall t_0 \in \mathfrak{T}_i, \forall \mathbf{I}(.) \in \mathcal{J}^{\xi}$,. It follows that $\boldsymbol{\Upsilon}$ is the unique invariant set relative to the motions of the system (3.12) in $\mathfrak{N}\left(t_0;\mathbf{I};\boldsymbol{\Upsilon}\right)$, $\forall t_0 \in \mathfrak{T}_i, \forall \mathbf{I}(.) \in \mathcal{J}^{\xi}$. This proves the necessity of the condition 3). From $\mathfrak{N}\left(t_0;\mathbf{I};\boldsymbol{\Upsilon}\right) \equiv \mathcal{D}_T\left(t_0;\mathbf{I};\boldsymbol{\Upsilon}\right)$ it follows that $\mathfrak{T}_0$ is the interval of the existence of the system motions $\mathbf{Z}\left(t;t_0;\mathbf{Z}_0;\mathbf{I}\right)$ for every $\left(t,t_0,\mathbf{Z}_0\right) \in \mathfrak{T}_0 \times \mathfrak{T}_i \times \mathfrak{N}\left(t_0;\mathbf{I};\boldsymbol{\Upsilon}\right)$, $\forall \mathbf{I}(.) \in \mathcal{J}^{\xi}$, due to Definition 215. Let $W\left(.;\Upsilon\right) \in L\left[\mathfrak{T}_0,\mathfrak{T}_i,\mathfrak{S},\mathcal{J}^{\xi},\Upsilon\right]$, Definition 373, be an arbitrarily selected positive definite decrescent function with respect to Υ on $\mathfrak{T}_i \times \mathfrak{N}\left(t;\mathbf{I};\boldsymbol{\Upsilon}\right)$. Therefore, there is $\mu > 0$ such that there exists a solution $V\left(.;\Upsilon\right)$ to the equations (24.5) and (24.6), which is continuous in $\left(t,\mathbf{Z}\right) \in \mathfrak{T}_i \times \mathfrak{N}_{\mu}\left(\Upsilon\right)$ and satisfies (24.7). Hence,

$$\left|V\left(t,\mathbf{Z};\mathbf{I};\boldsymbol{\Upsilon}\right)\right| < \infty, \ \forall\left(t,\mathbf{Z}\right) \in \mathfrak{T}_i \times Cl\mathfrak{N}_{\mu}\left(\Upsilon\right), \forall \mathbf{I}(.) \in \mathcal{J}^{\xi}. \tag{D.4}$$

Let $\beta \in]1,\infty[$ and $\zeta \in \mathfrak{R}^{+}$ be such that

$$\mathfrak{N}_{\beta}\left(\Upsilon\right) \cap \mathfrak{N}_{\mu}\left(\Upsilon\right) \cap \mathfrak{S}\left(t,t_0;\Upsilon\right) \supset \mathfrak{W}_{\zeta}\left(t;\mathbf{I};\boldsymbol{\Upsilon}\right), \ \forall t \in \mathfrak{T}_i, \forall \mathbf{I}(.) \in \mathcal{J}^{\xi}. \tag{D.5}$$

Existence of such β and ζ is guaranteed by positive definiteness of $W\left(.;\Upsilon\right)$ on $\mathfrak{T}_i \times \mathfrak{N}\left(t;\mathbf{I};\boldsymbol{\Upsilon}\right)$ and by the fact that $\mathfrak{S}\left(t,t_0;\Upsilon\right)$ is a neighborhood of $\mathbf{Z} = \mathbf{0}_K$. Let $t_0 \in \mathfrak{T}_i$ be arbitrary and $\tau \in \mathfrak{R}^{+}$, $\tau = \tau\left(t_0,\mathbf{Z}_0,\zeta;\mathbf{I};W\right)$, be such that for any $\mathbf{Z}_0 \in \mathfrak{N}\left(t_0;\mathbf{I};\boldsymbol{\Upsilon}\right)$ the following condition holds:

$$\mathbf{Z}\left(t;t_0;\mathbf{Z}_0;\mathbf{I}\right) \in Cl\mathfrak{W}_{\zeta}\left(t;\mathbf{I};\boldsymbol{\Upsilon}\right), \ \forall t \in \left[t_0+\tau,\infty\right[, \forall \mathbf{I}(.) \in \mathcal{J}^{\xi}. \tag{D.6}$$

Such τ exists due to Definition 215, Definition 220, $\mathbf{Z}_0 \in \mathfrak{N}\left(t_0;\mathbf{I};\boldsymbol{\Upsilon}\right)$ and the fact that $\mathcal{D}_{ST}\left(t_0;\mathbf{I};\boldsymbol{\Upsilon}\right) \equiv \mathfrak{N}\left(t_0;\mathbf{I};\boldsymbol{\Upsilon}\right)$. Notice that $\mathbf{Z}_0 \in \mathfrak{N}\left(t_0;\mathbf{I};\boldsymbol{\Upsilon}\right)$ also yields

$$\rho\left[\mathbf{Z}\left(\infty;t_0;\mathbf{Z}_0;\mathbf{I}\right),\Upsilon\right] = 0, \ \text{ i.e., } \mathbf{Z}\left(\infty;t_0;\mathbf{Z}_0;\mathbf{I}\right) = \mathbf{0}_K, \forall \mathbf{I}(.) \in \mathcal{J}^{\xi}. \tag{D.7}$$

Let (24.5) be integrated from $t \in \mathfrak{T}_0$ to ∞,

$$\begin{aligned}&V\left[\infty,\mathbf{Z}\left(\infty;t_0;\mathbf{Z}_0;\mathbf{I}\right);\mathbf{I};\boldsymbol{\Upsilon}\right] - V\left[t,\mathbf{Z}\left(t;t_0;\mathbf{Z}_0;\mathbf{I}\right);\mathbf{I};\boldsymbol{\Upsilon}\right] \\ &\quad = -\int_t^{\infty} W\left[\sigma,\mathbf{Z}\left(\sigma;t_0;\mathbf{Z}_0;\mathbf{I}\right);\mathbf{I};\boldsymbol{\Upsilon}\right]d\sigma, \\ &\forall\left(t,\mathbf{Z}\right) \in \mathfrak{T}_0 \times \mathfrak{N}\left(t_0;\mathbf{I};\boldsymbol{\Upsilon}\right), \forall \mathbf{I}(.) \in \mathcal{J}^{\xi}.\end{aligned} \tag{D.8}$$

Now, (24.6) and (D.7) reduce (D.8) to

$$\begin{aligned}&V\left[t,\mathbf{Z}\left(t;t_0;\mathbf{Z}_0;\mathbf{I}\right);\mathbf{I};\boldsymbol{\Upsilon}\right] = \int_t^{\infty} W\left[\sigma,\mathbf{Z}\left(\sigma;t_0;\mathbf{Z}_0;\mathbf{I}\right);\mathbf{I};\boldsymbol{\Upsilon}\right]d\sigma = \\ &= \int_t^{t_0+\tau} W\left[\sigma,\mathbf{Z}\left(\sigma;t_0;\mathbf{Z}_0;\mathbf{I}\right);\mathbf{I};\boldsymbol{\Upsilon}\right]d\sigma + \int_{t_0+\tau}^{\infty} W\left[\sigma,\mathbf{Z}\left(\sigma;t_0;\mathbf{Z}_0;\mathbf{I}\right);\mathbf{I};\boldsymbol{\Upsilon}\right]d\sigma, \\ &\forall\left(t,\mathbf{Z}_0\right) \in \mathfrak{T}_0 \times \mathfrak{N}\left(t_0;\mathbf{I};\boldsymbol{\Upsilon}\right), \forall \mathbf{I}(.) \in \mathcal{J}^{\xi}.\end{aligned} \tag{D.9}$$

Invariance of $\mathcal{D}_T(t_0;\mathbf{I};\boldsymbol{\Upsilon})$ on $\mathfrak{T}_i$ with respect to the system motions (Lemma 1) $\mathfrak{S}(t,t_0;\Upsilon)\supseteq\mathcal{D}_T(t_0;\mathbf{I};\boldsymbol{\Upsilon})\equiv\mathfrak{N}(t_0;\mathbf{I};\boldsymbol{\Upsilon})$ (D.3), continuity of $\mathbf{Z}(t;t_0;\mathbf{Z}_0;\mathbf{I})$ in $(t;t_0;\mathbf{Z}_0)\in\mathfrak{T}_0\times\mathfrak{T}_i\times\mathfrak{S}(t,t_0;\Upsilon),\forall\mathbf{I}(.)\in\mathcal{J}^{\xi}$, (due to (i-b) of weak smoothness property 368), continuity, positive definiteness, and decrescency of $W(.;\Upsilon)$ relative to Υ on $\mathfrak{T}_i\times\mathfrak{S}(t;t_0;\Upsilon)$, the definition of τ, (D.6), and compactness of $[t,t_0+\gamma]$ for any $\gamma\in\mathfrak{R}_+$ imply:

$$\left|\int_t^{t_0+\gamma}W\left[\sigma,\mathbf{Z}\left(\sigma;t_0;\mathbf{Z}_0;\mathbf{I}\right);\mathbf{I};\boldsymbol{\Upsilon}\right]d\sigma\right|<\infty,$$
$$\forall\left(\gamma,t_0,\mathbf{Z}_0\right)\in\mathfrak{R}_+\times\mathfrak{T}_i\times\mathfrak{N}\left(t_0;\mathbf{I};\boldsymbol{\Upsilon}\right),\ \forall\mathbf{I}(.)\in\mathcal{J}^{\xi}. \tag{D.10}$$

Now, (D.4)-(D.6), (D.9), and (D.10) for $\gamma=\tau$ yield

$$\left|V\left[t,\mathbf{Z}\left(t;t_0;\mathbf{Z}_0;\mathbf{I}\right);\mathbf{I};\boldsymbol{\Upsilon}\right]\right|<\infty,$$
$$\forall\left(t,t_0,\mathbf{Z}_0\right)\in\mathfrak{T}_0\times\mathfrak{T}_i\times\mathfrak{N}\left(t_0;\mathbf{I};\boldsymbol{\Upsilon}\right),\forall\mathbf{I}(.)\in\mathcal{J}^{\xi}. \tag{D.11}$$

Let us replace t by t_0 and $\mathbf{Z}_0$ by $\mathbf{Z}$ in (D.11). Then,

$$\left|V\left(t,\mathbf{Z};\mathbf{I};\boldsymbol{\Upsilon}\right)\right|<\infty,\ \forall\left(t,\mathbf{Z}\right)\in\mathfrak{T}_i\times\mathfrak{N}\left(t;\mathbf{I};\boldsymbol{\Upsilon}\right),\forall\mathbf{I}(.)\in\mathcal{J}^{\xi}. \tag{D.12}$$

Continuity of $W(.;\Upsilon)$ on $\mathfrak{T}_i\times\mathfrak{S}(t;t_0;\Upsilon)$, $W(.;\Upsilon)\in L\left[\mathfrak{T}_0,\mathfrak{T}_i,\mathfrak{S},\mathcal{J}^{\xi},\Upsilon\right]$, Definition 373, $\mathfrak{S}(t,t_0;\Upsilon)\supseteq\mathfrak{N}(\mathfrak{T}_i;\mathbf{I})(t)$, (D.9), and (D.12) prove

$$V\left(t,\mathbf{Z};\mathbf{I};\boldsymbol{\Upsilon}\right)\in\mathfrak{C}\left[\mathfrak{T}_i\times\mathfrak{N}\left(t;\mathbf{I};\boldsymbol{\Upsilon}\right)\right]=\mathfrak{C}\left[\mathfrak{T}_i\times\mathcal{D}_{ST}\left(t_0;\mathbf{I};\boldsymbol{\Upsilon}\right)\right],$$
$$\forall\mathbf{I}(.)\in\mathcal{J}^{\xi}. \tag{D.13}$$

Invariance of $\mathcal{D}_T(t_0;\mathbf{I};\boldsymbol{\Upsilon})$, (a-1) of Lemma 1), $\mathcal{D}_T(t;\mathbf{Z}_d;\mathbf{I})\equiv\mathfrak{N}(t;\mathbf{I};\boldsymbol{\Upsilon})$, (D.3), continuity of $\mathbf{Z}(t;t_0;\mathbf{Z}_0;\mathbf{I})$ in $(t,t_0,\mathbf{Z}_0)\in\mathfrak{T}_0\times\mathfrak{T}_i\times\mathcal{D}_T(t_0;\mathbf{I};\boldsymbol{\Upsilon}),\forall\mathbf{I}(.)\in\mathcal{J}^{\xi}$, positive definiteness, and decrescency of $W(.;\Upsilon)$ relative to Υ on $\mathfrak{T}_0\times\mathfrak{T}_i\times\mathfrak{S}(t;t_0;\Upsilon)$, $W(.;\Upsilon)\in L\left[\mathfrak{T}_0,\mathfrak{T}_i,\mathfrak{S},\mathcal{J}^{\xi},\Upsilon\right]$, (24.7), Definition 373, $\mathfrak{S}(t,t_0;\Upsilon)\supseteq\mathfrak{N}(t;\mathbf{I};\boldsymbol{\Upsilon})$, the definition of τ (D.6), and compactness of $[t,t_0+\tau]$ guarantee the existence of $\zeta_k(.;\Upsilon):\mathfrak{R}^K\times\mathcal{J}^{\xi}\longrightarrow\mathfrak{R}$, $k=1,2$, $\zeta_1(\mathbf{Z};\mathbf{I};\boldsymbol{\Upsilon})\in\mathfrak{C}\left[\mathfrak{N}_M(\mathfrak{T}_i;\mathbf{I};\boldsymbol{\Upsilon})\right],\forall\mathbf{I}(.)\in\mathcal{J}^{\xi}$, and $\zeta_2(\mathbf{Z};\mathbf{I};\boldsymbol{\Upsilon})\in\mathfrak{C}\left[\mathfrak{N}_m(\mathfrak{T}_i;\mathbf{I};\boldsymbol{\Upsilon})\right],\forall\mathbf{I}(.)\in\mathcal{J}^{\xi}$, where

$$\mathfrak{N}_M\left(\mathfrak{T}_i;\mathbf{I};\boldsymbol{\Upsilon}\right)=\cup\left[\mathfrak{N}\left(t;\mathbf{I};\boldsymbol{\Upsilon}\right):t\in\mathfrak{T}_i\right],\ \forall\mathbf{I}(.)\in\mathcal{J}^{\xi},$$
$$\mathfrak{N}_m\left(\mathfrak{T}_i;\mathbf{I};\boldsymbol{\Upsilon}\right)=\cap\left[\mathfrak{N}\left(t;\mathbf{I};\boldsymbol{\Upsilon}\right):t\in\mathfrak{T}_i\right],\ \forall\mathbf{I}(.)\in\mathcal{J}^{\xi},$$

and $\psi_k(.;\Upsilon):\mathfrak{R}^K\times\mathcal{J}^{\xi}\longrightarrow\mathfrak{R}$, $k=1,2$, such that

$$0<\zeta_1\left(\mathbf{Z}_0;\mathbf{I};\boldsymbol{\Upsilon}\right)\leq\int_t^{t_0+\tau}\psi_1\left[\mathbf{Z}\left(\sigma;t_0;\mathbf{Z}_0;\mathbf{I}\right);\mathbf{I};\boldsymbol{\Upsilon}\right]d\boldsymbol{\sigma},$$
$$\forall\left(t,t_0,\mathbf{Z}\right)\in\mathfrak{T}_0\times\mathfrak{T}_i\times\left[\mathfrak{N}\left(t_0;\mathbf{I};\boldsymbol{\Upsilon}\right)\backslash Cl\mathfrak{N}_{\mu}\left(\Upsilon\right)\right],\forall\mathbf{I}(.)\in\mathcal{J}^{\xi}, \tag{D.14}$$

$$\infty>\zeta_2\left(\mathbf{Z}_0;\mathbf{I};\boldsymbol{\Upsilon}\right)\geq\int_t^{t_0+\tau}\psi_2\left[\mathbf{Z}\left(\sigma;t_0;\mathbf{Z}_0;\mathbf{I}\right);\mathbf{I};\boldsymbol{\Upsilon}\right]d\boldsymbol{\sigma},$$
$$\forall\left(t,t_0,\mathbf{Z}\right)\in\mathfrak{T}_0\times\mathfrak{T}_i\times\left[\mathfrak{N}_m\left(\mathfrak{T}_i;\mathbf{I};\boldsymbol{\Upsilon}\right)\backslash Cl\mathfrak{N}_{\mu}\left(\Upsilon\right)\right],\forall\mathbf{I}(.)\in\mathcal{J}^{\xi}, \tag{D.15}$$

and

$$\psi_1\left(\mathbf{Z};\mathbf{I};\boldsymbol{\Upsilon}\right)\in\mathfrak{C}\left[\mathfrak{N}_M\left(\mathfrak{T}_i;\mathbf{I};\boldsymbol{\Upsilon}\right)\right],\forall\mathbf{I}(.)\in\mathcal{J}^{\xi},$$
$$\psi_2\left(\mathbf{Z};\mathbf{I};\boldsymbol{\Upsilon}\right)\in\mathfrak{C}\left[\mathfrak{N}_m\left(\mathfrak{T}_i;\mathbf{I};\boldsymbol{\Upsilon}\right)\right],\forall\mathbf{I}(.)\in\mathcal{J}^{\xi}, \tag{D.16}$$

$$\psi_k\left(\mathbf{Z};\mathbf{I};\boldsymbol{\Upsilon}\right)=0\Longleftrightarrow\mathbf{Z}\in\partial\boldsymbol{\Upsilon},\ \forall\mathbf{I}\in\mathcal{J}^{\xi},\ k=1,2, \tag{D.17}$$

$$\psi_1(\mathbf{Z};\mathbf{I};\mathbf{\Upsilon}) > 0,\ \forall \mathbf{Z} \in [\mathfrak{N}_M(\mathfrak{T}_i;\mathbf{I};\mathbf{\Upsilon}) \backslash \mathfrak{D}], \forall \mathbf{I}(.) \in \mathcal{J}^\xi,$$
$$\psi_2(\mathbf{Z};\mathbf{I};\mathbf{\Upsilon}) > 0,\ \forall \mathbf{Z} \in [\mathfrak{N}_m(\mathfrak{T}_i;\mathbf{I};\mathbf{\Upsilon}) \backslash \mathfrak{D}], \forall \mathbf{I}(.) \in \mathcal{J}^\xi, \tag{D.18}$$

$$\psi_1(\mathbf{Z};\mathbf{I};\mathbf{\Upsilon}) \leq W(t,\mathbf{Z};\mathbf{I};\mathbf{\Upsilon}),\ \forall (t,\mathbf{Z}) \in \mathfrak{T}_i \times \mathfrak{N}_M(\mathfrak{T}_i;\mathbf{I};\mathbf{\Upsilon}), \forall \mathbf{I}(.) \in \mathcal{J}^\xi, \tag{D.19}$$
$$\psi_2(\mathbf{Z};\mathbf{I};\mathbf{\Upsilon}) \geq W(t,\mathbf{Z};\mathbf{I};\mathbf{\Upsilon}),\ \forall (t,\mathbf{Z}) \in \mathfrak{T}_i \times \mathfrak{N}_m(\mathfrak{T}_i;\mathbf{I};\mathbf{\Upsilon}), \forall \mathbf{I}(.) \in \mathcal{J}^\xi. \tag{D.20}$$

Such functions $\psi_k(.;\Upsilon)$, $k = 1,2$, exist due to decrescency and positive definiteness of $W(.;\mathbf{\Upsilon})$ relative to the set Υ on the set product $\mathfrak{T}_0 \times \mathfrak{T}_i \times \mathfrak{S}(t;t_0;\Upsilon)$,

$$W(.;\Upsilon) \in L\left[\mathfrak{T}_0, \mathfrak{T}_i, \mathfrak{S}, \mathcal{J}^\xi, \Upsilon\right]$$

and $\mathfrak{S}(t;t_0;\Upsilon) \supseteq \mathfrak{N}(t;\mathbf{I};\mathbf{\Upsilon})$. They can be of the form

$$\psi_k(\mathbf{Z};\mathbf{I};\mathbf{\Upsilon}) = g_k[\rho(\mathbf{Z},\mathbf{\Upsilon});\mathbf{I};\mathbf{\Upsilon}], k = 1,2,$$

together with $g_k(.;\mathbf{I};\mathbf{\Upsilon})$ in the class $\mathcal{K}$, $g_k(.;\mathbf{I};\mathbf{\Upsilon}) \in \mathcal{K}$ for every $\mathbf{I}(.) \in \mathcal{J}^\xi$. Let $\vartheta_k(.;\Upsilon) : \mathfrak{R}^K \times \mathcal{J}^\xi \longrightarrow \mathfrak{R}$, $k = 1,2$, obey (D.21) through (D.23):

$$\vartheta_k(\mathbf{Z};\mathbf{I};\mathbf{\Upsilon}) \in \mathfrak{C}\left(\mathfrak{R}^K\right) \ and\ \vartheta_k(\mathbf{Z};\mathbf{I};\mathbf{\Upsilon}) = 0 \Longleftrightarrow \mathbf{Z} \in \partial\mathbf{\Upsilon},$$
$$\forall \mathbf{I}(.) \in \mathcal{J}^\xi,\ k = 1,2, \tag{D.21}$$

$$0 < \vartheta_1(\mathbf{Z};\mathbf{I};\mathbf{\Upsilon}) \leq \left\{ \begin{array}{c} \zeta_1(\mathbf{Z};\mathbf{I};\mathbf{\Upsilon}), \\ \forall \mathbf{Z} \in [\mathfrak{N}_M(\mathfrak{T}_i;\mathbf{I};\mathbf{\Upsilon}) \backslash Cl\mathfrak{N}_\mu(\Upsilon)], \\ \omega_\mu(\mathbf{Z};\mathbf{I}),\ \forall \mathbf{Z} \in [Cl\mathfrak{N}_\mu(\Upsilon) \backslash \Upsilon], \\ \forall \mathbf{I}(.) \in \mathcal{J}^\xi, \end{array} \right\} \tag{D.22}$$

$$\vartheta_2(\mathbf{Z};\mathbf{I};\mathbf{\Upsilon}) \geq \left\{ \begin{array}{c} \left\langle \begin{array}{c} \zeta_2(\mathbf{Z};\mathbf{I};\mathbf{\Upsilon}) + \omega_\mu(\mathbf{Z}_\tau;\mathbf{I};\mathbf{\Upsilon}), \\ \mathbf{Z}_\tau = \mathbf{Z}(\tau;t;\mathbf{Z};\mathbf{I}),\ \forall (t,\mathbf{Z}) \in \\ \in \mathfrak{T}_i \times [\mathfrak{N}_m(\mathfrak{T}_i;\mathbf{I};\mathbf{\Upsilon}) \backslash Cl\mathfrak{N}_\mu(\Upsilon)], \end{array} \right\rangle, \\ \forall \mathbf{I}(.) \in \mathcal{J}^\xi, \\ \omega_\mu(\mathbf{Z};\mathbf{I}), \\ \forall \mathbf{Z} \in [Cl\mathfrak{N}_\mu(\Upsilon) \backslash \mathfrak{D}], \forall \mathbf{I}(.) \in \mathcal{J}^\xi, \end{array} \right\} \tag{D.23}$$

where $\omega_\mu(.;\mathbf{\Upsilon})$ is defined by (24.7). Now, (24.7), (D.9), positive definiteness of $W(.;\mathbf{\Upsilon})$ relative to Υ on $\mathfrak{T}_0 \times \mathfrak{T}_i \times \mathfrak{S}(t;t_0;\Upsilon)$, the uniqueness of the invariant set Υ in $\mathfrak{S}(t;t_0;\Upsilon)$, $\forall (t,t_0) \in \mathfrak{T}_0 \times \mathfrak{T}_i$, and (D.14) through (D.23) yield the following for $(t_0, \mathbf{Z}_0) = (t, \mathbf{Z})$:

$$\vartheta_1(\mathbf{Z};\mathbf{I};\mathbf{\Upsilon}) \leq V(t,\mathbf{Z};\mathbf{I}), \forall (t,\mathbf{Z}) \in \mathfrak{T}_i \times \mathfrak{N}(t;\mathbf{I};\mathbf{\Upsilon}), \forall \mathbf{I}(.) \in \mathcal{J}^\xi, \tag{D.24}$$

$$V(t,\mathbf{Z};\mathbf{I}) \leq \vartheta_2(\mathbf{Z};\mathbf{I};\mathbf{\Upsilon}),\ \forall (t,\mathbf{Z}) \in \mathfrak{T}_i \times \mathfrak{N}_m(\mathfrak{T}_i;\mathbf{I};\mathbf{\Upsilon}), \forall \mathbf{I}(.) \in \mathcal{J}^\xi, \tag{D.25}$$

$$V(t,\mathbf{Z};\mathbf{I};\mathbf{\Upsilon}) = 0 \Longleftrightarrow \mathbf{Z} \in \partial\mathbf{\Upsilon},\ \forall t \in \mathfrak{T}_i, \forall \mathbf{I}(.) \in \mathcal{J}^\xi, \tag{D.26}$$

From $W(.;\Upsilon) \in L\left[\mathfrak{T}_0, \mathfrak{T}_i, \mathfrak{S}, \mathcal{J}^\xi, \Upsilon\right]$, (24.6), (D.13), and (D.24) through (D.26) it follows that a solution function $V(.;\Upsilon)$ to (24.5), (24.6) is decrescent, positive definite, hence continuous, relative to Υ on $\mathfrak{T}_0 \times \mathfrak{T}_i \times \mathfrak{N}(t;\mathbf{I};\mathbf{\Upsilon}), \forall \mathbf{I}(.) \in \mathcal{J}^\xi$. Let it be assumed that there exist two such solutions $V_1(.)$ and $V_2(.)$ to (24.5), (24.6). Hence,

$$V_1(t_0,\mathbf{Z}_0;\mathbf{I};\mathbf{\Upsilon}) - V_2(t_0,\mathbf{Z}_0;\mathbf{I};\mathbf{\Upsilon})$$
$$= \int_{t_0}^{\infty} \left\{ \begin{array}{c} W[\sigma,\mathbf{Z}_1(\sigma;t_0;\mathbf{Z}_0;\mathbf{I});\mathbf{I};\mathbf{\Upsilon}] - \\ -W[\sigma,\mathbf{Z}_2(\sigma;t_0;\mathbf{Z}_0;\mathbf{I});\mathbf{I};\mathbf{\Upsilon}] \end{array} \right\} d\sigma,$$
$$\forall (t,t_0,\mathbf{Z}_0) \in \mathfrak{T}_0 \times \mathfrak{T}_i \times \mathfrak{N}(t_0;\mathbf{I};\mathbf{\Upsilon}), \forall \mathbf{I}(.) \in \mathcal{J}^\xi. \tag{D.27}$$

Uniqueness of the motions $\mathbf{Z}_2(.;t_0;\mathbf{Z}_0;\mathbf{I})$, $\forall (t,t_0,\mathbf{Z}_0) \in \mathfrak{T}_0 \times \mathfrak{T}_i \times \mathfrak{N}(t_0;\mathbf{I};\mathbf{\Upsilon})$, $\forall \mathbf{I}(.) \in \mathcal{J}^\xi$ (weak smoothness property 368), $\mathfrak{S}(t_0;t_0;\mathbf{\Upsilon}) \supseteq \mathfrak{N}(t_0;\mathbf{I};\mathbf{\Upsilon})$, and uniqueness of $W(t,\mathbf{Z};\mathbf{I};\mathbf{\Upsilon})$

for every $(t, t_0, \mathbf{Z}) \in \mathfrak{T}_0 \times \mathfrak{T}_i \times \mathfrak{S}(t; t_0; \Upsilon)$, $\forall \mathbf{I}(.) \in \mathcal{J}^{\xi}$, [due to positive definiteness of $W(.; \Upsilon)$ on $\mathfrak{S}(t; t_0; \Upsilon)$], imply

$$\int_{t_0}^{\infty} \{W[\sigma, \mathbf{Z}_1(\sigma; t_0; \mathbf{Z}_0; \mathbf{I}); \mathbf{I}; \mathbf{\Upsilon}] - W[\sigma, \mathbf{Z}_2(\sigma; t_0; \mathbf{Z}_0; \mathbf{I}); \mathbf{I}; \mathbf{\Upsilon}]\} d\sigma$$

$$= \int_{t_0}^{\infty} \{W[\sigma, \mathbf{Z}(\sigma; t_0; \mathbf{Z}_0; \mathbf{I}); \mathbf{I}; \mathbf{\Upsilon}] - W[\sigma, \mathbf{Z}(\sigma; t_0; \mathbf{Z}_0; \mathbf{I}); \mathbf{I}; \mathbf{\Upsilon}]\} d\sigma = 0,$$

$$\forall (t, t_0, \mathbf{Z}_0) \in \mathfrak{T}_0 \times \mathfrak{T}_i \times \mathfrak{N}(t_0; \mathbf{I}; \mathbf{\Upsilon}), \forall \mathbf{I}(.) \in \mathcal{J}^{\xi}.$$

This and (D.27) prove

$$V_1(t_0, \mathbf{Z}_0; \mathbf{I}; \mathbf{\Upsilon}) \equiv V_2(t_0, \mathbf{Z}_0; \mathbf{I}; \mathbf{\Upsilon}).$$

The function $V(.; \Upsilon)$ is the unique solution to (24.5), (24.6). This completes the proof of the necessity of the condition 4-a-i). If $\partial\mathfrak{N}(t_0; \mathbf{I}; \mathbf{\Upsilon}) \neq \phi$ then let $t_0 \in \mathfrak{T}_i$ and $\mathbf{I}(.) \in \mathcal{J}^{\xi}$ be arbitrary and $\mathbf{Z}_k$, $k = 1, 2, ...$, be a sequence converging to U, $\mathbf{Z}_k \longrightarrow U$ as $k \longrightarrow \infty$, $\mathbf{Z}_k \in \mathfrak{N}(t_0; \mathbf{I}; \mathbf{\Upsilon})$, for all $k = 1, 2, ...$, and $U \in \partial\mathfrak{N}(t_0; \mathbf{I}; \mathbf{\Upsilon})$. Let $\zeta \in \mathfrak{R}^+$ be arbitrarily chosen so that $\mathfrak{N}(\mathfrak{T}_i; \mathbf{I}; \mathbf{\Upsilon}) \supset \mathfrak{W}_{\zeta}(t; \mathbf{I}; \mathbf{\Upsilon})$ for all $t \in \mathfrak{T}_i, \forall \mathbf{I}(.) \in \mathcal{J}^{\xi}$. Such ζ exists because $W(.; \Upsilon)$ is positive definite relative to Υ on $\mathfrak{S}(t; t_0; \Upsilon)$, $\mathfrak{N}(\mathfrak{T}_i; \mathbf{I}; \mathbf{\Upsilon}) \subseteq \mathfrak{S}(t; t_0; \Upsilon)$, $\forall t \in \mathfrak{T}_i$, and defines $\text{Cl}\mathfrak{W}_{\zeta}(t; \mathbf{I}; \mathbf{\Upsilon})$, and because $\mathfrak{N}(\mathfrak{T}_i; \mathbf{I}; \mathbf{\Upsilon})$ is a neighborhood of $\mathbf{\Upsilon}$. Let τ_k, $\tau_k = \tau_k(t_0; \mathbf{Z}_k; \zeta; \mathbf{I}; \mathbf{\Upsilon}) \in \mathfrak{R}_+$, be the first instant satisfying (D.28):

$$\mathbf{Z}(t; t_0; \mathbf{Z}_0; \mathbf{I}) \in \text{Cl}\mathfrak{W}_{\zeta}(t; \mathbf{I}; \mathbf{\Upsilon}), \ \forall t \in \ [t_0 + \tau_k, \infty[\times \mathfrak{T}_i, \forall \mathbf{I}(.) \in \mathcal{J}^{\xi}. \tag{D.28}$$

The existence of such τ_k is ensured by $\mathbf{Z}_k \in \mathfrak{N}(t_0; \mathbf{I}; \mathbf{\Upsilon})$, $\mathfrak{N}(t; \mathbf{I}; \mathbf{\Upsilon}) = \mathcal{D}_T(t; \mathbf{I}; \mathbf{\Upsilon})$ and by the fact that $\cap [\mathfrak{W}_{\varsigma}(t; \mathbf{I}; \mathbf{\Upsilon}) : t \in \mathfrak{T}_i]$ is a neighborhood of $\mathbf{\Upsilon}$ due to the decrescency of $W(.; \Upsilon)$ relative to Υ on $\mathfrak{T}_0 \times \mathfrak{T}_i \times \mathfrak{N}(t; \mathbf{I}; \mathbf{\Upsilon}), \forall \mathbf{I}(.) \in \mathcal{J}^{\xi}$ (Proposition 304). Continuity of $\mathbf{Z}(t; t_0; \mathbf{Z}_0; \mathbf{I})$ in $(t; t_0; \mathbf{Z}_0) \in \mathfrak{T}_0 \times \mathfrak{T}_i \times \mathfrak{S}(t; t_0; \Upsilon)$, $\forall \mathbf{I}(.) \in \mathcal{J}^{\xi}$, (weak smoothness property 368), $\mathfrak{S}(t_0; t_0; \Upsilon) \supseteq \mathcal{D}_T(t_0; \mathbf{I}; \mathbf{\Upsilon}) = \mathfrak{N}(t_0; \mathbf{I}; \mathbf{\Upsilon})$, positive invariance of $\mathcal{D}_T(t; \mathbf{I}; \mathbf{\Upsilon})$, [a) of Lemma 1], the fact that $\cap [\mathcal{D}_T(t; \mathbf{I}; \mathbf{\Upsilon}) : t \in \mathfrak{T}_i] = \mathcal{D}_{TU}(\mathfrak{T}_i; \mathbf{I}; \mathbf{\Upsilon})$ is a neighborhood of $\mathbf{\Upsilon}$, (b) of Definition 216), and $\mathbf{Z}_k \in \mathfrak{N}(t_0; \mathbf{I}; \mathbf{\Upsilon})$ imply

$$\tau_k \longrightarrow \infty \text{ as } k \longrightarrow \infty. \tag{D.29}$$

Let $m \in \{1, 2, ...\}$ be such that $\mathbf{Z}_k \in \{\mathfrak{N}(t_0; \mathbf{I}; \mathbf{\Upsilon}) \backslash \text{Cl}\mathfrak{W}_{\zeta}(t_0; \mathbf{I})\}$ for all $k = m, m + 1, ...$, and $\mathbf{Z}_k \longrightarrow \partial\mathfrak{N}(t_0; \mathbf{I}; \mathbf{\Upsilon})$ as $k \longrightarrow \infty$, for every $\mathbf{I}(.) \in \mathcal{J}^{\xi}$. Such $\mathbf{Z}_k$ exists because $\mathfrak{N}(t_0; \mathbf{I}; \mathbf{\Upsilon}) \equiv \mathcal{D}_T(t_0; \mathbf{I}; \mathbf{\Upsilon})$ is open [a-2) of Lemma 1] and $\mathfrak{N}(t_0; \mathbf{I}; \mathbf{\Upsilon}) \supset \text{Cl}\mathfrak{W}_{\zeta}(t_0; \mathbf{I})$.

Let χ be defined by

$$\chi = \min \{W(t, \mathbf{Z}; \mathbf{I}; \mathbf{\Upsilon}) : (t, \mathbf{Z}) \in \mathfrak{T}_i \times [\mathfrak{S}(t; t_0; \Upsilon) \backslash \mathfrak{W}_{\zeta}(t; \mathbf{I}; \mathbf{\Upsilon})]\},$$
$$\forall \mathbf{I}(.) \in \mathcal{J}^{\xi}. \tag{D.30}$$

Because $W(.; \Upsilon) \in L[\mathfrak{T}_0, \mathfrak{T}_i, \mathfrak{S}, \mathcal{J}^{\xi}, \Upsilon]$ then $\chi \in \mathfrak{R}^+$. Therefore, (D.9), (D.28), (D.30), and the definitions of χ and τ_k yield $V(t_0, \mathbf{Z}_k; \mathbf{I}; \mathbf{\Upsilon}) \geq \chi\tau_k$, $\forall t_0 \in \mathfrak{T}_i$, $\forall \mathbf{I}(.) \in \mathcal{J}^{\xi}$, which together with (D.29) prove the necessity of the condition 4-a-ii). The conditions under 4-b) follow from 4-a) due to (24.5), (24.6), (24.13), (24.14), and (24.20)-(24.24). This completes the proof of the necessity part.

Sufficiency. Let all the conditions of the theorem statement hold. Because the function $V(.; \Upsilon)$ is the solution to (24.5), (24.6), [or, $V(.; \Upsilon)$ is the solution to (24.13), (24.14)], and it is positive definite and decrescent relative to Υ on $\mathfrak{T}_i \times \mathfrak{N}(t; \mathbf{I}; \mathbf{\Upsilon})$, for every $\mathbf{I}(.) \in \mathcal{J}^{\xi}$, because the function $W(.; \Upsilon) \in L[\mathfrak{T}_0, \mathfrak{T}_i, \mathfrak{S}, \mathcal{J}^{\xi}, \Upsilon]$, and/or, $W(.; \Upsilon) \in E[\mathfrak{T}_0, \mathfrak{T}_i, \mathfrak{S}, \mathcal{J}^{\xi}, \Upsilon]/$ is also positive definite and decrescent relative to Υ on $\mathfrak{T}_i \times \mathfrak{N}(t; \mathbf{I}; \mathbf{\Upsilon})$, then the motions of the system (3.12) exhibit uniform stable tracking of Υ on $\mathfrak{T}_0 \times \mathfrak{T}_i \times \mathcal{J}^{\xi}$, The proof of this statement is the proof of Theorem 364) when $\mathbf{z} = \mathbf{0}_K$, equivalently of $\mathbf{Z}_d(t)$, is replaced by the set

Υ, all positive definite and/or decrescent functions are relative to Υ, and all hyperballs are replaced by the corresponding neighborhoods of Υ. Hence, $\mp$ has both $\mathcal{D}_T(t_0;\mathbf{I};\mathbf{\Upsilon})$ at every $t_0 \in \mathfrak{T}_i$ and $\mathcal{D}_{TU}(\mathfrak{T}_i;\mathbf{I};\mathbf{\Upsilon})$, (Definition 215, Definition 216, Definition 220, and Definition 221). In order to show that both $\mathfrak{N}(t_0;\mathbf{I};\mathbf{\Upsilon}) \equiv \mathcal{D}_T(t_0;\mathbf{I};\mathbf{\Upsilon})$ and $\mathfrak{N}(\mathfrak{T}_i;\mathbf{I};\mathbf{\Upsilon}) \equiv \mathcal{D}_{TU}(\mathfrak{T}_i;\mathbf{I};\mathbf{\Upsilon})$ we proceed as follows. The condition (ii) of strong smoothness property 370 guarantees $\mathcal{D}_T(t_0;\mathbf{I};\mathbf{\Upsilon}) \subseteq \mathfrak{S}(t_0;t_0;\Upsilon)$, $\forall[t_0,\mathbf{I}(.)] \in \mathfrak{T}_i \times \mathcal{J}^\xi$. Let $[t_0,\mathbf{I}(.)] \in \mathfrak{T}_i \times \mathcal{J}^\xi$ be arbitrary and fixed. If $\partial\mathfrak{N}(t_0;\mathbf{I};\mathbf{\Upsilon}) = \phi$ then $\mathfrak{N}(t_0;\mathbf{I};\mathbf{\Upsilon}) = \mathfrak{R}^K$. Hence, $\mathcal{D}_T(t_0;\mathbf{I};\mathbf{\Upsilon}) \subseteq \mathfrak{N}(t_0;\mathbf{I};\mathbf{\Upsilon})$ is then possible. If $\mathcal{D}_T(t_0;\mathbf{I};\mathbf{\Upsilon}) \subset \mathfrak{N}(t_0;\mathbf{I};\mathbf{\Upsilon})$ then $\partial\mathcal{D}_T(t_0;\mathbf{I};\mathbf{\Upsilon}) \cap \mathfrak{N}(t_0;\mathbf{I};\mathbf{\Upsilon}) \neq \phi$ that implies $V(t_0,\mathbf{Z};\mathbf{I}) \longrightarrow \infty$ [because the function $V(.;\Upsilon)$ is the solution to (24.5), (24.6), as shown in the proof of the necessity], which contradicts the condition 4-a,i) because of $\mathfrak{N}(t_0;\mathbf{I};\mathbf{\Upsilon}) = \mathfrak{R}^K$. This implies $\partial\mathcal{D}_T(t_0;\mathbf{I};\mathbf{\Upsilon}) \cap \mathfrak{N}(t_0;\mathbf{I};\mathbf{\Upsilon}) = \partial\mathcal{D}_T(t_0;\mathbf{I};\mathbf{\Upsilon}) \cap \mathfrak{R}^K = \phi$. Because $\mathcal{D}_T(t_0;\mathbf{I};\mathbf{\Upsilon})$ is an open connected neighborhood of $\mathbf{\Upsilon}$ and Υ is a compact connected set then $\mathcal{D}_T(t_0;\mathbf{I};\mathbf{\Upsilon}) = \mathfrak{R}^K$; i.e., $\mathcal{D}_T(t_0;\mathbf{I};\mathbf{\Upsilon}) = \mathfrak{N}(t_0;\mathbf{I};\mathbf{\Upsilon})$. Let it be now supposed that $\partial\mathfrak{N}(t_0;\mathbf{I};\mathbf{\Upsilon}) \neq \phi$. i.e., $\mathfrak{N}(t_0;\mathbf{I};\mathbf{\Upsilon}) \subset \mathfrak{R}^K$. If we assume now $\partial\mathcal{D}_T(t_0;\mathbf{I};\mathbf{\Upsilon}) = \phi$, then $\mathcal{D}_T(t_0;\mathbf{I};\mathbf{\Upsilon}) = \mathfrak{R}^K$. It implies $\partial\mathfrak{N}(t_0;\mathbf{I};\mathbf{\Upsilon}) \cap \mathcal{D}_T(t_0;\mathbf{I};\mathbf{\Upsilon}) \neq \phi$. This and the condition 4-a,ii) show that there is a set $L \subseteq \partial\mathfrak{N}(t_0;\mathbf{I};\mathbf{\Upsilon}) \cap \mathcal{D}_T(t_0;\mathbf{I};\mathbf{\Upsilon})$ such that $V(t_0,\mathbf{Z};\mathbf{I}) \longrightarrow \infty$ as $\mathbf{Z} \longrightarrow L \subseteq \partial\mathfrak{N}(t_0;\mathbf{I};\mathbf{\Upsilon}) \cap \mathcal{D}_T(t_0;\mathbf{I};\mathbf{\Upsilon})$, which is impossible because the function $V(.;\Upsilon)$ is the unique solution of (24.5), (24.6) that is continuous on $\mathfrak{T}_i \times \mathcal{D}_T(t_0;\mathbf{I};\mathbf{\Upsilon})$, as shown in the details in the proof of the necessity. Assume $\partial\mathcal{D}_T(t_0;\mathbf{I};\mathbf{\Upsilon}) = \phi$ fails. Let $\partial\mathcal{D}_T(t_0;\mathbf{I};\mathbf{\Upsilon}) \neq \phi$ be considered. If $\partial\mathcal{D}_T(t_0;\mathbf{I};\mathbf{\Upsilon}) \cap \partial\mathfrak{N}(t_0;\mathbf{I};\mathbf{\Upsilon}) = \phi$ then either $\mathcal{D}_T(t_0;\mathbf{I};\mathbf{\Upsilon}) = \mathfrak{N}(t_0;\mathbf{I};\mathbf{\Upsilon})$ or $\mathcal{D}_T(t_0;\mathbf{I};\mathbf{\Upsilon}) \subset \mathfrak{N}(t_0;\mathbf{I};\mathbf{\Upsilon})$ or $\mathcal{D}_T(t_0;\mathbf{I};\mathbf{\Upsilon}) \supset \mathfrak{N}(t_0;\mathbf{I};\mathbf{\Upsilon})$ because both are open connected neighborhoods of $\mathbf{\Upsilon}$ and their boundaries are nonempty. The last two cases are impossible due to positive definiteness of the function $V(.;\Upsilon)$ on $\mathfrak{T}_i \times \mathfrak{N}(t;\mathbf{I};\mathbf{\Upsilon})$ and its construction via the equations (24.5), (24.6) as shown above. If $\partial\mathcal{D}_T(t_0;\mathbf{I};\mathbf{\Upsilon}) \cap \partial\mathfrak{N}(t_0;\mathbf{I};\mathbf{\Upsilon}) \neq \phi$ then either $\partial\mathcal{D}_T(t_0;\mathbf{I};\mathbf{\Upsilon}) = \partial\mathfrak{N}(t_0;\mathbf{I};\mathbf{\Upsilon})$, which implies $\mathcal{D}_T(t_0;\mathbf{I};\mathbf{\Upsilon}) = \mathfrak{N}(t_0;\mathbf{I};\mathbf{\Upsilon})$, or $\partial\mathcal{D}_T(t_0;\mathbf{I};\mathbf{\Upsilon}) \cap \mathfrak{N}(t_0;\mathbf{I};\mathbf{\Upsilon}) \neq \phi$ and/or $\mathcal{D}_T(t_0;\mathbf{I};\mathbf{\Upsilon}) \cap \partial\mathfrak{N}(t_0;\mathbf{I};\mathbf{\Upsilon}) \neq \phi$. If $\partial\mathcal{D}_T(t_0;\mathbf{I};\mathbf{\Upsilon}) \cap \mathfrak{N}(t_0;\mathbf{I};\mathbf{\Upsilon}) \neq \phi$ then it means that the function $V(.;\Upsilon)$ blows up (to ∞) on $\mathfrak{N}(t_0;\mathbf{I};\mathbf{\Upsilon})$, which contradicts its continuity on $\mathfrak{N}(t_0;\mathbf{I};\mathbf{\Upsilon})$ due to the condition 4-a,2. If $\mathcal{D}_T(t_0;\mathbf{I};\mathbf{\Upsilon}) \cap \partial\mathfrak{N}(t_0;\mathbf{I};\mathbf{\Upsilon}) \neq \phi$ then it means that $V(.;\Upsilon)$ blows up on $\mathcal{D}_T(t_0;\mathbf{I};\mathbf{\Upsilon})$; that is impossible due to (D.13) because $V(.;\Upsilon)$ is generated by (24.5), (24.6). Hence, $\partial\mathcal{D}_T(t_0;\mathbf{I};\mathbf{\Upsilon}) = \partial\mathfrak{N}(t_0;\mathbf{I};\mathbf{\Upsilon})$ that implies $\mathcal{D}_T(t_0;\mathbf{I};\mathbf{\Upsilon}) = \mathfrak{N}(t_0;\mathbf{I};\mathbf{\Upsilon})$, which holds as the overall result. Now, $\mathfrak{N}(\mathfrak{T}_i;\mathbf{I};\mathbf{\Upsilon}) = \cap[\mathfrak{N}(t_0;\mathbf{I};\mathbf{\Upsilon}) : t_0 \in \mathfrak{T}_i]$ (the condition of the theorem statement) and the conditions b) of Definition 221 and Definition 216 imply $\mathcal{D}_{TU}(\mathfrak{T}_i;\mathbf{I};\mathbf{\Upsilon}) \equiv \mathfrak{N}(\mathfrak{T}_i;\mathbf{I};\mathbf{\Upsilon})$. Positive definiteness of $W(.;\Upsilon)$ on $\mathfrak{S}(t;t_0;\Upsilon)$, $W(.;\Upsilon) \in L[\mathfrak{T}_0,\mathfrak{T}_i,\mathfrak{S},\mathcal{J}^\xi,\Upsilon]$, the equation (24.5), $\mathfrak{N}(t;\mathbf{I};\mathbf{\Upsilon}) \subseteq \mathfrak{S}(t;t_0;\Upsilon)$, $\forall t \in \mathfrak{T}_i$, the condition 4-a,i) and a) of Lemma 1 imply

$$V[t_0+\tau,\mathbf{Z}(t_0+\tau;t_0;\mathbf{Z}_0;\mathbf{I});\mathbf{I};\mathbf{\Upsilon}] \leq V(t_0;\mathbf{Z}_0;\mathbf{I};\mathbf{\Upsilon})$$
$$-\xi[\varsigma;W;V;\mathfrak{N}(\mathfrak{T}_i;\mathbf{I};\mathbf{\Upsilon});\mathfrak{T}_i;\Upsilon]\,\tau(t_0;\mathbf{Z}_0;\varsigma;W;\mathbf{I};\mathbf{\Upsilon}),\ \forall\mathbf{I}(.) \in \mathcal{J}^\xi,$$

where $\varsigma \in \mathfrak{R}^+$ is arbitrarily small,

$$\xi(\varsigma;W;V;\mathfrak{N}(\mathfrak{T}_i;\mathbf{I};\mathbf{\Upsilon});\mathfrak{T}_i;\mathbf{I};\mathbf{\Upsilon})$$

$$= \min\{V(t;\mathbf{Z};\mathbf{I};\mathbf{\Upsilon}) : (t;\mathbf{Z}) \in \mathfrak{T}_i \times [\mathfrak{N}(\mathfrak{T}_i;\mathbf{I};\mathbf{\Upsilon}) \backslash \mathfrak{V}_\psi(\mathfrak{T}_i;\mathbf{I};\mathbf{\Upsilon})]\} \in \mathfrak{R}^+,$$
$$\forall\mathbf{I}(.) \in \mathcal{J}^\xi,\ \psi = \varphi_1(\zeta;\Upsilon),\ \varphi_1(.;\Upsilon) \in \mathcal{K},$$

$$\mathfrak{V}_\psi(\mathfrak{T}_i;\mathbf{I};\mathbf{\Upsilon}) = \cap[\mathfrak{V}_\psi(t;\mathbf{I};\mathbf{\Upsilon}) : t \in \mathfrak{T}_i],\ \forall\mathbf{I}(.) \in \mathcal{J}^\xi,$$
$$\varphi_1[\rho(\mathbf{Z},\mathbf{\Upsilon});\Upsilon] \leq V(t;\mathbf{Z};\mathbf{I};\mathbf{\Upsilon}),\ \forall(t;\mathbf{Z}) \in \mathfrak{T}_i \times \mathfrak{N}(t;\mathbf{I};\mathbf{\Upsilon}),\ \forall\mathbf{I}(.) \in \mathcal{J}^\xi,$$

$$\varphi_2[\rho(\mathbf{Z},\mathbf{\Upsilon});\Upsilon] \geq V(t;\mathbf{Z};\mathbf{I};\mathbf{\Upsilon}),\ \forall(t;\mathbf{Z}) \in \mathfrak{T}_i \times \mathfrak{N}(\mathfrak{T}_i;\mathbf{I};\mathbf{\Upsilon}),\ \forall\mathbf{I}(.) \in \mathcal{J}^\xi,$$
$$\varphi_2(.;\Upsilon) \in \mathcal{K},$$

so that

$$\tau\left(t_0;\mathbf{Z}_0;\varsigma;W;\mathbf{I};\boldsymbol{\Upsilon}\right)\leq\left[\varphi_2\left[\rho\left(\mathbf{Z}_0,\Upsilon\right);\Upsilon\right]-\varphi_1\left(\zeta;\Upsilon\right)\right]$$
$$\bullet\,\xi^{-1}\left[\varsigma;W;V;\mathfrak{N}\left(\mathfrak{T}_i;\mathbf{I};\boldsymbol{\Upsilon}\right);\mathfrak{T}_i;\mathbf{I};\Upsilon\right],\ \forall\mathbf{I}\left(.\right)\in\mathcal{J}^{\xi},$$

$$\sup\left[\tau_m\left(t_0;\mathbf{Z}_0;\varsigma;W;\mathbf{I};\boldsymbol{\Upsilon}\right):t_0\in\mathfrak{T}_i\right]$$
$$\leq\left[\varphi_2\left[\rho\left(\mathbf{Z}_0,\Upsilon\right);\Upsilon\right]-\varphi_1\left(\zeta;\Upsilon\right)\right]\xi^{-1}\left[\varsigma;W;V;\mathfrak{N}\left(\mathfrak{T}_i;\mathbf{I};\boldsymbol{\Upsilon}\right);\mathfrak{T}_i;\mathbf{I}\right]<\infty,$$
$$\forall\mathbf{Z}_0\in\mathfrak{N}\left(\mathfrak{T}_i;\mathbf{I};\boldsymbol{\Upsilon}\right),\ \forall\mathbf{I}\left(.\right)\in\mathcal{J}^{\xi},$$

and, therefore, the conditions under (b) of Definition 221 and Definition 216 are satisfied. This completes the proof of the sufficiency of the conditions 1-4,a). The sufficiency of the conditions 1-3 and 4b) is implied by the sufficiency of 1-4a) and (24.20)-(24.24), which completes the proof ■

D.4 Proof of Theorem 364

Proof. Proof of Theorem 380.The following replacements in the proof D.3 (of Theorem 379) transform it into the proof of Theorem 380:

$$\text{“}W\left(t,\mathbf{z};\mathbf{i}\right)\text{” into “}W\left(t,\mathbf{z};\mathcal{J}^{\xi};\Upsilon\right)\text{”},$$

$$\text{“}\mathbf{z}_d\left(t\right)\equiv\mathbf{0}_K\text{ on }\mathfrak{T}_0\times\mathfrak{T}_i,\forall\mathbf{i}\left(.\right)\in\mathcal{J}^{\xi}\text{” into “}\boldsymbol{\Upsilon}\text{ on }\mathfrak{T}_0\times\mathfrak{T}_i\times\mathcal{J}^{\xi}\text{”},$$
$$\text{“tracking on }\mathfrak{T}_0\times\mathfrak{T}_i,\forall\mathbf{i}\left(.\right)\in\mathcal{J}^{jj}\text{” into “tracking on }\mathfrak{T}_0\times\mathfrak{T}_i\times\mathcal{J}^{\xi}\text{”},$$
$$\text{“on }\mathfrak{T}_0\times\mathfrak{T}_i,\forall\mathbf{i}\left(.\right)\in\mathcal{J}^{\xi}\text{” into “on }\mathfrak{T}_0\times\mathfrak{T}_i\times\mathcal{J}^{\xi}\text{”},$$
$$\text{“}\forall t_0\in\mathfrak{T}_i,\ \forall\mathbf{i}\left(.\right)\in\mathcal{J}^{\xi}\text{” into “}\forall\left[t_0,\mathbf{I}\left(.\right)\right]\in\mathfrak{T}_i\times\mathcal{J}^{\xi}\text{”},$$
$$\text{“}\mathcal{D}_{STU}\left(\mathfrak{T}_i;\mathbf{z}_d;\mathbf{i}\right)\text{” into “}\mathcal{D}_{STU}\left(\mathfrak{T}_i;\mathbf{z}_d;\mathcal{J}^{\xi};\Upsilon\right)\text{”},$$
$$\text{“}\mathcal{D}_{TU}\left(\mathfrak{T}_i;\mathbf{z}_d;\mathbf{i}\right)\text{” into “}\mathcal{D}_{TU}\left(\mathfrak{T}_i;\mathbf{z}_d;\mathcal{J}^{\xi};\Upsilon\right)\text{”},$$

$$\text{“}\mathfrak{T}_i\times\left[\mathcal{D}_T\left(t_0;\mathbf{z}_d;\mathbf{i}\right)\backslash\mathfrak{S}\left(t,t_0\right)\right],\forall\mathbf{i}\left(.\right)\in\mathcal{J}^{\xi}\text{” into}$$
$$\text{“}\mathfrak{T}_i\times\left[\mathcal{D}_T\left(t_0;\mathbf{I};\boldsymbol{\Upsilon}\right)\backslash\mathfrak{S}\left(t;t_0;\Upsilon\right)\right]\times\mathcal{J}^{\xi}\text{”},$$

$$\text{“}\mathfrak{N}\left(\mathfrak{T}_i;\mathbf{i}\right)=\cap\left[\mathfrak{N}\left(t_0;\mathbf{i}\right):t_0\in\mathfrak{T}_i\right]\text{” into}$$
$$\text{“}\mathfrak{N}\left(\mathfrak{T}_i;\mathcal{J}^{\xi};\Upsilon\right)=\cap\left[\mathfrak{N}\left(t_0;\mathbf{i};\boldsymbol{\Upsilon}\right):\left[t_0,(.)\right]\in\mathfrak{T}_i\times\mathcal{J}^{\xi}\right]\text{”}$$
$$\text{“}\mathfrak{N}\left(\mathfrak{T}_i;\mathbf{i}\right)\text{” into “}\mathfrak{N}\left(\mathfrak{T}_i;\mathcal{J}^{\xi};\Upsilon\right)\text{”},$$

$$\text{“}\left|v\left(t,\mathbf{z};\mathbf{i}\right)\right|<\infty,\ \forall\left(t,\mathbf{z}\right)\in\mathfrak{T}_i\times Cl\mathfrak{B}_{\mu},\forall\mathbf{i}\left(.\right)\in\mathcal{J}^{\xi}\text{” into}$$
$$\text{“}\left|v\left(t,\mathbf{z};\mathcal{J}^{\xi};\Upsilon\right)\right|<\infty,\ \forall\left[t,\mathbf{z},\mathbf{i}\left(.\right)\right]\in\mathfrak{T}_i\times Cl\mathfrak{N}_{\mu}\left(\Upsilon\right)\times\mathcal{J}^{\xi}\text{”},$$
$$\text{“}v\left(t,\mathbf{z};\mathbf{i}\right)\text{” into “}V\left(t,\mathbf{z};\mathcal{J}^{\xi};\Upsilon\right)\text{”},$$

$$\text{“}\mathfrak{W}_{\zeta}\left(t;\mathbf{i}\right)\text{” into “}\mathfrak{W}_{\zeta}\left(t;\mathcal{J}^{\xi};\Upsilon\right)\text{”},$$

$$\text{“}\mathfrak{N}_M\left(\mathfrak{T}_i;\mathbf{i}\right)=\cup\left[\mathfrak{N}\left(t;\mathbf{i}\right):t\in\mathfrak{T}_i\right]\text{” into}$$
$$\text{“}\mathfrak{N}_M\left(\mathfrak{T}_i;\mathbf{I};\boldsymbol{\Upsilon}\right)=\cup\left[\mathfrak{N}\left(t;\mathbf{I};\boldsymbol{\Upsilon}\right):\left[t,\mathbf{I}\left(.\right)\right]\in\mathfrak{T}_i\times\mathcal{J}^{\xi}\right]\text{”}$$
$$\text{“}\mathfrak{N}_m\left(\mathfrak{T}_i;\mathbf{i}\right)=\cap\left[\mathfrak{N}\left(t;\mathbf{i}\right):t\in\mathfrak{T}_i\right]\text{” into}$$
$$\text{“}\mathfrak{N}_m\left(\mathfrak{T}_i;\mathbf{I};\boldsymbol{\Upsilon}\right)=\cap\left[\mathfrak{N}\left(t;\mathbf{I};\boldsymbol{\Upsilon}\right):\left[t,\mathbf{I}\left(.\right)\right]\in\mathfrak{T}_i\times\mathcal{J}^{\xi}\right]\text{”}.$$

These substitutions explain clearly other necessary substitutions ■

D.5 Proof of Theorem 368

Proof. Proof of Theorem 384.

This proof relies largely on the proof D.3 of Theorem 379. Let the system (23.1) possess weak smoothness property 368.

Necessity. Let the system (23.1) exhibit uniform stable tracking of $\mathbf{z}_d(t) \equiv \mathbf{0}_K$ on $\mathfrak{T}_0 \times \mathfrak{T}_i$, $\forall \mathbf{i}(.) \in \mathcal{J}^j$. Definition 221 and Definition 220 imply that the system has the domain $\mathcal{D}_{STU}(\mathfrak{T}_i; \mathbf{z}_d; \mathbf{i})$ of the uniform stable tracking on $\mathfrak{T}_0 \times \mathfrak{T}_i$, $\forall \mathbf{i}(.) \in \mathcal{J}^j$, and domain $\mathcal{D}_{ST}(t_0; \mathbf{z}_d; \mathbf{i})$ at every $t_0 \in \mathfrak{T}_i$ of the stable tracking on $\mathfrak{T}_0 \times \mathfrak{T}_i$, $\forall \mathbf{i}(.) \in \mathcal{J}^j$. The preceding definitions, Definition 215 and Definition 216 show that the system also has the uniform tracking domain $\mathcal{D}_{TU}(\mathfrak{T}_i; \mathbf{z}_d; \mathbf{i})$ on $\mathcal{J}^j$ and the instantaneous tracking domain $\mathcal{D}_T(t_0; \mathbf{z}_d; \mathbf{i})$ at every $t_0 \in \mathfrak{T}_i$ on $\mathfrak{T}_0 \times \mathfrak{T}_i$, $\forall \mathbf{i}(.) \in \mathcal{J}^j$, so that (D.1) holds. $\mathfrak{N}(t_0; \mathbf{i})$ be defined by (D.3):

$$\mathfrak{N}(t_0; \mathbf{i}) \equiv \mathcal{D}_T(t_0; \mathbf{z}_d; \mathbf{i}) \tag{D.31}$$

so that $\mathfrak{S}(t_0, t_0) \supseteq \mathfrak{N}(t_0; \mathbf{i})$ and $\mathfrak{N}(\mathfrak{T}_i; \mathbf{i}) = \mathcal{D}_{TU}(\mathfrak{T}_i; \mathbf{z}_d; \mathbf{i})$. Hence, $\mathfrak{N}(t_0; \mathbf{i})$ is an open continuous connected neighborhood of $\mathbf{z} = \mathbf{0}_K$ on $\mathfrak{T}_0 \times \mathfrak{T}_i$, $\forall \mathbf{i}(.) \in \mathcal{J}^j$, and $\mathfrak{N}(\mathfrak{T}_i; \mathbf{i}) = \mathcal{D}_{TU}(\mathfrak{T}_i; \mathbf{z}_d; \mathbf{i})$ is a connected neighborhood of $\mathbf{z} = \mathbf{0}_K$ (Proof D.3). Let $w(.) \in L\left[\mathfrak{T}_0, \mathfrak{T}_i, \mathfrak{R}^K, \mathcal{J}^j\right]$, (or, $w(.) \in E\left[\mathfrak{T}_0, \mathfrak{T}_i, \mathfrak{R}^K, \mathcal{J}^j\right]$), Definition 373, be an arbitrarily selected positive definite decrescent function on $\mathfrak{T}_i \times \mathfrak{R}^K$. From now on we should repeat the necessity part of the proof D.3.

Sufficiency. Let the conditions 1) through 4-a) hold. Then the conditions 1) through 4) of Theorem 379 are satisfied for $\mathfrak{S}(t, t_0)$ replaced by $\mathfrak{R}^K$ ■

D.6 Lemma 415

Proof. Let all the conditions of the lemma statement hold.

Necessity. Let $\mathbf{r}^{\alpha-1}(t) \neq \mathbf{0}_{\alpha\rho}$, $t \in [t_0, \tau]$. Because

$$\left[\text{grad}_{tr} V(t, \mathbf{r}^{\alpha-1})\right]^T \widetilde{Q}^{-1}\left[t, \mathbf{r}^{\alpha-1}(t)\right] P(t) \in \mathfrak{R}^{1\times r}$$

then (30.3) guarantees

$$\det \left\langle \begin{array}{c} \left\{\left[grad_{tr} V(t, \mathbf{r}^{\alpha-1})\right]^T \widetilde{Q}^{-1}\left[t, \mathbf{r}^{\alpha-1}(t)\right] P(t)\right\} \\ \bullet \left\{\left[grad_{tr} V(t, \mathbf{r}^{\alpha-1})\right]^T \widetilde{Q}^{-1}\left[t, \mathbf{r}^{\alpha-1}(t)\right] P(t)\right\}^T \end{array} \right\rangle \neq 0,$$
$$\forall (t, \mathbf{r}^{\alpha-1} \neq \mathbf{0}_{\alpha\rho}) \in \mathfrak{T} \times \mathfrak{R}^{\alpha\rho}, \tag{D.32}$$

We can set (22.16) repeated as (D.33):

$$V^{(1)}(t, \mathbf{r}^{\alpha-1}; \mathbf{i}) = \frac{\partial V(t, \mathbf{r}^{\alpha-1})}{\partial t}$$
$$+ \left[\frac{\partial V(t, \mathbf{r}^{\alpha-1})}{\partial r_1} .. \frac{\partial V(t, \mathbf{r}^{\alpha-1})}{\partial r_\rho} .... \frac{\partial V(t, \mathbf{r}^{\alpha-1})}{\partial r_1^{(\alpha-2)}} .. \frac{\partial V(t, \mathbf{r}^{\alpha-1})}{\partial r_\rho^{(\alpha-2)}} \right] \frac{d\mathbf{r}^{\alpha-2}}{dt}$$
$$+ \left\langle \begin{array}{c} \left[\frac{\partial V(t, \mathbf{r}^{\alpha-1})}{\partial r_1^{(\alpha-1)}} .... \frac{\partial V(t, \mathbf{r}^{\alpha-1})}{\partial r_\rho^{(\alpha-1)}} \right] \\ \bullet\, \widetilde{Q}^{-1}\left[t, \mathbf{r}^{\alpha-1}(t)\right] \left\{ \widetilde{\mathbf{h}}\left[t, \mathbf{i}^{\xi}(t)\right] - \widetilde{\mathbf{q}}\left[t, \mathbf{r}^{\alpha-1}(t)\right] \right\} \end{array} \right\rangle . \tag{D.33}$$

into a compact form

$$V^{(1)}(t, \mathbf{r}^{\alpha-1}; \mathbf{i}) = \psi(t, \mathbf{r}^{\alpha-1})$$
$$+ \left[\text{grad}_{tr} V(t, \mathbf{r}^{\alpha-1})\right]^T \widetilde{Q}^{-1}\left[t, \mathbf{r}^{\alpha-1}(t)\right] \left\{ \widetilde{\mathbf{h}}\left[t, \mathbf{i}^{\xi}(t)\right] - \widetilde{\mathbf{q}}\left[t, \mathbf{r}^{\alpha-1}(t)\right] \right\} \tag{D.34}$$

For the plant (23.1) the equation (D.34) becomes:

$$V^{(1)}(t,\mathbf{r}^{\alpha-1};\mathbf{d}) = \psi(t,\mathbf{r}^{\alpha-1})$$
$$+\left[\operatorname{grad}_{tr}V(t,\mathbf{r}^{\alpha-1})\right]^{T}\widetilde{Q}^{-1}\left[t,\mathbf{r}^{\alpha-1}(t)\right]\left\{\begin{array}{c} E(t)\,\widetilde{\mathbf{e}}_{d}\left[\mathbf{d}^{\eta}(t)\right] \\ +\,P(t)\,\widetilde{\mathbf{e}}_{u}\left[\mathbf{u}^{\mu}(t)\right]-\widetilde{\mathbf{q}}\left[t,\mathbf{r}^{\alpha-1}(t)\right]\end{array}\right\}. \quad \text{(D.35)}$$

In order for control to ensure $V^{(1)}(t,\mathbf{r}^{\alpha-1}) = -W(t,\mathbf{r}^{\alpha-1};\mathbf{d})$ where $W(t,\mathbf{r}^{\alpha-1};\mathbf{d})$ is a positive definite function for every $\mathbf{d}(.) \in \mathfrak{D}$ it is necessary and sufficient that

$$\psi(t,\mathbf{r}^{\alpha-1}) + \left[\operatorname{grad}_{tr}V(t,\mathbf{r}^{\alpha-1})\right]^{T}\widetilde{Q}^{-1}\left[t,\mathbf{r}^{\alpha-1}(t)\right]$$
$$\bullet\left\{\begin{array}{c} E(t)\,\widetilde{\mathbf{e}}_{d}\left[\mathbf{d}^{\eta}(t)\right] \\ +\,P(t)\,\widetilde{\mathbf{e}}_{u}\left[\mathbf{u}^{\mu}(t)\right]-\widetilde{\mathbf{q}}\left[t,\mathbf{r}^{\alpha-1}(t)\right]\end{array}\right\} = -W(t,\mathbf{r}^{\alpha-1};\mathbf{d}), \quad \text{(D.36)}$$

i.e.

$$\left[\operatorname{grad}_{tr}V(t,\mathbf{r}^{\alpha-1})\right]^{T}\widetilde{Q}^{-1}\left[t,\mathbf{r}^{\alpha-1}(t)\right]P(t)\,\widetilde{\mathbf{e}}_{u}\left[\mathbf{u}^{\mu}(t)\right] = -\psi(t,\mathbf{r}^{\alpha-1})$$
$$-\,W(t,\mathbf{r}^{\alpha-1};\mathbf{d}) + \left[\operatorname{grad}_{tr}V(t,\mathbf{r}^{\alpha-1})\right]^{T}\widetilde{Q}^{-1}\left[t,\mathbf{r}^{\alpha-1}(t)\right]\left\{\begin{array}{c} \widetilde{\mathbf{q}}\left[t,\mathbf{r}^{\alpha-1}(t)\right] \\ -\,E(t)\,\widetilde{\mathbf{e}}_{d}\left[\mathbf{d}^{\eta}(t)\right]\end{array}\right\}.$$

This equation and (D.32) transform (D.36) into

$$\widetilde{\mathbf{e}}_{u}\left[\mathbf{u}^{\mu}(t)\right] = \left\{\left[\operatorname{grad}_{tr}V(t,\mathbf{r}^{\alpha-1})\right]^{T}\widetilde{Q}^{-1}\left[t,\mathbf{r}^{\alpha-1}(t)\right]P(t)\right\}^{\mathrm{T}}$$
$$\bullet\left\langle\begin{array}{c}\left\{\left[\operatorname{grad}_{tr}V(t,\mathbf{r}^{\alpha-1})\right]^{T}\widetilde{Q}^{-1}\left[t,\mathbf{r}^{\alpha-1}(t)\right]P(t)\right\} \\ \bullet\left\{\left[\operatorname{grad}_{tr}V(t,\mathbf{r}^{\alpha-1})\right]^{T}\widetilde{Q}^{-1}\left[t,\mathbf{r}^{\alpha-1}(t)\right]P(t)\right\}^{\mathrm{T}}\end{array}\right\rangle^{-1}$$
$$\bullet\left\{\begin{array}{c} -\psi(t,\mathbf{r}^{\alpha-1}) - W(t,\mathbf{r}^{\alpha-1};\mathbf{d}) + \left[\operatorname{grad}_{tr}V(t,\mathbf{r}^{\alpha-1})\right]^{T}\widetilde{Q}^{-1}\left[t,\mathbf{r}^{\alpha-1}(t)\right] \\ \bullet\left\{\widetilde{\mathbf{q}}\left[t,\mathbf{r}^{\alpha-1}(t)\right]-E(t)\,\widetilde{\mathbf{e}}_{d}\left[\mathbf{d}^{\eta}(t)\right]\right\}\end{array}\right\},$$

which is (30.6) if $\mathbf{r}^{\alpha-1}(t) \neq \mathbf{0}_{\alpha\rho}$, $t \in [t_0,\tau]$. This proves the necessity of (30.6) for $\mathbf{r}^{\alpha-1}(t) \neq \mathbf{0}_{\alpha\rho}$, $t \in [t_0,\tau]$.

Let $\tau \in \mathfrak{T}_0$ obey (30.5): $\mathbf{r}^{\alpha-1}(\tau) = \mathbf{0}_{\alpha\rho}$. Positive definiteness of $V(.)$ and 4) of Assumption 429 together with (30.5) imply $\mathbf{r}^{\alpha-1}(t) = \mathbf{0}_{\alpha\rho}$ that further implies $\mathbf{r}^{\alpha}(t) = \mathbf{0}_{\alpha\rho}$ for every $t \in [\tau,\infty[$. Then (23.1) becomes

$$\widetilde{\mathbf{q}}(t,\mathbf{0}_{\alpha\rho}) = \widetilde{\mathbf{q}}\left[t,\mathbf{r}^{\alpha}(t)\right] = E(t)\,\widetilde{\mathbf{e}}_{d}\left[\mathbf{d}^{\eta}(t)\right] + P(t)\,\widetilde{\mathbf{e}}_{u}\left[\mathbf{u}^{\mu}(t)\right],\ \forall t \in [\tau,\infty[,$$

which becomes

$$\widetilde{\mathbf{e}}_{u}\left[\mathbf{u}^{\mu}(t)\right] = P^{T}(t)\left[P(t)\,P^{T}(t)\right]^{-1}\left\{\begin{array}{c}\widetilde{\mathbf{q}}\left[t,\mathbf{r}^{\alpha}(t)\right]- \\ -E(t)\,\widetilde{\mathbf{e}}_{d}\left[\mathbf{d}^{\eta}(t)\right]\end{array}\right\},\ \forall t \in [\tau,\infty[,$$

due to $r \geq \rho$ and rank $P(t) = \rho$ for all $t \in \mathfrak{T}_0$. This proves the necessity of (30.6) if $\mathbf{r}^{\alpha-1}(\tau) = \mathbf{0}_{\alpha\rho}$.

Sufficiency. Let $\mathbf{r}^{\alpha-1}(t) \neq \mathbf{0}_{\alpha\rho}$, $t \in [t_0,\tau]$. Let us replace $\widetilde{\mathbf{e}}_{u}\left[\mathbf{u}^{\mu}(t)\right]$ by the right-hand side of (30.6) into (D.36). The result is $V^{(1)}(t,\mathbf{r}^{\alpha-1}) = -W(t,\mathbf{r}^{\alpha-1};\mathbf{d})$, which proves sufficiency of (30.6) if $\mathbf{r}^{\alpha-1}(t) \neq \mathbf{0}_{\alpha\rho}$,

Let then $\mathbf{r}^{\alpha-1}(\tau) = \mathbf{0}_{\alpha\rho}$ and let (30.6) hold. Then (23.1) becomes

$$\widetilde{Q}\left[t,\mathbf{r}^{\alpha-1}(t)\right]\mathbf{r}^{(\alpha)}(t) + \widetilde{\mathbf{q}}\left[t,\mathbf{r}^{\alpha-1}(t)\right] = E(t)\,\widetilde{\mathbf{e}}_{d}\left[\mathbf{d}^{\eta}(t)\right]$$
$$+\,P(t)\,P^{T}(t)\left[P(t)\,P^{T}(t)\right]^{-1}\left\{\begin{array}{c}\widetilde{\mathbf{q}}\left[t,\mathbf{r}^{\alpha}(t)\right] \\ -\,E(t)\,\widetilde{\mathbf{e}}_{d}\left[\mathbf{d}^{\eta}(t)\right]\end{array}\right\} =$$
$$= E(t)\,\widetilde{\mathbf{e}}_{d}\left[\mathbf{d}^{\eta}(t)\right] + \widetilde{\mathbf{q}}\left[t,\mathbf{r}^{\alpha}(t)\right] - E(t)\,\widetilde{\mathbf{e}}_{d}\left[\mathbf{d}^{\eta}(t)\right],\ \forall t \in [\tau,\infty[;$$

i.e.,

$$\widetilde{Q}\left(t,\mathbf{0}_{\alpha\rho}\right)\mathbf{r}^{(\alpha)}(t)=\mathbf{0}_{\rho},\ \ \forall t\in[\tau,\infty[.$$

The nonsingularity of $\widetilde{Q}\left(t,\mathbf{0}_{\alpha\rho}\right)$ for all $t\in\mathfrak{T}_0$ reduces the last equation into

$$\mathbf{r}^{(\alpha)}(t)=\mathbf{0}_{\rho},\ \ \forall t\in[\tau,\infty[.$$

This and $\mathbf{r}^{\alpha-1}\left(\tau\right)=\mathbf{0}_{\alpha\rho}$ imply

$$\mathbf{r}^{\alpha}(t)=\mathbf{0}_{\rho},\ \ \forall t\in[\tau,\infty[.$$

Hence, both $V\left(t,\mathbf{r}^{\alpha-1}\right)=0$ and $W\left(t,\mathbf{r}^{\alpha-1}\right)=0$, $\forall t\in[\tau,\infty[$. These results imply

$$V^{(1)}\left(t,\mathbf{r}^{\alpha-1}\right)=0=W\left(t,\mathbf{r}^{\alpha-1}\right),\forall t\in[\tau,\infty[.$$

This completes the proof ■

Part XI

USED LITERATURE

Bibliography

[1] A. B. Açìkmeşe and M. Corles, "Robust output tracking for uncertain/nonlinear systems subject to almost constant disturbances," *Automatica*, Vol. 38, pp. 1919-1926, 2002.

[2] J. L. M. Aguilar, R. A. García, and C. E. D'Attellis, "Exact linearization of nonlinear systems: trajectory tracking with bounded control and state constraints," *Int. J. Control*, Vol. 65, No. 3, pp. 455 – 467, 1996.

[3] T. Ahmed-Ali and F. Lamnabhi-Lagarrigue, "Tracking control of nonlinear systems with disturbance attenuation," *C. R. Acad. Sci.*, Paris, t. 325, Série I, pp. 329-338, 1997.

[4] V. M. Alekseev, *Optimal Control*, New York: Springer, 2013.

[5] G. Ambrosino, G. Celentano, and F. Garofalo, "Robust model tracking control for a class of nonlinear plants," *IEEE Trans. on Automatic Control*, Vol. AC-30, No. 3, pp. 275-279, 1985.

[6] B. D. O. Anderson, "Stability of control systems with multiple nonlinearities," *J. Franklin Institute*, Vol. 282, No. 3, pp. 155-160, 1966.

[7] B. D. O. Anderson and J. B. Moore, "Construction of Liapunov functions for nonstationary systems containing non-inertial nonlinearities," (in Russian), *Avotmatika i Telemehanika*, No. 5, pp. 14-21, 1972.

[8] H. A. Antosiewcz, "A survey of Liapunov's second method," in *Contributions to the Theory of Nonlinear Oscillations*, Ed. S. Lesfchetz, Vol. IV, Princeton N. J: Princeton University Press, 1958.

[9] H. A. Antosiewcz, "Recent contributions to Lyapunov's second method," *Colloques Internationaux sur les Vibrations Forces dans les Systemes Non-lineaires*, pp. 29-37, September 1964.

[10] P. J. Antsaklis and O. R. Gonzalez, "Compensator Structure and Internal Models in Tracking and Regulation," *Proceedings of the 23rd IEEE Conference on Decision and Control*, Las Vegas, Nevada, pp. 1-2, December 12-14, 1984

[11] P. J. Antsaklis and A. N. Michel, *Linear Systems*, New York: McGraw Hill, 1997.

[12] G. Arienti, C. Sutti, and G. P. Szegö, "On the numerical construction of Lyapunov function," *Preprints of the IV IFAC Congress*, Warszawa, Poland, pp. 3-20, 1969.

[13] M. A. Arteaga, "Tracking control of flexible robot arms with nonlinear observer," *Automatica,* Vol. 36, pp. 1329-1337, 2000.

[14] M. A. Arteaga and B. Siciliano, "On tracking control of flexible robot arms," *IEEE Trans. on Automatic Control*, Vol. 45, No. 3, pp. 520-527, March 2000.

[15] M. A. Athans and P. L. Falb, *Optimal Control*, New York: McGraw–Hill, 1966, 1994, 2007.

[16] M. Athanassiades, "Bang-bang control for tracking systems," *IRE Trans. Automatic Control*, Vol. AC-7, No. 3, pp. 77-78, 1962.

[17] E.-W. Bai and Y.-F. Huang, "Variable gain parameter estimation algorithms for fast tracking and smooth steady state," *Automatica*, Vol. 36, pp. 1001-1008, 2000.

[18] J. A. Ball, P. Kachroo, and A. J. Krener, "H_∞ tracking control for a class of nonlinear systems," *IEEE Trans. Automatic Control*, Vol. 44, No. 6, pp. 1202-1206, June 1999.

[19] A. Balluchi and A. Bicchi, "Necessary and sufficient conditions for robust perfect tracking under variable structure control," *Int. J. Robust and Nonlinear Control*, Vol. 13, pp. 141-151, 2003.

[20] Ye. A. Barbashin, *Introduction to the Theory of Stability*, (in Russian), Moscow: Nauka, 1967.

[21] Ye. A. Barbashin, *The Liapunov Functions* (in Russian), Moscow: Nauka, 1970.

[22] Ye. A. Barbashin and N. N. Krasovskii, "On the stability of motion in the large," (in Russian), *Dokl. Akad. Nauk SSSR*, Vol. 86, No. 3, pp. 453-456, 1952.

[23] Ye. A. Barbashin and N. N. Krasovskii, "On the existence of Liapunov functions in the case of asymptotic stability in the whole," (in Russian), *Prikl. Mat. Meh.*, Vol. XVIII, pp. 345-350, 1954.

[24] S. Barnett, "Some topics in algebraic systems theory: A survey," in *Recent Mathematical Developments in Control*, Ed. D. J. Bell, New York: Academic Press, pp. 323-344, 1973.

[25] S. Barnett and C. Storey, *Matrix Methods in Stability Theory*, London: Nelson, 1970.

[26] **Y.** Bar-Shalom, "Tracking methods in a multitarget environment," *IEEE Trans. Automatic Control*, Vol. AC-23, No. 4, pp.618-626, August 1978.

[27] Y. Bar-Shalom, L. J. Campo and P. B. Luh, "From Receiver Operating Characteristic to System Operating Characteristic: Evaluation of a Track Formation System," *IEEE Trans. Automatic Control*, Vol. AC-35, No. 2, pp. 172-179, February 1990.

[28] Y. Bar-Shalom and T. E. Fortmann, *Tracking and Data Association*, Boston: Academic Press, 1988.

[29] Y. Bar-Shalom and X. R. Li, *Estimation and Tracking, Principles, Techniques and Software*, Norwood, MA: Artech House, 1993.

[30] Y. Bar-Shalom and X. R. Li, *Multitarget-Multisensor Tracking, Principles and Techniques*, Storrs, CT: YBBS Publishing, 1995.

[31] Y. Bar-Shalom, X. R. Li and T. Kirubarajan, *Estimation with Applications to Tracking and Navigation*, New York: Wiley-Interscience, 2001.

[32] Y. Bar-Shalom, P. K Willet, and X. Tian, *Tracking, and Data Fusion*, Storrs, CT: YBBS Publishing, 2011.

[33] G. Bartolini and A. Ferrara, "Multi-input sliding mode control of a class of uncertain nonlinear systems," *IEEE Trans. Automatic Control*, Vol. 41, No. 11, pp. 1662-1666, November 1996.

[34] G. Bartolini, A. Ferrara, V. I. Utkin and T. Zolezzi, "A control vector simplex approach to variable structure control of nonlinear systems," *Int. J. Robust and Nonlinear Control*, Vol. 7, pp. 321-335, 1997.

[35] S. Battilotti and L. Lanari L., "Tracking with disturbance attenuation for rigid robots," *1996 IEEE International Conference on Robotics and Automation*, Minneapolis, Minnesota, pp. 1578-1583, April 1996.

[36] R. Bellman, *Dynamic Programming*, Mineola, NY: Dover, 2003.

[37] R. Bellman, *Stability Theory of Differential Equations*, New York: McGraw Hill, 1953.

[38] R. Bellman, "Vector Lyapunov functions," *J.S.I.A.M. Control*, Ser. A, Vol. 1, No.1, pp. 32-34, 1962.

[39] R. F. Berg, "Estimation and prediction for maneuvering target trajectories," *IEEE Trans. Automatic Control*, Vol. 28, No. 3, pp. 294-304, March, 1983.

[40] L. D. Berkovitz, *Optimal Control Theory*, New York: Springer Verlag, 2010.

[41] L. D. Berkovitz and N. G. Medhin, *Nonlinear Optimal Control*, Taylor & Francis-CRC, Boca Raton, FL, 2013.

[42] J. E. Bertram and P. E. Sarachik, "On optimal computer control," *Proc. First International Congress of the Federation of Automatic Control*, London: Butterworths, pp. 419-422, 1961.

[43] G. Besançon, "Global output feedback tracking control for a class of Lagrangian systems," *Automatica*, Vol. 36, pp. 1915-1921, 2000.

[44] J. T. Betts, *Practical Methods for Optimal Control Using Nonlinear Programming*, Philadelphia: SIAM Press, 2001.

[45] N. P. Bhatia, "Attraction and nonsaddle sets in dynamical systems," *J. Differential Equations*, Vol. 8, No. 2, 1970.

[46] N. P. Bhatia, "On asymptotic stability in dynamical systems," *Math. Systems Theory*, Vol. 1, No. 2, pp. 113-127, 1967.

[47] N. P. Bhatia and A. C. Lazer, "On global weak attractors in dynamical systems," in: J. K. Hale, J. P. LaSalle, Eds., *Differential Equations and Dynamical Systems*, New York: Academic Press, pp. 321-325, 1967.

[48] N. P. Bhatia and G. P. Szegö, *Dynamical Systems: Stability Theory and Applications*, Berlin: Springer-Verlag, 1967.

[49] S. P. Bhattacharyya, "Frequency domain conditions for disturbance rejection," *IEEE Trans. Automatic Control*, Vol. AC-25, No. 6, 1211-1213, December 1980.

[50] S. P. Bhattacharyya, "Transfer function conditions for output feedback disturbance rejection," *IEEE Trans. Automatic Control*, Vol. AC-27, No. 4, pp. 974-977, August 1982.

[51] S. P. Bhattacharyya, A. C. del Nero Gomes and J. W. Howze, "The structure of robust disturbance rejection," *IEEE Trans. Automatic Control*, Vol. AC-28, No. 9, 874-881, September 1983.

[52] S. S. Blackman, *Multiple-Target Tracking with radar Applications*, Norwood, MA: Artech House, 1999.

[53] S. S. Blackman and R. Popoli, *Design and Analysis of Modern Tracking Systems*, Norwood, MA: Artech House, 1999.

[54] J. H. Blakelock, *Automatic control of aircrafts and missiles*, New York: Wiley, 1991.

[55] V. Boltyanski, H. Martini and V. Soltan, *Geometric Methods and Optimization Problems*, New York: Springer, 2014.

[56] P. Borne, G. Dauphin-Tanguy, J.-P. Richard, F. Rotella, and I. Zambettakis, *Commande et Optimisation des Processus*, Paris: Éditions TECHNIP, 1990.

[57] R. K. Brayton and C.H. Tong, "Constructive stability and asymptotic stability of dynamical systems," *IEEE Trans. Circuits and Systems*, Vol. CAS-21, No. 1, pp. 1121-1130, 1980.

[58] R. W. Brockett and M. D. Mesarović, "The reproducibility of multivariable systems," *J. Math. Analysis Applications*, 1, pp. 548-563, 1965.

[59] W. L. Brogan, *Modern Control Theory*, New York: Quantum, 1974.

[60] J. R. Broussard and M. J. O'Brien, "Feedforward control to track the output of a forced model," *IEEE Trans. Automatic Control*, Vol. AC-25, No. 4, pp. 851-953, 1979.

[61] G. S. Brown and D. P. Campbell, *Principles of Servomechanisms*, New York: Wiley, 1948.

[62] A. E. Bryson, and Y. Ho, *Applied Optimal Control: Optimization, Estimation and Control*, (Revised Printing), New York: Wiley, 1975.

[63] A. E. Bryson, and Y. C. Ho, *Applied Optimal Control.* Washington, DC: Hemisphere, 1975.

[64] R. Bulirsch, A. Miele, J. Stoer, K. H. Well, Eds., *Optimal Control*, Basel: Springer Basel AG, 1993.

[65] F. Bullo and R. M. Murray, "Trajectory tracking for fully actuated mechanical systems," *Proc. European Control Conference*, Tu-E K2 pp. 1-6, Brussels, Belgium, 1-4 July, 1997.

[66] T. A. Burton, "On the construction of Lyapunov functions," *SIAM J. Appl. Math.*, Vol. 17, No. 6, pp. 1078-1085, 1969.

[67] D. Bushaw, "Stabilities of Lyapunov and Poisson types," *SIAM Review*, Vol. 11, No. 2, pp. 214-225, 1969.

[68] J. B. D. Cabrera and K. S. Narendra, "Issues in the Application of Neural Networks for Tracking Based on Inverse Control," *IEEE Trans. Automatic Control*, Vol. 44, No. 11, pp. 2007-2027, November 1999.

[69] L. Cai and Z. Zhao, "A new Control Method for Uncertain Nonlinear Systems Tracking Control," *1996 IEEE International Conference on Robotics and Automation*, Minneapolis, Minnesota, pp. 933-938, April 1996.

[70] A. Cavallo and G. De Maria, "A sliding manifold approach to the feedback control of rigid robots," *Int. J. Robust and Nonlinear Control*, Vol. 6, pp. 501-516, 1996.

[71] L. Cesari, *Asymptotic Behavior and Stability Problems in Ordinary Differential Equations*, Berlin: Springer, 1959.

[72] Y.-C. Chang, "Robust tracking control for nonlinear MIMO systems via fuzzy approaches," *Automatica*, Vol. 36, pp. 1535-1545, 2000.

[73] S. S. L. Chang, *Synthesis of Optimum Control Systems*, New York: McGraw–Hill, 1961.

[74] B. Chatterjec, "Nonlinear feedback in servo systems," *IRE Trans. Automatic Control*, Vol. AC-5, No. 4, pp. 329-330, 1960.

[75] C.-T. Chen, *Linear System Theory and Design*, New York, NY, USA: Holt, Rinehart and Winston, Inc., 1984.

[76] Y.-C. Chen and S. Chang, "Output tracking design of affine nonlinear plant via variable structure system," *IEEE Trans. Automatic Control*, Vol. 37, No. 11, pp. 1823-1828, November 1992.

[77] C.-T. Chen and G. H. Hostetter, "Design of two-input one output compensators to achieve asymptotic tracking," *Int. J. Control*, Vol. 46, No. 6, pp. 1883-1887, 1987.

[78] C. S. Chen and E. Kinnen, "Construction of Lyapunov functions," *J. Franklin Inst.*, Vol. 289, No. 2, pp. 133-146, 1970.

[79] Y.-C. Chen, P.-L. Lin and S. Chang, "Design of output tracking via variable structure system: for plants with redundant inputs," *IEE Proc.-D*, Vol. 139, No. 4 pp. 421-428, July 1992,.

[80] B. Chen and J. K. Tugnait, "Tracking of multiple maneuvering targets in clutter using IMM/JPDA filtering and fixed-lag smoothing," *Automatica*, Vol. 37, pp. 239-249, 2001.

[81] X. P. Cheng and R. V. Patel, "Neural network based tracking control of a flexible macro-micro manipulator system," *Neural Networks*, Vol. 16, pp. 271-286, 2003.

[82] H. Chestnut and R. W. Mayer, *Servomechanisms and Regulating System Design*, New York: Wiley, 1955.

[83] N. G. Chetaev, "On stable trajectories of dynamics," (in Russian), Kazan State University, Kazan, USSR, *Mathematics*. 1, Vol. 91, no. 4, pp. 3–8, 1931.

[84] N. G. Chetaev, *Stability of Motion* (in Russian), Moscow: OGIZ, 1946. English translation Oxford: Pergamon Press, 1961.

[85] P. S. M. Chin, "A general method to derive Lyapunov functions for non-linear systems," *Int. J. Control*, Vol. 44, No. 22, pp. 381-393, 1986.

[86] P. S. M. Chin, "Generalized integral method to derive Lyapunov functions for non-linear systems," *Int. J. Control*, Vol. 46, No. 22, pp. 933-943, 1987.

[87] S.-I. Cho and I.-J. Ha, "A learning approach to tracking in mechanical systems with friction," *IEEE Trans. Automatic Control*, Vol. 45, No. 1, pp. 111-116, January 2000.

[88] P. D. Christofides and S. K. Spurgeon, "Robust output tracking using a sliding-mode controller/observer scheme," *Int. J. Control*, Vol. 64, No. 5, pp. 967 -983, 1996.

[89] P. D. Christofides, A. R. Teel and P. Daoutidis, "Robust semi-global output tracking for nonlinear singularly perturbed systems," *Int. J. Control*, Vol. 65, No. 4, pp. 639-666, 1996.

[90] D. Chwa, "Sliding-mode tracking control of nonholonomic wheeled mobile robots in polar coordinates," *IEEE Trans. Control Systems Technology*, Vol. 12, No. 4, pp. 637-644, July 2004.

[91] D. H. Chyung, "Robust tracking controller for time delay systems," *28th Conference on Decision and Control*, pp. 1425-1426, 1989.

[92] E. A. Coddington and N. Levinson, *Theory of Ordinary Differential Equations*, New York: McGraw Hill, 1955.

[93] W. A. Coppel, *Stability and Asymptotic Behaviour of Differential Equations*, Boston: D. C. Heatch, 1965.

[94] J. L. Corne and N. Rouche, "Attractivity of closed sets proved by using a family of Liapunov functions," *J. Differential Equations*, Vol.13, No. 2, pp. 231-246, 1973.

[95] T. R. Crossley, "Model design of continuous-time tracking systems," *Int. J. Control*, Vol. 25, No. 1, pp. 153-163, 1977.

[96] F. E. Daum, "Bounds on performance for multiple target tracking," *IEEE Trans. Automatic Control*, Vol. 35, No. 4, pp. 443-446, 1990.

[97] F. E. Daum, "Bounds on track purity for multiple target tracking," *Proc. 28th Conference on Decision and Control*, Tampa, Florida, pp. 1423-1424, December 1989.

[98] F. E. Daum and R. J. Fitzgerald, "Array radar tracking," *IEEE Trans. Automatic Control*, Vol. 28, No. 3, pp. 269-282, March, 1983.

[99] G. Dauphin-Tanguy and Lj. T. Grujić, "Asymptotic stability via energy and power. Part II: bond-graph bridging for non-linear systems," *Proc. IFAC Conf. System, Structure and Control*, Nantes, France, pp. 96-101, 1995.

[100] R. Davies, C. Edwards and S. K. Spurgeon, "Robust tracking with a sliding mode," in *Variable Structure and Lyapunov Control*, A. S. I. Zinober Ed., London: Springer Verlag, pp. 51-73, 1994.

[101] J. H. Davis and R. M. Hirschorn, "Tracking control of a flexible robot link," *IEEE Trans. Automatic Control*, Vol. 33, No. 3, pp. 238-248, March 1998.

[102] E. J. Davison, "The robust decentralized control of a servomechanism problem," *IEEE Trans. Automatic Control*, Vol. AC-21, No. 1, 14-24, 1976.

[103] E. J. Davison, "The robust decentralized control of a servomechanism problem for composite systems with input-output interconnections," *IEEE Trans. Automatic Control*, Vol. AC-24, No. 4, pp. 325-327, 1979.

[104] E. J. Davison, "The robust decentralized servomechanism problem with extra stabilizing control agents," *IEEE Trans. Automatic Control*, Vol. AC-22, No. 3, 256-258, 1977.

[105] E. J. Davison and I. Ferguson, "The design of controllers for the multivariable robust servomechanism problem using parameter optimization methods," *IEEE Trans. Automatic Control*, Vol. AC-26, No. 1, 93-110, 1981.

[106] E. J. Davison and A. Goldenberg, "Robust control of a general servomechanism problem: The servo-compensator," *Automatica*, 11, 461-471, 1975.

[107] E. J. Davison and E. Kurak, "A computational method for determining quadratic Lyapunov functions for non-linear systems," *Automatica*, Vol. 7, pp. 627-636, 1971.

[108] E. J. Davison and P. Patel, "Application of the robust servomechanism controller to systems with periodic tracking/disturbance signals," *Int. J. Control*, Vol. 47, No. 1, pp. 111-127, 1988.

[109] E. J. Davison and B. M. Scherzinger, "Perfect control of the robust servomechanism problem," *IEEE Trans. Automatic Control*, Vol. AC-32, No. 8, pp.689-702, August 1987.

[110] J. J. D'Azzo and C. H. Houpis, "Design of tracking systems using output feedback," Chapter 20 in *Linear Control System Analysis & Design*, New York: McGraw Hill, pp. 660-665, 1988.

[111] L. Debnath, *Integral Transformations and Their Applications*, Boca Raton, FL: CRC Press, 1995.

[112] J. S. Demetry and H. A. Titus, "Adaptive tracking of Maneuvering targets," *IEEE Trans. Automatic Control*, Vol. AC-13, No. 6, 749-750, 1968.

[113] B. P. Demidovich, *Lectures on Mathematical Stability Theory* (in Russian), Moscow: Nauka, 1967.

[114] A. Denker and K. Ohnishi, "Robust tracking control of mechatronic arms," *IEEE/ASME Trans. Mechatronics*, Vol. 1, No. 2, pp. 181-188, June 1996.

[115] C. A. Desoer and C.-A. Lin, "Tracking and disturbance rejection of MIMO nonlinear systems with PI controller," *IEEE Trans. Automatic Control*, Vol. AC-30, No. 9, pp. 861-867, 1985.

[116] C. A. Desoer and Y. T. Wang, "The robust non-linear servomechanism problem ," *Int. J. Control*, Vol. 29, No. 5, pp. 803-828, 1979.

[117] S. Devasia, D. Chen, and B. Paden, "Nonlinear inversion-based output tracking," *IEEE Trans. Automatic Control*, Vol. 41, No. 7, pp. 930-942, July 1996.

[118] S. Di Gennaro, "Output attitude tracking for flexible spacecraft," *Automatica*, Vol. 38, pp. 1719-1726, 2002.

[119] W. E. Dixon, D. M. Dawson, E. Zergeroglu and F. Zhang, "Robust tracking and regulation control for mobile robots," *Int. J. Robust and Nonlinear Control*, Vol. 10, pp. 199-216, 2000.

[120] M. Doğruel, Ü. Özgüner, and S. Drakunov, "Sliding mode control in discrete-state and hybrid systems," *IEEE Trans. Automatic Control*, Vol. 41, No. 3, pp. 414-419, March 1996.

[121] C. Edward and S. K. Spurgeon, "Robust output tracking using a sliding mode controller / observer scheme," *Int. J. Control*, Vol. 64, No. 5, pp. 967-983, 1996.

[122] C. Edwards and S. K. Spurgeon, "Sliding mode output tracking with application to a multivariable high temperature furnace problem," *Int. J. Robust and Nonlinear Control*, Vol. 7, pp. 337-351, 1997.

[123] B. Etkin, *Dynamics of Flight-Stability and Control*, New York: John Wiley, 1982.

[124] E. Fabian and W. M. Wonham, "Decoupling and disturbance rejection," *IEEE Trans. Automatic Control*, Vol. AC-20, No. 4, 399-401, 1975.

[125] F. Fallside and M. R. Patel, "Control engineering applications of V. I. Zubov's construction procedure for Lyapunov functions," *IEEE Trans. Automatic Control*, Vol. AC-10, No. 2, 220 -222, 1965.

[126] L. Fang, P. J. Antsaklis, L. Montestruque, B. McMickell, M. Lemmon, Y. Sun, H. Fang, I. Koutroulis, M. Haenggi, M. Xie, and X. Xie, "A wireless dead reckoning pedestrian tracking system," *ACM Workshop on Applications of Mobile Embedded Systems*, Boston, MA, pp. 1 -3, 2004.

[127] L. Fang and P. J. Antsaklis, "Decentralized formation tracking of multi-vehicle systems with consensus-based controllers," Chapter 15 in *Advances in Unmanned Aerial Vehicles; State of the Art and the Road to Autonomy,* Ed. K. P. Valavanis, Berlin: Springer, pp. 455-471, 2007.

[128] L. Fang, P. Antsaklis, "Decentralized formation tracking of multi-vehicle systems with nonlinear dynamics," *14th Mediterranean Conference on Control and Automation, (MED '06)*, Universitá Politecnica delle Marche, Ancona, pp. 1-6, June 28-30, 2006.

[129] A. A. Fel'Dbaum, *Optimal Control Systems*, Salt Lake City: Academic Press, 1965.

[130] P. M. G. Ferreira, "Tracking with sensor failures," *Automatica*, Vol. 38, pp. 1621-1623, 2002.

[131] G. M. Fikhtengol'c, *Course on Differential and Integral Calculus* (in Russian), Moscow: Fizmatgiz, 1958.

[132] I. Flügge-Lotz and C. F. Taylor, "Synthesis of a nonlinear control system," (Paper was presented at the West Coast Conference sponsored by IRE, August 1955), *IRE Trans. Automatic Control*, Vol. 1, No. 1, pp.3-9, May 1956.

[133] T. E. Fortmann, Y. Bar-Shalom, M. Scheffe, and S.Gelfand, "Detection thresholds for tracking in clutter-A connection between estimation and signal processing," *IEEE Trans. Automatic Control*, Vol. AC-30, No. 3, pp. 221-228, 1985.

[134] T. I. Fossen, *Guidance and Control of Ocean Vehicles*, Chichester: John Wiley & Sons, 1994.

[135] T. I. Fossen and J. G. Balchen, "Modeling and non-linear self-tuning robust trajectory control of an autonomous underwater vehicle," *Modeling, Identification and Control*, Vol. 9, No. 4, pp. 165-177, 1988.

[136] G. F. Franklin, "Design of ripple-free multivariable robust servomechanisms," *IEEE Trans. Automatic Control*, Vol. AC-31, No. 7, pp. 661-664, 1986.

[137] R. A. Freeman and P. V. Kokotović, "Tracking controllers for systems linear in the unmeasured states," *Automatica*, Vol. 32, No. 5, pp. 735-746, 1996.

[138] L.-C. Fu and T.-L. Liao, "Globally stable robust tracking of nonlinear systems using variable structure control and with an application to a robotic manipulator," *IEEE Trans. Automatic Control*, Vol. 35, No. 12, pp. 1345-1350, December 1990.

[139] F. R. Gantmacher, *The Theory of Matrices*, Vol. 1, New York: Chelsea, 1974.

[140] F. R. Gantmacher, *The Theory of Matrices*, Vol. 2, New York: Chelsea, 1974.

[141] R. A. García and C. E. D'Attellis, "Trajectory tracking in nonlinear systems via nonlinear reduced-order observers," *Int. J. Control*, Vol. 62, No. 3, pp. 685-715, 1995.

[142] E. Gershon, U. Shaked and I. Yaesh, "H_∞ tracking of linear continuous-time systems with stochastic uncertainties and prev," *Int. J. Robust and Nonlinear Control*, Vol. 14, No. 7, pp. 607-626, 2004.

[143] E. G. Gilbert, "Controllability and observability in multivariable control systems," *SIAM J. Control*, Vol. 1, pp. 128-151, 1963.

[144] D. N. Godbole and S. S. Sastry, "Approximate decoupling and asymptotic tracking for MIMO systems," *IEEE Trans. Automatic Control*, Vol. 40, No. 3, pp. 441-450, March, 1995.

[145] O. R. González and P. J. Antsaklis, "Internal models in regulation, stabilization and tracking," *Proceedings of 28th Conference on Decision and Control*, Tampa, Florida, pp. 1343-1348, 1989.

[146] G. A. Gordon and S. J. Oh, "A theory of tracking for a dynamic radar scatterer ensemble," *IEEE Trans. Automatic Control*, Vol. AC-18, No. 2, pp. 91-97, 1973.

[147] T. J. Greattinger and B. H. Krogh, "On the computation of reference signal constraints for guaranteed tracking performance," *Automatica*, Vol. 18, No. 6, pp. 1125-1141, 1992.

[148] J. W. Grizzle, M. D. Di Benedetto and F. Lamnabhi-Lagarrigue, "Necessary conditions for asymptotic tracking in nonlinear systems," *IEEE Trans. Automatic Control*, Vol. AC-39, No. 9, pp.1782-1794, September 1994.

[149] M. Gruber, "Path integrals and Lyapunov functionals," *IEEE Trans. Automatic Control*, Vol. AC-14, No. 5, 465-478, 1969.

[150] Lj. T. Grujić, "Adaptive tracking control for a class of plants with uncertain parameters and non-linearities," *Int. J. Adaptive Control and Signal Processing*, Vol. 2, pp. 49-71, 1988.

[151] Lj. T. Grujić, "Algebraic conditions for absolute tracking control of continuous-time Lurie systems," *Proc. Conference on Linear Algebra in Signals, Systems, and Control*, (Boston, Massachussetts, August 12-14, 1986), published as *Linear Algebra in Signals, Systems, and Control*, Eds. B. N. Datta, C. R. Johnson, M. A. Kaashoek, R. J. Plemmons, and E. D. Sontag, Philadelphia: SIAM, pp. 535-555, 1988.

[152] Lj. T. Grujić, "Algebraic conditions for absolute tracking control of Lurie systems," *Int. J. Control*, Vol. 48, No. 2, pp. 729-754, 1988.

[153] Lj. T. Grujić, "Algorithms for CAD of continuous-time non-stationary non-linear tracking systems via the output-space," Anaheim, CA: *ACTA Press*, pp. 58-61, 1985.

[154] Lj. T. Grujić, "Algorithms for CAD of discrete-time non-stationary non-linear tracking systems via the state-space," Anaheim, CA: *ACTA Press*, pp. 135-138, 1985.

[155] Lj. T. Grujić, "A physical principle and consistent Lyapunov methodology: time-invariant nonlinear systems," *Proc. International Conference on Advances in Systems, Signals, Control and Computers*, 1, Durban, South Africa, pp. 42-50, 1998.

[156] Lj. T. Grujić, "Complete exact solution to the Lyapunov stability problem: Time-varying nonlinear systems with differentiable motions," *Nonlinear Analysis, Theory, Methods & Applications*, Vol. 22, No. 8, pp. 971-981, 1994.

[157] Lj. T. Grujić, "Concepts of stability domains" (in Serbo-Croatian), Zagreb (Croatia): *Automatika*, Vol. 26, No. 1-2, pp. 5-10, 1985.

[158] Lj. T. Grujić, "Consistent Lyapunov methodology for time-invariant nonlinear systems," *Avtomatika i Telemehanika* (in Russian), No. 12, December, pp. 35-73, 1997.

[159] Lj. T. Grujić (Ly. T. Gruyitch), "Continuous time control systems," *Lecture notes for the course "DNEL4CN2: Control Systems,* "Durban: Department of Electrical Engineering, University of Natal, South Africa, 1993.

[160] Lj. T. Grujić, "Exact both construction of Lyapunov function and asymptotic stability domain determination," *Proc. 1993 IEEE/SMC International Conference on System, Man and Cybernetics*, Le Touquet, France, Vol. I, pp. 331-336, October 17-20, 1993.

[161] Lj. T. Grujić, "Exact determination of a Lyapunov function and the asymptotic stability domain," *Int. J. Systems Sci.*, Vol. 23, No. 11, pp. 1871-1888, 1992.

[162] Lj. T. Grujić, "Exact solutions for asymptotic stability: Non-linear systems," *Int. J. Non-Linear Mechanics*, Vol. 30, No. 1, 45-56, 1995.

[163] Lj. T. Grujić, "Exponential quality of time-varying dynamical systems: Stability and tracking," in *Advances in Nonlinear Dynamics-Stability and Control: Theory, Methods and Applications*, Eds. S. Sivasundaram and A. A. Martynyuk, Amsterdam: Gordon and Breach Science, 5, pp. 51-61, 1997.

[164] Lj. T. Grujić, "Lyapunov-like solutions for stability problems of the most general Lurie-Postnikov systems," *Int. J. Systems Sci.*, Vol. 12, No. 7, pp. 813-833, 1981.

[165] Lj. T. Grujić, "Natural trackability and control: Multiple time scale systems," *Preprints of the IFAC-IFIP-IMACS Conference: Control of Industrial Systems*, 2, Pergamon, Elsevier, London, pp. 111-116, 1997.

[166] Lj. T. Grujić, "Natural trackability and control: Multiple time scale systems," *IFAC Conference: Control of Industrial Systems*, (Eds. Lj. T. Grujić, P. Borne, A. El Moudni and M. Ferney), Vol. 2, Pergamon, Elsevier, London, pp. 669-674, 1997.

[167] Lj. T. Grujić, "Natural trackability and control: Perturbed robots," *Preprints IFAC-IFIP-IMACS Conference: Control of Industrial Systems*, 3, Belfort, France, pp. 691-696, 1997.

[168] Lj. T. Grujić, "Natural trackability and control: perturbed robots," *Proc. IFAC Conference: Control of Industrial Systems*, (Ed's. Lj. T. Grujić, P. Borne, A. El Moudni and M. Ferney), Vol. 3, London: Pergamon, Elsevier, pp. 1641-1646, 1997.

[169] Lj. T. Grujić, "Natural trackability and tracking control of robots," *IMACS-IEEE-SMC Multiconference CESA'96: Symposium on Control, Optimization and Supervision*, Vol. 1, Lille, France, pp. 38-43, 1996.

[170] Lj. T. Grujić, "New approach to asymptotic stability: Time-varying nonlinear systems," *Int. J. of Mathematics and Mathematical Sciences*, Vol. 20, No. 2, pp. 347-366, 1997.

[171] Lj. T. Grujić, "New Lyapunov method based methodology for asymptotic stability," *Proc. IFAC Conference on System Structure and Control*, Nantes, France, pp. 536-541, July 5-7, 1995.

[172] Lj. T. Grujić, "New Lyapunov methodology and exact construction of a Lyapunov function: Exponential stability," *Problems of the Nonlinear Analysis in Engineering Systems*, Kazan, Russia, Vol. 1, pp. 9-16, 1995.

[173] Lj. T. Grujić, "Non-linear singularly perturbed tracking systems," *Proc. AMSE Conference on Modelling and Simulation*, Paris, pp. 116-123, 1982.

[174] Lj. T. Grujić, "Novel Lyapunov stability methodology for nonlinear systems: Complete solutions," *Nonlinear Analysis, Theory, Methods & Applications*, Vol. 30, No. 8, pp. 5315-5325, 1997.

[175] Lj. T. Grujić, "Novel development of Lyapunov stability of motion," *Int. Journal of Control*, Vol. 22, No. 4, pp. 525-549, 1975.

[176] Lj. T. Grujić, "On absolute stability and Aizerman conjecture," Automatica, Vol. 17, No. 2, pp. 335-349, 1981.

[177] Lj. T. Grujić, "On general solutions of non-linear tracking for stationary systems," *AI 83 IASTED Symposium*, Lille, pp. 49-53, 1983.

[178] Lj. T. Grujić, "On large-scale systems stability," *Computing and computers for control systems*, Eds. P. Borne et al., J. C. Baltzer AG, Scientific, IMACS, pp. 201-206, 1989.

[179] Lj. T. Grujić, "On non-linear tracking domain estimates: Discrete-time systems," *AI 83 IASTED Symposium*, Lille, pp. 59-63, 1983.

[180] Lj. T. Grujić, "On non-linear tracking domain estimates: Continuous-time systems," *AI 83 IASTED Symposium*, Lille, pp. 65-66, 1983.

[181] Lj. T. Grujić, "On non-linear tracking phenomena and problems," *AI 83 IASTED Symposium*, Lille, 1983, pp. 45-48.

[182] Lj. T. Grujić, "On solutions to Lyapunov stability problems," *Facta, Universitatis, Series: Mechanics, Automatic Control and Robotics*, University of Nish, Serbia, Yugoslavia, Vol. 1, No. 2, pp. 121-138, 1992.

[183] Lj. T. Grujić, "On the non-linear tracking systems theory: I-Phenomena, concepts and problems via output space," *Automatika*, Zagreb, Vol. 27, No. 1-2, pp. 3-8, 1986.

[184] Lj. T. Grujić, "On the non-linear tracking systems theory: II-Phenomena, concepts and problems via state space," *Automatika*, Zagreb, Vol. 27, No. 1-2, pp. 9-16, 1986.

[185] Lj. T. Grujić, "On the non-linear tracking systems theory: III-Liapunov-like approach via the output-space: Continuous-time," *Automatika*, Zagreb, Vol. 27, No. 3-4, pp. 99-104, 1986.

[186] Lj. T. Grujić, "On the non-Linear tracking systems theory: IV-Liapunov-like approach via the state-space: Continuous-time," *Automatika*, Zagreb, Vol. 27, No. 3-4, pp. 105-116, 1986.

[187] Lj. T. Grujić, "On the non-Linear tracking systems theory: V-Liapunov-like approach via the output-space: Discrete-time," *Automatika*, Zagreb, Vol. 27, No. 5-6, pp. 197-202, 1986.

[188] Lj. T. Grujić, "On the non-Linear tracking systems theory: VI-Liapunov-like approach via the state-space: Discrete-time," *Automatika*, Zagreb, Vol. 27, No. 5-6, pp. 203-211, 1986.

[189] Lj. T. Grujić, "On the theory and synthesis of non-linear non-stationary tracking singularly perturbed systems," *Control Theory and Advanced Technology*, Tokyo: MITA Press, Vol. 4, No. 4, pp. 395-409, 1988.

[190] Lj. T. Grujić, "On the theory of nonlinear systems tracking with guaranteed performance index bounds: Application to robot control," *Proc. 1989 IEEE Int. Conference on Robotics and Automation*, Scottsdale, Arizona: IEEE-Computer Society Press, Vol. 3, pp. 1486-1490, May 14-19, 1989.

[191] Lj. T. Grujić, "On the tracking problem for nonlinear systems," in *Applied Control*, New York: Marcel Dekker, pp. 325-343, 1993.

[192] Lj. T. Grujić, "On the tracking theory with a prespecified quality," (in Serbo-Croatian), *Zastava*, Vol. VIII, No. 28-29, pp. 18-22, Oct. 1990.

[193] Lj. T. Grujić, "On tracking control of singularly perturbed systems," *Proc. IMACS-IFAC Symposium: "Modelling and Simulation for Control of Lumped and Distributed Parameter Systems*," Villeneuve d'Ascq, pp. 565-568, June 1-3, 1986.

[194] Lj. T. Grujić, "On tracking domain estimates of large-scale systems," *AI 83 IASTED Symposium*, Lille, pp. 55-57, 1983.

[195] Lj. T. Grujić, "On tracking domains of continuous-time non-linear control systems," *Proc. 1982 American Control Conference*, AACC-IEEE, New York, 1982, pp. 670-674. Also: *R.A.I.R.O. Automatique / Systems Analysis and Control*, 16, No. 4, 1982, pp. 311-327.

[196] Lj. T. Grujić, "Phenomena, concepts and problems of automatic tracking: continuous-time stationary non-linear systems with variable inputs," (in Serbo-Croatian), *Proc. First Int. Seminar "AUTOMATON and ROBOT*, "USAUM Srbije i "OMO, "Belgrade, pp. 307-330, 1985.

[197] Lj. T. Grujić, "Phenomena, concepts and problems of automatic tracking: discrete-time stationary non-linear systems with variable inputs," (in Serbo-Croatian), *Proc. First Int. Seminar "AUTOMATON and ROBOT*, "USAUM Srbije i "OMO, "Belgrade, pp. 401-422, 1985.

[198] Lj. T. Grujić, "Sets and singularly perturbed systems," *Systems Science* (Wroclaw, Poland), Vol. 5, No. 4, pp. 327-338, 1979.

[199] Lj. T. Grujić, "Solutions to Lyapunov stability problems: nonlinear systems with continuous motions," *Int. J. Math. Math. Sci.*, Vol. 17, No.3, pp. 587-596, 1994.

[200] Lj. T. Grujić, "Solutions to Lyapunov stability problems: Nonlinear systems with differentiable motions," *Proc. 13th IMACS World Congress on Computation and Applied Mathematics*, Vol. 3, pp. 1228-1231, 1991, and in: *Computational and Applied Mathematics II: Differential Equations*, Eds. W.F Ames and P.J. van der Houwen, Amsterdam: Elsevier, pp. 39-47, 1992.

[201] Lj. T. Grujić, "Solutions to Lyapunov stability problems: nonlinear systems with differentiable motions," *Int. J. Systems Science*, Vol. 23, No. 11, pp. 1874-1888, 1992.

[202] Lj. T. Grujić, "Solutions to Lyapunov stability problems: nonlinear systems with globally differentiable motions," in *The Lyapunov Functions Method and Applications*, Ed. P. Borne and V. Matrosov, J.C. Baltzer AG, Scientific, IMACS, pp. 19-27, 1990.

[203] Lj. T. Grujić, "Solutions to Lyapunov stability problems of sets: nonlinear systems with differentiable motions," *Proc. 13th IMACS World Congress on Computation and Applied Mathematics*, Dublin, Vol. 3, pp. 1228-1231, and in *Int. J. Math. Math. Sci.*, Vol. 17, No.1, pp.103-112, 1994.

[204] Lj. T. Grujić, "Solutions to Lyapunov stability problems: time-invariant systems," *Proc. 14th IMACS World Congress on Computation and Applied Mathematics*, Vol. 1, pp. 203-205, 1994.

[205] Lj. T. Grujić, "Solutions to Lyapunov stability problems via O-uniquely bounded sets," *Control-Theory and Advanced Technology*, Tokyo, Vol. 10, No. 4, Part 2, pp. 1069-1091, 1995.

[206] Lj. T. Grujić, "Stability and instability of product sets," *Systems Science*, Vol. 3, No. 1, pp. 14-31, 1977.

[207] Lj. T. Grujić, "Stability Domains of General and Large-Scale Systems," *Proc. IMACS/IFAC Symposium on Modelling and Simulation for Control and Distributed Parameter Systems*, Eds. P. Borne and S. G. Tzafestas, Elsevier Science Publishers B. V. (North Holland), IMACS, pp. 317-327, 1987.

[208] Lj. T. Grujić, "Stability domains of general and large-scale stationary systems," in *Applied Modelling and Simulation of Technological Systems*, Institut Industriel du Nord (Lille, France), Vol. 1, pp. 267-272, 1986.

[209] Lj. T. Grujić, *Stability of Large-Scale Systems*, D. Sci. dissertation, Belgrade, Faculty of Mechanical Engineering, University of Belgrade, defended 1972, published 1974.

[210] Lj. T. Grujić, "Stability versus tracking in automatic control systems," (in Serbo-Croatian), *Proc. JUREMA 29*, Part 1, Zagreb, pp. 1-4, 1984.

[211] Lj. T. Grujić, "State-space domains of singularly perturbed systems," *Proc. 12th World Congress on Scientific Computation*, IMACS, Paris, pp. 83-91, July 18-22, 1988.

[212] Lj. T. Grujić, "Synthesis of automatic tracking systems and the output-space: continuous-time stationary non-linear systems with variable inputs," (in Serbo-Croatian), *Proc. First Int. Seminar "AUTOMATON and ROBOT*, "USAUM Srbije i "OMO, "Belgrade, pp. 331-370, 1985.

[213] Lj. T. Grujić, "Synthesis of automatic tracking systems and the state-space: continuous-time stationary non-linear systems with variable inputs" (in Serbo-Croatian), *Proc. First Int. Seminar "AUTOMATON and ROBOT*, "USAUM Srbije i "OMO, "Belgrade, pp. 371-400, 1985.

[214] Lj. T. Grujić, "Synthesis of automatic tracking systems and the output-space: discrete-time stationary non-linear systems with variable inputs" (in Serbo-Croatian), *Proc. First Int. Seminar "AUTOMATON and ROBOT*, "USAUM Srbije i "OMO, "Belgrade, pp. 423-448, 1985.

[215] Lj. T. Grujić, "Synthesis of automatic tracking systems and the state-space: discrete-time stationary non-linear systems with variable inputs," (in Serbo-Croatian), *Proc. First Int. Seminar "AUTOMATON and ROBOT*, "USAUM Srbije and "OMO, "Belgrade, pp. 449-476, 1985.

[216] Lj. T. Grujić, "The necessary and sufficient conditions for the exact construction of a Lyapunov function and the asymptotic stability domain," *Proc. 39th IEEE Conference on Decision and Control*, Brighton, Vol. 3, pp. 2885-2888, 1991.

[217] Lj. T. Grujić, "Time-varying continuous nonlinear systems: Uniform asymptotic stability," *Int. J. Systems Science*, Vol. 26, No. 5, pp. 1103-1127, 1995; "Corrigendum, "Ibid, Vol. 27, No. 7, p. 689, 1996.

[218] Lj. T. Grujić, "Tracking analysis for non-stationary non-linear systems," Proc. *1986 IEEE Int. Conference on Robotics and Automaton*, San Francisco, Vol. 2, pp. 713-721, 1986.

[219] Lj. T. Grujić, "Tracking control obeying prespecified performance index," *Proc. 12th World Congress on Scientific Computation*, IMACS, Paris, pp. 332-336, July 18-22, 1988, also in: *Computing and Computers for control systems*, Ed. P. Borne et al., J.C. Baltzer AG, Scientific, IMACS, pp. 229-233, 1989.

[220] Lj. T. Grujić, "Tracking with prespecified index limits: Control synthesis for non-linear objects," *Proc. II Int. Seminar and Symposium: "AUTOMATON and ROBOT,* "SAUM and IEE, Belgrade, pp. S-20-S-52, 1987.

[221] Lj. T. Grujić, "Tracking versus stability: Theory," *Proc. 12th World Congress on Scientific Computation*, IMACS, Paris, pp. 319-327, July 18-22, 1988, also in *Computing and Computers for Control Systems*, Ed. P. Borne et al., J.C. Baltzer AG, Scientific, IMACS, pp. 165-173, 1989.

[222] Lj. T. Grujić, "Uniquely bounded sets," presented at *1st Conference on Mathematics at the Service of Man*, Barcelona, Spain, July 11-16, 1977; published at *Proc. 1st Conference on Mathematics at the Service of Man,* Univ. Politec., Barcelona, Vol I, pp. 145-161, 1980.

[223] Lj. T. Grujić, "Uniquely bounded sets and nonlinear systems," *Proc. 1978 IEEE Conference on Decision and Control*, San Diego, California, Vol. I, pp. 325-333, 1979.

[224] Lj. T. Grujić and G. Dauphin-Tanguy, "Asymptotic stability via energy and Power. Part I: New Lyapunov methodology for nonlinear systems," *Proc. IFAC Conference on System Structure and Control*, Nantes, France, pp. 548-553, July 5-7, 1995.

[225] Lj. T. Grujić and Z. Janković, "Synthesis of tracking control for a plane motion," *Proc. Second Conference on Systems, Automatic Control and Measurement* (in Serbo-Croatian), SAUM, Belgrade, pp. 477-492, 1986.

[226] Lj. T. Grujić and Z. Janković, "Synthesis of tracking control for a process," *Proc. Second Conference on Systems, Automatic Control and Measurement*, (in Serbo-Croatian), SAUM, Belgrade, pp. 305-324, 1986.

[227] Lj. T. Grujić, A. A. Martynyuk, and M. Ribbens-Pavella, *Large-Scale Systems under Structural and Singular Perturbations* (in Russian, Kiev: Naukova Dumka 1984), Berlin: Springer Verlag, 1987.

[228] Lj. T. Grujić and A. N. Michel, "Discrete-time mathematical modeling of neural networks under pure structural variations," *Proc. 13th IMACS World Congress on Computational and Applied Mathematics*, Dublin, Ireland, Vol. 3, pp. 1256,1257, July 22-26, 1991.

[229] Lj. T. Grujić and A. N. Michel, "Exponential stability and trajectory bounds of neural networks under structural variations," *Proc. 29th IEEE Conference on Decision and Control*, Honolulu, Hawaii, Vol. 3, pp. 1713-1718, December 5-7, 1990.

[230] Lj. T. Grujić and A. N. Michel, "Modeling and qualitative analysis of continuous-time neural networks under pure structural variations," *Mathematics and Computers in Simulation*, North-Holland-Elsevier, Vol. 40, pp. 523-533, 1996.

[231] Lj. T. Grujić and A. N. Michel, "Qualitative analysis of neural networks under structural perturbations," *Proc. 1990 IEEE Intern. Symp. on Cyrc. and Syst.*, New Orleans, Vol. 1, pp. 391-394, May 1-3, 1990.

[232] Lj. T. Grujić and W. P. Mounfield, Jr., "Natural tracking control of linear systems," *Proc. 13th IMACS World Congress on Computation and Applied Mathematics*, Eds. R. Vichnevetsky and J. J. H. Miller, Trinity College, Dublin, Ireland, Vol. 3, pp. 1269-1270, July 22-26, 1991.

[233] Lj. T. Grujić and W. P. Mounfield, "Natural tracking control of linear systems," in *Mathematics of the Analysis and Design of Process Control*, Ed. P. Borne, S.G. Tzafestas and N.E. Radhy, Elsevier Science B. V., IMACS, pp. 53-64, 1992.

[234] Lj. T. Grujić and W. P. Mounfield, *Natural Tracking Controller*, US Patent No 5,379,210, Jan. 3, 1995.

[235] Lj. T. Grujić and W. P. Mounfield, "Natural tracking PID process control for exponential tracking," *American Inst. of Chemical Engineers J.*, 38, No. 4, pp. 555-562, 1992.

[236] Lj. T. Grujić and W. P. Mounfield, "PD-Control for stablewise tracking with finite reachability time: Linear continuous time MIMO systems with state-space description," *Int. J. Robust and Nonlinear Control*, England, Vol. 3, pp. 341-360, 1993.

[237] Lj. T. Grujić and W. P. Mounfield, "PD natural tracking control of an unstable chemical reaction," *Proc. 1993 IEEE Int. Conference on Systems, Man and Cybernetics*, Le Touquet, Vol. 2, pp. 730-735, 1993.

[238] Lj. T. Grujić and Mounfield W. P., "PID natural tracking control of a robot: theory," *Proc. 1993 IEEE Int. Conference on Systems, Man and Cybernetics*, Le Touquet, Vol. 4, pp. 323-327, 1993.

[239] Lj. T. Grujić and W. P. Mounfield, "Ship roll stabilization by natural tracking control: Stablewise tracking with finite reachability time," *Proc. 3rd IFAC Workshop on Control Applications in Marine Systems*, Trondheim, Norway, pp. 202-207, 10-12 May, 1995.

[240] Lj. T. Grujić and W. P. Mounfield, "Stablewise tracking with finite reachability time: Linear time-invariant continuous-time MIMO systems," *Proc. 31st IEEE Conference on Decision and Control*, Tucson, Arizona, pp. 834-839, 1992.

[241] Lj. T. Grujić and W. P. Mounfield, Jr., "Tracking control of time-invariant linear systems described by IO differential equations," *Proc. 30th IMACS Conference on Decision and Control*, Brighton, England, Vol. 3, pp. 2441-2446, December 11-13, 1991.

[242] Lj. T. Grujić and Z. Novaković, "Feedback principle via Liapunov function concept rejects robot control synthesis drawback," *Proc. Sixth IASTED Int. Symposium on Modelling*, Identification and Control, Grindelwald, pp. 227-230, February 17-20, 1987.

[243] Lj. T. Grujić and Z. Novaković, "Output robot control: Stablewise tracking with a requested reachability time," *Proc. IEEE Int. Workshop on Intelligent Robots and Systems*, Tokyo, pp. 85-90, Oct. 31-Nov. 2, 1988.

[244] Lj. T. Grujić and Z. Novaković, "Theory of robust adaptive exponential tracking control. Application to robots without using inverse mechanics," *Proc. Fifth Yale Workshop on Applications of Adaptive Systems Theory*, Center for Systems Science, Yale University, pp. 237-243, May 20-22, 1987.

[245] Lj. T. Grujić and B. Porter, "Continuous-time tracking systems incorporating Lur'e plants with single non-linearities," *Int. J. Systems Science*, Vol. 11, No. 2, pp. 177-189, 1980.

[246] Lj. T. Grujić and B. Porter, "Discrete-time tracking systems incorporating Lur'e plants with multiple non-linearities," *Int. J. Systems Science*, Vol. 11, No. 12, pp. 1505-1520, 1980.

[247] Lj. T. Grujić and Z. Ribar, "Synthesis of time-varying Lurie structurally variable tracking systems," *Proc. 13th IMACS World Congress on Computation and Applied Mathematics*, pp. 1267-1268, July 22-26, 1991.

[248] Lj. T. Grujić and M. Ribbens-Pavella, "Asymptotic stability of large-scale systems: Part 1: domain estimations," *Elec. Power and Energy Systems*, Vol. 1, No. 3, pp. 151-157, 1979.

[249] Lj. T. Grujić and D. D. Šiljak, "Asymptotic stability and instability of large-scale systems," *IEEE Trans. Automatic Control*, Vol. AC-18, No. 6, pp. 636-645, December 1973.

[250] Lj. T. Grujić and D. D. Šiljak, "On stability of discrete composite systems," *IEEE Trans. Automatic Control*, Vol. AC-18, No. 5, pp. 522-524, October 1973.

[251] Lj. T. Grujić and D. D. Šiljak, "Stability of large-scale systems with stable and unstable subsystems," *Proc. 1972 Joint Automatic Control Conference*, Stanford, California, pp. 550-555, 1972.

[252] Lj. T. Gruyitch, *Advances in the Linear Dynamic Systems Theory. Time-Invariant Continuous-Time Systems*, Tamarac, Fl.: Llumina press, 2013.

[253] Ly. T. Gruyitch, "Aircraft natural control synthesis: Vector Lyapunov function approach," *Actual Problems of Airplane and Aerospace Systems: Processes, Models, Experiments*, Vol. 2, No. 6, Kazan, Russia, and Daytona Beach, FL, pp. 1-9, 1998.

[254] Ly. T. Gruyitch, *Conduite des systèmes*, Lecture Notes: Notes de cours SY 98, Belfort: University of Technology Belfort-Montbeliard, 2000, 2001.

[255] Lj. T. Gruyitch, "Consistent Lyapunov methodology for exponential stability: PCUP approach," in *Advances in Stability Theory at the End of the 20th Century*, Ed. A. A. Martynyuk, London: Taylor and Francis, pp. 107-120, 2003.

[256] Ly. T. Gruyitch, "Consistent Lyapunov methodology: Non-differentiable non-linear systems," *Nonlinear Dynamics and Systems Theory*, Vol. 1, No. 1, pp. 1-22, 2001.

[257] Ly. T. Gruyitch, "Consistent Lyapunov methodology, time-varying nonlinear systems and sets," *Nonlinear Analysis, Theory and Applications*, Vol. 39, pp. 413-446, 2000.

[258] Ly. T. Gruyitch, *Contrôle commande des processus industriels*, Lecture Notes: Notes de cours SY 51 , Belfort: University of Technology Belfort-Montbeliard, 2002, 2003.

[259] Ly. T. Gruyitch, *Einstein's Relativity Theory. Correct, Paradoxical, and Wrong*, Victoria: Trafford, Canada, 2006.

[260] Ly. T. Gruyitch, "Exponential stabilizing natural tracking control of robots: theory," *Proc. Third ASCE Specialty Conference on Robotics for Challenging Environments*, held in Albuquerque, New Mexico, (Eds. Laura A. Demsetz, Raymond H. Bryne, and John P. Wetzel), Reston, Virginia: American Society of Civil Engineers (ASCE), pp. 286-292, April 26-30, 1998.

[261] Ly. T. Gruyitch, *Galilean-Newtonean Rebuttal to Einstein's Relativity Theory,* Cambridge: Cambridge International Science Publishing, 2015.

[262] Ly. T. Gruyitch, "Gaussian generalisations of the relativity theory fundaments with applications," *Proc. VII Int. Conference: Physical Interpretations of Relativity Theory,* Ed. M. C.Duffy, British Society for the Philosophy of Science, London, pp.125-136, September 15-18, 2000.

[263] Ly. T. Gruyitch, "Global natural $\theta-$tracking control of Lagrangian systems," *Proc. American Control Conference*, San Diego, California, pp. 2996-3000, June 1999.

[264] Ly. T. Gruyitch, "Natural control of robots for fine tracking," *Proc. 38th Conference on Decision and Control*, Phoenix, Arizona, pp. 5102-5107, December 1999.

[265] Ly. T. Gruyitch, "Natural tracking control synthesis for lagrangian systems," V International Seminar on Stability and Oscillations of Nonlinear Control Systems, *Russian Academy of Sciences*, Moscow, pp. 115-120, June 3-5, 1998.

[266] Ly. T. Gruyitch, "New development of vector Lyapunov functions and airplane control synthesis," Chapter 7 in *Advances in Dynamics and Control*, Ed. S. Sivasundaram, Boca Raton, Fl.: Chapman & Hall/CRC, pp. 89-102, 2004.

[267] Ly. T. Gruyitch, "On tracking theory with embedded stability: control duality resolution," *Proc. 40th IEEE Conference on Decision and Control*, Orlando, Fl., pp. 4003-4008, December 2001.

[268] Ly. T. Gruyitch, "Physical continuity and uniqueness principle. Exponential natural tracking control," *Neural, Parallel & Scientific Computations*, 6, pp. 143-170, 1998.

[269] Ly. T. Gruyitch, "Robot global tracking with finite vector reachability time," *Proc. European Control Conference*, Karlsruhe, Germany, Paper # 132, pp. 1-6, August 31-September 3 1999.

[270] Ly. T. Gruyitch, "Robust prespecified quality tracking control synthesis for 2D systems," *Proc. Int. Conference on Advances in Systems, Signals, Control and Computers*, 3, Durban, South Africa, pp. 171-175, 1998.

[271] Ly. T. Gruyitch, *Systèmes d'asservissement industriels*, Lecture Notes: Notes de cours SY 40, Belfort : Universite de Technologie de Belfort-Montbeliard, 2001.

[272] Ly. T. Gruyitch, *Time and Consistent Relativity. Physical and Mathemaical Fundamentals,* Waretown, NJ: Apple Academic Press and Ontario, Canada: Oakville, 2015.

[273] Ly. T. Gruyitch, *Time. Fields, Relativity, and Systems*, Coral Springs, Fl.: Llumina, 2006.

[274] Ly. T. Gruyitch, *Time and Time Fields. Modeling, Relativity, and Systems Control*, Victoria, Canada: Trafford, 2007.

[275] Ly. T. Gruyitch, "Time and uniform relativity theory fundaments," *Problems of Nonlinear Analysis in Engineering Systems,* 7, N° 2(14), Kazan, Russia, pp. 1-29, 2001.

[276] Ly. T. Gruyitch, "Time, relativity and physical principle : Generalizations and applications," *Proc. V Int. Conference: Physical Interpretations of Relativity Theory*, Ed. M. C. Duffy, pp. 134-170, London, 11-14 September, 1998; (also in: *Nelinijni Koluvannya*, Vol. 2, No. 4, pp. 465-489, Kiev, Ukraine, 1999).

[277] Ly. T. Gruyitch, "Time, systems, and control: Qualitative properties and methods," Chapter 2 in *Stability and Control of Dynamical Systems with Applications*, Eds. D. Liu and P. J. Antsaklis, Boston: Birkhâuser, pp. 23-46, 2003.

[278] Ly. T. Gruyitch, "Time, systems and control," invited, submitted and accepted paper has had 50 pages, its abstract was published in *Abstracts of the Papers of the VIII Int. seminar "Stability and Oscillations of Nonlinear Control Systems,* "Ed. V. N. Thai, Moscow: IPU RAN, ISBN 5-201-14972-3, June 2-4, 2004.

[279] Ly. T. Gruyitch, *Tracking Control of Linear Systems*, Boca Raton, Fl.: CRC Press, 2013.

[280] Ly. T. Gruyitch, "Vector Lyapunov function synthesis of aircraft control," *Proc. INPAA-98: Second Int. Conference on Nonlinear Problems in Aviation & Aerospace*, Ed. Seenith Sivasundaram, ISBN: 0 9526643 1 3, Cambridge UK: European Conference Publications, Vol. 1, pp. 253-260, 1999.

[281] Ly. T. Gruyitch and A. Kökösy, "Conceptual development of vector Lyapunov functions and control synthesis for mechanical systems," *CD Proc. 17th IMACS World Congress*, paper # T2-I-0572, Paris, pp. 1-8, July 11-15, 2005.

[282] Ly. T. Gruyitch and W. Pratt Mounfield, Jr., "Absolute output natural tracking control: MIMO Lurie systems," *Proc. 14th Triennial World Congress*, Beijing, P. R. China, Pergamon-Elsevier Science, Vol. C, pp. 389-394, July 5-9, 1999.

[283] Ly. T. Gruyitch and W. Pratt Mounfield, Jr., "Constrained natural tracking control algorithms for bilinear DC shunt wound motors," *Proc. 40th IEEE Conference on Decision and Control*, Orlando, Fl., pp. 4433-4438, December 2001.

[284] Ly. T. Gruyitch and W. Pratt Mounfield, Jr., "Elementwise stablewise tracking with finite reachability time: linear time-invariant continuous-time MIMO systems," *Int. J. Systems Science*, Vol. 33, No.4, pp. 277-299, 2002.

[285] Ly. T. Gruyitch and W. P. Mounfield, Jr., "Robust elementwise exponential tracking control: IO linear systems," *Proc. 36th IEEE Conference on Decision and Control,* San Diego, California, pp. 3836-3841, December 1997.

[286] Ly. T. Gruyitch and W. Pratt Mounfield, Jr., "Stablewise absolute output natural tracking control with finite reachability time: MIMO Lurie systems," *CD Rom Proc. 17th IMACS World Congress*, Invited session IS-2 : Tracking theory and control of nonlinear systems, Paris, France, pp. 1-17, July 11-15, 2005; *Mathematics and computers in simulation*, Vol. 76, pp. 330-344, 2008.

[287] L. Gruyitch, J-P. Richard, P. Borne and J-C. Gentina, *Stability Domains*, Boca Raton, Fl.: Chapman&Hall/CRC, 2004.

[288] J. Guldner and V. I. Utkin, "Sliding mode control for gradient tracking and robot navigation using artificial potential fields," *IEEE Trans. Robotics and Automation*, Vol. 11, No. 2, pp. 247-254, April 1995.

[289] S. C. Gupta and R. J. Solem, "Accurate error analysis of a satellite tracking control loop system," *Int. J. Control*, Vol. 2, No. 6, pp. 539-549, 1965.

[290] P.-O. Gutman and M. Velger, "Tracking targets with unknown process noise variance using adaptive Kalman filtering," *Proc. 27th Conference on Decision and Control*, Austin, Texas, pp. 869-874, December 1988.

[291] I.J.Ha and E. G. Gilbert, "Robust tracking in nonlinear systems," *IEEE Trans. Automatic Control*, Vol. AC-32, No. 9, pp. 763-771, 1987.

[292] W. Hahn, *Stability of Motion*, New York: Springer-Verlag, 1967.

[293] O. Hájek, "Compactness and asymptotic stability," *Math. Systems Theory*, Vol. 4, No. 2, pp. 154-159, 1970.

[294] O. Hájek, "Ordinary and asymptotic stability of noncompact sets," *J. Differential Equations*, Vol. 11, No. 1, pp. 49-65, 1972.

[295] A. Halanay, *Differential Equations*, New York: Academic Press, 1966.

[296] J. K. Hale, *Ordinary Differential Equations*, New York: Wiley-Interscience, 1969.

[297] C. C. Hang and J. A. Chang, "An algorithm for constructing Lyapunov functions based on the variable gradient method," *IEEE Trans. Automatic Control*, Vol. AC-15, No. 4, 510-512, 1970.

[298] S. Hara and T. Sugie, "Independent parametrization of two-degree-of-freedom compensators in general robust tracking systems," *IEEE Trans. Automatic Control*, Vol. 33, No. 1, pp. 59-67, 1988.

[299] C. R. Hargraves, and S. W. Paris, "Direct trajectory optimization using nonlinear programming and collocation," *J. Guidance, Control, Dynamics*, Vol. 10, No. 4., 1987, pp. 338–342.

[300] P. Hartmann, *Ordinary Differential Equations*, New York: Wiley, 1964.

[301] F. Hausdorf, *Grundzüge der Mengenlehre*, New York: Chelsea, 1949.

[302] M. L. J. Hautus, "Controllability and observability conditions of linear autonomous systems," *Proc. Nederland Academy of Science: Mathematics*, Ser. A, Vol. 72, pp. 443-448, 1969.

[303] M. A. Henson and D. E. Seborg, "An internal model control strategy for nonlinear systems," *AIChE J.*, Vol. 37, No. 7, pp. 1065-1081, 1991.

[304] J. R. Hewit and C. Storey, "Comparison of numerical methods in stability analysis," *Int. J. Control*, Vol. 10, No. 6, pp. 687-701, 1969.

[305] J. R. Hewit and C. Storey, "Numerical application of Szegö's method for constructing Lyapunov functions," *IEEE Trans. Automatic Control*, Vol., AC-14, No. 1, pp. 106-108, 1969.

[306] J. R. Hewit and C. Storey, "Optimization of the Zubov and Ingwerson methods for constructing Lyapunov functions," *Electronics Letters*, Vol. 3, No. 5, pp. 211-212, 1967.

[307] R. M. Hirschorn, "Output tracking in multivariable nonlinear systems," *IEEE Trans. Automatic Control*, Vol., AC-26, No. 6, pp. 593-595, 1981.

[308] R. M. Hirschorn, "Singular sliding-mode control," *IEEE Trans. Automatic Control*, Vol. 46, No. 2, pp. 276-285, February, 2001.

[309] R. M. Hirschorn and J. H. Davis, "Global output tracking for nonlinear systems," *SIAM J. Control and Optimization*, Vol. 26, No. 6, pp. 1321-1330, 1988.

[310] R. Hirschorn and J. Davis, "Output tracking for nonlinear systems with singular points," *SIAM J. Control and Optimization*, Vol. 25, No. 3, pp. 547-527, May 1987.

[311] J. G. Hocking and G. S. Young, *Topology*, Reading, MA: Addison-Wesley, 1961.

[312] L. Hong, "Multirate interacting multiple model filtering for target tracking using multirate models," *IEEE Trans. Automatic Control*, Vol. 44, No. 7, pp. 1326-1340, July 1999.

[313] L. Hsu, "Smooth sliding control of uncertain systems based on a prediction error," *Int. J. Robust and Nonlinear Control*, Vol. 7, pp. 353-372, 1997.

[314] J. Huang, "Asymptotic tracking of a nonminimum phase nonlinear system with nonhyperbolic zero dynamics," *IEEE Trans. Automatic Control*, Vol. 45, No. 3, pp. 542-546, June 2000.

[315] P.-Y. Huang and B.-S. Chen, "Robust tracking of linear MIMO time-varying systems," *Automatica*, Vol. 30, No. 5, pp. 817-830, 1994.

[316] A. Huaux, "On the construction of Lyapunov functions," *IEEE Trans. Automatic Control*, Vol., AC-12, No. 4, pp. 465-466, 1967.

[317] R. O. Hughes, "Optimal control of Sun tracking solar concentrators," *J. Dynamic Systems, Measurement, Control*, Vol. 101, No. 2, pp. 157-161, 1979.

[318] P. C. Hughes, *Spacecraft attitude dynamics*, New York: Wiley, 1986.

[319] D. G. Hull, *Optimal Control Theory for Applications*, New York: Springer Verlag, 2003.

[320] K. J. Hunt, "General polynomial solution to the optimal feedback/feedforward stochastic tracking problem," *Int. J. Control*, Vol. 48, No. 3, pp. 1057-1073, 1988.

[321] C.-L. Hwang, Y.-M. Chen and C. Jan, "Trajectory tracking of large-displacement piezoelectric actuators using a nonlinear observer-based variable structure control," *IEEE Trans. Control Systems Technology*, Vol. 13, No. 1, pp. 56-66, January 2005.

[322] D. R. Ingwerson, "A modified Lyapunov method for nonlinear stability analysis," *IEEE Trans. Automatic Control*, Vol., AC-6, pp. 199-210, 1961.

[323] A. Isidori, *Nonlinear Control Systems*, Berlin: Springer Verlag, 1995

[324] A. Isidori and C. I. Byrnes, "Output regulation of nonlinear systems," *IEEE Trans. Automatic Control*, Vol. 35, No. 2, pp. 131-140, 1990.

[325] B. de Jager and F. Veldpaus, "Multivariable H∞ tracking control of a mono-cycle," *Proc. European Control Conference*, WE-M, pp. 1-6, 1-4 July, 1997.

[326] E. Jarzębowska, *Model-Based Tracking Control of Nonlinear Systems*, Boca Raton, Fl.: Chapman and Hall/CRC Press, 2012.

[327] S. Jayasuriya, "Multivariable disturbance rejection tracking controllers for systems with slow and fast modes," *Trans. of the ASME*, Vol. 109, pp. 364-369, 1987.

[328] S. Jayasuriya, "Robust tracking for a class of uncertain linear systems," *Int. J. Control*, Vol. 45, No. 3, pp. 875-892, 1987.

[329] S. Jayasuriya and C.-D. Kee, "Circle-type criterion for synthesis of robust tracking controllers," *Int. J. Control*, Vol. 48, No. 3, pp. 865-886, 1988.

[330] L. Jetto, "Ripple-free tracking problem," *Int. J. Control*, Vol. 50, No. 1, pp. 349-359, 1989.

[331] Z.-P. Jiang and I. Kanellakopoulos, "Global output-feedback tracking for benchmark nonlinear system," *IEEE Trans. Automatic Control*, Vol. 45, No. 5, pp. 1023-1027, May 2000.

[332] Z.-P. Jiang and H. Nijmeijer, "A recursive technique for tracking control of nonholonomic systems in chained form," *IEEE Trans. Automatic Control*, Vol. 44, No. 2, pp. 265-279, June 1999.

[333] M. Kabuka, E. McVey, and P. Shionoshita, "An adaptive approach to video tracking," *IEEE Journal of Robotics and Automation*, Vol. 4, No. 2, pp. 228-236, April, 1988.

[334] P. Kachroo and M. Tomizuka, "Chattering reduction and error convergence in the sliding-mode control of a class of nonlinear systems," *IEEE Trans. Automatic Control*, Vol. 41, No. 7, pp. 1063-1068, July 1996.

[335] T. Kaczorek, "Dead-beat servo problem for 2-dimensional linear systems," *Int. J. Control*, Vol. 37, No. 6, pp. 1349-1353, 1983.

[336] R. E. Kalman, "Algebraic structure of linear dynamical systems, I. The module of Σ," *Proc. National Academy of Science: Mathematics*, USA NAS, Vol. 54, pp. 1503-1508, 1965.

[337] R. E. Kalman, "Canonical structure of linear dynamical systems," *Proc. National Academy of Science: Mathematics*, USA NAS, Vol. 48, pp. 596-600, 1962.

[338] R. E. Kalman, "Mathematical description of linear dynamical systems," *J.S.I.A.M. Control*, Ser. A, Vol. 1, No. 2, pp. 152-192, 1963.

[339] R. E. Kalman, "On the general theory of control systems," *Proc. First Int. Congress on Automatic Control*, pp. 481-491, London: Butterworth, 1960.

[340] R. E. Kalman and J. E. Bertram, "Control system analysis and design via the 'second method' of Lyapunov," part I, *Trans. ASME: J. Basic Eng.*, Vol. 82, pp. 371-393, 1960.

[341] R. E. Kalman, P. L. Falb and M. A. Arbib, *Topics in Mathematical System Theory*, New York: Mc Graw-Hill, 1969.

[342] R. E. Kalman, Y. C. Ho, and K. S. Narendra, "Controllability of linear dynamical systems," *Contributions to Differential Equations*, Vol. 1, No. 2, pp. 189-213, 1963.

[343] M. H. Khammash, "Robust Steady-State Tracking," *IEEE Trans. Automatic Control*, Vol. 40, No. 11, pp. 1872-1880, November 1995.

[344] J. Kieffer, A. J. Cahill and M. R. James, "Robust and accurate time-optimal path-tracking control for robot manipulators," *IEEE Trans. Robotics and Automation*, Vol. 13, No. 6, pp. 880-890, December 1997.

[345] Y.-H. Kim and I.-J. Ha, "Asymptotic state tracking in a class of nonlinear systems via learning-based inversion," *IEEE Trans. Automatic Control*, Vol. 45, No. 11, pp. 2001-2027, November 2000.

[346] E. Kinnen and C. S. Chen, "Lyapunov functions derived from auxiliary exact differential equations," *Automatica*, Vol. 4, pp. 195-204, 1968.

[347] D. E. Kirk, *Optimal Control Theory: An Introduction.* Englewood Cliffs, N. J.: Prentice-Hall 1970, Dover republication 2004.

[348] D. L. Kleinman and T. R. Perkins, "Modeling human performance in a time-varying anti-aircraft tracking loop," *IEEE Trans. Automatic Control*, Vol. AC-19, No. 4, pp. 297-306, 1974.

[349] D. E. Koditschek, "Application of a new Lyapunov function to global adaptive attitude tracking," *Proc. 27th IEEE Conference on Decision and Control*, pp. 63-68, 1988.

[350] A. Kojima and S. Ishijima, "H_∞ preview tracking in output feedback setting," *Int. J. Robust and Nonlinear Control*, Vol. 14, pp. 627-641, 2004.

[351] A. Kökösy, *Poursuite Pratique de Systemes de commande Automatique des Robots Industriels*, Ph. D. dissertation, Belfort, France: University of Belfort-Montbeliard, 1999.

[352] A. Kőkősy, "Practical tracking with settling time: Bounded control for robot motion," *Proc. 14th IFAC Triennial World Congress*, Beijing, P. R. China, C-2a-11-4, pp. 377-382, 1999.

[353] A. Kőkősy, "Practical tracking with settling time: Criteria and algorithms," *Proc. VI Int. SAUM Conference on Systems, Automatic Control and Measurement*, Nish, Serbia, Yugoslavia, pp. 296-301, September 28-30, 1998.

[354] A. Kökösy, "Practical tracking with vector settling and vector reachability time," *Proc. IFAC Workshop on Motion Control*, Grenoble, France, pp. 297-302, Sept. 21-23, 1998.

[355] A. Kökösy, "Robot control: Practical tracking with reachability time," *Proc. Int. Conference on System, Signals, Control, Computers*, Vol. 3, Durban, RSA, pp. 186-190, 1998.

[356] S. Köksal S. and V. Lakshmikantham, "Higher derivatives of Lyapunov functions and cone-valued Lyapunov functions," *Nonlinear Analysis, Theory, Methods & Applications*, Vol. 26, No. 9, pp. 1555-1564, 1996.

[357] J. C. Kollodge and J. A. Sand, "Advanced star tracker design using the charge injection device," *Automatica*, Vol. 20, No. 6, pp. 787-791, 1984.

[358] N. N. Krasovskii, *Some Problems of the Theory of Stability of Motion*, in Russian, Moscow: FIZMATGIZ, 1959.

[359] N. N. Krasovskii, *Stability of Motion*, Stanford: Stanford University Press, 1963.

[360] N. J. Krikelis and E. G. Papadopoulos, "An optimal design approach for tracking problems and its assessment against classical controllers," *Int. J. Control*, Vol. 36, No. 2, pp. 249-265, 1982.

[361] B. C. Kuo, *Automatic Control Systems*, Englewood Cliffs, NJ: Prentice-Hall, 1967

[362] B. C. Kuo, *Automatic Control Systems*, Englewood Cliffs, NJ: Prentice-Hall, 1987.

[363] Y. H. Ku and N. N. Puri, "On Liapunov functions of high order non-linear systems," *J. Franklin Inst.*, Vol. 276, No. 5, pp. 349-364, 1963.

[364] K. Kuratowskii, *Topology*, Vol. 1, London: Academic Press, 1966.

[365] H. Kwakernaak and R. Sivan, *Linear Optimal Control Systems*, New York: Wiley-Interscience, 1972.

[366] W. H. Kwon, "Receding horizon tracking control as a predictive control and its stability properties," *Int. J. Control*, Vol. 50, No. 5, pp. 1807-1824, 1989.

[367] G. S. Ladde, V. Lakshmikantham, and S. Leela, "Conditionally asymptotically invariant sets and perturbed systems," *Annali di Matematica Pura ed Applicata*, Ser. IV, XCIV, pp. 33-40, 1972.

[368] G. S. Ladde and S. Leela, "Analysis of invariant sets," *Annali di Matematica Pura ed Applicata*, Ser. IV, XCIV, pp. 283-289, 1972.

[369] G. S. Ladde and S. Leela, "Global results and asymptotically self-invariant sets," *Rendiconti Della Classe di Scienze Fisiche, Matematiche e Naturali*, Academia Nazionale dei Lincei, Ser. VIII LIV, No. 3, pp. 321-327, 1973.

[370] V. Lakshmikantham, "Vector Lyapunov functions and conditional stability," *J. Mathematical Analysis and Applications*, pp. 368-377, 1975.

[371] V. Lakshmikantham and S. Leela, "Asymptotically self-invariant sets and conditional stability," in *Differential Equations and Dynamical Systems*, Eds. J. K. Hale, J. P. LaSalle, New York: Academic Press, pp. 363-369, 1967.

[372] V. Lakshmikantham and S. Leela, *Differential and Integral Inequalities*, New York: Academic Press, 1969.

[373] J. P. LaSalle, *The Stability of Dynamical Systems*, Philadelphia: Society for Industrial and Applied Mathematics, 1976.

[374] J. P. LaSalle and S. Lefschetz, *Stability by Liapunov's Direct Method*, New York: Academic Press, 1961.

[375] H. Lauer, R. Lesnick and L. E. Matson, *Servomechanism Fundamentals*, New York, McGraw-Hill, 1947.

[376] D. V. Lazitch, *Analysis and Synthesis of Practical Tracking Automatic Control* (in Serb), D. Sci. Dissertation, Faculty of Mechanical Engineering, University of Belgrade, Belgrade Serbia, 1995.

[377] D. V. Lazitch, "Uniform exponential practical automatic control tracking," (in Serb), *Proc. of the V Conference on Systems, Automatic Control and Measurement (SAUM)*, Novi Sad, pp. 68-70, October 2-3, 1995.

[378] D. V. Lazitch, "Uniform practical automatic control tracking," (in Serb), *Proc. V Conference on Systems, Automatic Control and Measurement (SAUM)*, Novi Sad, pp. 53-57, October 2-3, 1995.

[379] D. V. Lazitch, "Uniform practical automatic control tracking with the vector reachability time," (in Serb), *Proc. V Conference on Systems, Automatic Control and Measurement (SAUM)*, Novi Sad, pp. 63-67, October 2-3, 1995.

[380] D. V. Lazitch, "Uniform practical automatic control tracking with the vector settling time," (in Serb), *Proc. of the V Conference on Systems, Automatic Control and Measurement (SAUM)*, Novi Sad, pp. 58-62, October 2-3, 1995.

[381] E. B. Lee and L. Markus, *Foundations of Optimal Control Theory*, London: John Wiley & Sons, 1967.

[382] E. Lefeber, A. Robertsson, and H. Nijmeijer, "Linear controllers for exponential tracking of systems in chained-form," *Int. J. Robust and Nonlinear Control*, Vol. 10, No. 4, pp. 243-263, 2000.

[383] S. Lefschetz, *Stability of Nonlinear Control Systems*, New York: Academic Press, 1965.

[384] W. Leighton, "On the construction of Lyapunov functions for certain autonomous nonlinear differential equations," in *Contributions to Differential Equations*, Ed. J. P. LaSalle, New York: Wiley, Vol. II, pp. 367-383, 1963.

[385] Leviner M. D. and Dawson D. M., "Position and force tracking control of rigid-link electrically driven robots actuated by switched reluctance motors," *Int. J. Systems Science*, Vol. 26, No. 8, pp. 1479-1500, 1995.

[386] X. R. Li, "Tracking in clutter with strongest neighbor measurements-Part I: Theoretical analysis," *IEEE Trans. Automatic Control*, Vol. 43, No. 11, pp. 1560-1578, November 1998.

[387] E. H. M. Lim and J. K. Hedrick, "Lateral and longitudinal vehicle control coupling for automated vehicle operation," *Proc. American Control Conference*, Vol. 5, pp. 3676-3680, 1999.

[388] C,-M. Lin and T.-D. Meng, "Simultaneous deadbeat tracking control of two plants," *Int. J. Control*, Vol. 67, No. 6, pp. 921-931, 1997.

[389] C. T. Liou and C. T. Yang, "Guaranteed cost control of tracking problems with large plant uncertainty," *Int. J. Control*, Vol. 45, No. 6, pp. 2161-2171, 1987.

[390] S. Lipschutz, *Set Theory and Related Topics*, New York: McGraw Hill, 1998.

[391] P.-T. Liu and P. L. Bongiovanni, "On a passive vehicle tracking problem and max-minimization," *IEEE Trans. Automatic Control*, Vol. AC-28, No. 2, pp. 269-304, 1983.

[392] G. Liu and A. A. Goldenberg, "Uncertainty decomposition-based robust control of robot manipulators," *IEEE Trans. Control Systems Technology*, Vol. 4, No. 4, 384-393, July 1996.

[393] J.-S. Liu and K. Yuan, "On tracking control for affine nonlinear systems by sliding mode," *Systems and Control Letters*, Vol. 13, pp. 439-443, 1989.

[394] A. Loria, "Global tracking control of one degree of freedom Euler-Lagrange system without velocity measurements," *European J. of Control*, Vol. 2, pp. 144-151, 1996.

[395] R. Lozano, "Independent tracking and regulation adaptive control with forgetting factor," *Automatica*, Vol. 18, No. 4, pp. 455-459, 1982.

[396] J. C. Lozier, "A steady state approach to the theory of saturable servo systems," *IRE Trans. Automatic Control*, Vol. 1, No. 1, pp.19-39, May 1956.

[397] J. C. Lozier, J. A. Norton, and M. Iwama, "The servo system for telestar antena positioning," *Automatica*, Vol. 2, pp. 129-149, 1965.

[398] P. Lu, "Tracking control of nonlinear systems with bounded controls and control rates," *Automatica*, Vol. 33, No. 6, pp. 1199-1202, 1997.

[399] D. G. Luenberger, "Tracking of goal seeking vehicles," *IEEE Trans. Automatic Control*, Vol. 13, No. 1, pp. 74-77, February 1968.

[400] A. I. Lurie and E. W. Rozenwasser, "On methods for generating Lyapunov functions in the theory of nonlinear regulating systems," *Proc. 1st IFAC Congress: Theory of Continuous Systems-Special Mathematical Problems*, Moscow: Academy of Sciences USSR, pp. 709-717, 1961.

[401] A. M. Lyapunov, *The General Problem of Stability of Motion* (in Russian), Kharkov Mathematical Society, Kharkov, 1892; in *Academician A. M. Lyapunov: Collected Papers*, U.S.S.R. Academy of Science, Moscow, II, pp. 5-263, 1956. French translation: "Problème général de la stabilité du mouvement," *Ann. Fac. Toulouse*, 9, pp. 203-474; also in: *Annals of Mathematics Study*, No. 17, Princeton University Press, 1949. English translation: *Int. J. Control*, 55, pp. 531-773, 1992; also the book, London: Taylor and Francis, 1992.

[402] L. A. MacColl, *Fundamental Theory of Servomechanisms*, New York: D. Van Nostrand, 1945.

[403] L. Magni and R. Scattolini, "On the solution of the tracking problem for non-linear system with MPC," *Int. J. System Science*, Vol. 36, No. 8, pp. 477-484, 2005.

[404] R. P. S. Mahler, *Statistical Multisource-Multitarget Tracking Information Fusion*, Norwood, MA: Artech House, 2007.

[405] I. G. Malkin, *Motion Stability Theory* (in Russian), Moscow: Nauka, 1968.

[406] I. G. Malkin, "On the question of the reciprocal Liapunov's theorem on asymptotic stability of control systems" (in Russian), *Avtom. i Telemeh.*, No. 1, pp. 188-191, 1954.

[407] L. Marconi and A. Isidori, "Mixed internal model-based and feedforward control for robust tracking in nonlinear systems," *Automatica*, Vol. 36, pp. 993-1000, 2000.

[408] R. Marino and P. Tomei, "Robust adaptive state-feedback tracking for nonlinear systems," *IEEE Trans. Automatic Control*, Vol. 43, No. 1, pp. 84-89, January 1998.

[409] R. Marino and P. Tomei, "Nonlinear output tracking with almost disturbance decoupling," *IEEE Trans. Automatic Control*, Vol. 44, No. 1, pp. 18-28, January 1999.

[410] S. G. Margolis and W. G. Vogt, "Control engineering applications of V. I. Zubov's construction procedure for Liapunov functions," *IEEE Trans. Automatic Control*, Vol. AC-8, No. 2, pp. 104-113, 1963.

[411] R. Marino and S. Nicosia, "Hamiltonian-type Lyapunov functions," *IEEE Trans. Automatic Control*, Vol. AC-28, No. 11, pp. 1055-1057, 1983.

[412] A. A. Martynyuk and R. Gutowski, *Integral Inequalities and Stability of Motion* (in Russian), Kiev: Naukova dumka, 1979.

[413] J. L. Massera, "On Liapunov's conditions of stability," *Ann. Math.*, Vol. 50, pp. 705-721, 1949.

[414] J. L. Massera, "Contributions to stability theory," *Ann. of Math.*, Vol. 64, pp. 182-206, 1956.

[415] V. M. Matrosov, "To the theory of stability of motion" (in Russian), *Prikl. Math. Mekh.* Vol. 26, No. 5, pp. 885-895, 1962.

[416] V. M. Matrosov, "Vector Lyapunov functions in the analysis of nonlinear interconnected systems," *Proc. Symp. Mathematica*, Bologna, Vol. 6, pp. 209-242, 1971.

[417] V. M. Matrosov, *Vector Lyapunov Function Method: Analysis of Dynamical Properties of Nonlinear Systems* (in Russian), Moscow: FIZMATLIT, 2001.

[418] G. P. Matthews and R. A. DeCarlo, "Decentralized tracking for a class of interconnected nonlinear systems using variable structure control," *Automatica*, Vol. 24, No. 2, pp.187-193, 1988.

[419] V. I. Matyukhin, "Motional stability of manipulator robots in a decomposition mode" (in Russian), *Avtomatika i Telemekhanika*, No. 3, pp. 33-44, March 1989.

[420] V. I. Matyukhin, "Stability of the motions of manipulators under persistent perturbations" (in Russian), *Avtomatika i Telemekhanika*, No. 11, pp. 124-134, November 1993.

[421] V. I. Matyukhin, "Strong stability of motions of mechanical systems" (in Russian), *Avtomatika i Telemekhanika*, Vol. 57, No. 1, pp. 28-44, January 1996.

[422] V. I. Matyukhin, and E. S. Pyatnitskii, "Controlling the motion of robot manipulators by decomposition taking into account the actuator dynamics" (in Russian), *Avtomatika i Telemekhanika*, No. 9, pp. 67-81, September 1989.

[423] H. Mayeda and S. Miyoshi, "Robust control of tracking problem with internal stability for linear structured system," *IFAC Control Science and Technology 8th Triennial World Congress*, pp. 1195-1201, 1981.

[424] F. Mazenc and L. Praly, "Asymptotic tracking of a state reference for systems with a feedforward structure," *Proc. European Control Conference*, Brussels, WE-A, A3, 1-4 July, 1997.

[425] E. J. McShane, *Integration*, Princeton, NJ: Princeton University Press, 1944.

[426] A. N. Michel and C. J. Herget, *Algebra and Analysis for Engineers and Scientists*, Boston: Birkhäuser, 2007.

[427] A. N. Michel, L. Hou, and D. Liu, *Stability of Dynamical Systems. Continuous, Discontinuous and Discrete Systems*, Boston-Basel-Berlin: Birkhäuser, 2008.

[428] A. N. Michel and Lj. T. Grujić, "Mathematical modeling of continuous-time neural networks under pure structural variations," *Proc. 13th IMACS World Congress on Computational and Applied Mathematics*, Dublin, Ireland, Vol. 3, pp. 1254,1255, July 22-26, 1991.

[429] A. N. Michel and Lj. T. Grujić, "Qualitative analysis of continuous-time neural networks under pure structural variations," *Proc. 13th IMACS World Congress on Computational and Applied Mathematics*, Dublin, Ireland, Vol. 3, pp. 1260,1261, July 22-26, 1991.

[430] A. N. Michel and R. K. Miller, *Qualitative Analysis of Large-Scale Dynamical Systems*, New York: Academic Press, 1977.

[431] A. N. Michel, R. K. Miller, and B. H. Nam, "Stability analysis of interconnected systems using computer generated Lyapunov functions," *IEEE Trans. Circuits and Systems*, Vol. CAS-29, No. 7, pp. 431-440, 1982.

[432] A. N. Michel, B. H. Nam, and V. Vittal, "Computer generated Lyapunov functions for interconnected systems: improved results with application to power systems," *IEEE Trans. Circuits and Systems*, Vol. CAS-31, No. 2, pp. 189-198, 1984.

[433] R. K. Miller and A. N. Michel, *Ordinary Differential Equations*, New York: Academic Press, 1982.

[434] D. E. Miller and E. J. Davison, "The self-tuning robust servomechanism problem," *IEEE Trans. Automatic Control*, Vol. AC-34, No. 5, pp. 511-523, May 1989.

[435] B. R. Milojković and L. T. Grujić, *Automatic control*, in Serbo-Croatian, Belgrade: Faculty of Mechanical Engineering, 1977.

[436] B. R. Milojković and L. T. Grujić, *Automatic control*, in Serbo-Croatian, Belgrade: Faculty of Mechanical Engineering, 1981.

[437] J.K. Mills and A. A. Goldenberg, "Robust control of robotic manipulators in the presence of dynamic parameter uncertainty," *Trans. of the ASME: J. Dynamic Systems, Measurement, and Control*, Vol. 111, pp. 444-451, September 1989.

[438] N. Minamide, "Design of a deadbeat adaptive tracking system," *Int. J. Control*, Vol. 39, No. 1, pp. 63-81, 1984.

[439] D. Mitra, "Tracking by band-limited inputs," *Int. J. Control*, Vol. 10, No. 2, pp. 221-226, 1969.

[440] D. Mitra and H. C. So, "Existence conditions for L_1 Lyapunov functions for a class of nonautonomous systems," *IEEE Trans. Circuit Theory*, Vol. CT-19, No. 6, pp. 594-598, 1972.

[441] S. Mitrović, "Passive Tracking of a Moving Acoustic Source," *Prepr. IFAC 5th Triennal World Congress*, Budapest, Hungary, pp. 1063-1068, 1984.

[442] E. Mosca, A. Casavola and L. Giarre, "Minimax LQ stochastic tracking and servo problems," *IEEE Trans. Automatic Control*, Vol. 35, No. 1, pp. 95-97, January 1990.

[443] E. Mosca and L. Giarre, "Minimax LQ Stochastic Tracking and Disturbance Rejection Problems," *Proc. 28th Conference on Decision and Control*, Tampa, Florida, pp. 1473-1476, December 1989.

[444] W. P. Mounfield, Jr. and Lj. T. Grujić, "High-gain natural tracking control of linear systems," *Proc. 13th IMACS World Congress on Computation and Applied Mathematics*, Eds. R. Vichnevetsky and J. J. H. Miller, Trinity College, Dublin, Ireland, Vol. 3, pp. 1271-1272, July 22-26, 1991.

[445] W. P. Mounfield, Jr. and Lj. T. Grujić, "High-gain natural tracking control of time-invariant systems described by IO differential equations," *Proc. Conference on Decision and Control*, Brighton, England, pp. 2447-2452, 1991.

[446] W. P. Mounfield and Lj. T. Grujić, "High-gain PI control of an aircraft lateral control system," *Proc. 1993 IEEE Int. Conference on Systems, Man and Cybernetics*, Le Touquet, Vol. 2, pp. 736-741, 1993.

[447] W. P. Mounfield, Jr. and Lj. T. Grujić, "High-gain PI natural tracking control for exponential tracking of linear MIMO systems with state-space description," *Int. J. Control*, Vol. 25, No. 11, pp. 1793-1817, 1994.

[448] W. P. Mounfield, Jr. and Lj. T. Grujić, "High-gain PI natural tracking control for exponential tracking of linear single-output systems with state-space description," *RAIRO-Automatique, Productique, Informatique Industrielle (APII)*, Vo. 26, pp. 125-146, 1992.

[449] W. P. Mounfield and Lj. T. Grujić, "Natural tracking control for exponential tracking: Lateral high-gain PI control of an aircraft system with state-space description," *Neural, Parallel & Scientific Computations*, Vol. 1, No. 3, pp. 357-370, 1993.

[450] W. P. Mounfield and Lj. T. Grujić, "PID-Natural tracking control of a robot: Application," *Proc. 1993 IEEE Int. Conference on Systems, Man and Cybernetics*, Le Touquet, Vol. 4, pp. 328-333, 1993.

[451] W. P. Mounfield and Lj. T. Grujić, "Robust natural tracking control for multi-zone space heating systems," *Proc. 14th IMACS World Congress*, Vol. 2, 841-843, 1994.

[452] W. P. Mounfield, Jr. and Ly. T. Gruyitch, "Control of aircrafts with redundant control surfaces: stablewise tracking control with finite reachability time," *Proc. Second Int. Conference on Nonlinear Problems in Aviation and Aerospace*, European Conference Publishers, Cambridge, Vol. 2, 1999, pp. 547-554, 1999.

[453] W. P. Mounfield, Jr. and Ly. T. Gruyitch, "Elementwise stablewise finite reachability time natural tracking control of robots," *Proc. 14th Triennial World Congress*, Beijing, P. R. China, Pergamon-Elsevier Science, Vol. B, pp. 31-36, July 5-9, 1999.

[454] T. Nagaraja and V. V. Chalam, "Generation of Lyapunov function-a new approach," *Int. J. Control*, Vol. 19, No. 4, pp. 781-787, 1974.

[455] K. S. Narendra and J. H. Taylor, *Frequency Domain Criteria for Absolute Stability*, New York: Academic Press, 1973.

[456] D. B. Nauparac, "Analysis of the tracking theory on the real electrohydraulic servosystem" (in Serb), M. Sci. Thesis, *Faculty of Mechanical Engineering*, University of Belgrade, Belgrade Serbia, 1993.

[457] N. Neditch/Nedić, *Synthesis of a computer controlled hydraulic servomechanism for a precise realization of industrial robots trajectories*, PhD. dissertation (in Serb), Cathedra for automatic control, Faculty of Mechanical Engineering, University of Belgrade, Belgrade, Serbia, (partially prepared at the Department of Mechanical Engineering, Clemson University, Clemson, SC, USA), May 18, 1987.

[458] N. N. Nedić and D. H. Pršić, "Pneumatic and hydraulic time varying desired motion control using natural tracking control," *SAI-Avtomatika i Informatika '2000*, Sofia, Vol. 3, pp. 37-40, 24 -26 ocktomvri, 2000.

[459] N. N. Nedić and D. H. Pršić, "Pneumatic position control using natural tracking law," *IFAC Workshop on Trends in Hydraulic and Pneumatic Components and Systems*, Chicago, pp. 1-13, Nov. 8-9, 1994.

[460] N. N. Nedić and D. H. Pršić, "Time variable speed control of pump controlled hydraulic motor using natural tracking control," *IFAC-IFIP-IMACS Int. Conference on Control of Industrial Systems*, Belfort, pp. 197-202, May 20-22, 1997.

[461] V. V. Nemytskii and V. V. Stepanov, *Qualitative Theory of Differential Equations*, Princeton: Princeton University Press, 1960.

[462] I. Newton, *Mathematical Principles of Natural Philosophy*-Book I. The Motion of Bodies, William Benton, Publisher, Encyclopaedia Britannica, Chicago, (first publication: 1687) 1952.

[463] M.-L. Ni and Y. Chen, "Decentralized stabilization and output tracking of large-scale uncertain systems," *Automatica*, Vol. 32, No. 7, pp. 1077-1080, 1996.

[464] S. Nicosia and P. Tomei, "Tracking control with disturbance attenuation for robot manipulators," *Int. J. Adaptive Control and Signal Processing*, Vol. 10, pp. 443-449, 1996.

[465] S. Nikosia and P. Tomei, "A tracking controller for flexible joint robots using only link position feedback," *IEEE Trans. Automatic Control*, Vol. 40, No. 5, pp. 885-890, May 1995.

[466] H. Nijmeijer and A. J. van der Schaft, *Nonlinear Dynamical Control Systems*, 3rd edition, New York: Springer Verlag, 1996.

[467] Z. Novaković and Lj. T. Grujić, "Robot control: Stablewise tracking with reachability time under load variations," *Proc. Sixth IASTED Int. Symposium on Modelling, Identification and Control*, Grindelwald, pp. 223-226, February 17-20, 1987.

[468] K. Ogata, *State Space Analysis of Control Systems*, Englewood Cliffs, NJ: Prentice Hall, 1967.

[469] K. Ogata, *State Space Analysis of Control Systems*, Englewood Cliffs, NJ: Prentice Hall, 1970.

[470] J. J. Olsen, "Unified' equations of motion for flexible aircraft and some special solutions," *Proc. Second Int. Conference on Nonlinear Problems in Aviation and Aerospace*, Ed. S. Sivasundaram, Cambridge, UK: European Conference Publications, Vol. 2, pp. 571-580, 1999.

[471] M. Pachter, P. R. Chandler, and M. Mears, "Reconfigurable tracking control with saturation," *J. Guidance, Control and Dynamics*, Vol. 18, No. 5, pp. 1016-1022, 1995.

[472] Z. Pan and T. Başar, "Adaptive controller design for tracking and disturbance attenuation parametric strict-feedback nonlinear systems," *IEEE Trans. Automatic Control*, Vol. 43, No. 8, pp. 1066-1083, August 1998.

[473] P. C. Parks and A. J. Pritchard, "On the construction and use of Lyapunov functionals," *Proc. 4th IFAC Congress*, pp. 59-76, 1969.

[474] V. Parra-Vega, S. Arimoto, Y.-H. Liu, G. Hirzinger, and P. Akella, "Dynamic sliding PID control for tracking of robot manipulators: Theory and experiments," *IEEE Trans. Robotics and Automation*, Vol. 19, No. 6, pp. 967-976, December 2003.

[475] K. M. Passino, "Disturbance rejection in nonlinear systems: examples," *IEEE Proceedings*, Vol. 136, No. 6, pp. 317-323, 1989.

[476] H. Peng. and M. Tomizuka, "Preview control for vehicle lateral guidance in highway automation," *ASME J. Dynamic Systems, Measurement and Control*, Vol. 115, No. 4, pp. 678 -686, 1993.

[477] R. S. Pindyck, "An application of the linear quadratic tracking problem to economic stabilization policy," *IEEE Trans. Automatic Control*, Vol. AC-17, No. 3, pp. 287-300, June 1972.

[478] J. B. Plant, Y. T. Chan, and D. A. Redmond ,"A discrete tracking control law for nonlinear plants," *IFAC Control Science and Technology*, 8th Triennial World Congress, Kyoto, Japan, 2, pp. 55-60, 1981.

[479] H. Poincaré, "Sur le courbes définies par une équation différentielle," *Journal de Mathématiques*, série 3, No. 7, pp. 375-422, 1881-1882.

[480] L. S. Pontryagin, *Ordinary Differential Equations*, (in Russian), Moscow: Nauka, 1970.

[481] L. S. Pontryagin, *The Mathematical Theory of Optimal Processes*, New York: Gordon and Breach, 1986.

[482] L.S. Pontryagin, V.G.Boltyanski, R.S.Gamkrelidze and E.F.Mishchenko, *The Mathematical Theory of Optimal Processes*, New York: Gordon and Breach, 1986.

[483] B. Porter, "Fast-sampling tracking systems incorporating Lur'e plants with multiple nonlinearities," *Int. J. Control*, Vol. 34, No. 2, pp. 333-344, 1981.

[484] B. Porter, "High-gain tracking systems incorporating Lur'e plants with multiple nonlinearities," *Int. J. Control*, Vol. 34, No. 2, pp. 345-358, 1981.

[485] B. Porter, "High-gain tracking systems incorporating Lur'e plants with multiple switching nonlinearities," *VIII IFAC World Congress*, Kyoto, Session 3, pp. I-72 to I-77, 1981.

[486] B. Porter, T. R. Crossley and A. Bradshaw, "Synthesis of disturbance-rejection controllers for linear multivariable continuous-time systems," *Israel Journal of Technology*, Vol. 13, pp. 25-30, 1975.

[487] B. Porter and Lj. T. Grujić, "Continuous-time tracking systems incorporating Lur'e plants with multiple non-linearities," *Int. J. Systems Science*, Vol. 11, No. 7, pp. 827-840, 1980.

[488] B. Porter and Lj. T. Grujić, "Discrete-time tracking systems incorporating Lur'e plants with single non-linearities," *Third IMA Conference on Control Theory*, Academic Press, London, pp. 115-133, 1981.

[489] H. M. Power and R. J. Simpson, *Introduction to Dynamics and Control*, London: McGraw-Hill (UK), 1978.

[490] H.-P. Preuss, "Perfect steady-state tracking and disturbance rejection by constant state feedback," *Int. J. Control*, Vol. 35, No. 1, pp. 75-94, 1982.

[491] F. D. Priscoli, "Sufficient conditions for robust tracking in nonlinear systems," *Int. J. Control*, Vol. 67, No. 5, pp. 825-836, 1997.

[492] D. Prshitch and N, Neditch, "Pneumatic cylinder control by using natural control," (in Serb), *HIPNEF'93*, Belgrade, pp. 127-132, 1993.

[493] E. S. Pyatnitsky, "To problem of control black box of mechanical nature," *Problems of Nonlinear Analysis in Engineering Systems*, Kazan, Russia, Vol. 2, No. 4, pp. 1-10, 1996.

[494] K. V. Ramachandra, *Kalman Filtering Techniques for Radar Tracking*, New York: Marcel Dekker, 2000.

[495] Z. Qu and D.M. Dawson, *Robust Tracking Control of Robot Manipulators*, Piscataway, NJ: IEEE Press, 1995.

[496] D. B. Reid, "An algorithm for tracking multiple targets," *IEEE Trans. Automatic Control*, Vol. AC-24, No. 6, pp.843-854, December 1979.

[497] R. Reiss and G. Geiss, "The construction of Lyapunov functions," *IEEE Trans. Automatic Control*, Vol. AC-8, No. 4, pp. 382-383, May 1963.

[498] Z. Retchkiman, "The Problem of Output Tracking for Nonlinear Systems in the Presence of Singular Points," *Proc. American Control Conference*, Vol. 3, pp. 2975-2979, San Francisco, CA, June 2-4, 1983.

[499] Z. Retchkiman, J. Alvarez, and R. Castro, "Asymptotic output tracking through singular points for nonlinear systems: Stability, disturbance rejection and robustness," *Int. J. Robust and Nonlinear Control*, Vol. 5, pp. 553 -572, 1995.

[500] Ribar Z., "Natural tracking control in process industry," *Preprints of the IFAC-IFIP-IMACS Conference: Control of Industrial Systems*, Vol. 3, pp. 185-190, Belfort, France, May 20-22, 1997.

[501] Z. B. Ribar, D. V. Lazic, M. R. Jovanovic, "Application of practical exponential tracking in fluid transportation industry," *Proc. 14th Int. Conference on Material Handling and Warehousing*, Belgrade, Serbia, pp. 5.65-5.70, December 11-12, 1996.

[502] Z. Ribar, R. Jovanović, and D. Sekulić, "Fuzzy control of a hydraulic servosystem based on practical tracking algorithms," *Proc. Bath Workshop on Power Transmission and Motion Control*, Eds. C. R. Burrows and K. A. Edge, London: Professional Engineering Publishing, 2000, pp. 73-87.

[503] Z. B. Ribar, M. R. Yovanovitch, R. Z. Yovanovitch, "Application of practical linear tracking in process industry" (in Serb), *Proc. XLI Conference ETRAN*, Zlatibor, Serbia, Notebook 1, pp. 444-447, June 3-6, 1997.

[504] M. Ribbens-Pavella, "Critical survey of transient stability studies of multimachine power systems by Lyapunov's direct method," *Proc. 9th Allerton Conf. Circ. and Syst. Theory*, pp. 151-167, 1971.

[505] M. Rios-Bolívar, A. S. I. Zinober and H. Sira-Ramírez, "Dynamical adaptive sliding mode output tracking control of a class of nonlinear systems," *Int. J. Robust and Nonlinear Control*, Vol. 7, pp. 387-405, 1997.

[506] N. Rouche, P. Habets and M. Laloy, *Stability Theory by Liapunov's Direct Method*, New York: Springer-Verlag, 1977.

[507] R. Saeks and J. Murray, "Feedback system design: the tracking and disturbance rejection problems," *IEEE Trans. Automatic Control*, Vol. AC-26, No. 3, 203-217, 1981.

[508] T. Sadeghi and M. Wozny, "An optimal proportional-plus-integral / tracking control law for aircraft applications," *IEEE Trans. Automatic Control*, Vol. 29, No. 9, pp. 827-829, September 1984.

[509] A. Sage and C. C. White, III, *Optimum Systems Control*, Englewood Cliffs, NJ: Prentice-Hall, 1977.

[510] W. E. Schmitendorf, "Methods for obtaining robust tracking control laws," *Automatica*, Vol. 23, No. 5, pp. 675-677, 1987.

[511] D. G. Schultz, "The generation of Lyapunov functions," in *Advances in Control Systems*, Ed. C. Leondes, New York: Academic Press, Vol. 2, pp. 1-64, 1965.

[512] D. G. Schultz and J. E. Gibson, "The variable method for generating Lyapunov functions," *IEEE Trans. Appl. Ind.*, Vol. 81, pp. 203-210, 1962.

[513] U. Shaked and C. E. de Souza, "Continuous-time tracking problems in an H_∞ setting: A game theory approach," *IEEE Trans. Automatic Control*, Vol. 40, No. 5, pp. 841-852, May 1995.

[514] H. M. Shertukde and Y. Bar-Shalom, "Tracking of crossing targets with forward looking infrared imaging Sensors," *Proc. 28th Conference on Decision and Control*, Tampa, Florida, pp. 1409-1416, December 1989.

[515] I. A. Shkolnikov, Y. B. Shtessel, "Tracking in a class of nonminimum-phase systems with nonlinear internal dynamics via sliding mode control using method of system center," *Automatica*, Vol. 38, pp. 837-842, 2002.

[516] Y. B. Shtessel, "Nonlinear nonminimum phase output tracking via dynamic sliding manifolds," *J. Franklin Inst.*, 335B, No. 5, pp. 841-850, 1998.

[517] Y. B. Shtessel, "Nonlinear output tracking in conventional and dynamic sliding manifolds," *IEEE Trans. Automatic Control*, Vol. 42, No. 9, pp. 1282-1286, September 1997.

[518] C. Silvestre, A. Pascoal and I. Kaminer, "On the design of gain-scheduled trajectory tracking controllers," *Int. J. Robust and Nonlinear Control*, Vol. 12, pp. 797-839, 2002.

[519] J. J. Slotine and S. S. Sastry, "Tracking control of non-linear systems using sliding surfaces, with application to robot manipulators," *Int. J. Control*, Vol. 38, No. 2, pp. 465-492, 1983.

[520] H. Sonia, W. Perruquetti, and P. Borne, "A new sliding mode controller for multivariable nonlinear systems," *Proc. 1996 IEEE Int. Conference on Systems, Man and Cybernetics*, Vol. 2, pp. 917-922, Beijing, China, October 14-17, 1996.

[521] Y. D. Song, "Adaptive motion tracking control of robot manipulators-Non-regresssor based approach," *Int. J. Control*, Vol. 63, No. 1, pp. 41-54, 1996.

[522] E. D. Sontag, *Mathematical Control Theory: Deterministic Finite Dimensional Systems*, Piscataway, NY: Springer, 1998.

[523] M. Spivak, *Calculus on Manifolds*, New York: W. A. Benjamin, 1965.

[524] J. T. Spooner and K. M. Passino, "Adaptive control off a class of decentralized nonlinear systems," *IEEE Trans. Automatic Control*, Vol. 41, No. 2, pp. 280 -284, February 1996.

[525] S. K. Spurgeon and X. Y. Lu, "Output tracking using dynamic sliding mode techniques," *Int. J. Robust and Nonlinear Control*, Vol. 7, pp. 407-427, 1997.

[526] R. F. Stengel, *Optimal Control and Estimation*, Mineola, NY: Dover, 1994.

[527] Y. Stepanenko and J. Yuan, "Robust adaptive control of a class of nonlinear mechanical systems with unbounded and fast-varying uncertainties," *Automatica*, Vol. 28, No. 2, pp. 265-276, 1992.

[528] L. D. Stone, C. A. Barlow, and T. L. Corwin, *Bayesian Multiple Target Tracking*, Norwood, MA: Artech House, 1999.

[529] M. Y. Stoychitch, *Practical Tracking of Digital Control Systems* (in Serb), D. Sci. Dissertation, Faculty of Mechanical Engineering, University of Banya Luka, Banya Luka, Republic of Serb, 2004.

[530] T. Sugie and M. Vidyasagar, "Further results on the robust tracking problem in two-degree-of-freedom control systems," *Systems & Control Letters*, Vol. 13, pp. 101-108, 1989.

[531] G. P. Szegö, "New methods for constructing Lyapunov functions for time invariant control systems," *Proc. 2nd IFAC World Congress*, London: Butterworths, pp. 584-589, 1964.

[532] D. D. Šiljak, *Large-Scale Dynamic Systems: Stability and Structure*, New York: North Holland, 1978.

[533] D. D. Šiljak, *Nonlinear Systems*, New York: Wiley, 1969.

[534] D. D. Šiljak, *Stability of Control Systems* (in Serbo-Croatian), Belgrade: Faculty of Electrical Engineering, 1974.

[535] A. I. Talkin, "Adaptive servo tracking," *IRE Trans. Automatic Control*, Vol. 6, No. 2, pp.167-172, May 1961.

[536] S. Tarbouriech, C. Pittet, and C. Burgat, "Output tracking problem for systems with input saturations via nonlinear integrating actions," *Int. J. Robust and Nonlinear Control*, Vol. 10, pp. 489-512, 2000.

[537] M. Tarokh, "A decentralized nonlinear three-term controller for manipulator trajectory tracking," *Proc. 1996 IEEE Int. Conference on Robotics and Automation*, pp. 3683-3688, Minneapolis, April 1996.

[538] B. O. S. Teixeira, M. A. Santillo, R. S. Erwin and D. S. Bernstein, "Spacecraft tracking using sampled-data Kalman filters," *IEEE Control Systems Magazine*, Vol. 28, No. 4, pp.78-94, August 2008.

[539] H. T. Toivonen and J. Pensar, "A worst-case approach to optimal tracking control with robust performance," *Int. J. Control*, Vol. 65, No. 1, pp. 17-32, 1996.

[540] D. E. Torfs, R. Vuerinckx, J. Swevers and J. Schoukens, "Comparison of two feed-forward design methods aiming at accurate trajectory tracking of the end point of a flexible robot arm," *IEEE Trans. Control Systems Technology*, Vol. 6, No. 1, pp. 2-14, January 1998.

[541] J. Tsiniaas and J. Karafyllis, "ISS property for time-varying systems and application to partial-static feedback stabilization and asymptotic tracking," *IEEE Trans. Automatic Control*, Vol. 44, No. 11, pp. 2179-2184, November 1999.

[542] A. Vanelli and M. Vidyasagar, "Maximal Lyapunov functions and domains of attraction for autonomous nonlinear systems," *Automatica*, Vol. 21, No. 1, pp. 69-80, 1985.

[543] R. Vinter, *Optimal Control*, New York: Springer, 2010.

[544] E. T. Wall, "A synthesis of Lyapunov functions for non-linear time-varying control systems," *Int. J. Sys. Science*, Vol. 4, No. 4, pp. 565-575, 1973.

[545] E. T. Wall, "A topological approach to the generation of Lyapunov functions," *Acta Technica SCAV*, Vol. 2, pp. 159-177, 1968.

[546] E. T. Wall, "The generation of Lyapunov functions in control theory by an energy matric algorithm," *Proc. 1968 J. A. C. C.*, pp. 172-179, 1968.

[547] E. T. Wall and M. L. Moe, "An energy metric algorithm for the generation of Lyapunov functions," *IEEE Trans. Automatic Control*, Vol. AC-13, pp. 121-122, 1968.

[548] E. T. Wall and M. L. Moe, "Generation of Lyapunov functions for time-varying nonlinear systems," *IEEE Trans. Automatic Control*, Vol. AC-14, p. 211, 1969.

[549] S.-S. Wang, B-S Chen, "Simultaneous deadbeat tracking controller synthesis," *Int. J. Control*, Vol. 44, No.6, pp. 1579-1586, 1986.

[550] P. E. Wellstead and P. Zanker, "Servo self-tuners," *Int. J. Control*, Vol. 30, No. 1, pp. 27-36, 1979.

[551] J. T. Wen and K. Kreutz, "Globally stable control laws for attitude maneuver problem: Tracking control and adaptive control," *Proc. 27th Conference on Decision and Control*, pp. 69-74, 1988.

[552] J. T. Wen, K. Kreutz-Delgado, and D. S. Bayard, "Lyapunov function-based control laws for revolute robot arms: Tracking control, robustness, and adaptive control," *IEEE Trans. Automatic Control*, Vol. 37, No. 2, pp. 231-237, February 1992.

[553] J. C. West, *Textbook of Servomechanisms*, London: English Universities Press, 1953.

[554] D. von Wissel, R. Nikoukhah, F. Delebecque, P.-A. Bliman, and M. Sorine, "Output trajectory tracking for mechanical systems with dry friction: A DPC approach," *Proc. European Control Conference*, Brussels, TH-E, G6, 1-4 July, 1997.

[555] W. A. Wolovich, *Linear Multivariable Systems*, New York: Springer-Verlag, 1974.

[556] W. M. Wonham, "Tracking and regulation in linear multivariable systems," *SIAM J. Control*, Vol. 11, No. 3, pp. 424-437, 1973.

[557] A. R. Woodyatt, M. M. Seron, J. S. Freudenberg, and R. H. Middleton, "Cheap control tracking performance," *Int. J. Robust and Nonlinear Control*, Vol. 12, pp. 1253-1273, 2002.

[558] J.-X. Xu and T. Zhu, "Dual-scale direct learning control of trajectory tracking for a class of nonlinear uncertain systems," *IEEE Trans. Automatic Control*, Vol. 44, No. 10, pp. 1884-1888, October 1999.

[559] H. Yamane and B. Porter, "Synthesis of limit-tracking incorporating linear multivariable plants," *Int. J. Systems Science*, Vol. 25, No. 12, pp. 2095-2111, 1994.

[560] J.-M. Yang and J.-H. Kim, "Sliding mode control for trajectory tracking nonholonomic wheeled mobile robots," *IEEE Trans. Robotics and Automation*, Vol. 15, No. 3, pp. 578-587, June 1999.

[561] T. Yoshikawa and T. Sugie, "Analysis and synthesis of tracking systems considering sensor dynamics," *Int. J. Control*, Vol. 41, No. 4, pp. 961-971, 1985.

[562] T. Yoshizawa, "Asymptotic behaviour of solutions of ordinary differential equations near sets," *Proc. Int. Symp. on Nolinear Oscillations*, Kiev, Ukraine, Vol. 1, pp. 213-225, 1963.

[563] T. Yoshizawa, "Eventual properties and quasi-asymptotic stability of a non-compact set, "*Funkcialaj Ekvacioj*, Vol. 8, No. 2, pp. 25-90, 1966.

[564] T. Yoshizawa, "Some notes on stability of sets and perturbed system," *Funkcialaj Ekvacioj*, Vol. 6, No. 1, pp. 1-11, 1964.

[565] T. Yoshizawa, *Stability Theory by Lyapunov's Second Method*, Tokyo: Mathematical Society of Japan, 1966.

[566] K. Youcef-Toumi and O. Ito, "A time delay controller for systems with unknown dynamics," *J. Dynamic Systems, Measurement, and Control*, Vol. 112, pp. 133-142, 1990.

[567] K. Youcef-Toumi and O. Ito, "Controller design for systems with unknown nonlinear dynamics," *Proc. 1987 American Control Conference*, Minneapolis, MN, Vol. 2, pp. 836-844, June 10-12, 1990.

[568] M. R. Yovanovitch, *Practical Tracking Automatic Control of the Axial Piston Hydraulic Motors* (in Serb), M. Sci. Thesis, Faculty of Mechanical Engineering, University of Belgrade, Belgrade Serbia, 1998.

[569] R. Zh. Yovanovitch, *Fuzzy Logic Based Realization of Control Systems* (in Serb), M. Sci. Thesis, Faculty of Mechanical Engineering, University of Belgrade, Belgrade, Serbia, 1999.

[570] R. Zh. Yovanovitch, *Fuzzy Tracking Control Algorithms of Electrohydraulic Servosystems* (in Serb), D. Sci. Dissertation, Faculty of Mechanical Engineering, University of Belgrade, Belgrade, Serbia, 2011.

[571] R. Zh. Yovanovitch and Z. B. Ribar, "Fuzzy practical exponential tracking of an electrohydraulic servosystems," Faculty of Mechanical Engineering (FME), Belgrade, Serbia, *FME Trans.*, Vol. 39, pp. 9-15, 2011.

[572] W.-S. Yu and Y.-H. Chen, "Decoupled variable structure control design for trajectory tracking on mechatronic arms," *IEEE Trans. Control Systems Technology*, Vol. 13, No. 5, pp. 798-806, September 2005.

[573] S. Y. Zhang and C. T. Chen, "Design of compensators for robust tracking and disturbance rejection," *IEEE Trans. Automatic Control*, Vol. AC-30, No. 7, 684-687, July 1985.

[574] J. Zhao and I. Kanellakopoulos, "Flexible backstepping design for tracking and disturbance attenuation," *Int. J. Robust and Nonlinear Control*, Vol. 8, pp. 331-348, 1998.

[575] M. Zhihong, A. P. Paplinski, and H. R. Wu, "A Robust MIMO terminal sliding mode control scheme for rigid robotic manipulators," *IEEE Trans. Automatic Control*, Vol. 39, No. 12, pp. 2464-2469, December 1994.

[576] Y.-S. Zhong, "Robust output tracking control of SISO plants with multiple operating points and with parametric and unstructured uncertainties," *Int. J. Control*, Vol. 75, No. 4, pp. 219-241, 2002.

[577] Y. Zhu, D. Dawson, T. Burg, and J. Hu, "A cheap output feedback tracking controller with robustness: The RLFJ problem," *Proc. 1996 IEEE International Conference on Robotics and Automation*, Minneapolis, pp. 939-944, April 1996.

[578] V. I. Zubov, *Methods of A. M. Liapunov and Their Applications* (in Russian), Leningrad: Leningrad Gos. University, (English translation: Groningen: P. Noordhoff Ltd., 1964).

Part XII

INDEXES

Author Index

Subject Index